Lecture Notes in Computer Science 6841

Commenced Publication in 1973
Founding and Former Series Editors:
Gerhard Goos, Juris Hartmanis, and Jan van Leeuwen

Phillip Rogaway (Ed.)

Advances in Cryptology – CRYPTO 2011

31st Annual Cryptology Conference
Santa Barbara, CA, USA, August 14-18, 2011
Proceedings

Springer

Volume Editor

Phillip Rogaway
University of California
Department of Computer Science
Davis, CA 95616, USA
E-mail: rogaway@cs.ucdavis.edu

ISSN 0302-9743 e-ISSN 1611-3349
ISBN 978-3-642-22791-2 e-ISBN 978-3-642-22792-9
DOI 10.1007/978-3-642-22792-9
Springer Heidelberg Dordrecht London New York

Library of Congress Control Number: 2011932695

CR Subject Classification (1998): E.3, G.2.1, F.2.1-2, D.4.6, K.6.5, C.2, J.1

LNCS Sublibrary: SL 4 – Security and Cryptology

Typesetting: Camera-ready by author, data conversion by Scientific Publishing Services, Chennai, India

Printed on acid-free paper

Springer is part of Springer Science+Business Media (www.springer.com)

Preface

CRYPTO 2011, the 31st Annual International Cryptology Conference, was held August 14–18 on the campus of the University of California, Santa Barbara. The event was sponsored by the International Association for Cryptologic Research (the IACR) in cooperation with the UCSB Computer Science Department and the IEEE Computer Society's Technical Committee on Security and Privacy.

We received 230 submissions, a new record, of which 43 were accepted for publication. With one pair of papers merged, these proceedings contain the revised versions of 42 papers.

There were also two invited talks. On Monday, Ron Rivest delivered the 2011 IACR Distinguished Lecture. On Wednesday, Roger Dingledine spoke about Tor, a widely used system for online anonymous communication. For Tuesday afternoon, traditionally left free, Shai Halevi graciously offered a three-hour tutorial on Fully Homomorphic Encryption. That evening, Dan Bernstein and Tanja Lange chaired the traditional rump session.

I have tried to assemble a technical program not only strong, but also balanced. Efforts in this direction included selection of a particularly large and broad Program Committee (PC), and a Call for Papers explicitly indicating receptiveness to cryptographic topics not routinely appearing at recent CRYPTOs. I encouraged PC members to focus on the positive aspects of submissions. When it came time to vote on second-round accepts, partitioning the papers into topical categories may also have helped.

For the Best Paper Award, the PC overwhelmingly selected "Computer-Aided Security Proofs for the Working Cryptographer," by Gilles Barthe, Benjamin Grégoire, Sylvain Heraud, and Santiago Zanella Béguelin. The Committee praised the work for its broad appeal, its connections to programming language, and its potential impact.

Papers were reviewed in the customary way, double-blind, with non-PC contributions generally receiving three or more reviews, and PC contributions getting four or more. I encouraged (anonymized) questions from PC members to authors, and ended up relaying several tens of such messages. Throughout the review process I tried to treat each submission as its authors' well-loved child, never as a three-digit number in need of categorization.

I would like to most sincerely thank the authors of submissions—both those who did and who did not get their papers in. Contributing research from all corners of the earth, it is the fine work of the authors that makes a conference like ours possible and worthwhile.

My deepest appreciation goes out to the PC. I find something wonderful and touching about so many busy and brilliant people putting in enormous amounts of time to perform so thankless and difficult a service. I was repeatedly impressed by the dedication, integrity, knowledge, and extraordinary technical skills of so many on our PC. A list of PC members appears after this note.

The external reviewers play a key role in assessing the submissions, and are heartily thanked for their contribution. A list of external reviewers likewise appears after this note. My apologies in advance for any errors or omissions.

I would like to thank Tom Shrimpton, the General Chair, for working closely with me and handling the myriad of matters associated to putting on a great conference. Rei Safavi-Naini served double duty as both PC member and Junior Chair. I kept in close touch with John Benaloh, my IACR point of contact, who could always be counted on for timely information and feedback. I repeatedly got invaluable and frank advice from Tal Rabin, the CRYPTO 2010 Program Chair. Shai Halevi wrote, explained, and maintained the superb *websubrev* software on which we conducted our business. Alfred Hofmann and his colleagues at Springer saw to the timely production of this volume. Finally, Bongkotrattana Lailert afforded me the time and space needed to do this piece of work as well as I possibly could, smilingly accepting her and Banlu's exile to distant lands.

In closing, I would like to acknowledge something that all old-timers know, but which, as authors, we may sometimes fail to internalize: that there's an awful lot of randomness in the paper-selection process. Wonderful papers sometimes get rejected; mediocre papers sometimes get in. After serving as PC Chair I am more convinced than ever that it is fundamentally wrong to feel much of anything when any particular paper one submits does or doesn't make the cut. I hope that, over a period of years, important papers do get in, and do get recognized as well.

Serving as a CRYPTO Program Chair is a big job, and it can be a stressful one as well. Yet, somehow, I feel like I have grown more than gray hairs with this job, and am happy to have taken it on.

June 2011 Phillip Rogaway

CRYPTO 2011

The 31st Annual International Cryptology Conference

Santa Barbara, California, USA
August 14–18, 2011

Sponsored by the
International Association of Cryptologic Research (IACR)

in cooperation with the
Computer Science Department of the University of California, Santa Barbara
and the
IEEE Computer Society's Technical Committee on Security and Privacy

General Chair

Thomas Shrimpton	Portland State University, USA

Program Chair

Phillip Rogaway	University of California, Davis, USA

Program Committee

Masayuki Abe	NTT, Japan
Michael Backes	Saarland University and MPI-SWS, Germany
Paulo Barreto	University of São Paulo, Brazil
Mihir Bellare	UC San Diego, USA
Alex Biryukov	University of Luxembourg
Dan Boneh	Stanford, USA
Jung Hee Cheon	Seoul National University, Korea
Jean-Sébastien Coron	University of Luxembourg
Marten van Dijk	RSA Labs and MIT/CSAIL, USA
Yevgeniy Dodis	New York University, USA
Orr Dunkelman	University of Haifa and Weizmann Institute, Israel
Serge Fehr	CWI, The Netherlands
Steven Galbraith	University of Auckland, New Zealand
Craig Gentry	IBM Research, USA
Louis Goubin	Université de Versailles, France
Vipul Goyal	Microsoft Research, India
Aggelos Kiayias	University of Connecticut, USA
Eike Kiltz	Ruhr-Universität Bochum, Germany
Anja Lehmann	IBM Zurich, Switzerland
Arjen Lenstra	EPFL, Switzerland
Stefan Mangard	Infineon Technologies, Germany

Program Committee (Continued)

Daniele Micciancio	UC San Diego, USA
Tal Moran	Harvard, USA
Chanathip Namprempre	Thammasat University, Thailand
Phong Nguyen	INRIA and ENS, France
Jesper Buus Nielsen	Aarhus University, Denmark
Rafael Pass	Cornell University, USA
Kenny Paterson	Royal Holloway, University of London, UK
Benny Pinkas	Bar Ilan University, Israel
Bart Preneel	Katholieke Universiteit Leuven, Belgium
Leonid Reyzin	Boston University, USA
Vincent Rijmen	Katholieke Universiteit Leuven, Belgium and TU Graz, Austria
Rei Safavi-Naini	University of Calgary, Canada
Andre Scedrov	University of Pennsylvania, USA
Adam Smith	Pennsylvania State University, USA
François-Xavier Standaert	UCL, Belgium
Stefano Tessaro	UC San Diego, USA
Bogdan Warinschi	University of Bristol, UK
Hoeteck Wee	Queens College, CUNY, USA

Advisory Members

Tal Rabin (CRYPTO 2010 Program Chair)	IBM Research, USA
Rei Safavi-Naini (CRYPTO 2012 Program Chair)	University of Calgary, Canada

External Reviewers

Michel Abdalla
Divesh Aggarwal
Hadi Ahmad
Mohsen Alimomeni
Joel Alwen
Elena Andreeva
Kazumaro Aoki
Gilad Asharov
Maxime Augier
Paul Baecher
Kfir Barhum
Alexandre Berzati
Gaëtan Bisson
Bruno Blanchet
Andrey Bogdanov
Alexandra Boldyreva
Joppe Bos
Charles Bouillaguet
Niek Bouman
Colin Boyd
Christina Brzuska
Jan Camenisch
Sebastien Canard
Ran Canetti
David Cash
Nishanth Chandran
Melissa Chase
Hao Chen
Alessandro Chiesa
Sherman S.M. Chow
Kai-Min Chung
Iwen Coisel
Cas Cremers
Dana Dachman-Soled
Ivan Damgård
Jean Paul Degabriele
Cécile Delerablée
Ante Derek
Claus Diem
Vivien Dubois
Maria Dubovitskaya
Léo Ducas
Andrej Dujella
Iwan Duursma
Stefan Dziembowski

Pooya Farshim
Sebastian Faust
Matthieu Finiasz
Dario Fiore
Marc Fischlin
Pierre-Alain Fouque
David Freeman
Georg Fuchsbauer
Eiichiro Fujisaki
Jakob Funder
Martin Gagne
David Galindo
Sanjam Garg
Peter Gaži
Ran Gelles
Rosario Gennaro
Benedikt Gierlichs
Henri Gilbert
Michael Goodrich
Thomas Gross
Jeffrey Guarente
Kil-Chan Ha
Robbert de Haan
Iftach Haitner
Shai Halevi
Sean Hallgren
Mike Hamburg
Safuat Hamdy
Goichiro Hanaoka
Danny Harnik
Carmit Hazay
Jens Hermans
Mathias Herrmann
Florian Hess
Martin Hirt
Viet Tung Hoang
Dennis Hofheinz
Susan Hohenberger
Peter Hoyer
Pavel Hubacek
Andreas Hülsing
Sebastiaan Indesteege
Yuval Ishai
Tibor Jager
Abhishek Jain
Dimitar Jetchev
Shaoquan Jiang
Antoine Joux
Pascal Junod
Seny Kamara
Bhavana Kanukurthi
Alexandre Karlov
Shiva Kasiviswanathan
Jonathan Katz
Marcel Keller
Jihye Kim
Minkyu Kim
Myungsun Kim
Sungwook Kim
Thorsten Kleinjung
Robin Künzler
Ralf Küsters
Soonhak Kwon
Taekyoung Kwon
Mario Lamberger
Hyung Tae Lee
Younho Lee
Gaëtan Leurent
Allison Lewko
Benoit Libert
Changlu Lin
Huijia Lin
Yehuda Lindell
Satya Lokam
Adriana López-Alt
Carolin Lunemann
Anna Lysyanskaya
Vadim Lyubashevsky
Mohammad Mahmoody
Alexander May
Marcel Medwed
Sebastian Meiser
Florian Mendel
Bart Mennink
Alexander Meurer
Petros Mol
Hart Montogomery
Kirill Morozov
Elchanan Mossel
Serban Nacu
Tomislav Nad
Michael Naehrig
Arvind Narayanan
Gregory Neven
Ivica Nikolic
Ryo Nishimaki
Kobbi Nissim
Peter Sebastian Nordholt
Adam O'Neill
Miyako Ohkubo
Tatsuaki Okamoto
Elisabeth Oswald
Onur Ozen
Pascal Paillier
Paolo Palmieri
Omkant Pandey
Valerio Pastro
Jacques Patarin
Arpita Patra
Thomas Peeters
Serdar Pehlivanoglu
Chris Peikert
Robin Pemantle
Olivier Pereira
Edoardo Persichetti
Christophe Petit
Krzysztof Pietrzak
David Pointcheval
Manoj Prabhakaran
Ananth Raghunathan
Francesco Regazzoni
Oded Regev
Tzachy Reinman
Renato Renner
Thomas Ristenpart
Matthieu Rivain
Guy Rothblum
Yannis Rouselakis
Arnab Roy
Nashad Safa
Louis Salvail
Juraj Sarinay
Sumanta Sarkar
Yu Sasaki
Christian Schaffner

Martin Schläffer
Thomas Schneider
Dominique Schröder
Gil Segev
Jae Hong Seo
Yannick Seurin
Siamak Shahandashi
Elaine Shi
Thomas Shrimpton
Marcos A. Simplicio Jr
Thomas Sirvent
William E. Skeith III
Arkadii Slinko
Nigel Smart
Fang Song
Martijn Stam
John Steinberger
Marc Stevens
Gabor Tardos
Aris Tentes
Enrico Thomae
Mehdi Tibouchi
Elmar Tischhauser
Tomas Toft
Nikos Triandopoulos
Tomasz Truderung
Wei-lung Tseng
Ashraful Tuhin
Yevgeniy Vahlis
Vinod Vaikuntanathan
Kerem Varıcı
Damien Vergnaud
Ivan Visconti
Huaxiong Wang
Meiqin Wang
Yongge Wang
Brent Waters
Gaven Watson
Benne de Weger
Ralf-Philipp Weinmann
Daniel Wichs
Steve Williams
Christopher Wolf
Jürg Wullschleger
Andy Yao
Sarah Zakarias
Hong-Sheng Zhou
Angela Zottarel

Table of Contents

Randomness and Its Use

Computer-Assisted Cryptographic Proofs

Outsourcing and Delegating Computation

Symmetric Cryptanalysis and Constructions

Secure Computation

Leakage and Side Channels

Quantum Cryptography

Lattices and Knapsacks

Invited Talk

Public-Key Encryption

Symmetric Schemes

Signatures

Oblivious Transfer and Secret Sharing

Multivariate and Coding-Based Schemes

Leftover Hash Lemma, Revisited

Boaz Barak[1], Yevgeniy Dodis[2], Hugo Krawczyk[3], Olivier Pereira[4], Krzysztof Pietrzak[5], François-Xavier Standaert[4], and Yu Yu[6]

[1] Microsoft Research New England
boaz@microsoft.com
[2] New York University
dodis@cs.nyu.edu
[3] IBM Research
hugo@ee.technion.ac.il
[4] Université Catholique de Louvain
{Olivier.Pereira,fstandae}@uclouvain.be
[5] CWI Amsterdam
pietrzak@cwi.nl
[6] East China Normal University
yuyu@yuyu.hk

Abstract. The famous *Leftover Hash Lemma* (LHL) states that (almost) universal hash functions are good randomness extractors. Despite its numerous applications, LHL-based extractors suffer from the following two limitations:

- **Large Entropy Loss:** to extract v bits from distribution X of min-entropy m which are ε-close to uniform, one must set $v \leq m - 2\log(1/\varepsilon)$, meaning that the entropy loss $L \stackrel{\text{def}}{=} m - v \geq 2\log(1/\varepsilon)$. For many applications, such entropy loss is too large.
- **Large Seed Length:** the seed length n of (almost) universal hash function required by the LHL must be at least $n \geq \min(u - v, v + 2\log(1/\varepsilon)) - O(1)$, where u is the length of the source, and must grow with the number of extracted bits.

Quite surprisingly, we show that both limitations of the LHL — large entropy loss and large seed — can be overcome (or, at least, mitigated) in various important scenarios. First, we show that entropy loss could be reduced to $L = \log(1/\varepsilon)$ for the setting of deriving secret keys for a wide range of cryptographic applications. Specifically, the security of these schemes with an LHL-derived key gracefully degrades from ε to at most $\varepsilon + \sqrt{\varepsilon 2^{-L}}$. (Notice that, unlike standard LHL, this bound is meaningful even when one extracts more bits than the min-entropy we have!) Based on these results we build a general *computational extractor* that enjoys low entropy loss and can be used to instantiate a generic key derivation function for *any* cryptographic application.

Second, we study the soundness of the natural *expand-then-extract* approach, where one uses a *pseudorandom generator* (PRG) to expand a short "input seed" S into a longer "output seed" S', and then use the resulting S' as the seed required by the LHL (or, more generally, by any randomness extractor). We show that, in general, the expand-then-extract approach is not sound if the Decisional Diffie-Hellman assumption is true. Despite that, we show that it is sound either: (1) when

P. Rogaway (Ed.): CRYPTO 2011, LNCS 6841, pp. 1–20, 2011.

extracting a "small" (logarithmic in the security of the PRG) number of bits; or (2) in minicrypt. Implication (2) suggests that the expand-then-extract approach is likely secure when used with "practical" PRGs, despite lacking a reductionist proof of security!

1 Introduction

The famous Leftover Hash Lemma [18] (LHL; see also [18] for earlier formulations) has found a huge number of applications in many areas of cryptography and complexity theory. In its simplest form, it states that universal hash functions [7] are good (strong) randomness extractors [31]. Specifically, if X is a distribution of min-entropy m over some space $\mathcal{X}$, $\mathcal{H}$ is a family of universal functions (see Definition 2) from $\mathcal{X}$ to $\{0,1\}^v$, and H is a random member of $\mathcal{H}$, then, *even conditioned on the "seed"* H, the statistical distance between $H(X)$ and the uniform distribution U_v on $\{0,1\}^v$ is bounded by $\sqrt{2^{-L}}$, where $L \stackrel{\mathsf{def}}{=} m - v$. The parameter L is defined as the *entropy loss* and it measures the amount of min-entropy "sacrificed" in order to achieve good randomness extraction. Thus, no application can tell apart the "extracted" randomness $H(X)$ from uniform randomness U_v, with advantage greater than $\varepsilon \stackrel{\mathsf{def}}{=} \sqrt{2^{-L}}$, even if the seed H is published (as long as H is independent of X).

The LHL is extremely attractive for many reasons. First, and foremost, it leads to simple and efficient randomness extractors, and can be used in many applications requiring good secret randomness. One such major setting is that of cryptographic key derivation, which is needed in many situations, such as privacy amplification [4], Diffie-Hellman key exchange [14,25], biometrics [11,5] and random number generators from physical sources of randomness [3,2]. Second, many simple functions, such as the inner product or, more generally, matrix-vector multiplication, are universal. Such elegant functions have nice algebraic properties which can be used for other reasons beyond randomness extraction (for a few examples, see [26,10,29]). Third, many simple and efficient constructions of (almost) universal hash functions are known [7,37,30,24], making LHL-based extractors the most efficient extractors to date. Finally, LHL achieves the optimal value of the entropy loss $L = m - v$ sufficient to achieve the desired statistical distance ε. Specifically, LHL achieves $L = 2\log(1/\varepsilon)$, which is known to be the the smallest possible entropy loss for *any* extractor [34].

Despite these extremely attractive properties, LHL-based extractors are not necessarily applicable or sufficient in various situations. This is primarily due to the following two limitations of the LHL: large entropy loss and large seed.

Large Entropy Loss. In theory, the entropy loss of $2\log(1/\varepsilon)$ might appear quite insignificant, especially in the asymptotic sense. However, in practical situations it often becomes a deal-breaker, especially when applied to the setting of key derivation. In this case the main question is to determine the smallest min-entropy value m sufficient to extract a v-bit key with security ε. Minimizing this value m, which we call *startup entropy*, is often of critical importance, especially in entropy constrained scenarios, such as Diffie-Hellman key exchange (especially on elliptic curves) or biometrics. For example, for the Diffie-Hellman key exchange, the value m corresponds to the size of the elliptic curve group, which directly affects efficiency. This is one of the reasons why statistical extractors are often replaced in practice with heuristic constructions based on cryptographic hash functions.

LARGE SEED. Another significant hurdle in the use of LHL comes from the fact that universal hash functions require long description, which means that LHL-based extractors have long seeds. Indeed, Stinson [37] showed that (perfectly) universal hash functions require the length of the seed to be linear in the length of the source X. More generally, even "good-enough" *almost* universal hash functions for LHL require seeds of length at least $\min(|X| - v, v + 2\log(1/\varepsilon)) - O(1)$ [37], and, thus, must grow with the number of extracted bits. This large seed length makes it inconvenient in many applications of extractors (e.g., [6,35,25]), including any use of extractors for derandomization, where one must be able to enumerate over all the seeds efficiently.

Large (and variable-length) seeds are also inconvenient for standardized cryptographic applications where fixed-size keys, independent of the size of inputs, are favored (as in the case of block ciphers or cryptographic hash functions). When extractors are used in cryptographic settings, seeds are viewed as keys and hence fixed-size seeds are very desirable. In applications of extractors, where the attacker is assumed to be sufficiently limited as to not make the source X dependent on the seed (e.g., when extracting keys from biometrics, physical measurements or in the Diffie-Hellman key exchange), one might consider fixing a good *public* seed, and use it repeatedly with a fast provably secure extractor. As said, this is not possible with universal hash functions as their seed length must grow with the length of X.[1]

OUR RESULTS. Quite surprisingly, we show that both limitations of the LHL — large entropy loss and large seed — can be overcome or, at least, mitigated in various important scenarios. We describe these results below.

1.1 Reducing the Entropy Loss

At first, reducing the entropy loss L might seem impossible since we already mentioned that any extractor must have entropy loss $L \geq 2\log(1/\varepsilon) - O(1)$ [34]. However, the impossibility is for general applications of extractors, where we must ensure that the extracted string R cannot be distinguished from random by *any* statistical test D. In contrast, when extractors are used to derive *cryptographic keys*, we only care about *limited types* of statistical tests D. Concretely, the tests that correspond to the security game between the attacker A and the challenger C. For example, when deriving the key for a signature scheme, the only tests we care about correspond to the attacker seeing several signatures and then outputting a

[1] In theory one can build (non-LHL-based) extractors where the length n of the seed H is roughly logarithmic in the length of the source X (see [16,36] and many references therein). However, the resulting constructions are mainly of theoretical value and lose the extreme simplicity and efficiency of LHL-based extractors.

new signature. Namely, we only care that the probability of a successful forgery does not suddenly become non-negligible when the secret key is obtained using an extractor instead of being random. And since the signature scheme is assumed to be secure with a truly random key, we can restrict our attention to a very restricted class of statistical tests which almost never output 1. Similar restrictions on the distinguisher naturally arise for other cryptographic primitives, which gives us hope that the lower bound of [34] might be overcome in such settings.

Generalized LHL and Applications. Indeed, we derive a tighter form of the LHL, called *generalized LHL* (see Theorem 1), which non-trivially depends on the type of distinguisher D we care about. Our improved bound contains a novel term informally measuring the *standard deviation* of the distinguisher's advantage (the standard LHL is a particular case where this term is bounded by 1). Applying this new bound to the analysis of cryptographic functions, we obtain much tighter bounds for the security of a wide class cryptographic applications. These include key derivation for *all* "unpredictability" applications, such as signatures, MACs, one-way functions, identification schemes, etc. More surprisingly, they also include key derivation for some prominent "indistinguishability" applications that include all stateless encryption schemes, both CPA- and CCA-secure and in the public- and symmetric-key settings, as well as weak pseudorandom functions. Specifically, in each of these cases, denote by ε the security of the cryptographic primitive (i.e., the best success probability or advantage of an attacker with certain resources) when keyed with a perfectly random v-bit key, and by ε' the corresponding security value when the key is derived from an imperfect m-bit entropy source via the LHL. We show (recall that $L = m - v$ represents the entropy loss):

$$\varepsilon' \leq \varepsilon + \sqrt{\varepsilon 2^{v-m}} = \varepsilon + \sqrt{\varepsilon 2^{-L}} \tag{1}$$

Comparing with Standard LHL. Let us first compare this bound with the regular $\varepsilon + \sqrt{2^{-L}}$ LHL bound. The latter required $L \geq 2\log(1/\varepsilon)$ to achieve the same type of security $O(\varepsilon)$ as with the ideal randomness. Using our improved bound, we show that only half of that amount, $L = \log(1/\varepsilon)$, already suffices. In fact, not only do we get improved bounds on the entropy loss L, but we also get meaningful security bounds for arbitrary values of L, even negative ones (when the entropy loss becomes *"entropy gain"*)! E.g., standard LHL does not give anything for $L \leq 0$, while we achieve significant $\varepsilon' \approx \sqrt{\varepsilon}$ security for $L = 0$ (no entropy loss!), and even start to "gracefully borrow" security from our application when we extract more bits than the min-entropy of the source, up to $L = -\log(1/\varepsilon)$ (i.e., $v = m + \log(1/\varepsilon)$).

Computational Extractor with Improved Loss. Although our improved bound, as stated, is not applicable to all cryptographic applications (the most important omission being pseudorandom functions and stream ciphers), in Section 3.2 we use our results to build *general-purpose* key derivation function for *any* (computationally-secure) cryptographic application, while providing the full entropy benefits derived from Equation (1). The scheme combines any LHL-based extractor with any (weak) pseudorandom function family.

1.2 Reducing the Seed Length

EXPAND-THEN-EXTRACT. A natural idea to reduce the seed length is to use a *pseudorandom generator* (PRG) to expand a short "input seed" S into a longer "output seed" S', and then use the resulting S' as the seed required by the LHL, or, more generally, by any randomness extractor. Let us call this natural approach *expand-then-extract.* Of course, as long as one hides the short S and uses the long S' as the public seed for the extractor, the extracted bits are pseudorandom. But is it possible to ensure the pseudorandomness of the extractor's output if the actual short seed S is made *public*? Had this been the case, we would obtain efficient LHL-based extractors with a short seed and, moreover, an extractor whose seed length is independent of the length of the input, as desired for the practical scenarios discussed earlier.

COUNTER-EXAMPLE. In Section 4.1 we show that the expand-then-extract approach will *not* work in general. We construct a simple PRG (which is secure under the *Decisional Diffie-Hellman* (DDH) assumption) and an extractor (which is a natural, perfectly universal hash function), where the output of the extractor — on any (e.g., even uniform) distribution — can be efficiently distinguished from random with probability close to 1, when given the short seed S used to generate the pseudorandom long seed S' for the extractor. Despite the above, we also show two positive results which nicely complement our counter-example.

EXTRACTING FEW BITS. First, in Section 4.2 we show that the expand-then-extract approach always works provided *the number of extracted bits v is "small"*. Here "small" means logarithmic in the security level of the PRG, which could range from $O(\log k)$ to $\Omega(k)$ (where k is the security parameter), depending on whether PRG is assumed to be polynomially or exponentially hard. Quite interestingly, in this case we do not even have to settle for pseudorandom bits: our small number v of extracted bits is actually *statistically* random, as long as the PRG is secure against circuits whose size is exponential in v. The intuition for this result comes from the fact that we can test, in time exponential in v, whether a given n-bit extractor seed s' is "good" or "bad" for our source X. We also know that most *random* long seeds $s' \leftarrow U_n$ must be good. Hence, by the PRG security, the same must be true for "most" *pseudorandom* seeds $s' \leftarrow \mathsf{Prg}(U_k)$, which is precisely what we need to show.

SECURITY IN minicrypt. Second, although our original counterexample is fairly simple and natural, it involves an assumption (DDH) from the "public-key world". In Section 4.3, we show, somewhat surprisingly, that such "public-key" type assumption is indeed *necessary* for any counter-example. We do this by showing that the expand-then-extract approach is sound in minicrypt [22] (i.e. in a hypothetical world where pseudorandom generators exist, but public-key cryptography does not). In particular, we construct a simple 2-message protocol (built from a PRG and an extractor) which constitutes a secure key-agreement protocol for any PRG/extractor combination for which the expand-then-extract approach is *insecure*. Since our protocol only has 2 messages, we even get semantically secure public-key encryption (PKE). Hence, since no such protocol/PKE exist in minicrypt, expand-then-extract must be secure.

Practical Interpretation. This leads to the following *practical interpretation* of our results indicating that using the expand-then-extract approach with common pseudorandom primitives, such as AES, is secure in spite of a lack of direct (reductionist) proof of security. Indeed, consider the expand-then-extract scheme implemented via AES (in some stream cipher mode). Our results show that, if this extraction scheme fails, then we have found *a public-key encryption scheme that is provable secure based on the security of AES as a block cipher!* Moreover, the resulting PKE has a very restrictive form, where the secret key is a PRG seed S, and the public-key is the PRG output $S' = \mathsf{Prg}(S)$. (E.g., in the case of AES, the public key is simply the evaluation of AES on several distinct points.) As we argue in Section 4.3, the existence of such a PKE appears to be extremely unlikely, and would be a major breakthrough given current state-of-the-art. Thus, our results give strong evidence that the expand-then-extract approach might be secure "in practice", — when used with "fast" ciphers (like AES), — despite being (generally) insecure "in theory"!

We also remark that all our results elegantly handle side information Z the attacker might have about our source X (as advocated by [11], such "average-case" extractors are very handy in cryptographic applications), and also generalize to the case of *almost* universal hash functions.

1.3 Related Work

Hast [17] also observed that for certain cryptographic applications, the relevant attackers correspond to restricted classes of distinguishers, which allowed him to obtain improved security bounds when the Goldreich-Levin hardcore bit [15] is used as a "computational" randomness extractor. This result is incomparable to ours. On the one hand, we consider general (multi-bit) LHL-based extractors and not just the single bit inner-product function (which is the form of the Goldreich-Levin predicate). On the other hand, Hast was working in the computational setting, and had to make an explicit reduction from the distinguisher to the predictor of the source X, which is not required in our setting.

We also mentioned the notion of *slice extractors* defined by Radhakrishnan and Ta-Shma [34], which limits the type of statistical tests to "rare distinguishers". To the best of our understanding, this definition was not motivated by applications, but rather was a convenient "parametrization" on a road to other results. Still, this notion roughly correspond to the setting of key derivation for authentication applications, when the attacker rarely succeeds. Interestingly, [34] showed a lower bound for the entropy loss of slice extractors (which was lower than that of general extractors), and matched this lower bound by an existential construction. As it turns out, our improved LHL immediately gives a *constructive* way to match this lower bound, showing that LHL-based extractors are optimal slice extractors in terms of the entropy loss. This connection is outlined in more detail in the full version of this paper [1].

In a very different (non-cryptographic) context of building hash tables, Mitzenmacher and Vahdan [27] also observed that improved bounds on the "entropy loss" could be obtained when the standard deviation of the "distinguisher" is much less than 1. In their setting the entropy loss was the minimum entropy required from the input stream to hash well, and the distinguisher was the characteristic function of a set of occupied buckets.

We note that our "win-win" result for the expand-then-extract approach is similar in spirit to several other "win-win" results [13,32,12,33], where a (hypothetical) adversary for one task is turned into a "surprising useful" protocol for a seemingly unrelated task. Among the above, the result most similar to ours is [13], where a PRG is used to expand the key for "forward secure storage", which is a concept related to "locally computable" extractors.

On the more practical side of our results, particularly in what refers to key derivation, it is worth mentioning the work of [9,25] that analyze constructions of key derivation functions (KDFs) based on cryptographic hash functions. These constructions do not use standard, generic assumptions, such as pseudorandomness, but build on specific modes of operations on their compression function f, and rely on dedicated, sometimes idealized, assumptions. Under such idealized assumptions, these schemes support situations where the KDF needs to be modeled as a random oracle, or where the source only has "unpredictability entropy" [21]. On the other hand, outside of the random oracle heuristics, much of the analysis of [9,25] studied sufficient conditions on compression function f and/or the source input distribution, under which cryptographic hash functions are "universal enough" so as to apply the standard LHL. As such, these analyses suffer the same drawbacks as any other LHL-based extractor. In particular, our results regarding the improved entropy loss for LHL-based extractors should carry over to improve the results of [9,25], while our results on the expand-then-extract approach could be viewed as partial justification of the heuristic where a fixed-description-length compression function is replaced by random in most (but not all) of the analyses of [9,25].

2 Standard Leftover Hash Lemma

NOTATION. For a set S, we let U_S denote the uniform distribution over S. For an integer $v \in \mathbb{N}$, we let U_v denote the uniform distribution over $\{0,1\}^v$, the bit-strings of length v. For a distribution or random variable X we write $x \leftarrow X$ to denote the operation of sampling a random x according to X. For a set S, we write $s \leftarrow S$ as shorthand for $s \leftarrow U_S$.

MIN-ENTROPY AND EXTRACTORS. The min-entropy of a random variable X is defined as $\mathbf{H}_\infty(X) \stackrel{\text{def}}{=} -\log(\max_x \Pr[X = x])$. In cryptographic applications, one often uses the average min-entropy of a random variable X conditioned on another random variable Z. This is defined as

$$\widetilde{\mathbf{H}}_\infty(X|Z) \stackrel{\text{def}}{=} -\log \mathbb{E}_{z\leftarrow Z}\left[\max_x \Pr[X=x|Z=z]\right] = -\log \mathbb{E}_{z\leftarrow Z}\left[2^{-\mathbf{H}_\infty(X|Z=z)}\right]$$

where $\mathbb{E}_{z\leftarrow Z}$ denotes the expected value over $z \leftarrow Z$, and measures the worst-case predictability of X by an adversary that may observe a correlated variable Z.

We denote with $\Delta_D(X,Y)$ the advantage of a circuit D in distinguishing the random variables X, Y: $\Delta_D(X,Y) \stackrel{\text{def}}{=} |\ \Pr[D(X)=1] - \Pr[D(Y)=1]\ |$. The *statistical distance* between two random variables X, Y is defined by

$$\mathsf{SD}(X,Y) \stackrel{\text{def}}{=} \frac{1}{2}\sum_x |\Pr[X=x] - \Pr[Y=x]| = \max_D \quad \Delta_D(X,Y)$$

where the maximum is taken over all (potentially computationally unbounded) D. Given side information Z, we write $\Delta_D(X,Y|Z)$ and $\mathsf{SD}(X,Y|Z)$ as shorthands for $\Delta_D((X,Z),(Y,Z))$ and $\mathsf{SD}((X,Z),(Y,Z))$, respectively.[2]

An extractor [31] can be used to extract uniform randomness out of a weakly-random value which is only assumed to have sufficient min-entropy. Our definition follows that of [11], which is defined in terms of conditional min-entropy.

Definition 1 (Extractors). *An efficient function* $\mathsf{Ext} : \mathcal{X} \times \{0,1\}^n \to \{0,1\}^v$ *is an (average-case, strong)* (m,ε)-extractor *(for space* $\mathcal{X}$*), if for all* X, Z *such that* X *is distributed over* $\mathcal{X}$ *and* $\widetilde{\mathbf{H}}_\infty(X|Z) \geq m$*, we get*

$$\mathsf{SD}(\ \mathsf{Ext}(X;S)\ ,\ U_v \mid (S,Z)) \leq \varepsilon$$

where $S \equiv U_n$ *denotes the coins of* Ext *(called the* seed*). The value* $L = m - v$ *is called the* entropy loss *of* Ext*, and the value* n *is called the* seed length *of* Ext.

Universal hashing and Leftover Hash lemma. We now recall the definition of universal-hashing [7,37] and the leftover-hash lemma [18], which states that universal hash functions are also good extractors.

Definition 2 (ρ-Universal Hashing). *A family* $\mathcal{H}$ *of (deterministic) functions* $h : \mathcal{X} \to \{0,1\}^v$ *is a called* ρ-universal hash family *(on space* $\mathcal{X}$*), if for any* $x_1 \neq x_2 \in \mathcal{X}$ *we have* $\Pr_{h\leftarrow\mathcal{H}}[h(x_1) = h(x_2)] \leq \rho$*. When* $\rho = 1/2^v$*, we say that* $\mathcal{H}$ *is* universal.

We can finally state the Leftover Hash Lemma (LHL). (Multiple versions of this lemma have appeared; we use the formulation of [38, Theorem 8.1], augmented by [11, Lemma 2.4] for the conditional entropy case; see [18] and references therein for earlier formulations.)

Lemma 1 (Leftover-Hash Lemma). *Assume that the family* $\mathcal{H}$ *of functions* $h : \mathcal{X} \to \{0,1\}^v$ *is a* $\frac{1+\gamma}{2^v}$*-universal hash family. Then the extractor* $\mathsf{Ext}(x;h) \stackrel{def}{=} h(x)$*, where* h *is uniform over* $\mathcal{H}$*, is an* (m,ε)*-extractor, where* $\varepsilon = \frac{1}{2}\cdot\sqrt{\gamma + \frac{2^v}{2^m}} = \frac{1}{2}\cdot\sqrt{\gamma + \frac{1}{2^L}}$ *(recall,* $L = m - v$ *is the entropy loss). In particular,* $\frac{1+3\varepsilon^2}{2^v}$*-universal hash functions yield* $(v + 2\log(1/\varepsilon), \varepsilon)$*-extractors.*

[2] Notice, $\Delta_D(X,Y|Z) \leq \mathbb{E}_{z\leftarrow Z}[\Delta_D(X|_{Z=z}, Y|_{Z=z})]$, but $\mathsf{SD}(X,Y|Z) = \max_D \mathbb{E}_{z\leftarrow Z}[\Delta_D(X|_{Z=z}, Y|_{Z=z})]$.

3 Reducing the Entropy Loss

As we mentioned, the entropy loss of $2\log(1/\varepsilon)$ is optimal when one is concerned with general distinguishers D [34]. As we show, in various cryptographic scenarios we only care about a somewhat restrictive class of distinguishers, which will allow us to reduce the entropy loss for such applications.

Below, we state a generalization of the LHL (Theorem 1), which will include a novel term measuring the *standard deviation* of the distinguisher's advantage, and then derive some useful special cases (Theorem 2). In Section 3.1, we then apply our tighter bound to derive improved entropy loss bounds for various cryptographic applications, including bounds for all authentication applications (Theorem 3) and some privacy applications,including chosen plaintext secure encryption (Theorem 4) and weak PRFs. In Section 3.2 we further extend our results to get a generic key derivation function with improved entropy loss for *any* computationally secure application, including stream ciphers and PRFs.

COLLISION PROBABILITY AND c-VARIANCE. Given a distribution Y, its collision probability is $\mathsf{Col}(Y) \stackrel{\text{def}}{=} \sum_y \Pr[Y = y]^2 \leq 2^{-\mathbf{H}_\infty(Y)}$. Given a joint distribution (Y, Z), we let $\mathsf{Col}(Y|Z) \stackrel{\text{def}}{=} \mathbb{E}_z[\mathsf{Col}(Y|Z = z)] \leq 2^{-\widetilde{\mathbf{H}}_\infty(Y|Z)}$. We also use the notation $(U_\mathcal{Y}, Z)$ to denote the probability distribution which first samples $(y, z) \leftarrow (Y, Z)$, but then replaces y by an independent, uniform sample from $U_\mathcal{Y}$. Finally, given a random variable W and a constant c, we define its *c-variance* as $\mathsf{Var}_c[W] \stackrel{\text{def}}{=} \mathbb{E}[(W - c)^2]$ and its *c-standard deviation* as $\sigma_c[W] \stackrel{\text{def}}{=} \sqrt{\mathsf{Var}_c[W]}$. When $c = \mathbb{E}[W]$, we recover the standard notion of variance and standard deviation, $\mathsf{Var}[W]$ and $\sigma[W]$, and notice that these are the smallest possible values for $\mathsf{Var}_c[W]$ and $\sigma_c[W]$. Still, we will later find it easier to use (slightly weaker) bounds obtained with specific values of $c \in \{0, \frac{1}{2}\}$.

We start with an useful lemma of independent interest which generalizes Lemma 4.5.1 of [27].

Lemma 2. *Assume (Y, Z) is a pair of correlated random variables distribution on a set $\mathcal{Y} \times \mathcal{Z}$. Then for any (deterministic) real-valued function $f : \mathcal{Y} \times \mathcal{Z} \to \mathbb{R}$ and any constant c, we have*

$$| \; \mathbb{E}[f(Y, Z)] - \mathbb{E}[f(U_\mathcal{Y}, Z)] \; | \;\; \leq \;\; \sigma_c[f(U_\mathcal{Y}, Z)] \cdot \sqrt{|\mathcal{Y}|\mathsf{Col}(Y|Z) - 1} \qquad (2)$$

The proof of this lemma can be found in the full version of this paper [1]. The useful feature of the bound given in Equation (2) comes from the fact that the value $\sigma_c[f(U_\mathcal{Y}, Z)]$ does not depend on the actual distribution Y used to replace the uniform distribution, while the value $\sqrt{|\mathcal{Y}|\mathsf{Col}(Y|Z) - 1}$ does not depend on the function f whose average we are trying to preserve (by using Y in place of $U_\mathcal{Y}$). In particular, we get the following corollary which will allow us to eventually get all our improved bounds.

Theorem 1 (Generalized LHL). *Let (X, Z) be some joint distribution over $\mathcal{X} \times \mathcal{Z}$, $\mathcal{H} = \{h : \mathcal{X} \to \{0, 1\}^v\}$ be a family of $\frac{1+\gamma}{2^v}$-universal hash functions, H be a random member of $\mathcal{H}$, and let $L \stackrel{\text{def}}{=} \widetilde{\mathbf{H}}_\infty(X|Z) - v$ be the entropy loss.*

Then, for any constant $c \in [0,1]$ and any (possibly probabilistic) distinguisher $D(r,h,z)$, we have

$$\Delta_D(H(X), U_v \mid (H,Z)) \le \sigma_c \left[\Pr_D[D(U_v,H,Z)=1]\right] \cdot \sqrt{\gamma + \frac{1}{2^L}} \tag{3}$$

The proof of the theorem can be found in the full version of this paper [1]. Equation (3) bounds the advantage of D in distinguishing real and extracted randomness using two terms. The second term (under the square root) depends on the universality of $\mathcal{H}$ and the entropy loss L (but not on D). The novel term is the c-standard deviation $\sigma_c[\Pr_D[D(U_v,H,Z)=1]]$ of D, which we will simply call c-standard deviation of D and denote $V(D,c)$. Intuitively, it measures how "concentrated" the expected output of D is to some value c in the "ideal setting" (when fed U_v rather than $H(X)$). We notice that for any D and any $c \in [0,1]$, the c-standard deviation $V(D,c) \le 1$. Plugging this trivial bound in Equation (3) removes the dependence on D, and (essentially)[3] gives us the statement of the standard LHL from Lemma 1. As mentioned, though, this forces the entropy loss to be at least $2\log(1/\varepsilon)$ to achieve security ε. Below, we show several special cases when we can upper bound $V(D,c)$ by roughly $\sqrt{\varepsilon}$, which means that the entropy loss L only needs to be roughly $\log(1/\varepsilon)$ (say, with perfectly universal $\mathcal{H}$) to achieve security ε.

Theorem 2. *Let (X,Z) be some joint distribution over $\mathcal{X} \times \mathcal{Z}$, $\mathcal{H} = \{h : \mathcal{X} \to \{0,1\}^v\}$ be a family of $\frac{1+\gamma}{2^v}$-universal hash functions, H be a random member of $\mathcal{H}$, $L \stackrel{def}{=} \widetilde{\mathbf{H}}_\infty(X|Z) - v$ be the entropy loss, and $D(r,h,z)$ be some (possibly probabilistic) distinguisher. Then, for each of the values ε defined in scenarios (a)-(c) below it holds:*

$$\Delta_D(H(X), U_v \mid (H,Z)) \le \sqrt{\varepsilon \cdot \left(\gamma + \frac{1}{2^L}\right)} \tag{4}$$

(a) Assume for some $c, \delta, \tau \in [0,1]$, $\varepsilon = \tau^2 + \delta$ and the following condition is satisfied:[4]

$$\Pr_{r \leftarrow U_v, h \leftarrow \mathcal{H}, z \leftarrow Z}[\ |\Pr_D[D(r,h,z)=1] - c| \ge \tau\ ] \le \delta \tag{5}$$

(b) Assume $\Pr[D(U_v,H,Z)=1] \le \varepsilon$ (where probability is taken over U_v, H, Z and the coins of D).

(c) For fixed r, h, and z, define the distinguisher $D'(r,h,z)$ as follows. First, make two independent samples $\tilde{d}, d \leftarrow D(r,h,z)$. Then, if $\tilde{d} = 1$, return d else return $(1-d)$. Assume further that $\Pr[D'(U_v,H,Z)=1] \le \frac{1}{2} + 2\epsilon$.

The proof of this Theorem is given in the full version [1]. In the following section, we demonstrate the use of Theorem 2 by concentrating on the important case of *key derivation* using LHL, where the value ε will essentially correspond to the "cryptographic security" of the application at hand.

[3] The exact bound claimed in Lemma 1 follows when $V(D,c) \le \frac{1}{2}$, which is true for $c = 1/2$.

[4] Note that the condition below implies $|\Pr[D(U_v,H,Z)=1] - c| \le \tau + \delta$.

3.1 Improved LHL for Key Derivation

Consider any cryptographic primitive P (e.g., signature, encryption, etc.), which uses randomness $R \in \{0,1\}^v$ to derive its secret (and, public, if needed) key(s). Without loss of generality, we can assume that R itself is the secret key. In the "ideal" setting, $R = U_v$ is perfectly uniform and independent from whatever side information Z available to the attacker. In the "real setting", the key owner has a randomness source, represented by a random variable X and possibly correlated with the attacker's side information Z. It then samples a universal hash function H using (fresh) *public* randomness and uses the extracted value $R = H(X)$ as its key. We would like to argue that if P is "ε-secure" in the ideal setting (against attackers with resources[5] less than T), then P is also "ε'-secure" in the real setting (against attackers with resources less than $T' \approx T$), where ε' is not much larger than ε. Of course, to have a hope of achieving this, $\mathcal{H}$ must be "universal-enough" and $L = \widetilde{\mathbf{H}}_\infty(X|Z) - v$ must "high-enough". To parameterize this, we will sometimes explicitly write (L, γ)-*real model* to denote the real model above, where $\mathcal{H}$ is $(1+\gamma)2^{-v}$-universal and $\widetilde{\mathbf{H}}_\infty(X|Z) \geq v + L$. We formalize this general setting as follows.

Abstract Security Games. We assume that the security of P is defined via an interactive game

between a probabilistic attacker $A(h,z)$ and a probabilistic challenger $C(r)$. Here one should think of h and z as particular values of the hash function and the side information, respectively, and r as a particular value used by the challenger in the key generation algorithm of P. We note that C only uses the secret key r and does not depend on h and z. In the ideal setting, where $r \leftarrow U_v$, the attacker A does not use the values h and z (and anyway the optimal values of h and z can be hardwired into A in the non-uniform model), yet, for notation convenience, we will still pass h and z to A even in the ideal setting.

At the end of the game, $C(r)$ outputs a bit b, where $b = 1$ indicates that the attacker "won the game". Since C is fixed by the definition of P (e.g., C runs the unforgeability game for signature or the semantic security game for encryption, etc.), we denote by $D_A(r,h,z)$ the (abstract) distinguisher which simulates the entire game between $A(h,z)$ and $C(r)$ and outputs the bit b, and by $\mathsf{Win}_A(r,h,z) = \Pr[D_A(r,h,z) = 1]$ the probability that $A(h,z)$ wins the game against $C(r)$. With this notation, the probability of winning the game in the "real setting" is given by the random variable $\mathsf{Win}_A(H(X),H,Z)$, and the same probability in the ideal setting becomes $\mathsf{Win}_A(U_v,H,Z)$. Moreover, the difference between these probabilities is simply the distinguishing advantage of D_A of telling apart real and extracted randomness when given H,Z:

$$|\mathsf{Win}_A(H(X),H,Z) - \mathsf{Win}_A(U_v,H,Z)| = \Delta_{D_A}(H(X),U_v \mid (H,Z)) \qquad (6)$$

As we justify next, to argue the security of P in the real setting assuming its security in the ideal setting, it is sufficient for us to argue that the above

[5] We use the word "resource" to include all the efficiency measures we might care about, such as running time, circuit size, number of oracle queries, etc.

distinguishing advantage is "small" for all legal attackers A. And since the security of P will usually restrict the power of attackers A (hence, also the power of abstract distinguishers D_A), we may use the results of Theorem 2 to argue better bounds on the entropy loss $L = \widetilde{\mathbf{H}}_\infty(X|Z) - v$.

Definition 3. *Let $c = 0$ for unpredictability applications P (signature, MAC, one-way function, etc.) and $c = \frac{1}{2}$ for indistinguishability applications P (encryption, pseudorandom function/permutation, etc.). Fix also the $(1+\gamma)2^{-v}$-universal hash family $\mathcal{H}$ and the joint distribution (X, Z) satisfying $\widetilde{\mathbf{H}}_\infty(X|Z) \geq v + L$, so that the real and the ideal model are well-defined.*

We say that P is (T, ε)-secure in the ideal model if for all attackers A with resources less than T, we have $\mathsf{Win}_A(U_v, H, Z) \leq c + \varepsilon$.

Similarly, P is (T', ε')-secure in the real model if for all attackers A have resources less than T', we have $\mathsf{Win}_A(H(X), H, Z) \leq c + \varepsilon'$.

Triangle inequality coupled with Equation (6) immediately yields the following Corollary.

Lemma 3. *Fix L and γ defining the real and the ideal models. Assume P is (T, ε)-secure in the ideal model, and for all attackers A with resources less than T' (where $T' \leq T$) we have $\Delta_{D_A}(H(X), U_v \mid (H, Z)) \leq \delta$. Then P is $(T', \varepsilon + \delta)$-secure in the (L, γ)-real model.*

We are now ready to apply Lemma 3 and Theorem 2 to various cryptographic primitives P. Below, we let $c \in \{0, \frac{1}{2}\}$ be the constant governing the security of P (0 for unpredictability and 1/2 for indistinguishability applications). Due to space constraint, we leave the application of part (a) of Theorem 2 to so called "strongly secure" primitives, where (for some τ and δ) any attacker has advantage more than τ on at most δ the fraction of keys r, to the full version [1] of the paper, and move directly to other applications which use parts (b) and (c) of Theorem 2. We also give several concrete examples in the full version [1].

Improved Bound for Unpredictability Applications. Recall, authentication applications correspond to $c = 0$, and include signature schemes, MACs, one-way functions/permutations, etc. In this case (T, ε)-security in the ideal model implies that for any T-bounded attacker A, $\mathbb{E}[\mathsf{Win}_A(U_v, H, Z)] \leq \varepsilon$. Recalling the definition of the abstract distinguisher D_A, this is the same as $\Pr[D_A(U_v, H, Z) = 1] \leq \varepsilon$, which is precisely the pre-condition for part (b) of Theorem 2. Thus, combining Equation (4) with Lemma 3, we immediate get:

Theorem 3. *Fix L and γ defining the real and the ideal models. Assume authentication primitive P (corresponding to $c = 0$) is (T, ε)-secure in the ideal model. Then P is (T, ε')-secure in the (L, γ)-real model, where*

$$\varepsilon' \leq \varepsilon + \sqrt{\varepsilon\left(\gamma + \frac{1}{2^L}\right)} \tag{7}$$

In particular, if $\gamma = 0$ *and* $L = \log(1/\varepsilon)$*, then* $\varepsilon' \leq 2\varepsilon$*. Moreover, when* $\gamma = 0$*, the security bound is meaningful even for* negative *entropy "loss"* $0 \geq L > -\log(1/\varepsilon)$*, when one extracts more bits than the min-entropy* $\widetilde{\mathbf{H}}_\infty(X|Z)$ *and "borrows the security deficit" from the ideal security of* P*.*

Intuitively, Theorem 3 uses the fact that for authentication applications one only cares about distinguishers which almost never output 1, since the attacker almost never forges successfully.

Improved Bound for Some Indistinguishability Applications. We now move to the more difficult case of indistinguishability applications, where $c = 1/2$. In general, we do not expect to outperform the standard LHL, as illustrated by the one-time pad example. Quite surprisingly, we show that the for a class of applications, including chosen plaintext attack (CPA) secure encryption, one can still get improved bounds as compared to the standard LHL. Specifically, as long as the primitive P allows the attacker A to "test" its success before playing the actual "challenge" from C, we still get significant savings. We start by defining the general type of security games where our technique applies.

BIT-GUESSING GAMES. As usual the game is played by the attacker $A = A(h, z)$ and the challenger $C(r)$. The game can have an arbitrary structure, except the winning condition is determined as follows. At some point A asks C for a "challenge". C flips a random bit $b \in \{0, 1\}$, and sends A a value $e = e(b, r)$. The game continues in an arbitrary way and ends with A making a guess b'. A wins if $b = b'$.

So far, this still includes all standard indistinguishability games, including the one-time pad. The extra assumption we make is the following. For any valid attacker A having resources less than T' there exists another valid attacker A' (having somewhat larger resources $T \geq T'$) such that:

(1) The execution between A' and $C(r)$ defines four bits $b, b', \tilde{b}, \tilde{b}'$, such that the joint distribution of $(b, b', \tilde{b}, \tilde{b}')$ is the same as two *independent* tuples (b, b') obtained when A runs with $C(r)$.
(2) The bits b and b' are precisely the secret bit of C and the guess of A', so that A' wins iff $b = b'$.
(3) A' learns if $\tilde{b} = \tilde{b}'$ before outputting b'.

We will call such indistinguishability games (T', T)*-simulatable*. In the full version [1], we show a general result stating improved bounds on entropy loss for any (T', T)-simulatable application, and also show that CPA-secure (public- or symmetric-key) encryption schemes are simulatable, where, as expected, the "resources" T are roughly doubled compared to T'. (Intuitively, the attacker can run the challenger in "his head" and see if it won.) In particular, we get the following theorem as a corollary of this general result:

Theorem 4. *Fix* L *and* γ *defining the real and the ideal models, and set* $\varepsilon' = \varepsilon + \sqrt{\varepsilon(\gamma + 2^{-L})}$.

Assume P is public-key encryption scheme which is ε-secure, in the ideal model, against attackers with running time $2t + t_{enc}$, where t_{enc} is the runtime of the encryption process. Then P is ε'-secure, in the (L, γ)-real model, against attackers with running time t.

Similarly, assume P is a symmetric-key encryption scheme which is ε-secure, in the ideal model, against attackers with running time $2t + O(1)$ and making $2q + 1$ encryption queries. Then P is ε'-secure, in the (L, γ)-real model, against attackers with running time t and making q encryption queries.

Limitations and Extensions. Unfortunately, several other indistinguishability primitives, such as pseudorandom generators, functions or permutations, do not appear to be simulatable. The problem seems to be in verifying the winning condition (condition (3) of simulatability), since this has to be done with respect to the actual secret key r not known to the attacker. For PRFs (or PRPs), it is tempting to solve the problem by using an equivalent definition, where the attacker can learn the value of PRF at any point, but then, as a challenge, must distinguish the value of the PRF at an *un-queried* point from random. Although this variant allows the attacker to check the winning condition during the first "virtual" run, it now creates a different problem in that the challenge point during the second "actual" run might have been queried during the first run, making such an attacker A' invalid.

Interestingly, the above "fix" works for a useful relaxation of PRFs, known as *weak PRFs* (wPRFs). Here the attacker only gets to see the values of the PRF at several *random points*, and has to distinguish a new value from random. Assuming a large enough input domain, the probability of collision between the PRF values revealed in the first first run and challenged in the second run, is negligible, which allows to complete the (valid) simulation. Similarly, it works for a slightly stronger relaxation of PRFs, known as *random-challenge* PRFs. As with wPRFs, A gets as the challenge a real-or-random evaluations of the PRF at a random point, but can additionally query the PRF at arbitrary points different from the challenge point. In Section 3.2 we show that wPRFs are all we need to apply our results to a generic key derivation function.

3.2 A Generic Key Derivation Function

So far we have discussed the applications of our generalized Leftover Hash Lemma and Theorem 2 to the derivation of cryptographic keys in entropy-constraint environments for specific applications. Although our analysis covers many such applications, it does not cover all, including PRFs and stream ciphers. In this section we make a simple observation which allows us to overcome this limitation and design a *generic* key derivation function (KDFs) which is (computationally) secure for any cryptographic application while still enjoying the same entropy loss savings. The idea is to compose universal hash functions a *weak* PRF (wPRF), where the random input to the wPRF now becomes part of the extractor seed, and use the fact that wPRFs fall under the class of *simulatable* applications as defined in Section 3.1.

Specifically, we define the KDF on the basis of a $(1+\gamma)/2^v$-universal hash family $\mathcal{H}$ with v-bit outputs and a wPRF F taking a k-bit input w and a v-bit key r, and outputting a v-bit output $y = F_r(w)$, as follows. The public seed s for the KDF Ext is a pair $s = (h, w)$, where h is a random universal function from $\mathcal{H}$ and w is a random element in the domain of F. We then define $\mathsf{Ext}(x, (h, w)) \stackrel{\text{def}}{=} F_{h(x)}(w)$, i.e., the initially extracted value $h(x)$ is used as a wPRF key, which is then used to evaluate F on w.

We also notice that if one needs to extract multiple keys for several applications, we can simply use the output of our computational KDF as a seed of a regular PRG or PRF, since such applications are now "covered".

Remark 1. For the case of deriving multiple keys, as above, we notice that the wPRF step is not needed provided *all* the keys are for cryptographic applications covered by our technique (i.e., strongly secure, unpredictable, or simulatable primitives). Namely, in such a case we can directly use the initially extracted key $H(X)$ as a seed for the (regular) PRF/PRG to derive all the required keys. This allows us to avoid increasing the seed length by k bits, and saves one application of wPRF. The proof of this claim follows by a simple hybrid argument (which we omit). In general, though, the wPRF-based solution is preferable, as it adds considerable generality at a relatively moderate cost.

4 Reducing the Seed Length

In this section we study the soundness of the natural expand-then-extract approach described in the Introduction, showing our negative result in Section 4.1 and our two positive results in Section 4.2 and Section 4.3.

NEGLIGIBLE AND FRIENDS. We use k to denote a security parameter. A function $\mu : \mathbb{N} \rightarrow [0,1]$ is *negligible* if for any $c > 0$ there is a k_0 such that $\mu(k) \leq 1/k^c$ for all $k \geq k_0$. To the contrary, μ is *non-negligible* if for some $c > 0$ we have $\mu(k) \geq 1/k^c$ for infinitely many k. Throughout, $\mathsf{negl}(k)$ denotes a negligible function in k.

A function $\tau(\cdot) : \mathbb{N} \rightarrow [0,1]$ is *overwhelming* if $1 - \tau(\cdot)$ is negligible. A function $\phi : \mathbb{N} \rightarrow [0,1]$ is *noticeable* if for some $c > 0$ there is an k_0 such that $\phi(k) \geq 1/k^c$ for all $k \geq k_0$. Note that non-negligible is not the same as noticeable. For example, $\mu(k) \stackrel{\text{def}}{=} k \bmod 2$ is non-negligible but not noticeable.

COMPUTATIONAL EXTRACTORS AND PRGS. Recall that with $\Delta_D(X, Y) \stackrel{\text{def}}{=} |\Pr[D(X) = 1] - \Pr[D(Y) = 1]|$ we denote the advantage of a circuit D in distinguishing the random variables X and Y. Let $\mathcal{D}_t$ denote the class of all probabilistic circuits of size t. With $\mathsf{CD}_t(X, Y) = \max_D \Delta_D(X, Y)$ we denote the *computational distance* of X and Y, here the maximum is over $D \in \mathcal{D}_t$. When $t = \infty$ in unbounded, we recover the notion of statistical distance $\mathsf{SD}(X, Y)$. When $X = X_k$ and $Y = Y_k$ are families of distributions indexed by the security parameter k, we will say that X and Y are computationally indistinguishable, denoted $X \approx_c Y$, if for every polynomial $t(.)$, $\mathsf{CD}_{t(k)}(X_k, Y_k) = \mathsf{negl}(k)$.

Definition 4 (Computational Extractor). *We say that an efficient function* $\mathsf{Ext} : \mathcal{X} \times \{0,1\}^n \to \{0,1\}^v$ *is an (average-case, strong)* **computational** *m-extractor (for space $\mathcal{X}$), if for all* efficiently samplable X, Z *such that X is distributed over $\mathcal{X}$ and $\widetilde{\mathbf{H}}_\infty(X|Z) \geq m$ (here $\mathcal{X}, n, v, m$ are all indexed by a security parameter k)*

$$(\mathsf{Ext}(X; S), S, Z) \approx_c (U_v, S, Z)$$

Definition 5 (Pseudorandom Generator). *A length increasing function* $\mathsf{Prg} : \{0,1\}^k \to \{0,1\}^n$ *is a pseudorandom generator (PRG) if*[6] $\mathsf{Prg}(U_k) \approx_c U_n$. *We also say that* Prg *is (T, δ)-secure if* $\mathsf{CD}_T(\mathsf{Prg}(U_k), U_n) \leq \delta$.

Computational Extractors with Short Seeds? We are ready to formalize our main question: *Is it safe to expand an extractor seed using a PRG?*

Hypothesis 1. *[Expand-then-Extract] If* Ext *is an $(m(k), \varepsilon(k))$-extractor with seed length $n(k)$ where $\varepsilon(.)$ is negligible and* $\mathsf{Prg} : \{0,1\}^k \to \{0,1\}^{n(k)}$ *is a pseudorandom generator, then* Ext′ *defined as*

$$\mathsf{Ext}'(x; s) = \mathsf{Ext}(x; \mathsf{Prg}(s))$$

is a computational m-extractor.

4.1 Counter-Example: Expanding Seeds Is Insecure in General

In this section we show that, unfortunately, Hypothesis 1 is wrong in general.

Theorem 5 (Hypothesis 1 wrong assuming DDH). *Under the DDH assumption, there exists a pseudorandom generator* $\mathsf{Prg}(.)$ *and a strong extractor* $\mathsf{Ext}(.;.)$ *(which is a perfectly universal hash function) such that* $\mathsf{Ext}'(x; s) \stackrel{\mathsf{def}}{=} \mathsf{Ext}(x; \mathsf{Prg}(s))$ *can be efficiently distinguished from uniform on any input distribution (i.e.* Ext′ *is not a computational extractor.)*

Proof (of Theorem 5). Let $\mathcal{G}$ be a prime order p cyclic group with generator g where the DDH problem is hard. Then $\mathsf{Prg} : \mathbb{Z}_p^3 \to \mathcal{G}^6$ defined as

$$\mathsf{Prg}(a, b, c) = (g^a, g^b, g^{ab}, g^{ac}, g^{bc}, g^{abc})$$

is a a secure pseudorandom generator [28]. Let $\mathsf{Ext} : \mathbb{Z}_p^3 \times \mathcal{G}^6 \to \mathcal{G}^2$ be

$$\mathsf{Ext}((x, y, z); (A, B, C, D, E, F)) = (A^x B^y C^z, D^x E^y F^z)$$

It is easy to see that Ext is a perfectly universal hash function from $\mathbb{Z}_p^3 \to \mathcal{G}^2$ (and, by Lemma 1, strong $(2\log p + 2\log(1/\varepsilon), \varepsilon)$-extractor). Now consider the distribution

$$[\mathsf{Ext}((x, y, z); \mathsf{Prg}(a, b, c)) \ , \ (a, b, c)] = [(g^{ax} g^{by} g^{abz}, g^{acx} g^{bcy} g^{abcz}) \ , \ (a, b, c)] \quad (8)$$

The distribution (8) is not pseudorandom as any tuple $(\alpha, \beta), (a, b, c) \in \mathcal{G}^2 \times \mathbb{Z}_p^3$ of the form (8) satisfies $\alpha^c = \beta$, which can be efficiently verified, while a random distribution will satisfy this relation with probability $1/p$.

[6] In order to avoid an extra parameter, we simply assume wlog that the seed length of our PRG is equal to the security parameter k.

4.2 Expanding Seeds Is Safe When Extracting Few Bits

By the following theorem, the expand-then-extract Hypothesis does hold, if the pseudorandom generator Prg used for expansion is sufficiently strong. The required hardness depends exponentially on the output length v of the extractor. The proof of the following Theorem is given in the full version of this paper [1].

Theorem 6. *Assume* $\mathsf{Ext} : \mathcal{X} \times \{0,1\}^n \to \{0,1\}^v$ *is a* (m, ε)-extractor *with running time* t_{Ext}, *and* $\mathsf{Prg} : \{0,1\}^k \to \{0,1\}^n$ *is a* $(T, \sqrt{\varepsilon})$*-pseudorandom generator, for some*

$$T \in O\left(2^{2v}(n+v)t_{\mathsf{Ext}}/\varepsilon\right) \tag{9}$$

Then $\mathsf{Ext}'(x;s) \stackrel{\text{def}}{=} \mathsf{Ext}(x;\mathsf{Prg}(s))$ *is a* $(m, 4\sqrt{\varepsilon})$-extractor. *In particular, if the running time of* Ext *is polynomial in* k, *its error* $\varepsilon(k) = \mathsf{negl}(k)$, *its output size* $v = O(\log k)$, *and* Prg *is secure against polynomial (in* k*) size distinguishers, then* Ext' *is an* (m, ε')*-extractor for* $\varepsilon'(k) = \mathsf{negl}(k)$.

4.3 Expanding Seeds Is Safe in Minicrypt

Before we can state the main result of this section, we need a few more definitions.

Bit-Agreement. *Bit-agreement* is a protocol between two efficient parties, which we refer to as Alice and Bob. They get the security parameter k in unary (denoted 1^k) as a common input and can communicate over an authentic channel. Finally, Alice and Bob output a bit b_{A} and b_{B}, respectively. The protocol has *correlation* $\epsilon = \epsilon(k)$, if for all k, $\Pr[b_{\mathsf{A}} = b_{\mathsf{B}}] \geq (1+\epsilon(k))/2$. Furthermore, the protocol has security $\delta = \delta(k)$, if for every efficient adversary Eve, which can observe the whole communication C, and for all k, $\Pr[\mathsf{Eve}(1^k, C) = b_{\mathsf{B}}] \leq 1 - \delta(k)/2$.

Key-Agreement & PKE. If $\epsilon(\cdot)$ and $\delta(\cdot)$ are overwhelming then such a protocol achieves *key-agreement.* Using parallel repetition and privacy amplification, it is known [20,19] that any protocol which achieves bit-agreement with noticeable correlation $\epsilon(\cdot)$ and overwhelming security $\delta(\cdot)$ can be turned into a key-agreement protocol, without increasing the number of rounds. A 2-message key-agreement protocol is equivalent to public-key encryption (PKE). The proof of the following Theorem is given in the full version of this paper [1].

Theorem 7 (Hypothesis 1 holds in minicrypt). *If there exists a secure pseudorandom generator* Prg *and a strong extractor* Ext *where* $\mathsf{Ext}'(.;.) \stackrel{\text{def}}{=} \mathsf{Ext}(\mathsf{Prg}(.);.)$ *is* <u>*not*</u> *a computational extractor, then the protocol from Figure 1 is a two-message bit-agreement protocol with noticeable correlation and overwhelming security (and thus implies PKE).*

Remark 2. In the above theorem not being a secure computational extractor means that there exists an efficient <u>uniform</u> D that can distinguish $\mathsf{Ext}(\mathsf{Prg}(.);.)$ with <u>noticeable</u> advantage (in the security parameter k). If $\mathsf{Ext}(\mathsf{Prg}(.);.)$ is only insecure against non-uniform adversaries, then also the resulting protocol (which uses D) will be non-uniform. If the distinguisher D only has non-negligible advantage (i.e. only works for infinitely many, but not all, security parameters k), then

Alice		Bob
$s \leftarrow U_k;\ s' \leftarrow \mathsf{Prg}(s)$	$\xrightarrow{s'}$	$(x, z) \leftarrow (X, Z)$
		$b_{\mathsf{B}} \leftarrow U_1$
		if $b_{\mathsf{B}} = 0$ then $r \leftarrow \mathsf{Ext}(x; s')$
		else if $b_{\mathsf{B}} = 1$ then $r \leftarrow U_v$
	$\xleftarrow{r, z}$	
$b_{\mathsf{A}} \leftarrow D(r, s, z)$		

Fig. 1. A bit agreement protocol from any Prg, Ext that constitute a counterexample (via distinguisher D) to Hypothesis 1

also the protocol will work for infinitely many k. This issue is inherent in win-win type results where an adversary is turned into a "useful" protocol [13,32,12,33]. It roots in the fact that in cryptography we usually put weak requirements for adversaries to be considered efficient (can be non-uniform and only have non-negligible advantage), whereas we usually require from practical algorithms to be uniform and secure for all (sufficiently large) security parameters.

We obtain the following corollary whose proof is immediate from Figure 1.

Corollary 1. *Assume* Prg *is a secure pseudorandom generator. Assume further that there exists no public-key encryption scheme (with non-negligible gap between security and decryption correctness) with pseudorandom ciphertexts, whose secret key is the seed of* Prg *and whose public key is the output of* Prg*. Then the expand-then-extract hypothesis is true for* Prg.

Discussion. Corollary 1 reduces the soundness of the expand-then-extract approach to the impossibility of constructing public-key encryption from the given PRG in a very particular way, where the ciphertexts are pseudorandom and, more importantly, the key-generation algorithm samples a random s and sets $sk = s, pk = \mathsf{Prg}(s)$. To the best of our knowledge, this impossibility assumption seems very likely for "practical" PRGs, such as AES. For example, not only there is no black-box construction of PKE (or key agreement) from a PRG alone, as shown by Impagliazzo and Rudich [23], but, in fact, it is entirely consistent with current knowledge that these two tasks are separable, in the sense that there is some computational model/complexity class (e.g., perhaps some extension of BQP or $AM \cap coAM$) that is powerful enough to break all public key schemes, but not powerful enough to break AES. If this is the case, then the AES-bases expand and extract scheme is secure with respect to all efficient input distributions and distingiuishers (even those that are based on public key tools such as factoring, lattices etc.). Moreover, we do not know any black-box construction of PKE from PRG and *any other* "cryptomania" assumption (like non-interactive zero-knowledge proofs, fully-homomorphic or identity-based encryption, etc.), where the public key of the PKE is simply the output of the PRG. To summarize, our results give strong evidence that the expand-then-extract approach is secure using any practical PRG (like AES), despite being (generally) insecure in theory.

Acknowledgements. We would like to thank Russell Impagliazzo and Ronen Shaltiel for useful discussions.

References

1. Barak, B., Dodis, Y., Krawczyk, H., Pereira, O., Pietrzak, K., Standaert, F.-X., Yu, Y.: Leftover hash lemma, revisited. Cryptology ePrint Archive, Report 2011/088 (2011), `http://eprint.iacr.org/`
2. Barak, B., Halevi, S.: A model and architecture for pseudo-random generation with applications to /dev/random. In: ACM CCS (2005)
3. Barak, B., Shaltiel, R., Tromer, E.: True Random Number Generators Secure in a Changing Environment. In: Walter, C.D., Koç, Ç.K., Paar, C. (eds.) CHES 2003. LNCS, vol. 2779, pp. 166–180. Springer, Heidelberg (2003)
4. Bennett, C.H., Brassard, G., Robert, J.-M.: Privacy amplification by public discussion. SIAM Journal on Computing 17(2), 210–229 (1988)
5. Boyen, X., Dodis, Y., Katz, J., Ostrovsky, R., Smith, A.: Secure Remote Authentication Using Biometric Data. In: Cramer, R. (ed.) EUROCRYPT 2005. LNCS, vol. 3494, pp. 147–163. Springer, Heidelberg (2005)
6. Canetti, R., Dodis, Y., Halevi, S., Kushilevitz, E., Sahai, A.: Exposure-resilient functions and all-or-nothing transforms. In: Preneel, B. (ed.) EUROCRYPT 2000. LNCS, vol. 1807, p. 453. Springer, Heidelberg (2000)
7. Carter, J.L., Wegman, M.N.: Universal classes of hash functions. Journal of Computer and System Sciences 18, 143–154 (1979)
8. Chevalier, C., Fouque, P.-A., Pointcheval, D., Zimmer, S.: Optimal randomness extraction from a diffie-hellman element. In: Joux, A. (ed.) EUROCRYPT 2009. LNCS, vol. 5479, Springer, Heidelberg (2009)
9. Dodis, Y., Gennaro, R., Håstad, J., Krawczyk, H., Rabin, T.: Randomness extraction and key derivation using the cbc, cascade and hmac modes. In: Franklin, M. (ed.) CRYPTO 2004. LNCS, vol. 3152, Springer, Heidelberg (2004)
10. Dodis, Y., Katz, J., Reyzin, L., Smith, A.: Robust fuzzy extractors and authenticated key agreement from close secrets. In: Dwork, C. (ed.) CRYPTO 2006. LNCS, vol. 4117, Springer, Heidelberg (2006)
11. Dodis, Y., Ostrovsky, R., Reyzin, L., Smith, A.: Fuzzy extractors: How to generate strong keys from biometrics and other noisy data. SIAM Journal on Computing 38(1), 97–139 (2008)
12. Dubrov, B., Ishai, Y.: On the randomness complexity of efficient sampling. In: STOC (2006)
13. Dziembowski, S.: On Forward-Secure Storage. In: Dwork, C. (ed.) CRYPTO 2006. LNCS, vol. 4117, pp. 251–270. Springer, Heidelberg (2006)
14. Gennaro, R., Krawczyk, H., Rabin, T.: Secure hashed diffie-hellman over non-ddh groups. In: Cachin, C., Camenisch, J.L. (eds.) EUROCRYPT 2004. LNCS, vol. 3027, Springer, Heidelberg (2004)
15. Goldreich, O., Levin, L.: A hard-core predicate for all one-way functions. In: STOC (1989)
16. Guruswami, V., Umans, C., Vadhan, S.: Unbalanced expanders and randomness extractors from parvaresh–vardy codes. J. ACM 56(4) (2009)
17. Hast, G.: Nearly one-sided tests and the goldreich?levin predicate. J. Cryptology 17(3), 209–229 (2004)

18. Håstad, J., Impagliazzo, R., Levin, L.A., Luby, M.: Construction of pseudorandom generator from any one-way function. SIAM Journal on Computing 28(4), 1364–1396 (1999)
19. Holenstein, T.: Key agreement from weak bit agreement. In: STOC (2005)
20. Holenstein, T.: Strengthening Key Agreement using Hard-Core Sets. PhD thesis, ETH Zurich, Zurich, Switzerland (2006)
21. Hsiao, C.-Y., Reyzin, L.: Finding Collisions on a Public Road, or Do Secure Hash Functions Need Secret Coins? In: Franklin, M. (ed.) CRYPTO 2004. LNCS, vol. 3152, pp. 92–105. Springer, Heidelberg (2004)
22. Impagliazzo, R.: A personal view of average-case complexity. In: Structure in Complexity Theory Conference, pp. 134–147 (1995)
23. Impagliazzo, R., Rudich, S.: Limits on the provable consequences of one-way permutations. In: STOC (1989)
24. Ishai, Y., Kushilevitz, E., Ostrovsky, R., Sahai, A.: Cryptography with constant computational overhead. In: STOC (2008)
25. Krawczyk, H.: Cryptographic Extraction and Key Derivation: The HKDF Scheme. In: Rabin, T. (ed.) CRYPTO 2010. LNCS, vol. 6223, pp. 631–648. Springer, Heidelberg (2010)
26. Maurer, U., Wolf, S.: Privacy amplification secure against active adversaries. In: Kaliski Jr., B.S. (ed.) CRYPTO 1997. LNCS, vol. 1294, Springer, Heidelberg (1997)
27. Mitzenmacher, M., Vadhan, S.P.: Why simple hash functions work: exploiting the entropy in a data stream. In: SODA (2008)
28. Naor, M., Reingold, O.: Number-theoretic constructions of efficient pseudo-random functions. In: FOCS (1997)
29. Naor, M., Segev, G.: Public-key cryptosystems resilient to key leakage. In: Halevi, S. (ed.) CRYPTO 2009. LNCS, vol. 5677, Springer, Heidelberg (2009)
30. Nevelsteen, W., Preneel, B.: Software Performance of Universal Hash Functions. In: Stern, J. (ed.) EUROCRYPT 1999. LNCS, vol. 1592, p. 24. Springer, Heidelberg (1999)
31. Nisan, N., Zuckerman, D.: Randomness is linear in space. Journal of Computer and System Sciences 52(1), 43–53 (1996)
32. Pietrzak, K.: Composition implies adaptive security in minicrypt. In: Vaudenay, S. (ed.) EUROCRYPT 2006. LNCS, vol. 4004, Springer, Heidelberg (2006)
33. Pietrzak, K., Sjödin, J.: Weak pseudorandom functions in minicrypt. In: Aceto, L., Damgård, I., Goldberg, L.A., Halldórsson, M.M., Ingólfsdóttir, A., Walukiewicz, I. (eds.) ICALP 2008, Part II. LNCS, vol. 5126, pp. 423–436. Springer, Heidelberg (2008)
34. Radhakrishnan, J., Ta-Shma, A.: Bounds for dispersers, extractors, and depth-two superconcentrators. SIAM Journal on Computing 13(1), 2–24 (2000)
35. Renner, R., Wolf, S.: Unconditional authenticity and privacy from an arbitrarily weak secret. In: Boneh, D. (ed.) CRYPTO 2003. LNCS, vol. 2729, Springer, Heidelberg (2003)
36. Shaltiel, R.: Recent developments in explicit constructions of extractors. Bulletin of the EATCS 77, 67–95 (2002)
37. Stinson, D.R.: Universal hashing and authentication codes. Designs, Codes, and Cryptography 4(4), 369–380 (1994)
38. Stinson, D.R.: Universal hash families and the leftover hash lemma, and applications to cryptography and computing. Journal of Combinatorial Mathematics and Combinatorial Computing 42, 3–31 (2002), `http://www.cacr.math.uwaterloo.ca/~dstinson/publist.html`

Random Oracle Reducibility

Paul Baecher and Marc Fischlin

Darmstadt University of Technology, Germany
www.minicrypt.de

Abstract. We discuss a reduction notion relating the random oracles in two cryptographic schemes A and B. Basically, the random oracle of scheme B reduces to the one of scheme A if *any* hash function instantiation of the random oracle (possibly still oracle based) which makes A secure also makes B secure. In a sense, instantiating the random oracle in scheme B is thus not more demanding than the one for scheme A. If, in addition, the standard cryptographic assumptions for scheme B are implied by the ones for scheme A, we can conclude that scheme B actually relies on weaker assumptions. Technically, such a conclusion cannot be made given only individual proofs in the random oracle model for each scheme.

The notion of random oracle reducibility immediately allows to transfer an uninstantiability result from an uninstantiable scheme B to a scheme A to which the random oracle reduces. We are nonetheless mainly interested in the other direction as a mean to establish hierarchically ordered random-oracle based schemes in terms of security assumptions. As a positive example, we consider the twin Diffie-Hellman (DH) encryption scheme of Cash et al. (Journal of Cryptology, 2009), which has been shown to be secure under the DH assumption in the random oracle scheme. It thus appears to improve over the related hashed ElGamal encryption scheme which relies on the random oracle model and the strong DH assumption where the adversary also gets access to a decisional DH oracle. As explained above, we complement this believe by showing that the random oracle in the twin DH scheme actually reduces to the one of the hashed ElGamal encryption scheme. We finally discuss further random oracle reductions between common signature schemes like GQ, PSS, and FDH.

Keywords: Random Oracle Model, Uninstantiability, Diffie Hellman, Encryption.

1 Introduction

Suppose you have a cryptographic scheme A which can be shown to be secure in the random oracle model [4] under some assumption $\mathbb{A}$, say, the RSA assumption. Assume furthermore that someone presents to you a scheme B for the same purpose which is also secure in the random oracle, but now under the potentially weaker assumption $\mathbb{B}$ like factoring. Clearly, if it was not for the random oracle, and scheme B would also improve over A in other relevant aspects like efficiency, then scheme B should be preferred. Unfortunately, the random oracle model introduces some undesirable uncertainty when simply following the strategy of picking the scheme with the weaker assumption.

P. Rogaway (Ed.): CRYPTO 2011, LNCS 6841, pp. 21–38, 2011.

Formally, proofs in the random oracle model (ROM) all rely on equally powerful random hash functions, but very often the *exact* requirements for the hash functions to conduct a security proof for a scheme remain unclear. This is all the more true since the random oracle in some schemes is uninstantiable in the sense that no efficient hash function can securely replace the random oracle [9]. For our example of schemes A and B above this means that scheme B may rely on a weaker assumption $\mathbb{B}$, but the actual requirements on the hash function may be much stronger than the ones for A. In the extreme, the hash function in scheme B may be uninstantiable, whereas the hash function for A may rely on a very mild cryptographic assumption like collision-resistance (albeit no proof has been found for this so far).

A natural approach to overcome the problem would be to determine the exact requirements on the hash function and to show that scheme B also relies on weaker assumptions for the hash function than scheme A. However, pinning down these properties of random oracles is often tedious and does not yield the desired result, especially since one would also need to show that the properties are necessary. One example are the hash function properties for OAEP, where Boldyreva and Fischlin [6,7] and later Kiltz et al. [19] gave necessary and, for much weaker security notions than IND-CCA, sufficient conditions on the hash function (in combination with further assumption about the underlying trapdoor permutation). None of these results, however, shows the desired kind of strong security. To complement these results, Kiltz and Pietrzak [20] claimed that for arbitrary trapdoor permutations the hash function in OAEP cannot be instantiated securely to derive IND-CCA security. The latter result is not known to be applicable to specific trapdoor permutations like RSA, though.

Random Oracle Reducibility. The strategy we suggest here is based on the classical reductionist approach to relate cryptographic assumptions: Show that any hash function H, ranging from efficient instantiations to random oracles, which makes scheme A secure under assumptions $\mathbb{A}$ also makes scheme B secure under assumptions $\mathbb{B}$. Technically, this seems to be too optimistic because hash functions in different schemes often cannot be used unchanged but rely on different domains, ranges etc. We thus allow for a "structural" transformation T^H of H for scheme B, possibly depending on the specific hash function. There are three possibilities to relate the hash functions in the schemes:

Definition 1 (Random Oracle Reducibility — Informally). *Let $\mathbb{A}$ and $\mathbb{B}$ be some sets of assumptions. A random oracle in scheme B strictly resp. strongly resp. weakly reduces to the random oracle in scheme A if for every hash function H there exists a transformation T such that*

$$\textit{strictly: } A^H \textit{ secure under } \mathbb{A} \Longrightarrow \textit{scheme } B^{T^H} \textit{ secure under } \mathbb{B}$$

$$\textit{strongly: } A^H \textit{ secure under } \mathbb{A} \Longrightarrow \begin{cases} \textit{scheme } B^{T^H} \textit{ secure under } \mathbb{A}\cup\mathbb{B} \\ \textit{and } B^{T^{H'}} \textit{ secure under } \mathbb{B} \textit{ for some } H' \end{cases}$$

$$\textit{weakly: } A^H \textit{ secure under } \mathbb{A} \Longrightarrow \textit{scheme } B^{T^H} \textit{ secure under } \mathbb{A}\cup\mathbb{B}$$

Several details are hidden in this informal definition, of course, e.g., what a "secure" scheme is, which properties the transformation must satisfy, or how cryptographic assumptions of hash function instantiations are dealt with. We fill in these details in the formal definition and keep it informally for now. We note, however, that the formal definition covers *any* type of hash function, i.e., both oracle-based ones, as well as keyed hash functions with succinct descriptions, or mixtures thereof. In particular, the security of scheme B in the strong case may only be known for a random oracle H'.

For identical assumptions $\mathbb{A} = \mathbb{B}$, or even if $\mathbb{A} \subseteq \mathbb{B}$, all three notions coincide. The difference can be best explained for the case $\mathbb{B} \subsetneq \mathbb{A}$, i.e., that the assumptions $\mathbb{A}$ are strictly stronger than $\mathbb{B}$. The strict notion can in this case be put informally as saying "Scheme B is strictly superior to scheme A in regard of the assumptions, even for the hash function." The strong and presumably more accessible approach can be described as "Scheme B is at least as good as scheme A in regard of the assumptions, but potentially superior." The weak case says that "Scheme B is at least as good as scheme A." In terms of security assumptions it seems that the strict and strong versions are the interesting ones (hence the names); the weak version does not provide any potential improvement concerning the assumption. We note that in the strong case often a security proof for scheme B in the random oracle model can be given without assuming security of A. We merely introduced the dependence via the prerequisite of the implication to make the notions comparable.

As a first sanity check note how previous uninstantiability results relate to either kind of definition. If the hash function security of B can be (weakly, strongly, or strictly) reduced to the hash function security of A and B turns out to be uninstantiable, then this also follows for scheme A (else T^H would be a valid instantiation for B). In this regard the reduction approach also allows to extend uninstantiability results without directly showing the ineffectiveness of efficient hash functions. Vice versa, any new result about secure instantiations of A would immediately transfer to B. Also, uninstantiability immediately implies that there are schemes A (allowing efficient instantiations) and B (being uninstantiable) such that the random oracle for B does not (even weakly) reduce to the one for scheme A.

Example: Hashed ElGamal Encryption. To show that the strong approach is applicable and the definition non-trivial we discuss the case of Hashed ElGamal encryption [1] and its chosen-ciphertext security proof under the strong Diffie-Hellman (DH) assumption in the random oracle model [12]. Here the strong DH assumption says that computing DH keys is infeasible even if given (restricted) access to a decisional DH oracle. Cash et al. [10] present a variant which can be shown to be CCA secure under the (regular) DH assumption in the random oracle model. This is a clear example of two schemes where the variant seems to improve over the original one in terms of assumption, but where this conclusion is technically not known to be sound because of the random oracle model.

The original hashed ElGamal encryption scheme encrypts a message under public key $X = g^x$ as (Y, c), where $Y = g^y$ and $c = \mathsf{Enc}(k, m)$ for the hashed Diffie-Hellman key $k = H(Y, X^y)$ and the symmetric encryption scheme

Enc. The variant in [10] instead computes two related ephemeral Diffie-Hellman keys from public keys $X_0 = g^{x_0}$ and $X_1 = g^{x_1}$, and derives a ciphertext (Y, c) for $Y = g^y$ and $c = \mathsf{Enc}(k, m)$ for $k = H(Y, X_0^y, X_1^y)$. We show (for a slight derivate of the scheme in [10]) that the random oracles can be strongly reduced to the one of hashed ElGamal. Ciphertexts in our variant are defined as

$$(Y, c, k_1), \quad \text{where } Y = g^y,\ c = \mathsf{Enc}(k_0, m) \text{ for } k_0 = H(Y, X_0^y),\ k_1 = H(Y, X_1^y),$$

i.e., we split the hashing into two evaluations, one for each public key part, and use the second key as a kind of confirmation that the first key is computed correctly. We can view this as the transformation

$$T^H(Y, Z_0, Z_1) = H(Y, Z_0)||H(Y, Z_1)$$

and where we use a special symmetric encryption scheme where the key part k_1 is output in clear.

We then prove that IND-CCA security for hashed ElGamal implies security of (our variant) of the twin DH scheme *for any hash function* H for the same assumptions that the hashed ElGamal is secure for. We also show that our variant is secure in the random oracle model assuming only the assumptions given in [10]. It follows that the random oracle in our scheme is strongly reducible to the one of hashed ElGamal.

Note that yet another hashed ElGamal scheme, related to the original scheme, has been shown to be uninstantiable [2]. The scheme differs in two important aspects from our scheme, though. First, their hashed ElGamal encryption does not use randomness and is thus deterministic. Second, the security notion considered in [2] is *IND-CCA-preservation* which gives the adversary simultaneously access to the algorithms of the public-key scheme and the symmetric scheme involving secret keys. In contrast, we use the standard notion of IND-CCA security for the hybrid (public-key) scheme.

We note that the security reduction for our variant to the underlying primitives like the Diffie-Hellman problem for random oracle H' is looser than the one in [10] in terms of concrete bounds.[1] At the same time our scheme relates the random oracle to the one in the original scheme. Of course, concreteness of security bounds is another important aspect, besides efficiency, when considering random oracle reducibility. In principle, it could be incorporated as an explicit requirement in the notion. We relinquish to do so because both aspects, tightness and efficiency, depend to some extend on the individual willingness to pay for the additional security guarantees through the random oracle reducibility.

Reductions for Signature Schemes. We give further examples of the applicability of the notion of random oracle reducibility by considering common signature schemes like Guillou-Quisquater [17] or PSS [5] and showing that the random oracle of a probabilistic version of FDH [11] (Full-Domain Hash) reduces to the random oracles in these schemes. However, note that FDH signatures are

[1] Note that both proofs are in the random oracle model where concrete bounds must be taken with a grain of salt anyway.

only known to be uninstantiable according to [14,13] for plain hash evaluations over the message (i.e., no randomness). The first result only applies to random trapdoor permutations (i.e., the result is not known to apply to RSA), and the second more recent result holds when RSA is treated as a black-box group. Any progress in terms of uninstantiability of FDH signatures to our probabilistic case would thus immediately allow to conclude that the Guillou-Quisquater signature scheme and the PSS scheme are uninstantiable. This would somehow extend the uninstantiability result of Goldwasser and Kalai [16] about general (and somewhat contrived) Fiat-Shamir schemes to the "more natural" species.

We discuss another random oracle reduction of (probabilistic) BLS signatures [8] to the Schnorr signature scheme [25]. In this case, however, we need a non-standard assumption to make the reduction work, namely, the knowledge of exponent assumption KEA1 [18,3] which roughly says that when complementing a value X to a Diffie-Hellman tuple $(X, Y, \mathrm{DH}(X, Y))$ one must know the discrete logarithm y of Y. For the random oracles this means that, if our version of the BLS scheme is uninstantiable, then so is the Schnorr signature scheme, or the KEA1 assumption is false.[2]

Some Words of Caution. Just as reductions between number-theoretic assumptions merely relate problems like factoring and RSA, but do not touch the question if RSA is really hard, a reduction for random oracles does not mean that scheme B, in and of itself, is secure (under assumptions $\mathbb{B}$) or that the hash function can be securely instantiated. The reduction only says that scheme B can be made as secure as scheme A in regard of the hash function. Since we do not put any formal prerequisite about the security of scheme A, which could thus be insecure, the reduction could potentially be trivial.

However, as for relating number-theoretic assumptions, where the stronger assumption is usually accompanied by some hardness analysis, scheme A typically comes with some form of security guarantee. Often, this is at least a security proof in the random oracle model, or sometimes for a relaxation thereof like non-programmable random oracles [24,15], traceable random oracles [23], or leaky random oracles [26]. The advantage of our approach is that it follows immediately that B can also be shown secure under the corresponding assumption about the hash function.

One caveat is that the transformed hash function T^H, unlike random oracles, obeys some structure, as the "split" evaluation in our ElGamal example. Hence, when instantiated with some efficient hash function h, scheme B could become insecure for the transformed hash function T^h, despite the reduction and a proof that T^H makes B secure for random oracle H. Noteworthy, at the same time B could be secure when instantiated with h directly, instead of going through the transformation T! We observe, however, that this is an inherent limitation of the random oracle model: It solely provides a heuristic which does not allow to conclude security under concrete instantiations. Our approach at least gives

[2] To be precise need the KEA1 assumption to hold even if one can get additional Schnorr signatures under the key X.

some confidence in the choice of the hash function in the sense that the security is at least as good as the one of another, hopefully well-examined scheme.

2 Random Oracle Reducibility

Hash Functions. We consider families $\mathcal{H}$ of hash functions H where it is understood that H is (not necessarily efficiently) sampleable from $\mathcal{H}$ according to a security parameter λ. It is thus also clear that a hash function H may have a restricted input or output length, depending on λ. We write $H \leftarrow \mathcal{H}(1^\lambda)$ for the sampling. For example, to model a random oracle we let $\mathcal{H}(1^\lambda)$ be the family of all functions with the specified domain and range and the sampling picks a random function from this set. In the sequel we usually simply identify the hash function $H(\cdot)$ with its description H itself. We assume that hash functions are deterministic in the sense that, once a hash function has been sampled, its behavior is fixed. A hash function family may rely on some cryptographic assumptions $\mathbb{H}$; in case of random oracles no assumption $\mathbb{H}$ is necessary as the sampling of a random function already provides all desirable security properties.

Given a hash function H for a scheme A we write A^H for the scheme where each party or algorithm gets oracle access to H. Furthermore, the hash function H may include a public description part which is then also given to all parties and algorithms as additional input. This public part may be for example the full description of H, or only parts thereof, e.g., if H is a hybrid between a random oracle and a keyed hash function. A hash function family is *efficient* if it follows the usual notion of an efficient keyed hash function, i.e., the sampling is efficient, a sampled function H is efficiently computable and entirely described through a public part.

Transformations. A hash function H used in a cryptographic scheme A may not be immediately applicable to another scheme B for the mere fact that the domain and range do not fit. We therefore "slot in" a transformation algorithm T in, such that scheme B then uses the hash function T^H (with the semantic that any algorithm or party gets public descriptions of T^H as additional input). We write $T^{\mathcal{H}}$ for the corresponding hash function family (described by sampling $H \leftarrow \mathcal{H}(1^\lambda)$ and evaluating T^H). Ideally, the transformation should only make structural modifications (like adapting the domain and range) and should be deterministic.

There is, however, one technical subtlety concerning the statefulness of transformations. Namely, we explicitly require that the transformation is stateless. The reason is that if we would allow stateful transformations, it is possible to construct particular contrived transformations that trivialize our notion, in the sense that some scheme B always reduces to any scheme A. To get a sense of the problem, consider arbitrary schemes A and B which are secure in the random oracle model. Let T denote the stateful transformation which ignores its oracle H and (efficiently) implements a random oracle via lazy sampling. Since B^{T^H} is then clearly secure, scheme B is —as per Definition 1— reducible to scheme

A, despite the arbitrary choice of the two schemes. A similar issue arises with instantiations: Suppose scheme B is now instantiable for some hash function family $\mathcal{H}$. Construct the transformation T which again ignores its oracle and instead initially samples a function $H' \leftarrow \mathcal{H}$ and answers subsequent queries according to H'. Again, scheme B remains secure and reduces to any scheme A. Both examples rely on a stateful transformation function —to answer consistently and to remember the choice of the hash function, respectively. Thus, in order to rule out such trivial cases, we ask that transformations are always stateless. This may seem overly restrictive at first glance, but, in fact, is easily justified because hash functions are inherently stateless entities.

One can nonetheless allow for rather general transformations, possibly even considering transformations which themselves rely on assumptions $\mathbb{T}$.

Security of Schemes. We consider security of schemes to be defined via a general notion of games, albeit our games also allow to state simulation-based security properties (by saying that the game returns 1 iff for any simulator there exists a successful distinguisher). As we can subsume several games like the ones for blindness and unforgeability for blind signature schemes into a single game, with corresponding sub games for which the adversary initially decides to mount the attack against, we consider a single game G only. We let $\mathsf{Adv}(\mathcal{A}, G)$ denote the *advantage* of adversary $\mathcal{A}$ playing game G, i.e., the adversary's success probability of winning the game. This makes Adv an implicit part of G. Here, in decisional games the advantage usually denotes the adversary's success probability minus the trivial guessing probability of $1/2$, and in computational games the advantage is usually the adversary's probability of computing a solution.

Analogously to hash functions, we write G^H for a security game in which all parties and algorithms get access to H in the same manner. It is understood that the choice of $H \leftarrow \mathcal{H}(1^\lambda)$ is part of the security game. We write $G^{\mathcal{H}}$ for a game for which a hash function is chosen from the family $\mathcal{H}$.

We envision security assumptions $\mathbb{A}$ for a scheme A as a set of elementary properties such as unforgeability of an underlying MAC or number-theoretic assumptions like the hardness of factoring. We can then apply common set operations and relations to assumptions in a well-defined way, e.g., $\mathbb{A} \cup \mathbb{B}$ comprises all assumptions stated in $\mathbb{A}$ and $\mathbb{B}$, and $\mathbb{B} \subseteq \mathbb{A}$ means that assumptions in $\mathbb{B}$ hold if $\mathbb{A}$ holds. This approach is too applicable for the hash function assumptions $\mathbb{H}$ and possibly the transformation assumptions $\mathbb{T}$. We also assume that assumptions are "opt-in", i.e., need to specified in the set, or else the assumption does not hold. Formally we can define this by considering a universe $\mathbb{U}$ of assumptions and say that any assumption in $\mathbb{U} \setminus \mathbb{A}$ is false.

Note that we keep the formal specifications of games and assumptions at a minimal level. This is possible as we later demand random oracle reducibility with respect to specific games and assumptions. It is thus up to the reduction statement to consider "reasonable" games and assumptions. We only need very limited syntactical requirements here and can, for example, even allow conflicting assumptions in $\mathbb{A}\cup\mathbb{B}$ (in which case, however, the claims usually become trivial).

Definition 2 (Game-based Security). *Let A denote a cryptographic scheme and $G_A^{\mathcal{H}}$ an associated security game for hash family $\mathcal{H}$. Scheme A is called* $G_A^{\mathcal{H}}$-secure under assumptions $\mathbb{A}$ *for hash family $\mathcal{H}$ relying on assumptions $\mathbb{H}$ if for any efficient adversary $\mathcal{A}$ we have that $\mathsf{Adv}(\mathcal{A}, G_A^{\mathcal{H}}) \approx 0$ is negligible in the security parameter, where the probability is over all random choices of the game (including the choice of the hash function), the algorithms, and the adversary.*

As an example consider the IND-CCA security game for an encryption scheme A (in the random oracle model), in which the game $G_A^{\mathcal{H}}$ proceeds in stages where $\mathcal{A}$ in the first phase receives a public key (in case of an asymmetric scheme) and gets access to a decryption oracle plus the random oracle, then outputs a pair of equal-length messages m_0, m_1 to receive a *single* challenge ciphertext of m_b for secret random bit b, and finally continues asking decryption queries except for the challenge ciphertext. The adversary wins if correctly predicts b, and the advantage of the adversary is the probability for a correct prediction minus 1/2. In the notation above an IND-CCA secure encryption scheme relying on some cryptographic assumption $\mathbb{A}$ is $G_A^{\mathcal{H}}$-secure under $\mathbb{A}$ for random oracle $\mathcal{H}$.

Random Oracle Reducibility. As explained in the introduction we introduce a weak, strong, and strict notion of random oracle reducibility:

Definition 3 (Random Oracle Reducibility). *Let A be a cryptographic scheme with security game G_A and assumptions $\mathbb{A}$, and let B be a cryptographic scheme with game G_B and assumptions $\mathbb{B}$. Then the random oracle in scheme B (strictly resp. strongly resp. weakly)* reduces *to the random oracle in scheme A if for every hash function family $\mathcal{H}$ relying on assumptions $\mathbb{H}$ there exists a stateless transformation T such that*

$$\text{strict: } A \text{ is } G_A^{\mathcal{H}}\text{-secure under } \mathbb{A} \Rightarrow B \text{ is } G_B^{T^{\mathcal{H}}}\text{-secure under } \mathbb{B}$$

$$\text{strong: } A \text{ is } G_A^{\mathcal{H}}\text{-secure under } \mathbb{A} \Rightarrow \begin{cases} B \text{ is } G_B^{T^{\mathcal{H}}}\text{-secure under } \mathbb{A} \cup \mathbb{B} \text{ and} \\ B \text{ is } G_B^{T^{\mathcal{H}'}}\text{-secure under } \mathbb{B} \text{ for some } \mathcal{H}' \\ \qquad\qquad\qquad\qquad\qquad \text{relying on } \mathbb{H}' \end{cases}$$

$$\text{weak: } A \text{ is } G_A^{\mathcal{H}}\text{-secure under } \mathbb{A} \Rightarrow B \text{ is } G_B^{T^{\mathcal{H}}}\text{-secure under } \mathbb{A} \cup \mathbb{B}$$

We say that $(B, G_B^{\mathcal{H}}, \mathbb{B})$ is (weakly or strongly or strictly) random oracle reducible *to $(A, G_A^{\mathcal{H}}, \mathbb{A})$. It is* polynomial-time (weakly or strongly or strictly) random oracle reducible *if it is random oracle reducible via (deterministic) stateless polynomial-time transformations T for any hash function family $\mathcal{H}$.*

We occasionally simply say that B is random oracle reducible (RO-reducible) to A if the games and assumptions are clear from the context.

Some remarks about the definition and variations follow:

– The above does not rule out trivial examples where scheme B actually relies on stronger assumptions $\mathbb{B}$ than scheme A, e.g., if $\mathbb{A}$ is a subset of $\mathbb{B}$. As explained in the introduction, the most interesting examples seem to be the ones where assumptions $\mathbb{B}$ are weaker than $\mathbb{A}$ or at least incomparable.

Occasionally, however, one may be interested in a scheme B which requires stronger assumptions $\mathbb{B}$ but which is more efficient (or has other desirable properties).

- We can devise stronger notions concerning the order of quantification for our reducibility notion. Above, the transformation can depend on the specific hash function family $\mathcal{H}$, and thus possibly specific properties of $\mathcal{H}$. One could alternatively demand that the transformation needs to be universal in the sense that it works for any $\mathcal{H}$.
- The above definition assumes that transformation T does not rely on additional assumptions. More generally, one could specify assumptions $\mathbb{T}$ and say that scheme B is secure under assumptions $\mathbb{B}' = \mathbb{B} \cup \mathbb{T}$.
- According to our syntax the adversary $\mathcal{B}$ in game G_B with the transformed random oracle would get access to T^H, but not H itself. This can be easily patched by letting the transformation T give direct access to H through a special query mode.

3 Basic Results

Relating the Reducibility Notions. We first show that strict reducibility implies strong reducibility which implies weak reducibility. The proof is rather syntactical and omitted for space reasons.

Proposition 1 (Strict $\Rightarrow$ Strong $\Rightarrow$ Weak Reducibility). *Let A be a cryptographic scheme with security game G_A and assumptions $\mathbb{A}$, and let B be a cryptographic schemes with game G_B and assumptions $\mathbb{B}$. If the random oracle in scheme B strictly reduces to the random oracle in scheme A, then it also strongly reduces to it. If it strongly reduces to it, then it also weakly reduces to it.*

We next discuss a scheme which supports a strong reduction, but not a strict one. Note that for $\mathbb{A} \subseteq \mathbb{B}$ this claim would be trivial because then the notions coincide. Instead, our separation example even holds for $\mathbb{B} \subsetneq \mathbb{A}$.

Proposition 2 (Strong $\not\Rightarrow$ Strict Reducibility). *There exists schemes A, B for games $G_A^{\mathcal{H}}$ and $G_B^{\mathcal{H}}$ and assumptions $\mathbb{A}, \mathbb{B}$ such that $\mathbb{B} \subsetneq \mathbb{A}$, and the random oracle of B strongly reduces to the one of scheme A, but not strictly.*

Proof. Let scheme A run two copies of Lamport's one-time signature scheme [21], one based on an alleged one-way function f, and the other one by using the given hash function (oracle). Verification checks if both signatures are valid. Let $G_A^{\mathcal{H}}$ be the standard unforgeability game for one-time signature schemes, and let $\mathbb{A}$ be the assumption that an underlying function f is really one-way. Let B and $G_B^{\mathcal{H}}$ be the same scheme and game, but let $\mathbb{B}$ be the empty set.

Consider the hash function family $\mathcal{H}$ which samples trivial functions $H : \{0,1\}^* \to \{0\}$ only and where $\mathbb{H}$ is empty. Then scheme A is still unforgeable if f is one-way, independently of the H-part of the signature. In contrast, B would be insecure under $\mathbb{B}$ and for the trivial hash function family, because, by assumption about the "minimalistic" approach for the set $\mathbb{B}$, the function f is not one-way then. Hence, the random oracle in B cannot be strictly reduced to the one in A.

Finally, note that for a hash function family $\mathcal{H}'$ which is one-way the signature scheme B becomes secure even under $\mathbb{B}$, because any forger would need to forge the one-time signature scheme for the hash function. At the same time, for any hash function family scheme B is secure under $\mathbb{A} \cup \mathbb{B}$. These two properties show that the random oracle in B strongly reduces to the one in A. □

For the next separation we further need to exclude contrived examples where the hash function assumptions $\mathbb{H}$ "makes up" for assumptions in $\mathbb{A} \setminus \mathbb{B}$ to make scheme B secure. We say that $\mathbb{H}$ is *non-interfering* with $\mathbb{A}$ and $\mathbb{B}$ iff $\mathbb{H} \cap (\mathbb{A} \setminus \mathbb{B}) = \emptyset$. In this case we say that the random oracle in scheme B reduces to the one in scheme A under non-interfering hash assumptions if reducibility holds for all hash function families $\mathcal{H}$ with non-interfering assumption $\mathbb{H}$.

Proposition 3 (Weak $\not\Rightarrow$ Strong Reducibility). *There exists schemes A, B for games $G_A^{\mathcal{H}}$ and $G_B^{\mathcal{H}}$ and assumptions $\mathbb{A}, \mathbb{B}$ such that $\mathbb{B} \subsetneq \mathbb{A}$, and the random oracle of B weakly reduces to the one of scheme A, but not strongly for non-interfering hash functions.*

Proof. Consider again Lamport's one-time signature scheme as scheme A, relying on a one-way function f (whose one-wayness is postulated in $\mathbb{A}$). The scheme ignores the hash function. Let $G_B^{\mathcal{H}}$ be again the unforgeability game for one-time signature schemes. Let B the same scheme with the same security game, but let $\mathbb{B}$ be empty.

Any hash function makes both schemes secure under assumptions $\mathbb{A} \cup \mathbb{B}$ such that the (irrelevant) random oracle of B weakly reduces to the one of A. Since B cannot be secure assuming only $\mathbb{B}$, because the hash function cannot include the assumption about the one-wayness of f by the non-interference, the scheme cannot strongly reduce the random oracle. □

Uninstantiability Implications. In this section we briefly show fundamental results about (un)instantiable random oracles. We define uninstantiability with respect to a very loose requirement on the assumptions, leaving it up to the reduction statement to consider only "standard" cryptographic assumptions in $\mathbb{A}$ and $\mathbb{B}$.

Definition 4 (Uninstantiability). *Let A be $G_A^{\mathcal{H}}$-secure under assumptions $\mathbb{A}$ for random oracle $\mathcal{H}$. Then the random oracle is* uninstantiable *for $G_A^{\mathcal{H}}$ and $\mathbb{A}$ if for any efficient hash function family $\mathcal{H}$ with assumption $\mathbb{H}$ the scheme A is not $G_A^{\mathcal{H}}$-secure under assumptions $\mathbb{A}$.*

Proposition 4 (B uninstantiable $\Rightarrow$ A uninstantiable). *Assume that scheme B with game $G_B^{\mathcal{H}}$ and assumptions $\mathbb{B}$ is (strictly resp. strongly resp. weakly) polynomial-time RO-reducible to scheme A for $G_A^{\mathcal{H}}$ and (true) assumptions $\mathbb{A}$. If B is uninstantiable for $G_B^{\mathcal{H}}$ under $\mathbb{B}$ (for strict reductions) resp. $\mathbb{A} \cup \mathbb{B}$ (for strong or weak reduction), then so is A for $G_A^{\mathcal{H}}$ and assumptions $\mathbb{A}$.*

The proof is again rather straightforward from the definitions. It is clear that this, vice versa, implies that any instantiability result about A transfers accordingly to B.

Given the uninstantiability notion we next note that there are schemes for which the random oracles are not (even weakly) reducible to each other:

Proposition 5 (Impossibility of Reducibility). *There exists schemes A, B for games $G_A^{\mathcal{H}}$ and $G_B^{\mathcal{H}}$ and (true) assumptions $\mathbb{A}, \mathbb{B}$ such that the random oracle of B does not support a weak or strong or strict polynomial-time reduction to the one of scheme A, even though B is secure in the random oracle model.*

The proof appears in the full version of the paper.

4 Example: Hashed ElGamal

In this section we show that the hash function in (a variant) the Twin Diffie-Hellman encryption scheme is RO-reducible to the hash function in hashed ElGamal. We remark that we are not aware if the original twin DH scheme allows the same reduction.

Hashed ElGamal. We first review the classical hashed ElGamal-encryption scheme as presented in [1]. This scheme, denoted by $A = (\mathsf{KGen}_A, \mathsf{Enc}_A, \mathsf{Dec}_A)$ is based on the Diffie-Hellman problem and uses a hash function H and a symmetric cipher $(\mathsf{Enc}, \mathsf{Dec})$. Specifically,

Construction 1 (Hashed ElGamal Encryption Scheme). *The hashed ElGamal encryption scheme $A = (\mathsf{KGen}_A, \mathsf{Enc}_A, \mathsf{Dec}_A)$ in the ROM is defined as follows:*

$\mathsf{KGen}_A(\lambda)$	$\mathsf{Enc}_A(pk, m)$	$\mathsf{Dec}_A(sk, Y, c)$
pick $(\mathcal{G}, g, q)$	$y \leftarrow \mathbb{Z}_q$; $Y \leftarrow g^y$	$Z \leftarrow Y^x$
$x \leftarrow \mathbb{Z}_q$; $X \leftarrow g^x$	$Z \leftarrow X^y$; $k \leftarrow H(Y, Z)$	$k \leftarrow H(Y, Z)$
$sk \leftarrow x$; $pk \leftarrow (\mathcal{G}, g, q, X)$	$c \leftarrow \mathsf{Enc}_k(m)$	$m \leftarrow \mathsf{Dec}_k(c)$
Return (sk, pk)	*Return* (Y, c)	*Return* m

Assuming that the symmetric cipher is secure against chosen ciphertext attacks[3] and that the strong Diffie-Hellman assumption holds (where the adversary has access to a restricted DH decisional oracle), it is proven in [12] that scheme A is secure against chosen ciphertext attacks if H is modeled as a random oracle. The milder ordinary DH assumption is not known to be sufficient to prove CCA security, since the attacker obtains a decision oracle through the decryption oracle here, such that some information about the key may be leaked.

Twin DH Scheme. Subsequently, Cash et al. [10] introduce the so-called strong twin DH assumption which holds if and only if the regular DH assumption holds. Their corresponding DH problems are equally hard but the twin case includes

[3] We always refer to attacks involving a single challenge only throughout the paper.

access to a decision oracle. This enables a clean security proof for a variant of the hashed ElGamal scheme, because the decryption oracle is not more powerful than the decision oracle in the strong twin DH case. Thus, the twin ElGamal scheme allows for milder number-theoretic assumptions while preserving CCA security.

However, the random oracle in the twin ElGamal scheme is used slightly differently than in the original scheme: Its domain is the set of group element triples, as opposed to tuples in the original scheme. While this is unproblematic in the ROM for hash functions $H : \{0,1\}^* \to \{0,1\}^\lambda$ with arbitrary input length, the implications for other security properties for instantiations are less clear. For example, it may be that the twin Diffie-Hellman scheme demands stronger properties from the hash function. We show via our notion of RO-reducibility that this is not the case, at least for our slight variation:

Construction 2 (Twin Diffie-Hellman Encryption Scheme). *The twin DH encryption scheme* $B = (\mathsf{KGen}_B, \mathsf{Enc}_B, \mathsf{Dec}_B)$ *in the ROM is defined as follows:*

$\mathsf{KGen}_B(\lambda)$	$\mathsf{Enc}_B(pk, m)$	$\mathsf{Dec}_B(sk, Y, c, k_1)$
pick $(\mathcal{G}, g, q)$	$y \leftarrow \mathbb{Z}_q;\ Y \leftarrow g^y$	$Z_0 \leftarrow Y^{x_0}$
$x_0 \leftarrow \mathbb{Z}_q;\ X_0 \leftarrow g^{x_0}$	$Z_0 \leftarrow X_0^y;\ k_0 \leftarrow H(Y, Z_0)$	$Z_1 \leftarrow Y^{x_1}$
$x_1 \leftarrow \mathbb{Z}_q;\ X_1 \leftarrow g^{x_1}$	$Z_1 \leftarrow X_1^y;\ k_1 \leftarrow H(Y, Z_1)$	$k_0 \leftarrow H(Y, Z_0)$
$sk \leftarrow (x_0, x_1)$	$c \leftarrow \mathsf{Enc}_{k_0}(m)$	$m \leftarrow \mathsf{Dec}_{k_0}(c)$
$pk \leftarrow (\mathcal{G}, g, q, X_0, X_1)$	*Return* (Y, c, k_1)	*if* $k_1 \neq H(Y, Z_1)$
Return (sk, pk)		*set* $m \leftarrow \perp$
		Return m

We can view the transformation $T^H : \mathcal{G}^3 \to \{0,1\}^{2\lambda}$ of the hash function $H : \mathcal{G}^2 \to \{0,1\}^\lambda$ as follows:

$$T^H(Y, Z_0, Z_1) = H(Y, Z_0)||H(Y, Z_1).$$

Splitting the actual encryption of the message into an encryption for one key half and where we output the other half in clear can then be seen as a special encryption scheme (with double-length keys).

RO-Reducibility. We first show that our twin DH scheme *weakly* reduces the random oracle to the one of the hashed ElGamal scheme for IND-CCA security, i.e., assuming the strong DH assumption. We discuss afterward that the scheme is also secure in the random oracle model assuming the regular DH assumption, implying that the reducibility is also strong:

Theorem 3. *Consider the hashed ElGamal encryption scheme for the IND-CCA security game and the assumptions* $\mathbb{A}$ *that the symmetric encryption scheme is IND-CCA secure and the strong DH assumption holds. Then the twin DH encryption scheme* B *with the IND-CCA security game and the assumptions* $\mathbb{B}$ *that the symmetric encryption scheme is IND-CCA secure and that the DH assumption holds, is strongly RO-reducible to the hashed ElGamal encryption scheme via* $T^H(Y, Z_0, Z_1) = H(Y, Z_0)||H(Y, Z_1)$.

The proof follows from the following two propositions.

Proposition 6. *Under the assumptions as in Theorem 3 the twin DH encryption scheme is weakly RO-reducible to the hashed ElGamal encryption scheme.*

Proof. Assume towards contradiction that there exists an algorithm $\mathcal{B}$ that breaks the CCA-security of B. We then describe an adversary $\mathcal{A}$ that breaks the CCA-security of A. This adversary essentially simulates the "second key half" of the scheme by itself.

Description. To initialize the simulation adversary $\mathcal{A}$ on input $(\mathcal{G}, g, q, X_0) = (\mathcal{G}, g, q, g^{x_0})$ chooses the other half of the secret key $x_1 \leftarrow \mathbb{Z}_q$ and calculates the corresponding public key $X_1 \leftarrow g^{x_1}$. Adversary $\mathcal{A}$ next runs adversary $\mathcal{B}$ with input $(\mathcal{G}, g, q, (X_0, X_1))$ and answers $\mathcal{B}$'s oracle queries as follows:

- First, $\mathcal{A}$ translates any hash query $H(A, B, C)$ from $\mathcal{B}$ into two queries to $\mathcal{A}$'s own hash oracle. More precisely, $\mathcal{A}$ answers an (A, B, C) query with $(H(A, B), H(A, C)) = T^H(A, B, C)$.
- In order to answer $\mathcal{B}$'s challenge query (m_0, m_1), the adversary submits (m_0, m_1) to his own challenge oracle and parses the corresponding ciphertext answer as (Y, c). It remains to compute the extra value by re-using the randomness Y obtained from the oracle. Adversary $\mathcal{A}$ thus computes $k_1 = H(Y, Y^{x_1}) = H(Y, X_1^y)$ and finally returns the ciphertext (Y, c, k_1) to $\mathcal{B}$.
- On a decryption query (Y, c, k_1) of $\mathcal{B}$ adversary $\mathcal{A}$ first checks if (Y, c) corresponds to the value in the challenge ciphertext, or if $k_1 \neq H(Y, Y^{x_1})$. If so, then $\mathcal{A}$ immediately returns $\perp$. Else $\mathcal{A}$ asks its own decryption oracle for the decryption m of (Y, c). To answer the query, it then returns m.
- Note also that we can grant $\mathcal{B}$ direct access to the H oracle. Adversary $\mathcal{A}$ would simply forward this query and hand back the answer.

When $\mathcal{B}$ eventually outputs a guess b then $\mathcal{A}$ outputs the same bit.

Analysis. The simulation is perfect in the following sense: $\mathcal{B}$ cannot submit a ciphertext (Y, c, k_1^*) to the decryption oracle (after receiving the challenge ciphertext (Y, c, k_1)) for $k_1^* \neq k_1$ which would decrypt correctly. Hence, $\mathcal{A}$ can reject such ciphertexts immediately and therefore only submits "pruned" ciphertexts to its decryption oracle which have never appeared before. Hence, $\mathcal{A}$, too, represents a successful attacker on the hashed ElGamal scheme if $\mathcal{B}$ is one for the twin DH scheme. Moreover, the advantages of both algorithms in their corresponding IND-CCA game are identical. □

To complete the proof for a strong reduction we finally show that our version is secure in the random oracle model:

Proposition 7. *The twin DH encryption scheme B with the assumptions $\mathbb{B}$ that the symmetric encryption scheme is IND-CCA secure and that the DH assumption holds, is IND-CCA-secure in the random oracle model.*

Proof. The proof is more involved that then one in [10], owned to the fact that the random oracle $H(X, Z_0, Z_1)$ in [10] ties together the twin DH tuples and that this property is required for the twin DH oracle. In contrast, in our scheme the pairs (X, Z_0) and (X, Z_1) are only loosely connected. We show that this loose connection can be made a strong one with multiple simulations of the adversary.

In a first step we can "normalize" an adversary $\mathcal{A}$ against IND-CCA of our twin DH scheme. First we may assume that $\mathcal{A}$ never makes a hash query twice. Second, we can assume that $\mathcal{A}$ never submits a tuple (Y_i, c_i, k_i) to the decryption oracle before receiving the challenge ciphertext (Y, c, k) where $Y_i = Y$. This decreases the adversary's success probability by a negligible amount D/q for the polynomial number D of $\mathcal{A}$'s decryption queries. Third, we can assume that adversary $\mathcal{A}$ never submits a decryption request (Y_i, c_i, k_i) such that, in case $Y_i \neq Y$, it has not queried the hash function about $(Y_i, Y_i^{x_1})$ for $Y_i \neq Y$ before. The loss is at most $D \cdot 2^{-\lambda}$ for this. Fourth, we assume that the adversary never submits (Y_i, c_i, k_i) to the decryption oracle where $Y_i = Y$ but $k_i \neq k$; such a query cannot be valid. Fifth, we assume that $X_0 \neq X_1$ which happens with probability $1 - 1/q$.

Taming Hash Queries. Consider a normalized adversary $\mathcal{A}$ against our twin DH scheme. We assume that $\mathcal{A}$ in addition to T^H also has direct access to the random oracle $H : \mathcal{G}^2 \to \{0,1\}^*$. In fact, we assume from now on that all algorithms, including the adversary and the scheme's algorithms, never call T^H, but use H to simulate T^H with two queries. Define the following event HASHQUERY that, during the IND-CCA attack, $\mathcal{A}$ at some point asks a query (Y, Z) to H such that Y appears in the challenge ciphertext, and $Z = Y^{x_0}$ or $Z = Y^{x_1}$ for the public key entries $X_0 = g^{x_0}$ and $X_1 = g^{x_1}$.

We show that the probability $\epsilon(\lambda)$ of event HASHQUERY must be negligible. Assume toward contradiction that this was not the case. We then show how to break the the twin DH problem (and thus the DH problem) via algorithm $\mathcal{B}$. Algorithm $\mathcal{B}$ receives a group description $(\mathcal{G}, g, q)$ and values Y, X_0, X_1 as input. It can also query a twin DH oracle about values (g^a, B_0, B_1) which outputs 1 iff $B_0 = X_0^a$ and $B_1 = X_1^a$. The values X_0, X_1 serve as the public key presented to $\mathcal{A}$, and Y will be placed in the challenge ciphertext.

Algorithm $\mathcal{B}$ runs $\mathcal{A}$'s attack by using the input data as the public key, and simulating the random oracle and decryption queries as follows:

- $\mathcal{B}$ will maintain a list $\mathcal{L}$ of tuples of the form (A, B, k) or $(\mathtt{dh}, A, X_b, k)$ where the former type corresponds to direct hash queries of $\mathcal{A}$ and the latter type to implicit hash queries. Initially, $\mathcal{B}$ sets $\mathcal{L} := \{(\mathtt{dh}, Y, X_0, k_0), (\mathtt{dh}, Y, X_1, k_1)\}$ for random values k_0, k_1 for the hash values to compute the challenge ciphertext (note that Y is already known at the outset).
- Whenever $\mathcal{A}$ makes a hash query (A, B) algorithm $\mathcal{B}$ first searches for an entry (A, C, k) in $\mathcal{L}$ such that (A, B, C) or (A, C, B) forms a correct twin DH tuple (under X_0, X_1). Since $X_0 \neq X_1$ only one case can happen. If found, and there exists an entry $(\mathtt{dh}, A, X_0, k)$ in $\mathcal{L}$ for the case (A, B, C) resp. $(\mathtt{dh}, A, X_1, k)$ for the case (A, C, B), then replace this entry by (A, B, k) in $\mathcal{L}$. In any other case, pick k at random and store (A, B, k) in $\mathcal{L}$. Return k.

– If $\mathcal{A}$ makes a decryption request (Y_i, c_i, k_i) then check whether $Y_i = Y$ or not. In case $Y_i = Y$ look up the entry $(\mathtt{dh}, Y, X_0, k_0)$ in $\mathcal{L}$ and use k_0 to decrypt c_i. (Note that, by assumption, k_1 must be correct.) Suppose $Y_i \neq Y$. Then, since the adversary is normalized, there must be an entry (Y_i, Z_1, k_1) in $\mathcal{L}$ already, caused by a hash query, where $Z_1 = Y_i^{x_1}$. (There cannot exist another entry (Y_i, Z_1, k_1') for $k_1' \neq k_1$ as hash queries never repeat.) Given (Y_i, Z_1, k_1) check for an entry (Y_i, Z_0, k_0) such that (Y_i, Z_0, Z_1) forms a valid twin DH tuple for X_0, X_1. If such an entry exist then use k_0 to decrypt c_i. If no such entry exist, check for a tuple $(\mathtt{dh}, Y_i, X_0, k_0)$ in $\mathcal{L}$ and use k_0 to decrypt. Else, pick a new value k_0, store $(\mathtt{dh}, Y_i, X_0, k_0)$ in $\mathcal{L}$, and use k_0 to decrypt. Return the decrypted message.

To prepare the challenge ciphertext $\mathcal{B}$ uses the previously chosen values k_0, k_1 placed in $\mathcal{L}$, also picks one of the two messages m_0, m_1 at random, and returns $(Y, \mathsf{Enc}(k_0, m_b), k_1)$.

If $\mathcal{A}$ finishes algorithm $\mathcal{B}$ records all entries (A, B) in $\mathcal{L}$ with $A = Y$ and now reruns the above procedure, with the same group but for re-randomized data $Y' = Y^s$, $X_0' := X_a^{s_a}$, $X_1' := X_{1-a}^{s_{1-a}}$ for random $s, s_0, s_1 \leftarrow \mathbb{Z}_q^*$ and random bit a. Every other random choice is based on fresh randomness. Any query (A, B, C) to the twin DH oracle in this second run is first transformed into $(A, B^{1/s_0}, C^{1/s_1})$ for $a = 0$ resp. $(A, C^{1/s_0}, B^{1/s_1})$ for $a = 1$. At the end, $\mathcal{B}$ transforms all pairs (A', B') in the list $\mathcal{L}$ of the second run by computing $((A')^{1/s}, (B')^{1/s_0})$ and $((A')^{1/s}, (B')^{1/s_1})$, effectively doubling the number of pairs. Sieve to keep only those with first element Y. Run on all combinations of the two (sieved) lists the twin DH oracle to find a solution (Y, Z_0, Z_1) to the twin DH problem.

Analysis. The maintenance of the hash list $\mathcal{L}$ provides a more fine-grained implementation of how a random oracle would behave: Since any decryption query for $Y_i \neq Y$ must already contain a corresponding entry $(Y_i, Y_i^{x_1}, k_1)$ by assumption, we can check via the twin DH oracle if we already have a matching entry (Y_i, Z_0, k_0). If not, we generate a fresh random string and store the implicit representation $(\mathtt{dh}, Y_i, X_0, k_0)$ in $\mathcal{L}$, and will later carefully check if a hash query for $Y_i^{x_0}$ is made (in which case we update the entry in $\mathcal{L}$ and re-use the value k_0).

As for $\mathcal{B}$'s success probability, we call a group $(\mathcal{G}, g, q)$ good if $\mathcal{A}$'s success probability conditioned on this group exceeds $\epsilon/2$. By an averaging argument a group is good with probability at least $\epsilon/2$. Hence, given such a good group, and the fact that $\mathcal{B}$ provides a perfect simulation, $\mathcal{B}$ obtains a valid entry (Y, Y^{x_0}) or (Y, Y^{x_1}) with probability at least $\epsilon/2$ in the first run. The same applies in the second run where the re-randomization is correctly undone for each twin DH oracle query. With probability $1/2$ algorithm $\mathcal{B}$ then obtains matching values (Y, Y^{x_0}) and (Y, Y^{x_1}) because the order bit a in the second run is information-theoretically hidden from $\mathcal{A}$. Overall, and neglecting the minor loss due to normalization of $\mathcal{A}$, algorithm $\mathcal{B}$ thus solves the twin DH problem with probability at least $\epsilon^3/16$. By assumption this is still non-negligible.

Conditioning on the adversary not making bad hash queries, it is now easy to give a reduction to the IND-CCA security of the symmetric cipher (with the attacker against the symmetric scheme providing all the public-key operations itself). □

5 Reductions among Signature Schemes

In this section we briefly outline a few more applications of our notion. Specifically, we give three relations among signature schemes including the Guillou-Quisquater (GQ) signature scheme [17] which we reduce to a probabilistic version of FDH, the PSS signature scheme [5] which we also reduce to a probabilistic FDH variation, and finally a reduction from Schnorr signatures [25] to a (probabilistic version of) BLS signatures [8].

GQ ⇒ FDH. We first consider the RSA-based Guillou-Quisquater identification scheme and its derived signature scheme via the Fiat-Shamir heuristic [17]. For public key $pk = (X, N, e)$ and secret key x with $X = x^e \bmod N$ the signer computes a signature as (R, y) for random $R = r^e \bmod N$, and where $y = r^c x \bmod N$ for $c = H(pk, R, m)$. A probabilistic full-domain hash (FDH) RSA signature scheme with signatures of the form (R, σ) for $\sigma = (H(pk, R, m))^d \bmod N$ is (strictly) random oracle reducible to the aforementioned Guillou-Quisquater scheme via the transformation $T^H(pk, R, m) = R^{H(pk,R,m)} X \bmod N$ for any type of forgery attack under the RSA assumption. The reason is that any Guillou-Quisquater signature for H can be seen as a FDH signature for $T^H : \{0,1\}^* \to \mathbb{Z}_N^*$, and any successful forgery for the FDH scheme for T^H is vice versa a valid forgery for the Guillou-Quisquater scheme.

PSS ⇒ FDH. The reduction of another probabilistic version of FDH to the PSS signature scheme is similar to the GQ case. Consider FDH signatures $(T^H(r, m))^d \bmod N$ for the PSS-encoding $T^H(r, m) = \texttt{str2int}(0||w||r^*||H_2(w))$ for $w = H_0(r, m)$ and $r^* r \oplus = H_1(w)$. Here, H_0, H_1, H_2 are hash functions derived from H as in the PSS scheme. Then any successful attack on FDH with hash function T^H easily yields a forgery against PSS with hash function H. Hence, PSS allows a strict random oracle reduction to the probabilistic version of FDH under the RSA assumption for any type of forgery attack.

Schnorr ⇒ BLS. Consider a probabilistic version of the BLS signature scheme [8], where signatures are of the form $\sigma = (R, H(R, X, m)^x)$ for randomness R, message m, private key x and public key $X = g^x$. Verification is performed analogously to the original scheme via a pairing computation. We argue that the Schnorr signature scheme (recall that a signature there is of the form $\sigma = (c, r + cx \bmod q)$ for public key $x = g^x$, $R = g^r$, and $c = H(R, m)$) is (strictly) random oracle reducible to the BLS version via the transformation $T^H(R, X, m) = RX^{H(R,m)}$. This holds assuming the discrete logarithm assumption and under an augmented version of the KEA1 assumption [18,3] which states that, for any adversary $\mathcal{A}$ which for input a description of the group, g, X, and with access to a Schnorr signing oracle under key X and a hash function oracle, outputs a pair (Y, Y^x), there exists an adversary $\mathcal{A}'$ which, on the same input and with access to the same oracles, outputs y with $X^y = Y^x$. The probability that $\mathcal{A}$ succeeds, but $\mathcal{A}'$ does not, must be negligible for all $\mathcal{A}$.

Suppose now that there exists some successful adversary $\mathcal{B}$ against our version of BLS. Construct adversary $\mathcal{A}$ against the Schnorr scheme as follows. Whenever $\mathcal{B}$ makes some query m, adversary $\mathcal{A}$ forwards this query to its own signing oracle. It uses the answer (c, y) to calculate $h = X^y$, computes $R = g^y X^{-c}$ (such that $H(R, m) = c$) and finally answers $\mathcal{B}$'s query with (R, h). This simulates a correct signature since $\mathcal{B}$ expects R and $T^H(R, X, m)^x = (RX^{H(R,m)})^x = (g^y)^x = X^y = h$. It remains to construct a Schnorr forgery from $\mathcal{B}$'s forgery, denoted by (m^*, R^*, Z^*). To this end we note that, under the augmented KEA1 assumption, for $\mathcal{A}$ (running $\mathcal{B}$ as a subroutine) outputting $Y^* = T^H(R^*, X, m^*)$ and $Z^* = (Y^*)^x$ for the valid forgery (m^*, R^*, Z^*), there must exist an adversary $\mathcal{A}'$ returning y^* with $Z^* = X^{y^*}$. This must be true with non-negligible probability, because $\mathcal{A}$ succeeds with non-negligible probability, and otherwise the augmented KEA1 assumption would be false. Hence, there exists an adversary which creates a valid forgery $(m^*, H(R^*, m^*), y^*)$ for the Schnorr scheme with non-negligible probability.

Acknowledgments. We thank the anonymous reviewers for valuable comments, especially Mihir Bellare. We also thank Anja Lehmann, Adam O'Neill, and Tom Ristenpart for listening to this idea and providing feedback at an early stage. Both authors are supported by grants Fi 940/2-1 and Fi 940/4-1 of the German Research Foundation (DFG). This work was also supported by CASED (`www.cased.de`).

References

1. Abdalla, M., Bellare, M., Rogaway, P.: The oracle diffie-hellman assumptions and an analysis of DHIES. In: Naccache, D. (ed.) CT-RSA 2001. LNCS, vol. 2020, pp. 143–158. Springer, Heidelberg (2001)
2. Bellare, M., Boldyreva, A., Palacio, A.: An uninstantiable random-oracle-model scheme for a hybrid-encryption problem. In: Cachin, C., Camenisch, J.L. (eds.) EUROCRYPT 2004. LNCS, vol. 3027, pp. 171–188. Springer, Heidelberg (2004)
3. Bellare, M., Palacio, A.: The knowledge-of-exponent assumptions and 3-round zero-knowledge protocols. In: Franklin, M. (ed.) CRYPTO 2004. LNCS, vol. 3152, pp. 273–289. Springer, Heidelberg (2004)
4. Bellare, M., Rogaway, P.: Random oracles are practical: A paradigm for designing efficient protocols. In: ACM CCS 1993, pp. 62–73 (1993)
5. Bellare, M., Rogaway, P.: The exact security of digital signatures - how to sign with RSA and rabin. In: Maurer, U.M. (ed.) EUROCRYPT 1996. LNCS, vol. 1070, pp. 399–416. Springer, Heidelberg (1996)
6. Boldyreva, A., Fischlin, M.: Analysis of random oracle instantiation scenarios for OAEP and other practical schemes. In: Shoup, V. (ed.) CRYPTO 2005. LNCS, vol. 3621, pp. 412–429. Springer, Heidelberg (2005)

7. Boldyreva, A., Fischlin, M.: On the security of OAEP. In: Lai, X., Chen, K. (eds.) ASIACRYPT 2006. LNCS, vol. 4284, pp. 210–225. Springer, Heidelberg (2006)
8. Boneh, D., Lynn, B., Shacham, H.: Short signatures from the Weil pairing. Journal of Cryptology 17(4), 297–319
9. Canetti, R., Goldreich, O., Halevi, S.: The random oracle methodology, revisited (preliminary version). In: 30th ACM STOC, pp. 209–218
10. Cash, D., Kiltz, E., Shoup, V.: The twin DiffieHellman problem and applications. Journal of Cryptology 22(4), 470–504
11. Coron, J.S.: On the exact security of full domain hash. In: Bellare, M. (ed.) CRYPTO 2000. LNCS, vol. 1880, pp. 229–235. Springer, Heidelberg (2000)
12. Cramer, R., Shoup, V.: Design and analysis of practical public-key encryption schemes secure against adaptive chosen ciphertext attack. SIAM Journal on Computing 33(1), 167–226 (2003)
13. Dodis, Y., Haitner, I., Tentes, A.: On the (in)security of rsa signatures. Cryptology ePrint Archive, Report 2011/087 (2011), http://eprint.iacr.org/
14. Dodis, Y., Oliveira, R., Pietrzak, K.: On the generic insecurity of the full domain hash. In: Shoup, V. (ed.) CRYPTO 2005. LNCS, vol. 3621, pp. 449–466. Springer, Heidelberg (2005)
15. Fischlin, M., Lehmann, A., Ristenpart, T., Shrimpton, T., Stam, M., Tessaro, S.: Random oracles with(out) programmability. In: Abe, M. (ed.) ASIACRYPT 2010. LNCS, vol. 6477, pp. 303–320. Springer, Heidelberg (2010)
16. Goldwasser, S., Kalai, Y.T.: On the (in)security of the Fiat-Shamir paradigm. In: 44th FOCS, pp. 102–115. IEEE Computer Society Press, Los Alamitos
17. Guillou, L.C., Quisquater, J.-J.: A practical zero-knowledge protocol fitted to security microprocessor minimizing both transmission and memory. In: Günther, C.G. (ed.) EUROCRYPT 1988. LNCS, vol. 330, pp. 123–128. Springer, Heidelberg (1988)
18. Hada, S., Tanaka, T.: On the existence of 3-round zero-knowledge protocols. In: Krawczyk, H. (ed.) CRYPTO 1998. LNCS, vol. 1462, pp. 408–423. Springer, Heidelberg (1998)
19. Kiltz, E., O'Neill, A., Smith, A.: Instantiability of RSA-OAEP under chosen-plaintext attack. In: Rabin, T. (ed.) CRYPTO 2010. LNCS, vol. 6223, pp. 295–313. Springer, Heidelberg (2010)
20. Kiltz, E., Pietrzak, K.: On the security of padding-based encryption schemes – or – why we cannot prove OAEP secure in the standard model. In: Joux, A. (ed.) EUROCRYPT 2009. LNCS, vol. 5479, pp. 389–406. Springer, Heidelberg (2009)
21. Lamport, L.: Constructing digital signatures from a one-way function. Technical Report SRI-CSL-98, SRI International Computer Science Laboratory
22. Maurer, U.M., Renner, R., Holenstein, C.: Indifferentiability, impossibility results on reductions, and applications to the random oracle methodology. In: Naor, M. (ed.) TCC 2004. LNCS, vol. 2951, pp. 21–39. Springer, Heidelberg (2004)
23. Naito, Y., Yoneyama, K., Wang, L., Ohta, K.: How to confirm cryptosystems security: The original merkle-damgård is still alive! In: Matsui, M. (ed.) ASIACRYPT 2009. LNCS, vol. 5912, pp. 382–398. Springer, Heidelberg (2009)
24. Nielsen, J.B.: Separating random oracle proofs from complexity theoretic proofs: The non-committing encryption case. In: Yung, M. (ed.) CRYPTO 2002. LNCS, vol. 2442, pp. 111–126. Springer, Heidelberg (2002)
25. Schnorr, C.P.: Efficient signature generation by smart cards. Journal of Cryptology 4(3), 161–174 (1991)
26. Yoneyama, K., Miyagawa, S., Ohta, K.: Leaky random oracle (extended abstract). In: Baek, J., Bao, F., Chen, K., Lai, X. (eds.) ProvSec 2008. LNCS, vol. 5324, pp. 226–240. Springer, Heidelberg (2008)

Time-Lock Puzzles in the Random Oracle Model

Mohammad Mahmoody[1], Tal Moran[2], and Salil Vadhan[2,⋆]

[1] Department of Computer Science, Cornell University
mohammad@cs.cornell.edu
http://www.cs.cornell.edu/~mohammad/
[2] School of Engineering and Applied Sciences and
Center for Research on Computation and Society, Harvard University
{talm,salil}@seas.harvard.edu
http://seas.harvard.edu/~{talm,salil}/

Abstract. A time-lock puzzle is a mechanism for sending messages "to the future". The sender publishes a puzzle whose solution is the message to be sent, thus hiding it until enough time has elapsed for the puzzle to be solved. For time-lock puzzles to be useful, generating a puzzle should take less time than solving it. Since adversaries may have access to many more computers than honest solvers, massively parallel solvers should not be able to produce a solution much faster than serial ones.

To date, we know of only one mechanism that is believed to satisfy these properties: the one proposed by Rivest, Shamir and Wagner (1996), who originally introduced the notion of time-lock puzzles. Their puzzle is based on the serial nature of exponentiation and the hardness of factoring, and is therefore vulnerable to advances in factoring techniques (as well as to quantum attacks).

In this work, we study the possibility of constructing time-lock puzzles in the random-oracle model. Our main result is negative, ruling out time-lock puzzles that require more parallel time to solve than the total work required to generate a puzzle. In particular, this should rule out black-box constructions of such time-lock puzzles from one-way permutations and collision-resistant hash-functions. On the positive side, we construct a time-lock puzzle with a linear gap in parallel time: a new puzzle can be generated with one round of n parallel queries to the random oracle, but n rounds of *serial* queries are required to solve it (even for massively parallel adversaries).

1 Introduction

In this paper we revisit the subject of "timed-release crypto" based on "time-lock puzzles". The goal of timed-release crypto, introduced by May [22], is to encrypt a message in such a way that it will be readable at some specified time in the future (even without additional help from the sender), but not before then.

In addition to the basic use of "sending messages to the future", there are many other potential uses of timed-release crypto. Rivest, Shamir and Wagner [25] suggest, among other uses, delayed digital cash payments, sealed-bid auctions and key escrow. Boneh and Naor [9] define timed commitments and timed signatures and show that they can be used for fair contract signing, honesty-preserving auctions and more.

⋆ Supported by NSF grant CNS-0831289.

P. Rogaway (Ed.): CRYPTO 2011, LNCS 6841, pp. 39–50, 2011.

A natural approach to building a timed-release crypto scheme is the use of time-lock puzzles: puzzles that take a prespecified amount of time to solve (which should be significantly longer than the time to generate the puzzle). Intuitively, using the solution of a time-lock puzzle as the key to an encryption scheme would force anyone wanting to decrypt the message to perform the computation for the time required to solve the puzzle. By tuning the difficulty of the solution according to the time we would like the message to remain secure, we can ensure that decryption will take at least that amount of time.

Inverting a (suitably weak) one-way function seems like an obvious candidate for a time-lock puzzle. However, as Rivest et al. observed, for many uses a generic one-way function would not suffice. This is because a potential adversary may have access to much larger computational resources than an honest party. Even if the processors available to the adversary are not be significantly faster than those available to the honest parties, it is reasonable to assume that a well-funded adversary could have access to many more processors (that could be used in parallel). Thus, we require that time-lock puzzles be "essentially sequential" in nature: having many parallel processors should not give a large advantage over a single processor in solving the puzzle.

The puzzles proposed in [25] are based on the conjecture that exponentiation (modulo an RSA integer) is such a task. In particular, if the factorization of the modulus is not known, the best known method for exponentiation is repeated squaring (which is conjectured to be essentially sequential). Given the factors of the modulus, however, there is a shortcut that allows the exponentiation to be performed much faster (so that the puzzles can be generated efficiently). Thus, there seems to be a super-polynomial gap between the work required to generate the puzzle and the parallel time required to solve it (for a polynomial number of parallel processors).

To the best of our knowledge, this construction of time-lock puzzles is the only one currently known that is resistant to parallel attack. The construction of Boneh and Naor [9] uses essentially the same idea. This leads to the natural question of whether we can construct time-lock puzzles based on other assumptions, preferably weaker and more general ones.

Biham, Goren and Ishai [5] suggest an additional motivation as well: obtaining (weak) key-agreement protocols based on one-way functions that resist quantum attack. They show that in the classical world there do exist weak key-agreement protocols based on one-way functions (of exponential strength) that force an adversary to work in time quadratic in the time of the honest parties, based on a variant of Merkle puzzles [23]. However, both their construction and the original Merkle puzzles are vulnerable to quantum attack via Grover's search algorithm [20]. Biham et al. note that Grover's speedup only applies to *parallel* search, and leave as an open problem whether such puzzles exist that are resistant to parallel attack (and thus, potentially, to quantum attack as well).

In this paper, we study time-lock puzzles in the *random oracle model*. In the random oracle model, we assume all parties have access to an oracle, H, modeled as a random function. In a real implementation, the random oracle is usually "instantiated" with a cryptographic hash function. We assume the adversary in this model is computationally unbounded, and measure the difficulty of the time-lock puzzle by the number of queries the adversary is required to make to the random oracle in order to solve it.

The random oracle model is interesting for several reasons. First, negative results in this model rule out "black-box" constructions from one-way permutations and collision-resistant hash functions (since a random function is collision-resistant and indistinguishable from a random permutation using only a small number of queries; see e.g., [21,19,3] for details). Second, most "natural" protocols that have been proven secure in the random oracle model appear to be secure in practice as well (even though some "artificial" protocols are secure in this model but insecure for *any* explicit instantiation of the random oracle [11]), and constructing a protocol in this model is sometimes a first step towards constructing a provably-secure protocol in the plain model (e.g., the first efficient IBE scheme was proven secure in the random oracle model [8], after which constructions were found in the standard model as well [12,6,7]). Finally, the random oracle model is much simpler to analyze than models that incorporate computational complexity, and better understanding the problem in this setting may give insight into the complexity-theoretic case.

We can think of a time-lock puzzle generator as a randomized oracle algorithm f. The output of $f^H(r_A)$ (where r_A is the random input and H the random oracle) is a pair $(M, \mathcal{V})$: the puzzle M and a solution validator $\mathcal{V}$. The solver, given M, must output a solution x such that $\mathcal{V}(x) = 1$. When a time-lock puzzle has a single solution, such as when it is used to hide an encrypted message, $\mathcal{V}$ just compares its input to that constant value. In general, however, $\mathcal{V}$ may perform more complex verification and our negative results hold even when $\mathcal{V}$ is not efficient (note that $\mathcal{V}$ does not have access to the random oracle). For this to be a good time-lock puzzle, we would like f to be easy to compute but moderately hard to solve, even for a parallel adversary. More precisely, if we can compute $(M, \mathcal{V}) \leftarrow f^H(r_A)$ using n queries to H, we would like f to satisfy:

Completeness

- There exists a (randomized) polynomial-time algorithm g (the honest solver) that solves puzzles generated by f: with high probability (over the random coins of f and g and the random oracle H), if we generate $(M, \mathcal{V}) \leftarrow f^H(r_A)$ and $x \leftarrow g^H(r_B, M)$ then $\mathcal{V}(x) = 1$. We use the shorthand notation $[(M, \mathcal{V}) \leftarrow f^H(r_A); x \leftarrow g^H(r_B, M); \mathcal{V}(x) = 1]$ to denote the event that the puzzle was generated as described above, the solver was run and its solution was valid. We denote by $m > n$ the number of queries g makes to H. m measures the difficulty for the honest solver and should be moderately larger than n, e.g., a large polynomial in n.

Soundness

- Any algorithm that solves f and makes up to $q \gg m$ queries to H must use at least $m' \approx m$ levels of adaptivity. For example, we might take $q = n^{\omega(1)}$ and $m' = m/2$. The number of levels of adaptivity measures the complexity for a parallelized adversary; this requirement means that unless the adversary makes a very large number of queries, using parallelism won't give it an advantage over the honest solver.

1.1 Our Results

Time-lock puzzles with large difficulty gap are impossible. Our main result is a negative one. We show that for every time-lock puzzle there exists a parallel adversary that can solve the puzzle in no more time than it takes to generate and makes only polynomially more queries to the random oracle than the best honest (serial) solver. Thus, constructing time-lock puzzles with a "gap" between the work of the puzzle generator and the parallel time of the solver cannot be done in the random-oracle model.

Concretely, we prove two similar theorems but with incomparable parameters. The first provides an adversary that makes an optimal number of parallel query rounds, but may require super-polynomial time to run, even if the honest solver is efficient. Our second theorem gives a much simpler adversary construction that runs in polynomial time if the honest solver does, but has slightly worse adaptivity.

Formally, we prove the following two theorems:

Theorem 1 (Optimally Adaptive but Inefficient Adversary). *Let f be an oracle algorithm that makes at most n queries to a random oracle H and g an oracle algorithm that makes at most $m > n$ queries to H. If*

$$\Pr\left[(M, \mathcal{V}) \leftarrow f^H(r_A); x \leftarrow g^H(r_B, M); \mathcal{V}(x) = 1\right] \geq 1 - \nu$$

(i.e., when a puzzle is randomly generated after which the solver g is executed, its output is a valid solution with probability at least $1-\nu$ over the random coins r_A, r_B and H) then for all $\varepsilon \in (0, 1)$ there exists an adversary, Ivy, that makes $\tilde{O}(nm/\varepsilon)$ queries to H, uses only n levels of adaptivity and satisfies $\Pr\left[(M, \mathcal{V}) \leftarrow f^H(r_A); x \leftarrow h^H(r_I, M); \mathcal{V}(x) = 1\right] \geq 1 - \nu - \varepsilon$ (where r_I is the variable denoting the random coins used by Ivy).

Theorem 2 (Efficient but Non-Optimal Adversary). *Let f be an oracle algorithm that makes at most n queries to a random oracle H and g an oracle algorithm that makes at most $m > n$ queries to H. If $\Pr\left[(M, \mathcal{V}) \leftarrow f^H(r_A); x \leftarrow g^H(r_B, M); \mathcal{V}(x) = 1\right] \geq 1 - \nu$ then for all $\varepsilon \in (0, 1)$ there exists a deterministic adversary Javier (denoted by J) who makes at most nm/ε queries to H, uses only n/ε levels of adaptivity and satisfies $\Pr\left[(M, \mathcal{V}) \leftarrow f^H(r_A); x \leftarrow J^H(M); \mathcal{V}(x) = 1\right] \geq 1 - \nu - \varepsilon$. Moreover, the running time of J is $O(n/\varepsilon)$ times the running time of g.*

Some intuition for the proofs of both theorems, as well as the full proof of Theorem 2, can be found in Section 2. Due to space considerations, the proof of Theorem 1 is deferred to the full version.

By combining Theorem 1 with the result of [4], we partially resolve the open question of Biham et al. [5] by showing that every key-agreement protocol in the random-oracle model can be broken by a parallel attack that makes polynomially many queries to the random-oracle: Biham et al. were interested in whether there exist key-agreement protocols that resist quantum attack, but as a step towards this goal asked the question of whether there exist such protocols secure against a parallel *classical* attacker, and specifically whether such protocols could be based on random functions.

Corollary 1. *Let Π be a two-party protocol in the random oracle model such that when executing Π the two parties Alice and Bob make at most n queries each and their outputs* *are identical with probability at least* $1 - \nu$. *Then for every* $0 < \varepsilon < 1$ *there exists an adversary that, given the public transcript of the protocol, outputs a value that agrees with Bob's output with probability* $1 - \nu - \varepsilon$ *using* $2n$ *levels of adaptivity and making* $\tilde{O}(n^3/\varepsilon^3)$ *total queries to* H.

Proof. Let $f^H(r_A, r_B) = (M, \mathcal{V})$, where M is the complete public transcript of Π when Alice uses the random coins r_A and Bob the random coins r_B. Define the corresponding solution validator to be

$$\mathcal{V}(x) = \begin{cases} 1 & \text{if } x \text{ is Bob's output in the execution of } \Pi \text{ in } f^H(r_A, r_B) \\ 0 & \text{otherwise} \end{cases} .$$

By the result of [4], there exists an adversary that makes at most $O(n^2/\varepsilon^2)$ queries to H and outputs a "correct" solution to this puzzle (i.e., an output that agrees with Bob) with probability $1 - \nu - \varepsilon/2$. Think of this adversary as the solver, g. By Theorem 1, this implies that there exists a solver that succeeds with probability $1 - \nu - \varepsilon$, makes $\tilde{O}((2n/\varepsilon) \cdot n^2/\varepsilon^2) = \tilde{O}(n^3/\varepsilon^3)$ total queries to H and uses only $2n$ levels of adaptivity (the $2n$ is because the total number of queries made by f is bounded by the total number of queries made by both Alice and Bob).

A time-lock puzzle with a linear gap in parallel time. Although our negative results rule out "strong" time-lock puzzles, they still leave open the possibility for a weaker version: one that can be generated with n parallel queries to the oracle but requires n rounds of adaptive queries to solve.

In a positive result, we show that such a puzzle can indeed be constructed. More formally, we prove:

Theorem 3. *Let* k *be a security parameter. There exist oracle functions* f *and* g *that satisfy:*

1. (Efficiency) $(M, \mathcal{V}_M) \leftarrow f^H(k, r)$ *can be computed using* n *parallel (non-adaptive) queries to* H.
2. (Completeness) $x \leftarrow g^H(k, M)$ *can be computed using* n *serial (adaptive) queries to* H *and the output of* g *always satisfies* $\mathcal{V}_M(x) = 1$ *(*g *is deterministic).*
3. (Soundness) *For every oracle function* h *that makes less than* n *serial rounds of queries to* H *and poly*(k) *queries overall to* H *in total,* $\Pr\left[(M, \mathcal{V}_M) \leftarrow f^H(k, r); x \leftarrow h^H(k, r_J, M); \mathcal{V}_M(x) = 1\right] = neg(k)$ *(where* neg *is some negligible function in* k*).*

The idea behind the construction is to force the solver to make sequential queries by "encrypting" each successive query with the result of an oracle call on its predecessor. The full construction and a sketch of its security proof appear in Section 3.

1.2 Related Work

Timed-Release Crypto Constructions. The notion of timed-release crypto was introduced by May [22]. May's proposal was to publish an encrypted message and distribute the decryption key between several trusted agents. The agents would be instructed to publish their shares of the key at a specified future date. Rivest, Shamir and Wagner [25] introduced the idea of using time-lock puzzles instead of requiring a sender to trust an external entity and also developed May's "trusted-agent" approach, suggesting a scheme where the trusted agents' storage does not grow with the number of timed-release messages (as it does in May's scheme).

These two approaches, one based on puzzles and the other on trusted agents, have remained the basis of all new timed-release crypto schemes that we know of. There have been many improvements in the agent-based approach, focusing on reducing interaction between the agents and the users, achieving various verifiability and privacy properties ([8,13,14], among others). On the other hand, to the best of our knowledge, all existing time-lock puzzle constructions (that are resistant to parallel attack) are based on the problem originated by Rivest et al., namely that of exponentiation modulo an RSA integer.

Puzzles. The term "puzzle" to describe a cryptographic construction that is "meant to be broken" was first used by Merkle in the context of key agreement protocols [23]. Merkle's key-exchange protocol allows two users to exchange a key by solving a single puzzle, while forcing an adversary to solve multiple puzzles in order to discover it. The protocol does not require much structure from the puzzles, and can be instantiated with black-box use of one-way functions. The computation gap between the honest users and the adversary is quadratic in Merkle's scheme: if an honest user requires $O(N)$ time to recover the key, an adversary can recover it in $O(N^2)$ time.

Barak and Mahmoody showed that this is essentially optimal [4], improving a previous result by Impagliazzo and Rudich [21]. Both of these works give an upper bound for the computation gap of arbitrary key-exchange protocols in the random oracle model (including protocols that require multiple rounds of interaction between the two honest parties). Our work considers only one-message protocols, but bounds the *parallel* complexity of the adversary (in contrast to [21,4], who analyze the complexity of serial adversaries).

Puzzles have also been proposed as proof-of-work mechanisms for controlling spam and preventing denial-of-service-attacks. The idea was first introduced by Dwork and Naor [17], and was developed in multiple subsequent works [1,2,16,18]. Rivest and Shamir even suggest one variant for use as a micropayment system [24].

One major difference between these types of puzzles and those we consider in this work is that resistance to parallel attack is not as critical: for example, an adversary generating spam messages can always parallelize at the message level rather than by attacking a specific puzzle. Proofs-of-work, on the other hand, must be resistant to amortization (solving one puzzle should not help in solving others), whereas this is usually not a concern for time-lock puzzles.

For both types of puzzles, it is still important to take into account the gap between the computational capability of an honest user and that of the adversary. Abadi, Burrows, Manasse and Wobbler suggest basing the difficulty of a puzzle on memory access time [1], under the assumption that this has less variance among users than CPU speed. In a subsequent work, Dwork, Goldberg and Naor [16] construct such a function in the random oracle model that uses "pointer-chasing" in a large random table. This has a very similar flavor to our time-lock puzzle construction in Section 3, although the goal is somewhat different and the analysis focuses on bounding memory accesses (to the table) rather than layers of adaptivity or queries to the random oracle.

2 Negative Results for Time-Lock Puzzles

In this section we give the intuition behind the proofs of our main results (Theorems 1 and 2) as well as the full proof of Theorem 2.

Following the seminal work of Impagliazzo and Rudich [21] on key agreement protocols in the random oracle model, our adversaries (for both theorems) attempt to find all *intersection queries* between the puzzle-generator f (Alice) and the solver g (Bob) — all queries made by both Alice and Bob. If successful, the adversary can then simulate an honest solver without asking additional queries. The novelty of our work is that we care not just about the total number of queries made by the adversary, but about the number of levels of adaptivity.

Both adversary constructions work in rounds, and query the random oracle only at the end of a round. Our aim is to reduce the total number of rounds (this is the "adaptivity level" of the adversary). The constructions differ in how they choose which queries to ask in each round, and in the corresponding proofs that the adversary succeeds in learning all of the intersection queries with high enough probability.

2.1 Intuition for Theorem 1

For the proof of Theorem 1, we use ideas (and a construction) of Impagliazzo and Rudich [21] (and later improvements by Barak and Mahmoody [4]), but modified to minimize the number of query rounds used by the adversary. Our attacker, Ivy, selects her queries to the random oracle in n rounds (n is the number of queries made by Alice). In round j, Ivy computes a set of *heavy* queries on which she will query the oracle at the end of the round. Heavy queries are those that have a high probability (given Ivy's view) of having been made by Alice, where "high" is a parameter that depends on n, m (the number of queries made by the honest solver) and ε (the probability with which Ivy is allowed to fail).

The intuition for why Ivy's attack works is that, as long as Bob has not hit any of Alice's "private" queries (those not made by Ivy), Bob doesn't know any more than Ivy about Alice's view. Thus, any private query must be "light" conditioned on his view. By definition, the probability that Bob hits a light query is small. We can then take a union bound over all of Bob's queries, and conclude that the total probability that Bob hits a private query is small.

Unfortunately, the intuition above isn't entirely correct: even querying an index that was not queried by Alice may give Bob information about Alice's view: the fact that Alice *didn't* query a particular index. We observe, however, that this is the *only* information about Alice's view that Bob can gain from making a non-intersection query. Thus, if we condition on Bob not having made any private intersection queries so far, our intuition still holds.

The main technical difficulty in the proof is making sure that Ivy can ask many queries in parallel, in order to bound the number of rounds of adaptivity. Our solution to this is to have Ivy condition her probability space after each query on the event that this query was *not an intersection query* (rather than on the response to the query). Since this event does not require Ivy to query the oracle in order to compute the new query probabilities, she can ask multiple parallel queries in each round. Loosely speaking, Ivy's query strategy ensures that if there are any remaining heavy queries, then one of her queries will be an intersection query with high probability. Since the number of intersection queries can be at most n, within n rounds Ivy can ensure that there are no remaining heavy queries.

Note that in order to find heavy queries, Ivy uses her unbounded computational power (e.g., if Alice queries the oracle on an index i and sends Bob an encryption of i, the index i is heavy conditioned on Ivy's view, but Ivy may have to break the encryption to find it).

2.2 Proof of Theorem 2

Javier, the adversary constructed in the proof of Theorem 2, works by running the honest solver in each round, but replacing its queries to the random oracle with a simulated oracle (so no queries to the real oracle are made during an execution). After the simulation, Javier updates the simulated oracle by querying the real oracle (in parallel) on every index that was queried during the simulated execution. The main idea in the proof is that, since the puzzle generator asks only n queries, there can be at most n rounds in which the simulated execution "hits" an intersection query that was not already known to Javier. In the remaining rounds, Javier does know all the intersection queries, and hence the simulated solver will behave just like the real honest solver (and output a correct solution to the puzzle with the same probability). Formally:

Proof (of Theorem 2). The adversary Javier follows Alg. 1. In the algorithm description, $Q_i(J)$ is the set of queries Javier made to H up to (but not including) round i, while $Q(B_i)$ is the set of queries the simulated Bob made to H_i in Javier's i^{th} round.

Algorithm 1. Javier's query algorithm on message M and oracle H with parameter ε

Randomly choose $i^* \in [n/\varepsilon]$.
for $i \in \{1, \dots, i^*\}$ **do**
  Run Bob to get: $x_i \leftarrow g^{H_i}(r_B, M)$ where H_i is an oracle that answers any query $x \in Q(J)$ the same as H does, and H_i answers any new query $q \notin Q(J)$ uniformly at random. Note that to run $g^{H_i}(r_B, M)$ we do not need to ask any new query to H because all the answers to queries in $Q(J)$ are already known and the rest are answered at random.
  Query H on all indices in $Q(B_i) \setminus Q_i(J)$ where $Q(B_i)$ is the queries Bob made to H_i.
Output x_{i^*}.

The total number of queries made by Javier is at most nm/ε and Javier's running time is $O(n/\varepsilon)$ times the running time of Bob. It remains to show that the probability that Javier's output is accepted by the solution validator is at least $1 - \nu - \varepsilon$.

Denote the event that Javier's output is accepted by the solution validator:

$$\text{Success} \overset{def}{=} (M, \mathcal{V}) \leftarrow f^H(r_A) \wedge x_{i^*} \leftarrow g^{H_{i^*}}(r_B, M) \wedge \mathcal{V}(x_{i^*}) = 1 \; .$$

Call a round i *good* if Javier did not ask any new intersection queries in round i (i.e., $Q(B_i) \cap Q(A) \subseteq Q_i(J)$). Denote Good_i the event that round i was good. Since Alice asks at most n queries, there can be at most n rounds that are not good. Thus, $\Pr[\text{Good}_{i^*}] \geq 1 - \varepsilon$. Note that if Good_{i^*} holds, the tuple $(f^H(r_A), g^{H_{i^*}}(r_B, M))$ is distributed identically to $(f^H(r_A), g^H(r_B, M))$, because as long as Bob's queries in round i^* were not queried by Alice, H and H_{i^*} both choose their answers at random and independently of all previous queries and answers. Therefore, for an arbitrary event E defined over the joint view of $f^H(r_A)$ and that of Javier till the end of round i^*, to know the quantity $\Pr[E \wedge \text{Good}_{i^*}]$ it does not matter weather we use the oracle H or H_{i^*} in round i^*, and the probabilities will remain the same. By misusing the notation, we also use Good_{i^*} to refer to the similar event when Javier uses the oracle H in his simulation of Bob in round i^*. Thus, we finally conclude:

$$\begin{aligned}
\Pr[\text{Success}] &\geq \Pr[\text{Success} \wedge \text{Good}_{i^*}] \\
&= \Pr\left[(M, \mathcal{V}) \leftarrow f^H(r_A); x_{i^*} \leftarrow g^{H_{i^*}}(r_B, M); \mathcal{V}(x_{i^*}) = 1 \wedge \text{Good}_{i^*}\right] \\
&= \Pr\left[(M, \mathcal{V}) \leftarrow f^H(r_A); x \leftarrow g^H(r_B, M); \mathcal{V}(x) = 1 \wedge \text{Good}_{i^*}\right] \\
&\geq \Pr\left[(M, \mathcal{V}) \leftarrow f^H(r_A); x \leftarrow g^H(r_B, M); \mathcal{V}(x) = 1\right] - (1 - \Pr[\text{Good}_{i^*}]) \\
&\geq 1 - \nu - \varepsilon.
\end{aligned}$$

3 A Time-Lock Puzzle with a Linear Difficulty Gap

In this section we give the construction and proof for Theorem 3.

In the description below, we omit the security parameter k: the security parameter is only used to determine the range of the random oracle — we assume w.l.o.g. that $H(q)$ returns k bits (if the random oracle returns fewer bits, we can interpret a query $H(q)$ as concatenation of multiple queries (e.g., $H(kq) \odot H(kq+1) \odot \cdots \odot H(kq+k-1)$). To further simplify notation, our definition of f only generates the message M. The (implicit) solution validator $\mathcal{V}_M$ checks whether its input is equal to f's input (our soundness proof is slightly stronger — we show that no adversary making less than n serial rounds of queries to H can find *any* valid preimage of M under f).

We define the puzzle-generating function f to be:

$$f^H(x_0, \ldots, x_n) \stackrel{def}{=} (x_0, H(x_0) \oplus x_1, \ldots, H(x_{n-1}) \oplus x_n)$$

(where the input is interpreted as $n+1$ k-bit query indices).

The honest solver g inverts f by running Algorithm 2:

Proof (Sketch for Theorem 3). By inspection, f can be computed with n non-adaptive queries: the values $H(x_1), \ldots, H(x_n)$ can be obtained in parallel. The correctness of the honest inverter (Alg. 2) and the fact that it uses n serial queries is also easy to see.

The main part of the proof is to show that every inverter making poly(k) queries to H needs to use at least n rounds of adaptive queries. To prove this, we first claim that any algorithm that outputs x_{i+1} with non-negligible probability must query H on

Algorithm 2. Honest solver g on input $M = (M_0, \ldots, M_n)$ and oracle H

$x_0 \leftarrow M_0$ // x_0 is not "encrypted".
for $i \in \{1, \ldots, n\}$ **do**
  $x_i \leftarrow H(x_{i-1}) \oplus M_i$ // "decrypting" x_i requires an oracle query on index x_{i-1}.
Output $(x_0, \ldots, x_n)$

x_i. This is because, even taken together, the value of $f^H(x_0, \ldots, x_n)$, the values of $\{x_0, \ldots, x_n\} \setminus \{x_{i+1}\}$ and the responses of the random oracle on all queries except x_i give no information (in the information-theoretic sense) about x_{i+1}. Thus, the probability that an algorithm outputs x_{i+1} without querying H on x_i is negligible in k (the output size of H). Note that this remains true if we allow the algorithm to output a polynomial number of guesses for x_{i+1}.

Now, consider an algorithm h making multiple rounds of queries to the oracle H, such that in each round the indices queried depend only on the responses from previous rounds. We can think of h as also outputting the indices it queries in each round (and the total number of indices output by h is polynomial in k). If h correctly inverts f on input $M = f^H(x_0, \ldots, x_n)$, it must output x_n at some round (since f is injective). By induction (and using the reasoning above), the probability that h first outputs (queries) x_i and x_{i+1} in the same round is negligible (since we showed it must query x_i before x_{i+1}). Therefore, the algorithm must use at least n rounds of adaptivity.

3.1 Increasing the Computation/Communication Ratio

Note that while our positive construction of a time-lock puzzle in the random-oracle model is optimal with respect to query complexity, the description of a puzzle that requires n adaptive queries to solve is also linear in n. When the cost of communication is comparable to an oracle query, simply communicating the puzzle takes $O(n)$ time, negating the benefit of parallel queries. We improve this ratio arbitrarily by replacing each oracle call with d composed calls (i.e., each call querying the oracle on the response to the previous call). This will increase both the (parallel) generation and solution time by a factor of d without changing the size of the puzzle description. Formally, let $H^{(1)} \stackrel{def}{=} H$ and for $d > 1$ let $H^{(d)}(q) \stackrel{def}{=} H(H^{(d-1)}(q))$. Then the function $f^{H^{(d)}}(x_0, \ldots, x_n)$ can be computed with d rounds of n non-adaptive queries, and the soundness condition from Theorem 3 holds with the increased parameters:

Claim. For every oracle function h that makes less than dn serial rounds of queries to H and poly(k) queries overall to H in total,

$$\Pr\left[(M, \mathcal{V}_M) \leftarrow f^{H^{(d)}}(k, r); x \leftarrow h^H(k, r_J, M); \mathcal{V}_M(x) = 1\right] = \text{neg}(k)$$

(where neg is some negligible function in k).

Proof (Sketch). The main idea is that any algorithm that outputs $H^{(i)}(x)$ with non-negligible probability must query H on $H^{(i-1)}(x)$ (otherwise the algorithm has no information about $H^{(i)}(x)$). By induction, it follows that an algorithm that makes only a polynomial number (in k) of queries to H needs d adaptive rounds to compute $H^{(d)}(x)$. Composing this idea with the induction in the proof of Theorem 3, we get the required parameters.

4 Discussion and Open Questions

The most obvious open question relating to time-lock puzzles is finding constructions based on assumptions other than the difficulty of factoring. Although this work rules out black-box constructions (with a super-constant gap) from one-way permutations and collision-resistant hash functions, we have no reason to believe that time-lock puzzles based on other concrete problems (e.g., lattice-based problems) do not exist. Extending our approach to other general assumptions (e.g., trapdoor permutations) is also an interesting open problem.

One of the motivations for looking at time-lock puzzles in the random-oracle model is the search for puzzles that are resistant to quantum attack. In this direction there still remains work to be done: on the positive side, our construction may not be secure against adversaries with quantum access to the random oracle (e.g., Dagdelen et al. show protocols that are secure in the random oracle model but can be broken by attackers with quantum access to the random oracle [15]). On the other hand, when the honest parties are quantum, the lower bound question is still open as well (Brassard and Salvail [10] and, independently, Biham et al [5], give a quantum version of Merkle puzzles that require the adversary to make $n^{3/2}$ queries in order to recover the shared key, but do not prove optimality).

Acknowledgements. We thank the anonymous reviewers for their helpful comments and suggestions.

References

1. Abadi, M., Burrows, M., Manasse, M.S., Wobber, T.: Moderately hard, memory-bound functions. ACM Trans. Internet Techn. 5(2), 299–327 (2005)
2. Back, A.: Hashcash — a denial of service counter-measure (2002), `http://www.hashcash.org/papers/hashcash.pdf`
3. Barak, B., Mahmoody, M.: Lower bounds on signatures from symmetric primitives. In: FOCS 2007, pp. 680–688. IEEE Computer Society, Los Alamitos (2007)
4. Barak, B., Mahmoody-Ghidary, M.: Merkle puzzles are optimal — an $O(n^2)$-query attack on any key exchange from a random oracle. In: Halevi, S. (ed.) CRYPTO 2009. LNCS, vol. 5677, pp. 374–390. Springer, Heidelberg (2009)
5. Biham, E., Goren, Y.J., Ishai, Y.: Basing weak public-key cryptography on strong one-way functions. In: Canetti, R. (ed.) TCC 2008. LNCS, vol. 4948, pp. 55–72. Springer, Heidelberg (2008)
6. Boneh, D., Boyen, X.: Efficient selective-id secure identity-based encryption without random oracles. In: Cachin, C., Camenisch, J.L. (eds.) EUROCRYPT 2004. LNCS, vol. 3027, pp. 223–238. Springer, Heidelberg (2004)
7. Boneh, D., Boyen, X.: Secure identity based encryption without random oracles. In: Franklin, M. (ed.) CRYPTO 2004. LNCS, vol. 3152, pp. 443–459. Springer, Heidelberg (2004)

8. Boneh, D., Franklin, M.K.: Identity-based encryption from the weil pairing. SIAM J. Comput. 32(3), 586–615 (2003)
9. Boneh, D., Naor, M.: Timed commitments. In: Bellare, M. (ed.) CRYPTO 2000. LNCS, vol. 1880, pp. 236–254. Springer, Heidelberg (2000)
10. Brassard, G., Salvail, L.: Quantum merkle puzzles. In: ICQNM, pp. 76–79. IEEE Computer Society, Los Alamitos (2008)
11. Canetti, R., Goldreich, O., Halevi, S.: The random oracle methodology, revisited. J. ACM 51(4), 557–594 (2004)
12. Canetti, R., Halevi, S., Katz, J.: A forward-secure public-key encryption scheme. In: Biham, E. (ed.) EUROCRYPT 2003. LNCS, vol. 2656, pp. 255–271. Springer, Heidelberg (2003)
13. Cathalo, J., Libert, B., Quisquater, J.-J.: Efficient and non-interactive timed-release encryption. In: Qing, S., Mao, W., López, J., Wang, G. (eds.) ICICS 2005. LNCS, vol. 3783, pp. 291–303. Springer, Heidelberg (2005)
14. Di Crescenzo, G., Ostrovsky, R., Rajagopalan, S.: Conditional oblivious transfer and timed-release encryption. In: Stern, J. (ed.) EUROCRYPT 1999. LNCS, vol. 1592, pp. 74–89. Springer, Heidelberg (1999)
15. Dagdelen, O., Fischlin, M., Lehmann, A., Schaffner, C.: Random oracles in a quantum world. Cryptology ePrint Archive, Report 2010/428 (2010), http://eprint.iacr.org/2010/428.pdf
16. Dwork, C., Goldberg, A., Naor, M.: On memory-bound functions for fighting spam. In: Boneh, D. (ed.) CRYPTO 2003. LNCS, vol. 2729, pp. 426–444. Springer, Heidelberg (2003)
17. Dwork, C., Naor, M.: Pricing via processing or combatting junk mail. In: Brickell, E.F. (ed.) CRYPTO 1992. LNCS, vol. 740, pp. 139–147. Springer, Heidelberg (1993)
18. Dwork, C., Naor, M., Wee, H.: Pebbling and proofs of work. In: Shoup, V. (ed.) CRYPTO 2005. LNCS, vol. 3621, pp. 37–54. Springer, Heidelberg (2005)
19. Gennaro, R., Gertner, Y., Katz, J., Trevisan, L.: Bounds on the efficiency of generic cryptographic constructions. SIAM J. Comput. 35(1), 217–246 (2005)
20. Grover, L.K.: A fast quantum mechanical algorithm for database search. In: STOC 1996, pp. 212–219. ACM, New York (1996)
21. Impagliazzo, R., Rudich, S.: Limits on the provable consequences of one-way permutations. In: STOC 1989, pp. 44–61. ACM, New York (1989)
22. May, T.C.: Timed-release crypto (February 1993), http://www.hks.net/cpunks/cpunks-0/1460.html
23. Merkle, R.C.: Secure communications over insecure channels. Commun. ACM 21(4), 294–299 (1978)
24. Rivest, R.L., Shamir, A.: Payword and micromint: Two simple micropayment schemes. In: Lomas, M. (ed.) Security Protocols 1996. LNCS, vol. 1189, pp. 69–87. Springer, Heidelberg (1997)
25. Rivest, R.L., Shamir, A., Wagner, D.A.: Time-lock puzzles and timed-release crypto. Technical Report MIT/LCS/TR-684, MIT (February 1996)

Physically Uncloneable Functions in the Universal Composition Framework

Christina Brzuska, Marc Fischlin, Heike Schröder, and Stefan Katzenbeisser

Darmstadt University of Technology
Center for Advanced Security Research Darmstadt

Abstract. Recently, there have been numerous works about hardware-assisted cryptographic protocols, either improving previous constructions in terms of efficiency, or in terms of security. In particular, many suggestions use Canetti's universal composition (UC) framework to model hardware tokens and to derive schemes with strong security guarantees in the UC framework. In this paper, we augment this approach by considering Physically Uncloneable Functions (PUFs) in the UC framework. Interestingly, when doing so, one encounters several peculiarities specific to PUFs, such as the intrinsic non-programmability of such functions. Using our UC notion of PUFs, we then devise efficient UC-secure protocols for basic tasks like oblivious transfer, commitments, and key exchange. It turns out that designing PUF-based protocols is fundamentally different than for other hardware tokens. For one part this is because of the non-programmability. But also, since the functional behavior is unpredictable even for the creator of the PUF, this causes an asymmetric situation in which only the party in possession of the PUF has full access to the secrets.

1 Introduction

Cryptographic protocols which simultaneously satisfy high efficiency demands as well as strong security requirements (like composable security), are scarce. One recent trend in this regard is to use the potential of hardware components like signature cards [20], one-time programs [14], standard smart cards [19], or even more complex tokens [21]. Most of these hardware-assisted protocols actually achieve security in Canetti's universal composition (UC) framework [4] and thus provide strong security guarantees.

1.1 Physically Uncloneable Functions

In this paper, we consider another type of hardware component which recently gained a lot of attention because of the irresistible progress in their realization: Physically Uncloneable Functions (PUFs) [27,26]. Basically, a PUF is a noisy device derived through a complex physical manufacturing process such that the behavior of the PUF is hard to clone. The PUF itself can be evaluated by a physical stimulus (aka. challenge) on which it provides a noisy response.

P. Rogaway (Ed.): CRYPTO 2011, LNCS 6841, pp. 51–70, 2011.

Modeling PUFs appropriately is a highly non-trivial task. Most importantly, there are different types of PUFs with different (physical) properties. Furthermore, there does not seem to be a general agreement upon common security properties of PUFs, even for a single type (e.g., whether a PUF is one-way or not, or if the output is pseudorandom). See [29,23] for more information. We thus consider a very minimalistic model which basically says that only the party in possession of a PUF can evaluate it by sending some stimulus to the PUF and observing the output, and where learning outputs for some stimuli does not facilitate the task of predicting the function's output for other stimuli.

There have been several approaches to define PUFs cryptographically, see [27,17,12,2,29,11,30]. However, these definitions usually are either rather informal, or follow the more stringent game-based approach, but stipulate uncloneability and tamper-resistance as an external property "outside of the game". A recent exception is the work of Armknecht et al. [1] which provides a game-based definition for uncloneability on a physical level. In the UC world, such features are more handy to specify. We hence follow previous approaches for other token-based protocols to model PUFs formally in the UC framework, exposing several peculiarities for this kind of hardware.

1.2 PUFs and the UC Framework

The UC Framework. The UC framework supports an easy modeling of tamper-proof hardware tokens via ideal functionalities. Roughly, the ideal functionality captures the abstract security properties of the token, and one considers a hybrid world in which real-world protocols and parties also have access to this ideal functionality (and thus the token). This is the approach which has been used extensively in the literature [21,19,24,16,15] and which we also use to model PUFs in the UC framework, in particular, to model restricted access depending on possession of the token or uncloneability.

In its original form the hybrid model supports the decomposition of cryptographic tasks into basic building blocks and to conclude security of protocols which are composed out of such building blocks. Loosely speaking, Canetti's composition theorem—or, actually a corollary of a more general statement—says that, if a protocol $\pi^{\mathcal{F}}$ UC-securely realizes some functionality $\mathcal{G}$ in the hybrid world with efficient functionality $\mathcal{F}$, and some protocol ρ UC-securely realizes $\mathcal{F}$, then the composed protocol π^{ρ} which invokes ρ whenever π would call $\mathcal{F}$, also UC-securely realizes $\mathcal{G}$.

PUFs in the UC Framework. Our ideal functionality for PUFs only allows the party in possession to stimulate it in order to retrieve a response, thus ensuring restricted access. Uncloneability is enforced through unpredictability. Parties can hand over the PUF to other parties. During transition, we allow the adversary temporary access before the PUF reaches the recipient. This models the classical example of PUF-augmented credit cards, sent via postal service, which are read out before getting delivered. As in other works about hardware based tokens, we assume some kind of tamper-evidence in the sense that the receiver can later verify the authenticity and integrity of the PUF. We note that this need not be ensured by the PUF technology itself. One may also consider reliable delivery (in which case the adversary may have read-only access during the manufacturing process).

Our ideal functionality covers different kinds of PUF technologies and comprises even PUFs with small input or output size (in which case unpredictability should be understood relative to the small output size). We note that for designing secure *protocols*, the intermediate access of the adversary also necessitates that the challenge space of the PUF is super-polynomial; else the adversary could clone the PUF easily. This domain requirement may currently not be true for all kinds of PUF technology; we comment on this in the full version of the paper.

The usage of hardware components in the UC context, especially of PUFs, causes several unpleasant side effects, though. At foremost, PUFs are not known to be implementable by probabilistic polynomial-time (PPT) Turing machines; the manufacturing process seems to be inherently based on physical properties. Hence, while the claims in the hybrid model are technically sound, any realization in practice through actual PUFs leaves a gap in the security claim of the composed protocol, as, strictly speaking, the composition theorem only applies to *probabilistic polynomial-time computable* functionalities $\mathcal{F}$. Fortunately, Canetti [4] proves the composition theorem to hold for a broader class of interactive Turing machines, and we sketch in the full version of this paper that the same holds for PUFs.

The uninstantiability of PUFs through efficient algorithms causes another issue when it comes to complex cryptographic protocols. For any PUF-based protocol relying on further cryptographic assumptions like the hardness of computing discrete logarithms, the assumption would need to hold relative to the additional computational power given through PUFs. That is, the underlying problem must be hard to solve even for attackers with "more than probabilistic polynomial-time power". It is therefore advantageous to avoid additional cryptographic assumptions in protocols and provide solutions with statistical security.

Non-Programmability. For PUFs, another aspect is the intrinsic *non-programmability* of these tokens: Even the manufacturer usually has no control over the functional behavior of the PUF. Hence, the ability of the ideal-world simulator to adapt the outcome of a PUF measurement adaptively, as guaranteed when modeling the PUF through an ideal functionality in the hybrid world, appears to be exceedingly optimistic. A similar observation has been made by Nielsen [25] about the (non-)programmability of random oracles in the UC framework. Roughly, Nielsen takes away the ability of the simulator to program the random oracle by giving the environment direct access to the random oracle. To support the argument in favor of non-programmable PUFs we also note that for random oracles it is straightforward to program consistently given a partial view of the function for other values, namely, by providing independent random values; for PUFs this is less clear since one would need to take the (not necessarily efficiently computable) conditional distribution of the specific PUF type into account.

We adopt Nielsen's approach and augment the environment's ability by giving it also access to the concrete PUF instantiation used in a protocol. Unlike in the case of the publicly available random oracle, though, the environment can only access this PUF when it is in possession of the adversary, i.e., we assume that a PUF, once in possession of the user, can only be accessed by this user. This corresponds to an honest user who prevents further unauthorized access. In a stronger version one could also allow further "uncontrolled" interaction between the environment, i.e., other protocols, and the PUF even when in possession of the honest user. This would somehow correspond to a permanently shared PUF functionality in the GUC model [5]. However, many advantages of deploying PUFs for designing efficient protocols would then disappear. With the restriction on temporary access we can still devise efficient solutions, e.g., circumventing impossibility results for UC commitment schemes in the plain model [6] and for GUC commitment schemes in the common reference string model [5].

1.3 PUF-Based Protocols in the UC Framework

We finally exemplify the usability of our PUF modeling by presenting PUF-based protocols for three classical areas: oblivious transfer (OT), commitments, and key exchange. Our protocols are UC-secure in the hybrid world (where we grant the environment access to the PUF instantiation as described above), and typically require only a few operations besides PUF evaluations. In particular, all protocols require only sending one party a token in the first step. The protocols do not rely on additional cryptographic assumptions, except for authenticated channels.

Designing PUF-based protocols is not just a matter of adopting other token-based solutions. One reason is clearly the non-programmability property which is usually not stipulated for other tokens (cf. [21,14]). In fact, most protocols take advantage of the ability to adapt the token's outputs on the fly. But more importantly, the main difference between PUFs and other tokens is that PUFs are by nature even unpredictable for the manufacturer. It follows that only the party in possession of the PUF has full access to the secrets; other parties may only draw from a small set of previously sampled values. In comparison, for the wrapper tokens [21], for example, the creator still knows the program placed inside the token, and the token holder can fully access this program in a black-box way. Hence, both parties somehow share a complete view of the secret. For PUFs the situation is rather "asymmetric".

Our oblivious transfer protocol bears some similarity to a PUF-based protocol of Rührmair [28]. His protocol, however, has a high round complexity due to an interactive hashing step. Still, [28] points out that, using symmetry of oblivious transfer [32] in the sense that one can change the roles of sender and receiver, one obtains an oblivious transfer protocol in which the other party sends the PUF. We confirm that this symmetry also holds in the UC setting.

Designing a UC-secure commitment scheme with the help of our PUFs turns out to be quite challenging. The non-programmability of our PUFs inhibits equivocality, a property which allows to adapt committed values appropriately, and which is usually required for such commitments [6]. We therefore use our PUF-based oblivious transfer protocol to derive a UC-secure bit commitment scheme. Interestingly, while the standard construction of commitment schemes out of OT [9] uses cut-and-choose techniques with a linear number of oblivious transfers, our transformation does not add any significant overhead. It only needs a single execution of the OT protocol and one extra message. We were not able to trace this idea back to any previous work.

A noteworthy aspect is that, while our OT protocol only withstands static corruptions, our derived commitment scheme is secure in presence of adaptive corruptions. The reason is that for commitments, in contrast to OT, the receiver does not obtain any external input; the values used in the OT sub protocol are chosen internally. This facilitates the simulation of the receiver's side. Hence, if we use our transformation we derive an adaptively secure commitment protocol from a concrete statically secure OT protocol!

Finally, our key exchange protocol follows the folklore approach of using the PUF to transport the key, only that our protocol is stated and formalized in the UC framework. That is, the sender samples some challenge/response pairs, sends the PUF, and later reveals a challenge to the receiver who recovers the image with the help of the PUF. Both parties use the images as the key, after applying a fuzzy extractor for error correction and smoothing the output. It is clear that for a key exchange protocol where only one party sends a PUF, some additional, one-sided authentication mechanism is required. Else the adversary with temporary access to the PUF could impersonate the honest sender.

All our protocols allow to re-use the PUF for multiple executions. By the unpredictable nature of PUFs, however, it is clear that the number of executions must be fixed in advance and must be known to the parties: The sender, once having sent the PUF, cannot access the PUF anymore and must thus challenge the PUF before sufficiently often, unless the PUF is frequently exchanged or further PUF tokens are sent. Note that in this case, attacks such as described in [31] will also be covered by the security proof. An interesting feature of PUFs is that, unlike other hardware tokens (e.g., [21]), protocols using PUFs are automatically secure against reset attacks because they implement (noisy) functions.

2 Physically Uncloneable Functions

A Physically Uncloneable Function (PUF) is a source of randomness that is implemented by a physical system. Roughly speaking, the randomness of PUFs relies on uncontrollable manufacturing variations during their fabrication. For PUF evaluation, the physical system is queried with a stimulus, usually called *challenge*. The device then produces a physical output, which is usually referred to as *response*. A pair of a stimulus and an output is called a *challenge/response pair* (CRP). Furthermore, a PUF, being a physical system, might not necessarily implement a mathematical function, i.e., querying the PUF twice on the same challenge may yield distinct responses. However, we require such "noise" to be bounded so that the two responses are closely related in terms of distance.

2.1 Defining PUFs

A PUF-family $\mathcal{P}$ consists of two (not necessarily efficient) algorithms Sample and Eval. The index sampling algorithm Sample which obtains as input the security parameter and returns as output an index id of the PUF family corresponds to the PUF fabrication process. The evaluation algorithm Eval takes as input a challenge c, evaluates the PUF on c, and generates as output the corresponding response r.

Note that we require the challenge space to be equal to a full set of strings of a certain length. For some classes of PUFs, this is naturally satisfied, for example arbiter PUFs and SRAM PUFs For others types this can be achieved through appropriate encoding, as for angles in optical PUFs.

Definition 1 (Physically Uncloneable Functions). *Let rg indicate the dimension of the range of the PUF responses of PUF-family, and let d_{noise} be a bound on the PUF's noise. A pair $\mathcal{P} = (\mathsf{Sample}, \mathsf{Eval})$ is a family of (rg, d_{noise})-PUFs if it satisfies the following properties:*

Index Sampling. *Let $\mathcal{I}_\lambda$ be an index set. The sampling algorithm* Sample *outputs, on input the security parameter 1^λ, an index $\mathsf{id} \in \mathcal{I}_\lambda$. We do not require that the index sampling can be done efficiently. Each index $\mathsf{id} \in \mathcal{I}_\lambda$ corresponds to a set $\mathcal{D}_{\mathsf{id}}$ of distributions. For each challenge $c \in \{0,1\}^\lambda$, $\mathcal{D}_{\mathsf{id}}$ contains a distribution $\mathcal{D}_{\mathsf{id}}(c)$ on $\{0,1\}^{rg(\lambda)}$. We do* not *require that $\mathcal{D}_{\mathsf{id}}$ has a short description or an efficient sampling algorithm.*

Evaluation. *The evaluation algorithm* Eval *gets as input a tuple $(1^\lambda, \mathsf{id}, c)$, where $c \in \{0,1\}^\lambda$. It outputs a response $r \in \{0,1\}^{rg(\lambda)}$ according to distribution $\mathcal{D}_{\mathsf{id}}(c)$. It is* not *required that* Eval *is a PPT algorithm.*

Bounded Noise. *For all indices $\mathsf{id} \in \mathcal{I}$, for all challenges $c \in \{0,1\}^\lambda$, we have that when running $\mathsf{Eval}(1^\lambda, \mathsf{id}, c)$ twice, then the Hamming distance of any two outputs r_1, r_2 of the algorithm is smaller than $d_{\mathsf{noise}}(\lambda)$.*

Instead of $\mathcal{D}_{\mathsf{id}}(c)$, we usually write $\mathsf{PUF}_{\mathsf{id}}(c)$. Moreover, if misunderstandings are unlikely to occur, we write $\mathcal{D}(c)$ instead of $\mathcal{D}_{\mathsf{id}}(c)$ and PUF instead of $\mathrm{PUF}_{\mathsf{id}}$. Finally, we usually write rg instead of $rg(\lambda)$ and $\mathcal{I}$ instead of $\mathcal{I}_\lambda$.

2.2 Security of PUFs

Various security properties of PUFs have been introduced in the literature (see [1,23,29] for overviews) such as unpredictability, uncloneability, bounded noise, uncorrelated outputs, one-wayness, and tamper-evidence. We give a detailed analysis of these properties in the full version of this paper as well as the relation to our security notions. The main security properties of PUFs are *uncloneability* and *unpredictability*. Unpredictability is covered via an entropy condition on the PUF distribution. This condition also implies mild forms of uncloneability as well as uncorrelated outputs. Moreover, one usually requires that tampering with PUFs can be detected easily, the idea being that a user does not use the PUF anymore after detecting it has been tampered with. Our UC-functionality will cover this property implicitly, as we permit the adversary black-box access to the PUF and the choice of delivering the PUF or not. Tampering with the PUF is treated as not delivering it. For an explicit treatment of tamper-evidencen.

We will now turn to our main security definition of PUFs, namely the unpredictability. The behavior of the PUF on input a challenge c should be unpredictable, i.e., have some significant amount of intrinsic entropy, even if the PUF has been measured before on several challenge values. Here, (*conditional*) *min-entropy* is the main tool. It indicates the residual min-entropy on a response value for a challenge c, when one has already measured the PUF on (not necessarily different) challenges $c_1, ..., c_\ell$ before. Since the random responses are not under adversarial control we can look at the residual entropy for the answer to r by taking the (weighted) average over all possible response values $r_1, ..., r_\ell$. Demanding that a PUF has a certain *average* min-entropy [10] is weaker than asking for all possible responses $r_1, ..., r_\ell$, that the residual entropy remains above a certain level. This weaker requirement suffices for our purposes. However, as the challenges c are chosen by the adversary, we ask the average min-entropy to be high for all challenges and defined by the maximal probability of a possible response r.

Definition 2 (Average Min-Entropy). *The average min-entropy of* $\mathsf{PUF}(c)$ *conditioned on the measurements of challenges* $\mathcal{C} = (c_1, \ldots, c_\ell)$ *is defined by*

$$\begin{aligned}&\tilde{H}_\infty(\mathsf{PUF}(c)|\mathsf{PUF}(\mathcal{C}))\\ &\quad := -\log\left(\mathbb{E}_{r_i \leftarrow \mathsf{PUF}(c_i)}\left[\max_r \Pr[\mathsf{PUF}(c) = r | r_1 = \mathsf{PUF}(c_1), \ldots, r_\ell = \mathsf{PUF}(c_\ell)]\right]\right)\\ &\quad := -\log\left(\mathbb{E}_{r_i \leftarrow \mathsf{PUF}(c_i)}\left[2^{-H_\infty(\mathsf{PUF}(c)|r_1=\mathsf{PUF}(c_1),\ldots,r_\ell=\mathsf{PUF}(c_\ell))}\right]\right),\end{aligned}$$

where the probability is taken over the choice of id *from* $\mathcal{I}$ *and the choice of possible PUF responses on challenge c. The term* $\mathsf{PUF}(\mathcal{C})$ *denotes a sequence of random variables* $\mathsf{PUF}(c_1), ..., \mathsf{PUF}(c_\ell)$ *each corresponding to an evaluation of the PUF on challenge* c_k.

We occasionally also write $\tilde{H}_\infty(\mathsf{PUF}(c)|\mathcal{C})$ as an abbreviation for $\tilde{H}_\infty(\mathsf{PUF}(c)|\mathsf{PUF}(\mathcal{C}))$. We now turn to our definition of unpredictability, which is derived from the notion of unpredictability for random variables.

Definition 3 (Unpredictability). *A* (rg, d_{noise})*-PUF family* $\mathcal{P} = (\mathsf{Sample}, \mathsf{Eval})$ *for security parameter* 1^λ *is* $(d_{\mathsf{min}}(\lambda), m(\lambda))$*-unpredictable if for any* $c \in \{0,1\}^\lambda$ *and any challenge list* $\mathcal{C} = (c_1, \ldots, c_\ell)$*, one has that, if for all* $1 \leq k \leq \ell$ *the Hamming distance satisfies* $\mathsf{dis}_{\mathsf{ham}}(c, c_k) \geq d_{\mathsf{min}}(\lambda)$*, then the average min-entropy satisfies* $\tilde{H}_\infty(\mathsf{PUF}(c)|\mathsf{PUF}(\mathcal{C})) \geq m(\lambda)$*. Such a PUF-family is called a* $(rg, d_{\mathsf{noise}}, d_{\mathsf{min}}, m)$*-PUF family.*

Note that one could also define a computational version of unpredictability via computational min-entropy (aka. HILL entropy, named after [18]) where the entropy is defined via the entropy of computationally indistinguishable random variables. All proofs considered in this paper carry through when replacing statistical by computational min-entropy; we nonetheless use the statistical variant for sake of simplicity. As explained in the introduction, indistinguishability for defining computational min-entropy then needs to be considered with respect to distinguishers that have PUF power (see also full version), and not with respect to mere PPT algorithms.

Also, one could define unpredictability in terms of a game where an efficient adversary, after seeing some challenge/response pairs, tries to predict the response for another challenge which is not within close distance to the previous queries (see full version); the success probability should then be negligible. Clearly, the PUF would need super-logarithmic min-entropy in the above sense to make it unpredictable according to this game, but the lower bound on the entropy would vary with the adversary. We do not take this approach because the fuzzy extractors, which are necessary to eliminate the noise of a PUF, usually need a *fixed* lower bound on the min-entropy in order to be applicable. Also, while it is easy to incorporate a distributional property as above into the ideal functionality, using game-based properties to specify abstract and ideal security requirements appears to be very peculiar.

2.3 PUFs and Fuzzy Extractors

By nature, PUF evaluation is noisy, so that same stimuli results in closely related but different outputs. We use fuzzy extractors of Dodis et al. [10] to convert noisy, high-entropy measurements of PUFs into reproducible random values.

An (m, ℓ, t, ϵ)-fuzzy extractor consists of a pair of algorithms $(\mathsf{Gen}, \mathsf{Rep})$. The generation algorithm Gen takes as input a noisy measurement w and generates as output an ℓ-bit secret st together with helper data p. The helper data can be stored publicly, since it does not reveal information about the secret: as long as the measurement contains m bits of min-entropy the secret has statistical distance ϵ to the uniform distribution, even if given p. The helper data is later used to reproduce the same secret st from related measurements within a certain distance t according to some metric space.

We now determine parameters to combine a PUF and the fuzzy extractor in order to achieve almost uniformly random values. Let λ be the security parameter. We let the parameters of the fuzzy extractor depend on the parameters of the PUF. Assume that we have a $(rg(\lambda), d_{\mathsf{noise}}(\lambda), d_{\mathsf{min}}(\lambda), m(\lambda))$-PUF family with d_{min} being in the order of $o(\lambda/\log\lambda)$. We now determine the corresponding parameters for the fuzzy extractor as follows. Let $\ell(\lambda) := \lambda$ be the length parameter for value st. Let $\epsilon(\lambda)$ be a negligible function and let $t(\lambda) = d_{\mathsf{noise}}(\lambda)$. For each λ, let $(\mathsf{Gen}, \mathsf{Rep})$ be a $(m(\lambda), \ell(\lambda), t(\lambda), \epsilon(\lambda))$-fuzzy extractor. The metric space $\mathcal{M}$ is $\{0,1\}^{rg(\lambda)}$ with Hamming distance $\mathsf{dis}_{\mathsf{ham}}$.[1]

[1] Note that such fuzzy extractors only exist if $rg(\lambda)$ and $m(\lambda)$ are sufficiently large. In order to achieve this, several PUFs can be combined. When combining two PUFs of the same family, rg gets doubled and so does m. Thus, if there are PUFs with $m(\lambda)$ being non-negligible they can be combined to a useful PUF-family — even if a PUF-family has *less than one bit* entropy, it still can be combined to obtain a good PUF-family with outputs which has high entropy of many bits. Thus, we may assume that the PUF has corresponding parameters.

Definition 4. *If a PUF and a fuzzy extractor* $(\mathsf{Gen}, \mathsf{Rep})$ *satisfy the above requirements, then they are said to have* matching parameters.

If a PUF and a fuzzy extractor have matching parameters, then the following properties of a well-spread domain, extraction independence and response consistency hold. A formal proof is given in the full version of this paper.

Well-Spread Domain: For all polynomials $p(\lambda)$ and all sets of challenges $c_1, ..., c_{p(\lambda)}$, the probability of a random challenge to be within distance smaller d_{min} of any of the c_k is negligible.

Extraction Independence: For all challenges $c_1, ..., c_{p(\lambda)}$, it holds that the PUF evaluation on a challenge c with $\mathsf{dis}(c_k, c) > d_{\mathsf{min}}$ for all $1 \leq k \leq p(\lambda)$ and subsequent application of Gen yields an almost uniform value st even for those who observe p.

Response Consistency: The fuzzy extractor maps two evaluations of the same PUF to the same random string, i.e., if PUF is measured on challenge c twice and returns r and r', then for $(st, p) \leftarrow \mathsf{Gen}(r)$, one has $st \leftarrow \mathsf{Rep}(r', p)$.

3 Universally Composable Security and PUFs

We model PUFs in the universal composition framework introduced by Canetti in [4]. Note that we use, among other things, well-studied UC basics, such as authenticated message transmissions.

3.1 Modeling PUFs in UC

In the following we propose an ideal functionality $\mathcal{F}_{\mathsf{PUF}}$ that will model PUFs. The functionality is presented in Figure 1 and handles the following operations: (1) a party P_i is allocated a PUF; (2) P_i can query the PUF; (3) P_i gives the PUF to another party P_j who can also query the device; (4) an adversary can query the PUF during transition.

The functionality $\mathcal{F}_{\mathsf{PUF}}$ maintains a list $\mathcal{L}$ of tuples $(\mathtt{sid}, P_i, \mathsf{id}, \tau)$ where $\mathtt{sid}$ is the (public) session identifier and id is the (internal) PUF-identifier, essentially describing the output distribution. Note that the PUF itself does not use $\mathtt{sid}$. The element $\tau \in \{\mathsf{trans}(P_j), \mathsf{notrans}\}$ denotes whether the PUF is in transition to P_j. For $\mathsf{trans}(P_j)$, indicating that the PUF is in transition to P_j, the adversary is able to query the PUF. In turn, if it is set to notrans then only the possessing party can query the PUF.

The PUF functionality $\mathcal{F}_{\mathsf{PUF}}$ is indexed by the PUF parameters $(rg, d_{\mathsf{noise}}, d_{\mathsf{min}}, m)$ and gets the security parameter λ in unary encoding as additional input. It is required to satisfy the bounded noise property for $d_{\mathsf{noise}}(\lambda)$ and the unpredictability property for $(d_{\mathsf{min}}(\lambda), m(\lambda))$. This enforces that the outputs obey the basic entropic requirements of PUFs (analogously to the requirement for the random oracle functionality to produce random and independent outputs). We write $\mathcal{F}_{\mathsf{PUF}}$ and $\mathcal{F}_{\mathsf{PUF}}(rg, d_{\mathsf{noise}}, d_{\mathsf{min}}, m)$ interchangeably.

$\mathcal{F}_{\mathsf{PUF}}(rg, d_{\mathsf{noise}}, d_{\mathsf{min}}, m)$ receives as initial input a security parameter 1^λ and runs with parties $P_1, ..., P_n$ and adversary $\mathcal{S}$.

- Whenever a party P_i writes $(\mathsf{init}_{\mathsf{PUF}}, \mathtt{sid}, P_i)$ on the input tape of $\mathcal{F}_{\mathsf{PUF}}$ then $\mathcal{F}_{\mathsf{PUF}}$ checks whether $\mathcal{L}$ already contains a tuple $(\mathtt{sid}, *, *, *, *)$:
  - ⋆ If this is the case then turn into the waiting state.
  - ⋆ Else, draw $\mathsf{id} \leftarrow \mathsf{Sample}(1^\lambda)$ from the PUF family. The functionality $\mathcal{F}_{\mathsf{PUF}}$ puts the following tuple in $\mathcal{L}$: $(\mathtt{sid}, \mathsf{id}, P_i, *, \mathsf{notrans})$ and writes $(\mathsf{initialized}_{\mathsf{PUF}}, \mathtt{sid})$ on the communication input tape of P_i.
- Whenever a party P_i writes $(\mathsf{eval}_{\mathsf{PUF}}, \mathtt{sid}, P_i, c)$ on $\mathcal{F}_{\mathsf{PUF}}$'s input tape then $\mathcal{F}_{\mathsf{PUF}}$ first checks, if there exists a tuple $(\mathtt{sid}, \mathsf{id}, P_i, \mathsf{notrans})$ in $\mathcal{L}$:
  - ⋆ If this is not the case then turn into the waiting state.
  - ⋆ Else, run $r \leftarrow \mathsf{Eval}(1^\lambda, \mathsf{id}, c)$ and write $(\mathsf{eval'ed}_{\mathsf{PUF}}, \mathtt{sid}, c, r)$ on P_i's communication input tape.
- Whenever a party P_i sends $(\mathsf{handover}_{\mathsf{PUF}}, \mathtt{sid}, P_i, P_j)$ to $\mathcal{F}_{\mathsf{PUF}}$ then $\mathcal{F}_{\mathsf{PUF}}$ first checks, if there exists a tuple $(\mathtt{sid}, *, P_i, \mathsf{notrans})$ in $\mathcal{L}$:
  - ⋆ If this is not the case then turn into the waiting state.
  - ⋆ Else, modify the tuple $(\mathtt{sid}, \mathsf{id}, P_i, \mathsf{notrans})$ to the updated tuple $(\mathtt{sid}, \mathsf{id}, \perp, \mathsf{trans}(P_j))$. Write $\mathsf{invoke}_{\mathsf{PUF}}(\mathtt{sid}, P_i, P_j)$ on $\mathcal{S}$'s communication input tape to indicate that a $\mathsf{handover}_{\mathsf{PUF}}$ occurs between P_i and P_j.
- Whenever the adversary writes $(\mathsf{eval}_{\mathsf{PUF}}, \mathtt{sid}, \mathcal{S}, c)$ on the input tape of $\mathcal{F}_{\mathsf{PUF}}$ then $\mathcal{F}_{\mathsf{PUF}}$ first checks, if $\mathcal{L}$ contains a tuple $(\mathtt{sid}, \mathsf{id}, \perp, \mathsf{trans}(*))$:
  - ⋆ If this is not the case then turn into the waiting state.
  - ⋆ Else, run $r \leftarrow \mathsf{Eval}(1^\lambda, \mathsf{id}, c)$ and return $(\mathsf{eval'ed}_{\mathsf{PUF}}, \mathtt{sid}, c, r)$ to $\mathcal{S}$.
- Whenever the adversary writes $(\mathsf{ready}_{\mathsf{PUF}}, \mathtt{sid}, \mathcal{S})$ on $\mathcal{F}_{\mathsf{PUF}}$'s input tape then $\mathcal{F}_{\mathsf{PUF}}$ searches for a tuple $(\mathtt{sid}, \mathsf{id}, \perp, \mathsf{trans}(P_j))$ in $\mathcal{L}$:
  - ⋆ If such a tuple does not exist then turn into the waiting state.
  - ⋆ Else, modify the tuple $(\mathtt{sid}, \mathsf{id}, \perp, \mathsf{trans}(P_j))$ to the updated tuple $(\mathtt{sid}, \mathsf{id}, P_i, \mathsf{notrans})$. Write the message $(\mathsf{handover}_{\mathsf{PUF}}, \mathtt{sid}, P_i)$ on P_j's communication input tape and store the tuple $(\mathsf{received}_{\mathsf{PUF}}, \mathtt{sid}, P_i)$.
- Whenever the adversary sends $(\mathsf{received}_{\mathsf{PUF}}, \mathtt{sid}, P_i)$ to $\mathcal{F}_{\mathsf{PUF}}$, $\mathcal{F}_{\mathsf{PUF}}$ checks if a tuple $(\mathsf{received}_{\mathsf{PUF}}, \mathtt{sid}, P_i)$ has been stored. If so, it writes this tuple to the communication input tape of P_i. Else, $\mathcal{F}_{\mathsf{PUF}}$ turns into the waiting state.

Fig. 1. The ideal functionality $\mathcal{F}_{\mathsf{PUF}}$ for PUFs

Also note that our definition requires that a PUF is somehow certified. That is, the adversary cannot replace a PUF sent to an honest party by a fake token including some "software emulation"; the adversary can only measure the PUF when in transition. The receiver can verify the constitution and authenticity of the received hardware. Our functionality also implies that the sender knows when the PUF has been delivered to the receiver. Relaxing this requirement is delicate as the adversary could then still be in possession of the PUF. Formally, delivery confirmation can be ensured by having the receiver send an acknowledgment message (via an authenticated channel).

3.2 Non-programmability

As explained in the introduction we envision a non-programmable version of PUFs. The functionality above, if used in the standard way within the hybrid model, would be programmable, though, because the environment would not have direct access (even if the PUF is in possession of the adversary). One way to enforce non-programmability is to switch to the extended UC (EUC) model [5] where all parties, including the environment, share the above functionality.

The PUF could then also be evaluated by the environment in which case the simulator is informed about the challenge and response.

To simplify we linger within the basic UC framework and instead allow the environment to dispatch special PUF queries to the adversary/simulator. This query needs to be answered faithfully by forwarding it to a genuine PUF instance, and the response is handed back to the environment. Put differently, we put some restriction on the how the simulator behaves, formally giving a UC-security proof which would transfer to the EUC model.

4 Oblivious Transfer with PUFs

In a 1-out-of-2 oblivious transfer (OT) protocol the sender possesses two secrets $\mathsf{s}_0, \mathsf{s}_1$ and the receiver holds a selection bit $b \in \{0, 1\}$, thereby choosing one of the two secrets. A 1-out-of-2 OT-protocol assures that at the end of the protocol execution, the receiver learns the secret s_b, but nothing about s_{1-b}, and the sender does not learn anything about the selection bit b.

Oblivious Transfer is a widely used cryptographic primitive for many cryptographic applications [22,9,13]. However, in many of those applications a bottleneck of OT is the computational requirements since, for instance, several public key operations are necessary. We here show how to avoid the number of public key operations by adopting hardware. In the following, we recall the oblivious transfer ideal functionality and then provide a PUF-based oblivious transfer protocol. As noted in the introduction, we envision a scenario in which the PUF is used multiple times. In the plain UC model, however, a fresh PUF would need be sent for each OT execution. An alternative would be to switch to the joint-state theorem (JUC) [8] for the UC framework. However, JUC applies a transformation to the original protocol, and if a single session of a PUF protocol requires to hand over a PUF once, the JUC transformation would also require a handover per session. Nothing would be gained. Thus, we define and analyze multi-session protocols instead of the more common one-session protocols.

4.1 The Oblivious Transfer Ideal Functionality

1-out-of-2 oblivious transfer is an interaction between a sender P_i and a receiver P_j where the environment $\mathcal{Z}$ provides P_i with two inputs $\mathsf{s}_0, \mathsf{s}_1$ and P_j with an input bit b. As soon as both parties provided their inputs (and the simulator $\mathcal{S}$ allows delivery), the ideal functionality returns the secret s_b to the receiver. The ideal functionality for oblivious transfer $\mathcal{F}_{\mathsf{OT}}$ is given in Figure 2. We stress that this functionality only supports static corruption and can be used a bounded number of times, and only by the parties which have exchanged the PUF. Each execution will be accompanied by a unique sub session identifier `ssid`.

4.2 Oblivious Transfer Scheme

In Figure 3, we provide an oblivious transfer protocol. For simplicity of exposition, we use the following notation. For a possibly empty set $\mathcal{C}$ we let

$\mathcal{F}_{\mathrm{OT}}$ is parameterized by an integer N and receives as input a security parameter 1^λ, and runs with parties $P_1, ..., P_n$ and adversary $\mathcal{S}$. The functionality initially sets $(n, S, R) = (1, \perp, \perp)$.
In the following, the functionality ignores any input if $n > N$, or if $n > 1$ and $(S, R) \neq (P_i, P_j)$ for the parties' identities (P_i, P_j) in the input. Else,

- Whenever P_i writes $(\mathsf{send}_{\mathrm{OT}}, \mathtt{sid}, \mathtt{ssid}, P_i, P_j, (\mathsf{s}_0, \mathsf{s}_1))$ with $\mathsf{s}_0, \mathsf{s}_1 \in \{0,1\}^\lambda \cup \{\perp\}$ on $\mathcal{F}_{\mathrm{OT}}$'s input tape, $\mathcal{F}_{\mathrm{OT}}$ stores $(\mathsf{send}_{\mathrm{OT}}, \mathtt{sid}, \mathtt{ssid}, P_i, P_j, (\mathsf{s}_0, \mathsf{s}_1))$ and writes $(\mathsf{send}_{\mathrm{OT}}, \mathtt{sid}, \mathtt{ssid}, P_i, P_j)$ to the communication input tape of $\mathcal{S}$. The functionality increments n to $n + 1$ and stores $(S, R) = (P_i, P_j)$ if $n = 2$ now.
- Whenever P_j writes $(\mathsf{choose\text{-}secret}_{\mathrm{OT}}, \mathtt{sid}, \mathtt{ssid}, P_i, P_j, b)$ on the input tape of $\mathcal{F}_{\mathrm{OT}}$, the functionality $\mathcal{F}_{\mathrm{OT}}$ stores this tuple and writes $(\mathsf{choose\text{-}secret}_{\mathrm{OT}}, \mathtt{ssid}, \mathtt{sid}, P_i, P_j)$ on the input tape of $\mathcal{S}$.
- When $\mathcal{S}$ writes $(\mathsf{deliver}_{\mathrm{OT}}, \mathtt{sid}, \mathtt{ssid}, P_i, P_j)$ on $\mathcal{F}_{\mathrm{OT}}$'s input communication tape then $\mathcal{F}_{\mathrm{OT}}$ checks if tuples $(\mathsf{send}_{\mathrm{OT}}, \mathtt{sid}, \mathtt{ssid}, P_i, P_j, (\mathsf{s}_0, \mathsf{s}_1))$ and $(\mathsf{choose\text{-}secret}_{\mathrm{OT}}, \mathtt{sid}, \mathtt{ssid}, P_i, P_j, b)$ have been stored. If so, write $(\mathsf{deliver}_{\mathrm{OT}}, \mathtt{sid}, \mathtt{ssid}, P_i, P_j, \mathsf{s}_b)$ on the input communication tape of P_j.

Fig. 2. The ideal functionality for oblivious transfer adapted from [4]

Sender P_i	session sid	Receiver P_j
		$(\mathsf{init}_{\mathsf{PUF}}, \mathtt{sid}_0, P_i, \lambda)$
		$k = 1, ..., N$: $c_k \leftarrow \{0,1\}^\lambda$
		$r_k \leftarrow (\mathsf{eval}_{\mathsf{PUF}}, \mathtt{sid}_0, P_i, c_k)$
$\mathcal{C} := \emptyset$	$\xleftarrow{(\mathsf{handover}_{\mathsf{PUF}}, \mathtt{sid}_0, P_i, P_j)}$	$\mathcal{L} := (c_1, r_1, ..., c_\ell, r_\ell)$, $\mathcal{C} := \emptyset$
	Repeat at most N times with fresh ssid	
Input: $\mathsf{s}_0, \mathsf{s}_1 \in \{0,1\}^\lambda$, sid		Input: $b \in \{0,1\}$, sid
$x_0, x_1 \xleftarrow{\$} \{0,1\}^\lambda$	$\xrightarrow{(\mathsf{send}_{\mathsf{auth}}, \mathtt{sid}, \mathtt{ssid}, P_i, P_j, (x_0, x_1))}$	Draw $(c, r) \xleftarrow{\$} \mathcal{L}$
		$v := c \oplus x_b$, $c' := c \oplus x_0 \oplus x_1$
$\mathsf{dis}(v \oplus x_0, \mathcal{C}) > d_{\mathsf{min}}$?	$\xleftarrow{(\mathsf{send}_{\mathsf{auth}}, \mathtt{sid}, \mathtt{ssid}, P_i, P_j, v)}$	$\mathsf{dis}(c, \mathcal{C}) > d_{\mathsf{min}}$?
$\mathsf{dis}(v \oplus x_1, \mathcal{C}) > d_{\mathsf{min}}$?		$\mathsf{dis}(c', \mathcal{C}) > d_{\mathsf{min}}$?
Add $v \oplus x_0$, $v \oplus x_1$ to $\mathcal{C}$		Add c, c' to $\mathcal{C}$
$r'_0 \leftarrow (\mathsf{eval}_{\mathsf{PUF}}, \mathtt{sid}_0, P_i, v \oplus x_0)$		Delete (c, r) in $\mathcal{L}$
$r'_1 \leftarrow (\mathsf{eval}_{\mathsf{PUF}}, \mathtt{sid}_0, P_i, v \oplus x_1)$		
$(st_0, p_0) \leftarrow \mathsf{Gen}(r'_0)$		
$(st_1, p_0) \leftarrow \mathsf{Gen}(r'_1)$		
$\mathsf{S}_0 := \mathsf{s}_0 \oplus st_0$,	$\xrightarrow{(\mathsf{send}_{\mathsf{auth}}, \mathtt{sid}, \mathtt{ssid}, P_i, P_j, (\mathsf{S}_0, p_0, \mathsf{S}_1, p_1))}$	$st'_b \leftarrow \mathsf{Rep}(r, p_b)$
$\mathsf{S}_1 := \mathsf{s}_1 \oplus st_1$		$\mathsf{s}_b = \mathsf{S}_b \oplus st'_b$

Fig. 3. Oblivious transfer scheme with PUFs

$\mathsf{dis}(c, \mathcal{C}) > d_{\mathsf{min}}$ denote the check that each element c_i in $\mathcal{C}$ satisfies the bound $\mathsf{dis}(c, c_i) > d_{\mathsf{min}}$. If not, we assume that the corresponding party aborts. Also, when interacting with the PUF (functionality), we simply write for example $r \leftarrow (\mathsf{eval}_{\mathsf{PUF}}, \mathtt{sid}_0, P_i, c)$ to denote the fact that, for a call $(\mathsf{eval}_{\mathsf{PUF}}, \mathtt{sid}_0, P_i, c)$ the functionality has replied with $(\mathsf{eval'ed}_{\mathsf{PUF}}, \mathtt{sid}_0, c, r)$. Here, $\mathtt{sid}_0$ is the session identifier for $\mathcal{F}_{\mathsf{PUF}}$, as opposed to sid and ssid for the oblivious transfer protocol.

We note that the protocol does not achieve perfect completeness in the sense that executions between honest parties may fail. The probability for this is negligible, though. This follows straightforwardly again from the fact that the domain is well-spread: All (at most polynomial) challenges are independent random values such that one is within small distance of the others with negligible probability only. If all challenges are sufficiently far apart, the receiver always obtains the correct value.

We now sketch the security arguments for the OT-protocol in Figure 3, i.e., at the end of the OT protocol (1) a malicious sender learns nothing about the bit b and (2) a malicious receiver learns only the secret s_b and remains oblivious about s_{1-b}. For case (1), the receiver chooses the challenge c at random. Thus, $v = c \oplus x_b$ hides x_b information-theoretically and thus also b. We now consider case (2). For simplicity, assume that $b = 0$. Then, the sender shall remain oblivious about any information about s_1. If st_1 looks uniform to the sender, then s_1 is information-theoretically hidden. If the fuzzy extractor and the PUF have matching parameters (see Definition 4), then with overwhelming probability this is the case, as — due to the well-spread domain property (see Subsection 2.3) — the probability that the receiver queried the PUF on values c_k with $\mathsf{dis}_{\mathsf{ham}}(c_k, v \oplus x_1) < d_{\mathsf{min}}$ is negligible, and the checks on the sender side about list $\mathcal{L}$ provided that the sender does not reveal PUF responses to critical challenges.

Theorem 1. *Assuming that* $(\mathsf{Gen}, \mathsf{Rep})$ *is a* (m, ℓ, t, ϵ)*-fuzzy generator and that* $\mathsf{PUF} = (\mathsf{Sample}, \mathsf{Eval})$ *is a PUF-family with matching parameters (see Definition 4), then protocol* PUFOT *securely realizes the functionality* $\mathcal{F}_{\mathsf{OT}}$ *in the* $\mathcal{F}_{\mathsf{PUF}}$*-hybrid model.*

Security holds in a statistical sense, i.e., the environment's views in the two worlds are statistically close. This remains true for unbounded algorithms $\mathcal{A}, \mathcal{S}$, and $\mathcal{Z}$, as long as the number of PUF evaluations is polynomially bounded. The proof is delegated to the full version of the paper.

4.3 Oblivious Transfer with Sender-PUF

Our OT-protocol requires the receiver to send a PUF to the sender. Sometimes it may be desirable to have the sender prepare the PUF, though. This can be achieved by switching the roles of the sender and the receiver via the protocol by Wolf and Wullschleger [32], but at the expense of having to run linear many OT executions for strings of length λ. This is unavoidable since the receiver in an OT-protocol just enters a bit such that, when acting as a sender, it can only transmit a single bit. In this protocol the sender of the outer OT-protocol acts as a receiver in the inner OT-protocol, thus sending the PUF.

The protocol in [32] requires only a single round of additional communication. It is UC-secure in the $\mathcal{F}_{\mathsf{OT}}$-hybrid world and inherits the security properties (statistical vs. computational security, and adaptive vs. static corruptions). With a linear overhead [3] and another extra round of communication one can then get an OT-protocol for strings, which is also UC-secure in the $\mathcal{F}_{\mathsf{OT}}$-hybrid world for bit-functionality $\mathcal{F}_{\mathsf{OT}}$. The final protocol is now a UC-secure OT-protocol for strings with linear many calls to $\mathcal{F}_{\mathsf{OT}}$, a few extra rounds, and inheriting all security characteristics from $\mathcal{F}_{\mathsf{OT}}$.

5 PUF-Based Commitment Scheme

A commitment scheme is a two-party protocol between a sender and a receiver where the sender (also called committer) first sends a disguised version of the value to the receiver such that, later, only this value can be revealed. More precisely, a commitment scheme allows the committer to compute to a value msg a pair $(\mathsf{com}, \mathsf{decom})$ such that com reveals nothing about the value msg but using the pair $(\mathsf{com}, \mathsf{decom})$ one can open msg. Moreover it should be infeasible to find a value decom' such that $(\mathsf{com}, \mathsf{decom}')$ reveals $\mathsf{msg}' \neq \mathsf{msg}$.

5.1 The Commitment Scheme Ideal Functionality

In the UC world, the commitment scheme is realized by the (bounded) functionality $\mathcal{F}_{\mathsf{com}}$ as follows: $\mathcal{F}_{\mathsf{com}}$ receives an input $(\mathsf{commit}, \mathtt{sid}, \mathtt{ssid}, \mathsf{msg})$ from some committer P_i where msg is the value committed to. After verifying the validity of the session identifier $\mathtt{sid}$, $\mathcal{F}_{\mathsf{com}}$ records the value msg. Subsequently, the functionality lets both the receiver P_j and the adversary $\mathcal{S}$ know that the committer has committed to some value by computing a public delayed output $(\mathsf{receipt}, \mathtt{sid}, \mathtt{ssid})$ and sending it to P_j (this phase is called the commitment phase).

To initiate the decommitment phase, the committer P_i sends $(\mathsf{open}, \mathtt{sid}, \mathtt{ssid})$ to the functionality $\mathcal{F}_{\mathsf{com}}$. Thereupon, $\mathcal{F}_{\mathsf{com}}$ checks if there exists a value msg; if so, the functionality computes a public delayed output $(\mathsf{open}, \mathtt{sid}, \mathtt{ssid}, \mathsf{msg})$ and sends it to P_j. When the adversary corrupts the committer by sending $(\mathsf{corrupt\text{-}committer}, \mathtt{sid}, \mathtt{ssid})$ to $\mathcal{F}_{\mathsf{com}}$, the functionality reveals the recorded value msg to the adversary $\mathcal{S}$. Furthermore, if the $\mathsf{receipt}$ value was not yet delivered to P_j, then $\mathcal{F}_{\mathsf{com}}$ allows the adversary to modify the committed value. This is in order to deal with adaptive corruptions. The ideal functionality for commitment schemes $\mathcal{F}_{\mathsf{com}}$ is given in Figure 4.

$\mathcal{F}_{\mathsf{com}}$ is parameterized by an integer N and runs with parties P_i, P_j, and adversary $\mathcal{S}$. It initially sets $(n, S, R) = (1, \perp, \perp)$.
The functionality ignores any commit-input if $n > N$, or if $n > 1$ and $(S, R) \neq (P_i, P_j)$ for the parties' identities in the input. Else,

- Upon receiving input $(\mathsf{commit}, \mathtt{sid}, \mathtt{ssid}, P_i, P_j, \mathsf{msg})$ from party P_i, $\mathcal{F}_{\mathsf{com}}$ proceeds as follows:
  - ⋆ Records msg, generate a public delayed output $(\mathsf{receipt}, \mathtt{sid}, \mathtt{ssid})$, and send the output to P_j. Increment n to $n+1$ and store $(S, R) = (P_i, P_j)$ if now $n = 2$.
- Upon receiving input $(\mathsf{open}, \mathtt{sid}, \mathtt{ssid})$ from party P_i, $\mathcal{F}_{\mathsf{com}}$ proceeds as follows:
  - ⋆ If a value msg is recorded, generate a public delayed output $(\mathsf{open}, \mathtt{sid}, \mathtt{ssid}, \mathsf{msg})$ and send it to P_j.
  - ⋆ Otherwise, do nothing.
- Upon receiving the input $(\mathsf{corrupt\text{-}committer}, \mathtt{sid}, \mathtt{ssid})$ from the adversary $\mathcal{S}$, $\mathcal{F}_{\mathsf{com}}$ proceeds as follows:
  - ⋆ Send the value msg to $\mathcal{S}$.
  - ⋆ If $\mathcal{S}$ provides a value msg' and the $\mathsf{receipt}$ output was not yet written on P_j's tape, then $\mathcal{S}$ can change the recorded value to msg'.

Fig. 4. The ideal functionality for commitment schemes adapted from [4]

5.2 PUF-Based Commitment Scheme

We now provide a *universal* transformation from OT-protocols to bit commitment schemes which—to our knowledge—has not been considered so far. Previous transformations [22,9] rely on cut-and-choose and require linear many executions of the OT-protocol. Our transformation only requires a single additional message to be sent after executing the OT-protocol. The main idea of the protocol in Figure 5 is to inverse the roles of the sender and the receiver. The OT-protocol transfers two secrets, and the committer only learns one of them, namely the one corresponding to its secret bit b. This secret is then used to open the commitment.

The main idea is to inverse the roles of the sender and the receiver. The commitment protocol uses the OT-protocol as a building block or, more precisely, since we work in the UC framework, the corresponding ideal OT-functionality. Consider a commitment scheme with OT-sender P_i and OT-receiver P_j. Then, P_j is the committer and has a secret bit b which it submits to the OT-functionality. The receiver P_i draws two sufficiently long random strings s_0 and s_1 which it submits to the OT-functionality. The OT-functionality then provides P_j with the secret s_b. This terminates the commitment phase.

In the opening phase, the committer P_j sends the pair (b, s_b). The receiver P_i then checks whether s_b matches the b-th secret. The protocol is binding, as the OT-functionality does not allow to modify the secret bit b and the secret s_{1-b} is statistically hidden from P_i. Thus, P_i can determine s_{1-b} only with negligible probability. The protocol is hiding, as the OT-functionality does not reveal information about the bit b to P_i.

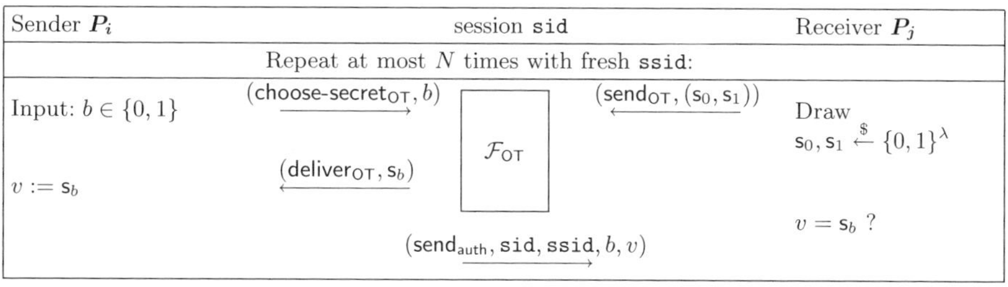

Fig. 5. Commitment scheme with $\mathcal{F}_{\mathsf{OT}}$

Theorem 2. *The commitment protocol in Figure 5 securely UC-realizes the functionality $\mathcal{F}_{\mathsf{com}}$ in the $\mathcal{F}_{\mathsf{OT}}$-hybrid model.*

If functionality $\mathcal{F}_{\mathsf{OT}}$ is replaced by some OT-protocol, then the derived commitment protocol basically inherits the characteristics of the OT-protocol. That is, it is secure against adaptive corruptions if OT is, and it is statistically secure if OT is. Remarkably, we show in the next section that our PUF-based OT-protocol, while being only statically secure, makes the commitment scheme even adaptively secure.

We merely provide a proof sketch for Theorem 2 here. Note that in the case where both users are honest, only the modeling of the final message needs to be taken into consideration. The simulator learns the secret bit b from the commitment functionality $\mathcal{F}_{\mathsf{com}}$. It then draws a random string v from $\{0,1\}^\lambda$ and sends $(\mathsf{send}_{\mathsf{auth}}, \mathtt{sid}, \mathtt{ssid}, b, v)$. If the receiver is dishonest, then it provides $\mathcal{F}_{\mathsf{OT}}$ with two secrets $\mathsf{s}_0, \mathsf{s}_1$. The simulator lets the sender provide a random bit b' to the simulated $\mathcal{F}_{\mathsf{OT}}$. It receives back the secret $\mathsf{s}_{b'}$. In the opening phase, $\mathcal{S}$ learns the (real) secret bit b from the commitment functionality and simulates the final protocol message as $(\mathsf{send}_{\mathsf{auth}}, \mathtt{sid}, \mathtt{ssid}, b, \mathsf{s}_b)$. If the sender is corrupt then it provides the simulated $\mathcal{F}_{\mathsf{OT}}$ with a secret bit b. The simulator creates two random strings $\mathsf{s}_0, \mathsf{s}_1$ and passes them to the simulated $\mathcal{F}_{\mathsf{OT}}$ which passes s_b to the receiver. The simulator commits to the sender's bit b in the ideal world. If the sender sends a message $(\mathsf{send}_{\mathsf{auth}}, \mathtt{sid}, \mathtt{ssid}, b, v)$ then $\mathcal{S}$ checks if $\mathsf{s}_b = v$. If so, it instructs $\mathcal{F}_{\mathsf{com}}$ to open the commitment. All simulations are perfect.

5.3 Adaptively Secure Commitments

Consider the concrete commitment protocol where we plug in our OT-protocol from the previous section into the abstract scheme above (and work in the $\mathcal{F}_{\mathsf{PUF}}$-hybrid model instead of the $\mathcal{F}_{\mathsf{OT}}$-hybrid model then), then we observe the following: In the commitment phase (i.e., the OT-phase), the message sent by the OT-receiver (the commitment sender) is statistically independent from its secret input: the OT-receiver merely sends a single uniformly random message.

For the OT-sender, this is not the case: When having access to the PUF, one can extract both secrets from the mere transcript of the protocol. This enables the simulator $\mathcal{S}$ to derive both secrets from the protocol, as it accesses the PUF. It can thus provide the simulated committer with open messages for both bit values. As the remaining part of the committer's state merely consists in challenge/response-pairs, the simulator can thus provide genuine internal state.

6 Key Exchange with PUFs

In a key exchange (ke) protocol two parties interact over an insecure network to establish a common secret key κ. This common secret key can then be used to build a secure channel or to ensure confidentiality of transmitted data.

6.1 The Key Exchange Ideal Functionality

The main idea of the key exchange ideal functionality $\mathcal{F}_{\mathsf{ke}}$ is the following: if both parties are honest, the functionality provides them with a common random value which is invisible to the adversary. If one of them is corrupted, though, the adversary determines the session key entirely thus modeling the participation of a corrupted party. The definition of the key exchange functionality $\mathcal{F}_{\mathsf{ke}}$ is depicted in Figure 6, adapted from [7].

$\mathcal{F}_{\mathsf{ke}}$ is parameterized by an integer N and receives as input a security parameter 1^λ, and runs with parties $P_1, ..., P_n$ and adversary $\mathcal{S}$. $\mathcal{F}_{\mathsf{ke}}$ obtains a list of corrupt parties. It initially sets $(n, S, R) = (1, \perp, \perp)$.

Ignore any $\mathsf{establish\text{-}session_{ke}}$-input if $n > N$, or if $n > 1$ and $(S, R) \neq (P_i, P_j)$ for the parties' identities in the input. Else,

- When a message $(\mathsf{establish\text{-}session_{ke}}, \mathtt{sid}, \mathtt{ssid}, P_i, P_j)$ is written on $\mathcal{F}_{\mathsf{ke}}$'s input tape by a party P_i. Then $\mathcal{F}_{\mathsf{ke}}$ stores the tuple $(\mathsf{establish\text{-}session_{ke}}, \mathtt{sid}, \mathtt{ssid}, P_i, P_j)$ (and refuses if there already is a tuple $(\mathsf{establish\text{-}session_{ke}}, \mathtt{sid}, \mathtt{ssid}, P_j, P_i)$ or a tuple $(\mathsf{establish\text{-}session_{ke}}, \mathtt{sid}, \mathtt{ssid}, P_i, P_j)$). $\mathcal{F}_{\mathsf{ke}}$ outputs $(\mathsf{establish\text{-}session_{ke}}, \mathtt{sid}, \mathtt{ssid}, P_i, P_j)$ to the adversary $\mathcal{S}$. If both users are honest then draw a random value κ from $\{0,1\}^\lambda$ and store the messages $(\mathsf{deliver_{ke}}, \mathtt{sid}, \mathtt{ssid}, \kappa, P_i)$ and $(\mathsf{deliver_{ke}}, \mathtt{sid}, \mathtt{ssid}, \kappa, P_j)$. Increment n to $n+1$.
- When $\mathcal{S}$ writes $(\mathsf{choose\text{-}value_{ke}}, \mathtt{sid}, \mathtt{ssid}, P_i, P_j, \kappa)$ on $\mathcal{F}_{\mathsf{ke}}$'s input tape then check whether there is a message $(\mathsf{establish\text{-}session_{ke}}, \mathtt{sid}, \mathtt{ssid}, P_i, P_j)$ or a message $(\mathsf{establish\text{-}session_{ke}}, \mathtt{sid}, \mathtt{ssid}, P_j, P_i)$ and whether at least one of the users P_i and P_j is corrupt. If so, store the messages $(\mathsf{deliver_{ke}}, \mathtt{sid}, \mathtt{ssid}, \kappa, P_i)$ and $(\mathsf{deliver_{ke}}, \mathtt{sid}, \mathtt{ssid}, \kappa, P_j)$.
- $\mathcal{S}$ writes $(\mathsf{deliver_{ke}}, \mathtt{sid}, \mathtt{ssid}, P_i)$ on $\mathcal{F}_{\mathsf{ke}}$'s input communication tape. Check if a tuple $(\mathsf{deliver_{ke}}, \mathtt{sid}, \mathtt{ssid}, \kappa, P_i)$ is stored. If so, write $(\mathsf{deliver_{ke}}, \mathtt{sid}, \mathtt{ssid}, \kappa, P_i)$ to P_i's input tape and delete $(\mathsf{deliver_{ke}}, \mathtt{sid}, \mathtt{ssid}, \kappa, P_i)$. Else, do nothing.

Fig. 6. The key exchange ideal functionality adapted from [7]

6.2 Minimal Requirements

We present a key exchange protocol in Section 6.3 which sends a PUF in a setup phase. Afterwards, a single message per protocol execution is sent via a unidirectional authenticated channel. As mentioned in the introduction, it is desirable to circumvent the use of complexity-theoretic assumptions. However, for practical reasons, PUF transfers should also be minimized. If only a single PUF transfer occurs, then the assumption of a unidirectional authenticated channel cannot be dropped: The sender of the PUF measured the PUF several times and sent it to the receiver. The adversary can query the PUF during its transition. If the sender does not have any further secret information for authentication, then the adversary can can carry out the same computations as the sender. Thus, the protocol cannot be secure against impersonation attacks. In the following, we use the standard bidirectional $\mathcal{F}_{\mathsf{auth}}$ functionality. Deriving corresponding unidirectional definitions is straightforward.

6.3 PUF-Based Key Exchange Scheme

Intuitively, our key exchange protocol proceeds as follows. In an enrollment phase, a server issues a PUF, measures for a set of randomly chosen challenges the corresponding responses, and finally ensures a noisy-free PUF measurement by generating for each response r a fuzzy extractor secret st from a set of random secrets as well as a corresponding helper data p. The server then sends the PUF to the client. Upon finishing the enrollment phase the server broadcasts a randomly chosen challenge c including its helper data p to the client and sets $\kappa = st$ to obtain the protocol key. The client evaluates the PUF on the challenge c, computes the corresponding fuzzy secret st due to the helper data p, and obtains the protocol key by setting $\kappa = st$. Consequently, both parties use the fuzzy extractor secret st as their common protocol key κ. We again note that

Server P_i		Client P_j
$(\mathsf{init}_{\mathsf{PUF}}, \mathtt{sid}, P_i, \lambda)$		
Repeat N times:		
$r \leftarrow (\mathsf{eval}_{\mathsf{PUF}}, \mathtt{sid}, P_i, c)$		
$(st, p) \leftarrow \mathsf{Gen}(r)$		
add (c, r, st, p) to $\mathcal{L}$	$\xrightarrow{(\mathsf{handover}_{\mathsf{PUF}}, \mathtt{sid}, P_i, P_j)}$	
	Repeat at most N times	
pick $(c, r, st, p) \leftarrow \mathcal{L}$		
remove the entry from $\mathcal{L}$	$\xrightarrow{(\mathsf{send}_{\mathsf{auth}}, \mathtt{sid}, \mathtt{ssid}, P_j, (c, p))}$	$r' \leftarrow (\mathsf{eval}_{\mathsf{PUF}}, \mathtt{sid}, P_j, c)$
$\kappa = st$		$st \leftarrow \mathsf{Rep}(r', p)$
		$\kappa = st$

Fig. 7. Key exchange scheme with PUFs

the sender is informed about the point in time when the receiver is in possession of the PUF.

Theorem 3. *Protocol* PUFKE *securely realizes functionality* $\mathcal{F}_{\mathsf{ke}}$ *in the* $\mathcal{F}_{\mathsf{PUF}}$*-hybrid model.*

In the following we merely provide a proof sketch for Theorem 3. The idea is that, for an honest sender, the simulator can easily emulate the setup phase by simply querying the PUF honestly. The simulator simply reveals these samples step by step. If the receiver is honest then, due to the well-spread domain, the adversary will most likely not have queried the PUF about any of the sampled values during the transition phase, such that all the derived keys are statistically indistinguishable from random. It follows that the simulation is statistically close to an actual protocol execution. Finally note that, if one of the parties in the key exchange protocol is corrupt, then the simulator can easily set the key to one of the obtained PUF measurements (after running the fuzzy extractor).

Acknowledgments. We thank the anonymous reviewers for valuable comments. Marc Fischlin is supported by the Emmy Noether Program Fi 940/2-1 of the German Research Foundation (DFG). This work was supported by CASED (`http://www.cased.de`).

References

1. Armknecht, F., Maes, R., Sadeghi, A.-R., Standaert, F.-X., Wachsmann, C.: A formal foundation for the security features of physical functions. To appear at IEEE S&P (2011)
2. Armknecht, F., Maes, R., Sadeghi, A.-R., Sunar, B., Tuyls, P.: Memory Leakage-Resilient Encryption Based on Physically Unclonable Functions. In: Matsui, M. (ed.) ASIACRYPT 2009. LNCS, vol. 5912, pp. 685–702. Springer, Heidelberg (2009)
3. Brassard, G., Crépeau, C., Robert, J.-M.: Information theoretic reductions among disclosure problems. In: FOCS, pp. 168–173. IEEE, Los Alamitos (1986)

4. Canetti, R.: Universally composable security: A new paradigm for cryptographic protocols. In: FOCS, pp. 136–145 (2001)
5. Canetti, R., Dodis, Y., Pass, R., Walfish, S.: Universally Composable Security with Global Setup. In: Vadhan, S.P. (ed.) TCC 2007. LNCS, vol. 4392, pp. 61–85. Springer, Heidelberg (2007)
6. Canetti, R., Fischlin, M.: Universally composable commitments. In: Kilian, J. (ed.) CRYPTO 2001. LNCS, vol. 2139, pp. 19–40. Springer, Heidelberg (2001)
7. Canetti, R., Krawczyk, H.: Universally composable notions of key exchange and secure channels. In: Knudsen, L.R. (ed.) EUROCRYPT 2002. LNCS, vol. 2332, pp. 337–351. Springer, Heidelberg (2002)
8. Canetti, R., Rabin, T.: Universal Composition with Joint State. In: Boneh, D. (ed.) CRYPTO 2003. LNCS, vol. 2729, pp. 265–281. Springer, Heidelberg (2003)
9. Crépeau, C.: Equivalence between Two Flavours of Oblivious Transfers. In: Pomerance, C. (ed.) CRYPTO 1987. LNCS, vol. 293, pp. 350–354. Springer, Heidelberg (1988)
10. Dodis, Y., Ostrovsky, R., Reyzin, L., Smith, A.: Fuzzy extractors: How to generate strong keys from biometrics and other noisy data. SIAM J. Comput. 38, 97–139 (2008)
11. Frikken, K.B., Blanton, M., Atallah, M.J.: Robust Authentication Using Physically Unclonable Functions. In: Samarati, P., Yung, M., Martinelli, F., Ardagna, C.A. (eds.) ISC 2009. LNCS, vol. 5735, pp. 262–277. Springer, Heidelberg (2009)
12. Gassend, B., van Dijk, M., Clarke, D.E., Torlak, E., Devadas, S., Tuyls, P.: Controlled physical random functions and applications. ACM Trans. Inf. Syst. Secur. 10(4) (2008)
13. Goldreich, O., Micali, S., Wigderson, A.: How to play any mental game or a completeness theorem for protocols with honest majority. In: STOC, pp. 218–229. ACM, New York (1987)
14. Goldwasser, S., Kalai, Y.T., Rothblum, G.N.: One-Time Programs. In: Wagner, D. (ed.) CRYPTO 2008. LNCS, vol. 5157, pp. 39–56. Springer, Heidelberg (2008)
15. Goyal, V., Ishai, Y., Mahmoody, M., Sahai, A.: Interactive Locking, Zero-Knowledge PCPs, and Unconditional Cryptography. In: Rabin, T. (ed.) CRYPTO 2010. LNCS, vol. 6223, pp. 173–190. Springer, Heidelberg (2010)
16. Goyal, V., Ishai, Y., Sahai, A., Venkatesan, R., Wadia, A.: Founding Cryptography on Tamper-Proof Hardware Tokens. In: Micciancio, D. (ed.) TCC 2010. LNCS, vol. 5978, pp. 308–326. Springer, Heidelberg (2010)
17. Guajardo, J., Kumar, S.S., Schrijen, G.-J., Tuyls, P.: FPGA Intrinsic PUFs and Their Use for IP Protection. In: Paillier, P., Verbauwhede, I. (eds.) CHES 2007. LNCS, vol. 4727, pp. 63–80. Springer, Heidelberg (2007)
18. Håstad, J., Impagliazzo, R., Levin, L.A., Luby, M.: A pseudorandom generator from any one-way function. SIAM J. Comput. 28(4), 1364–1396 (1999)
19. Hazay, C., Lindell, Y.: Constructions of truly practical secure protocols using standard smartcards. In: ACM CCS, pp. 491–500. ACM, New York (2008)
20. Hofheinz, D., Unruh, D., Müller-Quade, J.: Universally composable zero-knowledge arguments and commitments from signature cards. Tatra Mt. Math. Pub., 93–103 (2007)
21. Katz, J.: Universally Composable Multi-party Computation Using Tamper-Proof Hardware. In: Naor, M. (ed.) EUROCRYPT 2007. LNCS, vol. 4515, pp. 115–128. Springer, Heidelberg (2007)
22. Kilian, J.: Founding cryptography on oblivious transfer. In: STOC, pp. 20–31. ACM, New York (1988)

23. Maes, R., Verbauwhede, I.: hysically Unclonable Functions: a Study on the State of the Art and Future Research Directions, section 1. Towards Hardware-Intrinsic Security. Springer, Heidelberg (2010)
24. Moran, T., Segev, G.: David and Goliath Commitments: UC Computation for Asymmetric Parties Using Tamper-Proof Hardware. In: Smart, N.P. (ed.) EUROCRYPT 2008. LNCS, vol. 4965, pp. 527–544. Springer, Heidelberg (2008)
25. Nielsen, J.B.: Separating random oracle proofs from complexity theoretic proofs: The non-committing encryption case. In: Yung, M. (ed.) CRYPTO 2002. LNCS, vol. 2442, pp. 111–126. Springer, Heidelberg (2002)
26. Pappu, R., Recht, B., Taylor, J., Gershenfeld, N.: Physical one-way functions. Science 297, 2026–2030 (2002)
27. Pappu, R.S.: Physical One-Way Functions. Phd thesis, Massachusetts Institut of Technology (2001)
28. Rührmair, U.: Oblivious Transfer Based on Physical Unclonable Functions. In: Acquisti, A., Smith, S.W., Sadeghi, A.-R. (eds.) TRUST 2010. LNCS, vol. 6101, pp. 430–440. Springer, Heidelberg (2010)
29. Rührmair, U., Sölter, J., Sehnke, F.: On the foundations of physical unclonable functions. Cryptology ePrint Archive, Report 2009/277 (2009)
30. Sadeghi, A.-R., Visconti, I., Wachsmann, C.: Enhancing RFID Security and Privacy by Physically Unclonable Functions. Towards Hardware-Intrinsic Security. Springer, Heidelberg (2010)
31. Rührmair, C.J.U., Algasinger, M.: An attack on puf-based session key exchange and a hardware-based countermeasure: Erasable pufs. In: Proc. Financial Cryptoghraphy (2011)
32. Wolf, S., Wullschleger, J.: Oblivious Transfer Is Symmetric. In: Vaudenay, S. (ed.) EUROCRYPT 2006. LNCS, vol. 4004, pp. 222–232. Springer, Heidelberg (2006)

Computer-Aided Security Proofs for the Working Cryptographer*

Gilles Barthe[1], Benjamin Grégoire[2], Sylvain Heraud[2], and Santiago Zanella Béguelin[1]

[1] IMDEA Software Institute, Madrid, Spain
[2] INRIA Sophia Antipolis-Méditerranée, France

Abstract. We present EasyCrypt, an automated tool for elaborating security proofs of cryptographic systems from proof sketches—compact, formal representations of the essence of a proof as a sequence of games and hints. Proof sketches are checked automatically using off-the-shelf SMT solvers and automated theorem provers, and then compiled into verifiable proofs in the CertiCrypt framework. The tool supports most common reasoning patterns and is significantly easier to use than its predecessors. We argue that EasyCrypt is a plausible candidate for adoption by working cryptographers and illustrate its application to security proofs of the Cramer-Shoup and Hashed ElGamal cryptosystems.

Keywords: Provable security, verifiable security, game-based proofs, Cramer-Shoup cryptosystem, ElGamal encryption.

1 Introduction

The game-playing technique [8, 17, 20] is an established methodology for structuring cryptographic proofs. Its essence lies in giving precise mathematical descriptions, referred to as games, of the interaction between adversaries and oracle systems. Proofs are organized as sequences of games, starting from a game that represents a security goal (e.g. indistinguishability against chosen-ciphertext attacks), and proceeding to games that represent security assumptions (e.g. Decision Diffie-Hellman) by successive transformations that can be shown to preserve, or alter only slightly the overall security. In a typical step in a game-based proof the goal is to relate the probability of an event A in a game G to the probability of a possibly different event A' in a game G'. For example, the goal may be to establish an inequality of the form $\Pr[G : A] \leq \Pr[G' : A'] + \Delta$, where Δ is an arithmetic expression that depends on the number of oracle queries made by an adversary. The prevailing practice for proving the validity of such proof steps is to use standard mathematical tools, which interleave reasoning about the semantics of games with information-theoretic or arithmetical arguments.

* Partially funded by European Project FP7-256980 NESSoS, French project ANR SESUR-012 SCALP, Spanish project TIN2009-14599 DESAFIOS 10, and Madrid Regional project S2009TIC-1465 PROMETIDOS.

P. Rogaway (Ed.): CRYPTO 2011, LNCS 6841, pp. 71–90, 2011.

In the code-based approach to the game-playing technique [8,17] games are cast as probabilistic algorithms. The adoption of programming idioms allows to give precise definitions of games, and paves the way for applying programming language methods to justify proof steps rigorously. As anticipated by their proponents, code-based game-playing proofs are amenable to formal verification, and a number of tools provide support for building them. CryptoVerif [11] is a tool for conducting security proofs in a game-based setting in which games are modeled as processes and transitions are justified by means of process-algebraic concepts such as bisimulations. One strength of CryptoVerif, apart from being the first tool to have supported game-based proofs, is that it applies both to protocols and primitives; it has been successfully applied to verify Kerberos [10] and the Full-Domain Hash (FDH) signature scheme [12]. CertiCrypt [6] is another framework that allows for the interactive construction of game-based proofs in the Coq proof assistant [22]. One specificity of CertiCrypt is that proofs can be verified independently and automatically by a small trustworthy checker; it has been successfully applied to verify prominent cryptographic constructions, including OAEP [5], FDH [24], and zero-knowledge protocols [7].

While the developments based on CryptoVerif and CertiCrypt make a convincing case that computer-aided cryptographic proofs are indeed plausible, neither tool has reached a wide audience among cryptographers. In [5], we contrast the high guarantees given by CertiCrypt with the effort and expertise required to build machine-checked proofs, and conclude that cryptographers are unlikely to adopt verifiable security in its current form. In this sense, it can be considered that CryptoVerif and CertiCrypt only provide a partial realization of Halevi's programme of systematically building computer-aided cryptographic proofs [17].

The thesis of this article is that verifiable security can dramatically benefit from automation using state-of-the-art verification technology, and that verifiable game-based proofs can be constructed with only a moderate effort. The thesis is realized with the presentation of EasyCrypt, an automated tool that builds machine-checked proofs from *proof sketches*, which offer a machine-processable representation of the essence of a security proof. We argue that EasyCrypt is significantly easier to use than previous tools, making an important step towards the adoption of computer-aided security proofs by working cryptographers and hence towards fulfilling Halevi's programme. To substantiate our claim, we present computer-aided proofs of security of Hashed ElGamal encryption and the Cramer-Shoup cryptosystem.

EasyCrypt adopts the principled approach mandated by CertiCrypt to conduct game-based proofs and imposes a clear separation between program verification and information-theoretic reasoning. Transitions between games are justified in two steps: first, one proves logical relations between the games using probabilistic Relational Hoare Logic (pRHL); second, one applies information-theoretic reasoning to derive claims about the probability of events from pRHL judgments. We provide for each step highly effective mechanisms that build upon a combination of off-the-shelf and purpose-specific tools. Specifically, EasyCrypt implements an automated procedure that computes for any pRHL judgment a set of sufficient conditions for its validity, known as verification conditions. The outstanding feature of this procedure, and the key to the effectiveness of EasyCrypt, is that verification conditions are expressed in the language of first-order logic, without any mention of probability, and can be discharged automatically by state-of-the-art tools such as SMT solvers and theorem provers. The verification condition generator is *proof-producing*, in the sense that it generates Coq files that can be machine-checked using the CertiCrypt framework. Moreover, the connection to CertiCrypt makes it possible to benefit from the expressivity and flexibility of a general-purpose proof assistant for advanced verification goals that fall out of the scope of automated techniques. Additionally, EasyCrypt implements an automated mechanism for proving claims about probability. The mechanism combines some elementary rules to compute (bounds on) probabilities of events—e.g. the probability of a uniformly sampled element to belong to a list—with rules to derive (in)equalities between probabilities of events in games from judgments in pRHL. The combination of these tools with other more mundane features such as a limited form of specification inference for procedures provides substantial leverage towards making verifiable security practical and makes EasyCrypt a plausible candidate for adoption by working cryptographers.

2 Introductory Example: Hashed ElGamal Encryption

This section illustrates the application of EasyCrypt to a proof of IND-CPA security of Hashed ElGamal encryption in the Random Oracle Model. The example serves to introduce the notion of proof sketch and to give the reader an idea of the input that the tool expects. It also allows for a preliminary comparison between EasyCrypt and CertiCrypt. We refer the reader to [4] for a proof of the same result in CertiCrypt.

Hashed ElGamal is a variant of ElGamal encryption that does not require plaintexts to be elements of a group. Instead, plaintexts are bitstrings of a certain length k and group elements are mapped into bitstrings using a hash function $H : \mathcal{G} \rightarrow \{0,1\}^k$. Let $\mathcal{G}$ be a multiplicative cyclic group of order q with generator g. Formally, the scheme is defined by the following triple of algorithms:

$$\begin{array}{lcl} \mathcal{KG}() & \stackrel{\text{def}}{=} & x \xleftarrow{\$} \mathbb{Z}_q;\ \mathsf{return}\ (g^x, x) \\ \mathcal{E}(\alpha, m) & \stackrel{\text{def}}{=} & y \xleftarrow{\$} \mathbb{Z}_q;\ h \leftarrow H(\alpha^y);\ \mathsf{return}\ (g^y, h \oplus m) \\ \mathcal{D}(x, (\beta, \zeta)) & \stackrel{\text{def}}{=} & h \leftarrow H(\beta^x);\ \mathsf{return}\ (\zeta \oplus h) \end{array}$$

The security of Hashed ElGamal can be reduced to the Computational Diffie-Hellman (CDH) assumption on the underlying group family. This is the assumption that it is hard to compute g^{xy} given g^x and g^y where x and y are uniformly random elements in $\mathbb{Z}_q$. To match the existing proof in CertiCrypt, we exhibit a reduction to the LCDH assumption, the *set* version of the CDH assumption—the reduction from LCDH to CDH is immediate.

Figure 1 shows the sequence of games used to justify the security reduction. This is an essential part of the proof sketch that is input to EasyCrypt, and which is composed of five ingredients:[1]

1. Type, constant and operator declarations, which introduce the objects manipulated by the scheme. In this case, they include a type for elements of the cyclic group $\mathcal{G}$, constants representing the length of messages k, the order of the group q and a generator g, and operators denoting the group law and exponentiation, and exclusive or on bitstrings;
2. Axioms, which capture mathematical properties of these objects, and are used by automated tools to check the validity of the proof sketch. We use axioms to state properties of the group law and exponentiation, and the exclusive or operator;
3. Game definitions, where adversaries are specified as abstract procedures with access to oracles. In all games in the figure the hash function H is modeled as a random oracle and the adversary is represented as two procedures $\mathcal{A}_1$ and $\mathcal{A}_2$ that share state. The procedures representing the adversary are given access to a wrapper $H_{\mathcal{A}}$ for the hash oracle that just stores queries in a list $\boldsymbol{L}_{\mathcal{A}}$ before forwarding them to H:

$$\begin{array}{lcl} H(x) & \stackrel{\text{def}}{=} & \text{if } x \notin \text{dom}(\boldsymbol{L}) \text{ then } h \stackrel{\$}{\leftarrow} \{0,1\}^k; \boldsymbol{L}[x] \leftarrow h \text{ end if}; \text{return } \boldsymbol{L}[x] \\ H_{\mathcal{A}}(x) & \stackrel{\text{def}}{=} & \boldsymbol{L}_{\mathcal{A}} \leftarrow x :: \boldsymbol{L}_{\mathcal{A}};\ m \leftarrow H(x);\ \text{return } m \end{array}$$

4. Judgments in pRHL. The general form of judgments is $\models \mathsf{G}_1 \sim \mathsf{G}_2 : \Psi \Rightarrow \Phi$, where G_1 and G_2 are games, and the pre-condition Ψ and the post-condition Φ are relations on program memories (memories map program variables to values). Pre- and post-conditions are first-order formulae built from relational expressions, in which language expressions are tagged with $\langle 1 \rangle$ or $\langle 2 \rangle$ to denote their interpretation in the first or second game. We often consider equivalence of memories on a set of variables X; we use $=_X$ as a shorthand for the formula $\forall x \in X.\ x\langle 1 \rangle = x\langle 2 \rangle$;
5. Claims about probability, built from probability quantities (the probability of an event in a game), arithmetic operators, and mathematical relations (e.g. $=, <, \leq$). The final statement that expresses the overall security guarantee brought by the proof sketch is usually a claim that upper bounds the probability of adversary success in an initial attack game in terms of the probabilities of one or more adversaries breaking security assumptions.

We briefly comment on the sequence of games in Figure 1. The first and last games encode the IND-CPA and LCDH experiments, respectively. We obtain G_1 by inlining the key generation and encryption procedures in the initial game and rearranging instructions so that random choices are made upfront. We prove that games IND-CPA and G_1 yield identical distributions on the result of the game (denoted by the keyword res). We deduce from this that the probability of the event $b = b'$ is the same in both games.

[1] The first two are omitted from the figure. We include an extract of the actual input file for reference in Appendix A.

Game IND-CPA :
$(\alpha, x) \leftarrow \mathcal{KG}()$;
$(m_0, m_1) \leftarrow \mathcal{A}_1(\alpha)$;
$b \xleftarrow{\$} \{0,1\}$;
$(\beta, \gamma) \leftarrow \mathcal{E}(\alpha, m_b)$;
$b' \leftarrow \mathcal{A}_2(\beta, \gamma)$;
return $(b = b')$

Game G_1 :
$x \xleftarrow{\$} \mathbb{Z}_q;\ \alpha \leftarrow g^x$;
$y \xleftarrow{\$} \mathbb{Z}_q;\ \hat{y} \leftarrow \alpha^y$;
$(m_0, m_1) \leftarrow \mathcal{A}_1(\alpha)$;
$b \xleftarrow{\$} \{0,1\}$;
$h \leftarrow H(\hat{y})$;
$b' \leftarrow \mathcal{A}_2(g^y, h \oplus m_b)$;
return $(b = b')$

$\models \mathsf{IND\text{-}CPA} \sim \mathsf{G}_1 : \mathsf{true} \Rightarrow =_{\{\mathsf{res}\}}$
$\Pr[\mathsf{IND\text{-}CPA} : b = b'] = \Pr[\mathsf{G}_1 : b = b']$

Game G_1 :
$x \xleftarrow{\$} \mathbb{Z}_q;\ \alpha \leftarrow g^x$;
$y \xleftarrow{\$} \mathbb{Z}_q;\ \hat{y} \leftarrow \alpha^y$;
$(m_0, m_1) \leftarrow \mathcal{A}_1(\alpha)$;
$b \xleftarrow{\$} \{0,1\}$;
$h \leftarrow H(\hat{y})$;
$b' \leftarrow \mathcal{A}_2(g^y, h \oplus m_b)$;
return $(b = b')$

Game G_2 :
$x \xleftarrow{\$} \mathbb{Z}_q;\ \alpha \leftarrow g^x$;
$y \xleftarrow{\$} \mathbb{Z}_q;\ \hat{y} \leftarrow \alpha^y$;
$(m_0, m_1) \leftarrow \mathcal{A}_1(\alpha)$;
$b \xleftarrow{\$} \{0,1\}$;
$h \xleftarrow{\$} \{0,1\}^k$;
$b' \leftarrow \mathcal{A}_2(g^y, h \oplus m_b)$;
return $(b = b')$

$\models \mathsf{G}_1 \sim \mathsf{G}_2 : \mathsf{true} \Rightarrow (\hat{y} \in \boldsymbol{L}_\mathcal{A})\langle 1 \rangle \leftrightarrow (\hat{y} \in \boldsymbol{L}_\mathcal{A})\langle 2 \rangle \wedge \big((\hat{y} \notin \boldsymbol{L}_\mathcal{A})\langle 1 \rangle \rightarrow =_{\{\mathsf{res}\}}\big)$
$|\Pr[\mathsf{G}_1 : b = b'] - \Pr[\mathsf{G}_2 : b = b']| \leq \Pr[\mathsf{G}_2 : \hat{y} \in \boldsymbol{L}_\mathcal{A}]$

Game G_2 :
$x \xleftarrow{\$} \mathbb{Z}_q;\ \alpha \leftarrow g^x$;
$y \xleftarrow{\$} \mathbb{Z}_q;\ \hat{y} \leftarrow \alpha^y$;
$(m_0, m_1) \leftarrow \mathcal{A}_1(\alpha)$;
$b \xleftarrow{\$} \{0,1\}$;
$h \xleftarrow{\$} \{0,1\}^k$;
$b' \leftarrow \mathcal{A}_2(g^y, h \oplus m_b)$;
return $(b = b')$

Game G_3 :
$x \xleftarrow{\$} \mathbb{Z}_q;\ \alpha \leftarrow g^x$;
$y \xleftarrow{\$} \mathbb{Z}_q;\ \hat{y} \leftarrow \alpha^y$;
$(m_0, m_1) \leftarrow \mathcal{A}_1(\alpha)$;
$\gamma \xleftarrow{\$} \{0,1\}^k$;
$b' \leftarrow \mathcal{A}_2(g^y, \gamma)$;
$b \xleftarrow{\$} \{0,1\}$;
return $(b = b')$

$\models \mathsf{G}_2 \sim \mathsf{G}_3 : \mathsf{true} \Rightarrow =_{\{\mathsf{res}, \hat{y}, \boldsymbol{L}_\mathcal{A}\}}$
$\Pr[\mathsf{G}_2 : b = b'] = \Pr[\mathsf{G}_3 : b = b'] = 1/2$ $\quad$ $\Pr[\mathsf{G}_2 : \hat{y} \in \boldsymbol{L}_\mathcal{A}] = \Pr[\mathsf{G}_3 : \hat{y} \in \boldsymbol{L}_\mathcal{A}]$

Game G_3 :
$x \xleftarrow{\$} \mathbb{Z}_q;\ \alpha \leftarrow g^x$;
$y \xleftarrow{\$} \mathbb{Z}_q;\ \hat{y} \leftarrow \alpha^y$;
$(m_0, m_1) \leftarrow \mathcal{A}_1(\alpha)$;
$\gamma \xleftarrow{\$} \{0,1\}^k$;
$b' \leftarrow \mathcal{A}_2(g^y, \gamma)$;
$b \xleftarrow{\$} \{0,1\}$;
return $(b = b')$

Game LCDH :
$x \xleftarrow{\$} \mathbb{Z}_q;\ y \xleftarrow{\$} \mathbb{Z}_q$;
$L \leftarrow B(g^x, g^y)$;
return $(g^{xy} \in L)$

Adversary $\mathcal{B}(\alpha, \beta)$:
$(m_0, m_1) \leftarrow \mathcal{A}_1(\alpha)$;
$\gamma \xleftarrow{\$} \{0,1\}^k$;
$b' \leftarrow \mathcal{A}_2(\beta, \gamma)$;
return $\boldsymbol{L}_\mathcal{A}$

$\models \mathsf{G}_3 \sim \mathsf{LCDH} : \mathsf{true} \Rightarrow (\hat{y} \in \boldsymbol{L}_\mathcal{A})\langle 1 \rangle \leftrightarrow \mathsf{res}\langle 2 \rangle$
$\Pr[\mathsf{G}_3 : \hat{y} \in \boldsymbol{L}_\mathcal{A}] = \Pr[\mathsf{LCDH} : g^{xy} \in L]$

$$\left|\Pr[\mathsf{IND\text{-}CPA} : b = b'] - \tfrac{1}{2}\right| \leq \Pr[\mathsf{LCDH} : g^{xy} \in L]$$

Fig. 1. Proof sketch of Hashed ElGamal security

In game G_2 we substitute the value $H(\hat{y})$ used to compute the challenge ciphertext by a uniformly chosen value. This only makes a difference if $\mathcal{A}_1$ queries $\hat{y}$ to H, and this happens with the same probability in either game. Thus, the difference in the probability of any event in these games is bounded by the probability of $\hat{y} \in \boldsymbol{L}_{\mathcal{A}}$ in G_2. This can be seen as a semantic variant of the Fundamental Lemma of Game-Playing; the logic allows to dispense with the code instrumentation needed to apply the syntactic counterpart of the lemma.

The transition from G_2 to G_3 uses a code transformation known as *optimistic sampling*: instead of sampling h and defining a value γ as $h \oplus m_b$, we sample γ and define $h = \gamma \oplus m_b$; we then remove the definition of h as dead code. This transformation is proven admissible within the logic and removes the dependency of the adversary's output from the challenge bit b.

The final transition performs the reduction to LCDH by exhibiting an adversary $\mathcal{B}$ that uses $\mathcal{A}$ as a sub-procedure and for which the semantics of games LCDH and G_3 coincide. Finally, from the preceding claims, the advantage of $\mathcal{A}$ can be bounded by the probability of $\mathcal{B}$ in solving LCDH. The resulting proof sketch is about 250 lines long, about 5 times shorter than the proof in CertiCrypt reported in [4]—and arguably much simpler and close to a pen-and-paper proof.

3 An Overview of EasyCrypt

Programming Language. Games are modeled as programs in a typed, probabilistic, procedural, imperative language. Types include Booleans, integers, bitstrings, pairs, lists, maps, and user-defined types. Expressions are built from variables and operators in the usual way; for instance, Boolean-valued operators include the usual connectives, equality, list membership, arithmetic comparisons. The commands of the language are defined by the following grammar:

$$
\begin{array}{rll}
\mathcal{I} ::= & \mathcal{V} \leftarrow \mathcal{E} & \text{assignment} \\
| & \mathcal{V} \overset{\$}{\leftarrow} \mathcal{DE} & \text{random sampling} \\
| & \mathsf{if}\ \mathcal{E}\ \mathsf{then}\ \mathcal{C}\ \mathsf{else}\ \mathcal{C} & \text{conditional} \\
| & \mathcal{V} \leftarrow \mathcal{P}(\mathcal{E}, \dots, \mathcal{E}) & \text{procedure call} \\
\mathcal{C} ::= & \mathsf{skip} & \text{nop} \\
| & \mathcal{I};\ \mathcal{C} & \text{sequence}
\end{array}
$$

where $\mathcal{V}$ is a set of variables, $\mathcal{P}$ is a set of procedures, and $\mathcal{DE}$ is a set of distribution expressions. For the purpose of this article, distribution expressions are restricted to uniform distributions over specific domains, for instance integers in $\mathbb{Z}_q$ or (non-neutral) elements of some group $\mathcal{G}$. Adversaries are modeled as abstract procedures with an interface that specifies the oracles they may query.

Games can be given a semantics as memory distribution transformers, in the style of [6]. Formally, memories are well-typed mappings from variables to values, and the semantics of a game G is a function, denoted $[\![\mathsf{G}]\!]$, that returns for an initial memory m the (sub-)distribution on final memories resulting from executing G in m. Given an initial memory m and an event A (a Boolean expression), we let $\Pr[\mathsf{G}, m : A]$ denote the probability of A w.r.t. the distribution $[\![\mathsf{G}]\!]\ m$; we simply write $\Pr[\mathsf{G} : A]$ when the initial memory is not relevant.

Relational Judgments. Pre- and post-conditions in pRHL judgments are first-order formulae built from relational expressions. Relational expressions are arbitrary Boolean expressions over logical variables and program variables tagged with $\langle 1\rangle$, $\langle 2\rangle$; the only restriction is that logical variables may only appear quantified. By abuse of notation, we write $e\langle i\rangle$ for the expression e in which all variables have been tagged with $\langle i\rangle$. Let b stand for an arbitrary Boolean expression over tagged and logical variables, then logical formulae are defined by the following grammar:

$$\Psi, \Phi ::= b \mid \neg\Phi \mid \Psi \wedge \Phi \mid \Psi \vee \Phi \mid \Psi \rightarrow \Phi \mid \Psi \leftrightarrow \Phi \mid (\Phi) \mid \forall x.\ \Phi \mid \exists x.\ \Phi$$

A logical formula is interpreted as a relation on program memories. For example, the formula $x\langle 1\rangle + y\langle 2\rangle \leq z\langle 1\rangle$ is interpreted as the relation

$$R = \{(m_1, m_2) \mid m_1(x) + m_2(y) \leq m_1(z)\}$$

A pRHL judgment $\models \mathsf{G}_1 \sim \mathsf{G}_2 : \Psi \Rightarrow \Phi$ is valid iff for any pair of initial memories m_1, m_2 satisfying the pre-condition Ψ, the distributions $[\![\mathsf{G}_1]\!]\ m_1$ and $[\![\mathsf{G}_2]\!]\ m_2$ satisfy the lifting of post-condition Φ, $([\![\mathsf{G}_1]\!]\ m_1)\, \mathcal{L}(\Phi)\, ([\![\mathsf{G}_2]\!]\ m_2)$. The lifting of a relation to a distribution is defined as a max-cut min-flow problem, in the style of [18]. Formally, let μ_1 be a probability distribution on a set A and μ_2 a probability distribution on a set B. We define the lifting $\mu_1\, \mathcal{L}(R)\, \mu_2$ of a relation $R \subseteq A \times B$ to μ_1 and μ_2 as follows:[2]

$$\exists \mu : \mathcal{D}(A \times B).\ \pi_1(\mu) = \mu_1 \wedge \pi_2(\mu) = \mu_2 \wedge \forall (a, b) : A \times B.\ \mu(a, b) > 0 \implies a\ R\ b$$

where the projections $\pi_1(\mu)$ and $\pi_2(\mu)$ of μ are defined as

$$\pi_1(\mu)(a) \stackrel{\text{def}}{=} \sum_{b \in B} \mu(a, b) \qquad \pi_2(\mu)(b) \stackrel{\text{def}}{=} \sum_{a \in A} \mu(a, b)$$

Claims about probability can be derived from valid relational judgments by means of the following rules:

$$\frac{m_1\ \Psi\ m_2 \qquad \models \mathsf{G}_1 \sim \mathsf{G}_2 : \Psi \Rightarrow \Phi \qquad \Phi \rightarrow (A\langle 1\rangle \leftrightarrow B\langle 2\rangle)}{\Pr[\mathsf{G}_1, m_1 : A] = \Pr[\mathsf{G}_2, m_2 : B]}\ [\text{PrEq}]$$

$$\frac{m_1\ \Psi\ m_2 \qquad \models \mathsf{G}_1 \sim \mathsf{G}_2 : \Psi \Rightarrow \Phi \qquad \Phi \rightarrow (A\langle 1\rangle \rightarrow B\langle 2\rangle)}{\Pr[\mathsf{G}_1, m_1 : A] \leq \Pr[\mathsf{G}_2, m_2 : B]}\ [\text{PrLe}]$$

Automated Proofs of Relational Judgments Most practical verification tools adopt a similar methodology: a weakest precondition (wp) calculus is used to compute from a program and its specification a set of sufficient conditions, known as verification conditions, and these conditions are discharged by automated

[2] For the clarity of presentation, we assume that A and B are discrete and cast our definitions using the usual representation of distributions. However, the tool builds on a monadic representation of distributions, as in [6].

tools. Extending the methodology to the logic pRHL is a significant challenge, for two reasons: first, generating verification conditions for a relational program logic is an open topic of research, and second, there is no prior application of the methodology to procedural nor probabilistic programs.

There are at least two natural strategies for defining a wp calculus in a relational setting. The calculus can either operate on both games in lockstep, or else it can operate on each game separately, in the style of self-composition [2]. Both strategies are incomplete: the lockstep wp calculus fails on programs that are not structurally equivalent, whereas self-composition fails to handle random assignments and adversary calls. In order to circumvent these limitations, EasyCrypt implements an alternative approach that mixes both strategies:

1. Calls to non-adversary procedures are eliminated from the games by successive inlining their definitions. In the absence of recursion, the transformation terminates successfully and only adversary calls remain;
2. Random assignments are moved upfront. The resulting code consists of a sequence of random assignments followed by deterministic code, possibly with adversary calls;
3. A relational weakest precondition calculus is applied to the deterministic fragment of the game, using relational specifications to deal with adversary calls. Each adversary specification induces a proof obligation, expressed as a pRHL judgment, on the oracles in its interface. Self-composition is applied to verify the code of oracles with respect to these pRHL judgments. This results in a judgment of the form

$$\models x_1 \stackrel{\$}{\leftarrow} T_1; \ldots x_l \stackrel{\$}{\leftarrow} T_l \sim y_1 \stackrel{\$}{\leftarrow} U_1; \ldots y_n \stackrel{\$}{\leftarrow} U_n : \Psi \Rightarrow \Phi$$

4. A mapping $f : T_1 \times \cdots \times T_l \rightarrow U_1 \times \cdots \times U_n$ is selected, and used to generate the verification condition $\Phi \Rightarrow_f \Psi$, defined as[3]

$$\forall m_1\ m_2\ t_1 \ldots t_l\ .\ m_1\ \Psi\ m_2 \implies m_1 \left\{\vec{t}/\vec{x}\right\}\ \Phi\ m_2 \left\{f(t_1, \ldots, t_l)/\vec{y}\right\}$$

   Under specific conditions on f, see [23], the validity of $\Phi \Rightarrow_f \Psi$ entails the validity of the corresponding pRHL judgment. In practice, it is generally sufficient to require that f is a 1-1 mapping, and taking f as the identity function works most of the time. However, in some cases other mappings must be used. For example, to prove the equivalence between games G_2 and G_3 in the proof of Hashed ElGamal described in the previous section, it is necessary to prove a judgment like the following:

$$\models h \stackrel{\$}{\leftarrow} \{0,1\}^k;\ \gamma \leftarrow h \oplus m_b \sim \gamma \stackrel{\$}{\leftarrow} \{0,1\}^k;\ h \leftarrow \gamma \oplus m_b : =_{\{m_b\}} \Rightarrow =_{\{h,\gamma\}}$$

   The wp will stop after computing the weakest precondition for the deterministic fragment of the two programs, yielding

$$\models h \stackrel{\$}{\leftarrow} \{0,1\}^k \sim \gamma \stackrel{\$}{\leftarrow} \{0,1\}^k : =_{\{m_b\}} \Rightarrow (h\langle 1\rangle = \gamma\langle 2\rangle \oplus m_b\langle 2\rangle)$$

[3] The memory $m_1 \left\{\vec{t}/\vec{x}\right\}$ maps x_i to t_i for $i = 1 \ldots l$ and y to $m_1(z)$ for $z \notin \{x_1 \ldots x_l\}$. Likewise, $m_2 \left\{f(t_1, \ldots, t_l)/\vec{y}\right\}$ is the memory that maps y_i to $\pi_i(f(t_1, \ldots, t_l))$ for $i = 1 \ldots n$ and z to $m_2(z)$ for $z \notin \{y_1 \ldots y_n\}$.

This equivalence is proved in EasyCrypt by providing the bijective function $f(x) = x \oplus m_b$ as a witness. The fact that f is bijective is established automatically since f is idempotent. In the general case this is proved by providing also the inverse mapping.

5. Since $\Phi \Rightarrow_f \Psi$ is a first-order formula, its validity can be established by off-the-shelf tools. In order to target multiple tools, EasyCrypt generates its verification conditions in the intermediate format of the Why tool [16]. We then use the Simplify prover [15] and the alt-ergo SMT solver [13] to discharge the conditions (although many others provers are supported, including interactive theorem provers such as Coq).

Verification condition generation is incomplete (in the logical sense), and would fail on pRHL judgments where games perform calls to adversaries in a different order. Pleasingly, the strategy is extremely effective in practice—so that we have found no need to implement alternatives for dealing with programs not handled by our approach.

A Mechanized Probabilistic Relational Hoare Logic. EasyCrypt implements a simple tactic language to prove the validity of judgments using rules of the logic and program transformations. The tactics allow the application of two-sided rules, which require that the two commands of a judgment have the same shape, and one-sided rules, which operate on only one of the games in a judgment. All language constructs admit both one-sided and two-sided rules, except for random assignments and adversary calls, for which only two-sided rules exist.

The lack of one-sided rules for random assignments and adversary calls limits the applicability of the logic: e.g., it cannot relate the programs $x \xleftarrow{\$} X; y \leftarrow \mathcal{A}(z)$ and $y \leftarrow \mathcal{A}(z); x \xleftarrow{\$} X$, because instructions are executed in a different order. To mitigate this limitation, EasyCrypt implements program transformations for code motion, allowing to swap instructions that are independent. Moreover, EasyCrypt implements tactics for inlining procedure calls and eagerly/lazily sample random values. Basic tactics can be combined using tacticals to increase automation. The tactic language provides the necessary infrastructure for making most components of EasyCrypt proof-producing, as discussed below.

Reasoning about Failure Events. Game-based proofs often include steps in which it is argued that two games G_1 and G_2 behave identically unless a designated failure event F occurs. Such transitions are justified using the so-called Fundamental Lemma [20, 8], which allows to bound the difference between the probability of an event A in game G_1 and a possibly different event B in game G_2 by the probability of F in either game. Although a syntactical characterization of this lemma is often used, in which the failure event is represented by a Boolean flag in the code of the games, we state a more general version of the lemma using relational logic.

Lemma 1 (Fundamental Lemma). *Let* G_1*,* G_2 *be two games and* A, B*, and* F *be events such that*

$$\models \mathsf{G}_1 \sim \mathsf{G}_2 : \Psi \Rightarrow (F\langle 1\rangle \leftrightarrow F\langle 2\rangle) \wedge (\neg F\langle 1\rangle \rightarrow (A\langle 1\rangle \leftrightarrow B\langle 2\rangle))$$

Then, if $m_1 \ \Psi \ m_2$*,*

1. $\Pr[\mathsf{G}_1, m_1 : A \wedge \neg F] = \Pr[G_2, m_2 : B \wedge \neg F]$,
2. $|\Pr[\mathsf{G}_1, m_1 : A] - \Pr[\mathsf{G}_2, m_2 : B]| \leq \Pr[\mathsf{G}_1, m_1 : F] = \Pr[\mathsf{G}_2, m_2 : F]$

The hypothesis of the lemma can be checked using the pRHL prover. The key to proving the validity of the judgment is finding an appropriate specification for adversaries. EasyCrypt infers for each adversary call $x \leftarrow \mathcal{A}(\vec{e})$ a relation Θ and checks the validity of the judgment

$$\begin{aligned} \models \mathcal{A} \sim \mathcal{A} : (\neg F\langle 1\rangle \wedge \neg F\langle 2\rangle \wedge =_{\mathsf{args}(\mathcal{A})} \wedge \Theta) \Rightarrow \\ (F\langle 1\rangle \leftrightarrow F\langle 2\rangle) \wedge \left(\neg F\langle 1\rangle \rightarrow =_{\{\mathsf{res}\}} \wedge \Theta\right) \end{aligned}$$

where $\mathsf{args}(\mathcal{A})$ denotes the set of formal parameters of $\mathcal{A}$. This in turn, requires inferring and checking similar specifications for oracles. Although these heuristically inferred specifications suffice in most cases, the user can choose to prove their own specifications for one or more oracles or adversaries when needed, leaving the tool to infer the rest.

Computing Probabilities. EasyCrypt can prove claims about the probability of events in games using properties of probability (e.g. inclusion-exclusion principle), arithmetic laws, and the rules [PrEq] and [PrLe] above, which allow deriving probability claims from valid relational judgments. We also implement a simple mechanism for computing probability bounds. This mechanism can establish, for instance, that the probability that a value uniformly chosen from a set T is equal to an arbitrary expression is $1/|T|$, or the probability it belongs to a list of n values is at most $n/|T|$.

Generating Verifiable Evidence. EasyCrypt implements a compiler that turns proof sketches into Coq files that are compatible with the CertiCrypt framework and can be verified using the type checker of Coq. The compiler serves two purposes: first, it significantly increases confidence in proof sketches by producing independently verifiable proofs, and providing means of checking the consistency of the set of axioms used in a proof sketch. Second, it opens the possibility to conduct in a general-purpose proof assistant proof steps that fall out of the scope of automated methods.

We briefly describe the workings of the compiler. The declarations, definitions of games, and axioms of a proof sketch admit an immediate translation into CertiCrypt. The recommended practice is to prove the axioms used by EasyCrypt in CertiCrypt. In most cases, the axioms already exist in CertiCrypt, or are simple consequences of proven facts. Then, using the proof-producing option of the pRHL prover, all judgments of a proof sketch are compiled into pRHL derivations in CertiCrypt. Finally, the compiler generates for each claim in a proof sketch a Coq lemma that may need to be completed manually with justifications of the probability reasoning performed by EasyCrypt.

4 Advanced Application: Cramer-Shoup Cryptosystem

The Cramer-Shoup cryptosystem is a public-key encryption scheme based on ElGamal encryption that gained fame for being the first efficient asymmetric encryption scheme to be proven secure against adaptive chosen-ciphertext attacks under standard assumptions—the length of ciphertexts is just twice the length of ElGamal ciphertexts. Given a cyclic group (family) $\mathcal{G}$ of order q and a keyed hash function $\{H_k : \mathcal{G}^3 \to \mathbb{Z}_q\}_{k \in K}$ mapping triples of group elements into integers in $\mathbb{Z}_q$, key generation, encryption, and decryption are defined as follows:

$\mathcal{KG}() \stackrel{\text{def}}{=}$
$g, \hat{g} \stackrel{\$}{\leftarrow} \mathcal{G} \setminus \{1\};$
$x_1, x_2, y_1, y_2, z_1, z_2 \stackrel{\$}{\leftarrow} \mathbb{Z}_q;\ k \stackrel{\$}{\leftarrow} K;$
$e \leftarrow g^{x_1} \hat{g}^{x_2};$
$f \leftarrow g^{y_1} \hat{g}^{y_2};$
$h \leftarrow g^{z_1} \hat{g}^{z_2};$
$pk \leftarrow (k, g, \hat{g}, e, f, h);$
$sk \leftarrow (k, g, \hat{g}, x_1, x_2, y_1, y_2, z_1, z_2);$
return (pk, sk)

$\mathcal{E}((k, g, \hat{g}, e, f, h), m) \stackrel{\text{def}}{=}$
$u \stackrel{\$}{\leftarrow} \mathbb{Z}_q;\ a \leftarrow g^u;\ \hat{a} \leftarrow \hat{g}^u;\ c \leftarrow h^u \cdot m;$
$v \leftarrow H_k(a, \hat{a}, c);\ d \leftarrow e^u \cdot f^{uv};$
return $(a, \hat{a}, c, d)$

$\mathcal{D}((k, g, \hat{g}, x_1, x_2, y_1, y_2, z_1, z_2), (a, \hat{a}, c, d)) \stackrel{\text{def}}{=}$
$v \leftarrow H_k(a, \hat{a}, c);$
if $d = a^{x_1 + v y_1} \cdot \hat{a}^{x_2 + v y_2}$ then
  return $c / (a^{z_1} \cdot \hat{a}^{z_2})$
else return $\perp$

We prove that the Cramer-Shoup cryptosystem is secure against adaptive chosen-ciphertext attacks (IND-CCA secure) in the standard model assuming the DDH problem is hard in the underlying group family and the hash function H is target collision-resistant (i.e., universal one-way).

Definition 1 (Target Collision-Resistance). *Let $\{H_k : A \to B\}_{k \in K}$ be a keyed family of hash functions. The advantage of an adversary $\mathcal{C}$ against the target collision-resistance of H is defined as*

$$\mathbf{Adv}^{\mathcal{C}}_{\mathsf{TCR}} \stackrel{def}{=} \Pr[\mathsf{TCR} : H_k(x) = H_k(y) \wedge x \neq y]$$

where the experiment TCR is defined by means of the following game:

$$\textbf{Game } \mathsf{TCR} : x \leftarrow \mathcal{C}_1();\ k \stackrel{\$}{\leftarrow} K;\ y \leftarrow \mathcal{C}_2(k)$$

Definition 2 (CCA-advantage). *Let $(\mathcal{KG}, \mathcal{E}, \mathcal{D})$ be an asymmetric encryption scheme. The CCA-advantage of an adversary $\mathcal{A}$ limited to $q_{\mathcal{D}}$ decryption queries against the adaptive chosen-ciphertext security of the scheme is defined as*

$$\mathbf{Adv}^{\mathcal{A}}_{\mathsf{CCA}}(q_{\mathcal{D}}) \stackrel{def}{=} \left| \Pr[\mathsf{IND\text{-}CCA} : b = b'] - \frac{1}{2} \right|$$

where the experiment IND-CCA is defined by means of the following game:

Game *IND-CCA* :
$(pk, \boldsymbol{sk}) \leftarrow \mathcal{KG}();$
$(m_0, m_1) \leftarrow \mathcal{A}_1(pk);$
$b \stackrel{\$}{\leftarrow} \{0, 1\};$
$\gamma^* \leftarrow \mathcal{E}(pk, m_b);\ \gamma^*_{\mathsf{def}} \leftarrow$ true;
$b' \leftarrow \mathcal{A}_2(\gamma^*);$
return $(b = b')$

Oracle $\mathcal{D}_{\mathcal{A}}(\gamma)$:
if $|\boldsymbol{L}_{\mathcal{D}}| < q_{\mathcal{D}} \wedge \neg(\gamma^*_{\mathsf{def}} \wedge \gamma = \gamma^*)$ then
  $\boldsymbol{L}_{\mathcal{D}} \leftarrow \gamma :: \boldsymbol{L}_{\mathcal{D}};$
  return $\mathcal{D}(\boldsymbol{sk}, \gamma)$
else return $\perp$

Theorem 1 (Security of Cramer-Shoup). *Let $\mathcal{A}$ be an adversary against the IND-CCA security of Cramer-Shoup limited to $q_\mathcal{D}$ decryption queries. Then, there exists an algorithm $\mathcal{B}$ for solving the DDH problem in $\mathcal{G}$ and an adversary $\mathcal{C}$ against the target collision-resistance of the hash function H such that*

$$\mathbf{Adv}^{\mathcal{A}}_{\mathsf{CCA}}(q_\mathcal{D}) \leq \mathbf{Adv}^{\mathcal{B}}_{\mathsf{DDH}} + \mathbf{Adv}^{\mathcal{C}}_{\mathsf{TCR}} + \frac{q_\mathcal{D}^4}{q^4} + \frac{q_\mathcal{D}+2}{q}$$

Figures 2-4 show a proof sketch of the above theorem in EasyCrypt. The proof follows closely the one presented in [17]; we give only a high-level description here. Game G_1 in the figure is obtained directly from the IND-CCA game instantiated for Cramer-Shoup by inlining the definitions of the key generation and encryption procedures, propagating assignments, and replacing expressions by equivalent ones. We observe that all verification conditions that ensure the validity of this transformation can be discharged automatically using an SMT solver. This surpasses Halevi's expectations [17], who suggested this transformation be split in three steps so that it could be handled by an automated tool.

We then build a DDH distinguisher $\mathcal{B}$ such that the output distribution on the value of $(b = b')$ is identical in games DDH_0 (where $\mathcal{B}$ receives valid DDH triples) and G_1, on the one hand, and in games DDH_1 (where $\mathcal{B}$ receives random triples) and G_2, on the other. In addition, we instrument the decryption oracle in G_2 to raise a flag **bad** whenever $\mathcal{A}$ queries for the decryption of a valid ciphertext with $\log_a \hat{a} \neq \log_g \hat{g}$. We then show using our semantic characterization of the Fundamental Lemma that the difference in the probability of $(b = b')$ in this game and in game G_3, where $\mathcal{D}$ rejects such ciphertexts, is bounded by the probability of **bad** in the latter game. We also change the way e, f and h are computed in a semantics-preserving way. Up to this point, by the triangular inequality we have

$$|\Pr[\mathsf{IND\text{-}CCA} : b = b'] - \Pr[\mathsf{G}_3 : b = b']| \leq \mathbf{Adv}^{\mathcal{B}}_{\mathsf{DDH}} + \Pr[\mathsf{G}_3 : \mathbf{bad}]$$

The next game in the sequence, G_4, removes the dependency of the adversary's output from bit b by choosing uniformly r and setting $c = g^r$. This requires to be able to compute z_2 from $\log_g(c) = uz + (u - u')wz_2 + \log_g(m_b)$, which is not possible if $u = u'$, but this happens only with probability $1/q$. We use again the semantic formulation of the Fundamental Lemma to bound the difference in the probability of $(b = b')$ between G_3 and G_4 by $1/q$. After straightforward information-theoretic reasoning we get

$$|\Pr[\mathsf{IND\text{-}CPA} : b = b'] - 1/2| \leq \mathbf{Adv}^{\mathcal{B}}_{\mathsf{DDH}} + 2/q + \Pr[\mathsf{G}_4 : \mathbf{bad} \wedge u \neq u']$$

We can now move most of the code of the game before the call to $\mathcal{A}_1$. This in turn allows to make d random by uniformly choosing $r' = \log_g(d)$ and defining x_2 in terms of it, rather than the other way around. Since now the game computes the challenge ciphertext in advance, we can instrument $\mathcal{D}$ to raise a flag $\mathbf{bad}_1$ when the challenge is queried during the first phase of the game. Note that at this point the challenge ciphertext is a 4-tuple of uniformly random elements,

Game G_1 :
$g, \hat{g} \xleftarrow{\$} \mathcal{G} \setminus \{1\}; x_1, x_2, y_1, y_2, z_1, z_2 \xleftarrow{\$} \mathbb{Z}_q;$
$k \xleftarrow{\$} K;$
$e \leftarrow g^{x_1}\hat{g}^{x_2};\ f \leftarrow g^{y_1}\hat{g}^{y_2};\ h \leftarrow g^{z_1}\hat{g}^{z_2};$
$(m_0, m_1) \leftarrow \mathcal{A}_1(k, g, \hat{g}, e, f, h);\ b \xleftarrow{\$} \{0,1\};$
$u \xleftarrow{\$} \mathbb{Z}_q;\ a \leftarrow g^u;\ \hat{a} \leftarrow \hat{g}^u;$
$c \leftarrow a^{z_1} \cdot \hat{a}^{z_2} \cdot m_b;$
$v \leftarrow H_k(a, \hat{a}, c);\ d \leftarrow a^{x_1+vy_1} \cdot \hat{a}^{x_2+vy_2};$
$\gamma^* \leftarrow (a, \hat{a}, c, d);\ \gamma^*_{\mathsf{def}} \leftarrow$ true;
$b' \leftarrow \mathcal{A}_2(\gamma^*)$; return $(b = b')$

Oracle $\mathcal{D}(a, \hat{a}, c, d)$:
if $|\boldsymbol{L}_\mathcal{D}| < q_\mathcal{D} \wedge \neg(\gamma^*_{\mathsf{def}} \wedge (a, \hat{a}, c, d) = \gamma^*)$
then
  $\boldsymbol{L}_\mathcal{D} \leftarrow \gamma :: \boldsymbol{L}_\mathcal{D};$
  $v \leftarrow H_k(a, \hat{a}, c);$
  if $d = a^{x_1+vy_1} \cdot \hat{a}^{x_2+vy_2}$ then
    return $c/(a^{z_1} \cdot \hat{a}^{z_2})$
  else return $\perp$
else return $\perp$

$\models \mathsf{G}_1 \sim \mathsf{DDH}_0$: true $\Rightarrow =_{\{\mathsf{res}\}}$ $\qquad \Pr[\mathsf{G}_1 : b = b'] = \Pr[\mathsf{DDH}_0 : b = b']$

Game DDH_0 DDH_1 :
$g \xleftarrow{\$} \mathcal{G} \setminus \{1\}; x \xleftarrow{\$} \mathbb{Z}_q^*;\ y \xleftarrow{\$} \mathbb{Z}_q;$
$z \leftarrow xy$ | $z \xleftarrow{\$} \mathbb{Z}_q$;
return $\mathcal{B}(g, g^x, g^y, g^z)$
Adversary $\mathcal{B}(g, \hat{g}, a, \hat{a})$:
$x_1, x_2, y_1, y_2, z_1, z_2 \xleftarrow{\$} \mathbb{Z}_q;\ k \xleftarrow{\$} K;$
$e \leftarrow g^{x_1}\hat{g}^{x_2};\ f \leftarrow g^{y_1}\hat{g}^{y_2};\ h \leftarrow g^{z_1}\hat{g}^{z_2};$
$(m_0, m_1) \leftarrow \mathcal{A}_1(k, g, \hat{g}, e, f, h);\ b \xleftarrow{\$} \{0,1\};$
$c \leftarrow a^{z_1} \cdot \hat{a}^{z_2} \cdot m_b;$
$v \leftarrow H_k(a, \hat{a}, c);\ d \leftarrow a^{x_1+vy_1} \cdot \hat{a}^{x_2+vy_2};$
$\gamma^* \leftarrow (a, \hat{a}, c, d);\ \gamma^*_{\mathsf{def}} \leftarrow$ true;
$b' \leftarrow \mathcal{A}_2(\gamma^*)$; return $(b = b')$

Oracle $\mathcal{D}(a, \hat{a}, c, d)$:
if $|\boldsymbol{L}_\mathcal{D}| < q_\mathcal{D} \wedge \neg(\gamma^*_{\mathsf{def}} \wedge (a, \hat{a}, c, d) = \gamma^*)$
then
  $\boldsymbol{L}_\mathcal{D} \leftarrow \gamma :: \boldsymbol{L}_\mathcal{D};$
  $v \leftarrow H_k(a, \hat{a}, c);$
  if $d = a^{x_1+vy_1} \cdot \hat{a}^{x_2+vy_2}$ then
    return $c/(a^{z_1} \cdot \hat{a}^{z_2})$
  else return $\perp$
else return $\perp$

$\models \mathsf{DDH}_1 \sim \mathsf{G}_2$: true $\Rightarrow =_{\{\mathsf{res}\}}$ $\qquad \Pr[\mathsf{DDH}_1 : b = b'] = \Pr[\mathsf{G}_2 : b = b']$

Game G_2 :
$g \xleftarrow{\$} \mathcal{G} \setminus \{1\};\ w \xleftarrow{\$} \mathbb{Z}_q^*;\ \hat{g} \leftarrow g^w;$
$u, u' \xleftarrow{\$} \mathbb{Z}_q;\ a \leftarrow g^u;\ \hat{a} \leftarrow \hat{g}^{u'};$
$x_1, x_2, y_1, y_2, z_1, z_2 \xleftarrow{\$} \mathbb{Z}_q;\ k \xleftarrow{\$} K;$
$e \leftarrow g^{x_1}\hat{g}^{x_2};\ f \leftarrow g^{y_1}\hat{g}^{y_2};\ h \leftarrow g^{z_1}\hat{g}^{z_2};$
$(m_0, m_1) \leftarrow \mathcal{A}_1(k, h, \hat{g}, e, f, h);\ b \xleftarrow{\$} \{0,1\};$
$c \leftarrow a^{z_1} \cdot \hat{a}^{z_2} \cdot m_b;$
$v \leftarrow H_k(a, \hat{a}, c);\ d \leftarrow a^{x_1+vy_1} \cdot \hat{a}^{x_2+vy_2};$
$\gamma^* \leftarrow (a, \hat{a}, c, d);\ \gamma^*_{\mathsf{def}} \leftarrow$ true;
$b' \leftarrow \mathcal{A}_2(\gamma^*)$;
return $(b = b')$

Oracle $\mathcal{D}(a, \hat{a}, c, d)$:
if $|\boldsymbol{L}_\mathcal{D}| < q_\mathcal{D} \wedge \neg(\gamma^*_{\mathsf{def}} \wedge (a, \hat{a}, c, d) = \gamma^*)$
then
  $\boldsymbol{L}_\mathcal{D} \leftarrow \gamma :: \boldsymbol{L}_\mathcal{D};\ v \leftarrow H_k(a, \hat{a}, c);$
  if $\hat{a} = a^w$ then ;
    if $d = a^{x_1+vy_1} \cdot \hat{a}^{x_2+vy_2}$ then
      return $c/(a^{z_1} \cdot \hat{a}^{z_2})$
    else return $\perp$
  elsif $d = a^{x_1+vy_1} \cdot \hat{a}^{x_2+vy_2}$ then
    bad $\leftarrow$ true; return $c/(a^{z_1} \cdot \hat{a}^{z_2})$
  else return $\perp$
else return $\perp$

Fig. 2. Proof sketch of the IND-CCA security of the Cramer-Shoup cryptosystem

therefore, the probability of $\mathbf{bad}_1$ is bounded by $(q_\mathcal{D}/q)^4$—this is achieved by means of an intermediate game, not shown in the figure, that stores the 4 components of queried ciphertexts in different lists, and by independently bounding the probability of each component of the challenge appearing in the corresponding list. Hence, we have

$$\Pr[\mathsf{G}_4 : \mathbf{bad} \wedge u \neq u'] \leq \Pr[\mathsf{G}_5 : \mathbf{bad} \wedge u \neq u'] + (q_\mathcal{D}/q)^4$$

Game G_3 :
$g \xleftarrow{\$} \mathcal{G} \setminus \{1\};\ w \xleftarrow{\$} \mathbb{Z}_q^*;\ \hat{g} \leftarrow g^w;\ k \xleftarrow{\$} K;$
$x, x_2 \xleftarrow{\$} \mathbb{Z}_q;\ x_1 \leftarrow x - wx_2;\ e \leftarrow g^x;$
$y, y_2 \xleftarrow{\$} \mathbb{Z}_q;\ y_1 \leftarrow y - wy_2;\ f \leftarrow g^y;$
$z, z_2 \xleftarrow{\$} \mathbb{Z}_q;\ z_1 \leftarrow z - wz_2;\ h \leftarrow g^z;$
$(m_0, m_1) \leftarrow \mathcal{A}_1(k, h, \hat{g}, e, f, h);\ b \xleftarrow{\$} \{0, 1\};$
$u, u' \xleftarrow{\$} \mathbb{Z}_q;\ a \leftarrow g^u;\ \hat{a} \leftarrow \hat{g}^{u'};$
$c \leftarrow a^{z_1} \cdot \hat{a}^{z_2} \cdot m_b;$
$v \leftarrow H_k(a, \hat{a}, c);\ d \leftarrow a^{x_1+vy_1} \cdot \hat{a}^{x_2+vy_2};$
$\gamma^* \leftarrow (a, \hat{a}, c, d);\ \gamma^*_{\mathsf{def}} \leftarrow$ true;
$b' \leftarrow \mathcal{A}_2(\gamma^*)$; return $(b = b')$

Oracle $\mathcal{D}(a, \hat{a}, c, d)$:
if $|\boldsymbol{L}_\mathcal{D}| < q_\mathcal{D} \land \neg(\gamma^*_{\mathsf{def}} \land (a, \hat{a}, c, d) = \gamma^*)$
then
  $\boldsymbol{L}_\mathcal{D} \leftarrow \gamma :: \boldsymbol{L}_\mathcal{D};\ v \leftarrow H_k(a, \hat{a}, c);$
  if $\hat{a} = a^w$ then
    if $d = a^{x+vy}$ then return c/a^z
    else return $\bot$
  elsif $d = a^{x_1+vy_1} \cdot \hat{a}^{x_2+vy_2}$ then
    bad $\leftarrow$ true; return $\bot$
  else return $\bot$
else return $\bot$

$\models \mathsf{G}_3 \sim \mathsf{G}_4 : \mathsf{true} \Rightarrow (u = u')\langle 1\rangle \leftrightarrow (u = u')\langle 2\rangle \land \big((u \neq u')\langle 1\rangle \rightarrow =_{\{\mathsf{res},\mathbf{bad}\}}\big)$
$\Pr[\mathsf{G}_4 : b = b'] = 1/2 \qquad |\Pr[\mathsf{G}_3 : b = b'] - \Pr[\mathsf{G}_4 : b = b']| \leq \Pr[\mathsf{G}_3 : u = u'] = 1/q$

Game G_4 :
$g \xleftarrow{\$} \mathcal{G} \setminus \{1\};\ w \xleftarrow{\$} \mathbb{Z}_q^*;\ \hat{g} \leftarrow g^w;\ k \xleftarrow{\$} K;$
$x, x_2 \xleftarrow{\$} \mathbb{Z}_q;\ x_1 \leftarrow x - wx_2;\ e \leftarrow g^x;$
$y, y_2 \xleftarrow{\$} \mathbb{Z}_q;\ y_1 \leftarrow y - wy_2;\ f \leftarrow g^y;$
$z \xleftarrow{\$} \mathbb{Z}_q;\ h \leftarrow g^z;$
$u, u' \xleftarrow{\$} \mathbb{Z}_q;\ a \leftarrow g^u;\ \hat{a} \leftarrow \hat{g}^{u'};$
$r \xleftarrow{\$} \mathbb{Z}_q;\ c \leftarrow g^r;$
$v \leftarrow H_k(a, \hat{a}, c);\ d \leftarrow a^{x_1+vy_1} \cdot \hat{a}^{x_2+vy_2};$
$(m_0, m_1) \leftarrow \mathcal{A}_1(k, h, \hat{g}, e, f, h);\ b \xleftarrow{\$} \{0, 1\};$
$\gamma^* \leftarrow (a, \hat{a}, c, d);\ \gamma^*_{\mathsf{def}} \leftarrow$ true;
$b' \leftarrow \mathcal{A}_2(\gamma^*)$; return $(b = b')$

Oracle $\mathcal{D}(a, \hat{a}, c, d)$:
if $|\boldsymbol{L}_\mathcal{D}| < q_\mathcal{D} \land \neg(\gamma^*_{\mathsf{def}} \land (a, \hat{a}, c, d) = \gamma^*)$
then
  $\boldsymbol{L}_\mathcal{D} \leftarrow \gamma :: \boldsymbol{L}_\mathcal{D};\ v \leftarrow H_k(a, \hat{a}, c);$
  if $\hat{a} = a^w$ then
    if $d = a^{x+vy}$ then return c/a^z
    else return $\bot$
  elsif $d = a^{x_1+vy_1} \cdot \hat{a}^{x_2+vy_2}$ then
    bad $\leftarrow$ true; return $\bot$
  else return $\bot$
else return $\bot$

$\models \mathsf{G}_4 \sim \mathsf{G}_4' : \mathsf{true} \Rightarrow (u = u')\langle 1\rangle \leftrightarrow (u = u')\langle 2\rangle \land \big((u \neq u')\langle 1\rangle \rightarrow =_{\{\mathbf{bad}\}}\big)$
$\models \mathsf{G}_4' \sim \mathsf{G}_5 : \mathsf{true} \Rightarrow =_{\{\mathbf{bad}_1\}} \land \big(\neg\mathbf{bad}_1\langle 1\rangle \rightarrow =_{\{\mathbf{bad},u,u'\}}\big)$
$\Pr[\mathsf{G}_4 : \mathbf{bad} \land u \neq u'] \leq \Pr[\mathsf{G}_5 : \mathbf{bad} \land u \neq u'] + (q_\mathcal{D}/q)^4$

Fig. 3. Proof sketch of the IND-CCA security of the Cramer-Shoup cryptosystem

The decryption oracle in game G_5 also raises a flag $\mathbf{bad}_2$ when a valid ciphertext with $H_k(a, \hat{a}, c) = H_k(g^u, \hat{g}^{u'}, g^r)$ is queried. Since this leads to a collision, we can build an adversary $\mathcal{C}$ against the TCR of H such that its success probability is lower bounded by the probability of $\mathbf{bad}_2$ being raised in G_5. Thus,

$$\Pr[\mathsf{G}_5 : \mathbf{bad} \land u \neq u'] \leq \mathbf{Adv}^{\mathcal{C}}_{\mathsf{TCR}} + \Pr[\mathsf{G}_5 : \mathbf{bad} \land u \neq u' \land \neg\mathbf{bad}_2]$$

The proof concludes by showing that the probability in G_5 of **bad** being set while $\mathbf{bad}_2$ is not is bounded by $q_\mathcal{D}/q$. This is done by reformulating the test under which $\mathbf{bad}_2$ is set so that it does not depend on x_1, x_2, y_1, y_2. Therefore, the probability of this test succeeding in any decryption query (under the condition that $u \neq u'$) is the probability of the adversary guessing a random value in the group, at most $q_\mathcal{D}/q$ summing over all queries. The bound in the statement follows.

Game $\boxed{\mathsf{G}_4'}$ G_5 :	**Oracle** $\mathcal{D}(a,\hat{a},c,d)$:

$g \xleftarrow{\$} \mathcal{G}\setminus\{1\};\ w \xleftarrow{\$} \mathbb{Z}_q^*;\ \hat{g} \leftarrow g^w;\ k \xleftarrow{\$} K;$
$u,u' \xleftarrow{\$} \mathbb{Z}_q;\ a \leftarrow g^u;\ \hat{a} \leftarrow \hat{g}^{u'};$
$y,y_2 \xleftarrow{\$} \mathbb{Z}_q;\ y_1 \leftarrow y - wy_2;\ f \leftarrow g^y;$
$x \xleftarrow{\$} \mathbb{Z}_q;\ e \leftarrow g^x;\ r' \xleftarrow{\$} \mathbb{Z}_q;\ d \leftarrow g^{r'};$
$x_2 \leftarrow (r' - u(x+vy))/(w(u'-u)) - vy_2;$
$x_1 \leftarrow x - wx_2;\ z \xleftarrow{\$} \mathbb{Z}_q; h \leftarrow g^z;$
$r \xleftarrow{\$} \mathbb{Z}_q;\ c \leftarrow g^r;$
$v \leftarrow H_k(a,\hat{a},c);\ \gamma^* \leftarrow (a,\hat{a},c,d);$
$(m_0,m_1) \leftarrow \mathcal{A}_1(k,h,\hat{g},e,f,h);$
$\gamma^*_{\mathsf{def}} \leftarrow$ true; $b' \leftarrow \mathcal{A}_2(\gamma^*);$ return $(b = b')$
	if $\|\boldsymbol{L}_\mathcal{D}\| < q_\mathcal{D} \wedge \neg\gamma^*_{\mathsf{def}} \wedge (a,\hat{a},c,d) = \gamma^*$
	then $\mathbf{bad}_1 \leftarrow$ true;
	if $\|\boldsymbol{L}_\mathcal{D}\| < q_\mathcal{D} \wedge (\boxed{\neg\gamma^*_{\mathsf{def}} \vee}(a,\hat{a},c,d) \neq \gamma^*)$
	then $\boldsymbol{L}_\mathcal{D} \leftarrow \gamma :: \boldsymbol{L}_\mathcal{D};\ v \leftarrow H_k(a,\hat{a},c);$
	  if $\hat{a} = a^w$ then
	    if $d = a^{x+vy}$ then return c/a^z
	    else return $\perp$
	  elsif $d = a^{x_1+vy_1} \cdot \hat{a}^{x_2+vy_2}$ then
	    $\mathbf{bad} \leftarrow$ true;
	    if $v = H_k(g^u, \hat{g}^{u'}, g^r)$ then
	      $\mathbf{bad}_2 \leftarrow$ true
	  else return $\perp$
	else return $\perp$

$\models \mathsf{G}_5 \sim \mathsf{TCR} : \mathsf{true} \Rightarrow \mathbf{bad}_2\langle 1\rangle \rightarrow \mathsf{res}\langle 2\rangle$
$\Pr[\mathsf{G}_5 : \mathbf{bad} \wedge u \neq u'] \leq$
$\Pr[\mathsf{TCR} : H_k(m_0) = H_k(m_1) \wedge m_0 \neq m_1] + \Pr[\mathsf{G}_5 : \mathbf{bad} \wedge u \neq u' \wedge \neg\mathbf{bad}_2]$

Game TCR :	**Oracle** $\mathcal{D}(a,\hat{a},c,d)$:

$m_0 \leftarrow \mathcal{C}_1(\,);\ k \xleftarrow{\$} K;\ m_1 \leftarrow \mathcal{C}_2(k);$
return $(H_k(m_0) = H_k(m_1) \wedge m_0 \neq m_1)$
Adversary $\mathcal{C}_1()$:
$g \xleftarrow{\$} \mathcal{G}\setminus\{1\};\ w \xleftarrow{\$} \mathbb{Z}_q^*;\ \hat{g} \leftarrow g^w;$
$u,u' \xleftarrow{\$} \mathbb{Z}_q;\ a \leftarrow g^u;\ \hat{a} \leftarrow \hat{g}^{u'};$
$r \xleftarrow{\$} \mathbb{Z}_q;\ c \leftarrow g^r;$ return $(a,\hat{a},c)$
Adversary $\mathcal{C}_2(k)$:
$r',x,y,z \xleftarrow{\$} \mathbb{Z}_q;$
$d \leftarrow g^{r'}; e \leftarrow g^x; f \leftarrow g^y;\ h \leftarrow g^z;$
$y_2 \xleftarrow{\$} \mathbb{Z}_q;\ y_1 \leftarrow y - wy_2;\ \hat{k} \leftarrow k;$
$v \leftarrow H_k(a,\hat{a},c);$
$x_2 \leftarrow (r' - u(x+vy))/(w(u'-u)) - vy_2;$
$x_1 \leftarrow x - wx_2;$
$(m_0,m_1) \leftarrow \mathcal{A}_1(h,\hat{g},e,f,h);$
$\gamma^* \leftarrow (a,\hat{a},c,d);\ b' \leftarrow \mathcal{A}_2(\gamma^*);$ return $\hat{m}$
	if $\|\boldsymbol{L}_\mathcal{D}\| < q_\mathcal{D} \wedge (a,\hat{a},c,d) \neq \gamma^*$ then
	  $\boldsymbol{L}_\mathcal{D} \leftarrow \gamma :: \boldsymbol{L}_\mathcal{D};$
	  $v \leftarrow H_{\hat{k}}(a,\hat{a},c);$
	  if $\hat{a} = a^w$ then
	    if $d = a^{x+vy}$ then return c/a^z
	    else return $\perp$
	  elsif $d = a^{x_1+vy_1} \cdot \hat{a}^{x_2+vy_2}$ then
	    if $v = H_{\hat{k}}(g^u, \hat{g}^{u'}, g^r)$ then
	      $\hat{m} \leftarrow (a,\hat{a},c);$
	      return $\perp$
	  else return $\perp$
	else return $\perp$

Fig. 4. Proof sketch of the IND-CCA security of the Cramer-Shoup cryptosystem

5 Limitations and Extensions

EasyCrypt is in its early stages of development; we briefly comment on some of its main limitations and possible extensions:

- Programming language: in comparison with CertiCrypt, the language of EasyCrypt lacks loops, recursive procedures, and drawing from skewed distributions. We do not see the need for extending the current language with recursive procedures. In contrast, we believe that more general forms for sampling and bounded loops are useful and foresee no specific difficulty in

adding them to the language (note that annotating loops with invariants may be required for verification condition generation);
- Verifiable evidence: EasyCrypt only generates partial verifiable evidence. As there is currently no SMT solver that generates Coq proofs, the verification conditions are admitted in order to make the output derivations checkable by the Coq proof assistant. Making SMT solvers proof-producing is an active subject of research [21], and advances towards this goal shall benefit immediately to EasyCrypt;
- Computation of probability: EasyCrypt generates proof skeletons for claims about probability rather than fully machine-checked proofs. While it is entirely feasible to extend the compiler for justifying more reasonings, a more principled solution would require a tool that can symbolically compute the probability of an event in a distribution.

Further research into the theory of cryptographic proofs, in the line of [3], is needed to broaden the scope of applications and effectiveness of EasyCrypt. Essential goals include providing a formal account of useful reasoning principles, such as rewinding arguments or coin-fixing, and notions, such as statistical distance, that have not yet been considered in our setting.

There remain ample opportunities to apply methods from programming languages and formal verification to computer-aided cryptographic proofs. We mention two exciting avenues for improving automation in EasyCrypt. The first avenue is to improve our mechanism for inferring relational specifications of adversaries: there is a large body of knowledge on inferring invariants, and it would be beneficial to transpose them to our setting. More speculatively, program synthesis could be used to discover part of the sequence of games needed to conclude a proof, and to build adversaries that justify reductions to cryptographic assumptions. Both specification inference and program synthesis rely on verification condition generation and SMT solving, hence the basic blocks for such an investigation are in place.

Finally, Halevi [17] stresses that "the usefulness of (a) tool will depend crucially on the willingness of the customers (in this case the cryptographic community) to use it", and suggests on this account that an appropriate user interface will be a crucial component of the tool. We fully adhere to his view, and see building such an interface as an important objective for further work.

5.1 Comparison with CertiCrypt

Table 1 compares CertiCrypt and EasyCrypt on various security proofs formalized in both systems. Times are measured on a 2.8GHz Intel Core 2 Duo processor with 4GB of RAM under Mac OS X 10.6.7. For comparison, we show the size and checking time of CertiCrypt proofs extracted from EasyCrypt proof sketches. This is not an altogether fair comparison, because extracted proofs assume as axioms proof obligations checked by automated provers. As an experiment, we completed interactively the extracted proof of security of ElGamal encryption,

Table 1. Comparison of proof size and checking time between CertiCrypt and EasyCrypt

	CertiCrypt		EasyCrypt		Extracted	
	Lines	Time	Lines	Time	Lines	Time
ElGamal (IND-CPA)	565	45s	190	12s	1130	23s
Hashed ElGamal (IND-CPA)	1255	1m05s	243	33s	1772	41s
Full-Domain Hash (EF-CMA)	2035	5m46s	509	1m26s	2724	1m11s
Cramer-Shoup (IND-CCA)	n/a	n/a	1637	5m12s	5504	3m14s
OAEP (IND-CPA)	2451	3m27s	n/a	n/a	n/a	n/a
OAEP (IND-CCA)	11162	37m32s	n/a	n/a	n/a	n/a

thus obtaining a full proof verifiable under Coq. The resulting proof is 1173 long (meaning that only 43 lines are needed to prove in Coq the proof obligations checked by automated provers) and takes 25s to check.

6 Conclusion

Computer-aided verification of cryptographic protocols in the symbolic model is an established field of research: robust tools are available and have been used successfully to analyze realistic protocols (e.g. [1,9,14,19]). In contrast, there is little prior work on computer-aided cryptographic proofs in the computational model. The importance of such proofs was suggested independently by Bellare and Rogaway [8] and, more explicitly, by Halevi [17], who convincingly argues that they can be viewed as the "natural next step along the way of viewing cryptographic proofs as a sequence of probabilistic games". To date, there are two main tools for computer-aided cryptographic proofs: CertiCrypt, which favors generality and verifiable proofs, and CryptoVerif, which favors automation. We have presented EasyCrypt, a new tool which provides the first flexible and automated framework for building machine-checkable cryptographic proofs, and illustrated its use through computer-aided security proofs of Hashed ElGamal encryption in the Random Oracle Model and the Cramer-Shoup cryptosystem in the standard model. These examples demonstrate that proofs in EasyCrypt are significantly easier and faster to build than in any previous tool, while providing guarantees similar to CertiCrypt. Overall, we believe that EasyCrypt makes an important step towards the adoption of computer-aided proofs by working cryptographers.

Acknowledgments. We are grateful to Daniel Hedin and Anne Pacalet for their participation in the initial phases of the project, to Yassine Lakhnech and David Pointcheval for useful discussions, and to the anonymous reviewers for their insightful comments.

References

1. Backes, M., Maffei, M., Unruh, D.: Computationally sound verification of source code. In: 17th ACM Conference on Computer and Communications Security, CCS 2010, pp. 387–398. ACM, New York (2010)

2. Barthe, G., D'Argenio, P., Rezk, T.: Secure information flow by self-composition. In: 17th IEEE Workshop on Computer Security Foundations, CSFW 2004, pp. 100–114. IEEE Computer Society, Washington (2004)
3. Barthe, G., Daubignard, M., Kapron, B., Lakhnech, Y.: Computational indistinguishability logic. In: 17th ACM Conference on Computer and Communications Security, CCS 2010, pp. 375–386. ACM, New York (2010)
4. Barthe, G., Grégoire, B., Heraud, S., Zanella Béguelin, S.: Formal certification of ElGamal encryption. A gentle introduction to CertiCrypt. In: Degano, P., Guttman, J., Martinelli, F. (eds.) FAST 2008. LNCS, vol. 5491, pp. 1–19. Springer, Heidelberg (2009)
5. Barthe, G., Grégoire, B., Lakhnech, Y., Zanella Béguelin, S.: Beyond provable security verifiable IND-CCA security of OAEP. In: Kiayias, A. (ed.) CT-RSA 2011. LNCS, vol. 6558, pp. 180–196. Springer, Heidelberg (2011)
6. Barthe, G., Grégoire, B., Zanella Béguelin, S.: Formal certification of code-based cryptographic proofs. In: 36th ACM SIGPLAN-SIGACT Symposium on Principles of Programming Languages, POPL 2009, pp. 90–101. ACM, New York (2009)
7. Barthe, G., Hedin, D., Zanella Béguelin, S., Grégoire, B., Heraud, S.: A machine-checked formalization of Sigma-protocols. In: 23rd IEEE Computer Security Foundations Symposium, CSF 2010, pp. 246–260. IEEE Computer Society, Los Alamitos (2010)
8. Bellare, M., Rogaway, P.: The security of triple encryption and a framework for code-based game-playing proofs. In: Vaudenay, S. (ed.) EUROCRYPT 2006. LNCS, vol. 4004, pp. 409–426. Springer, Heidelberg (2006)
9. Bhargavan, K., Fournet, C., Gordon, A.D.: Modular verification of security protocol code by typing. In: 37th ACM SIGPLAN-SIGACT Symposium on Principles of Programming Languages, POPL 2010, pp. 445–456. ACM, New York (2010)
10. Blanchet, B., Jaggard, A.D., Scedrov, A., Tsay, J.K.: Computationally sound mechanized proofs for basic and public-key Kerberos. In: 15th ACM Conference on Computer and Communications Security, CCS 2008, pp. 87–99. ACM, New York (2008)
11. Blanchet, B.: A computationally sound mechanized prover for security protocols. In: 27th IEEE Symposium on Security and Privacy, S&P 2006, pp. 140–154. IEEE Computer Society, Los Alamitos (2006)
12. Blanchet, B., Pointcheval, D.: Automated security proofs with sequences of games. In: Dwork, C. (ed.) CRYPTO 2006. LNCS, vol. 4117, pp. 537–554. Springer, Heidelberg (2006)
13. Conchon, S., Contejean, E., Kanig, J., Lescuyer, S.: CC(X): Semantic combination of congruence closure with solvable theories. Electronic Notes in Theoretical Computer Science 198(2), 51–69 (2008)
14. Cremers, C.: The scyther tool: Verification, falsification, and analysis of security protocols. In: Gupta, A., Malik, S. (eds.) CAV 2008. LNCS, vol. 5123, pp. 414–418. Springer, Heidelberg (2008)
15. Detlefs, D., Nelson, G., Saxe, J.B.: Simplify: A theorem prover for program checking. Tech. Rep. HPL-2003-148, HP Laboratories Palo Alto (2003)
16. Filliâtre, J.C.: The WHY verification tool: Tutorial and Reference Manual Version 2.28 (2010), `http://why.lri.fr`
17. Halevi, S.: A plausible approach to computer-aided cryptographic proofs. Cryptology ePrint Archive, Report 2005/181 (2005)
18. Jonsson, B., Yi, W., Larsen, K.G.: Probabilistic extensions of process algebras. In: Bergstra, J., Ponse, A., Smolka, S. (eds.) Handbook of Process Algebra, pp. 685–710. Elsevier, Amsterdam (2001)

19. Paulson, L.C.: The inductive approach to verifying cryptographic protocols. J. of Comput. Secur. 6(1-2), 85–128 (1998)
20. Shoup, V.: Sequences of games: a tool for taming complexity in security proofs. Cryptology ePrint Archive, Report 2004/332 (2004)
21. Stump, A.: Proof checking technology for satisfiability modulo theories. Electr. Notes Theor. Comput. Sci. 228, 121–133 (2009)
22. The Coq development team: The Coq Proof Assistant Reference Manual Version 8.3 (2010), http://coq.inria.fr
23. Zanella Béguelin, S.: Formal Certification of Game-Based Cryptographic Proofs. Ph.D. thesis, Ecole Nationale Supérieure des Mines de Paris – Mines ParisTech (2010)
24. Zanella Béguelin, S., Grégoire, B., Barthe, G., Olmedo, F.: Formally certifying the security of digital signature schemes. In: 30th IEEE Symposium on Security and Privacy, S&P 2009, pp. 237–250. IEEE Computer Society, Los Alamitos (2009)

A Input File for the Proof of Security of Hashed ElGamal

The following is an extract taken from the EasyCrypt input file corresponding to the proof of IND-CPA security of Hashed ElGamal described in Section 2:

```
type group

cnst q    : int
cnst g    : group
cnst k    : int
cnst zero : bitstring{k}

type skey    = int
type pkey    = group
type key     = skey * pkey
type message = bitstring{k}
type cipher  = group * bitstring{k}

op (*)  : group, group → group                          = mul
op (^)  : group, int → group                            = pow
op (^^) : bitstring{k}, bitstring{k} → bitstring{k}     = xor

axiom pow_mul  : ∀(x:int, y:int). { (g^x)^y = g^(x*y) }
axiom xor_comm : ∀(x:bitstring{k}, y:bitstring{k}). {(x^^y) = (y^^x)}

...

adversary A1(pk:pkey) : message * message { group → message}
adversary A2(pk:pkey) : bool              { group → message}

game INDCPA = {
  var L  : (group, bitstring{k}) map
  var LA : group list

  fun H(x:group) : message = {
    var h : message = {0,1}^k;
    if (¬in_dom(x,L)) { L[x] = h; };
    return L[x];
  }

  fun H_A(x:group) : message = {
    var m : message;
    LA = x :: LA;
    m = H(x);
    return m;
```

```
  }

  ...

  abs A1 = A1 {H_A}
  abs A2 = A2 {H_A}

  fun Main() : bool = {
    var sk : skey;
    var pk : pkey;
    var m0, m1 : message;
    var c : cipher;
    var b, b' : bool;

    L = empty_map();
    LA = [];
    (sk,pk) = KG();
    (m0,m1) = A1(pk);
    b = {0,1};
    c = Enc(pk, b ? m0 : m1);
    b' = A2(c);
    return (b = b');
  }
}

game G1 = INDCPA
  var y' : group
  where Main = {
    var m0, m1 : message;
    var c : cipher;
    var b, b' : bool;
    var x, y : int;
    var hy : message;
    var α : group;

    L = empty_map();
    LA = [];
    x = [0..q-1]; α = g^x;
    y = [0..q-1]; y' = α^y;
    (m0,m1) = A1(α);
    b = {0,1};
    hy = H(y');
    b' = A2((g^y, hy ^^ (b ? m0 : m1)));
    return (b = b');
  }

equiv Fact1 : INDCPA.Main ~ G1.Main : {true} ==> ={res}
 inline KG, Enc; derandomize;
 auto inv ={L,LA};
 pop⟨2⟩ 1; repeat rnd; trivial;;
save;;

claim Pr1 : INDCPA.Main[res] = G1.Main[res] using Fact1
...
```

Optimal Verification of Operations on Dynamic Sets

Charalampos Papamanthou[1], Roberto Tamassia[1], and Nikos Triandopoulos[2,3]

[1] Brown University, Providence RI, USA
[2] RSA Laboratories, Cambridge MA, USA
[3] Boston University, Boston MA, USA

Abstract. We study the design of protocols for *set-operation verification*, namely the problem of cryptographically checking the correctness of outsourced set operations performed by an untrusted server over a dynamic collection of sets that are owned (and updated) by a trusted source. We present new authenticated data structures that allow any entity to *publicly* verify a proof attesting the correctness of primitive set operations such as *intersection*, *union*, *subset* and *set difference*. Based on a novel extension of the security properties of *bilinear-map accumulators* as well as on a primitive called *accumulation tree*, our protocols achieve *optimal* verification and proof complexity (i.e., only proportional to the size of the query parameters and the answer), as well as *optimal* update complexity (i.e., constant), while incurring no extra asymptotic space overhead. The proof construction is also efficient, adding a *logarithmic* overhead to the computation of the answer of a set-operation query. In contrast, existing schemes entail high communication and verification costs or high storage costs. Applications of interest include efficient verification of keyword search and database queries. The security of our protocols is based on the *bilinear q-strong Diffie-Hellman* assumption.

1 Introduction

Providing integrity guarantees in third-party data management settings is an active area of research, especially in view of the growth in usage of cloud computing. In such settings, verifying the correctness of outsourced computations performed over remotely stored data becomes a crucial property for the trustworthiness of cloud services. Such a verification process should incur minimal overheads to the clients or otherwise the benefits of computation outsourcing are dismissed; ideally, computations should be verified without having to locally rerun them or to utilize too much extra cloud storage.

In this paper, we study the verification of outsourced operations on general sets and consider the following problem. Assuming that a *dynamic collection* of m sets $S_1, S_2, \ldots, S_m$ is remotely stored at an untrusted server, we wish to *publicly* verify basic operations on these sets, such as *intersection*, *union* and *set difference*. For example, for an intersection query of t sets specified by indices $1 \leq i_1, i_2, \ldots, i_t \leq m$, we aim at designing techniques that allow any client to cryptographically check the correctness of the returned answer $\mathsf{I} = S_{i_1} \cap S_{i_2} \cap \ldots \cap S_{i_t}$. Moreover, we wish the verification of any set operation to be *operation-sensitive*, meaning that the resulting complexity depends *only on the (description and outcome of the) operation, and not on the sizes of the involved sets*. That is, if $\delta = |\mathsf{I}|$ is the answer size then we would like the verification cost

P. Rogaway (Ed.): CRYPTO 2011, LNCS 6841, pp. 91–110, 2011.

to be proportional to $t+\delta$, and independent of m or $\sum_i |S_i|$; note that work at least proportional to $t + \delta$ is needed to verify any such query's answer. Applications of interest include *keyword search* and *database* queries, which boil down to set operations.

Relation to verifiable computing. Recent works on *verifiable computing* [1,12,16] achieve operation-sensitive verification of general functionalities, thus covering set operations as a special case. Although such approaches clearly meet our goal with respect to optimal verifiability, they are inherently inadequate to meet our other goals with respect to *public verifiability* and *dynamic updates*, both important properties in the context of outsourced data querying. Indeed, to outsource the computation as an encrypted circuit, the works in [1,12,16] make use of some secret information which is also used by the verification algorithm, thus effectively supporting only one verifier; instead, we seek for schemes that allow *any client* (knowing only a public key) to query the set collection and verify the returned results. Also, the description of the circuit in these works is fixed at the initialization of the scheme, thus effectively supporting no updates in the outsourced data; instead, we seek for schemes that are dynamic. In other scenarios, but still in the secret-key setting, protocols for general functionalities and polynomial evaluation have recently been proposed in [11] and [6] respectively.

Aiming at both publicly verifiable and dynamic solutions, we study set-operation verification in the model of *authenticated data structures* (ADSs). A typical setting in this model, usually referred to as the *three-party* model [36], involves protocols executed by three participating entities. A trusted party, called *source*, owns a data structure (here, a collection of sets) that is replicated along with some cryptographic information to one or more untrusted parties, called *servers*. Accordingly, *clients* issue data-structure queries to the servers and are able to verify the correctness of the returned answers, based only on knowledge of public information which includes a *public key* and a *digest* produced by the source (e.g., the root hash of a Merkle tree).[1] Updates on the data structure are performed by the source and appropriately propagated by the servers. Variations of this model include: (i) a *two-party* variant (e.g., [30]), where the source keeps only a small state (i.e., only a digest) and performs both the updates/queries and the verifications—this model is directly comparable to the model of verifiable computing; (ii) the *memory checking* model [7], where read/write operations on an array of memory cells is verified—however, the absence of the notion of proof computation in memory checking (the server is just a storage device) as well as the feature of public verifiability in authenticated data structures make the two models fundamentally different.[2]

Achieving operation-sensitive verification. In this work, we design authenticated data structures for the verification of set operations in an *operation-sensitive* manner, where the proof and verification complexity depends only on the description and outcome of the operation and not on the size of the involved sets. Conceptually, this property is similar to the property of *super-efficient verification* that has been studied in certifying algorithms [21] and certification data structures [19,37], which is achieved as well as in the context of verifiable computing [1,12,16], where an answer can be verified with complexity asymptotically less than the complexity required to produce it. Whether the

[1] Conveying the trust clients have in the source, the authentic digest is assumed to be publicly available; in practice, a time-stamped and digitally signed digest is outsourced to the server.

[2] Indeed, memory checking might require *secret* memory, e.g., as in the PRF construction in [7].

above optimality property is achievable for set operations (while keeping storage linear) was posed as an open problem in [23]. We close this problem in the affirmative.

All existing schemes for set-operation verification fall into the following two rather straightforward and highly inefficient solutions. Either short proofs for the answer of every possible set-operation query are precomputed allowing for optimal verification at the client at the cost of exponential storage and update overheads at the source and the server—an undesirable trade-off, as it is against storage outsourcing. Or integrity proofs for all the elements of the sets involved in the query are given to the client who locally verifies the query result: in this case the verification complexity can be linear in the problem size—an undesirable feature, as it is against computation outsourcing.

We achieve optimal verification by departing from the above approaches as follows. We first reduce the problem of verifying set operations to the problem of *verifying the validity of some more primitive relations on sets*, namely *subset containment* and *set disjointness*. Then for each such primitive relation we employ a corresponding cryptographic primitive to optimally verify its validity. In particular, we extend the *bilinear-map accumulator* to optimally verify subset containment (Lemmas 1 and 4), inspired by [32]. We then employ the extended Euclidean algorithm over polynomials (Lemma 5) in combination with subset containment proofs to provide a novel optimal verification test for set disjointness. The intuition behind our technique is that disjoint sets can be represented by polynomials mutually indivisible, therefore there exist other polynomials so that the sum of their pairwise products equals to one—this is the test to be used in the proof. Still, transmitting (and processing) these polynomials is bandwidth (and time) prohibitive and does not lead to operation-sensitive verification. Bilinearity properties, however, allow us to compress their coefficients in the exponent and, yet, use them meaningfully, i.e., compute an internal product. This is why although a conceptually simpler RSA accumulator [5] would yield a mathematically sound solution, a bilinear-map accumulator [28] is essential for achieving the desired complexity goal.

We formally describe our protocols using an *authenticated data structure scheme* or *ADS scheme* (Definition 1). An ADS scheme consists of algorithms $\{\mathsf{genkey}, \mathsf{setup}, \mathsf{update}, \mathsf{refresh}, \mathsf{query}, \mathsf{verify}\}$ such that: (i) genkey produces the secret and public key of the system; (ii) on input a plain data structure D, setup initializes the authenticated data structure $\mathsf{auth}(D)$; (iii) *having* access to the secret key, update computes the updated digest of $\mathsf{auth}(D)$; (iv) *without having* access to the secret key, refresh updates $\mathsf{auth}(D)$; (v) query computes cryptographic proofs $\Pi(q)$ for answers $\alpha(q)$ to data structure queries q; (vi) verify processes a proof Π and an answer α and either *accepts* or *rejects*. Note that neither query nor verify have access to the secret key, thus modeling computation outsourcing and public verifiability. An ADS scheme must satisfy certain correctness and security properties (Definitions 2 and 3). We note that protocols in both the three-party and the two-party models can be realized via an ADS scheme.

Our main result, Theorem 1, presents the first ADS scheme to achieve *optimal verification* of the set operations *intersection*, *union*, *subset* and *set difference*, as well as *optimal updates* on the underlying collection of sets. Our scheme is proved secure under the bilinear extension of the q-strong Diffie-Hellman assumption (see, e.g., [8]).

Table 1. Asymptotic access and group complexities of various ADS schemes for intersection queries on $t = O(1)$ sets in a collection of m sets with answer size δ. Here, M is the sum of sizes of all the sets and $0 < \epsilon < 1$ is a constant. Also, all sizes of the intersected or updated sets are $\Theta(n)$, $|\Pi|$ denotes the size of the proof, and CR stands from "collision resistance".

	setup	update, refresh	query	verify, $\|\Pi\|$	**assumption**
[23,38]	$m + M$	$\log n + \log m$	$n + \log m$	$n + \log m$	Generic CR
[26]	$m + M$	$m + M$	n	n	Strong RSA
[29]	$m^t + M$	m^t	1	δ	Discrete Log
this work	$m + M$	1	$n \log^3 n + m^\epsilon \log m$	δ	Bilinear q-Strong DH

Complexity model. To explicitly measure complexity of various algorithms with respect to number of primitive cryptographic operations, without considering the dependency on the security parameter, we adopt the complexity model used in memory checking [7,14], which has been only implicitly used in ADS literature. The *access complexity* of an algorithm is defined as the number of memory accesses performed during its execution on the authenticated data structure that is stored in an indexed memory of n cells.[3] E.g., a Merkle tree [24] has $O(\log n)$ update access complexity since the update algorithm needs to read and write $O(\log n)$ memory cells of the authenticated data structure, each cell storing exactly one hash value. The *group complexity* of a data collection (e.g., proof or ADS group complexity) is defined as the number of elementary data objects (e.g., hash values or elements in $\mathbb{Z}_p$) contained in this collection. Note that although the access and group complexities are respectively related to the time and space complexities, the former are in principle smaller than the latter. This is because time and space complexities are counting number of bits and are always functions of the security parameter which, in turn, is always $\Omega(\log n)$. Therefore time and space complexities are always $\Omega(\log n)$, whereas access and group complexities can be $O(1)$. Finally, whenever it is clear from the context, we omit the terms "access" and "group".

Related work. The great majority of authenticated data structures involve the use of cryptographic hashing [2,7,18,20,23,27,39] or other primitives [17,31,32] to hierarchically compute over the outsourced data one or more digests. Most of these schemes incur verification costs that are proportional to the time spent to produce the query answer, thus they are not operation sensitive. Some bandwidth-optimal and operation-sensitive solutions for verification of various (e.g., range search) queries appear in [2,19].

Despite the fact that privacy-related problems for set operations have been extensively studied in the cryptographic literature (e.g., [9,15]), existing work on the integrity dimension of set operations appears mostly in the database literature. In [23], the importance of coming up with an operation-sensitive scheme is identified. In [26], possibly the closest in context work to ours, set intersection, union and difference are authenticated with linear costs. Similar bounds appear in [38]. In [29], a different approach is taken: In order to achieve operation-sensitivity, expensive pre-processing and

[3] We use the term "access complexity" instead of the "query complexity" used in memory checking [7,14] to avoid ambiguity when referring to algorithm query of the ADS scheme. We also require that each memory cell can store up to $O(\text{poly}(\log n))$ bits, a word size used in [7,14].

exponential space are required (answers to all possible queries are signed). Finally, related to our work are non-membership proofs, both for the RSA [22] and the bilinear-map [3,13] accumulators. A comparison of our work with existing schemes appears in Table 1.

2 Preliminaries

We denote with k the security parameter and with $\text{neg}(k)$ a negligible function.[4]

The bilinear-map accumulator. Let $\mathbb{G}$ be a cyclic multiplicative group of prime order p, generated by element $g \in \mathbb{G}$. Let also $\mathcal{G}$ be a cyclic multiplicative group of the same order p, such that there exists a pairing $e : \mathbb{G} \times \mathbb{G} \rightarrow \mathcal{G}$ with the following properties: (i) Bilinearity: $e(P^a, Q^b) = e(P, Q)^{ab}$ for all $P, Q \in \mathbb{G}$ and $a, b \in \mathbb{Z}_p$; (ii) Non-degeneracy: $e(g, g) \neq 1$; (iii) Computability: For all $P, Q \in \mathbb{G}$, $e(P, Q)$ is efficiently computable. We call $(p, \mathbb{G}, \mathcal{G}, e, g)$ a tuple of bilinear pairing parameters, produced as the output of a probabilistic polynomial-time algorithm that runs on input 1^k.

In this setting, the bilinear-map accumulator [28] is an efficient way to provide short *proofs of membership* for elements that belong to a set. Let $s \in \mathbb{Z}_p^*$ be a randomly chosen value that constitutes the trapdoor in the scheme. The accumulator primitive accumulates elements in $\mathbb{Z}_p - \{s\}$, outputting a value that is an element in $\mathbb{G}$. For a set of elements $\mathcal{X}$ in $\mathbb{Z}_p - \{s\}$ the accumulation value $\text{acc}(\mathcal{X})$ of $\mathcal{X}$ is defined as

$$\text{acc}(\mathcal{X}) = g^{\prod_{x \in \mathcal{X}}(x+s)} .$$[5]

Value $\text{acc}(\mathcal{X})$ can be constructed using $\mathcal{X}$ and $g, g^s, g^{s^2}, \ldots, g^{s^q}$ (through polynomial interpolation), where $q \geq |\mathcal{X}|$. Subject to $\text{acc}(\mathcal{X})$ each element in $\mathcal{X}$ has a succinct membership proof. More generally, the *proof of subset containment* of a set $\mathcal{S} \subseteq \mathcal{X}$—for $|\mathcal{S}| = 1$, this becomes a membership proof—is the *witness* $(\mathcal{S}, \mathsf{W}_{\mathcal{S},\mathcal{X}})$ where

$$\mathsf{W}_{\mathcal{S},\mathcal{X}} = g^{\prod_{x \in \mathcal{X}-\mathcal{S}}(x+s)} . \tag{1}$$

Subset containment of $\mathcal{S}$ in $\mathcal{X}$ can be checked through relation $e(\mathsf{W}_{\mathcal{S},\mathcal{X}}, g^{\prod_{x \in \mathcal{S}}(x+s)}) \stackrel{?}{=} e(\text{acc}(\mathcal{X}), g)$ by any verifier with access only to public information. The security property of the bilinear-map accumulator, namely that computing fake but verifiable subset containment proofs is hard, can be proved using the *bilinear* q-strong Diffie-Hellman assumption, which is *slightly* stronger than the q-strong Diffie-Hellman assumption [8].[6]

Assumption 1 (Bilinear q-strong Diffie-Hellman assumption). *Let k be the security parameter and $(p, \mathbb{G}, \mathcal{G}, e, g)$ be a tuple of bilinear pairing parameters. Given the elements $g, g^s, \ldots, g^{s^q} \in \mathbb{G}$ for some s chosen at random from $\mathbb{Z}_p^*$, where $q = \text{poly}(k)$, no probabilistic polynomial-time algorithm can output a pair $(a, e(g, g)^{1/(a+s)}) \in \mathbb{Z}_p \times \mathcal{G}$, except with negligible probability* $\text{neg}(k)$.

---

[4] Function $f : \mathbb{N} \rightarrow \mathbb{R}$ is $\text{neg}(k)$ if and only if for any nonzero polynomial $p(k)$ there exits N such that for all $k > N$ it is $f(k) < 1/p(k)$.

[5] $\prod_{x \in S_i}(x + s)$ is called *characteristic polynomial* of set S_i in the literature (e.g., see [25]).

[6] However, the plain q-strong Diffie-Hellman assumption [28] suffices to prove just the *collision resistance* of the bilinear-map accumulator.

We next prove the security of subset witnesses by generalizing the proof in [28]. Subset witnesses also appeared (independent of our work but without a proof) in [10].

Lemma 1 (Subset containment). *Let k be the security parameter and $(p, \mathbb{G}, \mathcal{G}, e, g)$ be a tuple of bilinear pairing parameters. Given the elements $g, g^s, \ldots, g^{s^q} \in \mathbb{G}$ for some s chosen at random from $\mathbb{Z}_p^*$ and a set of elements $\mathcal{X}$ in $\mathbb{Z}_p - \{s\}$ with $q \geq |\mathcal{X}|$, suppose there is a probabilistic polynomial-time algorithm that finds $\mathcal{S}$ and W such that $\mathcal{S} \nsubseteq \mathcal{X}$ and $e(\mathsf{W}, g^{\prod_{x \in \mathcal{S}}(x+s)}) = e(\mathsf{acc}(\mathcal{X}), g)$. Then there is a probabilistic polynomial-time algorithm that breaks the bilinear q-strong Diffie-Hellman assumption.*

Proof. Suppose there is a probabilistic polynomial-time algorithm that computes such a set $\mathcal{S} = \{y_1, y_2, \ldots, y_\ell\}$ and a fake witness W. Let $\mathcal{X} = \{x_1, x_2, \ldots, x_n\}$ and $y_j \notin \mathcal{X}$ for some $1 \leq j \leq \ell$. This means that

$$e(\mathsf{W}, g)^{\prod_{y \in \mathcal{S}}(y+s)} = e(g, g)^{(x_1+s)(x_2+s)\ldots(x_n+s)} .$$

Note that $(y_j + s)$ does not divide $(x_1 + s)(x_2 + s)\ldots(x_n + s)$. Therefore there exist polynomial $Q(s)$ (computable in polynomial time) of degree $n - 1$ and constant $\lambda \neq 0$, such that $(x_1 + s)(x_2 + s)\ldots(x_n + s) = Q(s)(y_j + s) + \lambda$. Thus we have

$$e(\mathsf{W}, g)^{(y_j+s)\prod_{1 \leq i \neq j \leq \ell}(y_i+s)} = e(g, g)^{Q(s)(y_j+s)+\lambda} \Rightarrow$$

$$e(g, g)^{\frac{1}{y_j+s}} = \left[e(\mathsf{W}, g)^{\prod_{1 \leq i \neq j \leq \ell}(y_i+s)} e(g, g)^{-Q(s)}\right]^{\lambda^{-1}} .$$

Thus, this algorithm can break the bilinear q-strong Diffie-Hellman assumption. □

Tools for polynomial arithmetic. Our solutions use (modulo p) polynomial arithmetic. We next present two results that are extensively used in our techniques, contributing to achieve the desired complexity goals. The first result on polynomial interpolation is derived using an FFT algorithm (see Preparata and Sarwate [34]) that computes the DFT in a finite field (e.g., $\mathbb{Z}_p$) for arbitrary n and performing $O(n \log n)$ field operations. We note that an n-th root of unity is not required to exist in $\mathbb{Z}_p$ for this algorithm to work.

Lemma 2 (Polynomial interpolation with FFT [34]). *Let $\prod_{i=1}^{n}(x_i + s) = \sum_{i=0}^{n} b_i s^i$ be a degree-n polynomial. The coefficients $b_n \neq 0, b_{n-1}, \ldots, b_0$ of the polynomial can be computed with $O(n \log n)$ complexity, given $x_1, x_2, \ldots, x_n$.*

Lemma 2 refers to an efficient process for computing the coefficients of a polynomial, given its roots $x_1, x_2, \ldots, x_n$. In our construction, we make use of this process a numbers of times, in particular, when, given some values $x_1, x_2, \ldots, x_n$ to be accumulated, an untrusted party needs to compute $g^{(x_1+s)(x_2+s)\ldots(x_n+s)}$ without having access to s. However, access to $g, g^s, \ldots, g^{s^n}$ (part of the public key) is allowed, and therefore computing the accumulation value boils down to a polynomial interpolation.

We next present a second result that will be used in our verification algorithms. Related to *certifying algorithms* [21], this result states that if the vector of coefficients $\mathbf{b} = [b_n, b_{n-1}, \ldots, b_0]$ is claimed to be correct, then, given the vector of roots $\mathbf{x} = [x_1, x_2, \ldots, x_n]$, with high probability, vector $\mathbf{b}$ can be certified to be correct with complexity asymptotically less than $O(n \log n)$, i.e., without an FFT computation from scratch. This is achieved with the following algorithm:

Algorithm $\{\texttt{accept}, \texttt{reject}\} \leftarrow \mathsf{certify}(\mathbf{b}, \mathbf{x}, \mathsf{pk})$: The algorithm picks a random $\kappa \in \mathbb{Z}_p^*$. If $\sum_{i=0}^{n} b_i \kappa^i = \prod_{i=1}^{n}(x_i + \kappa)$, then the algorithm accepts, else it rejects.

Lemma 3 (Polynomial coefficients verification). *Let* $\mathbf{b} = [b_n, b_{n-1}, \ldots, b_0]$ *and* $\mathbf{x} = [x_1, x_2, \ldots, x_n]$. *Algorithm* $\mathsf{certify}(\mathbf{b}, \mathbf{x}, \mathsf{pk})$ *has* $O(n)$ *complexity. Also, if* $\texttt{accept} \leftarrow \mathsf{certify}(\mathbf{b}, \mathbf{x}, \mathsf{pk})$, *then* $b_n, b_{n-1}, \ldots, b_0$ *are the coefficients of the polynomial* $\prod_{i=1}^{n}(x_i + s)$ *with probability* $\Omega(1 - \mathsf{neg}(k))$.

Authenticated data structure scheme. We now define our *authenticated data structure scheme* (ADS scheme), as well as the correctness and security properties it must satisfy.

Definition 1 (ADS scheme). *Let* D *be* any *data structure that supports queries* q *and updates* u. *Let* $\mathsf{auth}(D)$ *denote the resulting authenticated data structure and* d *the digest of the authenticated data structure, i.e., a constant-size description of* D. *An ADS scheme* $\mathcal{A}$ *is a collection of the following six probabilistic polynomial-time algorithms:*

1. $\{\mathsf{sk}, \mathsf{pk}\} \leftarrow \mathsf{genkey}(1^k)$: *On input the security parameter* k, *it outputs a secret key* sk *and a public key* pk;
2. $\{\mathsf{auth}(D_0), d_0\} \leftarrow \mathsf{setup}(D_0, \mathsf{sk}, \mathsf{pk})$: *On input a (plain) data structure* D_0 *and the secret and public keys, it computes the authenticated data structure* $\mathsf{auth}(D_0)$ *and the respective digest* d_0 *of it;*
3. $\{D_{h+1}, \mathsf{auth}(D_{h+1}), d_{h+1}, \mathsf{upd}\} \leftarrow \mathsf{update}(u, D_h, \mathsf{auth}(D_h), d_h, \mathsf{sk}, \mathsf{pk})$: *On input an update* u *on data structure* D_h, *the authenticated data structure* $\mathsf{auth}(D_h)$, *the digest* d_h, *and the secret and public keys, it outputs the updated data structure* D_{h+1} *along with the updated authenticated data structure* $\mathsf{auth}(D_{h+1})$, *the updated digest* d_{h+1} *and some relative information* upd;
4. $\{D_{h+1}, \mathsf{auth}(D_{h+1}), d_{h+1}\} \leftarrow \mathsf{refresh}(u, D_h, \mathsf{auth}(D_h), d_h, \mathsf{upd}, \mathsf{pk})$: *On input an update* u *on data structure* D_h, *the authenticated data structure* $\mathsf{auth}(D_h)$, *the digest* d_h, *relative information* upd *(output by* update*), and the public key, it outputs the updated data structure* D_{h+1} *along with the updated authenticated data structure* $\mathsf{auth}(D_{h+1})$ *and the updated digest* d_{h+1};
5. $\{\Pi(q), \alpha(q)\} \leftarrow \mathsf{query}(q, D_h, \mathsf{auth}(D_h), \mathsf{pk})$: *On input a query* q *on data structure* D_h, *the authenticated data structure* $\mathsf{auth}(D_h)$ *and the public key, it returns the answer* $\alpha(q)$ *to the query, along with a proof* $\Pi(q)$;
6. $\{\texttt{accept}, \texttt{reject}\} \leftarrow \mathsf{verify}(q, \alpha, \Pi, d_h, \mathsf{pk})$: *On input a query* q, *an answer* α, *a proof* Π, *a digest* d_h *and the public key, it outputs either* $\texttt{accept}$ *or* $\texttt{reject}$.

Let $\{\texttt{accept}, \texttt{reject}\} \leftarrow \mathsf{check}(q, \alpha, D_h)$ be an algorithm that decides whether α is a correct answer for query q on data structure D_h (check is not part of the definition of an ADS scheme). There are two properties that an ADS scheme should satisfy, namely *correctness* and *security* (intuition follows from signature schemes definitions).

Definition 2 (Correctness). *Let* $\mathcal{ASC}$ *be an ADS scheme* $\{\mathsf{genkey}, \mathsf{setup}, \mathsf{update}, \mathsf{refresh}, \mathsf{query}, \mathsf{verify}\}$. *We say that the ADS scheme* $\mathcal{ASC}$ *is* correct *if, for all* $k \in \mathbb{N}$, *for all* $\{\mathsf{sk}, \mathsf{pk}\}$ *output by algorithm* genkey, *for all* D_h, $\mathsf{auth}(D_h)$, d_h *output by one invocation of* setup *followed by polynomially-many invocations of* $\mathsf{refresh}$, *where* $h \geq 0$, *for all queries* q *and for all* $\Pi(q), \alpha(q)$ *output by* $\mathsf{query}(q, D_h, \mathsf{auth}(D_h), \mathsf{pk})$, *with all but negligible probability, whenever algorithm* $\mathsf{check}(q, \alpha(q), D_h)$ *outputs* $\texttt{accept}$, *so does algorithm* $\mathsf{verify}(q, \Pi(q), \alpha(q), d_h, \mathsf{pk})$.

Definition 3 (Security). *Let $\mathcal{ASC}$ be an ADS scheme* {genkey, setup, update, refresh, query, verify}, *k be the security parameter, $\nu(k)$ be a negligible function and* {sk, pk} $\leftarrow$ genkey(1^k). *Let also* Adv *be a probabilistic polynomial-time adversary that is only given* pk. *The adversary has unlimited access to all algorithms of $\mathcal{ASC}$, except for algorithms* setup *and* update *to which he has only oracle access. The adversary picks an initial state of the data structure D_0 and computes D_0,* auth(D_0), *d_0 through oracle access to algorithm* setup. *Then, for $i = 0, \ldots, h =$* poly(k), Adv *issues an update u_i in the data structure D_i and computes D_{i+1},* auth(D_{i+1}) *and d_{i+1} through oracle access to algorithm* update. *Finally the adversary picks an index $0 \leq t \leq h+1$, and computes a query q, an answer α and a proof Π. We say that the ADS scheme $\mathcal{ASC}$ is* secure *if for all $k \in \mathbb{N}$, for all* {sk, pk} *output by algorithm* genkey, *and for any probabilistic polynomial-time adversary* Adv *it holds that*

$$\Pr\left[\begin{array}{l}\{q, \Pi, \alpha, t\} \leftarrow \mathsf{Adv}(1^k, \mathsf{pk});\ \texttt{accept} \leftarrow \mathsf{verify}(q, \alpha, \Pi, d_t, \mathsf{pk});\\ \qquad\qquad\qquad\qquad\qquad\ \ \texttt{reject} \leftarrow \mathsf{check}(q, \alpha, D_t).\end{array}\right] \leq \nu(k)\,. \quad (2)$$

3 Construction and Algorithms

In this section we present an ADS scheme for set-operation verification. The underlying data structure for which we design our ADS scheme is called *sets collection*, and can be viewed as a generalization of the *inverted index* [4] data structure.

Sets collection. The *sets collection* data structure consists of m sets, denoted with $S_1, S_2, \ldots, S_m$, each containing elements from a universe $\mathcal{U}$. Without loss of generality we assume that the universe $\mathcal{U}$ is the set of nonnegative integers in the interval $[m+1, p-1] - \{s\}$,[7] where p is k-bit prime, m is the number of the sets in our collection that has bit size $O(\log k)$, k is the security parameter and s is the trapdoor of the scheme (see algorithm genkey). A set S_i does not contain duplicate elements, however an element $x \in \mathcal{U}$ can appear in more than one set. Each set is sorted and the total space needed is $O(m + M)$, where M is the sum of the sizes of the sets.

In order to get some intuition, we can view the sets collection as an *inverted index*. In this view, the elements are pointers to documents and each set S_i corresponds to a term w_i in the dictionary, containing the pointers to documents where term w_i appears. In this case, m is the number of terms being indexed, which is typically in the hundreds of thousands, while M, bounded from below by the number of documents being indexed, is typically in the billions. Thus, the more general terms "elements" and "sets" in a sets collection can be instantiated to the more specific "documents" and "terms".

The operations supported by the sets collection data structure consist of *updates* and *queries*. An update is either an *insertion* of an element into a set or a *deletion* of an element from a set. An update on a set of size n takes $O(\log n)$ time. For simplicity, we assume that the number m of sets does not change after updates. A query is one of the following standard set operations: (i) *Intersection:* Given indices $i_1, i_2, \ldots, i_t$, return set $\mathsf{I} = S_{i_1} \cap S_{i_2} \cap \ldots \cap S_{i_t}$; (ii) *Union:* Given indices $i_1, i_2, \ldots, i_t$, return set $\mathsf{U} = S_{i_1} \cup S_{i_2} \cup \ldots \cup S_{i_t}$; (iii) *Subset query:* Given indices i and j, return true if

[7] This choice simplifies the exposition; however, by using some collision-resistant hash function, universe $\mathcal{U}$ can be set to $\mathbb{Z}_p - \{s\}$.

$S_i \subseteq S_j$ and `false` otherwise; (iv) *Set difference:* Given indices i and j, return set $\mathsf{D} = S_i - S_j$. For the rest of the paper, we denote with δ the size of the answer to a query operation, i.e., δ is equal to the size of I, U, or D. For a subset query, δ is $O(1)$.

We next detail the design of an ADS scheme $\mathcal{ASC}$ for the *sets collection* data structure. This scheme provides protocols for verifying the integrity of the answers to set operations in a dynamic setting where sets evolve over time through updates. The goal is to achieve optimality in the communication and verification complexity: a query with t parameters and answer size δ should be verified with $O(t+\delta)$ complexity, and at the same time query and update algorithms should be efficient as well.

3.1 Setup and Updates

We describe an ADS scheme $\mathcal{ASC} = \{\mathsf{genkey}, \mathsf{setup}, \mathsf{update}, \mathsf{refresh}, \mathsf{query}, \mathsf{verify}\}$ for the sets collection data structure and we prove that its algorithms satisfy the complexities of Table 1. We begin with the algorithms that are related to the setup and the updates of the authenticated data structure.

Algorithm $\{\mathsf{sk}, \mathsf{pk}\} \leftarrow \mathsf{genkey}(1^k)$**:** Bilinear pairing parameters $(p, \mathbb{G}, \mathcal{G}, e, g)$ are picked and an element $s \in \mathbb{Z}_p^*$ is chosen at random. Subsequently, an one-to-one function $h(\cdot) : \mathbb{G} \rightarrow \mathbb{Z}_p^*$ is used. This function simply outputs the bit description of the elements of $\mathbb{G}$ according to some canonical representation of $\mathbb{G}$. Finally the algorithm outputs $\mathsf{sk} = s$ and $\mathsf{pk} = \{h(\cdot), p, \mathbb{G}, \mathcal{G}, e, g, \mathbf{g}\}$, where vector $\mathbf{g}$ contains values

$$\mathbf{g} = \left[g^s, g^{s^2}, \ldots, g^{s^q}\right],$$

where $q \geq \max\{m, \max_{i=1,\ldots,m}\{|S_i|\}\}$. The algorithm has $O(1)$ access complexity.

Algorithm $\{D_0, \mathsf{auth}(D_0), d_0\} \leftarrow \mathsf{setup}(D_0, \mathsf{sk}, \mathsf{pk})$**:** Let D_0 be our initial data structure, i.e., the one representing sets $S_1, S_2, \ldots, S_m$. The authenticated data structure $\mathsf{auth}(D_0)$ is built as follows. First, for each set S_i its accumulation value $\mathsf{acc}(S_i) = g^{\prod_{x \in S_i}(x+s)}$ is computed (see Section 2). Subsequently, the algorithm picks a constant $0 < \epsilon < 1$. Let T be a tree that has $l = \lceil 1/\epsilon \rceil$ levels and m leaves, numbered $1, 2, \ldots, m$, where m is the number of the sets of our sets collection data structure. Since T is a *constant-height* tree, the degree of any internal node of it is $O(m^\epsilon)$. We call such a tree an *accumulation tree*, which was originally introduced (combined with different cryptography) in [32]. For each node of the tree v, the algorithm recursively computes the *digest* $d(v)$ of v as follows. If v is a leaf corresponding to set S_i, where $1 \leq i \leq m$, the algorithm sets $d(v) = \mathsf{acc}(S_i)^{(i+s)}$; here, raising value $\mathsf{acc}(S_i)$ to exponent $i+s$, under the constraint that $i \leq m$, is done to also accumulate the index i of set S_i (and thus prove that $\mathsf{acc}(S_i)$ refers to S_i). If node v is not a leaf, then

$$d(v) = g^{\prod_{w \in \mathcal{N}(v)}(h(d(w))+s))}, \quad (3)$$

where $\mathcal{N}(v)$ denotes the set of children of node v. The algorithm outputs all the sets S_i as the data structure D_0, and all the accumulation values $\mathsf{acc}(S_i)$ for $1 \leq i \leq m$, the tree T and all the digests $d(v)$ for all $v \in T$ as the authenticated data structure $\mathsf{auth}(D_0)$. Finally, the algorithm sets $d_0 = d(r)$ where r is the root of T, i.e., d_0 is the digest of the authenticated data structure (defined similarly as in a Merkle tree).[8] The access complexity of the algorithm is $O(m + M)$ (for postorder traversal of T and computation of $\mathsf{acc}(S_i)$), where $M = \sum_{i=1}^{m} |S_i|$. The group complexity of $\mathsf{auth}(D_0)$ is also $O(m + M)$ since the algorithm stores one digest per node of T, T has $O(m)$ nodes and there are M elements contained in the sets, as part of $\mathsf{auth}(D_0)$.

Algorithm $\{D_{h+1}, \mathsf{auth}(D_{h+1}), d_{h+1}, \mathsf{upd}\} \leftarrow \mathsf{update}(u, D_h, \mathsf{auth}(D_h), d_h, \mathsf{sk}, \mathsf{pk})$**:** We consider the update "insert element $x \in \mathcal{U}$ into set S_i" (note that the same algorithm could be used for element deletions). Let v_0 be the leaf node of T corresponding to set S_i. Let $v_0, v_1, \ldots, v_l$ be the path in T from node v_0 to the root of the tree, where $l = \lceil 1/\epsilon \rceil$. The algorithm initially sets $d'(v_0) = \mathsf{acc}(S_i)^{(x+s)}$, i.e., it updates the accumulation value that corresponds to the updated set (note that in the case where x is deleted from S_i, the algorithm sets $d'(v_0) = \mathsf{acc}(S_i)^{(x+s)^{-1}}$). Then the algorithm sets

$$d'(v_j) = d(v_j)^{(h(d'(v_{j-1}))+s)(h(d(v_{j-1}))+s)^{-1}} \text{ for } j = 1, \ldots, l\,, \tag{4}$$

where $d(v_{j-1})$ is the current digest of v_{j-1} and $d'(v_{j-1})$ is the updated digest of v_{j-1}.[9] All these newly computed values (i.e., the new digests) are stored by the algorithm. The algorithm then outputs the new digests $d'(v_{j-1})$, $j = 1, \ldots, l$, along the path from the updated set to the root of the tree, as part of information upd. Information upd also includes x and $d'(v_l)$. The algorithm also sets $d_{h+1} = d'(v_l)$, i.e., the updated digest is the newly computed digest of the root of T. Finally the new authenticated data structure $\mathsf{auth}(D_{h+1})$ is computed as follows: in the current authenticated data structure $\mathsf{auth}(D_h)$ that is input of the algorithm, the values $d(v_{j-1})$ are overwritten with the new values $d'(v_{j-1})$ ($j = 1, \ldots, l$), and the resulting structure is included in the output of the algorithm. The number of operations performed is proportional to $1/\epsilon$, therefore the complexity of the algorithm is $O(1)$.

Algorithm $\{D_{h+1}, \mathsf{auth}(D_{h+1}), d_{h+1}\} \leftarrow \mathsf{refresh}(u, D_h, \mathsf{auth}(D_h), d_h, \mathsf{upd}, \mathsf{pk})$**:** We consider the update "insert element $x \in \mathcal{U}$ into set S_i". Let v_0 be the node of T corresponding to set S_i. Let $v_0, v_1, \ldots, v_l$ be the path in T from node v_0 to the root of the tree. Using the information upd, the algorithm sets $d(v_j) = d'(v_j)$ for $j = 0, \ldots, l$, i.e., it updates the digests that correspond to the updated path. Finally, it outputs the updated sets collection as D_{h+1}, the updated digests $d(v_j)$ (along with the ones that belong to the nodes that are not updated) as $\mathsf{auth}(D_{h+1})$ and $d'(v_l)$ (contained in upd) as d_{h+1}.[10] The algorithm has $O(1)$ complexity as the number of performed operations is $O(1/\epsilon)$.

3.2 Authenticity of Accumulation Values

So far we have described the authenticated data structure $\mathsf{auth}(D_h)$ that our ADS scheme $\mathcal{ASC}$ will use for set-operation verifications. Overall, $\mathsf{auth}(D_h)$ comprises a set

---

[8] Digest $d(r)$ is a "secure" succinct description of the set collection data structure. Namely, the accumulation tree protects the integrity of values $\mathsf{acc}(S_i)$, $1 \leq i \leq m$, and each accumulation value $\mathsf{acc}(S_i)$ protects the integrity of the elements contained in set S_i.

[9] Note that these update computations are efficient because update has access to secret key s.

[10] Note that information upd is not *required* for the execution of $\mathsf{refresh}$, but is rather used for efficiency. Without access to upd, algorithm $\mathsf{refresh}$ could compute the updated values $d(v_j)$ using polynomial interpolation, which would have $O(m^\epsilon \log m)$ complexity (see Lemma 2).

of m accumulation values $\mathsf{acc}(S_i)$, one for each set S_i, $i = 1, \ldots, m$, and a set of $O(m)$ digests $d(v)$, one for each internal node v of the accumulation tree T. Our proof construction and verification protocols for set operations (described in Section 3.3) make use of the accumulation values $\mathsf{acc}(S_i)$ (subject to which subset-containment witnesses can be defined), and therefore it is required that the authenticity of each such value can be verified. Tree T serves this exact role by providing short correctness proofs for each value $\mathsf{acc}(S_i)$ stored at leaf i of T, this time subject to the (global) digest d_h stored at the root of T. We next provide the details related to proving the authenticity of $\mathsf{acc}(S_i)$.

The *correctness proof* Π_i of accumulation value $\mathsf{acc}(S_i)$, $1 \leq i \leq m$, is a collection of $O(1)$ bilinear-map accumulator witnesses (as defined in Section 2). In particular, Π_i is set to be the ordered sequence $\Pi = (\pi_1, \pi_2, \ldots, \pi_l)$, where π_j is the pair of the digest of node v_{j-1} and a witness that authenticates v_{j-1}, subject to node v_j, in the path $v_0, v_1, \ldots, v_l$ defined by leaf v_0 storing accumulation value $\mathsf{acc}(S_i)$ and the root v_l of T. Conveniently, π_j is defined as $\pi_j = (\beta_j, \gamma_j)$, where

$$\beta_j = d(v_{j-1}) \text{ and } \gamma_j = \mathsf{W}_{v_{j-1}(v_j)} = g^{\prod_{w \in \mathcal{N}(v_j) - \{v_{j-1}\}} (h(d(w))+s)} . \tag{5}$$

Note that π_j is the witness for a subset of *one* element, namely $h(d(v_{j-1}))$ (recall, $d(v_0) = \mathsf{acc}(S_i)^{(i+s)}$). Clearly, pair π_j has group complexity $O(1)$ and can be constructed using polynomial interpolation with $O(m^\epsilon \log m)$ complexity, by Lemma 2 and since v_j has degree $O(m^\epsilon)$. Since Π_i consists of $O(1)$ such pairs, we conclude that the proof Π_i for an accumulation value $\mathsf{acc}(S_i)$ can be constructed with $O(m^\epsilon \log m)$ complexity and has $O(1)$ group complexity. The following algorithms $\mathsf{queryTree}$ and $\mathsf{verifyTree}$ are used to formally describe the construction and respectively the verification of such correctness proofs. Similar methods have been described in [32].

Algorithm $\{\Pi_i, \alpha_i\} \leftarrow \mathsf{queryTree}(i, D_h, \mathsf{auth}(D_h), \mathsf{pk})$**:** Let $v_0, v_1, \ldots, v_l$ be the path of T from the node storing $\mathsf{acc}(S_i)$ to the root of T. The algorithm computes Π_i by setting $\Pi_i = (\pi_1, \pi_2, \ldots, \pi_l)$, where $\pi_j = (d(v_{j-1}), \mathsf{W}_{v_{j-1}(v_j)})$ and $\mathsf{W}_{v_{j-1}(v_j)}$ is given in Equation 5 and computed by Lemma 2. Finally, the algorithm sets $\alpha_i = \mathsf{acc}(S_i)$.

Algorithm $\{\texttt{accept}, \texttt{reject}\} \leftarrow \mathsf{verifyTree}(i, \alpha_i, \Pi_i, d_h, \mathsf{pk})$**:** Let the proof be $\Pi_i = (\pi_1, \pi_2, \ldots, \pi_l)$, where $\pi_j = (\beta_j, \gamma_j)$. The algorithm outputs $\texttt{reject}$ if one of the following is true: (i) $e(\beta_1, g) \neq e(\alpha_i, g^i g^s)$; or (ii) $e(\beta_j, g) \neq e\left(\gamma_{j-1}, g^{h(\beta_{j-1})} g^s\right)$ for some $2 \leq j \leq l$; or (iii) $e(d_h, g) \neq e\left(\gamma_l, g^{h(\beta_l)} g^s\right)$. Otherwise, it outputs $\texttt{accept}$.

We finally provide some complexity and security properties that hold for the correctness proofs of the accumulated values. The following result is used as a building block to derive the complexity of our scheme and prove its security (Theorem 1).

Lemma 4. *Algorithm* $\mathsf{queryTree}$ *runs with* $O(m^\epsilon \log m)$ *access complexity and outputs a proof of* $O(1)$ *group complexity. Moreover algorithm* $\mathsf{verifyTree}$ *has* $O(1)$ *access complexity. Finally, for any adversarially chosen proof* Π_i *(*$1 \leq i \leq m$*), if* $\texttt{accept} \leftarrow \mathsf{verifyTree}(i, \alpha_i, \Pi_i, d_h, \mathsf{pk})$*, then* $\alpha_i = \mathsf{acc}(S_i)$ *with probability* $\Omega(1 - \mathsf{neg}(k))$.

3.3 Queries and Verification

With the correctness proofs of accumulation values at hand, we complete the description of our scheme $\mathcal{ASC}$ by presenting the algorithms that are related to the construction and verification of proofs attesting the correctness of set operations. These proofs are efficiently constructed using the authenticated data structure presented earlier, and they have optimal size $O(t+\delta)$, where t and δ are the sizes of the query parameters and the answer. In the rest of the section, we focus on the detailed description of the algorithms for an *intersection* and a *union* query, but due to space limitations, we omit the details of the *subset* and the *set difference* query. We note, however, that the treatment of the subset and set difference queries is analogous to that of the intersection and union queries.

The parameters of an intersection or a union query are t indices $i_1, i_2, \ldots, i_t$, with $1 \leq t \leq m$. To simplify the notation, we assume without loss of generality that these indices are $1, 2, \ldots, t$. Let n_i denote the size of set S_i ($1 \leq i \leq t$) and let $N = \sum_{i=1}^{t} n_i$. Note that the size δ of the intersection or union is always $O(N)$ and that operations can be performed with $O(N)$ complexity, by using a generalized merge.

Intersection query. Let $\mathsf{I} = S_1 \cap S_2 \cap \ldots \cap S_t = \{y_1, y_2, \ldots, y_\delta\}$. We express the correctness of the set intersection operation by means of the following two conditions:

$$\textbf{Subset condition: } \mathsf{I} \subseteq S_1 \wedge \mathsf{I} \subseteq S_2 \wedge \ldots \wedge \mathsf{I} \subseteq S_t\,; \tag{6}$$

$$\textbf{Completeness condition: } (S_1 - \mathsf{I}) \cap (S_2 - \mathsf{I}) \cap \ldots \cap (S_t - \mathsf{I}) = \emptyset\,. \tag{7}$$

The completeness condition in Equation 7 is necessary since set I must contain *all* the common elements. Given an intersection I, and for every set S_j, $1 \leq i \leq t$, we define the degree-n_j polynomial

$$P_j(s) = \prod_{x \in S_j - \mathsf{I}} (x+s)\,. \tag{8}$$

The following result is based on the extended Euclidean algorithm over polynomials and provides our core verification test for checking the correctness of set intersection.

Lemma 5. *Set I is the intersection of sets $S_1, S_2, \ldots, S_t$ if and only if there exist polynomials $q_1(s), q_2(s), \ldots, q_t(s)$ such that $q_1(s)P_1(s) + q_2(s)P_2(s) + \ldots + q_t(s)P_t(s) = 1$, where $P_j(s)$, $j = 1, \ldots, t$, are defined in Equation 8. Moreover, the polynomials $q_1(s), q_2(s), \ldots, q_t(s)$ can be computed with $O(N \log^2 N \log\log N)$ complexity.*

Using Lemmas 2 and 5 we next construct efficient proofs for both conditions in Equations 6 and 7. In turn, the proofs are directly used to define the algorithms query and verify of our ADS scheme $\mathcal{ASC}$ for *intersection* queries.

Proof of subset condition. For each set S_j, $1 \leq j \leq t$, the *subset witnesses* $\mathsf{W}_{\mathsf{I},j} = g^{P_j(s)} = g^{\prod_{x \in S_j - \mathsf{I}}(x+s)}$ are computed, each with $O(n_j \log n_j)$ complexity, by Lemma 2. (Recall, $\mathsf{W}_{\mathsf{I},j}$ serves as a proof that I is a subset of set S_j.) Thus, the total complexity for computing all t required subset witnesses is $O(N \log N)$, where $N = \sum_{i=1}^{t} n_i$.[11]

Proof of completeness condition. For each $q_j(s)$, $1 \leq j \leq t$, as in Lemma 5 satisfying $q_1(s)P_1(s) + q_2(s)P_2(s) + \ldots + q_t(s)P_t(s) = 1$, the *completeness witnesses* $\mathsf{F}_{\mathsf{I},j} = g^{q_j(s)}$ are computed, by Lemma 5 with $O(N \log^2 N \log\log N)$ complexity.

[11] This is because $\sum n_j \log n_j \leq \log N \sum n_j = N \log N$.

Algorithm $\{\Pi(q), \alpha(q)\} \leftarrow \mathsf{query}(q, D_h, \mathsf{auth}(D_h), \mathsf{pk})$ **(Intersection):** Query q consists of t indices $\{1, 2, \ldots, t\}$, asking for the intersection I of $S_1, S_2, \ldots, S_t$. Let $\mathsf{I} = \{y_1, y_2, \ldots, y_\delta\}$. Then $\alpha(q) = \mathsf{I}$, and the proof $\Pi(q)$ consists of the following parts.

1. *Coefficients* $b_\delta, b_{\delta-1}, \ldots, b_0$ of polynomial $(y_1 + s)(y_2 + s) \ldots (y_\delta + s)$ that is associated with the intersection $\mathsf{I} = \{y_1, y_2, \ldots, y_\delta\}$. These are computed with $O(\delta \log \delta)$ complexity (Lemma 2) and they have $O(\delta)$ group complexity.
2. *Accumulation values* $\mathsf{acc}(S_j)$, $j = 1, \ldots, t$, which are associated with sets S_j, along with their respective *correctness proofs* Π_j. These are computed by calling algorithm $\mathsf{queryTree}(j, D_h, \mathsf{auth}(D_h), \mathsf{pk})$, for $j = 1, \ldots, t$, with $O(tm^\epsilon \log m)$ total complexity and they have $O(t)$ total group complexity (Lemma 4).
3. *Subset witnesses* $\mathsf{W}_{\mathsf{I},j}$, $j = 1, \ldots, t$, which are associated with sets S_j and intersection I (see proof of subset condition). These are computed with $O(N \log N)$ complexity and have $O(t)$ total group complexity (Lemma 2).
4. *Completeness witnesses* $\mathsf{F}_{\mathsf{I},j}$, $j = 1, \ldots, t$, which are associated with polynomials $q_j(s)$ of Lemma 5 (see proof of completeness condition). These are computed with $O(N \log^2 N \log \log N)$ complexity and have $O(t)$ group complexity (Lemma 5).

Algorithm $\{\texttt{accept}, \texttt{reject}\} \leftarrow \mathsf{verify}(q, \alpha, \Pi, d_h, \mathsf{pk})$ **(Intersection):** Verifying the result of an intersection query includes the following steps.

1. First, the algorithm uses the coefficients $\mathbf{b} = [b_\delta, b_{\delta-1}, \ldots, b_0]$ and the answer $\alpha(q) = \{y_1, y_2, \ldots, y_\delta\}$ as an input to algorithm $\mathsf{certify}(\mathbf{b}, \alpha(q), \mathsf{pk})$, in order to certify the validity of $b_\delta, b_{\delta-1}, \ldots, b_0$. If certify outputs `reject`, the algorithm also outputs `reject`.[12] This step has $O(\delta)$ complexity (Lemma 3).
2. Subsequently, the algorithm uses the proof Π_j to verify the correctness of $\mathsf{acc}(S_j)$, by running algorithm $\mathsf{verifyTree}(j, \mathsf{acc}(S_j), \Pi_j, d_h, \mathsf{pk})$ for $j = 1, \ldots, t$. If, for some j, verifyTree running on $\mathsf{acc}(S_j)$ outputs `reject`, the algorithm also outputs `reject`. This step has $O(t)$ complexity (Lemma 4).
3. Next, the algorithm checks the subset condition:[13]

$$e\left(\prod_{i=0}^{\delta} \left(g^{s^i}\right)^{b_i}, \mathsf{W}_{\mathsf{I},j}\right) \stackrel{?}{=} e\left(\mathsf{acc}(S_j), g\right), \text{ for } j = 1, \ldots, t. \tag{9}$$

   If, for some j, the above check on subset witness $\mathsf{W}_{\mathsf{I},j}$ fails, the algorithm outputs `reject`. This step has $O(t + \delta)$ complexity.
4. Finally, the algorithm checks the completeness condition:

$$\prod_{j=1}^{t} e\left(\mathsf{W}_{\mathsf{I},j}, \mathsf{F}_{\mathsf{I},j}\right) \stackrel{?}{=} e(g, g). \tag{10}$$

   If the above check on the completeness witnesses $\mathsf{F}_{\mathsf{I},j}$, $1 \leq j \leq t$, fails, the algorithm outputs `reject`. Or, if this relation holds, the algorithm outputs `accept`, i.e., it accepts $\alpha(q)$ as the *correct* intersection. This step has $O(t)$ complexity.

[12] Algorithm certify is used to achieve optimal verification and avoid an $O(\delta \log \delta)$ FFT computation from scratch.

[13] Group element $\prod_{i=0}^{\delta} g^{s^i b_i} = g^{(y_1+s)(y_2+s)\ldots(y_\delta+s)}$ is computed once with $O(\delta)$ complexity.

Note that for Equation 10, it holds $\prod_{j=1}^{t} e(\mathsf{W}_{\mathsf{I},j}, \mathsf{F}_{\mathsf{I},j}) = e(g,g)^{\sum_{j=1}^{t} q_j(s)P_j(s)} = e(g,g)$ when all the subset witnesses $\mathsf{W}_{\mathsf{I},j}$, all the completeness witnesses $\mathsf{F}_{\mathsf{I},j}$ and all the sets accumulation values $\mathsf{acc}(S_j)$ have been computed *honestly*, since $q_1(s)P_1(s) + q_2(s)P_2(s) + \ldots + q_t(s)P_t(s) = 1$. This is a required condition for proving the correctness of our ADS scheme, as defined in Definition 2. We continue with the description of algorithms query and verify for the union query.

Union query. Let $\mathsf{U} = S_1 \cup S_2 \cup \ldots \cup S_t = \{y_1, y_2, \ldots, y_\delta\}$. We express the correctness of the set union operation by means of the following two conditions:

$$\textbf{Membership condition:} \quad \forall y_i \in \mathsf{U}\ \exists j \in \{1,2,\ldots,t\} : y_i \in S_j\,; \tag{11}$$

$$\textbf{Superset condition:} \quad (\mathsf{U} \supseteq S_1) \wedge (\mathsf{U} \supseteq S_2) \wedge \ldots \wedge (\mathsf{U} \supseteq S_t)\,. \tag{12}$$

The superset condition in Equation 12 is necessary since set U must exclude none of the elements in sets $S_1, S_2, \ldots, S_t$. We formally describe algorithms query and verify of our ADS scheme $\mathcal{ASC}$ for *union* queries.

Algorithm $\{\Pi(q), \alpha(q)\} \leftarrow \mathsf{query}(q, D_h, \mathsf{auth}(D_h), \mathsf{pk})$ **(Union):** Query q asks for the union U of t sets $S_1, S_2, \ldots, S_t$. Let $\mathsf{U} = \{y_1, y_2, \ldots, y_\delta\}$. Then $\alpha(q) = \mathsf{U}$ and the proof $\Pi(q)$ consists of the following parts. **(1)** *Coefficients* $b_\delta, b_{\delta-1}, \ldots, b_0$ of polynomial $(y_1 + s)(y_2 + s)\ldots(y_\delta + s)$ that is associated with the union $\mathsf{U} = \{y_1, y_2, \ldots, y_\delta\}$. **(2)** *Accumulation values* $\mathsf{acc}(S_j)$, $j = 1, \ldots, t$, which are associated with sets S_j, along with their respective *correctness proofs* Π_j, both output of algorithm $\mathsf{queryTree}(j, D_h, \mathsf{auth}(D_h), \mathsf{pk})$. **(3)** *Membership witnesses* W_{y_i,S_k} of y_i, $i = 1, \ldots, \delta$ (see Equation 1), which prove that y_i belongs to *some* set S_k, $1 \leq k \leq t$, and which are computed with $O(N \log N)$ total complexity and have $O(\delta)$ total group complexity (Lemma 2). **(4)** *Subset witnesses* $\mathsf{W}_{S_j,\mathsf{U}}$, $j = 1, \ldots, t$, which are associated with sets S_i and union U and prove that U is a superset of S_j, $1 \leq k \leq t$, and which are computed with $O(N \log N)$ total complexity and have $O(t)$ total group complexity (Lemma 2).

Algorithm $\{\texttt{accept}, \texttt{reject}\} \leftarrow \mathsf{verify}(q, \alpha, \Pi, d_h, \mathsf{pk})$**: (Union):** Verifying the result of a union query includes the following steps. **(1)** First, the algorithm uses $\mathbf{b} = [b_\delta, b_{\delta-1}, \ldots, b_0]$ and the answer $\mathsf{U} = \alpha(q) = \{y_1, y_2, \ldots, y_\delta\}$ as an input to algorithm $\mathsf{certify}(\mathbf{b}, \alpha(q), \mathsf{pk})$, in order to certify the validity of $b_\delta, b_{\delta-1}, \ldots, b_0$. **(2)** Subsequently, the algorithm uses the proofs Π_j to verify the correctness of $\mathsf{acc}(S_j)$, by using algorithm $\mathsf{verifyTree}(j, \mathsf{acc}(S_j), \Pi_j, d_h, \mathsf{pk})$ for $j = 1, \ldots, t$. If the verification fails for at least one of $\mathsf{acc}(S_j)$, the algorithm outputs `reject`. **(3)** Next, the algorithm verifies that each element y_i, $i = 1, \ldots, \delta$, of the reported union belongs to some set S_k, for some $1 \leq k \leq t$ ($O(\delta)$ complexity). This is done by checking that relation $e(\mathsf{W}_{y_i,S_k}, g^{y_i} g^s) = e(\mathsf{acc}(S_k), g)$ holds for all $i = 1, \ldots, \delta$; otherwise the algorithm outputs `reject`. **(4)** Finally, the algorithm verifies that all sets specified by the query are *subsets* of the union, by checking the following conditions:

$$e\left(\mathsf{W}_{S_j,\mathsf{U}}, \mathsf{acc}(S_j)\right) \stackrel{?}{=} e\left(\prod_{i=0}^{\delta} \left(g^{s^i}\right)^{b_i}, g\right), \text{ for } j = 1, \ldots, t.$$

If any of the above checks fails, the algorithm outputs `reject`, otherwise, it outputs `accept`, i.e., U is accepted as the *correct* union.

Subset and set difference query. For a subset query (positive or negative), we use the property $S_i \subseteq S_j \Leftrightarrow \forall y \in S_i, y \in S_j$. For a set difference query we use the property

$$\mathsf{D} = S_i - S_j \Leftrightarrow \exists F : F \cup \mathsf{D} = S_i \wedge F = S_i \cap S_j \,.$$

The above conditions can both be checked in an operation-sensitive manner using the techniques we have presented before. We now give the main result in our work.

Theorem 1. *Consider a collection of m sets $S_1, \ldots, S_m$ and let $M = \sum_{i=1}^{m} |S_i|$ and $0 < \epsilon < 1$. For a query operation involving t sets, let N be the sum of the sizes of the involved sets, and δ be the answer size. Then there exists an ADS scheme $\mathcal{ASC} = \{\mathsf{genkey}, \mathsf{setup}, \mathsf{update}, \mathsf{refresh}, \mathsf{query}, \mathsf{verify}\}$ for a sets collection data structure D with the following properties: (1) $\mathcal{ASC}$ is correct and secure according to Definitions 2 and 3 and based on the bilinear q-strong Diffie-Hellman assumption; (2) The access complexity of algorithm (i) genkey is $O(1)$; (ii) setup is $O(m + M)$; (iii) update is $O(1)$ outputting information upd of $O(1)$ group complexity; (iv) refresh is $O(1)$; (3) For all queries q (intersection/union/subset/difference), constructing the proof with algorithm query has $O(N \log^2 N \log\log N + tm^{\epsilon} \log m)$ access complexity, algorithm verify has $O(t+\delta)$ access complexity and the proof $\Pi(q)$ has $O(t+\delta)$ group complexity; (4) The group complexity of the authenticated data structure $\mathsf{auth}(D)$ is $O(m + M)$.*

4 Security, Protocols and Applications

In this section we give an overview of the security analysis of our ADS scheme, describe how it can be employed to provide verification protocols in the three-party [36] and two-party [30] authentication models, and finally discuss some concrete applications.

Security proof sketch. We provide some key elements of the security of our verification protocols focusing on set intersection queries. The security proofs of the other set operations share similar ideas. Let D_0 be a sets collection data structure consisting of m sets $S_1, S_2, \ldots, S_m$,[14] and consider our ADS scheme $\mathcal{ASC} = \{\mathsf{genkey}, \mathsf{setup}, \mathsf{update}, \mathsf{refresh}, \mathsf{query}, \mathsf{verify}\}$. Let k be the security parameter and let $\{\mathsf{sk}, \mathsf{pk}\} \leftarrow \mathsf{genkey}(1^k)$. The adversary is given the public key pk, namely $\{h(\cdot), p, \mathbb{G}, \mathcal{G}, e, g, g^s, \ldots, g^{s^q}\}$, and unlimited access to all the algorithms of $\mathcal{ASC}$, except for setup and update to which he only has oracle access. The adversary initially outputs the authenticated data structure $\mathsf{auth}(D_0)$ and the digest d_0, through an oracle call to algorithm setup. Then the adversary picks a polynomial number of updates u_i (e.g., insertion of an element x into a set S_r) and outputs the data structure D_i, the authenticated data structure $\mathsf{auth}(D_i)$ and the digest d_i through oracle access to update. Then he picks a set of indices $q = \{1, 2, \ldots, t\}$ (wlog), all between 1 and m and outputs a proof $\Pi(q)$ and an answer $\mathcal{I} \neq \mathsf{I} = S_1 \cap S_2 \cap \ldots \cap S_t$ which is rejected by check as incorrect. Suppose the answer $\alpha(q)$ contains d elements. The proof $\Pi(q)$ contains (i) *Some* coefficients

[14] Note here that since the sets are picked by the adversary, we have to make sure that no element in any set is equal to s, the trapdoor of the scheme (see definition of the bilinear-map accumulator domain). However, this event occurs with negligible probability since the sizes of the sets are polynomially-bounded and s is chosen at random from a domain of exponential size.

$b_0, b_1, \ldots, b_d$; (ii) *Some* accumulation values acc_j with *some* respective correctness proofs Π_j, for $j = 1, \ldots, t$; (iii) *Some* subset witnesses W_j with *some* completeness witnesses F_j, for $j = 1, \ldots, t$ (this is, what algorithm verify expects for input).

Suppose verify accepts. Then: (i) By Lemma 3, $b_0, b_1, \ldots, b_d$ are *indeed* the coefficients of the polynomial $\prod_{x \in \mathcal{I}}(x + s)$, except with negligible probability; (ii) By Lemma 4, values acc_j are *indeed* the accumulation values of sets S_j, except with negligible probability; (iii) By Lemma 1, values W_j are *indeed* the subset witnesses for set $\mathcal{I}$ (with reference to S_j), i.e., $\mathsf{W}_j = g^{P_j(s)}$, except with negligible probability; (iv) However, $P_1(s), P_2(s), \ldots, P_t(s)$ are not coprime since $\mathcal{I}$ is incorrect and therefore $\mathcal{I}$ cannot contain *all* the elements of the intersection. Thus the polynomials $P_1(s), P_2(s), \ldots, P_t(s)$ (Equation 8) have at least one common factor, say $(r + s)$ and it holds that $P_j(s) = (r + s)Q_j(s)$ for some polynomials $Q_j(s)$ (computable in polynomial time), for all $j = 1, \ldots, t$. By the verification of Equation 10 (completeness condition), we have

$$e(g, g) = \prod_{j=1}^{t} e(\mathsf{W}_j, \mathsf{F}_j) = \prod_{j=1}^{t} e\left(g^{P_j(s)}, \mathsf{F}_j\right) = \prod_{j=1}^{t} e\left(g^{(r+s)Q_j(s)}, \mathsf{F}_j\right)$$
$$= \prod_{j=1}^{t} e\left(g^{Q_j(s)}, \mathsf{F}_j\right)^{(r+s)} = \left(\prod_{j=1}^{t} e\left(g^{Q_j(s)}, \mathsf{F}_j\right)\right)^{(r+s)} .$$

Therefore we can derive an $(r + s)$-th root of $e(g, g)$ as

$$e(g, g)^{\frac{1}{r+s}} = \prod_{j=1}^{t} e\left(g^{Q_j(s)}, \mathsf{F}_j\right) .$$

This means that if the intersection $\mathcal{I}$ is incorrect and all the verification tests are satisfied, we can derive a polynomial-time algorithm that outputs a bilinear q-strong Diffie-Hellman challenge $(r, e(g, g)^{1/(r+s)})$ for an element r that is a common factor of the polynomials $P_1(s), P_2(s), \ldots, P_t(s)$, which by Assumption 1 happens with probability $\mathsf{neg}(k)$. This concludes an ouline of the proof strategy for the case of intersection.

Protocols. As mentioned in the introduction, our ADS scheme $\mathcal{ASC}$ can be used by a verification protocol in the three-party model [36]. Here, a trusted entity, called source, owns a sets collection data structure D_h, but desires to outsource query answering, in a trustworthy (verifiable) way. The source runs genkey and setup and outputs the authenticated data structure $\mathsf{auth}(D_h)$ along with the digest d_h. The source subsequently signs the digest d_h, and it outsources $\mathsf{auth}(D_h)$, D_h, the digest d_h and its signature to some untrusted entities, called servers. On input a data structure query q (e.g., an intersection query) sent by clients, the servers use $\mathsf{auth}(D_h)$ and D_h to compute proofs $\Pi(q)$, by running algorithm query, and they return to the clients $\Pi(q)$ and the signature on d_h along with the answer $a(q)$ to q. Clients can verify these proofs $\Pi(q)$ by running algorithm verify (since they have access to the signature of d_h, they can verify that d_h is authentic). When there is an update in the data structure (issued by the source), the source uses algorithm update to produce the new digest d'_h to be used in next verifications, while the servers update the authenticated data structure through refresh.

Additionally, our ADS scheme $\mathcal{ASC}$ can also be used by a *non-interactive* verification protocol in the two-party model [30]. In this case, the source and the client coincide, i.e., the client issues both the updates and the queries, and it is required to keep only constant state, i.e., the digest of the authenticated data structure. Whenever there is an update by the client, the client retrieves a verifiable, constant-size portion of the authenticated data structure that is used for locally performing the update and for computing the new local state, i.e., the new digest. A non-interactive two-party protocol that uses an ADS scheme for a data structure D is directly comparable with the recent protocols for verifiable computing [1,12,16] for the functionalities offered by the data structure D, e.g., computation of intersection, union, etc. Due to space limitations, we defer the detailed description of these protocols to the full version of the paper.

Applications. First of all, our scheme can be used to verify *keyword-search queries* implemented by the *inverted index* data structure [4]: Each term in the dictionary corresponds to a set in our sets collection data structure which contains all the documents that include this term. A usual text query for terms m_1 and m_2 returns those documents that are included in both the sets that are represented by m_1 and m_2, i.e., their intersection. Moreover, the derived authenticated inverted index can be efficiently updated as well. However, sometimes in keyword searches (e.g., keyword searches in the email inbox) it is desirable to introduce a "second" dimension: For example, a query could be "return emails that contain terms m_1 and m_2 and which were received between time t_1 and t_2", where $t_1 < t_2$. We call this variant a *timestamped keyword-search*. One solution for verifying such queries could be to embed a timestamp in the documents (e.g., each email message) and have the client do the filtering locally, after he has verified—using our scheme—the intersection of the sets that correspond to terms m_1 and m_2. However, this approach is not operation-sensitive: The intersection can be bigger than the set output after the local filtering, making this solution inefficient. To overcome this inefficiency, we can use a *segment-tree* data structure [35], verifying in this way timestamped keyword-search queries efficiently with $O(t \log r + \delta)$ complexity, where r is the total number of timestamps we are supporting. This involves building a binary tree T on top of sets of messages sent at certain timestamps and requiring each internal node of T be the union of messages stored in its children. Finally, our method can be used for verifying equi-join queries over relational tables, which boil down to set intersections.

5 Conclusion

In this paper, we presented an authenticated data structure for the optimal verification of set operations. The achieved efficiency is mainly due to new, extended security properties of accumulators based on pairing-based cryptography. Our solution provides two important properties, namely *public verifiability* and *dynamic updates*, as opposed to existing protocols in the verifiable computing model that provide *generality* and *secrecy*, but *verifiability in a static, secret-key setting* only.

A natural question to ask is whether outsourced verifiable computations with *secrecy* and *efficient dynamic updates* are feasible. Analogously, it is interesting to explore whether other specific functionalities (beyond set operations) can be optimally and publicly verified. Finally, according to a recently proposed definition of optimality [33], our construction is *nearly* optimal: verification and updates are optimal, but not queries. It is interesting to explore whether an *optimal* authenticated sets collection data structure exists, i.e., one that asymptotically matches the bounds of the plain sets collection data structure, reducing the query time from $O(N \log^2 N)$ to $O(N)$.

Acknowledgments. This work was supported in part by the U.S. National Science Foundation under grants CNS–1012060 and CNS–1012798 and by the Kanellakis Fellowship and the Center for Geometric Computing at Brown University, the RISCS Center at Boston University and NetApp. The authors thank Michael T. Goodrich for useful discussions. The views in this paper do not necessarily reflect the views of the sponsors.

References

1. Applebaum, B., Ishai, Y., Kushilevitz, E.: From secrecy to soundness: Efficient verification via secure computation. In: Abramsky, S., Gavoille, C., Kirchner, C., Meyer auf der Heide, F., Spirakis, P.G. (eds.) ICALP 2010. LNCS, vol. 6198, pp. 152–163. Springer, Heidelberg (2010)
2. Atallah, M.J., Cho, Y., Kundu, A.: Efficient data authentication in an environment of untrusted third-party distributors. In: Int. Conference on Data Engineering (ICDE), pp. 696–704 (2008)
3. Au, M.H., Tsang, P.P., Susilo, W., Mu, Y.: Dynamic universal accumulators for DDH groups and their application to attribute-based anonymous credential systems. In: Fischlin, M. (ed.) CT-RSA 2009. LNCS, vol. 5473, pp. 295–308. Springer, Heidelberg (2009)
4. Baeza-Yates, R., Ribeiro-Neto, B.: Modern Information Retrieval. Addison-Wesley Publishing Company, Reading (1999)
5. Bellare, M., Micciancio, D.: A new paradigm for collision-free hashing: Incrementality at reduced cost. In: Fumy, W. (ed.) EUROCRYPT 1997. LNCS, vol. 1233, pp. 163–192. Springer, Heidelberg (1997)
6. Benabbas, S., Gennaro, R., Vahlis, Y.: Verifiable delegation of computation over large datasets. In: Int. Cryptology Conference, CRYPTO (2011)
7. Blum, M., Evans, W.S., Gemmell, P., Kannan, S., Naor, M.: Checking the correctness of memories. Algorithmica 12(2/3), 225–244 (1994)
8. Boneh, D., Boyen, X.: Short signatures without random oracles and the SDH assumption in bilinear groups. J. Cryptology 21(2), 149–177 (2008)
9. Boneh, D., Waters, B.: Conjunctive, subset, and range queries on encrypted data. In: Vadhan, S.P. (ed.) TCC 2007. LNCS, vol. 4392, pp. 535–554. Springer, Heidelberg (2007)
10. Canard, S., Gouget, A.: Multiple denominations in E-cash with compact transaction data. In: Sion, R. (ed.) FC 2010. LNCS, vol. 6052, pp. 82–97. Springer, Heidelberg (2010)
11. Chung, K.-M., Kalai, Y., Liu, F.-H., Raz, R.: Memory delegation. In: Int. Cryptology Conference, CRYPTO (2011)
12. Chung, K.-M., Kalai, Y., Vadhan, S.: Improved delegation of computation using fully homomorphic encryption. In: Rabin, T. (ed.) CRYPTO 2010. LNCS, vol. 6223, pp. 483–501. Springer, Heidelberg (2010)
13. Damgård, I., Triandopoulos, N.: Supporting non-membership proofs with bilinear-map accumulators. Cryptology ePrint Archive, Report 2008/538 (2008), `http://eprint.iacr.org/`
14. Dwork, C., Naor, M., Rothblum, G.N., Vaikuntanathan, V.: How efficient can memory checking be? In: Reingold, O. (ed.) TCC 2009. LNCS, vol. 5444, pp. 503–520. Springer, Heidelberg (2009)

15. Freedman, M.J., Nissim, K., Pinkas, B.: Efficient private matching and set intersection. In: Cachin, C., Camenisch, J.L. (eds.) EUROCRYPT 2004. LNCS, vol. 3027, pp. 1–19. Springer, Heidelberg (2004)
16. Gennaro, R., Gentry, C., Parno, B.: Non-interactive verifiable computing: Outsourcing computation to untrusted workers. In: Rabin, T. (ed.) CRYPTO 2010. LNCS, vol. 6223, pp. 465–482. Springer, Heidelberg (2010)
17. Goodrich, M.T., Tamassia, R., Hasic, J.: An efficient dynamic and distributed cryptographic accumulator. In: Chan, A.H., Gligor, V.D. (eds.) ISC 2002. LNCS, vol. 2433, pp. 372–388. Springer, Heidelberg (2002)
18. Goodrich, M.T., Tamassia, R., Schwerin, A.: Implementation of an authenticated dictionary with skip lists and commutative hashing. In: DARPA Information Survivability Conference and Exposition II (DISCEX II), pp. 68–82 (2001)
19. Goodrich, M.T., Tamassia, R., Triandopoulos, N.: Super-efficient verification of dynamic outsourced databases. In: Malkin, T. (ed.) CT-RSA 2008. LNCS, vol. 4964, pp. 407–424. Springer, Heidelberg (2008)
20. Goodrich, M.T., Tamassia, R., Triandopoulos, N.: Efficient authenticated data structures for graph connectivity and geometric search problems. Algorithmica 60(3), 505–552 (2011)
21. Kratsch, D., McConnell, R.M., Mehlhorn, K., Spinrad, J.P.: Certifying algorithms for recognizing interval graphs and permutation graphs. In: Symposium on Discrete Algorithms (SODA), pp. 158–167 (2003)
22. Li, J., Li, N., Xue, R.: Universal accumulators with efficient nonmembership proofs. In: Katz, J., Yung, M. (eds.) ACNS 2007. LNCS, vol. 4521, pp. 253–269. Springer, Heidelberg (2007)
23. Martel, C., Nuckolls, G., Devanbu, P., Gertz, M., Kwong, A., Stubblebine, S.G.: A general model for authenticated data structures. Algorithmica 39(1), 21–41 (2004)
24. Merkle, R.C.: A certified digital signature. In: Brassard, G. (ed.) CRYPTO 1989. LNCS, vol. 435, pp. 218–238. Springer, Heidelberg (1990)
25. Minsky, Y., Trachtenberg, A., Zippel, R.: Set reconciliation with nearly optimal communication complexity. IEEE Transactions on Information Theory 49(9), 2213–2218 (2003)
26. Morselli, R., Bhattacharjee, S., Katz, J., Keleher, P.J.: Trust-preserving set operations. In: Int. Conference on Computer Communications, INFOCOM (2004)
27. Naor, M., Nissim, K.: Certificate revocation and certificate update. In: USENIX Security Symposium, pp. 217–228 (1998)
28. Nguyen, L.: Accumulators from bilinear pairings and applications. In: Menezes, A. (ed.) CT-RSA 2005. LNCS, vol. 3376, pp. 275–292. Springer, Heidelberg (2005)
29. Pang, H., Tan, K.-L.: Authenticating query results in edge computing. In: Int. Conference on Data Engineering (ICDE), pp. 560–571 (2004)
30. Papamanthou, C., Tamassia, R.: Time and space efficient algorithms for two-party authenticated data structures. In: Qing, S., Imai, H., Wang, G. (eds.) ICICS 2007. LNCS, vol. 4861, pp. 1–15. Springer, Heidelberg (2007)
31. Papamanthou, C., Tamassia, R.: Cryptography for efficiency: Authenticated data structures based on lattices and parallel online memory checking. In: Cryptology ePrint Archive, Report 2011/102 (2011), `http://eprint.iacr.org/`
32. Papamanthou, C., Tamassia, R., Triandopoulos, N.: Authenticated hash tables. In: Int. Conference on Computer and Communications Security (CCS), pp. 437–448 (2008)
33. Papamanthou, C., Tamassia, R., Triandopoulos, N.: Optimal authenticated data structures with multilinear forms. In: Joye, M., Miyaji, A., Otsuka, A. (eds.) Pairing 2010. LNCS, vol. 6487, pp. 246–264. Springer, Heidelberg (2010)
34. Preparata, F.P., Sarwate, D.V.: Computational complexity of Fourier transforms over finite fields. Mathematics of Computation 31(139), 740–751 (1977)

35. Preparata, F.P., Shamos, M.I.: Computational Geometry: An Introduction. Springer, New York (1985)
36. Tamassia, R.: Authenticated data structures. In: Di Battista, G., Zwick, U. (eds.) ESA 2003. LNCS, vol. 2832, pp. 2–5. Springer, Heidelberg (2003)
37. Tamassia, R., Triandopoulos, N.: Certification and authentication of data structures. In: Alberto Mendelzon Workshop on Foundations of Data Management (2010)
38. Yang, Y., Papadias, D., Papadopoulos, S., Kalnis, P.: Authenticated join processing in outsourced databases. In: Int. Conf. on Management of Data (SIGMOD), pp. 5–18 (2009)
39. Yiu, M.L., Lin, Y., Mouratidis, K.: Efficient verification of shortest path search via authenticated hints. In: Int. Conference on Data Engineering (ICDE), pp. 237–248 (2010)

Verifiable Delegation of Computation over Large Datasets*

Siavosh Benabbas[1], Rosario Gennaro[2], and Yevgeniy Vahlis[3]

[1] University of Toronto
siavosh@cs.toronto.edu
[2] IBM Research
rosario@us.ibm.com
[3] Columbia University
evahlis@cs.columbia.edu

Abstract. We study the problem of computing on large datasets that are stored on an untrusted server. We follow the approach of *amortized verifiable computation* introduced by Gennaro, Gentry, and Parno in CRYPTO 2010. We present the first practical verifiable computation scheme for high degree polynomial functions. Such functions can be used, for example, to make predictions based on polynomials fitted to a large number of sample points in an experiment. In addition to the many non-cryptographic applications of delegating high degree polynomials, we use our verifiable computation scheme to obtain new solutions for verifiable keyword search, and proofs of retrievability. Our constructions are based on the DDH assumption and its variants, and achieve adaptive security, which was left as an open problem by Gennaro *et al* (albeit for general functionalities).

Our second result is a primitive which we call a *verifiable database* (VDB). Here, a weak client outsources a large table to an untrusted server, and makes retrieval and update queries. For each query, the server provides a response and a proof that the response was computed correctly. The goal is to minimize the resources required by the client. This is made particularly challenging if the number of update queries is unbounded. We present a VDB scheme based on the hardness of the subgroup membership problem in composite order bilinear groups.

1 Introduction

This paper presents very efficient protocols that allow a computationally weak client to securely outsource some computations over very large datasets to a powerful server. Security in this context means that the client will receive an assurance that the computation performed by the server is correct, with the optional property that the client will be able to hide some of his data from the server.

The problem of securely outsourcing computation has received widespread attention due to the rise of *cloud computing:* a paradigm where businesses lease computing resources from a service (the *cloud provider*) rather than maintain

* A full version of this paper is available at http://eprint.iacr.org/2011/132

P. Rogaway (Ed.): CRYPTO 2011, LNCS 6841, pp. 111–131, 2011.

their own computing infrastructure [2,57]. A crucial component of secure cloud computing is a mechanism that enforces the integrity and correctness of the computations done by the provider.

Outsourced computations are also increasingly relevant due to the proliferation of mobile devices, such as smart phones and netbooks, computationally weak devices which might off-load heavy computations, e.g., a cryptographic operation or a photo manipulation, to a network server. Here too, a proof of the correctness of the result might be desirable if not necessary.

A crucial requirement in all of these cases is that the computation invested by the (weak) client in order to verify the result of the server's work must be substantially smaller than the amount of computation required to perform the work to begin with. Indeed if that was not the case, the client could perform the computation on its own without interacting with the server! It is also desirable to keep the server's overhead as small as possible: in other words the computation of the server to provide both the result and a correctness proof to the client should be as close as possible to the amount of work needed to simply compute the original function (otherwise, the server, which might provide this service to many clients, may become overwhelmed by the computational load).

This paper initiates a line of research about efficient protocols for verifiable computation of specific functions, in our case the evaluation of polynomials derived from very large datasets. Most of the prior work (reviewed below) has focused on generic solutions for arbitrary functions. So while in "general" the problem we are considering has been solved, by focusing on specific computations we are able to obtain much more efficient protocols. This is similar to the way research over secure multiparty computation has evolved: following generic protocols for the evaluation of arbitrary functions [60,29,9,17], there has been a twenty-plus year effort to come up with efficient distributed protocols for specific computations encountered in practical applications (e.g. the entire work on threshold cryptography [21], or protocols on set intersection and pattern matching such as [34]).

Our Results. This paper focuses on the evaluation of polynomials derived from very large datasets. While the computations themselves are simple, it's the magnitude of data that prevents the client (who cannot even store the entire data) to perform them by itself. In our protocols the client will initially store the data at the server (with the option of encrypting it for confidentiality, if desired), with some authenticating information. The client will only keep a short secret key. Later, every time the client requests the value of a computation over the data, the server will compute the result and return it together with an *authentication code*, which the client will be able to quickly verify with the secret key. This description shows that our problem naturally fits into the *amortized* model for outsourced computation introduced in [27]: the client performs a one-time computationally expensive phase (in our case storing the data with its authentication information) and then quickly verifies the results provided by the server.

Our protocols are very efficient. The computation of the authentication data is comparable to encrypting the file using the ElGamal encryption scheme (i.e.

roughly 2 exponentiations per data block). Verification takes at most a logarithmic (in the number of blocks) number of exponentiations under the DDH Assumption. Additionally, we present a faster protocol (which requires only a *single* exponentiation to verify the result) which is secure under a decisional variant of the Strong Diffie Hellman Assumption in single groups[1].

An immediate application of our results is the ability to verifiably outsource computations to make predictions based on polynomials fitted to a large number of sample points in an experiment.

In the second part of our paper, we present an extension to our protocols, which allows the client to efficiently *update* the data (and its associated authentication information) stored at the server. We also present applications of our protocols to the problems of verifiable *keyword search* (the client stores a large database with the server and it queries if a specific keyword appears in it) and secure *proofs of retrievability* (the client checks that the file stored with the server is indeed retrievable) [50,36].

VERIFIABLE DELEGATION OF POLYNOMIALS. The basis of all our protocols is verifiable delegation of polynomials. Assume the client has a polynomial $P(\cdot)$ of large degree d, and it wants to compute the value $P(x)$ for an arbitrary inputs x. In our basic solution the client stores the polynomial in the clear with the server as a vector $\boldsymbol{c}$ of coefficients in $\mathbb{Z}_p$. The client also stores with the server a vector $\boldsymbol{t}$ of group elements of the form $g^{ac_i+r_i}$ where $a \in_{\mathrm{R}} \mathbb{Z}_p$ and r_i is the ith-coefficient of a polynomial $R(\cdot)$ of the same degree as $P(\cdot)$. When queried on input x the server returns $y = P(x)$ and $t = g^{aP(x)+R(x)}$ and the client accepts y iff $t = g^{ay+R(x)}$.

If $R(\cdot)$ was a random polynomial, then we can prove that this is a secure delegation scheme in the sense of [27]. However checking that $t = g^{ay+R(x)}$ would require the client to perform computation polynomial in the degree of $P(\cdot)$ – the exact work that we set out to avoid. The crucial point, therefore, is how to perform this verification fast, in time which is independent, or at the very least sublinear in the degree of $P(\cdot)$. We do that by defining $r_i = F_K(i)$ where F is a pseudo-random function (PRF in the following) with a special property which we call *closed form efficiency*. The property is that given the polynomial $R(\cdot)$ defined by the r_i coefficients, the value $R(x)$ (for any input x) can be computed very efficiently (sub-linearly in d) by a party who knows the secret key K for the PRF. Since F is a PRF, the security of the scheme is not compromised (as F is indistinguishable from a random function), and the closed form efficiency of F will allow the client to verify the result in time sub-linear in the degree of the polynomial.

We generalize our result for PRFs with other types of closed form efficiency, which yield efficient and secure delegation protocols not only for single-variable polynomials of degree d, but also for multivariate polynomials with total degree d or of degree d in each variable. We have several different variations of PRFs: the least efficient one is secure under the Decisional Diffie-Hellman assumption, while more efficient ones require a decisional variant of the Strong DH assumption.

[1] See e.g., [19] for a survey of the strong DH family of assumptions.

Adaptivity: One of the main questions to remain open after the work of GGP [27] is whether we can achieve verifiable delegation even if the malicious server knows whether the verifier accepted or rejected the correctness proof of the value computed by the server. Indeed, the GPV scheme becomes insecure if the server learns this single bit of information after each proof is sent to the verifier. Our constructions are the first to achieve adaptive security in the amortized setting.

Privacy: Our solution allows the client to preserve the secrecy of the polynomial stored with the server, by encrypting it with an additively homomorphic encryption scheme. In this case the server returns an encrypted form of y which the client will decrypt.

Keyword Search: The applications to keyword search without updates is almost immediate. Consider a text file $F = \{w_1, \ldots, w_\ell\}$ where w_i are the words contained in it. Encode F as the polynomial $P(\cdot)$ of degree ℓ such that $P(w_i) = 0$. To make this basic solution efficiently updatable we use a variation of the polynomial delegation scheme which uses bilinear maps. We also present a generic, but less efficient way to make *any* static keyword search protocol updatable which might be of independent interest.

Proof of Retrievability: Again the application of our technique is quite simple. The client encodes the file as a polynomial $F(x)$ of degree d (each block representing a coefficient), and delegates the computation of $F(x)$ to the server. The proof of retrievability consists of the client and the server engaging in our verifiable delegation protocol over a random point r: the client sends r and the server returns the value $F(r)$ together with a proof of its correctness. The client accepts if it accepts the proof that $F(r)$ is the correct value.

VERIFIABLE DATABASES WITH EFFICIENT UPDATES. In the second part of our paper we study the problem of verifiable databases, where a resource constrained client wishes to store an array DB on a server, and to be able to retrieve the value at any cell $DB[i]$, and to update the database by assigning $DB[i] \leftarrow v$ for a new value v. The goal is to achieve this functionality with an additional guarantee that if a server attempts to tamper with the data, the tampering will be detected when the client queries the database.

Simple solutions (based on Message Authentication Codes or Signature Schemes) exist for the restricted case where the database is static – i.e. the client only needs to retrieve data, but does not modify the database. One example is to have the client sign each pair (index,value) that is sent to the server. Clearly, if no updates are performed, the server has no choice but to return the correct value for a given cell. However, the problem becomes significantly harder when efficient updates are needed. One solution is for the client to just keep track of all the changes locally, and apply them as needed, but this contradicts our goal of keeping client state and workload as small as possible. On a high level, the main technical difficulty stems from the fact that the client must revoke any authenticating data that the server has for the previous value of the updated cell. This issue has been addressed in the line of works on cryptographic accumulators [16,47,53], and, using different techniques, in the authenticated datastructures literature [48,41,52,59].

We present a verifiable database delegation scheme based on the hardness of the subgroup membership problem in composite order bilinear groups (this assumption was originally introduced in [13]). Our solution allows the client to query any location of the database, and verify the response in time that is independent of the size of the database. The main advantage of our construction is that it allows the client to insert and delete values, as well as update the value at any cell by sending a single group element to the server after retrieving the current value stored in the cell. Prior solutions either rely on non-constant size assumptions (such as variants of the Strong Diffie-Hellman assumption [23,15]), require expensive generation of primes for each operation (in the worst case), or require expensive "re-shuffling" procedures to be performed once in a while on the data. On the other hand, our construction works in the private key setting, whereas some prior solutions allow public verification (e.g., [16,47]).

Roadmap. The rest of the paper is organized as follows. In Section 2 we define the security assumptions used in the paper. Readers interested in the precise definition of Verifiable Computation and its security can find them in Section 3. In Section 4 we introduce our notation of Algebraic Pseudorandom Functions which are the main building block of our constructions. In Section 5 we show how to securely delegate polynomial evaluations to an untrusted server using Algebraic PRFs. In the full version of the paper [7] we show how to use delegation of polynomials to implement verifiable databases, and to obtain new constructions of Proofs of Retrievability.

1.1 Related Work

As mentioned above our work follows the paradigm introduced in [27] which is also adopted in [20,3]. The protocols described in those papers allow a client to outsource the computation of an arbitrary function (encoded as a Boolean circuit) and use fully homomorphic encryption (i.e. [28]) resulting in protocols of limited practical relevance. Our protocols on the other hand work for only a very limited class of computations (mostly polynomial evaluations) but are very efficient and easily implementable in practice.

The previous schemes based on fully homomorphic encryption also suffer from the following drawback: if a malicious server tries to cheat and learns if the client has accepted or rejected its answer, then the client must repeat the expensive pre-processing stage. The only alternative way to deal with this problem proposed in these papers is to protect this bit of information from the server (which is a very strong assumption to make). Somewhat surprisingly our scheme remains secure even if a cheating server learns the acceptance/rejection bit of the client, without any need to repeat the pre-processing stage. This is not only conceptually interesting, but also a very practical advantage.

There is a large body of literature, prior to [27], that investigates the problem of verifiably outsourcing the computation of an arbitrary functions (we refer to [27] for an exhaustive list of citations). This problem has attracted the attention of the Theory community, starting from the work on Interactive Proofs [5,31], efficient arguments based on probabilistically checkable proofs (PCP) [37,38], CS

Proofs [43] and the *muggles proofs* in [30]. However in PCP-based schemes, the client must store the large data in order to verify the result and therefore these solutions might not be applicable to our setting.

This problem has also been studied by the Applied Security community, with solutions which are practical but whose security holds under some very strong assumptions on the behavior of the adversary. For example, solutions based on audit (e.g. [46,6]) which typically assume many clients, and require a fraction of them to recompute some of the results provided by the server, but are secure only under the assumption that bad clients do not collude. Another approach is to use secure co-processors (e.g. [56,61]) which "sign" the computation as correct, under the assumption that the adversary cannot tamper with the processor. Finally, other trust models have been considered. The area of authenticated data structures [58,41,54] aims to provide delegation solutions when the owner of the data is decoupled from the client. In this scenario, the owner maintains a large state, and acts as a trusted third party, but delegates his data to an untrusted server that can be queried by weak clients.

For the specific case of outsourcing expensive cryptographic operations, Chaum and Pedersen in [18], describe protocols to allow a client to verify the behavior of a piece of hardware placed on the client's device by a service provider such as a bank. Hohenberger and Lysyanskaya formalize this model [35], and present protocols for the computation of modular exponentiations (arguably the most expensive step in public-key cryptography operations). Their protocol requires the client to interact with *two* non-colluding servers. Other work targets specific function classes, such as one-way function inversion [32].

The application of secure keyword search over a stored file can be handled using *zero-knowledge sets*, [44] which however does not allow for an easy way to update the file. Our protocol for keyword search combines ideas from our polynomial delegation scheme with some machinery inspired by zero-knowledge sets, to obtain a protocol that allows for efficient updates and other additional desirable properties (see full version [7]).

The problem of proof of retrievability was first posed in [50,36], and subsequent protocols include [4,55,22]. A proof of retrievability protocol usually goes like this: after storing a (potentially large) file with the server, the client issues a *query* to receive an assurance that the file is still correctly stored. The server computes an answer based on the query and the file, and finally the client performs some verification procedure on the answer. All of these protocols incur a substantial storage overhead for the server (since the file is stored using an erasure code) and, except for [22], require communication which is quadratic in the security parameter. The protocol in [22] has linear communication complexity but it requires *both* the server and the client to work in time proportional to the size of the file. Our solution achieves linear communication complexity in the security parameter and is very efficient for the client (as its work is sublinear in the size of the file).

Our verifiable database construction is closely related to Memory Checkers (see e.g. [10,26,1,24,50]). However, our setting differs from the memory checking setting in that we allow the server to be an arbitrary algorithm, whereas a memory checker interacts with a RAM (an oracle that accepts store/retrieve queries). In this context, our construction would yield a memory checker with poor performance since it would require the checker to issue a number of queries that is linear in the size of the memory. In contrast, we focus on optimizing the communication and the work of the client when the server can perform arbitrary computation on its data. Our construction requires the server to perform a linear amount of work to answer one type of queries (update/retrieve), while the other type of queries requires only a constant amount of work. Finally, we note that the work on accumulators [16,47,53] and authenticated data structures [48,41,52,59] can be used to construct verifiable databases with similar efficiency under different assumptions.

2 Assumptions

In this work we rely on the following assumptions about computational groups.

Decisional Diffie Hellman. The standard Decisional Diffie-Hellman Assumption (DDH) is defined as follows. For every PPT distinguisher A there exists a negligible function $neg(\cdot)$ such that for all n,

$$|\Pr[A(1^n, g, g^x, g^y, g^{xy}) = 1] - \Pr[A(1^n, g, g^x, g^y, g^z) = 1]| \leq neg(n)$$

where g is a generator of a group $\mathbb{G}$ of order p where p is a prime of length approximately n, and $x, y, z \in_R \mathbb{Z}_p$.

Strong Diffie Hellman. The strong Diffie-Hellman family of assumptions allows an adversary to obtain group elements $g, g^x, g^{x^2}, \ldots, g^{x^d}$, and requires the adversary to compute or distinguish a related group element from a random one. Computational variants of the problem appeared as early as the work of Mitsunari *et al* [45]. More recently, bilinear versions of the assumptions, starting with the works of Boneh and Boyen [11,12], were used in several applications (e.g. [23,15]). Boneh and Boyen gave a proof of the bilinear assumptions in the generic group model. In one of our constructions, we achieve high efficiency by relying on a decisional version of the strong DH assumption in single groups.

The d-SDDH assumption is stated as follows. For every PPT distinguisher A there exists a negligible function $neg(\cdot)$ such that for all n,

$$|\Pr[A(1^n, g, g^x, g^{x^2}, \ldots, g^{x^d}) = 1] - \Pr[A(1^n, g, g^{x_1}, g^{x_2}, \ldots, g^{x_d}) = 1]| \leq neg(n)$$

where g is a generator of a group $\mathbb{G}$ of order p where p is a prime of length approximately n, and $x, x_1, \ldots, x_d \in_R \mathbb{Z}_p$.

Subgroup membership assumption in composite order bilinear groups. The subgroup membership assumption in composite order bilinear groups first appeared in [13], and has seen many recent applications in the areas of Identity Based Encryption (IBE), Hierarchical IBE, and others [25,13,8]. The assumption we rely on (for our verifiable database delegation scheme) is the following.

For every PPT distinguisher A there exists a negligible function $neg(\cdot)$ such that for all n,

$$|\Pr[A(1^n, g_1g_2, u_2, (g_1g_2)^x) = 1] - \Pr[A(1^n, g_1g_2, u_2, u_2^x) = 1]| \leq neg(n)$$

where $\mathbb{G}$ is a group of order $N = p_1p_2$ where p_1 and p_2 are primes of length approximately n, $\mathbb{G}_1$ and $\mathbb{G}_2$ are subgroups of $\mathbb{G}$ of orders p_1 and p_2 respectively, $g_1 \in_{\mathrm{R}} \mathbb{G}_1$, $g_2, u_2 \in_{\mathrm{R}} G_2$, and $x \in_{\mathrm{R}} \mathbb{Z}_N$.

In addition, we require the existence of an efficiently computable pairing $e : \mathbb{G} \times \mathbb{G} \to \mathbb{G}_T$ where $\mathbb{G}_T$ is a group of order N. We shall make use of the following property of pairings over composite order groups: for $g_1 \in \mathbb{G}_1$ and $g_2 \in \mathbb{G}_2$, $e(g_1, g_2) = 1_{\mathbb{G}_T}$. This property holds for every bilinear pairing over composite order groups (as shown e.g. in [39]).

Bilinear sub-group projection assumption. In the analysis of our verifiable database scheme we first show the security of the scheme based on the following new assumption. We then apply Lemma 1 (given below) to obtain a reduction to the subgroup membership problem. The Bilinear Sub-Group Projection Assumption (BSGP) is stated as follows: for every PPT adversary A, there exists a negligible function $neg(\cdot)$ such that for all n,

$$\Pr[A(1^n, (g_1g_2), (h_1h_2), u_2) = e(g_1, h_1)] \leq neg(n)$$

where $\mathbb{G}$ is a group of order $N = p_1p_2$ where p_1 and p_2 are primes of length approximately n, $\mathbb{G}_1$ and $\mathbb{G}_2$ are subgroups of $\mathbb{G}$ of orders p_1 and p_2 respectively, $g_1, h_1 \in_{\mathrm{R}} \mathbb{G}_1$, and $g_2, h_2, u_2 \in_{\mathrm{R}} G_2$. The following lemma shows that the BSGP assumption is implied by the standard sub-group membership assumption in composite order bilinear groups.

Lemma 1. *The subgroup membership assumption in composite order bilinear groups reduces to the BSGP assumption.*

The proof of the lemma, as well as its application to delegation of data structures, is given in the full version of this paper [7].

3 Verifiable Computation

A verifiable computation scheme is a two-party protocol between a *client* and a *server*. The client chooses a function and an input which he provides to the server. The latter is expected to evaluate the function on the input and respond with the output together with a *proof* that the result is correct. The client then verifies that the output provided by the worker is indeed the output of the function computed on the input provided.

The goal of a verifiable computation scheme is to make such verification very efficient, and particularly much faster than the computation of the function itself. We adopt the *amortized* model of Gennaro *et al.* [27]: for each function F, the client is allowed to invest a one-time expensive computational effort (comparable to the effort to compute F itself) to produce a public/secret key pair, which he will use to efficiently (e.g. in linear-time) verify the computation of F by the server on many inputs.

We prove our results in a stronger version of the [27] definition of verifiable computation scheme. As we discussed in the Introduction, the main difference is that in our protocols the server is allowed to learn if the client accepts or rejects the output of a particular computation (in [27] and following works in the amortized model, leaking this bit of information to the server would help him cheat in the following executions).

We refer the reader to the full version of this paper [7] for the precise definition of a verifiable computation scheme.

4 Algebraic Pseudorandom Functions

Our main technical tool is a new way of viewing pseudo-random functions (PRF) with algebraic properties to achieve efficient verification of server certificates in the delegation setting. Intuitively, we rely on the fact that certain pseudo-random functions (such as the Naor-Reingold PRF [49]) have outputs that are members of an abelian group, and that certain algebraic operations on these outputs can be computed significantly more efficiently if one possesses the key of the pseudo-random function that was used to generate them. In this section we present an abstraction of the said property, and several constructions achieving different trade-offs between the types of functions that can be efficiently evaluated given the key, and the assumption that is needed to guarantee pseudo-randomness.

An algebraic pseudorandom function (PRF) consists of algorithms $\mathcal{PRF} = \langle \mathsf{KeyGen}, F, \mathsf{CFEval} \rangle$ where KeyGen takes as input a security parameter 1^n and a parameter $m \in \mathbb{N}$ that determines the domain size of the PRF, and outputs a pair $(K, param) \in \mathcal{K}_n$, where $\mathcal{K}_n$ is the key space for security parameter n. K is the secret key of the PRF, and $param$ encodes the public parameters. F takes as input a key K, public parameters $param$, an input $x \in \{0,1\}^m$, and outputs a value $y \in Y$, where Y is some set determined by $param$.

We require the following properties:

- *(Algebraic)* We say that $\mathcal{PRF}$ is algebraic if the range Y of $F_K(\cdot)$ for every $n \in \mathbb{N}$ and $(K, param) \in \mathcal{K}_n$ forms an abelian group. We require that the group operation on Y be efficiently computable given $param$. We are going to use the multiplicative notation for the group operation.
- *(Pseudorandom)* $\mathcal{PRF}$ is pseudorandom if for every PPT adversary A, and every polynomial $m(\cdot)$, there exists a negligible function $neg : \mathbb{N} \to \mathbb{N}$, such that for all $n \in \mathbb{N}$:

$$|\Pr[A^{F_K(\cdot)}(1^n, param) = 1] - \Pr[A^{R(\cdot)}(1^n, param) = 1]| \leq neg(n)$$

  where $(K, param) \leftarrow_{\mathrm{R}} \mathsf{KeyGen}(1^n, m(n))$, and $R : \{0,1\}^m \to Y$ is a random function.

- *(Closed form efficiency)* Let N be the order of the range sets of F for security parameter n. Let $z = (z_1, \ldots, z_l) \in (\{0,1\}^m)^l$, $k \in \mathbb{N}$, and efficiently computable $h : \mathbb{Z}_N^k \to \mathbb{Z}_N^l$ with $h(x) =< h_1(x), \ldots, h_l(x) >$. We say that (h, z) is *closed form efficient* for $\mathcal{PRF}$ if there exists an algorithm $\mathsf{CFEval}_{h,z}$ such that for every $x \in \mathbb{Z}_N^k$,

$$\mathsf{CFEval}_{h,z}(x, K) = \prod_{i=1}^{l} [F_K(z_i)]^{h_i(x)}$$

and the running time of CFEval is polynomial in n, m, k but sublinear in l. When $z = (0, \ldots, l)$ we will omit it from the subscript, and write $\mathsf{CFEval}_h(x, K)$ instead.

The last condition (which distinguishes our notion from traditional PRFs) allows to compute a "weighted product" of l PRF values much more efficiently than by computing the l values separately and then combining them. Indeed, given *param*, h, x, and $F_K(z)$, one can always compute the value $\prod_{i=1}^{l}[F_K(z_i)]^{h_i(x)}$ in time linear in l (this follows from the algebraic property of the PRF). The purpose of the closed form efficiency requirement is therefore to capture the existence of a more efficient way to compute the same value given the secret key K.

Note that closed form efficiency can be defined for PRFs over arbitrary input spaces. In particular, it is a non-trivial condition to attain even when the input space is polynomial in the security parameter[2]. In the constructions needed for our delegation scheme, this will be the case.

4.1 Small Domain Algebraic PRFs from Strong DDH

Construction 1. *Let $\mathcal{G}$ be a computational group scheme. The following construction $\mathcal{PRF}_1$ is an algebraic PRF with polynomial sized domains.*

$\mathsf{KeyGen}(1^n, m)$: *Generate a group description $(p, g, \mathbb{G}) \leftarrow_R \mathcal{G}(1^n)$. Choose $k_0, k_1 \in_R \mathbb{Z}_p$. Output param $= (m, p, g, \mathbb{G})$, $K = (k, k')$.*

$F_K(x)$: *Interpret x as an integer in $\{0, \ldots, D = 2^m\}$ where D is polynomial in n. Compute and output $g^{k_0 k_1^x}$.*

Closed form efficiency for polynomials. We now show an efficient closed form for $\mathcal{PRF}_1$ for polynomials of the form

$$p(x) = F_K(0) + F_K(1)x + \cdots + F_K(d)x^d$$

[2] When the input space is polynomial in the security parameter traditional PRFs exist unconditionally: if the input space has ℓ elements $\{x_1, \ldots, x_\ell\}$, define the key as ℓ random values $y_1, \ldots, y_\ell$ and $F_K(x_i) = y_i$. Notice however that this function does not have closed-form efficiency.

where $d \leq D$. Let $h : \mathbb{Z}_p \to \mathbb{Z}_p^{d+1}$, be defined as $h(x) \stackrel{\mathsf{def}}{=} (1, x, \dots, x^d)$. Then, we can define

$$\mathsf{CFEval}_h(x, K) \stackrel{\mathsf{def}}{=} g^{\frac{k_0(1-k_1^{d+1}x^{d+1})}{1-k_1 x}}$$

Let us now write the $\prod_{i=0}^{d}[F_K(z_i)]^{h_i(x)}$ where $(z_0, \dots, z_d) = (0, \dots, d)$:

$$\prod_{i=0}^{d}[F_K(z_i)]^{h_i(x)} = \prod_{i=0}^{d}[g^{k_0 k_1^i}]^{x^i} = g^{k_0 \sum_{i=0}^{d} k_1^i x^i}$$

Applying the identity $\sum_{i=0}^{d} k_0 k_1^i x^i = \frac{k_0(1-(k_1 x)^{d+1})}{1-k_1 x}$ we obtain the correctness of $\mathsf{CFEval}_h(x)$.

Theorem 1. *Suppose that the D-Strong DDH assumption holds. Then, $\mathcal{PRF}_1$ is a pseudorandom function.*

Proof. The input to the reduction is a description $(p, g, \mathbb{G})$ of a group, and a challenge $t_1, \dots, t_d$ where t_i is either a random member of $\mathbb{G}$, or $g^{k_1^i}$, and $k_1 \in \mathbb{Z}_p$ is randomly chosen once for the entire challenge. The reduction then chooses $k_0 \in_{\mathrm{R}} \mathbb{Z}_p$, and computes the function $H(i) = t_i^{k_0}$ for $0 \leq i \leq d$. Clearly, H is a random function if the t_i are random, and is equal to $F_K(\cdot)$ for $K = (k_0, k_1)$ if the t_i are determined by k_1.

Construction 2. *Let $\mathcal{G}$ be a computational group scheme. We define $\mathcal{PRF}_{2,d}$, for $d \in \mathbb{N}$, as follows:*

$\mathsf{KeyGen}(1^n, m)$: *Generate a group description $(p, g, \mathbb{G}) \leftarrow_R \mathcal{G}(1^n)$. Choose $k_0, k_1, \dots, k_m \in_R \mathbb{Z}_p$. Output param $= (m, p, g, \mathbb{G})$, $K = (k_0, k_1, \dots, k_m)$.*

$F_K(x)$: *Interpret x as a vector $(x_1, \dots, x_m) \in \{0, \dots, d\}^m$. Compute and output $g^{k_0 k_1^{x_1} \cdots k_m^{x_m}}$.*

Closed form for m-variate polynomials of total degree at most d. We describe an efficient closed form for $\mathcal{PRF}_{2,d}$ for computing polynomials of the form

$$p(x_1, \dots, x_m) = \sum_{\substack{i_1, \dots, i_m \\ i_1 + \dots + i_m \leq d}} F_K(i_1, \dots, i_m) x_1^{i_1} x_2^{i_2} \cdots x_m^{i_m}.$$

Let $h : \mathbb{Z}_p^m \to \mathbb{Z}_p^l$, where $l = \binom{m+d}{d}$, be defined as

$$h(x_1, \dots, x_m) \stackrel{\mathsf{def}}{=} \left(\binom{d}{i_1 \dots, i_m} \prod_{j=1}^{m} x_j^{i_j} \right)_{i_1 + \dots + i_m \leq d}$$

Let $z = [z_1, \dots, z_l] = [(i_1, \dots, i_m)]_{i_1 + \dots + i_m \leq d} \in \mathbb{Z}_d^{m \times l}$. We can now define

$$\mathsf{CFEval}_{h,z}(x_1, \dots, x_m, K) \stackrel{\mathsf{def}}{=} g^{k_0(1 + k_1 x_1 + \dots + k_m x_m)^d}$$

Correctness follows by algebraic manipulation and is given in the full version of this paper.

Theorem 2. *Let $d \in \mathbb{N}$, and suppose that the d-Strong DDH assumption holds. Then, $\mathcal{PRF}_{2,d}$ is a pseudorandom function.*

We refer the reader to the full version of this paper [7] for the proof of Theorem 2.

Remark 1. It is interesting to note that the Naor-Reingold PRF is a special case of Construction 2 obtained by setting $d = 1$. Therefore, our construction provides a tradeoff between the security assumption and the size of the key of the PRF: to operate on binary inputs of length n our construction requires $n/\log_2(d+1)$ elements of $\mathbb{Z}_p$ in the key.

Remark 2. One can change the above construction so that it becomes slightly less efficient but secure under the *standard DDH* assumption. We call this modified version $\mathcal{PRF}_{4,d}$ and refer the reader to the full version for its exact definition. The reader can also get a glimpse of this PRF's parameters in Table 1.

4.2 Small Domain Algebraic PRFs from DDH

Construction 3. *Let $\mathcal{G}$ be a computational group scheme. We define $\mathcal{PRF}_3$ as follows:*

$\mathsf{KeyGen}(1^n, m)$: *Generate a group description $(p, g, \mathbb{G}) \leftarrow_R \mathcal{G}(1^n)$. Choose k_0, $k_{1,1}, \ldots, k_{1,s}, \ldots, k_{m,1}, \ldots, k_{m,s} \in_R \mathbb{Z}_p$. Output param $= ((m, s), p, g, \mathbb{G})$, $K = (k_0, k_{1,1}, \ldots, k_{1,s}, \ldots, k_{m,1}, \ldots, k_{m,s})$.*

$F_K(x)$: *Interpret $x = (x_1, \ldots, x_m)$ with each $x_i = [x_{i,1}, \ldots, x_{i,s}]$ as an s-bit string. Compute and output $g^{k_0 k_{1,1}^{x_{1,1}} \cdots k_{1,s}^{x_{1,s}} \cdots k_{m,1}^{x_{m,1}} k_{m,s}^{x_{m,s}}}$.*

Closed form for polynomials of degree d in each variable. We describe an efficient closed form for $\mathcal{PRF}_3$ for computing polynomials of the form

$$p(x_1, \ldots, x_m) = \sum_{i_1, \ldots, i_m \leq d} F_K(i_1, \ldots, i_m) x_1^{i_1} \cdots x_m^{i_m}$$

where the PRF F is initialized with m and $s = \lceil \log d \rceil$. Let $h : \mathbb{Z}_p^m \to \mathbb{Z}_p^l$, where $l = md$, be defined as $h(x_1, \ldots, x_m) = (x_1^{i_1} \cdots x_m^{i_m})_{i_1, \ldots, i_m \leq d}$. Let $z = [z_1, \ldots, z_l] = [(i_1, \ldots, i_m)]_{i_1, \ldots, i_m \leq d}$ then

$$\mathsf{CFEval}_{h,z}(x_1, \ldots, x_m, K) \stackrel{\mathsf{def}}{=} g^{k_0 \prod_{j=1}^{m} (1+k_{j,1}x_j)(1+k_{j,2}x_j^2) \cdots (1+k_{j,s}x_j^{2^s})}$$

Correctness follows directly by expanding the expression in the exponent.

Remark 3. Note that for $m = 1$ we obtain an alternative construction for single-variable polynomials of degree d. Below we prove that Construction 3 is a PRF under the DDH Assumption. Therefore compared to Construction 1, this construction relies on a weaker assumption (DDH vs. D-strong DDH). However the efficiency of the closed form computation in Construction 1 is better: constant vs. $O(\log d)$ in Construction 3. Jumping ahead this will give us two alternative ways to delegate the computation of a single-variable polynomial of degree d with the following tradeoff: either one assumes a weaker assumption (DDH) but verification of the result will take $O(\log d)$ time, or one assumes a stronger assumption to obtain constant verification time.

Closed form for 1-out-of-2 multivariate polynomials of degree 1. We now consider polynomials of the form

$$p(x_1, y_1, \ldots, x_m, y_m) = \sum_{s \in \{0,1\}^m} F_K(s) x_1^{s_1} y_1^{1-s_1} \cdots x_m^{s_m} y_m^{1-s_m}$$

In such polynomials, each monomial contains exactly one of x_i and y_i for $1 \leq i \leq m$. We initialize the PRF F with m and $s = 1$ (and for simplicity we drop the double subscript and denote the key $k_{i,1}$ as k_i). Specifically, let $h : \mathbb{Z}_p^{2m} \to \mathbb{Z}_p^l$, where $l = 2^m$, be defined as $h(x_1, y_1, \ldots, x_m, y_m) = (x_1^{s_1} y_1^{1-s_1} \cdots x_m^{s_m} y_m^{1-s_m})_{s \in \{0,1\}^m}$. We can then define

$$\mathsf{CFEval}_h(x_1, y_1, \ldots, x_m, y_m) \stackrel{\mathsf{def}}{=} g^{k_0(x_1+k_1 y_1)\cdots(x_m+k_m y_m)}$$

Correctness is straightforward by expanding the expression in the exponent. The proof of the following theorem was given in [49];

Theorem 3. *[49] Suppose that the DDH assumption holds for $\mathcal{G}$. Then, $\mathcal{PRF}_3$ is a pseudorandom function.*

5 Verifiable Delegation of Polynomials

The basic idea of our construction is the following. The client stores the polynomial $P(\cdot)$ in the clear with the server as a vector $\boldsymbol{c}$ of coefficient in $\mathbb{Z}_p$. The client also stores with the server a vector $\boldsymbol{t}$ of group elements of the form $g^{ac_i+r_i}$ where $a \in_{\mathrm{R}} \mathbb{Z}_p$ and r_i is the ith-coefficient of a polynomial $R(\cdot)$ of the same degree as $P(\cdot)$. When queried on input x the server returns $y = P(x)$ and $t = g^{aP(x)+R(x)}$ and the client accepts y iff $t = g^{ay+R(x)}$.

If $R(\cdot)$ was a random polynomial, then our proof below shows that this is a secure delegation scheme. However checking that $t = g^{ay+R(x)}$ would require the client to evaluate the random polynomial $R(\cdot)$, which is just as inefficient as evaluating the original polynomial $P(\cdot)$. Moreover, the client would have to store a long description of a random polynomial that is as long as the original polynomial[3] $P(\cdot)$. The crucial point, therefore, is how to make this computation fast. We do that by defining $r_i = F_K(i)$ for an algebraic PRF that has a closed form efficient computation for polynomials, such as the ones described in the previous

[3] Alternatively, the client could generate the coefficients of R using a standard PRF, thereby avoiding storing a large polynomial. However, this would still require the client to recompute all the coefficients of R each time a verification needs to be performed.

Table 1. Parameters of different protocols for verifiable delegation of polynomials. Numbers inside the parenthesis show the number of group operations. Columns starting with "C." are the client's requirements and the ones starting with "S." are the server's. In each case the server's query runtime (resp. space requirements) is asymptotically the same as evaluating (resp. storing) the polynomial. Note that $(\frac{n+1}{\sqrt{d+1}})^d \le \binom{n+d}{d} \le (n+d)^d$, and in particular for constant d it is $\Theta(n^d)$.

Polynomial Type	Setup	C. Query	S. Query	Assumption	$\mathcal{PRF}$
1-variable, degree d	$O(d)$	$O(1)^4$ (1)	$O(d)$	d-Strong DDH	$\mathcal{PRF}_1$
n-variable, variable degree d	$O((d+1)^n)$	$O(n \log d)$ (1)	$O((d+1)^n)$	DDH	$\mathcal{PRF}_3$
n-variable, total degree d	$O(\binom{n+d}{d})$	$O(n \log d)$ (1)	$O(\binom{n+d}{d})$	d-Strong DDH	$\mathcal{PRF}_{2,d}$
n-variable, total degree d	$O((n+1)^d)$	$O(nd)$ (1)	$O((n+1)^d)$	DDH	$\mathcal{PRF}_{4,d}$

section. Since F is a PRF, the security of the scheme is not compromised, and the closed form efficiency of F will allow the client to verify the result in time sub-linear in the degree of the polynomial.

The result is described in general form, using algebraic PRFs. It therefore follows that we obtain efficient and secure delegation protocols not only for single-variable polynomials of degree d, but also for multivariate polynomials with total degree d or of degree d in each variable. The relevant parameters of the resulting protocols for each of these cases can be seen in Table 1.

Finally at the end of the section we show how to protect the privacy of the polynomial, by encrypting it with an homomorphically additive encryption scheme.

5.1 Construction Based on Algebraic PRFs

We describe a verifiable delegation scheme for functions of the form $f_{\boldsymbol{c},h}(\boldsymbol{x}) = \langle h(\boldsymbol{x}), \boldsymbol{c}\rangle$, where $\boldsymbol{c}$ is a (long) vector of coefficients, $\boldsymbol{x}$ is a (short) vector of inputs, and h expands $\boldsymbol{x}$ to a vector of the same length of $\boldsymbol{c}$. Our construction is generic based on any algebraic PRF that has closed form efficiency relative to h.

Protocol Delegate-Polynomial($\boldsymbol{c}$)

KeyGen($\boldsymbol{c}, n$): Generate $(K, param) \leftarrow_{\mathrm{R}} \mathsf{KeyGen}(1^n, \lceil \log d \rceil)$. Parse $\boldsymbol{c}$ as a vector $\boldsymbol{c} = (c_0, \ldots, c_d) \in \mathbb{Z}_p^{d+1}$. Let $\mathbb{G}$ be the range group of F_K, and let g be a generator for that group. Compute $g_i \leftarrow F_K(i)$ for $0 \le i \le d$, choose $a \in_{\mathrm{R}} \mathbb{Z}_p$, and set $\boldsymbol{t} = [t_0, \ldots, t_d] \leftarrow (g_0 g^{ac_0}, \ldots, g_d g^{ac_d})$. Output $PK \leftarrow (param, \boldsymbol{c}, \boldsymbol{t})$, and $SK \leftarrow (K, a)$.

ProbGen($\boldsymbol{SK}, \boldsymbol{x}$): Output $(\sigma_x, \tau_x) = (\boldsymbol{x}, \boldsymbol{x})$.

Compute($\boldsymbol{PK}, \sigma_x$): Parse PK as $(param, \boldsymbol{c}, \boldsymbol{t})$, $\boldsymbol{c}$ as $c_0, \ldots, c_d$, and σ_x as $\boldsymbol{x}$. Compute $\boldsymbol{w} \leftarrow h(\boldsymbol{x}) = [h_0(\boldsymbol{x}), \ldots, h_d(\boldsymbol{x})] \in \mathbb{Z}_p^{d+1}$, $y \leftarrow \sum_{i=0}^{d} c_i h_i(\boldsymbol{x})$, and $t \leftarrow \prod_{i=0}^{d} t_i^{h_i(\boldsymbol{x})}$. Output $\nu_x = (y, t)$.

Verify($\boldsymbol{SK}, \tau_x, \nu_x$): Parse SK as (K, a), τ_x as $\boldsymbol{x}$, and ν_x as $(y, t) \in \mathbb{Z}_p \times \mathbb{G}$. Compute $z \leftarrow \mathsf{CFEval}_h(\boldsymbol{x}, K)$, and accept if $t \stackrel{?}{=} z \cdot g^{a \cdot y}$. Otherwise, reject.

[4] The client needs to do two exponentiations.

CORRECTNESS. The correctness of the above scheme follows straightforwardly from the correctness of CFEval for the algebraic PRF F.

The security analysis of the above scheme, as well as an extension allowing the client to preserve the privacy of the polynomial, are given in the full version of the paper [7].

6 Verifiable Database Queries with Efficient Updates

We have shown a general framework that allows any resource constrained client to verifiably delegate large polynomials to a server. As we have already mentioned in the introduction, this immediately gives a verifiable delegation solution to many natural practical applications (such as prediction using fitted polynomials). In this Section we present an application of our techniques to the problem of efficient verification of the result to queries posed to a dynamic database. In other words the client stores a database with the server together with some authentication informatiom. It then needs to be able to efficiently verify that the results of its queries are correct, and also to efficiently update the database and its associated authenticator. The techniques we developed for delegation of polynomials are at the basis of the solution we present, which however requires other novel and interesting technical ideas.

The protocol uses ideas borrowed from our polynomial verification scheme: the authenticator for every database entry can be efficiently reconstructed by the client using a PRF with closed form efficiency. However as we pointed out in the Introduction the main challenge comes with the updates: the client must revoke any authenticating data that the server has for the previous value of the updated cell. We deal with this problem by "masking" the authenticator with another closed-form efficient PRF. This "mask" can be efficiently removed by the client to perform the authentication and can also be efficiently updated so that old masked values cannot be reused. The technical details are somewhat involved and are described below.

HANDLING LARGE PAYLOADS. For simplicity we consider databases of the form (i, v_i) where i is the index and v_i the payload data. The construction we describe below allows data values to be only polynomially large (in the security parameter). Before proceeding to describe our protocol, we show a simple transformation that allows us to support databases with arbitrary payload sizes.

On a high level, we will use a small payload protocol, such as the one described below, to store a database of the form (i, s_i) where s_i is a counter that counts the number of times index i has been updated. The server will also store the MAC of the tuple (i, s_i, v_i) where v_i is the (possibly large) payload.

When the client queries i, it will receive the value s_i through the verifiable database protocol. The security of this protocol will guarantee to the client that s_i is correct. Then the server will also answer with v_i and the MAC on (i, s_i, v_i) and the client will accept v_i if the MAC is correct. To update index i, the client will first query i and retrieve the current tuple (i, s_i, v_i). It will then update s_i on the verifiable database by setting $s'_i = s_i + 1$. This can be done since s_i is bounded by the running time of the client and therefore polynomial. Finally it will store the new v_i' together with a MAC on (i, s_i', v_i'). Since s_i will only ever increase, the server will not be able to re-use old MACs once an index has been updated.

Therefore for now we will focus on databases with small (polynomial) payloads. The protocol is described in detail below.

Relation to Merkle Trees. Merkles trees [42] are a standard technique for efficient authentication of data. Each element is represented as a leaf of a binary tree. The internal nodes contain hashes of their two children, and the owner of the data keeps only the root, which is essentially a hash of the entire tree. Retrieval queries can now be answered and verified in logarithmic time: the server simply sends the hashes along the path from the root to the desired leaf (along with any hashes within distance 1 of the path), and the client uses these values to compute the hash at the root of the tree. The client then checks that the stored hash value of the root is equal to the recomputed hash. Updating the tree is also quite efficient – only the hashes along the path to the updated leaf must be recomputed. In comparison, our scheme requires the client to perform a constant amount of work both during retrieval and updates, while the server must choose one of the two types of queries where he will do a linear amount of work (the other type of queries requires a constant amount of work from the server as well).

Our protocol. We give a fully detailed protocol in the full version [7]. Here, we present a high level overview of the approach. The basic tools that we use are computational bilinear groups of composite order. In this setting, a pair of groups $\mathbb{G}, \mathbb{G}_T$ are generated, along with a pairing $e : \mathbb{G} \times \mathbb{G} \to \mathbb{G}_T$. Here each of the groups are of some order $N = p_1 p_2$ where p_1 and p_2 are large primes. The cryptographic assumption is that it is infeasible to distinguish a random member of $\mathbb{G}$ (or $\mathbb{G}_T$) from a random member of the subgroup of order p_1 or p_2. Such groups have recently been used to solve several important open problems in the areas of identity based encryption, leakage resilient cryptography, and other related problems (see e.g. [13]).

The basic approach of our construction (leaving some details to the full description below) can be described as follows: each entry (i, v_i) in the database (where i is an index and v_i is data) is encoded as a composite group element of the form

$$t_i = g_1^{r_i + a v_i} g_2^{w_i}.$$

Here, g_1 and g_2 are generators of the subgroups $\mathbb{G}_1$ and $\mathbb{G}_2$ of $\mathbb{G}$, and the values r_i, w_i are generated using a pseudo-random function. To retrieve the value of the database at index i, we will have the server compute (given keys that we shall describe in a moment) the value

$$t = g_1^{r_i + a v_i} g_2^{\sum_i w_i}.$$

Forgetting (for now) how the server should compute these values, the client can easily strip off the $\mathbb{G}_2$ masking by keeping the single group element $g_2^{\sum_i w_i}$ in his private key. It is now easy to see that if we replace the r_i's with random values, then our scheme is secure before any updates are performed. This follows from the fact that each entry in the database is MAC'ed with an information theoretically secure MAC (the $\mathbb{G}_2$ part hasn't played a role so far), and so the server must return the correct value in the $\mathbb{G}_1$ part of each entry. The difficulty is in allowing updates that do not require the client to change his keys for the pseudo-random functions, which in turn would require the server to obtain new MACs for all the entries in the database.

A naive solution to change the value of index i from v_i to v_i' can be for the client to send to the server a new encoding $g_1^{r_i+bv_i}g_2^{w_i}$. However, the server can then easily recover the MAC keys r_i and a by dividing the new group element that he receives during the update by the previous encoding that he already has. Our solution is therefore to randomize the new encoding by having the client send

$$t_x' = g_1^{r_i+a\delta}g_2^{w_i'},$$

where $\delta = v_i' - v_i$, and w_i' is a new pseudorandom value (generated by using a counter). Intuitively, this allows the client to send t_x' as an update token that the server can multiply into his existing group element t_i to obtain $g_1^{r_i+av_i'}g_2^{w_i+w_i'}$. Notice that the $\mathbb{G}_1$ part is a MAC of the value v_i' using the same key that was previously used to MAC v_i. We show, relying on the subgroup membership assumption, that the random mask $g_2^{w_i+w_i'}$ effectively makes the MAC in the $\mathbb{G}_1$ of the token indistinguishable from a new MAC using fresh keys. We now arrive at the problem of allowing the server to compute the value t, which requires stripping the $\mathbb{G}_1$ part of all the tokens except the token that corresponds to index i, without compromising security. We achieve this by issuing to the server random group elements $\hat{t}_1$ from $\mathbb{G}$, and $\hat{t}_0$ from $\mathbb{G}_2$. The server then computes the response to query i as

$$t = e(t_i, \hat{t}_1)\prod_{j\neq i} e(t_j, \hat{t}_0).$$

A remaining technical issue is the fact the in the above discussion we haven't mentioned anything about how the client should remember the new masked value w_i' after an update. Our solution is to compute it pseudo-randomly as $F_k(i, s_i)$ where s_i is a counter that is incremented with each update and is stored together with the payload v_i. This guarantees that a fresh pseudo-random value is used after each update, which in turn allows us to substitute the pseudo-random w_i's by random ones in the security analysis.

Acknowledgments. Rosario Gennaro's research was sponsored by US Army Research laboratory and the UK Ministry of Defence and was accomplished under Agreement Number W911NF-06-3-0001. The views and conclusions contained herein are those of the authors and should not be interpreted as representing the official policies, either expressed or implied, of the US Army Research Laboratory, the US Government, the UK Ministry of Defence, or the UK Government.

The US and UK Governments are authorized to reproduce and distribute reprints of this work for Government purposes, notwithstanding any copyright notation hereon.

References

1. Ajtai, M.: The invasiveness of off-line memory checking. In: STOC, pp. 504–513 (2002)
2. Amazon Elastic Compute Cloud, http://aws.amazon.com/ec2
3. Applebaum, B., Ishai, Y., Kushilevitz, E.: From Secrecy to Soundness: Efficient Verification via Secure Computation. In: Abramsky, S., Gavoille, C., Kirchner, C., Meyer auf der Heide, F., Spirakis, P.G. (eds.) ICALP 2010. LNCS, vol. 6198, pp. 152–163. Springer, Heidelberg (2010)
4. Ateniese, G., Burns, R., Curtmola, R., Herring, J., Kissner, L., Peterson, Z., Song, D.: Provable data possession at untrusted stores. In: ACM CCS, pp. 598–609 (2007)
5. Babai, L.: Trading group theory for randomness. In: Proceedings of the ACM Symposium on Theory of Computing (STOC), pp. 421–429. ACM, New York (1985)
6. Belenkiy, M., Chase, M., Erway, C.C., Jannotti, J., Küpçü, A., Lysyanskaya, A.: Incentivizing outsourced computation. In: Proceedings of the Workshop on Economics of Networked Systems (NetEcon), pp. 85–90. ACM, New York (2008)
7. Benabbas, S., Gennaro, R., Vahlis, Y.: Verifiable Delegation of Computation over Large Datasets. Cryptology ePrint Archive, Report 2011/132 (2011), http://eprint.iacr.org/
8. Bellare, M., Waters, B., Yilek, S.: Identity-based encryption secure against selective opening attack. In: Ishai, Y. (ed.) TCC 2011. LNCS, vol. 6597, pp. 235–252. Springer, Heidelberg (2011)
9. Ben-Or, M., Goldwasser, S., Wigderson, A.: Completeness theorems for non-cryptographic fault-tolerant distributed computation. In: STOC, pp. 1–10 (1988)
10. Blum, M., Evans, W., Gemmell, P., Kannan, S., Naor, M.: Checking the correctness of memories. In: 32nd Annual IEEE Symposium of Foundations of Computer Science (FOCS 1991), pp. 90–99 (1991)
11. Boneh, D., Boyen, X.: Efficient selective-ID secure identity-based encryption without random oracles. In: Cachin, C., Camenisch, J.L. (eds.) EUROCRYPT 2004. LNCS, vol. 3027, pp. 223–238. Springer, Heidelberg (2004)
12. Boneh, D., Boyen, X.: Short signatures without random oracles and the sdh assumption in bilinear groups. J. Cryptology 21(2), 149–177 (2008)
13. Boneh, D., Goh, E.-J., Nissim, K.: Evaluating 2-DNF formulas on ciphertexts. In: Kilian, J. (ed.) TCC 2005. LNCS, vol. 3378, pp. 325–341. Springer, Heidelberg (2005)
14. Boneh, D., Montogomery, H., Raghunathan, A.: Algebraic pseudorandom functions with improved efficiency from the augmented cascade. In: Proc. of ACM CCS 2010 (2010)
15. Boneh, D., Sahai, A., Waters, B.: Fully collusion resistant traitor tracing with short ciphertexts and private keys. In: Vaudenay, S. (ed.) EUROCRYPT 2006. LNCS, vol. 4004, pp. 573–592. Springer, Heidelberg (2006)
16. Camenisch, J.L., Lysyanskaya, A.: Dynamic accumulators and application to efficient revocation of anonymous credentials. In: Yung, M. (ed.) CRYPTO 2002. LNCS, vol. 2442, pp. 61–76. Springer, Heidelberg (2002)

17. Chaum, D., Crepeau, C., Damgard, I.: Multiparty unconditionally secure protocols. In: STOC, pp. 11–19 (1988)
18. Chaum, D., Pedersen, T.: Wallet databases with observers. In: Brickell, E.F. (ed.) CRYPTO 1992. LNCS, vol. 740, pp. 89–105. Springer, Heidelberg (1993)
19. Cheon, J.H.: Security analysis of the strong diffie-hellman problem. In: Vaudenay, S. (ed.) EUROCRYPT 2006. LNCS, vol. 4004, pp. 1–11. Springer, Heidelberg (2006)
20. Chung, K.-M., Kalai, Y., Vadhan, S.P.: Improved Delegation of Computation Using Fully Homomorphic Encryption. In: Rabin, T. (ed.) CRYPTO 2010. LNCS, vol. 6223, pp. 483–501. Springer, Heidelberg (2010)
21. Desmedt, Y.: Threshold Cryptography. In: Encyclopedia of Cryptography and Security (2005)
22. Dodis, Y., Vadhan, S.P., Wichs, D.: Proofs of Retrievability via Hardness Amplification. In: Reingold, O. (ed.) TCC 2009. LNCS, vol. 5444, pp. 109–127. Springer, Heidelberg (2009)
23. Dodis, Y., Yampolskiy, A.: A verifiable random function with short proofs and keys. In: Vaudenay, S. (ed.) PKC 2005. LNCS, vol. 3386, pp. 416–431. Springer, Heidelberg (2005)
24. Dwork, C., Naor, M., Rothblum, G.N., Vaikuntanathan, V.: How Efficient Can Memory Checking Be? In: Reingold, O. (ed.) TCC 2009. LNCS, vol. 5444, pp. 503–520. Springer, Heidelberg (2009)
25. Freeman, D.M.: Converting pairing-based cryptosystems from composite-order groups to prime-order groups. In: Gilbert, H. (ed.) EUROCRYPT 2010. LNCS, vol. 6110, pp. 44–61. Springer, Heidelberg (2010)
26. Gemmell, P., Naor, M.: Codes for Interactive Authentication. In: Stinson, D.R. (ed.) CRYPTO 1993. LNCS, vol. 773, pp. 355–367. Springer, Heidelberg (1994)
27. Gennaro, R., Gentry, C., Parno, B.: Non-interactive verifiable computing: Outsourcing computation to untrusted workers. In: Rabin, T. (ed.) CRYPTO 2010. LNCS, vol. 6223, pp. 465–482. Springer, Heidelberg (2010)
28. Gentry, C.: Fully homomorphic encryption using ideal lattices. In: Proceedings of the ACM Symposium on the Theory of Computing (STOC) (2009)
29. Goldreich, O., Micali, S., Wigderson, A.: How to play ANY mental game. In: STOC, pp. 218–229 (1987)
30. Goldwasser, S., Kalai, Y.T., Rothblum, G.N.: Delegating computation: interactive proofs for muggles. In: Proceedings of the ACM Symposium on the Theory of Computing (STOC) (2008)
31. Goldwasser, S., Micali, S., Rackoff, C.: The knowledge complexity of interactive proof-systems. SIAM Journal on Computing 18(1), 186–208 (1989)
32. Golle, P., Mironov, I.: Uncheatable distributed computations. In: Naccache, D. (ed.) CT-RSA 2001. LNCS, vol. 2020, p. 425. Springer, Heidelberg (2001)
33. Hall, W.E., Jutla, C.S.: Parallelizable authentication trees. In: Preneel, B., Tavares, S. (eds.) SAC 2005. LNCS, vol. 3897, pp. 95–109. Springer, Heidelberg (2006)
34. Hazay, C., Lindell, Y.: Efficient Protocols for Set Intersection and Pattern Matching with Security Against Malicious and Covert Adversaries. J. Cryptology 23(3), 422–456 (2010)
35. Hohenberger, S., Lysyanskaya, A.: How to securely outsource cryptographic computations. In: Kilian, J. (ed.) TCC 2005. LNCS, vol. 3378, pp. 264–282. Springer, Heidelberg (2005)
36. Juels, A., Kaliski Jr., B.S.: Pors: proofs of retrievability for large files. In: ACM Conference on Computer and Communications Security, pp. 584–597 (2007)

37. Kilian, J.: A note on efficient zero-knowledge proofs and arguments (extended abstract). In: Proceedings of the ACM Symposium on Theory of Computing (STOC), pp. 723–732. ACM, New York (1992)
38. Kilian, J.: Improved efficient arguments (preliminary version). In: Proceedings of the International Cryptology Conference on Advances in Cryptology, pp. 311–324. Springer, London (1995)
39. Lewko, A., Rouselakis, Y., Waters, B.: Achieving leakage resilience through dual system encryption. In: Ishai, Y. (ed.) TCC 2011. LNCS, vol. 6597, pp. 70–88. Springer, Heidelberg (2011)
40. Lewko, A.B., Waters, B.: Efficient pseudorandom functions from the decisional linear assumption and weaker variants. In: Proceedings of the 16th ACM Conference On Computer and Communications Security, CCS 2009, pp. 112–120. ACM, New York (2009)
41. Martel, C., Nuckolls, G., Devanbu, P., Gertz, M., Kwong, A., Stubblebine, S.G.: A general model for authenticated data structures. Algorithmica 39(1), 21–31 (2004)
42. Merkle, R.C.: A Digital Signature Based on a Conventional Encryption Function. In: Pomerance, C. (ed.) CRYPTO 1987. LNCS, vol. 293, pp. 369–378. Springer, Heidelberg (1988)
43. Micali, S.: CS proofs (extended abstract). In: Proceedings of the IEEE Symposium on Foundations of Computer Science (1994)
44. Micali, S., Rabin, M.O., Kilian, J.: Zero-Knowledge Sets. In: FOCS, pp. 80–91 (2003)
45. Mitsunari, S., Sakai, R., Kasahara, M.: A new traitor tracing. IEICE Transactions on Fundamentals of Electronics, Communications and Computer Sciences 85(2), 481–484 (2002)
46. Monrose, F., Wyckoff, P., Rubin, A.: Distributed execution with remote audit. In: Proceedings of ISOC Network and Distributed System Security Symposium, NDSS (February 1999)
47. Nguyen, L.: Accumulators from bilinear pairings and applications. In: Menezes, A. (ed.) CT-RSA 2005. LNCS, vol. 3376, pp. 275–292. Springer, Heidelberg (2005)
48. Naor, M., Nissim, K.: Certificate revocation and certificate update. In: USENIX Security, pp. 17–17 (1998)
49. Naor, M., Reingold, O.: Number-theoretic constructions of efficient pseudo-random functions. J. ACM 51, 231–262 (2004)
50. Naor, M., Rothblum, G.N.: The Complexity of Online Memory Checking. In: FOCS, pp. 573–584 (2005)
51. Paillier, P.: Public-key cryptosystems based on composite degree residuosity classes. In: Stern, J. (ed.) EUROCRYPT 1999. LNCS, vol. 1592, pp. 223–238. Springer, Heidelberg (1999)
52. Papamanthou, C., Tamassia, R.: Time and space efficient algorithms for two-party authenticated data structures. In: Qing, S., Imai, H., Wang, G. (eds.) ICICS 2007. LNCS, vol. 4861, pp. 1–15. Springer, Heidelberg (2007)
53. Papamanthou, C., Tamassia, R., Triandopoulos, N.: Authenticated hash tables. In: CCS, pp. 437–448 (October 2008)
54. Papamanthou, C., Tamassia, R., Triandopoulos, N.: Optimal Authentication of Operations on Dynamic Sets. Cryptology ePrint Archive, Report 2010/455 (2010), http://eprint.iacr.org/
55. Shacham, H., Waters, B.: Compact Proofs of Retrievability. In: Pieprzyk, J. (ed.) ASIACRYPT 2008. LNCS, vol. 5350, pp. 90–107. Springer, Heidelberg (2008)

56. Smith, S., Weingart, S.: Building a high-performance, programmable secure coprocessor. Computer Networks (Special Issue on Computer Network Security) 31, 831–960 (1999)
57. Sun Utility Computing, `http://www.sun.com/service/sungrid/index.jsp`
58. Tamassia, R.: Authenticated data structures. In: Di Battista, G., Zwick, U. (eds.) ESA 2003. LNCS, vol. 2832, pp. 2–5. Springer, Heidelberg (2003)
59. Tamassia, R., Triandopoulos, N.: Certification and authentication of data structures. In: Proc. Alberto Mendelzon Workshop on Foundations of Data Management, Cite-seer (2010)
60. Yao, A.: Protocols for secure computations. In: FOCS, p. 1982
61. Yee, B.S.: Using Secure Coprocessors. PhD thesis, Carnegie Mellon University (1994)

Secure Computation on the Web: Computing without Simultaneous Interaction

Shai Halevi[1], Yehuda Lindell[2,⋆], and Benny Pinkas[2,⋆,⋆⋆]

[1] IBM T.J. Watson Research Center
shaih@alum.mit.edu
[2] Bar-Ilan University
lindell@biu.ac.il, benny@pinkas.net

Abstract. Secure computation enables mutually suspicious parties to compute a joint function of their private inputs while providing strong security guarantees. However, its use in practice seems limited. We argue that one of the reasons for this is that the model of computation on the web is not suited to the type of communication patterns needed for secure computation. Specifically, in most web scenarios clients independently connect to servers, interact with them and then leave. This rules out the use of secure computation protocols that require that *all* participants interact simultaneously.

We initiate a study of secure computation in a client-server model where each client connects to the server *once* and interacts with it, without any other client necessarily being connected at the same time. We point out some inherent limitations in this model and present definitions that capture what can be done. We also present a general feasibility result and several truly practical protocols for a number of functions of interest. All our protocols are based on standard assumptions, and we achieve security both in the semi-honest and malicious adversary models.

1 Introduction

Web-servers are a dominant communication medium in today's society. Some examples include users of social networks that communicate by sending messages to the web-servers of their network to "write on the wall" of their friends (and these servers distribute the messages to the intended recipients), program committees that use web-based systems to share their reviews and discussions, readers that participate in on-line polls on newspaper web sites, bidders engaging in on-line auctions, voters using web-based election systems, and so on. In many cases, direct interaction between users is impossible simply because users are off line most of the time. In almost all systems today, the web-server serves not only as a communication medium but also as a trusted party. It receives all the information from the users and does all the processing, and it is trusted by the users to only use their information as needed for the application (or as

⋆ Supported by the European Research Council as part of the ERC project LAST.
⋆⋆ Supported in part by the Israel Science Foundation (grant No. 860/06).

P. Rogaway (Ed.): CRYPTO 2011, LNCS 6841, pp. 132–150, 2011.

specified in the "privacy policy" of the web site). This may be appropriate in some cases, but there are many cases where there is no reason for users to trust the server or each other, and indeed many cases where this trust was found to be unjustified in retrospect. (For example see [1].)

A natural approach toward rectifying this problem is to use cryptographic techniques for eliminating trusted parties. Indeed, the last three decades saw a very significant body of work within the cryptography research community (going under the general name of *secure multi-party computation*), devoted to finding various ways of transforming systems that rely on trusted parties into systems that do not need them (see, e.g., [2, Ch. 7] for an overview).

In fact, with client-side processing in Web 2.0 we now have a huge mass of parties with serious computing platforms and conflicting interests, all wishing to interact with each other to perform some joint tasks. This seems to offer the perfect setting for mass deployment of secure multi-party computation, but in reality such mass deployment has not happened. Some of the reasons are related to practical issues with browser technology (e.g., clients cannot verify that they run the right program), but here we focus on a more cryptographic reason; specifically, the fact that our current multi-party protocols seem incompatible with the communication patterns of today's web applications. Much of the work on secure multi-party computation assumes that all parties remain on-line throughout the computation, and most solutions also rely on strong communication primitives like broadcast. In contrast, clients on the web connect in an ad hoc manner via a server at different times, and typically do not communicate with each other. We thus ask whether one can eliminate the need for the web-server being a trusted party, even in this setting of loosely connected parties that are off line most of the time.

Beyond the practical interest that we discussed above, addressing multi-party computation in this model is also of significant theoretical interest. It is not at all clear that theoretically meaningful secure computation can be achieved in a setting where each party carries out a single interaction with an untrusted server at a different time (either in the semi-honest or the malicious settings). The power of this model is therefore a natural theoretical question to consider.

We note at the outset that a naive approach using fully homomorphic encryption [3,4] does not solve the problem of secure computation in our setting. This is due to the fact that although each party can encrypt its input and the computation can be done homomorphically, there is still the need to decrypt the final ciphertexts while preventing decryption of the intermediate ciphertexts.

1.1 Our Contributions

We initiate a study of secure computation with loosely connected parties. We define security, and observe that in this setting it is not always possible to achieve the same level of security as in the standard setting of secure computation. We formalize what can be achieved in this model, and then present theoretical and practical constructions, for both the cases of semi-honest and malicious adversaries. Our constructions all rely on standard assumptions (like the DDH assumption) and are in the standard model. The only exception is that for our practical construction in the case of malicious adversaries, we use random oracles in order to obtain practical non-interactive zero-knowledge via the Fiat-Shamir paradigm [5].

We begin by considering a very basic setting of a server and n parties, denoted $P_1, P_2, \ldots, P_n$. Each party P_i has an input x_i, and the parties wish to jointly evaluate a function $f(x_1, \ldots, x_n)$ (e.g., the sum of the inputs, or their maximum value), such that the server learns the output value. To simplify the exposition, consider the case where the parties talk to the server in order, first party P_1, then party P_2, all the way up to party P_n, and if everyone cooperates then after talking to them all the server should be able to learn the output value.

We stress that although our basic model assumes a pre-set order, many of the protocols that we describe allow the parties to interact with the server in an arbitrary order, which need not be set up in advance. However, all our protocols assume that the clients connect to the server sequentially, removing this requirement is an interesting open problem.

Consider first the case of semi-honest parties. It is easy to see that even in this model protocols *cannot* always provide the same privacy guarantees as standard secure function evaluation protocols (SFE). For example, if the last $n-i$ parties collude with the server, then they can always evaluate the residual function $g_i^{\vec{x}}(z_{i+1}, \ldots, z_n) \stackrel{\text{def}}{=} f(x_1, \ldots, x_i, z_{i+1}, \ldots, z_n)$ on as many inputs $(z_{i+1}, \ldots, z_n)$ as they like. This is due to the fact that these last $n-i$ parties must have the capability of computing $f(x_1, \ldots, x_i, x_{i+1}, \ldots, x_n)$ for every possible vector of their inputs $x_{i+1}, \ldots, x_n$. Furthermore, since the first i parties are no longer involved, nothing prevents the last $n-i$ parties from just rerunning the rest of the protocol many times with different inputs $z_{i+1}, \ldots, z_n$.

We formalize the inherent "leakage" in this model by introducing the concept of a *one-pass decomposition* of a function: A decomposition of an n-input function $f(x_1, x_2, \ldots, x_n)$ is a vector of functions $\{f_i(y_{i-1}, x_i) : i = 1, \ldots, n\}$, such that for all inputs $x_1, \ldots, x_n$ it holds that $f(x_1, x_2, \ldots, x_n) = f_n(\cdots f_2(f_1(x_1), x_2) \ldots, x_n)$. Here y_i represents the *intermediate result* based on the inputs of parties P_1 through P_i and y_0 is defined as the empty string. One can see that every protocol for computing f in our model corresponds to some (possibly randomized) decomposition of f, roughly because we can think of y_i as the state of the server after interacting with party P_i. However, as we will see, not all decompositions are equal; some are better than others (and some are incomparable). We therefore break up the problem of secure computation in this model into (a) finding a "good" decomposition of the given function f, and (b) devising a protocol to securely compute a given decomposition.

Good decompositions. Although every function f can be decomposed as described above, some decompositions are more "interesting" or "natural" than others. A trivial example is that any function f can be decomposed by setting the functions $f_1, \ldots, f_{n-1}$ to all be the identity function and then setting $f_n = f$. A more interesting example is that the sum function, $f(x_1, \ldots, x_n) = \sum_i x_i$, can be decomposed by letting the f_i's be the partial sums, $f_i(y_{i-1}, x_i) = y_{i-1} + x_i$.

Clearly, the decomposition of the sum function using partial sums is much better than its decomposition using the identity functions, since it reveals much less information to the adversary (in the case of a corrupted server and corrupted party P_n the adversary learns all the inputs when the identity function is used, in contrast to a partial sum only).

We are particularly interested in "*minimum-disclosure*" decompositions of f, where $y_i = f_i(\cdots)$ carries no more information about the inputs $x_1, \ldots, x_i$ than the truth-table of the residual function $g_i^{\vec{x}}$ from above. For example, it is easy to see that for the sum function, having the f_i's be the partial sums is indeed a minimum-disclosure decomposition, because given $x_{i+1}, \ldots, x_n$ and the output y_n it is possible to compute the partial sum y_i. In Section 2 we define this notion of minimum-disclosure decompositions and describe many functions that have efficient minimum-disclosure decompositions. Then in Section 4.1 we describe practical protocols for securely computing some of these decompositions (in a PKI model). The functions that we can handle in this fashion include all the symmetric functions on small domains (and also some other functions). Thus, for example, we construct a practical protocol for computing a referendum, as privately as is possible in our model.

Securely computing any decomposition. Given a specific decomposition of f (that codifies the "leakage" that we are willing to tolerate while computing f in our model), what does it mean for a protocol to securely compute this decomposition? In keeping with the intuition that y_i represents the partial result up to party P_i, we set out to formalize the requirement that these partial results are the only thing that can be learned by the bad parties.

First, observe that many of the intermediate results y_i's can be hidden from the corrupted parties. For example, if parties P_1, P_2 and P_3 are honest then we expect the partial results y_1 and y_2 to remain hidden, even if a dishonest P_4 learns y_3. In fact our formal definition requires a little more: A protocol is said to securely compute a given decomposition of f if the only partial result that it leaks is the one after *the last honest party*. Namely, the view of any set of adversarial parties can be simulated knowing only the value $y_i = f_i(\ldots)$, where i is the index of the last honest party. Furthermore, if the server is honest, then nothing but the output of f is revealed. (We remark that a weaker definition where bad parties can learn all the y_i's for which party P_{i+1} is dishonest, is essentially equivalent to the notion of i-Hop homomorphic encryption from [6].)

In Section 5 we consider the task of devising a protocol to securely compute a particular given decomposition of a function f. Using re-randomizable garbled circuits similar to Gentry et al. [6] we show that under the DDH assumption any efficient decomposition of f can be securely computed in our model (if a PKI is available). Our treatment simplifies the techniques from [6], in that we use re-randomizable garbled circuits only in conjunction with re-randomizable encryption (whereas [6] needed also re-randomizable OT). We also strengthen the construction from [6] slightly in order to deal with malicious parties. See Section 5 for more details about these points.

1.2 Some Related Work

Some of the techniques that we use are similar to those used in the work of Harnik et al. [7]. In that work they considered a multi-party computation settings where the inputs of parties are incorporated one at a time, with the goal of minimizing the number of OTs that are needed every time a new input is received. In particular our protocols for symmetric functions are reminiscent of their tables method.

Another related work is that of Choi et al. [8]. They considered a setting where the parties can interact in a setup phase before receiving their inputs, and then they want to minimize online communication while maintaining full security. Their results are not applicable in our model, however, since, as we explained, full security cannot be obtained in our model (and this remains true even given an interactive setup phase).

2 One-Pass Decompositions

Throughout the text we denote the number of parties (not counting the server) by n, and the security parameter by m. For an integer n we denote $\mathbb{Z}_n = \{0, 1, \ldots, n-1\}$ and $[n] = \{1, 2, \ldots, n\}$. In the text we also refer to randomized functions which can be viewed as distributions over deterministic functions all with the same domain and range.

Definition 1 (Decomposition). *Let $f : D^n \to R$ be an n-variable function (from domain D to range R). A deterministic* one-pass decomposition *of f is a sequence of functions $f_1 : D \to \{0,1\}^*$, $f_i : \{0,1\}^* \times D \to \{0,1\}^*$ for $i = 2, 3, \ldots, n-1$, and $f_n : \{0,1\}^* \times D \to R$ such that for all $x_1, \ldots, x_n \in D$, it holds that $f(x_1, x_2, \ldots, x_n) = f_n(\cdots f_2(f_1(x_1), x_2) \cdots, x_n)$.*

A randomized one-pass decomposition of f is a sequence of n randomized functions with the same domains and ranges as above, such that the equality above holds with overwhelming probability (in the implicit security parameter).

Below we will omit the "one-pass" qualifier and just call this sequence of functions a decomposition. We often also omit the distinction between deterministic and randomized decompositions. Given a decomposition $\bar{f} = \langle f_1, \ldots, f_n \rangle$, we denote by $\tilde{f}_i$ the concatenation of the first i functions,

$$\tilde{f}_i(x_1, x_2, \ldots, x_i) \stackrel{\text{def}}{=} f_i(\cdots f_2(f_1(x_1), x_2) \cdots, x_i). \tag{1}$$

2.1 Minimum-Disclosure Decompositions

As was mentioned above, some decompositions are better than others and some functions have efficient decompositions that are "as good as possible" (in that they do not leak anything beyond the ability to compute the residual functions g_i), while others do not. Fix an n-input function f and n particular inputs

$x_1, \ldots, x_n$. For all $i = 0, \ldots, n$ we denote by $g_i^{\vec{x}}$ the "residual function" with the first i variables fixed. That is, for $\vec{x} = \langle x_1, \ldots, x_n \rangle$, define

$$g_i^{\vec{x}}(z_{i+1}, \ldots, z_n) \stackrel{\text{def}}{=} f(x_1, \ldots, x_i, z_{i+1}, \ldots, z_n). \tag{2}$$

(In particular $g_0^{\vec{x}} = f$ and $g_n^{\vec{x}}$ is the constant function $g_n^{\vec{x}}(\cdot) = f(x_1, \ldots, x_n)$.) As we explained above, any decomposition of f must at least leak the ability to compute $g_i^{\vec{x}}$ on all residual input vectors $z_{i+1}, \ldots, z_n$. A minimum-disclosure decomposition is one that does not leak anything else. Namely, for all i it is possible to compute the output of the composition of the first i functions $f_1, \ldots, f_i$, given only *oracle access* to the residual function $g_i^{\vec{x}}(\cdot)$.

Definition 2 (Minimum-Disclosure). *A decomposition $\bar{f}$ is* minimum disclosure *if there exists a probabilistic black box simulator S such that for every vector of inputs $\vec{x} = \langle x_1, \ldots, x_n \rangle$ of total length ℓ and every $i \in [n]$, $S^{g_i^{\vec{x}}(\cdot)}(\ell, n, i)$ runs in time polynomial in $\ell + n$, and the output of $S^{g_i^{\vec{x}}(\cdot)}(\ell, n, i)$ equals $\tilde{f}_i(x_1, \ldots, x_i)$, except with negligible probability.*[1]

We stress that not all functions have efficient minimum-disclosure decompositions,[2] as is stated in the following theorem.

Theorem 1. *If one-way functions exist, then there are functions that do not have efficient minimum-disclosure decompositions.*

The theorem is proved in the full version of the paper [9]. Roughly, a decomposition is minimum-disclosure only when the residual functions g_i are efficiently learnable. Hence, a pseudorandom function f : $\mathsf{Seeds} \times \mathsf{Inputs} \to \mathsf{Outputs}$ (when viewed as a two-input function $f(s, x)$) does not have an efficient minimum-disclosure decomposition. In the full version we also include a discussion about functions with incomparable decompositions.

2.2 Some Functions with Minimum-Disclosure Decompositions

The sum function. Perhaps the simplest example is the sum function over a group: $f(x_1, \ldots, x_n) = \sum_{j=1}^{n} x_j$. In this case the decomposition into partial sums $f_i(y_{i-1}, x_i) = y_{i-1} + x_i$ is clearly minimum disclosure. Indeed, we have $\tilde{f}_i(x_1, \ldots, x_i) = \sum_{j=1}^{i} x_j$, and the simulator S can simply query $g_i^{\vec{x}}(0, \ldots, 0)$ and return the answer that it gets: $g_i^{\vec{x}}(0, \ldots, 0) = f(x_1, \ldots, x_i, 0, \ldots, 0) = \sum_{j=1}^{i} x_i$.

Selection functions. Other illustrating examples of functions with minimum-disclosure decompositions are selection functions. Consider first the selection function with index at the end, $f(x_1, \ldots, x_{n-1}, j) = x_j$. Here we can see that the trivial decomposition, where for $i < n$ we have f_i = identity and for $i = n$

[1] For randomized functionalities we require that $\{S^{g_i^{\vec{x}}(\cdot)}(\ell, n, i)\} \stackrel{c}{\equiv} \{\tilde{f}_i(x_1, \ldots, x_i)\}$.

[2] The residual truth table of a function is always minimum disclosure; however, it may be exponentially large.

we have $f_n = f$, is minimum disclosure. This is because given oracle access to $g_i^{\vec{x}}$ for any $i < n$, the simulator can just query it with varying inputs of the selection variable j, thus getting all the inputs $x_1, \ldots, x_i$.

On the other hand, consider the selection function with index at the beginning, $f(j, x_2, \ldots, x_n) = x_j$. Here a minimum disclosure decomposition would maintain a value and a state bit (wait/done), such that when the state is wait then the value is j, and when the state is done then the value is x_j. To see that this is indeed minimum disclosure, notice that given access to $g_i^{\vec{x}}$ the simulator can test if the selection index j is larger than i, e.g., by testing if $g_i^{\vec{x}}$ gives different values on $\langle 0, 0, \ldots, 0\rangle$ and $\langle 1, 1, \ldots, 1\rangle$. If $j > i$ then the simulator can find j by testing which is the input that $g_i^{\vec{x}}$ depends on, and if $j < i$ the simulator can output x_j (which is the output of $g_i^{\vec{x}}$ on every input).

Binary symmetric functions. An n-input binary symmetric function takes n bits as input, and the output depends only on the number of 1's in the input (i.e., the Hamming weight). Some examples include the AND, OR, PARITY, and MAJORITY functions. We note that the truth table of a binary symmetric function has an efficient representation: we just list for every $0 \le j \le n$ the output of f on inputs with Hamming-weight j. Thus, the truth table is of length $n+1$ rather than of length 2^n. We also note that for a binary symmetric function f and input $\vec{x}$, all the corresponding $g_i^{\vec{x}}$'s are also binary symmetric functions, and moreover the truth table of $g_{i+1}^{\vec{x}}$ can be computed from the value of x_i and the truth table of $g_i^{\vec{x}}$. Specifically, for $x_i = 0$ the truth table of $g_{i+1}^{\vec{x}}$ is obtained from that of $g_i^{\vec{x}}$ by removing the last row, and for $x_i = 1$ the truth table of $g_{i+1}^{\vec{x}}$ is obtained by removing the first row from that of $g_i^{\vec{x}}$.

For a binary symmetric function f, consider the decomposition that outputs at every step i the truth table of $g_i^{\vec{x}}$. The above observations imply that this decomposition is efficient, and it is minimum disclosure since it is easy to compute the truth table of a symmetric function given oracle access to that function.

Symmetric functions over other domains. The observations from above can be extended to symmetric functions over other domains. We assume without loss of generality that the domain is $\mathbb{Z}_c = \{0, 1, \ldots, c-1\}$ for some integer c. An n-input symmetric function over $\mathbb{Z}_c$ is one where permuting the inputs does not affect the output. In other words, the output depends only on how many of the inputs assume what value of the domain. This type of function is common for statistical measurement, including functions like SUM, AVERAGE, MEDIAN, MAJORITY, MAXIMUM and more.

The truth table for a symmetric function over $\mathbb{Z}_c$ can be expressed using a single row for all the inputs that have exactly j_0 inputs of value 0, j_1 inputs of value 1, and so on up to j_{c-2} inputs of value $c-2$ and $j_{c-1} = n - \sum_{i=0}^{c-2} j_i$ inputs of value $c-1$. That is, we have a row in the truth table for every c-vector of non-negative integers $\langle j_0, j_1, \ldots, j_{c-1}\rangle$ that sum up to n, so we have a total of $\binom{n+c-1}{n}$ rows. Hence the truth table is of polynomial-size $O(n^c)$ for any constant c. Moreover, in this case we again have the properties that all the $g_i^{\vec{x}}$'s are symmetric, and the truth table of $g_{i+1}^{\vec{x}}$ can be computed efficiently from the value of x_i and the truth table of $g_{i+1}^{\vec{x}}$ (see the full version for more details).

Also similarly to the binary case, when the truth table has polynomial size then it can be constructed efficiently given only oracle access to the function, hence the functions that output at every step i the truth table of $g_i^{\vec{x}}$ constitute a minimum-disclosure decomposition of the original symmetric function f.

3 Server-Based One-Pass Protocols

All our protocols are staged in the PKI model. Namely, where each party knows the public keys of all other parties, and each honest party knows the private key corresponding to its own public key.

A server-based one-pass protocol for n clients and a server is a sequence of n two-party protocols, $\bar{\pi} = \langle \pi_1, \ldots, \pi_n \rangle$, which are carried out sequentially with π_i being a two-party protocol between the server and the ith client P_i. The output of the protocol $\bar{\pi}$ is defined as the output of the server after the last protocol π_n. Below we denote the clients by $P_1, P_2, \ldots, P_n$ and the server by P_{n+1}. We denote the joint outputs of an adversary $\mathcal{A}$ and server P_{n+1} after a real execution of $\bar{\pi}$ with inputs $\vec{x} = (x_1, \ldots, x_n)$, vector of public/private key-pairs $\vec{kp}$, auxiliary input z to $\mathcal{A}$, corrupted parties $I \subseteq [n+1]$, and security parameter m, by $\text{REAL}_{\bar{\pi},\mathcal{A}(z),I}(\vec{x}, \vec{kp}, 1^m)$.

Securely computing a decomposition. We define security via the ideal/real paradigm in the stand-alone setting with static corruptions. In the ideal world, there is an additional trusted party that carries out the computation for the parties. In our setting, the trusted party receives the input of all clients and the identities of corrupted parties, and sends to the server the function output as well as any information that is inherently learned in our model (based on who is corrupted). Note that the ideal model is defined for a function *decomposition* $\bar{f}$. (It is not necessary to include f since $\bar{f}$ fully determines f.)

In the ideal world of the *semi-honest model*, the output that is given to the server is always the value of the function $f(x_1, \ldots, x_n)$ on the given inputs of all the clients. In addition, if the server is corrupted, then the trusted party sends it the value $\tilde{f}_i(x_1, \ldots, x_i) = f_i(\cdots, f_2(f_1(x_1), x_2)\cdots, x_i)$ where i is the index of the last honest party. We denote the outputs of a semi-honest ideal-world adversary $\mathcal{S}$ and server P_{n+1} after an ideal execution with inputs $\vec{x} = (x_1, \ldots, x_n)$, auxiliary input z to $\mathcal{S}$, corrupted parties $I \subseteq [n+1]$, and security parameter m, by $\text{IDEAL}^{\text{SH}}_{\bar{f},\mathcal{S}(z),I}(\vec{x}, z, 1^m)$.

The ideal-world of the malicious model is exactly the same, except that corrupted clients may send any arbitrary inputs to the trusted party, not necessarily the ones from their input. By convention, if a client sends input $\perp$, then the output of the function is defined to be $\perp$ (representing an aborted execution). The joint output here is denoted $\text{IDEAL}^{\text{MAL}}_{\bar{f},\mathcal{S}(z),I}(\vec{x}, z, 1^m)$.

Definition 3 (Securely Computing a Decomposition). *Let f be an n-input function and let $\bar{f} = \langle f_1, \ldots, f_n \rangle$ be a decomposition of f. A server-based one-pass protocol $\bar{\pi}$* securely computes the decomposition *$\bar{f}$ in the semi-honest (resp.*

malicious) model, if for every non-uniform probabilistic polynomial-time semi-honest (resp. malicious) adversary $\mathcal{A}$ in the real world, there exists a non-uniform probabilistic polynomial-time adversary $\mathcal{S}$ for the semi-honest (resp. malicious) ideal world, such that for all $\vec{x} \in (\{0,1\}^*)^n$ and $z \in \{0,1\}^*$

$$\left\{\text{IDEAL}_{\bar{f},\mathcal{S}(z),I}(\vec{x}, 1^m)\right\} \stackrel{c}{\equiv} \left\{\text{REAL}_{\bar{\pi},\mathcal{A}(z),I}(\vec{x}, \vec{kp}, 1^m)\right\}$$

where the key-pairs $\vec{kp}$ are chosen as described above.

We stress that if the server is honest, then in all cases nothing is learned by the adversary. When the function has a minimum-disclosure decomposition and a protocol that realizes that decomposition, then that protocol is called *optimally-private.*

Definition 4 (Optimally-Private). *Let f be an n-input function. We say that $\bar{\pi}$ is an* optimally-private server-based one-pass protocol for computing f *if there exists a minimum-disclosure decomposition $\bar{f}$ of f such that $\bar{\pi}$ securely computes $\bar{f}$ in the semi-honest (resp. malicious) model.*

4 Practical Optimal Protocols

In Section 5 we show that *any* decomposition can be securely computed given a public-key infrastructure, under the DDH assumption. As a corollary we obtain that any function that has a minimum-disclosure decomposition can be computed with optimal privacy. However, this construction is far from being practical; even for simple functions and semi-honest adversaries, it requires computing hundreds of exponentiations per gate. In this section, we present highly efficient protocols for specific examples from Section 2.2. These protocol are truly practical and could be implemented, for example, in a conference program committee review site in order to carry out secure voting. (With only a few tens of users, the solution that provides security in the presence of malicious adversaries would only require a few seconds of computation by each client and the server.) In the full version we describe protocols for the other functions from Section 2.2.

4.1 Protocols for Symmetric Functions

We begin by showing how to securely compute any binary symmetric function, based on the truth-table decomposition described in Section 2.2.

The Semi-Honest Case. Recall that symmetric functions have a concise truth table of size $n+1$, that the minimum-disclosure decomposition for functions of this class consists of the truth table of the $g_i^{\vec{x}}$'s, and that computing the next truth table is carried out by removing the first or last row of the current truth table. Intuitively, our protocol works by having the first client P_1 encrypt each entry of the truth table iteratively (in a layered, or onion like, structure) under all parties' public keys. Then, each party in turn removes the encryption under its public key, and removes the first row of the truth table if its input is 0, or the last row of the truth table if its input is 1. After the last party, the table contains just one row which is encrypted under the server's key.

This solution is not quite enough, however. For example, a collusion of P_1 and P_3 can learn P_2's exact input (irrespective of whether or not the server is corrupted). To see this, observe that P_1 generates all the ciphertexts. In addition, it can see all the ciphertexts received by P_3 after P_2 decrypts its layer of encryption. Hence, given P_3's view P_1 can determine if P_2 removed the first or the last row of the table.

We solve this problem by using rerandomizable public-key encryption. Loosely speaking, this means that given an encryption $c = E_{pk}(x)$ and the public key pk it is possible to generate an equivalent encryption $c' = E_{pk}(x)$ with *independent* randomness. We stress that the rerandomization must work on all layers of the (onion-type) encryption. The requirements here are therefore different from the standard notion. Let $E_{pk}(x;r)$ denote an encryption of x using randomness r, and let $E_{pk_1,\ldots,pk_{n+1}}(x;r_1,\ldots,r_{n+1}) = E_{pk_1}(\cdots E_{pk_{n+1}}(x;r_{n+1})\cdots;r_1)$ denote a layered encryption starting with the encryption of x under pk_{n+1} with randomness r_{n+1} and re-encrypting under each pk_i in turn, using randomness r_i. For shorthand, we write $\bar{E}_{\vec{pk}}(x;\vec{r})$ where $\vec{pk} = (pk_1,\ldots,pk_{n+1})$ and $\vec{r} = (r_1,\ldots,r_{n+1})$.[3] We define:

Definition 5. *A public-key scheme (G,E,D) is* layer rerandomizable *if there exists a procedure $\mathcal{R}$ such that for every $x \in \{0,1\}^*$ and every $\vec{r} \in (\{0,1\}^*)^n$,*

$$\left\{\vec{pk}\ ,\bar{E}_{\vec{pk}}(x;\vec{r}),\ \bar{E}_{\vec{pk}}(x;\vec{s})\right\} \equiv \left\{\vec{pk},\ \bar{E}_{\vec{pk}}(x;\vec{r}),\ \mathcal{R}(\vec{pk},\bar{E}_{\vec{pk}}(x;\vec{r}))\right\}$$

where $\vec{pk} = (pk_1,\ldots,pk_n)$ is such that all the pk_i's are in the range of G, and $\vec{s} \in_R (\{0,1\}^*)^n$ is a vector of uniformly distributed random strings.

We stress that the definition requires the rerandomization to work for *all* randomness $\vec{r}$ (even randomness that is "badly chosen"). However, it is assumed that all the public keys are "legitimate" in that they are in the range of G. Layer rerandomizability can be obtained from any additively homomorphic encryption scheme. Namely, define an initial layered encryption of x by

$$\bar{E}_{\vec{pk}}(x;\vec{r}) \stackrel{\text{def}}{=} \langle E_{pk_1}(x_1;r_1),\ldots,E_{pk_n}(x_n;r_n)\rangle$$

where $x_1,\ldots,x_n$ are chosen at random under the constraint that $\oplus_{j=1}^n x_j = x$. A *jth step layered encryption* of x is defined as

$$\bar{E}^j_{\vec{pk}}(x;\vec{r}) \stackrel{\text{def}}{=} \left\langle x_1,\ldots,x_j,E_{pk_{j+1}}(x_{j+1};r_{j+1}),\ldots,E_{pk_n}(x_n;r_n)\right\rangle$$

Rerandomization works by adding to the x_i's random δ_i's that sum up to zero, and then rerandomizing each ciphertext separately, under the appropriate key.

[3] Below we abuse these notations somewhat, denoting by $\bar{E}_{\vec{pk}}(x;\vec{r})$ a procedure that encrypts x under all the public keys but not necessarily in an onion fashion.

In addition, it is possible to decrypt in layers by having each party decrypt its ciphertext in turn and pass on the decrypted value along with the rest. Namely, the jth party transforms a $(j-1)$th level layered encryption to a jth level layered encryption.

A more efficient layer rerandomizable encryption scheme can be constructed from El Gamal. Let $\mathbb{G}$ be a group of prime order q with generator G. Then, for public-key $h = G^\alpha$ and $E_{pk}(x) = (G^r, h^r \cdot x)$, define $\mathcal{R}(pk, \langle u, v \rangle) = \langle u \cdot G^s, v \cdot h^s \rangle$, where $s \in_R \mathbb{Z}_q$. Observe that for $u = G^r, v = h^r \cdot x$ it follows that $\mathcal{R}(pk, u, v) = (G^{r+s}, h^{r+s} \cdot x)$, which is distributed identically to an encryption of x under an independent random string $r' = r + s \bmod q$.

In order to make this layer rerandomizable without increasing the size of the ciphertext, we define layered encryption as follows. Each party P_i has an El Gamal public-key $h_i = G^{\alpha_i}$ relative to the same group $(\mathbb{G}, q, g)$ as before. However, an encryption of x under the public keys $h_1, \ldots, h_n$ is defined to be $(G^r, (H_{1,n})^r \cdot x)$, where $H_{1,n} = \prod_{j=1}^n h_j = G^{\sum_{j=1}^n \alpha_j}$. In general, we define $H_{i,n} = \prod_{j=i}^n h_j = G^{\sum_{j=i}^n \alpha_j}$. It remains to show how P_i "decrypts" under its key h_i and rerandomizes the result. Given (u, v) where $u = G^r$ and $v = (H_{i,n})^r \cdot x$, party P_i decrypts by computing $u' = u$ and $v' = v \cdot u^{-\alpha_i}$. This works because taking $u = G^r$ and $v = x \cdot (H_{i,n})^r$ we have that

$$v \cdot u^{-\alpha_i} = x \cdot (H_{i,n})^r \cdot (G^r)^{-\alpha_i} = x \cdot \left(G^{\sum_{j=i}^n \alpha_j}\right)^r \cdot \left(G^{-\alpha_i}\right)^r = x \cdot \left(G^{\sum_{j=i+1}^n \alpha_j}\right)^r$$

and so (u', v') is a valid encryption of x with randomness r, under public key $H_{i+1,n}$. Rerandomization is then carried out as described above, using public-key $H_{i+1,n}$. That is, we compute $u'' = u' \cdot G^s$ and $v'' = v' \cdot (H_{i+1,n})^s$.

Returning to symmetric functions, in our protocol we will now use layer rerandomizable encryption to encrypt the lines of the truth table, and each party in turn will decrypt its own layer, remove either the first or last row from the table, rerandomize and then send back to the server.

Theorem 2. *Let f be a binary symmetric function. If the encryption scheme (G, E, D) is layer rerandomizable, and all honest parties' public keys are generated honestly using G, then the protocol above is an optimally-private server-based one-pass protocol for computing f, in the presence of semi-honest adversaries. Moreover, it is secure even if the semi-honest adversary can choose the randomness for the protocol in an arbitrary manner.*

Proof (sketch). We separately prove the case that P_{n+1} is corrupted and the case that it is not. If P_{n+1} is not corrupted, then it suffices to prove that it obtains correct output and that the adversary's view can be simulated without any help from the trusted party. Correctness is immediate from the construction. The view of the adversary can be simulated since everything is encrypted under the key of the honest server. Specifically, every time an honest party P_i is supposed to carry out its interaction with the server, construct a brand new truth table $\mathcal{C}_i$ which contains $n - i + 1$ encryptions of 0 under the public-keys $pk_{i+1}, \ldots, pk_{n+1}$, in turn. The fact that this is indistinguishable from a real execution follows directly from the hiding property of encryptions, and the rerandomizability property.

Next, we consider the case that the server P_{n+1} is corrupted, and $1 \le i \le n$ is the index of the last honest party. In this case, the simulator $\mathcal{S}$ is given the value $y_i = \tilde{f}_i(x_1, \ldots, x_n)$, which in this case is the appropriate partial truth table. The simulation is the same as before for every iteration up to and including $i-1$. In the ith iteration, $\mathcal{S}$ simulates the message sent by the honest P_i by encrypting under the public keys $pk_{i+1}, \ldots, pk_{n+1}$ the partial truth table that it received from the trusted party. As before, the output distribution of the adversary is indistinguishable from a real execution (note that the last simulated message is actually identical to in a real execution; the difference comes from prior ones which are all encryptions of 0 instead of the real partial truth table). □

In the protocol above, using the El-Gamal-based rerandomizable encryption, each party computes less than $3n$ exponentiations, so the total number of exponentiations if at most $3n^2$. Hence this protocol could be practical for a large (but not huge) number of parties, perhaps even for n in the thousands. We remark however that the parties must work sequentially, and this may be a limitation if n is too large. Also, in this concrete instantiation the parties can connect and interact with the server in any order, which is an important property for practical implementation and deployment.

The Malicious Case. Since the semi-honest protocol is secure for any random coins used by the dishonest parties, it is enough to add signatures so that corrupt parties and/or server cannot modify the messages sent by previous parties, and (non-interactive) zero-knowledge proofs of good behavior to obtain security in the malicious model. We describe the resulting protocol in the full version. For our concrete El Gamal implementation, all these proofs can be made efficient since they can all be reduced to compound statements about equality of discrete logarithms, and these can be made non-interactive using the Fiat-Shamir transformation in the random-oracle model. In particular, each party needs to compute $O(n^2)$ exponentiations, we estimate that running the protocol with $n = 100$ parties will take just a few minutes per party.

Symmetric Functions Over Larger Domains. In the full version we also show that the protocols for binary symmetric functions extend to symmetric functions over any domain $\mathbb{Z}_c$, where the complexity grows as $n^{O(c)}$. Hence we get efficient protocols for any constant c.

4.2 Selection Functions

In this section, we construct an optimally-private protocol for the selection function $f(j, x_2, \ldots, x_n) = x_j$; i.e., where the selector is first. As we have seen in Section 2.2, the disclosure in this case is the least. Specifically, if the last honest party is after the selected party, then the only thing learned by the server is the selected value and not even its position. Otherwise, the position is learned, but nothing else. (Note that hiding the position is really the only interesting issue in this function, since otherwise it can be trivially solved by having the selector first announce who is selected and next having the selected party send its value.)

The semi-honest case. Our protocol is similar to the following 1-out-of-N (semi-honest) oblivious transfer protocol, using additively homomorphic encryption: The receiver, who wants to get the jth value, generates N ciphertexts, all encrypting 1 except the jth that encrypts a 0. Using the additive-homomorphism, the sender multiplies the ciphertexts by random scalars (a different random number for each ciphertext) and then adds its value x_i to the ith ciphertext. When the receiver decrypts, it gets the jth value intact and all other values are random.

Our setting is a little more complicated than the OT setting, since (a) the inputs are split between parties $P_2, \ldots, P_n$ rather than all belonging to one sender, and (b) the receiver in our case is the server P_{n+1}, while the selection index j is known to the first party P_1. The latter concern is handled by choosing an encryption scheme with plaintext space much larger than the domain of inputs to the parties. Now with high probability the jth entry will be the only one in the domain of inputs, so the server can identify it. To handle the first concern we will use a mix-net-like construction (using a layer-rerandomizable encryption), with each party shuffling the ciphertexts so that the following parties cannot tell which ciphertext came from what party. (Also, we use El Gamal which is multiplicative rather than additive-homomorphic, and so we modify the underlying OT protocol accordingly.)

In more detail, P_1 with selector input j prepares a vector of El Gamal ciphertexts, all encrypting the group generator G except for the jth entry that encrypts the group element 1. The ith ciphertext in this vector is encrypted under the compound El Gamal public key $H_{i,n+1} = \prod_{t=i}^{n+1} h_t$. (When using a generic layer-rerandomizable encryption, the ith ciphertext is encrypted onion-style under the public keys of parties i though $n+1$.) We call this vector the "initial ciphertexts" and denote it by $\mathcal{I}$. During the protocol the initial ciphertexts will be passed unchanged, and the parties use them to process another vector of ciphertexts that contain the actual values. We call that other vector of ciphertexts the "work ciphertexts", and denote it by $\mathcal{W}$.

Each party P_i ($i \geq 2$) receives the initial ciphertexts $\mathcal{I}$ and a vector $\mathcal{W}_{i-1}$ of $i-2$ ciphertexts. The ciphertexts in $\mathcal{W}_{i-1}$ are all encrypted under $H_{i,n+1}$. P_i takes the ith ciphertext from $\mathcal{I}$ (which is also encrypted under $H_{i,n+1}$), uses the multiplicative homomorphism of El Gamal to raise the plaintext inside it to a random power in $\mathbb{Z}_q$, then uses the homomorphism again to multiply the plaintext by its input x_i. It inserts the resulting ciphertext to $\mathcal{W}_{i-1}$, thus getting a vector of $i-1$ ciphertexts which we denote by $\mathcal{W}'_i$. P_i then peels off its layer of encryption (resulting in ciphertexts under $H_{i+1,n+1}$), randomly permutes the ciphertexts and re-randomizes them, thus obtaining a new vector of ciphertexts $\mathcal{W}_i$, which P_i sends back to the server.

After all the parties have participated, the server has a vector of "work ciphertexts" $\mathcal{W}_n$, encrypted under the public key of the server $H_{n+1} = h_{n+1}$. The server decrypts this vector, and if the corresponding plaintext vector has a single element from the input domain of the protocol then the server outputs that element. A pseudocode description of this protocol (described using a generic additively homomorphic encryption layer-rerandomizable) can be found in Protocol 3.

PROTOCOL 3 (Semi-Honest Optimal Protocol for the Selection Function)

- **Inputs:** Party P_1 has an index j ($2 \leq j \leq n$), and each party P_i ($2 \leq i \leq n$) has a private input x_i, its own private key sk_i, and a vector of public keys $(pk_2, \ldots, pk_n, pk_{n+1})$.
- **The protocol:**
  1. *First party instructions:*
     (a) For every $i = 2, \ldots, n$, $i \neq j$, P_1 computes $c_i = \bar{E}_{pk_i, \ldots, pk_{n+1}}(1)$. For $i = j$, P_1 computes $c_j = \bar{E}_{pk_j, \ldots, pk_{n+1}}(0)$.
     (b) P_1 sends the vector of initial ciphertexts $\mathcal{I} = (c_2, \ldots, c_n)$ to the server P_{n+1}.
  2. *Interaction of clients $P_2, \ldots, P_n$ with server.* For $i = 2, \ldots, n$:
     (a) P_{n+1} sends P_i the initial ciphertexts $\mathcal{I}$, and a vector $\mathcal{W}_{i-1}$ of $i-2$ ciphertexts, encrypted under $\vec{\mathsf{pk}}_i = (pk_i, \ldots, pk_{n+1})$. (For $i = 2$, $\mathcal{W}_1$ is empty.)
     (b) P_i extracts the ith ciphertext from $\mathcal{I}$, $c_i = \mathcal{I}[i]$ (encrypting a bit $b_i \in \{0, 1\}$ under $\vec{\mathsf{pk}}_i$.) It chooses a random number r_i from the plaintext space and uses the additive-homomorphic property of the encryption to compute an encryption of $r_i \cdot b_i + x_i$, using $c_i = E_{\vec{\mathsf{pk}}_i}(b_i)$, r_i and x_i.
     (c) P_i adds c_i' to the vectors $\mathcal{W}_{i-1}$ (thus receiving a vector of $i-1$ ciphertexts under $(pk_i, \ldots, pk_{n+1})$) and decrypts a layer of all of these ciphertexts using its secret key sk_i; denote the result by $\mathcal{W}_i'$.
     (d) P_i permutes the ciphertexts in $\mathcal{W}_i'$ and rerandomizes all of them using the public keys $pk_{i+1}, \ldots, pk_{n+1}$. Denoting the result by $\mathcal{W}_i$, P_i sends $\mathcal{W}_i$ back to the server.
  3. *Concluding the computation:* Upon receiving the encrypted vector $\mathcal{W}_n$ (of length $n-1$) from P_n, the server P_{n+1} decrypts all the ciphertext using its secret key sk_{n+1}. If the corresponding plaintext vector includes a single element from the input space then the server outputs that plaintext (otherwise it outputs '?').

Using similar arguments as in the binary symmetric case, we have that Protocol 3 is optimally-private in the presence of semi-honest adversaries, if the encryption schemes used is additively homomorphic and layer rerandomizable, and has plaintext space which is super-polynomially larger than the input space for the protocol.

The malicious case. As above, in this case we need to have the parties sign on their messages and prove that they behaved honestly. This can be achieved using similar techniques as those described above.

5 Securely Computing any Decomposition

In this section, we present a basic feasibility result regarding the possibility of securely computing an arbitrary given decomposition in our model. For this we use *re-randomizable garbled circuits* that were introduced by Gentry et al. for the purpose of multi-Hop homomorphic encryption [6]. (Below we call this the GHV construction.) Roughly, each party P_i receives from the server a garbled circuit encoding $\tilde{f}_{i-1}(x_1, \ldots, x_{i-1})$, adds its input to generate a garbled circuit for $\tilde{f}_i(x_1, \ldots, x_i)$, then re-randomizes this garbled circuit (so as to hide x_i from colluding parties $i-1$ and $i+1$) and sends the result back to the server.

The main problem that arises is that in our setting we do not want the server to be able to evaluate all the $\tilde{f}_i$'s. More specifically, if i is the index of the last honest party then we do not want the adversary to be able to evaluate $\tilde{f}_j$ for any $j < i$. (In contrast, in the setting of multi-Hop homomorphic encryption if party P_{i+1} is dishonest then the adversary can evaluate $\tilde{f}_i$.)

To solve this we again use layered re-randomizable encryption: instead of giving the parties the input labels for the garbled circuit, we give them only the encryption of these input labels, encrypted under all the keys of the parties that did not participate yet. Each party peels of its layer of encryption and re-randomizes the result, hence the server learns the input label only after all the (honest) parties decrypted their layers, and it cannot evaluate the circuit earlier.

We note that the layered re-randomizable encryption is intertwined with the garbled-circuit construction, since each party has to be able to transform the encryption of the inputs of one garbled circuit into "freshly random" encryption of the inputs to a re-randomized version of the garbled circuit. Recall that in the GHV construction the labels on the wires are balanced bit-strings (with half 0s and half 1s), and re-randomizing a circuit is done by bitwise permuting the labels. Hence we use bit-wise encryption (to handle the permutation) where ciphertexts can be re-randomized (to hide the correlation to the previous circuit).

We mention that the original construction from [6] is secure only in the semi-honest model. In particular a malicious party can choose "bad labels" to wires to foil re-randomization, by choosing the two labels on a wire with a very small (or very large) Hamming distance. We thus modify the construction slightly and require that the Hamming distance between the two labels be exactly half their length. This turns the GHV construction into one that works for any adversarial coins in the semi-honest model, so we can add (non-interactive) zero-knowledge proofs and get resilience against malicious adversaries.

5.1 Our Construction, Semi-Honest Model

As described above, we obtain security in our model by augmenting the GHV construction with encryption of the input labels. Differently from Gentry et al., we do not use oblivious transfer to encode the input of the first party but instead have that party encrypt the labels corresponding to its input bits with El Gamal. (We note that the same simplification could also be used in the context of multi-Hop homomorphic encryption.)

In more detail, our construction works in the PKI model, where each party P_i has a secret key sk_i and a corresponding public key $\mathsf{pk}_i = \mathsf{pk}(\mathsf{sk}_i)$, and every party knows the public keys of all other parties. In the description below we assume that these are all keys for El Gamal encryption, namely we have $\mathsf{sk}_i = \alpha_i \in \mathbb{Z}_q$ and $\mathsf{pk}_i - G^{\alpha_i}$ where G is a generator in an order-q group $\mathbb{G}$ in which DDH is hard.

The protocol. Let $\langle f_1, \ldots, f_n \rangle$ be a given decomposition that we want to implement. Namely, we want a protocol where the view of any set of cooperating semi-honest parties can be simulated knowing only the value $y_i = \tilde{f}_i(x_1, \ldots, x_i) = f_i(\ldots f_1(x_1), \ldots, x_i)$, where i is the index of the last honest party (i.e., the last party not in the set of corrupted parties).

Throughout the computation, we maintain the invariant that before interacting with party P_i the server has a garbled circuit of the function $\tilde{f}_{i-1}(x_1, \ldots, x_{i-1})$ and an encryption of all the labels corresponding to the inputs bits in $x_1, \ldots, x_{i-1}$, where the encryption is with respect to the public keys of the remaining parties $\mathsf{pk}_i, \ldots, \mathsf{pk}_n, \mathsf{pk}_{n+1}$ (pk_{n+1} is the key of the server.)

In more detail, let Λ_{i-1} be a garbled circuit that the server has before talking to party P_i (where Λ_0 is the empty garbled circuit with no inputs). To slightly simplify notations we assume that all the inputs x_i are exactly t-bit long, and let x_{ij} denote the jth bit of x_i, i.e. $x_i = x_{i1}x_{i2}\ldots x_{it}$. Hence Λ_{i-1} has $(i-1)t$ input wires, and there are two ℓ-bit labels associated with each input wire. We denote the 0 and 1 labels associated with the jth input wire of the ith party by $L0_{ij}, L1_{ij}$, respectively.

Below we also denote by σ_{ij}^k the kth bit of the label corresponding to the input bit x_{ij}. That is, if $x_{ij} = 0$ then $(\sigma_{ij}^1 \ldots \sigma_{ij}^\ell) = L0_{ij}$ and if $x_{ij} = 1$ then $(\sigma_{ij}^1 \ldots \sigma_{ij}^\ell) = L1_{ij}$. Hence before talking to party P_i the server should have encryptions of all the bits $\sigma_{i'j}^k$ for $i' < i$, $j = 1, \ldots, t$ and $k = 1, \ldots, \ell$. Specifically, let H_i be the compounded public key of parties i through $n+1$, namely $H_i \stackrel{\text{def}}{=} \prod_{j=i}^{n+1} h_i$. Then for each bit $\sigma_{i'j}^k$ with $i' < i$, $j \le t$ and $k \le \ell$, the server has an El Gamal encryption of $\sigma_{i'j}^k$ relative to public key H_i, i.e., a pair of the form $(G^r, G^{\sigma_{ij}^k} \cdot {H_i}^r)$. (Of course, the exponents r in all these ciphertexts are chosen independently.)

The *i*th party. The ith party has its input $x_i = (x_{i1} \ldots x_{it})$, its secret key α_i and the public keys of the parties after it, $h_{i+1}, \ldots, h_n, h_{n+1}$. It receives from the server the garbled circuit Λ_{i-1} corresponding to $\tilde{f}_{i-1}$, and the encryption of all the bits σ_{ij}^k relative to the compounded public key H_i. Recall that $\tilde{f}_i$ is an extension of $\tilde{f}_{i-1}$ via $f_i(y_{i-1}, x_i)$, namely

$$\tilde{f}_i(x_1, \ldots, x_{i-1}, x_i) \;=\; f_i(\; \underbrace{\tilde{f}_{i-1}(x_1, \ldots, x_{i-1})}_{y_{i-1}} \;, x_i).$$

Hence party P_i can extend the garbled circuit Λ_{i-1} corresponding to $\tilde{f}_{i-1}$ into a garbled circuit Λ_i corresponding to $\tilde{f}_i$, using the output labels of Λ_{i-1} as input labels for the wires of y_{i-1} and choosing new input labels for the wires of x_i.

That is, party P_i builds the Yao circuit for $\tilde{f}_i$, choosing random labels for all wires except for the input wires corresponding to the output of $\tilde{f}_{i-1}$; the garbled labels on the input wires are taken as the output labels for the wires of the received circuit. Thus, the two circuits are composed into one.

Next, party P_i uses its secret key α_i to convert all the El Gamal ciphertexts relative to H_i into encryption of the same bits relative to H_{i+1}. Namely, given a ciphertext $(u = G^r, v = G^\sigma \cdot {H_i}^r)$, Party P_i computes $v' = v/u^{\alpha_i}$ and outputs the ciphertexts (u, v'). This is indeed an encryption of the bit σ with respect to H_{i+1}, since $H_{i+1} = H_i/h_i = H_i/G^{\alpha_i}$ and therefore

$$v' = \frac{v}{u^{\alpha_i}} = \frac{G^\sigma \cdot {H_i}^r}{G^{r\alpha_i}} = G^\sigma \cdot \left(\frac{H_i}{G^{\alpha_i}}\right)^r = G^\sigma \cdot {H_{i+1}}^r.$$

Party P_i also encrypts the bits σ_{ij}^k of the labels corresponding to all of its input bits x_{ij}, relative to the compounded public key H_{i+1}.

At this point Party P_i holds the complete state as needed for the next step of the computation, and it only remains to re-randomize this state (so as to hide x_i). To this end, Party P_i applies the re-randomization procedure to the garbled circuit Λ_i to get a new garbled circuit Λ'_i. This includes in particular choosing a random permutation π_{ij} for the wire of every input bit x_{ij}. Party P_i permutes accordingly the vector of El Gamal ciphertexts for the bits on that wire $(\sigma_{ij}^1 \dots \sigma_{ij}^\ell)$, thus obtaining an encryption of the new input label for this wire. (All these encryptions are relative to the compound public key H_{i+1}.) Finally it re-randomizes these encryptions by choosing for each ciphertext a new exponent r' and replacing the pair $\langle u = G^r,\ v = G^\sigma \cdot H_{i+1}^r \rangle$ with $u' = u \cdot G^{r'} = G^{r+r'}$ and $v' = v \cdot H_{i+1}^{r'} = G^\sigma \cdot H_{i+1}^{r+r'}$. Party P_i sends Λ'_i and all the ciphertexts (in order) to the server, and the server is now ready for party P_{i+1}.

The server. After the interaction with the last party n, the server has a garbled circuit for the function $\tilde{f}_n = f$, and encryption of the input labels corresponding to all the input bits of all the parties, relative to the public key $H_{n+1} = h_{n+1}$. Since the server knows the secret key α_{n+1} corresponding to h_{n+1}, it can decrypt all these ciphertexts and recover the label on each input wire. The server then evaluates the garbled circuit and obtains the result $f(x_1, \dots, x_n)$, as needed. The proof of the following theorem can be found in the full version [9].

Theorem 3. *For any decomposition $\bar{f} = \langle f_1, \dots, f_n \rangle$, the protocol from Section 5.1 is a server-based one-pass protocol that securely computes $\bar{f}$ in the semi-honest model, even if the dishonest parties can choose arbitrary random coins for the protocol.*

5.2 The Malicious Model

As we saw in Theorem 3, the security of the semi-honest protocol holds even if dishonest parties are allowed to choose their coins arbitrarily (rather than at random). Thus, to achieve security in the presence of malicious adversaries, we have each party prove that it followed the instructions of the protocol relative to *some* input and set of random coins. This proof must be non-interactive and verified by all subsequent parties. This requires a common reference string (or perhaps re-use of the available PKI). In order for us to extract the inputs used in the ideal-model simulation, the proof also has to be a proof of knowledge. One option for this is to use a universally composable non-interactive system of zero-knowledge proofs of knowledge, using enhanced trapdoor permutations [10].

In addition, we need to ensure that if the server is corrupted, then it does not modify any of the constructions carried out by the honest parties. This can be achieved using digital signatures (and having the signing key be part of the public-key infrastructure).

Theorem 4. *Assume the existence of enhanced trapdoor permutations and that DDH holds. Then, for any decomposition $\bar{f} = \langle f_1, \ldots, f_n \rangle$, there exists a server-based one-pass protocol that securely computes $\bar{f}$ in the malicious model, with a public-key infrastructure and in the common reference string model.*

6 Extensions and Open Problems

In this work we considered a very simple setting of a server and n clients that all know about each other (and in particular have each other's public keys), and where the order in which the clients connect to the server is pre-set. Our practical solutions for symmetric functions extend also to the "first come first serve" setting with no pre-set order, but still require all parties to know about each other. In addition, all our solutions are sequential, they all rely heavily on the fact that client i completes the interaction with the server before client $i+1$ begins. Allowing concurrency between clients is a very interesting open problem and may be crucial for a large number of clients.

Another possible extension deals with functions that have natural "projections" on any subset of their variables. (For example, for the AVERAGE function, it is natural to talk about the average of any subset of the variables.) In this case, it may be desirable that the server be able to compute the function value as soon as at least t of the n clients connected to it.[4] Although it may be possible to replace the onion-like encryption in our protocols with encryption in a t-out-of-n manner, it seems nontrivial to do it in such a way that will still not allow a subset of t parties to decrypt the entire transcript of the protocol.

Another very interesting extension is when we have a large universe of clients that do not have each other's public keys, and we want to compute some function as soon as n of them connect to the server (e.g., polling). In this case it may be reasonable to assume that the clients all share some system parameters, and maybe even that each client has some secret key for the system, so perhaps tools from identity-based cryptography can be used here.

[4] In general, if we have a decomposition of f then we can think of $\tilde{f}_t(x_1, \ldots, x_t)$ as the projection of f on the first t variables. Computing $\tilde{f}_i$ may or may not be desirable, depending on the application.

Finally, we point out that if we can have each of the parties connect twice to the server (rather than once), then our protocols can be used for achieving the standard notion of privacy for secure computation. Indeed, instead of computing the original n-input function $f(x_1, \ldots, x_n)$, we set up a protocol for computing the extended $2n$-input function that depends only the first n inputs $\hat{f}(x_1, \ldots, x_n, x_{n+1}, \ldots, x_{2n}) = f(x_1, \ldots, x_n)$. We consider a decomposition of $\hat{f}$ where the intermediate value after the nth input is $f(x_1, \ldots, x_n)$, design a protocol to realize it, and let party P_i play the role of both parties i and $i + n$ in this protocol. With this protocol, as soon as one of the parties is honest we have that the intermediate result after "the last honest party" in the protocol is $f(x_1, \ldots, x_n)$. Hence the view of the corrupted parties can be simulated knowing only this value.

Acknowledgments. We thank the CRYPTO 2011 reviewers for their many helpful comments.

References

1. A Face Is Exposed for AOL Searcher No. 4417749 (The New York Times) (August 9, 2006), http://www.nytimes.com/2006/08/09/technology/09aol.html
2. Goldreich, O.: Foundations of Cryptography, Basic Applications, vol. 2. Cambridge University Press, Cambridge (2004)
3. Rivest, R., Adleman, L., Dertouzos, M.: On data banks and privacy homomorphisms. In: Foundations of Secure Computation, pp. 169–177. Academic Press, London (1978)
4. Gentry, C.: Fully homomorphic encryption using ideal lattices. In: 41st ACM Symposium on Theory of Computing – STOC 2009, pp. 169–178. ACM, New York (2009)
5. Fiat, A., Shamir, A.: How to Prove Yourself: Practical Solutions to Identification and Signature Problems. In: Odlyzko, A.M. (ed.) CRYPTO 1986. LNCS, vol. 263, pp. 186–194. Springer, Heidelberg (1987)
6. Gentry, C., Halevi, S., Vaikuntanathan, V.: i-Hop Homomorphic Encryption and Rerandomizable Yao Circuits. In: Rabin, T. (ed.) CRYPTO 2010. LNCS, vol. 6223, pp. 155–172. Springer, Heidelberg (2010), Full version available online at http://eprint.iacr.org/2010/145
7. Harnik, D., Ishai, Y., Kushilevitz, E.: How Many Oblivious Transfers Are Needed for Secure Multiparty Computation? In: Menezes, A. (ed.) CRYPTO 2007. LNCS, vol. 4622, pp. 284–302. Springer, Heidelberg (2007)
8. Choi, S.G., Elbaz, A., Malkin, T., Yung, M.: Secure Multi-party Computation Minimizing Online Rounds. In: Matsui, M. (ed.) ASIACRYPT 2009. LNCS, vol. 5912, pp. 268–286. Springer, Heidelberg (2009)
9. Halevi, S., Lindell, Y., Pinkas, B.: Secure Computation on the Web: Computing without Simultaneous Interaction, http://eprint.iacr.org/2011/157
10. De Santis, A., Di Crescenzo, G., Ostrovsky, R., Persiano, G., Sahai, A.: Robust Non-interactive Zero Knowledge. In: Kilian, J. (ed.) CRYPTO 2001. LNCS, vol. 2139, pp. 566–598. Springer, Heidelberg (2001)

Memory Delegation*

Kai-Min Chung[1,**], Yael Tauman Kalai[2], Feng-Hao Liu[3], and Ran Raz[4]

[1] Department of Computer Science, Cornell University, Ithaca, NY, USA
chung@cs.cornell.edu

[2] Microsoft Research New England, Cambridge MA, USA
yael@microsoft.com

[3] Department of Computer Science, Brown University, Providence RI, USA
fenghao@cs.brown.edu

[4] Department of Mathematics and Computer Science, Weizmann Institute of Science, Rehovot, Israel
ran.raz@weizmann.ac.il

Abstract. We consider the problem of delegating computation, where the delegator doesn't even know the input to the function being delegated, and runs in time significantly smaller than the input length.

For example, consider the setting of *memory delegation*, where a delegator wishes to delegate her entire memory to the cloud. The delegator may want the cloud to compute functions on this memory, and prove that the functions were computed correctly. As another example, consider the setting of *streaming delegation*, where a stream of data goes by, and a delegator, who cannot store this data, delegates this task to the cloud. Later the delegator may ask the cloud to compute statistics on this streaming data, and prove the correctness of the computation. We note that in both settings the delegator must keep a (short) certificate of the data being delegated, in order to later verify the correctness of the computations. Moreover, in the streaming setting, this certificate should be computed in a streaming manner.

We construct both memory and streaming delegation schemes. We present non-interactive constructions based on the (standard) delegation scheme of Goldwasswer *et. al.* [GKR08]. These schemes allow the delegation of any function computable by an $\mathcal{L}$-uniform circuit of low depth (the complexity of the delegator depends linearly on the depth). For memory delegation, we rely on the existence of a polylog PIR scheme, and for streaming, we rely on the existence of a fully homomorphic encryption scheme.

We also present constructions based on the CS-proofs of Micali. These schemes allow the delegation of any function in **P**. However, they are interactive (i.e., consists of 4 messages), or are non-interactive in the Random Oracle Model.

1 Introduction

The problem of delegating computation considers a scenario where one party, the *delegator*, wishes to delegate the computation of a function f to another

* A full version of this paper can be found on [CKLR11].

** Supported by US-Israel BSF grant 2006060 and NSF grant CNS-0831289.

P. Rogaway (Ed.): CRYPTO 2011, LNCS 6841, pp. 151–168, 2011.

party, the *worker*. The challenge is that the delegator may not trust the worker, and thus it is desirable to have the worker "prove" that the computation was done correctly. Obviously, verifying this proof should be easier than doing the computation.

This concept of "outsourcing" computation received a lot of attention in recent years, partly due to the increasing interest in cloud computing, where the goal is to outsource all the computational resources to a (possibly untrusted) "cloud". There are several reasons why the client (or delegator) may not trust the cloud, and thus would like to receive proofs for the correctness of the computation. For example, the cloud may have an incentive to return incorrect answers. Such an incentive may be a financial one, if the real computation requires a lot of work, whereas computing incorrect answers requires less work and is unlikely to be detected by the client. Moreover, in some cases, the applications outsourced to the cloud may be so critical that the delegator wishes to rule out accidental errors during the computation.

In order to ensure that the worker (or the cloud) performed the computation correctly, we would like the worker to *prove* this to the delegator. Of course, it is essential that the time it takes to verify the proof is significantly smaller than the time needed to actually run the computation. At the same time, the running time of the worker carrying out the proof should also be reasonable — comparable to the time it takes to do the computation.

The problem of delegating computation has been studied excessively (see Section 1.2 for an overview on previous work). However, most previous work on delegation allow the delegator to run in time polynomial in the input size, as long as this runtime is significantly smaller than the time it takes to do the computation. For example, when delegating the computation of a function f that runs in time T and has inputs of size n, typically the desired runtime of the delegator is $\mathrm{poly}(n, \log T)$ and the desired runtime of the worker is $\mathrm{poly}(T)$.

In this work, we want the delegator to run in time that is even smaller than the input size n. Namely, we don't allow the delegator even to read the input! At first, this requirement may seem unreasonable and unachievable. So, let us start by motivating this requirement with two examples.

Memory delegation. Suppose that Alice would like to store all her memory in the cloud. The size of her memory may be huge (for example, may include all the emails she ever received). Moreover, suppose she doesn't trust the cloud. Then, every time she asks the cloud to carry out some computation (for example, compute how many emails she has received from Bob during the last year), she would like the answer to be accompanied by a proof that indeed the computation was done correctly. Note that the input to these delegated functions may be her entire memory, which can be huge. Therefore, it is highly undesirable that Alice runs in time that is proportional to this input size. More importantly, Alice doesn't even hold on to this memory anymore, since she delegated it to the cloud.

Thus, in a memory delegation scheme, a delegator delegates her entire memory to the cloud, and then may ask the could to compute functions of this memory, and expects the answers to be accompanied by a proof. Note that in order to verify the correctness of these proofs, the delegator must save some short certificate of her memory, say a certificate of size polylog(n), where n is the memory size. The proofs should be verifiable very efficiently; say, in time polylog(n, T), where T is the time it takes to compute the function. Moreover, Alice should be able to update her memory efficiently.

Streaming delegation. Suppose that there is some large amount of data that is streaming by, and suppose that a user, Alice, wishes to save this data, so that later on she will be able to compute statistics on this data. However, Alice's memory is bounded and she cannot store this data. Instead, she wishes to delegate this to the cloud. Namely, she asks the cloud to store this streaming data for her, and then she asks the cloud to perform computation on this data. As in the case of memory delegation, in order to later verify the correctness of these computations, Alice must save some short certificate of this streaming data. As opposed to the setting of memory delegation, here the certificate should be computed (and updated) in a streaming manner.

The settings of memory delegation and streaming delegation are quite similar. In both settings Alice asks the cloud to store a huge object (either her memory or the streaming data). There are two main differences between the two: (1) In the setting of streaming delegation, the certificates and updates must be computed in a streaming manner. Thus, in this sense, constructing streaming delegation schemes may be harder than constructing memory delegation schemes. Indeed, our streaming delegation scheme is more complicated than our memory delegation scheme, and proving soundness in the streaming setting is significantly harder than proving soundness in the memory setting. (2) In the setting of streaming delegation, the memory is updated by simply adding elements to it. This is in contrast to the setting of memory delegation, where the memory can be updated in arbitrary ways, depending on the user's needs. However, in the memory setting, we allow the delegator to use the help of the worker when updating her certificate (or secret state), whereas in the streaming setting we require that the delegator updates her certificate on her own. The reason for this discrepancy, is that in the memory setting the delegator may not be able to update her certificate on her own, since she may want to update her memory in involved ways (such as, erase all emails from Bob). On the other hand, in the streaming setting, it seems essential that the delegator updates her certificate on her own, since in this setting the data may be streaming by very quickly, and there may not be enough time for the delegator and worker to interact during each update.

1.1 Our Results

We construct both memory delegation and streaming delegation schemes. The memory delegation scheme consists of an offline phase, where the delegator D delegates her memory $x \in \{0,1\}^n$ to a worker W. This phase is non-interactive, where the delegator sends a single message, which includes her memory content x to the worker W. The runtime of both the delegator and the worker in the offline phase is poly(n), where n is the memory size. At the end of this phase, the delegator saves a short certificate σ of her memory, which she will later use when verifying delegation proofs.

The streaming delegation scheme, on the other hand, doesn't have such an offline phase. In the streaming setting, we consider the scenario where at each time unit t a bit x_t is being streamed. The delegator starts with some secret state (or certificate) σ_0, and at time unit $t+1$ she uses her secret state σ_t and the current bit x_{t+1} being streamed, to efficiently update her secret state from σ_t to σ_{t+1}.

In both settings, each time the delegator D wants the worker W to compute a function $f(x)$, they run a delegation protocol, which we denote by Compute(f). The memory delegation scheme also has an Update protocol, where the delegator D asks the worker W to update her memory and to help her update her secret state σ. The latter can be thought of as a delegation request, and the guarantees (in term of runtime and communication complexity) are similar to the guarantees of the Compute protocol.

In the streaming setting, the delegator updates her secret state on her own in time polylog(N), where N is an upper bound on the length of the stream. Namely, the update function, that takes as input a certificate σ_t and a bit x_{t+1}, and outputs a new certificate σ_{t+1}, can be computed in time polylog(N).

We present two memory and streaming delegation protocols. The first are non-interactive (i.e, Compute(f) consists of two messages, the first sent by the delegator and the second sent by the worker). They are based on the non-interactive version of the delegation protocol of Goldwasser *et. al.* [GKR08, KR09], denoted by GKR (though are significantly more complicated than merely running GKR). As in GKR, the efficiency of the delegator depends linearly on the depth of the circuit being delegated. Our second memory and streaming delegation protocols are interactive (i.e., Compute(f) consists of four messages). These schemes are based on CS-proofs of Micali [Mic94], and allow for efficient delegation of all functions in **P**.

In what follows, we state our theorems formally. However, due to the lack of space, we refer the reader to the full version of this paper [CKLR11] for the formal definition of a memory delegation scheme and a streaming delegation scheme.

Theorem 1 (Memory Delegation). *Assume the existence of a poly-log PIR scheme, and assume the existence of a collision resistant hash family. Let $\mathcal{F}$ be the class of all $\mathcal{L}$-uniform poly-size boolean circuits. Then there exists a non-interactive (2-message) memory delegation scheme* mDel, *for delegating any function $f \in \mathcal{F}$. The delegation scheme,* mDel *has the following properties, for security parameter k.*

- *The scheme has perfect completeness and negligible (reusable) soundness error.*
- *The delegator and worker are efficient in the offline stage; i.e., both the delegator and the worker run in time* poly(k, n).

– *The worker is efficient in the online phase. More specifically, it runs in time* poly(k, S) *during each* Compute(f) *and* Update(f) *operation, where* S *is the size of the* $\mathcal{L}$*-uniform circuit computing* f*. The delegator runs in time* poly(k, d) *during each* Compute(f) *and* Update(f) *operation, where* d *is the depth of the* $\mathcal{L}$*-uniform circuit computing* f.[1]

In particular, assuming the existence of a poly-logarithmic PIR scheme, and assuming the existence of a collision resistent hash family, we obtain a memory delegation scheme for $\mathcal{L}$-uniform **NC** computations, where the delegator D runs in time *poly-logarithmic* in the length of the memory.

Theorem 2 (Streaming Delegation). *Let* k *be a security parameter, and let* N *be a parameter (an upper bound on the length of the stream). Let* $\mathcal{F}$ *be the class of all* $\mathcal{L}$*-uniform poly-size boolean circuits. Assume the existence of a fully-homomorphic encryption scheme secure against* poly(N)*-size adversaries. Then there exists a streaming delegation scheme* $\mathsf{sDel}_{\mathcal{F}}$ *for* $\mathcal{F}$ *with the following properties.*

– $\mathsf{sDel}_{\mathcal{F}}$ *has perfect completeness and negligible reusable soundness error.*
– D *updates her secret state in time* polylog(N)*, per data item.*
– *In the delegation protocol, when delegating a function* $f \in \mathcal{F}$ *computable by an* $\mathcal{L}$*-uniform circuit of size* S *and depth* d*, the delegator* D *runs in time* poly$(k, d, \log N)$*, and the worker* W *runs in time* poly$(k, \log N, S)$.

In particular, assuming the existence of a fully-homomorphic encryption scheme secure against adversaries of size poly(N), we obtain a streaming delegation scheme for $\mathcal{L}$-uniform **NC** computations, where the delegator D runs in time *poly-logarithmic* in the length of data stream.

Remark. We note that the property we needed from the GKR protocol is that the verifier does not need to read the entire input in order to verify, but rather only needs to access a single random point in the low-degree extension of the input. (We refer the reader to Section 2.1 for the definition and properties of a low-degree extension.) We note that the CS-proof delegation scheme of Micali [Mic94], for delegating the computation of (uniform) Turing machines, also has the property that verification can be done by only accessing a few random points in the low-degree extension of the input, assuming the underlying PCP is a PCP of Proximity [BSGH+05].

Indeed using this delegation scheme, we get a memory delegation scheme and a streaming delegation scheme for all of **P**. Using this scheme, the Compute(f) protocol is interactive (i.e., it is a 4-message protocol). The runtime of the delegator is polylog(T) and the runtime of the worker is poly(T), where T is the runtime of the Turing machine computing the function f.[2] Furthermore, the memory delegation scheme relies only on the existence of a collision resistant hash family, without the need of a poly-log PIR scheme.

[1] Thus, for every constant $c \in \mathbb{N}$, if we restrict the depth of f to be at most k^c, then the delegator is considered efficient.

[2] We assume that $T \geq n$.

Theorem 3 (Interactive Memory Delegation). *Assume the existence of a collision resistant hash family. Then there exists a memory delegation scheme* mDel*, for delegating any function computable by a polynomial-time Turning machine. The delegation scheme,* mDel *has the following properties, for security parameter k.*

- *The scheme has perfect completeness and negligible (reusable) soundness error.*
- *The delegator and worker are efficient in the offline stage; i.e., both the delegator and the worker run in time* $\text{poly}(k, n)$.
- *The worker is efficient in the online phase. More specifically, it runs in time* $\text{poly}(k, T)$ *during each* Compute(f) *and* Update(f) *operation, where T is an upper-bound on the running time of f. The delegator runs in time* $\text{poly}(k, \log T)$ *during each* Compute(f) *and* Update(f) *operation.*
- *Both* Compute(f) *and* Update(f) *operations consist of* 4 *message exchanges.*

Theorem 4 (Interactive Streaming Delegation). *Let k be a security parameter, and let N be a parameter (an upper bound on the length of the stream). Let $\mathcal{F}$ be the class of all functions computable by a polynomial-time Turning machine. Assume the existence of a fully-homomorphic encryption scheme secure against* $\text{poly}(N)$*-size adversaries. Then there exists a streaming delegation scheme* $\mathsf{sDel}_{\mathcal{F}}$ *for $\mathcal{F}$ with the following properties.*

- $\mathsf{sDel}_{\mathcal{F}}$ *has perfect completeness and negligible reusable soundness error.*
- D *updates her secret state in time* $\text{polylog}(N)$*, per data item.*
- *In the delegation protocol, when delegating a function $f \in \mathcal{F}$ computable in time T, the delegator* D *runs in time* $\text{poly}(k, \log N, \log T)$*, and the worker* W *runs in time* $\text{poly}(k, \log N, T)$*. The delegation protocol consists of* 4 *message exchanges.*

We note that in the Random Oracle Model (ROM) [BR97], the delegation scheme of Micali is non-interactive. This yields a non-interactive memory delegation scheme and a non-interactive streaming delegation scheme, for delegating all functions in **P**, in the ROM.

Due to the lack of space, we focus on our results using the GKR delegation protocol, and refer the reader to the full version of this paper [CKLR11] for the details on our results using the CS-delegation protocol. However, we note that the techniques and proofs are essentially the same in both cases.

1.2 Previous Work

Various delegation protocols have been proposed in the literature. Some provide delegation protocols that are sound against any cheating worker, whereas others provide delegation protocols that are secure only against computationally bounded cheating worker (i.e., arguments as opposed to proofs). Some of these protocols are interactive, whereas others are non-interactive. We survey some of these results below, however, we emphasize that in all these solutions, the delegator runs in time that is (at least) linear in the input size, and thus do not apply to our settings of memory delegation or streaming delegation.

Interactive proofs. The celebrated IP=PSPACE Theorem [LFKN92, Sha92] yields interactive proofs for any function f computable in polynomial space, with a verifier (delegator) running in polynomial time. Thus, the IP=PSPACE protocol can be seen as a delegation protocol for languages in PSPACE $\setminus$ **P**. However, the complexity of the prover (worker) is only bounded by polynomial space (and hence exponential time). This theorem was refined and scaled down in [FL93] to give verifier complexity $\text{poly}(n, s)$ and prover complexity $2^{\text{poly}(s)}$ for functions f computable in time T and space s, on inputs of length n. Note that the prover complexity is still super-polynomial in T, even for computations that run in the smallest possible space, namely $s = O(\log T)$.

The prover complexity was recently improved by Goldwasser et al. [GKR08] to $\text{poly}(T, 2^s)$, which is $\text{poly}(T)$ when $s = O(\log T)$. More generally, Goldwasser et al. [GKR08] give interactive proofs for computations of small *depth* d (i.e. parallel time). For these, they achieve prover complexity $\text{poly}(T)$ and verifier complexity $\text{poly}(n, d, \log T)$. (This implies the result for space-bounded computation because an algorithm that runs in time T and space s can be converted into one that runs in time $\text{poly}(T, 2^s)$ and depth $d = O(s^2)$.) However, if we do not restrict to computations of small space or depth, then we cannot use interactive proofs. Indeed, any language that has an interactive proof with verifier running time (and hence communication) T_V can be decided in space $\text{poly}(n, T_V)$.

Interactive arguments. Interactive arguments [BCC88] (aka computationally sound proofs [Mic00]) relax the soundness condition to be computational. Namely, instead of requiring that no prover strategy whatsoever can convince the verifier of a false statement, we instead require that no computationally feasible prover strategy can convince the verifier of a false statement. In this model, Kilian [Kil92] and Micali [Mic00] gave constant-round protocols with prover complexity $\text{poly}(T, k)$ and verifier complexity $\text{poly}(n, k, \log T)$ (where k is the security parameter), assuming the existence of collision-resistant hash functions [BG02].

Toward non-interactive Solutions. This possibility of efficient non-interactive arguments was suggested by Micali [Mic00], who showed that non-interactive arguments with prover complexity $\text{poly}(T, k)$ and verifier complexity $\text{poly}(n, k, \log T)$ are possible in the Random Oracle Model (the oracle is used to eliminate interaction a la Fiat–Shamir [FS86]). Heuristically, one might hope that by instantiating the random oracle with an appropriate family of hash functions, we could obtain a non-interactive solution to delegating computation: first the delegator (or a trusted third party) chooses and publishes a random hash function from the family, and then, the proofs are completely non-interactive (just one message from the prover to the verifier). However, the Random Oracle Heuristic is known to be unsound in general [CGH04] and even in the context of Fiat–Shamir [Bar01, GK03]. Thus, despite extensive effort, the existence of efficient non-interactive arguments remains a significant open problem in complexity and cryptography.

There has been some recent progress in reducing the amount of interaction needed. Using a transformation of Kalai and Raz [KR09], the GKR delegation protocol [GKR08] can be converted into a 2-message argument (assuming the existence of single-server private-information retrieval (PIR) schemes). However, like the interactive proofs of [GKR08], this solution applies only to small-depth computations, as the verifier's complexity grows linearly with the depth.

Very recently, Gennaro, Gentry, and Parno [GGP10], and the followup work of Chung, Kalai, and Vadhan [CKV10], gave a 2-message delegation scheme for arbitrary functions. However, these constructions have an offline phase, where the delegator invests time $\text{poly}(T, k)$ and computes a *secret state* (T is the time it takes to compute the function, and k is the security parameter). In the online phase, the delegator's running time is reduced to $\text{poly}(n, k, \log T)$ for an input of length n, and the worker's complexity is $\text{poly}(T, k)$. Thus, the delegator's large investment in the offline phase can be amortized over many executions of the online phase to delegate the computation of f on many inputs. Their online phase is not completely non-interactive, but rather consists of two messages. However, in many applications, two messages will be necessary anyway, as the delegator may need to communicate the input x to the worker.

We remark that one main drawback of these works [GGP10, CKV10] is that soundness is only guaranteed as long as the adversarial worker does not learn whether the delegator accepted or rejected the proofs.

In another followup work, Applebaum, Ishai, and Kushilevitz [AIK10] also consider the offline/online setting, but focus on efficient solutions for one-time delegation (i.e., the online phase can only be executed one time). They also consider the case when the delegation functions are represented as arithmetic circuits.

PCPs and MIPs. The MIP=NEXP Theorem [BFL91] and its scaled-down version by Babai et al. [BFLS91] yield multi-prover interactive proofs and probabilistically checkable proofs for time T computations with a prover running in time $\text{poly}(T)$ and a verifier running in time $\text{poly}(n, \log T)$, exactly as we want. However, using these for delegation require specialized communication models — either 2 non-communicating provers, or a mechanism for the prover to give the verifier random access to a long PCP (of length $\text{poly}(T)$) that cannot be changed by the prover during the verification.

Streaming Interactive Proofs. Recently, Cormode, Thaler, and Yi [CTY10] considered streaming interactive proofs, which is a strengthening of interactive proofs where the input is given to the verifier in a streaming manner and the verifier is restricted to have sub-linear (ideally, poly-logarithmic) space. They observed that both the GKR protocol [GKR08] and universal arguments [BG02] can be modified to yield efficient streaming interactive proofs/arguments.

Streaming interactive proofs are closely related to streaming delegation. The main difference is that streaming interactive proofs correspond to *one-time* streaming delegation, whereas in our streaming delegation model, the delegator is allowed to delegate as many computations to the worker as she want. Indeed, the GKR protocol is also the starting point of our construction of streaming delegation scheme, and the main effort is to make the scheme *reusable*.

2 Preliminaries

2.1 Low Degree Extension

Let $\mathbb{H}$ be an extension field of $\mathbb{GF}[2]$, and let $\mathbb{F}$ be an extension field of $\mathbb{H}$ (and in particular, an extension field of $\mathbb{GF}[2]$), where $|\mathbb{F}| = \text{poly}(|\mathbb{H}|)$.[3] We always assume that field operations can be performed in time that is poly-logarithmic in the field size. Fix an integer $m \in \mathbb{N}$. In what follows, we define the low degree extension of an n-element string $(w_0, w_1, \ldots, w_{n-1}) \in \mathbb{F}^n$ with respect to $\mathbb{F}, \mathbb{H}, m$, where $n \leq |\mathbb{H}|^m$.

Fix $\alpha : \mathbb{H}^m \rightarrow \{0, 1, \ldots, |\mathbb{H}|^m - 1\}$ to be any (efficiently computable) one-to-one function. In this paper, we take α to be the lexicographic order of $\mathbb{H}^m$. We can view $(w_0, w_1, \ldots, w_{n-1})$ as a function $W : \mathbb{H}^m \rightarrow \mathbb{F}$, where

$$W(z) = \begin{cases} w_{\alpha(z)} & \text{if } \alpha(z) < n, \\ 0 & \text{otherwise.} \end{cases} \tag{1}$$

A basic fact is that there exists a unique extension of W into a function $\tilde{W} : \mathbb{F}^m \rightarrow \mathbb{F}$ (which agrees with W on $\mathbb{H}^m$; i.e., $\tilde{W}|_{\mathbb{H}^m} \equiv W$), such that $\tilde{W}$ is an m-variate polynomial of degree at most $|\mathbb{H}| - 1$ in each variable. Moreover, as is formally stated in the proposition below, the function $\tilde{W}$ can be expressed as

$$\tilde{W}(t_1, \ldots, t_m) = \sum_{i=0}^{n-1} \tilde{\beta}_i(t_1, \ldots, t_m) \cdot w_i,$$

where each $\tilde{\beta}_i : \mathbb{F}^m \rightarrow \mathbb{F}$ is an m-variate polynomial, that depends only on the parameters $\mathbb{H}, \mathbb{F}$, and m (and is independent of w), of size $\text{poly}(\mathbb{H}, m)$ and degree $|\mathbb{H}| - 1$ in each variable.

The function $\tilde{W}$ is called the *low degree extension* of $w = (w_0, w_1, \ldots, w_{n-1})$ with respect to $\mathbb{F}, \mathbb{H}, m$, and is denoted by $\text{LDE}_w^{\mathbb{F},\mathbb{H},m}$. We omit the index of $\mathbb{F}, \mathbb{H}, m$ when the context is clear. Also, sometimes we use $\tilde{W}$ for simplicity.

Proposition 1. *There exists a Turing machine that takes as input an extension field $\mathbb{H}$ of $\mathbb{GF}[2]$,*[4] *an extension field $\mathbb{F}$ of $\mathbb{H}$, and integer m. The machine runs in time* $\text{poly}(|\mathbb{H}|, m)$ *and outputs the unique $2m$-variate polynomial $\tilde{\beta} : \mathbb{F}^m \times \mathbb{F}^m \rightarrow$*

[3] Usually, when doing low degree extensions, $\mathbb{F}$ is taken to be an extension field of $\mathbb{GF}[2]$, and $\mathbb{H}$ is simply a subset of $\mathbb{F}$ (not necessarily a subfield). In this work, following the work of [GKR08], we take $\mathbb{H}$ to be a subfield. However, all that is actually needed is that it is of size 2^ℓ for some $\ell \in \mathbb{N}$.

[4] Throughout this work, when we refer to a machine that takes as input a field, we mean that the machine is given a short (poly-logarithmic in the field size) description of the field, that permits field operations to be computed in time that is poly-logarithmic in the field size.

$\mathbb{F}$ of degree $|\mathbb{H}|-1$ in each variable (represented as an arithmetic circuit of degree $|\mathbb{H}|-1$ in each variable), such that for every $w = (w_0, w_1, \ldots, w_{n-1}) \in \mathbb{F}^n$, where $n \leq |\mathbb{H}|^m$, and for every $z \in \mathbb{F}^m$,

$$\tilde{W}(z) = \sum_{p \in \mathbb{H}^m} \tilde{\beta}(z,p) \cdot W(p),$$

where $W : \mathbb{H}^m \rightarrow \mathbb{F}$ is the function corresponding to $(w_0, w-1, \ldots, w_{n-1})$ as defined in Equation (1), and $\tilde{W} : \mathbb{F}^m \rightarrow \mathbb{F}$ is its low degree extension (i.e., the unique extension of $W : \mathbb{H}^m \rightarrow \mathbb{F}$ of degree at most $\mathbb{H}-1$ in each variable).

Moreover, $\tilde{\beta}$ can be evaluated in time $\text{poly}(|\mathbb{H}|, m)$. Namely, there exists a Turing machine that runs in time $\text{poly}(|\mathbb{H}|, m)$ that takes as input parameters $\mathbb{H}, \mathbb{F}, m$ (as above), and a pair $(z,p) \in \mathbb{F}^m \times \mathbb{F}^m$, and outputs $\tilde{\beta}(z,p)$.

Corollary 1. *There exists a Turing machine that takes as input an extension field $\mathbb{H}$ of $\mathbb{GF}[2]$, an extension field $\mathbb{F}$ of $\mathbb{H}$, an integer m, a sequence $w = (w_0, w_1, \ldots, w_{n-1}) \in \mathbb{F}^n$ such that $n \leq |\mathbb{H}|^m$, and a coordinate $z \in \mathbb{F}^m$. It runs in time $n \cdot \text{poly}(|\mathbb{H}|, m)$, and outputs the value $\tilde{W}(z)$, where $\tilde{W}$ is the unique low-degree extension of w (with respect to $\mathbb{H}, \mathbb{F}, m$).*

2.2 Delegation Schemes

In recent years, as cloud computing is gaining popularity, there have been many attempts to construct efficient delegation schemes. Loosely speaking, a delegation scheme is a protocol between a delegator D and a worker W, where the delegator asks the worker to do some computation, and prove that he indeed did the computation correctly. Typically, a delegation scheme is with respect to a class of functions $\mathcal{F}$, and the requirement is that on input (f, x) where $f \in \mathcal{F}$ and x is in the domain of f, the worker outputs $f(x)$, along with a proof (which may be interactive or non-interactive). The requirement is that the worker runs in time that is polynomial in the size of f (when representing f as a circuit), and the delegator runs in time that is significantly shorter than the size of f (as otherwise, it would simply compute $f(x)$ on its own). In this work, we use the 2-message delegation protocol of [GKR08], which in turn uses a round reduction technique from [KR09]. The protocol has the following guarantees.

Theorem 5. *[GKR08, KR09] Assume the existence of a poly-logarithmic PIR scheme. Let k be the security parameter, and let $\mathcal{F}$ be the family of functions computable by $\mathcal{L}$-space uniform boolean circuits of size $S(n)$ and depth $d(n) \geq \log S(n)$. Then, there exists a delegation protocol for $\mathcal{F}$ with the following properties.*

1. *The worker runs in time $\text{poly}(S, k)$ and the delegator runs in time $n \cdot \text{poly}(k, d(n))$.*
2. *The protocol has perfect completeness and soundness $s \leq \frac{1}{2}$ (can be made arbitrarily small), where soundness is against any cheating worker of size $\leq 2^{k^3}$.*

3. *The protocol consists of two messages, with communication complexity* $d(n) \cdot \text{poly}(k, \log S)$. *Moreover, the first message sent by the delegator depends only on her random coin tosses, and is independent of the statement being proved.*
4. *If the delegator is given oracle access to the low-degree extension of* x, *rather than being given the input* x *itself, then it runs in time* $\text{poly}(k, d(n))$, *and the protocol still has all the properties described above, assuming the parameters* $\mathbb{H}, \mathbb{F}, m$ *of the low-degree extension satisfy the following:*

$$|\mathbb{H}| = \theta(d \cdot \log n),\ m = \theta\left(\frac{\log n}{\log d}\right),\ |\mathbb{F}| = \text{poly}(|\mathbb{H}|)$$

where poly *is a large enough polynomial.*[5] *Moreover, the delegator queries the low-degree extension of* x *at a single point, which is uniformly random (over his coin tosses).*

Throughout this paper, we denote this protocol by GKR.

2.3 Merkle Tree Commitments

Definition 1. *Let* $h : \{0,1\}^k \times \{0,1\}^k \to \{0,1\}^k$ *be a hash function. A Merkle tree commitment of a sting* $x \in \{0,1\}^n$ *w.r.t.* h, *denoted by* $T_h(x)$, *is a* k*-bit string, computed as follows: The input* x *is partitioned into* $m = \lceil n/k \rceil$ *blocks* $x = (B_1, \ldots, B_m)$, *each block of size* k. *These blocks are partitioned into pairs* (B_{2i-1}, B_{2i}), *and the hash function* h *is applied to each pair, resulting in* $m/2$ *blocks. Then, again these* $m/2$ *blocks are partitioned into pairs, and the hash function* h *is applied to each of these pairs, resulting with* $m/4$ *blocks. This is repeated* $\log m$ *times, resulting in a binary tree of hash values, until one block remains. This block is* $T_h(x)$.

3 Overview of Our Constructions

In what follows we present a high-level overview of our memory and streaming delegation schemes. In this extended abstract, we focus on our non-interactive constructions, based on the GKR delegation schemes, and only present the high-level overview of these constructions. We refer the reader to the full version of this paper [CKLR11] for a formal presentation of our constructions and analysis.

3.1 Overview of Our Memory Delegation Scheme

The starting point of this work is the observation of Goldwasswer *et. al.* [GKR08], that their delegation protocol can be verified *very* efficiently (in time sub-linear in the input size), if the delegator has oracle access to the low-degree extension of the input x. Moreover, as observed by [GKR08], the delegator needs to access this low-degree extension LDE_x at a single point z, which depends only on the random coin tosses of the delegator.

[5] The larger poly is, the smaller the soundness becomes.

This observation immediately gives rise to a memory delegation scheme with *one-time* soundness: The delegator's secret state will be $(z, \mathrm{LDE}_x(z))$. Then, she will use this secret state in order to verify computation using the GKR protocol. As was argued by Goldwasswer *et. al.*, this indeed works if the delegator runs the delegation protocol *once*. However, the soundness crucially relies on the fact that the delegator's secret state is indeed secret, and if the delegator uses this state more than once, then soundness breaks completely.

One idea, following the idea of Gennaro *et. al.* [GGP10], is to use a fully homomorphic encryption (FHE) scheme to encrypt all the communication, in order to hide the secret state. This indeed works if the worker does not learn whether the delegator accepts or rejects his proofs. However, if the worker does learn the verdict of the delegator, then there are known attacks that break soundness.

In the streaming setting, we follow this approach, and we succeed in overcoming this problem, and construct a scheme that is sound even if the worker does learn the verdict of the delegator. We could follow this approach in the memory delegation setting as well. However, for several reasons, we choose to take a different approach. First, the approach above relies on the existence of an FHE scheme, whereas our memory delegation scheme relies on the existence of a poly-logarithmic PIR scheme, arguably a more reasonable assumption. Second, the approach above results with the delegator having a secret state, whereas in our memory delegation scheme, the state of the delegator is public. Finally, the construction and proof of the memory delegation scheme is simpler.

In our approach, instead of having $(z, \mathrm{LDE}_x(z))$ as the delegator's secret state, the delegator keeps a tree-commitment of the entire LDE_x as her secret state (see Section 2.3 for the definition of a tree-commitment). Namely, she chooses a random hash function h from a collision-resistant hash family, and keeps $(h, T_h(\mathrm{LDE}_x))$ as her state. In addition to giving the worker her memory x, she also gives him the hash function h. We stress that her state is not secret, which makes the proof of security significantly simpler than that in the streaming setting (where the delegator's state is secret).

Very roughly speaking, when the delegator wishes to delegate the computation of a function f, they execute $\mathsf{Compute}(f)$ by simply running the (non-interactive) delegation protocol $\mathrm{GKR}(f)$. Recall that at the end of the GKR protocol the delegator needs to verify the value of $\mathrm{LDE}_x(r)$ for a random r. However, she doesn't have x, since it was delegated to the prover, and all she has is the state $(h, T_h(\mathrm{LDE}_x))$. So, rather than computing the value of $\mathrm{LDE}_x(r)$ on her own, the worker will reveal this value, by sending the augmented path in the Merkle tree corresponding to the leaf r.

Unfortunately the high-level description given above is a gross oversimplification of our actual scheme, and there are several technical issues that complicate matters. We elaborate on these in Section 3.3.

We mention that when the delegator wishes to update her memory from x to $g(x)$, she needs to update her secret state from $(h, T_h(\mathrm{LDE}_x))$ to $(h, T_h(\mathrm{LDE}_{g(x)}))$.[6]

[6] Actually, for technical reasons she will need to choose a fresh hash function $h' \leftarrow \mathcal{H}$ during each Update. We discard this technical issue here.

However, she cannot perform this operation on her own, since she does not have x. Instead she will delegate this computation to the worker, by requesting a $\mathsf{Compute}(g')$ operation, where $g'(x) = T_h(\mathrm{LDE}_{g(x)})$.

3.2 Overview of Our Streaming Delegation Scheme

Our streaming delegation scheme is similar to our memory delegation scheme described above, and the main difference is in the way the certificate is generated and updated, and in the way the worker reveals the value $\mathrm{LDE}_x(r)$.

Generating and updating the certificate. Recall that in the memory delegation scheme, the certificate of the delegator D consists of a tree-commitment to the low-degree extension of her memory x. Namely, her certificate is $(h, T_h(\mathrm{LDE}_x))$, where h is a collision resistant hash function. Note that this certificate cannot be updated in a streaming manner, since any change to x changes the low-degree extension LDE_x almost everywhere.

Instead, in the streaming setting, we replace the tree commitment with an "*algebraic commitment*", which has the property that it can be updated efficiently when new data items arrive. The resulting certificate is a random point in the low-degree extension of the stream x; i.e., $(z, \mathrm{LDE}_x(z))$ for a random point z. This certificate is efficiently updatable, if we assume some upper-bound N on the size of the stream, and we take parameters $\mathbb{H}, \mathbb{F}, m$ of the low-degree extension, such that

$$|\mathbb{H}| = \mathrm{polylog}(N),\ m = \theta\left(\frac{\log N}{\log\log N}\right),\ |\mathbb{F}| = \mathrm{poly}(|\mathbb{H}|) \tag{2}$$

(this follows from Proposition 1).

As in the memory delegation scheme, at the end of each delegation protocol, the delegator needs to verify the value of $\mathrm{LDE}_x(r)$ at a random point r. In the memory delegation scheme this was done using a Reveal protocol where the worker reveals the augmented path of the leaf r in the Merkle tree-commitment of LDE_x. In the streaming setting, the Reveal protocol is totally different, since the delegator cannot compute the tree-commitment of LDE_x. Unfortunately, unlike in the memory delegation scheme, in the streaming setting constructing a *reusable* and *sound* reveal protocol is highly non-trivial.

The Reveal protocol. Our starting point is a basic reveal protocol Reveal_1 described in Figure 1. Note that the soundness of Reveal_1 relies on the secrecy of the certificate σ. Namely, assuming that W does not know the point z, it is not hard to see, by the Schwartz-Zippel Lemma, that an adversarial worker can cheat with probability at most $d/|\mathbb{F}|$, where d is the (total) degree of LDE_x.

However, note that the Reveal_1 protocol is not reusable. Suppose that D uses the above reveal protocol to learn the value of LDE_x on two random points $s, s' \in \mathbb{F}^m$. From the two executions, an adversarial worker W^* receives two lines $\ell_{s,z}$ and $\ell_{s',z}$, and can learn the secret point z by taking the intersection of the two lines. Once W^* learns z, W^* can easily cheat by returning any polynomial p^* that agrees with LDE_x only on point z but disagrees on the remaining points.

Reveal$_1$ protocol: D stores a *secret* state $\sigma = (z, \mathrm{LDE}_x(z))$, where $x \in \{0,1\}^N$ and z is a random point in $\mathbb{F}^m$, and wants to learn the value of $\mathrm{LDE}_x(s)$ from W.

- D sends to W the line ℓ_{sz} that passes through the points s and z. More specifically, D chooses two random points $\alpha_1, \alpha_2 \leftarrow \mathbb{F}$, and defines $\ell_{s,z}$ to be the line that satisfies $\ell_{s,z}(\alpha_1) = z$ and $\ell_{s,z}(\alpha_2) = s$.
- W returns a univariate polynomial $p : \mathbb{F} \to \mathbb{F}$, which is the polynomial LDE_x restricted to the line $\ell_{s,z}$ (i.e., $p = \mathrm{LDE}_x|_{\ell_{s,z}}$).
- D checks whether $p(\alpha_1) = \mathrm{LDE}_x(z)$, and if so accepts the value $p(\alpha_2) = \mathrm{LDE}_x(s)$. Otherwise, she rejects.

Fig. 1. Reveal$_1$ protocol

As observed by Gennaro *et. al.* [GGP10], a natural way to protect the secret point z, is to run the above Reveal protocol under a fully-homomorphic encryption (FHE) scheme. Namely, D generates a pair of keys $(\mathsf{pk}, \mathsf{sk})$ for a FHE $(\mathrm{Gen}, \mathrm{Enc}, \mathrm{Dec}, \mathrm{Eval})$, and sends pk and an encrypted line $\hat{\ell}_{s,z} = \mathrm{Enc}_{\mathsf{pk}}(\ell_{s,z})$ to W, who can compute the polynomial $p = \mathrm{LDE}_x|_\ell$ homomorphically under the encryption. Indeed, by the semantic security of FHE, an adversarial worker W* cannot learn any information from D's message $\hat{\ell}_{s,z}$. This indeed makes the protocol reusable provided that W* does not learn the decision bits of D, as proved in [GGP10, CKV10].

However, since the decision bit of D can potentially contain one bit information about the secret point z, it is not clear that security holds if W* learns these decision bits. In fact, for both of the delegation schemes of [GGP10, CKV10], which use FHE to hide the delegator D's secret state, there are known attacks that learn the whole secret state of D bit-by-bit from D's decision bits.

Fortunately, we are able to show that a variant of the Reveal$_1$ protocol described in Figure 2 is reusable even if W* learns the decision bits of D. The main difference between Reveal$_1$ and Reveal$_2$ is that in Reveal$_2$, the delegator D uses a random *two-dimensional* affine subspace instead of a line, and uses an FHE to mask the entire protocol.

We prove that no efficient adversarial W* can learn useful information about the secret point z from the Reveal$_2$ protocol. We note that the proof of the above statement is highly non-trivial, and is one of the main technical difficulties in this work. Informally, we first prove a "leakage-resilient lemma", which claims that the ciphertext $\hat{S}_{rz}$ and the decision bit b of D (which depend on the strategy of W^*) do not give too much information about S_{rz} to W*. In other words, the random subspace $S_{s,z}$ still has high (pseudo-)entropy from the point of view of W*. Then we use an *information-theoretic* argument to argue that a random point z in a sufficiently random (with high entropy) subspace $S_{s,z}$ is *statistically close* to a random point in $\mathbb{F}^m$, which implies that W* does not learn useful information about z.

The Field Size. Recall that by Schwartz-Zippel Lemma, an adversarial worker can cheat with probability at most $d/|\mathbb{F}|$, where d is the (total) degree of LDE_x.

Reveal$_2$ protocol: D stores a secret state $\sigma = (z, \mathrm{LDE}_x(z))$, where $x \in \{0,1\}^N$ and z is a random point in $\mathbb{F}^m$, and wants to learn the value of $\mathrm{LDE}_x(s)$ from W.

- D does the following.
  1. Generate a pair of keys $(\mathsf{pk}, \mathsf{sk}) \leftarrow \mathrm{Gen}(1^k)$ for a fully homomorphic encryption scheme FHE.
  2. Choose a random two-dimensional affine subspace $S_{s,z} \subset \mathbb{F}^m$ that contains the points s and z. More specifically, choose two random points $\alpha_1, \alpha_2 \leftarrow \mathbb{F}^2$ and let $S_{s,z} \subset \mathbb{F}^m$ be a random two-dimensional affine subspace that satisfies $S_{s,z}(\alpha_1) = z$ and $S_{s,z}(\alpha_2) = s$.
  3. Send $\hat{S}_{s,z} \leftarrow \mathrm{Enc}_{\mathsf{pk}}(S_{s,z})$ and pk to W.
- W homomorphically computes the two-variate polynomial $p = \mathrm{LDE}_x|_{S_{s,z}}$ under the FHE (denote the resulting ciphertext $\hat{p}$), and sends $\hat{p}$ to D.
- D decrypts and checks whether $p(\alpha_1) = \mathrm{LDE}_x(z)$, and if so accepts the value $p(\alpha_2) = \mathrm{LDE}_x(s)$.

Fig. 2. Protocol Reveal$_2$

Recall that in our setting of parameters:

$$|\mathbb{H}| = \mathrm{polylog}(N), \; m = \theta\left(\frac{\log N}{\log\log N}\right), \; |\mathbb{F}| = \mathrm{poly}(|\mathbb{H}|).$$

Thus, a cheating worker can cheat (and more importantly, obtain information about the secret z) with probability $d/|\mathbb{F}| = O(1/\mathrm{polylog}(N))$, which is not low enough.

The idea is to reduce the cheating probability to negligible by simply increasing the field size to be super-polynomial. However, we cannot increase the field size in the GKR protocol, since it will increase the complexity of the worker. Instead, we use an extension field $\tilde{\mathbb{F}}$ of $\mathbb{F}$, of super-polynomial size, only in the certificate and the Reveal protocol, but run the GKR protocols as before. Namely, the secret state is $\sigma = (z, \mathrm{LDE}^{\tilde{\mathbb{F}},\mathbb{H},m}(z))$ where $z \leftarrow \tilde{\mathbb{F}}^m$, The GKR protocol is run exactly as before with the parameters $(\mathbb{H}, \mathbb{F}, m)$.

3.3 Additional Technicalities

The high-level description given above (in Sections 3.1 and 3.2) is a gross oversimplification of our actual schemes, and there are several technical issues that complicate matters.

Recall that in the overview above, we claimed that $\mathsf{Compute}(f)$ merely runs GKR, in addition to a Reveal protocol which helps the delegator verify the GKR protocol.[7] There are several technical reasons why this actually does not work. In what follows, we explain what are the main technical problems with this simple idea, and we give the highlevel idea of how to overcome these problems.

[7] The Reveal protocol in the memory setting is totally different from the Reveal protocol in the streaming setting.

1. The first technicality (the easiest one to deal with), is that the GKR delegation scheme does not have a negligible soundness error. In our setting, especially in the setting of memory delegation, it is very important to have negligible soundness. The reason is that if the soundness is non-negligible, then a cheating worker may cheat in the update procedure of the memory delegation scheme (which is also being delegated). The problem is that if a worker cheats even once in an update procedure, all soundness guarantees are mute from that point on. So, we really need the soundness error to be negligible. In order to reduce the soundness error, we will run the GKR protocol in parallel u times (for any parameter u such that $1/2^u = \mathsf{ngl}(k)$, where k is the security parameter). We denote the u-fold parallel repetition of GKR by $\mathrm{GKR}^{(u)}$. As a result the worker will need to reveal to u random points in the low-degree extension: $\mathrm{LDE}_x(r_1), \ldots, \mathrm{LDE}_x(r_u)$.
2. The second technical point is more subtle. In the offline stage, when the delegator computes the tree commitment $T_h(\mathrm{LDE}_x)$, she needs to choose the parameters $\mathbb{H}, \mathbb{F}, m$ for the low-degree extension. The typical choice for these parameters is:

$$|\mathbb{H}| = \mathrm{polylog}(n),\ |\mathbb{F}| = \mathrm{poly}(|\mathbb{H}|),\ m = O\left(\frac{\log n}{\log\log n}\right),$$

where $n = |x|$. When delegating the computation of a function f, the worker and delegator run $\mathrm{GKR}^{(u)}(f)$ and need to verify $\mathrm{LDE}_x(r_i) = v_i$ for random points $r_1, \ldots, r_u$. However, here the parameters of the low-degree extension LDE_x depend on the depth d of the circuit computing f. Namely, looking at the parameters given in [GKR08] (see Theorem 5), the parameters of the low-degree extension are

$$|\mathbb{H}'| = \theta(d \cdot \log n),\ m' = \theta\left(\frac{\log n}{\log d}\right),\ |\mathbb{F}'| = \mathrm{poly}(|\mathbb{H}'|).$$

Therefore, the worker cannot simply execute the Reveal protocols of the memory delegation or the streaming delegation. In the memory setting, the tree commitment is w.r.t. parameters $\mathbb{H}, \mathbb{F}, m$ whereas the delegator needs to verify $\mathrm{LDE}_x^{\mathbb{F}',\mathbb{H}',m'}(r_i) = v_i$. In the streaming setting, the secret state of the delegator is $(z, \mathrm{LDE}_x^{\mathbb{F},\mathbb{H},m}(z))$, as opposed to $(z, \mathrm{LDE}_x^{\mathbb{F}',\mathbb{H}',m'}(z))$, thus the Reveal protocol described in Section 3.2 doesn't work.
We get around this technical problem by delegating the functions $g_{r_i}(x) \triangleq \mathrm{LDE}_x^{\mathbb{H}',\mathbb{F}',m'}(r_i)$. Luckily, these functions can be computed by a poly-size circuit of depth at most $\log^2 n$, assuming the delegated function f is of poly-size (see Proposition 1). We delegate the computation of each of these g_{r_i} using $\mathrm{GKR}^{(u)}$ to ensure negligible soundness. Thus, finally the worker will need to reveal to u^2 points in LDE_x (u points for each g_{r_i}).[8]

[8] We note that there are several ways to improve efficiency, such as thinking of $(g_{r_1}, \ldots, g_{r_u})$ as one function. However, for the sake of simplicity of exposition, we focus on the simplest (rather than most efficient) solution.

3. The final technical difficulty is that all these algorithms need to run in parallel, since we want our final delegation schemes to be non-interactive (i.e., to consist of only two messages). Typically, there is no problem in running several two-message protocols in parallel [BIN97, CHS05]. However, in our case, the delegator uses a common *secret* input in these protocols. Namely, the delegator uses secret randomness $r_1, \ldots, r_u \in (\mathbb{F}')^{m'}$ in the parallel repetition of the delegation protocol GKR(f) which ends with her needing to verify that $\mathrm{LDE}_x^{\mathbb{H}',\mathbb{F}',m'}(r_i) = v_i$ for every $i \in [u]$. In addition she uses these same r_i's in the delegation protocols GKR(g_{r_i}). Moreover, at the end of each of the GKR(g_{r_i}) protocols, the delegator needs to verify that $\mathrm{LDE}_x^{\mathbb{H},\mathbb{F},m}(z_{i,j}) = w_{i,j}$ for random points $z_{i,1}, \ldots, z_{i,u} \in \mathbb{F}^m$. Finally, they also run a reveal protocol for each $z_{i,j}$, denoted by Reveal($z_{i,j}$).
We note that the protocol GKR(f) (resp. GKR(g)) is not sound if the r_i's (resp. $z_{i,j}$'s) are a priori known to the worker. To ensure that soundness still holds even if we run all these algorithms in parallel, we mask parts of the delegator's message using a PIR scheme or an FHE scheme, and then we claim that the soundness error remains negligible. To this end, we use a "parallel composition lemma", which roughly states that if a set of protocols $\Pi_1, \ldots \Pi_t$ are executed in parallel, and the verifiers use the same common private randomness p in all these protocols, then the soundness remains if the messages of the verifiers hide this common secret randomness p.

Acknowledgments. We are very grateful to Shai Halevi for collaborating with us in the initial phase of this work, and to Salil Vadhan for several helpful discussions.

References

[AIK10] Applebaum, B., Ishai, Y., Kushilevitz, E.: From secrecy to soundness: Efficient verification via secure computation. In: Abramsky, S., Gavoille, C., Kirchner, C., Meyer auf der Heide, F., Spirakis, P.G. (eds.) ICALP 2010. LNCS, vol. 6198, pp. 152–163. Springer, Heidelberg (2010)

[Bar01] Barak, B.: How to go beyond the black-box simulation barrier. In: FOCS, pp. 106–115 (2001)

[BCC88] Brassard, G., Chaum, D., Crépeau, C.: Minimum disclosure proofs of knowledge. Journal of Computer and System Sciences 37(2), 156–189 (1988)

[BFL91] Babai, L., Fortnow, L., Lund, C.: Non-deterministic exponential time has two-prover interactive protocols. Computational Complexity 1, 3–40 (1991)

[BFLS91] Babai, L., Fortnow, L., Levin, L.A., Szegedy, M.: Checking computations in polylogarithmic time. In: STOC, pp. 21–31 (1991)

[BG02] Barak, B., Goldreich, O.: Universal arguments and their applications. In: Proceedings of the 17th Annual IEEE Conference on Computational Complexity, pp. 194–203 (2002)

[BIN97] Bellare, M., Impagliazzo, R., Naor, M.: Does parallel repetition lower the error in computationally sound protocols? In: FOCS, pp. 374–383 (1997)

[BR97] Bellare, M., Rogaway, P.: Minimizing the use of random oracles in authenticated encryption schemes. In: ICICS, pp. 1–16 (1997)

[BSGH+05] Ben-Sasson, E., Goldreich, O., Harsha, P., Sudan, M., Vadhan, S.P.: Short pcps verifiable in polylogarithmic time. In: IEEE Conference on Computational Complexity, pp. 120–134 (2005)

[CGH04] Canetti, R., Goldreich, O., Halevi, S.: The random oracle methodology, revisited. Journal of the ACM 51(4), 557–594 (2004)

[CHS05] Canetti, R., Halevi, S., Steiner, M.: Hardness amplification of weakly verifiable puzzles. In: Kilian, J. (ed.) TCC 2005. LNCS, vol. 3378, pp. 17–33. Springer, Heidelberg (2005)

[CKLR11] Chung, K.-M., Kalai, Y.T., Liu, F.-H., Raz, R.: Memory delegation. Cryptology ePrint Archive, Report 2011/273 (2011), `http://eprint.iacr.org/`

[CKV10] Chung, K.-M., Kalai, Y., Vadhan, S.P.: Improved delegation of computation using fully homomorphic encryption. In: Rabin, T. (ed.) CRYPTO 2010. LNCS, vol. 6223, pp. 483–501. Springer, Heidelberg (2010)

[CTY10] Cormode, G., Thaler, J., Yi, K.: Verifying computations with streaming interactive proofs. Technical Report TR10-159, ECCC Report (2010)

[FL93] Fortnow, L., Lund, C.: Interactive proof systems and alternating time-space complexity. Theoretical Computer Science 113(1), 55–73 (1993)

[FS86] Fiat, A., Shamir, A.: How to prove yourself: Practical solutions to identification and signature problems. In: Odlyzko, A.M. (ed.) CRYPTO 1986. LNCS, vol. 263, pp. 186–194. Springer, Heidelberg (1987)

[GGP10] Gennaro, R., Gentry, C., Parno, B.: Non-interactive verifiable computing: Outsourcing computation to untrusted workers. In: Rabin, T. (ed.) CRYPTO 2010. LNCS, vol. 6223, pp. 465–482. Springer, Heidelberg (2010)

[GK03] Goldwasser, S., Kalai, Y.T.: On the (in)security of the fiat-shamir paradigm, pp. 102–113 (2003)

[GKR08] Goldwasser, S., Kalai, Y.T., Rothblum, G.N.: Delegating computation: interactive proofs for muggles. In: STOC, pp. 113–122 (2008)

[Kil92] Kilian, J.: A note on efficient zero-knowledge proofs and arguments (extended abstract). In: STOC, pp. 723–732 (1992)

[KR09] Kalai, Y.T., Raz, R.: Probabilistically checkable arguments. In: Halevi, S. (ed.) CRYPTO 2009. LNCS, vol. 5677, pp. 143–159. Springer, Heidelberg (2009)

[LFKN92] Lund, C., Fortnow, L., Karloff, H.J., Nisan, N.: Algebraic methods for interactive proof systems. J. ACM 39(4), 859–868 (1992)

[Mic94] Micali, S.: Cs proofs (extended abstracts). In: FOCS, pp. 436–453 (1994)

[Mic00] Micali, S.: Computationally sound proofs. SIAM J. Comput. 30(4), 1253–1298 (2000)

[Sha92] Shamir, A.: IP = PSPACE. Journal of the ACM 39(4), 869–877 (1992)

Automatic Search of Attacks on Round-Reduced AES and Applications

Charles Bouillaguet, Patrick Derbez, and Pierre-Alain Fouque

ENS, CNRS, INRIA, 45 rue d'Ulm, 75005 Paris, France
{charles.bouillaguet,patrick.derbez,pierre-alain.fouque}@ens.fr

Abstract. In this paper, we describe versatile and powerful algorithms for searching guess-and-determine and meet-in-the-middle attacks on byte-oriented symmetric primitives. To demonstrate the strengh of these tool, we show that they allows to automatically discover new attacks on round-reduced AES with very low data complexity, and to find improved attacks on the AES-based MACs Alpha-MAC and Pelican-MAC, and also on the AES-based stream cipher LEX. Finally, the tools can be used in the context of fault attacks. These algorithms exploit the algebraically simple byte-oriented structure of the AES. When the attack found by the tool are practical, they have been implemented and validated.

1 Introduction

Since the introduction of the AES in 2001, it has been questioned whether its simple algebraic structure could be exploited by cryptanalysts. Soon after its publication as a standard [30], Murphy and Robshaw showed in 2002 an interesting algebraic property: the AES encryption process can be described only with simple algebraic operations in $GF(2^8)$ [29]. Such a result paved the way for multivariate algebraic techniques [13,11] since the AES encryption function can be described by a very sparse overdetermined multivariate quadratic system over $GF(2)$. However, so far this approach has not been so promising [28,12], and the initial objective of this simple structure, providing good security protections against differential and linear cryptanalysis, has been fulfilled.

Recently, much attention has been devoted to the AES block cipher as a by-product of the NIST SHA-3 competition. The low diffusion property of the key schedule has been used to mount several related-key attacks [6,5,3,27] and differential characteristic developed for hash functions have been used to also improve single-key attacks [20]. In order to improve these attacks, new automatic tools have been designed to automatically search either related-key attacks or collision attacks on byte-oriented block ciphers [7] or AES-based hash functions [27].

In this paper, we look at the security of round-reduced versions of the AES block cipher in a practical security model, in continuity with [8]. The adversary knows a *very small number* of plaintext/ciphertext pairs, one or two, and his goal is to recover the secret key. Studying reduced-round versions of AES is motivated by the proliferation, these last years, of many AES-based primitives for hashing

P. Rogaway (Ed.): CRYPTO 2011, LNCS 6841, pp. 169–187, 2011.

or authentication, such as the Grøstl, ECHO, Shavite, LANE hash functions, the LEX [1] stream cipher, or the Alpha-MAC [14] and Pelican-MAC [15] message authentication codes. A possible explanation of this fancy is that the AES enjoys very interesting security properties against statistical attacks. Namely, two rounds achieve full diffusion, and there exist very good differential and linear lower bounds for the best differential on four rounds [26,25,24]. Consequently, for some applications such as hashing and authentication where the adversary has little or no access to the internal state, the full ten AES rounds may be overkill, and some designers proposed to use less rounds for more efficiency. In these applications, the adversary has less control over the AES than in the usual block-cipher setting, and has access to *a very few* number of plaintext/ciphertext pairs. For example, in the LEX stream cipher [2], only a quarter of the state is leaked at each round and to generate the next 32 bits of keystream, only one round of AES is performed. Furthermore, in some particular attacks, such as side-channel attacks, only a small number of rounds of the cipher needs to be studied [31,4]. In the latter scenario, the adversary does not know plaintext/ciphertext pairs, but that some difference in intermediate states results in two different ciphertexts. Finally, in symmetric cryptanalysis, statistical attacks usually use distinguishers on a small number of rounds and then, extend these distinguishers to more rounds. Consequently, it is important to search the best attack in this model.

Related Work. In this security model, statistical attacks may be not the best possible attacks, since they usually require many pairs with specific input difference and algebraic attacks seem to be more well-suited. However, such attacks using either SAT solvers or Gröbner basis algorithms [29,10], have never been able, so far, to endanger even very reduced versions of the AES even though its structure exhibits some algebraic properties. These attacks encode the problem into a system of equations, then feeds the equations to a generic, sometimes off-the-shelf equation solver, such as a SAT-solver or a Gröbner basis algorithm. The main obstacle in these approaches is the S-box, that only admits "bad" representations (for instance, it is a high degree polynomial over the AES finite field), and increases the complexity of the equations, even though low degree implicit equations may also exist.

Our tools, instead of using pre-existing generic equations solvers, first run a search for an *ad hoc* solver tailored for the equations to solve, build it, and then run it to obtain the actual solutions. They can be applied to systems of linear equations containing a non-linear permutation of the field, such as an S-box. Our idea is to consider the S-box as a black box permutation. We only use few properties of this function and our attacks works for *any instantiation* of the S-box. This approach is reminiscent of the ideas used in [27] by Khovratovich *et al.* where similar systems of linear equations are written to describe a hash function, and where additional constraints enforce the message and chaining value to follow a certain truncated differential characteristic inside the function. Solving the equations would then yield a collision. The basic strategy for finding a message pair conforming a differential characteristic consists in exhaustively trying values for the variables and checking if the constraints are satisfied. In order to speed up the collision search, they propose to look for a maximum-sized set of variables that could take freely any values without violating the constraints. To this end, they use linear algebra, and essentially consider x and $S(x)$ to be independent variables, to find such maximum set using a greedy strategy. During the search of a conforming message pair, the free variables can take all the possible values while the value of the other variables are deduced from the free ones. Consequently, the search avoids trying bad values for the latter variables which improves the probabilistic trial stage. The algorithm in [27] is however limited in that when the greedy strategy aborts, no other solutions are explored.

Our Techniques and Results. Our tools try to find attacks automatically by searching some classes of guess-and-determine and meet-in-the-middle attacks. They take as input a system of equations that describes the cryptographic primitive and some constraints on the plaintext and ciphertext variables for example. Then, it solves the equations by first running a (potentially exponential) search for a customized solver for the input system. Then, the solver is run, and the solutions are computed.

We describe two tools. Our preliminary tool uses a depth-first branch-and-bound search to find "good" guess-and-determine attacks. It has been (covertly) used to generate some of the attacks found in [8], and outperformed human cryptanalyst in several occasions. However, the class of attack searched for by this preliminary tool is quite restricted, and it fails to take into account important differential properties of the S-box. Our second, more advanced tool, allows to find more powerful attacks, such as Meet-in-the-Middle attacks. For instance, it automatically exploits the useful fact that an input and output difference on the S-box determine almost uniquely the actual input and output values. The algorithmic techniques used by this tool are reminiscent of the Buchberger algorithm [9]. The results found by these algorithms are summarized in tables 1 and 2.

We improve many existing attacks in the "very-low data complexity" league. For instance, we find a certificational attack on 4 full AES rounds using just a single known plaintext, and a practical attack on the same 4 full AES rounds with 4 chosen plaintexts. We also look at AES-based primitives. We independently discovered (along with [21]) the best known attack on Pelican-MAC, and automatically rediscover the best attacks on Alpha-MAC and LEX. We also used our tool to find a new, faster, attack on LEX. Lastly, we improve the efficiency of the state-recovery part of the Piret-Quisquater fault attack against the full AES. While it required 2^{32} elementary operations, it now takes about one second on a laptop.

Organization of the Paper. In section 2, we describe how the equations are constructed given the AES description and how we represent them. Then, we present our preliminary guess-and-determine attack finder in section 3 and then a more advanced tool that finds meet-in-the-middle attacks in section 4. Finally, in section 5, we show four different attacks that were automatically found by the previous tool.

Table 1. Summary of our Proposed Attacks on AES-128

Attacks on round reduced version of the AES-128						
		This paper		Previous Best Attacks		
#Rounds	Data	Time	Memory	Time	Memory	Ref.
1	1 KP	2^{32}	2^{16}	2^{48}	1	[19]
1.5	1 KP	2^{56}	1			
1.5	2 KP	2^{24}	2^{16}			
2	1 KP	2^{64}	2^{48}	2^{80}	1	[8] ⋆
2	2 KP	2^{32}	2^{24}	2^{48}	1	[8]
2	2 CP	2^{8}	2^{8}	2^{28}	1	[8]
2.5	1 KP	2^{88}	2^{88}			
2.5	2 KP	2^{80}	2^{80}			
2.5	2 CP	2^{24}	2^{16}			
3	1 KP	2^{96}	2^{96}	2^{120}	1	[8] ⋆
3	2 CP	2^{16}	2^{8}	2^{32}	1	[8]
4	1 KP	2^{120}	2^{120}			
4	2 CP	2^{80}	2^{80}	2^{104}	1	[8]
4	4 CP	2^{32}	2^{24}			
4.5	1 KP	2^{120}	2^{120}			

KP — Known plaintext, CP — Chosen plaintext,
Time complexity is measured in encryption units unless mentioned otherwise.
Memory complexity is measured approximately
⋆ : previously published, but found with these tools
"r.5 rounds" — r full rounds and the final round

Table 2. Summary of our Proposed Attacks on Primitives based on AES

Attacks on Primitives based on AES						
Primitive	Complexity			G & D Part		References
	Data	Time	Memory	Time	Memory	
Pelican-MAC	$2^{85.5}$ queries	$2^{85.5}$	$2^{85.5}$			[32]
Pelican-MAC	2^{64} queries	2^{64}	2^{64}	2^{32}	2^{24}	Sect. 5.2
Alpha-MAC	2^{65} queries	2^{64}	2^{64}	2^{32}	2^{16}	[32] †
LEX	$2^{36.3}$ bytes	2^{112}	2^{36}			[17]
LEX	2^{40} bytes	2^{100}	2^{64}	2^{80}	1	[18]
LEX	$2^{36.3}$ bytes	2^{96}	2^{80}	2^{64}	2^{64}	Full version
LEX	2^{50} bytes	2^{80}	2^{48}	2^{16}	2^{8}	Full version
AES-128	1 fault	2^{32}	2^{32}	2^{32}	2^{32}	[31]
AES-128	1 fault	2^{24}	2^{16}	2^{24}	2^{16}	Sect. 5.3

Time complexity is measured in encryption units unless mentioned otherwise.
Memory complexity is measured approximately
† : the tools can find automatically a comparable attack

2 Preliminaries

Let $\mathbb{F}_{256}$ denote the finite field with 256 elements used in the AES. We denote the Sbox of the `SubBytes` transformation by $S : \mathbb{F}_{256} \rightarrow \mathbb{F}_{256}$. In this paper we only consider the 128-bit version of the AES. Keys, plaintext, ciphertext and internal states of the cipher are represented by 4×4 matrices over $\mathbb{F}_{256}$. In such a matrix, we use the following numbering of bytes: byte zero is the top-left corner, the first column is made of bytes 0-3, while the last column is made of bytes 12-15, with byte 15 in the bottom-right corner. We also denote by $M[\bullet, j]$ the j-th column of M. In r-round AES, a master key, K_0 is expanded into r round keys, $K_1, \ldots, K_r$ by a key-schedule algorithm[1] which is described by the following equations:

$$KS_i : \begin{cases} K_i[\bullet, j] + K_i[\bullet, j-1] + K_{i-1}[\bullet, j] = 0, & j = 1, 2, 3 \\ K_i[0] + K_{i-1}[0] + S\left(K_{i-1}[13]\right) + \mathtt{RCON}_i = 0 \\ K_i[1] + K_{i-1}[1] + S\left(K_{i-1}[14]\right) = 0 \\ K_i[2] + K_{i-1}[2] + S\left(K_{i-1}[15]\right) = 0 \\ K_i[3] + K_{i-1}[3] + S\left(K_{i-1}[12]\right) = 0 \end{cases}$$

An AES round performs the following sequence of operations: `SubBytes`, `ShiftRows`, `MixColumns`, and a round subkey addition. We refer the reader to the AES specification for more details [30]. We denote by X_i the internal state entering round i (*i.e.*, before `SubBytes`), while Y_i and W_i denote the internal state before and after the `MixColumns` operation, respectively. The master key is XORed to the plaintext before entering the first round. The process is summarized by these equations, where MC denote the `MixColumns` matrix:

$$R_i : \begin{cases} W_i + MC \times \begin{pmatrix} S\left(X_i[0]\right) & S\left(X_i[4]\right) & S\left(X_i[8]\right) & S\left(X_i[12]\right) \\ S\left(X_i[5]\right) & S\left(X_i[9]\right) & S\left(X_i[13]\right) & S\left(X_i[1]\right) \\ S\left(X_i[10]\right) & S\left(X_i[14]\right) & S\left(X_i[2]\right) & S\left(X_i[6]\right) \\ S\left(X_i[15]\right) & S\left(X_i[3]\right) & S\left(X_i[7]\right) & S\left(X_i[11]\right) \end{pmatrix} = 0 \\ X_{i+1} + W_i + K_{1+i} = 0 \end{cases}$$

It is straightforward to form the system of equations $\mathbb{E}$ describing the full encryption process along with the key schedule: we just have to concatenate some KS_i's and some R_i's (without forgetting the initial key addition). Since the right-hand-side of all these equations are zero, we stop representing them from now on.

Let us denote by $\mathcal{V}(X)$ the vector space spanned by $1, x, S(x)$ for all $x \in X$, for any set of variables X. If we denote by $\mathbb{X}$ the set of all key and internal state variables then the cipher equations can be seen as a subspace of $\mathcal{V}(\mathbb{X})$. We also introduce the notation $\mathcal{S}(\mathbb{E})$ to denote the set of solutions of a system of equations $\mathbb{E}$.

[1] Note that the AES-256 has a different key-schedule.

3 A preliminary Tool for Guess-And-Determine Attacks

Confronted with a system of equations in $\mathcal{V}(\mathbb{X})$ (possibly describing a cryptographic problem), the most naive way to obtain its solutions consists in enumerating all the variables and retaining only the combination that satisfy all the equations. However, equations in $\mathcal{V}(\mathbb{X})$ are such that, in a given equation, once all the terms but one are known then the last one can be found efficiently. This enables more or less efficient *guess-and-determine* techniques to solve the equations. In a cryptographic setting, guess-and-determine attacks are often found when data is very scarce, and statistic attacks are therefore impossible. Guess-and-determine attacks can be more or less sophisticated, but the simplest ones typically take the following form:

```
for all values of some part of the (unknown) internal state do
    Compute the full internal state
    Retrieve the secrets
    Try to regenerate available data using secrets
    if match available data then return secrets
end for
```

The difficulty in finding such an attack is to find which parts of the internal state to enumerate, and to find how to recover the rest. In this section, we present a *Preliminary Tool* that finds such attacks automatically. It takes as input a system of equations in $\mathcal{V}(\mathbb{X})$ and a set $\mathbb{K}_0 \subset \mathbb{X}$ of initially *known* variables—these are the variables corresponding to the available data, for instance the plaintext, the ciphertext, the keystream, etc. The Preliminary Tool returns a C++ function (the "solver") that enumerates its solutions (using negligible memory), along with the exact number of elementary operations it performs.

This Preliminary Tool has for instance been used to find *one known plaintext* attacks against 1, 1.5, 2, 2.5 and 3 rounds of AES. Some of these attacks have been published in [8]. While performing the research that lead to the publication [8], the Preliminary Tool (which was designed for the occasion) improved on the best results found by well-known human cryptanalysts. For instance, prior to the publication of [8], the best attack on one (full) round of AES was a guess-and-determine attack with complexity 2^{48} guessed described in [19]. This preliminary tool found in less than a second an attack with 5 guesses and generated its implementation: the C++ file is available on the web page of the first author.

Knowledge Propagation. The core idea of the Preliminary Tool is quite simple: if there is a linear combination of the equations in which the values of all terms are known except one, then the value of this last term can be determined efficiently.

When applied to the AES, this simple procedure automatically harnesses the simple and clean algebraic structure of the cipher. It automatically exploits the linear relations existing in the key-schedule, as well as the `MixColumns` property:

if $y = \texttt{MixColumns}(x)$ then knowlege of any four bytes in (x, y) is sufficient to recover the remaining four in a unique way.

An "algebraic" Point of View. The acquisition of further knowledge, either by "guessing" or "determining" has an algebraic effect on the equations. Let $\mathbb{K} \subset \mathbb{X}$ be a set of variables whose value is known. If we substituted the values of known variables into the original equations $\mathbb{E}$, we would get a system with less variables. In fact, this reduced system is essentially the quotient space of $\mathbb{E}$ by $\mathcal{V}(\mathbb{K})$: starting from an equation $f \in \mathbb{E}$, its equivalence class $[f]$ in the quotient contains a representative where all the variables in $\mathbb{K}$ have disappeared. Alternatively, the variable x can be deduced from $\mathbb{K}$ if either $[x]$ or $[S(x)]$ belong to the quotient. We will write $x \in \textsc{Propagate}(\mathbb{K})$ when it is the case.

3.1 Automatic Search for a Minimal Number of Guesses

Given a set of "known" variables $\mathbb{K} = \mathbb{K}_0$, we may propagate knowledge and obtain the value of new variables, yielding a new set of known variables $\mathbb{K}_1$. But it may turn out that new variables may again be obtained from $\mathbb{K}_1$. We therefore define the function $\textsc{Propagate}^*(X)$ which returns the least fixed point of $\textsc{Propagate}$ containing X.

A guess-and-determine solver has been found as soon as we have found a set $\mathbb{G}$ of "guesses" such that $\textsc{Propagate}^*(\mathbb{G}) = \mathbb{X}$. In that case, we will say that $\mathbb{G}$ is *sufficient*. The problem thus comes down to automatically find a sufficient set of minimal size.

The process of exhaustively searching such a guess-and-determine attack can be seen as the exploration of a DAG whose nodes are sets of variables. The starting node is the set $\mathbb{K}_0$, and the terminal node is $\mathbb{X}$. For any set of variables X, and any $y \notin X$ there is an edge $X \xrightarrow{y} X \cup \{y\}$, meaning that we may always choose to enumerate y to gain knowledge. Finally, for any set of variables X, there is an edge $X \rightarrow \textsc{Propagate}^*(X)$, symbolizing the fact that we may propagate knowledge.

In this setting, the objective of the Preliminary Tool is to find a path from $\mathbb{K}$ to $\mathbb{X}$ traversing a small (if not the smallest) number of "guess" edges. Indeed, the cost of the resulting attack is exponential in the number of traversed "guess edges". The problem is that the size of the DAG is exponential in the number of variables.

The search works in a depth-first branch-and-bound fashion reminiscent of the DPLL procedure implemented in many SAT-solvers. The pseudo-code of the search procedure is shown in Figure 1. The function $\textsc{Explore}(\mathbb{K}, \mathbb{G}, \mathbb{B})$ returns a minimal set of variables to guess in order to be able to recover the entire internal state. Here $\mathbb{K}$ denotes the set of *currently known* variables (*i.e.*, the current node of the DAG), $\mathbb{G}$ denotes the set of variables that have been guessed so far, and $\mathbb{B}$ denotes the set of variables that have been guessed in the best known solution. This implicit assumption is that $|\mathbb{G}| < |\mathbb{B}|$, and that the result of explore has cardinality smaller than or equal to $\mathbb{B}$. To find the best solution, just run $\textsc{Explore}(\mathbb{K}_0, \emptyset, \mathbb{X})$.

```
function Explore(𝕂, 𝔾, 𝔹)
    if 𝕂 = 𝕏 then return 𝔾
    if 𝕂 → Propagate*(𝕂) then
        return Explore(Propagate*(𝕂), 𝔾, 𝔹)
    if |𝔾| = |𝔹| − 1 then return 𝔹
    for all x ∈ FilterGuesses(𝕂) do
        recursive ← Explore(𝕂 ∪ {x}, 𝔾 ∪ {x}, 𝔹)
        if |recursive| < 𝔹 then 𝔹 ← recursive
        if |𝔾| = |𝔹| − 1 then return 𝔹
    end for
    return 𝔹
end function
```

Fig. 1. Pseudo-code of the Preliminary Tool

In order to speed-up the search procedure, we used several *pruning strategies* that remove "guess" edges from the DAG without modifying its reachability properties.

Local Pruning. In simple words, if we need to guess a new variable, and if guessing x allows to deduce y, then it is useless to guess y instead of x. More formally, we see that if $y \in \text{Propagate}^*(\mathbb{K} \cup \{x\})$, then:

$$\text{Propagate}^*(\mathbb{K} \cup \{y\}) \subseteq \text{Propagate}^*(\mathbb{K} \cup \{x\})$$

A reasonable pruning strategy is to consider only the candidate guesses that are not "subsumed" by any other.

Global Pruning. A somewhat surprising consequence of the fact that Propagate^* is *monotonic* brings in a powerful result, enabling us to further discard some bad guesses.

Lemma 1. *Let $V \subsetneq \mathbb{X}$ be an* insufficient *set of variables, and let $\mathbb{G} \subseteq \mathbb{X}$ be a sufficient set of variables. Then:*

$$\mathbb{G} \cap (\mathbb{X} - \text{Propagate}^*(V)) \neq \emptyset$$

If $\mathbb{G}$ denotes a sufficient set of minimal size, then Lemma 1 gives us *a priori* knowledge on $\mathbb{G}$, and it enables to choose the first guess of the search procedure in $\mathbb{X} - \text{Propagate}^*(V)$ without risking to throw the best solution away. It is also possible to use lemma 1 at any point of the search, but then V must be chosen to be a superset of the currently known variables (otherwise we may not learn anything).

The problem remains to find the biggest possible sets V of variables such that $\text{Propagate}^*(V) \neq \mathbb{X}$. At each step, there is a different tradeoff to make between pruning and exploring the DAG. In any case, a simple greedy heuristic—add to V the variable x that minimizes the size of $\text{Propagate}^*(V \cup \{x\})$—already give interesting results.

3.2 Limitations

The main limitation of this approach is that it completely fails to take into account the differential properties of the S-box. For instance, it cannot exploit the fact that when the input and output differences of the S-box are fixed and non-zero, then at most 4 possible input values are possible. Therefore, this approach alone does not bring useful result when more than one plaintext is available. However, it can be used as a sub-component in a more complex technique. We now move on to describe a generalization of this technique that allows to find more powerful attacks.

4 A Tool for Meet In-The-Middle Attacks

The equations describing the AES enjoy an interesting and important property. Let us consider a partition of the set of variables, $\mathbb{X} = \mathbb{X}_1 \cup \mathbb{X}_2$. Then any equation $f \in \mathbb{E}$ can be written $f = f_1 + f_2$, with $f_1 \in \mathcal{V}(\mathbb{X}_1)$ and $f_2 \in \mathcal{V}(\mathbb{X}_2)$. In some sense, these equations are *separable*. We will see that this allows a recursive "meet-in-the-middle" approach.

4.1 Solving Subsystems Recursively

The simple algebraic structure of the equations allows us to efficiently extract from a system $\mathbb{E}$ a *subsystem* containing only certain variables (say $\mathbb{X}_1$), by simply computing the vector space intersection $\mathbb{E} \cap \mathcal{V}(\mathbb{X}_1)$. In the sequel we will denote it by $\mathbb{E}(\mathbb{X}_1)$. We note that a solution of $\mathbb{E}$ is also a solution of $\mathbb{E}(\mathbb{X}_1)$, for any $\mathbb{X}_1 \subsetneq \mathbb{X}$, but that the converse is not true in general.

Now let us be given a partition $\mathbb{X} = \mathbb{X}_1 \cup \mathbb{X}_2$ and two *black-box solvers* $\mathcal{A}_1$ and $\mathcal{A}_2$ that find all the solutions of $\mathbb{E}(\mathbb{X}_1)$ and $\mathbb{E}(\mathbb{X}_2)$. The two sub-solvers $\mathcal{A}_1$ and $\mathcal{A}_2$ can be used to find the solutions $\mathcal{S}$ of the full problem $\mathbb{E}$. An obvious way would be to compute the solutions $\mathcal{S}_1$ of $\mathbb{E}(\mathbb{X}_1)$ and $\mathcal{S}_2$ of $\mathbb{E}(\mathbb{X}_2)$, and to test all the solutions in the Cartesian product $\mathcal{S}_1 \times \mathcal{S}_2$. This would require about $|\mathcal{S}_1| \cdot |\mathcal{S}_2|$ evaluations of the equations.

However, it is possible to do better. Firstly, we observe that the vectors in $\mathcal{S}_1 \times \mathcal{S}_2$ automatically satisfy the equations in $\mathbb{E}(\mathbb{X}_1) + \mathbb{E}(\mathbb{X}_2)$. Therefore we first compute a supplementary of $\mathbb{E}(\mathbb{X}_1) + \mathbb{E}(\mathbb{X}_1)$ inside $\mathbb{E}$ (let us denote it by $\mathcal{M}$). The solutions of $\mathbb{E}$ are in fact the elements of $\mathcal{S}_1 \times \mathcal{S}_2$ satisfying the equations of $\mathcal{M}$. This already makes less constraints to check. Second, sieving the elements satisfying these constraints can be done in roughly $|\mathcal{S}_1| + |\mathcal{S}_2|$ operations, using variable separation and a table. Let $(f_i)_{1 \leq i \leq n}$ be a basis of $\mathbb{E}$, and $f_i = g_i + h_i$ with $g_i \in \mathcal{V}(\mathbb{X}_1)$ and $h_i \in \mathcal{V}(\mathbb{X}_2)$. If the values of all the variables in $\mathbb{X}_1$ (resp. $\mathbb{X}_2$) are available, then the g_i's (resp. h_i) may be evaluated. We denote by G (resp. H) the function that evaluates all the g_i on its input. We build two tables:

$$L_1 \longleftarrow \{(G(x_1), x_1) \mid x_1 \text{ solution of } \mathbb{E}(\mathbb{X}_1)\}$$
$$L_2 \longleftarrow \{(H(x_2), x_2) \mid x_2 \text{ solution of } \mathbb{E}(\mathbb{X}_2)\}$$

Then, the solutions of $\mathbb{E}$ are the pairs (x, y) for which there exist a z such that $(z, x) \in L_1$ and $(z, y) \in L_2$. They can be identified efficiently by various methods (sorting the tables, using a hash index, etc.). We have just combined $\mathcal{A}_1$ and $\mathcal{A}_2$ to form a new solver, $\mathcal{A} = \mathcal{A}_1 \bowtie \mathcal{A}_2$, that enumerates the solutions $\mathcal{S}$ of $\mathbb{E}$.

Note that the guess-and-determine attacks discussed in the previous section form a particular case of this more general framework. They can be described by a recursive combination where $\mathbb{X}_2$ always contain a single variable.

Complexity of the Combination. Given two sub-solvers $\mathcal{A}_1$ and $\mathcal{A}_2$, the complexity and the properties of $\mathcal{A}_1 \bowtie \mathcal{A}_2$ are easy to determine. Let us denote by $T(\mathcal{A})$ the running time of $\mathcal{A}$, by $M(\mathcal{A})$ its memory consumption, by $V(\mathcal{A})$ the set of variables occurring in the corresponding equations, and by $\mathcal{S}(\mathcal{A})$ the set of solutions it outputs. The number of operations performed by the combination is the sum of the number of operations produced by the sub-solvers, plus the number of solutions (the time required to scan the tables, namely $|\mathcal{S}_1| + |\mathcal{S}_2|$, is in the worst case of the same order as the running time of the two sub-solvers). However, we use the following approximation

$$T(\mathcal{A}_1 \bowtie \mathcal{A}_2) = \max\big(T(\mathcal{A}_1), T(\mathcal{A}_2), |\mathcal{S}(\mathbb{E}(V(\mathcal{A}_1) \cup V(\mathcal{A}_2)))|\big)$$

It is possible to store only the smallest table, and to enumerate the content of the other "on the fly", while looking for a collision. This reduces the memory complexity to the maximum of the memory complexity of the sub-solvers, and the size of the smaller table. This yields:

$$M(\mathcal{A}_1 \bowtie \mathcal{A}_2) = \max\Big\{M(\mathcal{A}_1), M(\mathcal{A}_2), \min\Big(|\mathcal{S}(\mathcal{A}_1)|, |\mathcal{S}(\mathcal{A}_2)|\Big)\Big\}$$

Heuristic Assumption on the Number of Solutions. Evaluating the complexity of a given (possibly recursive) combination requires evaluating the number of solutions of various sub-systems. This is a difficult problem in general, and in order to be able to quickly evaluate the properties of a combination, we use the following *heuristic assumption* : if $\mathcal{S}_1$ are the solutions of $\mathbb{E}(\mathbb{X}_1)$, then $|\mathcal{S}_1| \approx 2^{8(|\mathbb{X}_1| - \dim \mathbb{E}(\mathbb{X}_1))}$. This heuristic assumption introduces a risk of failure, or of wrong estimation of the complexity. To protect ourselves against this risk, we have tried, when possible, to implement the solvers and check if this assumption holds.

4.2 Automatic Search for Recursive Combinations of Solvers

Given a system of equations, we would like to build an efficient solver by breaking the problem down to smaller and smaller subsystems, recursively generating efficient sub-solver for the sub-problems and combining them back.

Note that $\mathbb{E}(\{x\})$ cannot be further broken down, and is a "base case" of the decomposition, which is dealt with by a "base solver". We can safely assume that $\mathbb{E}(\{x\}) = 0$, since otherwise, for a maximum cost of 2^8, one can determine x uniquely (according to our hypothesis) and add it to the set of known variables.

Combining base solvers in various ways yields *solving trees* of various shapes. It is often possible to construct several solving trees that solve the same problem in different ways, and sometimes more or less efficiently.

Comparing Solvers. We therefore want to be able to compare solvers in a meaningful way. We want $\mathcal{A}_1 \succeq \mathcal{A}_2$ if $\mathcal{A}_1$ is overally more interesting (works faster, finds solution of a bigger system). We also want the order relation to be compatible with the combination operation (*i.e.*, $\mathcal{A}_1 \succeq \mathcal{A}_2$ implies $\mathcal{A}_1 \bowtie \mathcal{A}_3 \succeq \mathcal{A}_2 \bowtie \mathcal{A}_3$). We thus define:

$$\mathcal{A}_1 \succeq \mathcal{A}_2 \Longleftrightarrow T(\mathcal{A}_1) \leq T(\mathcal{A}_2), V(\mathcal{A}_1) \supseteq V(\mathcal{A}_2), |S(\mathcal{A}_1)| \leq |S(\mathcal{A}_2)|$$

The equivalence relation induced by this order carries an interesting meaning: if $\mathcal{A}_1 \succeq \mathcal{A}_2$ and $\mathcal{A}_2 \succeq \mathcal{A}_1$, then $\mathcal{A}_1$ and $\mathcal{A}_2$ offer essentially the same functionality. The equivalence relation is also compatible with the combination operation. We observe that given a set of variables $\mathbb{X}_1$, there can be only one maximal solver (up to equivalence) for $\mathbb{E}(\mathbb{X}_1)$. Thus, our objective is now clearly identified: find a maximal (*i.e.*, the best) solver for $\mathbb{E}$.

Exhaustive Search for the Best Recursive Solver. The procedure ExhaustiveSearch on fig. 2 computes the set of all maximal solvers for all sub-systems of a given system of equations $\mathbb{E}$. In particular, it will construct a maximal solver for $\mathbb{E}$ itself. The algorithm is reminiscent of (and inspired by) the Buchberger algorithm for Gröbner bases [9]. The complexity of this algorithm seems difficult to evaluate. It depends on the equations, and on the order in which the combinations are performed. In any case, the size of its ouput is upper bounded by $2^{|\mathbb{X}|}$ (because it will return only one maximal solver for each subset of $\mathbb{X}$). The parameter T_{up} allows the user to enforce an upper-bound on the time complexity of the generated solvers (by discarding the others). For small values of T_{up}, this may for instance allow to prove the non-existence of recursive solvers with complexity lower than a threshold. The running time of the exhaustive search also gets smaller with lower values of T_{up}.

In practice, what dominates the execution of this algorithm is the computation of the dimension of the combination C, and the bookkeeping required to update G. A nice improvement is to use the Propagate* function from section 3: each time a new solver C is constructed, we check whether $V(C)$ is stable by Propagate*. If not, we combine it with the base solvers in Propagate*$(V(C)) - V(C)$, thus improving it without increasing its running time. We also have the following theorem which allows us to reduce the size of the search space and to *refine* solvers :

Theorem 1. *Let $\mathbb{X}$ be a set of variables, $x \in \mathbb{X}$ and $\mathcal{A}$ an optimal solver for $\mathbb{E}(\mathbb{X} - \{x\})$. If $\dim \mathbb{E}(\mathbb{X} - \{x\}) = \dim \mathbb{E}(\mathbb{X}) - 1$ then $\mathcal{A} \bowtie \{x\}$ is an optimal solver for $\mathbb{E}(\mathbb{X})$.*

```
function Add-Reduce(G, A)
    if there exist A' ∈ G such that A' ⪰ A then return G
    return {A} ∪ {A' ∈ G | A ⋡ A'}
end function

function ExhaustiveSearch(E, T_up)
    G ← Base Solvers for E (one for each variable)
    repeat
        G' ← G
        for all pairs (A_1, A_2) ∈ G' do
            C ← A_1 ⋈ A_2
            if T(C) ≤ T_up then G ← Add-Reduce(G, C)
        end for
    until G = G'
    return G
end function
```

Fig. 2. Exhaustive Search for a good recursive solver

Randomized Search. The complexity of the exhaustive search is inherently exponential, and exploring the whole space might not be feasible. In that case, a non-exhaustive randomized search might find good results, without offering the guarantee that they are the best possible. The procedure RandomizedSearch on fig. 3 shows a possible randomized search that we have found to give good results. The idea is again quite simple: at each step, we choose a random set of variables Y, we build a solver for $\mathbb{E}(Y)$, and if it is not subsumed by any previously known solver, we include it in the current solver list, and we try to combine it with all the solvers we know. It would make sense to choose Y with some care, for instance using the pruning strategies discussed in section 3.

There are many possible other ways to perform such a randomize search: Choose the size of the random subsets of $\mathbb{X}$ according to some distribution, periodically restart the procedure, periodically flush "bad" solvers from G, run the exhaustive search for a while, fill G, then switch to randomized search, etc.

4.3 Usage

When an interesting solver for $\mathbb{E}$ is found by the search procedure, it is not particularly complicated to recursively generate a C++ implementation thereof (*i.e.*, a function that takes as input the "known" variables, and returns the solutions of the system of equations), or a text file that describes which variables to enumerate, which tables to join, in a nearly human-readable language.

5 Applications

In this section, we show several attacks that were found by the tool of section 4. The attacks were found completely automatically. The human intervention consisted in writing down the right equations, which sometimes required

```
function RandomizedSearch(𝔼, T_up)
    G ← ∅
    loop
        Y ← random subset of 𝕏, of size T_up
        (Z_1, Z_2, ...) ← Propagate*(Y)
        B ← BaseSolver(Z_1) ⋈ BaseSolver(Z_2) ⋈ ...
        G ← Add-Reduce(G, B)
        for all 𝒜 ∈ G do
            C ← 𝒜 ⋈ B
            if T(C) > T_up then drop C
            if V(C) = 𝕏 then return C
            L ← Add-Reduce(G, C)
        end for
    end loop
end function
```

Fig. 3. Randomized Search for a good recursive solver

some knowledge of the primitive (for instance, to choose a sparse input difference for Pelican-MAC, or to use a 3-collision for LEX). The tool re-discovered attacks equivalent to the best published results on LEX and Alpha-MAC. We also used it to find a better attack on LEX (which is not included in this paper due to lack of space, but is present in the full version). This illustrates that the tool can be a useful research assistant, allowing the cryptanalyst to quickly test a global idea ("let's use a 3-collision against LEX"), while the tool takes care of the tedious, nasty and delicate details. When possible, we implemented these attacks (either manually or using code generated by the tools), and checked them using a reference implementation of the AES.

These attacks that we improve upon are all relatively recent. We also improve on the Piret-Quisquater fault attack against the AES, an older result that had a suboptimal state recovery procedure. The tool automatically found a better one.

5.1 Improved Attacks on Reduced-Round Rijndael

We present a new key-recovery attack with a negligible complexity about 2^8 encryptions. This is a significant improvement of the best previous attack published in [8] with a complexity about of 2^{32} encryptions, demonstrating the power of our tool. The adversary asks for the encryption of two plaintexts which differ only in four bytes composing one column. The attack relies on Lemma 2 which cleverly uses the linearity in the key-schedule of the AES.

Lemma 2. *For all $i, j \geq 1$ we have the following equation :*

$$MC\left(Y_{i-1}[\bullet, j] + Y_i[\bullet, j-1] + Y_i[\bullet, j]\right) = X_i[\bullet, j] + X_{i+1}[\bullet, j-1] + X_{i+1}[\bullet, j]$$

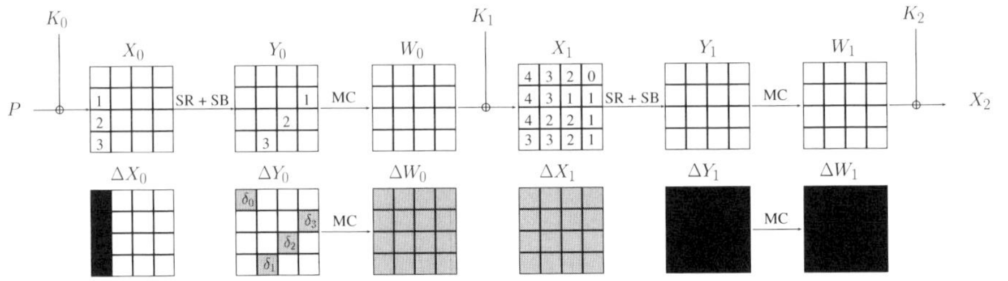

Fig. 4. Gray squares denote the presence of a difference. Black squares denote a known difference.

We denote by δ_i the non-zero byte of column $\Delta Y_0[\bullet, i]$. We begin by constructing, in table form, the inverse of the following functions:

- $T_1 : Y_0[1,3] \longmapsto \delta_3$
- $T_2 : Y_0[2,2] \longmapsto \delta_2$
- $T_3 : Y_0[3,1] \longmapsto \delta_1$
- $T_{ij} : X_1[i,j] \longmapsto \delta_j,\ i+j \neq 3$

Then, for each possible value of $X_1[0,3]$, we perform following steps :

1-a Get δ_3 and, using $T_{\bullet 3}$ and T_1, get $X_1[\bullet, 3]$ and $Y_0[1,3]$.
1-b Compute $X_1[1,2]$ by applying lemma 2.
2-a Get δ_2 and, using $T_{\bullet 2}$ and T_2, get $X_1[\bullet, 2]$ and $Y_0[2,2]$.
2-b Compute $X_1[2,1]$ by applying lemma 2.
3-a Get δ_1 and, using $T_{\bullet 1}$ and T_3, get $X_1[\bullet, 1]$ and $Y_0[3,1]$.
3-b Compute $X_1[3,0]$ by applying lemma 2.
4 Get δ_0 and, using $T_{\bullet 0}$, get $X_1[\bullet, 0]$.
5 Compute K_2 and check whether it is correct.

We have implemented and tested this attack. On average, there are $2^{8.65}$ candidates for K_2, which is very close to our hypothesis.

We can easily extend the attack to three rounds. The adversary simply asks for the encryption of two plaintexts which differ only in one byte and guesses the corresponding byte on K_0. The configuration is the same as before and we can apply the previous attack. This gives a new attack with a time complexity of about 2^{16} encryptions and negligible memory requirement.

5.2 Improved Forgery Attacks on Pelican-MAC

The best published attacks against Alpha-MAC and Pelican-MAC is [32]. For Alpha-MAC, after having found an internal collision (this requires 2^{65} queries), the internal state is recovered with a guess-and-determine attack that makes about 2^{64} simple operations. For Pelican-MAC, an impossible differential attack recovers the internal state with data and time complexity $2^{85.5}$.

The general idea we exploit is to find a single collision in the internal state, found by injecting message blocks following a fixed truncated differental characteristic. Then, the state recovery problem is encoded in equations and given to the tool. It must be noted that an attack with the same global complexity has been independently found time by Dunkelman, Keller and Shamir [21], using an impossible differential. The "state-recovery" phase presented here is faster though.

Pelican-MAC. We now present a new attack against Pelican-MAC, with time and data complexity 2^{64}. We pick an arbitrary message block M_1 and query the MAC with 2^{64} random two-block messages $M_1 \parallel M_2$, and store the (message,tag) pair in a table. Then, we query the MAC on $(M_1 + \Delta_i) \parallel (M_2')$, where Δ_i is zero everywhere except on the first byte, and M_2' is random. When a collision is found, we know that the pair of internal states follows the differential characteristic of figure 5 (there could be accidental difference cancellations with small probability though).

We then wrote down the state-recovery problem as a system of equations: two unknown states with a known one-byte difference yields two unknown states with a known (full) difference. The tool of section 4 quickly found an attack that runs in time and space about 2^{32} (the attack with 2^{24} in memory is much more complicated to describe), and which is summarized by fig. 5. The key observation (which the tool found all by itself) is that if α, β, γ and δ denote the differences in Y_1, then the differences in X_2 are:

$$\Delta X_2 = \begin{pmatrix} 02\alpha & \beta & \gamma & 03\delta \\ \alpha & \beta & 03\gamma & 02\delta \\ \alpha & 03\beta & 02\gamma & \delta \\ 03\alpha & 02\beta & \gamma & \delta \end{pmatrix}$$

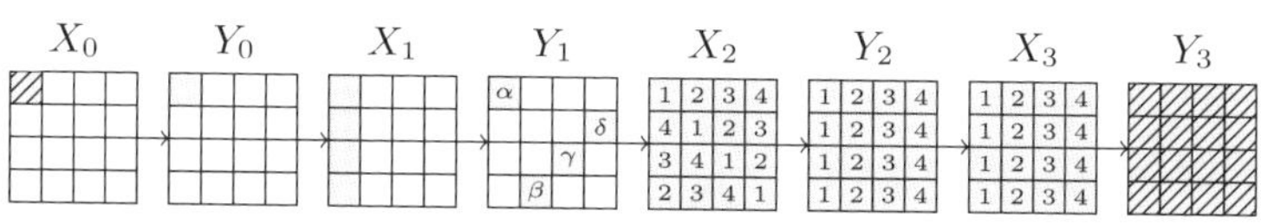

Fig. 5. Gray squares denote the presence of a difference. Hatched squares denote a known difference.

We extracted a description of the attack from the tool's output. It proceeds as follows:

1-a Guess bytes 0-3 of X_3. The corresponding values in X_3' can be found thanks to the known difference in Y_3.

1-b Partially decrypt in the second round to get suggestions for α, β, γ and δ.
1-c Store bytes $0 - 3$ of X_3 in a hash table $\mathcal{T}_0$ indexed by $(\alpha, \beta, \gamma, \delta)$
2 Repeat the process with the second column of X_3. Store bytes $4 - 7$ of X_3 in a table $\mathcal{T}_1$ indexed by $(\alpha, \beta, \gamma, \delta)$.
3 Repeat the process with the third and fourth column of X_3. Build tables $\mathcal{T}_2$ and $\mathcal{T}_3$
4 Enumerate $(\alpha, \beta, \gamma, \delta)$. Look-up $\mathcal{T}_0, \mathcal{T}_1, \mathcal{T}_2$ and $\mathcal{T}_3$ and retrieve the parts of X_3 corresponding to $(\alpha, \beta, \gamma, \delta)$, if present.
5 if $(\alpha, \beta, \gamma, \delta)$ occurs in the 4 tables, then we get a complete suggestion for X_3. Decrypt 3 rounds and recover X_0. Check if the input difference is right.

We implemented the state-recovery part of the attack (the collision-finding would not be feasible in practice for us) and validated it experimentally. The number of tested candidates is consistent with the expected number (2^{32}).

Alpha-MAC. Obviously, we cannot overally improve on the attack of [32], since finding the collision dominates the running time. However, it is noteworthy that the tool found a state-recovery procedure that requires only 2^{32} elementary operations and memory, when the first input message difference contains only one active byte. This is much more efficient than its counterpart in [32].

5.3 Improvement to the Piret-Quisquater Fault Attack

In the Piret-Quisquater fault attack [31], an unknown difference is introduced in byte 0 of the internal state X_7. The adversary observes the output difference, and recovers the secret key in time 2^{32}. Here, we show an improved procedure (found by the Tool) working in time 2^{24} and memory 2^{16}. Let us denote by δ the difference $S(X_0[0,0]) + S(X'_0[0,0])$. For the sake of simplicity, we describe the attack assuming that the final `MixColumns` operation has *not* been removed.

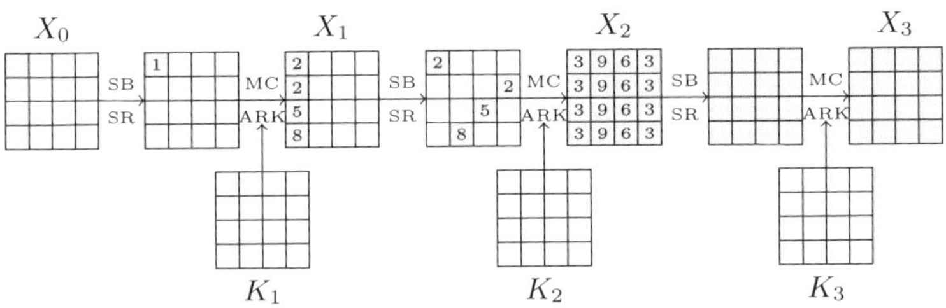

Fig. 6. Fault attack against the AES. Gray square indicates the presence of a difference.

The attack can be replayed without it, but some details become significantly messier. The attack makes use of the following non-trivial observation, that we extracted from the rool's output:

Lemma 3. *i) $X_1[1]$ can be deduced from $X_2[\bullet, 0]$ and $X_2[\bullet, 3]$*
ii) $X_1[2]$ can be deduced from $X_2[\bullet, 0], X_2[\bullet, 2]$ and $X_2[\bullet, 3]$
iii) $X_1[3]$ can be deduced from $X_2[\bullet, 0], X_2[\bullet, 1]$ and $X_2[\bullet, 3]$

1. Guess the difference in $X_1[0, 0]$
2. Guess the actual value of $X_1[0, 0]$ and $X_1[1, 0]$
3. Compute the difference in $X_2[\bullet, 0]$ and $X_2[\bullet, 3]$, then the actual values.
4. Use lemma 3, item *i* to filter the guesses of step 3. Only 2^{16} out of 2^{24} should pass the test.
5. Guess the actual value of $X_1[2, 0]$
6. Compute the difference in $X_2[\bullet, 2]$, then the actual values.
7. Use lemma 3, item *ii* to filter the guesses of step 5. Only 2^{16} should pass.
8. Guess the actual value of $X_1[3, 0]$
9. Compute the difference in $X_2[\bullet, 1]$, then the actual values.
10. Use lemma 3, item *iii* to filter the guesses of step 8. Only 2^{16} should pass.
11. At this point we should have 2^{16} candidates for $(X_1[\bullet, 0], X_1'[\bullet, 0])$. From those, X_2 can be reconstructed entirely, as well as K_3. Simply test all the candidates.

We implemented this attack and validated it in practice. It terminates in a couple of seconds on a laptop. In particular, we could check that the actual number of tested candidates was consistent with the expected number.

References

1. Biryukov, A.: The Design of a Stream Cipher LEX. In: Biham, E., Youssef, A.M. (eds.) SAC 2006. LNCS, vol. 4356, pp. 67–75. Springer, Heidelberg (2007)
2. Biryukov, A.: Design of a New Stream Cipher—LEX. In: Robshaw, M.J.B., Billet, O. (eds.) New Stream Cipher Designs. LNCS, vol. 4986, pp. 48–56. Springer, Heidelberg (2008)
3. Biryukov, A., Dunkelman, O., Keller, N., Khovratovich, D., Shamir, A.: Key Recovery Attacks of Practical Complexity on AES-256 Variants with up to 10 Rounds. In: [22], pp. 299–319
4. Biryukov, A., Khovratovich, D.: Two New Techniques of Side-Channel Cryptanalysis. In: Paillier, P., Verbauwhede, I. (eds.) CHES 2007. LNCS, vol. 4727, pp. 195–208. Springer, Heidelberg (2007)
5. Biryukov, A., Khovratovich, D.: Related-Key Cryptanalysis of the Full AES-192 and AES-256. In: Matsui, M. (ed.) ASIACRYPT 2009. LNCS, vol. 5912, pp. 1–18. Springer, Heidelberg (2009)
6. Biryukov, A., Khovratovich, D., Nikolic, I.: Distinguisher and Related-Key Attack on the Full AES-256. [23], 231–249
7. Biryukov, A., Nikolic, I.: Automatic Search for Related-Key Differential Characteristics in Byte-Oriented Block Ciphers: Application to AES, Camellia, Khazad and Others. [22], 322–344

8. Bouillaguet, C., Derbez, P., Dunkelman, O., Keller, N., Fouque, P.A.: Low Data Complexity Attacks on AES. Cryptology ePrint Archive, Report 2010/633 (2010), http://eprint.iacr.org/
9. Buchberger, B.: Ein Algorithmus zum Auffinden der Basiselemente des Restklassenringes nach einem nulldimensionalen Polynomideal. PhD thesis, University of Innsbruck (1965)
10. Buchmann, J., Pyshkin, A., Weinmann, R.-P.: A Zero-Dimensional Gröbner Basis for AES-128. In: Robshaw, M.J.B. (ed.) FSE 2006. LNCS, vol. 4047, pp. 78–88. Springer, Heidelberg (2006)
11. Cid, C.: Some Algebraic Aspects of the Advanced Encryption Standard. [16], 58–66
12. Cid, C., Leurent, G.: An Analysis of the XSL Algorithm. In: Roy, B. (ed.) ASIACRYPT 2005. LNCS, vol. 3788, pp. 333–352. Springer, Heidelberg (2005)
13. Courtois, N.T., Pieprzyk, J.: Cryptanalysis of Block Ciphers with Overdefined Systems of Equations. In: Zheng, Y. (ed.) ASIACRYPT 2002. LNCS, vol. 2501, pp. 267–287. Springer, Heidelberg (2002)
14. Daemen, J., Rijmen, V.: A New MAC Construction ALRED and a Specific Instance ALPHA-MAC. In: Gilbert, H., Handschuh, H. (eds.) FSE 2005. LNCS, vol. 3557, pp. 1–17. Springer, Heidelberg (2005)
15. Daemen, J., Rijmen, V.: The Pelican MAC Function. Cryptology ePrint Archive, Report 2005/088 (2005), http://eprint.iacr.org/
16. Dobbertin, H., Rijmen, V., Sowa, A. (eds.): AES 2005. LNCS, vol. 3373. Springer, Heidelberg (2005)
17. Dunkelman, O., Keller, N.: A New Attack on the LEX Stream Cipher. In: Pieprzyk, J. (ed.) ASIACRYPT 2008. LNCS, vol. 5350, pp. 539–556. Springer, Heidelberg (2008)
18. Dunkelman, O., Keller, N.: Cryptanalysis of the Stream Cipher LEX (2010), http://www.ma.huji.ac.il/~nkeller/Crypt-jour-LEX.pdf
19. Dunkelman, O., Keller, N.: The effects of the omission of last round's mixcolumns on aes. Inf. Process. Lett. 110(8-9), 304–308 (2010)
20. Dunkelman, O., Keller, N., Shamir, A.: Improved Single-Key Attacks on 8-Round AES-192 and AES-256. In: Abe, M. (ed.) ASIACRYPT 2010. LNCS, vol. 6477, pp. 158–176. Springer, Heidelberg (2010)
21. Dunkelman, O., Keller, N., Shamir, A.: Alred blues: New attacks on aes-based mac's. Cryptology ePrint Archive, Report 2011/095 (2011), http://eprint.iacr.org/
22. Gilbert, H. (ed.): EUROCRYPT 2010. LNCS, vol. 6110. Springer, Heidelberg (2010)
23. Halevi, S. (ed.): CRYPTO 2009. LNCS, vol. 5677. Springer, Heidelberg (2009)
24. Keliher, L.: Refined Analysis of Bounds Related to Linear and Differential Cryptanalysis for the AES. [16], 42–57
25. Keliher, L., Meijer, H., Tavares, S.: Improving the Upper Bound on the Maximum Average Linear Hull Probability for Rijndael. In: Vaudenay, S., Youssef, A.M. (eds.) SAC 2001. LNCS, vol. 2259, pp. 112–128. Springer, Heidelberg (2001)
26. Keliher, L., Meijer, H., Tavares, S.: New Method for Upper Bounding the Maximum Average Linear Hull Probability for SPNs. In: Pfitzmann, B. (ed.) EUROCRYPT 2001. LNCS, vol. 2045, pp. 420–436. Springer, Heidelberg (2001)
27. Khovratovich, D., Biryukov, A., Nikolic, I.: Speeding up Collision Search for Byte-Oriented Hash Functions. In: Fischlin, M. (ed.) CT-RSA 2009. LNCS, vol. 5473, pp. 164–181. Springer, Heidelberg (2009)

28. Monnerat, J., Vaudenay, S.: On Some Weak Extensions of AES and BES. In: López, J., Qing, S., Okamoto, E. (eds.) ICICS 2004. LNCS, vol. 3269, pp. 414–426. Springer, Heidelberg (2004)
29. Murphy, S., Robshaw, M.J.B.: Essential Algebraic Structure within the AES. In: Yung, M. (ed.) CRYPTO 2002. LNCS, vol. 2442, pp. 1–16. Springer, Heidelberg (2002)
30. NIST: Advanced Encryption Standard (AES), FIPS 197. Technical report, NIST (November 2001)
31. Piret, G., Quisquater, J.-J.: A Differential Fault Attack Technique against SPN Structures, with Application to the AES and KHAZAD. In: Walter, C.D., Koç, Ç.K., Paar, C. (eds.) CHES 2003. LNCS, vol. 2779, pp. 77–88. Springer, Heidelberg (2003)
32. Yuan, Z., Wang, W., Jia, K., Xu, G., Wang, X.: New Birthday Attacks on Some MACs Based on Block Ciphers. [23], 209–230

How to Improve Rebound Attacks

María Naya-Plasencia*

FHNW, Windisch, Switzerland

Abstract. Rebound attacks are a state-of-the-art analysis method for hash functions. These cryptanalysis methods are based on a well chosen differential path and have been applied to several hash functions from the SHA-3 competition, providing the best known analysis in these cases. In this paper we study rebound attacks in detail and find for a large number of cases that the complexities of existing attacks can be improved.

This is done by identifying problems that optimally adapt to the cryptanalytic situation, and by using better algorithms to find solutions for the differential path. Our improvements affect one particular operation that appears in most rebound attacks and which is often the bottleneck of the attacks. This operation, which varies depending on the attack, can be roughly described as *merging* large lists. As a result, we introduce new general purpose algorithms for enabling further rebound analysis to be as performant as possible. We illustrate our new algorithms on real hash functions.

Keywords: hash functions, SHA-3 competition, rebound attacks, algorithms.

1 Introduction

The rebound attack is a recent technique introduced in [13] by Mendel *et al.* It was conceived to analyze AES-like hash functions (like Grøstl [7] in [14,8,16], Echo [2] in [14,8,18], Whirlpool [1] in [11]). A rebound attack is composed of two parts: the inbound phase and the outbound phase. The aim of the inbound phase is to find, at a low cost, a large number of pairs of values that satisfy a part of a differential path that would be very expensive to satisfy in a probabilistic way. The outbound phase then uses these values to perform an attack.

This technique has been applied to other algorithms with inner permutations which are not AES-like; for instance it has been applied to JH [21] (reduced to 22 rounds) in [17] and Luffa [4] (reduced to 7 rounds) in [10]; both of those hash functions use Sboxes of size 4×4 and have a linear part in which the mixing is done in a very different way than in the AES. The hash function LANE [9], which includes several AES states, each treated by the AES round transformation, and a different transformation for mixing these states has also been analysed in [12,22] using rebound attacks.

* Supported by the National Competence Center in Research on Mobile Information and Communication Systems (NCCR-MICS), a center of the Swiss National Science Foundation under grant number 5005-67322.

P. Rogaway (Ed.): CRYPTO 2011, LNCS 6841, pp. 188–205, 2011.

In these cryptanalysis results, the rebound attack technique needs to be refined and adapted to each case, but all of them follow the same scheme: first find a differential path, then find solutions verifying this differential path. This paper focuses on optimizing the latter part. In all the previously mentioned cryptanalysis, that part involves enumerating, from a very large set of possible candidates represented as a cross product of lists, all those that verify a given relation. We call this operation *"merging" the lists*. The merging problem can be described more formally as follows.

Merging problem with respect to t: Let t be a Boolean function taking N k-bit words as input, *i.e.* $t : (\{0,1\}^k)^N \rightarrow \{0,1\}$. Let $L_1, \ldots, L_N$ be N given lists of k-bit words drawn uniformly and independently at random from $\{0,1\}^k$. We assume that the probability over all N-tuples X in $L_1 \times \ldots \times L_N$ that $t(X) = 1$ is P_t. For any given function t and any given N-tuple of lists $(L_1, \ldots, L_N)$ the *merging problem* consists in finding the list $\mathcal{L}_{sol}$ of all $X \in L_1 \times \ldots \times L_N$ satisfying $t(X) = 1$. We call this operation *merging* the lists $L_1, \ldots, L_N$ to obtain $\mathcal{L}_{sol}$.

It is assumed that the image of a given input under t can be easily computed. In the following, the size of a list L is denoted by $|L|$. A brute force method for solving this problem therefore consists in enumerating all the $|L_1| \times \ldots \times |L_N|$ inputs, in computing t on all of them and in keeping the ones verifying $t = 1$. Note that, in the lack of any additional information on t, it is theoretically impossible to do better. However, in practice, the function t often has a set of properties which can be exploited to optimize this approach. We aim at reducing the number of candidates which have to be examined, in some cases by a preliminary sieving similar to the one used in [5]. This paper presents such optimization techniques, that, when applied to most of the rebound attacks published on the SHA-3 candidates, yield significant improvements in the overall time and/or memory complexities of the attack, as shown on Table 1. In this table we can see that we have considered the best existing attacks against four hash functions and the best rebound attack on a fifth (two of them are finalists and two are second-round candidates of the SHA-3 competition), where by best attack we denote the one on the highest number of rounds. We have been able to improve their complexities by scrutinizing the original attack and finding a more efficient algorithm for obtaining the solutions for the differential path. Most of the time the improvement relies on a better *merging* of the lists, and sometimes it is due to the use of more adequate conditions in the general algorithm. Let us recall here that the aim is to find *all* the N-tuples that verify $t = 1$ for a complex function t, which is significantly different from finding just *one* (or few) of them for a linear t such as in [20,19,6,3]. As in the previous rebound analysis, we will work throughout the paper with average values in the probabilistic cases.

In Section 2, we define *Problem 1* that corresponds to functions t with a particular form, and we propose three generic algorithms to solve it. These 3 algorithms have different optimal scenarios. Some examples of applications are given. In Section 3 we define *Problem 2* and propose the *stop-in-the-middle algorithms* for solving it. We also present two concrete algorithms in this family

Table 1. Improvements on best known attacks. The highlighted values are the improved complexities. For Luffa we consider the best known rebound attack where the complexities presented in the second row have already been obtained in [10] by a dedicated algorithm similar to our general approach.

Hash function	SHA3 Round	Best Known Analysis	Rounds / total	Previous Time	Previous Memory	Ref.	This paper Time	This paper Memory
JH	Final	semi-free-start coll.	16 / 42	2^{190}	2^{104}	[17]	$\mathbf{2^{97}}$	$\mathbf{2^{97}}$
JH		semi-free-start near coll.	22 / 42	2^{168}	$2^{143.70}$	[17]	$\mathbf{2^{96}}$	$\mathbf{2^{96}}$
Grøstl-256	Final*	(compr. function property)	10 / 10	2^{192}	2^{64}	[16]	$\mathbf{2^{182}}$	2^{64}
Grøstl-256		(internal permutation dist.)	10 / 10	2^{192}	2^{64}	[16]	$\mathbf{2^{175}}$	2^{64}
Grøstl-512		(compr. function property)	11 / 14	2^{640}	2^{64}	[16]	$\mathbf{2^{630}}$	2^{64}
ECHO-256	2nd	internal permutation dist.	8 / 8	2^{182}	2^{37}	[18]	$\mathbf{2^{151}}$	2^{67}
Luffa	2nd	semi-free-start coll.	7 / 8	2^{132}	$2^{68.8}$	[10]	$\mathbf{2^{112.9}}$	$2^{68.8}$
							$(\mathbf{2^{104}})$	(2^{102})
LANE-256	1st	semi-free-start coll.	6+3 / 6+3	2^{96}	2^{88}	[12]	$\mathbf{2^{80}}$	$\mathbf{2^{66}}$
LANE-512		semi-free-start coll.	8+4 / 8+4	2^{224}	2^{128}	[12]	2^{224}	$\mathbf{2^{66}}$

* The Grøstl analysis does not apply after the final round tweak.

applied to the scenarios of ECHO and LANE. In Section 4 we show briefly how applying these algorithms combined with an appropriate definition and decomposition of the problem in each case, allows us to improve the complexities of the best known rebound attacks on 5 SHA-3 candidates. Besides the results in Table 1, the main interest of this paper is to present a general framework for improving rebound attacks. We introduce several new algorithms that considerably improve the overall effectiveness when the attack needs to *merge* large lists. We provide a formal definition of the field of application of those algorithms, and describe them as a set of constraints on t, in hope that designers of rebound attacks will be able to easily *identify* scenarios where one of these algorithms, or variants, may be applied. This was motivated by our own research path, when we realized that a generalization of the techniques leveraged in specific cases allowed us to find similar improvements in almost all of the rebound attacks that we have studied so far.

2 When *t* is Group-Wise

In some cases we can considerably reduce the complexity of the *merging* problem by redefining it into a more concrete one. We consider here a very common case that will appear in many rebound scenarios, as we will later show with the examples. This case corresponds to a function t that can be decomposed in smaller functions. After introducing the general problem, we will illustrate it with an example. Though we preferred to state the problem in full generality for any possible N, in the concrete rebound examples that we studied, the number of lists N was either 2, 4 or 6. Also, the elements of each list can be decomposed in sets of small size s, where s is typically the size of the involved Sbox; and z is the number of such sets involved[1] in the function t.

[1] Sometimes, elements are only partially involved in t.

Problem 1: Let $L_1, \ldots, L_N$ be N lists of size $2^{l_1}, \ldots, 2^{l_N}$ respectively, where the elements are drawn uniformly and independently at random from $\{0,1\}^k$.

Let t be a Boolean function, $t : \left(\{0,1\}^k\right)^N \to \{0,1\}$ for which there exists $N' < N$, an integer z and some triples of functions $t_j : \{0,1\}^{2s} \to \{0,1\}$, $f_j : (\{0,1\}^k)^{N'} \to \{0,1\}^s$ and $f'_j : (\{0,1\}^k)^{(N-N')} \to \{0,1\}^s$ for $j = 1, \ldots, z$ such that, $\forall\, (\boldsymbol{x_1}, \ldots, \boldsymbol{x_N}) \in L_1 \times \ldots \times L_N$:

$$t(\boldsymbol{x_1}, \ldots, \boldsymbol{x_N}) = 1 \quad \Leftrightarrow \quad \begin{array}{l} \forall j = 1, \ldots, z, \\ \left\{ \begin{array}{l} t_j(v_j, v'_j) = 1 \\ \text{with } v_j = f_j(\boldsymbol{x_1}, \ldots, \boldsymbol{x_{N'}}) \\ \text{and } v'_j = f'_j(\boldsymbol{x_{N'+1}}, \ldots, \boldsymbol{x_N}) \end{array} \right. \end{array}$$

Let P_t be the probability that $t = 1$ for a random input.

Problem 1 consists in *merging* these N lists to obtain the set $\mathcal{L}_{sol}$, of size $P_t 2^{\sum_{i=1}^{N} l_i}$, of all N-tuples of $(L_1 \times \ldots \times L_N)$ verifying $t = 1$.

Reduction from N to 2: For any $N \geq 2$ *Problem 1* can be reduced to an equivalent and simplified problem with $N = 2$, *i.e.* merging two lists L_A and L_B, which consist of elements in $(\{0,1\}^s)^z$ corresponding to $\boldsymbol{x_A} = \boldsymbol{v} = (v_1, \ldots, v_z)$ and $\boldsymbol{x_B} = \boldsymbol{v'} = (v'_1, \ldots, v'_z)$, with respect to the function $\boldsymbol{x_A}, \boldsymbol{x_B} \mapsto \Pi_{j=1}^{z} t_j(v_j, v'_j)$. The reduction is performed as follows:

1. Build a table T_A^* of size $2^{\sum_{i=1}^{N'} l_i}$ storing each element $\boldsymbol{e_A} = (\boldsymbol{x_1}, \ldots, \boldsymbol{x_{N'}})$ of $L_1 \times \ldots \times L_{N'}$, indexed[2] by the value of $(f_1(\boldsymbol{e_A}), \ldots, f_z(\boldsymbol{e_A}))$, *i.e.* $(v_1, \ldots, v_z)$. Store the corresponding $(v_1, \ldots, v_z)$ in a list L_A. Note that several $\boldsymbol{e_A}$ may lead to the same value of $(v_1, \ldots, v_z)$.
2. Build a similar table T_B^* of size $2^{\sum_{i=N'+1}^{N} l_i}$ storing each element $\boldsymbol{e_B} = (\boldsymbol{x_{N'+1}}, \ldots, \boldsymbol{x_N})$ of $L_{N'+1} \times \ldots \times L_N$, indexed by $(f_1(\boldsymbol{e_B}), \ldots, f_z(\boldsymbol{e_B}))$, *i.e.* $(v'_1, \ldots, v'_z)$. Store $(v'_1, \ldots, v'_z)$ in a list L_B.
3. *Merge* L_A and L_B with respect to $\Pi_{j=1}^{z} t_j$ and obtain $\mathcal{L}_{sol}$.
4. Build $\mathcal{L}_{sol}^*$ by iterating over each pair $((v_1, \ldots, v_z), (v'_1, \ldots, v'_z))$ of $\mathcal{L}_{sol}$, and adding the set of all $(\boldsymbol{x_1}, \ldots, \boldsymbol{x_{N'}}, \boldsymbol{x_{N+1}}, \ldots, \boldsymbol{x_N}) \in T_A^*[(v_1, \ldots, v_z)] \times T_B^*[(v'_1, \ldots, v'_z)]$. $\mathcal{L}_{sol}^*$ is the solution to the original problem.

Let $2^{T_{\text{merge}}}, 2^{M_{\text{merge}}}$ be the time and memory complexities of step 3. The total time complexity of solving *Problem 1* is $\mathcal{O}(sz2^{\sum_{i=1}^{N'} l_i} + sz2^{\sum_{i=N'+1}^{N} l_i} + 2^{T_{\text{merge}}} + P_t 2^{\sum_{i=1}^{N} l_i})$ where the last term comes from the fact that only the N-tuples satisfying $t = 1$ are examined at step 4 because of the sieve applied at step 3. The proportion of such tuples is then P_t. The memory complexity[3] is $\mathcal{O}((zs + N'k)2^{\sum_{i=1}^{N'} l_i} + (zs + (N - N')k)2^{\sum_{i=N'+1}^{N} l_i} + 2^{M_{\text{merge}}} + P_t 2^{\sum_{i=1}^{N} l_i})$, where the last term appears only when the solutions must be stored.

---

[2] Here and in the following sections we can use standard hash tables for storage and lookup in constant time, since the keys are integers.

[3] The first two terms, corresponding to the storage of T_A^* and T_B^* could be avoided if they were the bottleneck by slightly increasing the time complexity by a factor of 2.

Using the brute force approach, $2^{T_{\text{merge}}}$ would be $2^{l_A+l_B}$ where 2^{l_A} (respectively 2^{l_B}) denotes the size of L_A (L_B), and $2^{M_{\text{merge}}}$ would be negligible. We present in the following sections some algorithms for solving *Problem 1* considering $N = 2$ with L_A and L_B, that provide better complexities than the brute force approach. Note that the roles of L_A and L_B are assigned by choice to obtain the best overall complexity. Those algorithms can be applied for obtaining a smaller $2^{T_{\text{merge}}}$ when $N > 2$.

2.1 Basic Algorithm for Solving *Problem 1*: Instant Matching

As s is typically very small we can enumerate the solutions (v_j, v'_j) of $t_j(v_j, v'_j) = 1$ and store them in tables T_j of size $\leq 2^{2s}$, indexed by v'_j. This costs $\mathcal{O}(z \cdot 2^{2s})$ in time and memory. We propose in Fig. 1 a first algorithm for solving *Problem 1*, which has lower complexity than the brute-force approach. Although being the simplest algorithm presented in this paper, it has not been applied in critical steps of some of the previously mentioned attacks, though it could yield significant improvements.

Require: Two lists L_A, L_B and a Boolean function t as described in *Problem 1*.
Ensure: The returned list $\mathcal{L}_{sol}$ will contain all elements of $L_A \times L_B$ verifying t.

1: **for** j from 1 to z **do**
2:     **for all** (v_j, v'_j) in $\{0,1\}^s \times \{0,1\}^s$ **do**
3:         **if** $t_j(v_j, v'_j) = 1$, **then** add v_j to $T_j[v'_j]$.
4: **for** each $(v'_1, \ldots, v'_z) \in L_B$ **do**
5:     Empty L_{aux}.
6:     **for** j from 1 to z **do**
7:         **if** $T_j[v'_j]$ is empty, **then** go to 4.
8:     Add all tuples $(v_1, \ldots, v_z)$ verifying $\forall j\ v_j \in T_j[v'_j]$ to L_{aux}.
9:     **for** each $(v_1, \ldots, v_z)$ in L_{aux} **do**
10:         **if** $(v_1, \ldots, v_z) \in L_A$ **then**
11:             Add $(v_1, \ldots, v_z, v'_1, \ldots, v'_z)$ to $\mathcal{L}_{sol}$.
12: Return $\mathcal{L}_{sol}$.

Fig. 1. Instant matching algorithm

Let 2^{-p_j} be the probability over all pairs (v_j, v'_j) that $t_j(v_j, v'_j) = 1$. The relationship between t and the $(t_j)_{1 \leq j \leq z}$ implies that $\sum_{j=1}^{z} p_j = -\log_2(P_t)$ where P_t is the probability that $t = 1$.

Let us determine the average size of L_{aux}. The average size of $T_j[v'_j]$ over all v'_j is 2^{s-p_j}. Then the average size of L_{aux} is $2^{zs - \sum_{j=1}^{z} p_j} = P_t 2^{zs}$. It follows that the time complexity of the algorithm is $\mathcal{O}(z2^s + zP_t 2^{l_B+zs})$ and is proportional to the product of the size of L_B by the average size of[4] L_{aux}. The memory complexity is $\mathcal{O}(z2^s + 2^{l_A} + 2^{l_B} + P_t 2^{l_A+l_B})$. In some cases, the last term can disappear, namely if we do not need to store the list $\mathcal{L}_{sol}$, but just use each solution as soon as it is obtained. The same way, the list L_B does not need to be stored, if it can be given on the fly.

[4] The cost of building and storing the lists $T_j[v'_j]$ is negligible.

We now describe a concrete example of application of the *instant-matching algorithm* in a case included in a particular rebound attack, improving its complexity. In the extended version of this paper [15] two more examples are provided in Appendix A, where it clearly appears that identifying and isolating the most appropriate problem (or problems) to solve is of major importance. These two last examples might help also to understand the role of f_j and f'_j.

Example 1: Application of the Instant Matching Algorithm. We use here a case presented in the analysis of JH [17] which is the attack on 8 rounds using one inbound when the dimension of a block of bits denoted by d is 4. Here we improve step 3 of the attack, which is also the bottleneck in time complexity. Two lists are given, L_A and L_B of size $2^{24.18}$ elements each. The aim of step 3 is to *merge* those lists, *i.e.* find all pairs $(\boldsymbol{v}, \boldsymbol{v'}) \in L_A \times L_B$ verifying 10 conditions on groups of $s = 4$ bits of $(\boldsymbol{v}, \boldsymbol{v'})$.

In [17] this is solved by exhaustive search, *i.e.* all possible pairs are examined and only the ones that verify the 10 conditions are kept, which has cost $2^{48.36}$. We can improve this complexity by applying the instant-matching algorithm: first, we notice that 6 out of these 10 conditions can be written as

$$t_j(v_j, v'_j) = 1, \forall j \in \{1, \ldots, 6\},$$

where variables v_j and v'_j represent groups of differences of 4 bits. The functions t_j return 1 when the linear function of JH, L, applied to v_j and v'_j produces 4 bits out of 8 without difference in the wanted positions. Those functions t_j can be computed directly by using a precomputed table of size 2^8.

This is an instance of *Problem 1* with the parameters: $z = 6$ (corresponding to the number of relations $t_1, \ldots, t_6$), and $p_j = 3.91\ \forall j$. Hence $P_t 2^{zs} = 2^{0.09 \cdot 6} = 2^{0.54} \simeq 1.45$. The instant-matching algorithm allows us to find all pairs satisfying these 6 conditions with a complexity of $2^{27.8}$ in time and no additional memory. We then obtain $2^{24.9}$ pairs of elements that pass the first 6 conditions. To complete step 3 of the attack, we evaluate the 4 remaining conditions for each pair, for a global complexity of $2^{24.9}$.

To summarize, we were able to resolve step 3 of the attack with a time complexity of about $2^{27.8}$, improving significantly the complexity of $2^{48.36}$ given in [17].

2.2 Solving Problem 1 When $P_t 2^{zs} > 2^{l_A}$: Gradual Matching

In Fig. 2 we present an algorithm for solving *Problem 1* that is useful in cases where the average size of L_{aux} exceeds the size of L_A, *i.e.*[5] $P_t 2^{zs} > 2^{l_A}$. In this case the instant-matching algorithm has a higher complexity than the exhaustive search. This is why here, instead of directly matching the z groups that appear in relation t, we will first match the $z' < z$ ones, and next, the $z - z'$ remaining

[5] When $P_t 2^{zs}$ is close to 2^{l_A} this algorithm might also outperform the instant-matching technique.

ones. We present here how to use one step of the *gradual-matching algorithm* for solving Problem 1. This algorithm reminds the method used in Example 1 where the problem is first solved with only 6 relations. But the difference is that the remaining $z - z'$ relations can also be written in the form needed for *Problem 1* and $P_t 2^{zs} > 2^{l_A}$. Let us suppose that we choose z' so that $z's < l_A$ (the best value for z' depends on the situation).

Require: Two lists L_A and L_B and a function t as described in *Problem 1.*
Ensure: List $\mathcal{L}_{sol} \subset L_A \times L_B$ of all elements verifying t.
for j from 1 to z **do**
  for all (v_j, v'_j) in $\{0,1\}^s \times \{0,1\}^s$ **do**
    if $t_j(v_j, v'_j) = 1$, **then** add v_j to $T_j[v'_j]$.
for each $\alpha = (\alpha_1, \ldots, \alpha_{z'})$ in $(\{0,1\}^s)^{z'}$ **do**
  Empty L_{aux}.
  Consider the sublist $L_B(\alpha)$ of all elements in L_B with $(v'_1, \ldots, v'_{z'}) = \alpha$.
  for each $(v_1, \ldots, v_{z'})$ in $T_1[\alpha_1] \times \ldots \times T_{z'}[\alpha_{z'}]$ **do**
    add $(v_1, \ldots, v_{z'})$ to L_{aux}.
  for each $\gamma = (\gamma_1, \ldots, \gamma_{z'})$ in L_{aux} **do**
    Consider the sublist $L_A(\gamma)$ of all elements of L_A with $(v_1, \ldots, v_{z'}) = \gamma$.
    Merge $L_A(\gamma)$ with $L_B(\alpha)$ with respect to $t' = \Pi_{j=z'+1}^{z} t_j$.
    Add the solutions to $\mathcal{L}_{sol}$.
Return $\mathcal{L}_{sol}$, containing about $P_t 2^{l_A + l_B}$ elements.

Fig. 2. Gradual matching algorithm

Let 2^{merge} be the time complexity of *merging* once lists $L_B(\alpha)$ and $L_A(\gamma)$ as defined in Fig. 2. Since their respective average sizes are $2^{l_A - z's}$ and $2^{l_B - z's}$ the complexity of the brute force is $2^{l_A + l_B - 2z's}$. It can be improved by using one of the proposed algorithms from this section but it cannot be smaller than the size of the resulting merged list, *i.e.* $2^{l_A + l_B - 2z's - \sum_{j=z'+1}^{z} p_j}$. Now the average size of L_{aux}[6] is $\mathcal{S} = 2^{z's - \sum_{j=1}^{z'} p_j}$. Then, the time complexity of this algorithm is $\mathcal{O}(z2^s + 2^{z's}(z' + \mathcal{S}2^{\text{merge}}))$. It is worth noticing that this complexity corresponds to $z'2^{z's} + 2^{l_A + l_B - \sum_{j=1}^{z'} p_j}$ when the intermediate lists are merged by the brute force algorithm and to $z'2^{z's} + P_t 2^{l_A + l_B}$ if they are merged by an optimal algorithm. The memory complexity is $\mathcal{O}(z2^s + 2^{l_A} + 2^{l_B} + \mathcal{S} + P_t 2^{l_A + l_B})$. Again, in some cases, the last term can disappear, if we do not need to store the list $\mathcal{L}_{sol}$, but just use the solutions on the fly.

2.3 Time-Memory Trade-Offs When $P_t 2^{zs} > 2^{l_A}$: Parallel Matching

The *parallel-matching algorithm* improves the time complexity of the gradual-matching by a time-memory trade-off and can be applied in the same situations.

[6] Here and in the previous section, there is no need for storing L_{aux}, as each element can be treated as soon as it is obtained, but these auxiliary lists are very useful for describing the complexities.

It is a generalization of an algorithm proposed in [10]. As the gradual-matching algorithm this algorithm first finds elements that verify $t_j = 1$ for $j \in \{1, \ldots, z'\}$ and then, for each of them, it checks if the remaining $(z - z')$ relations are also verified. However, in this algorithm, the matching of the z' relations is done in parallel for n and m relations, so that $z' = m + n$. The motivation of choosing different variables for n and m is showing that there is no need for them to be the same when applying the algorithm. We choose n so that $n < z$, $ns < l_A$

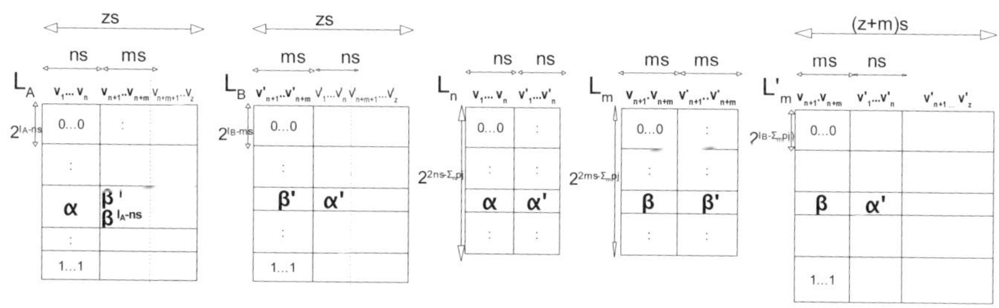

Fig. 3. Representation of the parallel-matching algorithm

and $ns < l_B$, and in the same way, we choose m $(n + m = z' \leq z)$. This algorithm will be explained with ordered lists, as it is more graphical and helps the understanding. However, since we can perform it with hash tables indexed by the values we want to have ordered, we do not need to take into account the logarithmic terms for ordering and searching in the final complexity. First we build the lists that we will use and that are represented in Fig. 3:

- We order the list L_A by the first n groups $(v_1, \ldots, v_n)$. L_A has $2^{l_A - sn}$ elements in average corresponding to a given value of these n groups.
- We order the list L_B by the next m groups $(v'_{n+1}, \ldots, v'_{n+m})$. L_B has $2^{l_B - sm}$ elements in average corresponding to a given value of these m groups.
- We build the list L_n of size $2^{2ns - \sum_{j=1}^{n} p_j}$ formed by all $(v_1, \ldots, v_n, v'_1, \ldots, v'_n)$ with $v_j \in T_j[v'_j]$ for all $1 \leq j \leq n$. All the elements from this list satisfy $t_j(v_j, v'_j) = 1$ for $j \in [1, \ldots, n]$.
- We build the list L_m of size $2^{2ms - \sum_{j=n+1}^{n+m} p_j}$ formed by all $(v_{n+1}, \ldots, v_{n+m}, v'_{n+1}, \ldots, v'_{n+m})$ with $v_j \in T_j[v'_j]$ for all $(n + 1) \leq j \leq (n + m)$. All the elements from this list satisfy $t_j(v_j, v'_j) = 1$ for $j \in [n + 1, \ldots, n + m]$.
- From L_m and L_B we build L'_m as follows: for each (β, β') in L_m, we add to L'_m all elements $(\beta, v'_1, \ldots, v'_z)$ of L_B such that $(v'_{n+1}, \ldots, v'_{n+m}) = \beta'$ and we store them ordered by the values of $(\beta, v'_1, \ldots, v'_n)$. The average size of L'_m is $2^{l_B + sm - \sum_{j=n+1}^{n+m} p_j}$. Then we perform the algorithm given in Fig. 4.

In total we find the $2^{l_A + l_B - \sum_{j=1}^{z} p_j}$ existing matches, with a complexity of $\mathcal{O}(2^{l_n} + 2^{l_m} + 2^{l_A + l_B - \sum_{j=1}^{n+m} p_j} + 2^{l_A + ns - \sum_{j=1}^{n} p_j} + 2^{l_B + ms - \sum_{j=n+1}^{m} p_j})$ in time

for each (α, α') in L_n **do**
  for each $(v_1, \ldots, v_z)$ in L_A with $(v_1, \ldots, v_n) = \alpha$ **do**
    if L'_m contains any element $(v_{n+1}, \ldots, v_{n+m}, v'_1, \ldots, v'_z)$ starting by $(v_{n+1}, \ldots, v_{n+m}, \alpha')$ **then**
      if $(v_1, \ldots, v_z, v'_1, \ldots, v'_z)$ satisfies the remaning $(z-n-m)$ conditions **then**
        Add $(v_1, \ldots, v_z, v'_1, \ldots, v'_z)$ to $\mathcal{L}_{sol}$.
Return $\mathcal{L}_{sol}$ containing about $P_t 2^{l_A + l_B}$ elements.

Fig. 4. Parallel matching algorithm

and $\mathcal{O}(2^{l_n} + 2^{l_m} + 2^{l_B} + 2^{l_B + ms - \sum_{j=n+1}^{m} p_j} + 2^{l_A + l_B - \sum_{j=1}^{z} p_j})$ in memory, where the last term corresponds to the storage of all solutions, not always needed. In this case, the storage of L_A is not necessary.

2.4 Example 2: Gradual Matching vs Parallel Matching

We are going to apply both previous algorithms to the analysis of Luffa presented in [10]. We are given two lists L_A and L_B of size 2^{67} and $2^{65.6}$. These lists contain elements formed by $z = 52$ groups of differences of $s = 4$ bits. List L_A contains the possible differences for the input of 52 Sboxes. List L_B contains the possible differences for the output of the same 52 Sboxes. For the j-th Sbox, the probability that one input difference can be associated to one output difference is $2^{-p_j} = 2^{-1.23}$. The average size of L_{aux} if we apply the instant-matching algorithm is then $P_t 2^{zs} = 2^{144.04}$. In this case t can be decomposed in 52 t_j, one per Sbox. So $t_j(v_j, v'_j) = 1$ if there exists $x \in \{0,1\}^s$ such that

$$\text{Sbox}(x) \oplus \text{Sbox}(x \oplus v_j) = v'_j.$$

The brute force algorithm for solving this problem has complexity of $2^{65.6+67} = 2^{132.6}$ in time and of $2^{68.8}$ in memory. If we apply the gradual-matching algorithm with $z' = 16$ we have $\mathcal{S} = 2^{44.32}$, and we obtain the $2^{68.8}$ solutions with a time complexity of $2^{112.9}$ and the same memory as before as no additional memory is needed. If instead we apply the parallel-matching algorithm with $m = n = 13$, we can obtain the solutions with a time complexity of 2^{104} and a memory complexity of 2^{102}. Different choices of parameters allow many other time-memory trade-offs, but we just show here the one that provides the lowest time complexity, and so the highest memory needs, for contrast with the gradual matching algorithm.

3 Stop-in-the-Middle Algorithms

In this section we present another case that allows to reduce the complexity of solving the basic problem. It is described in *Problem 2*. Then, we define the main lines of the *stop-in-the-middle algorithms*, that we use for solving *Problem 2*. Next, we present such an algorithm that solves *Problem 2* in the scenario of LANE-256. Then a more complex variant of this algorithm is applied to an

ECHO-256 scenario. We believe that, in particular, this kind of algorithms can be adapted and applied to functions that use several AES (like) states in parallel which are then merged at the end of each round.

In the following, we consider a permutation F from $\{0,1\}^{sk}$ to $\{0,1\}^{sk}$ and we assume that there exists a decomposition function ϕ (respectively ψ) of the input of F (respectively the output) in k elements of $\{0,1\}^s$. These two decompositions may be different. Then, instead of the original function F we will now focus on the function $f = \psi \circ F \circ \phi^{-1}$ which is a function over $(\{0,1\}^s)^k$ (see Fig. 5). In the following (u, w) denotes the word corresponding to the concatenation of the vectors u and w.

Problem 2: Let z_A and z_B be two integers less than or equal to k . Let L_A be a list of elements in $(\{0,1\}^s)^{z_A}$ and L_B be a list of elements on $(\{0,1\}^s)^{z_B}$. The *Problem 2* consists on finding all triples (a, b, c) with $a \in L_A$, $b \in L_B$ and $c \in L_C = (\{0,1\}^s)^k$ such that

$$f(c) \oplus f(c \oplus (a, 0^{s(k-z_A)})) = (b, 0^{s(k-z_B)}),$$

where there exists the function $F_1 : (\{0,1\}^s)^k \rightarrow (\{0,1\}^s)^k$ and some permutations of $\{0,1\}^s$, $g_1, \ldots, g_k$ and $h_1, \ldots, h_k$ over $\{0,1\}^s$ such that

$$f = H \circ F_1 \circ G$$

where

$$\begin{aligned} G : \quad & (\{0,1\}^s)^k \rightarrow (\{0,1\}^s)^k \\ & (x_1, \ldots, x_k) \rightarrow (g_1(x_1), \ldots, g_k(x_k)) \end{aligned}$$

and

$$\begin{aligned} H : \quad & (\{0,1\}^s)^k \rightarrow (\{0,1\}^s)^k \\ & (x_1, \ldots, x_k) \rightarrow (h_1(x_1), \ldots, h_k(x_k)) \end{aligned}$$

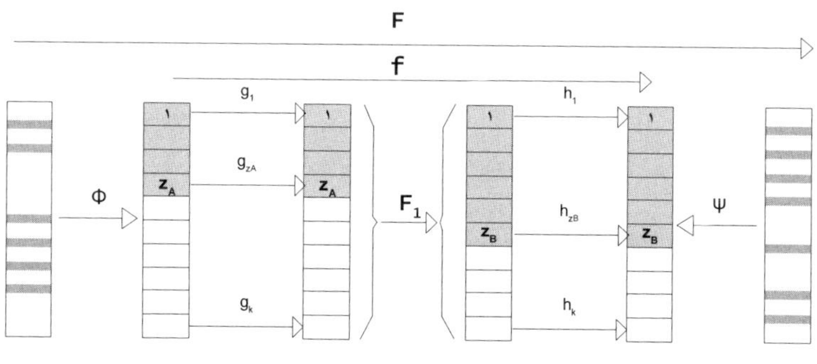

Fig. 5. Representation of F from *Problem 2*

for each b in L_B **do**
  for each $j \in [1, \ldots, z_B]$ **do**
    for each $y_j \in \{0,1\}^s$ **do**
      add $(h_j^{-1}(y_j), h_j^{-1}(y_j) \oplus h_j^{-1}(y_j \oplus b_j))$ to $L_{j,b}$.
for each a in L_A **do**
  for each $i \in [1, \ldots, z_A]$ **do**
    for each x_i in $\{0,1\}^s$ **do**
      add $(g_i(x_i), g_i(x_i) \oplus g_i(x_i \oplus a_i))$ to L_i.
  Using the previous lists L_i and $L_{j,b}$, match in the middle using F_1, *i.e.* construct $L_{aux} = \{(x, b_1, \ldots, b_{z_B}), x \in (\{0,1\}^s)^k\}$ such that $((F_1[g_1(x_1), \ldots, g_{z_A}(x_{z_A}), x^*)], F_1[g_1(x_1 \oplus a_1), \ldots, g_{z_A}(x_{z_A} \oplus a_{z_A}), x^*)]) = ((h_1^{-1}(y_1), \ldots, h_{z_B}^{-1}(y_{z_B}), y^*), (h_1^{-1}(y_1 \oplus b_1), \ldots, h_{z_B}^{-1}(y_{z_B} \oplus b_{z_B}), y^*))$ for some $x^* \in (\{0,1\}^s)^{k-z_A}$ and $y^* \in (\{0,1\}^s)^{k-z_B}$.
  for all $(x, b_1, \ldots, b_{z_B})$ in L_{aux} **do**
    if $b = (b_1, \ldots, b_{z_B}) \in L_B$ **then**
      add (a, b, x) to $\mathcal{L}_{sol}$.
Return $\mathcal{L}_{sol}$.

Fig. 6. General scheme of stop-in-the-middle algorithms

It is worth noting that we assume that both decompossitions ϕ and ψ have been chosen in an appropriate way such that the z_A words of a (respectively the z_B words of b) correspond to the first words of the input state (respectively of the output state). We call *stop-in-the-middle algorithms* those that solve *Problem 2* following the main general scheme described in Fig. 6. The associated complexities depend on the particular form of F_1, as we show in the next sections.

In the cases we have studied and that we detail below, the function f is formed by several AES transformations in parallel. We then expect $2^{l_A+l_B}$ solutions, as for each $a \in L_A$ and each $b \in L_B$ there exists one $c \in L_C$ so that the condition of *Problem 2* holds. The match-in-the-middle step is assumed to be simple due to the simple form of F_1 (typical functions F_1 are linear diffusion layers). For the same reason, L_{aux} can typically be written in a compact way, for example, in several independent lists.

3.1 Algorithm for Lane-256

Each lane of the internal state of Lane-256 is composed of two AES states. An AES state is a state of size 128 bits that can be seen as a 4×4 matrix of bytes. The AES transformations are noted: SB for SubBytes, SR for ShiftRows and MC for MixColumn. The transformation SC mixes the two AES states at the end of each round by interchanging their columns. We consider Fig. 7 that represents a part of the differential path used in [12]. In that attack it was treated as the merging of two inbounds and 2^{64} solutions were found with a complexity of 2^{96} in time and 2^{88} in memory. We consider the scheme represented in Fig. 7 where we have swapped lines and columns for a more easy intuitive understanding (so SR is applied to the columns and MC is applied to the lines).

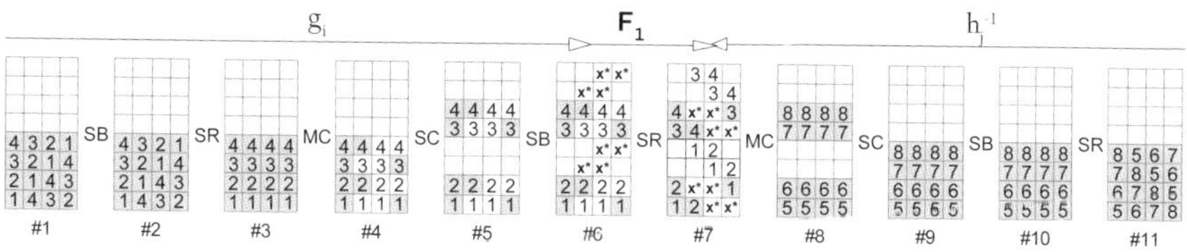

Fig. 7. Differential path associated to the first improvement on the LANE analysis

Using the example from [12], $l_A = 32$ and $l_B = 32$ and L_C is the list of all possible input values and needs to be neither stored nor computed. We consider that the input state (respectively the output state) of the function f presented in Fig. 7 is decomposed into eight 32-bit words (*i.e.* $s = 32$ and $k = 8$). The input differences and output differences that we consider in L_A and L_B correspond to the first $z_A = z_B = 4$ 32-bit words of the state. In Fig. 7 each one of the $4 + 4 = 8$ 32-bits active word corresponds to the four active bytes with the same number written on them (1 to 4 for the four active input words and 5 to 8 for the 4 active output words).

With the algorithm described in Fig. 8 we find the 2^{64} solutions with a complexity of 2^{66} in time and 2^{65} in memory. The time complexity associated to the studied path is $z_B 2^{l_B+32} + 2^{l_A+32}$. This comes from the fact that each L_i has average size 2^{16}. Then, $L_{5,6}$ and $L_{7,8}$ have size 2^{l_B+32}. Then the size of both L^0_{aux} and L^1_{aux} is 2^{l_B} since in each we keep the pairs of elements that match on 4 active bytes, and this happens with a probability of 2^{-64} (32 values and 32 differences); and the number of possible pairs is $2^{16+16+l_B+32}$. The memory complexity is $2^{l_B+32+1} + 2^{32+1} + 2^{l_A+l_B}$ for obtaining $2^{l_A+l_B}$ solutions. In [15] a detailed explanation on how this algorithm allows to considerably reduce the complexity of the LANE-256 semi-free-start collision presented in [12] is given, when applied jointly with other improvements concerning other steps of the attack.

3.2 Algorithm for ECHO-256

An ECHO-256 state is a state of size 2048 bits that can be seen as a 4×4 matrix of AES states. The ECHO operations BigSR, BigMC and BigSB are similar to the AES ones, but they operate on AES states instead of bytes. A SuperSbox is an Sbox defined by SR ∘ SB ∘ MC ∘ SR ∘ SB. Applied on an AES state, it can be seen as a 32×32 Sbox. We define a *SuperSbox set* as each one of the 4 (in the AES state) sets of bits that act as input and output of the SuperSbox. We define a *BigSuperSbox* as an Sbox defined by BigSR ∘ BigSB ∘ BigMC ∘ BigSR ∘ BigSB. Applied to ECHO it defines 4 sets of size 4 AES-states.

We consider Fig. 9, where each column represents the four AES states that form a BigSuperSbox at a certain state #i, for i from 1 to 13. Each possible differences in #1 in L_A consist of $z_A = 12$ 32-bit words and the possible

Require: Function f and lists L_A and L_B of differences in #1 and #11 respectively.
Ensure: List $\mathcal{L}_{sol} = \{(a, b, c)$ such that $f(c \oplus (a, 0^{s(k-z_A)})) \oplus f(c) = (b, 0^{s(k-z_B)})\}$.

for each b in L_B **do**
  for i from 5 to 8 **do**
    for each $y \in \{0,1\}^{32}$ **do**
      if $h_i^{-1}(y) \oplus h_i^{-1}(y \oplus b_i)$ has only the two bytes active (*see state #7*) **then**
        Store $(y, b_i, h_i^{-1}(y), h_i^{-1}(y \oplus b_i))$ in L_i, where the last two terms are truncated to the 2 active bytes.
  for each (y_5, b_5, u_5, w_5) from L_5 and (y_6, b_6, u_6, w_6) from L_6 **do**
    Add $(u_5, w_5, u_6, w_6, y_5, y_6, b_5, b_6)$ in $L_{5,6}$ indexed by u_5, w_5, u_6, w_6.
  for each (y_7, b_7, u_7, w_7) from L_7 and (y_8, b_8, u_8, w_8) from L_8 **do**
    Add $(u_7, w_7, u_8, w_8, y_7, y_8, b_7, b_8)$ in $L_{7,8}$ indexed by u_7, w_7, u_8, w_8.
  Empty L_5, L_6, L_7 and L_8.
for each a in L_A **do**
  for i from 1 to 4 **do**
    for each $x_i \in \{0,1\}^{32}$ **do**
      if $g_i(x_i) \oplus g_i(x_i \oplus a_i)$ has only the two bytes active (*see state #4*) **then**
        Store $(x_i, g_i(x_i), g_i(x_i \oplus a_i))$ in L_i, where the two last terms are truncated to the 2 active bytes.
  for i from 0 to 1 **do**
    for each $(x_{2i+1}, u_{2i+1}, w_{2i+1})$ in L_{2i+1} and $(x_{2i+2}, u_{2i+2}, w_{2i+2})$ in L_{2i+2} **do**
      if there exists an element in $L_{5+2i,6+2i}$ indexed by $(u_{2i+1}, w_{2i+1}, u_{2i+2}, w_{2i+2})$ **then**
        Add $(x_{2i+1}, x_{2i+2}, b_{5+2i}, b_{6+2i})$ to L_{aux}^i indexed by (b_{5+2i}, b_{6+2i}).
  for each (x_1, x_2, b_5, b_6) in L_{aux}^0 **do**
    for each (b_7, b_8) such that $(b_5, b_6, b_7, b_8) \in L_B$ **do**
      if there exists an element in L_{aux}^1 indexed by (b_7, b_8) **then**
        add $(a, (b_5, b_6, b_7, b_8), (x_1, x_2, x_3, x_4))$ to $\mathcal{L}_{sol}$.
Return $\mathcal{L}_{sol}$.

Fig. 8. Algorithm for solving two inbounds of LANE-256

differences in #13 consist of $z_B = 8$ 32-bit words, where L_B can be written as $L_B = L_{B_1} \times L_{B_2}$ with both L_{B_1} (associated to AES state B_1 in Fig. 9) and L_{B_2} (associated to AES state B_4) are subsets of $(\{0,1\}^{32})^4$ each of size 2^{32} (this is a particular case which has to be adapted in other cases). Finding solutions for this differential path with the previously mentioned conditions is a problem proposed in [18] and was solved in such a way that 2^{32} solutions could be found with a complexity of 2^{128} in time and 2^{37} in memory. We propose here a new algorithm that can solve it for obtaining 2^{64} solutions with the same time complexity and a memory of 2^{67}. Variants of our algorithm can be applied in several cases, like when the transition in #7 to #8 is from 2 active states to 3, or from 1 to 4 or from 4 to 1. Additionally we believe that it can improve the complexity of other future attacks on ECHO-256.

The list L_C contains all the possible values for the input state. This list needs to be neither computed nor stored. Here the aim is to find for each possible (a, b^1, b^2) in $L_A \times L_{B_1} \times L_{B_2}$ the associated c so that $f(c) \oplus f(c \oplus (a, 0^{s(k-z_A)}) =$

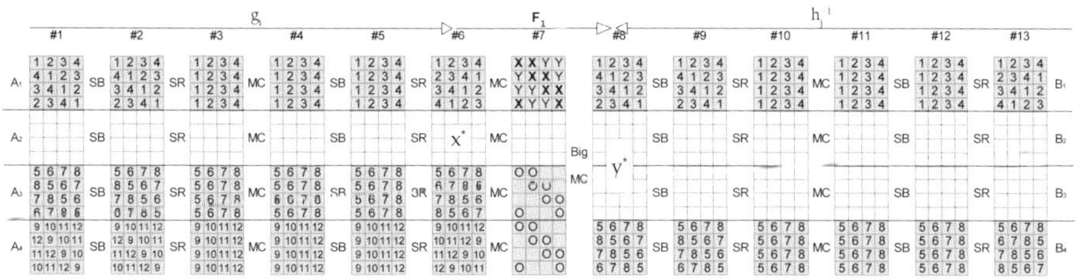

Fig. 9. Differential path on a BigSuperSbox of ECHO-256

$(b^1, b^2, 0^{s(k-z_B)})$. In Fig. 9 we can see how the function f can be written in the way requested by *Problem 2*. We omit the operation BigSR as it does not affect the states, as well as the round keys that are taken into account in the different g_i and h_j. For the sake of simplicity we consider in Fig. 9 the list L_B of possible differences before the last MC of the BigSuperSbox. This can be done by a simple transformation MC^{-1} of the differences in #B'. The grey bytes represent the bytes with differences. We can observe that, from #1 to #6 there are $z_A = 12$ independent active SuperSbox sets ($s = 32$), denoted in Fig. 9 by a number from 1 to 12. To each of these groups we can associate a difference from L_A and a value from L_C at state #1 and we can apply independently g_i, $i \in [1, \ldots, 12]$ to obtain the value and the difference of the group in #6. The same way, from #8 to #13 there are $z_B = 8$ independent active SuperSbox sets and the corresponding functions $h_i^{-1}, i \in [1, \ldots 8]$ that link state #13 with state #8. The function F_1 =MC∘BigMC takes a complete internal state in #6 and computes the corresponding state in #8. Let $f(x) = y$, and let $d_i^{\#7}$ the ith active diagonal in state #7. Without knowing the values of x^* nor of y^* represented in Fig. 9 we can still write the following equations that have to be verified, that are obtained from BigMC, and that are used in the algorithm:

$$2 \times d_i^{\#7} \oplus d_{i+4}^{\#7} \oplus d_{i+8}^{\#7} \oplus 9 \times d_i^{\#7} \oplus 3 \times d_{i+4}^{\#7} \oplus 6 \times d_{i+8}^{\#7} = h_i^{-1}(y_i) \oplus 3 \times h_{i+4}^{-1}(y_{i+4}) \quad (1)$$

for $i \in 1, \ldots, 4$ where the multiplication corresponds to the one in the definition of MC applied independently to each byte of the diagonal.

We consider that the input state (respectively the output state) of the function presented in Fig. 9 is decomposed into sixteen 32-bit words (*i.e.* $s = 32$ and $k = 16$). The input differences (respectively output differences) that we consider in L_A (L_B) correspond to the first $z_A = 12$ ($z_B = 8$) 32-bit words of the state. In Fig. 9 each one of the 12 (respectively 8) 32-bits active word from the input (respectively the output) corresponds to the four active bytes with the same number written on them (1 to 12 for the twelve active input words and 1 to 8 for the eight active output words).

Let V_X (V_Y, V_O respectively) be the values at the positions in #7 marked with an X (Y, O) and Δ_X (Δ_Y, Δ_O) their differences. Let $\Delta_{j'}^{\#r}$ be an auxiliary variable denoting the difference for the SuperSbox set j' in state $\#r$. The algorithm is described in Fig. 10.

Require: Function f, lists of differences L_A (in #1) and L_{B_1} and L_{B_2} (in #13).
Ensure: $\mathcal{L}_{sol} = \{(a, b^1, b^2, c)$, such that $f(c) \oplus f(c \oplus (a, 0^{s(k-z_A)})) = (b^1, b^2, 0^{s(k-z_B)})\}$

1: **for** j from 1 to 4 **do**
2:   **for** each $y_j \in \{0,1\}^{32}$ and **for** each b^1 from L_{B_1} **do**
3:     Store $(h_j^{-1}(y_j), h_j^{-1}(y_j) \oplus h_j^{-1}(y_j \oplus b_j^1))$ in $L^j_{\#8,b^1}$ (4×2^{32} *lists of size* 2^{32}).
4: **for** j from 5 to 8 **do**
5:   **for** each $y_j \in \{0,1\}^{32}$ and **for** each b^2 from L_{B_2} **do**
6:     Store $(h_j^{-1}(y_j), h_j^{-1}(y_j) \oplus h_j^{-1}(y_j \oplus b_j^2))$ in $L^j_{\#8,b^2}$ (4×2^{32} *lists of size* 2^{32}).
7: **for** each a in L_A **do**
8:   **for** i from 1 to 12 **do**
9:     **for** each $x_i \in \{0,1\}^{32}$ **do**
10:       Store $(g_i(x_i), g_i(x_i) \oplus g_i(x_i, a_i))$ in $L^i_{\#6}$.
11:   **for** Δ_X from 0 to $2^{64} - 1$ (*and not the 128 bits as done in [18]*) **do**
12:     Compute Δ_O and $\Delta^{\#8}_{j'}$ for $j' \in \{1, 2, 5, 6\}$ (*with BigMC and linear condit.*).
13:     **for** each b^1 in L_{B_1} and for $j = [1, 2]$ **do**
14:       Find an element in $L^j_{\#8,b^1}$ such that $h_j^{-1}(y_j) \oplus h_j^{-1}(y_j \oplus b_j^1) = \Delta^{\#8}_j$ and store $(h_1^{-1}(y_1), \Delta^{\#8}_1, h_2^{-1}(y_2), \Delta^{\#8}_2, b^1)$ in L_{aux_1}.
15:     **for** each b^2 in L_{B_2} and for $j = [5, 6]$ **do**
16:       Find an element in $L^j_{\#8,b^2}$ such that $h_j^{-1}(y_j) \oplus h_j^{-1}(y_j \oplus b_j^2) = \Delta^{\#8}_j$ and store $(h_5^{-1}(y_5), \Delta^{\#8}_5, h_6^{-1}(y_6), \Delta^{\#8}_6, b^2)$ in L_{aux_2}.
17:     **for** each $(h_1^{-1}(y_1), \Delta^{\#8}_1, h_2^{-1}(y_2), \Delta^{\#8}_2, b^1)$ in L_{aux_1} and **for** each $(h_5^{-1}(y_5), \Delta^{\#8}_5, h_6^{-1}(y_6), \Delta^{\#8}_6, b^2)$ in L_{aux_2} **do**
18:       Compute $V_1' = h_1^{-1}(y_1) \oplus 3 \times h_5^{-1}(y_5)$ and $V_2' = h_2^{-1}(y_2) \oplus 3 \times h_6^{-1}(y_6)$, and store $((h_1^{-1}(y_1), \Delta^{\#8}_1, h_2^{-1}(y_2), \Delta^{\#8}_2, b^1), (h_5^{-1}(y_5), \Delta^{\#8}_5, h_6^{-1}(y_6), \Delta^{\#8}_6, b^2))$ in a hash table T indexed by these (V_1', V_2').
19:     **for** Δ_Y from 0 to $2^{64} - 1$ **do**
20:       Determine by BigMC $\Delta^{\#8}_{j'}$ for $j' = 3, 4, 7, 8$; and $\Delta^{\#6}_j$ for $j \in [1, \ldots, 12]$.
21:       **for** i from 1 to 12 **do**
22:         Find the element from $L^i_{\#6}$ such that $g_i(x_i) \oplus g_i(x_i, a_i) = \Delta^{\#6}_i$.
23:       Compute with them by MC the values $d_i^{\#7}$ of the active diagonals in #7 and $V_j = 2 \times d_j^{\#7} \oplus d_{j+4}^{\#7} \oplus d_{j+8}^{\#7} \oplus 9 \times d_j^{\#7} \oplus 3 \times d_{j+4}^{\#7} \oplus 6 \times d_{j+8}^{\#7}$ for $j = 1, 2$.
24:       **if** there is an element such that $V_1' = V_1$ and $V_2' = V_2$ in T (*one on average, determines* b^1 *and* b^2) **then**
25:         Find $(h_{j'}^{-1}(y_{j'}), \Delta^{\#8}_{j'})$ from $L^{j'}_{\#8,b^1}$ for $j' = 3, 4$. This implies y_3 and y_4.
26:         Find $(h_{j'}^{-1}(y_{j'}), \Delta^{\#8}_{j'})$ from $L^{j'}_{\#8,b^2}$ for $j' = 7, 8$. This implies y_7 and y_8.
27:         **if** with these values of $(h_{j'}^{-1}(y_{j'}),\ j' = 3, 4, 7, 8$ and the ones obtained in step 22 of $g_i(x_i)$ for $i = 3, 4, 7, 8, 11, 12$, the equation (1) for $i = 3, 4$ derived from F_1 can be verified (*happens with a probability of* 2^{-64}) **then**
28:           The value x_* is determined. Add $(x_1, \ldots, x_{z_A}, x^*, a, b^1, b^2)$ to $\mathcal{L}_{sol}$
29: Return $\mathcal{L}_{sol}$, containing about 2^{64+l_A} elements.

Fig. 10. Algorithm for finding solutions for one ECHO BigSuperSbox

So the time complexity is $\mathcal{O}(z_B 2^{l_{B_1}+s} + z_B 2^{l_{B_2}+s} + z_A 2^s + 2^{l_A+64}(2^{l_{B_1}} + 2^{l_{B_2}} + 2^{l_{B_1}} 2^{l_{B_2}} + z_A 2^{64}))$. The memory complexity is $\mathcal{O}(z_B 2^{l_{B_1}+s} + z_B 2^{l_{B_2}+s} + 2^{l_{B_1}+l_{B_2}} + |\mathcal{L}_{sol}|)$. In the case of $l_A = 0$, we will obtain a complexity of 2^{129} in time and 2^{66} in memory for obtaining 2^{64} solutions. This algorithm proposes several trade-offs when changing the values of $|\Delta_X|$, and can be adapted for other forms of L_B.

4 How to Improve the Best Known Attacks on Five SHA-3 Candidates

In this section we enumerate briefly the main algorithms or ideas that we use to improve the best known attacks on the hash functions JH, Grøstl, ECHO, Luffa and LANE as shown on Table 1. In the full version of the paper [15] more detailed descriptions are provided.

- **JH:** To improve the complexities over the ones in [17] we use the instant-matching (as in Section 2.1) and gradual-matching algorithms as well as the fact that we do not merge the lists until we really have to (to keep lists of smaller sizes, with a smaller complexity).
- **Grøstl:** Instead of the initial lists used in [16], we can define them so that we erase the elements that for sure won't verify the outbound part. Having lists of smaller size translates to a smaller complexity.
- **ECHO:** Using conviniently the algorithm from Section 3.2 we provide better trade-offs improving the time complexity from [18].
- **Luffa:** The parallel-matching algorithm is applied in [10], improving the time complexity over the brute force merging method by increasing the memory requirements. If we apply instead the gradual-matching algorithm, the time complexity can still be better than the brute force one while the memory needs are not increased.
- LANE**:** In the cases of LANE-256 and LANE-512 several improvements are applied at different steps of the attacks from [12]. They use the instant-matching algorithm, as well as some more appropriate ways to formulate the problem, and the algorithms from Section 3.1 and from [15, App.B].

5 Conclusion

The main contribution of this paper is to propose several algorithms for solving the problem which constitutes the bottleneck of most rebound attacks, leading to improvements of the previously known complexities. We also highlight the importance of identifying the situations that could help improving the complexity of this type of attacks. This is often a difficult task due to the high technicality of the attacks and algorithms.

Finally, the previous contributions lead to improvements of most of the best known rebound attacks applied to the SHA-3 candidates JH, Grøstl, Luffa, ECHO-256 and LANE. It is important to point out that we just tried to improve the complexities of existing attacks. However, the work presented in this paper can be very useful for future rebound attacks, in particular we believe that the attacks on JH and on the compression function of ECHO can be improved (extending the number of rounds attacked) by exploiting the algorithms and ideas presented here. Finally, we believe that some of these algorithms, specially those of Section 2, will be applicable in other contexts besides rebound attacks.

Acknowledgements. The author would like to thank Willi Meier, Simon Knellwolf, Marine Minier, Thomas Peyrin, Martin Schläffer, Joana Treger and Fabien Viger for many helpful comments and discussions. A special mention is needed for Anne Canteaut and Andrea Röck for all the help and suggestions to improve this paper.

References

1. Barreto, P.S.L.M., Rijmen, V.: The Whirlpool Hashing Function (revised in 2003)
2. Benadjila, R., Billet, O., Gilbert, H., Macario-Rat, G., Peyrin, T., Robshaw, M., Seurin, Y.: Sha-3 proposal: ECHO. Submission to NIST (2009) (updated)
3. Camion, P., Patarin, J.: The knapsack hash function proposed at crypto 1989 can be broken. In: Davies, D.W. (ed.) EUROCRYPT 1991. LNCS, vol. 547, pp. 39–53. Springer, Heidelberg (1991)
4. Canniere, C.D., Sato, H., Watanabe, D.: Hash Function Luffa: Specification. Submission to NIST (2009) (Round 2)
5. Canteaut, A., Naya-Plasencia, M.: Structural weaknesses of permutations with low differential uniformity and generalized crooked functions. In: Finite Fields: Theory and Applications - Selected Papers from the 9th International Conference Finite Fields ans Applications. Contemporary Mathematics, vol. 518, pp. 55–71. AMS, Providence (2010), `http://www-rocq.inria.fr/secret/Maria.Naya_Plasencia/papers/canteaut-nayaplasencia.pdf`
6. Chose, P., Joux, A., Mitton, M.: Fast correlation attacks: An algorithmic point of view. In: Knudsen, L.R. (ed.) EUROCRYPT 2002. LNCS, vol. 2332, pp. 209–221. Springer, Heidelberg (2002)
7. Gauravaram, P., Knudsen, L.R., Matusiewicz, K., Mendel, F., Rechberger, C., Schläffer, M., Thomsen, S.S.: Grøstl – a SHA-3 candidate. Submitted to the SHA-3 competition, NIST (2008), `http://www.groestl.info`
8. Gilbert, H., Peyrin, T.: Super-sbox cryptanalysis: Improved attacks for AES-like permutations. In: Hong, S., Iwata, T. (eds.) FSE 2010. LNCS, vol. 6147, pp. 365–383. Springer, Heidelberg (2010)
9. Indesteege, S.: The Lane hash function. Submitted to the SHA-3 competition, NIST (2008), `http://www.cosic.esat.kuleuven.be/publications/article-1181.pdf`
10. Khovratovich, D., Naya-Plasencia, M., Röck, A., Schläffer, M.: Cryptanalysis of *Luffa* v2 Components. In: Biryukov, A., Gong, G., Stinson, D.R. (eds.) SAC 2010. LNCS, vol. 6544, pp. 388–409. Springer, Heidelberg (2011)
11. Lamberger, M., Mendel, F., Rechberger, C., Rijmen, V., Schläffer, M.: Rebound Distinguishers: Results on the Full Whirlpool Compression Function. In: Matsui, M. (ed.) ASIACRYPT 2009. LNCS, vol. 5912, pp. 126–143. Springer, Heidelberg (2009)
12. Matusiewicz, K., Naya-Plasencia, M., Nikolić, I., Sasaki, Y., Schläffer, M.: Rebound Attack on the Full Lane Compression Function. In: Matsui, M. (ed.) ASIACRYPT 2009. LNCS, vol. 5912, pp. 106–125. Springer, Heidelberg (2009)

13. Mendel, F., Rechberger, C., Schläffer, M., Thomsen, S.S.: The Rebound Attack: Cryptanalysis of Reduced Whirlpool and Grøstl. In: Dunkelman, O. (ed.) FSE 2009. LNCS, vol. 5665, pp. 260–276. Springer, Heidelberg (2009)
14. Mendel, F., Peyrin, T., Rechberger, C., Schläffer, M.: Improved cryptanalysis of the reduced grøstl compression function, ECHO permutation and AES block cipher. In: Jacobson Jr., M.J., Rijmen, V., Safavi-Naini, R. (eds.) SAC 2009. LNCS, vol. 5867, pp. 16–35. Springer, Heidelberg (2009)
15. Naya-Plasencia, M.: How to Improve Rebound Attacks. Cryptology ePrint Archive, Report 2010/607 (2010), http://eprint.iacr.org/2010/607.pdf (extended version)
16. Peyrin, T.: Improved Differential Attacks for ECHO and Grøstl. In: Rabin, T. (ed.) CRYPTO 2010. LNCS, vol. 6223, pp. 370–392. Springer, Heidelberg (2010)
17. Rijmen, V., Toz, D., Varıcı, K.: Rebound Attack on Reduced-Round Versions of JH. In: Hong, S., Iwata, T. (eds.) FSE 2010. LNCS, vol. 6147, pp. 286–303. Springer, Heidelberg (2010)
18. Sasaki, Y., Li, Y., Wang, L., Sakiyama, K., Ohta, K.: Non-full-active Super-Sbox Analysis: Applications to ECHO and Grøstl. In: Abe, M. (ed.) ASIACRYPT 2010. LNCS, vol. 6477, pp. 38–55. Springer, Heidelberg (2010)
19. Schroeppel, R., Shamir, A.: A $T=O(2^{n/2})$, $S=O(2^{n/4})$ algorithm for certain NP-complete problems. SIAM J. Comput. 10(3), 456–464 (1981)
20. Wagner, D.: A generalized birthday problem. In: Yung, M. (ed.) CRYPTO 2002. LNCS, vol. 2442, pp. 288–303. Springer, Heidelberg (2002)
21. Wu, H.: The hash function JH. Submission to NIST (2009) (updated), http://icsd.i2r.a-star.edu.sg/staff/hongjun/jh/jh_round2.pdf
22. Wu, S., Feng, D., Wu, W.: Cryptanalysis of the LANE hash function. In: Jacobson Jr., M.J., Rijmen, V., Safavi-Naini, R. (eds.) SAC 2009. LNCS, vol. 5867, pp. 126–140. Springer, Heidelberg (2009)

A Cryptanalysis of PRINTcipher: The Invariant Subspace Attack

Gregor Leander, Mohamed Ahmed Abdelraheem,
Hoda AlKhzaimi, and Erik Zenner

Technical University of Denmark, DK-2800 Kgs. Lyngby, Denmark
{G.Leander,M.A.Abdelraheem,H.Alkhzaimi,E.Zenner}@mat.dtu.dk

Abstract. At CHES 2010, the new block cipher PRINTcipher was presented as a light-weight encryption solution for printable circuits [15]. The best attack to date is a differential attack [1] that breaks less than half of the rounds. In this paper, we will present a new attack called *invariant subspace attack* that breaks the full cipher for a significant fraction of its keys. This attack can be seen as a weak-key variant of a statistical saturation attack. For such weak keys, a chosen plaintext distinguishing attack can be mounted in unit time. In addition to breaking PRINTcipher, the new attack also gives us new insights into other, more well-established attacks. We derive a truncated differential characteristic with a round-independent but highly key-dependent probability. In addition, we also show that for weak keys, strongly biased linear approximations exists for any number of rounds. In this sense, PRINTcipher behaves very differently to what is usually – often implicitly – assumed.

Keywords: Symmetric cryptography, block cipher, invariant subspace attack, truncated differentials, linear cryptanalysis, statistical saturation attack.

1 Introduction

Block ciphers are often said to be amongst the best understood subjects in the area of symmetric cryptography. Compared to – for example – stream ciphers and hash functions, the design of a secure block cipher is probably more straightforward. However, designing a secure block cipher that is at the same time very efficient is still challenging.

Incidentally, most recent block cipher proposals aim for the area of light-weight cryptography [3,5,13]. Light-weight cryptography provides building blocks for secure communication on extremely constrained devices. The constraints are mainly cost driven and result in highly limited computing power, chip area and/or power supply. It is an ongoing competition to design the most efficient block cipher for such devices. This competition resulted in more and more aggressive designs that often show two characteristics: (1) Innovative techniques are used to improve upon known ciphers, often leading to less standard and thus less well-understood designs. (2) The security margins that block ciphers are

P. Rogaway (Ed.): CRYPTO 2011, LNCS 6841, pp. 206–221, 2011.

traditionally equipped with are reduced as much as possible in order to optimize the cipher performance.

Unsurprisingly, this has led to a number of attacks against these newer designs [4,7,11]. In addition to constituting a break of the light-weight cipher in question, these attacks sometimes also have an additional quality: They improve our understanding of block ciphers in general. Note that an attack that breaks a light-weight cipher may be prevented by a conventional block cipher not by design, but by accident: Even though the attack was not even known by the time of designing the cipher, it may not pose a threat to the cipher simply because of the security margin.

In the following, we will present such a new attack called *invariant subspace attack* that breaks the block cipher PRINTcipher [15] proposed at CHES 2010. The best currently known analysis of PRINTcipher is a differential-style attack presented at FSE 2011 [1] that could be applied for less than half the rounds of PRINTcipher, making use of the full code book. Apart from breaking PRINTcipher and providing us with a new tool for attacking block ciphers, the invariant subspace attack also displays interesting relationships to other well-established attack techniques that increase our understanding of block cipher cryptanalysis in general.

1.1 Our Contribution

In this paper, detailed in Section 2.2, we present a new attack on PRINTcipher. In a nutshell, the attack is based on the observation that for PRINTcipher there exist cosets of subspaces of $\mathbb{F}_2^n$ that the round function maps to cosets of the same subspace. The exact coset is determined by the round key only. Now, if the round key is such that a coset gets mapped to itself, the fact that all round keys are identical in PRINTcipher (almost) immediately leads to the conclusion that for certain (weak) keys some affine subspaces are invariant under encryption. The round constants, mainly introduced to avoid slide attacks, do not prevent the attack as the round constants are included in the subspace. The principle of the attack is described in Section 2.1.

More particular, using this new attack technique, which we call (for obvious reasons) *invariant subspace attack*, we demonstrate the existence of 2^{52} weak keys (out of 2^{80}) for PRINTcipher-48 and 2^{102} weak keys (out of 2^{160}) for PRINTcipher-96. If a key is weak, our attack results in a distinguisher using less than 5 chosen plain- or ciphertexts. That is, even in the case of RFID-tags, where the amount of data available for a practical attack is strictly limited, our attacks apply. In a known plain- or ciphertext scenario the data complexity increases by a factor of 2^{16} (PRINTcipher-48) resp. 2^{32} (PRINTcipher-96).

Besides the low data complexity of the distinguisher, the attack technique has interesting relations to more established attacks which we like to highlight. Firstly, see Section 3, the invariant subspace attack implies a truncated differential attack, where the probability of the truncated differential characteristic is highly key-dependent. For a weak key, this probability is 2^{-16}, independent of the number of rounds – while for a non-weak key the probability is zero for any number of rounds greater or equal to two.

Secondly, the invariant subspace attack can be interpreted as a statistical saturation attack [7,8]. Here a weak key, together with a special choice of the fixed bits in a statistical saturation attack, leads to a maximal bias, independent of the number of rounds. Taking into account the close relation of statistical saturation attacks to multi-dimensional linear attacks, we show that the invariant subspace attack implies the existence of strongly biased linear approximations for weak keys, again independent of the number of rounds. Details can be found in Section 4.

It follows in particular that PRINTcipher is an example of a non-toy cipher where attacks do not behave as we usually expect them to. The probability of truncated differential characteristics, the bias for statistical saturation attacks, and the bias of linear hulls are extremely key-dependent. For a weak key, increasing the number of rounds up to the full number of rounds does not increase the security of the cipher with respect to these attacks.

1.2 Related Work

As already mentioned in the abstract, our attack can be seen as a weak key variant of statistical saturation attacks [7,8]. As the statistical saturation attack itself is a special case of partitioning cryptanalysis [12], so is our attack. Again, the main difference is that we make use of weak keys and for those keys the bias is maximal. More loosely our work is related to conditional cryptanalysis [2,14] in the sense that the truncated differential characteristic described in Section 3 is conditioned to certain key and message bits. Moreover, our attack can also be interpreted as an extreme case of a dynamic cube attack [11]. Here, the algebraic normal form of certain ciphertext bits becomes a constant when a weak key is used and certain message bits are fixed correctly.

2 The Invariant Subspace Attack

2.1 General Idea

Consider an n-bit block cipher with a round function E_k consisting of a key addition and an SP-layer

$$E : \mathbb{F}_2^n \to \mathbb{F}_2^n,$$

that is E_k is defined by $E_k(x) = E(x + k)$. Assume that the SP-layer E is such that there exists a subspace $U \subseteq \mathbb{F}_2^n$ and two constants $c, d \in \mathbb{F}_2^n$ with the property:

$$E(U + c) = U + d.$$

Then, given a (round) key $k = u + c + d$ with $u \in U$, the following holds:

$$E_k(U + d) = E((U + d) + (u + c + d)) = E(U + c) = U + d,$$

i.e. the round function maps the affine subspace $U + d$ onto itself. If all round keys are in $k \in U + (c + d)$ (in particular if a constant round key is used), then this property is iterative over an arbitrary number of rounds. This yields a very efficient distinguisher for a fraction of the keys. U should be as large as possible to increase this fraction. We call this new attack technique an *invariant subspace attack*. In the next section we show an example of how to apply it to the light-weight block cipher PRINTcipher.

2.2 Attack against PRINTcipher

Description of PRINTcipher. PRINTcipher is a block cipher proposed by Knudsen et al. at CHES 2010 [15]. It is a class of two SP-networks with a block size of $n = 48$ (resp. $n = 96$) bits, a key size of $l = 80$ (resp. $l = 160$) bit, and 48 (resp. 96) rounds. One round of PRINTcipher-48 is shown in Figure 1.

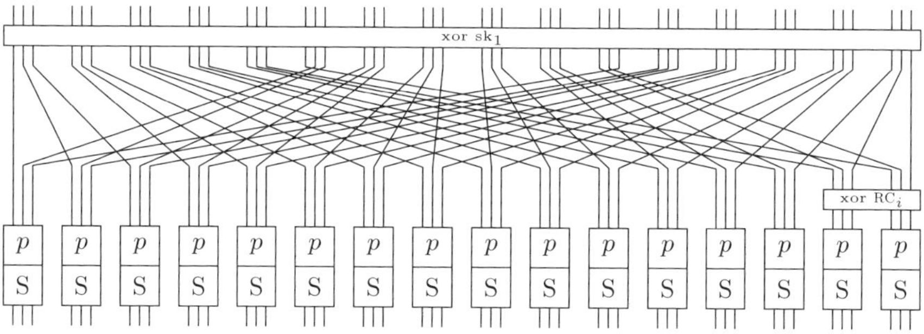

Fig. 1. One round of PRINTcipher-48 illustrating the bit-mapping between the 16 3-bit S-boxes from one round to the next. sk_1 denotes the xor key, p the permutation key, and RC_i the round counter.

PRINTcipher uses the same key for all rounds. It is split into two parts: The first n bits are used as an xor key, the remaining $l - n$ bits control the permutations p. In order to introduce differences between the rounds, a round counter RC_i is used which is generated by an LFSR (for details, see [15]). The other elements of the round function are defined as follows.

The **linear layer** consists of a bit permutation, where bit i of the current state is moved to bit position $P(i)$ where

$$P(i) = \begin{cases} 3i \bmod n - 1 & \text{for } 0 \leq i \leq n - 2, \\ n - 1 & \text{for } i = n - 1, \end{cases}$$

where $n \in \{48, 96\}$ is the block size.

Then the state bits are arranged in 16 (resp. 32) blocks of 3 bits each, which are permuted individually in the **permutation layer**. Out of 6 possible permutations on 3 bits, only four are valid permutations for PRINTcipher. Specifically, the three input bits $c_2||c_1||c_0$ are permuted to give the following output bits according to two key bits $a_1||a_0$.

nr.	$a_1\|\|a_0$	p
0	00	$c_2\|\|c_1\|\|c_0$
1	01	$c_1\|\|c_2\|\|c_0$
2	10	$c_2\|\|c_0\|\|c_1$
3	11	$c_0\|\|c_1\|\|c_2$

Finally, in the **non-linear layer**, each 3-bit block is processed by the same s-box, which is shown in the following table.

x	0	1	2	3	4	5	6	7
$S[x]$	0	1	3	6	7	4	5	2

An Attack on PRINTcipher. One interesting property of the PRINTcipher s-box is that a one bit difference in the input causes a one bit difference in the same bit in the output with probability 2/8. That is, there exists exactly one pair for each one bit input difference resulting in a one bit output difference (at the same position). More precisely, denoting by $*$ an arbitrary value in $\mathbb{F}_2$, the following holds for the PRINTcipher s-box:

$$\begin{array}{l}
\mathrm{S}(000) = 000 \\
\mathrm{S}(001) = 001
\end{array} \Leftrightarrow \mathrm{S}(00*) = 00*$$

$$\begin{array}{l}
\mathrm{S}(100) = 111 \\
\mathrm{S}(110) = 101
\end{array} \Leftrightarrow \mathrm{S}(1*0) = 1*1$$

$$\begin{array}{l}
\mathrm{S}(011) = 110 \\
\mathrm{S}(111) = 010
\end{array} \Leftrightarrow \mathrm{S}(*11) = *10$$

In addition, there exists a subset of s-boxes such that (1) two output bits of those s-boxes map onto two input bits of the same s-boxes in the next round and (2) the round-dependent RC_i is not involved (see Figure 2).

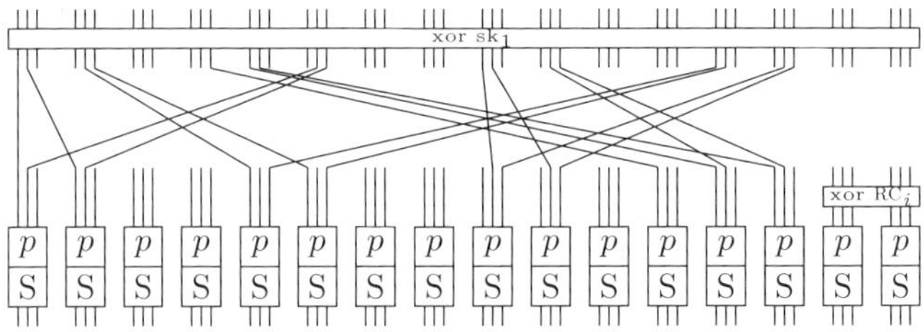

Fig. 2. A subset of PRINTcipher-48 s-boxes mapping onto itself

Now consider an xor-key sk_1 of the form

Xor key = `01* *11 *** *** 01* *11 *** *** 01* *11 *** *** 01* *11 *** ***`, and a permutation key with the following restrictions:

Perm. key = `0* 11 ** ** 10 01 ** ** 11 *0 ** ** *0 11 ** **`,

where again $*$ denotes an arbitrary value in $\mathbb{F}_2$. For those keys the following structural *iterative* one round property holds:

Start	00* *10 *** *** 00* *10 *** *** 00* *10 *** *** 00* *10 *** ***
Key xoring	01* *01 *** *** 01* *01 *** *** 01* *01 *** *** 01* *01 *** ***
Lin. layer	00* 11* *** *** 0*0 1*1 *** *** *00 *11 *** *** 00* 11* *** ***
RC	00* 11* *** *** 0*0 1*1 *** *** *00 *11 *** *** 00* 11* *** ***
Perm. layer	00* *11 *** *** 00* *11 *** *** 00* *11 *** *** 00* *11 *** ***
S-box layer	00* *10 *** *** 00* *10 *** *** 00* *10 *** *** 00* *10 *** ***

This property holds with probability one if both keys are of the above form. The fraction of those keys is $(1/2)^{16}$ for the XOR key and $(1/2)^{13}$ for the permutation key, meaning that the property is met for a fraction of $(1/2)^{29}$ of all keys. In other words, there exist 2^{51} weak keys of this form.

Thus, one can very efficiently check if a key of the above form is used by encrypting a few texts of the above form and check if the ciphertext is again of the same form. Given that the probability for false positives is $\approx 2^{-16}$, trial encrypting just a handful of selected plaintexts will uniquely identify such a weak key. If such a key is found, we do of course immediately have a distinguisher on PRINTCIPHER.

Invariant Subspace Description. Let us briefly rephrase the attack in terms of an invariant subspace attack. For this we fix a permutation key of the above form. Remember that the inner state at the beginning and the end of each round was

Start = `00* *10 *** *** 00* *10 *** *** 00* *10 *** *** 00* *10 *** ***`.

This means that the relevant subspace $U \subset \mathbb{F}_2^{48}$ is defined by

$$U = \{\texttt{00* *00 *** *** 00* *00 *** *** 00* *00 *** *** 00* *00 *** ***}\}, \quad (1)$$

and that the affine subspace is defined by any fixed vector d of the form

$$d = \texttt{00* *10 *** *** 00* *10 *** *** 00* *10 *** *** 00* *10 *** ***}. \quad (2)$$

Then for any fixed vector c of the form

$$c = \texttt{01* *01 *** *** 01* *01 *** *** 01* *01 *** *** 01* *01 *** ***}, \quad (3)$$

and any xor-key $k \in (U + c + d)$, the round function does indeed map $U + d$ onto itself.

2.3 Other Attack Profiles

In the following we describe other sets of weak keys for PRINTCIPHER-48 and similar ones for PRINTCIPHER-96.

Other Weak Keys for PRINTcipher-48. As it turns out, there are some more invariant subspaces that also can be used for PRINTcipher-48. They are all of the form

```
00* XXX *** 1*1 00* *10 *** *** 00* XXX *** 1*1 00* *10 *** ***,
```

where an 'X' marks a bit position where the attacker has to make an arbitrary assignment. Note that each position can be filled independently of the others. Thus, we have 2^6 possible plaintexts that we can work with, each of which targets another class of weak keys.

For each such assignment, the cipher behaves as follows:

Start (1)	`00* XXX *** 1*1 00* *10 *** *** 00* XXX *** 1*1 00* *10 *** ***`
Key xoring (2)	`0X* X01 *** X*1 01* *0X *** *** 0X* 001 *** X*X 01* *0X *** ***`
Lin. layer (3)	`00* XXX *** X*X 0*0 1*1 *** *** *00 XXX *** 10* 00* 11* *** ***`
RC (4)	`00* XXX *** X*X 0*0 1*1 *** *** *00 XXX *** 10* 00* 11* *** ***`
Perm. layer (5)	`00* XXX *** 1*0 00* *11 *** *** 00* XXX *** 1*0 00* *11 *** ***`
S-box layer (6)	`00* XXX *** 1*1 00* *10 *** *** 00* XXX *** 1*1 00* *10 *** ***`

The behaviour is best understood by traversing the cipher in the inverse direction, i.e. by starting from the end and then finding the key bits that ensure that all fixed bits in line (1) match their counterparts in line (6).

Let us start with the output of the s-box, i.e. line (6), and let the bit positions marked by 'X' be arbitrarily and independently fixed to either 0 or 1. Then going backwards through the s-box uniquely determines the bits in line (5). We then use a permutation key of the form

Perm. Key = `0* ** ** (00 or 11) 10 01 ** ** 11 ** ** 10 0* 11 ** **`

to obtain line (4), noting that 2^{-13} of all permutation keys meet this property. We then apply round counter and linear layer to obtain line (2). Now note that line (2) contains 22 bits that are fixed and that have to match the corresponding bits in line (1). Thus, 22 key bits of the xoring key are determined, meaning that 2^{-22} of all xoring keys are suitable for the attack.

Summing up, for each of the 2^6 possible assignments to the bits marked by 'X' in line (1) or (6), a fraction of exactly 2^{-35} keys are weak, meaning that in total, we have found another fraction of 2^{-29} weak keys that can be attacked by the above technique.

Analysis of PRINTcipher-96. As it turns out, the same attack can also be applied to PRINTcipher-96. Again, there are two types of weak keys. The first type is based on 32 active bits and is met by a fraction of 2^{-59} of all keys. The second type is based on 44 active bits and has an additional 12 freely chosable input bits. Each of the resulting 2^{12} inputs targets a fraction of 2^{-71} keys, meaning that this group, too, contains a fraction of 2^{-59} weak keys in total. The active bits for these weak keys are given in Table 1.

2.4 Protecting Against the Attack

The above attack against PRINTcipher is a special case of the general attack described in the beginning of the section, since the subspace is described by

Table 1. Subsets of active bits for PRINTcipher-96, grouped according to s-boxes

Subset	Bits
Subset 1	**Active input bits for linear layer:** (0 1) (4 5) (12 13) (16 17) (24 25) (28 29) (36 37) (40 41) (48 49) (52 53) (60 61) (64 65) (72 73) (76 77) (84 85) (88 89)
	Active output bits for linear layer: (0 2) (3 5) (12 13) (15 16) (25 26) (28 29) (36 38) (39 41) (48 49) (51 52) (61 62) (64 65) (72 74) (75 77) (84 85) (87 88)
Subset 2	**Active input bits for linear layer:** (0 1) (3 4 5) (9 11) (12 13) (16 17) (24 25) (27 28 29) (33 35) (36 37) (40 41) (48 49) (51 52 53) (57 59) (60 61) (64 65) (72 73) (75 76 77) (81 83) (84 85) (88 89)
	Active output bits for linear layer: (0 2) (3 4 5) (9 10) (12 13) (15 16) (25 26) (27 28 29) (33 35) (36 38) (39 41) (48 49) (51 52 53) (58 59) (61 62) (64 65) (72 74) (75 76 77) (81 82) (84 85) (87 88)

simply fixing some of its bits. In theory, describing the subspace by a set of linear equations is possible, opening for a wide range of attacks. The full potential of this generalized attack is yet to be determined.

As for the special case used against PRINTcipher, it is relatively easy to protect the design against the attack. Note that the list of attack profiles by fixing bits given here is complete, and that all attack profiles fix two of the bits 39-41 (PRINTcipher-48) resp. 87-89 (PRINTcipher-96). Thus, it would suffice to spread the round counter over the last three s-boxes, e.g. by assigning two counter bits to each s-box. This would destroy the only attack profiles available, at no extra hardware cost.

We also analysed the block cipher NOEKEON, which was proposed by Daemen et al. in 2000 [9]. NOEKEON is a 16-round block cipher with a constant round key, making it a particularly tempting target for the attack. However, as it turns out, the linear mixing layer of NOEKEON is much more resistant against the above type of attack. Here, the stronger round function (necessary for a cipher with only 16 rounds) works to the advantage of the cipher. As it turns out, even if there was no round counter involved in NOEKEON, the simple attack described above – i.e. where the subspace is defined by fixing certain bits – could not be applied. Whether or not the generalized attack has a better chance of succeeding remains yet to be determined.

3 Truncated Differential Attacks

As pointed out by Murphy in [18] the attack complexity for linear attacks is often wrongly stated in the literature. One of the reasons is that it is often easy to compute the average squared bias ϵ^2 when averaging over all keys. However, it is often stated that the average attack complexity is $\frac{\gamma}{\epsilon^2}$ for some small γ, which, in general, is wrong. In particular, the average complexity is formally infinite as soon as there exists a single key with no bias, while $\frac{\gamma}{\epsilon^2}$ is finite as soon as there exists a single key with a bias.

Now, to some extent the same is true for (truncated) differential attacks. A truncated differential characteristic on an n-bit block cipher can be, in general, described by a set of input and output differences. For $0 \leq i \leq r$ let $U_i \subset F_2^n$ and

$$U_i \xrightarrow{E_i} U_{i+1}$$

be a set of differential characteristics with probability p_i.

Assuming independent round keys the average probability, taken over all keys, of the truncated r-round differential characteristic

$$U_0 \xrightarrow{E_0} U_1 \xrightarrow{E_1} \cdots \xrightarrow{E_{r-1}} U_r$$

is $p = \prod_i p_i$. One normally assumes (cf. the hypothesis of stochastic equivalence in [16]) that for (almost) all keys it holds that $p_k \approx p$. Here p_k denotes the probability of the truncated differential characteristic for a fixed key k.

However, this may be highly incorrect. Indeed PRINTcipher is an example of the extreme opposite. We will show below that for PRINTcipher, the attack discussed in Section 2 implies the existence of a truncated differential characteristic such that

$$p_k \in \{2^{-16}, 0\},$$

for any number of rounds $r \geq 2$. Since a fraction of 2^{-29} of all keys is weak, the average probability over all keys is

$$p_{\text{av}} = 2^{-16} \cdot 2^{-29} = 2^{-45},$$

again noting that this holds for any number $r \geq 2$ of rounds. After introducing the invariant subspace attack, the existence of such truncated differential characteristics might not be so surprising, as one basically pays the price for following the characteristic only once. That is to say that pairs that follow the characteristic for two rounds automatically follow the characteristic for any number of rounds.

However, this disproves the naive assumption where multiplying the probabilities for the individual rounds yields an average attack complexity that tends to zero for an increasing number of rounds. Thus, not only is the assumption that all keys behave more or less similar wrong. Also, the assumption that the round keys are independent leads to a very wrong conclusion. Concluding this part, studying the average complexity does not reveal the actual behavior of PRINTcipher. On the contrary, PRINTcipher behaves completely opposite to what is usually assumed.

3.1 Rephrasing the Attack in Terms of Truncated Differentials

In this section, we will prove the above claims. To make the description easier, consider a PRINTcipher-48 version where we fix the permutation key to

```
00 11 00 00 10 01 00 00 11 00 00 00 00 11 00 00
```

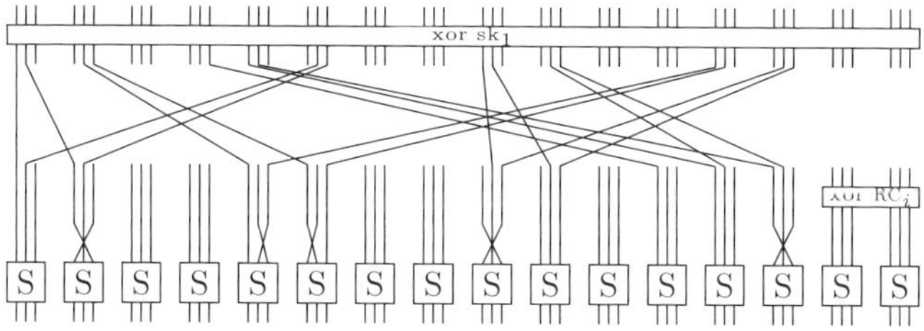

Fig. 3. One round of PRINTcipher with fixed permutation key. Only the bits that matter for the differential characteristic are shown in the linear layer.

One round of PRINTcipher with this key is given in Figure 3. Other weak permutation keys behave similarly.
Now, consider an r-round truncated differential characteristic[1] of the form

$$\alpha \overset{E_k}{\to} U' \overset{E_k}{\to} U \dots U \overset{E_k}{\to} U, \tag{4}$$

where α is given by

$\alpha =$ `000 100 011 101 000 100 001 100 000 000 001 110 001 000 101 110`,

and U' contains all vectors of the form

$U' = \{$`001 100 **1 1** 001 100 **1 1** 001 100 **1 1** 001 100 **1 1**`$\}$.

Finally, as in Section 2, U is defined by

$U = \{$`00* *00 *** *** 00* *00 *** *** 00* *00 *** *** 00* *00 *** ***`$\}$.

Note that $\alpha \in U' \subset U$. Given these definitions, we can prove the following theorem:

Theorem 1. *For a fixed (xor)-key k , denote the probability of the truncated differential characteristic given by Equation (4) by p_k. It holds that*

$$p_k = \begin{cases} 2^{-16} & \textit{if } k \textit{ is weak} \\ 0 & \textit{if } k \textit{ is not weak} \end{cases}$$

Here a (xor)-key k is weak if and only if it is of the form

$k =$ `01* *11 *** *** 01* *11 *** *** 01* *11 *** *** 01* *11 *** ***`.

[1] We emphasize that we deal with truncated differential characteristics and not with truncated differentials. In particular, for the characteristic we are using, for the corresponding differential one can expect a probability of 2^{-16} even for a random round function.

Table 2. This table shows how the differences of the truncated differential characteristic from Section 3 propagate through the round functions. The underlined values are the only positions where the difference propagation has a probability not equal to 1. That is, only those positions pose restrictions on the pairs satisfying the characteristic.

Input 1st Rnd.	000	100	011	101	000	100	001	100	000	000	001	110	001	000	101	110
After P-layer	001	001	001	100	010	010	001	100	100	100	001	100	001	001	001	100
After perm.key	001	100	001	100	001	100	001	100	001	100	001	100	001	100	001	100
S-box layer	001	100	**1	1**	001	100	**1	1**	001	100	**1	1**	001	100	**1	1**
Input 2st Rnd.	001	100	**1	1**	001	100	**1	1**	001	100	**1	1**	001	100	**1	1**
After P-layer	001	001	1**	1**	010	010	**1	**1	100	100	*1*	*1*	001	001	1**	1**
After perm.key	001	100	1**	1**	001	100	**1	**1	001	100	*1*	*1*	001	100	1**	1**
S-box layer	00*	*00	***	***	00*	*00	***	***	00*	*00	***	***	00*	*00	***	***
Input 3rd Rnd.	00*	*00	***	***	00*	*00	***	***	00*	*00	***	***	00*	*00	***	***
After P-layer	00*	00*	***	***	0*0	0*0	***	***	*00	*00	***	***	00*	00*	***	***
After perm.key	00*	*00	***	***	00*	*00	***	***	00*	*00	***	***	00*	*00	***	***
S-box layer	00*	*00	***	***	00*	*00	***	***	00*	*00	***	***	00*	*00	***	***

The idea behind this theorem is similar to results from [10]. Before we prove this result, we have to introduce some notation similar to that in [10]. Given a round function $E : \mathbb{F}_2^n \to \mathbb{F}_2^n$ excluding the initial key addition and two subsets $A, B \subseteq F_2^n$ we denote by

$$F_{(A,B)} = \{x \mid E(x) + E(x + \alpha) = \beta, \ \alpha \in A, \beta \in B\}$$

the set of input pairs satisfying the truncated differential $A \overset{E_k}{\to} B$.

As observed in [10] in many cases $F(A, B)$ is a coset of a subspace. This is also the case here. More precisely, we have the following lemma.

Lemma 1. *Let E be the round function of* PRINTcipher *excluding the inital key addition, α, U' and U as defined above. Then*

$$F_{(\{\alpha\},U')} = F_{(U',U)} = U + c,$$

where c is defined as in Equation (3).

Proof. This can be seen from Table 2, where it is shown how the differences propagate through the round functions. □

Now to prove Theorem 1, from the invariant subspace attack on PRINTcipher, as discussed in Section 2.2, we know that $E(U + c) = U + d$. It follows that for the good pairs for

$$\alpha \overset{E_k}{\to} U' \overset{E_k}{\to} U$$

the inputs to the second round are given by

$$(U + d + k) \ \cap \ (U + c) = \begin{cases} U + c & \text{if } k \in U + d + c \\ \emptyset & \text{if } k \neq U + d + c \end{cases}$$

This already proves that for a non-weak key the probability of the truncated differential characteristic given in the above theorem is 0. For a weak key it holds that $E_k(U + d) = U + d$ and therefore, for any x in $F_{a,b} + k = U + d$ the pair $(x, x + \alpha)$ fullfils the whole r round truncated differential characteristic. □

4 Statistical Saturation Attacks and Multidimensional Linear Attacks

The attack on PRINTcipher discussed in Section 2.2 is clearly strongly related to statistical saturation attacks as described in [7]. In this section, after briefly recalling some of the principles of statistical saturation attacks, we elaborate on the details of this relation. Maybe the most interesting finding here is that for PRINTcipher there exist strongly biased linear approximations for any number of rounds, if the key is weak in the sense of the invariant subspace attack. This result follows using a link between statistical saturation attacks and multidimensional linear attacks (see [17]). Understanding these strongly biased linear approximations by studying the linear hulls directly is an interesting problem that we leave open for further investigation.

4.1 Necessary Background Information

Notations. The canonical inner product on $\mathbb{F}_2^n$ is denoted by $\langle \cdot, \cdot \rangle$, i.e.

$$\langle (a_0, \dots, a_{n-1}), (b_0, \dots, b_{n-1}) \rangle := \sum_{i=0}^{n-1} a_i b_i.$$

We note that all linear forms, i.e. all linear functions $l : \mathbb{F}_2^n \to \mathbb{F}_2$, can be described as $\ell(x) = \langle a, x \rangle$ for a suitable $a \in \mathbb{F}_2^n$. Given a (vectorial Boolean) function $F : \mathbb{F}_2^n \to \mathbb{F}_2^m$ the *Fourier coefficient* of F at the pair $(a, b) \in \mathbb{F}_2^n \times \mathbb{F}_2^m$ is defined by

$$\widehat{F}(a, b) = \sum_{x \in \mathbb{F}_2^n} (-1)^{\langle b, F(x) \rangle + \langle a, x \rangle}.$$

The *bias* $\epsilon_F(a, b)$ of the linear approximation $\langle a, x \rangle$ of $\langle b, F(x) \rangle$ is defined as

$$\epsilon_F(a, b) := \frac{|\{x \mid \langle b, F(x) \rangle + \langle a, x \rangle = 0\}|}{2^n} - \frac{1}{2}.$$

The fundamental relation between the Fourier transformation of F and the bias of a linear approximation is given by

$$\epsilon_F(a, b) = \frac{\widehat{F}(a, b)}{2^{n+1}} \tag{5}$$

Given $F : \mathbb{F}_2^n \to \mathbb{F}_2^m$, the value used to determine the complexity of both multidimensional linear attacks and statistical saturation attacks is the capacity of F given by

$$\mathrm{Cap}(F) = \sum_{z \in \mathbb{F}_2^m} \frac{(2^{-n} \cdot |\{x \in F_2^n \mid F(x) = z\}| - 2^{-m})^2}{2^{-m}}.$$

Statistical Saturation Attacks. Let us first briefly recall some concepts from statistical saturation attacks. We refer to [7] for details. Given an encryption function

$$e : \mathbb{F}_2^n \to \mathbb{F}_2^n,$$

statistical saturation attacks study the distribution of e when some of its input bits are fixed. Up to a fixed bijective linear transformation before and after the cipher, we can restrict ourselves without loss of generality to the case where one fixes the first r bits in the inputs and considers only the first t bits of the output[2]. Thus we write

$$\begin{aligned} e : \mathbb{F}_2^r \times \mathbb{F}_2^s &\to \mathbb{F}_2^t \times \mathbb{F}_2^u \\ e(y,x) &= \left(e^{(1)}(y,x), e^{(2)}(y,x)\right), \end{aligned}$$

where $r + s = t + u = n$ and $e^{(1)}(y,x) \in \mathbb{F}_2^t$, $e^{(2)}(y,x) \in \mathbb{F}_2^u$. For convenience we denote by h_y the restriction of e by fixing the first r bits to y and considering only the first t bits of the output, that is

$$\begin{aligned} h_y : \mathbb{F}_2^s &\to \mathbb{F}_2^t \\ h_y(x) &= e^{(1)}(y,x). \end{aligned}$$

In a statistical saturation attack one considers the capacity of h_y, and the attack complexity is usually a constant times $1/\operatorname{Cap}(h_y)$. Computing this capacity is difficult in general. However, when averaging over all possible fixings y the following has been proven in [17]:

Theorem 2. *The average capacity in statistical saturation attacks where the average is taken over all possible fixations y is given by*

$$\overline{\operatorname{Cap}(h_y)} = 2^{-r} \sum_{y \in \mathbb{F}_2^r} \operatorname{Cap}(h_y) = 2^{-2n} \sum_{\substack{a \in \mathbb{F}_2^r \times \{0\} \\ b \in \mathbb{F}_2^t \times \{0\}, b \neq 0}} \left(\widehat{e}(a,b)\right)^2 \qquad (6)$$

4.2 On the Choice of the Values of the Fixed Bits

We now focus on the case where $r = t$, that is the number of fixed bits is the same as the number of bits considered at the output.

Assume a cipher is vulnerable to an invariant subspace attack. As for statistical saturation attacks, up to a fixed bijective linear transformation before and after the cipher, we can assume that, for a weak key, the affine subspace of the form $\{d\} \times \mathbb{F}_2^s$ is mapped to an affine subspace of the form $\{d\} \times \mathbb{F}_2^s$. It then follows immediately that (for a weak key) the function of the restriction h_y for $y = d$ is a constant, more precisely

$$h_d(x) = e^{(1)}(d,x) = d.$$

[2] This differs slightly from the notation in [17].

For the special choice of the values of the fixed bits the capacity is maximal. Hence for a weak key this special fixing of the bits leads to an optimal statistical saturation attack. Note that Theorem 2 does not reveal the existence of such extreme cases, as it only considers the average capacity of the restrictions.

While in an invariant subspace attack, given the subspace, the choice of the coset is crucial, for statistical saturation attacks the fixed bits are usually assigned with random values. As the invariant subspace attack on PRINTCIPHER does not imply that PRINTCIPHER is in general vulnerable to a statistical saturation attack, it does not come as a surprise that the experiments in [15] did not reveal any weakness of PRINTCIPHER with respect to those attacks.

4.3 On the Existence of Highly Biased Approximations

Theorem 2 was used to compute the average capacity using the Fourier coefficients. However, for us, the reciprocal is of interest as it implies the following corollary.

Corollary 1. *Assume an n-bit block cipher E_k is vulnerable to an invariant subspace attack, that is there exist a subspace U, a constant d and keys k such that*

$$E_k(U+d) = U+d.$$

Then, for those keys, there exist linear approximations with a bias ϵ such that

$$\epsilon \geq 2^{\dim(U)-n-1} - 2^{2(\dim(U)-n)-1}.$$

Proof. With the notation as in Section 4.2, h_d is a constant function. Thus $\mathrm{Cap}(h_d) = 2^r - 1$ and furthermore

$$\sum_{y \in \mathbb{F}_2^r} \mathrm{Cap}(h_y) \geq \mathrm{Cap}(h_d) = 2^r - 1.$$

Considering Equation (6) it follows that

$$\sum_{\substack{a \in \mathbb{F}_2^r \times \{0\} \\ b \in \mathbb{F}_2^t \times \{0\}, b \neq 0}} (\widehat{e}(a,b))^2 \geq 2^{2n}(1-2^{-r})$$

Lower bounding the maximal value by the average (and recalling that $r = t$), we compute

$$\max_{a, b \neq 0} (\widehat{e}(a,b))^2 \geq 2^{-2r} \sum_{\substack{a \in \mathbb{F}_2^r \times \{0\} \\ b \in \mathbb{F}_2^t \times \{0\}, b \neq 0}} (\widehat{e}(a,b))^2 \geq 2^{2n-2r}(1-2^{-r})$$

Thus there exists at least one Fourier coefficient such that

$$|\widehat{e}(a,b)| \geq 2^{n-r}\sqrt{1-2^{-r}} \geq 2^{n-r} - 2^{n-2r}$$

Applying identity (5) and remembering that $r = n - \dim U$, the theorem follows. □

Clearly, this Theorem is only interesting for the case where $\dim(U) > n/2$ as the existence of the stated approximations otherwise is trivial. For the case of PRINTCIPHER-48 we summarize the findings below

Corollary 2. *Given a weak key for any round $r \leq 48$ there exists at least one linear approximation for* PRINTCIPHER-*48 with bias at least* $2^{-17} - 2^{-33}$.

5 Conclusions

We have presented a new attack against iterative block ciphers named *invariant subspace attack* and demonstrated its validity by breaking PRINTCIPHER for a significant fraction of its keys. The presented *invariant subspace attack* shows that 2^{52} keys (out of 2^{80}) for PRINTCIPHER-48 and 2^{102} keys (out of 2^{160}) for PRINTCIPHER-96 are weak. In addition, we have shown the relationship of the *invariant subspace attack* to other classes of attacks such as truncated differential attack, multi-dimentional attack linear attack and statistical saturation attack. In doing this, we could provide an example for a truncated differential attack whose success probability is round-independent, disproving the common implicit assumptions that the total success probability is the product of the individual round probabilities and that the overall success probability against a cipher can be averaged over all keys. The probability of this truncated differential characteristic is 2^{-16} for weak keys and zero for non-weak keys given that the number of rounds is greater than or equal to two. Moreover, for PRINTCIPHER there are strongly biased linear approximations for any number of rounds, if a weak key is chosen. For example, there is at least one linear approximation for PRINTCIPHER-48 with bias at least 2^{-17}.

Open Questions and Future Work. The attack presented against PRINTCIPHER is a special case of the general *invariant subspace attack*. It should be evaluated whether the generalised attack provides even better results against PRINTCIPHER and other potentially vulnerable ciphers. Hence, the possibility of extending the presented distinguishing attack on weak keys classes into a key recovery attack is an open problem that needs to be further analysed. Understanding the strongly biased linear approximations by studying the linear hulls directly is another interesting open problem. We believe that it will increase our general understanding of linear hulls and how (very simple) key scheduling algorithms influence the distribution of biases.

References

1. Abdelraheem, M.A., Leander, G., Zenner, E.: Differential cryptanalysis of round-reduced PRINTCIPHER: Computing roots of permutations. In: Joux, A. (ed.) FSE 2011. LNCS, vol. 6733, pp. 1–17. Springer, Heidelberg (2011)

2. Ben-Aroya, I., Biham, E.: Differential cryptanalysis of Lucifer. Journal of Cryptology 9(1), 21–34 (1996)
3. Bogdanov, A., Knudsen, L.R., Leander, G., Paar, C., Poschmann, A., Robshaw, M.J.B., Seurin, Y., Vikkelsoe, C.: PRESENT: An Ultra-Lightweight Block Cipher. In: Paillier, P., Verbauwhede, I. (eds.) CHES 2007. LNCS, vol. 4727, pp. 450–466. Springer, Heidelberg (2007)
4. Bogdanov, A., Rechberger, C.: A 3-Subset Meet-in-the-Middle Attack: Cryptanalysis of the Lightweight Block Cipher KTANTAN. In: Biryukov, A., Gong, G., Stinson, D.R. (eds.) SAC 2010. LNCS, vol. 6544, pp. 229–240. Springer, Heidelberg (2011)
5. De Cannière, C., Dunkelman, O., Knežević, M.: KATAN and KTANTAN — A Family of Small and Efficient Hardware-Oriented Block Ciphers. In: Clavier, C., Gaj, K. (eds.) CHES 2009. LNCS, vol. 5747, pp. 272–288. Springer, Heidelberg (2009)
6. Cho, J.Y.: Linear Cryptanalysis of Reduced-Round PRESENT. In: Pieprzyk, J. (ed.) CT-RSA 2010. LNCS, vol. 5985, pp. 302–317. Springer, Heidelberg (2010)
7. Collard, B., Standaert, F.-X.: A Statistical Saturation Attack against the Block Cipher PRESENT. In: Fischlin, M. (ed.) CT-RSA 2009. LNCS, vol. 5473, pp. 195–210. Springer, Heidelberg (2009)
8. Collard, B., Standaert, F.-X.: Multi-trail Statistical Saturation Attacks. In: Zhou, J., Yung, M. (eds.) ACNS 2010. LNCS, vol. 6123, pp. 123–138. Springer, Heidelberg (2010)
9. Daemen, J., Peeters, M., van Assche, G., Rijmen, V.: Nessie proposal: NOEKEON (2000), `http://gro.noekeon.org/Noekeon-spec.pdf`
10. Daemen, J., Rijmen, V.: Plateau characteristics. Information Security, IET 1(1), 11–17 (2007)
11. Dinur, I., Shamir, A.: Breaking grain-128 with dynamic cube attacks. In: Joux, A. (ed.) FSE 2011. LNCS, vol. 6733, pp. 167–187. Springer, Heidelberg (2011)
12. Harpes, C., Massey, J.L.: Partitioning Cryptanalysis. In: Biham, E. (ed.) FSE 1997. LNCS, vol. 1267, pp. 13–27. Springer, Heidelberg (1997)
13. Hong, D., Sung, J., Hong, S.H., Lim, J.-I., Lee, S.-J., Koo, B.-S., Lee, C.-H., Chang, D., Lee, J., Jeong, K., Kim, H., Kim, J.-S., Chee, S.: HIGHT: A New Block Cipher Suitable for Low-Resource Device. In: Goubin, L., Matsui, M. (eds.) CHES 2006. LNCS, vol. 4249, pp. 46–59. Springer, Heidelberg (2006)
14. Knellwolf, S., Meier, W., Naya-Plasencia, M.: Conditional Differential Cryptanalysis of NLFSR-Based Cryptosystems. In: Abe, M. (ed.) ASIACRYPT 2010. LNCS, vol. 6477, pp. 130–145. Springer, Heidelberg (2010)
15. Knudsen, L.R., Leander, G., Poschmann, A., Robshaw, M.J.B.: PRINTcipher: A Block Cipher for IC-Printing. In: Mangard, S., Standaert, F.-X. (eds.) CHES 2010. LNCS, vol. 6225, pp. 16–32. Springer, Heidelberg (2010)
16. Lai, X., Massey, J.L., Murphy, S.: Markov Ciphers and Differential Cryptanalysis. In: Davies, D.W. (ed.) EUROCRYPT 1991. LNCS, vol. 547, pp. 17–38. Springer, Heidelberg (1991)
17. Leander, G.: On Linear Hulls, Statistical Saturation Attacks, PRESENT and a Cryptanalysis of PUFFIN. In: Paterson, K.G. (ed.) EUROCRYPT 2011. LNCS, vol. 6632, pp. 303–322. Springer, Heidelberg (2011)
18. Murphy, S.: The Effectiveness of the Linear Hull Effect. Technical report, RHUL-MA-2009-19 (2009)

The PHOTON Family of Lightweight Hash Functions

Jian Guo[1], Thomas Peyrin[2,⋆], and Axel Poschmann[2,⋆]

[1] Institute for Infocomm Research, Singapore
[2] Nanyang Technological University, Singapore
{ntu.guo,thomas.peyrin}@gmail.com, aposchmann@ntu.edu.sg

Abstract. RFID security is currently one of the major challenges cryptography has to face, often solved by protocols assuming that an on-tag hash function is available. In this article we present the PHOTON lightweight hash-function family, available in many different flavors and suitable for extremely constrained devices such as passive RFID tags. Our proposal uses a sponge-like construction as domain extension algorithm and an AES-like primitive as internal unkeyed permutation. This allows us to obtain the most compact hash function known so far (about 1120 GE for 64-bit collision resistance security), reaching areas very close to the theoretical optimum (derived from the minimal internal state memory size). Moreover, the speed achieved by PHOTON also compares quite favorably to its competitors. This is mostly due to the fact that unlike for previously proposed schemes, our proposal is very simple to analyze and one can derive tight AES-like bounds on the number of active Sboxes. This kind of AES-like primitive is usually not well suited for ultra constrained environments, but we describe in this paper a new method for generating the column mixing layer in a serial way, lowering drastically the area required. Finally, we slightly extend the sponge framework in order to offer interesting trade-offs between speed and preimage security for small messages, the classical use-case in hardware.

Keywords: lightweight, hash function, sponge function, AES.

1 Introduction

RFID tags are likely to be deployed widely in many different situations of everyday life and they represent a great business opportunity for various markets. However, this rising technology also provides new security challenges that the cryptography community has to handle. RFID tags can be used to fight product counterfeiting by authenticating them and on the other hand, we would also like to guarantee the privacy of the users.

These two security aspects have already been studied considerably and, interestingly, in most of the privacy-preserving RFID protocols proposed [3,20,23] a

⋆ The authors were supported in part by the Singapore National Research Foundation under Research Grant NRF-CRP2-2007-03.

P. Rogaway (Ed.): CRYPTO 2011, LNCS 6841, pp. 222–239, 2011.

hash function is required. Informally, such a primitive is a function that takes an arbitrary length input and outputs a fixed-size value. While no secret is involved in the computation, one would like that finding collisions (two distinct messages hashing to the same value) or (second)-preimages (a message input that hashes to a given challenge output value) is computationally intractable for an attacker. More precisely, for an n-bit ideal hash function we expect to perform $2^{n/2}$ and 2^n computations in order to find a collision and a (second)-preimage respectively. While not as mature as block-ciphers, the research on hash functions saw a rapid development lately, mainly due to the groundbreaking attacks on standardized primitives [37,35,36]. At the present time, most of the attention of the symmetric key cryptography academic community is focused on the SHA-3 competition organized by NIST [28], which should provide a potential replacement of the MD-SHA family.

In parallel, nice advances have also been made in the domain of lightweight symmetric key primitives in the last years. Protocol designers now have at disposal PRESENT[1], a 64-bit block-cipher with 80-bit key whose security has already been analyzed intensively and that can be as compact as 1075 GE [32]. Stream-ciphers are not outcast with implementations [19] with 80-bit security requiring about 1300 GE and 2600 GE reported for GRAIN and TRIVIUM respectively, two candidates selected in the final eSTREAM hardware portfolio. However, the situation is not as bright in the case of hash functions.

As already pointed out in [18] and echoed in, the community lacks very compact hash functions. Standardized primitives such as SHA-1 [26] or SHA-2 [27] are much too large to fit in very constrained hardware (5527 GE [29] and 10868 GE [18] for 80 and 128-bit aimed security respectively) and even compact-oriented proposals such as MAME require 8100 GE for 128-bit security. While hardware is an important criteria in the selection process, one can not expect the SHA-3 finalists to be much more compact. At the present time, all SHA-3 finalists require more than 12000 GE for 128-bit security (smaller versions of KECCAK that have not been submitted to the competition provide for example 64-bit security with 5090 GE). Note that a basic RFID tag may have a total gate count of anywhere from 1000-10000 gates, with only 200-2000 gates budgeted for security [22].

This compactness problem in hash algorithms is partly due to the fact that it widely depends on the memory registers required for the computation. Most hash functions proposed so far are software-oriented and output at least 256 bits in order to be out of reach of any generic collision search in practice. While such an output size makes sense where high level and long-term security are needed, RFID use-cases could bear much smaller security parameters. This is for example the path taken in, where the authors instantiate lightweight hash functions using literature-based constructions [21,31] with the compact block-cipher PRESENT. With SQUASH, Shamir proposed a compact keyed hash function inspired by the Rabin encryption scheme that processes short messages (at most

[1] Due to space limit, we omitted the references for many designs, interested readers are referred to [1] for an extended version of this article.

64-bit inputs) and that provides 64 bits of preimage security, without being collision resistant. At CHES 2010, the lightweight hash-function family ARMADILLO was proposed, but has recently been shown to present serious security weaknesses [11]. At the same conference, Aumasson *et al.* published the hash function QUARK, using sponge functions [4] as domain extension algorithm, and an internal permutation inspired from the stream-cipher GRAIN and the block-cipher KATAN [14]. Using sponge functions as operating mode is another step towards compactness. Indeed, classical n-bit hash function constructions like the MD-SHA family utilize a Merkle-Damgård [24,17] domain extension algorithm with a compression function h built upon an n-bit block-cipher E in Davies-Meyer mode ($h(CV, M) = E_M(CV) \oplus CV$), where CV stands for the chaining variable and M for the current message block. Avoiding any feed-forward like for sponge constructions saves a lot of memory registers at the cost of an invertible iterative process which induces a lower (second)-preimage security for the same internal state size. All in all, designers have to deal with a trade-off between security and memory requirements.

In this article, we describe a new hardware-oriented hash-function family: PHOTON. We chose to use the sponge functions framework in order to keep the internal memory size as low as possible. However, we extend this framework so as to provide very interesting trade-offs in hardware between preimage security and small messages hashing speed (small message scenario is a classical use-case and can be problematic for sponge functions because of their squeezing process that can be very slow in practice). The internal permutations of PHOTON can be seen as AES-like primitives especially derived for hardware: our columns mixing layer can be computed in a serial way while maintaining optimal diffusion properties. Overall, as shown in Table 2 in Section 4.3, not only PHOTON is easily the smallest hash function known so far, but it also achieves excellent area/throughput trade-offs.

In terms of security, it is particularly interesting to use AES-like permutations as we can fully leverage all the previous cryptanalysis performed on AES and on AES-based hash functions (again due to space limit we refer the reader to [1] for a detailed security analysis). Moreover, we can directly derive very simple bounds on the number of active Sboxes for 4 rounds of the permutation. These bounds being tight, we can confidently set an appropriate number of rounds that ensures a comfortable security margin.

2 Design Choices

In tag-based applications, one typically does not require high security primitives, such as a 512-bit output hash function. In contrary, 64 or 80-bit security is often appropriate considering the value of objects an RFID tag is protecting and the use cases. Moreover, a designer should use exactly the level that he expects from his primitive, so as to avoid any waste of area or computing power. This is the reason why we chose to precisely instantiate several security levels for PHOTON, ranging from 64-bit preimage resistance security to 128-bit collision resistance security.

2.1 Extended Sponge Functions

Sponge functions have been introduced by Bertoni *et al.* [4] as a new way of building hash functions from a fixed permutation (later more applications were proposed [7]). The internal state S, composed of the c-bit capacity and the r-bit bitrate, is first initialized with some fixed value. Then, after having appropriately padded and split the message into r-bit chunks, one simply and iteratively processes all r-bit message chunks by xoring them to the bitrate part of the internal state and then applying the $(c+r)$-bit permutation P. Once all message chunks have been handled by this absorbing phase, one successively outputs r bits of the final hash value by extracting r bits from the bitrate part of the internal state and then applying the permutation P on it (squeezing process).

When the internal permutation P is modeled as a randomly chosen permutation, a sponge function has been proven to be indifferentiable from a random oracle [5] up to $2^{c/2}$ calls to P. More precisely, for an n-bit sponge hash function with capacity c and bitrate r, when the internal primitive is modeled as a random permutation, one obtains $\min\{2^{n/2}, 2^{c/2}\}$ as collision resistance bound and $\min\{2^n, 2^{c/2}\}$ as (second)-preimage bound. However, in the case of preimage, there exists a gap between this bound and the best known generic attack[2]. Therefore, we expect the following complexities in the generic case:

- **Collision:** $\min\{2^{n/2}, 2^{c/2}\}$
- **Second-preimage:** $\min\{2^n, 2^{c/2}\}$
- **Preimage:** $\min\{2^n, 2^c, \max\{2^{n-r}, 2^{c/2}\}\}$

Moreover, sponge functions can be used as a Message Authentication Code with $MAC_K(M) = H(K||M)$, where $K \in \{0,1\}^k$ stands for the key and M for the message. It has been shown [8] that as long as the amount of message queries is limited to 2^a with $a \ll c/2$, then no attack better than exhaustive key search exists if $c \geq k + a + 1$.

Sponge functions seem a natural choice in order to minimize the amount of memory registers in hardware since they can offer speed/area/security tradeoffs. Indeed, the only memory required for the internal state is $c + r$ bits, while for a classical Davies-Meyer construction using an m-bit block cipher with a k-bit key input one needs to store $2m + k$ bits, out of which m bits are required for the feed-forward. For an equivalent ideal collision security level (thus setting $m = c = n$) and by minimizing the area (r and k are very small), the sponge function requires only about half of the memory. Note that if one looks for a perfectly (second)-preimage resistant hash function (up to the 2^n ideal bound), then it is required that $c \geq 2 \cdot n$ (which implies that the n-bit hash function built is indifferentiable from an n-bit random oracle anyway). In that particular case the sponge functions are not better than the Davies-Meyer construction in

[2] The 2^{n-r} term for preimage comes from the fact that in order to invert the hash function, the attacker will have to invert the squeezing process. The best known generic attack to solve this "multiblock constrained-input constrained-output problem" [6] requires 2^{n-r} computations.

terms of area requirements and therefore in this work we will not focus on this scenario. Instead, we will build hash functions that may have ideal resistance to collision, but not for (second)-preimage. The typical shape will be a capacity c equal to the hash output n and a very small bitrate r. This security/area trade-off, already utilized by the QUARK designers, will allow us to aim at extremly low area requirements, while maintaining security expectations very close to ideal.

In, the authors identify that in most RFID applications the user will not hash a large amount of data, *i.e.* in general less than 256 bits. Consider for example the *electronic product code (EPC)* number, which is a 96-bit string that is meant to identify globally any tag/product. In this particular case of small messages, sponge functions with a small bitrate r seem to be slow since one needs to call $(\lceil n/r \rceil - 1)$ times the internal permutation to complete the final squeezing process. This is for example the case with U-QUARK, that has a throughput of 1.47 kbps for very long messages which drops to 0.63 kbps for 96-bit inputs. On the other side, this "small messages" effect is reduced by the fact that having a small bitrate will reduce the amount of padding actually hashed (the padding simply consists in adding a "1" and as many "0" required to fill the last message block). Note that lightweight proposals based on classical Davies-Meyer construction that include the message length as suffix padding are also slow for small messages: DM-PRESENT-80 has a throughput of 14.63 kbps for very long messages which drops to 5.85 kbps for 96-bit inputs, because in the latter case many of the compression function calls are spent in order to handle padding blocks.

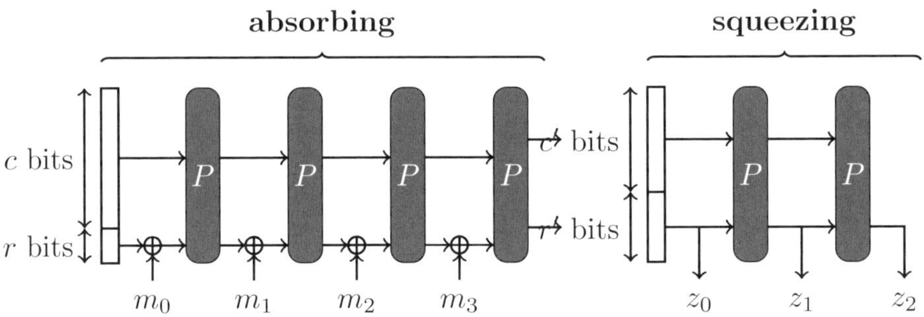

Fig. 1. The extended sponge framework, the domain extension algorithm used by the PHOTON hash-function family

In order to allow more flexibility about this issue, we propose to slightly extend the sponge framework by allowing the number r' of bits extracted during each iteration of the squeezing process to be different from the bitrate r[3] (see Figure 1). Increasing r' will directly reduce the time spent in the squeezing process, but might also reduce the preimage security. On the contrary, decreasing

[3] A recent work from Andreeva *et al.* [2] also independently proposed such an extension of the sponge model.

r' might improve the preimage bound at the cost of a speed drop for small messages. As long as the preimage security remains in an acceptable bound, this configuration can be interesting in many scenarios where only tiny inputs are to be hashed. More precisely, in this new model, the best known generic attacks require the following amount of computations:

- **Collision:** $\min\{2^{n/2}, 2^{c/2}\}$
- **Second-preimage:** $\min\{2^{n}, 2^{c/2}\}$
- **Preimage:** $\min\{2^{n}, 2^{c}, \max\{2^{n-r'}, 2^{c/2}\}\}$

Finally, in most tag-based applications the collision resistance is not a requirement, while only the one-wayness of the function must be ensured. However, as we previously explained, for lightweight scenarios the sponge construction does not maintain the (second)-preimage security at the full level of its capacity c. This is due to the output process of the sponge operating mode. Of course, performing a Davies-Meyer like feed-forward just after the final truncation would do the job, but that would also double the memory area required (which is precisely what we are trying to avoid). The nice trick of squeezing in the sponge functions framework permits to avoid any feed-forward while somehow rendering the process non-invertible, up to some extend (see multiblock constrained-input constrained-output problem in [6]). One solution to reach the full capacity preimage security would be to add one more squeezing iteration, thus increasing the output size of the hash by r' bits.[4] Then, the best known generic preimage attack for this $(n + r')$-bit hash function will run in $\min\{2^{n+r'}, 2^{c}, \max\{2^{n}, 2^{c/2}\}\} \geq 2^{n}$ when $c \geq n$ and one has to note that this hash output extension has no influence on the second-preimage resistance.

In this article, we will provide five sizes of internal permutations and one PHOTON flavor for each of them. The four biggest versions fit the classical sponge model and will ensure $2^{n/2}$ collision and second preimage resistance and 2^{n-r} concerning preimage. However, in order to illustrate the powerful trade-offs allowed by our extended model, the smaller PHOTON variant will have different input/output bitrates and an extended hash size. Using the five permutations defined in the next Section, one can derive its own PHOTON flavor depending on the collision / (second)-preimage / MAC security required, the maximal area and the maximal hash output size allowed. Note that the area required will only depend on the internal permutation chosen.

2.2 An AES-like Internal Permutation

We define an AES-like function to be a fixed key permutation P applied on an internal state of d^2 elements of s bits each, which can be represented as a $(d \times d)$ matrix. P is composed of N_r rounds, each containing four layers : AddConstants (AC), SubCells (SC), ShiftRows (ShR), and MixColumnsSerial (MCS). Informally, AddConstants simply consists in adding fixed values to the cells of the

[4] This generalization has been independently utilized by the QUARK designers in a revised version of their original article.

internal state, while SubCells applies an s-bit Sbox to each of them. ShiftRows rotates the position of the cells in each of the rows and MixColumnsSerial linearly mixes all the columns independently.

We chose to use AES-like permutations because they offer much confidence in the design strategy as one can leverage previous cryptanalysis works done on AES and on AES-like hash functions. Moreover, AES-like permutations allow to derive very simple proofs on the number of active Sboxes over four rounds of the primitive. More precisely, if the matrix underlying the MixColumnsSerial layer is Maximum Distance Separable (MDS), then one can immediately show that at least $(d+1)^2$ Sboxes will be active for any 4-round differential path [16]. This bound is tight, and we already know differential paths with only $(d+1)^2$ active Sboxes for four rounds (we will use them later for security analysis purposes). Moreover, note that the permutations we will design are fixed-key, so we naturally get rid of related-key attacks or any issue that might arise from the construction of a key-schedule [9,10].

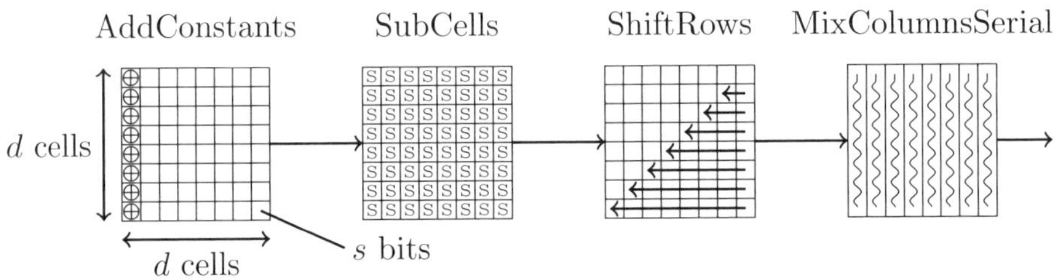

Fig. 2. One round of a PHOTON permutation

AddConstants. The constants have been chosen such that each of the N_r round computations are different, and such that the classical symmetry between columns in AES-like designs are destroyed (without the AddConstants layer, an input with all columns equal would maintain this property through any number of rounds). Also, the round constants can be generated by a combination of very compact Linear Feedback Shift Registers. For performance reasons, only the first column of the internal state is involved.

SubCells. Our choice of the Sboxes was mostly motivated by their hardware quality. 4-bit Sboxes can be very compact in hardware while the acceptable upper limit on the cell size is $s = 8$. We avoided to use an Sbox size s which is odd, because this leads to odd message block size or capacity when d is also odd. This leaves us with $s = 4, 6, 8$, but we also believe that reusing some already trusted and well analyzed components increases the confidence in the security of the scheme and saves a lot of time for cryptanalysts. Finally, we will use two types of Sboxes: the 4-bit PRESENT Sbox $\mathtt{SBOX_{PRE}}$ and the 8-bit AES Sbox $\mathtt{SBOX_{AES}}$ the latter being only utilized for high security levels (at least 128 bits of collision resistance). Note also that $s = 4, 8$ allows simpler and faster software implementations.

ShiftRows. The choice of the ShiftRows constants is very simple for PHOTON since our internal state is always a square of cells. Therefore, row i will classically be rotated by i positions to the left, i counts from 0.

MixColumnsSerial. The matrix underlying the AES MixColumns function is a circulant matrix with low hamming weight coefficients. Even if those coefficients and the irreducible polynomial used to create the Galois field for the AES MixColumns function have been chosen so as to improve the hardware footprint of the cipher, it can not be implemented in an extremely compact way. One of the main reason is that the byte-serial implementation of this function is not compact. Said in other words, if we write the AES MixColumns matrix as the composition of d operations each updating a single byte at a time in a serial way, then the coefficients of these d matrices will be very bad for small area implementations.

In order to solve this issue, we took the problem the other way round. Let A be the matrix that updates the last cell of the column vector with a linear combination of all of the vector cells and then rotates the vector by one position towards the top. Our new MixColumnsSerial layer will be composed of d applications of this matrix to the input column vector. More formally, let $X = (x_0, \ldots, x_{d-1})^T$ be an input column vector of MixColumnsSerial and $Y = (y_0, \ldots, y_{d-1})^T$ be the corresponding output. Then, we have $Y = A^d \times X$, where A is a $(d \times d)$ matrix of the form:

$$A = \begin{pmatrix} 0 & 1 & 0 & 0 & \cdots & 0 & 0 & 0 & 0 \\ 0 & 0 & 1 & 0 & \cdots & 0 & 0 & 0 & 0 \\ & \vdots & & & & & & \vdots & \\ 0 & 0 & 0 & 0 & \cdots & 0 & 0 & 0 & 1 \\ Z_0 & Z_1 & Z_2 & Z_3 & \cdots & Z_{d-4} & Z_{d-3} & Z_{d-2} & Z_{d-1} \end{pmatrix}$$

where coefficients $(Z_0, \ldots, Z_{d-1})$ can be chosen freely. We denote by $Serial(Z_0, \ldots, Z_{d-1})$ such a matrix. Of course, we would like the final matrix A^d to be MDS, so as to maintain as much diffusion as for the AES initial design strategy. For each square size d we picked during the design of PHOTON, we used MAGMA [12] to test all the possible values of $Z_0, \ldots, Z_{d-1}$ and picked the most compact candidate making A^d an MDS matrix. We also chose the irreducible polynomial with compactness as main criterion.

For design strategy comparison purposes, we can take as an example the AES case. By using our new mixing layer design method, we were able to find the matrix $A = Serial(1, 2, 1, 4)$ which gives the following MDS final matrix:

$$(A)^4 = \begin{pmatrix} 0 & 1 & 0 & 0 \\ 0 & 0 & 1 & 0 \\ 0 & 0 & 0 & 1 \\ 1 & 2 & 1 & 4 \end{pmatrix}^4 = \begin{pmatrix} 1 & 2 & 1 & 4 \\ 4 & 9 & 6 & 17 \\ 17 & 38 & 24 & 66 \\ 66 & 149 & 100 & 11 \end{pmatrix}$$

The smallest AES hardware implementation requires 2400 GE [25], for which 263 GE are dedicated to MixColumns. It is possible to implement MixColumns of AES in a byte-by-byte fashion, which requires only 81 GE to calculate one byte of the output column. However, since AES uses a circulant matrix, at least three additional 8-bit registers (144 GE), are required to hold the output, plus additional control logic, which increases the area requirements significantly. That is why [25] does not use a serial MixColumns, but rather processes one column at a time.

Please note that in general the choice of non-zero constants for any $d \times d$ MDS matrix on s-bit cells has only a minor impact of the area consumption, since a multiplication by x consists of w XOR gates, where w denotes the Hamming weight of the irreducible polynomial used. At the same time, $(d-1) \cdot s$ XOR gates are required to sum up the d individual terms of s bits each. It is no surprise, that multiplying with the constants above accounts for only 21.3 GE out of the 74 GE required. In fact, the efficiency of our approach lies in the shifting property of A, since this allows to re-use the existing memory with neither temporary storage nor additional control logic required.

All in all, using our approach would provide a tweaked AES cipher with the very same diffusion properties as the original one (the matrix being MDS), but that can fit in only 2210 GE, a total saving of around 8%. Moreover, for the deciphering process, a slightly modified hardware can be used in order to unroll the MixColumnsSerial, further reducing the area footprint of such a PHOTON-based cipher. One might think that the software implementations will suffer from this new layer. While our goal is to make a hardware-oriented primitive, we would like to remark that most AES software implementations are precomputed tables-based (applying both the Sbox and the MixColumns coefficients at the same time) and the very same method can be applied to PHOTON. This is confirmed by our first software implementations, whose benchmarks are given in Section 4.4.

3 The PHOTON Hash-Function Family

We describe in this section the PHOTON family of hash functions.[5] Each variant will be fully defined by its hash output size $64 \leq n \leq 256$, its input and its output bitrate r and r' respectively. Therefore we denote each function PHOTON-$n/r/r'$. The internal state size $t = (c + r)$ depends on the hash output size and can take only 5 distinct values: 100, 144, 196, 256 and 288 bits. As a consequence, we only have to define 5 internal permutations P_t, one for each internal state size.

In order to cover a wide spectrum of applications, we propose five different flavors of PHOTON, one for each internal state size: PHOTON-80/20/16, PHOTON-128/16/16, PHOTON-160/36/36, PHOTON-224/32/32 and PHOTON-256/32/32 will use internal permutations P_{100}, P_{144}, P_{196}, P_{256} and P_{288} respectively. Note that the first proposal is special in the sense that it is designed for the specific cases where 64-bit preimage security and 64-bit key MAC are considered to be

[5] An extended version of this paper including a more detailed description of PHOTON and test vectors can be found at the PHOTON website [1].

sufficient.[6] In contrary, the last proposal provides a high security level of 128-bit collision resistance, thus making it suitable for generic applications.

3.1 The Domain Extension Algorithm

The message M to hash is first padded by appending a "1" bit and as many zeros (possibly none) such that the total length is a multiple of the bitrate r and we can finally obtain l message blocks $m_0, \ldots, m_{l-1}$ of r bits each. The t-bit internal state S is initialized by setting it to the value $S_0 = IV = \{0\}^{t-24}||n/4||r||r'$, where $||$ denotes the concatenation and each value is coded on 8 bits. For implementation purposes, note that each byte is interpreted in big-endian form.

Then, as for the classical sponge strategy, at iteration i we absorb the message block m_i on leftmost part of the internal state S_i and then apply the permutation P_t, *i.e.* $S_{i+1} = P_t(S_i \oplus (m_i||\{0\}^c))$. Once all l message blocks have been absorbed, we build the hash value by concatenating the successive r'-bit output blocks z_i until we reach the appropriate output size n: $hash = z_0|| \ldots ||z_{l'-1}$, where l' denotes the number of squeezing iterations, that is $l' = \lceil n/r' \rceil - 1$. More precisely, z_i is the r' leftmost bits of the internal state S_{l+i} and we have $S_{l+i+1} = P_t(S_{l+i})$ for $0 \leq i < l'$. If the hash output size is not a multiple of r', one just truncates $z_{l'-1}$ to $n \bmod r'$ bits.

3.2 The Internal Permutations

We define here the internal permutations P_t, where $t \in \{100, 144, 196, 256, 288\}$. The internal state of the N_r-round permutation is viewed as a $(d \times d)$ matrix of s-bit cells and the corresponding values depending of t are given in Table 1. Note that we will always use a cell size of 4 bits, except for the largest version for which we use 8-bit cells, and that the number of rounds is always $N_r = 12$, whatever the value of t is. The internal state cell located at row i and column j is denoted $S[i,j]$ with $0 \leq i,j < d$.

One round is composed of four layers (see Figure 2): AddConstant (AC), SubCell (SC), ShiftRows (ShR) and MixColumnsSerial (MCS).

AddConstant. At round number v (starting the counting from 1), we first XOR a round constant $RC(v)$ to each cell $S[i,0]$ of the first column of the internal state. Then, we XOR distinct internal constants $IC_d(i)$ to each cell $S[i,0]$ of the same first column. Overall, for round v we have $S'[i,0] = S[i,0] \oplus RC(v) \oplus IC_d(i)$ for all $0 \leq i < d$. The round constants are

$$RC(v) = [1, 3, 7, 14, 13, 11, 6, 12, 9, 2, 5, 10].$$

[6] By sponge keying and using the security bound from [8], PHOTON-80/20/16 provides a secure 64-bit key MAC as long as the number of messages to be computed is lower than 2^{15}. For a secure 64-bit key MAC handling more messages (up to 2^{27}), one can for example go for a very similar PHOTON-80/8/8 version that also uses P_{100}. This version with capacity $c = 92$ would require the same area as PHOTON-80/20/16 but would be slower.

Table 1. The parameters of the internal permutations P_t, together with the internal constants IC_d, the irreductible polynomials and the Z_i coefficients for the MixColumnsSerial computation

	t	d	s	N_r	$IC_d(\cdot)$	irr. polynomial	Z_i coefficients
P_{100}	100	5	4	12	$[0,1,3,6,4]$	x^4+x+1	$(1,2,9,9,2)$
P_{144}	144	6	4	12	$[0,1,3,7,6,4]$	x^4+x+1	$(1,2,8,5,8,2)$
P_{196}	196	7	4	12	$[0,1,2,5,3,6,4]$	x^4+x+1	$(1,4,6,1,1,6,4)$
P_{256}	256	8	4	12	$[0,1,3,7,15,14,12,8]$	x^4+x+1	$(2,4,2,11,2,8,5,6)$
P_{288}	288	6	8	12	$[0,1,3,7,6,4]$	$x^8+x^4+x^3+x+1$	$(2,3,1,2,1,4)$

The internal constants depend on the square size d and on the row position i. They are given in Table 1.

SubCells. This layer simply applies an s-bit Sbox to each of the cells of the internal state, *i.e.* $S'[i,j] = \mathtt{SBOX}(S[i,j])$ for all $0 \leq i,j < d$. In the case of 4-bit cells, we use the PRESENT Sbox $\mathtt{SBOX}_{\mathtt{PRE}}$ while for the 8-bit cells case we use the AES Sbox $\mathtt{SBOX}_{\mathtt{AES}}$ [16].

ShiftRows. As for the AES, for each row i this layer rotates all cells to the left by i column positions. Namely, $S'[i,j] = S[i,(j+i) \bmod d]$ for all $0 \leq i,j < d$.

MixColumnsSerial. The final mixing layer is applied to each of the columns of the internal state independently. For each column j input vector $(S[0,j],\ldots,S[d-1,j])^T$, we apply d times the matrix $A_t = Serial(Z_0,\ldots,Z_{d-1})$. That is, for all $0 \leq j < d$: $(S'[0,j],\ldots,S'[d-1,j])^T = A_t^d \times (S[0,j],\ldots,S[d-1,j])^T$ where the coefficients $Z_0,\ldots,Z_{d-1}$ are given in Table 1. In the case of 4-bit cells, the irreducible polynomial we chose is x^4+x+1, while for the 8-bit case we chose the AES one, *i.e.* $x^8+x^4+x^3+x+1$. Note that all A_t^d matrices are Maximum Distance Separable.[7]

4 Performances and Comparison

Before we detail the hardware architectures and the optimizations done, we first describe the tools used. Finally we compare our results to previous work.

4.1 Design Flow

We used *Mentor Graphics ModelSimXE 6.4b* and *Synopsys DesignCompiler A-2007.12-SP1* for functional simulation and synthesis of the designs to the *Virtual Silicon* (VST) standard cell library *UMCL18G212T3* [34], which is based on the *UMC L180 0.18μm 1P6M* logic process with a typical voltage of 1.8 V. We used

[7] One could wonder why we did not propose a version with $d = 9$ and $s = 4$. The reason is that there is no matrix fulfilling the desired "serial MDS" properties for those parameters, whatever the irreducible polynomial chosen.

Synopsys Power Compiler version *A-2007.12-SP1* to estimate the power consumption of our ASIC implementations. For synthesis and for power estimation we advised the compiler to keep the hierarchy and use a clock frequency of 100 KHz. Note that the wire-load model used, though it is the smallest available for this library, still simulates the typical wire-load of a circuit with a size of around 10 000 GE.

4.2 Hardware Architectures

To substantiate our claims on the hardware efficiency of our PHOTON family, we have implemented the flavors specified in Section 3 in VHDL and simulated their post-synthesis performance. We designed two architectures: one is fully serialized, *i.e.* performing operations on one cell per clock cycle, and aims for the smallest area possible; the second one is a d times parallelization of the first architecture, thus performing operations on one row in one clock cycle, resulting in a significant speed-up. As can be seen in Figure 3, our serialized design consists of six modules: MCS, State, IO, AC, SC, and Controller.

IO allows to 1) initialize our implementation with an all '0' vector, 2) input the IV, 3) absorb message chunks, and 4) forward the output of the State module to the AC module without further modification. Instead of using two Multiplexer and an XOR gate, we used two NAND and one XOR gate thereby reducing the gate count required from $s \cdot 7.33$ to $s \cdot 4.67$ GE.

State comprises a $d \cdot d$ array of flip-flop cells storing s bits each. Every row constitutes a shift-register using the output of the last stage, *i.e.* column 0, as the input to the first stage (column $d-1$) of the same row and the next row. Using this feedback functionality ShiftRows can be performed in $d-1$ clock cycles with no additional hardware costs. Further, since MixColumnsSerial is performed on column 0, also a vertical shifting direction is required for this

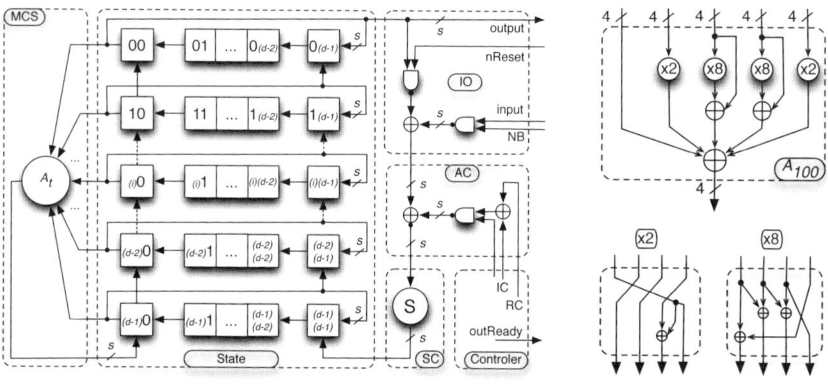

Fig. 3. Serial hardware architecture of PHOTON (left). As an example for its component A_t we also depict A_{100} with its sub-components (right).

column. Consequently, columns 0 and $d-1$ consist of flip-flop cells with two inputs (6 GE), while columns 1 to $d-2$ consist of flip-flop cells with only one input (4.67 GE). The overall gate count for this module is $s \cdot d \cdot ((d-2) \cdot 4.67 + 2 \cdot 6)$ GE and for all flavors it occupies the majority of the area required (between 65 and 77.5%).

`MCS` calculates the last row of A_t in one clock cycle. The result is stored in the `State` module, that is in the last row of column 0, which has been shifted upwards at the same time. Consequently, after d clock cycles the MixColumnsSerial operation is applied to an entire column. Then the whole state array is rotated by one position to the left and the next column is processed. In total $d \cdot (d+1)$ clock cycles are required to perform MCS. As an example of the hardware efficiency of MCS we depict A_{100} in the upper and its sub-components in the lower right part of Figure 3. Using our library, for a multiplication by 2, 4 and 8, we need 2.67 GE, 4.67 GE, and 7 GE when using the irreducible polynomial $x^4 + x + 1$, respectively. Therefore the choice of the coefficients has only a minor impact on the overall gate count, as the majority is required to sum up the intermediate results. For example, in the case of A_{100}, 56 out of 75.33 GE are required for the XOR sum. The gate counts for the other matrices are: 80 GE, 99 GE, 145 GE, and 144 GE for A_{144}, A_{196}, A_{256}, and A_{288}, respectively.

`AC` performs the AddConstant operation by XORing the sum of the round constant RC with the current internal constant IC. Furthermore, since AC is only applied to the first column, the input to the XNOR gate is gated with a NAND gate. Instead of using an AND gate in combination with an XOR gate, our approach allows to reduce the area required from $s \cdot 6.67$ to $s \cdot 6$ GE.

`SC` performs the SubCells operation and consists of a single instantiation of the corresponding Sbox. For $s = 4$ we used an optimized Boolean representation of the `PRESENT` Sbox, which only requires 22.33 GE and for $s = 8$ we used Canright's representation of the AES Sbox [15] which requires 233 GE. It takes $d \cdot d$ clock cycles to perform AddConstant and SubCells on the whole state.

`Controller` uses a Finite State Machine (FSM) to generate all control signals required. Furthermore, also the round constants and the internal constants are generated within this module, as their values are used for the transition conditions of the FSM. The FSM consists of one idle state, one state for the combined execution of AC and SC, $d-1$ states for ShR and two states for MCS (one for processing one column and another one to rotate the whole state to the left). Naturally, its gate count varies depending on d: 197 GE, 210 GE, 235 GE, and 254 GE for $d = 5, 6, 7, 8$, respectively.

4.3 Hardware Results and Comparison

We assume the message to be padded correctly and the IVs to be loaded at the beginning of the operation. Then it requires $d \cdot d + (d-1) + d \cdot (d+1)$ clock cycles to perform one round of the permutation P, resulting in a total latency of $12 \cdot (2 \cdot d \cdot (d+1) - 1)$ clock cycles. Table 2 compares our results to previous works, sorted after preimage and collision resistance levels. Area requirements are provided in

GE, while the latency is given in clock cycles for only the internal permutation P (or the internal block-cipher E), and the whole hash function H. Further metrics are Throughput in kbps and a Figure of Merit (FOM) proposed by. In order to have a comparison for a best case scenario and a real-world application, we provide the latter two metrics for 'long' messages (omitting any padding influences) and for 96-bit messages, where we do take padding into account. In particular this means that a 96-bit message is padded with "1" and as many "0"s as required. Furthermore Merkle-Damgård constructions need additional 64 bits to encode the message length. The parameters n, c, r and r' stand for the hash output size, the capacity, the input bitrate and the output bitrate respectively. Finally, the column "Pre" gives the claimed preimage resistance security and "Col" the claimed collision resistance security.

As can be seen, our proposals compete well in terms of area requirements, since they are 18% to 75% smaller compared to previous proposals with a similar preimage/collision resistance level. For a smaller area, the throughput of PHOTON variants is comparable to the QUARK proposals[8]. Alternatively, for a similar area, PHOTON variants are much faster than the QUARK proposals. This can be observed in the Figure of Merit column of the results Table. One could argue that the throughput of two proposals can not be compared because the security margin is not taken in account. However, we would like to emphasize that the security margin is very hard to measure as it greatly depends on the simplicity of the scheme, the amount of work spent by the cryptanalysts, etc. Unlike most of the lightweight hash functions proposed, in the case of PHOTON, we chose very simple to analyse internal permutations, thus directly leveraging the extensive analysis work already known for AES-like permutations. While 8 rounds over 12 of the internal permutations of PHOTON can be distinguished from a random permutation, we provide strong arguments that this is very unlikely to be much improved.

We did not include power figures in Table 2 for several reasons. First, the power consumption strongly depends on the technology used and cannot be compared between different technologies in a fair manner. Furthermore, simulated power figures strongly depend on the simulation method used, and the effort spent. Instead, we just briefly list the simulated power figures for our proposals here: 1.59, 2.29, 2.74, 4.01, and 4.55μW for serialized implementation of PHOTON-80/20/16, PHOTON-128/16/16, PHOTON-160/36/36, PHOTON-224/32/32, and PHOTON-256/32/32, respectively. The d-parallel implementations require 2.7, 3.45, 4.35, 6.5, and 8.38μW, respectively. This let us conclude that all PHOTON flavors seem to be suitable for ultra-constrained devices, such as passive RFID tags, which was one of our initial design goals.

[8] We synthesized the publicly available VHDL source code of U-QUARK using the same tool chain and ASIC library as for our proposals. The post-synthesis figures for U-QUARK are slightly higher than the previously published ones, *i.e.* 1400 GE instead of 1379 GE, which indicates that PHOTONs smaller footprint is not caused by a different tool chain. However, for comparison we took the previously available figures, which is in favour of QUARK.

Table 2. Overview of parameters, security level, and performance of several lightweight hash functions. Throughput and FOM figures have been derived at a clock frequency of 100 KHz. We marked by a * the preimage resistances of PHOTON-128/16/16, PHOTON-160/36/36, PHOTON-224/32/32 and PHOTON-256/32/32 in order to indicate that these PHOTON variants can achieve equal preimage resistance compared to its competitors by simply adding one more squeezing round. This will increase the hash output size n by r' bits and slightly reduce the throughput for small messages, while the area and the long message performances will remain the same.

Name	Ref.	Parameters				Security		Performance						
		n	c	r	r'	Pre	Col	Area [GE]	Latency [clk]		Throughput [kbps]		FOM [nb/clk/GE^2]	
									P/E	H	long	96-bit	long	96-bit
64-bit preimage resistance														
SQUASH	[38]	64	x	x	x	64	0	2646	31800	31800	0.2	0.15	0.29	0.14
DM-PRESENT-80		64	64	80	x	64	32	1600	547	547	14.63	5.85	57.13	19.04
DM-PRESENT-80		64	64	80	x	64	32	2213	33	33	242.42	96.67	495.01	165.00
DM-PRESENT-128		64	64	128	x	64	32	1886	559	559	22.90	8.59	64.37	32.19
DM-PRESENT-128		64	64	128	x	64	32	2530	33	33	387.88	145.45	605.98	302.99
KECCAK-f[200]		64	128	72	72	64	32	2520	900	900	8.00	5.33	12.6	8.4
PHOTON-80/20/16		80	80	20	16	64	40	865	708	3540	2.82	1.51	37.73	20.12
PHOTON-80/20/16		80	80	20	16	64	40	1168	132	660	15.15	8.08	111.13	59.27
64-bit collision resistance														
U-QUARK		136	128	8	8	128	64	1379	544	9248	1.47	0.61	7.73	3.20
U-QUARK		136	128	8	8	128	64	2392	68	1156	11.76	4.87	20.56	8.51
H-PRESENT-128		128	128	64	x	128	64	2330	559	559	11.45	5.72	21.09	10.54
H-PRESENT-128		128	128	64	x	128	64	4256	32	32	200.00	100.00	110.41	55.21
ARMADILLO2-B		128	128	64	x	128	64	4353	256	256	25.00	12.50	13.19	6.60
ARMADILLO2-B		128	128	64	x	128	64	6025	64	64	100.00	50.00	27.55	13.77
KECCAK-f[400]		128	256	144	144	128	64	5090	1000	1000	14.40	9.60	5.56	3.71
PHOTON-128/16/16		128	128	16	16	112*	64	1122	996	7968	1.61	0.69	12.78	5.48
PHOTON-128/16/16		128	128	16	16	112*	64	1708	156	1248	10.26	4.4	35.15	15.06
80-bit collision resistance														
D-QUARK		176	160	16	16	160	80	1702	704	7744	2.27	0.80	7.85	2.77
D-QUARK		176	160	16	16	160	80	2819	88	968	18.18	6.42	22.88	8.08
ARMADILLO2-C		160	160	80	x	160	80	5406	320	320	25.00	10.00	8.55	3.42
ARMADILLO2-C		160	160	80	x	160	80	7492	80	80	100.00	40.00	17.82	7.13
SHA-1	[29]	160	160	512	x	160	80	5527	344	344	148.84	27.91	48.72	9.14
PHOTON-160/36/36		160	160	36	36	124*	80	1396	1332	6660	2.70	1.03	13.87	5.28
PHOTON-160/36/36		160	160	36	36	124*	80	2117	180	900	20	7.62	44.64	17.01
112-bit collision resistance														
S-QUARK		256	224	32	32	224	112	2296	1024	8192	3.13	0.85	5.93	1.62
S-QUARK		256	224	32	32	224	112	4640	64	512	50.00	13.64	23.22	6.33
PHOTON-224/32/32		224	224	32	32	192*	112	1736	1716	12012	1.86	0.56	6.19	1.86
PHOTON-224/32/32		224	224	32	32	192*	112	2786	204	1428	15.69	4.71	20.21	6.06
128-bit collision resistance														
ARMADILLO2-E		256	256	128	x	256	128	8653	512	512	25.00	9.38	3.34	1.25
ARMADILLO2-E		256	256	128	x	256	128	11914	128	128	100.00	37.50	7.05	2.64
SHA-2	[18]	256	256	512	x	256	128	10868	1128	1128	45.39	8.51	3.84	0.72
PHOTON-256/32/32		256	256	32	32	224*	128	2177	996	7968	3.21	0.88	6.78	1.85
PHOTON-256/32/32		256	256	32	32	224*	128	4362	156	1248	20.51	5.59	10.78	2.94

Table 3. Software performances in cycles per byte of the PHOTON variants for long messages

PHOTON-80/20/16	PHOTON-128/16/16	PHOTON-160/36/36	PHOTON-224/32/32	PHOTON-256/32/32
95 c/B	156 c/B	116 c/B	227 c/B	157 c/B

4.4 Software Implementation

We give in Table 3 our software implementation performances for the PHOTON variants. The processor used for the benchmarks is an Intel(R) Core(TM) i7 CPU Q 720 clocked at 1.60GHz. For comparison purposes, we also benchmarked the speed of an AES permutation (without the key schedule) and a modified version of it with a serially computable MDS matrix instead (the 4×4 matrix A given in Section 2.2). As expected, the table-based implementations reach the same speed for both versions. We also benchmarked other lightweight hash function designs. QUARK reference code, very likely to be optimizable, runs at 8k, 30k and 22k cycles per byte for U-QUARK, D-QUARK and S-QUARK, respectively. The optimized PRESENT code runs at 90 cycles per byte, hence the estimate speed for DM-PRESENT-80, DM-PRESENT-128 and H-PRESENT-128 are 72, 45 and 90 cycles per byte, respectively.

5 Conclusion

We proposed PHOTON, the most lightweight hash-function family known so far, very close to the theoretical optimum. Our proposal is based on the well known AES design strategy, but we introduced a new mixing layer building method that perfectly fits small area scenarios. This allows us to directly leverage the extensive work done on AES and AES-like hash functions so as to provide good confidence in the security of our scheme. Due to page restrictions we refer to an extended version of this paper [1] for a detailed security analysis of PHOTON. Finally, PHOTON is not only the smallest hash function, but it also achieves excellent area/throughput trade-offs and we obtained very acceptable performances with simple software implementations.

Acknowledgement. The authors would like to thank the anonymous referees for their helpful comments. Also, we are very grateful to Dag Arne Osvik and AlpCode for providing an optimized Boolean representation of the PRESENT Sbox, to Jean-Philippe Aumasson for providing his cube testers source code and to Christina Boura for her help with zero-sum distinguishers.

References

1. The PHOTON Family of Lightweight Hash Functions, http://sites.google.com/site/photonhashfunction/
2. Andreeva, E., Mennink, B., Preneel, B.: The Parazoa Family: Generalizing the Sponge Hash Functions. Cryptology ePrint Archive, Report 2011/028 028 (2011)

3. Avoine, G., Oechslin, P.: A Scalable and Provably Secure Hash-Based RFID Protocol. In: PerCom Workshops, pp. 110–114. IEEE Computer Society, Los Alamitos (2005)
4. Bertoni, G., Daemen, J., Peeters, M., Assche, G.V.: Sponge functions. In: Ecrypt Hash Workshop 2007 (May 2007)
5. Bertoni, G., Daemen, J., Peeters, M., Assche, G.V.: On the Indifferentiability of the Sponge Construction. In: Paterson [30], pp. 181–197 (2011)
6. Bertoni, G., Daemen, J., Peeters, M., Assche, G.V.: Keccak specifications. Submission to NIST (2009) (Round 2)
7. Bertoni, G., Daemen, J., Peeters, M., Van Assche, G.: Sponge-Based Pseudo-Random Number Generators. In: Mangard, S., Standaert, F.-X. (eds.) CHES 2010. LNCS, vol. 6225, pp. 33–47. Springer, Heidelberg (2010)
8. Bertoni, G., Daemen, J., Peeters, M., Assche, G.V.: On the security of the keyed sponge construction. In: Leander, G., Thomsen, S. (eds.) SKEW (2011)
9. Biryukov, A., Khovratovich, D.: Related-Key Cryptanalysis of the Full AES-192 and AES-256. In: Matsui, M. (ed.) ASIACRYPT 2009. LNCS, vol. 5912, pp. 1–18. Springer, Heidelberg (2009)
10. Biryukov, A., Khovratovich, D., Nikolić, I.: Distinguisher and Related-Key Attack on the Full AES-256. In: Halevi, S. (ed.) CRYPTO 2009. LNCS, vol. 5677, pp. 231–249. Springer, Heidelberg (2009)
11. Blondeau, C., Naya-Plasencia, M., Videau, M., Zenner, E.: Cryptanalysis of ARMADILLO2. Cryptology ePrint Archive, Report 2011/160 (2011)
12. Bosma, W., Cannon, J., Playoust, C.: The Magma algebra system. I. The user language. J. Symbolic Comput. 24(3-4), 235–265 (1997); Computational Algebra and Number Theory, Londan (1993)
13. Brassard, G. (ed.): CRYPTO 1989. LNCS, vol. 435. Springer, Heidelberg (1990)
14. De Cannière, C., Dunkelman, O., Knežević, M.: KATAN and KTANTAN — A Family of Small and Efficient Hardware-Oriented Block Ciphers. In: Clavier, C., Gaj, K. (eds.) CHES 2009. LNCS, vol. 5747, pp. 272–288. Springer, Heidelberg (2009)
15. Canright, D.: A Very Compact S-Box for AES. In: Rao, J.R., Sunar, B. (eds.) CHES 2005. LNCS, vol. 3659, pp. 441–455. Springer, Heidelberg (2005), The HDL specification is available at the author's official webpage `http://faculty.nps.edu/drcanrig/pub/index.html`
16. Daemen, J., Rijmen, V.: The Design of Rijndael: AES - The Advanced Encryption Standard. Springer, Heidelberg (2002)
17. Damgård, I.: A Design Principle for Hash Functions. In: Brassard [13], pp. 416–427 (1989)
18. Feldhofer, M., Rechberger, C.: A Case Against Currently Used Hash Functions in RFID Protocols. In: Meersman, R., Tari, Z., Herrero, P. (eds.) OTM 2006 Workshops. LNCS, vol. 4277, pp. 372–381. Springer, Heidelberg (2006)
19. Good, T., Benaissa, M.: ASIC Hardware Performance. In: Robshaw, M.J.B., Billet, O. (eds.) New Stream Cipher Designs. LNCS, vol. 4986, pp. 267–293. Springer, Heidelberg (2008)
20. Henrici, D., Götze, J., Müller, P.: A Hash-based Pseudonymization Infrastructure for RFID Systems. In: SecPerU, pp. 22–27. IEEE Computer Society, Los Alamitos (2006)
21. Hirose, S.: Some Plausible Constructions of Double-Block-Length Hash Functions. In: Robshaw, M.J.B. (ed.) FSE 2006. LNCS, vol. 4047, pp. 210–225. Springer, Heidelberg (2006)

22. Juels, A., Weis, S.A.: Authenticating Pervasive Devices with Human Protocols. In: Shoup [33], pp. 293–308 (2005)
23. Lee, S.-M., Hwang, Y.J., Lee, D.H., Lim, J.I.: Efficient Authentication for Low-Cost RFID Systems. In: Gervasi, O., Gavrilova, M.L., Kumar, V., Laganá, A., Lee, H.P., Mun, Y., Taniar, D., Tan, C.J.K. (eds.) ICCSA 2005. LNCS, vol. 3480, pp. 619–627. Springer, Heidelberg (2005)
24. Merkle, R.C.: One Way Hash Functions and DES. In: Brassard [13], pp. 428–446 (1989)
25. Moradi, A., Poschmann, A., Ling, S., Paar, C., Wang, H.: Pushing the Limits: A Very Compact and a Threshold Implementation of the AES. In: Paterson [30]
26. National Institute of Standards and Technology. FIPS 180-1: Secure Hash Standard (April 1995), http://csrc.nist.gov
27. National Institute of Standards and Technology. FIPS 180-2: Secure Hash Standard (August 2002), http://csrc.nist.gov
28. National Institute of Standards and Technology. Announcing Request for Candidate Algorithm Nominations for a NewCryptographic Hash Algorithm (SHA-3) Family. Federal Register 27(212), 62212–62220 (November 2007), http://csrc.nist.gov/groups/ST/hash/documents/FR_Notice_Nov07.pdf (October 17, 2008)
29. O'Neill, M.: Low-Cost SHA-1 Hash Function Architecture for RFID Tags. In: Dominikus, S., Aigner, M. (eds.) RFIDSec (2008), http://events.iaik.tugraz.at/RFIDSec08/Papers/
30. Paterson, K.G. (ed.): EUROCRYPT 2011. LNCS, vol. 6632. Springer, Heidelberg (2011)
31. Peyrin, T., Gilbert, H., Muller, F., Robshaw, M.J.B.: Combining Compression Functions and Block Cipher-Based Hash Functions. In: Lai, X., Chen, K. (eds.) ASIACRYPT 2006. LNCS, vol. 4284, pp. 315–331. Springer, Heidelberg (2006)
32. Rolfes, C., Poschmann, A., Leander, G., Paar, C.: Ultra-Lightweight Implementations for Smart Devices – Security for 1000 Gate Equivalents. In: Grimaud, G., Standaert, F.-X. (eds.) CARDIS 2008. LNCS, vol. 5189, pp. 89–103. Springer, Heidelberg (2008)
33. Shoup, V. (ed.): CRYPTO 2005. LNCS, vol. 3621. Springer, Heidelberg (2005)
34. Virtual Silicon Inc. 0.18 μm VIP Standard Cell Library Tape Out Ready, Part Number: UMCL18G212T3, Process: UMC Logic 0.18 μm Generic II Technology: 0.18μm (July 2004)
35. Wang, X., Yin, Y.L., Yu, H.: Finding Collisions in the Full SHA-1. In: Shoup [33], pp. 17–36 (2005)
36. Wang, X., Yu, H.: How to Break MD5 and Other Hash Functions. In: Cramer, R. (ed.) EUROCRYPT 2005. LNCS, vol. 3494, pp. 19–35. Springer, Heidelberg (2005)
37. Wang, X., Yu, H., Yin, Y.L.: Efficient Collision Search Attacks on SHA-0. In: Shoup [33], pp. 1–16 (2005)
38. Zhilyaev, S.: Evaluating a new MAC for current and next generation RFID. Master's thesis, University of Massachusetts Amherst (2010), http://scholarworks.umass.edu/cgi/viewcontent.cgi?article=1477&context=theses

Perfectly-Secure Multiplication for Any $t < n/3$

Gilad Asharov[1,⋆], Yehuda Lindell[1,⋆], and Tal Rabin[2]

[1] Bar-Ilan University
asharog@cs.biu.ac.il, lindell@biu.ac.il
[2] IBM T.J. Watson Research
talr@us.ibm.com

Abstract. In the setting of secure multiparty computation, a set of n parties with private inputs wish to jointly compute some functionality of their inputs. One of the most fundamental results of information-theoretically secure computation was presented by Ben-Or, Goldwasser and Wigderson (BGW) in 1988. They demonstrated that any n-party functionality can be computed with *perfect security*, in the private channels model. The most technically challenging part of this result is a protocol for multiplying two shared values, with perfect security in the presence of up to $t < n/3$ malicious adversaries.

In this paper we provide a full specification of the BGW perfect multiplication protocol and prove its security. This includes one new step for the perfect multiplication protocol in the case of $n/4 \leq t < n/3$. As in the original BGW protocol, this protocol works whenever the parties hold univariate (Shamir) shares of the input values. In addition, we present a new multiplication protocol that utilizes *bivariate* secret sharing in order to achieve higher efficiency while maintaining a round complexity that is constant per multiplication. Both of our protocols are presented with full proofs of security.

1 Introduction

The groundbreaking BGW protocol [4] for perfectly-secure multiparty computation appeared over 20 years ago and has had a huge impact on our field; the importance of the result coupled with its elegant and ingenious techniques are the source of this great following. The BGW protocol enables a set of parties $P_1, \ldots, P_n$ to compute any functionality of their inputs while preserving security in the presence of up to $t < n/3$ malicious parties. The protocol is comprised of a few components; a method for verifiable secret sharing (VSS), a protocol for adding two secrets that are in shared form and a protocol for multiplying two secrets given in shared form. Despite its importance and the fact that over a thousand papers have built upon it, a full proof of its correctness has never appeared. In [1] we rectify this situation and present a proof of the BGW protocol, together with a new more efficient multiplication protocol presented here. In addition, we provide a full specification of the protocol. This includes a new

⋆ Supported by the European Research Council as part of the ERC project LAST.

P. Rogaway (Ed.): CRYPTO 2011, LNCS 6841, pp. 240–258, 2011.

step that is needed for the case of $n/4 \leq t < n/3$. In this extended abstract we focus on the question of perfect multiplication, including the new step needed for the BGW protocol and our new more efficient protocol.

BGW Perfect Multiplication. The aim of the BGW multiplication protocol is to have a set of parties compute a sharing of the product $a \cdot b$, given a sharing of the individual values a and b.[1] Let $a_1, \ldots, a_n$ denote the parties' shares of a, and let $b_1, \ldots, b_n$ denote their shares of b. The protocol works according to the following steps:

1. Each party P_i shares its shares a_i and b_i with all other parties. This is carried out in a way (using error correcting codes) that prevents a corrupted P_i from sharing a value $a'_i \neq a_i$ or $b'_i \neq b_i$.
2. Next, each P_i needs to distribute shares of the product of its shares $a_i \cdot b_i$ as follows. We focus on a fixed party P_i, and let $A(x)$ and $B(x)$ be the respective polynomials for the sharing of a_i and b_i from the previous step.
   (a) Party P_i computes polynomials $D_1, \ldots, D_t$ so that $C(x) = A(x) \cdot B(x) - \sum_{\ell=1}^{t} x^\ell \cdot D_\ell(x)$ is a degree-t polynomial with free coefficient $a_i \cdot b_i$. Note that since each polynomial D_ℓ is multiplied by x^ℓ, we have that the free coefficient of $C(x)$ is always $a_i \cdot b_i$ (i.e., $A(0) \cdot B(0)$). As shown in BGW, P_i can choose the polynomials $D_1, \ldots, D_t$ in a special way so as to cancel out all coefficients of degree higher than t, and to ensure that $C(x)$ is of degree-t exactly. We stress that if P_i uses "incorrect" polynomials, then $C(x)$ may be of degree higher than t.
   (b) Party P_i verifiably shares the polynomials $D_1, \ldots, D_t$ with all parties.
   (c) Each party computes its share of $C(x)$ based on its shares of a_i, b_i and the polynomials $D_1, \ldots, D_t$.
   (d) At this point, it is guaranteed that the parties hold shares of a polynomial with free coefficient $a_i \cdot b_i$ (as described in Step 2a above) and it remains to *verify that these shares define a polynomial of degree t* (and not a higher degree).
3. Once the above is completed for all P_i we have that all parties hold valid shares of all share products $a_1 \cdot b_1, \ldots, a_n \cdot b_n$. Given these subshares, it suffices for each party to carry out a local linear computation [14] with the result being that they obtain valid shares of $a \cdot b$, as required.

Verifying the Degree of the Polynomial. We examine how to carry out Step 2d above, that is, how to verify that the shares held by the parties define a degree-t polynomial rather than a polynomial of higher degree.

First we need to touch on a subtle point which is the source of the challenge of realizing the verification step. The question is what we mean when we say that the shares of the honest parties should define a polynomial of degree t (or less)? There is a clear distinction between two cases. The first is that given the set of $2t+1$ shares held by the honest parties, we wish to ensure that their shares all lie

[1] There are actually some subtleties in formally defining the multiplication functionality since the adversary can determine some of the points that the sharing polynomial goes through. Nevertheless, this is the basic idea.

on the same degree-t polynomial. If they do not, then we are willing to modify up to t of the honest parties' shares to achieve this goal. This is what typically happens in the verification step of VSS protocols; the dealer modifies the shares that it originally gave to some of the parties by broadcasting new shares. The second interpretation is that we need to verify that *all* of the $2t+1$ shares held by the honest parties at the onset lie on a single degree-t polynomial. If not, then they should be notified of this fact, and should not change their shares. This is the cause of some difficulty as it requires a mechanism to distinguish between honest and corrupt parties; in particular, to distinguish between a corrupt party who lies about its share and an honest party who has an incorrect share.

In the BGW multiplication step we are in the second case. The shares that the honest parties hold have been created via the computation in Step 2c. The correctness of the computation requires that the constant term of the polynomial defined by the honest parties' shares be $a_i \cdot b_i$. However, a corrupted P_i who does not share the D_ℓ's in an appropriate manner can cause the resulting polynomial $C(x)$ to be of degree higher than t. In this case, there are at least two subsets of honest parties of size $t+1$ such that the polynomials $f(x)$ and $f'(x)$ defined by their shares have different free coefficients. Thus at least one is not equal to $a_i \cdot b_i$.

We conclude that in order to carry out the verification required in the multiplication, we cannot use the verification strategy offered by known VSS protocols (in particular the one in BGW). This is because their strategy just guarantees that all parties output shares on a polynomial defined by *some* subset of $t+1$ of the honest parties' shares. Furthermore, any verification technique that provides only this guarantee cannot be used.

Therefore, a *new* verification protocol is needed that guarantees that the polynomial $C(x)$ is of degree-t *without* changing the value of the free coefficient of $C(x)$, i.e. by not changing the shares of the honest parties. Conceptually, this can be achieved by constructing a protocol that enables honest parties to prove that their share is incorrect, and by that prove that P_i has cheated. We stress that the VSS methodology does not achieve this property since in the case of inconsistencies the parties cannot know if the dealer or another party is cheating.

Our Results. In this paper, we focus on perfect multiplication in the presence of up to $t < n/3$ malicious parties. We present two methods for carrying out the verification, along with a complete specification and proof of security of the resulting multiplication protocols. The first protocol works whenever the parties have univariate Shamir shares [19] of the input values. Thus, it does not depend on any specific properties of the VSS method used to initially share the values. Furthermore, it is close in flavor to the original protocol of BGW. The second protocol that we present utilizes the additional information given in a *bivariate* polynomial sharing (as used in the verifiable secret sharing of BGW and Feldman [4,12]) in order to significantly improve the efficiency of each multiplication, while preserving a constant round complexity for a single multiplication. For example, we can completely eliminate the first step of the BGW multiplication protocol, which is to share the shares a_i and b_i. In addition to being more efficient, our resulting multiplication protocol is also significantly simpler. The communication complexity of our protocol in the worst case (i.e., when some parties behave maliciously) is $O(n^5)$ field elements over point-to-point channels and $O(n^4)$ field elements over a broadcast channel. This is in contrast to the first protocol (the original BGW protocol) which has worst-case communication complexity of $O(n^6)$ field elements over point-to-point channels and $O(n^6)$ field elements over a broadcast channel. We remark that in the case that no parties actually cheat, both of the protocols have communication complexity of only $O(n^4)$ field elements over point-to-point channels, and require no message broadcasts at all.

In summary, our two protocols are incomparable. The first protocol is less efficient but works with *any* VSS of Shamir shares, and not necessarily with VSS that is based on bivariate techniques. The second protocol is simpler and more efficient but works only when the parties also have additional information on the shares that is a byproduct of the bivariate-based VSS protocol.

An additional important contribution of this paper is that we provide *full proofs of security* of all of our protocols and subprotocols, under the standard definitions of security following the ideal/real model paradigm [5,11]. This includes the non-trivial definition of the ideal multiplication functionality and other subfunctionalities used. The full proof of the protocols in this paper together with a full proof of the entire BGW protocol (including the semi-honest case, the VSS protocol and more) appears in [1].

We also consider our work as addressing the question whether or not it is possible to construct protocols with round and communication complexity that are both low. Our second protocol takes the first step by reducing the communication complexity of BGW and [8] while maintaining constant round complexity per multiplication.

Concurrent Composition and Adaptive Security. Both of our protocols achieve *perfect security*, as in the original work of BGW. We stress that perfect security is not just a question of aesthetics, but rather provides a substantive advantage over protocols that are only proven statistically secure. First, in [18] it is shown that if a protocol is perfectly secure in the stand-alone setting and has a black-box straight-line simulator, then it is also secure under concurrent general composition, or equivalently, universal composition [6]. Second, it was shown in [7] that any protocol that is perfectly secure in the presence of malicious static adversaries under the definition of security of [10], is also secure in the presence of malicious adaptive corruptions. The additional requirements of the definition of [10] clearly hold for all BGW protocols and subprotocols. Thus, we obtain both adaptive security and universal composition for free.

Related Work. We compare our second protocol to those in the existing literature. The only other protocol for perfectly-secure multiplication for any $t < n/3$ that is constant round (and in particular does not depend on the number of participating parties) is that of Cramer et al. [8]. This protocol works in a different way to the BGW protocol, and has worst-case communication complexity of $O(n^5)$ field elements over point-to-point channels and $O(n^5)$ field elements over a broadcast channel, in contrast to $O(n^4)$ broadcasts in our protocol.

Furthermore, in the case that no parties actually cheat, the cost of [8] is $O(n^4)$ field elements over point-to-point channels and $O(n^3)$ field elements over a broadcast channel, in contrast to $O(n^4)$ field elements over point-to-point channels (and no broadcast) in our protocol.

There has been a considerable amount of work focused on improving the communication complexity of information-theoretic protocols using the player elimination technique [15,16,2,17,9,3]. This work culminated in linear communication complexity in [3], providing highly efficient protocols for achieving perfect secure computation. However, all of these works have round complexity that depends on the number of participating parties, and not just on the depth of the circuit being computed. This is inherent in the player elimination technique since every time cheating is detected, two players are eliminated and some computations are repeated by the remaining parties. Thus, this technique yields protocols that have round complexity of at least $\Omega(t)$. We remark that the round complexity of these protocol are actually higher; e.g., the round complexity of [15] is $O(d+n^2)$ where d is the depth of the circuit. Although in many cases player elimination would give a more efficient protocol, there are some cases where it would not; for example, when a low-depth circuit is computed by many parties. In addition, from a theoretical perspective the question of low round and communication complexity is an important one. These protocols are therefore incomparable.

2 Preliminaries and Tools

In this paper we will refer to a few functionalities which are described formally in the full version [1]. Here we give a brief description of these functionalities.

We use the following VSS functionality F_{VSS}. The dealer inputs a polynomial $f(x)$ of degree t, and the parties receive shares of that polynomial; i.e., party P_i receives $f(\alpha_i)$ where $\alpha_1, \ldots, \alpha_n$ are fixed elements in the finite field. The "verifiable" part is that if f is of degree greater than t, then the parties reject the dealer's shares and output $\perp$. Observe that the secret $s = f(0)$ is only implicitly defined in the functionality; it is however well defined.

The second functionality which we need is for sub-sharing of shares, denoted $F_{VSS}^{subshare}$. Informally speaking, the $F_{VSS}^{subshare}$ functionality is a way for a set of parties to verifiably give out shares of values that are themselves shares. Specifically, assume that the parties $P_1, \ldots, P_n$ hold values $f(\alpha_1), \ldots, f(\alpha_n)$, respectively, where f is a degree-t polynomial.[2] The goal is for each party to *share its share* $f(\alpha_i)$ with all other parties while ensuring that a malicious P_i shares its correct $f(\alpha_i)$ and not something else. The protocol for achieving this sub-sharing is highly non trivial, and involves n invocations of VSS plus the

[2] If not all of the points lie on a single degree-t polynomial, then no security guarantees are obtained. Formally, this is achieved by defining that in this case the functionality sends the inputs of the honest parties to the corrupted parties, and sets the output of the honest parties to be whatever the adversary desires. In this way, any protocol is secure in this "bad case". From now on we just ignore this case, since our functionalities are used only when this property is fulfilled.

transmission of $O(n^3)$ field elements over private channels. A full discussion of the complexity and the solution from BGW appear in the full version.

We denote by $I \subset [n]$ the indices of the (up to t) corrupted parties.

3 Verifying That a Shared Polynomial is of Degree t

As discussed in the introduction, in order to complete the BGW multiplication protocol we need a subprotocol that enables the parties to verify that the shares held by *all* the honest parties for $C(x)$, as computed in Step 2a of the BGW perfect multiplication described above, lie on the same degree-t polynomial. That is, the parties all hold shares of $A(x)$, $B(x)$, $D_1(x), \ldots, D_t(x)$ and they wish to verify that their shares of $A(x) \cdot B(x) - \sum_{\ell=1}^{t} x^\ell \cdot D_\ell(x)$ define a degree-t polynomial. In this section we show how to do this; the full multiplication protocol using this verification step appears in [1].

We carry out this verification step by first having the dealer share the polynomial $C'(x) = C(x)$ using F_{VSS}, and then having each party P_j verify that $C'(\alpha_j) = C(\alpha_j)$ (where $C'(\alpha_j)$ is its output from F_{VSS} and $C(\alpha_j)$ is the result of its computation based on its shares of $A(x), B(x)$ and $D_1(x), \ldots, D_t(x)$). If equality does not hold then the party complains. As we have explained, we cannot have the dealer change the share of a complaining party, but rather the party needs to "prove" that its share does not lie on the polynomial. This is achieved by having the parties run a subprotocol to check whether or not the complaint is legitimate. Note that the parties all hold shares of $C'(x)$ and $C(x)$ so in principle there is enough information to verify a complaint. However, care must be taken not to reveal more information than allowed, in case the complaint is false. If there is a legitimate complaint against the dealer then the computation halts. Otherwise, we are guaranteed that the degree-t polynomial $C'(x)$ shared using F_{VSS} agrees with the computed polynomial $C(x)$ on at least $2t + 1$ honest parties' shares. Since $C(x)$ is of degree at most $2t$ (recall that $C(x) = A(x) \cdot B(x) - \sum_{\ell=1}^{t} x^\ell \cdot D_\ell(x)$ where every $D_\ell(x)$ is guaranteed to be of degree at most t since it was shared using F_{VSS}), this implies that $C(x) = C'(x)$ and so it is actually of degree-t, with the desired free coefficient.

3.1 The Verification Protocol

The verification procedure is defined formally in Functionality 1.

We stress that $C'(x)$ was already shared using F_{VSS} and so is guaranteed to be of degree-t. Thus, the aim of the parties is to verify that the shared $C'(x)$ equals $A(x) \cdot B(x) - \sum_{\ell=1}^{t} x^\ell \cdot D_\ell(x)$. We also remark that the adversary always receives the corrupted parties' shares as part of the output, and receives all of the shares in the case that the honest parties' output is 0. The fact that the adversary always receives the corrupted parties' shares makes no difference since it already knows these shares in any setting where this functionality is used. However, this is needed for technical reasons in order to prove the security of our protocol (according to simulation-based definitions). Furthermore, the fact that it learns everything if the output is 0 makes no difference because when the shares of

FUNCTIONALITY 1. (The F_{vrfy} functionality)

1. F_{vrfy} receives the shares $\{\beta_j^A, \beta_j^B, \beta_j^{D_1}, \ldots, \beta_j^{D_t}, \beta_j^{C'}\}_{j \notin I}$ of honest parties.
2. Let $A, B, D_1, \ldots, D_t$ and C' be the polynomials that are defined from the shares $\beta_j^A, \beta_j^B, \beta_j^{D_1}, \ldots, \beta_j^{D_t}$, and $\beta_j^{C'}$, respectively; this assumes that the polynomials $A, B, D_1, \ldots, D_t, C'$ are all of degree-t (see Footnote 2). Functionality F_{vrfy} sends the corrupted parties' shares $\{A(\alpha_i), B(\alpha_i), D_1(\alpha_i), \ldots, D_t(\alpha_i), C'(\alpha_i)\}_{i \in I}$ to the adversary.
3. If $C'(x) = A(x) \cdot B(x) - \sum_{\ell=1}^{t} x^{\ell} \cdot D_{\ell}(x)$ then F_{vrfy} sends 1 to every party for output, otherwise it sends 0 to every party for output and sends $A(x), B(x), D_1(x), \ldots, D_t(x), C'(x)$ to the adversary.

$A, B, D_1, \ldots, D_t, C'$ are all dealt by an honest party this case never happens, and when they are dealt by a corrupted party then the adversary already knows all the shares anyway.

The Protocol. The protocol is very simple. Each party P_j locally computes $\beta_j^A \cdot \beta_j^B - \sum_{\ell=1}^{t} (\alpha_j)^{\ell} \cdot \beta_j^{D_\ell}$ and complains if the result does *not* equal $\beta_j^{C'}$. In such a case, all parties use F_{eval}^j described in Section 3.2 below to publicly reconstruct all the input shares of the complainant, *without* exposing the shares of the other participating parties. This enables all parties to determine whether or not the complaint was legitimate. We stress that public reconstruction does not reveal anything since if the complaint is legitimate and so the output is 0, everything is anyway allowed to be revealed to the adversary. Furthermore, if the complaint is not legitimate then the complainant is corrupt and all that is revealed are a corrupted party's shares. See Protocol 2 for full details.

PROTOCOL 2. (Securely computing F_{vrfy} in the F_{eval}^j-hybrid model)

- **Inputs:** Each party P_i holds shares $\beta_i^A, \beta_i^B, \beta_i^{D_1}, \ldots, \beta_i^{D_t}, \beta_i^{C'}$, all on the degree-$t$ polynomials $A, B, D_1, \ldots, D_t, C'$ (resp).
- **The protocol:**
  1. Each party P_i computes $\beta' = \beta_i^A \cdot \beta_i^B - \sum_{\ell=1}^{t} (\alpha_i)^{\ell} \cdot \beta_i^{D_\ell}$.
     If $\beta' \neq \beta_i^{C'}$, then P_i broadcasts $(\mathsf{complaint}, i)$.
  2. For every party P_j that broadcast $(\mathsf{complaint}, j)$ do the following:
     (a) Run $t+3$ invocations of F_{eval}^j: Each party P_i inputs $\beta_i^A, \beta_i^B, \beta_i^{D_1}, \ldots, \beta_i^{D_t}, \beta_i^{C'}$, respectively, in each of the invocations.
     (b) Let $\tilde{\beta}_j^A, \tilde{\beta}_j^B, \tilde{\beta}_j^{D_1}, \ldots, \tilde{\beta}_j^{D_t}$ and $\tilde{\beta}_j^{C'}$ be the respective outputs from the invocations.
     (c) If $\tilde{\beta}_j^{C'} \neq \tilde{\beta}_j^A \cdot \tilde{\beta}_j^B - \sum_{\ell=1}^{t} (\alpha_j)^{\ell} \cdot \beta_j^{D_\ell}$, then output 0 and halt.
  3. If the output was not already determined to be 0 then output 1.

Theorem 3. *Let $t < n/3$. Then, Protocol 2 t-securely computes the F_{vrfy} functionality in the F_{eval}^j-hybrid model, in the presence of a static malicious adversary.*

The proof of this theorem is implicit in [1] (in the proof of security of the F_{VSS}^{mult} functionality).

3.2 Complaint Verification – The F_{eval}^j Functionality

When an honest party P_j complains, this implies that $C'(\alpha_j) \neq A(\alpha_j) \cdot B(\alpha_j) - \sum_{\ell=1}^{t} (\alpha_j)^\ell \cdot D_\ell(\alpha_j)$. In order to verify whether this is a legitimate complaint we need to reconstruct all the input shares of P_j, i.e. $\beta_j^A, \beta_j^B, \beta_j^{D_1}, \ldots, \beta_j^{D_t}, \beta_j^{C'}$, without revealing anything else. Note, that each such value can be calculated from the values of the honest parties as they all define a polynomial of degree t. Thus, using this information we can "extract" the values of the complaining party. However, this must be done without revealing anything but the complainants shares. We begin by formally defining the functionality; the functionality is parameterized by an index j that determines which party's share is to be revealed; equivalently, at which point the shared polynomial is to be evaluated.

FUNCTIONALITY 4. (The F_{eval}^j functionality)

1. The F_{eval}^j functionality receives the inputs of the honest parties $\{\beta_i\}_{i \notin I}$. Let $f(x)$ be the unique degree-t polynomial determined by the points $\{(\alpha_i, \beta_i)\}_{i \notin I}$. (If not all the points lie on a single degree-t polynomial, then no security guarantees are obtained. See Footnote 2.)
2. The functionality F_{eval}^j sends the output pair $(f(\alpha_i), f(\alpha_j))$ to every party P_i $(i = 1, \ldots, n)$.

We remark that although each party P_i already holds $f(\alpha_i)$ as part of its input, we need the output to include it in order to simulate in the case that a corrupted party has incorrect input. This will not make a difference in its use, since $f(\alpha_i)$ is supposed to be known to P_i.

In this paper we provide two methods for computing F_{eval}^j that are dependent on the specific implementation that we use for the secret sharing. In the following we describe the first implementation that uses univariate polynomials. The second solution uses bivariate sharing and will be given in Section 4.

Background. The parties' inputs are a (row) vector $\vec{\beta} \stackrel{\text{def}}{=} (\beta_1, \ldots, \beta_n)$ where for every i it holds that $\beta_i = f(\alpha_i)$. Thus, the parties' inputs are computed by $\vec{\beta} = V_\alpha \cdot \vec{f}$, where V_α is the $n \times (t+1)$ Vandermonde matrix with $\alpha_1, \ldots, \alpha_n$ and $\vec{f}$ is the length $t+1$ (column) vector of coefficients of the polynomial $f(x)$. Let $\vec{\alpha}_j = (1, \alpha_j, (\alpha_j)^2, \ldots, (\alpha_j)^t)$ be the jth row of V_α. Then the output of the functionality is $f(\alpha_j) = \vec{\alpha}_j \cdot \vec{f}$. We have:

$$\vec{\alpha}_j \cdot \vec{f} = \vec{\alpha}_j \cdot (V_\alpha^{-1} \cdot V_\alpha) \cdot \vec{f} = (\vec{\alpha}_j \cdot V_\alpha^{-1}) \cdot \left(V_\alpha \cdot \vec{f}\right) = (\vec{\alpha}_j \cdot V_\alpha^{-1}) \cdot \vec{\beta}$$

where V_α^{-1} is the left inverse of V_α, of degree $(t+1) \times n$. Thus, there exists a vector of *constants* $(\vec{\alpha}_j \cdot V_\alpha^{-1})$ so that the inner product of this vector and the inputs yields the desired result. In other words, F_{eval}^j is just a linear function of the parties' inputs.

The Protocol. Since F^j_{eval} is simply a linear function of the parties' inputs, it can be computed by each party sharing its share and then locally computing the function on the subshares. The result is that each party P_i holds a share δ_i of a polynomial whose free coefficient is the result $f(\alpha_j)$. Thus, the parties can now simply send their δ_i shares to each other and reconstruct the resulting polynomial. This suffices for the semi-honest case. However, malicious parties may send incorrect shares and try to cheat. In order to prevent them from doing this, the $F_{VSS}^{subshare}$ functionality is used in order to share the shares; see Section 2. Then, the reconstruction in the last stage is carried out using Reed-Solomon decoding to correct any bad values sent by the malicious parties. This ensures that $t < n/3$ malicious parties cannot affect the result. See Protocol 5 for the full description.

PROTOCOL 5. (Computing F^j_{eval} in the $F_{VSS}^{subshare}$-hybrid model)

- **Inputs:** Each party P_i holds a value β_i; we assume that the points (α_i, β_i) for every honest P_i all lie on a single degree-t polynomial f (see the definition of F^j_{eval} above and Footnote 2)
- **The protocol:**
  1. The parties invoke the $F_{VSS}^{subshare}$ functionality with each party P_i using $\beta_i = f(\alpha_i)$ as its private input.
  2. At the end of this stage, each party P_i holds $g_1(\alpha_i), \ldots, g_n(\alpha_i)$, where all the $g_i(x)$ are of degree t, and for every i, $g_i(0) = \beta_i$.
  3. Each party P_i locally computes: $H(\alpha_i) = \sum_{\ell=1}^{n} \gamma_\ell \cdot g_\ell(\alpha_i)$, where $(\gamma_1, \ldots, \gamma_n) = \vec{\alpha}_j \cdot V_{\vec{\alpha}}^{-1}$. Each party P_i sends $H(\alpha_i)$ to every other party.
  4. Upon receiving $(\hat{H}(\alpha_1), \ldots, \hat{H}(\alpha_n))$, each party runs the Reed-Solomon decoding procedure and receives $(H(\alpha_1), \ldots, H(\alpha_n))$. It then reconstructs $H(x)$ and computes $H(0)$.
  5. Each party P_i outputs $(\beta_i, H(0))$.

Theorem 6. *Let $t < n/3$. Then, Protocol 5 t-securely computes the F^j_{eval} functionality in the $F_{VSS}^{subshare}$-hybrid model, in the presence of a static malicious adversary.*

The motivation behind the security of the protocol appears above, and a full proof of Theorem 6 appears in [1].

4 Efficient Multiplication Using Bivariate VSS

We present a new BGW-based protocol that is more efficient than the original BGW protocol. In a nutshell, this protocol uses the bivariate structure introduced by BGW for the purpose of VSS throughout the entire multiplication protocol. Hirt et al. [15] also observed that the use of the bivariate polynomial can offer efficiency improvements; however they do not utilize this to the fullest extent possible. In this section we will show how this approach enables us to completely avoid the use of $F_{VSS}^{subshare}$ and compute the other subprotocols for the multiplication procedure more efficiently. As we have discussed in Section 2, $F_{VSS}^{subshare}$ is expensive and so this is a significant improvement.

Recall that the original BGW multiplication protocol follows the invariant that each wire in the circuit is hidden by a random univariate polynomial $f(x)$ of degree-t, and the share of each party is a point $(\alpha_i, f(\alpha_i))$. Multiplication then works as follows:

1. *Subsharing* – $F_{VSS}^{subshare}$*:* Given shares $a_1, \ldots, a_n$ and $b_1, \ldots, b_n$ of values a and b, each party *shares its shares* to all other parties. This step is carried out using $F_{VSS}^{subshare}$, as described above.
2. *Multiplication of subshares* – F_{VSS}^{mult}*:* Each party P_i plays the role of dealer in a protocol for which the result is that all parties hold shares (with threshold t) of the product $a_i \cdot b_i$ of its initial shares a_i and b_i. This step uses the fact that all parties hold subshares of a_i, b_i as carried out in the previous section.
3. *Linear combination:* As described in [14], once the parties all hold shares of $a_1 \cdot b_1, \ldots, a_n \cdot b_n$, they can each carry out a local linear combination of their shares, with the result being that they hold shares $c_1, \ldots, c_n$ of $a \cdot b$.

In our proposed protocol, we have an analogous invariant (as in [15]): each wire in the circuit is hidden by a (random) *bivariate polynomial* $F(x, y)$ of degree-t in both variables. As a result, the share of each party is the pair of degree-t polynomials $(F(x, \alpha_i), F(\alpha_i, y))$. We note that in the BGW protocol the VSS sharing is carried out using a bivariate polynomial; however after the initial sharing the parties resort back to the shares of a univariate polynomial, by setting their shares for further computations to $F(\alpha_i, 0)$. In contrast, we will preserve the shares of the bivariate but at times will also use univariate polynomials.

In the full version of this paper [1] we define the functionalities required, including a redefinition of the VSS protocol of BGW where the output includes the bivariate shares. In addition, we show how to reconstruct and add shared secrets in this format, and provide the full details of the multiplication protocol and its proof. In what follows we focus on our new protocols and techniques. First, we will explain why the first step of computing $F_{VSS}^{subshare}$ is not needed when bivariate shares are maintained. Next, we describe a functionality that enables the conversion (or extension) of a univariate polynomial secret sharing into a bivariate secret sharing. We then use a combination of the above to securely compute the bivariate analogue of the complaint verification functionality F_{eval}^{j} (see Section 3.2). Finally, we use all of the above to construct a simpler and more efficient version of F_{VSS}^{mult}.

From here on, we assume that there are two secrets a and b that were shared amongst the parties, and we denote by $F_A(x, y)$ and $F_B(x, y)$ the bivariate polynomials that hide a and b, respectively. The shares of party P_i are defined to be the pairs of univariate polynomials $F_A(x, \alpha_i), F_A(\alpha_i, y)$ and $F_B(x, \alpha_i), F_B(\alpha_i, y)$, respectively.

4.1 $F_{VSS}^{subshare}$ for Free

As described above, in order to carry out the "multiplication of subshares" step, the parties need to each have shares of all the other parties' univariate shares. Thus, in the univariate case, the parties first run the $F_{VSS}^{subshare}$ protocol at the cost of n executions of VSS plus the transmission of $O(n^3)$ field elements. Our first important observation is that in the bivariate case the subshares of each share *are already distributed* among the parties. In order to see this, recall that each party P_i holds shares $F_A(x, \alpha_i), F_A(\alpha_i, y)$. Based on this, we can define the univariate "Shamir" sharing of a via the polynomial $f_a(x) \stackrel{\text{def}}{=} F_A(x, 0)$ as in the original BGW protocol; due to the properties of the bivariate sharing, $f_a(x)$ is a univariate polynomial of degree-t that hides a. Furthermore, since each party P_i holds the polynomial $F_A(\alpha_i, y)$, it can locally compute its share $a_i = F_A(\alpha_i, 0) = f_a(\alpha_i)$ on the univariate polynomial $f_a(x)$.

We now claim that for every i, it holds that all the other parties P_j actually already have univariate shares of a_i. These shares of a_i are defined via the polynomial $g_{a_i}(y) = F_A(\alpha_i, y)$. This is due to the fact that each P_j holds the polynomial $F_A(x, \alpha_j)$ and so can compute $a_i^j = F_A(\alpha_i, \alpha_j) = g_{a_i}(\alpha_j)$. Observe that by the definition of the bivariate polynomial $F_A(x, y)$, it holds that $g_{a_i}(y)$ is a degree-t univariate polynomial. Furthermore, $g_{a_i}(0) = F_A(\alpha_i, 0) = a_i$ and each $a_i^j = g_{a_i}(\alpha_j)$. In other words, the values a_i^j that are locally computed by each party P_j are valid univariate shares of a_i, which is the univariate share of P_i in the polynomial $f_a(x)$ that hides a. We conclude that all of the subshares that are computed via the $F_{VSS}^{subshare}$ functionality in the original BGW protocol can actually be locally computed by each party using the bivariate shares that they already obtained in the VSS stage. (Of course, these bivariate shares need to be maintained throughout the circuit computation phase; we show how this is achieved below.)

4.2 Transformation from Univariate to Bivariate – $\widetilde{F}_{extend}$

As we will show below (in Section 4.3) and as we have seen regarding $F_{VSS}^{subshare}$, it is possible to utilize the additional information provided by a bivariate secret sharing in order to obtain higher efficiency. However, some of the intermediate sharings used in the multiplication protocol are inherently univariate. Thus, we introduce a new functionality called[3] $\widetilde{F}_{extend}$ that enables a dealer to efficiently extend shares of a univariate polynomial $q(x)$ of degree-t to a sharing based on a bivariate polynomial $S(x, y)$ of degree-t in both variables, with the guarantee that $q(x) = S(x, 0)$. In the functionality definition, the dealer receives the polynomial $q(x)$ that is distributed via the inputs of the honest parties. Although in any use of the functionality this is already known to the dealer, we need it for technical reasons in the simulation when the dealer is corrupted. See Functionality 7 for a full definition (observe that the dealer has as input the univariate polynomial $q(x)$ and a bivariate polynomial $S(x, y)$ such that $S(x, 0) = q(x)$).

[3] By convention, we denote bivariate-sharing based functionalities with a tilde.

FUNCTIONALITY 7. (The Reactive Functionality $\widetilde{F}_{extend}$)

1. The $\widetilde{F}_{extend}$ functionality receives the shares of the honest parties $\{\beta_j\}_{j\notin I}$. Let $q(x)$ be the unique degree-t polynomial determined by the points $\{(\alpha_j,\beta_j)\}_{j\notin I}$. (If no such polynomial exists then no security is guaranteed; see Footnote 2.)
2. In case that the dealer is corrupted, $\widetilde{F}_{extend}$ sends $q(x)$ to the adversary.
3. $\widetilde{F}_{extend}$ receives $S(x,y)$ from the dealer. Then, it checks that $S(x,y)$ is of degree-t in both variables x,y, *and* $S(x,0)=q(x)$.
4. If both conditions hold, $\widetilde{F}_{extend}$ accepts the bivariate polynomial $S(x,y)$, and sends to each party P_j the pair of polynomials $(f_j(x),g_j(y))$ (which are $(S(x,\alpha_j),S(\alpha_j,y))$).
5. If either of the conditions do not hold, $\widetilde{F}_{extend}$ rejects the bivariate polynomial $S(x,y)$ and sends to each party P_j the value $\perp$.

The protocol that implements this functionality is simple and efficient, but the argument for its security is delicate. The dealer distributes shares of $S(x,y)$, using the bivariate VSS protocol (securely computing the bivariate VSS functionality $\widetilde{F}_{VSS}$,[4] described in the full version [1]). Each party receives shares $S(x,\alpha_i),S(\alpha_i,y)$, and checks that $S(\alpha_i,0)=q(\alpha_i)$. If not, it broadcasts a complaint. The parties accept the shares of $S(x,y)$ if and only if there are at most t complaints. A formal description of the protocol appears in the full version. Before proceeding to describe why this protocol securely computes $\widetilde{F}_{extend}$, we remark that its cost is just a single VSS invocation and at most $O(n)$ broadcasts.

We now give an intuitive argument as to why the protocol securely computes the functionality. First, assume that the dealer is honest. In this case, the dealer inputs a degree-t bivariate polynomial that satisfies $S(x,0)=q(x)$, as required. The bivariate VSS functionality $\widetilde{F}_{VSS}$ ensures that the honest parties receive the correct shares. Now, since the polynomial satisfies the requirement, none of the honest parties complain. As a result, at most t parties complain, and all the honest parties accept the new bivariate shares.

The case where the dealer is corrupted is more subtle. At first, it may seem possible that t honest parties receive inconsistent shares and broadcast a complaint, while the remaining $t+1$ honest parties receive consistent shares and remain silent (together with all the corrupted parties). In such a case, only t complaints would be broadcast and so the parties would accept the bivariate polynomial even though it is not consistent with the inputs of all honest parties. Fortunately, as we show, such a situation can actually never occur. This is due to the fact that the $\widetilde{F}_{VSS}$ functionality ensures that the bivariate polynomial that is distributed is of degree-t in both variables, and due to the fact that the inputs of the honest parties lie on a polynomial with degree-t. As we show in the proof [1], this implies that if there exists a set of $t+1$ honest parties for which the bivariate polynomial agrees with their inputs, then this bivariate polynomial

[4] This functionality receives a bivariate polynomial $S(x,y)$ and hands each party P_i shares $S(x,\alpha_i),S(\alpha_i,y)$ if and only if $S(x,y)$ is of degree t in both variables.

must satisfy $S(x, 0) = q(x)$. In other words, we prove that once $t + 1$ of the bivariate shares are consistent with the points of $t + 1$ of the honest parties, then *all* of the bivariate shares must be consistent with *all* of the honest parties' points.

4.3 Bivariate Complaint Verification – The $\widetilde{F}^k_{eval}$ Functionality

In order to deal with complaint verification as discussed in the beginning of Section 3, we define an analogous functionality to F^j_{eval} in the bivariate setting. That is, given a sharing of a bivariate polynomial $S(x, y)$ of degree-t in both variables, the parties wish to evaluate the bivariate polynomial on some point α_k, or equivalently to learn the pair of polynomials $S(x, \alpha_k)$, $S(\alpha_k, y)$. Here, however, the implementation of this functionality is much easier than the implementation of F^j_{eval} in the univariate setting (we use the index k here instead of j since the bivariate setting requires additional indices). The $\widetilde{F}^k_{eval}$ functionality is defined as follows:

FUNCTIONALITY 8. (The Functionality $\widetilde{F}^k_{eval}$)

1. The $\widetilde{F}^k_{eval}$ functionality receives from each honest party P_j the pair of degree-t polynomials $(f_j(x), g_j(y))$, for every $j \notin I$. Let $S(x, y)$ be the single bivariate polynomial with degree-t in both variables that satisfies $S(x, \alpha_j) = f_j(x)$, $S(\alpha_j, y) = g_j(y)$ for every $j \notin I$. (If no such $S(x, y)$ exists, then no security is guaranteed; see Footnote 2).
2. $\widetilde{F}^k_{eval}$ sends every party P_i the polynomials $(S(x, \alpha_k), S(\alpha_k, y))$.

The protocol computing $\widetilde{F}^k_{eval}$ is straightforward and very efficient. Given input $(f_i(x), g_i(y))$ (which under the assumption on the inputs as in Footnote 2 equals $S(x, \alpha_i)$, $S(\alpha_i, y)$), each party P_i sends $f_i(\alpha_k), g_i(\alpha_k)$ (equivalently, $S(\alpha_k, \alpha_i), S(\alpha_i, \alpha_k)$)) to every other party; broadcast is not needed for this. Once a party holds all the points $\{S(\alpha_j, \alpha_k)\}_{j\in[n]}$, it can reconstruct the polynomial $f_k(x) = S(x, \alpha_k)$, and likewise $g_k(y) = S(\alpha_k, y)$ from $\{S(\alpha_k, \alpha_j)\}_{j\in[n]}$. Since $S(x, y)$ is of degree-t in both variables, the polynomials $f_k(x) = S(x, \alpha_k)$ and $g_k(y) = S(\alpha_k, y)$ are both of degree-t, and thus each party can reconstruct the polynomials even if the corrupted parties sent incorrect values, by using Reed-Solomon decoding.

The simplicity and efficiency of this protocol demonstrates the benefits of the approach of utilizing the bivariate shares throughout the entire multiplication protocol.

4.4 The $\widetilde{F}^{mult}_{VSS}$ Functionality for Sharing a Product of Shares

As we have described in the Introduction and in the beginning of Section 4, the main step for achieving secure multiplication is a method for a party P_i to share the product of its shares $a_i \cdot b_i$, while preventing a corrupted P_i from sharing an incorrect value. In the univariate case, the parties use $F_{VSS}^{subshare}$ to first share their shares, and then use F_{VSS}^{mult} to distribute shares of the product of their shares. In this section, we revisit the multiplication for the bivariate case. In this case, the parties hold shares of univariate polynomials that hide a party P_i's shares a_i, b_i, exactly as in the univariate solution with functionality F_{VSS}^{mult}. We stress that in our case these shares are univariate (i.e. points on a polynomial) and not bivariate shares (i.e. univariate polynomials) since we are referring to the *subshares*. Nevertheless, as we have shown, these can be separately extended to bivariate sharings of a_i and b_i using $\widetilde{F}_{extend}$. Our goal with $\widetilde{F}_{VSS}^{mult}$ is for the parties to obtain a sharing of $a_i \cdot b_i$, by holding shares of a bivariate polynomial $C(x, y)$ whose constant term is the desired product.

For the sake of clarity and to reduce the number of indices, in this section we refer to a and b as the shares of P_i (and not the secret), and to a_j and b_j the univariate subshares that P_j holds of P_i's shares a and b. We also write the functionality and protocol with P_1 as the dealer (i.e., the party who has shares a and b and wishes to share $a \cdot b$); in the full multiplication, each party plays the dealer in turn. See Functionality 9 for a specification of this step.

FUNCTIONALITY 9. (The reactive $\widetilde{F}_{VSS}^{mult}$ functionality)

1. The $\widetilde{F}_{VSS}^{mult}$ functionality receives input shares (a_j, b_j) from every honest party P_j $(j \notin I)$.
2. $\widetilde{F}_{VSS}^{mult}$ computes the unique degree-t polynomials $A'(x)$ and $B'(x)$ such that $A'(\alpha_j) = a_j$ and $B'(\alpha_j) = b_j$ for every $j \notin I$ (if no such A' or B' exist, then see Footnote 2).
3. $\widetilde{F}_{VSS}^{mult}$ sends $(A'(x), B'(x))$ to the dealer P_1.
4. $\widetilde{F}_{VSS}^{mult}$ receives a bivariate polynomial $C(x, y)$ from P_1, and chooses $C^*(x, y)$ as follows:
   (a) If the input is the special symbol $*$, then $\widetilde{F}_{VSS}^{mult}$ chooses a random bivariate polynomial $C^*(x, y)$ of degree-t in both variables under the constraint that $C^*(0, 0) = A'(0) \cdot B'(0)$.
   (b) Else, if the input is a bivariate polynomial C such that $\deg(C) = t$ in both variables and $C(0, 0) = A'(0) \cdot B'(0)$, then $\widetilde{F}_{VSS}^{mult}$ sets $C^* = C$.
   (c) Otherwise, if either $\deg(C) > t$ or $C(0, 0) \neq A'(0) \cdot B'(0)$, then $\widetilde{F}_{VSS}^{mult}$ sets $C^*(x, y) = A'(0) \cdot B'(0)$ to be the constant polynomial equalling $A'(0) \cdot B'(0)$ everywhere.
5. $\widetilde{F}_{VSS}^{mult}$ sends $C^*(x, y)$ to the dealer P_1, and sends $(A'(\alpha_i), B'(\alpha_i), C^*(x, \alpha_i), C^*(\alpha_i, y))$ to P_i for every $i = 1, \ldots, n$.

The special input symbol $*$ is an instruction for the trusted party computing $\widetilde{F}_{VSS}^{mult}$ to choose the polynomial $C^*(x, y)$ determining the output shares uniformly at random.

We remark that although the dealing party P_1 is supposed to already have $A'(x), B'(x)$ as part of its input, and each party P_i is also supposed to already have $A'(\alpha_i)$ and $B'(\alpha_i)$ as part of its input, this information is provided as output in order to enable simulation in the case that the corrupted parties use incorrect inputs.

The Protocol. As in the univariate case, the protocol for implementing this functionality is based on the BGW method "(II) Verifying that $c = a \cdot b$", with the addition of complaint verification. In addition, here the parties will output bivariate and not univariate shares.

As described in the Introduction, the dealer chooses the univariate polynomials $D_1(x), \ldots, D_t(x)$ as instructed in BGW; see the full version for a detailed description of this. It then distributes them using bivariate polynomials that hide them. That is, in order to distribute a polynomial $D_i(x)$, the dealer selects a bivariate polynomial $D_i(x,y)$ uniformly at random under the constraint that $D_i(x,0) = D_i(x)$, and then shares it using the bivariate VSS functionality $\widetilde{F}_{VSS}$. This ensures that all the polynomials $D_1(x,0), \ldots, D_t(x,0)$ are of degree-t. In addition, this comes at no additional cost since the BGW VSS protocol anyway uses bivariate polynomials. At this point, each party holds shares of the univariate polynomials $A(x), B(x)$, and shares of the t bivariate polynomials $D_1(x,y), \ldots, D_t(x,y)$. From the construction (see the brief explanation in the introduction), the univariate polynomial defined by:

$$C'(x) = A(x) \cdot B(x) - \sum_{k=1}^{t} x^k \cdot D_k(x,0)$$

is a random polynomial with free coefficient $a \cdot b$, and each party P_i can locally compute its share $C'(\alpha_i)$ on this polynomial. However, as in the univariate case, if the dealer did not choose the polynomials $D_i(x,y)$ as instructed, the polynomial $C'(x)$ may not be of degree-t, and in fact can be any polynomial of degree $2t$ (but no more since all the polynomials were shared using VSS and so are of degree at most t). We must therefore check the degree of $C'(x)$.

At this point, the dealer chooses a random bivariate polynomial $C(x,y)$ of degree-t in both variables under the constraint that $C(x,0) = C'(x)$, and shares it using the bivariate VSS functionality $\widetilde{F}_{VSS}$. This guarantees that the parties hold shares of a degree-t bivariate polynomial $C(x,y)$. If this polynomial satisfies

$$C(x,0) = C'(x) = A(x) \cdot B(x) - \sum_{k=1}^{t} x^k \cdot D_k(x,0)$$

then $C(0,0) = A(0) \cdot B(0) = a \cdot b$, and we are done.

We therefore want to check that indeed $C(x,0) = C'(x)$. Each party P_i holds shares of the polynomial $C(x,y)$, and so it holds the univariate polynomials $C(x,\alpha_i)$, $C(\alpha_i,y)$. Moreover, it has already computed its share $C'(\alpha_i)$. Thus, it can check that $C(\alpha_i,0) = C'(\alpha_i)$. Since $C'(x)$ is of degree at most $2t$, and since $C(x,y)$ is of degree-t, then if this check passes for *all* of the $2t+1$ honest parties, we are guaranteed that $C(x,0) = C'(x)$, and so $C(0,0) = a \cdot b$. Thus, each party checks that $C(\alpha_i,0) = C'(\alpha_i)$, and if not it broadcasts a complaint. If there are more than t complaints, then it is clear that the dealer is corrupted. However, as in the univariate case, even when there are less than t complaints the dealer can be corrupted, and so the parties need to unequivocally verify each complaint.

The way the parties verify whether or not a complaint is false is similar to the univariate case, described in Section 3.1. That is, the parties evaluate each one of the polynomials $D_1, \ldots, D_t$, A, B, and C on the point of the complaining party. However, this time we use the bivariate evaluation functionality $\widetilde{F}^k_{eval}$ (see Section 4.3) instead of the univariate one F^j_{eval}. Observe that all of the polynomials $D_1, \ldots, D_t, C$ are bivariate and of degree-t, and so the bivariate $\widetilde{F}^k_{eval}$ can be used. In contrast, $A(x)$ and $B(x)$ are only univariate polynomials and so $\widetilde{F}_{extend}$ (see Section 4.2) is first used in order to distribute bivariate polynomial $A(x, y)$ and $B(x, y)$ that fit $A(x)$ and $B(x)$, respectively. Following this, $\widetilde{F}^k_{eval}$ can also be used for $A(x, y)$ and $B(x, y)$. Finally, after the parties receive all of the shares of the complaining party, they can check whether the complaint is true or false. In case of a true complaint, the parties reconstruct the original shares and set their output to be $a \cdot b$. See Protocol 11 for a full specification.

We have the following theorem, that is proven in the full version.

Theorem 10. *Let $t < n/3$. Then, Protocol 11 t-securely computes the $\widetilde{F}^{mult}_{VSS}$ functionality in the $(\widetilde{F}_{VSS}, \widetilde{F}^k_{eval}, \widetilde{F}_{extend})$-hybrid model, in the presence of a static malicious adversary.*

PROTOCOL 11. (Securely computing $\widetilde{F}^{mult}_{VSS}$)

- **Inputs:** The dealer P_1 holds degree-t polynomials $A(x)$ and $B(x)$. Each party P_i holds a pair of shares a_i and b_i such that $a_i = A(\alpha_i)$ and $b_i = B(\alpha_i)$.
- **The protocol:**
  1. *Dealing phase:*
     (a) The dealer P_1 defines the degree-$2t$ polynomial $D(x) = A(x) \cdot B(x)$; denote $D(x) = a \cdot b + \sum_{k=1}^{2t} d_k \cdot x^k$.
     (b) P_1 chooses t^2 values $\{r_{k,j}\}$ uniformly and independently at random from $\mathbb{F}$, where $k = 1, \ldots, t$, and $j = 0, \ldots, t-1$. For every $k = 1, \ldots, t$, the dealer defines the polynomial $D_k(x)$:
$$D_k(x) = \sum_{\ell=0}^{t-1} r_{k,\ell} \cdot x^\ell + \left(d_{k+t} - \sum_{j=k+1}^{t} r_{j,t+k-j} \right) \cdot x^t.$$
     (c) P_1 computes the polynomial:
$$C'(x) = D(x) - \sum_{k=1}^{t} x^k \cdot D_k(x).$$
     (d) P_1 chooses t random degree-t bivariate polynomials $D_1(x, y)$, $\ldots, D_t(x, y)$ under the constraint that $D_k(x, 0) = D_k(x)$ for every $k = 1, \ldots, t$. In addition, it chooses a random bivariate polynomial $C(x, y)$ of degree-t under the constraint that $C(x, 0) = C'(x)$.
     (e) P_1 invokes the $\widetilde{F}_{VSS}$ functionality as dealer with the following inputs $C(x, y)$, and $D_k(x, y)$ for every $k = 1, \ldots, t$.

Protocol for securely computing $\widetilde{F}_{VSS}^{mult}$ (continued):

2. Each party P_i works as follows:
   (a) If any of the shares it receives from $\widetilde{F}_{VSS}$ equal $\perp$ then P_i proceeds to the *reject phase.*
   (b) P_i computes $c'(i) \stackrel{\text{def}}{=} a_i \cdot b_i - \sum_{k=1}^{t} (\alpha_i)^k \cdot D_k(\alpha_i, 0)$. If $C(\alpha_i, 0) \neq c'(i)$, then P_i broadcasts $(\mathsf{complaint}, i)$; note that $C(\alpha_i, y)$ is part of P_i's output from $\widetilde{F}_{VSS}$ with $C(x, y)$.
   (c) If any party P_j broadcast $(\mathsf{complaint}, j)$ then go to the *complaint resolution phase.*
3. *Complaint resolution phase:*
   (a) P_1 chooses two random bivariate polynomials $A(x, y)$, $B(x, y)$ of degree t under the constraint that $A(x, 0) = A(x)$ and $B(x, 0) = B(x)$.
   (b) The parties invoke the $\widetilde{F}_{extend}$ functionality twice, where P_1 inserts $A(x, y)$, $B(x, y)$ and each party inserts a_i, b_i. If any one of the outputs is $\perp$ (in which case all parties receive $\perp$), P_i proceeds to *reject phase.*
   (c) The parties run the following for every $(\mathsf{complaint}, k)$ message:
      i. Run $t + 3$ invocations of $\widetilde{F}_{eval}^{k}$, with each party P_i inputting its shares of $A(x, y)$, $B(x, y)$, $D_1(x, y), \ldots, D_t(x, y)$, $C(x, y)$, respectively.
      Let $A(\alpha_k, y), B(\alpha_k, y), D_1(\alpha_k, y), \ldots, D_t(\alpha_k, y), C(\alpha_k, y)$ be the resulting shares (we ignore the dual shares $S(x, \alpha_k)$ for each polynomial).
      ii. If: $C(\alpha_k, 0) \neq A(\alpha_k, 0) \cdot B(\alpha_k, 0) - \sum_{\ell=1}^{t} \alpha_k^{\ell} D_\ell(\alpha_k, 0)$, proceed to the *reject phase.*
4. *Reject phase:*
   (a) Every party P_i sends a_i, b_i to all P_j. Party P_i defines the vector of values $\vec{a} = (a_1, \ldots, a_n)$ that it received, where $a_j = 0$ if it was not received at all. P_i sets $A'(x)$ to be the output of Reed-Solomon decoding on $\vec{a}$. Do the same for $B'(x)$.
   (b) Every party P_i sets $C(x, \alpha_i) = C(\alpha_i, y) = A'(0) \cdot B'(0)$; a constant polynomial.
5. *Outputs:* Every party P_i outputs $C(x, \alpha_i), C(\alpha_i, y)$. Party P_1 outputs $(A(x), B(x), C(x, y))$.

4.5 Wrapping Things Up – Perfectly-Secure Multiplication

Given bivariate shares of the input wires to a multiplication gate, the parties each in turn play the dealer in $\widetilde{F}_{VSS}^{mult}$. At the end of these executions, all parties hold bivariate shares of the product of all other parties shares (recall that a bivariate share is a pair of univariate polynomials). As in [14], the parties can obtain bivariate shares of the product of the input-wire values by just carrying out a local computation on their shares. This therefore concludes the multiplication protocol. A full description and proof appears in the full version.

Efficiency Analysis. A detailed efficiency analysis of the protocols appears in [1]. In short, our protocol utilizing the bivariate properties costs up to $O(n^5)$ field elements in private channels, together with $O(n^4)$ field elements in broadcast per multiplication gate in the case of malicious behavior. We remark that when no parties actively cheat, the protocol requires $O(n^4)$ field elements in private channels and no broadcast at all.

References

1. Asharov, G., Lindell, Y., Rabin, T.: A Full Proof of the Perfectly-Secure BGW Protocol and Improvements. Cryptology ePrint Archive, 2011/136 (2011)
2. Beerliová-Trubíniová, Z., Hirt, M.: Efficient Multi-party Computation with Dispute Control. In: Halevi, S., Rabin, T. (eds.) TCC 2006. LNCS, vol. 3876, pp. 305–328. Springer, Heidelberg (2006)
3. Beerliová-Trubíniová, Z., Hirt, M.: Perfectly-Secure MPC with Linear Communication Complexity. In: Canetti, R. (ed.) TCC 2008. LNCS, vol. 4948, pp. 213–230. Springer, Heidelberg (2008)
4. Ben-Or, M., Goldwasser, S., Wigderson, A.: Completeness Theorems for Non-Cryptographic Fault-Tolerant Distributed Computation. In: 20th STOC, pp. 1–10 (1988)
5. Canetti, R.: Security and Composition of Multiparty Cryptographic Protocols. Journal of Cryptology 13(1), 143–202 (2000)
6. Canetti, R.: Universally Composable Security: A New Paradigm for Cryptographic Protocols. In: 42nd FOCS, pp. 136–145 (2001)
7. Canetti, R., Damgård, I., Dziembowski, S., Ishai, Y., Malkin, T.: Adaptive versus Non-Adaptive Security of Multi-Party Protocols. Journal of Cryptology 17(3), 153–207 (2004)
8. Cramer, R., Damgård, I., Maurer, U.M.: General Secure Multi-party Computation from any Linear Secret-Sharing Scheme. In: Preneel, B. (ed.) EUROCRYPT 2000. LNCS, vol. 1807, pp. 316–334. Springer, Heidelberg (2000)
9. Damgård, I., Nielsen, J.B.: Scalable and Unconditionally Secure Multiparty Computation. In: Menezes, A. (ed.) CRYPTO 2007. LNCS, vol. 4622, pp. 572–590. Springer, Heidelberg (2007)
10. Dodis, Y., Micali, S.: Parallel Reducibility for Information-Theoretically Secure Computation. In: Bellare, M. (ed.) CRYPTO 2000. LNCS, vol. 1880, pp. 74–92. Springer, Heidelberg (2000)
11. Goldreich, O.: Foundations of Cryptography: Volume 2 – Basic Applications. Cambridge University Press, Cambridge (2004)
12. Feldman, P.: Optimal Algorithms for Byzantine Agreement. PhD thesis, Massachusetts Institute of Technology (1988)
13. Feldman, P., Micali, S.: An Optimal Probabilistic Protocol for Synchronous Byzantine Agreement. SIAM - Journal on Computing 26(4), 873–933 (1997)
14. Gennaro, R., Rabin, M.O., Rabin, T.: Simplified VSS and Fact-Track Multiparty Computations with Applications to Threshold Cryptography. In: 17th PODC, pp. 101–111 (1998)
15. Hirt, M., Maurer, U.M., Przydatek, B.: Efficient Secure Multi-party Computation. In: Okamoto, T. (ed.) ASIACRYPT 2000. LNCS, vol. 1976, pp. 143–161. Springer, Heidelberg (2000)
16. Hirt, M., Maurer, U.: Robustness for Free in Unconditional Multi-party Computation. In: Kilian, J. (ed.) CRYPTO 2001. LNCS, vol. 2139, pp. 101–118. Springer, Heidelberg (2001)

17. Hirt, M., Nielsen, J.B.: Robust Multiparty Computation with Linear Communication Complexity. In: Dwork, C. (ed.) CRYPTO 2006. LNCS, vol. 4117, pp. 463–482. Springer, Heidelberg (2006)
18. Kushilevitz, E., Lindell, Y., Rabin, T.: Information-Theoretically Secure Protocols and Security Under Composition. SIAM Journal on Computing 39(5), 2090–2112 (2010)
19. Shamir, A.: How to Share a Secret. Communications of the ACM 22(11), 612–613 (1979)

The IPS Compiler: Optimizations, Variants and Concrete Efficiency*

Yehuda Lindell, Eli Oxman, and Benny Pinkas

Dept. of Computer Science, Bar Ilan University, Ramat Gan, Israel
lindell@cs.biu.ac.il, eli.oxman@gmail.com, benny@pinkas.net

Abstract. In recent work, Ishai, Prabhakaran and Sahai (CRYPTO 2008) presented a new compiler (hereafter the IPS compiler) for constructing protocols that are secure in the presence of malicious adversaries without an honest majority, from protocols that are only secure in the presence of semi-honest adversaries. The IPS compiler has many important properties: it provides a radically different way of obtaining security in the presence of malicious adversaries with no honest majority, it is black-box in the underlying semi-honest protocol, and it has excellent asymptotic efficiency.

In this paper, we study the IPS compiler from a number of different angles. We present an efficiency improvement of the "watchlist setup phase" of the compiler that also facilitates a simpler and tighter analysis of the cheating probability. In addition, we present a conceptually simpler variant that uses protocols that are secure in the presence of covert adversaries as its basic building block. This variant can be used to achieve more efficient asymptotic security, as we show regarding black-box constructions of malicious oblivious transfer from semi-honest oblivious transfer. Finally, we analyze the IPS compiler from a *concrete efficiency* perspective and demonstrate that in some cases it can be competitive with the best efficient protocols currently known.

1 Introduction

In the setting of secure multiparty computation, a set of parties wish to jointly compute some function of their inputs while preserving security properties such as *privacy*, *correctness*, *independence of inputs*, and more. These properties must be preserved in the face of adversarial behavior. In this paper, we consider security in the presence of three types of adversaries. The two classic adversary models are those of *semi-honest* adversaries that follow the protocol specification exactly but attempt to learn more than they should, and *malicious* adversaries that can behave as they wish and as such can arbitrarily deviate from the protocol specification. A more recent model, called security in the presence of *covert*

* Research generously supported by the European Research Council as part of the ERC project LAST. The first author was also supported by the ISRAEL SCIENCE FOUNDATION (grant No. 781/07).

P. Rogaway (Ed.): CRYPTO 2011, LNCS 6841, pp. 259–276, 2011.

adversaries, guarantees that if a malicious adversary behaves in a way that enables it to break the protocol in some way, then it will be caught cheating by the honest parties with some probability ϵ [1].

There are two rather distinct settings for studying the problem of secure multiparty computation. In the first, it is assumed that a majority of the parties are honest. In such a case, it is possible to securely compute any efficient functionality with information-theoretic security [3,4] assuming private channels (and broadcast, for the case of $n/3 \leq t < n/2$ corrupted parties). In the second setting, any number of the parties may be corrupted; this case includes the important two-party setting where one party may be corrupted. In this case of no guaranteed honest majority, information-theoretic security cannot be achieved. Nevertheless, assuming the existence of oblivious transfer, which can be constructed from enhanced trapdoor permutations and homomorphic encryption, it has been shown that any efficient functionality can be securely computed without an honest majority [24,12].

Beyond proving an important theorem stating that any functionality can be securely computed without an honest majority, the construction of [12] shows how to *compile* any protocol that is secure in the presence of semi-honest adversaries into a protocol that is secure in the presence of malicious adversaries, using one-way functions alone. This result is therefore often referred to as the GMW compiler. Recently, a new compiler was presented by Ishai, Prabhakaran and Sahai (IPS) [16]. This compiler works in a completely different way to that of the GMW compiler. First, unlike GMW, it does not compile an m-party protocol for securely computing a functionality f in the presence of semi-honest adversaries into an m-party protocol for securely computing the same f in the presence of malicious adversaries. Rather, the IPS compiler uses m-party protocols that securely compute some basic operations (like addition and multiplication in a finite field) in the presence of semi-honest adversaries with no honest majority, in order to transform a multiparty protocol that securely computes f in the presence of malicious adversaries with an honest majority, to an m-party protocol that securely computes f in the presence of malicious adversaries with *no honest majority*. As a specific example, note that by setting $m = 2$ the IPS compiler can generate a two-party protocol for computing f that is secure against malicious adversaries, from two-party protocols for computing basic functionalities (secure against semi-honest adversaries), and a multiparty protocol for computing f (secure against malicious adversaries which can corrupt only a minority of the parties). Intriguingly, the IPS compiler utilizes the world of information-theoretic secure computation in order to achieve security in the setting of no honest majority, since the basic protocol computing f, to which the compiler is applied, can be defined in an information-theoretic setting.

The IPS compiler has a number of important properties. First, it is black box in the underlying constructions; see [13] for why this is important. Second, it provides a uniform approach to both the two-party and multiparty settings, like the GMW compiler. Third, in some settings, it has excellent asymptotic efficiency. This is due to the fact that in some cases (e.g., when an arithmetic circuit computing f is significantly smaller than a Boolean circuit computing f), information-theoretic protocols for the setting of an honest majority are much more efficient than computational protocols for the setting of no honest majority. The compiler can be applied to these more efficient protocols. This property has already been utilized to present protocols that have excellent theoretical, asymptotic efficiency [17]. Despite the above, it is unclear as to whether this approach can be used to achieve *concrete* efficiency for functionalities of interest [14].

1.1 The IPS Compiler

The compiler of [16] utilizes the following components in order to achieve the secure m-party computation of a functionality f, with no honest majority:

- A multiparty information-theoretic protocol π computing f with m clients who provide input and $n = O(m^2k)$ servers who carry out the computation (where k is a security parameter analyzed in our work), which is secure in the presence of a malicious adversary corrupting a minority of the *servers*. This is called the *outer protocol* by [16].
- m-party subprotocols and local client instructions for simulation of the server computation in π, that are secure as long as the adversary behaves in a semi-honest manner. Given the known techniques for information-theoretic secure multiparty computation, it suffices for example to use m-party subprotocols for securely computing additive shares of the product of shares (all other steps can be carried using local client instruction for generating shares, adding shares, and so on). We stress that these subprotocols need only be secure in the presence of semi-honest adversaries. These are called the *inner protocols* by [16].

The way that the compiler works is for the m real parties to run the information-theoretic protocol π by emulating the operations of the n servers in π. This emulation is carried out using the secure m-party subprotocols, run between the m real parties (clients), to compute the next step of all parties in π. Thus, the n servers running π are virtual and are emulated by the m real parties running the protocol. Observe that if the real adversary were to behave in a semi-honest manner in each subprotocol, then the overall computation would clearly be secure. This is due to the fact that the emulation of the n parties in π is carried out using a protocol that is *secure* in the presence of semi-honest adversaries. Thus, f is securely computed, as guaranteed by π. However, the adversary here may be *malicious*.

The magic in the IPS compiler is how to leverage the semi-honest security of the subprotocols in order to achieve security in the presence of malicious adversaries. The central observation is that in order for a malicious adversary to cheat, it must cheat in at least $n/2$ of the subprotocols. This is due to the fact that π is secure unless a *majority* of the servers, namely at least $n/2$ of them, behave maliciously. In order to prevent such cheating, the IPS compiler sets up *watchlists*, which enable the honest parties to verify that malicious parties are not cheating. These watchlists are generated as follows. Each real party (client) chooses the randomness that it will use when running the semi-honest subprotocol for each virtual server. Then, using oblivious transfer, each other client obtains k of the random strings of all other clients. Now, given the randomness that a party is supposed to use in the semi-honest subprotocol, it is possible to check that it is indeed behaving honestly. Furthermore, since oblivious transfer is used to obtain these strings, no client can know which semi-honest executions are being "watched". It is important to note, however, that it is not possible to raise the number of watchlists too high, because each time the randomness of a client (used with respect to some server) is watched, the internal state of the server is seen and that server is corrupted. It is shown in [16] that $n = O(k^2 m)$ servers are required in order to obtain security with a probability of cheating that is negligible in k. This number is important because it determines the number of servers in the information-theoretic protocol and thus its complexity.

In our presentation, we assume some familiarity with the IPS compiler. See the IPS papers for a description [16,17], or the brief tutorial of the construction in the full version of our paper.

1.2 Optimizations of the IPS Compiler

As will become clear in our concrete analysis of the efficiency of the compiler, see Sections 1.4 and 4, the number of servers n can become very large for some choices of parameters. This can have a severe effect on the efficiency of the protocol in a number of ways, one of these being the watchlist setup phase where $m^2 n$ executions of Rabin oblivious transfers must be carried out (each of these costing $\log n$ regular oblivious transfers; see Section 2 for a detailed explanation). This can therefore quickly become the bottleneck of the protocol. We therefore first devise a method for reducing the required number of servers to $n = O(mk)$ rather than $n = O(m^2 k)$, and second for setting up the watchlists in a way that costs an equivalent of $O(mn)$ regular oblivious transfers (rather than $O(m^2 n \log n)$ oblivious transfers). These optimizations are significant since they enable a better choice of parameters for the outer information-theoretic protocol. (Specifically, the best efficiency is obtained by using a protocol with many servers and a small fraction of corrupted parties; this enables the heavy use of the packed or multi-secret sharing methodology of [10].) Using our method, a smaller number of virtual servers can be corrupted and so a more efficient protocol can be used. We stress that when measuring concrete complexity, our new watchlist setup protocol is substantially more efficient, even for the two-party case where $m = 2$. An additional optimization is described in Section 2.3.

1.3 Variants of the IPS Compiler for Covert Adversaries

A simple compiler from covert to malicious security. We present an analog of the IPS compiler that uses subprotocols that are secure in the presence of covert adversaries instead of protocols that are secure in the presence of semi-honest adversaries. Recall that a protocol is secure in the presence of covert adversaries, with a deterrent parameter ϵ, if any cheating by an adversary is detected by the honest parties with probability at least ϵ [1]. (For our purposes, it is convenient to assume that $\epsilon = 1/2$.) The use of subprotocols with this level of security fits naturally with the IPS paradigm: the information-theoretic outer protocol is emulated using protocols that are secure in the presence of covert adversaries with deterrent $\epsilon = 1/2$. Then, if the adversary tries to cheat in k of the subprotocol executions, it will be caught except with probability 2^{-k}. Observe that there is no need for any watchlists. In addition, the analysis and proof of security of this compiler are extraordinarily straightforward, since the cheating probability can be measured exactly with ease.

Beyond being a significant conceptual simplification, the usage of covert protocols also enables us to use an information theoretic protocol with just $n = m+2k$ parties (tolerating up to k corruptions), rather than $O(mk)$ parties when using semi-honest security. Relying on the fact that protocols that are secure in the presence of covert adversaries with deterrent $\epsilon = 1/2$ are only about 2–3 times the cost of semi-honest protocols, this results in an *asymptotic* efficiency improvement over the original compiler (we stress, though, that by our concrete analysis, the original compiler of [16] will typically be more efficient for *concrete* parameters since the use of watchlists means that local computation by a party can be checked directly and need not be distributed).

An IPS compiler from semi-honest to covert security. We observe that the IPS compiler with some minor modifications can be used to obtain security in the presence of covert adversaries from protocols that are secure in the presence of semi-honest adversaries, via a black-box reduction. We show that it suffices to use $O(m)$ watchlists and an oblivious transfer protocol that is secure in the presence of covert adversaries to set up those watchlists. (We also show that covert oblivious transfer can be constructed at the cost of only a constant number of semi-honest oblivious transfers.) This answers a major open question left by the work of [6].

IPS compilation and covert adversaries. Based on the above, we have that the IPS paradigm significantly contributes to our understanding of security in the presence of covert adversaries, and enables us to position covert adversaries in their natural place between semi-honest and malicious adversaries with respect to protocol constructions. In addition, as we show, it is possible to obtain quantitative improvements using this methodology. Specifically, we obtain a fully black-box reduction from semi-honest OT and one-way functions to malicious OT, at the cost of just a linear number of semi-honest OT invocations (the previous reduction of this type requires a quadratic number of semi-honest OTs [13]).

1.4 The Concrete Efficiency of the IPS Protocol

On an *abstract level*, the IPS compiler provides an elegant and conceptually simple way of constructing protocols that are secure in the presence of malicious adversaries. However, the actual instantiation of a protocol using the IPS approach depends on many different parameters and choices, all having a significant effect on the concrete efficiency of the result. To start with, appropriate inner and outer protocols must be chosen, and these choices are interdependent.

This is due to the fact that the most efficient information-theoretic outer protocol may require more invocations of the inner protocol for computing multiplications (which may be more expensive than other operations) than a less efficient protocol, when judging efficiency in the standard information-theoretic setting. Thus, the cost of running the inner multiplication protocol must be traded off with the cost of other operations in the outer protocol. In addition, there may be outer protocols that can utilize different inner protocols and obtain higher efficiency.

Another parameter that must be chosen is the exact number of servers $n = O(mk)$. Observe that each corrupted client has effectively corrupted k servers, since the watchlists it obtains from the honest parties reveal the internal state of these servers. In addition, some additional servers may be corrupted by the client cheating in the inner protocols emulating these servers, and hoping that it does not get caught. Based on this, it is clear that $n > 2mk$ since $m - 1$ corrupted parties have already effectively corrupted $(m - 1)k$ servers, from the watchlists of the honest parties that they observe. However, how large should n be? A naive approach, which is to take n to be the smallest possible function of k, actually may have the opposite effect. For example, if $m = 2$ and $n = 4k$ then the corrupted party, who corrupts k servers through its watchlists, needs to cheat in k inner protocols in order to cheat in the outer In order to be concrete, let the number of clients m equal 2, and consider an outer protocol that tolerates any $t < n/2$ corruptions. We briefly analyze the difference between setting $n = 4k$ and $n = 3k$, where k is the number of watchlists viewed. In the case of $n = 4k$, the adversary needs to corrupt an additional k servers in order to have a dishonest majority of $2k$ servers, and this requires cheating in the simulation of at least k servers in the inner semi-honest protocols. The honest party does not detect this if its watchlists all fall in the other $3k$ (out of $4k$) servers. A rough analysis gives that the probability that the adversary succeeds in this case is $(3/4)^k$. In contrast, if $n = 3k$ then the adversary needs to corrupt an addition $k/2$ servers and so it successfully cheats with probability only $(2.5/3)^k = (5/6)^k$. Setting a fixed error of 2^{-40}, we have that when $n = 4k$ we need to set $k = 97$ and so $n = 388$, and when $n = 3k$ we need to set $k = 152$ and so $n = 456$. We therefore conclude that it is better to take $n = 4k$ than $n = 3k$ since this results in a lower number of servers n relative to the same cheating probability of 2^{-40}.

The above analysis relates to an outer protocol that tolerates any $t < n/2$ corruptions. However, the best protocols for this setting [7] use the packed secret sharing methodology of [10]. This methodology enables the effective multiplication of an entire block of shares using a single multiplication protocol as well as other efficiency improvements. Thus, large blocks can significantly lower the complexity of the protocol. However, the outer protocol must have the property that the number of corrupted parties that can be tolerated is upper bounded by the difference between the secret sharing threshold (which for [7] must be less than $n/4$) and the block size. This demonstrate the complexity of choosing good parameters since they are all interdependent. Observe that in our concrete analysis, we use k as the number of watchlists and not an independent security parameter; in the above asymptotic treatment these were the same.

2 Optimizations of the IPS Compiler

2.1 Efficient Watchlist Setup

As we have mentioned, the first step in the IPS compiler involves the setup of watchlists. Recall that the number of real parties is m and the number of virtual servers is n. Denote the real parties by $P_1, \ldots, P_m$. Technically, each party P_i chooses n random strings $r_i^1, \ldots, r_i^n$ (say, of length the security parameter k) and runs a two-party protocol with every other party P_j in which P_j receives k of the strings without P_i knowing which. The value r_i^ℓ is the random tape used by P_i in the semi-honest protocol simulating the ℓth server. Thus, any party knowing r_i^j can verify that P_i is running the protocol honestly.

The IPS setup. The method proposed in [16] is for each pair of parties P_i, P_j to run n executions of Rabin oblivious transfer [22]; in the ℓth execution the sender P_i inputs r_i^ℓ and the receiver P_j obtains this string with probability k/n, and obtains nothing otherwise. The result of this procedure is that the expected number of strings obtained by the receiver is $n \cdot k/n = k$, as required. As described in [16] it is possible to construct a single Rabin oblivious transfer with receipt probability k/n by running a single 1-out-of-n oblivious transfer, which in turn can be constructed using $\log n$ regular 1-out-of-2 oblivious transfers [19]. Thus, the cost of this setup phase is $n \log n$ oblivious transfers for P_i as sender and each P_j as receiver, with an overall of $m(m-1)n \log n$ oblivious transfers between all parties. With $n = O(m^2 k)$ as stated in [16] we have that the cost is approximately $O(m^4 k \log(m^2 k))$ oblivious transfers.

Concretely, this step can be carried out in two ways; one is to use the efficient extending of oblivious transfers of [15]. However, in the case of malicious oblivious transfer the oblivious transfer extension is not so efficient since it is based on the cut-and-choose methodology; in addition, it requires the use of the less standard assumption of correlation-robust hash functions. Alternatively, one can use a highly efficient OT protocol with $O(1)$ exponentiations per transfer [21] (the cost here is 11 exponentiations per transfer, with UC and stand-alone variants at essentially the same cost).

A new watchlist setup. We propose a new watchlist setup with the following properties. First, the method guarantees that each malicious party views the *same* k server watchlists for every honest party. This means that an adversary can examine the computation of k, rather than $(m-1)k$, servers. As we will see, this means that it suffices to set $n = O(mk)$ instead of $O(m^2 k)$. Second, it guarantees that each party views *exactly* k watchlists; this enables a tighter and more straightforward analysis of the probability that an adversary can cheat. Finally, we present a concrete protocol that gives a considerable efficiency improvement over the IPS setup, for both aforementioned implementation options.

Recall that the aim of the watchlist setup procedure is for each party to obtain k watchlists that it can view. Rather than achieving this by having each pair of parties run a separate procedure, we have the parties run a protocol for what we call multi-sender k-out-of-n oblivious transfer. This m-party functionality enables a receiver P_m to obtain k-out-of-n strings from $m-1$ different senders

$P_1, \ldots, P_{m-1}$ (each with a set of n strings). The functionality is formally defined in Figure 1.

Figure 1 (Multi-Sender k-out-of-n Oblivious Transfer $\mathcal{F}_n^k$).
Inputs: For every $j = 1, \ldots, n-1$, party P_j inputs a vector of n strings $(x_1^j, \ldots, x_n^j)$. The receiver P_m inputs a set of indices $I \subset [n]$ of size exactly k.
Outputs: If $|I| \neq k$ then all parties receive $\perp$ as output. Otherwise, for every $j = 1, \ldots, m-1$, party P_m receives the set $\{(i, x_i^j)\}_{i \in I}$ of k strings. Parties $P_1, \ldots, P_{m-1}$ receive no output.

We stress that the receiver is forced to use the same index set I for all sender parties. In the context of the IPS compiler, this means that each party can choose k servers for which it watches *all* parties. Now, let t be the number of corrupted real parties in the protocol. Then, the parties can together view $t \cdot k$ servers, which is equivalent to these servers being corrupted. In addition, the probability that the corrupted real parties can cheat in the semi-honest subprotocols for more than L servers equals the probability that none of the honest parties have any of these servers in their watchlists. A single honest party chooses k out of n watchlists and so the probability that it does *not* detect such cheating equals $\binom{n-L}{k} / \binom{n}{k}$.

Let $m = 2$ and let $n = 4k$. We have that the corrupted real party views k of the servers in its watchlist and needs to cheat in $L = k$ more in order to have corrupted at least half of the servers. The probability that it can do this without being detected is therefore $\binom{3k}{k}/\binom{4k}{k} < (3/4)^k$. Now consider the general case of m real parties and $n = 4mk$. We first analyze the case that $m-1$ parties are corrupted. Then, the adversary can view $(m-1)k$ different servers, and needs to corrupt $(m+1)k$ additional servers in order to have corrupted half of the servers. Thus, the probability that it goes undetected is $\binom{(3m-1)k}{k}/\binom{4mk}{k} < (3/4)^k$.

Thus, we conclude that it suffices to use $n = 4mk$ virtual servers rather than $O(m^2k)$. In addition, as is evident, the proof of this fact is straightforward. (We stress that as shown in Section 4, this is not necessarily the best way to choose the parameters for concrete efficiency. However, here we are dealing with asymptotic efficiency.)

We conclude with the following somewhat informal claim; it's proof follows from the analysis above and the proof of security of [16].

Claim 2. *The IPS compiler with the watchlist setup phase using the multi-sender k-out-of-n oblivious transfer functionality is secure with $n = O(mk)$ servers.*

2.2 Securely Realizing Multi-sender k-out-of-n Oblivious Transfer

A general protocol from oblivious transfer. It is possible to securely realize the multi-sender k-out-of-n oblivious transfer functionality in a straightforward way using committed oblivious transfer, which in turn can be constructed in a black-box way from any 1-out-of-2 oblivious transfer [5]. This construction has the advantage of preserving the IPS general structure of working under general assumptions and using any oblivious transfer protocol that is secure in the presence of malicious adversaries. Thus, we obtain the efficiency improvement regarding the number of servers and the simpler analysis, while remaining within the same framework. We note, however, that this strategy will not yield a concretely efficient watchlist setup.

A concrete protocol with greater efficiency. One of the aims in this paper, which is dealt with in detail in Section 4, is to consider the concrete efficiency of the IPS compiler. As such, in this section we present a highly-efficient protocol for securely computing the multi-sender k-out-of-n oblivious transfer functionality in the presence of malicious adversaries. The security of the protocol is based on the DDH assumption and requires just $O(n)$ exponentiations. Below, we compare the concrete efficiency of our watchlist setup method based on this protocol to the best-known concrete instantiations of the original IPS watchlist setup, and show that the efficiency improvement is dramatic. As we will see, the use of this protocol enables the use of significantly more servers than otherwise (in Section 4 it will become apparent why this is so important).

The protocol uses ideas from the cut-and-choose oblivious transfer protocol of [18]. The idea is for the receiver P_m to choose n pairs of group elements (a_i, b_i) so that relative to two fixed group elements g, h it holds that at most k out of the n tuples (g, h, a_i, b_i) are Diffie-Hellman tuples (and all the others are non-DH tuples). The receiver then broadcasts all of these pairs to the sending parties $P_1, \ldots, P_{m-1}$ and proves that at most k are DH tuples, as required. Following this, all of the sending parties send values with the property that P_m can obtain the ith string if and only if (g, h, a_i, b_i) is a DH tuple. This ensures that P_m receives k-out-of-n of the strings and no more. Furthermore, since the pairs are broadcast to all $P_1, \ldots, P_{m-1}$, it is guaranteed that P_m receives the ith string from every $P_1, \ldots, P_{m-1}$ if and only if (g, h, a_i, b_i) is a DH tuple. Thus, intuitively, the protocol securely computes the multi-sender k-out-of-n oblivious transfer functionality. See the full version of our paper for a full description and proof of security. The overall number of exponentiations in that protocol is $4n + (11n+k)(m-1)$. (For large m, this is approximately $11mn+km$ exponentiations. For the special case of $m = 2$ this comes to $15n + k$.)

A comparison of concrete efficiency. We now compare the concrete cost of running our watchlist setup protocol to the method of [16]. Recall that the IPS setup requires $m(m-1)n \log n$ oblivious transfers and this can be implemented using [21] at the cost of $11 \cdot m(m-1)n \log n$ exponentiations, or using the method of extending oblivious transfers of [15]. In contrast, our protocol requires $4n + (11n + k)(m - 1)$ exponentiations. (All exponentiations here are in any group in which the DDH assumption holds.)

For the sake of comparison and simplicity, assume that the same level of security is obtained using both setups and so the same values of k and n can be used. We compare this for two sets of concrete parameters given in Section 4.3, optimizing the performance of the inner protocol for the case of $m = 2$ parties.

For the AES-type circuit, we have $k = 207$ and $n = 1752$. Then, the cost of the original IPS setup is $m(m-1)n \log n = 2 \times 1752 \times 11 = 38544$ oblivious transfers. This can be implemented using [21] at a cost of 11 exponentiations per oblivious transfer, resulting in 423,984 exponentiations. Alternatively, using the method of extending oblivious transfers of [15], the cost is about 6,000 oblivious transfers (requiring 66,000 exponentiations) and approximately 2,500,000 hash function computations.[1] In contrast, our new setup costs $15n+k = 15\cdot1752+207 = 26,487$ exponentiations, which is much less (even than the solution using [15]). An even more illustrative example relates to the two settings of parameters given for the case of a circuit of size 30,000 in Section 4.3; for this we just directly use the extending oblivious transfer alternative. With the first choice of parameters, $n = 19554$ and $k = 729$, we have that the IPS setup costs 66,000 exponentiations and 38,700,000 hash operations, versus 294,039 exponentiations for our protocol. The other choice of parameters, of $n = 3362$ and $k = 292$, requires 66,000 exponentiations and 5,300,000 hash operations, versus 50,772 exponentiations for our protocol. Thus, using our setup protocol, it is still feasible to choose either of the optimal choices of parameters and tradeoff the cost for the rest of the protocol. In contrast, using the IPS setup, only the latter setting of k and n can be reasonably used.

2.3 Flexibility of the Outer Protocol

In the original analysis carried out by [16] and that discussed above, the outer protocol chosen is secure in the presence of a malicious adversary that can adaptively corrupt up to t servers, for an appropriately chosen t. However, the corruptions of the servers are actually of two distinct types. The up to $(m-1)k$ corruptions that are due to the fact that the adversary sees the watchlists of the honest parties, are actually *semi-honest* corruptions, meaning that the adversary sees the internal state of these servers but does not cause them to deviate from the protocol specification. In contrast, the corruptions that are due to the corrupted real parties cheating in the semi-honest server simulation (without being caught) are *malicious* corruptions. Thus, it is possible to use an outer protocol that provides hybrid security in the presence of t_1 malicious corruptions and t_2 semi-honest corruptions, for appropriate t_1 and t_2. This model has been studied, and it has been demonstrated that better resilience can be achieved [9].

In order to be concrete, we demonstrate this on the parameters discussed above in Section 2.1. When $m = 2$ and $n = 4k$, we have that the corrupted party views k of the servers and needs to actively corrupt k more. Thus, the outer protocol needs to be secure in the presence of k malicious servers and k semi-honest servers, rather than $2k$ malicious servers. This can significantly simplify the outer protocol and yield higher efficiency. (For example, the simple and efficient multiplication protocol of BGW [3] for the case of $t < n/4$ malicious corruptions can also be used in the case of $t_1 < n/6$ malicious corruptions together

[1] This calculation is based on a 128-bit seed, and setting the parameter σ of the extending OT scheme to be of size 44 in order to obtain an error of 2^{-40} in the oblivious transfer extension protocol; see [15, Figure 2].

with $t_2 < n/6$ semi-honest corruptions. Thus, although the overall number of corruptions is $t < n/3$, the simpler and more efficient multiplication protocol can be used. This can be heavily utilized in the setting of IPS compilation.)

3 IPS Variants Using Covert Adversaries

In this section we present variants of the IPS compiler in order to obtain security in the presence of malicious adversaries from security in the presence of covert adversaries, and security in the presence of covert adversaries from security in the presence of semi-honest adversaries. In addition, we show that this approach has a quantitative advantage regarding the black-box construction of malicious oblivious transfer from semi-honest oblivious transfer.

3.1 Secure Computation for Malicious from Covert Adversaries

We show an extraordinarily simple analog of the IPS compiler when the starting point is a protocol for secure computation in the presence of covert adversaries. As we have already discussed the idea behind our construction in Section 1, we proceed directly to the construction. We construct a protocol for computing a function f for m parties, where any number of them can be corrupt. The protocol is secure against malicious adversaries. The security parameter is denoted by k. The protocol uses the following tools.

Outer protocol: Let π be a multiparty (outer) protocol for m clients and $n = 2k$ servers, which is secure for any number of corrupted clients and as long as less than k servers are corrupted by an adaptive malicious adversary. π computes the function f where parties $P_1, \ldots, P_m$ provide input and receive output. For simplicity, the protocol π is such that all messages are sent over a broadcast channel, and every party broadcasts in every round

Server protocols: Let $\pi_1, \ldots, \pi_{m+n}$ be the instructions for the different parties in π. That is, the clients $P_1, \ldots, P_m$ run $\pi_1, \ldots, \pi_m$ and the ith server runs π_{m+i}. For the servers, namely for $i = 1, \ldots, n$, let $\mathcal{F}_{m+i}$ be the *reactive* ideal functionality computing π_{m+i}. Loosely speaking, $\mathcal{F}_i$ is a functionality that receives $n + m - 1$ inputs in each round and generates a single output; the $m + n - 1$ inputs are the values broadcast by all parties P_j for $j \neq i$ in the previous round and the output is the value that P_i should broadcast in this round. The exact description of the functionality $\mathcal{F}_i$ is more involved. This functionality is actually run by the m real parties. Thus, each party inputs a vector of length $m + n$ with the values broadcast in the previous round. The functionality then verifies that *all* vectors input by the m clients are identical. If not, it outputs $\perp$. If yes, it computes the next message that P_i would send and hands it to all the m clients.

Covert model functionality: We denote by $\mathcal{F}^{\epsilon}_{m+i}$ the functionality $\mathcal{F}_{m+i}$ in the covert model with deterrent ϵ. This means that we consider an ideal functionality that computes $\mathcal{F}_i$ with the additional instructions of the trusted party of the ideal model of the definition of covert adversaries; see [1].

Protocol 3 uses these tools to obtain security in the presence of a malicious adversary controlling an arbitrary number of the m parties. The protocol is defined in a hybrid model where the functionalities $\mathcal{F}^{\epsilon}_{m+1}, \ldots, \mathcal{F}^{\epsilon}_{m+n}$ are executed by a trusted party. Security is derived when these functionalities are instantiated by real protocols via standard composition theorems.

Protocol 3 (Security for Malicious in the Covert $\mathcal{F}^{\epsilon}_{m+i}$-Hybrid Model).

Inputs: Real parties $P_1, \ldots, P_m$ hold respective inputs $x_1, \ldots, x_m$

The protocol: For every round of protocol π, the parties $P_1, \ldots, P_m$ do:

1. Each party P_j $(1 \leq j \leq m)$ broadcasts the message that π instructs the client P_j to send in this round (using π_j), based on the messages from the last round.
2. For every $i = 1, \ldots, n$, each party P_j $(1 \leq j \leq m)$ sends to the ideal functionality $\mathcal{F}^{\epsilon}_{m+i}$ the vector of all messages broadcast in the previous round. (In the first round, the vector contains $m+n$ empty values λ.)
3. For every $i = 1, \ldots, n$, each party P_j $(1 \leq j \leq m)$ receives an output from $\mathcal{F}^{\epsilon}_{m+i}$. If the output is $\mathsf{corrupted}_\ell$ or abort_ℓ (see [1]), then P_j halts and outputs abort_ℓ. Otherwise, it records the output as the message "broadcast" by server P_{m+i} in this round.

Output: Each party P_j $(1 \leq j \leq m)$ outputs the value that π instructs client P_j to output.

We now state the security of Protocol 3. The proof appears in the full version of the paper; it is very straightforward, and this highlights the conceptual advantage of this alternative IPS compiler.

Theorem 4. *Let π be a protocol for m clients and $n = 2k$ servers that securely computes the m-party functionality f with abort, in the presence of an adaptive malicious adversary corrupting any number of clients and less than k of the servers, and let $\epsilon > 0$ be any constant. Then, Protocol 3 securely computes f with abort in the $\mathcal{F}^{\epsilon}_{m+1}, \ldots, \mathcal{F}^{\epsilon}_{m+n}$ hybrid model, in the presence of an adaptive malicious adversary corrupting any number of parties.*

3.2 Secure Computation for Covert from Semi-Honest Adversaries

We describe a black-box transformation from semi-honest protocols to covert protocols, using a covert oblivious transfer protocol. This result answers an open question left by the work of [6], which showed a similar transformation in the information-theoretic setting with an honest majority, but did not cover the case of a majority of corrupted parties. The construction is similar to the original construction of IPS [16] with two exceptions. First, only a small number of watchlists are used. Second, it suffices for us to use an oblivious transfer protocol with security for covert adversaries, rather than security for malicious adversaries, in order to set up the watchlists. (Oblivious transfer protocols with security for covert adversaries with constant ϵ can be constructed using $O(1)$ black-box invocations of semi-honest OT, as described in the full version of our paper.)

The protocol computes a function f for m parties, where any number of them can be corrupt. The security parameter is denoted by k.

Tools:

- Let π be a multiparty protocol for m clients and $n = 4m$ servers, which is secure for any number of corrupted clients and less than $n/2$ corrupted servers. As in Protocol 3, all messages of π are sent over a broadcast channel and every party broadcasts in every round. Furthermore, the clients $P_1, \ldots, P_m$ are the only ones who provide input and receive output.
- Let $\pi_1, \ldots, \pi_{m+n}$ be the instructions for the parties in π, and let $\mathcal{F}_{m+i}$ be the *reactive* ideal functionality computing π_{m+i}, for $i = 1, \ldots, n$. The functionality is as defined for Protocol 3.
- Let $\rho_{m+1}, \ldots, \rho_{m+n}$ be m-party protocols such that ρ_{m+i} securely computes $\mathcal{F}_{m+i}$ in the presence of semi-honest adversaries. Without loss of generality we assume that the random-tape of each party in each ρ_i is of length exactly k (a pseudorandom generator can be used if it is longer).

The compiler is described in Protocol 5.

Protocol 5 (Security for Covert from Semi-Honest).

Inputs: Real parties $P_1, \ldots, P_m$ hold respective inputs $x_1, \ldots, x_m$

The protocol:

1. *Phase 1 – set up watchlists:*
   (a) For every $j = 1, \ldots, m$, party P_j chooses a vector of $n = 4m$ random seeds $s_1^j, \ldots, s_{4m}^j \in \{0,1\}^k$. The parties all then run m multi-sender 1-out-of-n oblivious transfers that are secure in the presence of covert adversaries, so that each party P_j receives $\{s_{r_j}^i\}_{i=1}^m$ for m random indices $r_j \in_R \{1, \ldots, n\}$.
   (b) At the conclusion of this phase, each client P_j holds the following:
      i. A vector $\overline{s}_j = (s_1^j, \ldots, s_n^j)$ of random seeds chosen by P_j
      ii. A set of strings $\{s_{r_j}^i\}_{i=1}^m$ received by P_j from others
2. *Phase 2 – emulate π:* For every round of the $n = 4m$-party protocol π, the parties $P_1, \ldots, P_m$ work as follows:
   (a) Each party P_j ($1 \le j \le m$) broadcasts the message that π instructs the client P_j to send in this round, based on the previous messages.
   (b) For every $i = 1, \ldots, n$, each party P_j ($1 \le j \le m$) runs ρ_{m+i} with input the vector of all messages broadcast in the previous round, and using random-tape s_i^j.
   (c) Each party P_j checks its watchlists for the executions run in the previous step. Specifically, for every $\ell = 1, \ldots, m$ ($\ell \neq j$), party P_j verifies that party P_ℓ computed the next message according to ρ_{m+r_j} using the random tape $s_{r_j}^\ell$ and the messages broadcast. If no, then P_j outputs $\mathsf{corrupted}_\ell$ (signifying that P_ℓ cheated) and halts.
   (d) For every $i = 1, \ldots, n$, each P_j ($1 \le j \le m$) receives from ρ_{m+i} an output and records it as the message "broadcast" by P_i in this round.

Output: Each party P_j ($1 \le j \le m$) outputs the value that π instructs client P_j to output.

Security is stated by the following theorem (proved in the full version).

Theorem 6. *Let π be a protocol for m clients and $n = 4m$ servers that securely computes the m-party functionality f with abort, in the presence of an adaptive malicious adversary corrupting any number of clients and a minority of servers. Then, Protocol 5 securely computes f in the $\mathcal{F}_{m+1}, \ldots, \mathcal{F}_{m+n}$ (semi-honest) hybrid model, where $n = 4m$, in the presence of an adaptive covert adversary corrupting any number of corrupted parties, with ϵ-deterrence for $\epsilon > 1 - e^{-0.25}$.*

3.3 The Semi-honest Cost of Malicious Oblivious Transfer

In the full version of this paper, we use our methodology of IPS compilation via covert adversaries to prove the following theorem:

Theorem 7. *There exists a black-box reduction from bit oblivious transfer that is secure in the presence of malicious adversaries to one-way functions and $O(k)$ invocations of bit oblivious transfer that is secure for semi-honest adversaries.*

Previously, the best known such reduction required $O(k^2)$ invocations of semi-honest oblivious transfer [13].

4 The Concrete Efficiency of IPS

In this section, we describe our analysis of the *concrete* efficiency of the best IPS-type protocols. Due to the high level of abstraction in the IPS construction, its concrete complexity was completely unknown.

The protocol that we examined is based on sharing values using block secret sharing as in [10], in which ℓ values are encoded in a single polynomial. Thus, given blocks $a = (a_1, ..., a_\ell), b = (b_1, ..., b_\ell)$ which are shared using two polynomials of degree δ, addition results in a sharing of a polynomial of degree δ that hides the block $a+b = (a_1+b_1, ..., a_\ell+b_\ell)$, while multiplication results in sharing a polynomial of degree 2δ which hides the block $ab = (a_1b_1, ..., a_\ell b_\ell)$; as usual, a protocol is used to reduce the degree of the polynomial to δ.

An in-depth analysis of the protocol, described in the full version of the paper, reveals that the overall complexity of the protocol is dominated by the number of multiplications and the number of OTs (both the communication complexity and other computational operations are negligible in comparison to these). Our efficiency analysis will therefore present those two factors. The OTs are only needed for the inner semi-honest multiplication protocols,[2] and so the other building blocks will be analyzed only in terms of the number of multiplications. We emphasize that these OTs must only be secure against a semi-honest adversary, and not a malicious one.

[2] We remark that the OTs needed for setting up the watchlists (which must be secure against malicious adversaries) are also a factor. However, they depend only on the number of servers n and so can be considered at the end.

4.1 An Analysis of the Building Blocks

Secret sharing for blocks: This secret sharing scheme is a variant of Shamir's secret sharing [23], presented in [10]. Each polynomial encodes a block of ℓ values. The cost of sharing w elements among n servers, using blocks of size ℓ and polynomials of degree δ, is $(w/\ell)(\delta^2 + n\delta)$ multiplications.

Proving that shares lie on δ-degree polynomials: After sharing the secrets it must be proved that the shares are indeed encoded by z polynomials, each of degree δ (each polynomial is used to hide ℓ field elements; depending on the number of inputs, multiple polynomials must be used for the sharing). A protocol for such a proof is presented in [17]. It requires $\delta(z + n + k)$ multiplications.

Proving a replication pattern of shared blocks: The protocol requires parties to prove that certain shared blocks follow some replication pattern (namely that a certain output value is used as an input value for the next layer). The protocol we used is mentioned in the computation complexity analysis of [17]. The cost is $(4\delta)^2 + 4\delta n + 2((2\delta n + (2\delta)^2)u + (\delta n + \delta^2)v) + 2(n + k)(u + v)$ multiplications, where v is the number of input blocks (represented by polynomials of degree δ) and u is the number of output blocks (represented by polynomials of degree 2δ).

Semi-honest inner multiplication: For multiplication gates the parties run a semi-honest protocol for the functionality $(x_1, x_2) \mapsto (x_1 x_2 - r, r)$ for a random $r \in_R \mathcal{F}$. Six different protocols are presented for this functionality in [17]. We present the analysis for the most efficient protocols only (based on our concrete analysis for all options). The first protocol is based on packed Reed-Solomon encoding and is black-box in the field [17], and the second protocol is due to Glboa [11] and makes nonblack-box usage of the field and assumes standard bit representation of elements.

As is detailed in the full version of the paper, for a security parameter $s = 40$ giving error 2^{-40}, the packed Reed-Solomon encoding protocol costs 2734 multiplications and 16 1-out-of-2 OTs per inner multiplication, while the protocol of [11] uses 40 multiplications and 40 1-out-of-2 OTs. Namely, one protocol is more efficient in terms of OTs and the other is more efficient in terms of multiplications. For concrete numbers this phenomenon might present implementers with a real dilemma.

4.2 Instantiating the Parameters

In order to count concrete efficiency, the values of the different parameters must be set. We do not claim to have found the absolute optimal parameters, as the analysis of their effect on the overhead is very complex. We do present for each parameter the different considerations affecting the choice of its value, and eventually show that the protocol is comparable in its efficiency to other protocols from the literature and may be competitive in some settings.

The four main parameters that must be set are the degree of the polynomials δ, the block size ℓ, the number of corrupted parties tolerated t, and the number of servers n. Three out of the four different parameters, the degree, the block size and the corruption threshold, are tightly interconnected in that setting any two of them determines the third one. In addition, these three parameters are all chosen as a function of the number of servers n, and given their descriptions the actual concrete value of n is chosen (independently of the circuit). As we will see below, the determination of the degree δ is a straightforward choice. We then determine the block size based on the actual circuit being computed, thereby essentially setting the threshold.

The degree δ: Due to the replication proof protocol, it must hold that $\delta < n/4$. Other than that δ should be maximized, and we therefore set $\delta = n/4 - 1$ (for simplicity of notation, from here on we write $\delta = n/4$).

The block size ℓ: The block size has to be strictly smaller than the degree δ, but otherwise the larger the block size, the more efficient the outer protocol gets. This is because more multiplications are carried out together (note that there is no use in having a block size larger than the width of a layer in the circuit since this already upper bounds the number of multiplications that can be carried out together). However, the number of corrupted servers that the outer protocol can tolerate is $\delta - \ell$, thus the closer ℓ is to δ, the smaller the fraction of corrupt servers that can be tolerated. As a result, more servers are required in order to ensure that the probability of catching the adversary cheating in the server simulation does not go down.

The corruption threshold t: As shown in [10], up to $t < \delta - \ell$ corrupted servers receive no information about the secret block when using block secret sharing with degree δ and block size ℓ. Thus t is set to $\delta - \ell - 1$.

The number of servers n: Define $\tau = n/t$, and so $1/\tau$ is the ratio of servers that the adversary needs to corrupt in order to successfully cheat. Denote $n = O(mk) = a \cdot mk$ for some parameter a which should be chosen to minimize n. We have already observed in Section 1.4, however, that the naive approach of minimizing a is not optimal. An analysis reveals that for the two-party case the best choice is to take $a = \tau$, and thus to use $n = 2\tau k$ servers in the outer protocol.

4.3 Setting Concrete Values

We now show the concrete cost of IPS based on the results of our analysis regarding parameter instantiation. The choices of parameters are demonstrated for two different circuits, which clarify the dilemmas that arise in practice. As stated above the only parameter which we did not set independently of the circuit, but rather want to optimize for the concrete circuit, is the block size. We therefore calculated the number of operations required for a large range of block sizes. Our calculations are based on a combination of an analytic and numerical analysis of the parameters that yield a cheating probability of at most 2^{-40}.

The first example uses circuit parameters similar to the AES circuit of [8], assuming 2400 multiplication gates split over 100 layers. In this case minimal values for the number of OTs and the number of multiplications occur for the same block size of $n/73$. Remembering that $\delta = n/4$, the threshold ratio is $\tau = \frac{1}{(\delta-\ell)/n} \approx 4.231$. Setting $a = \tau$ as suggested above results in $n = 4.231 \cdot 2k$.

In order to get a cheating probability of 2^{-40}, each client must check $k = 207$ watchlists, and the number of servers is $n \approx 1752$.[3]

The number of OTs and multiplications which are required in this setting depends on the inner multiplication protocol that is used (based on Reed-Solomon codes, or on the protocol of [11]). The first choice results in approximately $5.5 \cdot 10^6$ OTs and $5.5 \cdot 10^9$ multiplications, while the latter choice requires approximately $13.8 \cdot 10^6$ OTs and $4.5 \cdot 10^9$ multiplications. The choice of the inner protocol is therefore not trivial, and depends on the properties of concrete implementations of the OT and multiplication primitives. Recall that multiplications are in a finite field of size 2^{40} and therefore elements fit in a single word of a modern 64-bit architecture, and can be done very efficiently. The OTs need only be secure against semi-honest adversaries, and so can be efficiently implemented using methods of extending OT as in [15]. Given these two observations, the run time of the protocol seems reasonable in comparison to that of other protocols providing security against malicious adversaries.

The second example is of a circuit of 30000 multiplication gates split over 10 layers, and results in another optimization dilemma. Setting the block size $\ell = n/5.7$ results in the minimal number of OTs, but minimizing the number of multiplications requires setting $\ell = n/13.1$. The actual numbers of operations are described below.

	n	k	RS OT	Gilboa OT	RS mult	Gilboa mult
$\boldsymbol{\ell = n/5.7}$	19554	729	$5.6 \cdot 10^6$	$11.1 \cdot 10^6$	$54 \cdot 10^9$	$53 \cdot 10^9$
$\boldsymbol{\ell = n/13.1}$	3362	292	$12 \cdot 10^6$	$31 \cdot 10^6$	$13 \cdot 10^9$	$11 \cdot 10^9$

The reason for this tradeoff is that the number of OTs is minimized when the block size can accommodate an entire layer in a single block, but this setting requires more servers (compared to a smaller block size), and so multi-point evaluation and interpolation become more expensive (as they depend on the number of servers), which results in an increased amount of multiplications. Observe also that setting $\ell = n/5.7$ results in a much larger number of servers which in turn affects the cost of the watchlist setup protocol. Plugging in the cost of our watchlist setup ($15n + k$ exponentiations), we have that when $\ell = n/5.7$ the setup cost is $294,039$ exponentiations, versus just $50,722$ when $\ell = n/13.1$. This cost may also weigh in as a factor.

We conclude that the IPS protocol may be competitive in some settings. We are currently implementing the protocol in order to empirically verify our analysis and conclusions.

References

1. Aumann, Y., Lindell, Y.: Security Against Covert Adversaries: Efficient Protocols for Realistic Adversaries. J. of Cryptology 23(2), 281–343 (2010)
2. Beaver, D.: Correlated Pseudorandomness and the Complexity of Private Computations. In: The 28th STOC, pp. 479–488 (1996)

[3] Note that the block size, of $n/73 = 24$, is equal to its maximum reasonable value, namely to the width of a layer of the circuit, which is also 24.

3. Ben-Or, M., Goldwasser, S., Wigderson, A.: Completeness Theorems for Non-Cryptographic Fault-Tolerant Distributed Computation. In: 20th STOC, pp. 1–10 (1988)
4. Chaum, D., Crépeau, C., Damgård, I.: Multi-party Unconditionally Secure Protocols. In: 20th STOC, pp. 11–19 (1988)
5. Crépeau, C., van de Graaf, J., Tapp, A.: Committed Oblivious Transfer and Private Multi-Party Computation. In: Coppersmith, D. (ed.) CRYPTO 1995. LNCS, vol. 963, pp. 110–123. Springer, Heidelberg (1995)
6. Damgård, I., Geisler, M., Nielsen, J.B.: From Passive to Covert Security at Low Cost. In: Micciancio, D. (ed.) TCC 2010. LNCS, vol. 5978, pp. 128–145. Springer, Heidelberg (2010)
7. Damgård, I., Ishai, Y.: Scalable Secure Multiparty Computation. In: Dwork, C. (ed.) CRYPTO 2006. LNCS, vol. 4117, pp. 501–520. Springer, Heidelberg (2006)
8. Damgård, I., Keller, M.: Secure Multiparty AES. In: Sion, R. (ed.) FC 2010. LNCS, vol. 6052, pp. 367–374. Springer, Heidelberg (2010)
9. Fitzi, M., Hirt, M., Maurer, U.M.: Trading Correctness for Privacy in Unconditional Multi-Party Computation. In: Krawczyk, H. (ed.) CRYPTO 1998. LNCS, vol. 1462, pp. 121–136. Springer, Heidelberg (1998)
10. Franklin, M.K., Yung, M.: Communication Complexity of Secure Computation (Extended Abstract). In: The 24th STOC, pp. 699–710 (1992)
11. Gilboa, N.: Two Party RSA Key Generation (Extended Abstract). In: Wiener, M. (ed.) CRYPTO 1999. LNCS, vol. 1666, pp. 116–129. Springer, Heidelberg (1999)
12. Goldreich, O., Micali, S., Wigderson, A.: How to Play any Mental Game – A Completeness Theorem for Protocols with Honest Majority. In: 19th STOC, pp. 218–229 (1987)
13. Haitner, I., Ishai, Y., Kushilevitz, E., Lindell, Y., Petrank, E.: Black-Box Constructions of Protocols for Secure Computation. SIAM Journal on Computing 40(2), 225–266 (2011)
14. Ishai, Y.: Personal Communication (2011)
15. Ishai, Y., Kilian, J., Nissim, K., Petrank, E.: Extending Oblivious Transfers Efficiently. In: Boneh, D. (ed.) CRYPTO 2003. LNCS, vol. 2729, pp. 145–161. Springer, Heidelberg (2003)
16. Ishai, Y., Prabhakaran, M., Sahai, A.: Founding Cryptography on Oblivious Transfer – Efficiently. In: Wagner, D. (ed.) CRYPTO 2008. LNCS, vol. 5157, pp. 572–591. Springer, Heidelberg (2008)
17. Ishai, Y., Prabhakaran, M., Sahai, A.: Secure Arithmetic Computation with No Honest Majority. In: Reingold, O. (ed.) TCC 2009. LNCS, vol. 5444, pp. 294–314. Springer, Heidelberg (2009)
18. Lindell, Y., Pinkas, B.: Secure Two-Party Computation via Cut-and-Choose Oblivious Transfer. In: Ishai, Y. (ed.) TCC 2011. LNCS, vol. 6597, pp. 329–346. Springer, Heidelberg (2011)
19. Naor, M., Pinkas, B.: Oblivious Transfer with Adaptive Queries. In: Wiener, M. (ed.) CRYPTO 1999. LNCS, vol. 1666, pp. 573–590. Springer, Heidelberg (1999)
20. Pinkas, B., Schneider, T., Smart, N.P., Williams, S.C.: Secure Two-Party Computation Is Practical. In: Matsui, M. (ed.) ASIACRYPT 2009. LNCS, vol. 5912, pp. 250–267. Springer, Heidelberg (2009)
21. Peikert, C., Vaikuntanathan, V., Waters, B.: A Framework for Efficient and Composable Oblivious Transfer. In: Wagner, D. (ed.) CRYPTO 2008. LNCS, vol. 5157, pp. 554–571. Springer, Heidelberg (2008)
22. Rabin, M.: How to Exchange Secrets by Oblivious Transfer. Tech. Memo TR-81, Aiken Computation Laboratory, Harvard U. (1981)
23. Shamir, A.: How to Share a Secret. Communications of the ACM 22(11), 612–613 (1979)
24. Yao, A.: How to Generate and Exchange Secrets. In: 27th FOCS, pp. 162–167 (1986)

$1/p$-Secure Multiparty Computation without Honest Majority and the Best of Both Worlds

Amos Beimel[1,⋆], Yehuda Lindell[2,⋆⋆], Eran Omri[2,⋆⋆], and Ilan Orlov[1,⋆]

[1] Dept. of Computer Science, Ben Gurion University
[2] Dept. of Computer Science, Bar Ilan University

Abstract. A protocol for computing a functionality is secure if an adversary in this protocol cannot cause more harm than in an ideal computation, where parties give their inputs to a trusted party which returns the output of the functionality to all parties. In particular, in the ideal model such computation is fair – all parties get the output. Cleve (STOC 1986) proved that, in general, fairness is not possible without an honest majority. To overcome this impossibility, Gordon and Katz (Eurocrypt 2010) suggested a relaxed definition – $1/p$-secure computation – which guarantees partial fairness. For two parties, they construct $1/p$-secure protocols for functionalities for which the size of either their domain or their range is polynomial (in the security parameter). Gordon and Katz ask whether their results can be extended to multiparty protocols.

We study $1/p$-secure protocols in the multiparty setting for general functionalities. Our main result is constructions of $1/p$-secure protocols that are resilient against *any* number of corrupt parties provided that the number of parties is constant and the size of the range of the functionality is at most polynomial (in the security parameter n). If less than $2/3$ of the parties are corrupt, the size of the domain is constant, and the functionality is deterministic, then our protocols are efficient even when the number of parties is $\log\log n$. On the negative side, we show that when the number of parties is super-constant, $1/p$-secure protocols are not possible when the size of the domain is polynomial. Thus, our feasibility results for $1/p$-secure computation are essentially tight.

We further motivate our results by constructing protocols with stronger guarantees: If in the execution of the protocol there is a majority of honest parties, then our protocols provide full security. However, if only a minority of the parties are honest, then our protocols are $1/p$-secure. Thus, our protocols provide the best of both worlds, where the $1/p$-security is only a fall-back option if there is no honest majority.

1 Introduction

A protocol for computing a functionality is secure if an adversary in this protocol cannot cause more harm than in an ideal computation, where parties give their

⋆ Generously supported by ISF grant 938/09 and by the Frankel Center for Computer Science.

⋆⋆ Generously supported by the European Research Council as part of the ERC project LAST, and by ISF grant 781/07.

P. Rogaway (Ed.): CRYPTO 2011, LNCS 6841, pp. 277–296, 2011.

inputs to a trusted party which, in turn, returns the output of the functionality to all parties. This is formalized by requiring that for every adversary in the real world, there is an adversary in the ideal world, called simulator, such that the output of the real-world adversary and the simulator are indistinguishable in polynomial time. Such security can be achieved when there is a majority of honest parties [13]. Secure computation is fair – all parties get the output. Cleve [7] proved that, in general, fairness is not possible without an honest majority.

To overcome the impossibility of [7], Gordon and Katz [18] suggested a relaxed definition – $1/p$-secure computation – which guarantees partial fairness. Informally, a protocol is $1/p$-secure if for every adversary in the real world, there is a simulator running in the ideal world, such that the output of the real-world adversary and the simulator cannot be efficiently distinguished with probability greater than $1/p$. For two parties, Gordon and Katz construct $1/p$-secure protocols for functionalities whose size of either their domain or their range is polynomial (in the security parameter). They also give impossibility results when both the domain and range are super-polynomial. Gordon and Katz ask whether their results can be extended to multiparty protocols. We give positive and negative answers to this question.

Previous Results. Cleve [7] proved that any protocol for coin-tossing without an honest majority cannot be fully secure; specifically, if the protocol has r rounds, then it is at most $1/r$-secure. Protocols with partial fairness, under various definitions and assumptions, have been constructed for coin-tossing [7, 8, 23, 3], for contract signing/exchanging secrets [5, 22, 10, 4, 9, 6], and for general functionalities [26, 11, 1, 14, 25, 12, 18]. We next describe the papers that are most relevant to our paper. Moran, Naor, and Segev [23] construct 2-party protocols for coin tossing that are $1/r$-secure (where r is the number of rounds in the protocol). Gordon and Katz [18] define $1/p$-security and construct 2-party $1/p$-secure protocols for every functionality whose size of either the domain or the range of the functionality is polynomial. Finlay, Beimel, Omri, and Orlov [3] construct multiparty protocols for coin tossing that are $O(1/r)$-secure provided that the fraction of corrupt parties is slightly larger than half. In particular, their protocol is $O(1/r)$-secure when the number of parties is constant and the fraction of bad parties is less than $2/3$.

Gordon et al. [15] showed that complete fairness is possible in the two party case for some functions. Gordon and Katz [17] showed similar results for the multiparty case. The characterization of the functions that can be computed with full fairness without honest majority is open. Gordon et al. [16] studied completeness for fair computations. Specifically, they showed a specific function that is complete for fair two-party computation; this function is also complete for $1/p$-secure two-party computation.

Ishai et al. [19] considered "best of two worlds" protocols. Such protocols should provide full security with an honest majority and some (weaker) security if there is only a minority of honest parties. They give positive and negative results for the existence of such protocols. We discuss some of their results below.

1.1 Our Results

We study $1/p$-secure protocols in the multiparty setting. We construct protocols for general functionalities that are $1/p$-secure against *any* number of corrupt parties provided that the number of parties is constant. Our protocols require that the size of the range of the (possibly randomized) functionality is at most polynomial in the security parameter. That is, we show the following feasibility result.

Theorem (Informal). *Let $\mathcal{F}$ be a (possibly randomized) functionality with a constant number of parties whose size of range is at most polynomial in the security parameter n. Then, for every polynomial $p(n)$ there is a $1/p(n)$-secure protocol for $\mathcal{F}$ tolerating any number of corrupt parties.*

Our results are the first general feasibility results for $1/p$-secure protocols in the multi-party setting, e.g., even for the case that there are 3 parties and two of them might be corrupt. We provide two additional protocols that are $1/p$-secure assuming that the fraction of corrupt parties is less than 2/3. These two protocols are more efficient than the protocols discussed above. Specifically, one of the protocols is $1/p$-secure even when the number of parties is $\log\log n$ (where n is the security parameter) provided that the functionality is deterministic and the size of the domain of inputs is constant.

The definition of $1/p$-security allows that with probability $1/p$ the outputs of the honest parties will be arbitrary, e.g., for a Boolean function the outputs can be non-Boolean. Some of our protocols are always correct, that is, they always return an output of the functionality with the inputs of the honest parties and some inputs for the corrupt parties. This correctness property is essential for the best of both worlds results described below.

We further motivate our results by constructing protocols with best of both worlds guarantees: If in the execution of the protocol there is a majority of honest parties, then our protocols provide full security. However, if only a minority of parties are honest, then our protocols are $1/p$-secure. The protocols succeed although they do not know in advance if there is an honest majority or not. Specifically, we show that

Theorem (Informal). *Let $\mathcal{F}$ be a functionality with a constant number of parties whose size of domain and range is at most polynomial in the security parameter n. Then, for every polynomial $p(n)$ there is a protocol for $\mathcal{F}$ tolerating any number of corrupt parties such that*

- *If there is an honest majority, then the protocol is fully secure.*
- *If there is no honest majority, then the protocol is $1/p(n)$-secure.*

Thus, the $1/p$-security guarantee can be considered as a fall-back option if there is no honest majority. Our protocols provide the best of both worlds, the world of honest majority where the known protocols (e.g., [13]) provide full security if there is an honest majority and provide no security guarantees if no such majority exists and the world of secure computation without honest majority.

In the latter world the security is either security-with-abort or $1/p$-security. These types of security are incomparable. Ishai et al. [19] proved that there is no general protocol which provides full security when there is an honest majority and security-with-abort without an honest majority. Thus, our protocols provide the best possible combination of both worlds.

Katz [21] presented a protocol, for any functionality $\mathcal{F}$, with full security when there is an honest majority, as well as $1/p$-security *with abort* for any number of corrupt parties. This result assumes a non-rushing adversary. In contrast, our protocols achieve a stronger security with a minority of honest parties and can handle the more realistic case of a rushing adversary. However, our protocols only work with a constant number of parties and a polynomial size domain.

To complete the picture, we prove interesting impossibility results. We show that, in general, when the number of parties is super-constant, $1/p$-secure protocols are not possible without honest majority when the size of the domain is polynomial. This impossibility result justifies the fact that in our protocols the number of parties is constant. We also show that, in general, when the number of parties is $\omega(\log n)$, $1/p$-secure protocols are not possible without honest majority even when the size of the domain is 2. The proof of the impossibility results is rather simple and follows from an impossibility result of [18]. Nevertheless, they show that our general feasibility results are almost tight.

Our impossibility results should be contrasted with the coin-tossing protocol of [3] which is an efficient $1/p$-secure protocol even when $m(n)$, the number of parties, is polynomial in the security parameter and the number of bad parties is $m(n)/2 + O(1)$. Our results show that these parameters are not possible for general $1/p$-secure protocols even when the size of the domain of inputs is 2.

The above mentioned impossibility results do not rule out that the best of two worlds results of Katz [21] can be strengthened by removing the restriction that the adversary is non-rushing. We show that this is impossible, that is, in general, when the number of parties is super-constant and the size of the domain is polynomial, there is no protocol that is fully secure with an honest majority and $1/p$-secure-with-abort without such a majority.

The ideas behind our protocols. Our protocols use ideas from the protocols of Gordon and Katz [18] and Beimel et al. [3], both of which generalize the protocol of Moran, Naor, and Segev [23]. In addition, our protocols introduce new ideas that are required to overcome challenges that did not occur in previous works, e.g., dealing with inputs (in contrast to the scenario of [3]) and dealing with a dishonest majority even after parties abort (in contrast to the scenario of [18]). In particular, in order to achieve resilience against any number of corrupt parties we introduce new techniques for hiding the round in which parties learn the output of an execution. Specifically, our protocols proceed in rounds, where in each round values are given to subsets of parties. There is a special round $i^\star$ in the protocol. Prior to round $i^\star$, the values given to a subset of parties are values that can be computed from the inputs of the parties in this subset; starting from round $i^\star$ the values are the "correct" output of the functionality. The values given to a subset are secret shared such that only if all parties in the subset cooperate they can reconstruct the value. Similar to the protocols of [23, 18, 3], the adversary can cause harm (e.g., bias the output of the functionality) only if it guesses $i^\star$; we show that in our protocols this probability is small and the protocols are $1/p$-secure.

In our protocols that are $1/p$-secure against a fraction of 2/3 corrupt parties (which are described in Section 4), if in some round many (corrupt) parties have aborted and there is a majority of honest parties among the active parties, then the set of active parties reconstructs the value given to this set in the previous round. The mechanism to secret share the values in this protocols is similar to [3], however, there are important differences in this sharing, as the sharing mechanism of [3] is not appropriate for $1/p$-secure computations of functionalities which depend on inputs. The fact that the protocol proceeds until there is an honest majority imposes some restrictions that imply that the protocol can tolerate only a fraction of 2/3 corrupt parties.

Our protocols that are $1/p$-secure against any number of corrupt parties (which are described in Section 5) take a different route. To describe the ideas of the protocol, we consider only the three-party case, where at most two parties are corrupt. In the protocol if one party aborts, then the remaining two parties execute a two-party protocol for the functionality. Again, this protocol proceeds in rounds, where in each round each party gets a value. If the party in the three-party protocol aborts after round $i^\star$, then all these values are the "correct" output of the functionality. To hide $i^\star$, also prior to $i^\star$, with some probability all these values must be equal. With the remaining probability, a new $i^\star$ is chosen with uniform distribution for the two-party protocol. In other words, in the two-party protocol prior to the original $i^\star$, with some probability, we chose a "fake" value of 1 for the new $i^\star$ of the two-party protocol.

Open Problems. In our impossibility results the size of the range is super-polynomial (in the security parameter). However, in all our protocols the size of the range is polynomial. It is open if there is an efficient $1/p$-secure protocol when the number of parties is not constant and the size of both the domain and range is polynomial. In our protocols, the number of rounds is double-exponential in the number of parties. Our impossibility results do not rule out that this double-exponential dependency can be improved.

The protocols of [18] are private – the adversary cannot learn any information on the inputs of the honest parties (other than the information that it can learn in the ideal world of computing $\mathcal{F}$). The adversary can only bias the output. Some of our protocols are provably not private (that is, the adversary can learn extra information). However, for other protocols, we do not know whether they are private. It is open if there are general multiparty $1/p$-secure protocols that are also private.

2 Background and the Model of Computation

A multi-party protocol with m parties is defined by m interactive probabilistic polynomial-time Turing machines $p_1, \ldots, p_m$. Each Turning machine, called party, has the security parameter 1^n as a joint input and a private input y_j. The computation proceeds in rounds. In each round, the active parties broadcast and receive messages on a common broadcast channel. The number of rounds in the protocol is expressed as some function $r(n)$ in the security parameter (typically, $r(n)$ is bounded by a polynomial). At the end of the protocol, the (honest) parties should hold a common value w (which should be equal to an output of a predefined functionality).

In this work we consider a corrupt, static, computationally-bounded (i.e., non-uniform probabilistic polynomial-time) adversary that controls some subset of parties. That is, before the beginning of the protocol, the adversary corrupts a subset of the parties and may instruct them to deviate from the protocol in an arbitrary way. The adversary has complete access to the internal states of the corrupted parties and fully controls the messages that they broadcast throughout the protocol. The honest parties follow the instructions of the protocol.

The parties communicate via a synchronous network, using only a broadcast channel. The adversary is rushing, that is, in each round the adversary sees the messages broadcast by the honest parties before broadcasting the messages of the corrupted parties for this round (thus, the broadcast messages of the corrupted parties can depend on the messages of the honest parties in the same round).

In this work we consider $1/p$-secure computation. Roughly speaking, we say that a protocol Π is $1/p$-secure if for every adversary $\mathcal{A}$ attacking Π in the real-world there is a simulator $\mathcal{S}$ running in the ideal-world, such that the global output of the real-world and the ideal-world executions cannot be distinguished with probability greater than $1/p$. The formal definitions of $1/p$-security and security with abort and cheat detection, which is a tool used in this paper, will be given in the full version of the paper.

3 Feasibility Results for $1/p$-Secure Multiparty Computation

In this section we state our main feasibility results. Our main result asserts that any functionality with a polynomial size range for a constant number of parties can be $1/p$-securely computed in polynomial time tolerating any number of corrupt (malicious) parties. We next formally state this result.

Theorem 1. *Let $\mathcal{F}$ be an m-party (possibly randomized) functionality. If enhanced trap-door permutations exist, and if m is constant and the size of the range $g(n)$ is bounded by a polynomial in the security parameter n, then for any polynomial $p(n)$ there is an $r(n)$-round $1/p(n)$-secure protocol computing $\mathcal{F}$ tolerating up to $m-1$ corrupt parties, where $r(n) = \left(p(n) \cdot g(n)\right)^{2^{O(m)}}$.*

The protocol that implies Theorem 1 for general m will appear in the full version of this paper. In this extended abstract we present, in Section 5, the 3-party version of this protocol tolerating up to 2 corrupt parties. In addition, for functionalities where the domain size is also bounded by a polynomial, we will present in the full version of this paper a protocol with somewhat stronger security properties. Using these stronger security, we can transform it into a protocol of the best of both worlds type (see Section 6.1 for details).

We give substantially better protocols secure against an adversary that may corrupt strictly less than two-thirds of the parties. Formally, we prove the following theorem.

Theorem 2. *Let $\mathcal{F}$ be an $m(n)$-party (possibly randomized) functionality. Let $t(n)$ be such that $m(n)/2 \leq t(n) < 2m(n)/3$. If enhanced trap-door permutations exist, then for any polynomial $p(n)$ the following hold:*

- *If $m(n)$ is constant (hence, $t = t(n)$ is constant) and the size of the range $g(n)$ is bounded by a polynomial, then there exists an $r(n)$-round $1/p(n)$-secure protocol computing $\mathcal{F}$ tolerating up to t corrupt parties, where $r(n) = (2p(n))^{2^t+1} \cdot g(n)^{2^t}$.*
- *If $\mathcal{F}$ is deterministic and the size of the domain $d(n)$ is bounded by a polynomial, then there exists an $r(n)$-round $1/p(n)$-secure protocol computing $\mathcal{F}$ tolerating up to $t(n)$ corrupt parties, where $r(n) = p(n) \cdot d(n)^{m(n) \cdot 2^{t(n)}}$, provided that $r(n)$ is bounded by a polynomial.*

The protocols that imply the results of Theorem 2 are presented in Section 4. As implied by the second item of Theorem 2, the round complexity of our protocol when $\mathcal{F}$ is deterministic has only a linear dependency on $p(n)$. Specifically, this protocol has polynomially many rounds even when the number of parties is $0.5 \log \log n$ provided that the functionality is deterministic and the size of the domain of inputs is constant.

4 Protocols with Less Than Two-Thirds Corrupt Parties

In this section we describe our protocols that are secure when the adversary corrupts strictly less than two thirds of the parties. We start with a protocol that assumes that either the functionality is deterministic and the size of the domain is polynomial, or that the functionality is randomized and both the domain and range of the functionality are polynomial. We then present a modification of the protocol that is $1/p$-secure for (possibly randomized) functionalities if the size of the range is polynomial (even if the size of the domain of $\mathcal{F}$ is not polynomial). The first protocol is more efficient for deterministic functionalities with polynomial-size domain. Furthermore, the first protocol has full correctness, while in the modified protocol, correctness is only guaranteed with probability $1 - 1/p$.

Following [23, 3], we present the first protocol in two stages. We first describe in Section 4.1 a protocol with a dealer and then in Section 4.2 present a protocol without this dealer. The goal of presenting the protocol in two stages is to simplify the understanding of the protocol and to enable us to prove the protocol in a modular way. In Section 4.3, we present a modification of the protocol which is $1/p$-secure if the size of the range is polynomial (even if the size of the domain of f is not polynomial).

4.1 The Protocol for Polynomial-Size Domain with a Dealer

In this section we assume that there is a special trusted on-line dealer, denoted T. This dealer interacts with the parties in rounds, sending messages on private channels. We assume that the dealer knows the set of corrupt parties. In Section 4.2, we show how to remove this dealer and construct a protocol without a dealer.

In our protocol the dealer sends in each round values to subsets of parties; the protocol proceeds with the normal execution as long as at least $t+1$ of the parties are still active. If in some round i, there are at most t active parties, then the active parties reconstruct the value given to them in round $i-1$, output this value, and halt. Following [21, 15, 23, 18, 3], the dealer chooses at random with uniform distribution a special round $i^\star$. Prior to this round the adversary gets no information and if the corrupt parties abort the execution prior to $i^\star$, then they cannot bias the output of the honest parties or cause any harm. After round $i^\star$, the output of the protocol is fixed, and also in this case the adversary cannot affect the output of the honest parties. The adversary can cause harm only if it guesses $i^\star$ and this happens with small probability.

In this extended abstract, we only give a verbal description of the protocol. This protocol is designed such that the dealer can be removed from it in Section 4.2. At the beginning of the protocol each party sends its input y_j to the dealer. The corrupted parties may send any values of their choice. Let $x_1, \ldots, x_m$ denote the inputs received by the dealer. If a corrupt party p_j does not send an input, then the dealer sets x_j to be a random value selected uniformly from the input domain X_n. In a preprocessing phase, the dealer T selects uniformly at random a special round $i^\star \in \{1, \ldots, r\}$. The dealer computes $w \leftarrow f_n(x_1, \ldots, x_m)$. Then, for every round $1 \leq i \leq r$ and every $L \subset \{1, \ldots, m\}$ such that $m - t \leq |L| \leq t$, the dealer selects an output, denoted σ_L^i, as follows (this output is returned by the parties in $Q_L = \{p_j : j \in L\}$ if the protocol terminates in round $i+1$ and Q_L is the set of the active parties):

Case I: $1 \leq i < i^\star$. For every $j \in L$ the dealer sets $\widehat{x}_j = x_j$ and for every $j \notin L$ it chooses $\widehat{x}_j$ independently with uniform distribution from the domain X_n; it computes the output $\sigma_L^i \leftarrow f_n(\widehat{x}_1, \ldots, \widehat{x}_m)$.

Case II: $i^\star \leq i \leq r$. The dealer sets $\sigma_L^i = w$.

The dealer T interacts with the parties in rounds, where in round i, for $1 \leq i \leq r$, there are of three phases:

The peeking phase. The dealer T sends to the adversary all the values σ_L^i such that all parties in Q_L are corrupted.

The abort and premature termination phase. The adversary sends to T the identities of the parties that abort in the current round. If there are less than $t+1$ active parties, then T sends σ_L^{i-1} to the active parties, where Q_L is the set of the active parties, where parties can also abort during this phase. The honest parties return this output and halt.

The main phase. If at least $t+1$ parties are active, T notifies the active parties that the protocol proceeds normally to the next round.

If after r rounds there are at least $t+1$ active parties, then T sends w to all active parties and the honest parties output this value.

Example 1. As an example, assume that $m=5$ and $t=3$. In this case the dealer computes a value σ_L^i for every set of size 2 or 3. Consider an execution of the protocol where p_1 aborts in round 4 and p_3 and p_4 abort in round 100. In this case, T sends $\sigma_{\{2,5\}}^{99}$ to p_2 and p_5, which return this output.

We next hint why for deterministic functionalities, an adversary can cause harm in the above protocol by at most $O(d^{O(1)}/r)$, where $d=d(n)$ is the size of the domain of the inputs and the number of parties, i.e., m, is constant. As in the protocols of [23, 18, 3], the adversary can only cause harm by causing the protocol to terminate in round $i^\star$. In our protocol, if in some round there are two values σ_L^i and $\sigma_{L'}^i$ that the adversary can obtain such that $\sigma_L^i \neq \sigma_{L'}^i$, then the adversary can deduce that $i < i^\star$. Furthermore, the adversary might have some auxiliary information on the inputs of the honest parties, thus, the adversary might be able to deduce that a round is not $i^\star$ even if all the values that it gets are equal. However, there are less than 2^t values that the adversary can obtain in each round (i.e., the values of subsets of the t corrupt parties of size at least $m-t$). We will show that for a round i such that $i < i^\star$, the probability that all these values are equal to a fixed value is $1/d^{O(1)}$ for a deterministic function f_n (for a randomized functionality this probability also depends on the size of the range). By [18, Lemma 2], this implies that the protocol is $d^{O(1)}/r$-secure.

4.2 Eliminating the Dealer of the Protocol

We eliminate the trusted on-line dealer in a few steps using a few layers of secret-sharing schemes. First, we change the on-line dealer, so that, in each round i, it shares the value σ_L^i of each subset Q_L among the parties of Q_L using a $|L|$-out-of-$|L|$ secret-sharing scheme – called *inner* secret-sharing scheme. As in protocol with the dealer (described in Section 4.1), the adversary is able to obtain information on σ_L^i only if it controls all the parties in Q_L. On the other hand, the honest parties can reconstruct σ_L^{i-1} (without the dealer), where Q_L is the set of active parties containing the honest parties. In the reconstruction, if an active (corrupt) party does not give its share, then it is removed from the set of active parties Q_L. This is possible since in the case of a premature termination an honest majority among the active parties is guaranteed (as further explained below).

Next, we convert the on-line dealer to an off-line dealer. That is, we construct a protocol in which the dealer sends only one message to each party in an initialization stage; the parties interact in rounds using a broadcast channel (without the dealer) and in each round i each party learns its shares of the ith round inner secret-sharing schemes. In each round i, each party p_j learns a share of σ_L^i in a $|L|$-out-of-$|L|$ secret-sharing scheme, for every set Q_L such that $j \in L$ and $m-t \leq |L| \leq t$ (that is, it learns the share of the inner scheme). For this purpose, the dealer computes, in a preprocessing phase, the appropriate shares for the inner secret-sharing scheme. For each round, the shares of each party p_j are then shared in a 2-out-of-2 secret-sharing scheme, where p_j gets one of the two shares (this share is a mask, enabling p_j to privately reconstruct its shares of the appropriate σ_L^i although messages are sent on a broadcast channel). All other parties get shares in a t-out-of-$(m-1)$ Shamir secret-sharing scheme of the other share of the 2-out-of-2 secret-sharing. We call the resulting secret-sharing scheme the *outer* $(t+1)$-out-of-m scheme (since t parties and the holder of the mask are needed to reconstruct the secret).

To prevent corrupt parties from cheating, by say, sending false shares and causing reconstruction of wrong secrets, every message that a party should send during the execution of the protocol is signed in the preprocessing phase (together with the appropriate round number and with the party's index). In addition, the dealer sends a verification key to each of the parties. To conclude, the off-line dealer gives each party the signed shares for the outer secret sharing scheme together with the verification key.

The protocol with the off-line dealer proceeds in rounds. In round i of the protocol, all parties broadcast their (signed) shares in the outer $(t+1)$-out-of-m secret-sharing scheme. Thereafter, each party can unmask the message it receives (with its share in the appropriate 2-out-of-2 secret-sharing scheme) to obtain its shares in the $|L|$-out-of-$|L|$ inner secret-sharing of the values σ_L^i (for the appropriate sets Q_L's to which the party belongs). If a party stops broadcasting messages or broadcasts improperly signs messages, then all other parties consider it as aborted. If $m-t$ or more parties abort, the remaining parties reconstruct the value of the set that contains all of them, i.e., σ_L^{i-1}. If the premature termination occurs in the first round, then the remaining active parties engage in a fully secure protocol (with honest majority) to compute f_n.

The use of the outer secret-sharing scheme with threshold $t+1$ plays a crucial role in eliminating the on-line dealer. On the one hand, it guarantees that an adversary, corrupting at most t parties, cannot reconstruct the shares of round i before round i. On the other hand, at least $m-t$ parties must abort to prevent the reconstruction of the outer secret-sharing scheme (this is why we cannot proceed after $m-t$ parties aborted). Furthermore, since $t \leq 2m/3$, when at least $m-t$ corrupt parties aborted, there is an honest majority. To see this, assume that at least $m-t$ corrupt parties aborted. Thus, at most $t-(m-t)=2t-m$ corrupt parties are active. There are $m-t$ honest parties (which are obviously active), therefore, as $2t-m < m-t$ (since $t < 2m/3$), an honest majority is achieved when at least $m-t$ parties abort. In this case we can execute a protocol with full security for the reconstruction.

Finally, we replace the off-line dealer by using a secure-with-abort and cheat-detection protocol computing the functionality computed by the dealer. This is done similarly to the preprocessing phase in [3], which in turn use the results of [24, 2]. Obtaining the outputs of this computation, an adversary is unable to infer any information regarding the input of honest parties or the output of the protocol (since it gets t shares of a $(t+1)$-out-of-m secret-sharing scheme). The adversary, however, can prevent the execution, at the price of at least one corrupt party being detected cheating by all other parties. In such an event, the remaining parties will start over without the detected cheating party. This goes on either until the protocol succeeds or there is an honest majority and a fully secure protocol computing f_n is executed.

Comparison with the multiparty coin-tossing protocol of [3]. Our protocol combines ideas from the protocols of [18, 3]. However, there are some important differences between our protocol and the protocol of [3]. In the coin-tossing protocol of [3], the bits σ_L^i are shared using a threshold scheme where the threshold is smaller than the size of the set Q_L. This means that a proper subset of Q_L containing corrupt parties can reconstruct σ_L^i. In coin-tossing this is not a problem since there are no inputs. However, when computing functionalities with inputs, such σ_L^i might reveal information on the inputs of honest parties in Q_L, and we share σ_L^i with threshold $|Q_L|$. As a result, we use more sets Q_L than in [3] and the bias of the protocol is increased (put differently, to keep the same security, we need to increase the number of rounds in the protocol). For example, the protocol of [3] has small bias when there are polynomially many parties and $t = m/2$. Our protocol is efficient only when there are constant number of parties. As explained in Section 7, this difference is inherent as a protocol for general functionalities with polynomially many parties and $t = m/2$ cannot have a small bias.

4.3 A $1/p$-Secure Protocol for Polynomial Range

Using an idea of [18], we modify our protocol so that it will have a small bias when the size of the range of the functionality $\mathcal{F}$ is polynomially bounded (even if $\mathcal{F}$ is randomized and has a big domain of inputs). The only modification is the way that each σ_L^i is chosen prior to round $i^\star$: with probability $1/(2p)$ we choose σ_L^i as a random value in the range of f_n and with probability $1 - 1/(2p)$ we choose it as in Case I described in Section 4.1. More formally, in the protocol with the dealer, in the preprocessing phase we replace Case I with the following step:

- For each $i \in \{1, \dots, i^\star - 1\}$ and for each $L \subseteq [m]$ s.t. $m - t \leq |L| \leq t$,
  - with probability $1/(2p)$, the dealer selects uniformly at random $z_L^i \in Z_n$ and sets $\sigma_L^i = z_L^i$.
  - with the remaining probability $1 - 1/(2p)$, the dealer chooses σ_L^i as in Case I described in Section 4.1.

Similar changes are made in the protocol without the dealer.

The idea why this change improves the protocol is that now the probability that all values held by the adversary are equal prior to round $i^\star$ is larger, thus, the probability that the adversary guesses $i^\star$ is smaller. This modification, however, can cause the honest parties to output a value that is not possible given their inputs, and, in general, we cannot simulate the case (which happens with probability $1/(2p)$) when the output is chosen with uniform distribution from the range.

5 The 3-party Protocol Tolerating Two Corrupt Parties

In this section we describe an r-round 3-party protocol tolerating two corrupt parties. Unlike Section 4, we directly describe our protocol without any dealer. The formal description of the 3-party protocol, Protocol $\mathrm{MPCFor3Protocol}_r$, appears in Figure 1 and Figure 2.

We next sketch the ideas of the protocol. As in all our protocols, we construct a protocol with two phases. The first phase is a preliminary phase in which the parties compute a given functionality (securely-with-abort with cheat detection). The output of this functionality for party p_j includes the messages that p_j broadcasts throughout the second phase – called the interaction phase. For simplicity of presentation, in the rest of the paper, we assume that in the interaction phase of the protocol the adversary is a *fail-stop* adversary. That is, all parties follow the protocol with one exception: the corrupt parties may abort the computation at any time. For our protocols, this assumption is without loss of generality, since in each round there is a small number of messages that each party can send. We have already demonstrated how to limit the adversary to aborts in this case by signing (in the preprocessing phase) any such possible message. Using this assumption, we can omit the discussion regarding signing of the messages.

In the preliminary phase of the protocol, for each round i and for each subset $L \subset \{1, 2, 3\}$ of size one or two a value σ_L^i is chosen similarly to the way the values in the protocols in Section 4 are chosen; this value is used by the parties $\{p_j : j \in L\}$ if the other party/parties abort in round $i + 1$. Specifically, there is a special round, called $i^\star$, chosen with uniform distribution from $\{1, \dots, r\}$. Prior to round $i^\star$, the values chosen for each subset depends only on the inputs of the subset: random inputs are chosen for the parties not in the subset and the function f_n is computed with the inputs of the subset and the random inputs for other party/parties. Starting in round $i^\star$, the value of each subset is the output w of f_n on the inputs of all parties.

If no party aborts during the protocol, then each party p_j outputs the value $\sigma_{\{j\}}^r$. If two corrupt parties abort in some round i, then the third party p_j outputs the value $\sigma_{\{j\}}^{i-1}$. The difficult case is when one party, say p_3, aborts in some round i. In this case one of the active parties p_1, p_2 might be corrupt. Thus, the parties execute a variant of the two-party r-round $O(1/r)$-secure protocol of [18] to compute f_n. Specifically, if $i \geq i^\star$, then in each round of the two-party protocol the parties get the value w (thus, an abort of p_3 after round $i^\star$ does not affect the output). If $i < i^\star$, then we would like to execute the following protocol: (1) a new special round $i_{\{1,2\},i}^\star$ is selected with uniform distribution from $\{1, \dots, r\}$, and (2) an r-round protocol is executed, where prior to round $i_{\{1,2\},i}^\star$ each party gets a value that depends only on its input and starting from round $i_{\{1,2\},i}^\star$, each party gets $\sigma_{\{1,2\}}^{i-1}$.

The protocol sketched above is flawed: suppose now that p_1, p_2 are corrupt. In each round i they can simulate the execution of the two-party protocol that they would have executed if p_3 has aborted. The first round i in which *all* the values

Inputs: Each party p_j holds a private input $y_j \in X_n$ and the joint input: the security parameter 1^n and the number of rounds $r = r(n)$.

Computing default values:

1. Set $w \leftarrow f_n(x_1, x_2, x_3)$ and select $i^\star \in \{1, \dots, r\}$ with uniform distribution.
2. For each $1 \leq i < i^\star$ and for each $j \in \{1, 2, 3\}$,
   (a) Set $L = \{1, 2, 3\} \setminus \{j\}$,
   (b) With probability $1/\sqrt{r}$, set $i^\star_{L,i} = 1$. With the remaining probability, select $i^\star_{L,i} \in \{1, \dots, r\}$ with uniform distribution.
   (c) Select uniformly at random $\widehat{z}_j \in X_n$, for each $\ell \in L$ set $\widehat{z}_\ell = x_\ell$, and set $\upsilon^i_L \leftarrow f_n(\widehat{z}_1, \widehat{z}_2, \widehat{z}_3)$.
   (d) For each $1 \leq i_2 < i^\star_{L,i}$ and for each $\ell \in L$,
      i. Set $\widehat{z}_\ell = x_\ell$ and for each $j \in \{1, 2, 3\} \setminus \{j\}$, select $\widehat{z}_j$ with uniform distribution from X_n.
      ii. Set $\sigma^i_{L,\ell,i_2} \leftarrow f_n(\widehat{z}_1, \widehat{z}_2, \widehat{z}_3)$.
   (e) For each $i^\star_{L,i} \leq i_2 \leq r$ and for each $\ell \in \{1, 2, 3\} \setminus \{j\}$, set $\sigma^i_{L,\ell,i_2} = \sigma^i_L$.
3. For each $i^\star \leq i < r$, for each $j \in \{1, 2, 3\}$, for each $1 \leq i_2 \leq r$, set $L = \{1, 2, 3\} \setminus \{j\}$ and for each $\ell \in L$ set $\sigma^i_{L,\ell,i_2} = w$.
4. For each $1 \leq i < i^\star$ and for each $j \in \{1, 2, 3\}$,
   (a) Set $\widehat{x}_j = x_j$ and for each $\ell \in \{1, 2, 3\} \setminus \{j\}$, select uniformly at random $\widehat{x}_\ell \in X_n$.
   (b) Set $\sigma^i_j \leftarrow f_n(\widehat{x}_1, \widehat{x}_2, \widehat{x}_3)$.
5. For each $i^\star \leq i \leq r$ and for each $j \in \{1, 2, 3\}$, set $\sigma^i_j = w$.

Computing messages:

1. For each $1 \leq i \leq r$, each $1 \leq i_2 \leq r$, each $L \subset \{1, 2, 3\}$ s.t. $|L| = 2$, and each $\ell \in L$, share σ^i_{L,ℓ,i_2} in a $|2|$-out-of-$|2|$ secret-sharing scheme for the parties $\{p_j : j \in L\}$.
   For each $j \in L$, let $S^i_{L,\ell,i_2,j}$ be the share of p_j of the secret σ^i_{L,ℓ,i_2}.
2. For each $1 \leq i \leq r$ and for each $j \in \{1, 2, 3\}$,
   (a) Set $m_{i,j} = \sigma^i_j \circ \left(S^i_{\{j,\ell\},\ell,i_2,j}\right)_{1 \leq i_2 \leq r; \ell \in \{1,2,3\} \setminus \{j\}}$.
   (b) Share $m_{i,j}$ in a $|3|$-out-of-$|3|$ secret-scheme. For each $\ell \in \{1, 2, 3\}$, let $S_{i,j,\ell}$ be the share of p_ℓ of the secret $m_{i,j}$.
   (c) For each $j \in \{1, 2, 3\}$ compute $M_{i,j} \leftarrow (S_{i,j,\ell})_{\ell \in \{1,2,3\} \setminus \{j\}}$.

Outputs: Each party p_j receives
- The messages $M_{1,j}, \dots, M_{r,j}$ that p_j broadcasts during the protocol.
- p_j's shares $S_{1,j,j}, \dots, S_{r,j,j}$ for reconstructing the values for the three party protocol.
- p_j's shares $S^i_{L,j,i_2,j}$ for each $1 \leq i \leq r$, each $1 \leq i_2 \leq r$, each $L \subset \{1, 2, 3\}$ s.t. $|L| = 2$, and $j \in L$ for reconstructing the values in the two-party protocol for $\{p_\ell : \ell \in L\}$.

Fig. 1. Functionality ShareGenFor3$_r$

Inputs: Each party p_j holds the private input $y_j \in X_n$ and the joint input: the security parameter 1^n and the number of rounds in the protocol $r = r(n)$.

Preliminary phase:

1. The parties execute a secure-with-abort and cheat-detection protocol computing Functionality ShareGenFor3_r. Each honest party p_j inputs y_j as its input for the functionality.
2. If an abort has occurred, that is, the output of the honest parties is "$\texttt{abort}_j$" for at least one index j, then, for each such index j, the remaining active parties mark p_j as inactive, i.e., set $D = D \cup \{j\}$ and execute the instructions for one or two active parties with $i = 1$.
3. Else (no party has aborted), denote $D = \emptyset$ and proceed to the first round.

Instructions for three active parties:

`In each round` $i = 1, \ldots, r$ `do:`

(a) Each party p_j broadcasts $M_{i,j}$.

(b) For every p_j that aborts, all parties mark p_j as inactive and the active parties execute the instructions for one or two active parties.

`At the end of round` r`:` Each active party p_j reconstructs the value σ_j^r, outputs it, and halts.

Instructions for two active parties indexed by L:

(a) If $i = 1$, then execute the two party protocol of [18] for the functionality $f_n(\cdot, \cdot, \cdot)$ in which, the input for the aborted party is selected uniformly at random from X_n and halt.

(b) Else, each active party p_j reconstructs $m_{i-1,j}$ and for each round $i_2 = 1, \ldots, r$ in the two-party protocol:

  i. Each active party p_j where $j \in L$ broadcasts the share $S^{i-1}_{L,\ell,i_2,j}$ where p_ℓ is the other active party.

  ii. If p_j aborts, then the remaining active party p_ℓ marks p_j as inactive and

   – If $i_2 > 1$, i.e., at least one round of the two-party protocol was competed, then, p_ℓ reconstructs $\sigma^{i-1}_{L,\ell,i_2-1}$, outputs it, and halts.

   – Else, if no round of the two-party protocol was competed, then, the active party p_ℓ reconstructs the value σ_ℓ^{i-1} from the three party protocol, outputs it, and halts.

(c) At the end of round r: Each active party p_j reconstructs the value $\sigma^{i-1}_{L,j,r}$, outputs it, and halts.

Instructions for one active party p_ℓ:

(a) Set $\widehat{x}_\ell = y_\ell$ and for every $j \in D$, select $\widehat{x}_j$ with uniform distribution from X_n.

(b) Set $w \leftarrow f_n(\widehat{x}_1, \widehat{x}_2, \widehat{x}_3)$, output w, and halt.

Fig. 2. The 3-party protocol MPCFor3_r for computing $\mathcal{F}$

they get in the simulated protocol are equal is $i^\star$. Thus, they can determine $i^\star$ and bias the output of the protocol with a high probability. To overcome this problem, we modify the way that $i^\star_{L,i}$ is chosen prior to round i: with probability $O(1/\sqrt{r})$ set $i^\star_{L,i} = 1$, and with the remaining probability choose it at random from $\{1, \ldots, r\}$. Notice that the simulated protocol in the case $i^\star_{L,i} - 1$ looks like the simulated protocols in rounds starting from $i^\star$, thus, the probability that the corrupted parties guess $i^\star$ is $O(\sqrt{r}/r) = O(1/\sqrt{r})$. However, a corrupt p_2 can bias the protocol by guessing $i^\star_{L,i} = 1$ and aborting in round 1 of the two-party protocol. This can cause an additional bias of at most $O(1/\sqrt{r})$. All together, the resulting protocol is $O(1/\sqrt{r})$-secure.

We next explain how the two-party protocol is executed. The two-party protocol of [18] has, again, two stages: a preliminary stage and an interaction phase. In our protocol, we have only one preliminary stage, in which all preliminary phases of the two-party protocols are executed simultaneously. That is, in the preliminary phase, for every round $1 \leq i \leq r$ and for every $L \subset \{1, 2, 3\}$ of size two, the preliminary phase of the two-party protocol of [18] is executed for L (using σ^i_L and $i^\star_{L,i}$). Let $(S^i_{L,j})_{j \in L}$ be the two outputs of the preliminary phase that should be given to the parties indexed by L. Each $S^i_{L,j}$ for $j \in L$ is shared using a 3-out-of-3 secret sharing scheme. The output of the preliminary phase of each party includes exactly one of these shares.

Later, in each interaction round i, for each $L \subset \{1, 2, 3\}$ of size two and for each $j \in L$, the parties p_k, where $k \neq j$, broadcast their shares of $S^i_{L,j}$. Thus, p_j obtains $S^i_{L,j}$ while the other two parties learn nothing on it. Now, if a party, say p_3, aborts in round i, parties p_1 and p_2 can execute the two party protocol of round $i - 1$ using $S^{i-1}_{\{1,2\},1}$ and $S^{i-1}_{\{1,2\},2}$ respectively.

In the above, we only sketched the protocols. The formal description of the functionality computed by the preliminary phase appears in Figure 1 and the protocol appears in Figure 2. The proof that the protocol is $1/p$-secure will appear in the full version of the paper. To construct a 3-party protocol for functionalities where the size of range is small we use the same trick used in Section 4.3: With some small probability a value given to a set is chosen from the range prior to $i^\star$ in the 3-party interaction and prior to $i^\star_{L,i}$ in the two parties' protocols. The m-party protocols tolerating up to $m - 1$ corrupt parties, uses the same ideas as our 3-party protocols. In a preliminary phase, $i^\star$ and values σ^i_L are chosen as above. If one party aborts in some round i, then the remaining $m - 1$ parties execute our $(m - 1)$-party protocol, where if $i \geq i^\star$ then it uses $i^\star_{L,i} = 1$, and if $i < i^\star$ then $i^\star_{L,i} = 1$ with some probability and $i^\star_{L,i}$ is random otherwise. In this $(m - 1)$-party protocol, if a party aborts the remaining $m - 2$ parties execute our $(m - 2)$-party protocol (again with its special round being set to 1 with some probability), and so on.

6 Best of Both Worlds – The $1/p$ Way

We study the question of whether or not it is possible to construct "best of both worlds" protocols, when the fall-back security guarantee is $1/p$-security or $1/p$-security-with-abort. We investigate whether protocols with these weaker notions of security are possible when full privacy cannot be guaranteed. In the full version of the paper, we construct protocols that guarantee full-security whenever less than half of the parties are corrupt, and $1/p$-security-with-abort otherwise. These protocols are simpler and more efficient than the protocols that guarantee fall-back $1/p$-(full)-security, which we describe below.

To construct the protocols that have fall-back $1/p$-(full)-security, we show in Section 6.1 how to transform $1/p$-secure protocols of a certain type into protocols that retain the same security for the case of no honest majority, while guaranteeing full-security whenever less than half of the parties executing the protocol are corrupt. Specifically, we will prove the following theorem.

Theorem 3. *Let $\mathcal{F}$ be an m-party (possibly randomized) functionality. If enhanced trap-door permutations exist, and if m is constant and the size of the domain $g(n)$ and the size of the range $g(n)$ are bounded by a polynomial in the security parameter n, then for any polynomial $p(n)$ there is an $r(n)$-round $1/p(n)$-secure protocol computing $\mathcal{F}$ tolerating up to $m-1$ corrupt parties and, in addition, guarantees full-security in the presence of an honest majority, where* $r(n) = 2 \cdot p(n)^{2^m} \cdot \left(d(n) \cdot g(n)\right)^{2^{O(m)}}$.

A similar theorem will appear in the full version of the paper for the result of applying the above transformation to the protocol implying the second item of Theorem 2. The resulting protocol has polynomially many rounds even when the number of parties is $\frac{1}{2}\log\log n$ provided that the functionality is deterministic and the size of the domain of inputs is constant.

6.1 Best of Both Worlds – The $1/p$-(full)-Security Variant

In this section we show how to transform $1/p$-secure protocols of a certain type into protocols that retain the security of the original protocol for the case of no honest majority, while guaranteeing full-security whenever less than half of the parties executing the protocol are corrupted. Intuitively, the transformation works if the original protocol has full security against a weaker adversary that can only abort at the beginning of each round (i.e., before seeing the messages of the honest parties for this round). Specifically, this transformation can be applied to all protocols in this paper that have full correctness (namely, the protocols that assume that the sizes of the domain and the range are polynomial). Note that protocols that do not have full correctness (at least for the case of honest majority) do not guarantee full-security for the case of honest majority. At the end of this section, we will hint why the resulting protocols guarantee the desired security notion. The full argument will appear in the full version of the paper.

The basic structure of protocols that can be transformed. For simplicity of presentation we first present our transformation for an (original) protocol with a certain structure. Consider an m-party protocol for computing a functionality $\mathcal{F}$ that has the following structure: The interaction starts with a preliminary phase in which the parties execute a secure-with-abort with cheat-detection protocol for computing the messages that the parties are to send in the next r interaction rounds; after this phase, each party p_j holds a (signed) message M_j^i for each round $1 \leq i \leq r$. In each interaction round i, each party p_j broadcasts the message M_j^i. Any failure of party p_j to broadcast the signed message as prescribed by the protocol is considered as an abort of p_j. The adversary can cause the protocol to prematurely terminate by instructing some $t_A < \lceil \frac{m}{2} \rceil$ corrupted parties to abort. Unless premature termination takes place, the protocol proceeds normally (that is, as long as less than t_A of parties have aborted). In the case of premature termination, the remaining parties engage in a protocol Π_{TERM} for agreeing on the output of the protocol, based on the view of the parties in the protocol so far. More specifically, the decision upon the output is based on the outputs of the (remaining) parties from the preliminary phase, on the messages broadcast until round $i-1$, and on the set of parties that have aborted D.

Indeed, all our protocols that were described in previous sections have the above structure. For the sake of being concrete, however, in the following we will describe the transformation as applied to the protocol of Section 4.2. In this protocol Π_{TERM} is a protocol for reconstructing the output that is always executed with a guaranteed honest majority.

The transformation. The core of the change is a mask we add to the messages of the parties in each round. This mask is shared in a $(\lfloor \frac{m}{2} \rfloor + 1)$-out-of-$m$ secret-sharing scheme. Hence, the messages of the parties disclose the original messages if and only if a majority of the parties work together to reconstruct the appropriate masks. Below we explain this change in more detail.

Denote by M_j^i the message that party p_j is instructed to broadcast in round i of the original protocol. That is, the output of party p_j from the preliminary phase of the original protocol includes the messages $M_j^1, \ldots, M_j^r$. In the preliminary phase of the new protocol a random string r_j^i will be selected for each party p_j and each round i, and the sequence $\hat{M}_j^1, \ldots, \hat{M}_j^r$ will be given to party p_j, where $\hat{M}_j^i = M_j^i \oplus r_j^i$. In addition, each party will also receive a share of r_j^i in a $(\lfloor \frac{m}{2} \rfloor + 1)$-out-of-$m$ Shamir secret-sharing scheme. The message $\hat{M}_j^i$ and the shares of its mask r_j^i are all signed.

Each interaction round of the original protocol is turned into a two-phased round in the new protocol. In the first phase, each party p_j broadcasts the message $\hat{M}_j^i$. In the second phase, the parties reconstruct all masks of round i by broadcasting all shares of masks r_j^i, for $1 \leq j \leq m$. If both phases are completed, then the parties have the same information as in the original protocol. Any failure of party p_j to broadcast the signed message as prescribed by the protocol is considered as an abort of p_j (including messages added by the transformation).

The adversary can cause the protocol to prematurely terminate only by instructing some $t_A < \lceil \frac{m}{2} \rceil$ corrupted parties to abort. We handle such premature termination in round i by instructing the parties to behave as if premature termination has occurred at the beginning or at the end of round i (i.e., at the beginning of round $i+1$). Specifically, if premature termination takes place before the reconstruction of the masks (in the second phase of round i) is completed, then the remaining parties will behave as if the original protocol was terminated at the beginning of round i. That is, they will engage in a protocol Π_{TERM} for agreeing on the output of the protocol, based on the messages broadcast until round $i-1$ and on the set of parties that have aborted D. Otherwise, if the reconstruction of the masks was completed before the abort, then the remaining parties will behave as if the original protocol was terminated at the beginning of round $i+1$.

The security of the new protocol. In the full version of this paper, we argue that applying the above transformation to any of the our protocols that assume that the domain and the range are polynomial, results in a protocol that is (i) fully secure against a malicious adversary that can corrupt any strict minority of the parties, and (ii) $1/p$-secure against a malicious adversary that can corrupt up to t parties. Furthermore, we will show that this is true for any protocol that has the structure defined above and, in addition, satisfies a few simple requirements.

We now give some intuition for why this is true if the transformation is applied to the protocol of Section 4.2. We need to consider two cases. In the case that at list half of the parties are malicious, it is quite straightforward to see that the adversary attacking the transformed protocol is not any more powerful than an adversary for the original protocol, since once the adversary sees the messages of the corrupted parties, the masks add no new information.

In the case of an honest majority, the shares of r_j^i that the corrupted parties see, do not reveal anything to the adversary as long as the shares of honest parties are not revealed (these shares are only revealed in the second phase of round i). Thus, if the adversary causes a premature termination during the first phase of round i, then it has no more information than is obtained in the original protocol (by an adversary corrupting the same subset of parties) until the beginning of round i. If it aborts after the first phase, then the honest parties will succeed in reconstructing the masks. Thus, the adversary is no more powerful then an adversary for the original protocol that can only abort at the beginning of each round. However, the security of the original protocol can only be violated if the adversary causes premature termination during round $i^{\star}$. Finally, the reconstruction is fully secure in the presence of an honest majority.

7 Impossibility of $1/p$-secure Computation with Non-constant Number of Parties

For deterministic functions, our protocols are efficient when the number of parties m is constant and the size of the domain or range is at most polynomial (in the security parameter n) or when the number of parties is $\log\log n$ and the size of the domain is constant. We show that, in general, there is no efficient protocol when the number of parties is $m(n) = \omega(1)$ and the size of the domain is polynomial and when $m(n) = \omega(\log n)$ and the size of the domain of each party is 2. That is, we prove the following two theorems.

Theorem 4. *For every $m(n) = \omega(\log n)$, there exists a deterministic $m(n)$-party functionality $\mathcal{F}'$ with domain $\{0,1\}$ that cannot be $1/p$-securely computed for $p \geq 2 + 1/\operatorname{poly}(n)$ without an honest majority.*

Theorem 5. *For every $m(n) = \omega(1)$, there exists a deterministic $m(n)$-party functionality $\mathcal{F}''$ with domain $\{0,1\}^{\log n}$ that cannot be $1/p$-securely computed for $p \geq 2 + 1/\operatorname{poly}(n)$ without an honest majority.*

7.1 Impossibility of Achieving "The Best of Both Worlds" for General Functionalities

Above we showed that $1/p$-secure computation is impossible in general when the number of parties is $m(n) = \omega(1)$ and the size of the domain is polynomial and when $m(n) = \omega(\log n)$ and the size of the domain of each party is 2. Since a "Best of Both Worlds" type protocol with fall-back $1/p$-security is in particular $1/p$-secure, the same impossibility results are implied for protocols of this type (i.e., guaranteeing full-security with an honest majority and $1/p$-security otherwise). We show that such protocols are impossible in general, even when allowing the fall-back security to be the weaker notion of $1/p$-security-with-abort. Hence, we show that the results discussed in Section 6 are somewhat optimal.

We start by showing in that for general functionalities (i.e., where both domains and both ranges may be super-polynomial), it is impossible to construct even 3-party protocols that simultaneously achieve full-security for the case of honest majority (i.e., at most one corrupted party) and $1/p$-security-with-abort with no honest majority. We then use this result to prove general impossibility results, that is, to prove the two following theorems:

Theorem 6. *For every $m(n) = \omega(\log n)$, there exists a deterministic $m(n)$-party functionality $\mathcal{F}'$ with domain $\{0,1\}$ that cannot be computed simultaneously guaranteeing full-security with an honest majority and $1/p$-security-with-abort for $p \geq 2 + 1/\operatorname{poly}(n)$ against an adversary controlling $\lfloor m(n)/2 \rfloor + 1$ parties.*

Theorem 7. *For every $m(n) = \omega(1)$, there exists a deterministic $m(n)$-party functionality $\mathcal{F}''$ with domain $\{0,1\}^{\log n}$ that cannot be computed simultaneously guaranteeing full-security with an honest majority and $1/p$-security-with-abort for $p \geq 2 + 1/\operatorname{poly}(n)$ against an adversary controlling $\lfloor m(n)/2 \rfloor + 1$ parties.*

References

[1] Beaver, D., Goldwasser, S.: Multiparty computation with faulty majority. In: 30th FOCS, pp. 468–473 (1989)

[2] Beaver, D., Micali, S., Rogaway, P.: The round complexity of secure protocols. In: 22nd STOC, pp. 503–513 (1990)

[3] Beimel, A., Omri, E., Orlov, I.: Protocols for multiparty coin toss with dishonest majority. In: Rabin, T. (ed.) CRYPTO 2010. LNCS, vol. 6223, pp. 538–557. Springer, Heidelberg (2010)

[4] Ben-Or, M., Goldreich, O., Micali, S., Rivest, R.: A fair protocol for signing contracts. In: 12th ICALP, pp. 43–52 (1985)

[5] Blum, M.: How to exchange (secret) keys. ACM Trans. Comput. Syst. 1(2), 175–193 (1983)
[6] Boneh, D., Naor, M.: Timed commitments. In: Bellare, M. (ed.) CRYPTO 2000. LNCS, vol. 1880, pp. 236–254. Springer, Heidelberg (2000)
[7] Cleve, R.: Limits on the security of coin flips when half the processors are faulty. In: 18th STOC, pp. 364–369 (1986)
[8] Cleve, R.: Controlled gradual disclosure schemes for random bits and their applications. In: Brassard, G. (ed.) CRYPTO 1989. LNCS, vol. 435, pp. 573–588. Springer, Heidelberg (1990)
[9] Damgård, I.: Practical and provably secure release of a secret and exchange of signatures. J. of Cryptology 8(4), 201–222 (1995)
[10] Even, S., Goldreich, O., Lempel, A.: A randomized protocol for signing contracts. CACM 28(6), 637–647 (1985)
[11] Galil, Z., Haber, S., Yung, M.: Cryptographic computation: Secure fault tolerant protocols and the public-key model. In: Pomerance, C. (ed.) CRYPTO 1987. LNCS, vol. 293, pp. 135–155. Springer, Heidelberg (1988)
[12] Garay, J.A., MacKenzie, P.D., Prabhakaran, M., Yang, K.: Resource fairness and composability of cryptographic protocols. In: Halevi, S., Rabin, T. (eds.) TCC 2006. LNCS, vol. 3876, pp. 404–428. Springer, Heidelberg (2006)
[13] Goldreich, O., Micali, S., Wigderson, A.: How to play any mental game. In: 19th STOC, pp. 218–229 (1987)
[14] Goldwasser, S., Levin, L.: Fair computation of general functions in presence of immoral majority. In: Menezes, A., Vanstone, S.A. (eds.) CRYPTO 1990. LNCS, vol. 537, pp. 77–93. Springer, Heidelberg (1991)
[15] Gordon, S.D., Hazay, C., Katz, J., Lindell, Y.: Complete fairness in secure two-party computation. In: 40th STOC, pp. 413–422 (2008)
[16] Gordon, D. S., Ishai, Y., Moran, T., Ostrovsky, R., Sahai, A.: On complete primitives for fairness. In: Micciancio, D. (ed.) TCC 2010. LNCS, vol. 5978, pp. 91–108. Springer, Heidelberg (2010)
[17] Gordon, S.D., Katz, J.: Complete fairness in multi-party computation without an honest majority. In: Reingold, O. (ed.) TCC 2009. LNCS, vol. 5444, pp. 19–35. Springer, Heidelberg (2009)
[18] Gordon, S.D., Katz, J.: Partial fairness in secure two-party computation. In: Gilbert, H. (ed.) EUROCRYPT 2010. LNCS, vol. 6110, pp. 157–176. Springer, Heidelberg (2010)
[19] Ishai, Y., Katz, J., Kushilevitz, E., Lindell, Y., Petrank, E.: On achieving the "best of both world" in secure multiparty computation. SIAM J. on Computing 40(1) (2011) (Journal version of [20, 21])
[20] Ishai, Y., Kushilevitz, E., Lindell, Y., Petrank, E.: On combining privacy with guaranteed output delivery in secure multiparty computation. In: Dwork, C. (ed.) CRYPTO 2006. LNCS, vol. 4117, pp. 483–500. Springer, Heidelberg (2006)
[21] Katz, J.: On achieving the "best of both worlds" in secure multiparty computation. In: 39th STOC, pp. 11–20 (2007)
[22] Luby, M., Micali, S., Rackoff, C.: How to simultaneously exchange a secret bit by flipping a symmetrically-biased coin. In: 24th FOCS, pp. 11–21 (1983)
[23] Moran, T., Naor, M., Segev, G.: An optimally fair coin toss. In: Reingold, O. (ed.) TCC 2009. LNCS, vol. 5444, pp. 1–18. Springer, Heidelberg (2009)
[24] Pass, R.: Bounded-concurrent secure multi-party computation with a dishonest majority. In: 36th STOC, pp. 232–241 (2004)
[25] Pinkas, B.: Fair secure two-party computation. In: Biham, E. (ed.) EUROCRYPT 2003. LNCS, vol. 2656, pp. 87–105. Springer, Heidelberg (2003)
[26] Yao, A.C.: How to generate and exchange secrets. In: 27th FOCS, pp. 162–167 (1986)

Leakage-Resilient Zero Knowledge

Sanjam Garg, Abhishek Jain*, and Amit Sahai*

UCLA
{sanjamg,abhishek,sahai}@cs.ucla.edu

Abstract. In this paper, we initiate a study of zero knowledge proof systems in the presence of side-channel attacks. Specifically, we consider a setting where a cheating verifier is allowed to obtain arbitrary bounded leakage on the *entire state* (including the witness and the random coins) of the prover *during the entire protocol execution*. We formalize a meaningful definition of *leakage-resilient zero knowledge* (LR-ZK) proof system, that intuitively guarantees that *the protocol does not yield anything beyond the validity of the statement and the leakage obtained by the verifier.*

We give a construction of LR-ZK interactive proof system based on standard general assumptions. To the best of our knowledge, this is the first instance of a cryptographic *interactive protocol* where the adversary is allowed to perform leakage attacks during the protocol execution on the *entire state* of honest party (in contrast, prior work only considered leakage *prior* to the protocol execution, or very limited leakage *during* the protocol execution). Next, we give an LR-NIZK proof system based on standard number-theoretic assumptions.

Finally, we demonstrate the usefulness of our notions by giving two concrete applications:

- We initiate a new line of research to relax the assumption on the "tamper-proofness" of hardware tokens used in the design of various cryptographic protocols. In particular, we give a construction of a universally composable multiparty computation protocol in the *leaky token model* (where an adversary in possession of a token is allowed to obtain arbitrary bounded leakage on the *entire state* of the token) based on standard general assumptions.
- Next, we give simple, generic constructions of *fully* leakage-resilient signatures in the bounded leakage model as well as the continual leakage model. Unlike the recent constructions of such schemes, we also obtain security in the "noisy leakage" model.

1 Introduction

Zero knowledge proof systems, introduced in the seminal work of Goldwasser, Micali and Rackoff [31], have proven fundamental to cryptography. Very briefly, a zero knowledge proof system is an interactive proof between two parties – a

* Research supported in part from a DARPA/ONR PROCEED award, NSF grants 0916574 and 0830803, a Xerox Foundation Award, a Google Faculty Research Award, an equipment grant from Intel, and an Okawa Foundation Research Grant.

P. Rogaway (Ed.): CRYPTO 2011, LNCS 6841, pp. 297–315, 2011.

prover, and a verifier – with the remarkable property that the verifier does not learn anything beyond the validity of the statement being proved. Subsequent to their introduction, zero knowledge proofs have been studied in various adversarial settings such as concurrency attacks [23], malleability attacks [22], to list a few, with very successful results. Over the years, zero knowledge proofs (and its various strengthened notions) have turned to be extremely useful, finding numerous applications in the design of various cryptographic protocols.

We note that the standard definition of zero knowledge proofs, like most classical security notions, assumes that an adversary is given only black-box access to the honest party algorithms. Unfortunately, over the last two decades, it has become increasingly evident that such an assumption may be unrealistic when arguing security in the real world where the physical implementation (e.g. on a smart card or a hardware token) of an algorithm is under attack. Motivated by such a scenario, in this paper, we initiate a study of zero knowledge proof systems in the presence of *side-channel attacks* [46,5,61,28,37]. Specifically, we study zero knowledge proofs in the intriguing setting where a cheating verifier, in addition to receiving a proof of some statement, is able to obtain arbitrary bounded leakage on the *entire state* (*including the witness and the random coins*) of the prover *during the entire protocol execution*. We note that while there has been an extensive amount of research work on leakage-resilient cryptography in the past few years, to the best of our knowledge, almost all prior work has either been on leakage resilient *primitives* such as encryption and signature schemes [24,2,59,21,4,56,44,18,25,3,45,10,19,20,48,51,9,47], or leakage-resilient (and tamper-resilient) *devices* [41,40,26,1], while very limited effort has been dedicated towards constructing leakage-resilient *interactive protocols*. To the best of our knowledge, the recent works on correlation extractors [39], and leakage-resilient identification and authenticated key agreement protocols [4,20,19] come closest to being considered in the latter category. However, we stress that in all these works, either leakage attacks are allowed only *prior* to the protocol execution, or very limited leakage is allowed *during* the protocol execution; in contrast, we consider the setting where the adversary can obtain leakage on the *entire state* of the honest party during the protocol execution.

We find it imperative to stress that handling leakage attacks on interactive protocols can be particularly challenging. On the one hand, for the leakage attacks to be meaningful, we would want to allow leakage on the secret state of the protocol participants. However, the state of a party typically includes a secret value (witness and random coins of the prover in the case of zero knowledge proofs) and any leakage on that secret value might immediately violate a security property (e.g., the zero knowledge property) of the protocol. Then, coming back to setting of zero knowledge proofs, it is not immediately clear how to even define "leakage-resilient zero knowledge."

How to define Leakage-Resilient Zero Knowledge? One possibility is to pursue an assumption such as *only computation leaks information* [53] (i.e., assuming that there is no leakage in the absence of computation). While this is a valuable and interesting approach, we note that this assumption is often problematic (e.g. cold-boot attacks [37]). In our work here, therefore, we do *not* make any such assumption. We seek a general definition maximizing the potential applicability of that definition to different application scenarios.

Another possibility is to allow a "leakage-free pre-processing phase" prior to the actual protocol execution, in an attempt to render the leakage attacks during the protocol useless. We note, however, that allowing pre-processing would limit the applicability of our notion. In particular, such a definition would be problematic for scenarios where the statement to be proven is generated "online" (thereby eliminating the possibility of pre-processing the witness "safely"). Furthermore, we give strong evidence that such an approach is unlikely to yield better guarantees than what we are able to achieve (see the full version for further discussion on this issue).

Indeed, our goal is to obtain a meaningful and appropriate definition of zero knowledge in the model where an adversarial verifier can obtain leakage on any content (state) of the prover machine at any time. We do not consider any "leakage-free" time-period; in particular, any pre-processing phase is subject to leakage as well. However, in such a setting, it is important to note that since the adversary could simply choose to leak on the *witness* (and no other prover state), the zero knowledge *simulator* must be able to obtain similar amount of leakage in order to perform correct simulation. We shall see that even with this limitation, our notion turns out to be both quite nontrivial to obtain and very useful in application scenarios.

Our Definition – Informally. To this end, we consider a definition of leakage-resilient zero knowledge that provides the intuitive guarantee that *the protocol does not yield anything beyond the validity of the statement and the leakage obtained by the adversary.* In other words, whatever an adversary "learns" from the protocol (with leakage) should be no more than what she can learn from only the leakage without running the protocol. To formalize the above intuition, as a first step, we consider a *leakage oracle* that gets as private input the witness of the honest prover; the zero knowledge simulator is then given access to such a leakage oracle. More concretely, we consider a parameter λ, and say that an interactive proof system is λ*-leakage-resilient zero knowledge* (LR-ZK) if for every cheating verifier, there exists a simulator with access to a leakage oracle (that gets the honest prover's witness as private input) that outputs a view of the verifier (indistinguishable from the real execution), with the following requirement. Let ℓ bits be an upper bound on the total amount of leakage obtained by the adversarial verifier. Then the simulator is allowed to obtain at most $\lambda \cdot \ell$ bits of leakage. (In the full version, we show that constructing an LR-ZK proof system with $\lambda < 1$ is in fact impossible.)

Applications of Our Definition. Now that we have a definition for LR-ZK proof system, one may question how meaningful it is. As we now discuss, the above definition indeed turns out to be very useful. Intuitively, our definition is appropriate for a scenario where a leakage-resilient primitive A is being used in conjunction with a zero knowledge proof system (where the proof system is used to prove some statement about A), in the design of another cryptographic protocol B. The reason for this is that our definition of LR-ZK allows us to directly reduce the leakage-resilience property of B on the leakage-resilience property of A.

As an application of our LR-ZK interactive proof system, we first construct a universally composable (UC) multiparty computation protocol in the *leaky token model* (which is a relaxation of the model of Katz [43] in that a malicious token user is now allowed to leak arbitrary bounded information on the *entire state* of the token). Very briefly, we use *leakage-resilient hard relations* [20] and hardware tokens that implement the prover algorithm of our LR-ZK proof system where we prove the validity of an instance of the hard relation; then the leakage on the state of the token can be easily "reduced" to leakage on (the witness corresponding to) an instance of the hard relation.

Next, we are able to extend the notion of LR-ZK to the non-interactive setting in a natural way. Then, as an application of LR-NIZKs, we give generic constructions of *fully* leakage-resilient (FLR) signature schemes (where leakage is allowed on the *entire state* as opposed to only the secret key). Very briefly, we use leakage-resilient hard relations in conjunction with "simulation-extractable" LR-NIZKs (see below); we are then able to reduce the leakage-resilience property of the signature scheme to that of the hard relation. We now summarize our results.

1.1 Our Results

We first study the possibility of constructing leakage-resilient zero knowledge protocols and obtain the following results:

- We construct a $(1+\epsilon)$-leakage-resilient zero knowledge interactive proof system (where ϵ is any positive constant) based on standard general assumptions (specifically, the existence of a statistically hiding commitment scheme that is public-coin w.r.t. the receiver). To the best of our knowledge, this is the first instance of a cryptographic *interactive protocol* where an adversary is allowed to obtain arbitrary bounded leakage on the *entire state* of the honest parties *during* the protocol execution.
- Next, we consider the non-interactive setting and show that any NIZK proof system with *honest prover state reconstruction* property [36] is an LR-NIZK proof system for $\lambda = 1$. As a corollary, we obtain an LR-NIZK proof system from [36] based on the decisional linear assumption.

We supplement our above positive results by proving the impossibility of constructing an LR-ZK proof (or argument) system for $\lambda < 1$. Then, as applications of leakage-resilient zero knowledge, we obtain the following results:

- We initiate a new line of research to relax the assumption on the "tamper-proofness" of hardware tokens used in the design of various cryptographic protocols. In particular, assuming semi-honest oblivious transfer, we give a construction of a universally composable (UC) multiparty computation protocol in the *leaky token model*, where the token user is allowed to obtain arbitrary bounded leakage on the *entire state* of the token. We stress that all prior works on designing cryptographic protocols using hardware tokens, including the work on UC secure computation [43,14,54,15], made the implicit assumption that the tokens are completely leakage-resilient.

- Next, we extend the notion of leakage-resilient NIZKs to incorporate the property of *simulation-extractability* [63,17] (also see [58] in the context of interactive proofs), in particular, the "true" variant [20]. We are then able to adapt the approach of Katz and Vaikuntanathan [44], and in particular, Dodis et al [20,19] (who use a leakage-resilient hard relation in conjunction with a true simulation-extractable NIZK argument system to construct leakage-resilient signatures) to the setting of *full leakage*. As a result, we obtain simple, generic constructions of *fully* leakage resilient signature schemes in the bounded leakage model as well as the continual leakage model. Similar to [20,19], our signature scheme inherits the leakage-resilience properties (and the leakage bounds) of the hard relation used in its construction.[1] In contrast to the recent constructions of FLR signature schemes by [51,9,47] in the standard model[2], our scheme is also secure in the *noisy leakage model* [56]. We supplement our result by showing that a true simulation-extractable leakage-resilient NIZK argument system is implied by the UC-NIZK of Groth et al. [36], which can be based on the decisional linear assumption.

1.2 Our Techniques

We now briefly discuss the main techniques used to obtain our positive results on leakage-resilient zero knowledge proof systems. Recall that our goal is to realize a definition where a cheating verifier does not learn anything from the protocol beyond the validity of the statement and the leakage information obtained from the prover. Further, recall that in our definition, simulator is given access to a leakage oracle that gets the honest prover's witness as private input and accepts leakage queries on the witness string. (In contrast, the verifier is allowed to make leakage queries on the entire state, including the witness and the random coins used by the prover thus far in the protocol execution.) Then, during the simulation, on receiving a leakage query from the verifier, our simulator attempts to convert it into a "valid" query to the leakage oracle. Now, note that the simulator may be cheating in the protocol execution (which is typically the case since it does not possess a valid witness); then, since the verifier can arbitrarily leak on both the witness and the random coins (which completely determine the actions of the prover thus far), at every point in the protocol execution, the simulator must find a way to "explain its actions so far". Note that this is reminiscent of *adaptive security* [7,11,13,50] in the context of secure computation

[1] As such, if we use the key pairs from the encryption scheme of Lewko et al [47] as a hard relation, then our signature scheme can tolerate leakage during the update process as well.

[2] Earlier, FLR signature schemes were constructed either only in the random oracle model [4,20,10], or were only "one-time" [44].

protocols. We stress, however, that adaptive security does not suffice to achieve the property of leakage-resilient zero knowledge in the interactive proofs setting, as we explain below.

Recall that the notion of adaptive security corresponds to the setting where an adversary is allowed to corrupt parties *during* the protocol execution (as opposed to static corruption, where the parties can only be corrupted *before* the protocol begins). Once a party is corrupted, the adversary learns the entire state (including the input and random coins) of that party. The adversary may choose to corrupt several parties (in the case of multi-party protocols) throughout the course of the protocol. The notion of adaptive security guarantees security for the remaining uncorrupted parties.

While adaptive corruption itself is not our focus, note that in our model, a cheating verifier may obtain leakage on the prover's state at *several points* during the protocol execution. Furthermore, the honest prover may not even be aware as to what was leaked. Our goal is to guarantee that the adversary does not learn anything beyond the leaked information. Then, in order to provide such a guarantee, note that our simulator must *continue to simulate the prover even after leakage happens*, in a way that is consistent with the leaked information even though it does not know the prover's witness or what information was leaked. In contrast, the simulator for adaptively secure protocols does *not* need to simulate a party once it is corrupted.[3] In summary, we wish to guarantee some security for the honest party even *after* leakage happens, while adaptive security does not provide any such guarantees. We stress that this difference is crucial, and explains why known techniques for achieving adaptive security do not suffice for our purposes. Nevertheless, as we explain below, adaptive security serves as a good starting point for our purpose.

Recall that the main challenge in the setting of adaptive security is that whenever an adversary chooses to corrupt a party, the simulator must be able to explain its random coins, in a way that is consistent with the party's input and the messages it generated so far in the protocol. The main technique for overcoming this challenge is to allow the simulator to *equivocate*. For our purposes, we will also make use of equivocation so that the leakage queries can be answered correctly by the simulator. However, since our simulator would need to simulate the prover even after leakage happens (without the knowledge of the prover's witness or the information that was leaked), we do not want this equivocation to interfere with the simulation of prover's messages. In other words we want to be able to simulate the prover's messages independent of what information is being leaked but still remain consistent with it. Our solution is to have two separate and independent ways of cheating at the simulator's disposal. It will use one way to cheat in the protocol messages and the second way is reserved for answering the leakage queries correctly. Furthermore, we would need to make sure that the simulator does not "step on its own toes" when using the two ways to cheat *simultaneously*.

[3] Indeed, for this reason, known adaptively secure ZK protocols are not leakage-resilient.

We now briefly discuss the actual construction of our protocol in order to illustrate the above ideas. We recall two well-known ways of constructing constant-round zero knowledge protocols – the Feige-Shamir [27] approach of using equivocal commitments (also used in adaptive security), and the Goldreich-Kahan [29] approach of requiring the verifier to commit to its challenges in advance. Now, armed with the intuition that our simulator will need two separate ways of cheating, we use both the above techniques *together*. Our protocol roughly consists of two phases: in the first phase, the verifier commits to a challenge string using a standard challenge-response based extractable commitment scheme (in a manner similar to [62]); in the second phase, we execute the Blum-Hamiltonicity protocol instantiated with an equivocal commitment scheme. While leakage during the first phase can be handled easily by our simulator, handling leakage during the second phase makes use of the ideas discussed above.

Unfortunately, although the above construction seems to satisfy most of our requirements, it fails on the following account. Recall that our goal is to obtain a leakage-resilient zero knowledge protocol with nearly optimal precision (i.e., $\lambda = 1 + \epsilon$) with respect to the leakage queries of the simulator. Now note that in the above construction, the simulator would need to extract the verifier's challenge in the first phase by means of rewinding before proceeding to phase two of the protocol. Then, depending upon the verifier's behavior, the simulator may need to perform several rewinds in order to succeed in extraction. Now, note that a cheating verifier may be able to make a *different* leakage query during each rewind, thus forcing our simulator to make a new query as well to its leakage oracle. As a result, depending upon the number of such rewinds, the total leakage obtained by the simulator may potentially become a polynomial factor of the leakage obtained by the adversary in a real execution.

In order to obtain a precision in the leakage queries of the simulator, we borrow techniques from the work on *precise zero knowledge* pioneered by Micali and Pass [52]. We remark that in the context of precise ZK, (for fundamental reasons of modeling) it is typically not possible to obtain a precision of almost 1. In our case, however, we are able to achieve a precision of $\lambda = 1 + \epsilon$ (where ϵ is any positive constant) with respect to the leakage queries of the simulator.

Finally, we note that in the case of non-interactive zero knowledge, since the simulator does not need to simulate any "future messages" after the leakage, we are indeed able to show that an adaptively secure NIZK is also a leakage-resilient NIZK. Specifically, we show that any NIZK with *honest prover state reconstruction* property, as defined by Groth et al. [36] (in the context of adaptive security), is also a leakage-resilient NIZK with $\lambda = 1$.

Related Work. In a very recent and exciting concurrent work, Canetti et al. consider a model of leakage in the context of UC protocols [8].

2 Leakage-Resilient Zero Knowledge: Interactive Case

We consider the scenario where a malicious verifier can obtain arbitrary bounded leakage on the entire state (including the witness and the random coins) of the prover during the protocol execution. We wish to give a meaningful definition of zero knowledge interactive proofs in such a setting. To this end, we first modify the standard model for zero knowledge interactive proof system in order to incorporate leakage attacks and then proceed to give our definition.

We model the prover P and the verifier V as interactive turing machines that have the ability to flip coins during the protocol execution (such that the random coins used by a party in any round are determined only at the beginning of that round). In order to incorporate leakage attacks, we allow a malicious verifier V^* to make *adaptive* leakage queries on the state of the prover during the protocol execution. A leakage query to the prover consists of an efficiently computable function f_i (described as a circuit), to which the prover responds with $f_i(\texttt{state})$, where $\texttt{state}$ is a variable that denotes the "current state" of the prover at any point during the protocol execution. The variable $\texttt{state}$ is initialized to the witness of the prover. At the completion of each step of the protocol execution (that corresponds to the prover sending a protocol message to the verifier), the random coins used by the prover in that step are appended to $\texttt{state}$. That is, $\texttt{state} := \texttt{state} \| r_i$, where r_i denote the random coins used by the prover in that step. The verifier may make any arbitrary polynomial number of such leakage queries during the protocol execution. Unlike prior works, we do not require an a-priori bound on the total leakage obtained by the verifier in order to satisfy our definition (described below). Nevertheless, in order for our definition to be meaningful, we note that the total leakage obtained by the verifier must be smaller than the witness size.

We model the zero knowledge simulator $\mathcal{S}$ as a PPT machine that has access to a leakage oracle $L_w^{k,\lambda}(\cdot)$ that is parameterized by the honest prover's witness w, a leakage parameter λ (see below), and the security parameter k. A query to the oracle consists of an efficiently computable function $f(\cdot)$, to which the oracle answers with $f(w)$. In order to bound the total leakage available to the simulator, we consider a parameter λ and require that if the verifier obtains ℓ bits of total leakage in the real execution, then the total leakage obtained by the simulator (from the leakage oracle) must be bounded by $\lambda \cdot \ell$ bits. Finally, we require that the view output by the simulator be computationally indistinguishable from the verifier's view in the real execution. We formalize this in the definition below.

Definition 1 (Leakage-Resilient Zero Knowledge). *An interactive proof system $\langle P, V \rangle$ for a language $\mathcal{L}$ with a witness relation $\mathcal{R}$ is said to be* λ-leakage-resilient zero knowledge *if for every* PPT *machine V^* that makes any arbitrary polynomial number of leakage queries on P's state (in the manner as described above) with ℓ bits of total leakage, there exists a* PPT *algorithm $\mathcal{S}$ that obtains at most $\lambda \cdot \ell$ bits of total leakage from a leakage oracle $L_w^{k,\lambda}(\cdot)$ (as defined above) such that for every $(x, w) \in \mathcal{R}$, every $z \in \{0,1\}^*$, $\text{view}_{V^*}(x, z)$ and $\mathcal{S}^{L_w^{k,\lambda}(\cdot)}(x, z)$ are computationally indistinguishable.*

Some observations on the above definition are in order.

Leakage parameter λ. Note that when $\lambda = 0$, no leakage is available to the simulator (as is the case for the standard zero knowledge simulator). In this case, our definition guarantees the standard zero knowledge property. It is not difficult to see that it is impossible to realize such a definition. In fact, as we show in the full version, it is impossible to realize the above definition for any $\lambda < 1$, where ϵ is any constant less than 1. On the other hand, in Section 2.1, we give a positive result for $\lambda = 1 + \epsilon$, where ϵ is any positive constant. The meaningfulness of our positive result stems from the observation that when λ is close to 1, very roughly, our definition guarantees that *a malicious verifier does not learn anything from the protocol beyond the validity of the statement being proved and the leakage obtained from the prover.*

Leakage-oblivious simulation. Note that in our definition of leakage resilient zero-knowledge, (apart from the total output length) there is no restriction on the nature of leakage queries that the simulator may make to the leakage oracle. Then, since the simulator has indirect access to the honest prover's witness (via the leakage oracle), it may simply choose to leak on the witness (regardless of the leakage queries of the verifier) in order to help with the simulation of protocol messages instead of using the leakage oracle to *only* answer the leakage queries of the verifier. We stress that this issue should not affect any potential application of leakage resilient zero-knowledge that one may think of. Nonetheless, we think that this is an important issue since it relates to the meaningfulness of the definition. To this end, we note that this issue can easily handled by putting a restriction on how the simulator accesses the leakage oracle. Specifically, we can model the interaction between the simulation and the oracle such that the simulator is not allowed to look at the oracle's responses to its queries. The simulator is still allowed to look at the leakage queries of the verifier, and use them to create new queries for the oracle; however, the oracle's responses are sent directly to the verifier and the simulator does not get to seem them. We call such simulators *leakage-oblivious*. We note that the simulator that we construct for our protocol $\langle P, V \rangle$ (described in the next subsection) is leakage-oblivious.[4]

2.1 Our Protocol

We now proceed to give our construction of a leakage-resilient zero knowledge interactive proof system as per Definition 1. Very roughly speaking, our protocol can be seen as a combination of Feige-Shamir [27] and Goldreich-Kahan [29], in that we make use of equivocal commitments from the prover's side, as well as require the verifier to commit to all its challenges in advance. Note that while either of the above techniques would suffice for standard simulation, interestingly, we need to use them *together* to help the simulator handle leakage queries from a cheating verifier. We now describe our protocol in more detail.

Let P and V denote the prover and verifier respectively. Our protocol $\langle P, V \rangle$ proceeds in three stages, described as follows. In *Stage 1*, V commits to its challenge and a large random string r' using a challenge-response based PRS [60] style preamble instantiated with a public-coin statistically hiding commitment

[4] Indeed, since we cannot rule out of obfuscation of arbitrary functionalities, we do not know how to obtain a formal proof without making the simulator leakage-oblivious.

scheme [57,38,16]. In *Stage 2*, P and V engage in coin-flipping (that was initiated in Stage 1 when V committed to r') to jointly compute a random string r. Finally, in *Stage 3*, P and V run k (where k denotes the security parameter) parallel repetitions of the 3-round Blum Hamiltonicity protocol, where P uses Naor's commitment scheme [55] to commit to the permuted graphs in the first round. Here, for each bit commitment i, P uses a different substring r_i (of appropriate length) of r as the first message of Naor's commitment scheme. Protocol $\langle P, V\rangle$ is described in Figure 1. Intuitively, the purpose of multiple challenge response slots in Stage 1 is to allow the simulator to extract the values committed by V^* with minimal use of the leakage oracle. With the knowledge of the extracted values, the simulator can force the output of the coin-flipping to a specific distribution of its choice. This, in turn, allows the simulator to convert Naor's commitment scheme into an *equivocal* commitment scheme during simulation.

Common Input: A k-vertex graph G.
Private Input to P: A Hamiltonian Cycle H in graph G.
Parameters: Security parameter 1^k, $n = \omega(\log(k))$, $t = 3k^4$, $\epsilon > 0$. Without loss of generality, we assume that $\frac{1}{\epsilon}$ is an integer.

Stage 1 (Commitment phase)

$V \rightleftarrows P$: Commit to a t-bit random string r' and $(\frac{n^2}{\epsilon})$-pairs of random shares $\left\{r'^{0}_{i,j}, r'^{1}_{i,j}\right\}_{i=1,j=1}^{i=\frac{n}{\epsilon},j=n}$ (such that $r'^{0}_{i,j} \oplus r'^{1}_{i,j} = r'$ for every $i \in [\frac{n}{\epsilon}], j \in [n]$) using a public-coin statistically hiding commitment scheme. Similarly commit to a k-bit random string ch and $(\frac{n^2}{\epsilon})$-pairs of random shares $\left\{ch^0_{i,j}, ch^1_{i,j}\right\}_{i=1,j=1}^{i=\frac{n}{\epsilon},j=n}$ (such that $ch^0_{i,j} \oplus ch^1_{i,j} = ch$ for every $i \in [\frac{n}{\epsilon}], j \in [n]$) using a public-coin statistically hiding commitment scheme.

Challenge-response slots: For every $i \in [\frac{n}{\epsilon}]$,

$P \to V$: Choose n-bit random strings $\alpha_i = \alpha_{i,1}, \ldots, \alpha_{i,n}$ and $\beta_i = \beta_{i,1}, \ldots, \beta_{i,n}$. Send α_i, β_i to V.

$V \to P$: For every $j \in [n]$, V^* decommits to $r'^{\alpha_{i,j}}_{i,j}$ and $ch^{\beta_{i,j}}_{i,j}$.

Stage 2 (Coin-flipping completion phase)

$P \to V$: Choose a t-bit random string r'' and send it to V.

$V \to P$: Decommit to r' and $r'^{0}_{i,j}, r'^{1}_{i,j}$ for every $i \in [\frac{n}{\epsilon}], j \in [n]$. Let $r = r' \oplus r''$.

Stage 3 (Blum Hamiltonicity protocol)

$P \to V$: Let $r = r_1, \ldots, r_{k^3}$, where $|r_i| = 3k$ for every $i \in [k^3]$. For every $i \in [k]$,

- Choose a random permutation π_i and prepare an isomorphic copy of G, denoted $G_i = \pi_i(G)$.
- For every $j \in [k^2]$, commit to bit b_j in the adjacency matrix of G_i using Naor's commitment scheme with $r_{i \times j}$ as the first message.

$V \to P$: Decommit to ch and $ch^0_{i,j}, ch^1_{i,j}$ for every $i \in [\frac{n}{\epsilon}], j \in [n]$.

$P \to V$: Let $ch = ch_1, \ldots, ch_k$. For each $i \in [k]$, if $ch_i = 0$, decommit to every edge in G_i and reveal the permutation π_i. Else, decommit to the edges in the Hamiltonian Cycle in G_i.

Fig. 1. Protocol $\langle P, V\rangle$

Theorem 1. *If public-coin statistically hiding commitment schemes exist, then the protocol $\langle P, V \rangle$, parameterized by ϵ, is a $(1+\epsilon)$-leakage-resilient zero knowledge proof system.*

We note that statistically hiding commitment schemes imply one-way functions, which in turn suffice for Naor's statistically binding commitment scheme used in our construction. In the interest of space, we give a proof of Theorem 1 in the full version.

3 Leakage-Resilient NIZK

We now turn our attention to the setting of non-interactive zero knowledge proof systems. We consider the scenario where a malicious verifier can obtain arbitrary leakage on the witness and the random coins used by an honest prover to generate the proof string. To model leakage attacks, we allow the cheating verifier to make adaptive leakage queries on the honest prover's witness and the random coins used to generate the proof string. A leakage query to the prover consists of an efficiently computable function f, to which the prover replies with $f(w\|r)$, where w and r denote the prover's witness and random coins respectively. It is easy to see that in the non-interactive proofs setting, a cheating verifier who is allowed multiple leakage queries enjoys no additional power than one who is allowed only one leakage query. Therefore, for simplicity of exposition, from now on, we only consider cheating verifiers who make only one leakage query. We note that our definition given below can be easily adapted to incorporate multiple leakage queries.

We model the zero knowledge simulator $\mathcal{S}$ as a PPT machine that has access to a leakage oracle $L_w^k(\cdot)$ that is parameterized by the honest prover's witness w and the security parameter k. (Unlike the interactive proofs setting, here we do not consider the leakage parameter λ for simplicity of exposition.) The leakage oracle accepts queries of the form f (where $f(\cdot)$ is an efficiently computable function) and outputs $f(w)$. In order to bound the total leakage available to the simulator, we require that if the verifier obtains ℓ bits of total leakage from the honest prover, then the total leakage obtained by the simulator (from the leakage oracle) must be bounded by ℓ bits.

We now setup some notation. Let $\mathcal{R}$ be an efficiently computable relation that consists of pairs (x, w), where x is called the statement and w is the witness. Let $\mathcal{L}$ denote the language consisting of statements in $\mathcal{R}$. Recall that a non-interactive proof system for a language $\mathcal{L}$ consists of a setup algorithm K, a prover P and a verifier V. The setup algorithm K generates a common reference string σ. The prover P takes as input (σ, x, w) and checks whether $(x, w) \in \mathcal{R}$; if so, it produces a proof string π, else it outputs `fail`. The verifier V takes as input (σ, x, π) and outputs 1 if the proof is valid, and 0 otherwise.

Definition 2 (LR-NIZK). *A non-interactive proof system (K, P, V) for a* PPT *relation $\mathcal{R}$ is said to be a* leakage-resilient NIZK *if there exists a simulator $\mathcal{S} = (\mathcal{S}_1, \mathcal{S}_2, \mathcal{S}_3)$ such that for all adversaries $\mathcal{A}$,*

$$\Pr[\sigma \leftarrow K(1^k) : \mathcal{A}^{PR(\sigma,\cdot,\cdot,\cdot)}(\sigma) = 1] \stackrel{c}{\equiv} \Pr[(\sigma, \tau) \leftarrow \mathcal{S}_1(1^k) : \mathcal{A}^{SR^{L_w^k(\cdot)}(\sigma,\tau,\cdot,\cdot,\cdot)}(\sigma) = 1],$$

where $PR(\sigma, x, w, f)$ *computes* $r \leftarrow \{0,1\}^{\ell_P(k)}$; $\pi \leftarrow P(\sigma, x, w; r)$; $y = f(w\|r)$ *and returns* (π, y), *while* $SR^{L_\lambda^k(\cdot)w}(\sigma, \tau, x, w, f)$ *computes* $r \leftarrow \{0,1\}^{\ell_S(k)}$; $\pi \leftarrow \mathcal{S}_2(\sigma, \tau, x; r)$; $f' \leftarrow \mathcal{S}_3(\sigma, \tau, x, r, f)$; $y \leftarrow L_w^k(f')$ *and returns* (π, y). *Here, the leakage query* f' *made to* $L_w^k(\cdot)$ *is such that its output length is no more than the output length of* f. *Both the oracles* PR *and* SR *output* `fail` *if* $(x, w) \notin \mathcal{R}$.

3.1 Our Result

We now claim that every NIZK proof system with the *honest prover state reconstruction property*[5] is in fact a leakage-resilient NIZK. An immediate corollary is that the Groth et al. [36] NIZK proof system is a leakage-resilient NIZK proof system. We refer the reader to the full version for a formal proof.

Theorem 2. *A NIZK proof system* (K, P, V) *for a relation* $\mathcal{R}$ *with honest prover state reconstruction is a leakage resilient NIZK for* $\mathcal{R}$.

4 Applications of Leakage-Resilient Zero Knowledge

4.1 UC with Leaky Tokens

Starting with the work of Goldreich and Ostrovsky on software protection [30], tamper-proof hardware tokens have been used for a variety of cryptographic tasks such as achieving universal composability [43,14,54,15], one-time-programs [32], unconditionally secure protocols [35,34], compilers for leakage-resilient computation [42,33], etc. To the best of our knowledge, all prior works using tamper-proof hardware tokens make the assumption that the tokens are completely leakage-resilient (i.e., a token does not leak any information to an adversary who is in possession of the token). Here, we start a new line of research to investigate whether it is possible to relax this assumption for various cryptographic tasks. In particular, we study the feasibility of doing universally composable secure computation using "leaky" tokens. We start with the tamper-proof hardware token model of Katz [43] and modify it appropriately to incorporate "bounded" leakage. Then, by making use of leakage-resilient hard relations [20] and our leakage-resilient zero knowledge proof system, we give a construction for a universally composable multi-party computation protocol in the leaky token model.

The Leaky Token Model. We first briefly recall the hardware token model of Katz [43]. In the model of [43], it is assumed that a party (referred to as the *creator*) can take some software code and "seal" it inside a tamper-proof hardware token; the party can then give this token to another party (referred to as the *user*), who can then access the embedded software in a black-box manner. This setup is

[5] Very briefly, this property (also known as *non-erasure zero knowledge*) requires that not only can we simulate an honest party making a proof, but also *how it constructed the proof* (i.e., create convincing randomness so that it looks like the simulated proof was constructed by an honest prover using this randomness). See [36] for a formal definition.

modeled by a "wrapper" functionality $\mathcal{G}_{wrap}$ that accepts two types of messages: the first type is used by a party P to "create" a hardware token (encapsulating an interactive protocol M) and to "send" this token to another party P'. On receiving the token, P' can interact with it in an arbitrary black-box manner. This is formalized by allowing P' to send messages of its choice to M via $\mathcal{G}_{wrap}$. Each time M is invoked, fresh random coins are chosen for M.

In order to incorporate leakage attacks, we consider a modified wrapper functionality $\mathcal{G}^{\ell}_{wrap}$ parametrized by a leakage-parameter ℓ that defines the "total" leakage available to a token user over all the executions of the token. More concretely, the new wrapper functionality $\mathcal{G}^{\ell}_{wrap}$ is defined in the same manner as $\mathcal{G}_{wrap}$, except that $\mathcal{G}^{\ell}_{wrap}$ accepts special `leak` queries (from the token user) that consist of a efficiently computable function $f_i : \{0,1\}^* \rightarrow \{0,1\}^{\ell_i}$ (described as a circuit), to which the functionality answers with $f(M, state)$, where M denotes the code of the interactive Turing machine encapsulated in the token and $state$ denotes the current state of M consisting of all the protocol messages received from the user and the random coins used so far by M in the current protocol execution. The token user can make any arbitrary polynomial number of such leakage queries over multiple protocol executions with M; we only require that the functions f_i be efficiently computable, and the total number of bits leaked (over all executions) is $\sum_i \ell_i = \ell$. We stress that by allowing leakage on M, we allow the token user to obtain leakage on any secret values hardwired into M. In the interest of space, we defer formal description of $\mathcal{G}^{\ell}_{wrap}$ to the full version.

UC Security via UC-Puzzles. In order to obtain our positive result, we build on the recent work of Lin et al. [49] which puts forward a unified framework for designing UC secure protocols from known setup assumptions [12,13,43,6]. Lin et al. observe that a general technique for constructing UC secure protocols is to have the simulator obtain a "trapdoor string" which is hard to compute for the adversary. This is formalized in the form of (two party) "UC-puzzle" protocols that enable the simulator to obtain such a trapdoor string (but prevent the adversary from doing so). Following the work of [49], the task of constructing UC secure protocols from any setup assumption reduces to the task of constructing a UC-puzzle in the hybrid model of the corresponding setup.[6] We obtain our positive result by following the same route. Specifically, we construct a "family of UC-puzzles" in the $\mathcal{G}^{\ell}_{wrap}$-hybrid model.

Our Protocol. Recall that in the hardware token model, each pair of parties in the system exchange hardware tokens with each other. Now consider a system with m parties $P_1, \ldots, P_m$. For each pair of parties (P_i, P_j), we will construct

[6] Very briefly, this is because Lin et al. show that a UC-puzzle can be used in conjunction with a strongly non-malleable witness indistinguishable protocol in order to construct a "concurrent simulation-sound" zero knowledge protocol with "UC simulation" property, which in turn is known to be sufficient to construct UC secure protocols (see e.g. [13]). We refer the reader to the full version for more details.

two different UC-puzzles in the $\mathcal{G}^{\ell}_{wrap}$ hybrid model, (a) one where P_i (resp., P_j) acts as the puzzle sender (resp., receiver) and (b) the other where the roles of P_i and P_j are reversed. This gives us a family of m^2 UC-puzzles.

We now describe the construction of a family of protocol and relation pairs $(\langle S_{ij}, R_{ij}\rangle, \mathcal{R}_{ij})$, where $i, j \in [m]$. Here the choice of notation is to highlight that party P_i (resp., P_j) plays the role of the sender (resp., receiver) in protocol $\langle S_{ij}, R_{ij}\rangle$. We will then prove that each pair $(\langle S_{ij}, R_{ij}\rangle, \mathcal{R}_{ij})$ is a UC-puzzle in the $\mathcal{G}^{\ell}_{wrap}$-hybrid model. In our construction, we will use a $(1+\epsilon)$-LR-ZK proof of knowledge system[7] as well as an ℓ'-leakage-resilient hard relation [20], where $\ell' = (1+\epsilon)\cdot\ell$.

Description of $\langle S_{ij}, R_{ij}\rangle$. The interactive Turing machine S_{ij}, when invoked with the inputs the identity of the sender P_i, the identity of the receiver P_j and the session id *sid*, proceeds as follows. It first checks whether this is the first time interacting with party P_j. If so, it first samples a pair (x, y) from an ℓ'-leakage resilient hard relation $\mathcal{R}_{\ell'}$ and then "creates" and "gives" P_j a token (by sending the "appropriate" message to $\mathcal{G}^{\ell}_{wrap}$), which encapsulates the interactive Turing machine M that gives a λ-LR-ZKPOK of the statement that there exists an x such that $(x, y) \in \mathcal{R}_{\ell'}$. To actually challenge P_j, S_{ij} simply sends y as the puzzle to the receiver.

The interactive Turing machine R_{ij}, on receiving y from S_{ij}, engages in an execution of our λ-LR-ZKPOK protocol with M (via $\mathcal{G}^{\ell}_{wrap}$) where M proves that there exists an x such that $(x, y) \in \mathcal{R}_{\ell'}$. An adversarial receiver R_{ij} may additionally send leakage queries $(\texttt{leak}, f)$ to $\mathcal{G}^{\ell}_{wrap}$, who responds with $f(M\|r)$ (where r denotes the random coins used by M "so far") as long as the total leakage (over all queries) is bounded by ℓ.

Description of $\mathcal{R}_{ij}$. The puzzle relation $\mathcal{R}_{ij}$ is simply $\{(x, y) | (x, y) \in \mathcal{R}_{\ell'}\}$. This completes the description of the pair $(\langle S_{ij}, R_{ij}\rangle, \mathcal{R}_{ij})$. We refer the reader to the full version for a proof of our claim that $(\langle S_{ij}, R_{ij}\rangle, \mathcal{R}_{ij})$ is a UC-puzzle in the $\mathcal{G}^{\ell}_{wrap}$-hybrid model.

4.2 Fully Leakage-Resilient Signatures

In this section, we give generic constructions of fully leakage-resilient (FLR) signature schemes in the bounded leakage model as well as the continual leakage model. In order to obtain our results, we will adapt the approach of Katz and Vaikuntnathan [44], and in particular, Dodis et al. [20,19] (who used leakage-resilient hard relations and tag-based *true simulation-extractable* (tSE) NIZK argument systems to construct "standard" leakage-resilient signature schemes) to the setting of *full leakage* (where the adversary can leak on the entire state, as opposed to only the secret key). Specifically, we first extend our notion of leakage-resilient NIZKs to incorporate the property of true simulation-extractability.

[7] We note here that the LR-ZK proof system discussed in Section 2.1 is not a *proof of knowledge*. However, it is easy to modify the construction to obtain a proof of knowledge system by using a *leakage-sound* zero knowledge proof system. We refer the reader to the full version for more details.

Then, by using a hard relation that is leakage-resilient in the bounded (resp., continual) leakage model along with our *true simulation-extractable leakage-resilient* (tSE-LR) NIZK argument system, we obtain FLR signatures in the bounded (resp., continual) leakage model. Somewhat interestingly, unlike the recent constructions of FLR signature schemes [51,9], our constructions are also secure in the noisy leakage model [56]. In interest of space, here we limit our discussion to the construction of an FLR signature scheme in the bounded leakage model. We refer the reader to the full version for further discussion on the continual leakage model and the noisy leakage model.

True Simulation-Extractable Leakage-Resilient NIZK. We first (informally) define tag-based tSE-LR-NIZK argument system and give a construction for the same. Let us first recall the notion of tSE NIZK, as defined in [20]. Very roughly, a NIZK proof system is true simulation extractable if there exists a PPT extractor which (when given an extraction trapdoor to the CRS) extracts a witness w^* from any proof π^* produced by an adversary $\mathcal{A}$ (using a tag $\mathtt{tag}^*$), even if $\mathcal{A}$ has previously seen some simulated proofs for other true statements (with different tags). Our notion of tSE-LR-NIZK extends the notion of tSE-NIZK by allowing the adversary to obtain (in addition to simulated proofs) leakage on the witness and randomness used to generate the simulated proofs.

A tag based tSE-LR-NIZK argument system $(\mathcal{K}, \mathcal{P}, \mathcal{V})$ follows directly from the adaptively secure UC-NIZK of Groth et al. [36]. The complete construction and proof is given in the full version.

Fully Leakage-Resilient Signatures in the Bounded Leakage Model. We will follow the definition of FLR signature schemes due to Boyle et al [9]. Very roughly, we say that a signature scheme is fully leakage-resilient in the bounded-leakage model if it is existentially unforgeable against any PPT adversary that can obtain polynomially many signatures over messages of her choice, as well as bounded leakage information on the secret key and the randomness used by the signing algorithm and the key generation algorithm) throughout the lifetime of the system. Due to space constraints, we omit the formal definition of FLR signatures from this manuscript. We now proceed to describe our construction. The security proof is deferred to the full version.

Our Construction. Let $\mathcal{R}_\ell$ be an ℓ-leakage-resilient hard relation with a PPT sampling algorithm $\textsc{kgen}(\cdot)$. Let (K, P, V) be a tag-based tSE-LR-NIZK argument system for a relation $\mathcal{R}$. The signature scheme (KeyGen, Sign, Verify) is described as follows.

- $\mathsf{KeyGen}(1^k, \ell)$: Sample $(x, y) \leftarrow \textsc{kgen}(1^k)$, $\sigma \leftarrow K(1^k)$. Output $sk = x$ and $pk = (\sigma, y)$.
- $\mathsf{Sign}_{sk}(m)$: Output $\Phi = \pi$, where $\pi \leftarrow P(\sigma, y, m, x)$. (Here m is the tag in the argument.)
- $\mathsf{Verify}_{pk}(m, \Phi)$: Output $V(\sigma, y, m, \Phi)$.

Theorem 3. *If $\mathcal{R}_\ell$ is an ℓ-leakage-resilient hard relation and (K, P, V) is a tag-based true simulation-extractable leakage-resilient NIZK argument system,* *then (*KeyGen, Sign, Verify*) is an ℓ-fully-leakage-resilient signature scheme in the bounded-leakage model.*

5 Conclusions

In this paper, we give definitions and constructions of leakage-resilient zero knowledge proof systems, where an adversarial verifier can obtain arbitrary bounded leakage on the secret state of the prover. It is natural to consider the (opposite) scenario where a malicious prover can obtain arbitrary leakage on the random coins of the verifier during the protocol execution. The question that we may ask is whether it is possible to construct interactive proofs that remain *sound* in such a scenario. Going even further, we can consider the question of constructing an interactive proof system that *simultaneously* satisfies the notions of "leakage-soundness" and leakage-resilient zero knowledge. In the full version, we give positive results for both these settings.

A natural question following our work is whether we can extend our notions and results to the setting of secure two-party computation. We address this in an upcoming work.

References

1. Ajtai, M.: Secure computation with information leaking to an adversary. In: STOC (2011)
2. Akavia, A., Goldwasser, S., Vaikuntanathan, V.: Simultaneous hardcore bits and cryptography against memory attacks. In: Reingold, O. (ed.) TCC 2009. LNCS, vol. 5444, pp. 474–495. Springer, Heidelberg (2009)
3. Alwen, J., Dodis, Y., Naor, M., Segev, G., Walfish, S., Wichs, D.: Public-key encryption in the bounded-retrieval model. In: Gilbert, H. (ed.) EUROCRYPT 2010. LNCS, vol. 6110, pp. 113–134. Springer, Heidelberg (2010)
4. Alwen, J., Dodis, Y., Wichs, D.: Leakage-resilient public-key cryptography in the bounded-retrieval model. In: Halevi, S. (ed.) CRYPTO 2009. LNCS, vol. 5677, pp. 36–54. Springer, Heidelberg (2009)
5. Anderson, R., Kuhn, M.: Tamper resistance: a cautionary note. In: WOEC (1996)
6. Barak, B., Canetti, R., Nielsen, J.B., Pass, R.: Universally composable protocols with relaxed set-up assumptions. In: FOCS (2004)
7. Beaver, D.: Adaptive zero knowledge and computational equivocation. In: STOC (1996)
8. Bitansky, N., Canetti, R., Halevi, S.: Leakage tolerant interactive protocols. Cryptology ePrint Archive, Report 2011/204 (2011)
9. Boyle, E., Segev, G., Wichs, D.: Fully leakage-resilient signatures. In: Paterson, K.G. (ed.) EUROCRYPT 2011. LNCS, vol. 6632, pp. 89–108. Springer, Heidelberg (2011)
10. Brakerski, Z., Kalai, Y.T., Katz, J., Vaikuntanathan, V.: Overcoming the hole in the bucket: Public-key cryptography resilient to continual memory leakage. In: FOCS (2010)
11. Canetti, R., Feige, U., Goldreich, O., Naor, M.: Adaptively secure multi-party computation. In: STOC (1996)

12. Canetti, R., Fischlin, M.: Universally composable commitments. In: Kilian, J. (ed.) CRYPTO 2001. LNCS, vol. 2139, p. 19. Springer, Heidelberg (2001)
13. Canetti, R., Lindell, Y., Ostrovsky, R., Sahai, A.: Universally composable two-party and multi-party secure computation. In: STOC (2002)
14. Chandran, N., Goyal, V., Sahai, A.: New constructions for UC secure computation using tamper-proof hardware. In: Smart, N.P. (ed.) EUROCRYPT 2008. LNCS, vol. 4965, pp. 545–562. Springer, Heidelberg (2008)
15. Damgård, I., Nielsen, J.B., Wichs, D.: Universally composable multiparty computation with partially isolated parties. In: Reingold, O. (ed.) TCC 2009. LNCS, vol. 5444, pp. 315–331. Springer, Heidelberg (2009)
16. Damgård, I., Pedersen, T.P., Pfitzmann, B.: On the existence of statistically hiding bit commitment schemes and fail-stop signatures. J. Cryptology (1997)
17. De Santis, A., Di Crescenzo, G., Ostrovsky, R., Persiano, G., Sahai, A.: Robust non-interactive zero knowledge. In: Kilian, J. (ed.) CRYPTO 2001. LNCS, vol. 2139, p. 566. Springer, Heidelberg (2001)
18. Dodis, Y., Goldwasser, S., Kalai, Y.T., Peikert, C., Vaikuntanathan, V.: Public-key encryption schemes with auxiliary inputs. In: Micciancio, D. (ed.) TCC 2010. LNCS, vol. 5978, pp. 361–381. Springer, Heidelberg (2010)
19. Dodis, Y., Haralambiev, K., López-Alt, A., Wichs, D.: Cryptography against continuous memory attacks. In: FOCS (2010)
20. Dodis, Y., Haralambiev, K., López-Alt, A., Wichs, D.: Efficient public-key cryptography in the presence of key leakage. In: Abe, M. (ed.) ASIACRYPT 2010. LNCS, vol. 6477, pp. 613–631. Springer, Heidelberg (2010)
21. Dodis, Y., Kalai, Y.T., Lovett, S.: On cryptography with auxiliary input. In: STOC (2009)
22. Dolev, D., Dwork, C., Naor, M.: Nonmalleable cryptography. SIAM J. Comput. (2000)
23. Dwork, C., Naor, M., Sahai, A.: Concurrent zero knowledge. In: STOC (1998)
24. Dziembowski, S., Pietrzak, K.: Leakage-resilient cryptography. In: FOCS (2008)
25. Faust, S., Kiltz, E., Pietrzak, K., Rothblum, G.N.: Leakage-resilient signatures. In: Micciancio, D. (ed.) TCC 2010. LNCS, vol. 5978, pp. 343–360. Springer, Heidelberg (2010)
26. Faust, S., Rabin, T., Reyzin, L., Tromer, E., Vaikuntanathan, V.: Protecting circuits from leakage: the computationally-bounded and noisy cases. In: Gilbert, H. (ed.) EUROCRYPT 2010. LNCS, vol. 6110, pp. 135–156. Springer, Heidelberg (2010)
27. Feige, U., Shamir, A.: Zero knowledge proofs of knowledge in two rounds. In: Brassard, G. (ed.) CRYPTO 1989. LNCS, vol. 435, pp. 526–545. Springer, Heidelberg (1990)
28. Gandolfi, K., Mourtel, C., Olivier, F.: Electromagnetic analysis: Concrete results. In: Koç, Ç.K., Naccache, D., Paar, C. (eds.) CHES 2001. LNCS, vol. 2162, p. 251. Springer, Heidelberg (2001)
29. Goldreich, O., Kahan, A.: How to construct constant-round zero-knowledge proof systems for NP. J. Cryptology (1996)
30. Goldreich, O., Ostrovsky, R.: Software protection and simulation on oblivious RAMs. J. ACM (1996)
31. Goldwasser, S., Micali, S., Rackoff, C.: The knowledge complexity of interactive proof-systems. In: STOC (1985)
32. Goldwasser, S., Kalai, Y.T., Rothblum, G.N.: One-time programs. In: Wagner, D. (ed.) CRYPTO 2008. LNCS, vol. 5157, pp. 39–56. Springer, Heidelberg (2008)

33. Goldwasser, S., Rothblum, G.N.: Securing computation against continuous leakage. In: Rabin, T. (ed.) CRYPTO 2010. LNCS, vol. 6223, pp. 59–79. Springer, Heidelberg (2010)
34. Goyal, V., Ishai, Y., Mahmoody, M., Sahai, A.: Interactive locking, zero-knowledge pCPs, and unconditional cryptography. In: Rabin, T. (ed.) CRYPTO 2010. LNCS, vol. 6223, pp. 173–190. Springer, Heidelberg (2010)
35. Goyal, V., Ishai, Y., Sahai, A., Venkatesan, R., Wadia, A.: Founding cryptography on tamper-proof hardware tokens. In: Micciancio, D. (ed.) TCC 2010. LNCS, vol. 5978, pp. 308–326. Springer, Heidelberg (2010)
36. Groth, J., Ostrovsky, R., Sahai, A.: Perfect non-interactive zero knowledge for NP. In: Vaudenay, S. (ed.) EUROCRYPT 2006. LNCS, vol. 4004, pp. 339–358. Springer, Heidelberg (2006)
37. Halderman, J.A., Schoen, S.D., Heninger, N., Clarkson, W., Paul, W., Calandrino, J.A., Feldman, A.J., Appelbaum, J., Felten, E.W.: Lest we remember: Cold boot attacks on encryption keys. In: USENIX Security Symposium (2008)
38. Halevi, S., Micali, S.: Practical and provably-secure commitment schemes from collision-free hashing. In: Koblitz, N. (ed.) CRYPTO 1996. LNCS, vol. 1109, pp. 201–215. Springer, Heidelberg (1996)
39. Ishai, Y., Kushilevitz, E., Ostrovsky, R., Sahai, A.: Extracting correlations. In: FOCS (2009)
40. Ishai, Y., Prabhakaran, M., Sahai, A., Wagner, D.: Private circuits II: Keeping secrets in tamperable circuits. In: Vaudenay, S. (ed.) EUROCRYPT 2006. LNCS, vol. 4004, pp. 308–327. Springer, Heidelberg (2006)
41. Ishai, Y., Sahai, A., Wagner, D.: Private circuits: Securing hardware against probing attacks. In: Boneh, D. (ed.) CRYPTO 2003. LNCS, vol. 2729, pp. 463–481. Springer, Heidelberg (2003)
42. Juma, A., Vahlis, Y.: Protecting cryptographic keys against continual leakage. In: Rabin, T. (ed.) CRYPTO 2010. LNCS, vol. 6223, pp. 41–58. Springer, Heidelberg (2010)
43. Katz, J.: Universally composable multi-party computation using tamper-proof hardware. In: Naor, M. (ed.) EUROCRYPT 2007. LNCS, vol. 4515, pp. 115–128. Springer, Heidelberg (2007)
44. Katz, J., Vaikuntanathan, V.: Signature schemes with bounded leakage resilience. In: Matsui, M. (ed.) ASIACRYPT 2009. LNCS, vol. 5912, pp. 703–720. Springer, Heidelberg (2009)
45. Kiltz, E., Pietrzak, K.: Leakage resilient elgamal encryption. In: Abe, M. (ed.) ASIACRYPT 2010. LNCS, vol. 6477, pp. 595–612. Springer, Heidelberg (2010)
46. Kocher, P.C.: Timing attacks on implementations of diffie-hellman, rsa, dss, and other systems. In: Koblitz, N. (ed.) CRYPTO 1996. LNCS, vol. 1109, pp. 104–113. Springer, Heidelberg (1996)
47. Lewko, A., Lewko, M., Waters, B.: How to leak on key updates. In: STOC (2011)
48. Lewko, A., Rouselakis, Y., Waters, B.: Achieving leakage resilience through dual system encryption. In: Ishai, Y. (ed.) TCC 2011. LNCS, vol. 6597, pp. 70–88. Springer, Heidelberg (2011)
49. Lin, H., Pass, R., Venkitasubramaniam, M.: A unified framework for concurrent security: universal composability from stand-alone non-malleability. In: STOC (2009)
50. Lindell, Y., Zarosim, H.: Adaptive zero-knowledge proofs and adaptively secure oblivious transfer. In: Reingold, O. (ed.) TCC 2009. LNCS, vol. 5444, pp. 183–201. Springer, Heidelberg (2009)
51. Malkin, T., Teranishi, I., Vahlis, Y., Yung, M.: Signatures resilient to continual leakage on memory and computation. In: EUROCRYPT (2011)

52. Micali, S., Pass, R.: Local zero knowledge. In: STOC (2006)
53. Micali, S., Reyzin, L.: Physically observable cryptography. In: Naor, M. (ed.) TCC 2004. LNCS, vol. 2951, pp. 278–296. Springer, Heidelberg (2004)
54. Moran, T., Segev, G.: David and goliath commitments: UC computation for asymmetric parties using tamper-proof hardware. In: Smart, N.P. (ed.) EUROCRYPT 2008. LNCS, vol. 4965, pp. 527–544. Springer, Heidelberg (2008)
55. Naor, M.: Bit commitment using pseudo-randomness (extended abstract). In: Brassard, G. (ed.) CRYPTO 1989. LNCS, vol. 435, pp. 128–136. Springer, Heidelberg (1990)
56. Naor, M., Segev, G.: Public-key cryptosystems resilient to key leakage. In: Halevi, S. (ed.) CRYPTO 2009. LNCS, vol. 5677, pp. 18–35. Springer, Heidelberg (2009)
57. Naor, M., Yung, M.: Universal one-way hash functions and their cryptographic applications. In: STOC (1989)
58. Pass, R., Rosen, A.: New and improved constructions of non-malleable cryptographic protocols. In: STOC (2005)
59. Pietrzak, K.: A leakage-resilient mode of operation. In: Joux, A. (ed.) EUROCRYPT 2009. LNCS, vol. 5479, pp. 462–482. Springer, Heidelberg (2009)
60. Prabhakaran, M., Rosen, A., Sahai, A.: Concurrent zero knowledge with logarithmic round-complexity. In: FOCS (2002)
61. Quisquater, J.J., Samyde, D.: Electromagnetic analysis (ema): Measures and counter-measures for smart cards. In: E-smart (2001)
62. Rosen, A.: A note on constant-round zero-knowledge proofs for NP. In: Naor, M. (ed.) TCC 2004. LNCS, vol. 2951, pp. 191–202. Springer, Heidelberg (2004)
63. Sahai, A.: Non-malleable non-interactive zero knowledge and adaptive chosen-ciphertext security. In: FOCS (1999)

A Comprehensive Evaluation of Mutual Information Analysis Using a Fair Evaluation Framework

Carolyn Whitnall and Elisabeth Oswald

University of Bristol, Department of Computer Science,
Merchant Venturers Building, Woodland Road, BS8 1UB, Bristol, UK
{carolyn.whitnall,elisabeth.oswald}@bris.ac.uk

Abstract. The resistance of cryptographic implementations to side-channel analysis is a matter of considerable interest to those concerned with information security. It is particularly desirable to identify the attack methodology (e.g. differential power analysis using correlation or distance-of-means as the distinguisher) able to produce the best results. Such attempts are complicated by the many and varied factors contributing to attack success: the device power consumption characteristics, an attacker's power model, the distinguisher by which measurements and model predictions are compared, the quality of the estimations, and so on. Previous work has delivered partial answers for certain restricted scenarios. In this paper we assess the effectiveness of mutual information-based differential power analysis within a generic and comprehensive evaluation framework. Complementary to existing work, we present several notions/characterisations of attack success with direct implications for the amount of data required. We are thus able to identify scenarios in which mutual information offers performance advantages over other distinguishers. Furthermore we observe an interesting feature—unique to the mutual information based distinguisher—resembling a type of stochastic resonance, which could potentially enhance the effectiveness of such attacks over other methods in certain noisy scenarios.

Keywords: side channel analysis, mutual information.

1 Introduction

Side-channel analysis (SCA) refers to a collection of cryptanalytic techniques for extracting secret information from the physical leakage of a device as it executes a cryptographic algorithm. Of the various types, one of the most popularly studied is differential power analysis (DPA); it involves applying some type of statistic (the *distinguisher*) to identify a correct hypothesis about (part of) the secret key from the set of all possible hypotheses about this key. Popular distinguisher choices are the Pearson correlation coefficient and the distance-of-means test. Mutual information (MI) measures the total dependency between two random variables, and was first proposed for use in DPA at CHES 2008 [6]. *A priori*

P. Rogaway (Ed.): CRYPTO 2011, LNCS 6841, pp. 316–334, 2011.

it was expected to display certain advantages over other distinguishers, loosely summarized by three (informal) conjectures:

1. By comprehensively exploiting *all* of the information contained within trace measurements it could have an efficiency advantage over existing side-channel distinguishers such as correlation (which measures linear dependencies only).
2. By capturing total dependency between the true device leakage and the modeled leakage it could prove effective in scenarios where an accurate model for the data-dependent leakage of the device is not known, thereby serving as a 'generic' distinguisher.
3. By natural extension to multivariate statistics it might be adapted to the context of higher-order attacks against (for example) protected implementations. Existing distinguishers operate on univariate data only and therefore require trace data to be pre-processed, resulting in loss of information.

Subsequent investigations such as [1,17,20,23] have found little evidence of the first two expectations being met in practice (there is rather more support for the third—see, for example, [1,5,17]). However, the literature has not been comprehensive in explaining why this might be. We must bear in mind that many factors influence DPA outcomes: not only the choice of distinguisher, but also the target intermediate function, the form of the data-dependent device leakage and how well this can be modeled, and the precision with which the distinguishing vector can be estimated using the resources and capabilities available. It is often unclear whether the observed underperformance of MI-based DPA is an inherent theoretical weakness of the distinguisher, a result of sub-optimal estimation procedures, or simply a failure to identify scenarios (i.e. combinations of target functions and power consumption patterns) where it offers a useful advantage: see Batina et al. [1] for an overview of these issues.

In this paper we introduce a framework for assessing and comparing DPA attacks in any given scenario on a theoretical basis, abstracting away from the problem of practical estimation. We use this to gain fresh insight into the findings of the existing literature and to clarify when and in what sense the *a priori* intuition regarding MI-based DPA *does* hold. Moreover, we are able to identify and describe attack scenarios in which MI-based DPA is theoretically successful whilst other distinguishers fail, or in which it displays a theoretic advantage large enough to potentially translate to a practical advantage. Further, we demonstrate that the (standardised) MI-based distinguishing vector exhibits the property of stochastic resonance as the noise levels in the power consumption vary. This feature, which is not shared by correlation-based DPA, could potentially be exploited to enhance MI-based attacks via noise injection.

In what follows, we first give the relevant preliminary information on DPA attacks, including details of particular distinguishers and a discussion of previous work in Sect. 2. In Sect. 3 we describe our methodology, whilst Sect. 4 reports on our findings as they relate to various attack scenarios. We conclude in Sect. 5.

2 DPA Attacks

We consider a 'standard DPA attack' scenario such as defined in [13]: The power consumption $\mathcal{L}$ of the target device depends on some internal value (or state) $f_{k^*}(x)$: a function of some part of the plaintext $x \in \mathcal{X}$, as well as some part of the secret key $k^* \in \mathcal{K}$. Hence, we have that $\mathcal{L} = L \circ f_{k^*}(x) + \varepsilon$, where L is some function which describes the data-dependent component and ε comprises the remaining power consumption which can be modeled as independent random noise. The attacker has N power measurements corresponding to encryptions of N known plaintexts $x_i \in \mathcal{X}$, $i = 1, \ldots, N$ and wishes to recover the secret key k^*. The attacker can accurately compute the internal values as they would be under each key hypothesis $\{f_k(x_i)\}_{i=1}^N$, $k \in \mathcal{K}$ and uses whatever information he possesses about the true leakage function L to construct a model M.

DPA exploits the fact that the modeled power traces corresponding to the correct key hypothesis should bear more resemblance to the true power traces than do the modeled traces corresponding to incorrect hypotheses. An attacker is thus concerned with quantifying and comparing the degree of similarity between the true and modeled traces for each key hypothesis. A range of comparison tools—'distinguishers'—are available, of which mutual information and Pearson's correlation coefficient are popular examples. We introduce these formally and examine them in more detail in the remaining parts of this section. We use the shorthands CPA and MIA to refer (respectively) to correlation-based and MI-based DPA attacks.

2.1 Reasoning about the Success and Efficiency of DPA Attacks

Previous work has made some progress towards providing meaningful and practically relevant definitions for the 'success' and 'efficiency' of DPA attacks. Standaert's work [21] formalised the notion of key-recovery success (and, correspondingly, success rate), which we adopt for our purposes here: The theoretic attack distinguisher is $\mathbf{D} = \{D(k)\}_{k\in\mathcal{K}} = \{D(L \circ f_{k^*}(X) + \varepsilon, M \circ f_k(X))\}_{k\in\mathcal{K}}$, where the plaintext input X takes values in $\mathcal{X}$ according to some known distribution (usually uniform). We say the attack is *theoretically successful* if $D(k^*) > D(k)\ \forall k \neq k^*$. We say it is *o-th order theoretically successful* if $\#\{k \in \mathcal{K} : D(k^*) \leq D(k)\} < o$.

However, in practice $\mathbf{D}$ must be estimated. Suppose we have observations corresponding to the vector of inputs $\mathbf{x} = \{x_i\}_{i=1}^N$, and write $\mathbf{e} = \{e_i\}_{i=1}^N$ to be the observed noise (i.e. drawn from the distribution of ε). Then the size $\#\mathcal{K}$ estimated vector is $\hat{\mathbf{D}}_N = \{\hat{D}_N(k)\}_{k\in\mathcal{K}} = \{\hat{D}_N(L \circ f_{k^*}(\mathbf{x}) + \mathbf{e}, M \circ f_k(\mathbf{x}))\}_{k\in\mathcal{K}}$. We then say the attack is *successful* if $\hat{D}_N(k^*) > \hat{D}_N(k)\ \forall k \neq k^*$ and *o-th order successful* if $\#\{k \in \mathcal{K} : \hat{D}_N(k^*) \leq \hat{D}_N(k)\} < o$.

Since we are particularly interested in the impact of L on attack outcomes, it is desirable to abstract away from the impact of noise, as well as from the estimation process. We define a distinguisher as *ideally successful* if it is theoretically successful in a noise-free scenario.

Ideal success thus depends on the target intermediate function, the form of the data-dependent device leakage L, the set $\mathcal{X}' \subseteq \mathcal{X}$ of plaintexts being encrypted, and the choice of power model and distinguisher. Theoretic success is further determined by the size and distribution of the noise ε whilst practical success depends additionally on the choice of estimator for the distinguisher and the number of trace measurements N. That is, given an attack which *theoretically* distinguishes the correct key (by a margin of a certain size), the practical outcome will be determined by whether or not an attacker has adequate resources to estimate $\hat{D}$ with sufficient precision to detect a difference of that size.

2.2 Distinguishers for DPA Attacks

Standaert *et al.* [20] provide a good overview of the many distinguishers that have been employed in the literature since DPA was first introduced in the late 1990s [9]. In this paper, we focus on mutual information and compare it with one other distinguisher of interest: Pearson's correlation coefficient.

In recent work, Mangard *et al.* [13] have shown that in the scenario of standard DPA attacks, the three most popular distinguishers, Pearson correlation, distance of means, and Bayes, are equally successful. Under additional, strong assumptions such that the MI can be estimated parametrically as a Gaussian mixture, they are even able to demonstrate a mapping between a correlation-based and an MI-based distinguisher. Our work relates to rather more general distributional assumptions.

Mutual Information. Mutual information measures, in bits, the total information shared between two random variables X and Y. It is most intuitively expressed in terms of entropies via Shannon's formula: $\mathrm{I}(X;Y) = H(X) - H(X|Y)$.[1]

Mutual information is a functional of probability distributions, and estimation is a much studied problem with no simple answers [3,8,14,19,22]. All estimators are biased, and further no 'ideal' estimator exists; different estimators perform differently depending on the underlying structure of the data.

The usual approach is to first estimate the underlying marginal and conditional densities and then to substitute these into Shannon's formula via a 'plug-in' estimator for discrete entropy. There are many different ways to estimate densities and the quality of the resulting estimator for MI is very sensitive to the methods and parameters chosen. If we have a good understanding of the underlying distributions we can fit a parametric model such as a Gaussian mixture (see Veyrat-Charvillon et al. [23]). However, since MIA has been proposed for use in scenarios where our usual assumptions do not hold we are generally more interested in nonparametric methods, which are somewhat sensitive to user approach and known to incur an overhead in terms of estimation costs. In practice,

[1] The original (but equivalent) definition is $\mathrm{I}(X;Y) = \sum_{y\in\mathcal{Y}}\sum_{x\in\mathcal{X}} p_{X,Y}(x,y)\log_2\left(\frac{p_{X,Y}(x,y)}{p_X(x)\,p_Y(y)}\right)$, where $p_{X,Y}$ is the joint probability density of X and Y and p_X, p_Y are the marginal densities.

due to the large sample space and small datasets we usually estimate the densities via an m-bin regularisation of the space. By an important data processing inequality[2] this means we are always estimating a lower bound on the mutual information—as the binning or mesh becomes finer the estimate approaches the true mutual information monotonically from below [14].

In security evaluations we often would like to be able to talk about the number of traces needed for an attack to be successful. This requires knowing the sampling distribution for the distinguisher under reasonable assumptions. Unfortunately, estimators for MI do not 'behave nicely' as do other statistics (such as the correlation coefficient—see below); in fact, there are no universal rates of convergence [14], so that whatever estimator we pick, we can always find a distribution for which the error vanishes arbitrarily slowly.

The relationship between the ideal MI and the theoretic MI in the presence of noise is complex (see, for example, [11]). In particular, whilst $\mathrm{I}(X+\varepsilon;Y) \leq \mathrm{I}(X;Y)$ (X, ε independent), nonetheless $\mathrm{I}(X;Y)-\mathrm{I}(X+\varepsilon;Y) \neq \mathrm{I}(X;Z)-\mathrm{I}(X+\varepsilon;Z)$. Thus, the elements of the theoretic MIA vector are differentially affected so that ideal outcomes do not directly generalise to theoretic outcomes in the presence of noise.

Pearson's Correlation Coefficient. Pearson's correlation coefficient measures the total linear dependency between two random variables X and Y. It is defined as $\rho(X,Y) = \frac{\mathrm{cov}(X,Y)}{\sigma_X \sigma_Y}$. It takes values from -1 to 1 and, as with mutual information, is zero whenever X and Y are independent. However, the converse is not true; namely, X and Y may be (non-linearly) dependent with a (linear) correlation of 0.

It is estimated from samples $\{x_i\}_{i=1}^N$, $\{y_i\}_{i=1}^N$ via the sample correlation coefficient: $r(X,Y) = \frac{\sum_{i=1}^N (x_i-\bar{x})(y_i-\bar{y})}{\sqrt{\sum_{i=1}^N (x_i-\bar{x})^2}\sqrt{\sum_{i=1}^N (y_i-\bar{y})^2}}$. This is a consistent estimator for $\rho(X,Y)$ and, moreover, is asymptotically unbiased and efficient if X and Y have a joint Normal distribution. Under the same assumptions, we can even approximate the sampling distribution which leads to 'nice' results such as the number of trace measurements required for attacks to be successful (see Chap. 6.4 of [12]).

The relationship between the ideal correlation and the theoretic correlation in the presence of noise is straightforward. In fact, as derived in Chap. 6.3 of [12], $\rho(L+\varepsilon, M_k) = \frac{\rho(L,M_k)}{\sqrt{1+\frac{\sigma_\varepsilon^2}{\mathrm{Var}(L)}}}$. Thus, the larger the noise, the more diminished are the correlations. But—crucially—the denominator does not depend on the key hypothesis; the theoretic distinguisher vector is thus scaled in such a way that the rankings and other *relative* features are preserved. This does not at all imply that *practical* CPA attacks are immune to noise: As the sample variance of the estimator increases, the number of traces required to reach a sufficient level of precision also increases (see Chap. 4 of [12]).

---

[2] $I(S(X);T(Y)) \leq I(X;Y)$ for any random variables X and Y and any functions S and T on the range of X and Y.

3 A Comprehensive Evaluation Framework

We compute and examine ideal/theoretic CPA and MIA vectors for a broad spectrum of possible leakage scenarios in unprofiled attacks where the true leakage L is unknown and modeled via the Hamming weight (HW) or the raw value (ID) of the target function output. For CPA, this is the same as assuming that the leakage is *proportional* to the HW or ID of the target, whereas for MIA this is the same as allowing the leakage to be *different* for each distinct HW or ID value, without any restriction on the nature of that dependency (for example, it needn't be a monotonic relationship). These vectors provide insight into the relative strengths and weaknesses of the distinguishers. We are particularly interested in finding scenarios where MIA has an ideal/theoretic advantage over CPA because we hope that a sufficiently large theoretic advantage would translate into a practical advantage. To do this we need to formulate an appropriate notion of "advantage".

An extremely desirable metric for security evaluation is the number of traces needed for an attack to be successful. We can compute this for a given estimator using the techniques of *statistical power analysis* [10], provided the sampling distribution can be approximated—but this is not achievable in general (see Sect. 2.2), besides which we are seeking to avoid estimator-specific comparisons. Our solution is to choose measures based on those characteristics of the theoretic vectors which have the greatest bearing on the trace efficiency of a practical attack:

1. *Correct key ranking*: The position of the correct key when ranked by distinguisher value. If the correct key is ranked joint first the *ranking order* is the number of keys sharing position 1, so that an attack with a ranking order of o is o^{th}-order theoretically successful as defined in Sect. 2.1. The relationship with practical efficiency is obvious: attacks which are not first-order successful will not be able to uniquely extract the correct key from *any* number of trace measurements (except by random chance).
2. *Average distinguishing score*: The number of standard deviations above (or below) the mean for the distinguisher value corresponding to the correct key. This matches the "DPA signal-to-noise ratio" described by [7] and indicates the sensitivity of the attack in isolating the correct key: A very sensitive attack may be able to succeed in practice with only a few trace measurements, as even imprecise estimates will detect a large difference. A theoretically 'unsuccessful' attack may still be able to return a small candidate subset containing the correct key if the average distinguishing power is high.
3. *Nearest-rival distinguishing score*: The distance from the 'nearest rival' (i.e. the difference between the correct key distinguisher value and the value for the highest ranked alternative), normalised by the standard deviation. This represents, more directly than the average distinguishing power, the margin to be detected by a practical attack.

By computing the above measures for uniformly drawn plaintexts $X \overset{unif.}{\leftarrow} \mathcal{X}$, we are able to compare theoretic behaviour of attacks when provided with full information. We propose to explore the sensitivity of attacks to restricted information by inspecting ideal/theoretic attack vectors for reduced subsets of the plaintext space. These vectors depend not only on the size but also on the composition of the input set; we cannot perform the computation exhaustively over the entire space of possible subsets (it is too large), but by repeated random draws of increasing size we can estimate the average support size needed for attack success. Thus we add the following measures as further clues to the "how many traces" problem:

5. *Average minimum support*: On average, the required support size of the input distribution for the attack to achieve o^{th}-order success (where o is the ranking order).
6. *Support required for $x\%$ success rate*: The support size for which the rate of success (of the appropriate order) is at least x per cent.

Our criteria are best viewed in conjunction with one another rather than in isolation, and trade-offs between them will interplay differently with practical considerations. For instance, a methodology which achieves only o^{th}-order success (where $o > 1$) might be preferable to one achieving 1st-order success if the distinguisher vector can be estimated more precisely and/or efficiently. Likewise, nearest-rival distinguishability may be more important than average minimum support in the presence of high noise.

In some parts of this study it is more desirable to measure the *average* behaviour of an attack in a class of scenarios than to describe results under a specific scenario. This is relevant, for example, when considering functions of sufficient arbitrariness that we cannot detail each case exhaustively. In such cases, as with the analysis of restricted input support, we estimate average behaviour by using randomly sampled examples (note that the distinguishing vectors themselves are still *computed*, not estimated).

We acknowledge that data complexity is not the only measure of cost and that considerations such as computational complexity also play a role in determining the practicality of an attack. A formal study is outside the scope of this paper, but we do try to comment where appropriate.

Ideal/Theoretic vs. Practical Attacks. Recall that we define theoretic (as well as ideal, i.e. noise-free) attacks to abstract away from the impact of the estimation process (and from noise). As such, theoretic outcomes depend on the target intermediate function, the device leakage (including how much noise is present), the set of plaintexts used as inputs, the attackers choice/knowledge about the power model, and the theoretical distinguisher (which is in this case the estimand). Practical outcomes depend on an additional, crucial factor, namely the estimator—the quality of which, and the sensitivity to the underlying population parameters and noise, will ultimately determine whether an observed ideal/theoretical advantage is translated into a real advantage in a practical attack.

We consider several outcome measures to allow for a nuanced analysis of the distinguisher qualities contributing to practical outcomes. For example, the notion of ranking order is needed in addition to correct key ranking because, whilst ties are highly unlikely in the estimated vectors arising from practical attacks, the underlying theoretic values may well rank keys equally. The approach of studying the distinguishing quality of the estimands separately from the qualities of the estimators is new and, as we will demonstrate in latter parts of the paper, provides fresh insight into the strengths and weaknesses of different distinguishers in practice.

4 Results

We now evaluate MIA and CPA distinguishers using the framework and considerations w.r.t. leakage models as spelled out before. For the sake of clarity and conciseness, we first show one detailed example (Hamming-weight leakage of a device implementing the DES algorithm), and then briefly report outcomes for some other leakage models. The choice for our focus is motivated by previous practical work which has focused on DES implementations [1], and the fact that DES is still used as predominant algorithm in the banking world. Note though that our framework could be used in the same way in a different context, and that the results of our evaluation of MI as a distinguisher are not strongly dependent on our specific choice.

4.1 Hamming-Weight Leakage

We begin with an ideal evaluation of MIA relative to CPA in the simplest and most popularly studied scenario: the first S-Box in a DES implementation (short: DES_{S1}) with a Hamming-weight (HW) leakage. As attacker power models we consider HW and the identity (ID) power model. For the sake of simplicity we use the following abbreviations: CPA(HW) as short-hand for correlation-based DPA with a HW power model, MIA(HW)/MIA(ID) as short-hand for MI-based DPA with a HW/ID power model, and MMIA for multivariate MI-based DPA. Using the notation as introduced before we first evaluate

$$\text{CPA(M)} : \{\rho(L \circ \text{DES}_{S1}(x, k^*), M \circ \text{DES}_{S1}(x, k))\}_{k \in \mathcal{K}}, \tag{1}$$

$$\text{MIA(M)} : \{\text{I}(L \circ \text{DES}_{S1}(x, k^*); M \circ \text{DES}_{S1}(x, k))\}_{k \in \mathcal{K}} \tag{2}$$

assuming that both the attacker's power model, as well as the device's power model is the Hamming weight, i.e. $L = M = HW$.

This is a scenario in which we expect CPA(HW) to perform well: the use of the true power model enables perfect prediction of the data-dependent leakage under the correct key hypothesis, whilst the choice of the S-Box as target ensures that the alternative hypotheses will each give rise to substantially different predictions (see [16]).

Figure 1 shows the ideal distinguisher values for a CPA(HW) and an MIA(HW) attack. Since the target function has the Equal Images under different Subkeys (EIS) property [18] and the plaintexts are assumed uniformly distributed, attack outcomes are key independent [13]: the correct hypothesis yields the same distinguisher value under any key, and only the arrangement of the remaining vector entries changes.

It is evident that both attacks are first-order successful by a clear margin, but that MIA(HW) has a substantial ideal advantage, with a nearest-rival distinguishability score of 5.61 compared with just 2.14 for CPA(HW). This simple result confirms that it must instead be a combination of the impact of noise and the relative efficiency of estimating the correlation coefficient which enables CPA to consistently outperform MIA in practical attacks with a good power model.

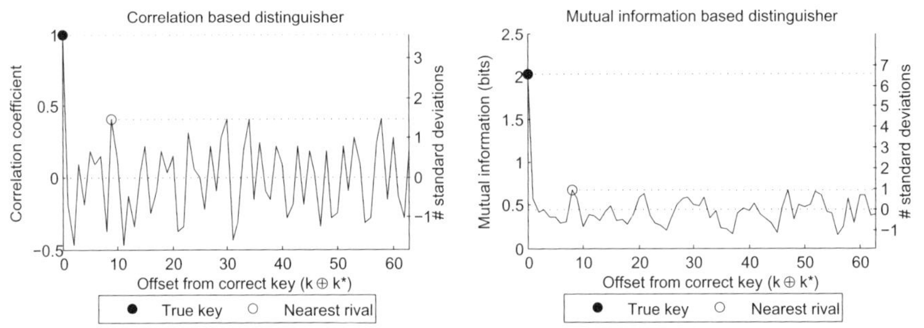

Fig. 1. Ideal distinguishing vectors using the HW power model against the output of the first DES S-Box

As a partial insight into the quantity of data needed we next look at the minimum input support size required for the distinguishers to approach their full ideal potential. The space of possible plaintext combinations is too large to explore exhaustively, so we look at the average behaviour of the attacks in repeated random draws from the plaintext space. We find that CPA is able to identify the correct key from a far smaller support than MIA, requiring just 6 inputs on average, and achieving 100% success with just 12, compared with an average of 8 and threshold of 14 for MIA. Note as well that even once a high ideal success rate is achieved, it may be that a broader support is required before MIA regains the distinguishing advantage it displays with respect to the full distribution.

We next investigate the enhancement of MIA via the incorporation of an additional data point in a multivariate attack on AddRoundKey (short: DES_{ARK}) and the first S-Box jointly. Figure 2 plots the ideal outcome[3]. First observe that the distinguisher values are greater in size (by a factor of about two) than that of the single point attack—that is, we *are* capturing a larger amount of information. However, the increase applies across the range of key hypotheses so does not

[3] Note that what we are proposing here is to use the mutual information between two *bivariate* variables; since joint entropy is well-defined this is entirely consistent with the formulation of MI described in Section 2.2. However, there are other notions of 'multivariate mutual information' which become more interesting and relevant in the context of higher-order attacks against protected implementations—see [1] for a full discussion.

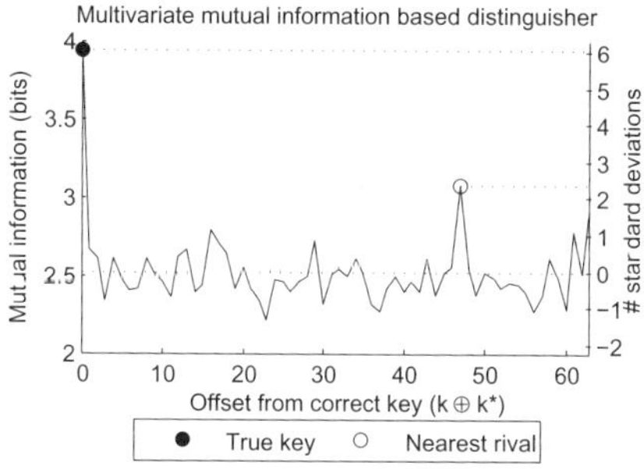

Fig. 2. Ideal MIA vector against the DES AddRoundKey and the first S-Box jointly

automatically raise the distinguishing power. In fact the true key is less strongly distinguished than in the attack against the S-Box alone: the nearest-rival distinguishability is reduced from 5.61 to 3.66. Moreover, the attack requires a larger input support—13 on average compared with 8 for MIA(HW).

Table 1 summarises outcomes for a wider selection of attacks, including MIA(ID): the proposed 'generic' attack of [6]. Unsurprisingly, in this first example where the leakage *is* proportional to the HW, MIA(ID) displays a disadvantage relative to MIA(HW). The generic capabilities of MIA will be of more relevance in leakage scenarios where the attacker is *not* able to correctly model the true leakage.

The attacks against AddRoundKey well illustrate the role of the target function: distinguishing power is greatly reduced in the case that incorrect key hypotheses give rise to outputs closely resembling the correct key outputs. Greater precision (and therefore a greater number of measured traces) will be required in order to detect a difference of this size in a practical attack, and moreover in the case of MIA there will remain an ambiguity between the true key k^* and its bitwise complement $\bar{k}^*$.

Table 1. Ideal strength of CPA and MIA attacks against DES with Hamming weight leakage

	AddRoundKey		**First S-Box**			**Multivariate**
DES with a HW leakage	**CPA (HW)**	**MIA (HW)**	**CPA (HW)**	**MIA (HW)**	**MIA (ID)**	**MMIA (HW)**
Correct key ranking (order)	1 (1)	1 (2)	1 (1)	1 (1)	1 (1)	1 (1)
Average score	2.45	4.48	3.61	6.59	6.35	6.04
Nearest-rival score	0.82	0.00	2.14	5.61	5.08	3.66
Average minimum support	6	9	6	8	16	13
Support required for 90% SR	8	11	8	11	19	15
Support required for 100% SR	11	15	12	14	22	21

Stochastic Resonance. We conclude this section by briefly considering the impact of (Gaussian) noise on theoretic outcomes. Figure 3 plots distinguishing scores against an increasing signal-to-noise ratio (SNR, defined as $\frac{\mathrm{var}(L \circ f_{k^*}(X))}{\mathrm{var}(\varepsilon)}$), confirming that (standardised) MIA outcomes are not constant. Moreover, the relationships are not monotonic: in each case there seems to be an optimal SNR at which the distinguishing scores reach a maximum, after which they diminish to that of the ideal (as depicted by the dashed lines). Such a phenomenon is a type of *stochastic resonance* [2], which can (in principle) occur in any nonlinear measurement system. Perhaps surprisingly, the required support sizes for both MIA(HW) and MIA(ID) match the ideal requirements and remain constant—though in general, such measures could also be subject to similar effects.

Recall, from Sect. 2.2, that by the properties of correlation, (standardised) CPA outcomes are unaffected by the level of noise. Hence the opportunity to enhance MIA (at least theoretically) by varying the noise is not available in the context of CPA.

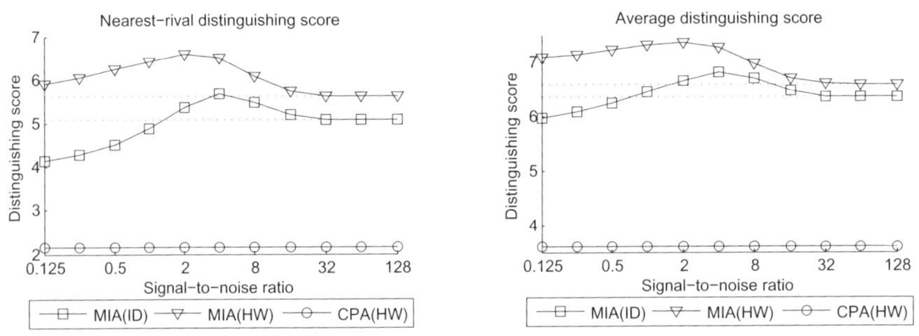

Fig. 3. The effect of Gaussian noise on HW and ID attacks against HW leakage of the first DES S-Box

4.2 Hamming-Distance Leakage

Whilst the Hamming weight model is very popular in the literature, Hamming distance leakage can be widely observed in practical devices using CMOS logic. Broadly speaking there are three scenarios which may be encountered. Firstly, the previous state is known to the attacker, in which case the attacks are equivalent to Hamming weight attacks. Secondly, the previous state is unknown to the attacker but fixed. Thirdly, the previous state is unknown to the attacker and can vary. The latter two scenarios are the focus of the following discussion.

Constant Reference State. Now let us suppose, as in [4], that the reference state is a constant but unknown machine word R. The device no longer leaks $\mathrm{L}(f_{k^*}(X))$ but rather $\mathrm{L}(R \oplus f_{k^*}(X))$.

First observe that no attack against a linear target function such as AddRoundKey can achieve first-order success, because the 'true key' values are perfectly replicated under an incorrect key hypothesis, namely $k^* \oplus R$. The power consumption for a plaintext X will be proportional to $\text{HD}((k^* \oplus X), R) = \text{HW}((k^* \oplus X) \oplus R) = \text{HW}((k^* \oplus R) \oplus X)$, so that when our hypothesis is $k = k^* \oplus R$ we get maximum correlation/MI (for both HW and ID models) and in fact the theoretical distinguishing vector is identical to that of a successful attack against HW leakage with a key of $k^* \oplus R$.

Targeting the S-Box avoids this predicament thanks to the high nonlinearity of the S-Box. In particular, there is no R' such that $\text{S-Box}(k^* \oplus X) \oplus R = \text{S-Box}((k^* \oplus R') \oplus X)\ \forall X \in \mathcal{X}$, so no incorrect key will produce the correct predictions. It remains to be seen whether the resemblance between the imperfect predictions (with naive power models) and the true power consumption remains strong enough for the correct key and weak enough for the alternative hypotheses for any sort of attack to be successful.

Ideal CPA(HW) succeeds precisely in those scenarios where the HW of the reference is 1 (or 0) and fails whenever it is 2 (see Table 2). Further, were we to use the absolute value of the correlation to distinguish (denoting this strategy as |CPA(HW)|) the resulting ideal attack would succeed whenever the HW of the reference state is 3 or 4; however, there is a substantial reduction in theoretic strength when the HW is 1 or 3, and for some reference states |CPA(HW)| requires almost the entire plaintext set to determine the correct key.

MIA(HW) succeeds in all scenarios and gains a considerable advantage both in terms of the ideal distinguishing scores with full information (nearest-rival scores are in the range of 3.6 to 4.5 for MIA(HW) but just 0.5-2.7 for |CPA(HW)|) and also in terms of the minimum input support required for success (on average, 14 to 15 for MIA(HW) compared with 17 to 18 for |CPA(HW)|). We can take advantage of the non-injectivity of the DES S-Box to launch generic MIA(ID) attacks. As the authors of [6] observed, these are essentially unaffected by a constant reference state so that the nearest-rival distinguishing score is always around 5 for MIA(ID) and average support requirement around 16. This means that when $R \in \{0000_{(2)}, 1111_{(2)}\}$ (i.e. L is the HW function) the generic attacks are less effective than the equivalent methods combined with a HW power model, but in all other reference state scenarios they gain an advantage. The consistency and ideal strength of these attacks might be sufficient to translate into a practical advantage—a possibility which we will investigate in a latter section.

We have thus shown that MIA applied with little consideration for or knowledge about the true leakage can be effective even when that leakage actually depends on an unknown reference state. CPA, applied equally blindly, is far less likely to yield a successful attack. However, Brier et al. [4] showed how to adapt it in order to determine R as an unknown of the problem in addition to $f_{k^*}(X) \oplus R$, which together reveals the secret key k^*. Whilst this simultaneous search process is more computationally costly than a standard CPA(HW) attack, MIA with an ID power model can itself be computationally costly in addition to the likely data

Table 2. Theoretical strength of CPA and MIA attacks against DES with Hamming distance leakage from a constant reference state

4 LSBs of reference state	CPA (HW)	|CPA| (HW)	MIA (HW)	MIA (ID)
Hamming weight 1				
Correct key ranking	1	1	1	1
Average score	2.04-4.05	2.56-4.94	5.48-5.97	5.81-6.46
Nearest-rival score	0.38-2.28	0.53-2.65	3.60-4.47	4.57-5.20
Average minimum support	17-25	20-34	14-15	16-17
Support required for 90% SR	31-49	33-53	20-22	19-20
Support required for 100% SR	40-59	44-61	28-32	21-24
Hamming weight 2				
Correct key ranking	27-32	54-63	1	1
Average score	0.00	-1.94-0.00	5.06-5.53	5.98-6.43
Nearest-rival score	-2.31-0.00	-5.62-0.00	3.05-3.16	4.49-5.42
Average minimum support	-	-	17-18	16-16
Support required for 90% SR	-	-	26-29	19-20
Support required for 100% SR	-	-	33-36	22
Hamming weight 3				
Correct key ranking	64	1	1	1
Average score	-2.44-0.00	2.56-4.94	5.48-5.97	5.81-6.46
Nearest-rival score	-4.58-0.00	0.53-2.65	3.60-4.47	4.57-5.20
Average minimum support	-	20-34	14-15	16-17
Support required for 90% SR	-	33-53	20-22	19-20
Support required for 100% SR	-	44-61	28-32	21-24
Hamming weight 4				
Correct key ranking	64	1	1	1
Average score	0.00	5.14	6.59	6.35
Nearest-rival score	0.00	3.56	5.61	5.08
Average minimum support	-	6	8	16
Support required for 90% SR	-	8	11	19
Support required for 100% SR	-	12	14	22

complexity overheads. Further work (and broader cost considerations) would be required to establish which of the two methods is most practical.

A Note on DRP logic. We observe an important and useful parallel between HD leakage and the expected behaviour of DPA-resistant dual-rail precharge (DRP) logic. In fact, an imperfect realisation of the logic style can be shown to exhibit data-dependent power consumption of a similar form to the HD from a constant reference state, enabling us to clarify its vulnerability to the 'generic' MIA(ID) attack described by Gierlichs et al. in [6].

DRP logic attempts to eradicate the data-dependency of the power consumption by making it equal in each clock cycle. This is achieved insofar as the capacitances of the complementary output wires in each logic gate can be balanced, a difficult feat in practice [15]. Suppose the ith bit of an m-bit word x is carried by a DRP logic gate driving two differential outputs with imperfectly balanced capacities (α_i, β_i), so that $\alpha_i = \beta_i + \gamma_i$. The power consumption of such a circuit can be shown to be equivalent to leakage scenarios with which we are more familiar, enabling us to comment on theoretical attack capabilities.

Let us initially consider the simplified case that both capacitances are the same throughout the circuit: $\beta_i = \beta$, $\alpha_i = \beta + \gamma$, $\forall i \in \{0, \dots, m-1\}$. Then the data-dependent leakage is proportional to:

$$\begin{aligned}\mathrm{HW}(x)\alpha + \mathrm{HW}(\bar{x})\beta &= \mathrm{HW}(x)(\beta + \gamma) + \mathrm{HW}(\bar{x})\beta \\ &= (\mathrm{HW}(x) + \mathrm{HW}(\bar{x}))\beta + \mathrm{HW}(x)\gamma \\ &= m\beta + \mathrm{HW}(x)\gamma\end{aligned}$$

The constant $m\beta$ is absorbed into the non-data-dependent component and we thus obtain the result that the leakage is proportional to the Hamming weight. Both CPA(HW) and MIA(HW) will be theoretically capable of returning the correct key; practical success will depend on ability and resources to estimate the distinguishing vectors with sufficient precision, in which case CPA(HW) is likely to have an advantage, as we have already seen.

Now suppose that the capacitances are the same throughout the circuit but that the order changes, i.e. so that some gates have capacitances (α, β) and others (β, α), where $\alpha = \beta + \gamma$. We can express this by introducing $R = (r_0, \dots, r_{m-1}) \in \{0,1\}^m$ such that gate i is (β, α) if $r_i = 1$ and (α, β) otherwise. Then the data-dependent leakage is:

$$\begin{aligned}\mathrm{HW}(x \oplus R)\alpha + \mathrm{HW}(x \oplus \bar{R})\beta &= \mathrm{HW}(x \oplus R)(\beta + \gamma) + \mathrm{HW}(x \oplus \bar{R})\beta \\ &= (\mathrm{HW}(x \oplus R) + \mathrm{HW}(x \oplus \bar{R}))\beta + \mathrm{HW}(x \oplus R)\gamma \\ &= m\beta + \mathrm{HW}(x \oplus R)\gamma\end{aligned}$$

That is, the data-dependent leakage is proportional to the Hamming distance from R, which equates to the scenario of a more conventional logic style (such as CMOS) consuming power proportional to the number of transitions from a constant, unknown reference state. We have already shown that MIA(ID) remains ideally successful against such leakage, whilst CPA(HW) is (depending on the state) either unsuccessful or greatly reduced in distinguishing power. This confirms that DRP logic gives rise to leakage scenarios under which first-order MIA(ID) could be useful, in particular, shedding light on the experimental result of [6].

In the most general case, the size of the capacitances and not just the direction of the differences may vary over the circuit. Suppose the gates corresponding to bits $i = 1, \dots, m$ have capacitances (α_i, β_i) such that $\alpha_i = \beta_i + \gamma_i$ where γ_i can be positive or negative. Letting $\mathbf{x} = (x_1, \ldots, x_m)$ and $\alpha = (\alpha_1, \ldots, \alpha_m)$, $\beta = (\beta_1, \ldots, \beta_m)$, $\gamma = (\gamma_1, \ldots, \gamma_m)$ we get a leakage function of $\mathbf{x} \cdot \alpha + (\mathbf{x} \oplus \mathbf{1}) \cdot \beta = (\mathbf{x} + \mathbf{x} \oplus \mathbf{1}) \cdot \beta + \mathbf{x} \cdot \gamma = \mathbf{1} \cdot \beta + \mathbf{x} \cdot \gamma$, so that the data-dependent power consumption is proportional to a weighted combination of the bits of $\mathbf{x}$, where the weights can take negative values. Further investigation is needed to establish the expected behaviour of our distinguishers as the relative weights become increasingly disproportionate.

Data-Dependent Reference State. We next investigate ideal performance against Hamming distance leakage allowing for R to take two or more different values depending on the plaintext, unknown to the attacker, but restricting it to be constant in repeated runs. In practice this could happen due to an incorrect implementation of a masking scheme.

In the (commonly studied) case of an 8-bit micro-controller, the reference states (or masks) take values in $\{0,1\}^8 = \{0, \ldots, 255\}$. Since our attacks on the first DES S-Box target 6-bit key portions, our plaintext inputs are drawn from $\{0,1\}^6 = \{0, \ldots, 63\}$—there could be up to 64 different input-dependent reference states. The number of possible ways that r reference states could be associated with the 64 input values is given by the Stirling number of the second kind: $\left\{ {64 \atop r} \right\} = \frac{1}{r!} \sum_{j=0}^{r} (-1)^{r-j} \binom{r}{j} j^{64}$, so it is no longer possible to exhaustively explore every scenario. Instead, we calculate the success rates in 1,000 random experiments for increasing numbers of different reference states, randomly assigned to approximately equal-sized subsets of the input space (see Figure 4)[4].

We find that MIA is much better able to succeed than |CPA|, particularly when provided with an ID power model—although even then it does not achieve 100% success for attacks with more than 2 different states and for more than 6 states success rates drop to below 50%. The success of |CPA(HW)| degrades rapidly; for attacks with about 20 different states it is no better than a random guess, whilst MIA(ID) and even MIA(HW) appear to retain some advantage over guessing.

Thus, when very little is known about the leakage an attacker may well be able to recover a great deal of information just by applying a 'blind' MIA—though even ideal success will be partially determined by chance, and the number of traces required for adequate estimation may be prohibitive. Such an approach may not be the best way of exploiting the available data: where resources permit, it may prove more effective or efficient to refine a CPA based approach (or similar), investing greater effort in understanding the leakage to begin with, perhaps through profiling.

[4] When the reference state is constant, only the 4 bits which are replaced by the S-Box output contribute to the data-dependent leakage whilst the contribution of the remaining bits is absorbed into the static component of the power consumption. However, when the state depends on the data in the manner described here, the contribution of the remaining bits *does* need to be taken into consideration as it becomes part of the data-dependent power consumption.

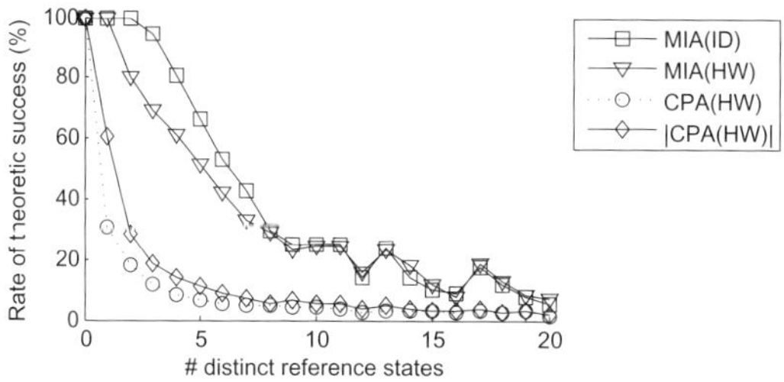

Fig. 4. Ideal success against the first DES S-Box in the presence of data-dependent reference states of length 8 bits, as the number of different states increases

4.3 Theoretical vs. Practical Success

We now return to a scenario which was identified as a candidate for MIA to hold an advantage over CPA in practice: Hamming-distance leakage from a reference state unknown to the attacker (taken to be $0100_{(2)}$ for the purposes of our example). We wish to investigate whether the observed ideal advantages generalise (theoretically) in the presence of noise and hence whether they can be translated into practical advantages. Figure 5 shows the impact of Gaussian noise on theoretic attack effectiveness, both in terms of nearest-rival distinguishability and in terms of the minimum support size required for first-order success. MIA(HW) distinguishability is not very robust to the addition of noise, even falling below that of CPA(HW). Moreover, there is a hefty penalty in terms of required support size. By contrast, MIA(ID) distinguishability is more robust and even exhibits some evidence of stochastic resonance-type behaviour, whilst required support size remains constant in the tested range.

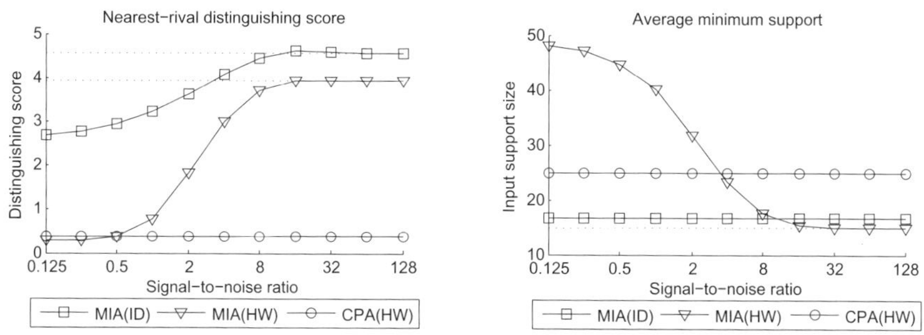

Fig. 5. Nearest-rival distinguishability and required support size of theoretic attacks against Hamming distance leakage (with a reference state of $0100_{(2)}$) for varying levels of Gaussian noise

Our simulated attacks use histogram-based estimators where bin counts are chosen equal to the cardinality of the power model domain, according to the

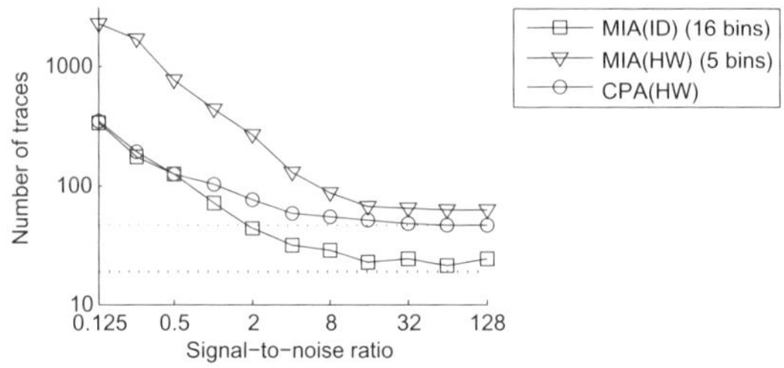

Fig. 6. Average number of traces required for key recovery in simulated practical attacks against Hamming-distance leakage (with a reference state of $0100_{(2)}$), for varying levels of Gaussian noise

heuristic which has emerged from the literature (see, for example, [1]). In a pure-signal scenario (see the dashed lines in Figure 6) the 5-bin estimator for MIA(HW) requires fewer traces than CPA(HW) to identify the correct key, but the introduction of even the smallest amount of noise incurs a burden so that across the tested range it is substantially less efficient. By contrast, the 16-bin estimator for MIA(ID) approaches the efficiency achieved in the pure-signal scenario as the SNR increases, and moreover substantially outperforms CPA(HW) once the SNR is at least 1. We have thus confirmed that—in this instance at least—ideal MIA advantages *can* be translated into practical advantages.

5 Conclusions

In this paper we have presented a framework for evaluating and comparing DPA methodologies on a like-for-like, ideal/theoretic basis. Our outcome measures allow for a nuanced assessment of the relative strengths and weaknesses of particular distinguishers as employed under different leakage scenarios. We have thus been able to compare MIA and CPA as abstracted away from the confounding problem of estimation, gaining valuable insight into the empirical results of existing literature which tends to focus on practical instantiations of the attacks. We have identified scenarios in which MIA offers a substantial theoretic advantage over CPA, and demonstrated that such theoretic advantages can be translated into practical advantages. Particular candidate scenarios for MIA to be useful arise when the leakage takes the form of the Hamming distance from an unknown reference state or in implementations using dual-rail precharge logic—and, in fact, we are able to demonstrate a relationship between these two cases. The generic capabilities of MIA are found to be an advantage as the HW model degrades relative to the true leakage, but multivariate extensions do not exhibit much if any advantage over univariate attacks in the first-order 'unprotected' setting. Lastly, we observe for the first time (to our knowledge) the noise-sensitivity of the (standardised) MIA distinguishing vector, which exhibits an effect which can be likened to stochastic resonance and which could possibly be exploited in certain noisy scenarios to enhance the distinguishing ability of

MIA attacks. This is a question for further research. Another open problem—persistently arising in the context of MIA—is that of finding estimators which most effectively translate theoretical advantages into practical ones.

Acknowledgements. The authors would like to thank Dave Cliff for pointing them towards the concept of stochastic resonance, and the anonymous referees for their comments. The first author of this paper has been funded via an EPSRC studentship. The second author has been supported by an EPSRC Leadership Fellowship I005226.

References

1. Batina, L., Gierlichs, B., Prouff, E., Rivain, M., Standaert, F.X., Veyrat-Charvillon, N.: Mutual Information Analysis: a Comprehensive Study. Journal of Cryptology 24, 269–291 (2011)
2. Benzi, R., Parisi, G., Sutera, A., Vulpiani, A.: Stochastic Resonance in Climatic Change. Tellus 34(1), 10–16 (1982)
3. Bonachela, J., Hinrichsen, H., Munoz, M.: Entropy Estimates of Small Data Sets. Journal of Physics A – Mathematical and Theoretical 41(20) (2008)
4. Brier, E., Clavier, C., Olivier, F.: Correlation Power Analysis with a Leakage Model. In: Joye, M., Quisquater, J.J. (eds.) CHES 2004. LNCS, vol. 3156, pp. 135–152. Springer, Heidelberg (2004)
5. Gierlichs, B., Batina, L., Tuyls, P., Preneel, B.: Mutual Information Analysis: a Generic Side-Channel Distinguisher. In: Oswald, E., Rohatgi, P. (eds.) CHES 2008. LNCS, vol. 5154, pp. 426–442. Springer, Heidelberg (2008)
6. Gierlichs, B., Batina, L., Tuyls, P., Preneel, B.: Mutual Information Analysis. In: Oswald, E., Rohatgi, P. (eds.) CHES 2008. LNCS, vol. 5154, pp. 426–442. Springer, Heidelberg (2008)
7. Guilley, S., Hoogvorst, P., Pacalet, R.: Differential Power Analysis Model and Some Results. In: Quisquater, J.J., Paradinas, P., Deswarte, Y., El Kalam, A. (eds.) Smart Card Research and Advanced Applications VI. IFIP, pp. 127–142. Springer, Boston (2004)
8. Hutter, M.: Distribution of Mutual Information. Advances in Neural Information Processing Systems 14, 399–406 (2002)
9. Kocher, P., Jaffe, J., Jun, B.: Differential Power Analysis. In: Wiener, M. (ed.) CRYPTO 1999. LNCS, vol. 1666, pp. 388–397. Springer, Heidelberg (1999)
10. Kreamer, H.C., Thiemann, S.: How many Subjects?: Statistical Power Analysis in Reasearch, 1st edn. Sage Publications Inc., Newbury Park (1987)
11. Madiman, M.: On the entropy of sums. In: 2008 IEEE Information Theory Workshop (2008)
12. Mangard, S., Oswald, E., Popp, T.: Power Analysis Attacks: Revealing the Secrets of Smart Cards. Springer, Heidelberg (2007)
13. Mangard, S., Oswald, E., Standaert, F.X.: One for all - all for one: Unifying standard DPA attacks. IET Information Security (to appear, 2011), preprint available from `http://eprint.iacr.org/2009/449`
14. Paninski, L.: Estimation of Entropy and Mutual Information. Neural Computation 15(6), 1191–1253 (2003)
15. Popp, T., Mangard, S.: Masked Dual-Rail Pre-charge Logic: DPA-Resistance Without Routing Constraints. In: Rao, J., Sunar, B. (eds.) CHES 2005. LNCS, vol. 3659, pp. 172–186. Springer, Heidelberg (2005)

16. Prouff, E.: DPA attacks and S-boxes. In: Gilbert, H., Handschuh, H. (eds.) FSE 2005. LNCS, vol. 3557, pp. 424–441. Springer, Heidelberg (2005)
17. Prouff, E., Rivain, M.: Theoretical and Practical Aspects of Mutual Information Based Side Channel Analysis. In: Abdalla, M., Pointcheval, D., Fouque, P.-A., Vergnaud, D. (eds.) ACNS 2009. LNCS, vol. 5536, pp. 499–518. Springer, Heidelberg (2009)
18. Schindler, W., Lemke, K., Paar, C.: A Stochastic Model for Differential Side Channel Cryptanalysis. In: Rao, J.R., Sunar, B. (eds.) CHES 2005. LNCS, vol. 3659, pp. 30–46. Springer, Heidelberg (2005)
19. Shiga, M., Yokota, Y.: An Optimal Entropy Estimator for Discrete Random Variables. In: Proceedings of the IJCNN, pp. 1280–1285. IEEE, New York (2005)
20. Standaert, F.X., Gierlichs, B., Verbauwhede, I.: Partition *vs.* Comparison Side-Channel Distinguishers: An Empirical Evaluation of Statistical Tests for Univariate Side-Channel Attacks against Two Unprotected CMOS Devices. In: Lee, P.J., Cheon, J.H. (eds.) ICISC 2008. LNCS, vol. 5461, pp. 253–267. Springer, Heidelberg (2009)
21. Standaert, F.X., Malkin, T.G., Yung, M.: A Unified Framework for the Analysis of Side-Channel Key Recovery Attacks. In: Joux, A. (ed.) EUROCRYPT 2009. LNCS, vol. 5479, pp. 443–461. Springer, Heidelberg (2009)
22. Treves, A., Panzeri, S.: The Upward Bias in Measures on Information Derived From Limited Data Samples. Neural Computation 7(2), 399–407 (1995)
23. Veyrat-Charvillon, N., Standaert, F.X.: Mutual Information Analysis: How, When and Why? In: Clavier, C., Gaj, K. (eds.) CHES 2009. LNCS, vol. 5747, pp. 429–443. Springer, Heidelberg (2009)

Key-Evolution Schemes Resilient to Space-Bounded Leakage*

Stefan Dziembowski[1], Tomasz Kazana[2], and Daniel Wichs[3]

[1] University of Warsaw and Sapienza University of Rome
[2] BioInfoBank Research Institute and University of Warsaw
[3] New York University

Abstract. Much recent work in cryptography attempts to build secure schemes in the presence of *side-channel leakage* or leakage caused by malicious software, like computer viruses. In this setting, the adversary may obtain some additional information (beyond the control of the scheme designer) about the internal secret state of a cryptographic scheme. Here, we consider key-evolution schemes that allow a user to evolve a secret-key K_1 via a *deterministic* function f, to get updated keys $K_2 = f(K_1), K_3 = f(K_2), \ldots$. Such a scheme is *leakage-resilient* if an adversary that can leak on the first i steps of the evolution process does not get any useful information about any future keys. For such schemes, one must assume some restriction on the *complexity* of the leakage to prevent *pre-computation attacks*, where the leakage on a key K_i simply pre-computes a future key K_{i+t} and leaks even a single bit on it.

Much of the prior work on this problem, and the restrictions made therein, can be divided into two types. Theoretical work offers rigor and provable security, but at the cost of having to make strong restrictions on the type of leakage and designing complicated schemes to make standard reduction-based proof techniques go through (an example of such an assumption is the "only computation leaks" axiom). On the other hand, practical work focuses on simple and efficient schemes, often at the cost of only achieving an intuitive notion of security without formal well-specified guarantees.

In this paper, we complement the two tracks via a middle-of-the-road approach. On one hand, we rely on the random-oracle model. On the other hand, we show that even in the random-oracle model, designing secure leakage-resilient schemes is susceptible to pitfalls. For example, just assuming that leakage "cannot evaluate the random oracle" can be misleading. Instead, we define a new model in which we assume that the "leakage" can be any arbitrary *space bounded* computation that can make random oracle calls itself. We connect the space-complexity of a computation in the random-oracle modeling to the *pebbling complexity* on graphs. Using this connection, we derive meaningful guarantees for relatively simple key-evolution constructions.

Our scheme is secure also against a large and natural class of active attacks, where an attacker can leak as well as tamper with the internals of a device. This is especially important if the key evolution is performed on a PC that can be attacked by a virus, a setting considered by prior work in the *bounded retrieval*

* The European Research Council has provided financial support to the first two authors of this paper under the European Community's Seventh Framework Programme (FP7/2007-2013) / ERC grant agreement no CNTM-207908.

P. Rogaway (Ed.): CRYPTO 2011, LNCS 6841, pp. 335–353, 2011.

model (BRM)). This paper provides the first scheme were the adversary in the BRM can also modify the data stored on the machine.

Keywords: graph pebbling, leakage-resilient cryptography, bounded-retrieval model.

1 Introduction

In the recent years, there has been a growing interest in the design of cryptographic schemes that are secure even if implemented on a devices that can leak information. The motivation for this research comes from the fact that, in practice, it is very hard to construct hardware that does not reveal any "extra" information about its internal data. In particular, leakage on the internals can often be obtained using *side-channel attacks*, that exploit physical phenomena such as electromagnetic radiation [36,30], timing [5], power consumption [29], acoustic emanations [38], and many others (see e.g. [33] for an overview). Another case in which the adversary may obtain leakage from cryptographic protocols is the situation when the protocols are implemented on PCs on which the adversary can install malicious software, like the computer viruses.

The first papers proposing algorithmic countermeasures against the side-channel attacks came from the practitioners' community (e.g. [7]). Later this area attracted a lot of attention also from the theoreticians, starting from the seminal papers of Ishai et al. [24], Micali and Reyzin [31], and Akavia et al. [1]. The theoretical countermeasures against the virus attacks are known under the name *bounded-retrieval model* [8,14].

In this paper we consider the following natural problem. Suppose a cryptographic key K_0 is stored on a device that leaks information. If leakage occurs continuously, then the adversary may obtain more and more information about the key, and eventually learn it entirely. A natural idea to prevent this from happening is to periodically update the key i.e. to repeatedly apply some *key evolution function* f to it, obtaining a sequence of keys $K_0, K_1, \ldots$, where each $K_{i+1} := f(K_i)$. The key evolution function should be constructed in such a way that the evolved key K_i used in period i is indistinguishable from uniform, even if the adversary can leak from the entire evolution process $K_1 \rightarrow K_2 \rightarrow \ldots \rightarrow K_{i-1}$. We will assume that the key evolution is deterministic (i.e. it does not depend on any external randomness) hence these keys K_i will be shared by synchronized devices, which evolve their keys simultaneously but independently (without communication). As we will see, the design of key-evolution functions is also intimately related to the design of leakage-resilient stream-ciphers, and pseudo-random generators. The problem of designing such primitives been studied before, both by the practitioners and by the theoreticians. Next, we look at several models of leakage-resilience and their results.

1.1 Leakage Resilient Key Evolution: Theory vs. Practice

Theoretical work on leakage-resilience usually included a formal model for reasoning about leakage. Usually, side-channel leakage is modeled as a family of functions where the attacker can choose a function from the family and learn the output of this function applied to the secret key. For example, a popular and powerful model of Akavia et al. [1], allows the adversary to compute arbitrary poly-time leakage functions of the internal secret key, subject only to the constraint that the amount of data retrieved (the output-length of the function) is bounded. Unfortunately, when it comes to key-evolution, it is clear that security *cannot* be achieved in this model, even if the adversary is restricted to leaking a *single bit*! In fact, just given the ability to leak on the initial key K_1, the adversarial leakage-function can *pre-compute* any future key K_i and output (say) the first bit of it. If the adversary can leak even 1 bit in several consecutive rounds, the adversary can eventually recover any future key K_i in full! This example (called a *key-precomputation attack* in [19]) shows that, when considering the security of the key-evolution schemes, the sole restriction that the output of the leakage function is bounded does not suffice. However, it is also easy to see that the leakage-functions used in the above counter-examples are extremely artificial, and are very unlikely to model natural side-channel attacks that occur in real life. Hence, it is natural to look for different (weaker) models for the leakage, that still cover all the *realistic* attacks, but in which key-evolution schemes may exist. We survey two such key-evolution schemes in their corresponding models of leakage. The two schemes come from two very different point views: one theoretical with an emphasis on models and proofs, and the other practical with an emphasis toward efficiency and simplicity. Therefore, it is interesting to compare their advantages and disadvantages.

The scheme of Dziembowski and Pietrzak [19]. On the theoretical side, [19] constructed a stream cipher (and, implicitly, a key evolution scheme) in a formal model called "only computation leaks information", first proposed by Micali and Reyzin [31] and refined in [19]. In this model, the internal memory of the device is separated into two or more segments, and all computation is divided into simpler sub-computations that access only some small subset of these segments. The assumption is that, during each sub-computation, the adversary can leak an arbitrary bounded-length function of only the memory segments accessed by the sub-computation. In other words, during any computational step, data can leak if and only if it is accessed. Pre-computation attacks can therefore be prevented, since the adversary can never get any global leakage of the entire state of the system needed to compute a future key.

The actual scheme of [19] (and a related scheme of [34]) uses an alternating structure with two memory segments accessed in alternating rounds. The main drawback of this model is that it relies on the highly controversial assumption that data which is not accessed cannot leak. Also, the security of solutions in this model is highly dependent on "implementation details" such what data is accessed when.

The scheme of Kocher [28]. On the practical side, Kocher [28] proposes a simple and efficient solution to just use a "sufficiently complicated" cryptographic hash function for the key-evolution function f. This scheme seems intuitively secure since it's unlikely that any natural and therefore "sufficiently simple" leakage could learn anything about $K_{i+1} = f(K_i)$ from K_i (or even from the entire key-evolution process used to derive K_i), if f is "sufficiently complicated". However, [28] does not offer any meaningful model in which to analyze the above intuition. The main idea is to assume that the adversarial leakage on the ith update $K_i \to K_{i+1}$ can be an *arbitrary function* of the entire state of that update subject to the constraints: (1) the leakage function cannot make any random-oracle calls, (2) the output-length of the leakage function is bounded to be just slightly smaller than the key-length $|K|$. At first, it may seem that constraint (1) offers a meaningful way of capturing "sufficiently simple" leakage. Unfortunately, it is not clear what this constraint means in practice, and can lead to counter-intuitive consequences, described next.

Since the amount of leakage tolerated by the scheme should be close to $|K|$, it is natural to try to increase $|K|$ if one would want to achieve more leakage. Of course, that means using a key-evolution function, and therefore hash function, with sufficiently large input-size and output-size. Assume we start with a compression function $H : \{0,1\}^{2\ell} \to \{0,1\}^{\ell}$, and we want to allow key-size $|K| = t\ell$ to allow for more leakage. The standard technique for domain-extension is the Merkle-Damgård transformation [9] (and variants of it in the indifferentiability framework) which gives a function $H' : \{0,1\}^* \to \{0,1\}^{\ell}$. Now, to increase the output-size, we can define $\tilde{H} : \{0,1\}^{t\ell} \to \{0,1\}^{t\ell}$ by $\tilde{H}(K) = H(H'(K)||1), \ldots, H(H'(K)||t)$. But, it is clear that, if one leaks the ℓ-bit value $H'(K_1)$ used as sub-component of the key-evolution computation $K_2 = \tilde{H}(K_1)$, then one can compute all future keys K_j and thus completely break the key-evolution scheme! Therefore, the scheme is not secure with respect to even ℓ bits of leakage when instantiated with real-world hash functions. Notice that the leakage-function does not perform any complicated computation; it just leaks several consecutive bits of the internal state, corresponding to $H'(K)$, which is computed as an intermediate value during the key-evolution computation.

So we see that, although the initial scheme of Kocher provides some intuitive leakage-resilience properties, one runs into pitfalls when trying to model and quantify them. In particular, by relying on the random-oracle model in an unintended way (assuming that simple functions cannot make random oracle calls), and assuming that leakage on a computation making random-oracle calls only gets the input and output of such calls, one reaches a model that doesn't correspond to reality in a meaningful way.

The scheme of Yu Yu et al. [40]. In a recent important work Yu Yu et al. [40] propose a practical scheme whose security is based on the assumptions that (1) the leakage functions cannot be chosen adaptively, and (2) the leakage function cannot evaluate the hash function (which is modeled as a random oracle). The security of the scheme that we construct in this paper does not require these assumptions. On the other hand, from the engineering point of view the assumptions made in [40] may look more attractive, since they are easier to verify empirically.

Other Models of Leakage-Resilience. We note that several other models of leakage resilience, with restricted leakage functions, have appeared in the literature. For example, [24] assumes leakage functions that leak individual wires from a circuit that performs a computation. Alternatively, Faust et al. [21] assume that leakage-function is an AC^0 circuit of the internal state. Several works consider computation that uses some small/simple leak-free components [22,21,23,26].

1.2 Our Model: Space-Bounded Leakage

In this paper we propose a method for the key evolution that combines the advantages of the Kocher's practical scheme (efficiency and simplicity of model, scheme) with some of the advantages of the theoretical scheme of [19] (provable security and scalability). On a high level, we restrict the class of leakage functions to ones that are bounded in the amount of "auxiliary work space" used during the computation of the output. In particular, this should model natural leakage which is unlikely to be sufficiently complex to require much space to compute. We will analyze a variant of the Kocher key-evolution scheme in the random-oracle model, but give all parties (including the leakage functions) the ability to compute the random oracle. In particular, we define a deterministic key-evolution function f that makes random-oracle calls to compute $K_{i+1} = f(K_i)$. On a high level, pre-computation attacks will not be possible because the space allowed to the leakage function is not sufficient to pre-compute K_{i+1} from K_i.

In greater detail, we model the leakage process on the key evolution as follows. We consider an adversary $\mathcal{A} = (\mathcal{A}_{small}, \mathcal{A}_{big})$ consisting of two parts. the adversary $\mathcal{A}_{big}$ corresponds to the external real-world attacker that tries to break the scheme, and $\mathcal{A}_{small}$ corresponds to the "space-bounded" device which is storing and evolving the secret key while leaking partial information to the external attacker $\mathcal{A}_{big}$. We do not assume anything about how the computation is implemented on the device, and thus allow the computation $\mathcal{A}_{small}$ itself to be adversarial. Initially, $\mathcal{A}_{small}$ is given the random starting key K_1. Since key evolution is deterministic, this will completely specify all future keys $K_2, K_3, \ldots$. Of course, we need somehow to "force" $\mathcal{A}_{small}$ to perform the key evolution (if $\mathcal{A}_{small}$ can completely "halt" the key evolution, then he can simply keep K_1 on $\mathcal{M}$ for a long time, and slowly retrieve it bit-by-bit). In the passive case we could simply assume that, in time period i, the key K_i is fully stored on the machine and therefore there is not enough memory on the machine to store much information about any of the prior keys from earlier time periods. However, if $\mathcal{A}_{small}$ is active then this assumption could be completely unreasonable as the attacker could just keep K_1 on the machine for arbitrarily many time periods and leak it entirely. Therefore, we will introduce a special procedure that we call Verify_i, and assume that this procedure is called in each time period i to ensure that the device is storing the full key K_i at that point in time.

During the entire key-evolution process, $\mathcal{A}_{small}$ can communicate with $\mathcal{A}_{big}$ and can perform arbitrary computation (in addition to / instead of computing K_i honestly) *including* the ability to make random oracle calls. We only make three restrictions: (1) the amount of data that $\mathcal{A}_{small}$ can send to $\mathcal{A}_{big}$ *in each period* is bounded (2) the space-complexity of $\mathcal{A}_{small}$ is bounded and not much larger then the space-complexity of the honest computation of f, (3) the number of oracle-queries made by all parties is polynomial (no other computational assumptions are made). (4) At the end of round i, the machine $\mathcal{A}_{small}$ stores the correct key K_i. We will allow unlimited communication in the other direction $\mathcal{A}_{big}$ to $\mathcal{A}_{small}$.

Let us elaborate on the restrictions in more detail. Restriction (1) models the fact that natural leakage is too simple to reveal too much data about any single computational step. Restriction (2) models that fact that the *complexity* of natural leakage functions is rather simple. In particular, the space-complexity of leaking on the internals of a computation should not be much larger then the amount of space actually used by the computation itself! This seems to be a rather conservative assumption. Lastly, restriction (3) models that all parties run in polynomial time, as is standard in cryptography, and restriction (4) models the fact that the device itself correctly computes the key K_i in round i.

Now that we have explained the model, let us give some intuition why key-evolution schemes are achievable in it. Firstly, note that, in round i, the adversary can (in principle) pre-compute any future key K_{i+t} in the space it is allotted. However, we will ensure that such computation would necessarily require the adversary to erase some data about K_i (since it cannot store all of K_i *and* compute K_{i+1} simultaneously with limited space) and hence it will be unable to satisfy the requirement that K_i is stored on the system at the end of round i.

1.3 Our Results

We construct a key-evolution function f that is secure in the model described above. Let c be the amount of bits that the adversary can retrieve in each round, and let s be the space that the adversary can use to compute the leakage function (including the $|K|$ bits needed to store the key K). We show that our scheme is secure as long as

$$4c + s \leq 3 \cdot |K|/2 \tag{1}$$

(cf. Theorem 1). Let us mention two applications of our construction.

Security against passive leakages. Firstly, suppose that the evolving key K_i is stored on some device (say, a smart-card) that may leak some information. Imagine that the device is used for (message or entity) authentication with a trusted server that has his own copy of K_i. Suppose the adversary can get a temporary access to the device, observe the process of key evolution and learn some partial information about the keys. At some later point the adversary looses access to the device. The properties of our function f will guarantee that the future keys are unknown to the adversary assuming that the leakage is bounded in the way described above. Since in this case the adversary is only passive, there is no need to perform the procedure Verify_i. It may look like the model described above is stronger than what we need for this application, since it seems too pessimistic to assume that the adversary fully controls how the keys K_i are computed on the device. It may seem tempting to consider a weaker model, where the computation is done is some honest way, and the adversary can apply the leakage functions during the evolution process. We believe that such a restriction would not make the proof simpler (while it would make the model more complicated). Moreover, going to the extreme and allowing the adversary to control the computation has the advantage that it protects us (to a certain extent) against implementation errors.

Security against active attacks in the BRM. The second application of our construction concerns the bounded-retrieval model (BRM) [8,15]. In this model one constructs schemes where the cryptographic key K is very large. The idea is that K can be stored on a PC that can be infected by viruses, and, as long as the virus does not retrieve a large portion of K, the scheme should remain secure. So far, all the work in the BRM considered only the passive attacks, where the virus was not allowed to modify the data on the machine. Now, consider the following problem: suppose we are using the BRM scheme for the session-key agreement [15,6] (where a pair of users share a secret key K), and we want to evolve the secret key K stored on the machine, so that in total over a long period of time we can tolerate a leakage of more than $|K|$ bits from the machine. We show that f can be used as such a key-evolution function. The details of the model are as follows. Suppose we store the keys K_i on the machine $\mathcal{M}$, and assume that the size of the local memory on $\mathcal{M}$ is s, and the amount of bits that can be retrieved in each key-evolution round is c, and c and s are such that (1) holds. Suppose $\mathcal{M}$ wants to authenticate to a trusted server that has his own copy of K_i. If there is a virus on $\mathcal{M}$ then we can even allow him to modify the data stored on $\mathcal{M}$. The restrictions that we impose are as follows. First, we assume that any computation that the virus performs has to be done within $\mathcal{M}$'s memory (of size s). Second, we somehow need to guarantee that the verification procedure Verify_i can be performed. We do it by assuming that $\mathcal{M}$ is equipped with a small tamper-free component $\mathcal{D}$ that can periodically check if the contents of the memory is "correct". For example, $\mathcal{D}$ could store the values of some hash function of $K_1, K_2, \ldots$,, and the Verify_i procedure would just consist of hashing the contents of the memory where K_i is supposed to be stored, and comparing the result with $H(K_i)$.

1.4 Some Implementation Details

In this paper we do not define formally the security of the concrete schemes (like the message authentication), since we are more interested in considering the key-evolution as an abstract procedure. Every key K_i will consist of $N-1$ blocks: $K_i = (K_i^0, \ldots, K_i^{N-1})$, each of the blocks being an output of a hash function modeled as a random oracle. Hence, we can simply say that K_i is secret if none of its blocks has every been calculated by the random oracle (in our model calculating such a block will correspond to "labeling" a vertex in some graph).

Of course, the key evolution scheme would be useless, if we could not use the evolved key in some other application. In other words, after each round i, the key evolution function should output some key κ_i, and in the formal model this κ_i should be given to $\mathcal{A}_{big}$ "for free". In our case we simply assume that κ_i is equal to one of the blocks of K_i. Note, that it does not require any modification of the model, since we can as well assume that κ_i is sent to $\mathcal{A}_{big}$ by $\mathcal{A}_{small}$.

The Verify_i procedure (that verifies the knowledge of K_i) can be implemented as follows: (1) the verifier (that knows K_i) sends a random value c to the device, and (2) the device replies with $v = \mathsf{MAC}(c, K_i)$ (where MAC is a tagging function of some message authentication code scheme, c is treated as a key for the MAC, and K_i is treated as a message), (3) the verifier checks if $v = \mathsf{MAC}(c, K_i)$, and if not then he aborts. Note, that we cannot hope for more than only verifying the correctness, since in the worst case an active adversary can anyway completely destroy the contents of the device. In our model we will not assume anything about how $\mathsf{MAC}(c, K_i)$ is computed by $\mathcal{A}_{small}$. For example, it will be possible that he partially "pre-computes" it before learning c. The only property of MAC that we use is that $\mathcal{A}_{small}$ can compute the value of $\mathsf{MAC}(c, K_i)$ only if each of the blocks of K_i were computed by him at some point earlier.

1.5 Organization

Our key-evolution function is defined using a special type of a graph, that we call a *tower graph*. The method of translating graphs into functions is described in Section 2. The tower-graphs and the function f are defined in Section 3. The main theorem is stated in Section 4, which also introduces most of the tools needed for the proof. The proof itself appears in Section 5.

1.6 Related Work

The theoretical countermeasures against the side-channel attacks were considered in [19,1,35,27,32,10,39,11,12,20,4,3]. The schemes in the Bounded Retrieval Model were constructed in [8,15,14,6,18,25,2]. We will use the technique called "graph pebbling" (cf. e.g. [37]), that was already used in cryptography in [13]. Some of our techniques (esp. those used in Sections 4.1 and 4.5) were introduced in [13] and recently extended in [17]. We note that, although our techniques are quite similar, the application is completely different (the main application of [17] is a construction of a scheme for the password-protected local storage). Unfortunately, it is impossible to use the theorems from [13,17] in a black-box way, and therefore we needed to non-trivially extend them. On a technical level, the main difference comes from the fact that in [17] the total amount of leakage was bounded globally, and in our paper we need to consider continual leakage over an unbounded number of round.

2 Random-Oracle Labeling of a Graph.

Let $G = (V, E)$ be a directed acyclic graph (DAG). A vertex v is a *child* of a vertex v' if there is an edge from v to v'. Let V_0 be the set of its *input vertices*, i.e. the vertices without children. A *labeling* of G is a function $\mathtt{label}(\cdot)$, which assigns values $\mathtt{label}(v) \in \{0,1\}^w$ to vertices $v \in V$. We call w the *label-length*. For any function $\mathcal{H} : \{0,1\}^* \rightarrow \{0,1\}^w$ and input-labels $K = (K_1, \ldots, K_N)$ with $K_i \in \{0,1\}^w$, we define the $(\mathcal{H}, K)$-labeling of G as follows:

- The labels of the N distinct input vertices $v_1, v_2, \ldots, v_N$ are given by $\mathtt{label}(v_i) \stackrel{\text{def}}{=} K_i$.
- The label of every other vertex v is defined recursively by

$$\mathtt{label}(v) \stackrel{\text{def}}{=} \mathcal{H}(\mathtt{label}(v_1), \cdots, \mathtt{label}(v_j), v)$$

  where $v_1, \cdots, v_j$ are the children of v.

A *random oracle labeling* of G is an $(\mathcal{H}, K)$-labeling of G where $\mathcal{H}$ is a random-function and K is chosen uniformly at random. For convenience, we also define $\mathtt{preLabel}(v) \stackrel{\text{def}}{=} (\mathtt{label}(v_1), \ldots, \mathtt{label}(v_j), v)$, where $v_1, \ldots, v_j$ are the children of v, so that $\mathtt{label}(v) = \mathcal{H}(\mathtt{preLabel}(v))$.

3 Our Key-Evolution Scheme

In this section we define our key-evolution function f. We start with defining a special type of graphs, that we call the "tower graphs". A graph $G = (V, E)$ is called an (N, M)-*tower graph* if $V = \{0, \dots, M-1\} \times \{0, \dots, N-1\}$ and $E = \{((i,j),(i+1,j)) : i \in \{0,1,\dots,\}, j \in \{0,\dots,N-1\}\} \cup \{((i,j),(i+1,(j-1) \bmod N) : i \in \{0,1,\dots\}, j \in \{0,\dots,N-1\}\}$ (cf. Figure in the appendix of extended version [16]). For $i = 0, \dots, t$ the set $V_i = \{(i,0), \dots, (i, N-1)\}$ is called the *ith line of G*. Let $V_{\geq i}$ denote the set $V_i \cup V_{i+1} \cup \cdots$. Note that the set of the input vertices of G is equal to V_0. We will say that an (infinite) graph G is an N-*tower graph* if it is an (N, ∞)-tower graph.

We are now ready to define f. If we fix a hash function $\mathcal{H}$ and label length w then the (N, M)-tower graph G defines a function $f : \{0,1\}^{Nw} \to \{0,1\}^{Nw}$ in the following way. On an input K the function f computes the $(\mathcal{H}, K)$-labeling of G and it outputs $(K'_1, \dots, K'_N)$, where each K_i is the label of $(M-1, i)$. The procedure for computing $f(K_0, \dots, K_{N-1})$ simply computes the labels bottom-up row-by-row in the following way:

- Set $(K^0_0, \dots, K^0_{N-1}) := (K_0, \dots, K_{N-1})$.
- For $j = 1, \dots, M-1$ do
  - For $i = 0, \dots, N-1$ do $K^j_i := \mathcal{H}(K^{j-1}_i, K^{j-1}_{i+1 \bmod N}, (i,j))$

Observe that the time needed to compute f is roughly equal to $N \cdot M$ times the time needed to compute $\mathcal{H}$, and the space needed to compute f is only slightly larger than the space needed to store K, since we can overwrite each $(K^j_0, \dots, K^j_{N-1})$ with $(K^{j+1}_0, \dots, K^{j+1}_{N-1})$ and hence re-use the space.

It is also easy to see that iterating the computation of f on the same input a times, i.e. computing $K' = f^a(K)$ can be seen as computing the labeling of an (N, aM)-tower graph, and in particular, if we want to evolve the key K^0 using the procedure $K^{i+1} = f(K^i)$ (for $i = 0, 1, \dots$) then we can look at it as a labeling the tower infinite N-tower graph, where the keys $K^1, K^2, \dots$ appear as labels of the lines $V_{1 \cdot M}, V_{2 \cdot M}, \dots$. We will call such lines the *round-switching lines*.

4 Games on Tower Graphs

We will show a connection between an adversary computing a "random oracle graph" and a pebbling game for the corresponding graph. A similar connection appears in [13] (and in [17], see Section 1.6 for more on relation between this work and [17]).

4.1 Model of Computation

Our main goal is to show that computing the labeling of a tower graph G requires a large amount of resources in the random-oracle model, and is therefore difficult. To do so, we must fix a model of computation in which we can make statements of the above form precise. Recall that we will usually consider an adversary that consists of two parts: a "space-bounded" component which gets access to the internals of an attacked device and has "bounded communication" to an external, and otherwise unrestricted, adversary.

We model such a adversary $\mathcal{A} = (\mathcal{A}_{big}, \mathcal{A}_{small})$ as a pair of interactive algorithms[1] with oracle-access to a random-oracle $\mathcal{H}(\cdot)$. Let M be some natural number that we will call the *round length*. While executing the algorithms the time is divided into rounds. Initially the computation is in a round 1. The adversary $\mathcal{A}_{small}$ is responsible for switching to next round. Namely: the round is changed to k when $\mathcal{A}_{small}$ calls special function $\texttt{nextRound}_k(\texttt{label}(a_1)\ldots\texttt{label}(a_n))$, where $\{a_1,\ldots,a_n\}$ is the kth *round-switching line* $V_{k\cdot M}$. A round k can be switched only to round $k+1$, in other words the order of the round-changing calls has to be $\texttt{nextRound}_1, \texttt{nextRound}_2, \ldots$. The period between the calls $\texttt{nextRound}_i$ and $\texttt{nextRound}_{i+1}$ will be called *the ith round*. The algorithm $\mathcal{A}_{big}$ will only be restricted in the number of oracle calls made. On the other hand, we impose the following additional restrictions on $\mathcal{A}_{small}$:

- s-bounded space: The total amount of space used by $\mathcal{A}_{small}$ is bounded by s. That is, we can accurately describe the entire configuration of $\mathcal{A}_{small}$ at any point in time using s bits.[2]
- c-bounded communication: The total number of outgoing bits communicated by $\mathcal{A}_{small}$ in each round is bounded by c.[3]

Note that these restrictions imply that he total number of outgoing bits communicated by $\mathcal{A}_{small}$ in every round is bounded by c and there is no global bound for communication. We use the notation $\mathcal{A}^{\mathcal{H}(\cdot)}(K) = \left(\mathcal{A}_{big}^{\mathcal{H}(\cdot)}() \leftrightarrows \mathcal{A}_{small}^{\mathcal{H}(\cdot)}(K)\right)$ to denote the interactive execution of $\mathcal{A}_{big}$ and $\mathcal{A}_{small}$, where $\mathcal{A}_{small}$ gets input K and both machines have access to the oracle $\mathcal{H}(\cdot)$. In particular, we will usually (only) care about the list of random-oracle calls made by $\mathcal{A}_{big}$ and $\mathcal{A}_{small}$ during such an execution. We say that an execution $\mathcal{A}^{H(\cdot)}(K)$ *labels* a vertex v, if a random-oracle call to $\texttt{preLabel}(v)$ is made by either $\mathcal{A}_{big}$ or $\mathcal{A}_{small}$. We are now ready to state our main theorem.

Theorem 1. *Let G be a N-tower graph and $\lambda > 0$. Suppose c, s and q are such that $\frac{4c+s+\lambda}{w-\log(q)} \leq N + N/2$, and let T be an arbitrary natural number. Let $\mathcal{A} = (\mathcal{A}_{big}, \mathcal{A}_{small})$ be an adversary with c-bounded communication and s-bounded storage that makes at most q queries to H. The probability p (taken over the choice of $(\mathcal{H}, K)$) that there exists $i = 1, \ldots, T-1$ such that $\mathcal{A}$ labels the line $V_{(i+1)\cdot M}$ of G in round i is at most*

$$q \cdot 2^{-w} + T \cdot 2^{1-\lambda} \tag{2}$$

The proof appears in Section 5. The necessary machinery is introduced in the next sections.

[1] Say ITMs, interactive RAMs, ... The exact model will not matter.

[2] This is somewhat different than standard space-complexity considered in complexity theory, even when we restrict the discussion to ITMs. Firstly, the configuration of $\mathcal{A}_{small}$ includes the value of *all* tapes, including the input tape. Secondly, it includes the current state that the machine is in and the position of all the tape heads.

[3] To be precise, we assume that we can completely describe the patters of outgoing communication of $\mathcal{A}_{small}$ using c bits. That is, $\mathcal{A}_{small}$ cannot convey additional information in when it sends these bits, how many bits are sent at a given time and so on.

4.2 Pebbling Games on Tower Graphs

We will consider a variant of the pebble game that we call the "red-black" pebble game over an N-tower graph $G = (V, E)$. Each vertex of the graph G can either be empty, contain a red pebble, contain a black pebble, or contain both types of pebbles. More precisely, if G is a tower graph, then a *pebbling configuration on* G is a function $\gamma : V \to \mathcal{P}(\{\mathtt{red}, \mathtt{black}\})$. Define $\mathit{Red}(\gamma) := \{v : \mathtt{red} \in \gamma(v)\}$, and $\mathit{Black}(\gamma) := \{v : \mathtt{black} \in \gamma(v)\}$. If $V' \subseteq V$ then define $\mathit{proj}(V') := (|V' \cap V_1|, \ldots, |V' \cap V_t|)$.

For a set $V' \subseteq V$ denote by $[V']$ the *closure of* V defined recursively as follows:

- if $v \in V'$ then $v \in [V']$,
- if all the children of v' are in $[V']$ then $v' \in [V']$.

An initial configuration γ_1 consists of (only) a black pebble placed on each input vertex of G. The game proceeds in steps where, in the ith step, the configuration γ_i is transformed into γ_{i+1} using one of the following four actions:

1. A red pebble can be placed on any vertex already containing a black pebble.
2. If both children of a vertex v have a red pebble on them, a red pebble can be placed on v.
3. If both children of v have *some* pebble on them (red or black), a black pebble can be placed on v.
4. A black pebble can be removed from any vertex.

A *pebbling game* is a sequence $\gamma_1 \to \gamma_2 \to \cdots \to_\ell$ of configurations. The game is — similarly to real computational model — divided into rounds. One starts a game in round 1. A round may be switched to u in a configuration γ_i if all vertices from a line $V_{u \cdot M}$ in γ_i are pebbled by some pebble (technically, a round is switched to u by issuing a request $\mathtt{nextRound}_u$). This switch is not obligatory when the specific row is pebbled. However, we require that the order of the request is $\mathtt{nextRound}_0, \mathtt{nextRound}_1, \ldots,$.

We define the *black-pebble complexity* of a round k of a pebbling game to be the maximum number of *black pebbles* on vertices in $V_{\geq k \cdot M}$ in use at any time of round k. More precisely if $\gamma_i \to \cdots \to \gamma_j$ are the configurations in round k. Then the black pebble complexity of k is equal to $\max_{\ell=i}^{j} |\mathit{Black}(\gamma_\ell) \cap V_{\geq k \cdot M}|$. If $\gamma_i, \ldots, \gamma_j$ are as above then the red pebble complexity of k is equal to the number of times in round k in which Step 1 was applied. For a parameter X a pebbling game is X*-bounded* if for every round k we have $2R_k + B_k < X$, where R_k and B_k denote the red- and the black-pebble complexities (resp.) of round k.

4.3 Auxiliary Lemmata

We need some auxiliary definitions and lemmas. For $(a_0, \ldots, a_t) \in \{0, \ldots, N\}^{t+1}$ define the *optimistic width of* $(a_0, \ldots, a_t)$ as: $\mathit{OptWidth}(a_0, \ldots, a_t) := (b_0, \ldots, b_t)$, where

- $b_0 := a_0$, and
- for every $i = 1, \ldots, t$ we set

$$b_i := \begin{cases} N & \text{if } b_{i-1} = N \\ \min(N, b_{i-1} - 1 + a_i) & \text{otherwise} \end{cases}$$

Intuitively, the idea is that if $(b_1, \ldots, b_t) := \mathit{OptWidth}(a_0, \ldots, a_t)$ then b_i's give an upper bound on the number of pebbles in the ith line of $[V']$ (for any $V' \subseteq V$), assuming that a_i is the number of pebbles in the ith line of V'. Formally, this is shown in the following lemma.

Lemma 1. *Take any set $V' \subseteq V$ and let $(a_0, \ldots, a_t) := \mathit{proj}([V'])$ and $(b_0, \ldots, b_t) := \mathit{OptWidth}(\mathit{proj}(V'))$. For every i we have that $a_i \leq b_i$.*

Proof. The proof goes by induction on $i = 0, \ldots, t$. Case $i = 0$ follows immediately from the fact that the closure operation does not change the configuration of the pebbles on the bottom row (V_0), and hence $a_0 = b_0$.

Now suppose the lemma holds for some i. The set of pebbles in the $(i+1)$st line of $[V']$ is equal to the sum of V'_{i+1} and the pebbles P that were derived (using the closure operation) from the pebbles in the ith line of $[V']$. By the induction hypothesis we get that the number of pebbles in the ith line of $[V']$ is at most b_{i-1}. Now, consider the case when $b_{i-1} \neq N$. From the definition of the closure operation it follows that $|P| \leq b_{i-1} - 1$. Therefore $|V'_{i+1} \cup P| \leq |V'_{i+1}| + |P| \leq a_i + b_{i-1} - 1$. Since the maximal value of $a_i + b_{i-1} - 1$ cannot be greater than N we get $|V'_{i+1} \cup P| \leq \min(N, a_i + b_{i-1} - 1)$.

The second case ($b_{i-1} = N$) follows easily from the fact that in this case $b_i = N$.

A sequence $(a_0, \ldots, a_t)$ is called *wide* if for some i we have $a_i = N$. A set $V' \subseteq V$ will be called *wide* if $\mathit{proj}([V'])$ is wide. We have the following simple observation.

Lemma 2. *For $U, W \subseteq V$ such that $U \cup W$ is not wide define $(a_0, \ldots, a_t) := \mathit{OptWidth}(\mathit{proj}([U \cup W]))$ and $(b_0, \ldots, b_t) := \mathit{OptWidth}(\mathit{proj}([W]))$. Then, for every i we have $b_i \leq a_i - |W|$. In other words: adding $|W|$ elements to U cannot increase the values on the coordinates in $\mathit{OptWidth}(\mathit{proj}([U]))$ by more than $|W|$ (as long as the resulting set $U \cup W$ is not wide).*

Proof (sketch). Suppose we add the elements of W to U one-by-one. From the definition of the closure operation it easily follows that adding one element cannot increase $\mathit{OptWidth}(\mathit{proj}([W]))$ by more than on each coordinate (as long as the resulting set is not wide). Hence the statement of the lemma follows.

A subgraph G' of a tower graph is a *pyramid graph* (cf. Fig. in Appendix in extended version [16]) if it is induced by the set of vertices: $\{(i + x \bmod N - 1, j + y \bmod N) : 0 \leq x + y \leq N - 1\}$ for some i and j. The vertex $(i + N, j)$ will be call the *root of* G'. We now have the following lemma whose proof appears in Appendix in extended version [16].

Lemma 3. *Consider a pebbling game for initially empty pyramid graph. If the root vertex is pebbled at the end of the game then there exist a configuration γ of the considered game with sum of red pebbles in the first row and the black pebbles is at least N.*

4.4 The Impossibility of Pebbling

Our goal is to show that — with some restrictions on red and black pebble complexity — it is impossible to pebble any vertex in $V_{\geq (u+2) \cdot M}$ in round u. Intuitively, it means that we cannot get any information about pebbles from any line $V_{\geq (u+2) \cdot M}$, before switching to round $u + 1$. More precisely, the following theorem holds:

Theorem 2. *Let N, T and X be arbitrary natural numbers such that*

$$X < \frac{3N}{2}.$$

Set $M := \frac{3N}{2}$. Suppose G is an N-tower graph. Then, for any X-bounded pebbling game for G with round length M and any configuration γ that belongs to the uth round, we have that in γ there are no pebbles on $V_{\geq(u+2)\cdot M}$.

Proof. In an execution of a pebbling game a pebble will be called *heavy* if it is a black pebble, or a red pebble placed on the graph using rule 1 (cf. Page 345). For a round u let γ_{i_u} be the last configuration of this round. We claim that in γ_{i_u} there is no pebble on $V_{\geq(u+2)\cdot M}$. Let A_u be the set of all pebbled vertices in the configuration γ_{i_u} and let Q_u be the set of all pebbles in A_u except of the black pebbles lying on the line $u \cdot M$. More precisely: $Q_u = A_u \setminus (A_u \cap V_{u \cdot M} \cap \mathit{Black}(\gamma))$. Set $Y_u := V_{\geq M \cdot u} \setminus V_{\geq M \cdot (u+1)}$

Lemma 4. *For every u we have:*

1. $A_u \cap V_{\geq(u+2)\cdot M} = \emptyset$
2. $\mathit{OptWidth}(\mathit{proj}([Q_u])) = (a_0, \ldots, a_{T\cdot M}) < (\underbrace{N, \ldots, N}_{u\cdot M}, \underbrace{N/2, \ldots, N/2}_{M}, 1, \ldots, 1),$

   in particular $[Q_u]$ is not wide.

After showing this we will be done with the proof since Point 1 of Lemma 4 clearly implies that Theorem 2 holds.

Proof (Proof of Lemma 4). Induction on $u = 1, 2, \ldots$. The base of the induction holds trivially since in the initial configuration only the bottom line is pebbled. Let us now assume the statement holds for some u. Now we prove the following claims for next round $u + 1$:

Claim. During the entire round $u + 1$ there must be at least $N/2$ heavy pebbles in the subgraph Y_u (i.e. the lines $u \cdot M, \ldots, (u + 1) \cdot M - 1$).

Proof. Let us consider any configuration γ from this round and denote the set of heavy pebbles in X in γ by P. At the end of the round every vertex on the $(u + 1)$st round-switching line $V_{(u+1)\cdot M}$ will contain a pebble. Therefore the closure $[P]$ of heavy pebbles P from the current configuration and the pebbles from the previous rounds Q_u need to contain whole line $V_{(u+1)\cdot M}$. This is because otherwise one would never be able to pebble $V_{(u+1)\cdot M}$ in the future (this follows easily from the definition of closure and the pebbling game). Hence $[P \cup Q_u]$ has to be wide and therefore (from Lemma 1) $\mathit{OptWidth}(P \cup Q_u$ also has to be wide. On the other hand, by the induction hypothesis we know that every coordinate of $\mathit{OptWidth}(Q_u)$ on positions $u \cdot M, \ldots$ is smaller than $N/2$. Now, by Lemma 2 adding to Q_u a set of cardinality $|P|$ cannot increase any of this coordinates by more than $|P|$. Hence $|P| \geq N/2$.

Claim. Through the whole round $u+1$ no vertex in $V_{\geq(u+2)\cdot M}$ is pebbled.

Proof. For the sake of contradiction assume that the claim is not true. So, we have a configuration γ from round $u+1$ with a pebble on some vertex v from set $V_{\geq(u+2)\cdot M}$. Denote by $(V', E')'$ the subgraph forming the pyramid graph with root in vertex v. From Lemma 3 we have that before γ there was a configuration γ' that: had b black pebbles on V' and had r red pebbles in bottom line of V' (which is the line $V_{\geq(u+2)\cdot M-N+1}$ of the tower graph) and $b+s \geq N$. However from Claim 4.4 there are at least $N/2$ heavy pebbles on Y_u, and Y_u is disjoint with the pyramid (V', E'). The number of all heavy pebbles is $A := B + R < (N-R) + (N/2)$. Therefore in the set $V_{\geq M\cdot(u+1)} \supset V'$ there are at most $(N-R)$ heavy pebbles. Since $b+s$ is bounded by the number of heavy pebbles so we have a contradiction with the fact that $b+s \geq N$.

Claim. $OptWidth(proj([Q_{u+1}])) = (a_0, \ldots, a_{T\cdot M}) < (\underbrace{N, \ldots, N}_{u+1\cdot M}, N/2, \ldots, N/2)$

Proof. Denote the configuration at the end of this round by γ and the set of heavy pebbles in γ by P. Let P' denote P without black-pebbled vertices from $(u+1)$th finishing line. From definition, we have $[Q_{u+1}] = [Q_u \cup P']$. In γ the $(u+1)$th finishing line is pebbled. There at most R red pebbles, so at least $N-R$ black pebbles are on this line. So $|P'|$ is at most $A - (N-R) = B + 2R - N < N/2$. Similarly as at the end of proof of the Claim 4.4 adding to Q_u a set of cardinality $|P'| < N/2$ cannot increase any coordinate of $OptWidth$ by more then $|P'|$. This finishes the proof.

Claims 4.4 and 4.4 prove inductive hypothesis for $u+1$. Hence we are done.

4.5 Connection between RO Labeling and the Pebbling Game

We now connect the random-oracle labeling of a tower graph G in our model of computation to the red-black pebbling game on G (a similar connection appeared recently in [17] and it is an extention of the technique from [13]). The idea is to show that from any execution of $\mathcal{A}$ (with space bounded by some s, and communication bounded by some c) we can construct some pebbling game for pebble game described before. Then, we show that (with high probability) this game respects the rules of the game and is (B, R)-bounded (for some B, R that will depend on c and s). We will then combine it with Theorem 2 to conclude that some specific oracle calls are impossible.

The main fact about the connection is that — in every round — the black-pebble complexity of the pebbling will correspond to the space-complexity of $\mathcal{A}_{small}$ and the red-pebble complexity corresponds to the communication-complexity of $\mathcal{A}_{small}$.

First, let us strictly define the method of translating an execution of $\mathcal{A}$ into a pebbling game. Let $\mathcal{H} : \{0,1\}^* \to \{0,1\}^w$ be a random oracle, and let $K = (K_1, \ldots, K_N)$ be a labeling of the input-vertices of G. For any algorithms $\mathcal{A} = (\mathcal{A}_{big}, \mathcal{A}_{small})$ we can use the execution $\mathcal{A}^{\mathcal{H}(\cdot)}(K) = \left(\mathcal{A}_{big}^{\mathcal{H}(\cdot)} \leftrightarrows \mathcal{A}_{small}^{\mathcal{H}(\cdot)}(K)\right)$ to construct a red-black pebbling of the graph G. In particular, we get a transcript listing all oracle calls made during its entire execution, and whether they were made by $\mathcal{A}_{small}$ or $\mathcal{A}_{big}$ and all `nextRound` calls made by $\mathcal{A}_{small}$.

We fix some terminology about the transcript. Given $(\mathcal{H}, K)$, we say that an oracle call of the form $\mathcal{H}(\texttt{label}_1, \texttt{label}_2, v)$ is *correct* if $(\texttt{label}_1, \texttt{label}_2, v) = \texttt{preLabel}(v)$. We call the children v_1, v_2 of v the input-vertices of the oracle call, and v is the output-vertex of the oracle call.

Using the transcript (along with the description of $\mathcal{H}, K$) we define the *ex-post-facto* pebbling of the graph G. We do so by processing the random-oracle calls and `nextRound` calls in the transcript one-by-one starting with the earliest one, and, for each call, we take the following steps:

Change round: If $\mathcal{A}_{small}$ calls `nextRound`, change round in the pebble game.

Place all necessary red pebbles: A vertex v is *red-necessary* if, looking at the *entire transcript* of all oracle calls, there exists some correct oracle call made by $\mathcal{A}_{big}$ with v as an input-vertex, which *precedes* all correct oracle calls made by $\mathcal{A}_{big}$ with v as an output-vertex. If the call is taken in kth round and $v \in V_{\geq k \cdot M}$ then we say that v is *k-red-necessary*.

Go through all red-necessary vertices v one-by-one and, for each one check if that has a black pebble,but no red pebble. If so, put red pebble on v.[4]

Delete all unnecessary black pebbles: A vertex v is *black-necessary* if it is *not* red-necessary and, *in the remainder of the transcript* of oracle-calls that have not yet been processed (including the current call), there exists some correct oracle call made by $\mathcal{A}_{small}$ with v as an input-vertex such that:

- In the *remainder of the transcript*, there is no *earlier* correct oracle call made by $\mathcal{A}_{small}$ with v as an output-vertex.
- In the *entire transcript*, there is no *earlier* correct oracle call made by $\mathcal{A}_{big}$ with v as an output-vertex.

Go through all vertices v which are *not* black-necessary but have a black pebble on them, one-by-one, and remove the black pebble.[5]

Process oracle call: If the current oracle call is correct and made by $\mathcal{A}_{small}$ (respectively $\mathcal{A}_{big}$) with output vertex v, we put a black (respectively red) pebble on v.

We notice that every vertex that is labeled by the execution of $\mathcal{A}^{\mathcal{H}(\cdot)}(K)$ gets a (red or black) pebble placed on it in the corresponding ex-post-facto pebbling (although, of course, this pebble may have been removed at some later point). Moreover, the order in which vertices get red/black pebbles corresponds to the order in which the oracle calls are made by $\mathcal{A}$.

[4] Note that the set of red-necessary vertices does not change throughout the process. Intuitively, these are the vertices whose labels must be communicated by $\mathcal{A}_{small}$ to $\mathcal{A}_{big}$ at some point in time, and correspondingly for which we need to take pebbling-action 1 to place a red pebble on them. We choose to take this action as early as legally possible, since it might allow us to remove related black pebbles early.

[5] Note that the set of black-necessary vertices can be different at different points in the process. Intuitively, at any point in time, a black-necessary vertex is one whose label must be stored in the memory of $\mathcal{A}_{small}$ since it will not be re-computed by $\mathcal{A}_{small}$ via oracle calls, it was never communicated to $\mathcal{A}_{big}$, nor will it be computed by $\mathcal{A}_{big}$ in time.

As mentioned before, we now show that, for any adversary $\mathcal{A} = (\mathcal{A}_{small}, \mathcal{A}_{big})$ which is space/communication bounded, and which makes a bounded number of oracle calls, the ex-post-facto pebbling is legal and has small space/communication complexity.

Theorem 3. *Let G be an N-tower graph. Let $\mathcal{A} = (\mathcal{A}_{big}, \mathcal{A}_{small})$ be any adversarial labeling game in our restricted model of computation. Let $(\mathcal{H}, K)$ define a random-oracle labeling of the graph G, with label-length w. Assume that $\mathcal{A}$ makes at most q random-oracle queries during the execution. Then, the ex-post-facto pebbling of G corresponding to an execution of $\mathcal{A}^{\mathcal{H}(\cdot)}(K)$ has the following properties (for any k):*

1. *It is a legal pebbling (i.e. follows the rules of the red-black pebbling game and changes round only when appropriate condition is hold) with probability $1 - \frac{q}{2^w}$ over the choice of $(\mathcal{H}, K)$.*
2. *Assuming that $\mathcal{A}_{small}$ has c-bounded communication and that in rounds $0, \ldots, k-2$ no vertices from $V_{\geq k \cdot M}$ were pebbled in the* ex-post-facto *game then, for any $\lambda \geq 0$ the red-pebble complexity of the round k is at most $\frac{2c+\lambda}{w-\log(q)}$ with probability $1 - 2^{-\lambda}$ over the choice of $(\mathcal{H}, K)$.*
3. *Assuming that $\mathcal{A}_{small}$ has s-bounded storage and c-bounded communication and that in rounds $0, \ldots, k-2$ no vertices from $V_{\geq k \cdot M}$ were pebbled then, for any $\lambda > 0$,* the sum *of the red-pebble complexity* and *the black-pebble complexity of the round k is at most $\frac{2c+s+\lambda}{w-\log(q)}$ with probability $1 - 2^{-\lambda}$ over the choice of $(\mathcal{H}, K)$.*

The proof appears in Appendix in extended version [16].

5 Proof of Theorem 1

Consider an execution of $\mathcal{A}$ and a corresponding ex-post-facto pebbling game $\mathcal{G}$. Let $\mathcal{X}$ denote an event that $\mathcal{G}$ is legal. For $i = 1, \ldots, T$ let B_i and R_i denote the respective black- and red-pebble complexities of round i in $\mathcal{G}$. Moreover, let $\mathcal{Y}_i$ denote the event that in the round i we have that (1) $B_i + 2R_i < N + N/2$, and (2) no vertex in $V_{(i+2) \cdot M}$ has been pebbled. Recall that every vertex that is labeled gets also pebbled. Therefore we have

$$\begin{aligned} 1 - p &\geq P(\mathcal{Y}_1 \wedge \cdots \wedge \mathcal{Y}_{T-2}) \\ &\geq P(\mathcal{X} \wedge \mathcal{Y}_1 \wedge \cdots \wedge \mathcal{Y}_{T-2}) \\ &= P(\mathcal{X}) \cdot P(\mathcal{Y}_1 | \mathcal{X}) \cdot P(\mathcal{Y}_2 | \mathcal{Y}_1 \wedge \mathcal{X}) \cdots P(\mathcal{Y}_{T-2} | \mathcal{Y}_1 \wedge \cdots \wedge \mathcal{Y}_{T-3} \wedge \mathcal{X}) \end{aligned} \quad (3)$$

Let us look at a term $P(\mathcal{Y}_i | \mathcal{Y}_1 \wedge \cdots \wedge \mathcal{Y}_{i-1} \wedge \mathcal{X})$. Suppose $\mathcal{Y}_1 \wedge \cdots \wedge \mathcal{Y}_{i-1} \wedge \mathcal{X}$ occurred. The events $\mathcal{Y}_1, \ldots \mathcal{Y}_{i-2}$ together imply that until round $i - 2$ no pebble has been placed on any vertex in $V_{\geq i \cdot M}$. Hence:

- $R_i \leq \frac{2c+\lambda}{w-\log(q)}$ with probability at least $1 - 2^{-\lambda}$ (from Part 2 of Theorem 3), and
- $B_i + R_i \leq \frac{2c+s+\lambda}{w-\log(q)}$ with probability at least $1 - 2^{-\lambda}$ (from Part 3 of Theorem 3).

Therefore we get that $B_i + 2R_i \leq \frac{4c+s+\lambda}{w\log(q)} \leq N + N/2$ with probability at least $1-2^{1-\lambda}$. Since the events $\mathcal{Y}_1, \ldots \mathcal{Y}_{i-1}$ also imply that $B_j + 2R_j \leq N + N/2$ holds for every $j < i$, therefore we can apply Theorem 2 and get that with probability $1 - 2^{1-\lambda}$ no pebble is put on any vertex in $V_{\geq(i+1)\cdot M}$ in round i. Since we also know that the pebbling is legal (because we assumed that $\mathcal{X}$ occurred), a vertex can be labeled by $\mathcal{A}$ only if it is pebbled. Hence $\mathcal{Y}_i$ holds (with probability at least $1 - 2^{1-\lambda}$). From Part 1 of Theorem 3 we have that $P(\mathcal{X}) \geq 1 - \frac{q}{2^w}$. Putting things together we get $P(\mathcal{Y}_i|\mathcal{Y}_1 \wedge \cdots \wedge \mathcal{Y}_{i-1} \wedge \mathcal{X}) \geq 1 - 2^{1-\lambda}$. Hence (3) is at least equal to

$$\begin{aligned}
&\left(1 - \frac{q}{2^w}\right) \cdot \left(1 - 2^{1-\lambda}\right)^{T-2} \\
&\geq \left(1 - q \cdot 2^{-w}\right) \cdot \left(1 - T \cdot 2^{1-\lambda}\right) \\
&\geq 1 - q \cdot 2^{-w} - T \cdot 2^{1-\lambda}
\end{aligned}$$

Therefore p is at most (2).

References

1. Akavia, A., Goldwasser, S., Vaikuntanathan, V.: Simultaneous hardcore bits and cryptography against memory attacks. In: Reingold, O. (ed.) TCC 2009. LNCS, vol. 5444, pp. 474–495. Springer, Heidelberg (2009)
2. Alwen, J., Dodis, Y., Wichs, D.: Leakage-resilient public-key cryptography in the bounded-retrieval model (2009), `http://eprint.iacr.org/`
3. Brakerski, Z., Goldwasser, S.: Circular and leakage resilient public-key encryption under subgroup indistinguishability (or: Quadratic residuosity strikes back) (2010)
4. Brakerski, Z., Kalai, Y.T., Katz, J., Vaikuntanathan, V.: Cryptography resilient to continual memory leakage (2010)
5. Brumley, D., Boneh, D.: Remote timing attacks are practical. Comput. Netw. (2005)
6. Cash, D., Ding, Y.Z., Dodis, Y., Lee, W., Lipton, R.J., Walfish, S.: Intrusion-resilient key exchange in the bounded retrieval model. In: Vadhan, S.P. (ed.) TCC 2007. LNCS, vol. 4392, pp. 479–498. Springer, Heidelberg (2007)
7. Chari, S., Jutla, C.S., Rao, J.R., Rohatgi, P.: Towards sound approaches to counteract power-analysis attacks. In: Wiener, M. (ed.) CRYPTO 1999. LNCS, vol. 1666, p. 398. Springer, Heidelberg (1999)
8. Di Crescenzo, G., Lipton, R.J., Walfish, S.: Perfectly secure password protocols in the bounded retrieval model. In: Halevi, S., Rabin, T. (eds.) TCC 2006. LNCS, vol. 3876, pp. 225–244. Springer, Heidelberg (2006)
9. Damgård, I.: A design principle for hash functions. In: Brassard, G. (ed.) CRYPTO 1989. LNCS, vol. 435, pp. 416–427. Springer, Heidelberg (1990)
10. Davì, F., Dziembowski, S., Venturi, D.: Leakage-resilient storage. In: Garay, J.A., De Prisco, R. (eds.) SCN 2010. LNCS, vol. 6280, pp. 121–137. Springer, Heidelberg (2010), `http://eprint.iacr.org/`
11. Dodis, Y., Goldwasser, S., Tauman Kalai, Y., Peikert, C., Vaikuntanathan, V.: Public-key encryption schemes with auxiliary inputs. In: Micciancio, D. (ed.) TCC 2010. LNCS, vol. 5978, pp. 361–381. Springer, Heidelberg (2010)

12. Dodis, Y., Haralambiev, K., Lopez-Alt, A., Wichs, D.: Cryptography against continuous memory attacks (2010)
13. Dwork, C., Naor, M., Wee, H.: Pebbling and proofs of work. In: Shoup, V. (ed.) CRYPTO 2005. LNCS, vol. 3621, pp. 37–54. Springer, Heidelberg (2005)
14. Dziembowski, S.: Intrusion-resilience via the bounded-storage model. In: Halevi, S., Rabin, T. (eds.) TCC 2006. LNCS, vol. 3876, pp. 207–224. Springer, Heidelberg (2006)
15. Dziembowski, S.: On forward-secure storage. In: Dwork, C. (ed.) CRYPTO 2006. LNCS, vol. 4117, pp. 251–270. Springer, Heidelberg (2006)
16. Dziembowski, S., Kazana, T., Wichs, D.: Key-evolution schemes resilient to space-bounded leakage (2011), http://eprint.iacr.org/
17. Dziembowski, S., Kazana, T., Wichs, D.: One-time computable self-erasing functions. In: Ishai, Y. (ed.) TCC 2011. LNCS, vol. 6597, pp. 125–143. Springer, Heidelberg (2011)
18. Dziembowski, S., Pietrzak, K.: Intrusion-resilient secret sharing. In: FOCS (2007)
19. Dziembowski, S., Pietrzak, K.: Leakage-resilient cryptography. In: FOCS (2008)
20. Faust, S., Kiltz, E., Pietrzak, K., Rothblum, G.N.: Leakage-resilient signatures. In: Micciancio, D. (ed.) TCC 2010. LNCS, vol. 5978, pp. 343–360. Springer, Heidelberg (2010)
21. Faust, S., Rabin, T., Reyzin, L., Tromer, E., Vaikuntanathan, V.: Protecting circuits from leakage: the computationally-bounded and noisy cases. In: Gilbert, H. (ed.) EUROCRYPT 2010. LNCS, vol. 6110, pp. 135–156. Springer, Heidelberg (2010)
22. Goldwasser, S., Kalai, Y.T., Rothblum, G.N.: One-time programs. In: Wagner, D. (ed.) CRYPTO 2008. LNCS, vol. 5157, pp. 39–56. Springer, Heidelberg (2008)
23. Goldwasser, S., Rothblum, G.N.: Securing computation against continuous leakage. In: Rabin, T. (ed.) CRYPTO 2010. LNCS, vol. 6223, pp. 59–79. Springer, Heidelberg (2010)
24. Ishai, Y., Sahai, A., Wagner, D.: Private Circuits: Securing Hardware against Probing Attacks. In: Boneh, D. (ed.) CRYPTO 2003. LNCS, vol. 2729, pp. 463–481. Springer, Heidelberg (2003)
25. Alwen, J., Dodis, Y., Wichs, D.: Leakage-resilient public-key cryptography in the bounded-retrieval model. In: Halevi, S. (ed.) CRYPTO 2009. LNCS, vol. 5677, pp. 36–54. Springer, Heidelberg (2009)
26. Juma, A., Vahlis, Y.: Protecting cryptographic keys against continual leakage. In: Rabin, T. (ed.) CRYPTO 2010. LNCS, vol. 6223, pp. 41–58. Springer, Heidelberg (2010)
27. Katz, J., Vaikuntanathan, V.: Signature schemes with bounded leakage resilience. In: Matsui, M. (ed.) ASIACRYPT 2009. LNCS, vol. 5912, pp. 703–720. Springer, Heidelberg (2009)
28. Kocher, P.: Design and validation strategies for obtaining assurance in countermeasures to power analysis and related attacks. In: NIST Physical Security Testing Workshop (2005)
29. Kocher, P.C., Jaffe, J., Jun, B.: Differential power analysis. In: Wiener, M. (ed.) CRYPTO 1999. LNCS, vol. 1666, p. 388. Springer, Heidelberg (1999)
30. Kuhn, M.G.: Compromising emanations: eavesdropping risks of computer displays. Technical Report UCAM-CL-TR-577 (2003)
31. Micali, S., Reyzin, L.: Physically observable cryptography (extended abstract). In: Naor, M. (ed.) TCC 2004. LNCS, vol. 2951, pp. 278–296. Springer, Heidelberg (2004)
32. Naor, M., Segev, G.: Public-key cryptosystems resilient to key leakage. In: Halevi, S. (ed.) CRYPTO 2009. LNCS, vol. 5677, pp. 18–35. Springer, Heidelberg (2009)
33. European Network of Excellence (ECRYPT). The side channel cryptanalysis lounge (retrieved on April 7, 2010), http://www.crypto.ruhr-uni-bochum.de/en_sclounge.html
34. Pietrzak, K.: A leakage-resilient mode of operation. In: Joux, A. (ed.) EUROCRYPT 2009. LNCS, vol. 5479, pp. 462–482. Springer, Heidelberg (2009)

35. Pietrzak, K.: A leakage-resilient mode of operation. In: Joux, A. (ed.) EUROCRYPT 2009. LNCS, vol. 5479, pp. 462–482. Springer, Heidelberg (2009)
36. Quisquater, J.-J., Samyde, D.: ElectroMagnetic analysis (EMA): Measures and countermeasures for smart cards. In: Attali, S., Jensen, T. (eds.) E-smart 2001. LNCS, vol. 2140, p. 200. Springer, Heidelberg (2001)
37. Savage, J.E.: Models of Computation: Exploring the Power of Computing. Addison Wesley, Reading (1997)
38. Shamir, A., Tromer, E.: Acoustic cryptanalysis. on nosy people and noisy machines. A webpage: `http://people.csail.mit.edu/tromer/acoustic/` (accessed on May 27, 2009)
39. Standaert, F.-X., Malkin, T.G., Yung, M.: A unified framework for the analysis of side-channel key recovery attacks. In: Joux, A. (ed.) EUROCRYPT 2009. LNCS, vol. 5479, pp. 443–461. Springer, Heidelberg (2009)
40. Yu, Y., Standaert, F.-X., Pereira, O., Yung, M.: Practical Leakage-Resilient Pseudorandom Generators. In: CCS: ACM Conference on Computer and Communications Security (2010) (to appear)

Generic Side-Channel Distinguishers: Improvements and Limitations

Nicolas Veyrat-Charvillon* and François-Xavier Standaert**

UCL Crypto Group, Université catholique de Louvain
Place du Levant 3, B-1348, Louvain-la-Neuve, Belgium
{nicolas.veyrat,fstandae}@uclouvain.be

Abstract. The goal of generic side-channel distinguishers is to allow key recoveries against any type of implementation, under minimum assumptions on the underlying hardware. Such distinguishers are particularly interesting in view of recent technological advances. Indeed, the traditional leakage models used in side-channel attacks, based on the Hamming weight or distance of the data contained in an implementation, are progressively invalidated by the increased variability in nanoscale electronic devices. In this paper, we consequently provide two contributions related to the application of side-channel analysis against emerging cryptographic implementations. First, we describe a new statistical test that is aimed to be generic and efficient when exploiting high-dimensional leakages. The proposed distinguisher is fully non-parametric. It formulates the leakage distributions using a copula and discriminates keys based on the detection of an "outlier behavior". Next, we provide experiments putting forward the limitations of generic side-channel analysis in advanced scenarios, where leaking devices are protected with countermeasures. Our results exhibit that all non-profiled attacks published so far can sometimes give a false sense of security, due to incorrect leakage models. That is, there exists settings in which an implementation is secure against such non-profiled attacks and can be defeated with profiling. This confirms that the evaluations of cryptographic implementations should always consider profiling, as a worst case scenario.

1 Introduction

Since the introduction of differential power analysis by Kocher, Jaffe and Jun in the late 1990s [13], physical attacks have become an important issue for the security of cryptographic devices. On the academic side, it gave rise to many exciting developments of new attacks, countermeasures and models for the evaluation (or, recently, proof) of physical security. On the industrial side, security against such attacks is now required to reach high certification levels for cryptographic products. Roughly speaking, side-channel attacks are usually classified

* Postdoctoral researcher supported by the Walloon region SCEPTIC project.
** Associate researcher of the Belgian Fund for Scientific Research (FNRS-F.R.S.).

P. Rogaway (Ed.): CRYPTO 2011, LNCS 6841, pp. 354–372, 2011.

as profiled or non-profiled [15]. Profiled attacks are typically useful in an evaluation context, where one can exploit devices with known keys in order to build precise leakage models, e.g. templates [5]. They can then be used in order to estimate evaluation metrics such as the mutual information, in order to capture a worst-case scenario [28]. By contrast, non-profiled attacks, on which we will focus in this paper, rather aim to capture the behavior of actual adversaries, who do not have a precise prior characterization of the devices they target. They are usually suboptimal from a data complexity point of view, and exploit the "on-the-fly" estimation of the leakage probability distributions (or their moments) in order to recover secret information. In general, the gap between these two scenarios can be large. Hence, profiled and non-profiled attacks are complementary and shed a different light on the security of embedded devices.

In brief, a non-profiled side-channel attack generally works by comparing key-dependent leakage models with actual measurements. If the attack is successful, the key candidate giving rise to the best comparison is the one manipulated by the target device. As a consequence, evaluating the performances of such a distinguisher can typically be done along two axes. On the one hand, attacks are expected to be efficient, meaning that they allow recovering keys with limited data (i.e. measurements), time and memory. On the other hand, attacks should also be generic, i.e. applicable against any type of device and (if possible) unsensitive to unprecise leakage models. A brief look at the state-of-the-art suggests that most previous works can be viewed as exploring the tradeoff between these two (usually contradictory) goals. For example, in the seminal paper of Crypto 1999, a "single-bit DPA" is performed using a simple difference-of-means test. This approach has limited efficiency, because the single-bit model implies a low SNR, and it cannot take advantage of any knowledge of the target device that may be available to the adversary. As a consequence, many "multiple bit" extensions have been proposed in the literature. Most prominently, the Correlation Power Analysis (CPA) introduced in 2004 works very efficiently in contexts where a device leaks according to a well known and linear (e.g. Hamming weight) model [2]. But it is also quite specific, and can end up to be completely ineffective if an unprecise leakage model is used. In order to avoid this model-dependency, a powerful solution is to build stochastic models "on-the-fly", as suggested by Schindler, Lemke and Paar in 2005 [26]. A possible drawback of stochastic models is their parametric nature. But as discussed in [6], such a linear regression-based approach gives excellent results in the context of first-order side-channel attacks against unprotected devices. Alternatives to stochastic models include the Mutual Information Analysis, introduced in 2008 [8], and tests such as the Cramér-von-Mises one, discussed in [31]. These last two distinguishers are quite generic, and can capture any type of leakage dependency. They are also non-parametric in the case of MIA (which still implies pdf estimation, e.g. with histograms or Kernels, hence requiring to select number of bins or Kernel bandwidth adequately [19]) and completely free of parameters for the Cramér-von-Mises test.

Interestingly, recent technological advances suggest that the genericity of side-channel distinguishers could be an important feature in the evaluation of future cryptographic devices. Indeed, as discussed in [24], the move towards nanoscale electronic circuits implies the apparition of new leakage functions, that strongly deviate from the traditional (Hamming weight, distance) assumptions. Also, the increasing device variability implies that each target implementation can be characterized by a different leakage model. Quite naturally, the situation turns out to be even nastier when moving to higher-order attacks against devices protected with masking [4,10]. Indeed, straightforward extensions of DPA and CPA require the introduction of a heuristic dimensionality reduction technique, usually denoted as the combination function in the literature [17]. But as discussed in [20,29], the selection of a good combination function is inherently dependent on the leakage function (i.e. the target device), and can only degrade the amount of information exploited by the distinguisher. Similarly, extensions of the stochastic model are direct in the profiled context [14], when masks are available during the profiling, but their application in a non-profiled scenario requires either to estimate pdf mixtures in an unsupervised manner (i.e. a problem for which we do not have systematic and efficient solutions), or to take advantage of some heuristic assumptions (e.g. using a combination function). In fact, only MIA directly generalizes to multivariate side-channel attacks, without requiring a combination function [1,7,19]. Given this attractive feature, it appears natural to investigate how such distinguishers can deal with advanced scenarios, mixing non-linear leakage functions and countermeasures like masking.

This paper presents two contributions in this direction. First, we propose a new test for side-channel analysis, aimed to be generic and efficient when exploiting high-dimensional leakages. For this purpose, we start from the observation that, in order to be generic, MIA selects the key candidate that maximizes the mutual information between an adversary's key-dependent leakage models and actual measurements. Our new distinguisher uses the alternative criterion to select the key candidate for which these leakage models deviate the most from a reference (e.g. uniform) distribution. Next, it has been observed in experiments on MIA that the generalization to multivariate attacks can be less efficient, because of the difficulty of estimating a multivariate pdf "on-the-fly", without specific assumptions [29]. Also, previously considered estimation methods such as using histograms, Kernels, or splines [30], generally require to tune a parameter, e.g. the number of bins in histograms, that directly impacts the efficiency of the attack. Hence, although MIA aims at genericity more than efficiency, it would be interesting to avoid such parameters, or to make their tuning as easy as possible. Our new distinguisher takes advantage of advanced statistical tools in order to mitigate these issues. For this purpose, we first apply a leakage transformation, exploiting copulas [18]. It projects the samples into a new space where their distribution (among each dimension) is uniform. Thanks to this transform, we base our distinguisher on the generic criterion of selecting the key candidate for which the model maximizes the deviation from uniform. Afterwards, we exploit distance sampling, i.e. we evaluate the distribution of the distance between two samples (conditioned on a leakage model), rather than the distribution of single samples. Distance sampling has interesting features in a multivariate setting, as it allows to avoid dealing with multivariate distributions directly (we rather evaluate the univariate distribution of a distance taken over several dimensions). As a result, our test is completely free of parameters and mainly requires to compute empirical cumulative distributions, for each dimension taken independently. Summarizing, it is pointed out in [34] that the efficiency loss of MIA is due to the problem of estimating the leakage distributions. The present paper complements this view and aims at making this estimation step easier.

Second, we propose different experiments in order to evaluate this new distinguisher. Namely, we investigated attacks against unprotected and masked S-box computations, in three different settings: naive Hamming weight simulations, real measurements of a 65 nanometer CMOS chip, Spice simulations of a dual-rail logic style. These examples are expected to be reflective of the variety of leakage functions that one can find in side-channel analysis. They highlight that the proposed generic test compares favorably with MIA, in all investigated scenarios. They also underline that, despite the generic nature of MIA and the new distinguisher, their application against modern electronic devices can be strongly affected by inaccurate leakage models, and that the impact of such imprecisions is amplified with countermeasures such as masking. In other words, even generic tests can be unsuccessful in certain contexts, unless preliminary assumptions (similar to profiling) are available to the adversary. We note that this last observation holds for all non-profiled distinguishers published so far. Hence, our results raise the question whether non-profiled attacks could be improved in order to deal with such critical contexts, or alternatively, whether these contexts can be formalized and used as a design criteria for new countermeasures. For now on, they at least confirm the importance of a profiled information theoretic analysis in the evaluation of leaking cryptographic devices.

2 Side-Channel Analysis

The next sections of the paper analyze the attack depicted in Figure 1, as described in [1]. That is, we consider a device performing several cryptographic computations $\mathsf{E}_k(p)$ on different plaintexts p drawn uniformly from the text space $\mathcal{P}$, using some fixed key k drawn uniformly from the key space $\mathcal{K}$. While computing $\mathsf{E}_k(P)$ (where P is a random variable over $\mathcal{P}$), the device will handle some intermediate values (defined as sensitive variables in [25]) that depend on the known input P and the unknown key k. In practice, the interesting sensitive variables in a DPA attack are the ones that only depend on an enumerable subkey s: we denote them as $V_{s,P}$. Anytime such a sensitive intermediate value is computed, the device generates some physical leakage, denoted as $Y_{k,P}$.

In order to perform a key recovery, an adversary first has to select a sensitive value. Given that this value only depends on a subkey s, he can then evaluate its result for the same plaintexts that have been used to generate $Y_{k,P}$ and all the possible subkey candidates $j \in \mathcal{S}$. It gives rise to different hypothetical values $V_{j,P}$. Afterwards, he uses a leakage model to map these values from their original space $\mathcal{V}$ towards a hypothetical leakage space $\mathcal{X}$. It is usually in this step that

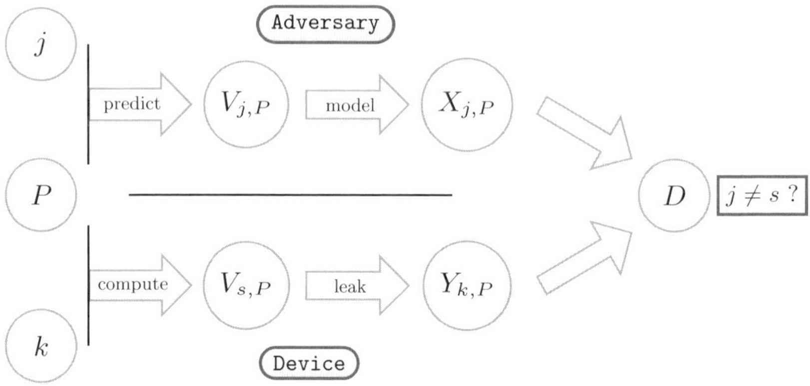

Fig. 1. Schematic illustration of a side-channel key recovery attack

engineering intuition can be exploited, if available. For example, a usual model that has been experimentally confirmed, e.g. in [15], is to take the Hamming weight of the values $V_{j,P}$. Such a model is justified by the dominating dynamic part of the power consumption in certain microelectronic devices. As a result, the adversary obtains $|\mathcal{S}|$ different models denoted as $X_{j,P}$, again corresponding to the different subkey candidates. Eventually, he uses a distinguisher D to compare the different models $X_{j,P}$ with the actual leakages $Y_{k,P}$. If the attack is successful, the best comparison result (*i.e.* the highest value of the distinguisher) should be obtained for the correct subkey candidate $j = s$. This procedure can then be repeated for different subkeys in order to eventually recover the full key.

3 The New Generic Test

A central problem in non-profiled side-channel analysis is to properly estimate the leakage distribution "on the fly" during an attack. Previous methods for this purpose typically range in two categories. A first class of distinguishers exploits specific assumptions about the target implementation, resulting in efficient attacks provided that these assumptions are fulfilled. Another class of distinguishers aims to avoid relying on assumptions, at the cost of a (hopefully small) efficiency loss. Quite naturally, the genericity of this second class of distinguishers essentially comes from that they try to completely characterize the leakage distribution (rather than its first- or second-order moments, typically). And its efficiency loss comes from the difficulty of the density estimation problem. For example, tools for estimating pdf usually rely on a good choice of parameters: number of modes in Gaussian mixtures, number of bins in histograms, bandwidth in Kernel estimators, to name a few. Also, these estimators strongly suffer from the curse of dimensionality: a 9-bin univariate histogram will typically require 81 bins in two dimensions. In the remainder of the section, we first describe a distinguisher, illustrated in Figure 2, that aims to limit these drawbacks. Next, we discuss its advantages and limitations.

3.1 Specification

The distinguisher is based on six main steps (one being optional).

Leakage space transform. In order to circumvent the problem of estimating the leakage distribution, our method first transforms the samples by means of copula. A copula simply applies the probability integral transform to every marginal variable, which renders the distribution of samples along each axis uniform. More precisely, for $Y = (Y_1, Y_2, \ldots, Y_d)$ a d-dimension random variable, the copula transformation gives a derived variable $Z = (Z_1, Z_2, \ldots, Z_d) = (F_1(Y_1), F_2(Y_2), \ldots, F_d(Y_d))$, where F_i is the cumulative distribution function of Y_i, defined by $F_i(y_i) = \Pr[Y_i \leq y_i]$. Interestingly, the Z_i have a uniform distribution by definition of the probability integral transform, but any dependency among the Y_i variables implies a corresponding dependency among the Z_i's. This is illustrated in Figure 2, where it is clearly seen that the marginal distribution $\Pr[Z = z]$ is uniform after reduction, while the conditional distributions remain easily distinguishable. For illustration, the figure represents the simple case of a single-bit leakage model. In practice, since the leakage density and its cumulative function are both unknown, we compute an empirical copula, i.e. a copula where the cumulative distributions are approximated by empirical distributions. For an n-sample set $y_1, \cdots y_n$ drawn from the distribution of a one-dimension random variable Y, the empirical cumulative distribution of a value y is given by $\hat{F}(y) = \frac{1}{n} \sum_i \mathrm{I}(y \leq y_i)$, where I is the indicator function. That is, it only requires to sort the samples and is equivalent to computing the quantile of y for that sample set. Thanks to the Glivenko-Cantelli theorem [3,9], we know that the empirical cumulative distribution function converges *almost surely* towards the common cumulative distribution function, uniformly over all y. That is:

$$\|\hat{F}_n - F\|_\infty \equiv \sup_{y \in \mathbb{R}} \left|\hat{F}_n(y) - F(y)\right| \xrightarrow{a.s.} 0. \tag{1}$$

An advantage of the empirical copula is that reduced samples only take quotient values: $0, \frac{1}{n}, \ldots, \frac{i}{n}, \ldots 1$, built from n real-valued leakages. Hence, it allows dealing with probability mass functions instead of densities, which simplifies computations. In general, the transform amounts to working on modified values $Z_{k,P}$ instead of the leakages $Y_{k,P}$, which have by construction a simple uniform distribution, but still retain the information contained in the original leakages.

Leakage partitioning. This step is common to all standard side-channel attacks. After transformation, the leakage samples are classified, based on predictions $X_{j,P}$ made up from the plaintexts and key hypotheses. Intuitively, the expectation is that predictions obtained from the correct key candidate will lead to a meaningful leakage partition, i.e. there will be a dependency between the categories x and the samples y they contain. By contrast, a wrong key hypothesis should give rise to random predictions, so that the categories only correspond to

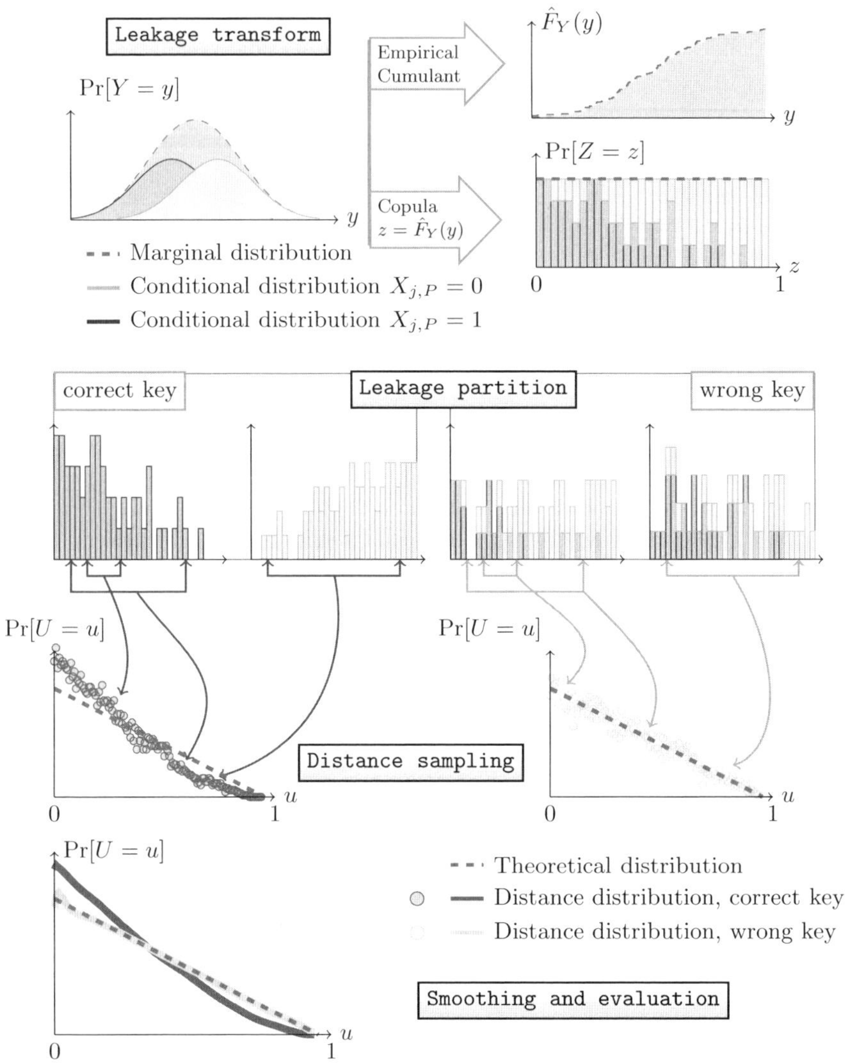

Fig. 2. Illustrated process of the distinguisher (1D)

a random shuffling of leakages from the sample set. Interestingly, thanks to the transformation step, a random sampling should tend towards a uniform distribution, as illustrated in the leakage partition step in Figure 2.

Feature selection and template building. Given the (uniform or not) distributions representing the key candidates, one can select a feature of these distributions, that properly captures possible non-uniformities. In the following, we will consider the distance between couples of samples for this purpose, which corresponds to the notion of spacings in statistics [21]. When dealing with multivariate leakages, we define spacings via the L_1 or Manhattan distance (instead of the Euclidean one), which avoids dealing with irrational values. The Manhattan distance between two samples z and z' is given by $d_1(z, z') = \sum_i |z_i - z'_i|$, where the sum is taken over the different dimensions of the samples. Next, since the marginal distribution of the leakages is known to be always uniform, we simply build a template for the distribution of the distance between two uniform samples. This template is precomputed once for all attacks, independently of the target implementation. Its shape is actually a spline, which ensure a degree of smoothness (see appendix A for details). We note that the interest of feature selection is not obvious in the univariate attack context of Figure 2. But it implies a dimensionality reduction that becomes convenient in a multivariate setting[1]. It also leads to an efficient solution for the estimation problem, as we now detail.

Estimation. This is the central step of the attack, in which we try to model the distributions corresponding to the different key candidates. For this purpose, one limitation of MIA was the need to estimate one conditional distribution per model value, for each key candidate. In other words, each of these distributions has to be estimated from only a part of the available samples. By contrast, our new test allows to estimate only one distribution, from all the available samples. This is possible because we know (again thanks to the copula) that the marginal distribution should be uniform for a wrong key candidate. Hence, we can sample the Manhattan distance for all partitions, and combine the results into a single probability mass function, which has to be consistent with uniformly drawn samples. Combined with the previous feature selection, it implies that one can estimate the distributions for each key candidate by iterating the next steps:

- Pick a random leakage z from the complete set of samples.
- Pick a different random leakage z', from the same model category as z.
- Compute their Manhattan distance $d_1(z, z')$.

Finally, we obtain the sampled probability mass function of the distance. This distance can only take $n \cdot d$ distinct values, with n the number of samples available and d the dimensionality of the leakages. This is possible because we use the Manhattan distance (i.e. there are as many possible distances as there are samples). Note that if there are m leakages corresponding to a model value $X = x$, the number of possible couples to sample scales as m^2. In our following experiments, it was always possible to test these couples exhaustively. But in attacks requiring millions of traces, exploiting Monte Carlo sampling would of course be an alternative to reduce the time complexity to more tractable values.

Smoothing. As can be seen from Figure 2, the distance histograms can be used to discriminate the different partitions, but they remain quite noisy. In order

[1] Distance sampling is reminiscent of the use of an absolute distance combining function in higher-order side-channel attacks [17]. The prior application of a copula allows us to give it a stronger foundation and to remove its device-dependent flavor.

to improve the signal-to-noise ratio of the pdf estimations, one straightforward solution is to apply a lowpass filter. Different solutions can be used for this purpose. A very generic one, that we applied in our experiments, is to use Kernel smoothing with an Epanechnikov function [12] (which is both the most efficient Kernel and the least costly to compute). The advantage of Kernel smoothing is to generalize easily to distributions with any number of dimensions. Its drawback is to introduce a parameter (the window size used in the smoothing). However, we note that this parameter is significantly easier to select than, e.g. the number of bins or Kernel bandwidth in the case of MIA, since we know exactly how the distribution of the wrong key candidates should look like. Namely, this distribution should correspond to the previously described uniform distance template. How to select this parameter adequately will be explained in the next subsection.

Evaluation. Finally, we compute the integrated square distance between the sample distributions of all the key candidates and the theoretical distance distribution (i.e. the uniform ditance template construted in step 3). The attack is successful if the correct key candidate corresponds to the largest deviation[2].

3.2 Pros and Cons

One important advantage of this new test is that it nicely extends to multivariate attacks. As illustrated in Figure 3 for a bivariate example (i.e. $d = 2$) and a Hamming weight leakage model, the leakage transformation is performed independently for each dimension. That is, we just have to compute d univariate empirical cumulative functions (rather than one d-dimensional distribution when applying MIA). The empirical cumulative functions are straightforward to compute and require no parameter at all. By applying the copula, the leakage samples are transformed in such a way that their marginal distribution along each dimension becomes uniform over $[0, 1]$. As a result, we can again discriminate key candidates by simply assuming that the leakage model generated with the good key candidate should lead to a partitioning for which the distance to uniform is large. On the negative side, the estimation of the distance distribution follows a multinomial distribution (a generalization of the binomial distribution to more than two categories), which implies that the sampled distribution tends to be noisy. The smoothing part that aims at reducing this noise requires to set a parameter that can be seen as the counterpart of the number of bins or Kernel bandwidth when directly trying to estimate the leakage pdf in MIA. However, this task is arguably easier in our new distinguisher, since we only need to detect departures from a well-characterized distribution. In practice, a Kernel smoothing with window size equal to 1 allows us to only retain the general features of the distance distribution, which is enough to detect departures from the uniform template. This is only suboptimal when dealing with very low-noise scenarios, where a smaller window size is enough to smooth out the estimated densities.

[2] As detailed in [31], different alternatives could be considered. A robust one would be to use the smoothed median of the distributions for all key hypotheses as a template.

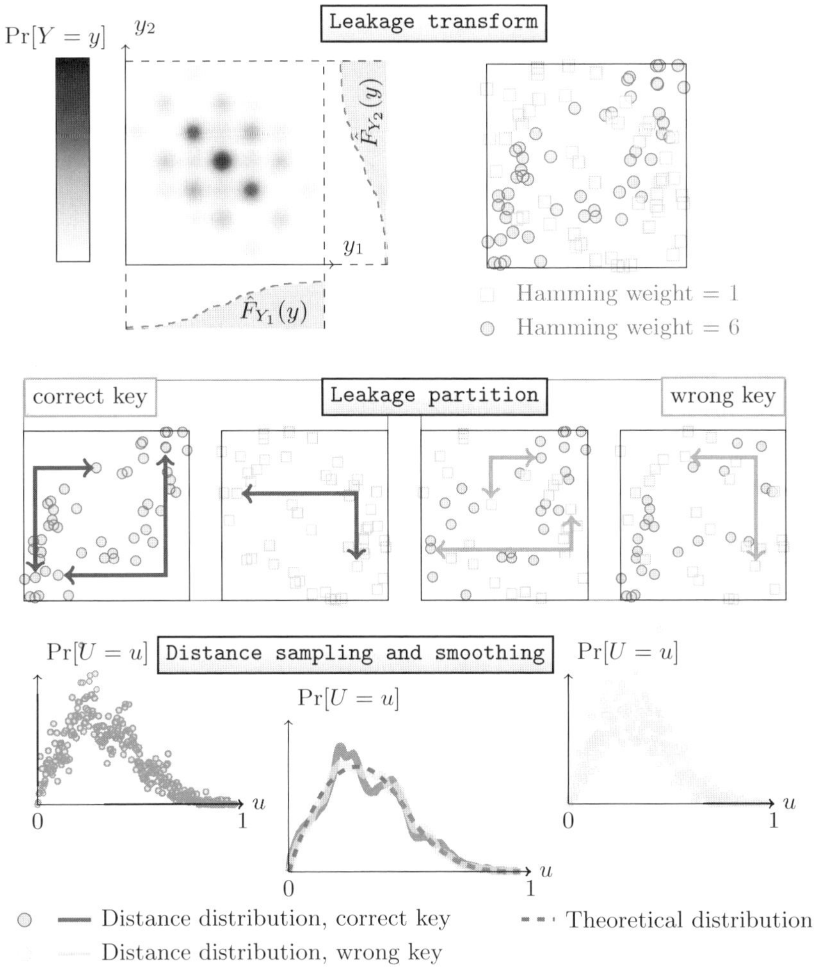

Fig. 3. Illustrated process of the distinguisher (2D)

4 Experiments

We now provide experiments, in order to verify the relevance of the previously introduced generic test and to compare its efficiency with other distinguishers used in side-channel analysis. For this purpose, we consider the following contexts.

1. Different target computations. We investigated three possible cases. In the first one, we target the leakage corresponding to the execution of an unprotected AES S-box, denoted as $v_{k,p} = \mathsf{S}(p \oplus k) \rightsquigarrow y_{k,p}$, where the adversary is provided with the plaintext p and leakage $y_{k,p}$. In the second one, we consider the execution of a masked AES S-box, denoted as $v^1_{k,p} = \mathsf{S}(p \oplus k) \oplus m \rightsquigarrow y^1_{k,p}$, $v^2_{k,p} = m \rightsquigarrow y^2_{k,p}$. In this case, the mask m is a uniformly random value and the adversary is provided with the plaintext p and leakages $y_{k,p}^1, y_{k,p}^2$. Finally, we consider the execution of a masked S-box where the the adversary only receives the plaintext p and a combination of the two leakage samples. Following the previous analyzes in [20,29], we used the normalized product between the samples, i.e. $\mathsf{C}(y_{k,p}^1, y_{k,p}^2) = (y_{k,p}^1 - \hat{\mathbf{E}}(Y_{k,P}^1)) \cdot (y_{k,p}^2 - \hat{\mathbf{E}}(Y_{k,P}^1))$, where $\hat{\mathbf{E}}$ denotes the sample mean operator.

2. Different leakage functions and target devices. We again analyzed three possible cases. In the first one, the leakages are simulated with a Hamming weight function, to which we added a Gaussian noise, with mean 0 and variance σ_n^2. Although not always realistic, the investigation of this setting is justified by the numerous works carried out under this assumption, as a reference. In the second case, we use the leakage measured from an S-box implemented in a 65 nanometer CMOS technology, running at 2MHz and sampled with a 1 Gsample/sec digital oscilloscope), previously analyzed in [24]. In the third case, the leakage of a dual-rail pre-charged S-box, implemented in the same 65 nanometer technology and using the logic style described in [11], was simulated with Spice. The details of this S-box are out of the scope of this paper, but it was selected as an example of leakage function that strongly deviates from the Hamming weight model. Both for the CMOS and the dual-rail S-boxes, we selected one single leakage sample per target operation. This selection is not supposed to be optimal, but was the same for all the investigated attacks, in order to allow fair comparisons.

3. Different distinguishers. First, correlation attacks based on Pearson's coefficient were applied, following the descriptions in [2]. Next, we performed MIA[3] with histogram-based pdf estimation, following [8]. The number of bins in histograms was selected according to Scott's rule of thumb [27]. Third, we used the stochastic approach first described in [26] and analyzed in the non-profiled setting by Doget et al. [6]. The goal of the stochastic approach is to approximate the S-box leakages with a linear function $\hat{\mathsf{L}}(j,p) = \sum \alpha_i \, g_i(j,p)$, where the coefficients α_i are determined by regression, and the $g_i(j,p)$'s correspond to the base functions used in the attack. Finally, we experimented our new generic test, with the window size in the Kernel smoothing step systematically set to one in the attacks (i.e. a version of the distinguisher completely free of parameters).

4. Different leakage assumptions. As detailed in Section 2, the non-profiled distinguishers studied in this paper need to rely on some preliminary assumptions on the leakage. For correlation attacks, MIA and the new test, we first evaluated the Hamming weight and identity leakage models (where one takes $x_{j,p}$ as the Hamming weight, or the 7 least significant bits of $v_{j,p}$, respectively). These are usual assumptions when performing a side-channel attack. However, as will be discussed in the remaining of the section, these models were not sufficient to perform successful key recoveries in all the investigated contexts. Hence, we additionally used a profiled leakage model in some experiments. Different solutions

[3] To avoid issues like described in [32], we performed a robust variant of MIA, by selecting the subkeys according to their mutual information bias, i.e. the distance between $\hat{\mathsf{I}}(X_{j,P}, Y_{k,P})$ and the median of this quantity, computed over all key candidates.

are possible for this purpose, e.g. exploiting templates [5]. In the following, we built model classes by grouping together transitions leading to similar leakage values (e.g. 9 such groups would appear for an 8-bit Hamming weight model, corresponding to the 9 possible weights). The model groups were built using a K-means clustering algorithm. Again, this selection is not supposed to be optimal, but to serve as a background to discuss generic distinguishers. As for the stochastic approach, one just needs to select the base vectors used in the adversary's predictions. We followed the classical strategy and used the target S-box output bits for this purpose (i.e. a 9-element basis, with 8 bits and a constant).

As our goal is to compare distinguishers, the evaluations we performed followed the security metrics in [28]. For each investigated context, we computed the success rate of the attacks, over a set of 100 to 500 independent experiments. The figures corresponding to these experiments have been reported in appendix. They allow a number of interesting observations that we now detail.

Observation 1. On different types of leakage functions. The different leakage functions considered imply very different constraints for the non-profiled adversaries. In the case of Hamming weight leakages (Figures 4 and 5 left), this function is purely linear, i.e. $\mathsf{L}(k,p) = \sum \alpha_i \ v_{j,p}[i]$, with $v_{j,p}[i]$ the ith bit of the S-box output and all α_i coefficients set to 1. In addition, as shown in [29], the bivariate distribution of the leakages conditioned on the key, for a masked S-box, is accurately characterized by the correlation between the samples $L^1_{k,P}$ and $L^2_{k,P}$ in this case. As a result, all investigated attacks are very efficient. By contrast, when moving to the analysis of real measurements on a 65nm chip (Figures 5 right and 6), the Hamming weight assumption becomes invalid, and the quadratic, cubic, ... terms in $\mathsf{L}(k,p) = \sum \alpha_i \ v_{j,p}[i] + \sum \beta_{i,j} \ v_{j,p}[i] \cdot v_{j,p}[j] + \ldots$ are non-negligible. As a result, attacks using this model are not successful anymore. Interestingly, the stochastic attack using a linear basis is still efficient, confirming the analysis in [24] that the linear terms of the leakage function are still significant. Surprisingly, the correlation attacks in Figure 6 suggests that a masked S-box may be easier to attack than an unmasked one, under a Hamming weight assumption. This is explained by the fact that actual leakages are not accurately predicted by Hamming weights in the unprotected case, whereas their inter-sample correlation remains informative in a second-order attack against a masked S-box. Eventually, the (simulated) dual-rail S-box is an example of implementation with completely non-linear leakages, as witnessed by the impossibility to perform a successful stochastic attack using a linear basis (see Figure 7).

Observation 2. On the limits of generic distinguishers and models. Generic distinguishers are expected to capture any type of leakage dependency. Still, they are dependent on the leakage model used to build the partitions in a side-channel attack. An interesting outcome of our experiments is that these distinguishers are in fact strongly affected by incorrect assumptions. For example, the Hamming weight leakage model does not lead to successful attacks, neither against the 65nm CMOS chip, nor against the dual-rail pre-charged one. More critically, the identity leakage model that is supposed to provide a generic way to target any implementation in [8], is not successful either in certain cases (Figures 5 right, 7). For Figures 5 right and 7 left, only models obtained through profiling lead to successful key recoveries with the new distinguisher. For Figure 7 right, only template attacks are successful. These limitations are due to the lack of relevance of the leakage models used by the adversary. They are in fact not new: already in 2005, Mangard et al. observed an implementation for which even single-bit leakage models were not accurate enough to perform a successful DPA [16]. More generally, and as also emphasized in [33], MIA-like distinguishers can naturally exploit identity models when applied to non-bijective S-boxes (as in the DES), because such S-boxes imply a meaningful partition by design. But the genericity of this model does not extend to bijective S-boxes (or other block cipher components), excepted if justified by specific implementation choices. In other words, there is no generic leakage model. As discussed in [31], even MIA requires a partitioning such that $\hat{\mathrm{I}}(X_{j,P}; Y_{k,P})$ is maximized for the correct key candidate. The extension of MIA in this paper faces a very similar requirement.

Observation 3. On the limits of non-profiled side-channel attacks. Another consequence of our experiments is to emphasize that, in the context of the dual-rail S-box (Figure 7), none of the non-profiled side-channel attacks could lead to successful key recoveries. Interestingly, the "on-the-fly" stochastic approach fails in this context, even when increasing the size of the basis (e.g. using not only linear, but quadratic, cubic, ... terms). The failure of a stochastic model using linear base vectors is easily explained by the strongly non-linear nature of the simulated leakages produced by the dual-rail S-box. The unsuccessful results with larger bases just derive from the fact that these large bases allow refining the model for all key candidates (i.e. not only the correct one). In fact, also in this case, it is important that the base vectors are justified by a reasonable physical intuition. For example, in case of linear leakage functions, the regression is easy for the correct key candidate (because provided with a good basis) and difficult for the wrong key candidates (because the regression essentially has to capture the non-linearity of a modified S-box S' such that $\mathsf{S}(x \oplus k_w) = \mathsf{S}'(x \oplus k_g)$, with k_g and k_w the good and a wrong key candidate). But as soon as the base vectors do not have a connection with the actual leakages, the advantage of the correct key candidate in building a good stochastic model vanishes[4]. When combining this dual-rail logic style with masking (in Figure 7 right), we see that only a bivariate template attack, similar to the ones in [29], allows recovering keys. In this respect, it is worth noting that a leakage model that is sound in an unprotected setting (e.g. the one based on clustering in Figure 7 left) does not translate into a sound model for the corresponding masked implementation (in Figure 7 right).

Observation 4. MIA versus the new test. Finally, our new distinguisher compares favorably to MIA in all the investigated experiments. We note that the efficiency of MIA could possibly be improved, by exploiting pdf estimation based on

[4] This limitation is only due to the application of the stochastic approach in a non-profiled scenario. In profiled attacks, the stochastic approach remains perfectly sound, as soon as provided with enough base vectors, just as a template attack.

Kernels, splines or parametric techniques [19]. However, more than the efficiency of the distinguisher, it is worth to notice that in certain settings, e.g. the masked 65nm CMOS S-boxes in Figures 5 right and 6 right, it allows exploiting a leakage assumption while MIA cannot. It is an open question to determine whether these experiments can be formally confirmed (e.g. are they due to identified limitations as in the example given in [32], Section 3) or are the result of measurement artifacts that would vanish with more intensive measurement efforts.

5 Conclusion and Open Problems

Generic distinguishers are a useful tool for evaluating leaking devices. In this paper, we first proposed a new and efficient generic test that is fully non-parametric, based on a natural discriminating criterion, and exploits state-of-the-art statistical tools. It can be useful in all scenarios where the previously introduced MIA shows significant advantages over other non-profiled distinguishers.

Next, we discussed the relevance of generic distinguishers in general. Based on experimental validation in different contexts, we put forward that such non-profiled attacks do not get rid of the need of sound assumptions during the partitioning step in a side-channel analysis. Similarly, when applied "on-the-fly", the stochastic approach of Schindler et al. is only sound when provided with meaningful based vectors, that may not be easy to guess for practical adversaries. This observation suggests that the gap between profiled and non-profiled side-channel attacks can be huge, when such assumptions are are not available. It re-emphasizes the need to consider two aspects in the security analysis of a leaking device, as advocated in [28]. First, the worst case security can only be evaluated with a profiled attack (e.g. using templates), and quantified with an information theoretic metric. Second, different types of non-profiled distinguishers can be compared with a security metric, in order to measure how efficiently they can take advantage of the available information. In this respect, generic tests bring an interesting alternative to more specific (e.g. correlation-based) statistical tools. But they are not immune to model imprecisions, and resisting such attacks is not sufficient to conclude that an implementation is secure.

Admittedly, the most critical examples we exhibit in this paper are based on simulations and practical implementations usually show linear dependencies in their leakages. Nevertheless, this discussion underlines that the theoretical limits of present non-profiled attacks have to be properly understood. It also leads to interesting questions for the design and analysis of secure implementations. First, as non-profiled distinguishers published in the literature seem specially affected by non-linear leakage functions, designing hardware logic styles with this criterion in mind appears as an interesting scope for further research. Preliminary experiments reported in [23] suggest that the DDSLL logic style may not be the most suitable for this purpose. Second, the partitioning step of generic distinguishers is made specially easy when non-bijective S-boxes (or components) are used. Hence, such S-boxes should be avoided when designing ciphers to be secured against physical attacks. In the same lines, targeting the output of MixColumn in an AES implementation could be an interesting topic for further investigation. Depending on the architectures (e.g. 8-bit software or 32-bit hardware), it could also lead to leakage models that are simple to exploit. Eventually, non-profiled security evaluations are typically misleading when randomization-based countermeasures such as masking are combined with the difficulty to make sound assumptions on the leakage model. Hence, developing new tools to get rid of this limitation, or showing that no such tools actually exist, is an important challenge for evaluating the security of future embedded cryptographic devices.

References

1. Batina, L., Gierlichs, B., Prouff, E., Rivain, M., Standaert, F.-X., Veyrat-Charvillon, N.: Mutual information analysis: a comprehensive study. J. Cryptology 24(2), 269–291 (2011)
2. Brier, E., Clavier, C., Olivier, F.: Correlation Power Analysis with a Leakage Model. In: Joye, M., Quisquater, J.-J. (eds.) CHES 2004. LNCS, vol. 3156, pp. 16–29. Springer, Heidelberg (2004)
3. Cantelli, F.P.: Sulla determinazione empirica della legge di probabilita. Giorn. Ist. Ital. 4, 421–424 (1933)
4. Chari, S., Jutla, C.S., Rao, J.R., Rohatgi, P.: Towards Sound Approaches to Counteract Power-Analysis Attacks. In: Wiener, M.J. (ed.) CRYPTO 1999. LNCS, vol. 1666, pp. 398–412. Springer, Heidelberg (1999)
5. Chari, S., Rao, J.R., Rohatgi, P.: Template attacks. In: Kaliski Jr., B.S., Koç, Ç.K., Paar, C. (eds.) CHES 2002. LNCS, vol. 2523, pp. 13–28. Springer, Heidelberg (2003)
6. Doget, J., Prouff, E., Rivain, M., Standaert, F.-X.: Univariate side channel attacks and leakage modeling. In: COSADE, Darmstadt, Germany, pp. 1–15 (February 2011)
7. Gierlichs, B., Batina, L., Preneel, B., Verbauwhede, I.: Revisiting higher-order DPA attacks. In: Pieprzyk, J. (ed.) CT-RSA 2010. LNCS, vol. 5985, pp. 221–234. Springer, Heidelberg (2010)
8. Gierlichs, B., Batina, L., Tuyls, P., Preneel, B.: Mutual Information Analysis. In: Oswald, E., Rohatgi, P. (eds.) CHES 2008. LNCS, vol. 5154, pp. 426–442. Springer, Heidelberg (2008)
9. Glivenko, V.: Sulla determinazione empirica della legge di probabilita. Giorn. Ist. Ital. 4, 92–99 (1933)
10. Goubin, L., Patarin, J.: DES and differential power analysis (the "duplication" method). In: Koç, Ç.K., Paar, C. (eds.) CHES 1999. LNCS, vol. 1717, pp. 158–172. Springer, Heidelberg (1999)
11. Hassoune, I., Macé, F., Flandre, D., Legat, J.-D.: Dynamic differential self-timed logic families for robust and low-power security ics. Integration, the VLSI Journal 40(3), 355–364 (2007)
12. Hastie, T., Tibshirani, R., Friedman, J.: The Elements of Statistical Learning, ch.6. Springer Series in Statistics. Springer New York Inc., New York (2001)

13. Kocher, P.C., Jaffe, J., Jun, B.: Differential power analysis. In: Wiener, M. (ed.) CRYPTO 1999. LNCS, vol. 1666, pp. 388–397. Springer, Heidelberg (1999)
14. Lemke-Rust, K., Paar, C.: Analyzing side channel leakage of masked implementations with stochastic methods. In: Biskup, J., López, J. (eds.) ESORICS 2007. LNCS, vol. 4734, pp. 454–468. Springer, Heidelberg (2007)
15. Mangard, S., Oswald, E., Popp, T.: Power Analysis Attacks: Revealing the Secrets of Smart Cards. Springer, Heidelberg (2007)
16. Mangard, S., Pramstaller, N., Oswald, E.: Successfully attacking masked aes hardware implementations. In: Rao, Sunar [22], pp. 157–171
17. Messerges, T.S.: Using second-order power analysis to attack dpa resistant software. In: Koç, Ç.K., Paar, C. (eds.) CHES 2000. LNCS, vol. 1965, pp. 238–251. Springer, Heidelberg (2000)
18. Nelsen, R.B.: An Introduction to Copulas, 1st edn. Lecture Notes in Statistics. Springer, Heidelberg (1998)
19. Prouff, E., Rivain, M.: Theoretical and practical aspects of mutual information based side channel analysis. In: Abdalla, M., Pointcheval, D., Fouque, P.-A., Vergnaud, D. (eds.) ACNS 2009. LNCS, vol. 5536, pp. 499–518. Springer, Heidelberg (2009)
20. Prouff, E., Rivain, M., Bevan, R.: Statistical analysis of second order differential power analysis. IEEE Trans. Computers 58(6), 799–811 (2009)
21. Pyke, R.: Spacings revisited. In: Proceedings of the Sixth Berkeley Symposium on Mathematical Statistics and Probability (Univ. California, Berkeley, Calif., 1970/1971), Vol. I: Theory of statistics, pp. 417–427. Univ. California Press, Berkeley (1972)
22. Rao, J.R., Sunar, B. (eds.): CHES 2005. LNCS, vol. 3659. Springer, Heidelberg (2005)
23. Renauld, M., Kamel, D., Standaert, F.-X., Flandre, D.: Scaling trends for dual rail logic styles (2011) (preprint)
24. Renauld, M., Standaert, F.-X., Veyrat-Charvillon, N., Kamel, D., Flandre, D.: A formal study of power variability issues and side-channel attacks for nanoscale devices. In: Paterson, K.G. (ed.) EUROCRYPT 2011. LNCS, vol. 6632, pp. 109–128. Springer, Heidelberg (2011)
25. Rivain, M., Dottax, E., Prouff, E.: Block ciphers implementations provably secure against second order side channel analysis. In: Nyberg, K. (ed.) FSE 2008. LNCS, vol. 5086, pp. 127–143. Springer, Heidelberg (2008)
26. Schindler, W., Lemke, K., Paar, C.: A stochastic model for differential side channel cryptanalysis. In: Rao, Sunar [22], pp. 30–46
27. Scott, D.W.: On optimal and data-based histograms. Biometrika 66(3), 605–610 (1979)
28. Standaert, F.-X., Malkin, T., Yung, M.: A unified framework for the analysis of side-channel key recovery attacks. In: Joux, A. (ed.) EUROCRYPT 2009. LNCS, vol. 5479, pp. 443–461. Springer, Heidelberg (2009)
29. Standaert, F.-X., Veyrat-Charvillon, N., Oswald, E., Gierlichs, B., Medwed, M., Kasper, M., Mangard, S.: The world is not enough: Another look on second-order dpa. In: Abe, M. (ed.) ASIACRYPT 2010. LNCS, vol. 6477, pp. 112–129. Springer, Heidelberg (2010)
30. Venelli, A.: Efficient entropy estimation for mutual information analysis using b-splines. In: Samarati, P., Tunstall, M., Posegga, J., Markantonakis, K., Sauveron, D. (eds.) WISTP 2010. LNCS, vol. 6033, pp. 17–30. Springer, Heidelberg (2010)

31. Veyrat-Charvillon, N., Standaert, F.-X.: Mutual information analysis: How, when and why? In: Clavier, C., Gaj, K. (eds.) CHES 2009. LNCS, vol. 5747, pp. 429–443. Springer, Heidelberg (2009)
32. Veyrat-Charvillon, N., Standaert, F.-X.: Generic side channel distinguishers: Improvements and limitations. Cryptology ePrint Archive, Report 2011/149 (2011), `http://eprint.iacr.org/`
33. Whitnall, C.: An information theoretic assessment of first-order mia. First year PhD report, University of Bristol (2010)
34. Whitnall, C., Oswald, E.: A comprehensive evaluation of mutual information analysis using a fair evaluation framework. In: To Appear In The Proceedings Of Crypto 2011, Santa Barbara, California, USA, August 2011, vol. xxxx, pp. yyy–zzz (2011)

A Results of Our Experiments

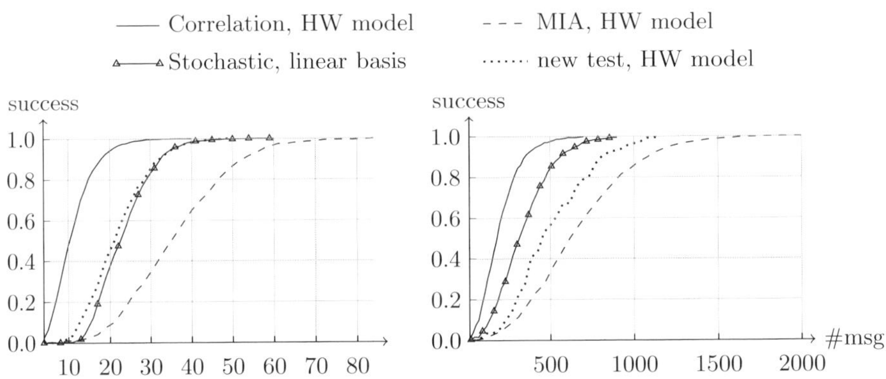

Fig. 4. Simulations with a Hamming weight leakage function, attacks against an unprotected S-box (left) and a masked S-box with combined samples (right)

B Building the Distance Templates

The distance templates are built by convolution of independent uniform random variables. That is, for one dimension, the probability of a spacing of length u is the probability that two random variables X and X' drawn from the n-valued discrete uniform distribution on the interval $[0, 1]$ (each value has probability $\frac{1}{n}$) will differ by an amount u. That is:

$$\Pr[U = u] = \sum_{x} \Pr[X = x] \cdot \Pr[X' = x \pm u]$$

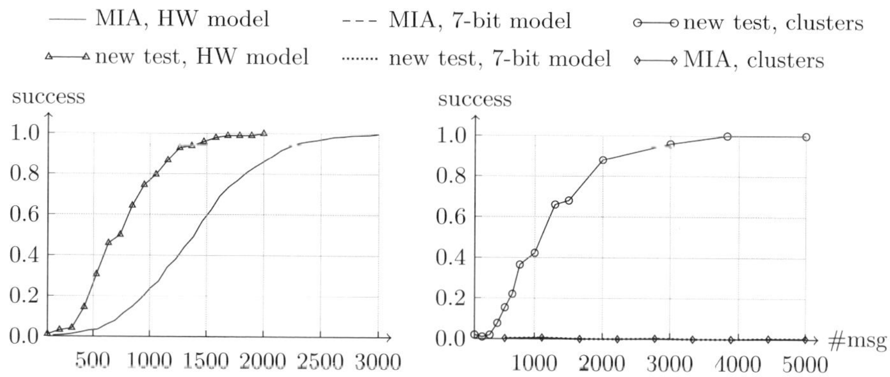

Fig. 5. Masked S-box with bivariate leakages. Attacks with simulated Hamming weight leakage function (left) and actual measurements on a 65nm CMOS chip (right).

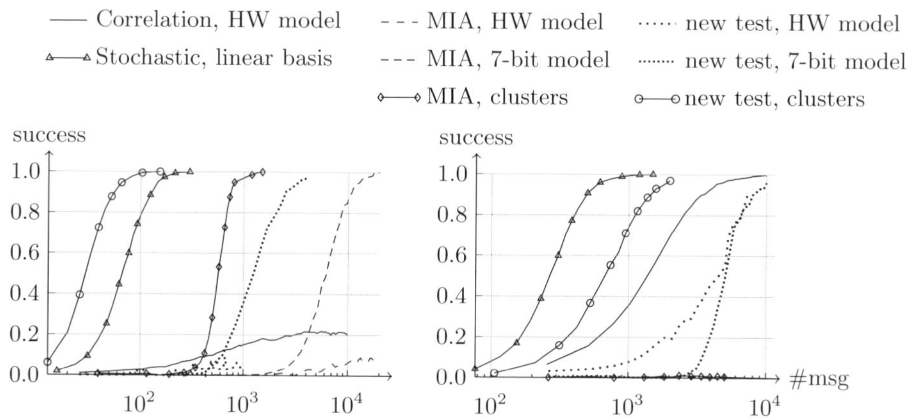

Fig. 6. Measurements on a 65nm CMOS chip, attacks against an unprotected S-box (left) and a masked S-box with combined samples (right)

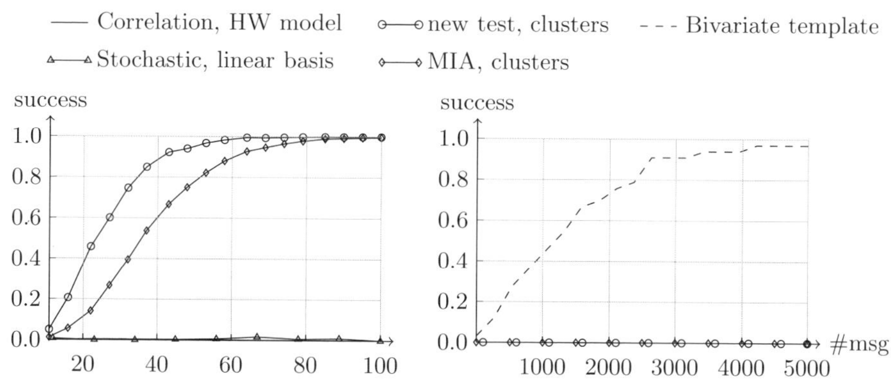

Fig. 7. Spice simulations of a 65nm dual-rail logic style, attacks against an unprotected S-box (left) and a masked S-box with bivariate samples (right)

In our specific case, since the discrete uniform distribution results from an integral transform of continuous variables, it is impossible to have a spacing of 0. The spacing distribution is therefore $\Pr[U = u] = \frac{2}{n-1} \times (1 - u)$ for $u > 0$, 0 otherwise. As one considers higher dimensions, using the Manhattan distance to extend the notion of spacings, the distribution becomes a convolution of spacings on one dimension:

$$\Pr[U = u] = \sum_{u_i} \prod_{i=1}^{d-1} \Pr[U_i = u_i] \cdot \Pr\left[U_d = \left(u - \sum_{i=1}^{d-1} u_i\right)\right],$$

where the U_i are spacings taken along dimension i. In the case of two dimensions, this formula simply gives :

$$\Pr[U = u] = \sum_{u_1} \Pr[U_1 = u_1] \cdot \Pr[U_2 = u - u_1],$$

which is the integral of two affine functions, therefore a piecewise cubic polynomial since U_1 and U_2 are only defined on the interval $[0, 1]$ while U ranges over $[0, 2]$. While it is possible to compute the sampling distribution analytically for higher dimensions, it quickly becomes cumbersome, and it is much more practical to build the distribution by composing lower-dimension distance histograms. This is done very efficient by following a method similar to the square-and-multiply algorithm, where for example the template for dimension 4 is built by convoluting the template for dimension 2 with himself.

Cryptography with Tamperable and Leaky Memory

Yael Tauman Kalai[1], Bhavana Kanukurthi[2,⋆], and Amit Sahai[3,⋆⋆]

[1] Microsoft Research
[2] Boston University
[3] University of California (UCLA)

Abstract. A large and growing body of research has sought to secure cryptographic systems against physical attacks. Motivated by a large variety of real-world physical attacks on memory, an important line of work was initiated by Akavia, Goldwasser, and Vaikuntanathan [1] where security is sought under the assumptions that: (1) *all memory is leaky*, and (2) leakage can be *an arbitrarily chosen (efficient) function* of the memory.

However, physical attacks on memory are not limited to leakagethrough side-channels, but can also include active *tampering* attacks through a variety of physical attacks, including heat and EM radiation. Nevertheless, protection against the analogous model for tampering – where (1) *all memory is tamperable*, and (2) where the tampering can be *an arbitrarily chosen (efficient) function* applied to the memory – has remained an elusive target, despite significant effort on tampering-related questions.

In this work, we tackle this question by considering a model where we assume that *both* of these pairs of statements are true – that all memory is both leaky and (arbitrarily) tamperable. Furthermore, we assume that this leakage and tampering can happen repeatedly and continually (extending the model of [10,7] in the context of leakage). We construct a signature scheme and an encryption scheme that are provably secure against such attacks, assuming that memory can be updated in a randomized fashion between episodes of tampering and leakage. In both schemes we rely on the linear assumption over bilinear groups.

We also separately consider a model where only continual and repeated tampering (but only bounded leakage) is allowed, and we are able to obtain positive results assuming only that "self-destruct" is possible, without the need for memory updates.

Our results also improve previous results in the continual leakage regime without tampering [10,7]. Whereas previous schemes secure against continual leakage (of arbitrary bounded functions of the secret key), could tolerate only $1/2 - \epsilon$ leakage-rate between key updates under the linear assumption over bilinear groups, our schemes can tolerate $1 - \epsilon$ leakage-rate between key updates, under the same assumption.

⋆ Research supported in part by CNS-0546614, CNS-0831281, CNS-1012910. Part of this work was done while visiting Microsoft Research, New England.

⋆⋆ Research supported in part from a DARPA/ONR PROCEED award, NSF grants 0916574 and 0830803, a Xerox Foundation Award, a Google Faculty Research Award, an equipment grant from Intel, and an Okawa Foundation Research Grant.

P. Rogaway (Ed.): CRYPTO 2011, LNCS 6841, pp. 373–390, 2011.

1 Introduction

A large and growing body of research has sought to secure cryptographic systems against physical attacks (e.g. [8,23,28,18,22,13,31,1,29,25,11,9,15,24,19]). Motivated by a large variety of real-world physical attacks on memory [26,27,30,21,20], an important line of work was initiated by Akavia, Goldwasser, and Vaikuntanathan [1] where security is sought under the assumptions that: (1) *all memory is leaky*, and (2) leakage can be *an arbitrarily chosen (efficient) function* of the memory[1].

However, physical attacks on memory are not limited to leakage through side-channels, but can also include active *tampering* attacks (see [20] and the references therein) through a variety of physical attacks, such as exposure to heat and EM radiation [17,33,32]. Nevertheless, protection against the analogous model for tampering – where (1) *all memory is tamperable*, and (2) where the tampering can be *an arbitrarily chosen (efficient) function* applied to the memory – has remained an elusive target, despite significant effort on tampering-related questions (e.g. [18,22,3,14,2])[2]. In this work, we tackle this question by considering a model where we assume that *both* of these pairs of statements are true – that all memory is both leaky and (arbitrarily) tamperable. Furthermore, we assume that this leakage and tampering can happen repeatedly and continually (extending the model of [10,7] in the context of leakage). We show strong positive results for cryptographic tasks including signatures and decryption, assuming that memory can be updated in a randomized fashion between episodes of tampering and leakage. We note that we only consider tampering with the memory and do not consider the question of tampering during the computation (which might occur, for instance, due to fault attacks [5,4,27]).

We also separately consider a model where only continual and repeated tampering (but bounded leakage) is allowed, and we are able to obtain positive results assuming only that "self-destruct" is possible, without the need for memory updates.

Background. Before explaining our results in greater detail, let us first recall the results of [10,7], which we strengthen. These results construct various cryptographic schemes (such as encryption and signature schemes), that remain secure even if the secret key is being continually leaked. In this continual leakage model, at each time period the adversary can make a *leakage query* L, where L is a poly-size circuit, and get back $L(sk)$. Clearly, in order to get security under continual leakage, one must bound the leakage function L in some way, since if, for example, L is the identity function, security is clearly breached. Both works [10,7] allow L to be any *shrinking* leakage function; *e.g.*, any L such that $|L(sk)| \leq 0.99|sk|$. More generally, they allow any leakage function L such that sk has "enough" min-entropy left conditioned on $L(sk)$. We note that several earlier continual leakage models were considered in the literature, such as the

[1] To prevent trivial attacks, the leakage function should either be significantly shrinking or leave the memory with enough entropy.

[2] We elaborate on these related works in Section 1.3.

one due to Micali and Reyzin [28] who consider the "only computation leaks information" assumption, and the one due to Ishai, Sahai and Wagner [23], who consider only leakage of individual bits produced during honest computation. We elaborate on these related works, as well as others, in Section 1.3.

As was argued in [7,10], since the leakage may be continual, and at any time period the adversary can learn *any* bounded poly-size function of the secret key, the secret-key must be periodically *updated*, since otherwise it will eventually be completely leaked. We emphasize that this should be done without modifying the public key, and these updates should be oblivious to all other users. In addition, *deletions* must be allowed, since otherwise the adversary can choose to (gradually) leak the initial secret-state from which everything can be derived. In such case, updating the state is useless because the adversary will leak from the original state and not the updated one.

We note that both these results rely on the linear assumption over bilinear groups. The work of [7] allow $1/2-\epsilon$ of the secret key to leak during each time period, whereas [10] allow only $1/3-\epsilon$ of the secret key to leak during each time period.[3] We note that in addition to our main results concerning tampering, we improve the leakage bounds of [10,7], even if we restrict our attention only to the regime of continual leakage (as opposed to continual leakage and tampering).

1.1 Our CTL Model

In this work, we extend the continual leakage model of [7,10], and consider not only continual leakage attacks, but also continual *tampering* attacks. Namely, we consider an adversary, that uses side channel attacks not only to continually leak information about the secret key, but also to continually tamper with the secret key. For example, consider an adversary that is given a signature card. The adversary may use various side channel attacks, such as timing attacks and power attacks, to continually *leak* information about the secret key stored in the card, but may also use various other physical attacks [20] to *tamper* with this secret key. For example, the adversary can extract information by causing mutations in the secret key, and then observing the input/output behavior of the tampered system (i.e., observing signatures of messages with respect to tampered secret keys).

More formally, our model generalizes the continual leakage model of [10,7], in that we allow the adversary at each time period to make any bounded leakage query L, *as well as any tampering query* T. Both types of queries are modeled as a poly-size circuit. After the adversary makes a tampering query T, the secret key sk is replaced with $T(sk)$. We call this model the *continual tampering and leakage* model, or the CTL model, for short. We note that we only allow tampering with the secret key and not the actual computation. We say that a scheme is secure in the CTL model if after continually leaking information about the secret keys, and continually tampering with them, the adversary cannot break security with respect to the *original* secret key; i.e., in the case of signature schemes, the

[3] We note that under the less standard SXDH assumption their leakage rate can increase to $1-\epsilon$ in [7] and $1/2-\epsilon$ in [10].

adversary cannot forge signatures with respect to the original verification key; in the case of encryption schemes, the adversary cannot break semantic security with respect to the original public key.

At first glance, the reader may wonder whether tampering attacks can be viewed as leakage attacks, and thus whether the CTL model is at all more general than the continual leakage model. The answer is that it is indeed more general, for the following two reasons. First, in leakage attacks there is always an assumed bound on the leakage size (within each time period, in the case of continual leakage). On the other hand, in a tampering attack, the adversary can mutate the key in an arbitrary manner, and receive signatures with this mutated key. Note that if such signatures reveal the entire mutated secret key (which is possible since the key is malformed), then this tampering attack is an illegal leakage attack (since this leakage is not bounded). The other (and perhaps more profound) reason why the CTL model seems much harder to deal with than the continual leakage model, is that in the CTL model the update procedure updates a *mutated* key, as opposed to an honestly generated key. Thus, we need to argue that the update procedure is still "effective" even if it is applied to mutated keys. We note that the schemes of [10,7] are not secure in the CTL model.

An additional assumption that we make is that we assume that a CRS is honestly chosen and hard-coded into the cryptographic system, not in memory but into the logic itself, so that the CRS is not tamperable. We stress that the CRS is fixed once and for all and does not depend on any secret information chosen later by the user, which is stored in memory. We note that to some extent, the hardware manufacturer must always be trusted to implement the correct algorithms in hardware. (For instance, we must trust that the manufacturer did not implement an algorithm that leaks secrets through specific subliminal channels.) Thus we are not asking a significantly higher level of trust from the user to assume that the manufacturer (or some outside entity with higher trust) honestly chose a CRS and that the manufacturer correctly hard-coded this CRS into the logic of the device. We leave the problem of eliminating the CRS, or exploring ways to reduce the level of trust needed in the manufacturer, as an important open problem.

1.2 Our Results

In what follows we present informal statements of our results.

Theorem 1. *Under the linear assumption in a prime order group G, there exists a encryption scheme that is semantically secure in the* CTL *model, tolerating a leakage of* $1-\epsilon$ *in each time period.*

Theorem 2. *Under the linear assumption in a prime order group G, there exists a signature scheme that is existentially unforgeable under adaptive chosen message attacks in the* CTL *model, tolerating a leakage of* $1-\epsilon$ *in each time period.*

Theorem 3. *Given a signature scheme that is existentially unforgeable in the bounded leakage model, there exists an efficient transformation to covert it into* *a signature scheme that is existentially unforgeable in the continual tampering and bounded leakage model.*[4]

The proofs of these theorems are deferred to the full version.

1.3 Previous Work

Work on tolerating leakage was initiated by Rivest and Boyko [34,6] in the context of increasing the cost of brute-force attacks on block ciphers and efficiency issues. Ideas there were applied to the context of leakage by numerous works on exposure-resilient cryptography [8,12,23]. These works consider simple leakage functions that reveal a subset of the bits of the secret key or the internal memory of the cryptographic device. This line of research culminated in the work of Ishai, Sahai and Wagner [23] who show how to "compile" any cryptographic algorithm into one that tolerates such types of leakage.

In contrast to these works, that consider leakage functions that act locally, the focus of later works has been on more powerful leakage functions that can perform some *global computation* on the secret key. Micali and Reyzin [28] proposed to construct and study *formal models* that capture *general* types of leakage. This study has led to two distinct strands of work, described below.

Bounded Leakage Models. This line of work considers the leakage model that allows the adversary to obtain the output of applying any efficiently computable function f, of his choice, to the secret key sk. This is allowed as long as the output $f(sk)$ "does not reveal the entire secret key". This latter condition has been formalized in many different ways, starting with the work of Akavia, Goldwasser and Vaikuntanathan [1] who restrict the leakage function f to have output length $\ell(|sk|) \ll |sk|$ and subsequently in the works of Naor and Segev [29] and Dodis, Kalai and Lovett [11]. Constructions of cryptosystems satisfying one or more of these definitions can be found in various works (for example [1,29,11,25,9]).

Continual Leakage Models. This line of work considers the case where the leakage is continual, i.e., a bounded amount of information about the secret key is leaked in each time period, but the overall leakage in the entire lifetime of the secret key is unbounded. It is easy to see that to guarantee any security in this model the secret key must necessarily be updated between time-periods. Micali and Reyzin [28] proposed to study security against continual leakage under the assumption that "only computation leaks information". In other words, information leakage may occur any time a computation takes place; however, the assumption is that the parts of memory that are not involved in the computation during a certain time-period, are not subject to leakage during that time-period. A number of works (for example [13,31,16,24,19]) design cryptographic schemes that are resilient to leakage under this assumption.

Very recently, Dodis, Haralambiev, Lopez-Alt and Wichs [10] and Brakerski, Kalai, Katz, and Vaikuntanathan [7] constructed cryptographic schemes (such

[4] For this result we need to assume that the signature card can self-destruct.

as signature schemes and encryption schemes) in the continual leakage model without this assumption that only computation leaks. Our main result builds upon the result of Brakerski *et al.*

Tampering models. Several models of security against tampering have been considered in the past, however all of them differ substantially from ours. The notion of algorithmic tamper-proof security was proposed by Gennaro, Lysyanskaya, Malkin, Micali, and Rabin [18] in which devices have *both* tamper-proof but fully leakable memory (which can be programmed with key information uniquely customized to each user), along with tamperable memory – which can be continually tampered with using arbitrary polynomial-time tampering functions. (In contrast, our model allows all memory to be arbitrarily tampered.) Security against related-key attacks was formalized by Bellare and Kohno [3] in the context of block-cipher security, but can be seen as relating to a model where the key can be tampered with. However, this line of work (see [2] and references therein) deals with limited kinds of tampering functions, such as functions that simply XOR the key with a fixed value or affine linear transformations. The recent work on non-malleable codes [14] also looks as such limited tampering functions. (In contrast, our model allows arbitrary polynomial-time tampering functions.) Finally, the work of Ishai, Prabhakaran, Sahai, and Wagner [22] considers still weaker tampering functions, which only allow individual bits to be toggled or set to a specific value. However, in contrast to our work, Ishai *et al.* also allow tampering to affect the computation, and not just the memory, which is beyond the scope of our work.

2 Overview

Notation. We use the following notational conventions. Bold uppercase denotes matrices ($\mathbf{X} \in \mathbb{Z}_p^{n\times k}$) and bold lowercase denotes vectors ($\mathbf{x} \in \mathbb{Z}_p^n$). All vectors are column vectors, row vectors are denoted by $\mathbf{x}^T$. In what follows, we let $\mathbb{G}$ be a multiplicative group of prime order p, and g be a generator of $\mathbb{G}$. We let $e : \mathbb{G} \times \mathbb{G} \to \mathbb{G}_T$ be a bilinear map. All our results assume that the linear assumption holds in $\mathbb{G}$.[5] For a matrix $\mathbf{X} \in \mathbb{Z}^{k\times n}$ (or a vector, as a special case), we let $g^{\mathbf{X}}$ denote a $k \times n$ matrix such that $(g^{\mathbf{X}})_{i,j} = g^{(\mathbf{X})_{i,j}}$.

Let X be a probability distribution over a domain S, we write $x \leftarrow X$ to indicate that x is sampled from the distribution X. For a set S, we write $x \leftarrow S$ to indicate that x is chosen uniformly from S. Two ensembles $X = \{X_k\}_k$, $Y = \{Y_k\}_k$ are said to be $\epsilon = \epsilon(k)$-*close* if the statistical distance between them is at most $\epsilon(k)$, this is also denoted by $X \stackrel{\epsilon}{\equiv} Y$.

As was mentioned in the introduction, our work builds on the work of [7], which (among other things) constructs encryption and signature schemes that were proven secure against continual leakage. Let us start by recalling some of the ideas in [7], which are relevant to this work. In both their encryption scheme

[5] Formally, $\mathbb{G} = \{\mathbb{G}_k\}_{k\in\mathbb{N}}$ where each group $\mathbb{G}_k$ is a group of prime order p, where p is a k bit prime, and where k is the security parameter.

and their signature scheme, the public key is of the form $g^{\mathbf{a}}$ and the secret key is of the form $g^{\mathbf{b}}$, where g is a generator of $\mathbb{G}$, and $\mathbf{a}, \mathbf{b} \leftarrow \mathbb{Z}_p^{\ell}$ such that $\mathbf{a} \cdot \mathbf{b} = 0$.[6] The secret key is updated by simply multiplying it by a random scalar in the exponent, i.e., updating it to be $g^{\alpha \cdot \mathbf{b}}$ where $\alpha \leftarrow \mathbb{Z}_p$. Thus, the updated secret keys are simply random elements in span($\mathbf{b}$) (in the exponent). To prove security against continual leakage, they rely on the following lemma, which states that random subspaces are resilient to continual leakage.

Lemma 1. *[7] Let $\mathbf{B} \leftarrow \mathbb{Z}_p^{\ell \times d}$ be a random subspace of dimension d where $\ell \geq d \geq 2$. Let $\mathbf{r} \leftarrow \mathbb{Z}_p^d$ and $\mathbf{u} \leftarrow \mathbb{Z}_p^{\ell}$. Then for any leakage function $L : \mathbb{Z}_p^{\ell} \rightarrow W$, chosen independent of $\mathbf{B}$,*

$$(\mathbf{B}, L(\mathbf{B} \cdot \mathbf{r})) \stackrel{\epsilon}{\equiv} (\mathbf{B}, L(\mathbf{u}))$$

as long as $|W| \leq p^{d-1} \cdot \epsilon^2$.

Intuitively, the above lemma says that leakage on random points taken from a random subspace is indistinguishable from leakage on random points taken from the whole space. In other words, leakage on random points of a subspace doesn't reveal *any* information about the subspace.

This lemma is used as follows: First, use the DDH assumption to claim that it is computationally hard to distinguish between the case that all the secret keys are indeed random in span($\mathbf{b}$) or are random in a random dimension-2 subspace $\mathbf{B} \subseteq \ker(\mathbf{a})$.[7] Therefore, if there exists an adversary that breaks security, then this adversary will also succeed in breaking the security if all the secret keys were random elements in $\mathbf{B}$ (in the exponent), as opposed to random elements in span($\mathbf{b}$). In this case, one can use Lemma 1 above, to claim that no information about the subspace $\mathbf{B}$ is revealed (information theoretically).

Suppose for the sake of simplicity, that the adversary that breaches security, actually finds a secret key (rather than merely forging a signature or breaking semantic security). Then, the fact that the random subspace $\mathbf{B}$ is (information theoretically) hidden, implies that with overwhelming probability the secret key $g^{\mathbf{b}'}$ that the adversary outputs is not in $\mathbf{B}$. If indeed $g^{\mathbf{b}'}$ is a valid secret key, i.e., $\mathbf{b}' \in \ker(\mathbf{a})$, and $\mathbf{b}'$ is not in the subspace $\mathbf{B}$, then one can use this to break the linear assumption.

---

6 Actually, using such keys, [7] get security under the SXDH assumption. To get security under the linear assumption, they take the public key to be $g^{\mathbf{A}}$ and the secret key to be $g^{\mathbf{B}}$, where $\mathbf{A} \leftarrow \mathbb{Z}_p^{2 \times \ell}$ and $\mathbf{B} \leftarrow \mathbb{Z}_p^{\ell \times 2}$ such that $\mathbf{A} \cdot \mathbf{B} = \mathbf{0}$. This results in a slightly more complicated scheme.

7 Note that they rely on the DDH assumption, and in addition, in order to verify or decrypt they need to assume the existence of a bilinear map. Thus, they need to rely on the SXDH assumption. However, by taking the public key to be *two* random vectors $\mathbf{A} \leftarrow \mathbb{Z}_p^{2 \times \ell}$ (in the exponent) and the secret key to be two vectors in $\ker(\mathbf{A})$ (in the exponent), they can get security based on the linear assumption over bilinear groups. For this, they use a more general form of Lemma 1.

This work. In this work, we allow the adversary to both leak and tamper with the secret key. Namely, the adversary can change the secret key from $g^{\mathbf{b}}$ to $T(g^{\mathbf{b}})$. Note that the above proof idea cannot handle tampering, since the update procedure updates a (tampered) secret key by raising it to the power of $\alpha \leftarrow \mathbb{Z}_p^\ell$, i.e., updates $T(g^{\mathbf{b}})$ to $T(g^{\mathbf{b}})^\alpha$. Since the tampering function T is adversarially chosen, we cannot rely on the DDH assumption anymore, and thus cannot use Lemma 1. Indeed, we cannot prove that the scheme of [7] is secure against tampering. Instead, we slightly modify their schemes.

Our schemes. We think of both $g^{\mathbf{a}}$ and $g^{\mathbf{b}}$ as public parameters (or crs), that may be shared among all users, and we assume that these parameters cannot be tampered with. As was mentioned in the introduction, the justification for this assumption is that these parameters are independent of the secret key, and can be thought of as generated in the manufacturing level, and hardwired into the card in a secure way.

Actually, to prove security under the linear assumption (rather than the SXDH assumption), we set the public parameters to be $g^{\mathbf{A}}$ and $g^{\mathbf{B}}$, where $\mathbf{A} \leftarrow \mathbb{Z}_p^{2\times\ell}$ and $\mathbf{B} \leftarrow \mathbb{Z}_p^{\ell\times 2}$ such that $\mathbf{A}\cdot\mathbf{B} = \mathbf{0}$. Then, we let the secret key be $g^{\mathbf{s}}$ where $\mathbf{s} \leftarrow \mathbb{Z}_p^\ell$, and the public key be $e(g,g)^{\mathbf{A}\cdot\mathbf{s}}$, where $e : \mathbb{G}\times\mathbb{G} \to \mathbb{G}_T$ is a bilinear map. The secret key $g^{\mathbf{s}}$ is updated by multiplying it by $g^{\mathbf{B}\cdot\mathbf{R}}$ where $\mathbf{R} \leftarrow \mathbb{Z}_p^{2\times 2}$. Note that this does not change the public key since

$$e(g,g)^{\mathbf{A}\cdot(\mathbf{s}+\mathbf{B}\cdot\mathbf{R})} = e(g,g)^{\mathbf{A}\cdot\mathbf{s}}.$$

Now, when the adversary tampers with the secret key, and converts it to $T(g^{\mathbf{s}})$, this tampered secret key is updated to be $T(g^{\mathbf{s}})\cdot g^{\mathbf{B}\cdot\mathbf{R}}$, and we can still use the linear assumption to claim that this is computationally indistinguishable from the case that the secret key is updated by adding to it (in the exponent) a random element from a random (independent) subspace $\mathbf{B}' \subseteq \ker(\mathbf{A})$.[8]

We would like to use Lemma 1 to argue that if indeed all the updates are done using $\mathbf{B}'$ (which is a random subspace in $\ker(\mathbf{A})$), then this subspace remains (information theoretically) hidden, even given all the leakages. If we could do so, then we would be in good shape, as before: For the sake of simplicity, suppose that an adversary that breaches security actually outputs a valid secret key (rather than a merely outputting a valid signature or breaking semantic security).[9] Then, this adversary outputs some $g^{\mathbf{s}'}$ such that $e(g,g)^{\mathbf{A}\mathbf{s}'} = e(g,g)^{\mathbf{A}\mathbf{s}}$, and thus $\mathbf{s}'-\mathbf{s} \in \ker(\mathbf{A})$. The fact that $\mathbf{B}'$ is information theoretically hidden implies that with overwhelming probability $\mathbf{s}'-\mathbf{s}$ is not in $\mathrm{span}(\mathbf{B}',\mathbf{B})$, and thus can be used to break the linear assumption.

Unfortunately, when trying to use Lemma 1, we face several technical difficulties. The main problem is the following: Note that we need some affine version of Lemma 1. Indeed, it is quite straightforward to prove that the affine version

[8] We defer the details of the actual encryption scheme and signature scheme to Sections 5 and 6, respectively, since they are not of relevance to the discussion here.

[9] Needless to say, this assumption significantly simplifies matters, and the actual proof contains several additional technicality that are hidden from this high level overview.

Lemma 1 is also true (while using Lemma 1 as a black box), assuming the affine element is *independent* of the (hidden) subspace $\mathbf{B}$. Namely, for any $\mathbf{s} \in \mathbb{Z}_p^\ell$

$$(\mathbf{B}, L(\mathbf{s} + \mathbf{B} \cdot \mathbf{r})) \stackrel{\epsilon}{\equiv} (\mathbf{B}, L(\mathbf{u}))$$

where $\mathbf{B} \leftarrow \mathbb{Z}_p^{\ell \times d}$, $\mathbf{r} \leftarrow \mathbb{Z}_p^d$ and $\mathbf{u} \leftarrow \mathbb{Z}_p^\ell$ (independent of $\mathbf{s}$). However, in our case the adversary, who controls the tampering function T, may cause the affine element $T(g^{\mathbf{s}+\mathbf{BR}})$ to depend in some (arbitrary) way on the subspace $\mathbf{B}$. In this case, we do not know how to prove an affine version of Lemma 1.

Nevertheless, we do succeed in proving an affine variant of Lemma 1 which is secure against tampering. This lemma (Lemma 2), does make use of Lemma 1 (in a black box manner), but requires overcoming several additional technical barriers. We present Lemma 2, which we think of as our main technical lemma, in Section 4, and defer its formal proof to the full version. We note that by updating the secret key using $g^{\mathbf{B}}$, which is part of the public parameters (rather than the secret key), doesn't only allow us to be resilient to tampering, but also improves the leakage bounds of [7]. Known schemes that were proven secure against continual leakage under the linear assumption [7,10], could resist at most $1/2 - \epsilon$ leakage rate between updates, whereas we can tolerate $1 - \epsilon$ leakage rate between updates.[10]

Roadmap. We define our CTL model in Section 3, followed by a formal statement of our main lemma in Section 4. A formal description of our encryption scheme and our signature scheme can be found in Sections 5 and 6 respectively. Our construction in the model of continual tampering with bounded leakage can be found in Section 7.

3 Our CTL Model

Our CTL model generalizes the continual memory leakage model of [7,10], to allow the adversary to (continually) tamper with the secret state, as well as (continually) leak. In what follows we define encryption scheme in the CTL model.

Encryption schemes in the CTL *model.* The definition of an encryption scheme in the CTL model is very similar to the definitions given in [7,10], except for the following difference. In our definition, we partition the key generation algorithm into two parts. The first part is the *public-parameter generation* algorithm, denoted by ppGen. This algorithm takes as input a security parameter 1^k and produces public parameters *pp*. These public parameters can be thought of as part of the verification key, however we choose to differentiate them from the verification key, since these are independent of the secret key, whereas the verification key does depend on the secret key. The reason for this distinction is that in our results, we do not allow tampering with these public parameters. The justification for this assumption

[10] We note that [7] could tolerate $1 - \epsilon$ leakage rate under the less standard SXDH assumption.

is that these parameters are independent of the secret key, and can be thought of as generated in the manufacturing level, and hardwired into the card in a secure way. We note that the same public parameters can be shared among all users. The second part is the key generation algorithm, denoted by Gen, which takes as input these public parameters *pp* and the security parameter 1^k, and generates a pair (sk, pk) of secret and public keys.

Formally, an encryption scheme in the CTL model consists of five PPT algorithms:

- The *public-parameter generation algorithm* ppGen takes as input the security parameter 1^k, and outputs public parameters *pp*. We denote this by $pp \leftarrow \mathsf{ppGen}(1^k)$.
- The *key-generation algorithm* Gen takes as input the security parameter 1^k and public parameters *pp*, and outputs a pair of secret and public keys (sk, pk). We denote this by $(sk, pk) \leftarrow \mathsf{Gen}(1^k, pp)$.
- *Encryption.* Takes as input the public parameters *pp*, a public-key *pk* and a message m in the message space $\mathcal{M}$, and outputs a ciphertext c. We denote this by $c \leftarrow \mathsf{Enc}_{pp,pk}(m)$.
- *Decryption.* Takes as input the public parameters *pp*, a secret-key *sk*, and a ciphertext c, and outputs a message m'. We denote this by $m' \leftarrow \mathsf{Dec}_{pp,sk}(c)$.
- *Key-update.* Takes as input the public parameters *pp* and a secret-key *sk*, and outputs an "updated" secret-key sk' such that $|sk'| = |sk|$. We denote this by $sk' \leftarrow \mathsf{Update}_{pp}(sk)$.

The correctness requirement is that for all $m \in \mathcal{M}$ and any polynomial t, setting $pp \leftarrow \mathsf{ppGen}(1^k)$, $(sk_0, pk) \leftarrow \mathsf{Gen}(1^k, pp)$ and then computing $sk_i \leftarrow \mathsf{Update}_{pp}(sk_{i-1})$ for $i \in [t]$, we have that for $c \leftarrow \mathsf{Enc}_{pp,pk}(m)$ and $m' \leftarrow \mathsf{Dec}_{pp,sk_t}(c)$, it holds that $m = m'$ with all but negligible probability (where the probability is over all the randomness in the experiment).

We next define semantic security in the CTL model. We use the definition from [7], and augment it by allowing for tampering queries. Formally, the leakage in this model is associated with three leakage parameters (ρ_G, ρ_U, ρ_M), where ρ_G bounds the leakage rate from the key-generation process, ρ_U bounds the leakage rate from the update process, and ρ_M is a "global" (relative) memory leakage bound that is enforced between key updates. For the sake of simplicity, we define security for the special case that $\rho_U = 0$. This simplifies the definition and allows for a cleaner exposition. The result of [7], can be used to show that *any* scheme that is secure against continual leakage, can tolerate $O(\log k)$ leakage from each update process, and thus our scheme can tolerate such leakage as well.

Definition 1. *An encryption scheme* $\mathcal{E} = (\mathsf{ppGen}, \mathsf{Gen}, \mathsf{Enc}, \mathsf{Dec}, \mathsf{Update})$ *is semantically secure in the* CTL *model with leakage rate* $(\rho_G, \rho_U, \rho_M) = (\rho_G, 0, \rho_M)$, *if any* PPT *adversary succeeds in the following game with probability* $1/2 + \text{negl}(k)$.

1. Initialize. *The forger specifies a circuit f such that $|f(r)| \leq \rho_G \cdot |r|$ for all r. The challenger chooses "secret randomness" r, generates $(sk_0, pk) = \mathsf{Gen}(1^k, pp; r)$, where pp are the public parameters generated by* ppGen, *sends $(pk, f(r))$ to the adversary, and sets $L := |f(r)|$.*

2. Tampering, leakage and updates. *The adversary makes queries of the following type:*
   - *Update queries* update. *The challenger sets* $sk \leftarrow \mathsf{Update}_{pp}(sk)$, *and sets* $L := 0$.
   - *Tampering queries* (tamper, T), *where* T *is a circuit. The challenger sets* $sk := T(sk)$.
   - *Leakage queries* (leak, f), *where* f *is a circuit. If* $L + |f(sk)| \leq \rho_M \cdot |sk|$ *then the challenger returns* $f(sk)$ *to the forger, and sets* $L := L + |f(sk)|$. *Otherwise, the challenger aborts.*
3. Challenge. *The adversary sends two messages* m_0, m_1 *to the challenger. The challenger flips a coin* $b \leftarrow \{0, 1\}$, *computes* $c \leftarrow \mathsf{Enc}_{pp,pk}(m_b)$, *and sends* c *to the adversary.*
4. Finish. *The adversary outputs a "guess"* $b' \in \{0, 1\}$.

The adversary **succeeds** *if* $b' = b$.

Signature schemes in the CTL *model.* The definition of signature scheme in the CTL model is similar to the definition of encryption scheme in the CTL model, except for the following two differences. First, a signature scheme in the CTL model is associated with an additional leakage parameter ρ_S, which captures leakage allowed from the signing process itself. Since our results cannot tolerate leakage from the signing process, throughout this work, we set $\rho_S = 0$. The second difference, which is more subtle, is the following: In the case of signature schemes we bound the number of tampering queries within each time period (however, the number of tampering queries overall is unbounded). The reason is that the adversary has access to a signing oracle, and the signatures themselves (w.r.t. tampered keys) leak some information about the tampered key. We defer the formal definition to the full version.

4 Main Lemma

Intuitively, the main lemma considers the setting where a random vector $\mathbf{s} \leftarrow \mathbb{Z}_p^\ell$ is chosen, and is being continually updated by adding to it either a random element from a small subspace, or a random element from a bigger subspace. In between every two updates the random vector can be tampered with, and partially leaked, many times, as long as the total amount of leakage is bounded. The claim is that it is hard to distinguish between the case where the updates were from a small subspace or from the large subspace, assuming the leakage functions take as input these vectors "in the exponent". More specifically, we consider a random matrix $\mathbf{A} \leftarrow \mathbb{Z}_p^{c \times \ell}$ consisting of c rows, for $2 \leq c < \ell$. The random vector $g^{\mathbf{s}}$ is either updated by adding to it a random element in $\ker(\mathbf{A})$ "in the exponent", or by adding to it a random element in $\mathrm{span}(\mathbf{B})$ "in the exponent", where $\mathbf{B} \in \mathbb{Z}_p^{\ell \times d}$ is a random d-dimensional subspace of $\ker(\mathbf{A})$. The former is called the *ideal* game, and the latter is called the *real* game, and the claim is that it is computationally hard to distinguish between the two, even given $\mathbf{B}$.

Lemma 2. *For any security parameter k, let $\mathbb{G}$ be a group of order p, where p is a k-bit prime, such that the linear assumption holds in $\mathbb{G}$. Let $\ell, c, d \in \mathbb{N}$ be polynomially bounded parameters such that $c \geq 2$, $4 + c \leq d \leq \ell - c$,[11] and let $t = \text{poly}(k)$. Then, for any* PPT *adversary $\mathcal{A}$,*

$$\mathsf{G}^{\mathcal{A}}_{\mathsf{real}}(p, \ell, c, d, t) \approx \mathsf{G}^{\mathcal{A}}_{\mathsf{ideal}}(p, \ell, c, d, t)$$

as long as each $L_i : \mathbb{G}^\ell \rightarrow W_i$ satisfies $|W_i| \leq p^{d-c-4}$, where $\mathsf{G}^{\mathcal{A}}_{\mathsf{real}}(p, \ell, c, d, t)$ and $\mathsf{G}^{\mathcal{A}}_{\mathsf{ideal}}(p, \ell, c, d, t)$ are defined below.

Experiment $\mathsf{G}^{\mathcal{A}}_{\mathsf{real}}(p, \ell, c, d, t)$. In $\mathsf{G}^{\mathcal{A}}_{\mathsf{real}}(p, \ell, c, d, t)$, a challenger $\mathcal{C}$ interacts with the adversary $\mathcal{A}$ as follows:

1. $\mathcal{C}$ chooses a random matrix $\mathbf{A} \leftarrow \mathbb{Z}_p^{c \times \ell}$ and a random vector $\mathbf{s} \leftarrow \mathbb{Z}_p^\ell$. It sets $\mathbf{v}_1 = g^{\mathbf{s}}$, and gives $\mathbf{A}, \mathbf{s}$ to $\mathcal{A}$.
2. $\mathcal{C}$ chooses a random d dimensional subspace $\mathbf{B} \leftarrow \mathbb{Z}_p^{\ell \times d}$ such that $\mathbf{A} \cdot \mathbf{B} = \mathbf{0}$ (i.e., each column in $\mathbf{B}$ is uniformly distributed in $\ker(\mathbf{A})$). For $i = 1$ to t, do:
   (a) $\mathcal{A}$ specifies a leakage queries L_i and a tampering function T_i.
   (b) $\mathcal{C}$ gives $L_i(\mathbf{v}_i)$ to $\mathcal{A}$. In addition he chooses $\mathbf{r}_i \leftarrow \mathbb{Z}_p^d$, sets $\mathbf{b}_i = \mathbf{B} \cdot \mathbf{r}_i$, and sets $\mathbf{v}_{i+1} = T_i(\mathbf{v}_i) \cdot g^{\mathbf{b}_i}$.
3. Finally, $\mathcal{C}$ gives the adversary $\mathcal{A}$ the subspace $\mathbf{B}$, and $\mathcal{A}$ outputs either 0 or 1.

Thus, during this experiment $\mathcal{A}$ gets

$$L_1(\mathbf{v}_1), L_2(\mathbf{v}_2), \ldots, L_t(\mathbf{v}_t),$$

where $\mathbf{v}_1 = g^{\mathbf{s}}$ and $\mathbf{v}_{i+1} = T_i(\mathbf{v}_i) \cdot g^{\mathbf{b}_i}$ for $1 \leq i \leq t - 1$.

Experiment $\mathsf{G}^{\mathcal{A}}_{\mathsf{ideal}}(p, \ell, c, d, t)$. This experiment is exactly the same as $\mathsf{G}^{\mathcal{A}}_{\mathsf{real}}(p, \ell, c, d, t)$ with the exception that $\mathbf{b}_i \leftarrow \ker(\mathbf{A})$. For the sake of clarity, we will use the notation $\mathbf{k}_i$ to denote updates in the ideal game (as opposed to $\mathbf{b}_i$ in the real game). That is, in the ideal game, $\mathcal{C}$ sets $\mathbf{v}_{i+1} = T_i(\mathbf{v}_i) \cdot g^{\mathbf{k}_i}$ in Step 2b above.

Thus, during this experiment $\mathcal{A}$ gets

$$L_1(\mathbf{v}_1), L_2(\mathbf{v}_2), \ldots, L_t(\mathbf{v}_t),$$

where $\mathbf{v}_1 = g^{\mathbf{s}}$ and $\mathbf{v}_{i+1} = T_i(\mathbf{v}_i) \cdot g^{\mathbf{k}_i}$ for $1 \leq i \leq t - 1$.

We defer the proof of Lemma 2 to the full version.

5 Encryption Scheme in the CTL Model

In this section, we construct an encryption scheme that is secure in the CTL model. Our encryption scheme $\mathcal{E} = (\mathsf{ppGen}, \mathsf{Gen}, \mathsf{Enc}, \mathsf{Dec}, \mathsf{Update})$ is defined in Figure 1.

[11] Think of c as a small constant (such as $c = 2$), think of ℓ as growing polynomially with the security parameter, and think of d as approaching ℓ (such as $d = \ell - c$). Such parameters will optimize our leakage bounds.

Encryption scheme $\mathcal{E}$

Public-parameter generation. The public-parameter generation algorithm ppGen takes as input the security parameter 1^k, and does as follows:

1. Choose at random $\mathbf{A} \leftarrow \mathbb{Z}_p^{2\times\ell}$ and $\mathbf{B} \leftarrow \mathbb{Z}_p^{\ell\times 2}$ such that $\mathbf{AB} = \mathbf{0}$ (i.e., both of the columns of $\mathbf{B}$ are randomly chosen in $\ker(\mathbf{A})$).

Output $pp = (g^{\mathbf{A}}, g^{\mathbf{B}})$

Key generation. The key generation algorithm Gen, takes as input the security parameter 1^k and public parameters $pp = (g^{\mathbf{A}}, g^{\mathbf{B}})$, and does as follows:

1. Choose a random a vector $g^{\mathbf{s}} \leftarrow \mathbb{G}^{\ell}$ (i.e., $\mathbf{s} \leftarrow \mathbb{Z}_p^{\ell}$).
2. Set $sk = g^{\mathbf{s}}$ and $pk = e(g,g)^{\mathbf{A}\cdot\mathbf{s}}$.

Output (sk, pk) as the secret and public key pair.

Encrypting. The encryption algorithm Enc, takes as input a message $b \in \{0,1\}$, public parameters $pp = (g^{\mathbf{A}}, g^{\mathbf{B}})$, and a public key $pk = e(g,g)^{\mathbf{A}\cdot\mathbf{s}}$, and does as follows:

1. If $b = 0$, choose a random vector $\mathbf{r} \leftarrow \mathbb{Z}_p^2$, and output $(\mathbf{c}_1, c_2) = (g^{\mathbf{r}^T\cdot\mathbf{A}}, e(g,g)^{\mathbf{r}^T\cdot\mathbf{A}\cdot\mathbf{s}})$.
2. If $b = 1$, choose a random vector $\mathbf{r} \leftarrow \mathbb{Z}_p^2$ and a random element $u \leftarrow \mathbb{Z}_p$, and output $(\mathbf{c}_1, c_2) = (g^{\mathbf{r}^T\cdot\mathbf{A}}, e(g,g)^{u})$.

Decrypting. The decryption algorithm Dec, takes as input a ciphertext $\mathbf{c} = (\mathbf{c}_1, c_2)$, public parameters $pp = (g^{\mathbf{A}}, g^{\mathbf{B}})$, and a secret key $sk = g^{\mathbf{s}}$, and does as follows:

1. If $e(\mathbf{c}_1, g^{\mathbf{s}}) = c_2$ then output $b' = 0$. Otherwise, output $b' = 1$.

Updates. The update procedure Update, takes as input public parameters $pp = (g^{\mathbf{A}}, g^{\mathbf{B}})$ and a secret key $sk = g^{\mathbf{s}}$, it chooses $\mathbf{u} \leftarrow \mathbb{Z}_p^2$ and outputs $sk = g^{\mathbf{s}+\mathbf{Bu}}$.

Fig. 1. Encryption scheme in the CTL model

Theorem 4. *Let $\mathbb{G}$ be a group of prime order p, with a bilinear map $e : \mathbb{G}\times\mathbb{G} \to \mathbb{G}_T$ and a generator g, and assume that the linear assumption holds in $\mathbb{G}$. Then the encryption scheme $\mathcal{E} = (\mathsf{ppGen}, \mathsf{Gen}, \mathsf{Enc}, \mathsf{Dec}, \mathsf{Update})$, described in Figure 1, is semantically secure in the* CTL *model, with leakage rate $(\rho_G, \rho_U, \rho_M) = (\frac{\ell-13}{\ell}, 0, \frac{\ell-13}{\ell})$.*

We defer the proof of this theorem, as well as those of theorems appearing in the subsequent sections, to the full version.

6 Signature Schemes in the CTL Model

In this section, we show how to convert any encryption scheme that is secure in the CTL model into a signature scheme that is secure in the CTL model. Our transformation follows the Katz-Vaikuntanathan paradigm [25] and uses the follows building blocks:

- An encryption scheme $\mathcal{E}_{\mathcal{TL}} = (\mathsf{ppGen}_{\mathcal{TL}}, \mathsf{Gen}_{\mathcal{TL}}, \mathsf{Enc}_{\mathcal{TL}}, \mathsf{Dec}_{\mathcal{TL}}, \mathsf{Update}_{\mathcal{TL}})$ that is semantically secure in the CTL model with leakage rate (ρ_G, ρ_U, ρ_M). We assume the encryption scheme $\mathcal{E}_{\mathcal{TL}}$ has a deterministic predicate T such that $T(pk, sk) = 1$ if and only if sk is a "valid" secret key corresponding to pk (i.e., if sk correctly decrypts ciphertexts encrypted using pk with overwhelming probability).
- An adaptive simulation-sound NIZK proof system $\Pi = (\ell, P, V, S = (S_1, S_2))$ (as defined by Sahai [35]) for the following language L:

$$L = \left\{(m, c, pk', pk) : \exists sk, r \text{ s.t. } c = \mathsf{Enc}_{pk'}(sk; r) \text{ and } T(pk, sk) = 1\right\}.$$

- An (ordinary) semantically secure encryption scheme $\mathcal{E} = (\mathsf{Gen}, \mathsf{Enc}, \mathsf{Dec})$ (without leakage or tampering).

Our signature scheme $\mathcal{S} = (\mathsf{ppGen}, \mathsf{Gen}, \mathsf{Sign}, \mathsf{Ver}, \mathsf{Update})$ is defined in Figure 2 and results in the following theorem:

Theorem 5. *The signature scheme* $\mathcal{S} = (\mathsf{ppGen}, \mathsf{Gen}, \mathsf{Sign}, \mathsf{Ver}, \mathsf{Update})$ *described in Figure 2 is existentially unforgeable under adaptive chosen message attacks in the* CTL *model, with leakage rate* $(\rho_G, \rho_U, \rho_S, \rho_M')$, *where* $\rho_M' = \rho_M - \frac{|vk|}{|sk|}$ *and* $\rho_S = 0$.

In particular, by using the encryption scheme described in Figure 1, we obtain a signature scheme $\mathcal{S} = (\mathsf{ppGen}, \mathsf{Gen}, \mathsf{Sign}, \mathsf{Ver}, \mathsf{Update})$ *that is existentially unforgeable under adaptive chosen message attacks in the* CTL *model, with leakage rate* $(\rho_G, \rho_U, \rho_S, \rho_M) = (\frac{\ell-13}{\ell}, 0, 0, \frac{\ell-15}{\ell})$.

7 Signature Scheme in the Continual Tampering and Bounded Leakage Model

In this section we show how to convert a signature scheme $\mathcal{S}$, that is secure in the bounded leakage model of [1], into a signature scheme that is secure in the continual tampering and bounded leakage model[12]. Our construction uses the following primitives as building blocks:

[12] To enable tampering in the bonded leakage model (where the secret key is never updated), we need to require that the circuit has a self-destruct mechanism i.e., it is possible for the circuit to erase the entire contents of memory. We defer these details to the full version.

Signature scheme $\mathcal{S}$

Public-parameter generation. The public-parameter generation algorithm ppGen takes as input a security parameter 1^k, and does the following:

1. Sample $pp \leftarrow \mathsf{ppGen}_{\mathcal{TL}}(1^k)$.
2. Sample a public key pk' for the (ordinary) encryption scheme $\mathcal{E}$.
3. sample a random string $\mathsf{crs} \leftarrow \{0,1\}^{\ell(k)}$.

Output (pp, pk', crs) as the public parameters.

Key generation. The key generation algorithm Gen, takes as input the security parameter 1^k and public parameters (pp, pk', crs), and generates $(sk, pk) \leftarrow \mathsf{Gen}_{\mathcal{TL}}(1^k, pp)$. Output sk as the secret key and pk as the verification key.

Signing. Given a message m, public parameters (pp, pk', crs), and a secret key sk, do the following:

1. Choose a random string r and compute $c = \mathsf{Enc}_{pk'}(sk; r)$.
2. Compute a proof π for the statement $(m, c, pk', pk) \in L$ w.r.t. the common random string crs, using (sk, r) as the witness, where

$$L = \left\{(m, c, pk', pk) : \exists sk, r \text{ s.t. } c = \mathsf{Enc}_{pk'}(sk; r) \text{ and } T(pk, sk) = 1\right\}.$$

   Namely, compute $\pi \leftarrow P_{\mathsf{crs}}((m, c, pk', pk), (sk, r))$.

Output $\sigma = (c, \pi)$ as a signature for m.

Verifying. To verify a signature $\sigma = (c, \pi)$ on a message m with respect to the verification key pk and the public parameters (pp, pk', crs), check whether π is a valid proof of the statement $(m, c, pk', pk) \in L$ with respect to the common random string crs.

Updates. The update procedure Update is identical to that of $\mathsf{Update}_{\mathcal{TL}}$. More specifically, it takes as input the public parameters (pp, pk, crs) and a secret key sk, and updates the secret key by computing $\mathsf{Update}_{\mathcal{TL}}(pp, sk)$.

Fig. 2. Signature scheme in the CTL model

- $\mathcal{S} = (\mathsf{Gen}, \mathsf{Sign}, \mathsf{Ver})$, a signature scheme secure in ρ-bounded leakage model, i.e., $\mathcal{S}$ is existentially unforgeable even if the adversary gets leakage $L(sk)$ of his choice such that $| L(sk) | \leq \rho \cdot |sk|$. We assume that $\mathcal{S}$ has the property that $sk \leftarrow \{0,1\}^{g(k)}$ and that there exists a PPT algorithm pkGen such that $vk \leftarrow \mathsf{pkGen}(sk)$ (as is satisfied by [25]).
- A pseudo-random generator $\mathsf{PRG} : \{0,1\}^k \to \{0,1\}^{g(k)}$.
- A single theorem adaptive "short" simulation sound (SS) NIZK POK [35] $\Pi = (\ell, P, V, S = (S_1, S_2), E)$ for the language $L = \{\psi : \exists s \text{ s.t. } \psi = \mathsf{PRG}(s)\}$, where the length of the proofs are polynomial in the witness length.

Signature scheme $\mathcal{S}'$

Public-parameter generation. The public-parameter generation algorithm ppGen' takes as input the security parameter 1^k, and does as follows.

1. Sample a string $\mathsf{crs} \leftarrow \{0,1\}^{\ell(k)}$.

Output $pp' = \mathsf{crs}$

Key generation. The key generation algorithm Gen', takes as input the security parameter 1^k and public parameters $pp' = \mathsf{crs}$, and does as follows:

1. Choose $s \leftarrow \{0,1\}^k$.
2. Set $sk = \mathsf{PRG}(s)$ and choose $vk \leftarrow \mathsf{pkGen}(sk)$.
3. Compute a short SS NIZK POK π for the statement $sk \in L$, w.r.t. crs, where
$$L = \{sk : \exists s' \text{ s.t. } sk = \mathsf{PRG}(s')\}$$
Namely, $\pi \leftarrow P(sk, s, \mathsf{crs})$.
4. Set $sk' = (sk, \pi)$.

Output (sk', vk) as the secret and verification key pair.

Signing. Given a message m, public parameter $pp' = \mathsf{crs}$, and a secret key $sk' = (sk, \pi)$, do the following:

1. Verify that π is a valid proof for the statement $sk \in L$ with respect to the common random string crs. Namely, check that $V(sk, \pi, \mathsf{crs}) = 1$. If not, self destruct.
2. Compute $\sigma \leftarrow \mathsf{Sign}_{sk}(m)$ (where Sign is the original leakage resilient signing algorithm).

Output σ as a signature for m.

Verifying. To verify a signature σ on a message m w.r.t. the verification key vk, run $\mathsf{Ver}_{vk}(m, \sigma)$ and output whatever it does.

Fig. 3. Signature scheme secure in the continual tampering and bounded leakage model

We convert $\mathcal{S}$ into $\mathcal{S}' = (\mathsf{ppGen}', \mathsf{Gen}', \mathsf{Sign}', \mathsf{Ver}')$, a signature scheme that is secure against continual tampering and bounded leakage, as depicted in Figure 3. This results in the following theorem:

Theorem 6. *The signature scheme $\mathcal{S}' = (\mathsf{ppGen}', \mathsf{Gen}', \mathsf{Sign}', \mathsf{Ver}')$, presented in Figure 3, is existentially unforgeable in the continual tampering and ρ'-bounded leakage model, where $\rho' = \left(\frac{\rho \cdot g(k)}{g(k)+\gamma(k)} - \frac{2k+\gamma(k)}{g(k)+\gamma(k)}\right)$.*

References

1. Akavia, A., Goldwasser, S., Vaikuntanathan, V.: Simultaneous hardcore bits and cryptography against memory attacks. In: Reingold, O. (ed.) TCC 2009. LNCS, vol. 5444, pp. 474–495. Springer, Heidelberg (2009)
2. Applebaum, B., Harnik, D., Ishai, Y.: Semantic security under related-key attacks and applications. Cryptology ePrint Archive, Report 2010/544 (2010), `http://eprint.iacr.org/`
3. Bellare, M., Kohno, T.: A theoretical treatment of related-key attacks: Rka-prps, rka-prfs, and applications. In: Biham, E. (ed.) EUROCRYPT 2003. LNCS, vol. 2656, pp. 491–506. Springer, Heidelberg (2003)
4. Biham, E., Shamir, A.: Differential fault analysis of secret key cryptosystems. In: Kaliski Jr., B.S. (ed.) CRYPTO 1997. LNCS, vol. 1294, pp. 513–525. Springer, Heidelberg (1997)
5. Boneh, D., DeMillo, R.A., Lipton, R.J.: On the importance of checking cryptographic protocols for faults. In: Fumy, W. (ed.) EUROCRYPT 1997. LNCS, vol. 1233, pp. 37–51. Springer, Heidelberg (1997)
6. Boyko, V.: On the security properties of OAEP as an all-or-nothing transform. In: Wiener, M. (ed.) CRYPTO 1999. LNCS, vol. 1666, pp. 503–518. Springer, Heidelberg (1999)
7. Brakerski, Z., Kalai, Y.T., Katz, J., Vaikuntanathan, V.: Overcoming the hole in the bucket: Public-key cryptography resilient to continual memory leakage. In: FOCS, pp. 335–359 (2010)
8. Canetti, R., Dodis, Y., Halevi, S., Kushilevitz, E., Sahai, A.: Exposure-resilient functions and all-or-nothing transforms. In: Preneel, B. (ed.) EUROCRYPT 2000. LNCS, vol. 1807, pp. 453–469. Springer, Heidelberg (2000)
9. Dodis, Y., Goldwasser, S., Tauman Kalai, Y., Peikert, C., Vaikuntanathan, V.: Public-key encryption schemes with auxiliary inputs. In: Micciancio, D. (ed.) TCC 2010. LNCS, vol. 5978, pp. 361–381. Springer, Heidelberg (2010)
10. Dodis, Y., Haralambiev, K., Lopez-Alt, A., Wichs, D.: Cryptography against continuous memory attacks. In: FOCS (2010)
11. Dodis, Y., Kalai, Y.T., Lovett, S.: On cryptography with auxiliary input. In: STOC, pp. 621–630 (2009)
12. Dodis, Y., Sahai, A., Smith, A.: On perfect and adaptive security in exposure-resilient cryptography. In: Pfitzmann, B. (ed.) EUROCRYPT 2001. LNCS, vol. 2045, pp. 301–324. Springer, Heidelberg (2001)
13. Dziembowski, S., Pietrzak, K.: Leakage-resilient cryptography. In: FOCS, pp. 293–302 (2008)
14. Dziembowski, S., Pietrzak, K., Wichs, D.: Non-malleable codes. In: ICS, pp. 434–452 (2010)
15. Faust, S., Kiltz, E., Pietrzak, K., Rothblum, G.N.: Leakage-resilient signatures. In: Micciancio, D. (ed.) TCC 2010. LNCS, vol. 5978, pp. 343–360. Springer, Heidelberg (2010)
16. Faust, S., Rabin, T., Reyzin, L., Tromer, E., Vaikuntanathan, V.: Protecting circuits from leakage: the computationally-bounded and noisy cases. In: Gilbert, H. (ed.) EUROCRYPT 2010. LNCS, vol. 6110, pp. 135–156. Springer, Heidelberg (2010)
17. Gandolfi, K., Mourtel, C., Olivier, F.: Electromagnetic analysis: Concrete results. In: Koç, Ç.K., Naccache, D., Paar, C. (eds.) CHES 2001. LNCS, vol. 2162, pp. 251–261. Springer, Heidelberg (2001)

18. Gennaro, R., Lysyanskaya, A., Malkin, T., Micali, S., Rabin, T.: Algorithmic tamper-proof (ATP) security: Theoretical foundations for security against hardware tampering. In: Naor, M. (ed.) TCC 2004. LNCS, vol. 2951, pp. 258–277. Springer, Heidelberg (2004)
19. Goldwasser, S., Rothblum, G.: How to play mental solitaire under continuous side-channels: A completeness theorem using secure hardware (2010) (manuscript)
20. Govindavajhala, S., Appel, A.W.: Using memory errors to attack a virtual machine. In: IEEE Symposium on Security and Privacy, pp. 154–165 (2003)
21. Alex Halderman, J., Schoen, S.D., Heninger, N., Clarkson, W., Paul, W., Calandrino, J.A., Feldman, A.J., Appelbaum, J., Felten, E.W.: Lest we remember: Cold boot attacks on encryption keys. In: USENIX Security Symposium, pp. 45–60 (2008)
22. Ishai, Y., Prabhakaran, M., Sahai, A., Wagner, D.: Private circuits II: Keeping secrets in tamperable circuits. In: Vaudenay, S. (ed.) EUROCRYPT 2006. LNCS, vol. 4004, pp. 308–327. Springer, Heidelberg (2006)
23. Ishai, Y., Sahai, A., Wagner, D.: Private circuits: Securing hardware against probing attacks. In: Boneh, D. (ed.) CRYPTO 2003. LNCS, vol. 2729, pp. 463–481. Springer, Heidelberg (2003)
24. Juma, A., Vahlis, Y.: Protecting cryptographic keys against continual leakage. In: Rabin, T. (ed.) CRYPTO 2010. LNCS, vol. 6223, pp. 41–58. Springer, Heidelberg (2010)
25. Katz, J., Vaikuntanathan, V.: Signature schemes with bounded leakage resilience. In: Matsui, M. (ed.) ASIACRYPT 2009. LNCS, vol. 5912, pp. 703–720. Springer, Heidelberg (2009)
26. Kocher, P.C.: Timing attacks on implementations of diffie-hellman, RSA, DSS, and other systems. In: Koblitz, N. (ed.) CRYPTO 1996. LNCS, vol. 1109, pp. 104–113. Springer, Heidelberg (1996)
27. Kocher, P.C., Jaffe, J., Jun, B.: Differential Power Analysis. In: Wiener, M. (ed.) CRYPTO 1999. LNCS, vol. 1666, pp. 388–397. Springer, Heidelberg (1999)
28. Micali, S., Reyzin, L.: Physically observable cryptography (extented abstract). In: Naor, M. (ed.) TCC 2004. LNCS, vol. 2951, pp. 278–296. Springer, Heidelberg (2004)
29. Naor, M., Segev, G.: Public-key cryptosystems resilient to key leakage. In: Halevi, S. (ed.) CRYPTO 2009. LNCS, vol. 5677, pp. 18–35. Springer, Heidelberg (2009)
30. Osvik, D.A., Shamir, A., Tromer, E.: Cache attacks and countermeasures: The case of AES. In: Pointcheval, D. (ed.) CT-RSA 2006. LNCS, vol. 3860, pp. 1–20. Springer, Heidelberg (2006)
31. Pietrzak, K.: A leakage-resilient mode of operation. In: Joux, A. (ed.) EUROCRYPT 2009. LNCS, vol. 5479, pp. 462–482. Springer, Heidelberg (2009)
32. Quisquater, J.-J., Samyde, D.: ElectroMagnetic analysis (EMA): Measures and counter-measures for smart cards. In: Attali, S., Jensen, T. (eds.) E-smart 2001. LNCS, vol. 2140, pp. 200–210. Springer, Heidelberg (2001)
33. Rao, J.R., Rohatgi, P.: Empowering side-channel attacks. Cryptology ePrint Archive, Report 2001/037 (2001), `http://eprint.iacr.org/`
34. Rivest, R.L.: All-or-nothing encryption and the package transform. In: Biham, E. (ed.) FSE 1997. LNCS, vol. 1267, pp. 210–218. Springer, Heidelberg (1997)
35. Sahai, A.: Non-malleable non-interactive zero knowledge and adaptive chosen-ciphertext security. In: FOCS, pp. 543–553 (1999)

Merkle Puzzles in a Quantum World

Gilles Brassard[1], Peter Høyer[2], Kassem Kalach[1],
Marc Kaplan[1], Sophie Laplante[3], and Louis Salvail[1]

[1] Département d'informatique et de recherche opérationnelle
Université de Montréal, C.P. 6128, Succursale Centre-ville
Montréal (QC), H3C 3J7 Canada
[2] Department of Computer Science, University of Calgary
2500 University Drive N.W., Calgary, AB, T2N 1N4 Canada
[3] LRI, Université Paris-Sud, 91400 Orsay, France

Abstract. In 1974, Ralph Merkle proposed the first unclassified scheme for secure communications over insecure channels. When legitimate communicating parties are willing to spend an amount of computational effort proportional to some parameter N, an eavesdropper cannot break into their communication without spending a time proportional to N^2, which is quadratically more than the legitimate effort. We showed in an earlier paper that Merkle's schemes are completely insecure against a quantum adversary, but that their security can be partially restored if the legitimate parties are also allowed to use quantum computation: the eavesdropper needed to spend a time proportional to $N^{3/2}$ to break our earlier quantum scheme. Furthermore, all previous *classical* schemes could be broken completely by the onslaught of a quantum eavesdropper and we conjectured that this is unavoidable.

We give two novel key agreement schemes in the spirit of Merkle's. The first one can be broken by a quantum adversary that makes an effort proportional to $N^{5/3}$ to implement a quantum random walk in a Johnson graph reminiscent of Andris Ambainis' quantum algorithm for the element distinctness problem. This attack is optimal up to logarithmic factors. Our second scheme is purely classical, yet it cannot be broken by a quantum eavesdropper who is only willing to expend effort proportional to that of the legitimate parties.

Keywords: Merkle Puzzles, Public Key Distribution, Quantum Cryptography.

1 Introduction

While Ralph Merkle was delivering the 2005 International Association for Cryptologic Research (IACR) Distinguished Lecture at the CRYPTO annual conference in Santa Barbara, describing his original unpublished 1974 scheme [15] for public key distribution (much simpler and more elegant than his subsequently published, yet better known, Merkle Puzzles [16]), one of us (Brassard) immediately realized that this scheme was totally insecure against an eavesdropper equipped with a quantum computer. The obvious question was: can Merkle's

P. Rogaway (Ed.): CRYPTO 2011, LNCS 6841, pp. 391–410, 2011.

idea be repaired and made secure again in our quantum world? The defining characteristics of Merkle's protocol are that (1) the legitimate parties communicate strictly through an authenticated classical channel on which eavesdropping is unrestricted and (2) a protocol is deemed to be *secure* if the cryptanalytic effort required of the eavesdropper to learn the key exchanged by the legitimate parties grows super-linearly with the legitimate work.

We partially repaired Merkle's scheme in Ref. [8] with a scheme in which the eavesdropper needed an amount of work in $\Omega(N^{3/2})$ to obtain the key established by quantum legitimate parties whose amount of work is in $O(N)$. This was not quite as good as the work in $\Omega(N^2)$ required by a *classical* eavesdropper against Merkle's original scheme, but significantly better than the work in $O(N)$ sufficient for a *quantum* eavesdropper against the same scheme. Two main questions were left open in Ref. [8]:

1. Can the quadratic security possible in a classical world be restored in our quantum world?
2. Is any security possible at all if the legitimate parties are purely classical, yet the eavesdropper is endowed with a quantum computer?

We give two novel key distribution protocols to address these issues. In the first protocol, the legitimate parties use quantum computers and classical authenticated communication to establish a shared key after $O(N)$ expected queries to two black-box random functions (which can be modelled with a single random oracle). We then give a nontrivial quantum cryptanalytic attack that uses a quantum random walk in a Johnson graph, much like Andris Ambainis' algorithm to solve the element distinctness problem [2], which allows a quantum eavesdropper to learn the key after $\Theta(N^{5/3})$ queries to the functions. Finally, we prove that our attack is optimal up to logarithmic factors. Therefore, we have not quite restored the quadratic security possible in a classical world, but we have made significant progress towards it.

Second, we give a *purely classical* protocol, in which the legitimate parties use classical communication and classical computation to establish a key after $O(N)$ calls to similar black-box random functions. We then attack this protocol with a quantum cryptanalytic algorithm that uses $\Theta(N^{13/12})$ queries to the functions. As unlikely as it may sound, this attack is optimal (up to logarithmic factors) and therefore it is not possible to break this purely classical protocol with a quantum attack that uses an amount of resource linear in the legitimate effort.

Before describing our new protocols (Sects. 3 and 4), quantum attacks against them (Sects. 3.1 and 4.1), proofs of optimality for those attacks (Sects. 3.2 and 4.2), and conjectures about the existence of even better schemes (Sect. 5), we begin with a review (lifted from Ref. [8]) of Merkle's original idea, its meltdown against a quantum eavesdropper, and our earlier partial quantum solution (Sect. 2). Some of the technical tools required by our quantum attacks are reviewed in the Appendix and new lower bound techniques are introduced.

2 Merkle's Original Scheme and How to Break and Partially Repair It

The first unclassified document ever written that pioneered public key distribution and public key cryptography was a project proposal written in 1974 by Merkle when he was a student in Lance Hoffman's CS244 course on Computer Security at the University of California, Berkeley [15]. Hoffman rejected the proposal and Merkle dropped the course but "kept working on the idea" and eventually published it as one of the most seminal cryptographic papers in the second half of the twentieth century [16]. Merkle's scheme in his published paper was somewhat different from his original 1974 idea, but both share the property that they "force any enemy to expend an amount of work which increases as the square of the work required of the two [legitimate] communicants" [16]. It took 35 years before Boaz Barak and Mohammad Mahmoody-Ghidary proved that this quadratic discrepancy between the legitimate and eavesdropping efforts are the best possible in a classical world [3].

In his IACR Distinguished Lecture[1], which he delivered at the CRYPTO '05 Conference in Santa Barbara, Merkle described from memory his first solution to the problem of secure communications over insecure channels. As a wondrous coincidence, he unsuspectingly opened up a box of old folders a mere three weeks after his Lecture and happily recovered his long-lost CS244 Project Proposal, together with comments handwritten by Hoffman [15]! To quote his original typewritten words:

```
Method 1:    Guessing.  Both sites guess at keywords.  These
             guesses are one-way encrypted, and transmitted to the
             other site.  If both sites should chance to guess at
             the same keyword, this fact will be discovered when
             the encrypted versions are compared, and this keyword
             will then be used to establish a communications link.
Discussion:  No, I am not joking.
```

In more modern terms, let f be a one-way permutation. In order to "one-way encrypt" x, as Merkle said in 1974, we assume that one can compute $f(x)$ in unit time for any given input x but that the only way to retrieve x given $f(x)$ is to try preimages and compute f on them until one is found that maps to $f(x)$. This is known as the *black-box* (or *oracle*) model. Hereinafter, in accordance with this model, efficiency is defined *solely* in terms of the number of calls to such black-box functions (there could be more than one). In the quantum case, these calls can be made in superposition of inputs. We also assume throughout this paper (as did Merkle) that an authenticated channel is available between the legitimate communicants, although this channel offers no protection against eavesdropping.

The "keywords" guessed at by "both sites" are random points in the domain of f. They are "one-way encrypted" by applying f to them. If there are N^2

[1] `www.iacr.org/publications/dl`.

points in the domain of f, it suffices to guess $O(N)$ keywords at each site before a variation on the birthday paradox makes it overwhelmingly likely that "both sites should chance to guess at the same keyword", which becomes their shared key. An eavesdropper who listens to the entire conversation has no other way to obtain this key than to invert f on the revealed common encrypted keyword. In accordance with the black-box model, this can only be done by trying on the average half the points in the domain of f before one is found that is mapped by f to the target value. This will require an expected number of calls to f in $\Omega(N^2)$, which is quadratic in the legitimate effort.

Shortly thereafter, Whitfield Diffie and Martin Hellman discovered a celebrated method for public-key distribution that makes the cryptanalytic effort *apparently* exponentially harder than the legitimate effort [10]. However, no proof is known that the Diffie-Hellman scheme is secure at all since it relies on the conjectured difficulty of extracting discrete logarithms, an assumption doomed to fail whenever quantum computers become available. In contrast, Merkle's approach offers provable quadratic security against any possible classical attack, under the sole assumption that f cannot be inverted by any other means than exhaustive search.

Next, we explain why Merkle's original proposal becomes completely insecure if the eavesdropper is capable of quantum computation (Merkle's published "puzzles" [16] are equally insecure). We then sketch a protocol from Ref. [8] that is not completely broken. This is be achieved by granting similar quantum computation capabilities to one of the legitimate communicating parties.

2.1 Quantum Attack and Partial Remedy

Let us now assume that function f can be computed quantum mechanically on a superposition of inputs. In this case, Merkle's original scheme is completely compromised by way of Grover's algorithm [11]. Indeed, this algorithm needs only $O(\sqrt{N^2}\,) = O(N)$ calls on f in order to invert it on any given point of its image, making the cryptanalytic task as easy (up to constant factors) as the legitimate key setup process. [2]

To remedy the situation, we allow the communicating parties to use quantum computers as well (actually, one of the parties will remain classical), and we increase the domain of f from N^2 to N^3 points. Instead of having both sites transmit one-way encrypted guesses to the other site, one site called Alice chooses N distinct random values $x_1, x_2, \ldots, x_N$ and transmits them, one-way encrypted by the application of f, to the other site called Bob. Let $Y = \{f(x_i) \mid 1 \leqslant i \leqslant N\}$ denote the set of encrypted keywords received by Bob, which becomes known

[2] If an unstructured search problem has t solutions among M candidates, Grover's algorithm [11], or more precisely its so-called BBHT generalization [6], can find one of the solutions after $O(\sqrt{M/t}\,)$ *expected* calls to a function that recognizes solutions among candidates. However, Theorem 4 of Ref. [7] implies that, whenever the number $t > 0$ is known, a solution can be found *with certainty* after $O(\sqrt{M/t}\,)$ calls to that function in the *worst case*. From now on, when we mention Grover's algorithm or BBHT, we really mean this improvement according to Ref. [7].

to the eavesdropper. Now, Bob defines Boolean function g on the same domain as f by

$$g(x) = \begin{cases} 1 \text{ if } f(x) \in Y \\ 0 \text{ otherwise}. \end{cases}$$

Out of N^3 points in the domain of f, there are exactly $t = N$ solutions to the problem of finding an x so that $g(x) = 1$. It suffices for Bob to apply the BBHT generalization [6] of Grover's algorithm [11], which finds such an x after $O(\sqrt{N^3/t}) = O(\sqrt{N^2}) = O(N)$ calls on g (and therefore on f). Bob sends back $f(x)$ to Alice, who knows the value of x because she was careful to keep her randomly chosen points. Therefore, it suffices of $O(N)$ calls on f by Alice and Bob for them to agree on key x.[3]

The eavesdropper, on the other hand, is faced with the need to invert f on a specific point of its image. Even with a quantum computer, this requires a number of calls on f proportional to the square root of the number of points in its domain [5], which is $\Omega(\sqrt{N^3}) = \Omega(N^{3/2})$. This is more effort than what is required of the legitimate parties, yet less than quadratically so, as would have been possible in a classical world. Even though we have avoided the meltdown of Merkle's original approach, the introduction of quantum computers available to all sides seems to be to the advantage of the codebreakers. Can we remedy this situation? Furthermore, is any security possible at all against a quantum computer if both legitimate parties are restricted to being purely classical? We address these two questions in the rest of this paper.

3 Improved Key Distribution Scheme

For any positive integer N, let $[N]$ denote the set of integers from 1 to N. We describe our novel key distribution protocol assuming the existence of *two* black-box random functions $f : [N^3] \rightarrow [N^k]$ and $g : [N^3] \times [N^3] \rightarrow [N^{k'}]$ that can be accessed in quantum superposition of inputs. Constants k and k' are chosen large enough so that there is no collision in the images of f and g, except with negligible probability. (For simplicity, we shall systematically disregard the possibility that such collisions might exist.) Notice that *a single binary random oracle* (which "implements" a random function from the integers to $\{0, 1\}$) could be used to define both functions f and g provided we disregard logarithmic factors in our analyses since $O(\log N)$ calls to the random oracle would suffice to compute f or g on any single input. For this reason, it is understood hereinafter that all our results are implicitly stated "up to logarithmic factors". As mentioned in the previous section, the only resource that we consider in our analyses of efficiency and lower bounds is the number of calls made to these functions or, equivalently, to the underlying binary random oracle.

---

[3] As we made clear already, we are only concerned in this paper by the number of calls made to black-box functions. Nevertheless, *if* we cared also about computational efficiency, Bob would sort the elements of Y in increasing order after receiving them from Alice so that he can quickly determine, given any $y = f(x)$, whether or not $y \in Y$, which is needed to compute function g.

Protocol 1.

1. Alice picks at random N distinct values $\{x_i\}_{i=1}^{N}$ with $x_i \in [N^3]$ and transmits the encrypted values $y_i = f(x_i)$ to Bob. Let X and Y denote $\{x_i \mid 1 \leqslant i \leqslant N\}$ and $\{y_i \mid 1 \leqslant i \leqslant N\}$, respectively. Note that Alice knows both X and Y, whereas Bob and the eavesdropper have immediate knowledge (i.e. without querying the black-box for function f) of Y only.
2. Bob finds the pre-images x and x' of *two* distinct random elements in Y. To find each one of them, he uses BBHT [6] to search for an x such that $\phi(x) = 1$, where $\phi : [N^3] \to \{0, 1\}$ is defined as follows:

$$\phi(x) = \begin{cases} 1 \text{ if } f(x) \in Y \\ 0 \text{ otherwise}. \end{cases}$$

   There are exactly N values of x such that $\phi(x) = 1$, out of N^3 points in the domain of ϕ. Therefore, Bob can find one such random x with $O(\sqrt{N^3/N}) = O(N)$ calls to function f. He needs to repeat this process twice in order to get both x and x'. (A small variation in function ϕ can be used the second time to make sure that $x' \neq x$).
3. Bob sends back $w = g(x, x')$ to Alice.
4. Because Alice had kept her randomly chosen set X, there are only N^2 candidate pairs $(x_i, x_j) \in X \times X$ such that $g(x_i, x_j)$ could equal w. Using Grover's algorithm, she can find the one pair (x, x') that Bob has in mind with $O(\sqrt{N^2}) = O(N)$ calls to function g.
5. The key shared by Alice and Bob is the pair (x, x').

All counted, Alice makes N calls to f in step 1 and $O(N)$ calls to g in step 4, whereas Bob makes $O(N)$ calls to f in step 2 and a single call to g in step 3. If the protocol is constructed over a binary random oracle, it will have to be called $O(N \log N)$ times since it takes $O(\log N)$ binary queries to compute either function on any given input.

3.1 Quantum Attack

All the obvious (and not so obvious) cryptanalytic attacks against this scheme, such as direct use of Grover's algorithm (or BBHT), or even more sophisticated attacks based on amplitude amplification [7], require the eavesdropper to call $\Omega(N^2)$ times functions f and/or g. Unfortunately, a more powerful attack based on the more recent paradigm of quantum walks in Markov chains [17] allows the eavesdropper to recover Alice and Bob's key (x, x') with an expected $O(N^{5/3})$ calls to f and $O(N)$ calls to g. This attack was inspired by Ambainis' quantum algorithm for element distinctness [2], which can find the unique pair (i, j) such that $c(i) = c(j)$ with $O(N^{2/3})$ expected queries to single-collision function c whose domain contains N elements (whereas all previous approaches based on Grover's algorithm and amplitude amplification [12,9] had required $\Omega(N^{3/4})$ queries).

Theorem 1. *There exists an eavesdropping strategy that outputs the pair (x, x') in Protocol 1 with $O(N^{5/3})$ expected quantum queries to functions f and g.*

Proof. In a nutshell, we apply Ambainis' algorithm for element distinctness with two modifications: (1) instead of looking for i and j such that $c(i) = c(j)$, we are looking for x and x' such that $g(x, x') = w$ and (2) instead of being able to get randomly chosen values in the image of c with a single call to oracle c per value, we need to get random elements of X by applying BBHT on the list Y, which requires $O(\sqrt{N^3/N}) = O(N)$ calls to oracle f per element. The second modification explains why the number of calls to f, compared to $O(N^{2/3})$ calls to c for element distinctness, is multiplied by $O(N)$. Hence, we need $O(N^{5/3})$ calls to function f. To determine the number of calls required to function g, however, we have to delve deeper into the eavesdropping algorithm.

The eavesdropping algorithm uses a quantum walk on a Johnson graph—see the Appendix for a review of this topic. Each node of the graph contains some number r (to be determined later) of distinct elements of X. We are looking for a node that contains the two elements x and x' such that $g(x, x') = w$, where w is the value announced by Bob in step 3 of the protocol. We apply Theorem 5 (Appendix) to analyse the cost of a quantum walk on this graph [2,17]. The set up cost S corresponds to finding r random elements of X. Since BBHT can be used to find one such element with $O(N)$ calls to f, and even to find an element of X guaranteed to be different from those already in the initial node (provided $k \ll N$, which it will be), $\mathsf{S} = O(rN)$ calls to f. The update cost U corresponds to finding one random element of X not already in the node, which is $\mathsf{U} = O(N)$ calls to f, again by BBHT. The checking cost C requires us to decide if there is a pair (x, x') of elements in the node such that $g(x, x') = w$, which can be done with $O(\sqrt{r^2}\,) = O(r)$ calls to g using Grover's algorithm since there are r^2 pairs of elements in the node. Putting it all together, the expected cryptanalytic cost is

$$\begin{aligned} &\mathsf{S} + O\left(\tfrac{N}{r}(\sqrt{r}\,\mathsf{U} + \mathsf{C})\right) \\ &= O\left((rN \text{ calls to } f) + \tfrac{N}{r}\left(\sqrt{r}(N \text{ calls to } f) + (r \text{ calls to } g)\right)\right) \\ &= O\left(rN + N^2/\sqrt{r}\right) \text{ calls to } f \ \ \textbf{and} \ \ O(N) \text{ calls to } g\,. \end{aligned}$$

To minimize the number of calls to f, we choose r so that $rN = N^2/\sqrt{r}$, which is $r = N^{2/3}$. It follows that a quantum eavesdropper is able to find the key (x, x') with an expected $O(rN) = O(N^{5/3})$ calls to f and $O(N)$ calls to g. □

Note that the use of Grover's algorithm in the checking step was not necessary to prove Theorem 1. Should this step be carried out classically, this would result in $\mathsf{C} = O(r^2)$ calls to g. The net result would be that the key is found after an expected $O(N^{5/3})$ calls to f and also $O(N^{5/3})$ calls to g.

3.2 Lower Bound

The proof that the quantum attack described above against our protocol is optimal proceeds in three steps.

1. We define a search problem reminiscent of element distinctness;
2. We prove a lower bound on the difficulty to solve this search problem; and
3. We reduce this search problem to the eavesdropping problem against our protocol. More precisely, we show that any attack on our key distribution scheme that would have a nonvanishing probability of success after $o(N^{5/3})$ calls to functions f and g could be turned into an algorithm capable of solving the search problem more efficiently than possible.

First, consider a function $c : [N] \to [N]$ so that there exists a single pair (i, j), $1 \leqslant i < j \leqslant N$, for which $c(i) = c(j)$. Ambainis' quantum algorithm for element distinctness [2] can find this pair with $O(N^{2/3})$ queries to function c and Scott Aaronson and Yaoyun Shi proved that this is optimal even for the decision version of this problem [1].

Now, consider a function $h : [N] \times [N^2] \to [N]'$, where $[N]'$ denotes $\{0\} \cup [N]$. The domain of this function is composed of N "buckets" of size N^2, where $h(i, \cdot)$ corresponds to the ith bucket, $1 \leqslant i \leqslant N$. In bucket i, all values of the function are 0 except for one single random $v_i \in [N^2]$ for which $h(i, v_i) = c(i)$:

$$h(i,j) = \begin{cases} c(i) & \text{if } j = v_i \\ 0 & \text{otherwise}. \end{cases}$$

It follows from the definitions of c and h that there is a single pair of distinct a and b in the domain of h such that $h(a) = h(b) \neq 0$. How difficult is it to find this pair given a black box for function h but no direct access to c?

Lemma 1. *Given h structured as above, finding the pair of distinct elements a and b in the domain of h such that $h(a) = h(b) \neq 0$ requires $\Omega(N^{5/3})$ quantum queries to h, except with vanishing probability.*

Proof. This problem can be modelled as the composition of element distinctness across buckets with finding the single non-zero entry in each bucket. It is therefore a special case of technical Lemma 5, stated in the Appendix, with parameters $\alpha = N$ (the number of buckets) and $\beta = N^2$ (the size of the buckets). It follows that finding the desired pair (a, b) requires

$$\Omega(\alpha^{2/3}\beta^{1/2}) = \Omega(N^{2/3}\sqrt{N^2}\,) = \Omega(N^{5/3})$$

quantum queries to h, except with vanishing probability. □

Consider now a slightly different search problem in which there are no buckets anymore, but there is an added coordinate in the image of the function: $h' : [N^3] \to [N]' \times [N]'$ is defined so that $h'(a) = (0, 0)$ on all but N randomly chosen points in its domain, namely $w_1, w_2, \ldots, w_N$. On these N points, $h'(w_i) = (i, c(i))$, where c is the function considered at the beginning of this section. We are required to find the unique pair of distinct a and b in $[N^3]$ such that $\pi_2(h'(a)) = \pi_2(h'(b)) \neq 0$, where "$\pi_2$" denotes the projection on the second coordinate (similarly for "π_1"). The lower bound on the earlier search problem concerning h implies directly the same lower bound on the new search problem concerning h' since any algorithm capable of solving the new problem can be used at the same cost to solve the earlier problem through randomization. In other words, the more structured version of the problem cannot be harder than the less structured one. The next Lemma formalizes the argument above.

Lemma 2. *Given h' structured as above, finding the pair of distinct elements a and b in the domain of h' such that $\pi_2(h'(a)) = \pi_2(h'(b)) \neq 0$ requires $\Omega(N^{5/3})$ quantum queries to h', except with vanishing probability.*

Proof. Define intermediary function $\tilde{h} : [N] \times [N^2] \to [N]' \times [N]'$ by

$$\tilde{h}(i,j) = \begin{cases} (i, h(i,j)) = (i, c(i)) & \text{if } h(i,j) \neq 0 \\ (0, h(i,j)) = (0,0) & \text{otherwise}. \end{cases}$$

It is elementary to reduce the search problem concerning h to the one concerning $\tilde{h}$ as well as the search problem concerning $\tilde{h}$ to the one concerning h'. Therefore, the lower bound concerning h given by Lemma 1 applies *mutatis mutandis* to h'. □

Finally, we show how to reduce the search problem concerning h' to the cryptanalytic difficulty for the eavesdropper to determine the key that Alice and Bob have established by using our protocol. This is the last step in proving the security of our scheme.

Theorem 2. *Any eavesdropping strategy that recovers the key (x, x') in protocol 1 requires a total of $\Omega(N^{5/3})$ quantum queries to functions f and g, except with vanishing probability.*

Proof. Consider any eavesdropping strategy $\mathcal{A}$ that listens to the communication between Alice and Bob and tries to determine the key (x, x') by querying black-box functions f and g. In fact, there are no Alice and Bob at all! Instead, there is a function $h' : [N^3] \to [N]' \times [N]'$ as described above, for which we want to solve the search problem by using unsuspecting $\mathcal{A}$ as a resource.

We start by supplying $\mathcal{A}$ with a completely fake "conversation" between "Alice" and "Bob": for sufficiently large k and k', we choose randomly N points $y_1, y_2, \ldots, y_N$ in $[N^k]$ and one point $w \in [N^{k'}]$ and we pretend that Alice has sent the y's to Bob and that Bob has responded with w. We also choose random functions $\hat{f} : [N^3] \to [N^k]$ and $\hat{g} : [N^3] \times [N^3] \to [N^{k'}]$, as well as a random Boolean $s \in \{\mathsf{true}, \mathsf{false}\}$. Note that the selection of $\hat{f}$ and $\hat{g}$ may take a lot of *time*, but this does not count towards the number of *queries* that will be made of function h', and our lower bound on the search problem concerns *only* this number of queries. The Boolean s indicates, when true (resp. false), that the fake "execution" is such that "Bob" has first picked x and then x' such that $x < x'$ (resp. $x' > x$). Both cases happen with probability $^1/_2$ in any real execution and for any public announcements Y and w. The value s will be used in the reduction to distinguish between $g(x, x')$ and $g(x', x)$ so that only $g(x, x')$ will be set to w.

Now, we wait for $\mathcal{A}$'s queries to f and g.

- When $\mathcal{A}$ asks for $f(i)$ for some $i \in [N^3]$, there are two possibilities.
  - If $h'(i) = (0,0)$, return $\hat{f}(i)$ to $\mathcal{A}$ as value for $f(i)$.
  - Otherwise, return $y_{\pi_1(h'(i))}$.
- When $\mathcal{A}$ asks for $g(i,j)$ for some $i, j \in [N^3]$, there are again two possibilities.
  - If $\pi_2(h'(i)) = \pi_2(h'(j)) \neq 0$ and either s is true and $i < j$ or s is false and $i > j$, return w as value for $g(i,j)$.
  - Otherwise, return $\hat{g}(i,j)$.

Suppose $\mathcal{A}$ happily returns the pair (i,j) for which it was told that $g(i,j) = w$, which is what a successful eavesdropper is supposed to do. This pair is in fact the answer to the search problem concerning h' since $g(i,j) = w$ implies that $\pi_2(h'(i)) = \pi_2(h'(j)) \neq 0$, except with the negligible probability that $\hat{g}(i',j') = w$ for some query (i',j') that $\mathcal{A}$ asks about g.

Queries asked by $\mathcal{A}$ concerning f and g are answered in the same way as they would be if f and g were two random functions consistent with the Y and w announced by Alice and Bob during the execution of a real protocol. To see this, remember that Y (subset of $[N^k]$) and w (element of $[N^{k'}]$) are uniformly picked at random in both the simulated and the real worlds. Moreover, the simulated function f is such that $f(i)$ is random when $h'(i) = (0,0)$. The remaining N output values are in Y, as expected by $\mathcal{A}$. On the other hand, the simulated function g is random everywhere except for one single input pair (i,j), $i \neq j$ for which $g(i,j) = w$, as it is also expected by $\mathcal{A}$. Therefore, $\mathcal{A}$ will behave in the environment provided by the simulation exactly as in the real world. Since we disregard the negligible possibility that g might not be be one-to-one, the reduction solves the search problem concerning h' whenever $\mathcal{A}$ succeeds in finding the key. Notice finally that each (new) question asked by $\mathcal{A}$ to either f or g translates to one or two questions actually asked to h'.

It follows that any successful cryptanalytic strategy that makes $o(N^{5/3})$ total queries to f and g would solve the search problem with only $o(N^{5/3})$ queries to function h', which is impossible, except with vanishing probability. This establishes the $\Omega(N^{5/3})$ lower bound on the cryptanalytic difficulty of breaking our key exchange protocol, again except with vanishing probability, which matches the upper bound provided by the explicit attack given in Sect. 3.1. □

4 Fully Classical Key Distribution Scheme

In this section, we revert to the original setting imagined by Merkle in the sense that Alice and Bob are now purely classical. However, we allow full quantum power to the eavesdropper. Recall that Merkle's original schemes [15,16] are completely broken in this context [8]. Is it possible to restore *some* security in this highly adversarial (and unfair!) scenario? The following purely classical key distribution protocol, which is inspired by our quantum protocol described in the previous section, provides a positive answer to this conundrum.

This time, black-box random functions f and g are defined on a smaller domain to compensate for the fact that classical Alice and Bob can no longer use Grover's algorithm. Specifically, $f : [N^2] \rightarrow [N^k]$ and $g : [N^2] \times [N^2] \rightarrow [N^{k'}]$, again with sufficiently large k and k' to avoid collisions in these functions, except with negligible probability (k and k' need not be the same here as in the previous section). As before, these two functions could be replaced by a single binary random oracle. For simplicity, we choose N to be a perfect square.

Protocol 2.

1. Alice picks at random N distinct values $\{x_i\}_{i=1}^N$ with $x_i \in [N^2]$ and transmits the encrypted values $y_i = f(x_i)$ to Bob. Let X and Y denote $\{x_i \mid 1 \leqslant i \leqslant N\}$ and $\{y_i \mid 1 \leqslant i \leqslant N\}$, respectively.
2. Bob finds the pre-images x and x' of two distinct random elements in Y. To find each one of them, he chooses random values in $[N^2]$ and applies f to them until one is found whose image is in Y. By virtue of the birthday paradox, he is expected to succeed after $O(\sqrt{N^2}) = O(N)$ calls to function f. *Until now this is identical to Merkle's original scheme*, except for the fact that Bob needs to find two elements of X rather than one.
3. Bob sends back $w = g(x, x')$ to Alice. In addition, he chooses $\sqrt{N} - 2$ random elements from $Y \setminus \{f(x), f(x')\}$ and he forms a set Y' of cardinality $\sqrt{N}$ by adding $f(x)$ and $f(x')$ to those elements. He sends the elements of Y' to Alice in increasing order of values.
4. Because Alice had kept her randomly chosen set X, she knows the preimages of each element of Y'. Let X' denote $\{x \in X \mid f(x) \in Y'\}$. By exhaustive search over all pairs of elements of X', Alice finds the one pair (x, x') such that $g(x, x') = w$.
5. The key shared by Alice and Bob is the pair (x, x').

All counted, Alice makes N calls to f in step 1 and at most N calls to g in step 4 because there are $\sqrt{N}\sqrt{N} = N$ pairs of elements of X' and one of them is the correct one. As for Bob, he makes an expected $O(N)$ calls to f in step 2 and a singe call to g in step 3. The total expected number of calls to f and g is therefore in $O(N)$ for both legitimate parties.

4.1 Quantum Attack

Theorem 3. *There exists an eavesdropping strategy that outputs the pair (x, x') in Protocol 2 with $O(N^{13/12})$ expected quantum queries to functions f and g.*

Proof. A quantum eavesdropper can set up a walk in a Johnson graph very similar to the one explained in Sect. 3.1, except that now the nodes in the graph contain some number r (to be determined later) of distinct elements of X' (rather than of X). The eavesdropper can find random elements of X' from his knowledge of Y' with an expected

$$O\left(\sqrt{N^2/\sqrt{N}}\right) = O(N^{3/4})$$

calls to f per element of X'. Therefore, $\mathsf{S} = O(rN^{3/4})$ calls to f, $\mathsf{U} = O(N^{3/4})$ calls to f and $\mathsf{C} = O(r)$ calls to g. Furthermore, δ is still $\Theta(1/r)$ but $\varepsilon = \Omega(r^2/N)$.

Putting it all together, the expected quantum cryptanalytic cost is

$$\begin{aligned} &\mathsf{S} + O\left(\tfrac{\sqrt{N}}{r}(\sqrt{r}\,\mathsf{U} + \mathsf{C})\right) \\ &= O\left((rN^{3/4} \text{ calls to } f) + \tfrac{\sqrt{N}}{r}\left(\sqrt{r}(N^{3/4} \text{ calls to } f) + (r \text{ calls to } g)\right)\right) \\ &= O\left(rN^{3/4} + N^{5/4}/\sqrt{r}\right) \text{ calls to } f \ \textbf{ and } \ O(\sqrt{N}\,) \text{ calls to } g\,. \end{aligned}$$

To minimize the number of calls to f, we choose r so that $rN^{3/4} = N^{5/4}/\sqrt{r}$, which is $r = N^{1/3}$. It follows that a quantum eavesdropper is able to find the key (x, x') with an expected $O(rN^{3/4}) = O(N^{13/12})$ calls to f and $O(\sqrt{N})$ calls to g. □

4.2 Lower Bound

The proof that it is not possible to find the key (x, x') with fewer than $\Omega(N^{13/12})$ calls to f and g, except with vanishing probability, follows the same lines as the lower bound proof in Sect. 3.2. It is therefore possible for purely classical Alice and Bob to agree on a shared key after calling f and g an expected number of times in the order of N whereas it is not possible, even for a *quantum* eavesdropper, to be privy of their secret with an effort in the same order, except with vanishing probability.

We refer the reader to Sect. 3 for the meaning of notation $[N]$ and to Sect. 3.2 for the definitions of projectors π_1, π_2, and the meaning of notation $[N]'$.

Consider a function $c : [\sqrt{N}\,] \rightarrow [\sqrt{N}\,]$ so that there is a single pair (i, j), $1 \leqslant i < j \leqslant \sqrt{N}$, for which $c(i) = c(j)$. Aaronson and Shi's lower bound [1] tells us that finding this pair requires $\Omega((\sqrt{N})^{2/3}) = \Omega(N^{1/3})$ calls to function c. Now, consider a function $h : [\sqrt{N}\,] \times [N^{3/2}] \rightarrow [\sqrt{N}\,]'$ where $h(i, \cdot)$ denotes the ith bucket, $1 \leqslant i \leqslant \sqrt{N}$. In bucket i, all values of the function are 0 except for one: there is a single random $v_i \in [N^{3/2}]$ such that $h(i, v_i) = c(i)$. It follows from the definitions of c and h that there is a single pair of distinct a and b in the domain of h such that $h(a) = h(b) \neq 0$.

Lemma 3. *Given h structured as above, finding the pair of distinct elements a and b in the domain of h such that $h(a) = h(b) \neq 0$ requires $\Omega(N^{13/12})$ quantum queries to h, except with vanishing probability.*

Proof. The proof is identical to the one for Lemma 1, *mutatis mutandis*. It is again a special case of Lemma 5, but with parameters $\alpha = \sqrt{N}$ (the number of buckets) and $\beta = N^{3/2}$ (the size of the buckets). It follows that finding the desired pair (a, b) requires

$$\Omega(\alpha^{2/3}\beta^{1/2}) = \Omega\left(\sqrt{N}^{\,2/3}\sqrt{N^{3/2}}\right) = \Omega(N^{13/12})$$

quantum queries to h, except with vanishing probability. □

Let $h' : [N^2] \to [\sqrt{N}\,]' \times [\sqrt{N}\,]'$ denote the unstructured version of the same search problem for h, defined the same way as in Sect. 3.2, *mutatis mutandis*. There is a single pair of distinct elements a and b such that $\pi_2(h'(a)) = \pi_2(h'(b)) \neq 0$. The problem of finding this pair is at least as difficult as finding the collision in h.

Lemma 4. *Given h' structured as above, finding the pair of distinct elements a and b in the domain of h' such that $\pi_2(h'(a)) = \pi_2(h'(b)) \neq 0$ requires $\Omega(N^{13/12})$ quantum queries to h', except with vanishing probability.*

It remains to show that the search problem concerning h' reduces to the cryptanalytic difficulty for the eavesdropper to determine the key established by Alice and Bob.

Theorem 4. *Any eavesdropping strategy that recovers the key (x, x') in protocol 2 requires a total of $\Omega(N^{13/12})$ quantum queries to functions f and g, except with vanishing probability.*

Proof. Consider any eavesdropping strategy $\mathcal{A}$ that listens to the communication between Alice and Bob and tries to determine the key (x, x') by querying the black-box functions f and g. As before, the reduction does not have access to Alice and Bob but instead, to a function $h' : [N^2] \to [\sqrt{N}\,]' \times [\sqrt{N}\,]'$ as described above and given as an oracle, for which we want to solve the search problem by using $\mathcal{A}$ as a resource.

We choose random functions $\hat{f} : [N^2] \to [N^k]$ and $\hat{g} : [N^2] \times [N^2] \to [N^{k'}]$, as well as a random Boolean $s \in \{\mathsf{true}, \mathsf{false}\}$, which has the same purpose as in the proof of Theorem 2. Let $\mathrm{Im}(\hat{f})$ denote the image of function $\hat{f}$. We then supply $\mathcal{A}$ with a fake "conversation" between "Alice" and "Bob": we choose randomly $\sqrt{N}$ points y'_1, y'_2,..., $y'_{\sqrt{N}}$ in $[N^k]$, $N - \sqrt{N}$ points $y_1, y_2, \ldots, y_{N-\sqrt{N}}$ in $\mathrm{Im}(\hat{f}) \subset [N^k]$, and one point $w \in [N^{k'}]$. We pretend that Alice has sent the list $Y = \{y_1, y_2, \ldots, y_{N-\sqrt{N}}\} \cup \{y'_1, y'_2, \ldots, y'_{\sqrt{N}}\}$ to Bob (in random order) and that Bob has responded with $Y' = \{y'_1, y'_2, \ldots, y'_{\sqrt{N}}\}$ (in increasing order) and w.

Now, we wait for $\mathcal{A}$'s queries to f and g.

- When $\mathcal{A}$ asks for $f(i)$ for some $i \in [N^2]$, there are two possibilities:
  - If $h'(i) = (0, 0)$, return $\hat{f}(i)$ to $\mathcal{A}$ as value for $f(i)$.
  - Otherwise, return $y'_{\pi_1(h'(i))}$.
- When $\mathcal{A}$ asks for $g(i, j)$ for some $i, j \in [N^2]$, there are two possibilities:
  - If $\pi_2(h'(i)) = \pi_2(h'(j)) \neq 0$ and either s is true and $i < j$ or s is false and $i > j$, return w as value for $g(i, j)$.
  - Otherwise, return $\hat{g}(i, j)$.

Suppose $\mathcal{A}$ happily returns the pair (i, j) for which it was told that $g(i, j) = w$, which is what a successful eavesdropper is supposed to do. This pair is in fact the answer to the search problem concerning function h'. Indeed, $g(i, j) = w$ for only the pair (i, j) for which $\pi_2(h'(i)) = \pi_2(h'(j)) \neq 0$, except with the negligible probability that $\hat{g}(i', j') = w$ for some query (i', j') that $\mathcal{A}$ asks about g. However, we need an additional condition for the reduction to create an environment identical to the real one: if $y \in Y$ then $h'(f^{-1}(y)) = (0,0)$. This is required for all elements in $Y \setminus Y'$ to be accessible when $\mathcal{A}$ is querying f in the reduction. Fortunately, it is easy to see that this condition is satisfied except with vanishing probability when k is large enough.

Provided this condition is satisfied, queries asked by $\mathcal{A}$ concerning f and g are answered in the same way as they would be if both f and g were random functions consistent with the Y, Y' and w announced by Alice and Bob during the execution of the protocol. To see this, remember that Y and Y' (subsets of $[N^k]$) and w (element of $[N^{k'}]$) are uniformly picked at random in both the simulated and the real worlds. Moreover, the simulated function f is such that $f(i)$ is random when $h'(i) = (0,0)$. Among these $N^2 - \sqrt{N}$ input values, there are exactly $N - \sqrt{N}$ output values in $Y \setminus Y'$, as expected by $\mathcal{A}$. The remaining $\sqrt{N}$ input values i also satisfy $f(i) \in Y'$ as it should be. On the other hand, the simulated function g is random everywhere except for one single input pair $(i,j), i \neq j$, for which $g(i,j) = w$, as it is also expected by $\mathcal{A}$. Therefore, $\mathcal{A}$ will behave in the environment provided by the simulation exactly as in the real case. Since we disregard the negligible possibility that g might not be be one-to-one, the reduction solves the search problem concerning h' whenever $\mathcal{A}$ succeeds in finding the key. Notice again that each (new) question asked by $\mathcal{A}$ to either f or g translates to one or two questions actually asked to h'.

It follows that any successful cryptanalytic strategy that makes $o(N^{13/12})$ total queries to f and g would solve the search problem with only $o(N^{13/12})$ queries to function h', which is impossible by Lemma 4, except with vanishing probability. This establishes the $\Omega(N^{13/12})$ lower bound on the cryptanalytic difficulty of breaking our key exchange protocol, which matches the upper bound provided by the explicit attack discussed in Sect. 4.1. □

5 Conclusion, Conjectures and Open Questions

We presented an improved protocol for quantum key distribution over a classical channel and the first secure protocol for classical key distribution against a quantum adversary. Is it possible that they are optimal? We conjecture that they are not.

Indeed, we have discovered two sequences of protocols Q_ℓ and C_ℓ for $\ell \geqslant 2$ (which we shall describe in a subsequent paper) with the following properties. In protocol Q_ℓ, a classical Alice exchanges a key with a quantum Bob after $O(N)$ accesses to a random oracle in such a way that our most efficient quantum eavesdropping strategy requires the eavesdropper to access the same random oracle $\Theta(N^{1+\frac{\ell}{\ell+1}})$ expected times. In protocol C_ℓ, purely classical Alice and Bob exchange a key after $O(N)$ accesses to a random oracle in such a way that our most efficient quantum eavesdropping strategy requires the eavesdropper to access the same random oracle $\Theta(N^{\frac{1}{2}+\frac{\ell}{\ell+1}})$ expected times.

Our attacks proceed by quantum walks in Johnson graphs similar to those exploited in the proofs of Theorems 1 and 3 to obtain optimal attacks against our protocols 1 and 2. If they are the best possible against our new protocols as well, then key distribution protocols à la Merkle can be arbitrarily as secure in our quantum world as they were in the whimsical classical world known to Merkle in 1974: arbitrarily close to quadratic security can be restored. The obvious open question is to prove the optimality of our attacks. It would also be interesting to find a quantum protocol that exactly achieves quadratic security... or better! Indeed, even though it has been proven in the classical case that quadratic security is the best that can be achieved [3], there is no compelling evidence yet that such a limitation exists in the quantum world.

If our quantum attacks against the classical protocols are optimal, classical Alice and Bob can exchange a secret key against a quantum eavesdropper with as good a security (in the limit) as it was known to be possible for *quantum* Alice and Bob before this work. The main open question would be to break the $\Omega(N^{3/2})$ barrier or prove that this is not possible.

Even though our protocols Q_ℓ and C_ℓ require classical Alice to access the random black-box function only N times, she has to work for a *time* in $\Theta(N^\ell)$ to complete her share of the protocol. (This could be reduced to $\Theta(N^{\ell/2})$ for Q_ℓ if both Alice and Bob used quantum computing capabilities, but this remains nonlinear as soon as $\ell \geqslant 3$.) Could similar protocols exist in which Alice would be efficient even outside the required calls to the black-box function?

Finally, our lower bounds prove that it is not possible for the eavesdropper to learn Alice and Bob's key (x, x'), except with vanishing probability, unless she queries the black-box functions significantly more than the legitimate parties. However, we have not addressed the possibility for the eavesdropper to obtain efficiently *partial information* about the key. We leave this important issue for further research.

Acknowledgements. We are grateful to Troy Lee and Mohammad Mahmoody-Ghidary for insightful discussions. G. B. is also grateful to Ralph Merkle for his most inspiring Distinguished Lecture at CRYPTO '05, which sparked this entire line of work.

G. B. is supported in part by Canada's Natural Sciences and Engineering Research Council of Canada (NSERC), the Institut transdisciplinaire d'informatique quantique (INTRIQ), the Canada Research Chair program, the Canadian Institute for Advanced Research (CIFAR) and the Quantum*Works* Network. P. H. is supported in part by NSERC, CIFAR, Quantum*Works*, and the Canadian Network Centres of Excellence for Mathematics of Information Technology and Complex Systems (MITACS). S. L. is supported in part by the European Union 7th framework program QCS, ANR Défis QRAC and ANR Jeune chercheur CRYQ. L. S. is supported in part by NSERC, Quantum*Works*, Fundamental Research on Quantum Networks and Cryptography (FREQUENCY) and INTRIQ.

References

1. Aaronson, S., Shi, Y.: Quantum lower bounds for the collision and the element distinctness problems. Journal of the ACM 51(4), 595–605 (2004)
2. Ambainis, A.: Quantum walk algorithm for element distinctness. SIAM Journal on Computing 37, 210–239 (2007)
3. Barak, B., Mahmoody-Ghidary, M.: Merkle puzzles are optimal — An $O(n^2)$–query attack on any key exchange from a random oracle. In: Halevi, S. (ed.) CRYPTO 2009. LNCS, vol. 5677, pp. 374–390. Springer, Heidelberg (2009)
4. Beals, R., Buhrman, H., Cleve, R., Mosca, M., de Wolf, R.: Quantum lower bounds by polynomials. Journal of the ACM 48(4), 778–797 (2001)
5. Bennett, C.H., Bernstein, E., Brassard, G., Vazirani, U.V.: Strengths and weaknesses of quantum computing. SIAM Journal on Computing 26(5), 1510–1523 (1997)
6. Boyer, M., Brassard, G., Høyer, P., Tapp, A.: Tight bounds on quantum searching. Fortschritte Der Physik 46, 493–505 (1998)
7. Brassard, G., Høyer, P., Mosca, M., Tapp, A.: Quantum amplitude amplification and estimation. In: Lomonaco Jr., S.J. (ed.) Quantum Computation and Quantum Information. Contemporary Mathematics, vol. 305, pp. 53–74. AMS, Providence (2002)
8. Brassard, G., Salvail, L.: Quantum Merkle puzzles. In: Proceedings of Second International Conference on Quantum, Nano, and Micro Technologies (ICQNM 2008), Sainte Luce, Martinique, pp. 76–79 (February 2008)
9. Buhrman, H., Dürr, C., Heiligman, M., Høyer, P., Magniez, F., Sántha, M., de Wolf, R.: Quantum algorithms for element distinctness (2000), `http://arxiv.org/abs/quant-ph/0007016v2`
10. Diffie, W., Hellman, M.E.: New directions in cryptography. IEEE Transactions on Information Theory 22(6), 644–654 (1976)
11. Grover, L.K.: Quantum mechanics helps in searching for a needle in a haystack. Physical Review Letters 79(2), 325–328 (1997)
12. Heiligman, M.: Finding matches between two databases on a quantum computer (2000), `http://arxiv.org/abs/quant-ph/0006136v1`
13. Høyer, P., Lee, T., Špalek, R.: Negative weights make adversaries stronger. In: Proceedings of 39th Annual Symposium on Theory of Computing (STOC), pp. 526–535 (June 2007), The complete version can be found at `http://arxiv.org/abs/quant-ph/0611054v2`
14. Lee, T., Mittal, R., Reichardt, B.W., Špalek, R.: An adversary for algorithms (2010), `http://arxiv.org/abs/1011.3020v1`
15. Merkle, R.: C.S. 244 Project Proposal (1974), Facsimile available at `http://www.merkle.com/1974`
16. Merkle, R.: Secure communications over insecure channels. Communications of the ACM 21(4), 294–299 (1978)
17. Sántha, M.: Quantum walk based search algorithms. In: Agrawal, M., Du, D.-Z., Duan, Z., Li, A. (eds.) TAMC 2008. LNCS, vol. 4978, pp. 31–46. Springer, Heidelberg (2008)

A Quantum Query Complexity

In our protocols, the work of the different parties is quantified by the number of queries made to black-box random functions, which can be modelled by a random oracle. In this Appendix, we review the main results from quantum query complexity that we used to prove our results and we sketch a new technical result that is needed for our lower-bound proofs.

Upper Bounds

Our attacks can be modelled as quantum walks on Johnson graphs. The graph $J(n, r)$ is an undirected graph in which each node contains some number r of distinct elements of $[n]$ and there is an edge between two nodes if and only if they differ by exactly two elements. Intuitively, we may think of "walking" from one node to an adjacent node by dropping one element and replacing it by another. The task is to find a specific k-subset of $[n]$. The nodes that contain this subset are called *marked*.

A random walk P on a Johnson graph can be quantized and the cost of the resulting quantum algorithm can be written as a function of S, U and C. These are the cost of setting up the quantum register in a state that corresponds to the stationary distribution, moving unitarily from one node to an adjacent node, and checking if a node is marked in order to flip its phase if it is, respectively.

Theorem 5. *[2,17] Let M be either empty, or the set of vertices that contain a fixed subset of constant size $k \leqslant r$. Then there is a quantum algorithm that finds, with high probability, the k-subset if M is not empty at an expected cost in the order of*

$$\mathsf{S} + \frac{1}{\sqrt{\varepsilon}}\left(\frac{1}{\sqrt{\delta}}\mathsf{U} + \mathsf{C}\right),$$

where $\delta = n/r(n-r)$ is the eigenvalue gap of the symmetric walk on $J(n, r)$ and $\varepsilon = \Omega(\frac{r^k}{n^k})$ is the probability that a node is marked.

Lower Bounds

The central technical part of our lower bound consists in analysing the complexity of a function closely related to the hardness of breaking the key distribution protocols. This function is obtained by composing element distinctness and a variant of the search problem.

Consider two integer parameters α and β and three functions $c : [\alpha] \to [\alpha]$, $v : [\alpha] \to [\beta]$ and $h : [\alpha] \times [\beta] \to [\alpha]'$ so that there exists a single pair (i, j), $1 \leqslant i < j \leqslant \alpha$, for which $c(i) = c(j)$, which is called a *collision*, and

$$h(i, j) = \begin{cases} c(i) & \text{if } j = v(i) \\ 0 & \text{otherwise}. \end{cases}$$

The task is to find the unique nonzero collision in h, having only access to a black-box that computes h. This can be thought of as *searching* among β possibilities for the sole nonzero $h(i,\cdot)$ for each i and then finding two of those *elements*, among α possibilities, that are not *distinct*. Our main technical lemma, below, gives a lower bound on the number of queries to h that are required.

Lemma 5. *Finding a nonzero collision in h, structured as above, requires $\Omega(\alpha^{2/3}\beta^{1/2})$ quantum queries to h, except with vanishing probability.*

A complete proof of this lemma will appear separately but we now proceed to sketch it. For technical reasons, it is more convenient to prove this lower bound for the related decision problem: we are given a function h of the type above, but it is either based on a function c that has a single collision (as above) or on a one-to-one function c (in which case h is collision-free, except for value 0 in its image). The task is to decide which is the case. Obviously, any algorithm that can solve the search problem with probability of success at least $p > 0$ can be used to solve the decision problem with error bounded by $\frac{1}{2} - \frac{p}{2}$: run the search algorithm; if a collision is found (and verified), output "collision", otherwise output either "collision" or "no collision" with equal probability after flipping a fair coin. It follows that any lower bound on the bounded-error decision problem applies equally well to the search problem.

Again for technical reasons, we shall change the notation in order to adapt it to the normal usage in the field of quantum query complexity. For instance, function $c : [\alpha] \rightarrow [\alpha]$ will be represented by an element of $[\alpha]^\alpha$. This makes it possible to think of the decision version of element distinctness as a Boolean function $\mathsf{ED} : [\alpha]^\alpha \rightarrow \{0,1\}$, although it is a partial function since there is a *promise* on the valid inputs to ED: Given α integers $(z_1,\ldots,z_\alpha) \in [\alpha]^\alpha$, the promise is that either all the elements are distinct; or that all the elements are distinct except two, say $z_i \neq z_j$. The goal is to decide which of the two cases occurs by making as few queries as possible to the function that returns z_i on input i.

Ambainis' element distinctness quantum algorithm [2] runs in $O(\alpha^{2/3})$ queries to the input, and Aaronson and Shi proved that this is optimal [1]. Although the lower bound was proven using the polynomial method [4], a recent theorem of Ref. [14] shows that the generalized adversary bound is tight. Since our proof of the lower bound is derived using the generalized adversary method [13], we may conclude that there exists an adversary bound for element distinctness.

We compose the element distinctness problem with α instances of a promise version of a Search problem, which we call $\mathsf{pSEARCH}$. $\mathsf{pSEARCH} : [\alpha]'^\beta \rightarrow [\alpha]$ is also a promise problem. On input $(a_1,\ldots a_\beta)$, the promise is that all but one of the numbers are zero. The goal is to find and output this non-zero number by making queries that take i as input and return a_i.

The composed function we study is $\mathsf{H} = \mathsf{ED} \circ \mathsf{pSEARCH}^\alpha$. We now restate Lemma 5 in its decision-problem version.

Lemma 6. *The quantum query complexity of* H *is in* $\Omega(\alpha^{2/3}\beta^{1/2})$.

The proof uses the generalized adversary method for quantum query complexity, which we briefly describe here. Suppose we want to determine the quantum query complexity of a function F. We will assign weights to pairs of inputs in such a way as to bring out how hard it is (in terms of number of queries) to distinguish these inputs apart from one another. The adversary lower bound is the worst ratio of the spectral norm of this matrix, which measures the overall progress necessary in order for the algorithm to be correct, to the spectral norms of a associated matrices, which measure the maximum amount of progress that can be achieved by making a single query.

Definition 1. *Fix a function* $\mathsf{F} : S^n \to T$*. A symmetric matrix* $\Gamma : S^n \times S^n \to \mathbb{R}$ *is an adversary matrix for* F *provided* $\Gamma[x, y] = 0$ *whenever* $\mathsf{F}(x) = \mathsf{F}(y)$*. Let* $D_i[x, y] = 1$ *if* $x_i \neq y_i$ *and 0 otherwise. The adversary bound of* F *using* Γ *is*

$$\mathsf{Adv}^{\pm}(\mathsf{F}; \Gamma) = \min_i \frac{\|\Gamma\|}{\|\Gamma \circ D_i\|},$$

where $\circ$ *denotes entrywise (or Hadamard) product, and* $\|A\|$ *denotes the spectral norm of* A *(which is equal to its largest eigenvalue). The adversary bound* $\mathsf{Adv}^{\pm}(\mathsf{F})$ *is the maximum, over all adversary matrices* Γ *for* F*, of* $\mathsf{Adv}^{\pm}(\mathsf{F}; \Gamma)$*.*

Since H is defined as the composition of ED and pSEARCH, one would like to apply a composition theorem for the generalized adversary method [13], which would say that if a function $\mathsf{H} = \mathsf{F} \circ \mathsf{G}^{\alpha}$, then $\mathsf{Adv}^{\pm}(\mathsf{H}) = \mathsf{Adv}^{\pm}(\mathsf{F})\mathsf{Adv}^{\pm}(\mathsf{G})$ (up to constant factors, which will no longer be mentioned). Unfortunately, the composition theorem of Ref. [13] requires the inner and outer functions to be Boolean, which is not the case here for the inner function. (In fact, the outer function does not need to be Boolean according Corollary 5.6 in Ref. [14], but there are no general results known to the authors that yield a similar theorem when the inner function is not Boolean.)

Nevertheless, we are still able to prove the lower bound using techniques from Ref. [13] . Although our inner function is not Boolean, it has a lot of structure, which turns out to be sufficient for the proof to go through, modulo some modifications, which we briefly sketch here. (For the full version of the composition theorem, we refer the reader to the `ArXiv` version of Ref. [13].)

Our goal is to construct an adversary matrix Γ_{H} that captures the difficulty of applying ED to α instances of pSEARCH. Recall that $\mathsf{Adv}^{\pm}$ is tight for query complexity, so we know that there exists an adversary matrix Γ_{ED} for which $\mathsf{Adv}^{\pm}(\mathsf{ED}; \Gamma_{\mathsf{ED}}) \geqslant \alpha^{2/3}$. We don't have an explicit expression for this matrix, let alone its spectral decomposition, but we know it exists.

For the inner pSEARCH problem, we construct an adversary matrix that we call Γ_{G} to keep consistent with the notation of Ref. [13]. We can analyse this matrix and give its explicit spectral decomposition. The block structure of the matrix and the form of its eigenvalues is key to proving the lower bound, so our proof does not hold for arbitrary non-Boolean inner functions g.

There are two main parts to the proof that $\mathsf{Adv}^{\pm}(\mathsf{H}) = \alpha^{2/3}\beta^{1/2}$. First we give a lower bound on $\|\Gamma_{\mathsf{H}}\|$, then we give an upper bound on $\|\Gamma_{\mathsf{H}} \circ D_i\|$, for each i.

Claim. $\|\Gamma_{\mathsf{H}}\| = \|\Gamma_{\mathsf{ED}}\| \|\Gamma_{\mathsf{G}}\|^{\alpha}$.

Proving this claim is the central and most technical part of the proof. In order to compute Γ_{H}'s spectral norm, we give its spectral decomposition. As in Ref. [13], we provide $(\alpha\beta)^\alpha$ eigenvectors and show that they form a basis. Our basis differs from that of Ref. [13], and is tailored to the properties we know about the spectral decomposition of Γ_{G}. We make essential use of the block structure of Γ_{G} and the fact that there are just two kinds of block: diagonal blocks, and off-diagonal blocks. (In the case of Boolean outputs, the same structure occurs, but there are just four blocks in that case, whereas we can handle many.)

Once we have the norm of Γ_{H}, we can use similar ideas to compute the value of $\|\Gamma_{\mathsf{H}} \circ D_i\|$. Because of the symmetry of H, it suffices to compute $\|\Gamma_{\mathsf{H}} \circ D_i\|$ for a fixed i. Fortunately, $\|\Gamma_{\mathsf{H}}\|$ and $\|\Gamma_{\mathsf{H}} \circ D_i\|$ share sufficient structure so that once the calculation of $\|\Gamma_{\mathsf{H}}\|$ is done, the calculation of $\|\Gamma_{\mathsf{H}} \circ D_i\|$ follows easily.

For any query i, we decompose it into the index p in which it occurs within x, and the index of the position queried within x_p. Then, D_i decomposes naturally into two parts, D_p and D_q.

Claim. $\forall i$ that decomposes into p, q, $\|\Gamma_{\mathsf{H}} \circ D_i\| = \|\Gamma_{\mathsf{ED}} \circ D_p\| \|\Gamma_{\mathsf{G}} \circ D_p\| \|\Gamma_{\mathsf{G}}\|^{\alpha-1}$.

Combining the two claims, we get

$$\mathsf{Adv}^{\pm}(\mathsf{H}; \Gamma_{\mathsf{H}}) = \mathsf{Adv}^{\pm}(\mathsf{ED})\mathsf{Adv}^{\pm}(\mathsf{pSEARCH}) = \mathsf{Q}(\mathsf{ED})\mathsf{Q}(\mathsf{OR}),$$

where Q denotes the quantum query complexity, and where the final inequality follows from the fact that OR is a special case of $\mathsf{pSEARCH}$. The lemma follows by using the known quantum query complexity lower bounds for $\mathsf{Q}(\mathsf{OR})$, which is in $\Omega(\beta^{1/2})$ [5], and for $\mathsf{Q}(\mathsf{ED})$, which is in $\Omega(\alpha^{2/3})$ [1].

Classical Cryptographic Protocols in a Quantum World

Sean Hallgren*, Adam Smith**, and Fang Song

Department of Computer Science and Engineering, Pennsylvania State University,
University Park, PA, U.S.A.

Abstract. Cryptographic protocols, such as protocols for secure function evaluation (SFE), have played a crucial role in the development of modern cryptography. The extensive theory of these protocols, however, deals almost exclusively with classical attackers. If we accept that quantum information processing is the most realistic model of physically feasible computation, then we must ask: what classical protocols remain secure against quantum attackers?

Our main contribution is showing the existence of classical two-party protocols for the secure evaluation of any polynomial-time function under reasonable computational assumptions (for example, it suffices that the learning with errors problem be hard for quantum polynomial time). Our result shows that the basic two-party feasibility picture from classical cryptography remains unchanged in a quantum world.

1 Introduction

Cryptographic protocols, such as protocols for secure function evaluation (SFE), have played a crucial role in the development of modern cryptography. Goldreich, Micali and Wigderson [25], building on the development of zero-knowledge (ZK) proof systems [27,26], showed that SFE protocols exist for any polynomial-time function under mild assumptions (roughly, the existence of secure public-key cryptosystems). Research into the design and analysis of such protocols is now a large subfield of cryptography; it has also driven important advances in more traditional areas of cryptography such as the design of encryption, authentication and signature schemes.

The extensive theory of these protocols, however, deals almost exclusively with classical attackers. However, given our current understanding of physics, quantum information processing is the most realistic model of physically feasible computation. It is natural to ask: *what classical protocols remain secure against quantum attackers?* In many cases, even adversaries with modest quantum computing capabilities, such as the ability to share and store entangled photon pairs, are not covered by existing proofs of security.

Clearly not all protocols are secure: we can rule out anything based on the computational hardness of factoring, the discrete log [43], or the principal ideal problem [28]. More subtly, the basic techniques used to reason about security may not apply in a

* Partially supported by National Science Foundation award CCF-0747274 and by the National Security Agency (NSA) under Army Research Office (ARO) contract number W911NF-08-1-0298.

** Partially supported by National Science Foundation award CCF-0747294.

P. Rogaway (Ed.): CRYPTO 2011, LNCS 6841, pp. 411–428, 2011.

quantum setting. For example, some information-theoretically secure two-prover ZK and commitment protocols are analyzed by viewing the provers as long tables that are fixed before queries are chosen by the verifier; quantum entanglement breaks that analysis and some protocols are insecure against colluding quantum provers (Crépeau *et al.*, [17]).

In the computational realm, *rewinding* is a key technique for basing the security of a protocol on the hardness of some underlying problem. Rewinding proofs consist of a mental experiment in which the adversary is run multiple times using careful variations of a given input. At first glance, rewinding seems impossible with a quantum adversary since running it multiple times might modify the entanglement between its internal storage and an outside reference system, thus changing the overall system's behavior.

In a breakthrough paper, Watrous [49] showed that a specific type of zero-knowledge proof (3-round, GMW-style protocols) can be proven secure using a rewinding argument tailored to quantum adversaries. Damgård and Lunemann [21] showed that a similar analysis can be applied to a variant of Blum's coin flipping protocol. Hallgren *et al.* [29] showed certain classical transformations from honest-verifier to malicious-verifier ZK can be modified to provide security against malicious quantum verifiers. Some information-theoretically secure classical protocols are also known to resist quantum attacks [15,5,23,47]. Finally, there is a longer line of work on protocols that involve quantum communication, dating back to the Bennett-Brassard key exchange paper. Overall, however, little is known about how much of the classical theory can be carried over to quantum settings. See "Related Work", below, for more detail.

1.1 Our Contributions

Our main contribution is showing the existence of classical two-party protocols for the secure evaluation of any polynomial-time function under reasonable computational assumptions (for example, it suffices that the learning with errors problem [42] be hard for quantum polynomial time). Our result shows that *the basic two-party feasibility picture from classical cryptography remains unchanged in a quantum world.* The only two-party general SFE protocols which had previously been analyzed in the presence of quantum attackers required quantum computation and communication on the part of the honest participants (e.g. [14,18]).

In what follows, we distinguish two basic settings: in the *stand-alone* setting, protocols are designed to be run in isolation, without other protocols running simultaneously; in *network* settings, the protocols must remain secure even when the honest participants are running other protocols (or copies of the same protocol) concurrently. Protocols proven secure in the *universal composability* (UC) model [11] are secure in arbitrary network settings, but UC-security is impossible to achieve in many settings.

Our contributions can be broken down as follows:

Classical Zero-knowledge Arguments of Knowledge Secure Against Quantum Adversaries. We construct a classical zero-knowledge argument of knowledge (ZKAoK) protocol that can be proven secure in our model. In particular it means that our construction is "witness-extendable" [33] in the sense that one can simulate an interaction with a malicious prover and simultaneously extracting a witness of the statement whenever the prover succeeds. Our construction overcomes a limitation of the proofs of knowledge recently analyzed by Unruh [46], where a simulator for the prover is not given, and thus it is unclear how to analyze security when using his proof of knowledge as a subprotocol. As in the classical case, our ZKAoK protocol is an important building block in designing general SFE protocols.

The main idea behind our construction is to have the prover and verifier first execute a weak coin-flipping protocol to generate a public key for a special type of encryption scheme. The prover encrypts his witness with respect to this public key and proves consistency of his ciphertext with the statement x using the ZK protocols analyzed by Watrous [49]. A simulator playing the role of the verifier can manipulate the coin-flipping phase to generate a public key for which she knows the secret key, thus allowing her to extract the witness without needing to rewind the prover. A simulator playing the role of the prover, on the other hand, cannot control the coin flip (to our knowledge) but can ensure that the public key is nearly random. If the encryption scheme satisfies additional, non-standard properties (that can be realized under widely used lattice-type assumptions), we show that the verifier's view can nonetheless be faithfully simulated. Lunemann and Nielsen [36] independently gave a similarly-flavored construction of ZKAoK for quantum adversaries; see "Related Work".

(More) General modeling of stand-alone security with quantum adversaries. We describe a security model for two-party protocols in the presence of a quantum attackers. Proving security in this model amounts to showing that a protocol for computing a function f behaves indistinguishably from an "ideal" protocol in which f is computed by a trusted third party, which we call the ideal functionality $\mathcal{F}$. Our model is a quantum analogue of the model of stand-alone security developed by Canetti [10] in the classical setting. It slightly generalizes the existing model of Damgård *et al.*[18] in two ways. First, our model allows for protocols in which the ideal functionalities that process quantum information (rather than only classical functionalities). Second, it allows for adversaries that take arbitrary quantum advice, and for arbitrary entanglement between honest and malicious players' inputs.

We also show a sequential modular composition theorem for protocols analyzed in our model. Roughly, it states that one can design protocols modularly, treating subprotocols as equivalent to their ideal versions when analyzing security of a high-level protocol. While the composition results of Damgaard *et al.* allow only for classical high-level protocols, our result holds for arbitrary quantum protocols.

Classical UC Protocols in a Quantum Context: Towards Unruh's Conjecture. We show that a large class of protocols which are UC-secure against computationally bounded classical adversaries are also UC-secure against quantum adversaries. In his recent paper, Unruh [47] showed that any classical protocol which is proven UC-secure against unbounded classical adversaries is also UC-secure against unbounded quantum adversaries. He conjectured (roughly, see [47] for the exact statement) that classical arguments of *computational* UC security should also go through as long as the underlying computational primitives are not easily breakable by quantum computers.

We provide support for this conjecture by describing a family of classical security arguments that go through verbatim with quantum adversaries. We call these arguments "simple hybrid arguments". They use rewinding neither in the simulation nor in any of the steps that show the correctness of simulation.[1]

Our observation allows us to port a general result of Canetti, Lindell, Ostrovsky and Sahai [13] to the quantum setting. We obtain the following: in the $\mathcal{G}_{ZK}$-hybrid model, where an ideal functionality $\mathcal{G}_{ZK}$ implementing ZKAoK is available, there exist classical protocols for the evaluation of any polynomial-time function f that are UC-secure against quantum adversaries under reasonable computational assumptions.

As an immediate corollary, we get a classical protocol that quantum UC-emulates ideal functionality $\mathcal{G}_{CF}$ for coin-flipping. Adapting ideas from [33], we also give a direct construction of coin-flipping from ZK. More interestingly, we can develop the converse by describing a simple classical protocol for ZKAoK that is UC-secure against quantum adversaries in the $\mathcal{G}_{CF}$-hybrid model (a.k.a the *common reference string model* where all participants have access to a common, uniformly distributed bit string). The "simple hybrid arguments" mentioned above do not suffice for proving the security of the UC-secure ZKAoK protocol. Specifically, one component of our protocol, a construction of a *witness-indistinguishable* proof system, needs a new proof of security. The basic strategy is still a hybrid argument, but its analysis requires breaking the space of possible executions into pieces (classically, this involves conditioning on complementary events; quantumly, this involves projecting onto orthogonal subspaces) and arguing that (a) the adversary cannot have a significant advantage in either piece and (b) the original state was a mixture, not a superposition, of the two pieces. This establishes the equivalence between $\mathcal{G}_{ZK}$ and $\mathcal{G}_{CF}$ in the UC model, which may be of independent interest, e.g., in simplifying protocol designs.

Implications. The modular composition theorem in our stand-alone model allows us to get the general feasibility result below by combining our stand-alone ZKAoK protocol and the UC-secure protocols in $\mathcal{G}_{ZK}$-hybrid model:

Under standard assumptions, *there exist* classical *SFE protocols in the plain model (without a shared random string) which are stand-alone-secure against static* quantum *adversaries*. This parallels the classic result of Goldreich, Micali and Wigderson [25].

The equivalence of zero-knowledge and coin-flipping functionalities in the UC model also gives rise to interesting implications. First, the availability of a common reference string suffices for implementing quantum UC-secure protocols. Secondly, given our stand-alone ZKAoK protocol, we get a quantum stand-alone coin-flipping protocol due to the aforementioned equivalence.

Independently of our work, Lunemann and Nielsen [36] obtained similar results to ours. See the discussion at the end of "Related Work".

1.2 Related Work

In addition to the previous work mentioned above, we expand here on three categories of related efforts.

---

[1] In general, it is hard to clearly define what it means for a security proof to "not use rewinding". It is not enough for the protocol to have a straight-line simulator, since the proof of the simulator's correctness might still employ rewinding. Simple hybrid arguments provide a clean, safe subclass of arguments that go through with quantum adversaries.

Composition Frameworks for Quantum Protocols. Systematic investigations of the composition properties of quantum protocols are relatively recent. Canetti's UC framework and Pfitzmann and Waidner's closely related *reactive functionality* framework were extended to the world of quantum protocols and adversaries by Ben-Or and Mayers [7] and Unruh [45,47]. These frameworks (which share similar semantics) provide extremely strong guarantees—security in arbitrary network environments. They were used to analyze a number of unconditionally secure quantum protocols (key exchange [6] and multi-party computation with honest majorities [5]). However, many protocols are not universally composable, and Canetti [11] showed that classical protocols cannot UC-securely realize even basic tasks such as commitment and zero-knowledge proofs without some additional setup assumptions such as a CRS or public-key infrastructure.

Damgård *et al.*[18], building on work by Fehr and Schaffner [23], proposed a general composition framework which applies only to secure quantum protocols of a particular form (where quantum communication occurs only at the lowest levels of the modular composition). As noted earlier, our model is more general and captures both classical and quantum protocols. That said, understanding the exact relationship between the models is delicate, and connected to basic questions in complexity theory such as the power of quantum advice (BQP/poly vs BQP/qpoly). We defer further discussion of this relationship to the full version.

Analyses of quantum protocols. The first careful proofs of security of quantum protocols were for key exchange (Mayers [37], Lo and Chau [35], Shor and Preskill [44], Beaver [2]). Research on quantum protocols for two-party tasks such as coin-flipping, bit commitment and oblivious transfer dates back farther [9,8] but many initially proposed protocols were insecure [37]. The first proofs of security of such protocols were based on computational assumptions [22,14]. They were highly protocol-specific and it was not known how well the protocols composed. The first proofs of security using the simulation paradigm were for information-theoretically-secure protocols for multi-party computations assuming a strict majority of honest participants [15,16,5]. Subsequently, a line of work on the *bounded quantum storage* model [20,19,23,48] developed tools for reasoning about specific types of composition of two-party protocols, under assumptions on the size of the adversary's quantum storage. Unruh's UC security work, mentioned above, was the first we are aware of that was sufficiently general to encompass classical and quantum protocols and generic composition.

Straight-line simulators and code-based games. As mentioned above, we introduce "simple hybrid arguments" to capture a class of straightforward security analyses that go through against quantum adversaries. Several formalisms have been introduced in the past to capture classes of "simple" security arguments. To our knowledge, none of them is automatically compatible with quantum adversaries. For example, *straight-line black-box simulators* [32] do not rewind the adversary nor use an explicit description of its random coins; however, it may be the case that rewinding is necessary to prove that the straight-line simulator is actually correct. In a different vein, the *code-based games* of Bellare and Rogaway [4] capture a class of hybrid arguments that can be encoded in a clean formal language; again, however, the arguments concerning each step of the hybrid may still require rewinding.

Independent Work: Lunemann and Nielsen [36]. Lunemann and Nielsen [36] independently obtained similar results to the ones described here, via a slightly different route. Specifically, they start by constructing a stand-alone coin-flipping protocol that is fully simulatable against quantum poly-time adversaries. Then they use the coin-flipping protocol to construct a stand-alone ZKAoK protocol, and finally by plugging into the GMW construction, they get quantum stand-alone-secure two-party SFE protocols as well. The computational assumptions in the two works are similar and the round complexities of the stand-alone SFE protocols are both polynomial in the security parameter. Our approach to composition is more general, however, leading to results that also apply (in part) to the UC model.

Organization. The rest of the paper is organized as follows: Section 2 reviews basic notations and definitions. In Section 3, we propose our quantum stand-alone security model. A quantum stand-alone-secure ZKAoK protocol is developed in Section 4. Section 5 studies a family of classical analysis that go through in the quantum UC model, and then Section 6 discusses equivalence of $\mathcal{G}_{ZK}$ and $\mathcal{G}_{CF}$. Finally in Section 7, we obtain, among other consequences, classical SFE that are quantum stand-alone-secure with no set-up assumptions. We conclude with future directions.

2 Preliminaries

For $m \in \mathbb{N}$, $[m]$ denotes the set $\{1, \ldots, m\}$. We use $n \in \mathbb{N}$ to denote a *security parameter*. The security parameter, represented in unary, is an implicit input to all cryptographic algorithms; we omit it when it is clear from the context. Quantities derived from protocols or algorithms (probabilities, running times, etc) should be thought of as functions of n, unless otherwise specified. A function $f(n)$ is said to be negligible if $f = o(n^{-c})$ for any constant c, and $negl(n)$ is used to denote an unspecified function that is negligible in n. We also use $poly(n)$ to denote an unspecified function $f(n) = O(n^c)$ for some constant c. Let $\mathbf{X} = \{X_n\}_{n\in\mathbb{N}}$ and $\mathbf{Y} = \{Y_n\}_{n\in\mathbb{N}}$ be two ensembles of binary random variables. We call $\mathbf{X}, \mathbf{Y}$ *indistinguishable*, denoted $\mathbf{X} \approx \mathbf{Y}$, if $|\Pr(X_n = 1) - \Pr(Y_n = 1)| \leq negl(n)$.

We assume the reader is familiar with the basic concepts of quantum information theory (see, e.g., [39]). We use a capital letter (e.g. X) to denote a quantum register and for each n, we use script letter (e.g., $\mathcal{X}(n)$) to denote the corresponding Hilbert space. Let $D(\mathcal{H})$ be the set of density operators acting on space $\mathcal{H}$. Let $\{\rho_n\}_{n\in\mathbb{N}}$ denote an ensemble of mixed states where $\rho_n \in D(\mathcal{H}_n)$ and $\mathcal{H}_n$ is a $poly(n)$-qubit space.

Quantum Machine Model. We adapt Unruh's machine model in [47] with minor changes. A *quantum interactive machine* (QIM) M is an ensemble of circuits $\{M_n\}_{n\in\mathbb{N}}$, for each value n of the security parameter. M operates on three registers: a state register S used for input and workspace; an output register O; and a network register N for communicating with other machines. We say the size (or running time) of M is $t(n)$, if there is a deterministic classical Turing machine that computes the description of M_n in time $t(n)$ on input 1^n. We say a machine is polynomial time if $t(n) = poly(n)$.

When two QIMs M and M' interact, their network register N is shared. The circuits M_n and M'_n are executed alternately. When three or more machines interact, the machines may share different parts of their network registers (for example, a private channel consists of a register shared between only two machines; a broadcast channel is a register shared by all machines). The order in which machines are activated may be either specified in advance (as in a synchronous network) or adversarially controlled.

A noninteractive quantum machine (referred to as QTM hereafter) is a QIM M with no network register that runs for only one round (for all n). This is equivalent to the *quantum Turing machine* model (see [50]). A classical interactive machine is a special case of a QIM, where the registers only store classical strings and all circuits are classical. Classical polynomial-time QIMs are equivalent to polynomial-time interactive Turing machines.

Indistinguishability of Quantum States. Recall Watrous's notion of indistinguishability of quantum states.

Definition 1. *((t,ϵ)-indistinguishable quantum states. [49, Definition 2]) We say two quantum state ensembles $\rho = \{\rho_n\}_{n\in\mathbb{N}}$ and $\eta = \{\eta_n\}_{n\in\mathbb{N}}$ are (t,ϵ)-quantum-indistinguishable, denoted $\rho \approx_Q^{t,\epsilon} \eta$, if for every $t(n)$-time QTM $\mathcal{Z}$ and every mixed state $\sigma_n \in \mathcal{W}(n)$, $\mathcal{W}(n)$ is a $t(n)$-qubit auxiliary system,*

$$|\Pr[\mathcal{Z}(\rho_n \otimes \sigma_n) = 1] - \Pr[\mathcal{Z}(\eta_n \otimes \sigma_n) = 1]| \leq \epsilon(n).$$

The states ρ and η are called *quantum computationally indistinguishable*, denoted $\rho \approx_Q \eta$, if for every $t(n) \leq poly(n)$, there exists a negligible $\epsilon(n)$ such that ρ_n and η_n are (t,ϵ)-indistinguishable. This definition subsumes classical distributions, since classical distributions can be represented by density matrices that are diagonal with respect to the standard basis.

Indistinguishability of quantum machines. Next we define indistinguishability of quantum interactive machines. Let $\mathcal{Z}, M$ be two QIMs, we denote $\langle \mathcal{Z}(\sigma), M\rangle$ as the process that $\mathcal{Z}$ with auxiliary input σ, interacts with M and finally $\mathcal{Z}$ outputs one classical bit 1 or 0.

Definition 2 ((t,ϵ)-indistinguishable QIMs). *We say two QIMs M_1 and M_2 are (t,ϵ)-*interactively indistinguishable, *denoted $M_1 \approx_I^{t,\epsilon} M_2$, if for any quantum $t(n)$-time interactive machine $\mathcal{Z}$ and every mixed state σ_n on $t(n)$ qubits, $\mathbf{X_1} \approx \mathbf{X_2}$, where $\mathbf{X}_i = \{\langle \mathcal{Z}(\sigma_n), M_i\rangle\}_{n\in\mathbb{N}}$ for $i = 1,2$. QIMs M_1 and M_2 are called* interactively indistinguishable, *denoted $M_1 \approx_I M_2$, if for every $t(n) \leq poly(n)$, there exists a negligible $\epsilon(n)$ such that M_1 and M_2 are (t,ϵ)-interactively indistinguishable.*

Finally we state the computational assumptions that we make in this work.

Assumption 1. *There exists a classical pseudorandom generator secure against quantum distinguishers.*

Based on this assumption and the construction of [38], we can obtain a statistically binding and quantum computationally hiding commitment scheme $(\mathbf{comm}, \mathbf{decom})$. All commitment scheme we use afterwards refers to this one. This assumption also suffices for Watrous's ZK proof system for any NP-language against quantum attacks.

Assumption 2. *There exists a* dense *classical public-key cryptosystem that is IND-CPA (chosen-plaintext attack) secure against quantum distinguishers. A public-key cryptosystem is dense if a valid public key is indistinguishable in quantum poly-time from a uniformly random string of the same length.*

Although it is likely that standard reductions would show that Assumption 2 implies Assumption 1, we chose to keep the assumptions separate because the instantiation one would normally use of the pseudorandom generator would not be related to the public-key system (instead, it would typically be based on a symmetric-key block or stream cipher). Both assumptions hold, for instance, assuming the hardness of *leaning with errors* (LWE) problem [42].

In one of our constructions (stand-alone ZKAoK), we need an encryption scheme that has one extra property than the one in Assumption 2.

Assumption 3. *There exists a dense classical public-key cryptosystem that is IND-CPA secure against quantum distinguishers. In addition, encryptions of two messages under a uniformly random string are statistically indistinguishable.*

Note that the *dense* property already implies encryptions under a random string are quantum computationally indistinguishable. Assumption 3 strengthens this requirement to be statistically indistinguishable. This allows "cheating" in the sense that if a ciphertext is generated under a uniformly random string, we can then claim it to be an encryption of an arbitrary message. This type of encryption scheme is sometimes called Meaningful/Meaningless encryption (e.g., see [31]). Again, the LWE assumption implies Assumption 3.

3 Quantum Stand-Alone Security and Modular Composition

In this section, we propose a stand-alone security model for two-party protocols in the presence of quantum attacks and show that modular composition holds in our model. Our definition can be viewed in two ways: either as a quantum analogue of Canetti's classical stand-alone model [10] or as a relaxed notion of a variant of Unruh's quantum UC security [47].

3.1 Security Definition

A two-party protocol Π consists of two quantum interactive machines $\mathbf{A}$ and $\mathbf{B}$. Two players Alice and Bob that execute Π are called honest if they run machines $\mathbf{A}$ and $\mathbf{B}$ respectively. An adversary in Π is one entity that corrupts some player and controls its behavior. We consider both *semi-honest* (a.k.a. *honest-but-curious*) and *malicious* adversaries. In the quantum setting, a semi-honest adversary runs the honest protocol *coherently*, that is, replacing measurements and classical operations with unitary equivalents.

We consider only *static* adversaries, which corrupt a set of players before the protocol execution starts, but do not perform further corruptions during the protocol execution.For ease of exposition, we merge the identities of an adversary and the party it corrupts. Machines run by an adversary are indicated by a ^ symbol (e.g., $\hat{\mathbf{B}}$).

Our definition of security follows the *simulation paradigm* where we compare two modes of execution called *real-world* and *ideal-world*. A *real-world* execution is an interaction between an honest player and a real-world adversary, e.g., $\mathbf{A}$ and $\hat{\mathbf{B}}$. In an *ideal world*, there is a trusted party that communicates with $\mathbf{A}_I$ and $\mathbf{B}_I$ (subscript I indicates entities in the ideal world) through private channels and completes the desired task. We model the trusted party as a quantum interactive machine, and call it an *ideal functionality* $\mathcal{F}$. For example, in the secure evaluation of a function f, an ideal functionality $\mathcal{F}$ would take inputs (x, y) from $\mathbf{A}_I$ and $\mathbf{B}_I$ respectively, compute $(f_A, f_B) = f(x, y)$ and give f_A to $\mathbf{A}_I$ and f_B to $\mathbf{B}_I$. We then say a protocol Π securely realizes a given task, formulated by an ideal functionality $\mathcal{F}$, if for any adversary $\hat{\mathbf{B}}$ attacking a real-world execution, there exists an ideal-world adversary $\hat{\mathbf{B}}_I$ emulating "equivalent" attacks in the ideal-world. Equivalent means on any input state the output states of the players in the real world and ideal world are indistinguishable.

To be more specific, in the real world we initialize $\mathrm{S}_{\mathbf{A}}$, $\mathrm{S}_{\hat{\mathbf{B}}}$ and an auxiliary register W with a quantum state $\sigma_n \in \mathcal{S}_{\mathbf{A}}(n) \otimes \mathcal{S}_{\hat{\mathbf{B}}}(n) \otimes \mathcal{W}(n)$. Then $\mathbf{A}$ and $\hat{\mathbf{B}}$ interact, and end up with a state $\sigma'_n \in \mathcal{O}_{\mathbf{A}}(n) \otimes \mathcal{O}_{\hat{\mathbf{B}}}(n) \otimes \mathcal{W}(n)$. Finally a QTM $\mathcal{Z}$, which we call an *environment*, takes σ'_n as input and outputs one classical bit. Abstractly, we treat $\mathbf{A}$ and $\hat{\mathbf{B}}$ collectively as a noninteractive machine $M_{\hat{\mathbf{B}}}$ with state space $\mathcal{S}_{\mathbf{A}} \otimes \mathcal{S}_{\hat{\mathbf{B}}}$ and output space $\mathcal{O}_{\mathbf{A}} \otimes \mathcal{O}_{\hat{\mathbf{B}}}$. Analogously, for each ideal world adversary $\hat{\mathbf{B}}_I$, we can model $\mathbf{A}_I$, $\hat{\mathbf{B}}_I$ and $\mathcal{F}$ as a single QTM $M_{\hat{\mathbf{B}}_I}$ with state space $\mathcal{S}_{\mathbf{A}_I} \otimes \mathcal{S}_{\hat{\mathbf{B}}_I}$ and output space $\mathcal{O}_{\mathbf{A}_I} \otimes \mathcal{O}_{\hat{\mathbf{B}}_I}$. Then let $\mathbf{EXEC}_{\Pi,\hat{\mathbf{B}},\mathcal{Z}} := \{\mathcal{Z}((M_{\hat{\mathbf{B}}} \otimes \mathbb{1}_{\mathcal{W}(n)})\sigma_n)\}_{n\in\mathbb{N}}$ and $\mathbf{IDEAL}_{\mathcal{F},\hat{\mathbf{B}}_I,\mathcal{Z}} := \{\mathcal{Z}((M_{\hat{\mathbf{B}}_I} \otimes \mathbb{1}_{\mathcal{W}(n)})\sigma_n)\}_{n\in\mathbb{N}}$ be the binary distribution ensembles of $\mathcal{Z}$'s output in the real-world execution and in the ideal-world execution respectively. See Fig. 3.1 for an illustration of real-world and ideal-world executions.

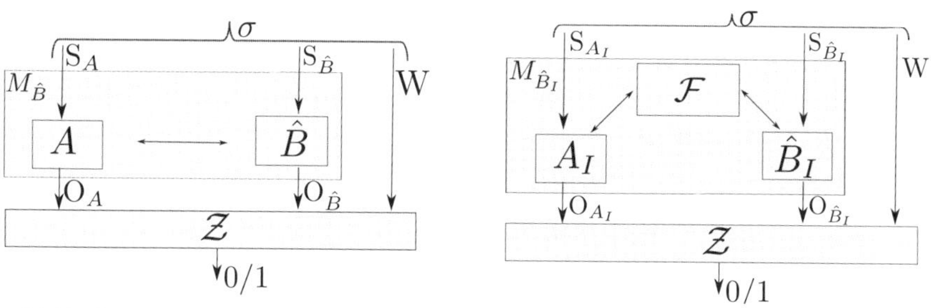

(a) Real-world execution $\mathbf{EXEC}_{\Pi,\hat{\mathbf{B}},\mathcal{Z}}$ (b) Ideal-world execution $\mathbf{IDEAL}_{\mathcal{F},\hat{\mathbf{B}}_I,\mathcal{Z}}$

Fig. 1. Real-world and Ideal-world Executions

Definition 3. *(Quantum Stand-alone Secure Emulation). Let $\mathcal{F}$ be a two-party functionality and let Π be a two-party protocol. We say Π quantum stand-alone-emulates $\mathcal{F}$, if for any poly-time QIM $\hat{\mathbf{B}}$, there is a poly-time QIM $\hat{\mathbf{B}}_I$, such that for any poly-time QTM $\mathcal{Z}$, and for any $\sigma = \{\sigma_n : \sigma_n \in \mathcal{S}_{\mathbf{A}}(n) \otimes \mathcal{S}_{\hat{\mathbf{B}}}(n) \otimes \mathcal{W}(n)\}_{n\in\mathbb{N}}$, $\mathbf{EXEC}_{\Pi,\hat{\mathbf{B}},\mathcal{Z}} \approx \mathbf{IDEAL}_{\mathcal{F},\hat{\mathbf{B}}_I,\mathcal{Z}}$.*

Remark. (I) Equivalently, the definition can be formulated as: for any $\hat{\mathbf{B}}$, there exists $\hat{\mathbf{B}}_I$, such that QTMs $M_{\hat{\mathbf{B}}}$ and $M_{\hat{\mathbf{B}}_I}$ are indistinguishable, as per Definition 2 restricting to non-interactive machines. (II) We focus on computational security in this work, and the model extends to information-theoretical setting straightforwardly. (III) We stress that

σ not only encodes the inputs to the players, but also contains auxiliary system $\mathcal{W}$ that might be entangled with the inputs and moreover serves as quantum advice to later assist $\mathcal{Z}$ in distinguishing the two worlds. There are other possible choices in the definition, e.g., disallowing auxiliary system $\mathcal{W}$ and only giving $\mathcal{Z}$ classical advice, which may give rise to variants that coincide with or subsume existing models. See the full version for a thorough discussion. (IV) For technical reasons, we require functionalities to be *well-formed* and protocols to be *nontrivial*, which are satisfied by all functionalities and protocols in our paper. (See [13, Sect.3] for details.) Aside from that, $\mathcal{F}$ could be as general as randomized, reactive, and evaluating quantum circuits, though in this work we concentrate on SFE of classical functions.

3.2 Modular Composition

It is common practice in the design of large protocols that we break a given task into subtasks, accomplish these subtasks and then use these modules as building blocks (subroutines) in a solution for the initial task. We formalize this paradigm by *hybrid models*[2]. A protocol in the *$\mathcal{G}$-hybrid model*, denoted $\Pi^{\mathcal{G}}$, has access to a trusted party that implements ideal functionality $\mathcal{G}$. As before, for each adversary $\hat{\mathbf{B}}_H$ (subscript H indicates entities in a hybrid model) in the $\mathcal{G}$-hybrid model, we can define $M_{\hat{\mathbf{B}}_H}$ and $\mathbf{EXEC}_{\Pi^{\mathcal{G}},\hat{\mathbf{B}}_H,\mathcal{Z}}$ likewise. Then we say $\Pi^{\mathcal{G}}$ quantum stand-alone-emulates ideal functionality $\mathcal{F}$ in the $\mathcal{G}$-hybrid model if $\mathbf{EXEC}_{\Pi^{\mathcal{G}},\hat{\mathbf{B}}_H,\mathcal{Z}} \approx \mathbf{IDEAL}_{\mathcal{F},\hat{\mathbf{B}}_I,\mathcal{Z}}$ for all poly-time QTMs $\mathcal{Z}$ and all σ.

Now suppose we have $\Pi_1^{\mathcal{G}}$ in the $\mathcal{G}$-hybrid model and a protocol Π_2 realizing $\mathcal{G}$. The operation of replacing an invocation of $\mathcal{G}$ with an invocation of Π_2 is done in the natural way: machines in Π_1 initialize machines in Π_2 and pause; machines in Π_2 execute Π_2 and generate outputs; then Π_1 resumes with these outputs. We denote the composed protocol $\Pi_1^{\Pi_2}$.

Theorem 1. *(Modular Composition Theorem) Let $\Pi_1^{\mathcal{G}}$ be a two-party protocol that quantum stand-alone-emulates $\mathcal{F}$ in the $\mathcal{G}$-hybrid model and let Π_2 be a two-party protocol that quantum stand-alone-emulates $\mathcal{G}$. Then the composed protocol $\Pi_1^{\Pi_2}$ quantum stand-alone-emulates $\mathcal{F}$.*

Remark. See the full version for its proof. It is easy to extend our analysis to a more general case where Π can invoke $\mathcal{G}$ multiple times and also access polynomially many ideal functionalities $(\mathcal{G}_1, \mathcal{G}_2, \ldots)$. However, we stress that at each round, only one functionality is invoked for at most once.

4 Quantum Stand-Alone-Secure ZK Arguments of Knowledge

A very important building block in cryptographic protocols is *Zero-Knowledge Arguments of Knowledge* (ZKAoK) for NP, formulated below as the ideal functionality $\mathcal{G}_{ZK}$. In this section we provide a construction that quantum stand-alone-emulates $\mathcal{G}_{ZK}$. Let

[2] In contrast, we call it a *plain model* if there are no trusted parties and no trusted setup assumptions like common reference string or public-key infrastructure, etc.

$L \in \text{NP}$ and let $R_L = \{(x,w) | w \text{ is a witness of } x\}$. Assume the length of the witness is bounded above by a polynomial $w(n)$.

Ideal Functionality $\mathcal{G}_{ZK}$: prover $\mathbf{P}_I$; verifier $\mathbf{V}_I$; NP-relation R_L

- Upon receiving (x,w) from $\mathbf{P}_I$, $\mathcal{G}_{ZK}$ verifies $(x,w) \stackrel{?}{\in} R_L$. If yes, it sends x to $\mathbf{V}_I$; otherwise it halts.

Notice that this is indeed an argument of knowledge, since the prover has to explicitly show a valid witness to the trusted party.

Let $\mathcal{E} =$ (**Gen**, **Enc**, **Dec**) be a cryptosystem as in Assumption 3.

ZKAoK Protocol Π_{ZK}

Phase 1

1. **V** chooses $a \leftarrow \{0,1\}^n$ at random, and sends **P** a commitment of a: $c = \mathbf{comm}(a)$.
2. **P** sends $b \leftarrow \{0,1\}^n$ to **V**.
3. **V** sends **P** string a.
4. **V** proves to **P** that c is indeed a commitment of a using Watrous's ZK protocol.
5. **P** and **V** set $pk = a \oplus b$ and interpret it as a public key.

Phase 2

1. **P**, holding an instance x and a witness w, encrypts w under pk. Let $e = \mathbf{Enc}_{pk}(w)$. **P** sends (x,e) to **V**.
2. **P** proves to **V** that e encodes a witness of x using Watrous's ZK protocol. **V** outputs x if it accepts in this ZK protocol. Otherwise it halts.

Theorem 2. *Protocol* Π_{ZK} *quantum stand-alone-emulates* $\mathcal{G}_{ZK}$.

The key idea lies in the inherent power of the simulator **S** of Watrous's ZK protocol. Namely, we can use **S** to generate a bogus proof that is indistinguishable from a real ZK proof run by a prover and a verifier, when we don't know a witness of a statement, or even when there isn't one, i.e., the statement is false. Specifically, an ideal-world $\hat{\mathbf{V}}_I$, receiving a true statement x from $\mathcal{G}_{ZK}$, needs to convince $\hat{\mathbf{V}}$ of the validity of x without knowing a witness. We do know that on true instances, i.e., the ciphertext e indeed encodes a witness w, **S** simulates a proof successfully by definition. The trouble then boils down to generating an encryption of w without knowing w. This might sound contradictory, but it is actually very natural. For instance, suppose a function f maps all strings to 0, then generating $f(r)$ without knowing r is trivial–just output 0! Our situation is more sophisticated, yet shares the same spirit. We need the fact that encryptions under a uniform string are statistically close. This implies, in particular, that encryption of any string under a uniform string pk, will coincide with $\mathbf{Enc}_{pk}(w)$ with high probability. In addition, if we let $\hat{\mathbf{V}}_I$ play an honest prover in Phase 1, the outcome pk will be guaranteed uniformly random. This shows how we handle corrupted verifiers.

On the other hand, in the case of a corrupted prover $\hat{\mathbf{P}}$, an ideal-world $\hat{\mathbf{P}}_I$ needs to extract a witness w from e when $\hat{\mathbf{P}}$ provides an accepting proof in Phase 2. The trick is that $\hat{\mathbf{P}}_I$ can use **S** to cheat in Phase 1 and force the outcome to be a real public key pk of which he knows a corresponding secret key sk, so that $\hat{\mathbf{P}}_I$ can decrypt e to recover w in the end. The difficulty is that $\hat{\mathbf{P}}_I$ wants to make $a = pk \oplus b$, but it has to commit to a before seeing b. It turns out we could commit to 0^n, and later run **S** on the *false* statement that $\mathbf{comm}(0^n)$ is a commitment of a. **S** must behave equally well as if it is given a true statement $(\mathbf{comm}(a), a)$, because otherwise **S** will break the hiding property of the commitment scheme. The formal proof can be found in the full version.

5 Classical Protocols with Quantum UC Security

In this section, we investigate classical protocols in the quantum Universal Composability (UC) model. We propose a framework, *simple hybrid arguments*, to capture a large family of classical security analyses that also go through against quantum adversaries (under reasonable computational assumptions). Applying our framework to the classical results of Canetti et el. [13], we get classical protocols that quantum UC-securely realize two-party SFE in the $\mathcal{G}_{ZK}$-hybrid model.

Universally Composable (UC) security, proposed in the classical context by Canetti [11], differs from the stand-alone definition of security in that the environment is allowed to be *interactive*: during the execution of the protocol, the environment may provide inputs and receive the outputs of the honest players, and exchange arbitrary messages with the adversary. In contrast, the environment in the stand-alone model runs only at the end of the protocol execution (and, implicitly, before the protocol starts, to prepare the inputs to all parties). UC-secure protocols enjoy a property called *general* (or *universal*) *composition*[3]: loosely speaking, the protocol remains secure even if it is run concurrently with an unbounded number of other arbitrary protocols (whereas proofs of security in the stand-alone model only guarantee security when only a single protocol at a time is running).

Earlier work on defining UC security and proving universal composition in the quantum setting appears in [7,45]. We will adapt the somewhat simpler formalism of Unruh [47]. Modulo a small change in Unruh's model (quantum advice, discussed below), our stand-alone model is exactly the restriction of Unruh's model to a *non-interactive* environment, that is one which is idle from the start to the finish of the protocol. [4]

We make one change to Unruh's model in order to be consistent with our earlier definitions and the work of Watrous on zero-knowledge [49]: we allow the environment to take quantum advice, rather than only classical advice. See the full version for details. This modification of Unruh's definition does not change the proof of the universal composition theorem:

[3] There is a distinction between UC security (a definition that may be satisfied by a specific protocol and ideal functionality) and universal composition (a property of the class of protocols that satisfy a security definition). Not all definitions that admit universal composition theorems are equivalent to UC security. See [30,34] for discussion.

[4] The only apparent difference in the models is that in the UC model, the environment runs for some time before the protocol starts to prepare inputs, while in Section 3.1 we simply quantify over all joint states σ of the honest players' and adversary's inputs and the auxiliary input W to the distinguisher. This difference is only cosmetic, though: the state σ can be taken to be the joint state of the outputs and internal memory of the environment at the time the protocol begins.

Theorem 3. *(Quantum UC Composition Theorem [47, Theorem 11]) Let Π_1, Π_2 and Π be quantum-polynomial-time protocols. Assume that Π_1 quantum UC-emulates Π_2. Then Π^{Π_1} quantum UC-emulates Π^{Π_2}.*

5.1 Classical Proofs for Quantum Adversaries: Simple Hybrid Argument

The goal of this section is to analyze a class of protocols, including the protocol of Canetti *et al.* [13] for two- and multi-party computation (referred to in the sequel as CLOS). These are classical protocols, proven secure in the classical UC model. We will show that these protocols remain secure in the presence of quantum adversaries as long as the underlying primitives (pseudorandom generators and a special kind of public-key encryption scheme) are secure against quantum adversaries. Specifically, we show:

Theorem 4. *Let $\mathcal{F}$ be a well-formed two-party functionality. Under Assumptions 1 and 2, there exists a nontrivial classical protocol that UC-emulates $\mathcal{F}$ in the $\mathcal{G}_{ZK}$-hybrid model in the presence of polynomial-time malicious, static* quantum *adversaries.*

To prove Theorem 4, we propose an abstraction that captures a family of classical security arguments in the UC model which remains valid in the quantum setting (as long as the underlying primitives are secure against quantum adversaries).

We use the term *experiment* loosely to describe a well-defined probability experiment which results in 0 or 1. The arguments described here could also be cast in the more stringent formalism of code-based games [4]; however, because the experiments we use are ultimately fairly simple, we have chosen a less formal exposition.

We'll use the following fact about UC-secure protocols, classical [11, Claim 10] and quantum [47, Lemma 10]: the adversary can be taken to be a "dummy" adversary, which simply relays messages faithfully to and from the environment without doing any actual processing. Because we will only discuss protocols with classical communication, we can assume w.l.o.g. that the adversary in our experiments is a known, classical machine; in particular, all quantum processing can be deferred to the environment. Note that ideal world adversaries will also be classical. Consequently, we can treat the process external to the environment as a whole, and view it as a classical interactive machine M. Namely, we let M describe the process $\langle \mathit{world}, \mathit{dummy\text{-}adv} \rangle$ or $\langle \mathit{world}, \mathit{simulator} \rangle$ where world is an ideal world, a real world or an execution in a hybrid model. (Recall that $\langle M_1, M_2 \rangle$ denotes the interaction between M_1 and M_2. It is itself an interactive machine whose inputs are the inputs expected by M_1 and M_2 together with messages expected by M_1 and M_2 from other entities. The outputs of $\langle M_1, M_2 \rangle$ are the outputs of M_1 and M_2 together with any messages sent by them to other entities.) Thus, all the experiments (real-world executions, ideal-world executions with simulators or without, executions in hybrid models, etc) we will analyze in this section have the form $\langle M, \mathcal{Z} \rangle$, where M is a classical interactive machine *which depends only on the protocol description* as we described above and $\mathcal{Z}$ is an adversarial environment.

Definition 4 (Simply related machines). *We say two QIMs M_a and M_b are (t, ϵ)-*simply related *if there is a classical time-t machine M and a pair of classical distributions D_a, D_b such that*

1. *$M(D_a) \equiv M_a$ (for two QIMs N_1 and N_2, we say $N_1 \equiv N_2$ if the two machines behave identically on all inputs, that is, if they can be described by the same circuits),*
2. *$M(D_b) \equiv M_b$, and*
3. *$D_a \approx_Q^{2t,\epsilon} D_b$.*

Definition 5 (Simple hybrid argument). *Two machines M_0 and M_ℓ are related by a* (t,ϵ)-simple hybrid argument of length ℓ *if there is a sequence of intermediate machines $M_1, M_2, ..., M_{\ell-1}$ such that each adjacent pair M_{i-1}, M_i of machines, $i = 1,\ldots,\ell$, is $(t, \frac{\epsilon}{\ell})$-simply related.*

Lemma 1. *For any t, ϵ and ℓ, if two machines are related by a (t,ϵ)-simple hybrid argument of length ℓ, then the machines are (t,ϵ)-interactively indistinguishable.*

Proofs of all the statements from this section are deferred to the full version.

Observation 5 (CLOS proof structure). *Except for the proof of security of protocol compilation from semi-honest to malicious adversaries, all the security proofs for static adversaries in CLOS consist of either (a) simple hybrid arguments with $t = poly(n)$ and $\epsilon = negl(n)$, or (b) applications of the UC composition theorem.*

Moreover, the underlying indistinguishable distributions in the CLOS arguments consist of either (i) switching between a real public key and a uniformly random string, (ii) changing the plaintext of an encryption, or (iii) changing the message in the commit phase of a commitment protocol.

From this observation, we get the corollary below, where $\mathcal{G}_{CP}$ denotes the "commit-and-prove" functionality of Canetti *et al.* [13, Figure 8].

Corollary 6 (CLOS—simple hybrids). *Under Assumptions 1 and 2,*

1. *In the $\mathcal{G}_{ZK}$-hybrid model, there is a nontrivial protocol that UC-emulates $\mathcal{G}_{CP}$ in the presence of polynomial-time* malicious *static* quantum *adversaries.*
2. *Let $\mathcal{F}$ be a well-formed two-party functionality. In the plain model, there is a protocol that UC-emulates $\mathcal{F}$ in the presence of polynomial-time* semi-honest *static* quantum *adversaries.*

It remains to discuss the proof of the security of the compiler from semi-honest to malicious adversaries in the $\mathcal{G}_{CP}$ model. The proof structure is only slightly different from the hybrid proofs above. Let Π be a protocol designed for the semi-honest model and let $Comp(\Pi)$ be the result of applying the CLOS compiler to Π to get a protocol in the (malicious) $\mathcal{G}_{CP}$-hybrid model. We use the following result from Canetti *et al.* [13]:

Proposition 7 (Canetti *et al.* [13, Proposition 8.1]). *Let Π be any real-world protocol designed for the semi-honest model. For every classical adversary $\hat{\mathbf{B}}$, there exists a classical adversary $\hat{\mathbf{B}}'$ with running time polynomial in that of $\hat{\mathbf{B}}$ such that the interaction of $\hat{\mathbf{B}}$ with honest players running $Comp(\Pi)$ in the $\mathcal{G}_{CP}$-hybrid model is* identical *to the interaction of $\hat{\mathbf{B}}'$ with Π in the semi-honest model; that is, $\langle Comp(\Pi), \hat{\mathbf{B}}\rangle \equiv \langle \Pi, \hat{\mathbf{B}}'\rangle$.*

Combining the previous proposition with the simpler arguments from CLOS (Corollary 6, above) we can prove Theorem 4. See the full version for further details.

6 Equivalence between $\mathcal{G}_{ZK}$ and $\mathcal{G}_{CF}$

In this section, we sketch the UC equivalence of zero-knowledge and coin-flipping in the quantum setting. The fact that coin-flipping can be realized in the $\mathcal{G}_{ZK}$ hybrid model follows from the general result of CLOS, discussed in the previous section. In the full version, we also give a direct construction of coin-flipping from ZK inspired by the parallel coin-flipping protocol of Lindell [33]. The direct construction relies only on the assumption of a quantum-secure PRG. More interestingly, we give a construction of $\mathcal{G}_{ZK}$ in the $\mathcal{G}_{CF}$-hybrid model which resists attacks by quantum adversaries.

Proposition 8. *1. Under Assumption 1, there is a constant-round protocol $\Pi_{CF}^{\mathcal{G}_{ZK}}$ that quantum UC-emulates $\mathcal{G}_{CF}$ in the $\mathcal{G}_{ZK}$-hybrid model.*

2. Under Assumptions 1 and 2, there is a constant-round protocol $\Pi_{ZK}^{\mathcal{G}_{CF}}$ that quantum UC-emulates $\mathcal{G}_{ZK}$ in the $\mathcal{G}_{CF}$-hybrid model.

This implies that in the stand-alone model, it suffices to construct a secure (simulatable) coin-flipping protocol to obtain secure SFE protocols for arbitrary functions. This gives a different avenue for constructing secure protocols, which might produce protocols that rely on assumptions weaker than (or incomparable to) those in our work, or that use fewer rounds. The related work of Lunemann and Nielsen [36] starts by constructing a coin-flipping protocol rather than a ZKAoK, though they rely on assumptions very similar to ours and have similar round complexity.

Our $\Pi_{ZK}^{\mathcal{G}_{CF}}$ protocol uses a standard transformation to get a ZKAoK from a witness-indistinguishable (WI) proof system in the CRS model. The main technical step in our analysis is showing that Blum's 3-round ZK protocol for Hamiltonian Cycle is in fact WI against a *malicious* quantum adversary. Our proof avoids rewinding, and is reminiscent of proofs that certain WI protocols can be composed concurrently. Details can be found in the full version.

7 Applications and Discussions

We first recap the results that we have obtained so far and derive a couple of straightforward yet important corollaries about two-party SFE in presence of quantum attacks.

1. Under Assumptions 1 and 2, for any well-formed two-party functionality $\mathcal{F}$, there is a classical protocol $\Pi^{\mathcal{G}_{ZK}}$ quantum UC-emulating $\mathcal{F}$ in the $\mathcal{G}_{ZK}$-hybrid model. (Theorem 4)
2. $\mathcal{G}_{ZK}$ and $\mathcal{G}_{CF}$ are equivalent in the quantum UC model. (Prop. 8)
3. There exists classical protocol Π_{ZK} that quantum stand-alone-emulates $\mathcal{G}_{ZK}$. (Theorem 2)

Applying modular composition theorem in the stand-alone model to item 1 and 3 we have:

Corollary 9. *For any well-formed classical two-party functionality $\mathcal{F}$, there exists a classical protocol Π that quantum stand-alone-emulates $\mathcal{F}$ with no set-up assumptions.*

Note that item 2 immediately implies equivalence of $\mathcal{G}_{ZK}$ and $\mathcal{G}_{CF}$ in the quantum stand-alone model. Combining with item 3 we get:

Corollary 10. *There exists a classical protocol Π_{CF} that quantum stand-alone-emulates $\mathcal{G}_{CF}$ with no set-up assumptions.*

Discussion. Our work suggests a number of straightforward conjectures. For example, it is likely that our techniques in fact apply to all the results in CLOS (multi-party, adaptive adversaries) and to corresponding results in the "generalized" UC model [12]. Essentially all protocols in the semi-honest model seem to fit the simple hybrids framework, in particular protocols based on Yao's garbled-circuits framework (e.g. [3]). It is also likely that existing proofs in security models which allow super-polynomial simulation (e.g., [40,41,1]) will carry through using a similar line of argument to the one here.

However, our work leaves open some basic questions: for example, can we construct constant-round ZK with negligible completeness and soundness errors against quantum verifiers? Watrous's technique does not immediately answer it since sequential repetition seems necessary in his construction to reduce the soundness error. A quick look at classical constant-round ZK (e.g., [24]) suggests that witness-indistinguishable proofs of knowledge are helpful. Is it possible to construct constant-round witness-extendable WI proofs of knowledge? Do our analyses apply to extensions of the UC framework, such the *generalized UC* framework of Canetti *et al.* [12]? Finally, more generally, which other uses of rewinding can be adapted to quantum adversaries? Aside from the original work by Watrous [49], Damgård and Lunemann [21] and Unruh [46] have shown examples of such adaption.

Acknowledgments. This work was informed by insightful discussions with many colleagues, notably Michael Ben-Or, Claude Crépeau, Ivan Damgård and Daniel Gottesman. Several of the results were obtained while A.S. was at the Institute for Pure and Applied Mathematics (IPAM) at UCLA in the fall of 2006. He gratefully acknowledges Rafi Ostrovksy and the IPAM staff for making his stay there pleasant and productive.

References

1. Barak, B., Sahai, A.: How to play almost any mental game over the net - concurrent composition via super-polynomial simulation. In: FOCS, pp. 543–552. IEEE, Los Alamitos (2005)
2. Beaver, D.: On deniability in quantum key exchange. In: Knudsen, L.R. (ed.) EUROCRYPT 2002. LNCS, vol. 2332, pp. 352–367. Springer, Heidelberg (2002)
3. Beaver, D., Micali, S., Rogaway, P.: The round complexity of secure protocols. In: STOC, pp. 503–513. ACM, New York (1990)
4. Bellare, M., Rogaway, P.: The security of triple encryption and a framework for code-based game-playing proofs. In: Vaudenay, S. (ed.) EUROCRYPT 2006. LNCS, vol. 4004, pp. 409–426. Springer, Heidelberg (2006)
5. Ben-Or, M., Crépeau, C., Gottesman, D., Hassidim, A., Smith, A.: Secure multiparty quantum computation with (only) a strict honest majority. In: FOCS, pp. 249–260. IEEE, Los Alamitos (2006)
6. Ben-Or, M., Horodecki, M., Leung, D.W., Mayers, D., Oppenheim, J.: The universal composable security of quantum key distribution. In: Kilian, J. (ed.) TCC 2005. LNCS, vol. 3378, pp. 386–406. Springer, Heidelberg (2005)

7. Ben-Or, M., Mayers, D.: General security definition and composability for quantum and classical protocols, arxiv:quant-ph/0409062v2 (September 2004)
8. Bennett, C.H., Brassard, G., Crépeau, C., Skubiszewska, M.-H.: Practical quantum oblivious transfer. In: Feigenbaum, J. (ed.) CRYPTO 1991. LNCS, vol. 576, pp. 351–366. Springer, Heidelberg (1992)
9. Brassard, G., Crépeau, C.: Quantum bit commitment and coin tossing protocols. In: Menezes, A., Vanstone, S.A. (eds.) CRYPTO 1990. LNCS, vol. 537, pp. 49–61. Springer, Heidelberg (1991)
10. Canetti, R.: Security and composition of multiparty cryptographic protocols. J. Cryptology 13(1), 143–202 (2000)
11. Canetti, R.: Universally composable security: A new paradigm for cryptographic protocols. In: FOCS, pp. 136–145. IEEE, Los Alamitos (2001)
12. Canetti, R., Dodis, Y., Pass, R., Walfish, S.: Universally composable security with global setup. In: Vadhan, S.P. (ed.) TCC 2007. LNCS, vol. 4392, pp. 61–85. Springer, Heidelberg (2007)
13. Canetti, R., Lindell, Y., Ostrovsky, R., Sahai, A.: Universally composable two-party and multi-party secure computation. In: STOC, pp. 494–503. ACM, New York (2002)
14. Crépeau, C., Dumais, P., Mayers, D., Salvail, L.: Computational collapse of quantum state with application to oblivious transfer. In: Naor, M. (ed.) TCC 2004. LNCS, vol. 2951, pp. 374–393. Springer, Heidelberg (2004)
15. Crépeau, C., Gottesman, D., Smith, A.: Secure multi-party quantum computation. In: STOC, pp. 643–652. ACM, New York (2002)
16. Crépeau, C., Gottesman, D., Smith, A.: Approximate quantum error-correcting codes and secret sharing schemes. In: Cramer, R. (ed.) EUROCRYPT 2005. LNCS, vol. 3494, pp. 285–301. Springer, Heidelberg (2005)
17. Crépeau, C., Salvail, L., Simard, J.-R., Tapp Classical, A.: quantum strategies for two-prover bit commitments. In: Quantum Information Processing, QIP (2006), `http://crypto.cs.mcgill.ca/~crepeau/PDF/CSST06.pdf`
18. Damgård, I., Fehr, S., Lunemann, C., Salvail, L., Schaffner, C.: Improving the security of quantum protocols via commit-and-open. In: Halevi, S. (ed.) CRYPTO 2009. LNCS, vol. 5677, pp. 408–427. Springer, Heidelberg (2009), Full version at arXiv:0902.3918v4
19. Damgård, I., Fehr, S., Salvail, L., Schaffner, C.: Secure identification and qkd in the bounded-quantum-storage model. In: Menezes, A. (ed.) CRYPTO 2007. LNCS, vol. 4622, pp. 342–359. Springer, Heidelberg (2007)
20. Damgård, I., Fehr, S., Salvail, L., Schaffner, C.: Cryptography in the bounded-quantum-storage model. SIAM J. Comput. 37(6), 1865–1890 (2008)
21. Damgård, I., Lunemann, C.: Quantum-secure coin-flipping and applications. In: Matsui, M. (ed.) ASIACRYPT 2009. LNCS, vol. 5912, pp. 52–69. Springer, Heidelberg (2009)
22. Dumais, P., Mayers, D., Salvail, L.: Perfectly concealing quantum bit commitment from any quantum one-way permutation. In: Preneel, B. (ed.) EUROCRYPT 2000. LNCS, vol. 1807, pp. 300–315. Springer, Heidelberg (2000)
23. Fehr, S., Schaffner, C.: Composing quantum protocols in a classical environment. In: Reingold, O. (ed.) TCC 2009. LNCS, vol. 5444, pp. 350–367. Springer, Heidelberg (2009)
24. Feige, U., Shamir, A.: Zero knowledge proofs of knowledge in two rounds. In: CRYPTO, pp. 526–544. Springer, Heidelberg(1990)
25. Goldreich, O., Micali, S., Wigderson, A.: How to play any mental game. In: STOC, pp. 218–229. ACM, New York (1987)
26. Goldreich, O., Micali, S., Wigderson, A.: Proofs that yield nothing but their validity for all languages in np have zero-knowledge proof systems. J. ACM 38(3), 691–729 (1991)
27. Goldwasser, S., Micali, S., Rackoff, C.: The knowledge complexity of interactive proof systems. SIAM J. Comput. 18, 186–208 (1989)

28. Hallgren, S.: Polynomial-time quantum algorithms for Pell's equation and the principal ideal problem. J. ACM 54(1), 1–19 (2007)
29. Hallgren, S., Kolla, A., Sen, P., Zhang, S.: Making classical honest verifier zero knowledge protocols secure against quantum attacks. In: Aceto, L., Damgård, I., Goldberg, L.A., Halldórsson, M.M., Ingólfsdóttir, A., Walukiewicz, I. (eds.) ICALP 2008, Part II. LNCS, vol. 5126, pp. 592–603. Springer, Heidelberg (2008)
30. Hofheinz, D., Unruh, D.: Simulatable security and polynomially bounded concurrent composability. In: Symposium on Security and Privacy, pp. 169–183. IEEE, Los Alamitos (2006)
31. Kol, G., Naor, M.: Games for exchanging information. In: STOC, pp. 423–432. ACM, New York (2008)
32. Kushilevitz, E., Lindell, Y., Rabin, T.: Information-theoretically secure protocols and security under composition. SIAM J. Comput. 39(5), 2090–2112 (2010)
33. Lindell, Y.: Parallel coin-tossing and constant-round secure two-party computation. J. Cryptology 16(3), 143–184 (2003)
34. Lindell, Y.: General composition and universal composability in secure multiparty computation. J. Cryptology 22(3), 395–428 (2009)
35. Lo, H.-K., Chau, H.F.: Unconditional security of quantum key distribution over arbitrarily long distances. Science 283(5410), 2050–2056 (1999)
36. Lunemann, C., Nielsen, J.B.: Fully simulatable quantum-secure coin-flipping and applications. In: Africacrypt (February 2011); arXiv:1102.0887
37. Mayers, D.: Unconditional security in quantum cryptography. J. ACM 48(3), 351–406 (2001)
38. Naor, M.: Bit commitment using pseudorandomness. J. Cryptology 4(2), 151–158 (1991)
39. Nielsen, M.A., Chuang, I.L.: Quantum Computation and Quantum Information. Cambridge University Press, Cambridge (2000)
40. Pass, R.: Simulation in quasi-polynomial time, and its application to protocol composition. In: Biham, E. (ed.) EUROCRYPT 2003. LNCS, vol. 2656, pp. 160–176. Springer, Heidelberg (2003)
41. Prabhakaran, M., Sahai, A.: New notions of security: achieving universal composability without trusted setup. In: STOC, pp. 242–251. ACM, New York (2004)
42. Regev, O.: On lattices, learning with errors, random linear codes, and cryptography. J. ACM 56(6) (2009); Preliminary version in STOC 2005
43. Shor, P.W.: Polynomial-time algorithms for prime factorization and discrete logarithms on a quantum computer. SIAM J. Comput. 26(5), 1484–1509 (1997)
44. Shor, P.W., Preskill, J.: Simple proof of security of the BB84 quantum key distribution protocol. Phys. Rev. Lett. 85(2), 441–444 (2000)
45. Unruh, D.: Simulatable security for quantum protocols, arXiv:quant-ph/0409125v2 (2004)
46. Unruh, D.: Quantum proofs of knowledge, IACR ePrint 2010/212 (April 2010)
47. Unruh, D.: Universally composable quantum multi-party computation. In: Gilbert, H. (ed.) EUROCRYPT 2010. LNCS, vol. 6110, pp. 486–505. Springer, Heidelberg (2010); arXiv:0910.2912v1
48. Unruh, D.: Concurrent composition in the bounded quantum storage model. In: Paterson, K.G. (ed.) EUROCRYPT 2011. LNCS, vol. 6632, pp. 467–486. Springer, Heidelberg (2011)
49. Watrous, J.: Zero-knowledge against quantum attacks. SIAM J. Comput. 39(1), 25–58 (2009); Preliminary version in STOC 2006
50. Yao, A.C.-C.: Quantum circuit complexity. In: FOCS, pp. 352–361. IEEE, Los Alamitos (1993)

Position-Based Quantum Cryptography: Impossibility and Constructions

Harry Buhrman[1,2,⋆], Nishanth Chandran[3,⋆⋆], Serge Fehr[1], Ran Gelles[3,⋆⋆], Vipul Goyal[4], Rafail Ostrovsky[3,⋆⋆⋆], and Christian Schaffner[2,1,†]

[1] Centrum Wiskunde & Informatica (CWI), The Netherlands
[2] University of Amsterdam, The Netherlands
[3] University of California (UCLA), CA, USA
[4] Microsoft Research, Bangalore, India

Abstract. The aim of position-based cryptography is to use the geographical position of a party as its only credential. In this work, we study position-based cryptography in the quantum setting.

We show that if collaborating adversaries are allowed to pre-share an arbitrarily large entangled quantum state, then position-verification, and as a consequence position-based cryptography in general, is *impossible* (also) in the quantum setting.

To this end, we prove that with the help of sufficient pre-shared entanglement, any non-local quantum computation, i.e., any computation that involves quantum inputs from two parties at different locations, can be performed *instantaneously* and *without any communication*, up to local corrections that need to be applied to the outputs. The latter can be understood in that the parties obtain their respective outputs "encrypted", where each corresponding encryption key is known by the opposite party. This result generalizes to any number of parties, and it implies that any non-local quantum computation can be performed using a *single* round of mutual communication (in which the parties exchange the encryption keys), and that any position-verification scheme can be broken, assuming sufficient pre-shared entanglement among the adversaries.

On the positive side, we show that for adversaries that are restricted to not share any entangled quantum states, secure position-verification is achievable. Jointly, these results suggest the interesting question whether secure position-verification is possible in case of a bounded amount of entanglement. Our positive result can be interpreted as resolving this question in the simplest case, where the bound is set to zero.

⋆ Supported by a NWO VICI grant and the EU 7th framework grant QCS.

⋆⋆ Supported in part by NSF grants 0716835, 0716389, 0830803, and 0916574.

⋆⋆⋆ Supported in part by IBM Faculty Award, Xerox Innovation Group Award, the Okawa Foundation Award, Intel, Teradata, DARPA, BSF grant 2008411, NSF grants 0716835, 0716389, 0830803, 0916574 and U.C. MICRO grant.

† Supported by a NWO VENI grant.

P. Rogaway (Ed.): CRYPTO 2011, LNCS 6841, pp. 429–446, 2011.

1 Introduction

1.1 Background

The goal of *position-based cryptography* is to use the geographical position of a party as its only "credential". For example, one would like to send a message to a party at a geographical position *pos* with the guarantee that the party can decrypt the message only if he or she is physically present at *pos*. The general concept of position-based cryptography was introduced by Chandran, Goyal, Moriarty and Ostrovsky [1]; certain specific related tasks have been considered before under different names (see below and Sect. 1.3).

A central task in position-based cryptography is the problem of *position-verification*. We have a *prover* P at position *pos*, wishing to convince a set of *verifiers* $V_0, \ldots, V_k$ (at different points in geographical space) that P is indeed at that position *pos*. The prover can run an interactive protocol with the verifiers in order to convince them. The main technique for such a protocol is known as distance bounding [2]. In this technique, a verifier sends a random nonce to P and measures the time taken for P to reply back with this value. Assuming that the speed of communication is bounded by the speed of light, this technique gives an upper bound on the distance of P from the verifier.

The problem of secure positioning has been studied before in the field of wireless security, and there have been several proposals for this task ([2,3,4,5,6,7,8,9]). However, [1] shows that there exists no protocol for secure positioning that offers security in the presence of *multiple colluding* adversaries. In other words, the set of verifiers cannot distinguish between the case when they are interacting with an honest prover at *pos* and the case when they are interacting with multiple colluding dishonest provers, none of which is at position *pos*. Their impossibility result holds even if one makes computational hardness assumptions, and it also rules out most other interesting position-based cryptographic tasks.

In light of the strong impossibility result, [1] considers a setting that assumes restrictions on the parties' storage capabilities, called the Bounded-Retrieval Model (BRM) in the full version of [1], and constructs secure protocols for position-verification and for position-based key exchange (wherein the verifiers, in addition to verifying the position claim of a prover, also exchange a secret key with the prover). While these protocols give us a way to realize position-based cryptography, the underlying setting is relatively hard to justify in practice.

This leaves us with the question: is there any other assumption or setting in which position-based cryptography is realizable?

1.2 Our Approach and Our Results

In this work, we study position-based cryptography in the *quantum* setting. To start with, let us briefly explain why moving to the quantum setting might be useful. The impossibility result of [1] relies heavily on the fact that an adversary can locally store all information he receives *and* at the same time share this information with other colluding adversaries, located elsewhere. Recall that the positive result of [1] in the BRM circumvents the impossibility result by assuming that an adversary *cannot* store all information he receives. By considering the quantum setting, one may be able to circumvent the impossibility result thanks to the following observation. If some information is encoded into a quantum state, then the above attack fails due to the no-cloning principle: the adversary can either store the quantum state or send it to a colluding adversary (or do something in-between, like store part of it), but *not both.*

However, this intuition turns out to be not completely accurate. Once the adversaries pre-share entangled states, they can make use of quantum teleportation [10]. Although teleportation on its own does not appear to immediate conflict with the above intuition, we show that, based on techniques by Vaidman [11], adversaries holding a large amount of entangled quantum states can perform *instantaneous nonlocal quantum computation*, which in particular implies that they can compute any unitary operation on a state shared between them, using only local operations and *one* round of classical mutual communication. Based on this technique, we show how a coalition of adversaries can attack and break any position-verification scheme.

Interestingly, sharing entangled quantum systems is vital for attacking the position-verification scheme. We show that there exist schemes that are secure in the information-theoretic sense, if the adversary is not allowed to pre-share or maintain entanglement. Furthermore, we show how to construct secure protocols for several position-based cryptographic tasks: position-verification, authentication, and key exchange.

This leads to an interesting open question regarding the amount of pre-shared entanglement required to break the positioning scheme: the case of a large amount of pre-shared states yields a complete break of any scheme while having no pre-shared states leads to information-theoretically secure schemes. The threshold of pre-shared quantum systems that keeps the system secure is yet unknown.

1.3 Related Work

To the best of our knowledge, quantum schemes for position-verification have first been considered by Kent in 2002 under the name of "quantum tagging". Together with Munro, Spiller and Beausoleil, a patent for an (insecure) scheme was filed for HP Labs in 2004 and granted in 2006 [12]. Their results have not appeared in the academic literature until 2010 [13]. In that paper, they describe several basic schemes and describe how to break them using teleportation-based attacks. They propose other variations (Schemes IV–VI in [13]) not suspect to their teleportation attack and leave their security as an open question. Our general attack presented here shows that these schemes are insecure as well.

Concurrent and independent of our work reported here and the work on quantum tagging described above, the approach of using quantum techniques for secure position-verification was proposed by Malaney [14,15]. However, the proposed scheme is merely claimed secure, and no rigorous security analysis is provided. As pointed out in [13], Malaney's schemes can also be broken by a teleportation-based attack. Chandran et al. have proposed and proved a secure quantum scheme for position-verification [16]. However, their proof implicitly assumed that the adversaries have no pre-shared entanglement; as shown in [13], their scheme also becomes insecure without this assumption.

In a subsequent paper [17], Lau and Lo use similar ideas as in [13] to show the insecurity of position-verification schemes that are of a certain (yet rather restricted) form, which include the schemes from [14,15] and [16]. Furthermore, they propose a position-verification scheme that resists their attack, and they conjecture it secure. While these protocols might be secure if the adversaries do not pre-share entanglement, our attack shows that all of them are insecure in general.

In a recent note [18], Kent considers a different model for position-based cryptography where the prover's position is *not* his only credential, but he is assumed to additionally share with the verifiers a classical key unknown to the adversary. In this case, quantum key distribution can be used to expand that key ad infinitum. This classical key stream is then used as authentication resource.

The idea of performing "instantaneous measurements of nonlocal variables" has been put forward by Vaidman [11] and was further investigated by Clark et al. [19]. The concept of instantaneous nonlocal quantum computation presented here is an extension of Vaidman's task. After the appearance and circulation of our work, Beigi and König [20] used the technique of port-based teleportation by Ishizaka and Hiroshima [21,22] to reduce the amount of entanglement required to perform instantaneous nonlocal quantum computation (from our double exponential) to exponential.

In [23], Giovannetti et al. show how to measure the distance between two parties by quantum cryptographic means so that only trusted people have access to the result. This is a different kind of problem than what we consider here, and the techniques used there are not applicable in our setting.

1.4 Our Attack and Our Schemes in More Detail

Position-Verification - A Simple Approach. Let us briefly discuss here the 1-dimensional case in which we have two verifiers V_0 and V_1, and a prover P at position *pos* that lies on the straight line between V_0 and V_1. Now, to verify P's position, V_0 sends a BB84 qubit $H^\theta|x\rangle$ to P, and V_1 sends the corresponding basis θ to P. The sending of these messages is timed in such a way that $H^\theta|x\rangle$ and θ arrive at position *pos* at the same time. P then has to measure the qubit in basis θ to obtain x, and immediately send x to both V_0 and V_1, who verify the correctness of x and if it has arrived "in time".

The intuition for this scheme is the following. Consider a dishonest prover $\hat{P}_0$ between V_0 and P, and a dishonest prover $\hat{P}_1$ between V_1 and P. (It is not too hard to see that additional dishonest provers do not help.) When $\hat{P}_0$ receives the BB84 qubit, she does not know yet the corresponding basis θ. Thus, if she measures it immediately when she receives it, then she is likely to measure it in the wrong basis and $\hat{P}_0$ and $\hat{P}_1$ will not be able to provide the correct x. However, if she waits until she knows the basis θ, then $\hat{P}_0$ and $\hat{P}_1$ will be too late in sending x to V_1 in time. Similarly, if she forwards the BB84 qubit to $\hat{P}_1$, who receives θ before $\hat{P}_0$ does, then $\hat{P}_0$ and $\hat{P}_1$ will be too late in sending x to V_0. It seems that in order to break the scheme, $\hat{P}_0$ needs to store the qubit until she receives the basis θ and at the same time send a copy of it to $\hat{P}_1$. But this is excluded by the no-cloning principle.

The Attack and Instantaneous Nonlocal Quantum Computation. The above intuition turns out to be wrong. Using pre-shared entanglement, $\hat{P}_0$ and $\hat{P}_1$ can perform quantum teleportation which enables them (in some sense) to act coherently on the complete state immediately upon reception. Combining this with the observation by Kent et al. [13] that the Pauli-corrections resulting from the teleportation commute with the actions of the honest prover in the above protocol shows that colluding adversaries can perfectly break the protocol.

Much more generally, we will show how to break *any* position-verification scheme, possibly consisting of multiple (and interleaved) rounds. To this end, we will show how to perform *instantaneous nonlocal quantum computation*. In particular, we prove that any unitary operation U acting on a composite system shared between players can be computed using only a single round of mutual classical communication. Based on ideas by Vaidman [11], the players teleport quantum states back and forth many times in a clever way, *without* awaiting the classical measurement outcomes from the other party's teleportations.

Position-Verification in the No-PE Model. On the other hand, the above intuition is correct in the *no pre-shared entanglement* (No-PE) model, where the adversaries are not allowed to have pre-shared entangled quantum states prior the execution the protocol, or, more generally, prior the execution of each round of the protocol in case of multi-round schemes. Even though this model may be somewhat unrealistic and artificial, analyzing protocols in this setting serves as stepping stone to obtaining protocols which tolerate adversaries who pre-share and maintain some *limited* amount of entanglement. But also, rigorously proving security in the restrictive (for the adversary) No-PE model is already non-trivial and requires heavy machinery. Our proof uses the *strong complementary information trade-off* (CIT) due to Renes and Boileau [24], and it guarantees that for any strategy, the success probability of $\hat{P}_0$ and $\hat{P}_1$ is bounded by approximately 0.89. By repeating the above simple scheme sequentially, we get a secure multi-round positioning scheme with exponentially small soundness error. We note that when performing sequential repetitions in the No-PE model, the adversaries must enter each round with no entanglement; thus, they are not allowed to generate entanglement in one round, store it, and use it in the next round(s).

Position-based authentication and key-exchange in the No-PE Model. Based on (sequential repetitions of) our position-verification scheme in the No-PE model, we can also construct schemes for position-based authentication and for position-based key-exchange, and prove their security in the No-PE model. Due to space limitation, these schemes and their analyses only appear in the full version of this paper [25].

2 Preliminaries

2.1 Notation and Terminology

We assume familiarity with the basic concepts of quantum information theory and refer to [26] for an excellent introduction; we merely fix some notation here.

Qubits. A *qubit* is a quantum system A with a 2-dimensional state space $\mathcal{H}_A = \mathbb{C}^2$. The *computational basis* $\{|0\rangle, |1\rangle\}$ (for a qubit) is given by $|0\rangle = \binom{1}{0}$ and $|1\rangle = \binom{0}{1}$, and the *Hadamard basis* by $H\{|0\rangle, |1\rangle\} = \{H|0\rangle, H|1\rangle\}$, where H denotes the 2-dimensional *Hadamard matrix*, which maps $|0\rangle$ to $(|0\rangle + |1\rangle)/\sqrt{2}$ and $|1\rangle$ to $(|0\rangle - |1\rangle)/\sqrt{2}$. The state space of an n-qubit system $A = A_1 \cdots A_n$ is given by the 2^n-dimensional space $\mathcal{H}_A = (\mathbb{C}^2)^{\otimes n} = \mathbb{C}^2 \otimes \cdots \otimes \mathbb{C}^2$.

Since we mainly use the above two bases, we can simplify terminology and notation by identifying the computational basis $\{|0\rangle, |1\rangle\}$ with the bit 0 and the Hadamard basis $H\{|0\rangle, |1\rangle\}$ with the bit 1. Hence, when we say that an n-qubit state $|\psi\rangle \in (\mathbb{C}^2)^{\otimes n}$ is measured in basis $\theta \in \{0,1\}^n$, we mean that the state is measured qubit-wise where basis $H^{\theta_i}\{|0\rangle, |1\rangle\}$ is used for the i-th qubit. As a result of the measurement, the string $x \in \{0,1\}^n$ is observed with probability $|\langle\psi|H^\theta|x\rangle|^2$, where $H^\theta = H^{\theta_1} \otimes \cdots \otimes H^{\theta_n}$ and $|x\rangle = |x_1\rangle \otimes \cdots \otimes |x_n\rangle$.

An important example of a 2-qubit state is the *EPR pair*, which is given by $|\Phi_{AB}\rangle = (|0\rangle|0\rangle + |1\rangle|1\rangle)/\sqrt{2} \in \mathcal{H}_A \otimes \mathcal{H}_B = \mathbb{C}^2 \otimes \mathbb{C}^2$ and has the following properties: if qubit A is measured in the computational basis, then a uniformly random bit $x \in \{0,1\}$ is observed and qubit B collapses to $|x\rangle$. Similarly, if qubit A is measured in the Hadamard basis, then a uniformly random bit $x \in \{0,1\}$ is observed and qubit B collapses to $H|x\rangle$.

Teleportation. The goal of teleportation is to transfer a quantum state from one location to another by only communicating classical information. Teleportation requires pre-shared entanglement among the two locations. To teleport a qubit Q in an arbitrary unknown state $|\psi\rangle$ from Alice to Bob, Alice performs a Bell-measurement on Q and her half of an EPR-pair, yielding a classical measurement outcome $k \in \{0,1,2,3\}$. Instantaneously, the other half of the corresponding EPR pair, which is held by Bob, turns into the state $\sigma_k^\dagger|\psi\rangle$, where $\sigma_0, \sigma_1, \sigma_2, \sigma_3$ denote the four Pauli-corrections $\{\mathbb{I}, X, Z, XZ\}$, respectively, and $\sigma^\dagger$ denotes the complex conjugate of the transpose of σ. The classical information k is then communicated to Bob who can recover the state $|\psi\rangle$ by performing σ_k on his EPR half. Note that the operator σ_k is Hermitian, thus $\sigma_k^\dagger = \sigma_k$.

3 Setup and the Task of Position Verification

3.1 The Security Model

We informally describe the model we use for the upcoming sections, which is a quantum version of the Vanilla (standard) model introduced in [1] (see there for a full description). We also describe our restricted model used for our security proof, that we call the *no pre-shared entanglement* (No-PE) model. We consider entities $V_0, \ldots, V_k$ called *verifiers* and an entity P, the (honest) *prover*. Additionally, we consider a coalition $\hat{P}$ of *dishonest provers* (or *adversaries*) $\hat{P}_0, \ldots, \hat{P}_\ell$. All entities can perform arbitrary quantum (and classical) operations and can communicate quantum (and classical) messages among them.

For our positive results, we consider a restricted model, which prohibits entanglement between the dishonest verifiers. Specifically, the *No-PE model* is such that the dishonest provers enter every new round of communication, initiated by the verifiers, with no pre-shared entanglement. That is, in every round, a dishonest prover can send an entangled quantum state only *after* it receives the verifier's message, and the dishonest provers cannot maintain such an entangled state in order to use it in the next round. As mentioned in the introduction, considering this simple (but possibly unrealistic) model may help us in obtaining protocols that are secure against adversaries with *limited* entanglement.

For simplicity, we assume that quantum operations and communication are noise-free; however, our results generalize to the more realistic noisy case, assuming that the noise is low enough. We require that the verifiers have a private and authenticated channel among themselves, which allows them to coordinate their actions by communicating before, during or after protocol execution. We stress however, that this does not hold for the communication between the verifiers and P: $\hat{P}$ has full control over the destination of messages communicated between the verifiers and P (both ways). This in particular means that the verifiers do not know per-se if they are communicating with the honest or a dishonest prover (or a coalition of dishonest provers).

The above model is now extended by incorporating the notion of *time* and *space*. Each entity is assigned an arbitrary fixed position *pos* in the d-dimensional space $\mathbb{R}^d$, and we assume that messages to be communicated travel at fixed velocity v (e.g. with the speed of light), and hence the time needed for a message to travel from one entity to another equals the Euclidean distance between the two (assuming that v is normalized to 1). This holds for honest and dishonest entities. We assume on the other hand that local computations take no time.

Finally, we assume that the verifiers have precise and synchronized clocks, so that they can coordinate exact times for sending off messages and can measure the exact time of a message arrival. We do not require P's clock to be precise or in sync with the verifiers. However, we do assume that P cannot be reset.

This model allows to reason as follows. Consider a verifier V_0 at position pos_0, who sends a challenge ch_0 to the (supposedly honest) prover claiming to be at position *pos*. If V_0 receives a reply within time $2d(pos_0, pos)$, where $d(\cdot,\cdot)$ is the Euclidean distance measure in $\mathbb{R}^d$ and thus also measures the time a message takes from one point to the other, then V_0 can conclude that he is communicating with a prover that is within distance $d(pos_0, pos)$.

We stress that in our model, the honest prover P has no advantage over the dishonest provers beyond being at its position *pos*. In particular, P does not share any secret information with the verifiers, nor can he per-se authenticate his messages by any other means.

Throughout the article, we require that the honest prover P is *enclosed* by the verifiers $V_0, \ldots, V_k$ in that the prover's position $pos \in \mathbb{R}^d$ lies within the tetrahedron, i.e., convex hull, $\text{Hull}(pos_0, \ldots, pos_k) \subset \mathbb{R}^d$ formed by the respective positions of the verifiers. Note that in this work we consider only *stand-alone security*, i.e., there exists only a single execution with a single honest prover, and we do not guarantee concurrent security.

3.2 Secure Position Verification

A position-verification scheme should allow a prover P at position $pos \in \mathbb{R}^d$ (in d-dimensional space) to convince a set of $k+1$ verifiers $V_0, \ldots, V_k$, who are located at respective positions $pos_0, \ldots, pos_k \in \mathbb{R}^d$, that he is indeed at position pos. We assume that P is enclosed by $V_0, \ldots, V_k$. We require that the verifiers jointly accept if an honest prover P is at position pos, and we require that the verifiers reject with "high" probability in case of a dishonest prover that is not at position pos. The latter should hold even if the dishonest prover consist of a *coalition* of collaborating dishonest provers $\hat{P}_0, \ldots, \hat{P}_\ell$ at arbitrary positions $apos_0, \ldots, apos_\ell \in \mathbb{R}^d$ with $apos_i \neq pos$ for all i. We refer to [1] for the general formal definition of the completeness and security of a position-verification scheme. In this article, we mainly focus on position-verification schemes of the following form:

Definition 1. *A 1-round* ***position-verification*** *scheme* $\mathsf{PV} = (\mathsf{Chlg}, \mathsf{Resp}, \mathsf{Ver})$ *consists of the following three parts. A challenge generator* Chlg*, which outputs a list of challenges* $(ch_0, \ldots, ch_k)$ *and auxiliary information* x*; a response algorithm* Resp*, which on input a list of challenges outputs a list of responses* $(x'_0, \ldots, x'_k)$*; and a verification algorithm* Ver *with* $\mathsf{Ver}(x'_0, \ldots, x'_k, x) \in \{0,1\}$*.*

PV *is said to have* **perfect completeness** *if* $\mathsf{Ver}(x'_0, \ldots, x'_k, x) = 1$ *with probability 1 for* $(ch_0, \ldots, ch_k)$ *and* x *generated by* Chlg *and* $(x'_0, \ldots, x'_k)$ *by* Resp *on input* $(ch_0, \ldots, ch_k)$*.*

The algorithms Chlg, Resp and Ver are used as described in Fig. 1 to verify the claimed position of a prover P. We clarify that in order to have all the challenges arrive at P's (claimed) location pos *at the same time*, the verifiers agree on a time T and each V_i sends off his challenge ch_i at time $T - d(pos_i, pos)$. As a result, all ch_i's arrive at P's position pos at time T. In Step 3, V_i receives x'_i *in time* if x'_i arrives at V_i's position pos_i at time $T + d(pos_i, pos)$. Throughout the article, we use this simplified terminology. Furthermore, we are sometimes a bit sloppy in distinguishing a party, like P, from its location pos.

We stress that we allow Chlg, Resp and Ver to be *quantum* algorithms and ch_i, x and x'_i to be quantum information. In our constructions, only ch_0 will actually be quantum; thus, we will only require quantum communication from V_0 to P, all other communication is classical. Also, in our constructions, $x'_0 = \ldots = x'_k$, and $\mathsf{Ver}(x'_0, \ldots, x'_k, x) = 1$ exactly if $x'_i = x$ for all i.

Definition 2. *A 1-round position-verification scheme* $\mathsf{PV} = (\mathsf{Chlg}, \mathsf{Resp}, \mathsf{Ver})$ *is called* ε**-sound** *if for any position* $pos \in \text{Hull}(pos_0, \ldots, pos_k)$*, and any coalition*

Common input to the verifiers: their respective positions $pos_0, \ldots, pos_k$, and P's (claimed) position pos.

0. V_0 generates a list of challenges $(ch_0, \ldots, ch_k)$ and auxiliary information x using Chlg, and sends ch_i to V_i for $i = 1, \ldots, k$.
1. Every V_i sends ch_i to P in such a way that all ch_i's arrive at the same time at P's position pos.
2. P computes $(x'_0, \ldots, x'_k) := \mathsf{Resp}(ch_0, \ldots, ch_k)$ as soon as all the ch_i's arrive, and he sends x'_i to V_i for every i.
3. The V_i's jointly accept if and only if all V_i's receive x'_i in time and $\mathsf{Ver}(x'_0, \ldots, x'_k, x) = 1$.

Fig. 1. Generic 1-round position-verification scheme

of dishonest provers $\hat{P}_0, \ldots, \hat{P}_\ell$ at arbitrary positions $apos_0, \ldots, apos_\ell$, all $\neq pos$, when executing the scheme from Fig. 1 the verifiers accept with probability at most ε. We then write PV^ε for such a protocol.

In order to be more realistic, we must take into consideration physical limitations of the equipment used, such as measurement errors, computation durations, etc. Those allow a dishonest prover which resides arbitrarily close to P to appear as if she resides at pos. Thus, we assume that all the adversaries are at least Δ-distanced from pos, where Δ is determined by those imperfections. For sake of simplicity, this Δ is implicit in the continuation of the paper.

4 Instantaneous Nonlocal Quantum Computation

In order to analyze the (in)security of position-verification schemes, we first address a more general task, which is interesting in its own right: *instantaneous nonlocal quantum computation*[1]. Consider the following problem, involving two parties Alice and Bob. Alice holds A and Bob holds B of a tripartite system ABE that is in some unknown state $|\psi\rangle$. The goal is to apply a known unitary transformation U to AB, but *without* using any communication, just by local operations. In general, such a task is clearly impossible, as it violates the non-signalling principle. The goal of instantaneous nonlocal quantum computation is to achieve almost the above but without violating non-signalling. Specifically, the goal is for Alice and Bob to compute, without communication, a state $|\varphi'\rangle$ that coincides with $|\varphi\rangle = (U \otimes \mathbb{I})|\psi\rangle$ up to *local* and *qubit-wise* operations on A and B, where $\mathbb{I}$ denotes the identity on E. Furthermore, these local and qubit-wise operations are determined by *classical* information that Alice and Bob obtain as part of their actions. In particular, if Alice and Bob share their classical information, which can be done with *one* round of simultaneous mutual communication,

[1] This is an extension of the task of "instantaneous measurement of nonlocal variables" introduced by Vaidman [11].

then they can transform $|\varphi'\rangle$ into $|\varphi\rangle = U|\psi\rangle$ by local qubit-wise operations. Following ideas by Vaidman [11], we show below that instantaneous nonlocal quantum computation, as described above, is possible if Alice and Bob share sufficiently many EPR pairs.

In the following, let $\mathcal{H}_A$, $\mathcal{H}_B$ and $\mathcal{H}_E$ be Hilbert spaces where the former two consist of n_A and n_B qubits respectively, i.e., $\mathcal{H}_A = (\mathbb{C}^2)^{\otimes n_A}$ and $\mathcal{H}_B = (\mathbb{C}^2)^{\otimes n_B}$. Furthermore, let U be a unitary matrix acting on $\mathcal{H}_A \otimes \mathcal{H}_B$. Alice holds system A and Bob holds system B of an arbitrary and unknown state $|\psi\rangle \in \mathcal{H}_{ABE} = \mathcal{H}_A \otimes \mathcal{H}_B \otimes \mathcal{H}_E$. Additionally, Alice and Bob share an arbitrary but finite number of EPR pairs.

Theorem 1. *For every unitary U and for every $\varepsilon > 0$, given sufficiently many shared EPR pairs, there exist local operations $\mathcal{A}$ and $\mathcal{B}$, acting on Alice's and Bob's respective sides, with the following property. For any initial state $|\psi\rangle \in \mathcal{H}_{ABE}$, the joint execution $\mathcal{A} \otimes \mathcal{B}$ transforms $|\psi\rangle$ into $|\varphi'\rangle$ and provides classical outputs k to Alice and ℓ to Bob, such that the following holds except with probability ε. The state $|\varphi'\rangle$ coincides with $|\varphi\rangle = (U \otimes \mathbb{I})|\psi\rangle$ up to local qubit-wise operations on A and B that are determined by k and ℓ.*

We stress that $\mathcal{A}$ acts on A as well as on Alice's shares of the EPR pairs, and the corresponding holds for $\mathcal{B}$. Furthermore, being equal up to local qubit-wise operations on A and B means that $|\varphi\rangle = (V^A_{k,\ell} \otimes V^B_{k,\ell} \otimes \mathbb{I})|\varphi'\rangle$, where $\{V^A_{k,\ell}\}_{k,\ell}$ and $\{V^B_{k,\ell}\}_{k,\ell}$ are fixed families of unitaries which act qubit-wise on $\mathcal{H}_A$ and $\mathcal{H}_B$, respectively. In our construction, the $V^A_{k,\ell}$ and $V^B_{k,\ell}$'s will actually be tensor products of one-qubit Pauli operators.

As an immediate consequence of Theorem 1, we get the following.

Corollary 1. *For every unitary U and for every $\varepsilon > 0$, given sufficiently many shared EPR pairs, there exists a nonlocal operation $\mathcal{AB}$ for Alice and Bob which consists of local operations and* one *round of mutual communication, such that for any initial state $|\psi\rangle \in \mathcal{H}_{ABE}$ of the tripartite system ABE, the joint execution of $\mathcal{AB}$ transforms $|\psi\rangle$ into $|\varphi\rangle = (U \otimes \mathbb{I})|\psi\rangle$, except with probability ε.*

For technical reasons, we will actually prove the following extension of Theorem 1, which is easily seen equivalent. The difference to Theorem 1 is that Alice and Bob are additionally given classical inputs: x to Alice and y to Bob, and the unitary U that is to be applied to the quantum input depends on x and y. In the statement below, x ranges over some arbitrary but fixed finite set $\mathcal{X}$, and y ranges over some arbitrary but fixed finite set $\mathcal{Y}$.

Theorem 2. *For every family $\{U_{x,y}\}$ of unitaries and for every $\varepsilon > 0$, given sufficiently many shared EPR pairs, there exist families $\{\mathcal{A}_x\}$ and $\{\mathcal{B}_y\}$ of local operations, acting on Alice's and Bob's respective sides, with the following property. For any initial state $|\psi\rangle \in \mathcal{H}_{ABE}$ and for every $x \in \mathcal{X}$ and $y \in \mathcal{Y}$, the joint execution $\mathcal{A}_x \otimes \mathcal{B}_y$ transforms the state $|\psi\rangle$ into $|\varphi'\rangle$ and provides classical outputs k to Alice and ℓ to Bob, such that the following holds except with probability ε. The state $|\varphi'\rangle$ coincides with $|\varphi\rangle = (U_{x,y} \otimes \mathbb{I})|\psi\rangle$ up to local qubit-wise operations on A and B that are determined by k and ℓ.*

The solution works by teleporting states back and forth in a clever way [11], but *without* communicating the classical outcomes of the Bell measurements, so that only local operations are performed. Thus, in the formal proof below, whenever we say that a state is teleported, this should be understood in this sense, i.e., the sender makes a Bell measurement resulting in some classical information, and the receiver takes his shares of the EPR pairs as the received state, but does/can not (yet) correct it.

Proof. To simplify notation, we assume that the joint state of A and B is pure, and thus we may ignore system E. However, all our arguments also hold in case the state of A and B is entangled with E.

Next, we observe that it is sufficient to prove Theorem 2 for the case where B is "empty", i.e., $\dim \mathcal{H}_B = 1$ and thus $n_B = 0$. Indeed, if this is not the case, then Alice and Bob can do the following. Bob first teleports B to Alice. Now, Alice holds $A' = AB$ with $n_{A'} = n_A + n_B$, and Bob's system has collapsed and thus Bob holds no quantum state anymore, only classical information. Then, they do the nonlocal computation, and in the end Alice teleports B back to Bob. The modification to the state of B introduced by teleporting it to Alice can be taken care of by modifying the set of unitaries $\{U_{x,y}\}$ accordingly (and making it dependent on Bob's measurement outcome, thereby extending the set $\mathcal{Y}$). Also, the modification to the state of B introduced by teleporting it back to Bob does not harm the requirement of the joint state being equal to $|\varphi\rangle = U_{x,y}|\psi\rangle$ up to local qubit-wise operations.

Hence, from now on, we may assume that B is "empty", and we write n for n_A. Next, we describe the core of how the local operations $\mathcal{A}_x$ and $\mathcal{B}_y$ work. To simplify notation, we assume that $\mathcal{X} = \{1, \dots, m\}$. Recall that Alice and Bob share (many) EPR pairs. We may assume that the EPR pairs are grouped into groups of size n; each such group we call a *teleportation channel*. Furthermore, we may assume that m of these teleportation channels are labeled by the numbers 1 up to m, and that another m of these teleportation channels are labeled by the numbers $m+1$ up to $2m$.

1. Alice teleports $|\psi\rangle$ to Bob, using the teleportation channel that is labeled by her input x. Let us denote her measurement outcome by $k_\circ \in \{0,1,2,3\}^n$.
2. For every $i \in \{1, \dots, m\}$, Bob does the following. He applies the unitary $U_{i,y}$ to the n qubits that make up his share of the EPR pairs given by the teleportation channel labeled by i. Then, he teleports the resulting state to Alice using the teleportation channel labeled by $m+i$. We denote the corresponding measurement outcome by $\ell_{\circ,i}$.
3. Alice specifies the n qubits that make up her share of the EPR pairs given by the teleportation channel labeled by $m+x$ to be the state $|\varphi'\rangle$.

Let us analyze the above. With probability $1/4^n$, namely if $k_\circ = 0 \cdots 0$, teleporting $|\psi\rangle$ to Bob leaves the state unchanged. In this case, it is easy to see that the resulting state $|\varphi'\rangle$ satisfies the required property of being identical to $|\varphi\rangle = U_{x,y}|\psi\rangle$ up to local qubit-wise operations determined by $\ell_{\circ,x}$, and thus determined by x and $\ell_\circ = (\ell_{\circ,1}, \ldots, \ell_{\circ,m})$. This proves the claim for the case where $\varepsilon \geq 1 - 1/4^n$.

We now show how to reduce ε. The crucial observation is that if in the above procedure $k_\circ \neq 0 \cdots 0$, and thus $|\varphi'\rangle$ is not necessarily identical to $|\varphi\rangle$ up to local qubit-wise operations, then

$$|\varphi'\rangle = V_{\ell_{\circ,x}} U_{x,y} V_{k_\circ} |\psi\rangle = V_{\ell_{\circ,x}} U_{x,y} V_{k_\circ} U_{x,y}^\dagger |\varphi\rangle ,$$

where $V_{\ell_{\circ,x}}$ and $V_{k_\circ}$ are tensor products of Pauli matrices. Thus, setting $|\psi'\rangle := |\varphi'\rangle$, $x' := (x, k_\circ)$ and $y' := (y, \ell_\circ)$, and $U'_{x',y'} := U_{x,y} V_{k_\circ} U_{x,y}^\dagger V_{\ell_{\circ,x}}$, the state $|\varphi\rangle$ can be written as $|\varphi\rangle = U'_{x',y'}|\psi'\rangle$. This means, we are back to the original problem of applying a unitary, $U'_{x',y'}$, to a state, $|\psi'\rangle$, held by Alice, where the unitary depends on classical information x' and y', known by Alice and Bob, respectively. Thus, we can re-apply the above procedure to the new problem instance. Note that in the new problem instance, the classical inputs x' and y' come from larger sets than the original inputs x and y, but the new quantum input, $|\psi'\rangle$, has the same qubit size, n. Therefore, re-applying the procedure will succeed with the same probability $1/4^n$.

As there is a constant probability of success in each round, re-applying the above procedure sufficiently many times to the resulting new problem instances guarantees that except with arbitrary small probability, the state $|\varphi'\rangle$ will be of the required form at some point (when Alice gets $k_\circ = 0 \cdots 0$). Say, this is the case at the end of the j-th iteration. Then, Alice stops with her part of the procedure at this point, keeps the state $|\varphi'\rangle$, and specifies k to consist of j and of her classical input into the j-th iteration (which consists of x and of the $k_\circ$'s from the prior $j-1$ iterations). Since Bob does not learn whether an iteration is successful or not, he has to keep on re-iterating up to some bound, and in the end he specifies ℓ to consist of the $\ell_\circ$'s collected over all the iterations. The state $|\varphi'\rangle$ then equals $|\varphi\rangle = U_{x,y}|\psi\rangle$ up to local qubit-wise operations that are determined by k and ℓ. □

Doing the maths shows that the number of EPR pairs needed by Alice and Bob in the scheme described in the proof is double exponential in $n_A + n_B$, the qubit size of the joint quantum system.

In recent subsequent work [20], Beigi and König have used a different kind of quantum teleportation by Ishizaka and Hiroshima [21,22] to reduce the amount of entanglement needed to to perform instantaneous nonlocal quantum computation to exponential in the qubit size of the joint quantum system. It remains an interesting open question whether such an exponentially large amount of entanglement is necessary.

5 Impossibility of Unconditional Position Verification

For simplicity, we consider the one-dimensional case, with two verifiers V_0 and V_1, but the attack can be generalized to higher dimensions and more verifiers.

We consider an arbitrary position-verification scheme in our model (as specified in Sect. 3.1). We recall that in this model, the verifiers must base their decision solely on *what* the prover replies and *how long* it takes him to reply, and the honest prover has no advantage over a coalition of dishonest provers beyond being at the claimed position[2]. Such a position-verification scheme may be of the form as specified in Fig. 1, but may also be made up of several, possibly interleaved, rounds of interaction between the prover and the verifiers.

For the honest prover P, such a general scheme consists of steps that look as follows. P holds a local quantum register R, which is set to some default value at the beginning of the scheme. In each step, P obtains a system A from V_0 and a system B from V_1, and V_0 and V_1 jointly keep some system E. Let $|\psi\rangle$ be the state of the four-partite system $ABRE$; it is determined by the scheme and by the step within the scheme we are focussing on. P then has to apply a fixed[3] known unitary transformation U to ABR, and send the (transformed) systems A and B back to V_0 and V_1 (and keep R). Note that after the transformation, the state of $ABRE$ is given by $|\varphi\rangle = (U \otimes \mathbb{I})|\psi\rangle$, where $\mathbb{I}$ is the identity acting on $\mathcal{H}_E$. For technical reasons, as in Sect. 4, it will be convenient to distinguish between classical and quantum inputs, and therefore, we let the unitary U depend on classical information x and y, where x has been sent by V_0 along with A, and y has been sent by V_1 along with B.

We now show that a coalition of two dishonest provers $\hat{P}_0$ and $\hat{P}_1$, where $\hat{P}_0$ is located in between V_0 and P and $\hat{P}_1$ is located in between V_1 and P, can perfectly simulate the actions of the honest prover P, and therefore it is impossible for the verifiers to distinguish between an honest prover at position *pos* and a coalition of dishonest provers at positions different from *pos*. The simulation of the dishonest provers perfectly imitates the *computation* as well as the *timing* of an honest P. Since in our model this information is what the verifiers have to base their decision on, the general impossibility of position-verification in our model follows.

Consider a step in the scheme as described above, but now from the point of view of $\hat{P}_0$ and $\hat{P}_1$. Since $\hat{P}_0$ is closer to V_0, he will first receive A and x; similarly, $\hat{P}_1$ will first receive B and y. We specify that $\hat{P}_1$ takes care of and maintains the local register R. If the step we consider here is the *first* step in the scheme, then the state of $ABRE$ equals $|\psi\rangle$, as in the case of an honest P. In order to have an invariant that holds for all the steps, we actually relax this statement and merely observe that the state of $ABRE$, say $|\psi'\rangle$, equals $|\psi\rangle$ up to local and qubit-wise operations on the subsystem R, determined by classical information $x_\circ$ and $y_\circ$, where $\hat{P}_0$ holds $x_\circ$ and $\hat{P}_1$ holds $y_\circ$. This invariant clearly holds for the first step in the scheme, when R is in some default state, and we will show that it also holds for the other steps.

By Theorem 2, it follows that without communication, just by instantaneous local operations, $\hat{P}_0$ and $\hat{P}_1$ can transform the state $|\psi'\rangle$ into a state $|\varphi'\rangle$ that coincides with $|\varphi\rangle = (U_{x,y} \otimes \mathbb{I})|\psi\rangle$ up to local and qubit-wise transformations on

---

[2] In particular, the prover does not share any secret information with the verifiers, differentiating our setting from models as described for example in [18].

[3] U is fixed for a fixed scheme and for a fixed step within the scheme, but of course may vary for different schemes and for different steps within a scheme.

A, B and R, determined by classical information k (known to $\hat{P}_0$) and ℓ (known to $\hat{P}_1$). Note that the initial state is not $|\psi\rangle$, but rather a state of the form $|\psi'\rangle = (V_{x_\circ,y_\circ} \otimes \mathbb{I})|\psi\rangle$, where $x_\circ$ is known to $\hat{P}_0$ and $y_\circ$ to $\hat{P}_1$. Thus, Theorem 2 is actually applied to the unitary $U'_{x',y'} = U_{x,y}V^\dagger_{x_\circ,y_\circ}$, where $x' = (x_\circ, x)$ and $y' = (y_\circ, y)$. Given $|\varphi'\rangle$ and k and ℓ, $\hat{P}_0$ and $\hat{P}_1$ can now exchange k and ℓ using *one* mutual round of communication and transform $|\varphi'\rangle$ into $|\varphi''\rangle$ that coincides with $|\varphi\rangle$ up to qubit-wise operations only on R, and send A to V_0 and B to V_1. It follows that the state of ABE and the time it took $\hat{P}_0$ and $\hat{P}_1$ for the computation and communication is identical to that of an honest P, i.e., $\hat{P}_0$ and $\hat{P}_1$ have perfectly simulated this step of the scheme.

Finally, we see that the invariant is satisfied, when moving on to the next step in the scheme, where $\hat{P}_0$ and $\hat{P}_1$ receive new A and B (along with new classical x and y) from V_0 and V_1, respectively. Even if this new round interleaves with the previous round in that the new A and B etc. arrive *before* $\hat{P}_0$ and $\hat{P}_1$ have finished exchanging (the old) k and ℓ, it still holds that the state of $ABRE$ is as in the case of honest P up to qubit-wise operations on the subsystem R. This implies that the above procedure works for all the steps and thus that $\hat{P}_0$ and $\hat{P}_1$ can indeed perfectly simulate honest P's actions throughout the whole scheme.

6 Secure Position-Verification in the No-PE Model

In this section we show the possibility of secure position-verification in the No-PE model. We consider the following basic 1-round position-verification scheme in the No-PE model, given in Fig. 2. It is based on the BB84 encoding.

We implicitly specify that parties abort if they receive any message that is inconsistent with the protocol, for instance (classical) messages with a wrong length, or different number of received qubits than expected, etc.

Theorem 3. *The 1-round position-verification scheme* $\mathsf{PV}^\varepsilon_{\text{BB84}}$ *from Fig. 2 is* ε*-sound with* $\varepsilon = 1 - \mathrm{h}^{-1}(\frac{1}{2})$*, in the No-PE model.*

The function $\mathrm{h} : [0,1] \to [0,1]$ denotes the *binary entropy function* defined as $\mathrm{h}(p) = -p\log(p) - (1-p)\log(1-p)$ for $0 < p < 1$ and as $\mathrm{h}(p) = 0$ for $p = 0$ or 1, and $\mathrm{h}^{-1} : [0,1] \to [0, \frac{1}{2}]$ denotes its inverse on the branch $0 \leq p \leq \frac{1}{2}$. A numerical calculation shows that $\mathrm{h}^{-1}(\frac{1}{2}) \geq 0.11$ and thus $\varepsilon \leq 0.89$. A particular attack for a dishonest prover $\hat{P}$, sitting in-between V_0 and P, is to measure

0. V_0 chooses two random bits $x, \theta \in \{0,1\}$ and privately sends them to V_1.
1. V_0 prepares the qubit $H^\theta|x\rangle$ and sends it to P, and V_1 sends the bit θ to P, so that $H^\theta|x\rangle$ and θ arrive at the same time at P.
2. When $H^\theta|x\rangle$ and θ arrive, P measures $H^\theta|x\rangle$ in basis θ to observe $x' \in \{0,1\}$, and sends x' to V_0 and V_1.
3. V_0 and V_1 accept if on both sides x' arrives in time and $x' = x$.

Fig. 2. Position-verification scheme $\mathsf{PV}^\varepsilon_{\text{BB84}}$ based on the BB84 encoding

the qubit $H^\theta|x\rangle$ in the *Breidbart* basis, resulting in an acceptance probability of $\cos(\pi/8)^2 \approx 0.85$. This shows that our analysis is pretty tight.

Proof. The proof uses several concepts of quantum information theory which are explained in more detail in the full version of this paper [25]. A key idea in this proof is the use of the *complementary information trade-off* (CIT) [24] (see also [27] for a generalization). In a form useful for us, CIT states that for any tri-partite state $|\psi_{AEF}\rangle \in \mathcal{H}_A \otimes \mathcal{H}_E \otimes \mathcal{H}_F$ with $\mathcal{H}_A = (\mathbb{C}^2)^{\otimes n}$, the following holds. If Θ is uniformly distributed in $\{0,1\}^n$ and X is the result of measuring A in basis Θ, then $\mathrm{H}(X|\Theta E) + \mathrm{H}(X|\Theta F) \geq n$, where H is the (conditional) von Neumann entropy.

In order to analyze the position-verification scheme it is convenient to consider an equivalent *purified* version, given in Fig. 3. The only difference between the original and the purified scheme is the preparation of the bit $H^\theta|x\rangle$. In the purified version, it is done by preparing $|\Phi_{AB}\rangle = (|0\rangle|0\rangle + |1\rangle|1\rangle)/\sqrt{2}$ and measuring A in basis θ. This changes the point in time when V_0 measures A, and the point in time when V_1 learns x. This, however, has no influence on the view of the (dishonest or honest) prover, nor on the joint distribution of θ, x and x', and thus neither on the probability that V_0 and V_1 accept. It therefore suffices to analyze the purified version.

0. V_0 and V_1 privately agree on a random bit $\theta \in \{0,1\}$.
1. V_0 prepares an EPR pair $|\Phi_{AB}\rangle \in \mathcal{H}_A \otimes \mathcal{H}_B$, keeps qubit A and sends B to P, and V_1 sends the bit θ to P, so that B and θ arrive at the same time at P.
2. When B and θ arrive, P measures B in basis θ to observe $x' \in \{0,1\}$, and sends x' to V_0 and V_1.
3. Only now, when x' arrives, V_0 measures A in basis θ to observe x, and privately sends x to V_1. V_0 and V_1 accept if on both sides x' arrives in time and $x' = x$.

Fig. 3. EPR version of $\mathsf{PV}^{\varepsilon}_{\mathrm{BB84}}$

We first consider security against two dishonest provers $\hat{P}_0$ and $\hat{P}_1$, where $\hat{P}_0$ is between V_0 and P and $\hat{P}_1$ is between V_1 and P. In the end we will argue that a similar argument holds for multiple dishonest provers on either side.

Since V_0 and V_1 do not accept if x' does not arrive in time and dishonest provers do not use pre-shared entanglement in the No-PE-model, any potentially successful strategy of $\hat{P}_0$ and $\hat{P}_1$ must look as follows. As soon as $\hat{P}_1$ receives the bit θ from V_1, she forwards (a copy of) it to $\hat{P}_0$. Also, as soon as $\hat{P}_0$ receives the qubit A, she applies an arbitrary quantum operation to the received qubit A (and maybe some ancillary system she possesses) that maps it into a bipartite state E_0E_1 (with arbitrary state space $\mathcal{H}_{E_0} \otimes \mathcal{H}_{E_1}$), and $\hat{P}_0$ keeps E_0 and sends E_1 to $\hat{P}_1$. Then, as soon as $\hat{P}_0$ receives θ, she applies some measurement (which may depend on θ) to E_0 to obtain $\hat{x}_0$, and as soon as $\hat{P}_1$ receives E_1, she applies some measurement (which may depend on θ) to E_1 to obtain $\hat{x}_1$, and both send $\hat{x}_0$ and $\hat{x}_1$ immediately to V_0 and V_1, respectively. We will now argue that the probability that $\hat{x}_0 = x$ *and* $\hat{x}_1 = x$ is upper bounded by ε as claimed.

Let $|\psi_{A\,E_0E_1}\rangle \in \mathcal{H}_A \otimes \mathcal{H}_{E_0} \otimes \mathcal{H}_{E_1}$ be the state of the tri-partite system $A\,E_0E_1$ after $\hat{P}_0$ has applied the quantum operation to the qubit B. It is important to realize that the state $|\psi_{A\,E_0E_1}\rangle$ is independent of θ. This is because $\hat{P}_0$ has to apply the quantum operation to B *before* learning θ.[4] Recall that x is obtained by measuring A in either the computational (if $\theta = 0$) or the Hadamard (if $\theta = 1$) basis. Writing x, θ, etc. as random variables X, Θ, etc., it follows from CIT that $\mathrm{H}(X|\Theta E_0) + \mathrm{H}(X|\Theta E_1) \geq 1$. Let Y_0 and Y_1 denote the classical information obtained by $\hat{P}_0$ and $\hat{P}_1$ as a result of measuring E_0 and E_1, respectively, with bases that may depend on Θ. By the well-known Holevo bound, it follows from the above that

$$\mathrm{H}(X|\Theta Y_0) + \mathrm{H}(X|\Theta Y_1) \geq 1 \; ,$$

therefore $\mathrm{H}(X|\Theta Y_i) \geq \frac{1}{2}$ for at least one $i \in \{0,1\}$. By Fano's inequality, we can conclude that the corresponding error probability $q_i = P[\hat{X}_i \neq X]$ satisfies $\mathrm{h}(q_i) \geq \frac{1}{2}$. It thus follows that the failure probability

$$q = P[\hat{X}_0 \neq X \vee \hat{X}_1 \neq X] \geq \max\{q_0, q_1\} \geq \mathrm{h}^{-1}(\frac{1}{2}) \; ,$$

and the probability of V_0 and V_1 accepting, $P[\hat{X}_0 = X \wedge \hat{X}_1 = X] = 1 - q$, is indeed upper bounded by ε as claimed. See full details in [25].

It remains to argue that more than two dishonest provers in the No-PE model cannot do any better. The reasoning is the same as above. Namely, in order to respond in time, the dishonest provers that are closer to V_0 than P must map the qubit A—possibly jointly—into a bipartite state E_0E_1 *without knowing* θ, and jointly keep E_0 and send E_1 to the dishonest provers that are "on the other side" of P (i.e., closer to V_1). Then, the reply for V_0 needs to be computed from E_0 and θ (possibly jointly by the dishonest provers that are closer to V_0), and the response for V_1 from E_1 and θ. Thus, it can be argued as above that the success probability is bounded by ε as claimed. □

The soundness error can be further reduced by sequentially repeating the scheme, assuming that the adversaries do not share entanglement at the beginning of each round. Also, the scheme can easily be extended to arbitrary dimension d. The idea is to involve additional verifiers $V_2, \ldots, V_d$ and have the basis θ secret-shared among $V_1, V_2, \ldots, V_d$.

7 Other Position-based Cryptographic Tasks

In the full version of this paper [25], we show the following additional results. Using a generic position-verification scheme, we construct a position-based *authentication* scheme, which ensures that a communicated message m originates from an entity P that is at some specific location. In combination with an off-the-shelf quantum-key-distribution (QKD) scheme, this results in a position-based

[4] We stress that this independency breaks down if $\hat{P}_0$ and $\hat{P}_1$ may start off with an entangled state, because then $\hat{P}_1$ can act on his part of the entangled state in a θ-dependent way, which makes the overall state dependent of θ.

key-distribution scheme, which enables the verifiers to exchange a cryptographic key K with the prover, with the guarantee that only the honest prover at location *pos* obtains K, but any adversary (or coalition of adversaries) not at location *pos* learns no information on K. Using our position-verification scheme in the No-PE model as underlying scheme, we obtain secure position-based *authentication* and position-based *key-distribution* schemes in the No-PE model.

8 Conclusion and Open Questions

We have proven a general impossibility result for position-based quantum cryptography, thereby showing the insecurity of several recently proposed schemes [13,14,15,16,17]. Our no-go result has already sparked subsequent work [20] about the *amount* of entanglement needed to break general position-verification schemes.

On the positive side, we have shown the existence of secure position-based quantum cryptographic schemes under the (strong) assumption that adversaries do not share any entanglement (prior to each round). An interesting open question is the existence of secure schemes under more relaxed and realistic assumptions, like in the bounded-quantum-storage model [28], where adversaries are limited in the number of qubits they can store reliably?

Acknowledgements. We thank Charles Bennett, Frédéric Dupuis and Louis Salvail for interesting discussions. HB would like to thank Sandu Popescu for explaining Vaidman's scheme and pointing [19] out to him.

References

1. Chandran, N., Goyal, V., Moriarty, R., Ostrovsky, R.: Position based cryptography. In: Halevi, S. (ed.) CRYPTO 2009. LNCS, vol. 5677, pp. 391–407. Springer, Heidelberg (2009)
2. Brands, S., Chaum, D.: Distance-bounding protocols. In: Helleseth, T. (ed.) EUROCRYPT 1993. LNCS, vol. 765, pp. 344–359. Springer, Heidelberg (1994)
3. Sastry, N., Shankar, U., Wagner, D.: Secure verification of location claims. In: WiSe 2003, pp. 1–10 (2003)
4. Vora, A., Nesterenko, M.: Secure location verification using radio broadcast. In: Higashino, T. (ed.) OPODIS 2004. LNCS, vol. 3544, pp. 369–383. Springer, Heidelberg (2005)
5. Bussard, L.: Trust Establishment Protocols for Communicating Devices. PhD thesis, Eurecom-ENST (2004)
6. Capkun, S., Hubaux, J.P.: Secure positioning of wireless devices with application to sensor networks. In: IEEE INFOCOM, 1917–1928 (2005)
7. Singelee, D., Preneel, B.: Location verification using secure distance bounding protocols. In: IEEE MASS'10 (2005)
8. Zhang, Y., Liu, W., Fang, Y., Wu, D.: Secure localization and authentication in ultra-wideband sensor networks. IEEE Journal on Selected Areas in Communications 24, 829–835 (2006)

9. Capkun, S., Cagalj, M., Srivastava, M.: Secure localization with hidden and mobile base stations. In: IEEE INFOCOM (2006)
10. Bennett, C.H., Brassard, G., Crépeau, C., Jozsa, R., Peres, A., Wootters, W.K.: Teleporting an unknown quantum state via dual classical and einstein-podolsky-rosen channels. Phys. Rev. Lett. 70(13), 1895–1899 (1993)
11. Vaidman, L.: Instantaneous measurement of nonlocal variables. Phys. Rev. Lett. 90(1), 010402 (2003)
12. Kent, A., Munro, W., Spiller, T., Beausoleil, R.: Tagging systems, US patent nr 2006/0022832 (2006)
13. Kent, A., Munro, B., Spiller, T.: Quantum tagging: Authenticating location via quantum information and relativistic signalling constraints, arXiv/quant-ph:1008.2147 (2010)
14. Malaney, R.A.: Location-dependent communications using quantum entanglement. Phys. Rev. A 81(4), 042319 (2010)
15. Malaney, R.A.: Quantum location verification in noisy channels, arXiv/quant-ph:1004.2689 (2010)
16. Chandran, N., Fehr, S., Gelles, R., Goyal, V., Ostrovsky, R.: Position-based quantum cryptography, arXiv/quant-ph:1005.1750 (2010)
17. Lau, H.K., Lo, H.K.: Insecurity of position-based quantum-cryptography protocols against entanglement attacks. Phys. Rev. A 83(1), 012322 (2011)
18. Kent, A.: Quantum tagging with cryptographically secure tags, arXiv/quant-ph:1008.5380 (2010)
19. Clark, S.R., Connor, A.J., Jaksch, D., Popescu, S.: Entanglement consumption of instantaneous nonlocal quantum measurements. New Journal of Physics 12(8), 083034 (2010)
20. Beigi, S., Koenig, R.: Simplified instantaneous non-local quantum computation with applications to position-based cryptography, arXiv/quant-ph:1101.1065 (2011)
21. Ishizaka, S., Hiroshima, T.: Asymptotic teleportation scheme as a universal programmable quantum processor. Phys. Rev. Lett. 101(24), 240501 (2008)
22. Ishizaka, S., Hiroshima, T.: Quantum teleportation scheme by selecting one of multiple output ports. Phys. Rev. A 79(4), 042306 (2009)
23. Giovannetti, V., Lloyd, S., Maccone, L.: Quantum cryptographic ranging. Journal of Optics B 4(4), 042319 (2002)
24. Renes, J., Boileau, J.: Conjectured strong complementary information tradeoff. Phys. Rev. Let. 103(2), 020402 (2009)
25. Buhrman, H., Chandran, N., Fehr, S., Gelles, R., Goyal, V., Ostrovsky, R., Schaffner, C.: Position-Based Quantum Cryptography: Impossibility and Constructions. Full version of this paper (2010), `http://arxiv.org/abs/1009.2490`
26. Nielsen, M.A., Chuang, I.L.: Quantum Computation and Quantum Information. Cambridge University Press, Cambridge (2000)
27. Berta, M., Christandl, M., Colbeck, R., Renes, J.M., Renner, R.: The uncertainty principle in the presence of quantum memory. Nature Physics (2010)
28. Damgård, I., Fehr, S., Salvail, L., Schaffner, C.: Cryptography in the bounded quantum-storage model. In: 46th Annual IEEE Symposium on Foundations of Computer Science (FOCS), pp. 449–458. IEEE, Los Alamitos (2005)

Analyzing Blockwise Lattice Algorithms Using Dynamical Systems

Guillaume Hanrot[1], Xavier Pujol[1], and Damien Stehlé[2]

[1] ÉNS Lyon, Laboratoire LIP (U. Lyon, CNRS, ENS Lyon, INRIA, UCBL), 46 Allée d'Italie, 69364 Lyon Cedex 07, France

[2] CNRS, Laboratoire LIP (U. Lyon, CNRS, ENS Lyon, INRIA, UCBL), 46 Allée d'Italie, 69364 Lyon Cedex 07, France

{guillaume.hanrot,xavier.pujol,damien.stehle}@ens-lyon.fr

Abstract. Strong lattice reduction is the key element for most attacks against lattice-based cryptosystems. Between the strongest but impractical HKZ reduction and the weak but fast LLL reduction, there have been several attempts to find efficient trade-offs. Among them, the BKZ algorithm introduced by Schnorr and Euchner [FCT'91] seems to achieve the best time/quality compromise in practice. However, no reasonable complexity upper bound is known for BKZ, and Gama and Nguyen [Eurocrypt'08] observed experimentally that its practical runtime seems to grow exponentially with the lattice dimension. In this work, we show that BKZ can be terminated long before its completion, while still providing bases of excellent quality. More precisely, we show that if given as inputs a basis $(\boldsymbol{b}_i)_{i\leq n} \in \mathbb{Q}^{n\times n}$ of a lattice L and a block-size β, and if terminated after $\Omega\left(\frac{n^3}{\beta^2}(\log n + \log\log \max_i \|\boldsymbol{b}_i\|)\right)$ calls to a β-dimensional HKZ-reduction (or SVP) subroutine, then BKZ returns a basis whose first vector has norm $\leq 2\nu_\beta^{\frac{n-1}{2(\beta-1)}+\frac{3}{2}} \cdot (\det L)^{\frac{1}{n}}$, where $\nu_\beta \leq \beta$ is the maximum of Hermite's constants in dimensions $\leq \beta$. To obtain this result, we develop a completely new elementary technique based on discrete-time affine dynamical systems, which could lead to the design of improved lattice reduction algorithms.

Keywords: Euclidean lattices, BKZ, lattice-based cryptanalysis.

1 Introduction

A (full-rank) n-dimensional lattice $L \subseteq \mathbb{R}^n$ is the set of integer linear combinations $\sum_{i=1}^n x_i \boldsymbol{b}_i$ of some linearly independent vectors $(\boldsymbol{b}_i)_{i\leq n}$. Such vectors are called a basis and we write $L = L[(\boldsymbol{b}_i)_i]$. Since L is discrete, it contains a shortest non-zero lattice vector, whose norm $\lambda_1(L)$ is called the lattice minimum. Computing such a vector given a basis is referred to as the (computational) Shortest Vector Problem (SVP), and is NP-hard under randomized reductions [1,12]. The complexities of the best known SVP solvers are no less than exponential [22,23,2,15] (the record is held by the algorithm from [22], with complexity $2^{2n+o(n)} \cdot \mathcal{P}oly(\log \max_i \|\boldsymbol{b}_i\|)$). Finding a vector reaching $\lambda_1(L)$

P. Rogaway (Ed.): CRYPTO 2011, LNCS 6841, pp. 447–464, 2011.

is polynomial-time equivalent to computing a basis of L that is reduced in the sense of Hermite-Korkine-Zolotarev (HKZ). The aforementioned SVP solvers can all be used to compute HKZ-reduced bases, in exponential time. On the other hand, bases reduced in the sense of Lenstra-Lenstra-Lovász (LLL) can be computed in polynomial time [16], but the first vector is only guaranteed to satisfy the weaker inequality $\|\boldsymbol{b}_1\| \leq (4/3+\varepsilon)^{\frac{n-1}{2}} \cdot \lambda_1(L)$ (for an arbitrary $\varepsilon > 0$). In 1987, Schnorr introduced time/quality trade-offs between LLL and HKZ [33]. In the present work, we propose the first analysis of the BKZ algorithm [36,37], which is currently the most practical such trade-off [40,8].

Lattice reduction is a popular tool in cryptanalysis [27]. For many applications, such as Coppersmith's method for computing the small roots of polynomials [5], LLL-reduction suffices. However, reductions of much higher quality seem required to break lattice-based cryptosystems. Lattice-based cryptography originated with Ajtai's seminal hash function [1], and the GGH and NTRU encryption schemes [9,14]. Thanks to its excellent asymptotic performance, provable security guarantees, and flexibility, it is currently attracting wide interest and developing at a steady pace. We refer to [21,31] for recent surveys. A major obstacle to the real-life deployment of lattice-based cryptography is the lack of a precise understanding of the limits of the best practical attacks, whose main component is the computation of strongly reduced lattice bases. This prevents from having a precise correspondence between specific security levels and practical parameters. Our work is a step towards a clearer understanding of BKZ, and thus of the best known attacks.

Strong lattice reduction has been studied for about 25 years (see among others [33,37,34,6,32,8,7]). From a theoretical perspective, the best known time/quality trade-off is due to Gama and Nguyen [7]. By building upon the proof of Mordell's inequality on Hermite's constant, they devised the notion of *slide reduction*, and proposed an algorithm computing slide-reduced bases: Given an arbitrary basis $B = (\boldsymbol{b}_i)_{i \leq n}$ of a lattice L, the slide-reduction algorithm finds a basis $(\boldsymbol{c}_i)_{i \leq n}$ of L such that

$$\|\boldsymbol{c}_1\| \leq ((1+\varepsilon)\gamma_\beta)^{\frac{n-\beta}{\beta-1}} \cdot \lambda_1(L), \tag{1}$$

within $\tau_{\texttt{slide}} := O\left(\frac{n^4}{\beta \cdot \varepsilon} \cdot \log \max_i \|\boldsymbol{b}_i\|\right)$ calls[1] to a β-dimensional HKZ-reduction algorithm and a β-dimensional (computational-)SVP solver, where $\gamma_\beta \approx \beta$ is the β-dimensional Hermite constant. If $L \subseteq \mathbb{Q}^n$, the overall cost of the slide-reduction algorithm is $\leq \mathcal{P}oly(n, \mathrm{size}(B)) \cdot \mathcal{C}_{\mathtt{HKZ}}(\beta)$, where $\mathcal{C}_{\mathtt{HKZ}}(\beta) = 2^{O(\beta)}$ is the cost of HKZ-reducing in dimension β. The higher β, the lower the achieved SVP approximation factor, but the higher the runtime. Slide reduction also provides a constructive variant of Minkowski's inequality, as (letting $\det L$ denote $\mathrm{vol}(\mathbb{R}^n/L)$):

$$\|\boldsymbol{c}_1\| \leq ((1+\varepsilon)\gamma_\beta)^{\frac{n-1}{2(\beta-1)}} \cdot (\det L)^{\frac{1}{n}}, \tag{2}$$

[1] The component $\frac{n^4}{\beta}$ of this upper bound is derived by adapting the results from [7] to our notations. A more thorough analysis leads to a smaller term.

From a practical perspective, however, slide reduction seems to be (significantly) outperformed by the BKZ algorithm [8]. BKZ also relies on a β-dimensional HKZ-reduction algorithm (resp. SVP-solver). The worst-case quality of the bases it returns has been studied in [34] and is comparable to that of the slide reduction algorithm. The first vector of the output basis $(\boldsymbol{c}_i)_{i<n}$ satisfies $\|\boldsymbol{c}_1\| \leq ((1+\varepsilon)\gamma_\beta)^{\frac{n-1}{\beta-1}} \cdot \lambda_1(L)$. Note that this bound essentially coincides with (1), except for large values of β. A bound similar to that of (2) also holds.[2] In practice, the quality of the computed bases seems much higher with BKZ than with the slide-reduction algorithm [8]. With respect to run-time, no reasonable bound is known on the number of calls to the β-dimensional HKZ reduction algorithm it needs to make before termination.[3] In practice, this number of calls does not seem to be polynomially bounded [8] and actually becomes huge when $\beta \geq 25$. Because of its large (and somewhat unpredictable) runtime, it is folklore practice to terminate BKZ before the end of its execution, when the solution of the problem for which it is used for is already provided by the current basis [38,24].

OUR RESULT. We show that if terminated within polynomially many calls to HKZ/SVP, a slightly modified version of BKZ (see Section 3) returns bases whose first vectors satisfy a slightly weaker variant of (2).

Theorem 1. *There exists*[4] $C > 0$ *such that the following holds for all* n *and* β. *Let* $B = (\boldsymbol{b}_i)_{i \leq n}$ *be a basis of a lattice* L*, given as input to the modified BKZ algorithm of Section 3 with block-size* β. *If terminated after* $\tau_{\mathsf{BKZ}} := C\frac{n^3}{\beta^2}\left(\log n + \log\log \max_i \frac{\|\boldsymbol{b}_i\|}{(\det L)^{1/n}}\right)$ *calls to an HKZ-reduction (or SVP solver) in dimension* β*, the output* $(\boldsymbol{c}_i)_{i \leq n}$ *is a basis of* L *that satisfies (with* $\nu_\beta \leq \beta$ *defined as the maximum of Hermite's constants in dimensions* $\leq \beta$*):*

$$\|\boldsymbol{c}_1\| \leq 2(\nu_\beta)^{\frac{n-1}{2(\beta-1)}+\frac{3}{2}} \cdot (\det L)^{\frac{1}{n}}.$$

If $L \subseteq \mathbb{Q}^n$*, then the overall cost is* $\leq \mathcal{P}oly(n, \mathrm{size}(B)) \cdot \mathcal{C}_{\mathsf{HKZ}}(\beta)$.

By using [18, p. 25], this provides an algorithm with runtime bounded by $\mathcal{P}oly(n, \mathrm{size}(B)) \cdot \mathcal{C}_{\mathsf{HKZ}}(\beta)$ that returns a basis whose first vector satisfies $\|\boldsymbol{c}_1\| \leq 4(\nu_\beta)^{\frac{n-1}{\beta-1}+3} \cdot \lambda_1(L)$, which is only slightly worse than (1). These results indicate that BKZ can be used to achieve essentially the same quality guarantees as slide reduction, within a number of calls to HKZ in dimension β that is no larger than that of slide reduction. Actually, note that τ_{BKZ} is significantly smaller than $\tau_{\texttt{slide}}$, in particular with a dependence with respect to $\max_i \|\boldsymbol{b}_i\|$ that is *exponentially* smaller. It may be possible to obtain a similar bound for the slide-reduction algorithm by adapting our analysis.

---

[2] In [8], the bound $\|\boldsymbol{c}_1\| \leq (\gamma_\beta)^{\frac{n-1}{2(\beta-1)}+\frac{1}{2}} \cdot (\det L)^{\frac{1}{n}}$ is claimed to hold, but without proof nor reference. We prove a (slightly) weaker bound, but we are able to improve it if γ_n is replaced by any linear function. See the appendix of the full version [10].

[3] A bound $(n\beta)^n$ is mentioned in [8]. For completeness, we give a proof of a similar result in the appendix of the full version [10].

[4] The constant C is used to absorb lower-order terms in n, and could be taken small.

To achieve our result, we use a completely new approach for analyzing lattice reduction algorithms. The classical approach to bound their runtimes was to introduce a quantity, sometimes called potential, involving the current Gram-Schmidt norms $\|\boldsymbol{b}_i^*\|$, which always strictly decreases every time some elementary step is performed. This technique was introduced by Lenstra, Lenstra and Lovász [16] for analyzing their LLL algorithm, and is still used in all complexity analyses of (variants of) LLL we are aware of. It was later adapted to stronger lattice reduction algorithms [33,6,32,7]. We still measure progress with the $\|\boldsymbol{b}_i^*\|$'s, but instead of considering a single scalar combining them all, we look at the *full vector* $(\|\boldsymbol{b}_i^*\|)_i$. More specifically, we observe that each call to HKZ within BKZ has the effect of applying an affine transformation to the vector $(\log \|\boldsymbol{b}_i^*\|)_i$: instead of providing a lower bound to the progress made on a "potential", we are then led to analyze a discrete-time dynamical affine system. Its fixed-points encode information on the output quality of BKZ, whereas its speed of convergence provides an upper bound on the number of times BKZ calls HKZ.

Intuitively, the effect of a call to HKZ on the vector $(\log \|\boldsymbol{b}_i^*\|)_{i \leq n}$ is to essentially replace β consecutive coefficients by their average. We formalize this intuition by making a specific assumption (see Section 4). Under this assumption, the execution of BKZ exactly matches with a dynamical system that we explicit and fully analyze. However, we cannot prove that this assumption is always correct (counter-examples can actually be constructed). To circumvent this difficulty, we instead consider the vector $\boldsymbol{\mu} = (\frac{1}{i}\sum_{j=1}^{i} \log \|\boldsymbol{b}_j^*\|)_{i \leq n}$. This amortization (also used in [11] for analyzing HKZ-reduced bases) allows us to *rigorously* bound the evolution of $\boldsymbol{\mu}$ by the orbit of a vector under another dynamical system. Since this new dynamical system happens to be a modification of the dynamical system used in the idealized model, the analysis performed for the idealized model can be adapted to the rigorous set-up.

This approach is likely to prove useful for analyzing other lattice reduction algorithms. As an illustration of its power, we provide two new results on LLL. First, we show that the SVP approximation factor $\sqrt{4/3}^{\,n-1}$ can be reached in polynomial time using only Gauss reductions. This is closely related to the question whether the "optimal LLL" (i.e., using LLL parameter $\delta = 1$) terminates in polynomial time [3,17]. Second, we give a LLL-reduction algorithm of bit-complexity $\mathcal{P}oly(n) \cdot \widetilde{O}(\text{size}(B))$. Such a complexity bound was only very recently achieved, with a completely different approach [29]. Note that close-by results on LLL have been concurrently and independently obtained by Schnorr [35].

Practical aspects. Our result is a (possibly pessimistic) worst-case quality bound on BKZ with early termination. In itself, this does not give a precise explanation of the practical behavior of BKZ. In particular, it does not explain why it outperforms slide reduction, but only why it does not behave significantly worse. However, this study illustrates the usefulness of early termination in BKZ: Much progress is done at the beginning of the execution, and quickly the basis quality becomes excellent; the rest of the execution takes much longer, for a significantly less dramatic quality improvement. This behavior is very clear in practice, as illustrated by Figure 1 of Section 2. Since most of the work performed by BKZ is completed within the first few calls to HKZ, it shows that the BKZ performance extrapolations used to estimate the hardness of cryptographic instances should focus only on the cost of a single call to HKZ and on the achieved basis quality after a few such calls. For instance, it indicates that the strategy (adopted, e.g., in [14,13]) consisting in measuring the full run-time of BKZ might be reconsidered.

Additionally, parts of the analysis might prove useful to better understand BKZ and devise reduction algorithms with improved practical time/quality trade-offs. In particular, the heuristic modelisation of BKZ as a discrete-time affine dynamical system suggests that the block of vectors on which HKZ-reduction is to be applied could be chosen adaptively, so that the system converges faster to its limit. It would not improve the output quality for BKZ, but it is likely to accelerate its convergence. Also, the second phase of BKZ, the one that takes longer but during which little progress is still made, could be understood by introducing some randomness in the model: most of the time, the norm of the first vector found by the HKZ-reduction sub-routine is around its expected value (a constant factor smaller than its worst-case bound), but it is significantly smaller every now and then. If such a model could predict the behavior of BKZ during its second phase, then maybe it would explain why it outperforms slide reduction. It might give indications on the optimal time for stopping BKZ with block-size β before switching to a larger block-size.

Notations. All vectors will be denoted in bold, and matrices in capital letters. If $\boldsymbol{b} \in \mathbb{R}^n$, the notation $\|\boldsymbol{b}\|$ will refer to its Euclidean norm. If $B \in \mathbb{R}^{n \times n}$, we define $\|B\|_2 = \max_{\|\boldsymbol{x}\|=1} \|B \cdot \boldsymbol{x}\|$ and we denote the spectral radius of B by $\rho(B)$. If B is a rational matrix, we define size(B) as the sum of the bit-sizes of the numerators and denominators of its entries. All complexity statements refer to elementary operations on bits. We will use the Landau notations $o(\cdot)$, $O(\cdot)$, $\widetilde{O}(\cdot)$ and $\Omega(\cdot)$. The notations $\log(\cdot)$ and $\ln(\cdot)$ respectively stand for the base 2 and natural logarithms.

2 Reminders

For an introduction to lattice reduction algorithms, we refer to [28].

Successive Minima. Let L be an n-dimensional lattice. Its i-th minimum $\lambda_i(L)$ is defined as the minimal radius r such that $\mathcal{B}(\mathbf{0}, r)$ contains $\geq i$ linearly independent vectors of L.

Hermite's constant. The n-dimensional Hermite constant γ_n is defined as the maximum taken over all lattices L of dimension n of the quantity $\frac{\lambda_1(L)^2}{(\det L)^{2/\dim(L)}}$. Let $\nu_n = \max_{k \leq n} \gamma_k$, an upper bound on γ_n which increases with n. Very few values of ν_n are known, but we have $\nu_n \leq 1 + \frac{n}{4}$ for all n (see [20, Re 2.7.5]).

Gram-Schmidt orthogonalisation. Let $(\boldsymbol{b}_i)_{i \leq n}$ be a lattice basis. Its Gram-Schmidt orthogonalization $(\boldsymbol{b}_i^*)_{i \leq n}$ is defined recursively by $\boldsymbol{b}_i^* = \boldsymbol{b}_i - \sum_{j<i} \mu_{i,j} \boldsymbol{b}_j^*$ with $\mu_{i,j} = \langle \boldsymbol{b}_i^*, \boldsymbol{b}_j^* \rangle / \|\boldsymbol{b}_j^*\|^2$ for $i > j$. The $\boldsymbol{b}_i^*$'s are mutually orthogonal. For $i \leq j$, we define $\boldsymbol{b}_j^{(i)}$ as the projection of $\boldsymbol{b}_j$ orthogonally to Span$(\boldsymbol{b}_k)_{k<i}$. Note that if L is an n-dimensional lattice, then $\det L = \prod_{i=1}^n \|\boldsymbol{b}_i^*\|$, for any basis $(\boldsymbol{b}_i)_{i \leq n}$ of L.

A few notions of reduction. Given a basis $(\boldsymbol{b}_i)_{i \leq n}$, we say that it is *size-reduced* if the Gram-Schmidt coefficients $\mu_{i,j}$ satisfy $|\mu_{i,j}| \leq 1/2$ for all $j < i \leq n$. We say that $(\boldsymbol{b}_i)_{i \leq n}$ is *δ-LLL-reduced* for $\delta \leq 1$ if it is size-reduced and the Lovász conditions $\delta \|\boldsymbol{b}_i^*\|^2 \leq \|\boldsymbol{b}_{i+1}^*\|^2 + \mu_{i+1,i}^2 \|\boldsymbol{b}_i^*\|^2$ are satisfied for all $i < n$. For any $\delta < 1$, a δ-LLL-reduced basis of a rational lattice L can be computed in polynomial time, given an arbitrary basis of L as input [16]. We say that $(\boldsymbol{b}_i)_{i \leq n}$ is HKZ-reduced if it is size-reduced and for all $i < n$, we have $\|\boldsymbol{b}_i^*\| = \lambda_1(L[(\boldsymbol{b}_j^{(i)})_{i \leq j \leq n}])$. An HKZ-reduced basis of a lattice $L \subseteq \mathbb{Q}^n$ can be computed in time $2^{2n+o(n)} \cdot \mathcal{P}oly(\text{size}(B))$, given an arbitrary basis B of L as input [22]. The following is a direct consequence of the definitions of the HKZ-reduction and Hermite constant.

Lemma 1. *For any HKZ-reduced basis* $(\boldsymbol{b}_i)_{i \leq n}$*, we have:* $\forall i < n$, $\|\boldsymbol{b}_i^*\| \leq \sqrt{\nu_{n-i+1}} \cdot (\prod_{j=i}^n \|\boldsymbol{b}_j^*\|)^{\frac{1}{n-i+1}}$.

The BKZ algorithm. We recall the original BKZ algorithm from [37] in Algorithm 1. BKZ was originally proposed as a mean of computing bases that are almost *β-reduced*. β-Reduction was proposed by Schnorr in [33], but without an algorithm for achieving it. The BKZ algorithm proceeds by iterating tours consisting of $n-1$ calls to a β-dimensional SVP solver called on the lattices $L[(\boldsymbol{b}_i^{(k)})_{k \leq i \leq k+\beta-1}]$. Its execution stops when no change occurs during a tour.

```
Input  : A (LLL-reduced) basis (b_i)_{i≤n}, a blocksize β and a constant δ < 1.
Output : A basis of L[(b_i)_{i≤n}].
repeat
    for k ← 1 to n − 1 do
        Find b such that ||b^(k)|| = λ_1(L[(b_i^(k))_{k≤i≤min(k+β−1,n)}]);
        if δ · ||b_k^*|| > ||b|| then
            LLL-reduce(b_1, ..., b_{k−1}, b, b_k, ..., b_{min(k+β,n)}).
        else
            LLL-reduce(b_1, ..., b_{min(k+β,n)}).
until no change occurs.
```

Algorithm 1. The Schnorr and Euchner BKZ algorithm

3 Terminating BKZ

In this article, we will not analyze the original BKZ algorithm, but we will focus on a slightly modified variant instead, which is given in Algorithm 2. It also performs BKZ tours, and during a tour it makes $n - \beta + 1$ calls to a β-dimensional HKZ-reduction algorithm. It fits more closely to what would be the simplest BKZ-style algorithm, aiming at producing a basis $(\boldsymbol{b}_i)_{i \leq n}$ such that the projected basis $(\boldsymbol{b}_i^{(k)})_{k \leq i \leq k+\beta-1}$ is HKZ-reduced for all $k \leq n - \beta + 1$.

Differences between the two variants of BKZ. The differences between the two algorithms are the following:

- In Algorithm 2, the execution can be terminated at the end of any BKZ tour.
- In the classical BKZ algorithm, the vector $\boldsymbol{b}$ found by the SVP solver is kept only if $\|\boldsymbol{b}^{(k)}\|$ is smaller than $\delta \cdot \|\boldsymbol{b}_k^*\|$. Such a factor $\delta < 1$ does not appear in Algorithm 2. It is unnecessary for our analysis to hold, complicates the algorithm, and leads to output bases of lesser quality.
- For each k within a tour, Algorithm 1 only requires an SVP solver while Algorithm 2 calls an HKZ-reduction algorithm, which is more complex. We use HKZ-reductions for the ease of the analysis. Our analysis would still hold if the loop was done for k from 1 to $n-1$ and if the HKZ-reductions were replaced by calls to any algorithm that returns bases whose first vector reaches the minimum (which can be obtained by calling any SVP solver, putting the output vector in front of the input basis and calling LLL to remove the linear dependency).
- Finally, to insert $\boldsymbol{b}$ in the current basis, Algorithm 1 performs an LLL-reduction. Indeed, applying LLL inside the projected block (i.e., to $\boldsymbol{b}^{(k)}, \boldsymbol{b}_k^{(k)}, \ldots, \boldsymbol{b}_{k+\beta-1}^{(k)}$) would be sufficient to remove the linear dependency while keeping $\boldsymbol{b}^{(k)}$ in first position, but instead it runs LLL from the beginning of the basis until the end of the next block to be considered (i.e., up to index $\min(k+\beta, n)$). This reduction is performed even if the block is already reduced and no vector is inserted. Experimentally, this seems to improve the speed of convergence of the algorithm by a small factor, but it does not seem easy to use our techniques to analyze this effect.

`Input` : A basis $(\boldsymbol{b}_i)_{i\leq n}$ and a blocksize β.
`Output` : A basis of $L[(\boldsymbol{b}_i)_{i\leq n}]$.
repeat
  for $k \leftarrow 1$ to $n-\beta+1$ **do**
    Modify $(\boldsymbol{b}_i)_{k\leq i\leq k+\beta-1}$ so that $(\boldsymbol{b}_i^{(k)})_{k\leq i\leq k+\beta-1}$ is HKZ-reduced;
    Size-reduce$(\boldsymbol{b}_1, \ldots, \boldsymbol{b}_n)$.
until no change occurs or termination is requested.

Algorithm 2. BKZ', the modified BKZ algorithm

On the practical behavior of BKZ. In order to give an insight on the practical behavior of BKZ and BKZ', we give experimental results on the evolution of the quantity $\frac{\|\boldsymbol{b}_1\|}{(\det L)^{1/n}}$ (the so-called Hermite factor) during their executions. The experiment corresponding to Figure 1 is as follows: We generated 64 knapsack-like bases [25] of dimension $n = 108$, with non-trivial entries of bit-length $100n$; Each was LLL-reduced using `fplll` [4] (with parameters $\delta = 0.99$ and $\eta = 0.51$); Then for each we ran NTL's BKZ [40] and an implementation of BKZ' in NTL, with blocksize 24. Figure 1 only shows the beginning of the executions. For both algorithms, the executions of about half the samples consisted in $\simeq 600$ tours, whereas the longest execution stopped after $\simeq 1200$ tours. The average value of $\frac{\|\boldsymbol{b}_1\|}{(\det L)^{1/n}}$ at the end of the executions was $\simeq 1.012$.

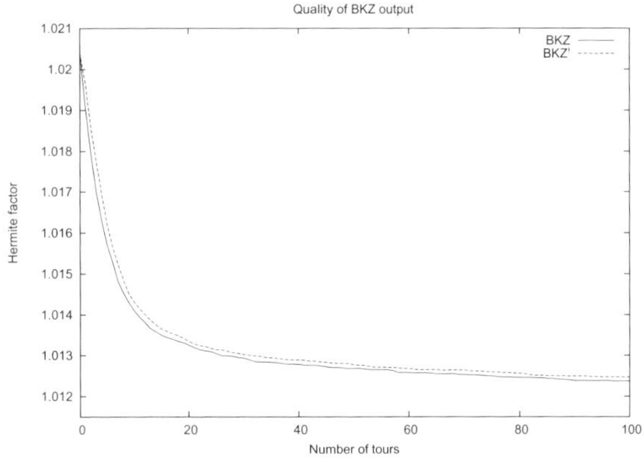

Fig. 1. Evolution of the Hermite factor $\frac{\|\boldsymbol{b}_1\|}{(\det L)^{1/n}}$ during the execution of BKZ and BKZ'

Cost of BKZ'. In order to bound the bit-complexities of BKZ and BKZ', it is classical to consider several cost components separately. In this article, we will focus on the number of tours. The number of calls to an SVP solver (for BKZ) or an HKZ-reduction algorithm (in the case of BKZ') is $\leq n$ times larger. A tour consists of efficient operations (LLL, size-reductions, etc) and of the more costly calls to SVP/BKZ. The cost of the SVP solver or the HKZ-reduction algorithm is often bounded in terms of the number of arithmetic operations it performs: For all known algorithms, this quantity is (at least) exponential in the block-size β. Finally, one should also take into account the bit-costs of the arithmetic operations performed to prepare the calls to SVP/HKZ, during these calls, and after these calls (when applying the computed transforms to the basis, and calling LLL or a size-reduction). These arithmetic costs are classically bounded by considering the bit-sizes of the quantities involved. They can easily be shown to be polynomial in the input bit-size, by relying on rational arithmetic and using standard tools from the analyses of LLL and HKZ [16,15]. It is likely that these costs can be lowered further by relying on floating-point approximations to these rational numbers, using the techniques from [26,30]. To conclude, the overall cost is upper bounded by $\mathcal{P}oly(n, \log\|B\|) \cdot 2^{O(\beta)} \cdot \tau$, where τ is the number of tours.

4 Analysis of BKZ' in the Sandpile Model

In this section, we (rigorously) analyze a *heuristic model of BKZ'*. In the following section, we will show how this analysis can be adapted to allow for a (rigorous) study of the *genuine BKZ'* algorithm.

We first note that BKZ' can be studied by looking at the way the vector $\boldsymbol{x} := (\log \|\boldsymbol{b}_i^*\|)_i$ changes during the execution, rather than considering the whole basis $(\boldsymbol{b}_i)_i$. This simplification is folklore in the analyses of lattice reduction algorithms, and allows for an interpretation in terms of sandpiles [19]. The study in the present section is heuristic in the sense that we assume the effect of a call to HKZ_β on $\boldsymbol{x}$ is determined by $\boldsymbol{x}$ only, in a deterministic fashion.

4.1 The Model and Its Dynamical System Interpretation

Before describing the model, let us consider the shape of a β-dimensional HKZ-reduced basis. Let $(\boldsymbol{b}_i)_{i\leq\beta}$ be an HKZ-reduced basis, and define $x_i = \log \|\boldsymbol{b}_i^*\|$. Then, by Lemma 1, we have:

$$\forall i \leq \beta,\ x_i \leq \frac{1}{2}\log \nu_{\beta-i+1} + \frac{1}{\beta-i+1}\sum_{j=i}^{\beta} x_j. \tag{3}$$

Our heuristic assumption consists in replacing these inequalities by equalities.

Heuristic Sandpile Model Assumption (SMA). We assume for any HKZ-reduced basis $(\boldsymbol{b}_i)_{i\leq\beta}$, we have $x_i = \frac{1}{2}\log \nu_{\beta-i+1} + \frac{1}{\beta-i+1}\sum_{j=i}^{\beta} x_j$ for all $i \leq \beta$, with $\boldsymbol{x} = (\log \|\boldsymbol{b}_i^*\|)_{i\leq\beta}$.

Under SMA, once $\sum_i x_i$ (i.e., $|\det(\boldsymbol{b}_i)_i|$) is fixed, an $\boldsymbol{x}$ of an HKZ-reduced basis is uniquely determined.

Lemma 2. *Let $(\boldsymbol{b}_i)_{i\leq\beta}$ be HKZ-reduced, $\boldsymbol{x} = (\log \|\boldsymbol{b}_i^*\|)_i$ and $\mathbb{E}[\boldsymbol{x}] = \sum_{i\leq\beta} \frac{x_i}{\beta}$. Then, under SMA, $x_\beta = \mathbb{E}[\boldsymbol{x}] - \Gamma_\beta(\beta-1)$ and:*

$$\forall i < \beta,\ x_i = \mathbb{E}[\boldsymbol{x}] - (\beta-i+1)\Gamma_\beta(i-1) + (\beta-i)\Gamma_\beta(i),$$

with $\Gamma_n(k) = \sum_{i=n-k}^{n-1} \frac{\log \nu_{i+1}}{2i}$ for all $0 \leq k < n$.

We now exploit SMA to interpret BKZ' as a discrete-time linear dynamical system. Let $(\boldsymbol{b}_i)_{i\leq n}$ be a lattice basis and $\boldsymbol{x} = (\log \|\boldsymbol{b}_i^*\|)_i$. Let $\beta \leq n$ be a blocksize and $\alpha \leq n-\beta+1$. When we apply an HKZ reduction algorithm to the projected sublattice $(\boldsymbol{b}_i^{(\alpha)})_{\alpha\leq i<\alpha+\beta-1}$, we obtain a new basis $(\boldsymbol{b}_i')_{i\leq n}$ such that (with $\boldsymbol{x}' = (\log \|\boldsymbol{b}_i'^*\|)_i$):

$$\sum_{i=\alpha}^{\alpha+\beta-1} x_i' = \sum_{i=\alpha}^{\alpha+\beta-1} x_i \text{ and } \forall i \notin [\alpha, \alpha+\beta-1],\ x_i' = x_i.$$

Under SMA, we also have:

$$\forall i \in [\alpha, \alpha+\beta-1],\ x_i' = \frac{1}{2}\log \nu_{\alpha+\beta-i} + \frac{1}{\alpha+\beta-i}\sum_{j=i}^{\alpha+\beta-1} x_j'.$$

By applying Lemma 2, we obtain $\boldsymbol{x}' = A^{(\alpha)} \cdot \boldsymbol{x} + \boldsymbol{g}^{(\alpha)}$, with:

$$A^{(\alpha)} = \begin{pmatrix} \ddots & & & & & \\ & 1 & & & & \\ & & \frac{1}{\beta} & \cdots & \frac{1}{\beta} & & \\ & & \vdots & \ddots & \vdots & & \\ & & \frac{1}{\beta} & \cdots & \frac{1}{\beta} & & \\ & & & & & 1 & \\ & & & & & & \ddots \end{pmatrix} \begin{matrix} \\ \\ (\alpha) \\ \\ (\alpha+\beta-1) \\ \\ \\ \end{matrix}$$

$$g_i^{(\alpha)} = \begin{cases} 0 \text{ if } i < \alpha \\ (\beta+\alpha-i-1)\Gamma_\beta(i-\alpha+1) - (\beta+\alpha-i)\Gamma_\beta(i-\alpha) \\ \qquad\qquad\qquad\qquad\qquad\qquad \text{if } i \in [\alpha, \alpha+\beta-2] \\ -\Gamma_\beta(\beta-1) \text{ if } i = \alpha+\beta-1 \\ 0 \text{ if } i \geq \alpha+\beta. \end{cases}$$

We recall that a *BKZ' tour* is the successive $(n-\beta+1)$ applications of an HKZ-reduction algorithm with $\alpha = 1, \ldots, n-\beta+1$ (in this order). Under SMA, the effect of a BKZ' tour on $\boldsymbol{x}$ is to replace it by $A\boldsymbol{x} + \boldsymbol{g}$ with $\boldsymbol{g} = \boldsymbol{g}^{(n-\beta+1)} + A^{(n-\beta+1)} \cdot (\boldsymbol{g}^{(n-\beta)} + A^{(n-\beta)} \cdot (\ldots))$ and:

$$A = A^{(n-\beta+1)} \cdot \ldots \cdot A^{(1)} = \begin{pmatrix} \frac{1}{\beta} & \cdots & \frac{1}{\beta} & & & \\ \frac{\beta-1}{\beta^2} & \cdots & \frac{\beta-1}{\beta^2} & \frac{1}{\beta} & & \\ \vdots & & \vdots & \ddots & \ddots & \\ \frac{(\beta-1)^{n-\beta}}{\beta^{n-\beta+1}} & \cdots & \frac{(\beta-1)^{n-\beta}}{\beta^{n-\beta+1}} & \cdots & \frac{\beta-1}{\beta^2} & \frac{1}{\beta} \\ \vdots & & \vdots & & \vdots & \vdots \\ \frac{(\beta-1)^{n-\beta}}{\beta^{n-\beta+1}} & \cdots & \frac{(\beta-1)^{n-\beta}}{\beta^{n-\beta+1}} & \cdots & \frac{\beta-1}{\beta^2} & \frac{1}{\beta} \end{pmatrix} \begin{matrix} \\ \\ \\ (n-\beta+1) \\ \\ (n) \end{matrix}.$$

(columns labelled (1) and (β))

We sum up the study of the discrete-time dynamical system $\boldsymbol{x} \leftarrow A \cdot \boldsymbol{x} + \boldsymbol{g}$ in the following Theorem. The solutions and speed of convergence respectively provide information on the output quality and runtime of BKZ' (under SMA). Overall, we have:

Theorem 2. *Under SMA, there exists $C > 0$ such that the following holds for all n and β. Let $(\mathbf{b}_i)_{i \leq n}$ be given as input to BKZ'$_\beta$ and L the lattice spanned by the $\mathbf{b}_i$'s. If terminated after $C\frac{n^2}{\beta^2}(\log n + \log\log \max_i \frac{\|\boldsymbol{b}_i^*\|}{(\det L)^{1/n}})$ tours, then the output $(\mathbf{c}_i)_{i \leq n}$ is a basis of L that satisfies $\|\boldsymbol{x} - \boldsymbol{x}^\infty\|_2 \leq 1$, where*

$x_i = \log \frac{\|\boldsymbol{c}_i^*\|}{(\det L)^{1/n}}$ *for all* i *and* $\boldsymbol{x}^\infty$ *is the unique solution of the equation* $\boldsymbol{x}^\infty = A \cdot \boldsymbol{x}^\infty + \boldsymbol{g}$ *with* $\mathbb{E}[\boldsymbol{x}^\infty] = 0$. *This implies that:*[5]

$$\|\boldsymbol{c}_1\| \leq 2(\nu_\beta)^{\frac{n-1}{2(\beta-1)}+\frac{3}{2}} \cdot (\det L)^{\frac{1}{n}}.$$

4.2 Solutions of the Dynamical System

Before studying the solutions of $\boldsymbol{x} = A \cdot \boldsymbol{x} + \boldsymbol{g}$, we consider the associated homogeneous system.

Lemma 3. *If* $A \cdot \boldsymbol{x} = \boldsymbol{x}$, *then* $\boldsymbol{x} \in \mathrm{span}(1, \ldots, 1)^T$.

It thus suffices to find one solution to $\boldsymbol{x} = A \cdot \boldsymbol{x} + \boldsymbol{g}$ to obtain all the solutions. We define $\overline{\boldsymbol{x}}$ as follows:

$$\overline{x}_i = \begin{cases} \frac{\beta}{2(\beta-1)} \log \nu_\beta + \frac{1}{\beta-1} \sum_{j=i+1}^{i+\beta-1} \overline{x}_j & \text{if } i \leq n-\beta \\ g_i^{(n-\beta+1)} & \text{if } i > n-\beta \end{cases}.$$

Lemma 4. *We have* $\overline{\boldsymbol{x}} = A \cdot \overline{\boldsymbol{x}} + \boldsymbol{g}$.

We now provide explicit lower and upper bounds for the coordinates of the solution $\overline{\boldsymbol{x}}$.

Lemma 5. *For all* $i \leq n - \beta + 1$, *we have* $\left(\frac{n-i}{\beta-1} - \frac{3}{2}\right) \log \nu_\beta \leq \overline{x}_i - \overline{x}_{n-\beta+1} \leq \frac{n-i}{\beta-1} \log \nu_\beta$.

We refer to [10] for proofs Lemmata 3, 4, 5.

As the set of solutions to $\boldsymbol{x} = A \cdot \boldsymbol{x} + \boldsymbol{g}$ is $\overline{\boldsymbol{x}} + \mathrm{Span}(1, \ldots, 1)^T$, the value of $\overline{\boldsymbol{x}}$ is only interesting up to a constant vector, which is why we bound $\overline{x}_i - \overline{x}_{n-\beta+1}$ rather than $\overline{x}_i$. In other words, since $\boldsymbol{x}^\infty$ of Theorem 1 is $\overline{\boldsymbol{x}} - (\mathbb{E}[\overline{\boldsymbol{x}}])_i$, the Lemma also applies to $\boldsymbol{x}^\infty$. It is also worth noting that the difference between the upper and lower bounds $\frac{3}{2} \log \nu_\beta$ is much smaller than the upper bound $\frac{n-i}{\beta-1} \log \nu_\beta$ (for most values of i). If we replace ν_β by β, then, via a tedious function analysis, we can improve both bounds so that their difference is lowered to $\frac{1}{2} \log \beta$. In the special case $\beta = 2$, the expression of $\overline{\boldsymbol{x}}$ is $\overline{x}_i = \overline{x}_n + (n-i) \log \nu_2$.

4.3 Speed of Convergence of the Dynamical System

The classical approach to study the speed of convergence (with respect to k) of a discrete-time dynamical system $\boldsymbol{x}_{k+1} := A_n \cdot \boldsymbol{x}_k + \boldsymbol{g}_n$ (where A_n and $\boldsymbol{g}_n$ are the n-dimensional values of A and $\boldsymbol{g}$ respectively) consists in providing an upper bound to the largest eigenvalue of $A_n^T A_n$. It is relatively easy to prove that it is 1 (note that A_n is doubly stochastic). We are to show that the second largest

[5] If we replace ν_β by a linear function that bounds it (e.g., $\nu_\beta \leq \beta$), then the constant $\frac{3}{2}$ may be replaced by $\frac{1-\ln 2}{2} + \varepsilon$ (with $\varepsilon > 0$ arbitrarily close to 0 and β sufficiently large).

singular value is $< 1 - \frac{\beta^2}{2n^2}$, and that this bound is sharp, up to changing the constant 1/2 and as long as $n - \beta = \Omega(n)$.

The asymptotic speed of convergence of the sequence $(A_n^k \cdot \boldsymbol{x})_k$ is in fact determined by the eigenvalue(s) of A_n of largest module[6] (this is the principle of the power iteration algorithm). However, this classical fact provides no indication on the dependency with respect to $\boldsymbol{x}$, which is crucial in the present situation. As we use the bound $\|A_n^k \cdot \boldsymbol{x}\| \leq \|A_n\|_2^{\frac{k}{2}} \cdot \|\boldsymbol{x}\|$, we are led to studying the largest singular values of $A_n^T A_n$.

We first explicit the characteristic polynomial χ_n of $A_n^T A_n$. The following lemma shows that it satisfies a second order recurrence formula.

Lemma 6. *We have* $\chi_\beta(t) = t^{\beta-1}(t-1)$, $\chi_{\beta+1}(t) = t^{\beta-1}(t-1)(t-\frac{1}{\beta^2})$ *and, for any* $n \geq \beta$*:*

$$\chi_{n+2}(t) = \frac{(2\beta(\beta-1)+1)t - 1}{\beta^2} \cdot \chi_{n+1}(t) - \left(\frac{\beta-1}{\beta}\right)^2 t^2 \cdot \chi_n(t).$$

This is used to study the roots of $\chi_n(t)$. The proof of the following result relies on several changes of variables to link the polynomials $\chi_n(t)$ to the Chebyshev polynomials of the second kind.

Lemma 7. *For any* $n \geq \beta \geq 2$*, the largest root of the polynomial* $\frac{\chi_n(t)}{t-1}$ *belongs to* $\left[1 - \frac{\pi^2\beta^2}{(n-\beta)^2},\ 1 - \frac{\beta^2}{2n^2}\right]$.

We refer to [10] for proofs of Lemmata 6 and 7.

Proof of Theorem 2. The unicity and existence of $\boldsymbol{x}^\infty$ come from Lemmata 3 and 4.

Let $(\boldsymbol{b}_i^{(k)})_{i \leq n}$ be the basis after k tours of the algorithm BKZ'$_\beta$ and $\boldsymbol{x}_i^{(k)} = \log \frac{\|\boldsymbol{b}_i^{(k)*}\|}{(\det L)^{1/n}}$. The definition of $\boldsymbol{x}^\infty$ and a simple induction imply that $\boldsymbol{x}^{(k)} - \boldsymbol{x}^\infty = A^k(\boldsymbol{x}^{(0)} - \boldsymbol{x}^\infty)$. Both $\boldsymbol{x}^{(0)}$ and $\boldsymbol{x}^\infty$ live in the subspace $\mathcal{E} := \mathrm{Span}(1, \ldots, 1)^\perp$, which is stabilized by A. Let us denote by $A_\mathcal{E}$ the restriction of A to this subspace. Then the largest eigenvalue of $A_\mathcal{E}^T A_\mathcal{E}$ is bounded in Lemma 7 by $\left(1 - \frac{\beta^2}{2n^2}\right)$. Taking the norm in the previous equation gives:

$$\begin{aligned} \|\boldsymbol{x}^{(k)} - \boldsymbol{x}^\infty\|_2 &\leq \|A_\mathcal{E}\|_2^{\frac{k}{2}} \cdot \|\boldsymbol{x}^{(0)} - \boldsymbol{x}^\infty\|_2 = \rho(A_\mathcal{E}^T A_\mathcal{E})^{k/2} \cdot \|\boldsymbol{x}^{(0)} - \boldsymbol{x}^\infty\|_2 \\ &\leq \left(1 - \frac{\beta^2}{2n^2}\right)^{k/2} \|\boldsymbol{x}^{(0)} - \boldsymbol{x}^\infty\|_2. \end{aligned}$$

The term $\|\boldsymbol{x}^{(0)} - \boldsymbol{x}^\infty\|_2$ is bounded by $\left(\log \frac{\max_i \|\boldsymbol{b}_i^*\|}{(\det L)^{1/n}}\right) n + n^{O(1)}$. Thus, there exists C such that $\|\boldsymbol{x}^{(k)} - \boldsymbol{x}^\infty\|_2 \leq 1$ when $k \geq C\frac{n^2}{\beta^2}(\log n + \log\log \max_i \frac{\|\boldsymbol{b}_i^*\|}{(\det L)^{1/n}})$.

[6] Which can also be proved to be $\leq 1 - c\beta^2/n^2$ for some constant c.

We now prove the last inequality of the theorem. By Lemma 5 and the fact that $\sum_{i=n-\beta+1}^{n} x_i^\infty \geq \beta x_{n-\beta+1}^\infty + \sum_{i=n-\beta+1}^{n}\left(\frac{\log \nu_\beta}{\beta-1}(n-i) - \frac{3}{2}\log \nu_\beta\right)$, we have:

$$\begin{aligned} x_1^\infty &\leq (n-1)\frac{\log \nu_\beta}{\beta - 1} - \frac{1}{n}\sum_{i=1}^{n}\left(\frac{\log \nu_\beta}{\beta-1}(n-i) \quad \frac{3}{2}\log \nu_\beta\right) \\ &= \left(\frac{n-1}{2(\beta-1)} + \frac{3}{2}\right)\log \nu_\beta. \end{aligned}$$

Using the inequality $x_1^{(k)} \leq x_1^\infty + 1$ and taking the exponential (in base 2) leads to the result. □

5 Analysis of BKZ'

We now show how the heuristic analysis of the previous section can be made rigorous. The main difficulty stems from the lack of control on the $\|\boldsymbol{b}_i^*\|$'s of an HKZ-reduced basis $(\boldsymbol{b}_i)_{i\leq\beta}$. More precisely, once the determinant and $\|\boldsymbol{b}_\beta^*\|$ are fixed, the $\|\boldsymbol{b}_i^*\|$'s are all below a specific curve (explicitly given in Lemma 2). However, if only the determinant is fixed, the pattern of the $\|\boldsymbol{b}_i^*\|$'s can vary significantly: as an example, taking orthogonal vectors of increasing norms shows that $\|\boldsymbol{b}_1^*\|$ (resp. $\|\boldsymbol{b}_\beta^*\|$) can be arbitrarily small (resp. large). Unfortunately, when applying HKZ within BKZ', it seems we only control the determinant of the HKZ-reduced basis of the considered block, although we would prefer to have an upper bound for each Gram-Schmidt norm individually. We circumvent this difficulty by *amortizing* the analysis over the $\|\boldsymbol{b}_i^*\|$'s: as observed in [11], we have a sharp control *on each average* of the first $\|\boldsymbol{b}_i^*\|$'s. For an arbitrary basis $B := (\boldsymbol{b}_i)_{i\leq n}$, we define $\mu_k^{(B)} = \frac{1}{k}\sum_{1\leq i\leq k}\log\|\boldsymbol{b}_i^*\|$, for $k \leq n$.

Lemma 8 ([11, Le. 3]). *If $B = (\boldsymbol{b}_i)_{i\leq\beta}$ is HKZ-reduced, then $\mu_k^{(B)} \leq \frac{\beta-k}{k}\log \Gamma_\beta(k) + \mu_\beta^{(B)}$ for all $k \leq \beta$.*

5.1 A Dynamical System for (Genuine) BKZ' Tours

We now reformulate the results of the previous section with the $\mu_i^{(B)}$'s instead of the $\log\|\boldsymbol{b}_i^*\|$'s. This amounts to a base change in the discrete-time dynamical system of Subsection 4.1. We define:

$$P = (\tfrac{1}{i}\mathbf{1}_{i\geq j})_{1\leq i,j\leq n},\ \widetilde{A} = PAP^{-1} \ \text{ and } \ \widetilde{\boldsymbol{g}} = P\cdot\boldsymbol{g}.$$

Note that $\boldsymbol{\mu}^{(B)} = P\cdot\boldsymbol{x}^{(B)}$, where $\boldsymbol{x}^{(B)} = (\log\|\boldsymbol{b}_i^*\|)_i$ and $\boldsymbol{\mu}^{(B)} = (\mu_i^{(B)})_i$.

Lemma 9. *Let B' be the basis obtained after a BKZ' tour given an n-dimensional basis B as input. Then $\boldsymbol{\mu}^{(B')} \leq \widetilde{A}\cdot\boldsymbol{\mu}^{(B)} + \widetilde{\boldsymbol{g}}$, where the inequality holds componentwise.*

5.2 Analysis of the Updated Dynamical System

Similarly to the analysis of the previous section, it may be possible to obtain information on the speed of convergence of BKZ' by estimating the eigenvalues of $\widetilde{A}^T \cdot \widetilde{A}$. However, the latter eigenvalues seem significantly less amenable to study than those of $A^T A$. The following lemma shows that we can short-circuit the study of the modified dynamical system. For a basis $B \in \mathbb{R}^{n \times n}$ given as input to BKZ'$_\beta$, we define $B^{[0]} = B$ and $B^{[i]}$ as the current basis after the i-th BKZ' tour. We also define $\boldsymbol{\mu}^\infty = P \cdot \boldsymbol{x}^\infty$.

Lemma 10. *Let $B \in \mathbb{R}^{n \times n}$ a basis given as input to BKZ'$_\beta$. Wlog we assume that $\mu_n^{(B)} = \mu_n^\infty$ (since $\mu_n^{(B)} = \frac{1}{n} \log |\det B|$, this can be achieved by multiplying B by a scalar). We have:*

$$\forall k \geq 0, \forall i \leq n, \quad \mu_i^{(B^{[k]})} \leq \mu_i^\infty + (1 + \log n)^{1/2} \cdot \left(1 - \frac{\beta^2}{2n^2}\right)^{k/2} \|\boldsymbol{x}^{(B^{[0]})} - \boldsymbol{x}^\infty\|_2.$$

Lemma 11. *There exists $C > 0$ such that the following holds for all integers $n \geq \beta$, and $\varepsilon \in (0,1]$. Let $(\boldsymbol{b}_i)_{i \leq n}$ be a basis of a lattice L, given as input to the modified BKZ' algorithm of Section 2 with block-size β. If terminated after $C \frac{n^3}{\beta^2} (\log \frac{n}{\varepsilon} + \log \log \max_i \frac{\|\boldsymbol{b}_i^*\|}{(\det L)^{1/n}})$ calls to an HKZ-reduction (resp. SVP solver) in dimension β, the output $(\boldsymbol{c}_i)_{i \leq n}$ is a basis of L that satisfies:*

$$\|\boldsymbol{c}_1\| \leq (1 + \varepsilon) \nu_\beta^{\frac{n-1}{2(\beta-1)} + \frac{3}{2}} \cdot (\det L)^{\frac{1}{n}}.$$

Theorem 1 corresponds to taking $\varepsilon = 1$ in Lemma 11. Also, when $\beta = 2$, using the explicit expression of $\boldsymbol{x}^\infty$ leads to the improved bound $\|\boldsymbol{c}_1\| \leq (1 + \varepsilon) \cdot (\nu_2)^{\frac{n-1}{2}} \cdot (\det L)^{\frac{1}{n}}$.

6 Applications to LLL-Reduction

In this section, we investigate the relationship between BKZ'$_2$ reduction and the notion of LLL-reduction [16]. Note that analogues of some of the results of this section have been concurrently and independently obtained by Schnorr [35].

Reminders on the LLL algorithm. The LLL algorithm with parameter δ proceeds by successive loop iterations. Each iteration has a corresponding index k, defined as the smallest such that $(\boldsymbol{b}_i)_{i \leq k}$ is not δ-LLL-reduced. The iteration consists in size-reducing $(\boldsymbol{b}_i)_{i \leq k}$ and then checking Lovász's condition $\delta \|\boldsymbol{b}_{k-1}^*\|^2 \leq \|\boldsymbol{b}_k^*\|^2 + \mu_{k,k-1}^2 \|\boldsymbol{b}_{k-1}^*\|^2$. If it is satisfied, then we proceed to the next loop iteration, and otherwise, we swap the vectors $\boldsymbol{b}_k$ and $\boldsymbol{b}_{k-1}$. Any such swap decreases the quantity $\Pi((\boldsymbol{b}_i)_i) = \prod_{i=1}^n \|\boldsymbol{b}_i^*\|^{2(n-i+1)}$ by a factor $\geq 1/\delta$ whereas it remains unchanged during size-reductions. Since $\Pi((\boldsymbol{b}_i)_i) \leq 2^{O(n^2 \operatorname{size}(B))}$ and since for any integer basis $\Pi((\boldsymbol{b}_i)_i)$ is an integer, this allows to prove termination within $O(n^2 \operatorname{size}(B))$ loop iterations when $\delta < 1$. When $\delta = 1$, we obtain the so-called *optimal LLL algorithm*. Termination can still be proven by using different arguments, but with a much larger bound $2^{\mathcal{P}oly(n)} \cdot \mathcal{P}oly(\operatorname{size}(B))$ (see [3,17]).

An iterated version of BKZ'$_2$. We consider the algorithm Iterated-BKZ'$_2$ (described in Algorithm 3) which given as input a basis $(\boldsymbol{b}_i)_{i\leq n}$ successively applies BKZ'$_2$ to the projected bases $(\boldsymbol{b}_i)_{i\leq n}$, $(\boldsymbol{b}_i^{(2)})_{2\leq i\leq n}$, ..., $(\boldsymbol{b}_i^{(n-1)})_{n-1\leq i\leq n}$. By using a quasi-linear time Gauss reduction algorithm (see [39,41]) as the HKZ$_2$ algorithm within BKZ'$_2$, Algorithm Iterated-BKZ'$_2$ can be shown to run in quasi-linear time.

```
Input  : A basis (b_i)_{i≤n} of a lattice L.
Output : A basis of L.
for k := 1 to n − 1 do
    Apply BKZ'_2 to the basis (b_i^{(k)})_{k≤i≤n};
    Let T be the corresponding transformation matrix;
    Update (b_i)_{i≤n} by applying T to (b_i)_{k≤i≤n}.
Return (b_i)_{i≤n}.
```

Algorithm 3. Iterated-BKZ'$_2$ Algorithm

Lemma 12. *Let B be a basis of an n-dimensional lattice, and $\varepsilon > 0$ be arbitrary. Then, using Algorithm Iterated-BKZ'$_2$, one can compute, in time $\mathcal{P}oly(n)\cdot \widetilde{O}(\text{size}(B))$, a basis $(\boldsymbol{b}'_i)_{i\leq n}$ such that*

$$\forall i\leq n,\ \|\boldsymbol{b}'^*_i\| \leq (1+\varepsilon)\left(\frac{4}{3}\right)^{\frac{n-i}{2}}\cdot\left(\prod_{j=i}^{n}\|\boldsymbol{b}'^*_i\|\right)^{\frac{1}{n-i+1}}. \tag{4}$$

A close analogue of the optimal LLL. Let $B = (\boldsymbol{b}_i)_{i\leq n}$ an integral basis output by Iterated-BKZ'$_2$. For $i \leq n$, we let p_i, q_i be coprime rational integers such that $\frac{p_i}{q_i} = \left(\frac{3}{4}\right)^{(n-i+1)(n-i)}\cdot\frac{\|\boldsymbol{b}_i{}^*\|^{2(n-i+1)}}{\prod_{j=i}^{n}\|\boldsymbol{b}_j{}^*\|^2}$. By (4), we know that $p_i/q_i \leq (1+\varepsilon)^{n-i+1}$. Note that p_i/q_i is a rational number with denominator $\leq 2^{O(n^2+\text{size}(B))}$. We can thus find a constant c such that, for all i, the quantity $|p_i/q_i - 1|$ is either 0 or $\geq 2^{-c(n^2+\text{size}(B))}$. Hence, if we choose $\varepsilon < \frac{1}{2n}\cdot 2^{-c(n^2+\text{size}(B'))}$, all the inequalities from (4) must hold with $\varepsilon = 0$. Overall, we obtain, in polynomial time and using only swaps and size-reductions, a basis for which (4) holds with $\varepsilon = 0$.

A quasi-linear time LLL-reduction algorithm. BKZ'$_2$ can be used to obtain a variant of LLL which given as input an integer basis $(\boldsymbol{b}_i)_{i\leq n}$ and $\delta < 1$ returns a δ-LLL-reduced basis of $L[(\boldsymbol{b}_i)_{i\leq n}]$ in time $\mathcal{P}oly(n)\cdot\widetilde{O}(\text{size}(B))$. First, we apply the modification from [18, p. 25] to a terminated BKZ'$_2$ so that the modified algorithm, when given as input an integer basis $(\boldsymbol{b}_i)_{i\leq n}$ and $\varepsilon > 0$, returns in time $\mathcal{P}oly(n)\cdot\widetilde{O}(\text{size}(B))$ a basis $(\boldsymbol{b}'_i)_{i\leq n}$ of $L[(\boldsymbol{b}_i)_{i\leq n}]$ such that $\|\boldsymbol{b}'_1\| \leq (1+\varepsilon)^2(4/3)^{n-1}\lambda_1(L)$. The complexity bound holds because the transformation from [18, p. 25] applies BKZ'$_2$ n times on bases whose bit-sizes are $\mathcal{P}oly(n)\cdot\widetilde{O}(\text{size}(B))$.

We iterate this algorithm n times on the projected lattices $(\boldsymbol{b}_i^{(k)})_{k\leq i\leq n}$ so that the output basis $(\boldsymbol{c}_i)_{i\leq n}$ of $L[(\boldsymbol{b}_i)_{i\leq n}]$ satisfies:

$$\forall i\leq n,\ \|\boldsymbol{c}_i{}^*\| \leq (1+\varepsilon)^2(4/3)^{n-i}\lambda_1(L[(\boldsymbol{b}_j^{(i)})_{i\leq j\leq n}]). \tag{5}$$

It follows from inequalities and the size-reducedness of $(\boldsymbol{c}_i)_{1\le i\le n}$ that $\text{size}(C) = \mathcal{P}oly(n) \cdot \text{size}(B)$.

We call δ-LLL' the successive application of the above algorithm based on BKZ'$_2$ and LLL with parameter δ. We are to prove that the number of loop iterations performed by δ-LLL is $\mathcal{P}oly(n)$.

Theorem 3. *Given as inputs a basis $B \in \mathbb{Z}^{n\times n}$ of a lattice L and $\delta < 1$, algorithm δ-LLL' algorithm outputs a δ-LLL-reduced basis of L within $\mathcal{P}oly(n) \cdot \widetilde{O}(\text{size}(B))$ bit operations.*

Proof. With the same notations as above, it suffices to prove that given as input $(\boldsymbol{c}_i)_{i\le n}$, algorithm δ-LLL terminates within $\mathcal{P}oly(n) \cdot \widetilde{O}(\text{size}(C))$ bit operations. Let $(\boldsymbol{c}'_i)_{i\le n}$ be the output basis. As size-reductions can be performed in time $\mathcal{P}oly(n) \cdot \widetilde{O}(\text{size}(C)))$, it suffices to show that the number of loop iterations of δ-LLL given $(\boldsymbol{c}_i)_{i\le n}$ as input is $\mathcal{P}oly(n)$. To do this, it suffices to bound $\frac{\Pi((\boldsymbol{c}_i)_{i\le n})}{\Pi((\boldsymbol{c}'_i)_{i\le n})}$ by $2^{\mathcal{P}oly(n)}$.

First of all, we have $\lambda_1(L[(\boldsymbol{c}_j^{(i)})_{i\le j\le n}]) \le \lambda_i(L)$, for all $i \le n$. Indeed, let $\boldsymbol{v}_1, \ldots, \boldsymbol{v}_i \in L$ be linearly independent such that $\max_{j\le i} \|\boldsymbol{v}_j\| \le \lambda_i(L)$; at least one of them, say $\boldsymbol{v}_1$, remains non-zero when projected orthogonally to Span $(\boldsymbol{c}_j)_{j<i}$. We thus have $\lambda_1(L[(\boldsymbol{c}_j^{(i)})_{i\le j\le n}]) \le \|\boldsymbol{v}_1\| \le \lambda_i(L)$. Now, using (5), we obtain:

$$\Pi((\boldsymbol{c}_i)_{i\le n}) = \prod_{i=1}^{n} \|\boldsymbol{c}_i^*\|^{2(n-i+1)} \le 2^{O(n^3)} \prod_{i=1}^{n} \lambda_i(L)^{2(n-i+1)}.$$

On the other hand, we have (see [16, (1.7)]) $\lambda_i(L) \le \max_{j\le i} \|\boldsymbol{c}'_j\| \le (\frac{1}{\sqrt{\delta - 1/4}})^{i-1} \|\boldsymbol{c}'^*_i\|$, for all $i \le n$. As a consequence, we have $\Pi((\boldsymbol{c}'_i)_{i\le n}) \ge 2^{-O(n^3)} \cdot \prod_{i=1}^{n} \lambda_i(L)^{2(n-i+1)}$. This completes the proof. □

Acknowledgments. We thank N. Gama and P. Q. Nguyen for explaining to us their bound on the number of tours of the original BKZ algorithm. We also thank C.-P. Schnorr for helpful discussions. The authors were partly supported by the LaRedA ANR grant and an ARC Discovery Grant DP110100628.

References

1. Ajtai, M.: Generating hard instances of lattice problems (extended abstract). In: Proc. of STOC, pp. 99–108. ACM, New York (1996)
2. Ajtai, M., Kumar, R., Sivakumar, D.: A sieve algorithm for the shortest lattice vector problem. In: Proc. of STOC, pp. 601–610. ACM, New York (2001)
3. Akhavi, A.: Worst-case complexity of the optimal LLL algorithm. In: Gonnet, G.H., Viola, A. (eds.) LATIN 2000. LNCS, vol. 1776, pp. 355–366. Springer, Heidelberg (2000)
4. Cadé, D., Pujol, X., Stehlé, D.: fplll-3.1, a floating-point LLL implementation, `http://perso.ens-lyon.fr/damien.stehle`

5. Coppersmith, D.: Small solutions to polynomial equations, and low exponent RSA vulnerabilities. Journal of Cryptology 10(4), 233–260 (1997)
6. Gama, N., Howgrave-Graham, N., Koy, H., Nguyên, P.Q.: Rankin's constant and blockwise lattice reduction. In: Dwork, C. (ed.) CRYPTO 2006. LNCS, vol. 4117, pp. 112–130. Springer, Heidelberg (2006)
7. Gama, N., Nguyen, P.Q.: Finding short lattice vectors within Mordell's inequality. In: Proc. of STOC, pp. 207–216. ACM, New York (2008)
8. Gama, N., Nguyen, P.Q.: Predicting lattice reduction. In: Smart, N.P. (ed.) EUROCRYPT 2008. LNCS, vol. 4965, pp. 31–51. Springer, Heidelberg (2008)
9. Goldreich, O., Goldwasser, S., Halevi, S.: Collision-free hashing from lattice problems. TR96-056 (1996), http://www.eccc.uni-trier.de/
10. Hanrot, G., Pujol, X., Stehlé, D.: Terminating BKZ. Cryptology ePrint Archive (2011), http://eprint.iacr.org/2011/198
11. Hanrot, G., Stehlé, D.: Improved analysis of Kannan's shortest lattice vector algorithm (extended abstract). In: Menezes, A. (ed.) CRYPTO 2007. LNCS, vol. 4622, pp. 170–186. Springer, Heidelberg (2007)
12. Haviv, I., Regev, O.: Tensor-based hardness of the shortest vector problem to within almost polynomial factors. In: Proc. of STOC, pp. 469–477. ACM, New York (2007)
13. Hirschhorn, P.S., Hoffstein, J., Howgrave-Graham, N., Whyte, W.: Choosing NTRUEncrypt parameters in light of combined lattice reduction and MITM approaches. In: Abdalla, M., Pointcheval, D., Fouque, P.-A., Vergnaud, D. (eds.) ACNS 2009. LNCS, vol. 5536, pp. 437–455. Springer, Heidelberg (2009)
14. Hoffstein, J., Pipher, J., Silverman, J.H.: NTRU: A ring-based public key cryptosystem. In: Buhler, J.P. (ed.) ANTS 1998. LNCS, vol. 1423, pp. 267–288. Springer, Heidelberg (1998)
15. Kannan, R.: Improved algorithms for integer programming and related lattice problems. In: Proc. of STOC, pp. 99–108. ACM, New York (1983)
16. Lenstra, A.K., Lenstra Jr., H.W., Lovász, L.: Factoring polynomials with rational coefficients. Math. Ann. 261, 515–534 (1982)
17. Lenstra Jr., H.W.: Flags and lattice basis reduction. In: Proceedings of the Third European Congress of Mathematics, vol. 1. Birkhäuser, Basel (2001) **1**
18. Lovász, L.: An Algorithmic Theory of Numbers, Graphs and Convexity. CBMS-NSF Regional Conference Series in Applied Mathematics. SIAM, Philadelphia (1986)
19. Madritsch, M. G., Vallée, B.: Modelling the LLL algorithm by sandpiles. In: López-Ortiz, A. (ed.) LATIN 2010. LNCS, vol. 6034, pp. 267–281. Springer, Heidelberg (2010)
20. Martinet, J.: Perfect Lattices in Euclidean Spaces. Springer, Heidelberg (2002)
21. Micciancio, D., Regev, O.: Lattice-based cryptography. In: Bernstein, D.J., Buchmann, J., Dahmen, E. (eds.) Post-Quantum Cryptography, pp. 147–191. Springer, Heidelberg (2009)
22. Micciancio, D., Voulgaris, P.: A deterministic single exponential time algorithm for most lattice problems based on Voronoi cell computations. In: Proc. of STOC, pp. 351–358. ACM, New York (2010)
23. Micciancio, D., Voulgaris, P.: Faster exponential time algorithms for the shortest vector problem. In: Proc. of SODA. ACM, New York (2010)
24. Nguyên, P.Q.: Cryptanalysis of the Goldreich-Goldwasser-Halevi cryptosystem from Crypto'97. In: Wiener, M. (ed.) CRYPTO 1999. LNCS, vol. 1666, pp. 288–304. Springer, Heidelberg (1999)

25. Nguyên, P.Q., Stehlé, D.: LLL on the average. In: Hess, F., Pauli, S., Pohst, M. (eds.) ANTS 2006. LNCS, vol. 4076, pp. 238–256. Springer, Heidelberg (2006)
26. Nguyen, P.Q., Stehlé, D.: An LLL algorithm with quadratic complexity. SIAM J. Comput. 39(3), 874–903 (2009)
27. Nguyên, P.Q., Stern, J.: The two faces of lattices in cryptology. In: Silverman, J.H. (ed.) CaLC 2001. LNCS, vol. 2146, pp. 146–180. Springer, Heidelberg (2001)
28. Nguyen, P.Q., Vallée, B. (eds.): The LLL Algorithm: Survey and Applications. Information Security and Cryptography. Springer, Heidelberg (2009)
29. Novocin, A., Stehlé, D., Villard, G.: An LLL-reduction algorithm with quasi-linear time complexity. To Appear in the Proceedings of STOC (2011), http://prunel.ccsd.cnrs.fr/ensl-00534899/en
30. Pujol, X., Stehlé, D.: Rigorous and efficient short lattice vectors enumeration. In: Pieprzyk, J. (ed.) ASIACRYPT 2008. LNCS, vol. 5350, pp. 390–405. Springer, Heidelberg (2008)
31. Regev, O.: The learning with errors problem. In: Invited Survey in CCC 2010 (2010), http://www.cs.tau.ac.il/~odedr/
32. Schnorr, C.P.: Progress on LLL and lattice reduction. In: [28]
33. Schnorr, C.P.: A hierarchy of polynomial lattice basis reduction algorithms. Theor. Comput. Science 53, 201–224 (1987)
34. Schnorr, C.P.: Block reduced lattice bases and successive minima. Combinatorics, Probability and Computing 3, 507–533 (1994)
35. Schnorr, C.P.: Accelerated slide- and LLL-reduction. Electronic Colloquium on Computational Complexity (ECCC) 11(50) (2011)
36. Schnorr, C.P., Euchner, M.: Lattice basis reduction: Improved practical algorithms and solving subset sum problems. In: Budach, L. (ed.) FCT 1991. LNCS, vol. 529, pp. 68–85. Springer, Heidelberg (1991)
37. Schnorr, C.P., Euchner, M.: Lattice basis reduction: improved practical algorithms and solving subset sum problems. Mathematics of Programming 66, 181–199 (1994)
38. Schnorr, C.P., Hörner, H.H.: Attacking the Chor-Rivest cryptosystem by improved lattice reduction. In: Guillou, L.C., Quisquater, J.-J. (eds.) EUROCRYPT 1995. LNCS, vol. 921, pp. 1–12. Springer, Heidelberg (1995)
39. Schönhage, A.: Fast reduction and composition of binary quadratic forms. In: Proceedings of the 1991 International Symposium on Symbolic and Algebraic Computation (ISSAC 1991), pp. 128–133. ACM, New York (1991)
40. Shoup, V.: NTL, Number Theory C++ Library, http://www.shoup.net/ntl/
41. Yap, C.K.: Fast unimodular reduction: planar integer lattices. In: Proceedings of the 1992 Symposium on the Foundations of Computer Science (FOCS 1992), pp. 437–446. IEEE Computer Society Press, Los Alamitos (1992)

Pseudorandom Knapsacks and the Sample Complexity of LWE Search-to-Decision Reductions*

Daniele Micciancio and Petros Mol

Department of Computer Science & Engineering,
University of California, San Diego
{daniele,pmol}@cs.ucsd.edu

Abstract. We study the pseudorandomness of bounded knapsack functions over arbitrary finite abelian groups. Previous works consider only specific families of finite abelian groups and 0-1 coefficients. The main technical contribution of our work is a new, general theorem that provides sufficient conditions under which pseudorandomness of bounded knapsack functions follows directly from their one-wayness. Our results generalize and substantially extend previous work of Impagliazzo and Naor (J. Cryptology 1996).

As an application of the new theorem, we give sample preserving search-to-decision reductions for the Learning With Errors (LWE) problem, introduced by (Regev, STOC 2005) and widely used in lattice-based cryptography. Concretely, we show that, for a wide range of parameters, m LWE samples can be proved indistinguishable from random just under the hypothesis that search LWE is a one-way function for the same number m of samples.

1 Introduction

The Learning With Errors (LWE) problem, introduced by Regev in [31], is the problem of recovering a secret n-dimensional integer vector $\mathbf{s} \in \mathbb{Z}_q^n$, given a collection of perturbed random equations $\mathbf{a}_i\mathbf{s} \approx b_i$ where $\mathbf{a}_i \in \mathbb{Z}_q^n$ is chosen uniformly at random and $b_i = \mathbf{a}_i\mathbf{s} + e_i$ for some small, randomly chosen error term e_i. In recent years, LWE has been used to substantially expand the scope of lattice based cryptography, yielding solutions to many important cryptographic tasks, including public key encryption secure against passive [31,20,29] and active attacks [30,28], (hierarchical) identity based encryption [14,10,1,2], digital signatures [14,10], oblivious transfer protocols [29], several forms of leakage resilient encryption [5,6,11,16], homomorphic encryption [13] and more. The versatility

* This research was supported in part by NSF under grants CNS-0831536 and CNS-0716790. Any opinions, findings, and conclusions or recommendations expressed in this material are those of the authors and do not necessarily reflect the views of the National Science Foundation. This is an extended abstract of the work. For the full version, see the authors' webpage.

P. Rogaway (Ed.): CRYPTO 2011, LNCS 6841, pp. 465–484, 2011.

of the LWE problem in the construction of a plethora of cryptographic applications is due in large part to its pseudorandomness properties: as proved in [31], if recovering (with high[1] probability) the secret $\mathbf{s}$ from the samples $(\mathbf{a}_i, \mathbf{a}_i\mathbf{s} + e_i)$ is computationally hard, then it is also hard to distinguish the LWE samples $(\mathbf{a}_i, \mathbf{a}_i\mathbf{s} + e_i)$ from uniformly random ones $(\mathbf{a}_i, b_i)$ where the $b_i \in \mathbb{Z}_q$ are chosen uniformly and independently at random. In other words, any efficient distinguisher (between the LWE and uniform distributions) can be turned into an inverter that recovers the secret $\mathbf{s}$, with only a polynomial slow-down.

On the theoretical side, cryptography based on LWE is supported by deep worst-case/average-case connections [31,28], showing that any algorithm that solves LWE (on the average) can be efficiently converted into a (quantum) algorithm to solve the (worst-case) hardest instances of several famous lattice approximation problems which are believed to be intractable, including approximating the minimum distance of a lattice within factors that grow polynomially in the dimension, and related problems [23]. It should be remarked that, while such proofs of security based on worst-case lattice assumptions provide a solid theoretical justification for the probability distributions used in LWE cryptography, they are hardly useful in practice: in order to get meaningful estimates on the hardness of breaking LWE cryptography, it is generally more useful and appropriate to conjecture the average-case hardness of solving LWE, and use that as a starting point. In fact, all recent work aimed at determining appropriate key sizes and security parameters [26,22,33] follows this approach, and investigates experimentally the concrete hardness of solving LWE on the average.

In light of that, LWE is best formulated as the problem of inverting the one-way function family (indexed by a random matrix $\mathbf{A} \in \mathbb{Z}_q^{m \times n}$, where m is the number of samples) that maps the secret $\mathbf{s}$ and error vector $\mathbf{e}$ to $\mathbf{Ax} + \mathbf{e}$. The search-to-decision reduction of [31] shows that if the LWE function family is one-way, then it is also a good pseudorandom generator. However, the reduction in [31] somehow hides a very important detail: the value of m for which the function is assumed to be one-way is much higher (still polynomially related) to the value of m for which the output of the function is pseudorandom. While theoretical results based on worst-case lattice problems are fairly insensitive to the value of m (i.e., the number of samples used in the LWE instance), this number becomes more important and relevant when considering concrete attacks on the average-case hardness of LWE.

For instance, recent algorithmic results [7], show that when the errors e_i are sufficiently small, the LWE problem can be solved in subexponential (or even polynomial) time, provided a sufficiently large (but still polynomial) number of samples is available. Therefore, for certain ranges of the parameters, the number of available samples can have a significant impact on the computational hardness of the LWE problem. Likewise, some lattice attacks perform better in practice when given many (typically $\omega(n)$) samples [26]. However, LWE-based encryption schemes (e.g., see [22]) typically expose only a small number of samples (say,

[1] Due to the self-reducibility properties of the LWE problem, here "high" can be interpreted in a variety of ways, ranging from "nonnegligible" to "very close to 1".

comparable to the dimension n of the LWE secret $\mathbf{s}$) during key generation and encryption. Fixing the number of available samples to a small value may significantly reduce the effectiveness of concrete attacks, and increase our confidence in the security of the schemes.

It should also be noted that when the number of available samples is above a certain threshold, one can efficiently generate an arbitrary number of additional samples [14,6,32], but at the cost of increasing the magnitude of the errors. So, for certain other ranges of the parameters the impact of increasing the number of samples may not be as critical as in [7]. Still, even in such situations, using a large number of samples comes at the price of lowering the quality of the samples, which can negatively impact the concrete security and performance of LWE-based cryptographic functions.

This motivates the following question: how big of a blow-up in the number of samples is required to prove the pseudorandomness of the LWE problem, based on the conjectured hardness of its search (secret recovery) version? The main result of this paper is that, perhaps surprisingly, in most common applications of LWE in cryptography, no such blow-up is necessary at all: there is a *sample preserving* reduction from solving the search LWE problem (with nonnegligible success probability) to the problem of distinguishing the LWE distribution from random (with nonnegligible advantage). At the core of our result is a general theorem about the pseudorandomness of the bounded knapsacks over arbitrary groups, that substantially extends previous results in the area and might be of independent interest.

Contributions and Applications. Let $(G, +)$ be a finite abelian group, and $\mathbf{g} = (g_1, \ldots, g_m) \in G^m$ a sequence of group elements chosen uniformly at random. The group elements $\mathbf{g}$ define a knapsack function $f_{\mathbf{g}}(\mathbf{x})$ that maps the vector $\mathbf{x} \in \mathbb{Z}^m$ to the group element $f_{\mathbf{g}}(\mathbf{x}) = \sum_i x_i g_i$. If the input $\mathbf{x}$ is restricted to vectors with small entries, then for a large variety of groups G, $f_{\mathbf{g}}$ is conjectured to be a one-way function family, i.e., a family of functions that are hard to invert on average when the key $\mathbf{g}$ is chosen uniformly at random. For example, when the input $\mathbf{x}$ is restricted to the set $\{0,1\}^m$ of binary vectors, inverting $f_{\mathbf{g}}$ is the famous subset-sum problem, which is conjectured to be hard to solve on average, and has been extensively studied in cryptography. In a classic paper [18], Impagliazzo and Naor showed that for some specific, but representative, choices of the group G, if the subset-sum function is one-way, then it is also a pseudorandom generator, i.e., it is computationally hard to distinguish $(\mathbf{g}, f_{\mathbf{g}}(\mathbf{x}))$ from a uniformly random element of G^{m+1}, when $\mathbf{g} \in G^m$ and $\mathbf{x} \in \{0,1\}^m$ are chosen uniformly at random. We generalize the results of [18] in two respects:

- We consider functions over arbitrary groups G. Only groups of the form $\mathbb{Z}_N$ were considered in [18], and for two specific (but representative) choices of N (prime and power of 2).
- We consider input coefficients x_i that take values from a set $\{0, \ldots, s\}$ (or $\{-s, \ldots, s\}$) for any (polynomially bounded) s. Moreover, we consider *arbitrary* input distributions. By contrast, the results in [18] hold for inputs $\mathbf{x}$ distributed *uniformly* with coefficients in $\{0,1\}$.

Both extensions are essential for the sample-preserving search-to-decision LWE reduction presented in Section 4.2, which requires the pseudorandomness of the knapsack function over vector groups $G = \mathbb{Z}_q^k$, and for inputs $\mathbf{x}$ following a nonuniform (Gaussian) distribution over a sufficiently large set $\{-s, \ldots, s\}$. Our main technical result (Theorem 2) shows that for any group G and input distribution $\mathcal{X}$, the output of the knapsack function is pseudorandom provided the following two conditions hold:

1. $f_{\mathbf{g}}$ is a one-way function family with respect to input distribution $\mathcal{X}$, and
2. certain folded versions of $f_{\mathbf{g}}$ (where both the key $\mathbf{g}$ and the output $f_{\mathbf{g}}(\mathbf{x})$ are projected onto a quotient group $G_d = G/dG$ for some $d \in \mathbb{Z}$,) have pseudorandom output.

The second condition above may seem to make the statement in the theorem vacuous, as it asserts the pseudorandomness of $f_{\mathbf{g}}$ assuming the pseudorandomness of (certain other versions of) $f_{\mathbf{g}}$. The power of the theorem comes from the fact that the quotient groups G_d considered are very small, so small that in many important settings the output of the folded knapsack function $f_{\mathbf{g}}(\mathbf{x}) \bmod dG$ is *statistically* close to uniform. As a technical tool, we provide upper bounds on the statistical distance between the distribution $(\mathbf{g}, f_{\mathbf{g}}(\mathbf{x})) \bmod dG$ and the uniform distribution over G_d^{m+1} (Lemma 4). We use these bounds to show that for many interesting groups and input distributions, the output of the folded knapsack function is statistically close to uniform. Therefore, as a corollary to the main theorem, we get that one-wayness of the bounded knapsack function implies that knapsacks are good pseudorandom generators. Specific groups and input distributions for which this holds include among others:

- Groups whose order contains only large prime factors, larger than the maximum value of the input coefficients. Cyclic groups with prime order and vector groups $\mathbb{Z}_p^k$ for prime p fall into this category. This result generalizes those in [18] from uniform binary input to arbitrary input distributions.
- Distributions that, when folded (modulo small divisors of the order of G,) maintain high entropy relative to the size of the quotient group G/dG. (See Theorem 6.) Groups of the form $G = \mathbb{Z}_{2^\ell}^k$ and uniform input distribution over $\mathbb{Z}_{2^i}^m$ for some $i < \ell$ satisfy this requirement. This parameter set is a very attractive choice in practice since both group operations and input sampling are particularly efficient and easy to implement.

Using the duality between LWE and the knapsack problem [35,25], we obtain sample preserving search-to-decision reductions for LWE for several interesting choices of the modulus q and input distribution, which include (among others):

- $q = 2$ and *any* error distribution. This directly proves the pseudorandomness of the well-known Learning Parity with Noise (LPN) problem, as already established in [9,19], but in a sample-preserving manner.
- prime q and *any polynomially bounded* error distribution.
- power-of-prime modulus $q = p^e$ for p large enough so that the error distribution is concentrated over $\{-(p-1)/2, \ldots, (p-1)/2\}$.
- $q = p^e$ for small prime p and uniform error distribution over $\mathbb{Z}_{p^i}$ $(i < e)$.

These results subsume (see below) several previous pseudorandomness results for LWE [31,6] and LPN [19] but with an important difference. While the proofs in [31,6,19] require that LWE (resp. LPN) is hard to solve for a very large number of samples, our reductions are *sample preserving*: the pseudorandomness of LWE (resp. LPN) holds, provided the same problem is one-way for the *same* number of samples. We remark that previous results are often phrased as reductions from solving the LWE search problem *with high probability*, to solving the LWE decision problem *with nonnegligible* advantage, combining the search-to-decision reduction and success probability amplification into a single statement. By contrast, our reduction shows how to solve the LWE search problem with *nonnegligible probability*. Our results subsume previous work in the sense that the LWE search problem can be solved with high probability by first invoking our reduction, and then amplifying the success probability using standard repetition techniques. Of course, any such success probability amplification would naturally carry the cost of a higher sample complexity. We remark that a close inspection of worst-case to average-case reductions for LWE [31,28] shows that these reductions directly support the conjecture that LWE is a *strong* one-way function. As already discussed, worst-case to average-case reductions do not provide quantitatively interesting results, and are best used as qualitative arguments to support the conjecture that certain problems are computationally hard on average. Under the standard conjecture that search LWE is a strong one-way function, the results in this paper offer a fairly tight, and sample preserving proof that LWE is also a good pseudorandom generator, which can be efficiently used for the construction of many other lattice based public key cryptographic primitives. By contrast, it is not known how to take advantage of the strong one-wayness of LWE within previous search-to-decision reductions, resulting in a major degradation of the parameters. Of course, if we change the complexity assumption, and as a starting point we use the worst-case hardness of lattice problems or the assumption that LWE is only a *weak* one-way function, then our reduction will also necessarily incur a large blow up in sample complexity through amplification, and lead to quantitatively uninteresting results.

2 Preliminaries

We use $\mathbb{N}, \mathbb{C}, \mathbb{T}$ for the sets of natural, complex and complex numbers of unit magnitude respectively. We use lower case for scalars, upper case for sets, bold lower case for vectors and bold upper case for matrices. We use calligraphic letters for probability distributions and (possibly randomized) algorithms. For $s \in \mathbb{N}$, the set of the first s nonnegative integers is denoted $[s] = \{0, 1, \ldots, s-1\}$.

2.1 Probability

We write $x \leftarrow \mathcal{X}$ for the operation of selecting x according to a probability distribution $\mathcal{X}$ or by running probabilistic algorithm $\mathcal{X}$. We use $\{(x, x') \mid x \leftarrow \mathcal{X}, x' \leftarrow \mathcal{X}\}$ to denote the probability distribution obtained by drawing two samples from $\mathcal{X}$ independently at random. For any probability distribution $\mathcal{X}$ over set X and any value $x \in X$, $\Pr\{x \leftarrow \mathcal{X}\}$ is the probability associated to x by distribution $\mathcal{X}$. The uniform distribution over a set A is denoted $\mathcal{U}(A)$, and the support of a distribution $\mathcal{X}$ is denoted $[\mathcal{X}] = \{x \in X \mid \Pr\{x \leftarrow \mathcal{X}\} > 0\}$. The *collision probability* of $\mathcal{X}$ is the probability $\mathsf{Col}(\mathcal{X}) = \Pr\{x = x' \mid x \leftarrow \mathcal{X}, x' \leftarrow \mathcal{X}\} = \sum_{x \in [\mathcal{X}]} \Pr\{x \leftarrow \mathcal{X}\}^2$ that two independent identically distributed samples from $\mathcal{X}$ take the same value. The *mode* of $\mathcal{X}$ is the probability of the most likely value, i.e. $\mathsf{mode}(\mathcal{X}) = \max_{x \in X} \Pr\{x \leftarrow \mathcal{X}\}$. It is easy to see that $\mathsf{Col}(\mathcal{X}) \leq \mathsf{mode}(\mathcal{X})$.

The *statistical distance* $\Delta(\mathcal{X}, \mathcal{Y})$ between distributions $\mathcal{X}$ and $\mathcal{Y}$, defined over the same set X, is the quantity $\frac{1}{2}\sum_{x \in X} |\Pr\{x \leftarrow \mathcal{X}\} - \Pr\{x \leftarrow \mathcal{Y}\}|$. The statistical distance satisfies $\Delta(f(\mathcal{X}), f(\mathcal{Y})) \leq \Delta(\mathcal{X}, \mathcal{Y})$ for any (possibly probabilistic) function f. Two distributions $\mathcal{X}, \mathcal{Y}$ are ϵ-*close* if $\Delta(\mathcal{X}, \mathcal{Y}) \leq \epsilon$. They are (t, ϵ)-*indistinguishable* if $\Delta(\mathcal{D}(\mathcal{X}), \mathcal{D}(\mathcal{Y})) \leq \epsilon$ for *any* probabilistic predicate $\mathcal{D} \colon X \to \{0,1\}$ (called the *distinguisher*) computable in time at most t. Otherwise, we say that $\mathcal{X}, \mathcal{Y}$ are (t, ϵ)-*distinguishable*. When $\mathcal{Y} = \mathcal{U}(X)$ is the uniform distribution, we use $\Delta_U(\mathcal{X}) = \Delta(\mathcal{X}, \mathcal{U}(X))$ as an abbreviation and say that $\mathcal{X}$ is ϵ-*random* (resp. (t, ϵ)-*pseudorandom*) if it is ϵ-close (resp. (t, ϵ)-close) to $\mathcal{U}(X)$.

Function families. A function family $(F, \mathcal{X})$ is a collection $F = \{f_i \colon X \to R\}_{i \in I}$ of functions indexed by $i \in I$ with common domain X and range R, together with a probability distribution $\mathcal{X}$ over the input set $X \supseteq [\mathcal{X}]$. For simplicity, in this paper we always assume that the set of functions is endowed with the *uniform* probability distribution $\mathcal{U}(F)$. Each function family $(F, \mathcal{X})$ naturally defines a probability distribution

$$\mathcal{F}(F, \mathcal{X}) = \{(f, f(x)) \mid f \leftarrow \mathcal{U}(F), x \leftarrow \mathcal{X}\} \tag{1}$$

obtained by selecting a function at random and evaluating it at a random input.

A function family $\mathcal{F} = (F, \mathcal{X})$ is called (t, ϵ)-*one-way* if there is no (probabilistic) algorithm $\mathcal{I}$ running in time at most t such that $\Pr\{f(x) = y \mid (f, y) \leftarrow \mathcal{F}(F, \mathcal{X}), x \leftarrow \mathcal{I}(f, y)\} \geq \epsilon$. In this paper it is convenient to use the related notion of "uninvertible function". A (t, ϵ)-*inverter* for a function family $(F, \mathcal{X})$ is a (probabilistic) algorithm $\mathcal{I}$ running in time at most t such that $\Pr\{x = y \mid f \leftarrow \mathcal{U}(F), x \leftarrow \mathcal{X}, y \leftarrow \mathcal{I}(f, f(x))\} \geq \epsilon$. If there exists a (t, ϵ)-inverter for a function family $(F, \mathcal{X})$, then we say that $(F, \mathcal{X})$ is (t, ϵ)-invertible. A function family such that there is no (t, ϵ)-inverter is called (t, ϵ)-*uninvertible*. In this paper, we deal with function families that are (almost) injective, i.e. with overwhelming probability over $f \leftarrow \mathcal{U}(F)$ and $x \leftarrow \mathcal{X}$, there exists no $x' \neq x$ such that $f(x) = f(x')$. When this is the case, then one-wayness and uninvertibility are equivalent notions. A (t, ϵ)-*pseudorandom generator family* is a function family $(F, \mathcal{X})$ such that the associated distribution (1) is (t, ϵ)-pseudorandom, i.e., it is (t, ϵ)-indistinguishable from the uniform distribution $\mathcal{U}(F \times R)$.

Asymptotics. We use n as a (security) parameter that controls all other parameters. Unless otherwise stated, any other parameter (say m) will be polynomially related to n. We use standard asymptotic notation $O(\cdot), \Omega(\cdot), o(\cdot), \omega(\cdot)$, etc.

We write $\mathsf{negl}(n)$ for the set of negligible functions and $\mathsf{poly}(n)$ for the set of polynomially bounded functions. In the asymptotic computational complexity setting, one often considers probability ensembles, i.e., sequences $\mathcal{X} = (\mathcal{X}_n)_{n\in\mathbb{N}}$ of probability distributions over possibly different sets $X_n \supseteq [\mathcal{X}_n]$. Two distributions ensembles $\mathcal{X} = (\mathcal{X}_n)_{n\in\mathbb{N}}$ and $\mathcal{Y} = (\mathcal{Y}_n)_{n\in\mathbb{N}}$ are *statistically close* (denoted $\mathcal{X} \overset{s}{\approx} \mathcal{Y}$) if $\mathcal{X}_n$ and $\mathcal{Y}_n$ are $\epsilon(n)$-close for some negligible function $\epsilon(n) = \mathsf{negl}(n)$. The ensembles $\mathcal{X}$ and $\mathcal{Y}$ are *computationally indistinguishable* (denoted $\mathcal{X} \overset{c}{\approx} \mathcal{Y}$) if $\mathcal{X}_n$ and $\mathcal{Y}_n$ are $(t(n), \epsilon(n))$-indistinguishable for $\epsilon(n) = \mathsf{negl}(n)$ and any $t(n) = \mathsf{poly}(n)$ under a uniform sequence $(\mathcal{D}_n \colon X_n \to \{0,1\})_{n\in\mathbb{N}}$ of distinguishers. Definitions for function families are also extended in the obvious way to function family ensembles $\mathcal{F} = (\mathcal{F}_n)_n$ in the asymptotic setting by taking $\epsilon(n) = \mathsf{negl}(n)$ and $t(n) = \mathsf{poly}(n)$, and considering uniform sequences of distinguishing algorithms. In particular, a function family ensemble $\mathcal{F} = (\mathcal{F}_n)_n$ is *one-way* if $\mathcal{F}_n$ is $(t(n), \epsilon(n))$-one-way for $\epsilon(n) = \mathsf{negl}(n)$ and any $t(n) = \mathsf{poly}(n)$. It is *pseudorandom* if the associated (asymptotic) distribution (1) is $(t(n), \epsilon(n))$-pseudorandom, i.e., it is $(t(n), \epsilon(n))$-indistinguishable from the uniform distribution $\mathcal{U}(F_n \times R_n)$.

Discrete Gaussian Distributions. Gaussian-like distributions play a central role in the Learning With Errors (LWE) problem. For each sample $(\mathbf{a}, b = \mathbf{a} \cdot \mathbf{s} + e)$, the distribution χ from which e is drawn, is a normal distribution over the integers. Below, we focus mainly on the *discrete* Gaussian distribution and provide bounds on its collision probability. Those bounds are used in establishing the search-to-decision reduction for LWE. Similar bounds can be established for the *discretized* Gaussian (defined in [31]).

The *Gaussian function* on $\mathbb{R}^m$ with parameter r and center $\mathbf{c}$ is defined as $\forall \mathbf{x} \in \mathbb{R}^m$, $\rho_{r,\mathbf{c}}(\mathbf{x}) = exp(-\pi\|\mathbf{x} - \mathbf{c}\|^2/r^2)$. The *discrete Gaussian distribution* over a countable set S is defined as

$$\forall \mathbf{x} \in S,\ \mathcal{D}_{S,r,\mathbf{c}} = \frac{\rho_{r,\mathbf{c}}(\mathbf{x})}{\sum_{\mathbf{y}\in S} \rho_{r,\mathbf{c}}(\mathbf{y})}$$

Here, we are interested in vectors $\mathbf{x} = (x_1, \ldots, x_m) \in \mathbb{Z}^m$ distributed according to $\mathcal{D}_{\mathbb{Z}^m,r}$ ($\mathbf{c} = \mathbf{0}$). In that case, each coordinate x_i of $\mathbf{x}$ is identically and independently distributed according to the 1-dimensional Gaussian $\mathcal{D}_{\mathbb{Z},r}$. For our search-to-decision reduction of LWE with discrete Gaussian error distribution, we need to consider the folded (1-dimensional) distribution $\mathcal{D}_{\mathbb{Z},r} \bmod d$. The following lemma bounds the collision probability of this distribution.

Lemma 1. *For any $r > 0$ and $d \in \mathbb{Z}$, we have $\mathsf{Col}(\mathcal{D}_{\mathbb{Z},r} \bmod d) \leq \frac{1}{r} + \frac{1}{d}$. Furthermore, if $r = d \cdot \omega(\sqrt{\log n})$, then $\mathsf{Col}(\mathcal{D}_{\mathbb{Z},r} \bmod d) \leq \frac{1}{d} + \mathsf{negl}(n)$.*

2.2 Groups and Knapsack Function Families

In this work, by group we always mean *finite abelian group*. We use additive notation for groups; 0_G is the *identity element*, $|G|$ is the *order* (size) of G and M_G its *exponent*, i.e. the smallest non-zero integer e such that $e \cdot g = 0_G$ for all $g \in G$. We use the dot product notation $\mathbf{x} \cdot \mathbf{y} = \sum_i x_i \cdot y_i$ both for the inner product of two vectors $\mathbf{x}, \mathbf{y} \in R^n$ with elements in a ring R, and also to take integer linear combinations $\mathbf{x} \in \mathbb{Z}^n$ of a vector $\mathbf{y} \in G^n$ with elements in an additive group. For $\mathbf{x} = (x_1, \ldots, x_n) \in R^n$ and $a \in R$, we also define $\mathbf{x} \cdot a = (x_1 \cdot a, \ldots, x_n \cdot a)$.

For any group G and (positive) integer d, we use G_d to denote the quotient group G/dG where dG is the subgroup $\{d \cdot g \mid g \in G\}$, in analogy with the usual notation $\mathbb{Z}_d = \mathbb{Z}/d\mathbb{Z}$ for the group of integers modulo d. Likewise, for an element $g \in G$, we use $g \bmod dG$ (or just $g \bmod d$) for the image of g under the natural homomorphism from G to G_d. For any integer vector $\mathbf{w} = (w_1, \ldots, w_r) \in \mathbb{Z}^r$, we write $\gcd_G(\mathbf{w}) = \gcd(w_1, \ldots, w_r, M_G)$ for the greatest common divisor of the elements of $\mathbf{w}$ and the group exponent.

Lemma 2. *For any group G and integer vector $\mathbf{w} \in \mathbb{Z}^r$, $\{\mathbf{w} \cdot \mathbf{g} \mid \mathbf{g} \leftarrow \mathcal{U}(G^r)\} = \mathcal{U}(\gcd_G(\mathbf{w}) \cdot G)$. In particular, $\Pr\left[\mathbf{w} \cdot \mathbf{g} = 0_G \mid \mathbf{g} \leftarrow \mathcal{U}(G^r)\right] = \frac{1}{|\gcd_G(\mathbf{w}) \cdot G|}$.*

Knapsack Families. For any group G and input distribution $\mathcal{X}$ over $\mathbb{Z}^m$, the knapsack family $\mathcal{K}(G, \mathcal{X})$ is the function family with input distribution $\mathcal{X}$ and set of functions $f_{\mathbf{g}} \colon [\mathcal{X}] \to G$ indexed by $\mathbf{g} \in G^m$ and defined as $f_{\mathbf{g}}(\mathbf{x}) = \mathbf{g} \cdot \mathbf{x} \in G$. We will often use $\mathbf{g}$ instead of $f_{\mathbf{g}}$ to describe a member function drawn from $\mathcal{K}(G, \mathcal{X})$. When $G, \mathcal{X}$ are clear from the context we will simply write $\mathcal{K}$. We often consider folded knapsack families $\mathcal{K}(G_d, \mathcal{X})$ over quotient groups G_d. For brevity, when G and $\mathcal{X}$ are clear from the context, we will write $\mathcal{K}_d$ instead of $\mathcal{K}(G_d, \mathcal{X})$. The following lemma shows that the distribution $\mathcal{F}(\mathcal{K}_d)$ associated to a folded knapsack function family is closely related to the distribution

$$\mathcal{F}_d(\mathcal{K}) = \{(\mathbf{g}, g + h) \mid (\mathbf{g}, g) \leftarrow \mathcal{F}(\mathcal{K}), h \leftarrow \mathcal{U}(d \cdot G)\}. \tag{2}$$

Lemma 3. *For any knapsack family $\mathcal{K}$ and $d \in \mathbb{Z}$, $\Delta_U(\mathcal{F}_d(\mathcal{K})) = \Delta_U(\mathcal{F}(\mathcal{K}_d))$. Also, $\mathcal{F}_d(\mathcal{K})$ is pseudorandom if and only if $\mathcal{F}(\mathcal{K}_d)$ is pseudorandom.*

For a group H such that $\mathcal{K}(H, \mathcal{X})$ compresses its input, Lemma 4 provides an upper bound on the statistical distance between $\mathcal{F}(\mathcal{K}(H, \mathcal{X}))$ and $\mathcal{U}(H^m \times H)$ by generalizing the Leftover Hash Lemma [17] to (non-necessarily universal) knapsack function families $\mathcal{K}(H, \mathcal{X})$ over arbitrary groups.

Lemma 4 (LHL, generalized). *For any finite abelian group H and integer d,*

$$\Delta_U(\mathcal{F}(\mathcal{K}(H, \mathcal{X}))) \leq \frac{1}{2} \sqrt{\sum_{1 < d \mid M} |H_d| \cdot \Pr\{\gcd_H(\mathbf{x} - \mathbf{y}) = d \mid \mathbf{x} \leftarrow \mathcal{X}, \mathbf{y} \leftarrow \mathcal{X}\}} \tag{3}$$

where M is the exponent of H and $d > 1$ ranges over all divisors of M.

2.3 Fourier Analysis and Learning

Fourier analysis has been used extensively in learning theory, especially in the context of learning functions defined over the boolean hypercube (see [21,8,27] for some examples). In Cryptography, two noteworthy examples are the

Kushilevitz-Mansour [21] formulation of the proof of the Goldreich-Levin [15] hard-core predicate for any one-way function and the proofs of hard-core predicates for several number-theoretic one-way functions by Akavia, Goldwasser and Safra [4].

Below we review some basic facts from Fourier analysis focusing on the discrete Fourier transform over finite abelian groups. We restrict the presentation to what is needed and refer the interested reader to [3,34] for more details.

Fourier Basics. Let H be a finite abelian group and $h_1, h_2 : H \to \mathbb{C}$ be functions from H to the complex numbers. The *inner product* of h_1 and h_2 is defined as

$$\langle h_1, h_2 \rangle = \mathop{\mathbb{E}}_{x \leftarrow \mathcal{U}(H)} \left[h_1(x)\overline{h_2(x)} \right] = \frac{1}{|H|} \sum_{x \in H} h_1(x)\overline{h_2(x)}$$

where $\bar{z}$ is the conjugate of $z \in \mathbb{C}$. The ℓ_2-norm and ℓ_∞-norm of h are defined as

$$\|h\|_2 = \sqrt{\langle h, h \rangle} \quad \text{and} \quad \|h\|_\infty = \max_{x \in H} |h(x)|.$$

The set of *characters* of H (denoted as $char(H)$) is the set of all the *homomorphisms* from H to the complex numbers of unit magnitude $\mathbb{T}$. Namely,

$$char(H) = \{\chi : H \to \mathbb{T} \mid \forall x, y \in H, \, \chi(x+y) = \chi(x) \cdot \chi(y)\}$$

When H is a *vector group*, i.e. $H \simeq \mathbb{Z}_k^\ell$, and $\boldsymbol{\alpha} = (\alpha_1, ..., \alpha_\ell) \in H$, then the character $\chi_{\boldsymbol{\alpha}} \colon H \to \mathbb{T}$ is defined as $\chi_{\boldsymbol{\alpha}}(\mathbf{x}) = (\omega_k)^{\sum_{i=1}^{\ell} \alpha_i x_i} = \omega_k^{\mathbf{x} \cdot \boldsymbol{\alpha}}$.

Fourier Transform. The *Fourier transform* of a function $h \colon H \to \mathbb{C}$ is the function $\widehat{h} \colon H \to \mathbb{C}$ defined as $\widehat{h}(\boldsymbol{\alpha}) = \langle h, \chi_{\boldsymbol{\alpha}} \rangle$. The Fourier transform measures the correlation of h with the characters in H.
The *energy* of a Fourier coefficient $\boldsymbol{\alpha}$ is defined as the square of its norm ($|\widehat{h}(\boldsymbol{\alpha})|^2$) while the *total energy* of h is defined as $\sum_{\boldsymbol{\alpha} \in H} |\widehat{h}(\boldsymbol{\alpha})|^2$. Parseval's identity says that $\sum_{\boldsymbol{\alpha} \in H} |\widehat{h}(\boldsymbol{\alpha})|^2 = \|h\|_2^2$.

Learning Heavy Fourier Coefficients. Let $\tau \in \mathbb{R}$, $\boldsymbol{\alpha} \in H$ and $h \colon H \to \mathbb{C}$ where H is a finite abelian group. Following the notation and terminology from [3], we say that $\boldsymbol{\alpha}$ is a τ-significant (or τ-heavy) Fourier coefficient of h if $|\widehat{h}(\boldsymbol{\alpha})|^2 \geq \tau$. The set of τ-significant Fourier coefficients of h is denoted by $Heavy_\tau(h)$, that is $Heavy_\tau(h) = \{\boldsymbol{\alpha} \in H \mid |\widehat{h}(\boldsymbol{\alpha})|^2 \geq \tau\}$. The following Theorem provides the conditions for learning heavy Fourier coefficients of functions defined over arbitrary finite groups and will be used in the proof of our main result.

Theorem 1. *(Significant Fourier Transform,[3, Theorem 3.3]) There exists a probabilistic algorithm ($\mathcal{SFT}$) that on input a threshold τ and given query access to a function $h \colon H \to \mathbb{C}$, returns all τ-heavy Fourier coefficients of h in time* $\mathsf{poly}(\log|H|, 1/\tau, \|h\|_\infty)$ *with probability*[2] *at least* $2/3$.

[2] The success probability is taken over the internal randomness of the $\mathcal{SFT}$ algorithm only, and can be amplified using standard repetition techniques. However, this is not needed in our context, so for simplicity we fix the success probability to 2/3.

3 Pseudorandomness of Knapsack Functions

In this section we establish the connection between the search and decision problems for families of bounded knapsack functions. The following theorem summarizes our main result.

Theorem 2 (Main). *Let $\mathcal{X}$ be a distribution over $[s]^m \subset \mathbb{Z}^m$ for some $s = \mathsf{poly}(n)$ and G be a finite abelian group. If $\mathcal{K}(G, \mathcal{X})$ is one-way and $\mathcal{K}(G_d, \mathcal{X})$ is pseudorandom for all $d < s$, then $\mathcal{K}(G, \mathcal{X})$ is pseudorandom.*

The proof of Theorem 2 makes use of the intermediate notion of (un)predictability defined below. Informally, for any $\ell \in \mathbb{N}$, a ℓ-predictor for a function family $(F, \mathcal{X})$ is a weak form of inverter algorithm that on input a function $f \in F$, a target value $f(\mathbf{x})$ and a query vector $\mathbf{r} \in \mathbb{Z}_\ell^m$, attempts to recover the value of $\mathbf{x} \cdot \mathbf{r} \pmod{\ell}$, rather than producing the entire input $\mathbf{x}$.

Definition 1. *For any $\ell \in \mathbb{N}$ and function family $(F, \mathcal{X})$ with domain $[\mathcal{X}] \subseteq \mathbb{Z}^m$, a* ℓ-predictor *for $(F, \mathcal{X})$ is a probabilistic algorithm $\mathcal{P}$ that on input $(f, y, \mathbf{r}) \in F \times R \times \mathbb{Z}_\ell^m$ outputs a value $\mathcal{P}(f, y, \mathbf{r}) \in \mathbb{Z}_\ell$ which is intended to be a guess for $\mathbf{x} \cdot \mathbf{r} \pmod{\ell}$. The error distribution of a predictor $\mathcal{P}$ is defined as*

$$\mathcal{E}_\ell(\mathcal{P}) = \{\mathbf{x} \cdot \mathbf{r} - \mathcal{P}(f, f(\mathbf{x}), \mathbf{r}) \pmod{\ell} \mid f \leftarrow \mathcal{U}(F), \mathbf{x} \leftarrow \mathcal{X}, \mathbf{r} \leftarrow \mathcal{U}(\mathbb{Z}_\ell^m)\}.$$

The bias *of a ℓ-predictor $\mathcal{P}$ is the quantity $\left|\sum_{k \in \mathbb{Z}_\ell} \Pr\{k \leftarrow \mathcal{E}_\ell(\mathcal{P})\} \cdot \omega_\ell^{-k}\right|$. If $\mathcal{P}$ runs in time t and has bias* at least *ϵ, we say that $\mathcal{P}$ is (t, ϵ)-biased. A function family $(F, \mathcal{X})$ that admits a (t, ϵ)-biased ℓ-predictor is* (t, ϵ, ℓ)-predictable.

The proof of Theorem 2 proceeds in two steps. In the first step (Lemma 5) we show that a certain (non-trivial) predictor for $\mathcal{K}$ implies a non-trivial inverter for $\mathcal{K}$. This step uses Fourier analysis and holds true for *any* function family (and not just for $\mathcal{K}$) with domain $[\mathcal{X}] \subseteq \mathbb{Z}^m$. In the second step (Proposition 2), we prove that if there exists a distinguisher for $\mathcal{K}(G, \mathcal{X})$, but no distinguisher for $\mathcal{K}(G_d, \mathcal{X})$ for small d, then there exists a predictor for $\mathcal{K}(G, \mathcal{X})$. This step is specific to knapsack families and depends on both the underlying group G and the distribution $\mathcal{X}$. The two steps combined yield Theorem 2. Sections 3.1 and 3.2 are devoted to each step of the reduction.

3.1 From Predictability to Invertibility

Proving that predictability implies invertibility is not specific to knapsack families. Rather, it holds for any function family $(F, \mathcal{X})$ with $F\colon X \to G$ where $X \subseteq \mathbb{Z}^m$ and G is a finite abelian group. Lemma 5 provides the conditions under which predictability implies invertibility.

Lemma 5. *Let $(F, \mathcal{X})$ be a function family with $[\mathcal{X}] \subseteq [s]^m \subset \mathbb{Z}^m$ for some $s = \mathsf{poly}(n)$. If $(F, \mathcal{X})$ is (t, ϵ, ℓ)-predictable for some $\ell > s$, then $(F, \mathcal{X})$ is $(\mathsf{poly}(n, \log \ell, 1/\epsilon) \cdot t, \frac{\epsilon}{3})$-invertible.*

Proof (Sketch). We use Fourier analysis and the $\mathcal{SFT}$ algorithm from Theorem 1. Let $\mathcal{P}$ be a ℓ-predictor for $\mathcal{F}$ that runs in time t. Roughly speaking, the inverter $\mathcal{I}$ on input $(f, f(\mathbf{x}))$ for some $f \leftarrow \mathcal{U}(F)$, simulates the execution of $\mathcal{SFT}$ in order to find $\mathbf{x}$. For every query $\mathbf{r} \in \mathbb{Z}_\ell^m$ issued by $\mathcal{SFT}$, $\mathcal{I}$ invokes $\mathcal{P}$ on appropriate input and sends back the result to $\mathcal{SFT}$. It turns out that, if the predictor $\mathcal{P}$ is (t, ϵ)-biased for some "sufficiently large" bias ϵ, then $\mathcal{I}$ simulates to $\mathcal{SFT}$ a (deterministic) function $h : \mathbb{Z}_\ell^m \to \mathbb{C}$ which is *highly correlated* with the character $\chi_{\mathbf{x}}(\cdot)$, that is, the function h $\mathcal{SFT}$ is given access to (through $\mathcal{I}, \mathcal{P}$) is such that $\widehat{h}(\mathbf{x})$ is "sufficiently heavy"[3] and therefore $\mathcal{SFT}$ will include $\mathbf{x}$ in the list it returns.

3.2 From Distinguishability to Predictability

We now proceed into proving that, under certain conditions, a distinguisher $\mathcal{D}$ for $\mathcal{K}(G, \mathcal{X})$ with noticeable distinguishing advantage implies a predictor for $\mathcal{K}(G, \mathcal{X})$ with noticeable bias. At a high level, the predictor works as follows: on input a modulus ℓ, function $\mathbf{g} \leftarrow \mathcal{U}(G^m)$, $y = \mathbf{g} \cdot \mathbf{x} \in G$ and $\mathbf{r} \in \mathbb{Z}_\ell^m$, it first makes a guess for the inner product $\mathbf{x} \cdot \mathbf{r} \bmod \ell$; it then uses that guess to modify the knapsack instance $(\mathbf{g}, y)$, and finally invokes the distinguisher $\mathcal{D}$ on the modified instance $(\mathbf{g}', y')$. To conclude, the output of $\mathcal{D}$ is used to determine whether the initial guess was correct or not. The same technique has been used by Impagliazzo and Naor in [18]. However, in the setting considered in [18] – subset-sum over a cyclic group of prime order[4] – the reduction is rather straightforward: if the guess for $\mathbf{x} \cdot \mathbf{r}$ is correct, then the modified knapsack instance $(\mathbf{g}', y')$ is distributed according to $\mathcal{F}(\mathcal{K}(G, \mathcal{X}))$, whereas if the guess is wrong, the distribution of $(\mathbf{g}', y')$ is (statistically close to) uniform. Therefore, a distinguisher with noticeable advantage implies almost immediately a 2-predictor with noticeable bias.

When considering general (not necessarily cyclic) abelian groups with possibly composite order and distributions $\mathcal{X}$ with $[\mathcal{X}] \not\subseteq \{0,1\}^m$, several technical difficulties arise. Unlike [18], if the guess for $\mathbf{x}\cdot\mathbf{r}$ is wrong, then the distribution of $(\mathbf{g}', y')$ can be statistically far from uniform. In fact, $(\mathbf{g}', y')$ can be distributed according to $\mathcal{F}_d(\mathcal{K}(G, \mathcal{X}))$ for any divisor d of the group exponent M_G. Depending on the order and structure of the underlying group, and the output distribution of the distinguisher $\mathcal{D}$ on the various auxiliary distributions $\mathcal{F}_d(\mathcal{K}(G, \mathcal{X}))$, the technical details of the reduction differ significantly. As a warm-up, we first state a weak form of our main Theorem.

Proposition 1. *If $\mathcal{K}(G, \mathcal{X})$ is (t, δ)-distinguishable from uniform for some noticeable δ, but $\mathcal{K}(G_d, \mathcal{X})$ is pseudorandom for all $d \le 2ms^2$ then there is a*

---

[3] In the context of polynomial time reductions "sufficiently high" and "sufficiently heavy" is to be interpreted as noticeable in the security parameter.

[4] [18] also consider cyclic groups with power-of-2 order but this makes their analysis only slightly more complicated.

poly(n)-bounded prime $p \geq s$ and a polynomial[5] $q(\cdot)$ such that $\mathcal{K}(G, \mathcal{X})$ is $(O(t+m), 1/q(n), p)$-predictable.

Even though Proposition 1 already gives search-to-decision reductions for some interesting families $\mathcal{K}$, it is not strong enough to establish Theorem 2 in its full generality. This is achieved in Proposition 2. Theorem 2 then follows directly if we combine Proposition 2 and Lemma 5.

Proposition 2. *If $\mathcal{K}(G, \mathcal{X})$ is (t, δ)-distinguishable from random for some noticeable δ, but $\mathcal{K}(G_d, \mathcal{X})$ is pseudorandom for all $d < s$, then there exists a polynomially bounded $d^* \geq s$ and polynomial $q(\cdot)$ such that $\mathcal{K}$ is $(O(t+m), 1/q(n), d^*)$-predictable.*

Proof. For simplicity, we write $\mathcal{K}$ (resp. $\mathcal{K}_d$) instead of $\mathcal{K}(G, \mathcal{X})$ (resp. $\mathcal{K}(G_d, \mathcal{X})$). We use $\mathcal{F}_d(\mathcal{K})$ (as in (2)) for all auxiliary distributions. For a distinguisher $\mathcal{D}$, $\mathsf{prob}_d^{\mathcal{D}} := \Pr[\mathcal{D}(\mathcal{F}_d(\mathcal{K})) = 1]$. Notice that $\Pr[\mathcal{D}(\mathcal{U}(G^m \times G)) = 1] = \mathsf{prob}_1^{\mathcal{D}}$ and $\Pr[\mathcal{D}(\mathcal{F}(\mathcal{K})) = 1] = \mathsf{prob}_{M_G}^{\mathcal{D}}$. The *distinguishing advantage* of $\mathcal{D}$ between distributions $\mathcal{F}_{d_1}(\mathcal{K})$ and $\mathcal{F}_{d_2}(\mathcal{K})$ is defined as $\mathsf{Adv}^{\mathcal{D}}(\mathcal{F}_{d_1}(\mathcal{K}), \mathcal{F}_{d_2}(\mathcal{K})) = |\mathsf{prob}_{d_1}^{\mathcal{D}} - \mathsf{prob}_{d_2}^{\mathcal{D}}|$. When one of the two distribution is $\mathcal{U}(G^m \times G)$, we write $\mathsf{Adv}_d^{\mathcal{D}}$ instead of $\mathsf{Adv}^{\mathcal{D}}(\mathcal{F}_d(\mathcal{K}), \mathcal{F}_1(\mathcal{K}))$. We often write $a \equiv_c b$ instead of $a \equiv b \pmod{c}$ and define $\delta_{ij} = 1$ if $i = j$ and 0 otherwise.

By hypothesis, there exists distinguisher $\mathcal{D}$ and polynomial $t(\cdot)$ such that $|\mathsf{Adv}_{M_G}^{\mathcal{D}}| \geq \frac{1}{t(n)}$ and $|\mathsf{Adv}_{d'}^{\mathcal{D}}| = \mathsf{negl}(n)\ \forall d' < s$. If $\mathsf{Adv}_{d'}^{\mathcal{D}} = \mathsf{negl}(n)\ \forall d' < 2ms^2$ then proof follows directly from Proposition 1. Else, there exists d with $s \leq d < 2ms^2$ (notice that since both s and m are polynomially bounded in n, so is d) and polynomial $w(\cdot)$ such that $|\mathsf{Adv}_d^{\mathcal{D}}| \geq \frac{1}{w(n)}$. Let d^* be the *smallest* divisor of d such that[6] $|\mathsf{Adv}_{d^*}^{\mathcal{D}}| \geq \frac{d^{*3}}{d^3 w(n)}$ (in particular, this implies that$|\mathsf{Adv}_{d'}^{\mathcal{D}}| < \frac{d'^3}{d^3 w(n)}$ for all $d' \mid d^*$). Since $\frac{d^{*3}}{d^3 w(n)} \geq \frac{1}{\mathsf{poly}(n)}$ and $|\mathsf{Adv}_{d'}^{\mathcal{D}}| = \mathsf{negl}(n)\ \forall d' < s$, it should be the case that $d^* \geq s$. Consider now the predictor $\mathcal{P}$ shown in Algorithm 1 ($\mathcal{P}$ tries to guess the inner product $\mathbf{r} \cdot \mathbf{x} \pmod{d^*}$).

It can be checked that, if $\mathcal{P}$'s guess for $\mathbf{r} \cdot \mathbf{x} \pmod{d^*}$ (line 1) is correct, then the input distribution to $\mathcal{D}$ (line 4) is exactly $\mathcal{F}_{d^*}(\mathcal{K})$. Otherwise the input distribution to $\mathcal{D}$ is $\mathcal{F}_{d'}(\mathcal{K})$ for some $d' \mid d^*$ with $d' < d^*$.

It only remains to compute the *bias* of $\mathcal{P}$. First notice that

$$\begin{aligned} Pr\{k \leftarrow \mathcal{E}_{d^*}(\mathcal{P})\} &= \Pr[\,guess \equiv_{d^*} v - k\,] \\ &= \sum_{j=0}^{d^*-1} \Pr\left[\,guess \equiv_{d^*} v - k \,\middle|\, c \equiv_{d^*} v - j\,\right] \Pr[\,c \equiv_{d^*} v - j\,] \\ &= \frac{1}{d^*} \sum_{j=0}^{d^*-1} \Pr\left[\,guess \equiv_{d^*} v - k \,\middle|\, c \equiv_{d^*} v - j\,\right] \qquad (4) \end{aligned}$$

[5] We only care about the predicting advantage being noticeable and do not seek to optimize it as a function of the distinguishing advantage. We simply mention that the success probability ϵ of the predictor is $\epsilon \geq \delta/4ms^2$.

[6] such a d^* always exists. Indeed d itself satisfies this condition and is a divisor of itself.

input : $(\mathbf{g}, y, \mathbf{r})$ // $y = \mathbf{g} \cdot \mathbf{x}$, $\mathbf{r} \leftarrow \mathcal{U}(\mathbb{Z}_{d^*}^m)$
output: $guess \in \mathbb{Z}_{d^*}$

Pick $c \leftarrow \mathcal{U}(\mathbb{Z}_{d^*})$;
Pick $g_1 \leftarrow \mathcal{U}(G), g_2 \leftarrow \mathcal{U}(G)$;
$\bar{\mathbf{g}} \leftarrow \mathbf{g} - \mathbf{r} \cdot g_1$ // $\mathbf{r} \cdot g_1 = (r_1 \cdot g_1, \ldots, r_m \cdot g_1)$;
Run $\mathcal{D}$ on input $(\bar{\mathbf{g}}, y - c \cdot g_1 + d^* \cdot g_2)$;
if $\mathcal{D}$ *returns 1* **then**
    $guess \leftarrow c$;
else
    $guess \leftarrow \mathcal{U}(\mathbb{Z}_{d^*} \setminus c)$;
end
return $guess$

Algorithm 1. Predictor for strong reduction (Proposition 2)

Conditioning on $\mathcal{D}$'s output and after doing some calculations, we get

$$Pr\{k \leftarrow \mathcal{E}_{d^*}(\mathcal{P})\} = \frac{1}{d^*} + \frac{1}{d^*}\mathsf{prob}^{\mathcal{D}}_{\gcd(k,d^*)} - \frac{1}{d^*(d^*-1)}\sum_{j \neq k} \mathsf{prob}^{\mathcal{D}}_{\gcd(j,d^*)}$$

which implies that

$$Pr\{k \leftarrow \mathcal{E}_{d^*}(\mathcal{P})\} - Pr\{1 \leftarrow \mathcal{E}_{d^*}(\mathcal{P})\} = \frac{\mathsf{prob}^{\mathcal{D}}_{\gcd(k,d^*)} - \mathsf{prob}^{\mathcal{D}}_{1}}{d^*-1} = \frac{\mathsf{Adv}^{\mathcal{D}}_{\gcd(k,d^*)}}{d^*-1}$$

Using this and the fact that $\sum_{k=0}^{d^*-1} \omega_{d^*}^{-k} = 0$ we get that

$$\begin{aligned} \left|\sum_{k=0}^{d^*-1} Pr\{k \leftarrow \mathcal{E}_{d^*}(\mathcal{P})\} \cdot \omega_{d^*}^{-k}\right| &= \frac{1}{d^*-1}\left|\sum_{k=0}^{d^*-1} \mathsf{Adv}^{\mathcal{D}}_{\gcd(k,d^*)}\omega_{d^*}^{-k}\right| \\ &\geq \frac{1}{d^*-1}\left[\left|\mathsf{Adv}^{\mathcal{D}}_{d^*}\right| - \sum_{k=1}^{d^*-1}\left|\mathsf{Adv}^{\mathcal{D}}_{\gcd(k,d^*)}\right|\right] \qquad (5) \end{aligned}$$

Next we bound $\sum_{k=1}^{d^*-1}\left|\mathsf{Adv}^{\mathcal{D}}_{\gcd(k,d^*)}\right|$. Define $\Phi(d^*, k) = \{1 \leq i < d^* : \gcd(i, d^*) = k\}$ and let[7] $\phi(d^*, k) = |\Phi(d^*, k)|$. Clearly $\phi(d^*, d') \leq \frac{d^*}{d'}$ $\forall d' \mid d^*$. So

$$\sum_{k=1}^{d^*-1}\left|\mathsf{Adv}^{\mathcal{D}}_{\gcd(k,d^*)}\right| \leq \sum_{\substack{d' \mid d^* \\ d' < d^*}} \phi(d^*, d')\left|\mathsf{Adv}^{\mathcal{D}}_{\gcd(k,d^*)}\right| \leq \frac{d^*}{d^3 w(n)} \sum_{\substack{d' \mid d^* \\ d' < d^*}} d'^2$$

[7] This is a generalization of Euler's totient function.

where in the last inequality we used the fact that for all proper divisors d' of d^*, $|\mathsf{Adv}^{\mathcal{D}}_{d'}| < \frac{d'^3}{d^3 w(n)}$. Replacing back in (5) we finally get

$$\left|\sum_{k=0}^{d^*-1} Pr\{k \leftarrow \mathcal{E}_{d^*}(\mathcal{P})\} \cdot \omega_{d^*}^{-k}\right| \geq \frac{1}{d^*-1}\left[\frac{d^{*3}}{d^3 w(n)} - \frac{d^*}{d^3 w(n)} \sum_{\substack{d' \mid d^* \\ d' < d^*}} d'^2\right]$$

$$\geq \frac{d^{*3}}{d^3(d^*-1)w(n)}\left(2 - \frac{\pi^2}{6}\right) \geq \frac{1}{q(n)}$$

for some polynomial $q(\cdot)$. In the last inequality we used the fact that for any $d \in \mathbb{N}$, $\sum_{r|d,r<d} r^2 \leq (\pi^2/6) \cdot d^2$.

4 Implications and Applications

Theorem 2 provides explicit criteria for checking if a knapsack family is pseudorandom. For a group G and input distribution $\mathcal{X}$, one needs only to check whether the folded families $\mathcal{K}_d = \mathcal{K}(G_d, \mathcal{X})$ are pseudorandom. As it turns out, for many choices of $(G, \mathcal{X})$, the folded knapsack functions $\mathcal{K}_d$ *compress* their input and map $\mathcal{X}$ to a distribution which is *statistically close* to uniform over G_d. More specifically, $\Delta_U(\mathcal{F}(\mathcal{K}(G_d, \mathcal{X}))) = \mathsf{negl}(n)$, and $\mathcal{K}(G_d, \mathcal{X})$ is pseudorandom in a strong statistical sense. Below, we provide some representative examples focusing on those that are most interesting in applications.

4.1 Specific Groups and Input Distributions

We start with groups G whose order does not contain any factors that are smaller than the maximum value the input can take, i.e. $[\mathcal{X}] \subseteq [s]^m$ and any prime factor of $|G|$ is at least as large as s. In this case, a direct interpretation of Theorem 2 reveals that one-wayness implies pseudorandomness for *any* input distribution.

Corollary 1. *Let p be the smallest prime factor of $|G|$ and $\mathcal{X}$ be such that $[\mathcal{X}] \subseteq [p]^m$. If $\mathcal{K}(G, \mathcal{X})$ is one-way, then it is also pseudorandom.*

Corollary 1 is already very powerful. For instance, in the standard subset sum problem we have $[\mathcal{X}] = \{0,1\}^m \subseteq [p]^m$ for any prime p. Therefore, Corollary 1 significantly generalizes the results from [18] and [12]. More specifically, it asserts that *any* knapsack family $\mathcal{K}(G, \mathcal{X})$ with $[\mathcal{X}] \subseteq \{0,1\}^m$ is pseudorandom provided it is one-way, for *any* abelian group G. Other interesting groups Corollary 1 is directly applicable to include groups with prime order, vector groups $\mathbb{Z}_p^k$ for prime p and generally groups $\mathbb{Z}_{p^e}^k$ where p is a prime such that $[\mathcal{X}] \subseteq [p]^m$.

For groups with small prime factors (smaller than s, where $[\mathcal{X}] \subseteq [s]^m$), the connection between one-wayness and pseudorandomness is more subtle: *search to decision* equivalence can be shown only for some input distributions and groups G. We summarize a few such examples focusing on *vector groups*, i.e. $G = \mathbb{Z}_q^k$ both for simplicity and because those groups are most interesting from a cryptographic viewpoint (see Section 4.2). Throughout, we assume $m - k = \omega(\log n)$.

For a vector group $G = \mathbb{Z}_q^k$ consider the folded knapsack function $\mathcal{K}_d = \mathcal{K}(G_d, \mathcal{X})$. First notice that $M_G = q$ and $dG = d\mathbb{Z}_q^k = \gcd(d,q) \cdot \mathbb{Z}_q^k$. By Theorem 2, proving pseudorandomness of $\mathcal{K}(G, \mathcal{X})$ amounts to proving that $\mathcal{K}_d$ are pseudorandom for all $d < s$ with $d \mid q$. In fact, below we study cases where $\mathcal{K}_d$ are *statistically* random, i.e. $\Delta_U(\mathcal{F}(\mathcal{K}(\mathbb{Z}_d^k, \mathcal{X}))) = \mathsf{negl}(n)$ for all divisors $d < s$ of q. Lemma 6 provides sufficient conditions for pseudorandomness expressed in terms of the statistical properties of $\mathcal{X}$ and the factorization of q. The statistical properties of $\mathcal{X}$ can be better expressed by defining the *d-folded* distribution $\mathcal{X}_d = \{\mathbf{x} \pmod d \mid \mathbf{x} \leftarrow \mathcal{X}\}$. Lemma 6 then requires that, for every "small" divisor of q, the d-folded distribution $\mathcal{X}_d$ has collision probability sufficiently smaller than a quantity that depends exclusively on d^k, the order of the quotient group $G_d = \mathbb{Z}_d^k$. The proof follows almost immediately from Theorem 2 and Lemma 4.

Lemma 6. *If $\mathcal{K} = \mathcal{K}(\mathbb{Z}_q^k, \mathcal{X})$ is one-way, $[\mathcal{X}] \subseteq [s]^m$ and $d^k \cdot \mathsf{Col}(\mathcal{X}_d) = \mathsf{negl}(n)$ for all $d \mid q$ with $d < s$, then $\mathcal{K}(\mathbb{Z}_q^k, \mathcal{X})$ is also pseudorandom.*

Below, we present 2 natural families of distributions which have small collision probability when "folded" over small d. Search to decision reductions for the corresponding knapsack families follow directly from Lemma 6 and the bounds on the collision probability of the two distributions. Lemmas 7 and 8 provide formal statements.

Uniformly Folded Distributions. For a given vector group G we say that a distribution $\mathcal{X}$ with $[\mathcal{X}] \subseteq [s]^m$ is *uniformly folded* with respect to G, if $\mathcal{X}_d = (\mathcal{X} \bmod d) \overset{s}{\approx} \mathcal{U}(\mathbb{Z}_d^m)$ for all $d < s$ such that $d \mid M_G$. When $G = \mathbb{Z}_q^k$, one such example is $\mathcal{X} = \mathcal{U}(\mathbb{Z}_q^m)$ or $\mathcal{X} = \mathcal{U}(\mathbb{Z}_{p^i}^m)$ when $q = p^e$ for some $e > i$.

Lemma 7. *If $\mathcal{K}(\mathbb{Z}_q^k, \mathcal{X})$ is one-way and $\mathcal{X}$ is uniformly folded with respect to $\mathbb{Z}_q^k$, (with $[\mathcal{X}] \subseteq [s]^m$ for $s = \mathsf{poly}(n)$), then it is also pseudorandom.*

Gaussian. Gaussian-like distributions are typically used for sampling the error in LWE-based cryptographic constructions. The following lemma establishes the search-to-decision reduction for knapsack families defined over $\mathbb{Z}_q^k$ and discrete Gaussian input distribution. Qualitatively similar results hold for *discretized* (rounded) Gaussians.

Lemma 8. *Let r be the Gaussian parameter with*[8] *$\omega(\log n) \le r \le \mathsf{poly}(n)$. If $\mathcal{K}(\mathbb{Z}_q^k, \mathcal{D}_{\mathbb{Z}^m,r})$ is one-way then it is also pseudorandom provided that either*
(a) q is prime or
(b) q is composite and there exists a function $\beta(n) = \omega(\sqrt{\log n})$ such that all divisors d of q lie outside the interval $[r/\beta(n), r \cdot \beta(n)]$.

---

[8] In typical instantiations, $r = \Omega(n^\theta)$ for some constant $\theta > 0$.

4.2 Applications to LWE

In this section, we show how our results for knapsack functions imply similar search-to-decision reductions for the Learning With Errors (LWE) problem with the interesting feature of being *sample-preserving*. Following existing LWE literature, we use n for the length of the secret vector $\mathbf{s}$, m for the number of samples, q for the modulus and χ for the error distribution. Let n, m, q be positive integers and χ a distribution with $[\chi] \subseteq \mathbb{Z}_q$. For a vector $\mathbf{s} \in \mathbb{Z}_q^n$, define the distribution

$$\mathcal{A}_{\mathbf{s},\chi} = \{(\mathbf{a}, \mathbf{a} \cdot \mathbf{s} + e) \mid \mathbf{a} \leftarrow \mathcal{U}(\mathbb{Z}_q^n), e \leftarrow \chi\}.$$

The LWE problem is the problem of recovering $\mathbf{s}$ given m samples from distribution $\mathcal{A}_{\mathbf{s},\chi}$. In its decisional version (DLWE), one is given m samples drawn independently at random either from $\mathcal{A}_{\mathbf{s},\chi}$ (for some secret $\mathbf{s}$) or from $\mathcal{U}(\mathbb{Z}_q^n \times \mathbb{Z}_q)$. The goal is to tell the two distributions apart with noticeable probability.

We are interested in reductions from LWE to DLWE that *preserve* all the parameters n, m, q, χ, including the *number of samples* m. Sample-preserving reductions are more naturally described using matrix notation for the LWE problem. Given a collection of m LWE samples $(\mathbf{a}_i, b_i) \leftarrow \mathcal{A}_{\mathbf{s},\chi}$, we can combine them in a matrix $\mathbf{A}$ having the vectors $\mathbf{a}_i$ as rows, and a column vector $\mathbf{b}$ with entries equal to b_i. That is, $\mathbf{b} = \mathbf{A}\mathbf{s} + \mathbf{e}$ where $\mathbf{e} \leftarrow \chi^m$. With this notation, we want to prove that any algorithm that distinguishes $(\mathbf{A}, \mathbf{b} = \mathbf{A}\mathbf{s} + \mathbf{e})$ from $\mathcal{U}(\mathbb{Z}_q^{m\times n} \times \mathbb{Z}_q^m)$ can be used to recover the secret $\mathbf{s}$. Notice that once the secret $\mathbf{s}$ has been recovered, one can also recover the error vector $\mathbf{e} = \mathbf{b} - \mathbf{A}\mathbf{s}$. So, we can equivalently define LWE as the problem of recovering both $\mathbf{s}$ and $\mathbf{e}$. This is exactly the problem of inverting the following function family.

Definition 2. *Let n, m, q and χ defined as above.* LWE(n, m, q, χ) *is the function family $(F, \mathcal{X})$ where $\mathcal{X} = \{(\mathbf{s}, \mathbf{e}) \mid \mathbf{s} \leftarrow \mathcal{U}(\mathbb{Z}_q^n), \mathbf{e} \leftarrow \chi^m\}$, and F is the set of functions $f_\mathbf{A}$ indexed by $\mathbf{A} \in \mathbb{Z}_q^{m\times n}$ and defined as $f_\mathbf{A}(\mathbf{s}, \mathbf{e}) = \mathbf{A}\mathbf{s} + \mathbf{e}$.*

The decision LWE is the problem of distinguishing $\mathcal{F}(\text{LWE}(n, m, q, \chi))$ from $\mathcal{U}(\mathbb{Z}_q^{m\times n} \times \mathbb{Z}_q^m)$. However, LWE$(n, m, q, \chi)$ is not a knapsack family. In order to apply the results from Section 3, we exploit the duality between the LWE problem and an associated knapsack function family described in the following lemmas.

Lemma 9. *For any*[9] *$n, m \geq n + \omega(\log n), q$ and χ, there is a polynomial time reduction from the problem of inverting* LWE(n, m, q, χ) *with probability ϵ, to the problem of inverting $\mathcal{K}(\mathbb{Z}_q^{m-n}, \chi^m)$ with probability $\epsilon' = \epsilon + \mathsf{negl}(n)$.*

Proof (Sketch). The transformation from the LWE problem into an equivalent knapsack problem requires that the matrix $\mathbf{A}$ be nonsingular, i.e., the rows of $\mathbf{A}$ generate $\mathbb{Z}_q^n$. When $\mathbf{A} \leftarrow \mathcal{U}(\mathbb{Z}_q^{m\times n})$, this is true except with probability at most $1/p^{m-n-1}$, where p is the smallest prime factor of q. So, for $m \geq n +$

[9] The requirement $m \geq n+\omega(\log n)$ is a standard assumption in the context of LWE, where typically $m \geq n + \Omega(n)$.

$\omega(\log n)$, $\Pr[\,\mathbf{A}\text{ singular}\,] = \mathsf{negl}(n)$. We can therefore assume $\mathbf{A}$ has been chosen at random, but conditioned on the property that it is nonsingular.

Consider now the set that contains all vectors $\mathbf{g}$ such that $\mathbf{gA} = 0 \pmod q$. Under the assumption that $\mathbf{A}$ is nonsingular, this set is generated by the rows of a matrix $\mathbf{G} \in \mathbb{Z}_q^{(m-n)\times m}$ that can be efficiently computed from $\mathbf{A}$ using linear algebra. We can further randomize $\mathbf{G}$ by left-multiplying it by a random unimodular matrix $\mathbf{U} \in \mathbb{Z}_q^{(m-n)\times(m-n)}$. Finally, if $\mathbf{A}$ is chosen at random among all nonsingular matrices, then this randomized $\mathbf{G}$ is also distributed uniformly at random among all matrices whose *columns* generate $\mathbb{Z}_q^{m-n}$. As before, the distribution of $\mathbf{G}$ is within negligible statistical distance from $\mathcal{U}(\mathbb{Z}_q^{(m-n)\times m})$, so we can treat the columns of $\mathbf{G}$ as random elements from the vector group $G = \mathbb{Z}_q^{m-n}$. Finally, we set $\mathbf{c} = \mathbf{Gb} = \mathbf{GAs} + \mathbf{Gc} = \mathbf{Ge}$, so the distribution $(\mathbf{G}, \mathbf{c})$ is *statistically close* to a *random instance* of the knapsack problem with group $G = \mathbb{Z}_q^{m-n}$ and input distributed according to the error distribution χ^m.

Lemma 10. *For any $n, m \geq n + \omega(\log n), q$ and χ, there is a polynomial time reduction from the problem of distinguishing $\mathcal{F}(\mathcal{K}(\mathbb{Z}_q^{m-n}, \chi^m))$ from uniform with advantage ϵ to the problem of distinguishing $\mathcal{F}(\mathrm{LWE}(n, m, q, \chi))$ from uniform with advantage $\epsilon' = \epsilon + \mathsf{negl}(n)$.*

Proof (Proof Sketch). The reduction reverses the steps taken to transform LWE into knapsack. We start from a pair $(\mathbf{G}, \mathbf{c})$. As before, we can assume that the columns of $\mathbf{G}$ generate $\mathbb{Z}_q^{m-n}$. Next, by linear algebra, we compute a matrix $\mathbf{A} \in \mathbb{Z}_q^{m\times n}$ whose columns generate the set of vectors $\mathbf{a}$ such that $\mathbf{Ga} = \mathbf{0} \pmod q$. As before, we can randomize $\mathbf{A}$ by right-multiplying it by a random unimodular matrix $\mathbf{U} \in \mathbb{Z}_q^{n\times n}$ to obtain $\mathbf{A}'$. We also map $\mathbf{c}$ to $\mathbf{A}'\mathbf{s}' + \mathbf{r}$ where $\mathbf{s}' \leftarrow \mathcal{U}(\mathbb{Z}_q^n)$ and $\mathbf{r}$ is a random solution to the equation $\mathbf{Gr} = \mathbf{c}$. It can be checked that this transformation maps the knapsack distribution $(\mathbf{G}, \mathbf{c} = \mathbf{Ge})$ to the LWE distribution $(\mathbf{A}', \mathbf{A}'\mathbf{s} + \mathbf{e})$ (with uniformly random $\mathbf{s}$), when $\mathbf{G}$ and $\mathbf{A}'$ are chosen at random subject to the constraint that they are nonsingular. The transformation also maps the uniform distribution to a (statistically close to) uniform distribution.

LWE: From Search to Decision. Sample-preserving search to decision reductions for LWE are immediately obtained combining the reductions from Lemma 9 and Lemma 10 with the results from Section 3 on $\mathcal{K}(\mathbb{Z}_q^{m-n}, \chi^m)$. Similarly to the knapsack case, the reductions *do not hold unconditionally*; rather they hold for specific, yet very broad, moduli q and error distributions χ. Below we give a general statement for the search to decision reduction parametrized by n, m, q and χ. Upon giving the statement, we provide specific instantiations of the error distribution χ and the modulus q for which the statement holds. Throughout, we assume that $m \geq n + \omega(\log n)$.

Proposition 3. *Assume there exists an efficient algorithm $\mathcal{D}$ that distinguishes between $\mathcal{F}(\mathrm{LWE}(n, m, q, \chi))$ and $\mathcal{U}(\mathbb{Z}_q^{m\times n} \times \mathbb{Z}_q^m)$ with noticeable advantage. Then there exists an efficient algorithm $\mathcal{I}$ that inverts $\mathrm{LWE}(n, m, q, \chi)$ with noticeable success probability.*

The following "assignments" provide examples of q and χ that make the above statement true.

- prime $q = \Theta(n^c)$ for *constant* c and $\chi = \mathcal{D}_{\mathbb{Z},r}$. The search to decision reduction of the corresponding bounded knapsack problem follows directly from Corollary 1. Setting q and χ as above is typical for instantiations of LWE-based cryptographic applications.
- $q = p^e$ for prime $p = \mathsf{poly}(n)$, and $\chi = \mathcal{D}_{\mathbb{Z},r}$ for "sufficiently narrow" standard deviation (more specifically, it is required that $r = o(\frac{p}{\log n})$). Again, the search to decision reduction of the bounded knapsack problem stems from Corollary 1. We note that this case provides a sample-preserving version of the search to decision reduction proved in [6].
- $q = p^e = \mathsf{poly}(n)$, with $\chi = \mathcal{U}(\mathbb{Z}_{p^i})$ for some $i < e$. Pseudorandomness of the knapsack instance stems directly from Lemma 7. Search to decision reduction for LWE with such noise distribution appears to be new; no such (even non-sample-preserving) reduction has previously appeared in the literature.

5 Open Problems

Our work leaves many interesting open questions. To start with, sample-preserving search to decision reductions for LWE with *bounded* noise as considered in this work, don't seem to extend to the unbounded noise regime, i.e. when each coefficient e_i of the error vector $\mathbf{e}$ of LWE is drawn from a set with *superpolynomial* size. We note that such search to decision reductions are known [28] but are *not* sampling preserving. These reductions rely heavily on a Chinese Reminder Theorem (CRT) approach: using a *perfect*[10] distinguisher, they first learn the secret modulo p_i with *overwhelming success probability* for each *polynomially bounded* prime factor p_i of the modulus q; they then use the CRT to recover the entire secret. In sample preserving reductions, where only an *imperfect* distinguisher can be afforded by the available number of samples, learning the secret modulo p_i can be performed in a much looser, list-decoding sense: the secret modulo p_i is included in the corresponding lists L_i but among possibly *many* other elements. And the only way to check which of the list elements corresponds to the secret modulo p_i seems to be by forming first the entire secret using CRT and then verifying that the result is the LWE secret. Thus, one has to solve superpolynomially many CRT instances before recovering the correct value of the secret. It would be nice to extend the list-decoding approach to work even in that case.

As an additional motivation, we mention that extending our sample preserving reductions to the unbounded error setting is likely to have implications to the search to decision equivalence of the newly introduced Ring LWE (R-LWE) problem [24]. R-LWE is an algebraic variant of LWE that leads to much more efficient constructions than standard LWE while still enjoying strong security

[10] By perfect here we mean a distinguisher with advantage almost 1. Getting a perfect distinguisher out of an imperfect one (one with only a nonnegligible advantage) is the main reason for the blowup in the number of samples the reduction consumes.

guarantees. Much like LWE with unbounded noise, existing search to decision reductions [24] decompose the secret (which is an element from a ring R) modulo $\mathfrak{q}_i$ where $\mathfrak{q}_i$ are prime ideal factors.

Our work also highlights the importance of understanding the hardness of LWE under various noise distributions. Current hardness proofs for search LWE [31] based on worst-case lattice problems rely on the noise following a Gaussian distribution. Can lattice-based hardness results for search LWE be extended to noise distributions other than Gaussian? Can we show similar lattice-based hardness results if the noise is distributed uniformly at random modulo 2^i? The latter case is very attractive from a practical viewpoint since arithmetic modulo 2 and sampling from uniform distributions can be implemented very efficiently.

References

1. Agrawal, S., Boneh, D., Boyen, X.: Efficient Lattice (H)IBE in the Standard Model. In: Gilbert, H. (ed.) EUROCRYPT 2010. LNCS, vol. 6110, pp. 553–572. Springer, Heidelberg (2010)
2. Agrawal, S., Boneh, D., Boyen, X.: Lattice Basis Delegation in Fixed Dimension and Shorter-Ciphertext Hierarchical IBE. In: Rabin, T. (ed.) CRYPTO 2010. LNCS, vol. 6223, pp. 98–115. Springer, Heidelberg (2010)
3. Akavia, A.: Learning Noisy Characters, Multiplication Codes and Hardcore Predicates. PhD thesis. MIT (February 2008)
4. Akavia, A., Goldwasser, S., Safra, S.: Proving Hard-Core Predicates Using List Decoding. In: FOCS, pp. 146–157 (2003)
5. Akavia, A., Goldwasser, S., Vaikuntanathan, V.: Simultaneous Hardcore Bits and Cryptography against Memory Attacks. In: Reingold, O. (ed.) TCC 2009. LNCS, vol. 5444, pp. 474–495. Springer, Heidelberg (2009)
6. Applebaum, B., Cash, D., Peikert, C., Sahai, A.: Fast Cryptographic Primitives and Circular-Secure Encryption Based on Hard Learning Problems. In: Halevi, S. (ed.) CRYPTO 2009. LNCS, vol. 5677, pp. 595–618. Springer, Heidelberg (2009)
7. Arora, S., Ge, R.: New algorithms for learning in presence of errors. In: ICALP (2011), `http://www.eccc.uni-trier.de/report/2010/066/`
8. Blum, A., Furst, M.L., Jackson, J.C., Kearns, M.J., Mansour, Y., Rudich, S.: Weakly Learning DNF and Characterizing Statistical Query Learning using Fourier Analysis. In: STOC, pp. 253–262 (1994)
9. Blum, A., Furst, M.L., Kearns, M. J., Lipton, R.J.: Cryptographic Primitives Based on Hard Learning Problems. In: Stinson, D.R. (ed.) CRYPTO 1993. LNCS, vol. 773, pp. 278–291. Springer, Heidelberg (1994)
10. Cash, D., Hofheinz, D., Kiltz, E., Peikert, C.: Bonsai Trees, or How to Delegate a Lattice Basis. In: Gilbert, H. (ed.) EUROCRYPT 2010. LNCS, vol. 6110, pp. 523–552. Springer, Heidelberg (2010)
11. Dodis, Y., Goldwasser, S., Tauman Kalai, Y., Peikert, C., Vaikuntanathan, V.: Public-Key Encryption Schemes with Auxiliary Inputs. In: Micciancio, D. (ed.) TCC 2010. LNCS, vol. 5978, pp. 361–381. Springer, Heidelberg (2010)
12. Fischer, J.-B., Stern, J.: An efficient pseudo-random generator provably as secure as syndrome decoding. In: Maurer, U.M. (ed.) EUROCRYPT 1996. LNCS, vol. 1070, pp. 245–255. Springer, Heidelberg (1996)
13. Gentry, C., Halevi, S., Vaikuntanathan, V.: A Simple BGN-Type Cryptosystem from LWE. In: Gilbert, H. (ed.) EUROCRYPT 2010. LNCS, vol. 6110, pp. 506–522. Springer, Heidelberg (2010)

14. Gentry, C., Peikert, C., Vaikuntanathan, V.: Trapdoors for Hard Lattices and New Cryptographic Constructions. In: STOC, pp. 197–206. ACM, New York (2008)
15. Goldreich, O., Levin, L.A.: A Hard-Core Predicate for All One-Way Functions. In: STOC, pp. 25–32 (1989)
16. Goldwasser, S., Kalai, Y.T., Peikert, C., Vaikuntanathan, V.: Robustness of the Learning with Errors Assumption. In: ICS (2010)
17. Impagliazzo, R., Zuckerman, D.: How to Recycle Random Bits. In: FOCS, pp. 248–253. IEEE Computer Society, Washington, DC, USA (1989)
18. Impagliazzo, R., Naor, M.: Efficient Cryptographic Schemes Provably as Secure as Subset Sum. J. Cryptology 9(4), 199–216 (1996)
19. Katz, J., Shin, J.S., Smith, A.: Parallel and Concurrent Security of the HB and HB^+ Protocols. J. Cryptology 23(3), 402–421 (2010)
20. Kawachi, A., Tanaka, K., Xagawa, K.: Multi-bit cryptosystems based on lattice problems. In: Okamoto, T., Wang, X. (eds.) PKC 2007. LNCS, vol. 4450, pp. 315–329. Springer, Heidelberg (2007)
21. Kushilevitz, E., Mansour, Y.: Learning Decision Trees Using the Fourier Sprectrum. In: STOC, pp. 455–464 (1991)
22. Lindner, R., Peikert, C.: Better Key Sizes (and Attacks) for LWE-Based Encryption. In: Kiayias, A. (ed.) CT-RSA 2011. LNCS, vol. 6558, pp. 319–339. Springer, Heidelberg (2011)
23. Lyubashevsky, V., Micciancio, D.: On bounded distance decoding, unique shortest vectors, and the minimum distance problem. In: Halevi, S. (ed.) CRYPTO 2009. LNCS, vol. 5677, pp. 577–594. Springer, Heidelberg (2009)
24. Lyubashevsky, V., Peikert, C., Regev, O.: On Ideal Lattices and Learning with Errors over Rings. In: Gilbert, H. (ed.) EUROCRYPT 2010. LNCS, vol. 6110, pp. 1–23. Springer, Heidelberg (2010)
25. Micciancio, D.: Duality in Lattice Based Cryptography. In: Public Key Cryptography (2010) (invited talk)
26. Micciancio, D., Regev, O.: Lattice-Based Cryptography. In: Post Quantum Cryptography, pp. 147–191. Springer Publishing Company, Heidelberg (2009)
27. Mossel, E., O'Donnell, R., Servedio, R.A.: Learning Juntas. In: STOC, pp. 206–212 (2003)
28. Peikert, C.: Public-Key Cryptosystems from the Worst-Case Shortest Vector Problem. In: STOC, pp. 333–342. ACM, New York (2009)
29. Peikert, C., Vaikuntanathan, V., Waters, B.: A Framework for Efficient and Composable Oblivious Transfer. In: Wagner, D. (ed.) CRYPTO 2008. LNCS, vol. 5157, pp. 554–571. Springer, Heidelberg (2008)
30. Peikert, C., Waters, B.: Lossy Trapdoor Functions and Their Applications. In: STOC, pp. 187–196. ACM, New York (2008)
31. Regev, O.: On lattices, learning with errors, random linear codes, and cryptography. Journal of ACM 56(6), 34 (2009); Preliminary version in STOC 2005
32. Regev, O.: The Learning with Errors Problem (Invited Survey). In: IEEE Conference on Computational Complexity, pp. 191–204 (2010)
33. Rückert, M., Schneider, M.: Estimating the Security of Lattice-based Cryptosystems. Technical Report 2010/137, IACR ePrint archive (2010)
34. Stefankovic, D.: Fourier Transform in Computer Science. Master's thesis, University of Chicago (October 2000)
35. Stehlé, D., Steinfeld, R., Tanaka, K., Xagawa, K.: Efficient Public Key Encryption Based on Ideal Lattices. In: Matsui, M. (ed.) ASIACRYPT 2009. LNCS, vol. 5912, pp. 617–635. Springer, Heidelberg (2009)

Tor and Circumvention: Lessons Learned*
(Abstract to Go with Invited Talk)

Roger Dingledine

The Tor Project

Tor is a free-software anonymizing overlay network that helps people around the world use the Internet in safety. Tor's 2500 volunteer relays carry almost 10Gb/s of traffic for several hundred thousand users each day.

While many in the research community know Tor as the primary fielded system in the anonymous communications literature [2], Tor has also played a central role in recent research on *blocking resistance*. That is, even if an anonymity system provides great anonymity, a government censor can render it moot by simply blocking the relays. In recent years we streamlined Tor's network communications to look more like ordinary SSL, and we introduced "bridge relays" that are harder for an attacker to find and block than Tor's public relays [1].

Tor played a key role in several Middle Eastern countries in early 2011. In this talk I'll walk the audience through how Iran used its Nokia DPI boxes to filter SSL flows that used Tor's original Diffie-Hellman parameter p; the surge in Tor traffic when Egypt blocked Facebook and the flatline when they unplugged the net; the continued bad news for Libya's Internet; and an intriguing trend in Saudi Arabia. I'll also cover current trends in China and Tunisia (not pictured).

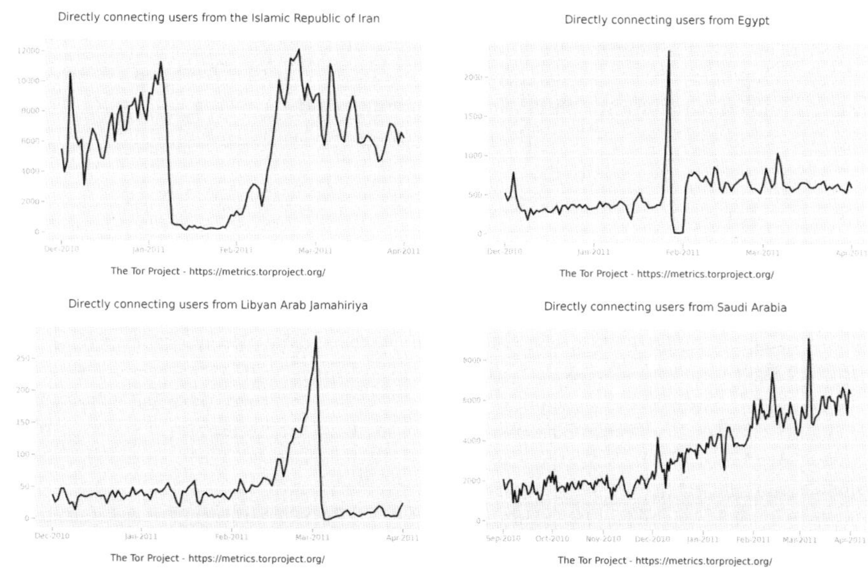

Fig. 1. Estimates of daily Tor clients connecting from each country

P. Rogaway (Ed.): CRYPTO 2011, LNCS 6841, pp. 485–486, 2011.

The data for these user graphs, along with historical Tor network data and ongoing performance statistics, are all available at https://metrics.torproject.org/. Our WECSR'10 paper [3] explains our aggregation techniques and why we think they're safe—we'd love for you to show us that we're wrong. Further, if you're working on Tor-related research, please talk to us (https://torproject.org/research) so we can explain what's available and help interpret your results.

Some open questions from the anonymity field. Here are a few examples of open anonymity and blocking-resistance problems:

1) How effective is the traffic correlation attack really? Tor's threat model assumes that an adversary who can see a traffic flow into the Tor network and the corresponding flow out of the Tor network can correlate them with high probability and low false positives. Recent results from Steven Murdoch [4] show confirmation attacks even when both sides only see a small sample of traffic on each side. But how quick can the attack actually be in practice, using how little traffic? Are there effective padding schemes to make correlation less effective?

2) For various diversity metrics (like entropy), how has the diversity of the Tor network changed over time? How robust is it to change or attack?[1]

3) How can we automatically recognize blocking events—when Tor relays are censored at a firewall by destination address or by traffic flow characteristics?

4) Clients who are censored from the public Tor relays can use private addresses to "bridge" into the public Tor network. What strategies should we use to give out these addresses such that legitimate users get enough addresses but adversaries can't learn too many?

5) How can we make it hard for censors to recognize Tor traffic flows by content (e.g. distinguishing Tor's handshake from other expected protocols) and by traffic characteristics (packet size, volume, and timing)? We need *obfuscation* metrics to let us anticipate which protocols will blend in better with background traffic or otherwise defeat deep packet inspection (DPI) algorithms.

References

1. Dingledine, R., Mathewson, N.: Design of a blocking-resistant anonymity system. Technical Report 2006-1, The Tor Project (November 2006)
2. Dingledine, R., Mathewson, N., Syverson, P.: Tor: The second-generation onion router. In: Proc. 13th USENIX Security Symposium (August 2004)
3. Loesing, K., Murdoch, S.J., Dingledine, R.: A Case Study on Measuring Statistical Data in the Tor Anonymity Network. In: Sion, R., Curtmola, R., Dietrich, S., Kiayias, A., Miret, J.M., Sako, K., Sebé, F. (eds.) RLCPS, WECSR, and WLC 2010. LNCS, vol. 6054, pp. 203–215. Springer, Heidelberg (2010)
4. Murdoch, S.J., Zieliński, P.: Sampled Traffic Analysis by Internet-Exchange-Level Adversaries. In: Borisov, N., Golle, P. (eds.) PET 2007. LNCS, vol. 4776, pp. 167–183. Springer, Heidelberg (2007)

[1] https://blog.torproject.org/blog/research-problem-measuring-safety-tor-network

Fully Homomorphic Encryption over the Integers with Shorter Public Keys

Jean-Sébastien Coron[1], Avradip Mandal[1], David Naccache[2], and Mehdi Tibouchi[1,2]

[1] Université du Luxembourg
{jean-sebastien.coron,avradip.mandal}@uni.lu
[2] École normale supérieure
{david.naccache,mehdi.tibouchi}@ens.fr

Abstract. At Eurocrypt 2010 van Dijk *et al.* described a fully homomorphic encryption scheme over the integers. The main appeal of this scheme (compared to Gentry's) is its conceptual simplicity. This simplicity comes at the expense of a public key size in $\tilde{\mathcal{O}}(\lambda^{10})$ which is too large for any practical system. In this paper we reduce the public key size to $\tilde{\mathcal{O}}(\lambda^7)$ by encrypting with a quadratic form in the public key elements, instead of a linear form. We prove that the scheme remains semantically secure, based on a stronger variant of the approximate-GCD problem, already considered by van Dijk *et al.*

We also describe the first implementation of the resulting fully homomorphic scheme. Borrowing some optimizations from the recent Gentry-Halevi implementation of Gentry's scheme, we obtain roughly the same level of efficiency. This shows that fully homomorphic encryption can be implemented using simple arithmetic operations.

1 Introduction

Fully Homomorphic Encryption. An encryption scheme is homomorphic if it supports operations on encrypted data. For example RSA is multiplicatively homomorphic since $c_1 = m_1^e \mod N$ and $c_2 = m_2^e \mod N$ yield the encryption of $m_1 \cdot m_2$ without using the private key.

Similarly, Paillier cryptosystem [12] is additively homomorphic because from $c_1 = g^{m_1} r^N \mod N^2$ and $c_2 = g^{m_2} s^N \mod N^2$ one can compute the encryption of $m_1 + m_2$.

In a breakthrough work Gentry described in 2009 the first encryption scheme that supports both addition and multiplication on ciphertexts, *i.e.* a fully homomorphic encryption scheme [5]. The construction proceeds by successive steps: First Gentry describes a "somewhat homomorphic" scheme that supports a limited number of additions and multiplications on ciphertexts. This is because every ciphertext has a noise component and any homomorphic operation applied to ciphertexts increases the noise in the resulting ciphertext. Once this noise reaches a certain threshold the resulting ciphertext does not decrypt correctly anymore; this limits the degree of the polynomial that can be applied to ciphertexts.

P. Rogaway (Ed.): CRYPTO 2011, LNCS 6841, pp. 487–504, 2011.

Secondly Gentry shows how to "squash" the decryption procedure so that it can be expressed as a low degree polynomial in the bits of the ciphertext and the secret key (equivalently a circuit of small depth). Then the breakthrough idea consists in evaluating this decryption polynomial not on the bits of the ciphertext and the secret key (as in regular decryption), but homomorphically on the *encryption* of those bits. Then instead of recovering the bit plaintext, one gets an encryption of this bit plaintext, *i.e.* yet another ciphertext for the same plaintext; see Figure 1 for an illustration. Now if the degree of the decryption polynomial is small enough, the resulting noise in this new ciphertext can be smaller than in the original ciphertext; this is called the "ciphertext refresh" procedure. Given two refreshed ciphertexts one can apply again the homomorphic operation (either addition or multiplication), which was not necessarily possible on the original ciphertexts because of the noise threshold. Using this "ciphertext refresh" procedure the number of permissible homomorphic operations becomes unlimited and we get a fully homomorphic encryption scheme.

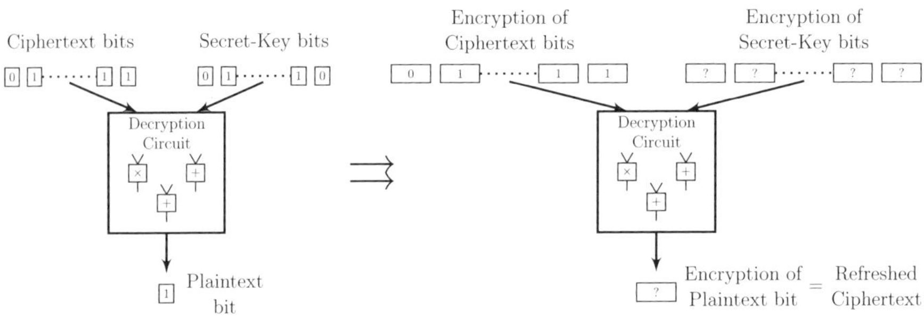

Fig. 1. The decryption circuit applied on the ciphertext bits and secret key bits (left), and the ciphertext refresh procedure with the decryption circuit applied homomorphically on the encryption of those bits (right).

The prerequisite for the "ciphertext refresh" procedure is that the degree of the polynomial that can be evaluated on ciphertexts exceeds the degree of the decryption polynomial (times two, since one must allow for a subsequent addition or multiplication of refreshed ciphertexts); this is called the "bootstrappability" condition. Once the scheme becomes bootstrappable it can be converted into a fully homomorphic encryption scheme by providing the encryption of the secret key bits inside the public key.

Based on Gentry's approach, two different fully homomorphic schemes are known: Gentry's scheme [5] based on ideal lattices and a scheme by van Dijk, Gentry, Halevi and Vaikuntanathan (DGHV) over the integers, that appeared at Eurocrypt 2010 [4].

Gentry's scheme and its implementations. Gentry described in [5] a somewhat homomorphic encryption scheme that is similar to GGH [7,14] over ideal lattices. To reduce the degree of the decryption polynomial, Gentry introduced the following transformation [5]: instead of using the original secret key, the decryption procedure uses a very sparse subset of values that adds up to the secret key; the full set of values is made part of the public key. To apply the new decryption procedure the original ciphertext must first be "expanded" using the full set of public values. This expanded ciphertext can then be decrypted with a low-degree polynomial in the bits of the new secret key (which are the characteristic vector of the sparse subset sum); this is called the "squashed decryption" procedure.

At PKC 2010 Smart and Vercauteren [16] made the first attempt to implement Gentry's scheme using a variant based on principal ideal lattices and requiring that the determinant of the lattice be a prime number. However the authors of [16] could not obtain a bootstrappable scheme because that would have required a lattice dimension of at least $n = 2^{27}$, whereas due to the prime determinant requirement they could not generate keys for dimensions $n > 2048$.

Gentry and Halevi described in [6] the first implementation of Gentry's scheme. The authors follow the same direction as Smart and Vercauteren, but for key generation they eliminate the requirement that the determinant is a prime. Additionally they present many clever optimizations. Four concrete parameter settings are provided, from a "toy" setting in dimension 512, to "small", "medium" and "large" settings of dimensions 2048, 8192 and 32768, respectively. For the "large" setting public key size is 2.3 Gigabytes. The authors of [6] report that for an optimized implementation on a high-end workstation, key generation takes 2.2 hours, encryption takes 3 minutes, and ciphertext refresh takes 30 minutes.

The DGHV fully homomorphic scheme over the integers. At Eurocrypt 2010, van Dijk, Gentry, Halevi and Vaikuntanathan described a fully homomorphic encryption scheme over the integers [4]. As in Gentry's scheme the authors first describe a somewhat homomorphic scheme supporting a limited number of additions and multiplications over encrypted bits. Then they apply Gentry's "squash decryption" technique to get a bootstrappable scheme and then Gentry's "ciphertext refresh" procedure (see Fig. 1) to get a fully homomorphic scheme.

The main appeal of the scheme (compared to the original Gentry's scheme) is its conceptual simplicity; namely all operations are done over the integers instead of ideal lattices. However the public-key was in $\tilde{\mathcal{O}}(\lambda^{10})$ which is too large for any practical system.

Our Contributions. In this paper we show how to reduce the public key size of the somewhat homomorphic scheme from $\mathcal{O}(\lambda^{10})$ down to $\mathcal{O}(\lambda^7)$. The idea consists in storing only a smaller subset of the public key and then generating the full public key on the fly by combining the elements in the small subset multiplicatively; we describe the new scheme in Section 3. In Section 4 we show that the new scheme is still semantically secure, but under a stronger variant of the approximate GCD assumption.

Our second contribution is to describe an implementation of the fully homomorphic DGHV scheme under our variant, using some of the optimizations from [6]. We use the refined analysis from [17] of the sparse subset sum problem; however we do not use the probabilistic decryption circuit from [17] because as in [6] the error probability is too high for our set of parameters. The main difficulty is to determine a secure set of concrete parameters; our approach is to implement the known attacks, measure their running time and extrapolate for large parameters; we can then fix the concrete parameters according to the desired level of security.

We obtain similar performances as the Gentry-Halevi implementation of Gentry's scheme [6]. More precisely we use four security levels inspired by the levels from [6] (though they may not be directly comparable due to different notions of "security bits"): "toy", "small", "medium" and "large", corresponding to 42, 52, 62 and 72 bits of security respectively. For "large" parameters, encryption and recryption take 3 minutes and 14 minutes respectively, with a public key size of 800 MBytes. Decryption is always close to instantaneous. This shows that fully homomorphic encryption can be implemented with a simple scheme.

2 The DGHV Scheme over the Integers

In this section we first recall the somewhat homomorphic encryption scheme published by van Dijk, Gentry, Halevi and Vaikuntanathan at Eurocrypt 2010 [4]. The scheme is based on a set of public integers: $x_i = p \cdot q_i + r_i$, $0 \leq i \leq \tau$, where the integer p is secret.

Notation. We use the same notation as in [4]. For a real number x, we denote by $\lceil x \rceil$, $\lfloor x \rfloor$ and $\lceil x \rfloor$ the rounding of x up, down, or to the nearest integer. For integers z, p we denote the reduction of z modulo p by $[z]_p$ with $-p/2 < [z]_p \leq p/2$. We also denote $[z]_p$ by $z \mod p$. We write $f(\lambda) = \tilde{\mathcal{O}}(g(\lambda))$ if $f(\lambda) = \mathcal{O}(g(\lambda) \log^k g(\lambda))$ for some $k \in \mathbb{N}$.

The scheme parameters. Given the security parameter λ, the following parameters are used:

- γ is the bit-length of the x_i's.
- η is the bit-length of secret key p.
- ρ is the bit-length of the noise r_i.
- τ is the number of x_i's in the public key.
- ρ' is a secondary noise parameter used for encryption.

For a specific η-bit odd integer p, we use the following distribution over γ-bit integers:

$$\mathcal{D}_{\gamma,\rho}(p) = \left\{ \textit{Choose } q \leftarrow \mathbb{Z} \cap [0, 2^\gamma/p), r \leftarrow \mathbb{Z} \cap (-2^\rho, 2^\rho) : \textit{ Output } x = q \cdot p + r \right\}$$

$\mathsf{KeyGen}(1^\lambda)$. Generate a random odd integer p of size η bits. For $0 \leq i \leq \tau$ sample $x_i \leftarrow \mathcal{D}_{\gamma,\rho}(p)$. Relabel so that x_0 is the largest. Restart unless x_0 is odd and $[x_0]_p$ is even. Let $pk = (x_0, x_1, \ldots x_\tau)$ and $sk = p$.

$\mathsf{Encrypt}(pk, m \in \{0,1\})$. Choose a random subset $S \subseteq \{1, 2, \ldots, \tau\}$ and a random integer r in $(-2^{\rho'}, 2^{\rho'})$, and output the ciphertext:

$$c = \left[m + 2r + 2 \sum_{i \in S} x_i \right]_{x_0}$$

$\mathsf{Evaluate}(pk, C, c_1, \ldots, c_t)$: given the circuit C with t input bits, and t ciphertexts c_i, apply the addition and multiplication gates of C to the ciphertexts, performing all the additions and multiplications over the integers, and return the resulting integer.

$\mathsf{Decrypt}(sk, c)$. Output $m \leftarrow (c \mod p) \mod 2$. Note that since $c \mod p = c - p \cdot \lfloor c/p \rceil$ and p is odd, one can compute instead: $m \leftarrow [c]_2 \oplus [\lfloor c/p \rceil]_2$.

This completes the description of the scheme. It is shown in [4] that the scheme is a somewhat homomorphic scheme and that it is semantically secure under the approximate-GCD assumption.

Definition 2.1 (Approximate GCD). *The (ρ, η, γ)-approximate-GCD problem is: For a random η-bit odd integer p, given polynomially many samples from $\mathcal{D}_{\gamma,\rho}(p)$, output p.*

Note that after one Mult operation $c \leftarrow c_1 \cdot c_2$ the ciphertext size doubles since there is no modular reduction involved. To reduce the ciphertext size after one Mult two techniques are described in [4]. The second and simpler technique consists in generating x_0 without noise, that is $x_0 = q_0 \cdot p$, and then reducing the ciphertext modulo x_0. The scheme is still semantically secure under the (stronger) approximate-GCD assumption with error-free x_0. While this problem seems easier to solve, as the adversary is given an exact multiple of p, no better attack is known against it than on the unmodified problem.

We recall in the full version of this paper [3] the constraints on the scheme parameters. To satisfy theses constraints the following parameter set is suggested in [4]: $\rho = \lambda$, $\rho' = 2\lambda$, $\eta = \tilde{\mathcal{O}}(\lambda^2)$, $\gamma = \tilde{\mathcal{O}}(\lambda^5)$ and $\tau = \gamma + \lambda$. The public key size is then $\tilde{\mathcal{O}}(\lambda^{10})$. In practice the size of the x_i's should be at least $\gamma = 2^{23}$ bits to prevent lattice attacks. The public key size is then at least 2^{46} bits, which is too large for any practical system.

3 Our Variant of the DGHV Scheme

3.1 Description

Our technique consists in working with integers x'_{ij} of the form $x'_{i,j} = x_{i,0} \cdot x_{j,1} \mod x_0$ for $1 \le i, j \le \beta$ where β is a new parameter. Then only 2β integers $x_{i,b}$ need to be stored in the public key in order to generate the $\tau = \beta^2$ integers x'_{ij} used for encryption. In other words we encrypt using a quadratic form in the public key elements instead of a linear form, which enables to reduce the public key size from τ down to roughly $2\sqrt{\tau}$ integers of γ bits.

Our technique requires to use an error-free x_0, that is $x_0 = q_0 \cdot p$, since otherwise the error would grow too large. Additionally for encryption we consider a linear combination of the $x'_{i,j}$ with coefficients in $[0, 2^\alpha)$ instead of bits; this enables to further reduce the public key size.

$\mathsf{KeyGen}(1^\lambda)$. Generate a random prime $p \in \cap[2^{\eta-1}, 2^\eta)$. Let $x_0 = q_0 \cdot p$ where q_0 is a random square free 2^λ-rough[1] integer in $[0, 2^\gamma/p)$. Generate integers $x_{i,b}$ for $1 \leq i \leq \beta$ and $b \in \{0, 1\}$:

$$x_{i,b} = p \cdot q_{i,b} + r_{i,b}, \quad 1 \leq i \leq \beta,\ 0 \leq b \leq 1 \tag{1}$$

where $q_{i,b}$ are random integers in $[0, q_0)$ and $r_{i,b}$ are integers in $(-2^\rho, 2^\rho)$. Let $sk = p$ and $pk = (x_0, x_{1,0}, x_{1,1}, \ldots x_{\beta,0}, x_{\beta,1})$.

$\mathsf{Encrypt}(pk, m \in \{0, 1\})$. Generate a random vector $\boldsymbol{b} = (b_{i,j})$ of size $\tau = \beta^2$ and with components in $[0, 2^\alpha)$. Generate a random integer r in $(-2^{\rho'}, 2^{\rho'})$. Output the ciphertext:

$$c = m + 2r + 2 \sum_{1 \leq i,j \leq \beta} b_{i,j} \cdot x_{i,0} \cdot x_{j,1} \mod x_0 \tag{2}$$

$\mathsf{Evaluate}$ and $\mathsf{Decrypt}$: same as in the original scheme, except that ciphertexts are reduced modulo x_0 after addition and multiplication.

3.2 Constraints on the Parameters

The first three constraints are the same as in the original DGHV scheme:

- $\rho = \omega(\log \lambda)$ to avoid brute force attack on the noise (see Section 6.1).
- $\eta \geq (2\rho + \alpha) \cdot \Theta(\lambda \log^2 \lambda)$ in order to support homomorphic operations for evaluating the "squashed decryption circuit" (see Section 5).
- $\gamma = \omega(\eta^2 \cdot \log \lambda)$ in order to thwart lattice-based attacks (see Section 6).
- $\alpha \cdot \beta^2 \geq \gamma + \omega(\log \lambda)$ for the reduction to approximate GCD (see Section 4).
- $\rho' = 2\rho + \alpha + \omega(\log \lambda)$ for the secondary noise parameter (see Section 4).

To satisfy these conditions we can still take $\rho = \lambda$, $\eta = \tilde{\mathcal{O}}(\lambda^2)$ and $\gamma = \tilde{\mathcal{O}}(\lambda^5)$ as in the original scheme, and we can take $\alpha = \lambda$, $\beta = \tilde{\mathcal{O}}(\lambda^2)$ and $\rho' = 4\lambda$. The main difference is that instead of having $\tau = \tilde{\mathcal{O}}(\lambda^5)$ integers x_i's, we now have only $2\beta = \tilde{\mathcal{O}}(\lambda^2)$ integers x_i. Hence the public key size becomes $\tilde{\mathcal{O}}(\lambda^7)$ instead of $\tilde{\mathcal{O}}(\lambda^{10})$. In Section 7.5 we describe concrete parameters in order to resist all known attacks.

Remark 3.1. It is possible to generate q_0 as a uniformly random square free 2^λ-rough integer of suitable size in probabilistic polynomial time: it suffices to generate a uniformly random number with known factorization [1] and try again if it has small or repeated factors. However, this makes key generation rather

[1] An integer is said to be a-rough when it does not contain prime factors smaller than a. Note that for $a > 2$ such integer must be odd.

unpractical. Alternatively, one can choose q_0 as the product of $(\gamma-\eta)/\lambda^2$ random primes, each of size λ^2 bits.[2] This is faster, but the security of the scheme then depends on a slightly more convoluted, though no less plausible, computational assumption, to account for the different key distribution.

3.3 Correctness

We refer to the full version of this paper [3] for the definition of correct homomorphic scheme with respect to a given circuit or circuit set. As in [4,5] we define a *permitted circuit* as one where for any $i \geq 1$ and any set of integer inputs all less than $\tau^i \cdot 2^{i(\rho'+2)}$ in absolute value, the generalized circuit's output has absolute value at most $2^{i(\eta-3-n)}$ with $n = \lceil \log_2(\lambda+1) \rceil$; we let $\mathcal{C}_\mathcal{E}$ be the set of permitted circuits. As in [4], we have (see proof in the full version of this paper [3]):

Lemma 3.1. *The scheme from above is correct for* $\mathcal{C}_\mathcal{E}$.

Remark 3.2. Since "fresh" ciphertexts output by Encrypt have noise at most $\tau \cdot 2^{\rho'+2}$, the ciphertext output by Evaluate applied to a permitted circuit has noise at most $2^{\eta-3-n} < p/(4(\lambda+1))$. A bound of $p/2$ would suffice to ensure correct decryption, but this stronger bound will be useful to prove the correctness of the bootstrappable version of this scheme later on.

Remark 3.3. The definition of a permitted circuit doesn't seem to give an easy criterion to determine whether a given computation is permitted. However, it is easy to give a sufficient condition on a multivariate polynomial f for the associated arithmetic circuit C to be permitted. If f is of degree d and if the sum of the absolute values of its coefficients is denoted by $\|f\|_1$, then $C \in \mathcal{C}_\mathcal{E}$ provided that:

$$d \leq \frac{\eta - 3 - n - \log \|f\|_1}{\rho' + 2 + 2\log \beta}$$

Following [4], we refer to such polynomials f as *permitted polynomials*, and denote the set of these polynomials by $\mathcal{P}_\mathcal{E}$.

4 Security of our Variant

4.1 Overview

In this section we show that our variant is still semantically secure, but under the (stronger) error-free approximate GCD assumption. Our security proof follows the same strategy as in [4]: show that an adversary breaking the scheme's semantic security can be converted into a LSB predictor for $z \bmod p$, where z is an integer such that $z \bmod p$ is small; this in turns enables to recover p in the approximate-GCD problem.

[2] The reason we choose λ^2-bit factors rather than λ is because factorization algorithms like ECM have a complexity subexponential in the size of factors, and can thus be used to extract λ-bit prime factors efficiently. In the implementation, to thwart this attack, it is safe to generate q_0 as a product of, say, 1000-bit primes.

For this one must show that given $c \leftarrow \mathsf{Encrypt}(pk, m)$, the distribution of $c' = [c+z]_{x_0}$ is essentially the same as $\mathsf{Encrypt}(pk, m')$ with $m' = m \oplus [z \bmod p]_2$. In [4] this is done by showing that the distribution of $\lfloor c/p \rceil = \sum_{i=1}^{\tau} b_i \cdot q_i$ where $\boldsymbol{b} \leftarrow \{0,1\}^\tau$ is statistically close to uniform in $\mathbb{Z}_{q_0}$. For this [4] applies the leftover hash lemma on the hash function family $h(\boldsymbol{b}) = \sum_{i=1}^{\tau} b_i \cdot q_i \mod q_0$ parametrized by the q_i's, which is clearly pairwise independent.

Similarly to prove the security of our variant we must apply the leftover hash lemma on the hash function family $h' : [0, 2^\alpha)^{\beta^2} \to \mathbb{Z}_{q_0}$ where:

$$h'(\boldsymbol{b}) = \sum_{1 \le i,j \le \beta} b_{i,j} \cdot q_{i,0} \cdot q_{j,1} \mod q_0$$

The main difficulty is to show that h' is (almost) pairwise independent; as shown below this requires to study the zeroes of the corresponding quadratic form. We note that our result might be of independent interest since it enables to construct a universal hash function with a small memory footprint.

4.2 Leftover Hash Lemma

A family $\mathcal{H}$ of hash functions $h : X \to Y$ is pairwise independent if for all $x \neq x'$ it holds that $\Pr_h[h(x) = h(x')] = 1/|Y|$. Since h' is not exactly pairwise independent we introduce a slightly more general definition:[3]

Definition 4.1. *A family $\mathcal{H}$ of hash functions $h : X \to Y$ is ε-pairwise independent if*

$$\sum_{x \neq x'} \left(\Pr_{h \leftarrow \mathcal{H}}[h(x) = h(x')] - \frac{1}{|Y|} \right) \le |X|^2 \cdot \frac{\varepsilon}{|Y|}$$

The following lemma is a straightforward generalization of the usual leftover hash lemma. We provide the proof in the full version of this paper [3].

Lemma 4.1 (Leftover hash lemma). *Let $\mathcal{H}$ be a family of ε-pairwise independent hash functions. Suppose that $h \leftarrow \mathcal{H}$ and $x \leftarrow X$ are chosen uniformly and independently. Then $(h, h(x))$ is $(\frac{1}{2}\sqrt{|Y|/|X| + \varepsilon})$-uniform over $\mathcal{H} \times Y$.*

4.3 Proof of Pairwise Independence

Let q be an integer. Let $\mathcal{H}$ be a hash function family from $\{0, \ldots, 2^\alpha - 1\}^{\beta \times \beta}$ to $\mathbb{Z}_q$. The members $h \in \mathcal{H}$ are associated to elements $q_{i,0}, q_{i,1}$ of $\mathbb{Z}_q$ for $1 \le i \le \beta$. For $\boldsymbol{b} \in \{0, \ldots, 2^\alpha - 1\}^{\beta \times \beta}$, we let:

$$h(\boldsymbol{b}) = \sum_{1 \le i,j \le \beta} b_{ij} q_{i,0} q_{j,1} \mod q$$

[3] Note that this is quite different from "ε-almost universal hash function families" in the sense of Wegman and Carter [19]. We need the collision probability $\Pr_{h \leftarrow \mathcal{H}}[h(x) = h(x')]$ to be at most $(1+\varepsilon)/|Y|$ on average, with negligible ε; $2/|Y|$ is not good enough.

Lemma 4.2. *For an odd prime integer q, the hash function family $\mathcal{H}$ is ε-pairwise independent, with:*

$$\varepsilon = \frac{1}{q} + \frac{\beta^2}{2^{\alpha\beta^2 - 2(\alpha+1)\beta}}$$

Proof. For each choice of $\boldsymbol{b} \neq \boldsymbol{b}'$, the probability $\Pr_{h \leftarrow \mathcal{H}}[h(\boldsymbol{b}) = h(\boldsymbol{b}')]$ can be expressed in terms of the number of zeros of a certain hyperbolic quadratic form in $\mathbb{Z}_q^{2\beta}$. More precisely let $A = (a_{ij})$ be the $\beta \times \beta$ matrix in $\mathbf{M}_\beta(\mathbb{Z}_q)$ given by $a_{ij} = b_{ij} - b'_{ij}$. We have:

$$\Pr_h[h(\boldsymbol{b}) = h(\boldsymbol{b}')] = \frac{1}{q^{2\beta}} \#\left\{(u_1, \ldots, u_\beta, v_1, \ldots, v_\beta) \in \mathbb{Z}_q^{2\beta} \ : \sum_{1 \le i,j \le \beta} a_{ij} u_i v_j = 0\right\}$$

Now the quadratic form $Q = \sum_{1 \le i,j \le \beta} a_{ij} u_i v_j$ has the matrix $\frac{1}{2}\begin{pmatrix} 0 & A \\ A^T & 0 \end{pmatrix}$, which is clearly conjugate to $\frac{1}{2}\begin{pmatrix} 0 & J \\ J & 0 \end{pmatrix}$ where J is the canonical row echelon form of A. It follows that Q is the orthogonal sum of r hyperbolic planes, with r the rank of A. Hence, its number of zeros is well-known (see e.g. [9, Theorem 6.32] for the non-degenerate case, from which the general case follows immediately):

$$\#\left\{(u_1, \ldots, u_\beta, v_1, \ldots, v_\beta) \in \mathbb{Z}_q^{2\beta} : \sum_{1 \le i,j \le \beta} a_{ij} u_i v_j = 0\right\} = q^{2\beta-1} + q^{2\beta-r} - q^{2\beta-r-1}$$

In particular, we get:

$$\Pr_h[h(\boldsymbol{b}) = h(\boldsymbol{b}')] - \frac{1}{q} \le \frac{1}{q^r}$$

This estimate is quite sufficient for our purposes, except in the case when $r = 1$. Therefore, we need to bound the number of pairs $(\boldsymbol{b}, \boldsymbol{b}')$ such that the corresponding matrix A is of rank 1. Noting that A has all its entries in $-2^\alpha + 1, \ldots, 2^\alpha - 1$, it is enough to bound the cardinality of the set U_α of matrices of rank 1 in $M_\beta(\mathbb{Z}_q)$ with entries in that interval.

To do so, note that a matrix of rank 1 with a nonzero upper-left entry is entirely determined by its first line and its first column. If the entries are in $\{-2^\alpha + 1, \ldots, 2^\alpha - 1\}$, this leaves $2^{\alpha+1} - 2$ choices for the upper-left entries and $(2^{\alpha+1} - 1)^{2\beta-2}$ choices for the remainder of the first line and the first column. Hence, there are less than $2^{2(\alpha+1)(\beta-1)}$ matrices in U_α with a nonzero upper-left entry (and usually much fewer, since not all first lines and first columns need to give rise to matrices with all their entries in the proper interval). The same argument can be applied to any other nonzero entry (i, j), leading to the coarse bound:

$$|U_\alpha| < \beta^2 \cdot 2^{2(\alpha+1)\beta}$$

Now, the number of pairs $(\boldsymbol{b}, \boldsymbol{b}')$ such that the corresponding matrix A is of rank 1 is at most $|X| \cdot |U_\alpha|$, since for any choice of $\boldsymbol{b}$, there are at most $|U_\alpha|$

possible values of $\boldsymbol{b}'$ such that A is in U_α. We can thus bound the value δ defined by:

$$\delta = \frac{|Y|}{|X|^2} \sum_{\boldsymbol{b} \neq \boldsymbol{b}'} \left(\Pr_h[h(\boldsymbol{b}) = h(\boldsymbol{b}')] - \frac{1}{|Y|} \right)$$

as required. Indeed:

$$\delta = \frac{q}{|X|^2} \sum_{\boldsymbol{b} \neq \boldsymbol{b}'} \left(\Pr_h[h(\boldsymbol{b}) = h(\boldsymbol{b}')] - \frac{1}{q} \right) \leq \frac{q}{|X|^2} \left(\sum_{\substack{\boldsymbol{b} \neq \boldsymbol{b}' \\ A \notin U_\alpha}} \frac{1}{q^2} + \sum_{\substack{\boldsymbol{b} \neq \boldsymbol{b}' \\ A \in U_\alpha}} \frac{1}{q} \right)$$

$$\leq \frac{q}{|X|^2} \left(\frac{|X|^2}{q^2} + \frac{|X| \cdot |U_\alpha|}{q} \right) \leq \frac{1}{q} + \frac{|U_\alpha|}{|X|} \leq \frac{1}{q} + \frac{\beta^2}{2^{\alpha\beta^2 - 2(\alpha+1)\beta}}$$

which concludes the proof. □

Corollary 4.1. *When q is a product of distinct primes greater than 2^α, the hash function family $\mathcal{H}$ is ε-pairwise independent, with:*

$$\varepsilon = \frac{\log q}{\log p} \left(\frac{e}{p} + \frac{\beta^2 \cdot 2^{(\log q)/(\log p)}}{2^{\alpha\beta^2 - 2(\alpha+1)\beta}} \right) \quad \textit{where } p \textit{ is the smallest prime factor of } q.$$

Proof. The proof is largely similar to the previous one. See the full version of this paper [3] for details.

4.4 Semantic Security

We are now ready to show that our variant is semantically secure under the (stronger) error-free approximate GCD assumption. The proof follows the same strategy as [4]; we refer to the full version of this paper [3] for the details. For two specific integers p and q_0, we define the modified distribution:

$$\mathcal{D}'_\rho(p, q_0) = \left\{ \textit{Choose } q \leftarrow [0, q_0), r \leftarrow \mathbb{Z} \cap (-2^\rho, 2^\rho) : \textit{ Output } x = q \cdot p + r \right\}$$

Definition 4.2 (Error-free approximate GCD). *The (ρ, η, γ)-error-free-approximate-GCD problem is: For a random η-bit prime integer p, given $x_0 = q_0 \cdot p$ where q_0 is a random square free 2^λ-rough integer in $[0, 2^\gamma/p)$, and polynomially many samples from $\mathcal{D}'_\rho(p, q_0)$, output p.*

Theorem 4.1. *Let $\mathcal{A}$ be an attacker with advantage ε against our variant encryption scheme with parameters $(\rho, \rho', \eta, \gamma, \tau = \beta^2)$ polynomial in the security parameter λ. There exists an algorithm $\mathcal{B}$ for solving the (ρ, η, γ)-error-free-approximate-GCD problem that succeeds with probability at least $\varepsilon/2$. The running time of $\mathcal{B}$ is polynomial in the running time of $\mathcal{A}$, λ and $1/\varepsilon$.*

5 Making the Scheme Fully Homomorphic

5.1 The Squashed Scheme

Gentry's transformation to "squash the decryption" consists in adding to the public key some extra information about the secret key and use this extra information to "post process" the ciphertext. Then the post-processed ciphertext can be decrypted by a decryption polynomial of small degree. This requires to introduce an additional complexity assumption, namely the sparse subset-sum assumption.

We follow the description of [4]. Three more parameters κ, θ and Θ are added. Concretely, one uses $\theta = \lambda$, $\kappa = \gamma + 2 + \lceil \log_2(\theta + 1) \rceil$, and $\Theta = \tilde{\mathcal{O}}(\lambda^3)$.[4] One adds to the public key a set $\boldsymbol{y} = \{y_1, \ldots, y_\Theta\}$ of rational numbers in $[0, 2)$ with κ bits of precision, such that there is a sparse subset $S \subset \{1, \ldots, \Theta\}$ of size θ with $\sum_{i \in S} y_i \simeq 1/p \mod 2$. The expanded ciphertext is computed using the y_i's. The secret key sk is replaced by the indicator vector of the subset S.

However adding Θ elements y_i each of size κ bits would give a public key of size $\Theta \cdot \kappa = \tilde{\mathcal{O}}(\lambda^8)$, instead of $\tilde{\mathcal{O}}(\lambda^7)$ in our variant. Therefore instead of storing the y_i's in the public key as in [4], we generate the y_i's using a pseudo-random generator[5] $f(\mathsf{se})$. Then only the seed se and y_1 need to be stored in the public key, and the other y_i's can be recovered during ciphertext expansion by applying $f(\mathsf{se})$ again. We obtain the following squashed scheme:

KeyGen. Generate $sk^* = p$ and pk^* as before. Set $x_p \leftarrow \lfloor 2^\kappa / p \rceil$, choose at random a Θ-bit vector $\boldsymbol{s} = (s_1, \ldots, s_\Theta)$ with Hamming weight θ with $s_1 = 1$, and let $S = \{i : s_i = 1\}$.

Initialize a system-wide pseudo-random number generator f with a random seed se, and use $f(\mathsf{se})$ to generate integers $u_i \in [0, 2^{\kappa+1})$ for $2 \le i \le \Theta$. Then, set u_1 such that $\sum_{i \in S} u_i = x_p \mod 2^{\kappa+1}$. Set $y_i = u_i / 2^\kappa$ and $\boldsymbol{y} = \{y_1, \ldots, y_\Theta\}$. Hence each y_i is a positive number smaller than two, with κ bits of precision after the binary point. Also, $[\sum_{i \in S} y_i]_2 = (1/p) - \Delta_p$ for some $|\Delta_p| < 2^{-\kappa}$.

Output the secret key $sk = \boldsymbol{s}$ and public key $pk = (pk^*, \boldsymbol{y})$.

Encrypt and Evaluate. Generate a ciphertext c^* as before. Then for $i \in \{1, \ldots, \Theta\}$ set $z_i \leftarrow [c^* \cdot y_i]_2$, keeping only $n = \lceil \log_2(\theta + 1) \rceil$ bits of precision after the binary point for each z_i. Output both c^* and $\boldsymbol{z} = (z_1, \ldots, z_\Theta)$.

Decrypt: Output $m \leftarrow [c^* - \lfloor \sum_i s_i z_i \rceil]_2$.

This completes the description of the scheme. Note that as in [6] we use $n = \lceil \log_2(\theta + 1) \rceil$ bits of precision, instead of $n = \lceil \log_2 \theta \rceil + 3$ in the original scheme. This enables to reduce the degree of the decryption polynomial.

---

[4] We use $\Theta = \tilde{\mathcal{O}}(\lambda^3)$ instead of $\Theta = \omega(\kappa \cdot \log \lambda)$ in [4] from a better analysis of the hardness of the SSSP problem (see Section 6.3).

[5] Note that f doesn't really need to be a cryptographically strong PRNG: all that is needed is that the sparse subset-sum problem remains hard when the subset is generated by f. Heuristically, this is a mild requirement. In our implementation, we use random numbers produced by the PRNG from the `glibc`.

In practice we will use $n = 4$. Note that for encryption we don't need to store all the y_i's in memory again; we can generate them one by one from the PRNG to compute $z_i \leftarrow [c^* \cdot y_i]_2$ with n bits of precision.

The proof of the following lemma is similar to the one in [4] (see the full version of this paper [3]), but we can handle a smaller precision n, as in [6], because in our scheme, ciphertext size does not grow in homomorphic operations.

Lemma 5.1. *The modified scheme is correct for the set $C(\mathcal{P}_\mathcal{E})$ of circuits that compute permitted polynomials.*

5.2 Bootstrapping

As in [4], one obtains that the scheme is bootstrappable. From Gentry's theorem we obtain homomorphic encryption schemes for circuits of any depth.

Theorem 5.1. *Let $\mathcal{E}$ be the scheme above, and let $D_\mathcal{E}$ be the set of augmented (squashed) decryption circuits. Then, $D_\mathcal{E} \subset C(\mathcal{P}_\mathcal{E})$.*

Proof. The proof is as in [4]. We provide a slightly different analysis. We consider the decryption equation:

$$m \leftarrow c^* - \left\lfloor \sum_{i=i}^{\Theta} s_i \cdot z_i \right\rceil \mod 2$$

where s_i are the secret key bits and z_i are rational numbers in $[0, 2)$ with n bits of precision after the binary point (therefore $n+1$ bits in total). We must express the decryption equation as a low degree polynomial in the bits s_i and the bits in z_i, i.e. a permitted polynomial.

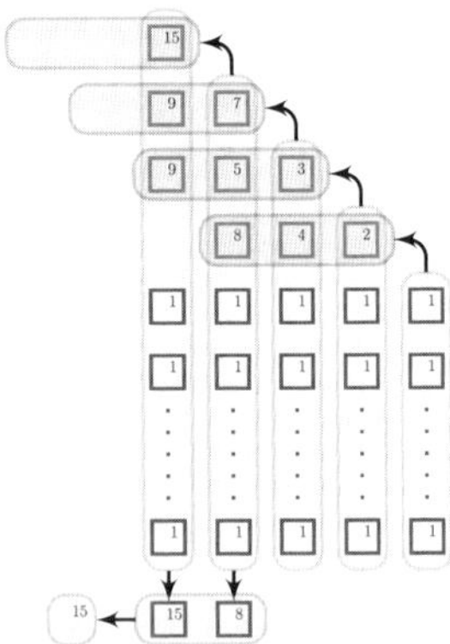

Fig. 2. Grade-school addition for Θ or $\theta = 15$ numbers of $n = 4$ bits of precision after the binary point. The numbers indicate the degree of each bit as a binary polynomial in the input bits.

For this one uses a simple grade-school addition of the numbers $a_i = s_i \cdot z_i$. As illustrated in Fig. 2 the bits of the a_i's are arranged in Θ rows and $n+1$ columns (one column before the binary point and n columns after). To see how this grade-school addition can be performed efficiently, first recall the following result from [4, §6.2].

Lemma 5.2. *Let* $\boldsymbol{b} = (b_1, b_2, \ldots, b_\Theta)$ *be any binary vector, and denote its Hamming weight by* W. *Write the binary digits of* W *as* $W = \overline{W_k \cdots W_1 W_0}^2$. *Then the* j*-th bit* W_j *of* W *can be expressed as a binary polynomial of degree exactly* 2^j *in the* b_i*'s, namely the* 2^j*-th elementary symmetric polynomial:*

$$W_j = \sum_{\substack{I \subset \{1,\ldots,\Theta\} \\ |I|=2^j}} \prod_{i \in I} b_i$$

Moreover, the bits $W_0, W_1, \ldots W_j$ *can be simultaneously computed by an arithmetic circuit of size* $2^j \cdot \Theta$.

That the W_j's are given by elementary symmetric polynomials is classical (see e.g. [2, Lemma 4]). Thus, to compute them, it suffices to find the top 2^j coefficients of the polynomial $(X - b_1)(X - b_2) \cdots (X - b_\Theta)$, which can be done iteratively with at most $2^j \cdot \Theta$ operations. We recall in the full version of this paper [3] the dynamic programming algorithm from [4].

Then, the procedure to compute

$$Q = \left\lfloor \sum_{k=1}^{\Theta} a_k \right\rceil$$

is as follows. We number the columns containing the a_k's from left to right as 0 (before the binary point), $-1, -2, \ldots, -n$.

As usual, grade-school addition starts from the rightmost column (column $-n$). Adding all Θ bits from that column produces a bit of result and a certain number of bits of carry. Since we are only interested in the $n+1$ least significant bits of the sum, we only need to keep track of the result and the first n carry bits: this amounts to computing the rightmost bits $W_0^{(-n)}, W_1^{(-n)}, \ldots, W_n^{(-n)}$ of the Hamming weight $W^{(-n)}$ of column $-n$, which can be done with at most $2^n \cdot \Theta$ multiplications according to the previous lemma.

Now, push carry bit $W_1^{(-n)}$ to column $-n+1$, carry bit $W_2^{(-n)}$ to column $-n+2$ and so on. We can then continue the grade-school addition process from column $-n+1$, where we only need to compute the result and $n-1$ carry bits, namely the bits $W_j^{(-n+1)}$ of the Hamming weight $W^{(-n+1)}$ of the column, including the possible carry bit from column $-n$. This amounts to at most $2^{n-1} \cdot (\Theta + 1)$ multiplications. When this is done, push the carry bits $W_1^{(-n+1)}, \ldots W_{n-1}^{(-n+1)}$ to columns $-n+2, -n+3, \ldots, 0$ respectively, move to the next column and continue as before. This is illustrated in Figure 2 for $n = 4$.

As shown in the full version of this paper [3], this can be done in $\mathcal{O}(\Theta \cdot \theta)$ multiplications, and the decryption polynomial is a polynomial f of the ciphertext bits and the secret key satisfying $d = \deg f = 2^{n+1}$ and $\|f\|_1 \leq 2\Theta$. In view of Remark 3.3, f is a permitted polynomial as long as $d \leq (\eta - 4 - n - \log_2 \Theta)/(\rho' + 2 + 2\log_2 \beta) \approx \eta/\rho'$ which is satisfied by choosing η according to the constraint in Section 3.2. □

6 Attacks

In this section we recall the known attacks. For each attack we provide an asymptotic analysis (as in [4]) and we also run the attacks in practice in order to derive concrete parameters for our implementation. We use four security levels inspired by the levels from [6]: "toy", "small", "medium" and "large", corresponding to 42, 52, 62 and 72 bits of security respectively. For security parameter λ we wish to ensure that the best attack requires at least 2^λ clock cycles on a standard PC.

Note that we use the SAGE [13] interface to the `fplll` lattice reduction package [15] which is to our knowledge the fastest publicly available. However any progress in LLL implementations will require an increase of our security parameters.

6.1 Brute Force Attack on the Noise

The easiest attack is the brute force attack on the noise in the public key. Given $x_0 = q_0 \cdot p$ and $x_1 = q_1 \cdot p + r_1$ with $|r_1| < 2^\rho$, one can guess r_1 and compute $\gcd(x_0, x_1 - r_1)$ to recover p. The state of the art algorithm for computing GCD's is the Stehlé-Zimmermann algorithm [18] with time complexity $\tilde{\mathcal{O}}(\gamma)$ for integers of γ bits. The attack complexity is then $2^\rho \cdot \tilde{\mathcal{O}}(\gamma)$. Therefore the attack is thwarted if $\rho = \omega(\log \lambda)$.

A better attack [11] consists in computing $p = \gcd(x_0, \prod_{i=-2^\rho}^{2^\rho}(x_1 - i) \ [x_0])$. Using fast multiplication the asymptotic complexity is also $2^\rho \cdot \tilde{\mathcal{O}}(\gamma)$. Experimentally this later attack is roughly 5 times faster. See the full version of this paper [3] for concrete parameters.

6.2 Approximate-GCD Attack on the Public Key

We do not consider Coppersmith's attack since as shown in [4] it does not apply for the range of parameters. We consider the attack based on Nguyen and Stern's orthogonal lattice [10] (see Section B.1 in [4]). One considers the first $t \leq \tau$ integers $x_i = p \cdot q_i + r_i$ and $x_0 = p \cdot q_0$. The attack builds the lattice L of row vectors orthogonal to $\boldsymbol{x} = (x_1, \ldots, x_t)$ modulo x_0 (see the full version of this paper [3] for more details). One must find a vector $\boldsymbol{u} \in L$ such that $\|u\|_\infty \leq 2^{\eta-\rho}$. From Minkowski's bound there exists a nonzero lattice vector of norm about $\sqrt{t} \cdot \det(L)^{1/t} \simeq 2^{\gamma/t}$, which gives the condition $t > \gamma/\eta$. However when the lattice dimension t is large, lattice reduction algorithms will not recover such a short vector but only an approximation.

As in [4] we use the following "rule of thumb" conjecture about lattice algorithms performance: there exists a constant μ such that for any k and any dimension t, one cannot find a $\mu^{t/k}$ approximation of the shortest vector in time smaller than 2^k. Since we must find a vector $\boldsymbol{u}$ such that $\|\boldsymbol{u}\|_\infty \leq 2^{\eta-\rho}$, we need better than a $2^{\eta-\rho}$ approximation of the shortest vector. To get a 2^η approximation (which is not quite enough to recover $\boldsymbol{u}$), from $t > \gamma/\eta$ the time required is then at least 2^k where $k = (\log_2 \mu)\gamma/\eta^2$. We recover the asymptotic condition $\gamma = \eta^2 \cdot \omega(\log \lambda)$. To obtain concrete parameters we have run some experiments with the LLL and BKZ-20 lattice reduction algorithms; see the full version of this paper [3].

6.3 Lattice Attack on the Sparse Subset-sum Problem

We use the refined analysis from [17] of the sparse subset sum problem. The attacker must solve the equation $\sum_{i=1}^{\Theta} s_i \cdot u_i = x_p \mod 2^\kappa$ where $\boldsymbol{s} = (s_1, \ldots, s_\Theta)$ is of small Hamming weight θ.

As shown in the full version of this paper [3] the lattice attack leads to a lattice L of determinant $\det L \simeq 2^{\Theta+\kappa} \simeq 2^{\Theta+\gamma}$. The lattice has a short vector of norm about $\sqrt{\Theta}$. From Minkowsky's bound we can expect that the norm of the second shortest vector is $\simeq (\det L)^{1/\Theta} \simeq 2^{\gamma/\Theta}$. Therefore to find the shortest vector we need better than a $2^{\gamma/\Theta}$ approximation. From the lattice "Rule of Thumb" conjecture with the previous notations the time required is then at least 2^k with $k = (\log_2 \mu)\Theta^2/\gamma$. Asymptotically the condition is therefore $\Theta^2 = \gamma \cdot \omega(\log \lambda)$. Therefore with $\gamma = \tilde{\mathcal{O}}(\lambda^5)$ we can take $\Theta = \tilde{\mathcal{O}}(\lambda^3)$. We refer to the full version of this paper [3] for concrete parameters; we also consider a birthday-like attack on the sparse subset-sum problem.

7 Implementation of the Fully Homomorphic Scheme

7.1 Recryption

Now that decryption can be expressed as an arithmetic circuit of low depth, it is possible to achieve bootstrapping, i.e. to publicly evaluate the decryption circuit homomorphically on a ciphertext, which produces another ciphertext for the same message, but with possibly less noise (a "recryption"). This process, which is Gentry's key idea [5] for achieving fully homomorphic encryption, is illustrated in Figure 1. The procedure that evaluates the decryption circuit homomorphically, called Recrypt, takes as input encryptions of the ciphertext bits, and encryptions of the secret key bits.

In the case of the DGHV scheme or of our variant, 0 and 1 are valid encryptions of themselves, so the ciphertext bits can be passed as is to the decryption circuit. However, encryptions of the secret key bits should also be made available publicly, so the key generation from §5.1 should be modified to include encryptions σ_i of the s_i's into the public key $pk = (pk^*, \boldsymbol{y}, \boldsymbol{\sigma})$. Then the Recrypt procedure is simply obtained by applying the decryption circuit to the ciphertext bits and the encrypted secret key bits.

Note that such ciphertexts σ_i are normally generated using the $x_{i,b}$'s from the public key, leading to σ_i's with noise of size ρ'. However since we are at key generation phase we can do better and let $\sigma_i = s_i + 2r'_i + 2p \cdot q'_i \mod x_0$ for $1 \le i \le \Theta$, for random integers q'_i modulo q_0 and random integers r'_i in $(-2^\rho, 2^\rho)$. The resulting ciphertexts σ_i have the same distribution as regular ciphertexts but with noise ρ instead of ρ'. Since $\rho < \rho'$ this enables to reduce the size η of p required to achieve bootstrappability.

For the refreshed ciphertext to decrypt correctly, its noise must be small enough, and in fact small enough that a multiplication operation between refreshed ciphertexts still decrypts correctly. The ciphertext bits are noise-free encryptions of themselves and the encrypted secret key bits contain ρ bits of noise, so one must have $d \cdot \rho < \eta/2$, where d is the degree of the decryption circuit discussed in the previous section (or in fact, only half that degree, because only the degree with respect to the secret key bits matters; the contribution in the ciphertext bits z_i can be ignored as they are used directly and without noise).

7.2 Optimization of the Decryption Circuit

We use the optimization from [4] which consists in representing the secret key $\boldsymbol{s}$ in θ boxes of $B = \Theta/\theta$ bits each, such that each box has a single 1-bit in it. This enables to obtain a grade-school addition algorithm that requires $\mathcal{O}(\theta^2)$ multiplications of bits instead of $\mathcal{O}(\Theta \cdot \theta)$. We refer to the full version of this paper [3] for the details. Note in particular that it results from the analysis that the degree of the decryption polynomial in the secret key bits is exactly θ. See also Fig. 2 for an illustration of the grade-school addition algorithm with $n = 4$.

7.3 Compression of Encrypted Secret Key Bits

The modification of the public key described previously, to accommodate for the Recrypt procedure, has the problem of increasing public key size significantly. Namely the vector $\boldsymbol{\sigma}$ in the enlarged public key consists of $\Theta = \tilde{\mathcal{O}}(\lambda^3)$ ciphertexts, each of size $\gamma = \tilde{\mathcal{O}}(\lambda^5)$, so we obtain a public key size of $\tilde{\mathcal{O}}(\lambda^8)$, instead of $\tilde{\mathcal{O}}(\lambda^7)$ in the basic scheme.

To alleviate this problem, an additional compression trick is described in [6]. Instead of generating the secret key as a single bit vector $\boldsymbol{s} = (s_1, \ldots, s_\Theta)$, one uses two bit vectors $\boldsymbol{s}^{(0)}$ and $\boldsymbol{s}^{(1)}$ of length $\sqrt{\Theta}$, and $\boldsymbol{s}$ is then recovered on the fly during decryption with $s_{i,j} = s_i^{(0)} \cdot s_j^{(1)}$. See the full version of this paper [3] for the details. This brings down the size of the encrypted secret key bits to about $\sqrt{\Theta} \cdot \gamma = \tilde{\mathcal{O}}(\lambda^{6.5})$. Note on the other hand that this increases the noise in $\boldsymbol{\sigma}$ by a factor of 2 since the $\sigma_{i,j}$ are obtained as products of two ciphertexts; this implies that to keep bootstrappability the size η of p must be doubled.

7.4 Smaller Dimension for Knapsack Encryption

From the previous section the size of the public key in the full scheme is now about $(\beta + \sqrt{\Theta}) \cdot \gamma$ bits. The conditions from Section 3.2 imply that we must have $\beta = \tilde{\mathcal{O}}(\lambda^2)$ to apply the leftover hash lemma. Since $\sqrt{\Theta} = \tilde{\mathcal{O}}(\lambda^{1.5})$ we have that β is the bottleneck. Therefore in practice we would like to use a smaller β, for which the leftover hash lemma would not apply but no attack would work.

This implies that we must consider a lattice attack against the knapsack sum in the encryption algorithm. The analysis is the same as in Section 6.3, with $\tau = \beta^2$ instead of Θ. This gives the asymptotic condition $\tau^2 = \gamma \cdot \omega(\log \lambda)$ which for $\alpha < \tau$ is weaker than the condition $\alpha \cdot \tau \geq \gamma + \omega(\log \lambda)$ necessary for the reduction to the approximate GCD problem. Under this condition we can take $\tau = \tilde{\mathcal{O}}(\gamma^3)$ instead of $\tau = \tilde{\mathcal{O}}(\gamma^4)$ and therefore $\beta = \tilde{\mathcal{O}}(\gamma^{1.5})$ instead of $\beta = \tilde{\mathcal{O}}(\gamma^2)$. The public key size is then $(\beta + \sqrt{\Theta}) \cdot \gamma = \tilde{\mathcal{O}}(\lambda^{6.5})$ instead of $\tilde{\mathcal{O}}(\lambda^7)$.

7.5 Concrete Parameters

From the analysis of the known attacks in the previous section we are now ready to derive the concrete parameters for the four levels of security. For all four levels we take $\theta = 15$. In this case the degree of the decryption polynomial is $2\theta = 30$ when using the degree-2 compression of the encryption of the secret key bits. Since we must allow for an additional multiplication after Recrypt, the total degree is $d = 4 \cdot \theta = 60$. To allow for some margin we take $\eta = (d+8)\rho = 68 \cdot \rho$. We obtain the parameters given in Table 1.

Table 1. Concrete parameters and corresponding timings, as measured using our implementation in Sage 4.5.3 [13] and GMP 4.3.2 [8], on a single core of a desktop computer with an Intel Core2 Duo E8500 CPU at 3.12 GHz. The public key is roughly $2(\beta + \sqrt{\Theta} + 1)\gamma$ bit long. Note that almost all the CPU time of key generation is spent in primality tests, to generate a rough q_0.

Parameters	λ	ρ	η	γ	β	Θ
Toy	42	16	1088	$1.6 \cdot 10^5$	12	144
Small	52	24	1632	$0.86 \cdot 10^6$	23	533
Medium	62	32	2176	$4.2 \cdot 10^6$	44	1972
Large	72	39	2652	$19 \cdot 10^6$	88	7897

Parameters	KeyGen	Encrypt	Expand	Decrypt	Recrypt	pk size
Toy	4.38 s	0.05 s	0.03 s	0.01 s	1.92 s	0.95 MB
Small	36 s	0.79 s	0.46 s	0.01 s	10.5 s	9.6 MB
Medium	5 min 9 s	10 s	8.1 s	0.02 s	1 min 20 s	89 MB
Large	43 min	2 min 57 s	3 min 55 s	0.05 s	14 min 33 s	802 MB

Acknowledgments. We would like to thank Phong Q. Nguyen, Nigel P. Smart and the CRYPTO referees for helpful comments. The work described in this paper has been supported in part by the European Commission through the ICT program under contract ICT-2007-216676 ECRYPT II.

References

1. Bach, E.: How to generate factored random numbers. SIAM J. Comput. 17, 179–193 (1988)
2. Boyar, J., Peralta, R., Pochuev, D.: On the multiplicative complexity of boolean functions over the basis $(\wedge, \oplus, 1)$. Theor. Comput. Sci. 235(1), 43–57 (2000)
3. Coron, J.S., Mandal, A., Naccache, D., Tibouchi, M.: Fully Homomorphic Encryption over the Integers with Shorter Public Keys, http://eprint.iacr.org
4. van Dijk, M., Gentry, C., Halevi, S., Vaikuntanathan, V.: Fully Homomorphic Encryption over the Integers. In: Gilbert, H. (ed.) EUROCRYPT 2010. LNCS, vol. 6110, pp. 24–43. Springer, Heidelberg (2010)
5. Gentry, C.: A fully homomorphic encryption scheme. Ph.D. thesis, Stanford University (2009), http://crypto.stanford.edu/craig
6. Gentry, C., Halevi, S.: Implementing Gentry's Fully-Homomorphic Encryption Scheme. In: Paterson, K.G. (ed.) EUROCRYPT 2011. LNCS, vol. 6632, pp. 129–148. Springer, Heidelberg (2011)
7. Goldreich, O., Goldwasser, S., Halevi, S.: Public-key cryptosystems from lattice reduction problems. In: Kaliski Jr., B.S. (ed.) CRYPTO 1997. LNCS, vol. 1294, pp. 112–131. Springer, Heidelberg (1997)
8. Grandlung, T., et al.: The GNU Multiple Precision arithmetic library, Version 4.3.2 (2010), http://gmplib.org
9. Lidl, R., Niederreiter, H.: Finite Fields. In: Encyclopedia of Mathematics and its Applications, vol. 20, Addison-Wesley, Reading (1983)
10. Nguyên, P.Q., Stern, J.: The Two Faces of Lattices in Cryptology. In: Silverman, J.H. (ed.) CaLC 2001. LNCS, vol. 2146, pp. 146–180. Springer, Heidelberg (2001)
11. Nguyen, P.Q.: Personal Communication
12. Paillier, P.: Public-key cryptosystems based on composite degree residuosity classes. In: Stern, J. (ed.) EUROCRYPT 1999. LNCS, vol. 1592, pp. 223–238. Springer, Heidelberg (1999)
13. Stein, W.A., et al.: Sage Mathematics Software (Version 4.5.3), The Sage Development Team (2010), http://www.sagemath.org
14. Micciancio, D.: Improving Lattice Based Cryptosystems Using the Hermite Normal Form. In: Silverman, J.H. (ed.) CaLC 2001. LNCS, vol. 2146, pp. 126–145. Springer, Heidelberg (2001)
15. Pujol, X., Stehlé, D., et al.: Fplll lattice reduction library, http://perso.ens-lyon.fr/xavier.pujol/fplll/
16. Smart, N.P., Vercauteren, F.: Fully Homomorphic Encryption with Relatively Small Key and Ciphertext Sizes. In: Nguyen, P.Q., Pointcheval, D. (eds.) PKC 2010. LNCS, vol. 6056, pp. 420–443. Springer, Heidelberg (2010)
17. Stehlé, D., Steinfeld, R.: Faster Fully Homomorphic Encryption. In: Abe, M. (ed.) ASIACRYPT 2010. LNCS, vol. 6477, pp. 377–394. Springer, Heidelberg (2010)
18. Stehlé, D., Zimmermann, P.: A binary recursive gcd algorithm. In: Buell, D.A. (ed.) ANTS 2004. LNCS, vol. 3076, pp. 411–425. Springer, Heidelberg (2004)
19. Wegman, M.N., Carter, J.L.: New hash functions and their use in authentication and set equality. Journal of Computer and System Sciences 22(3), 265–279 (1981)

Fully Homomorphic Encryption from Ring-LWE and Security for Key Dependent Messages

Zvika Brakerski[1] and Vinod Vaikuntanathan[2]

[1] Weizmann Institute of Science
zvika.brakerski@weizmann.ac.il
[2] Microsoft Research and University of Toronto
vinodv@cs.toronto.edu

Abstract. We present a somewhat homomorphic encryption scheme that is both very simple to describe and analyze, and whose security (quantumly) reduces to the worst-case hardness of problems on ideal lattices. We then transform it into a fully homomorphic encryption scheme using standard "squashing" and "bootstrapping" techniques introduced by Gentry (STOC 2009).

One of the obstacles in going from "somewhat" to full homomorphism is the requirement that the somewhat homomorphic scheme be circular secure, namely, the scheme can be used to securely encrypt its own secret key. For all known somewhat homomorphic encryption schemes, this requirement was not known to be achievable under any cryptographic assumption, and had to be explicitly assumed. We take a step forward towards removing this additional assumption by proving that our scheme is in fact secure when encrypting polynomial functions of the secret key.

Our scheme is based on the ring learning with errors (RLWE) assumption that was recently introduced by Lyubashevsky, Peikert and Regev (Eurocrypt 2010). The RLWE assumption is reducible to worst-case problems on ideal lattices, and allows us to completely abstract out the lattice interpretation, resulting in an extremely simple scheme. For example, our secret key is s, and our public key is $(a, b = as + 2e)$, where s, a, e are all degree $(n-1)$ integer polynomials whose coefficients are independently drawn from easy to sample distributions.

1 Introduction

Fully-homomorphic encryption is one of the most sought after goals of modern cryptography. In a nutshell, a fully homomorphic encryption scheme is an encryption scheme that allows evaluation of arbitrarily complex programs on encrypted data. The problem was first suggested by Rivest, Adleman and Dertouzos [36] back in 1978, yet the first plausible construction came thirty years later with the breakthrough work of Gentry in 2009 [14,15] (although, there has been partial progress in the meanwhile; see, e.g., [21,12,30,6]).

The cornerstone of Gentry's construction is the notion of a "somewhat homomorphic" encryption scheme – namely, an encryption scheme that allows evaluation of a class of functions below some complexity threshold. Specifically,

P. Rogaway (Ed.): CRYPTO 2011, LNCS 6841, pp. 505–524, 2011.

his construction of a somewhat homomorphic encryption scheme allows the homomorphic evaluation of any (arithmetic or Boolean) function whose polynomial representation has bounded degree. He then showed how to "bootstrap" from a sufficiently powerful somewhat homomorphic encryption scheme into a fully homomorphic encryption scheme. To construct a somewhat homomorphic encryption scheme, Gentry harnessed the power of *ideal lattices* – a sophisticated algebraic structure with many useful properties. Specifically, he was able to reduce the security of his *somewhat homomorphic encryption scheme* to the *worst-case hardness* of standard problems (such as the shortest vector problem) on ideal lattices [15].[1]

Gentry's construction is quite involved – the secret key, even in the private-key version of his scheme, is a short basis of a "random" ideal lattice. Generating pairs of public and secret bases with the right distributions appropriate for the worst-case to average-case reduction is technically quite complicated, and significant effort has been devoted recently to this issue [38,16]. We will present a scheme where key generation is simply sampling a random degree-$(n-1)$ polynomial with coefficients in $\mathbb{Z}_q$. Furthermore, all parts of our scheme can be described in elementary terms, with no reference to ideals.

A parallel line of work that utilizes ideal lattices in cryptography dates back to the NTRU cryptosystem [22]. The focus of this line of work is to use ideal lattices for *efficient cryptographic constructions*. The added structure of ideal lattices, compared to ordinary lattices, makes their representation more succinct and enables fast computation. Starting with the work of Micciancio [28], there has been an ongoing effort [31,23,32,25,24] to come up with very efficient constructions of various cryptographic primitives whose security can formally be reduced to the hardness of short-vector problems in ideal lattices. A recent work along these lines, which serves as an essential stepping stone for this work, is that of Lyubashevsky, Peikert and Regev [26].

Lyubashevsky et al. [26] present the *ring learning with errors* (RLWE) assumption, which is the "ring counterpart" of Regev's learning with errors assumption [34]. Roughly speaking, the assumption is that given polynomially many samples over a certain ring of the form $(a_i, a_i s + e_i)$, where s is a random "secret ring element", a_i's are uniformly random in the ring, and e_i are "small" ring elements, an adversary cannot distinguish this sequence of samples from random pairs of ring elements. They show that this simple to state assumption can be (very efficiently) reduced to the worst case hardness of short-vector problems on ideal lattices. They also construct a very efficient ring counterpart to Regev's [34] public-key encryption scheme, as well as a counterpart to the identity based encryption scheme of [17] (using the basis sampling techniques of [39]). The description of the scheme is very elegant since, as explained above, RLWE is stated without directly referring to lattices (similarly to the LWE assumption and ordinary lattices).

[1] The specific variant of the (approximate) shortest vector problem, as well as the specific approximation factor, are irrelevant for the current discussion.

A natural question that comes out of these two lines of work is whether one can get the best of both worlds, namely the expressive functionality on the one hand, and the simplicity and efficiency on the other. We show that indeed this can be done – we construct a somewhat homomorphic encryption scheme based on RLWE and thus inherit the simplicity and efficiency, as well as the worst case relation to ideal lattices. Furthermore, our scheme enjoys *key dependent message security* (KDM security, also known as "circular security") – namely, the scheme can securely encrypt polynomial functions (over an appropriately defined ring) of its own secret key. This property, while interesting in its own right,[2] carries special significance in the context of homomorphic encryption as we explain next.

All known constructions of fully homomorphic encryption employ a "bootstrapping" technique, which enforces the public key of the scheme to grow linearly with the maximal depth of evaluated circuits. This is a major drawback with regards to the usability and the efficiency of the scheme. However, the size of the public key can be made independent of the circuit depth if the somewhat homomorphic scheme can securely encrypt its own secret key. Achieving circular secure somewhat homomorphic encryption has been, thus, an interesting open problem[3] which we resolve in this paper. Unfortunately, the circular security we can prove is with respect to the representation of the secret key as a ring element, where bootstrapping requires circular security with respect to the bitwise representation of the secret key (to be precise: the bitwise representation of the "squashed" secret key). However, since prior to this work it was not known whether somewhat homomorphism can co-exist with *any* form of circular security, we view this property as a significant first step towards removing the above assumption.

We also show how to transform this into a fully homomorphic encryption scheme, following Gentry's blueprint of "squashing" and "bootstrapping".[4] Alternatively, applying techniques from a followup work [10], "squashing" can be avoided at the cost of relying on a "sparse" version of RLWE that is not known to reduce to worst case problems.

Lastly, we remark that our scheme is (additive) key-homomorphic, a property which recently found applications to achieving security against related-key attacks [3].

We elaborate more on the properties of our scheme below.

---

[2] In some ranges of parameters, we improve upon the best known based on any lattice assumption, see Section 1.1.

[3] Of course, one can just *assume* that some scheme is circular secure and hope that it is correct. This has indeed been the solution so far.

[4] Although our somewhat homomorphic encryption scheme assumes only the hardness of ring LWE (which can be based on the worst-case hardness of ideal lattice problems), the squashing step adds another assumption, namely the hardness of the sparse subset sum problem. This is completely analogous to what happens in Gentry's work.

1.1 Our Results and Techniques

We present a public-key encryption scheme under the polynomial LWE (PLWE) assumption, which is a simplified version of the aforementioned RLWE. We show that our scheme is both somewhat homomorphic and circular secure. The former means that bounded complexity functions can be evaluated on encrypted data. The latter means that non-trivial functions of the secret key (including the secret key itself) can be securely encrypted by our scheme. Finally, we show how fully homomorphic encryption can be obtained by bootstrapping, using "Gentry-style" squashing. We also mention how squashing can be traded for a sparse variant of PLWE using techniques from a follow-up work. Details follow.

The Assumption. We formally define the *polynomial learning with errors* assumption (PLWE), which is a simplified version of [26]'s RLWE assumption. We emphasize that PLWE is implicit in [26] and we just make it explicit. In particular, using the results of [26], the hardness of PLWE can be based on the worst-case hardness of ideal lattice problems.

In the standard parameter setting, we consider the polynomial ring $R_q \doteq \mathbb{Z}_q[x]/\langle x^n+1\rangle$, where n is a power of 2, namely the ring of all integer polynomials of degree $(n-1)$ and coefficients in $\mathbb{Z}_q$. Addition and multiplication over this ring are defined modulo (x^n+1, q). The PLWE assumption in this setting is that it is hard to distinguish polynomially many samples from the distribution $(a_i, a_i s + e_i)$ and the same number of samples from the distribution (a_i, u_i), where s, the a_i's and the u_i's are uniform in R_q and the e_i's are "noise polynomials" whose coefficients are sampled (independently) from a narrow Gaussian (which we refer to as the noise distribution χ). An important observation is that the assumption still holds if s is sampled from the noise distribution χ rather than the uniform distribution (this is the "Hermite normal form" of the assumption).

The resemblance to standard learning with errors is apparent, especially when noticing that the additive group of R_q and $\mathbb{Z}_q^n$ are isomorphic (as vector spaces, obviously multiplication in the latter is undefined). The new aspect of PLWE is the use of multiplication in R_q in the place of inner product, which results in a larger amount of pseudo-randomness generated per sample. Rather than obtaining just one element in $\mathbb{Z}_q$, as in standard LWE, we here obtain n such elements.

The PLWE problem, in some parameter settings, is reducible to the worst case hardness of "short vector problems" in ideal lattices. This is a straightforward consequence of [26]. Specifically, we require that q is a sub-exponential prime and that there is a sub-exponential gap between the q and the standard deviation of the Gaussian error (used to sample the coefficients of the e_i's described above). These parameters translate to the worst-case hardness of approximating the shortest vector problem to within a (slightly) sub-exponential approximation factor, using (slightly) sub-exponential algorithms. We note that the best (ideal) lattice algorithms run in time roughly $2^{n/k}$ to come up with a 2^k-approximation of shortest vectors (where k is a "tunable" parameter of the algorithm).

The Basic Scheme. Our somewhat homomorphic encryption scheme is so simple that the best way to present it is to spell it out. We first present the symmetric-key variant of the scheme and then explain how to transform it into a public-key scheme.

To generate the (symmetric) key for our scheme, we sample $s \stackrel{\$}{\leftarrow} \chi$ (in fact, if we only care about homomorphism and not KDM security, sampling $s \stackrel{\$}{\leftarrow} R_q$ is sufficient). Encryption is performed by sampling $a \stackrel{\$}{\leftarrow} R_q$ and $e \stackrel{\$}{\leftarrow} \chi$ and outputting the ciphertext $\mathbf{c} = (c_0, c_1)$ where $c_1 = -a$ and $c_0 = as + 2e + m$. The message m resides in the ring of polynomials with binary coefficients $R_2 = \mathbb{Z}_2[x]/\langle x^n + 1\rangle$ (which is isomorphic to $\{0,1\}^n$ but, as one might guess, has additional structure that will be used for homomorphism). To decrypt the ciphertext $\mathbf{c} = (c_0, c_1)$, one computes $c_0 + c_1 s \pmod 2$. Note that we slightly deviate from the standard notation for LWE based schemes for reasons that will be apparent below.

The correctness of the scheme is apparent, and security follows from the PLWE assumption by noting that $(a, as+2e)$ is indistinguishable from (a, u), where u is uniform (the additional factor of 2 is not a problem, since 2 is invertible in $\mathbb{Z}_q$). The ciphertext, thus, is indistinguishable from one that carries no information on the message.

A Public-Key Scheme. We obtain a public-key encryption scheme using a variant of the construction of [26]. Notice that in order to encrypt with our symmetric scheme, we only need the ability to generate pairs of the form $(a, as + 2e)$. We show that given one such pair, it is easy to re-randomize and generate as many of them as we want. Given $(a, b = as+2e)$, we consider the tuple $(a' = av+2e', b' = bv + 2e'')$, where $v, e' \stackrel{\$}{\leftarrow} \chi$, $e'' \stackrel{\$}{\leftarrow} \chi'$, where χ' is a noise distribution like χ, only with larger standard deviation. It holds that $b' = asv + 2(ev + e'') = a's + 2(ev + e'' - e's)$. If the standard deviation of e'' is sufficiently large, it holds that $b' \stackrel{s}{\approx} a's + 2e''$. In addition, a' is computationally indistinguishable from uniform (even for an adversary who knows s). Therefore (a', b') are computationally indistinguishable, even given the secret key, from an appropriately distributed pair that can be used for encryption. Security is not affected because the original pair (a, b) posted as the public key is computationally indistinguishable from being independent of s.

Somewhat Homomorphic Scheme. Achieving additive homomorphism is simple, via coordinate-wise addition: $\mathbf{c}_{\mathrm{add}} = \mathbf{c} + \mathbf{c}' = (c_0 + c_0', c_1 + c_1') = ((a + a')s + 2(e+e') + (m+m'), -(a+a'))$, which decrypts properly so long as the error does not "blow up". It is multiplicative homomorphism that requires careful handling. The intuition is that multiplying together the c_0 elements of 2 ciphertexts should create an element that depends on the product of the messages. Writing it down, we indeed see that $c_0 \cdot c_0' = -aa's^2 + (c_0 a' + c_0' a)s + 2(2ee' + em' + e'm) + mm'$, which almost looks like a legitimate ciphertext, except for the term $-aa's^2$, which contains a high power of s. The key observation is that we can make this ciphertext decryptable at the expense of adding an element to the ciphertext. Our new ciphertext will be $\mathbf{c}_{\mathrm{mult}} = (c_{\mathrm{mult},0}, c_{\mathrm{mult},1}, c_{\mathrm{mult},2})$, where $c_{\mathrm{mult},2} =$

c_1c_1', $c_{\mathrm{mult},1} = c_0c_1' + c_0'c_1$, $c_{\mathrm{mult},0} = c_0c_0'$. In other words, since we know that $m + 2e = c_0 + c_1s$ and $m' + 2e' = c_0' + c_1's$, then it holds that $(m + 2e) \cdot (m' + 2e') = (c_0 + c_1s) \cdot (c_0' + c_1's)$. We can open the parenthesis on the right hand side symbolically (without knowing s), and come up with $\mathbf{c}_{\mathrm{mult}}$ such that $(c_0 + c_1s) \cdot (c_0' + c_1's) = c_{\mathrm{mult},0} + c_{\mathrm{mult},1}s + c_{\mathrm{mult},2}s^2$. Note that all of the above can be computed from $\mathbf{c}, \mathbf{c}'$. To decrypt a 3 element ciphertext $\mathbf{c} = (c_0, c_1, c_2)$, the decryption process will be $c_0 + c_1s + c_2s^2 \pmod 2$. It is important to notice that ciphertexts of all lengths can be again added and multiplied, where addition results in a ciphertext of the maximal length of its operands and multiplication results in a ciphertext of length sum of operands minus 1.

The limiting factor on the number of homomorphic operations is the growth of the error term. We start with a sub-exponential ratio between the modulus q and the elements of e. In order to decrypt correctly, we need this ratio to be more than 2. Making the calculations (see Section 3), the total degree of the evaluated function (represented as a polynomial) needs to be less than n^ϵ for some constant ϵ (additions are relatively negligible).

The restriction on the error also limits the total length of a ciphertext: A decryptable ciphertext can have no more than n^ϵ elements (recall that the number of elements in a ciphertext and the degree of homomorphic operations are closely related). We denote this limit on the maximal degree by D.

KDM *Security.* The KDM properties of our scheme take after ideas from the work of Applebaum, Cash, Peikert and Sahai [2], who showed KDM security (w.r.t. linear functions) for Regev's LWE based scheme, and from the work of Malkin, Teranishi and Yung [27], who showed KDM security w.r.t. polynomials of the secret key (treated as integer) based on the decisional composite residuosity assumption.

To see that our scheme can encrypt non-trivial functions of its own secret key, consider the ciphertext $\mathbf{c} = (as + 2e + s, -a)$ which "looks like" an encryption of the secret key s.[5] If we define $a' = a + 1$, however, we have that $\mathbf{c} = (a's + 2e, -a' + 1)$. We notice that $(a', a's + 2e)$ is exactly a PLWE instance, so it is computationally indistinguishable from (a', u'), where u' is uniform. We have that $c \stackrel{c}{\approx} (u', -a' + 1)$, which is a completely uniform pair. This methodology is easy to extend to any linear function of s and for any polynomial number of ciphertexts.

We now revisit our previous claims that the aforementioned $\mathbf{c}$ "looks like" an encryption of s. In fact, s, drawn from χ, does not necessarily lie in the message space of the scheme, R_2, so the above statement is possibly meaningless! There are two ways to resolve this difficulty. One is to observe that choosing the parameters correctly, our c is statistically indistinguishable from an encryption of $s \pmod 2$, which is a non-trivial function of the secret key. Alternatively, and perhaps more satisfactory, is replacing the coefficient 2 in our scheme by a larger prime t, such that with all but negligible probability, $s \in R_t$. This enables achieving KDM security w.r.t. linear functions over the ring R_t.

[5] In fact, there is an important discrepancy between c and a legal encryption of s, but we will ignore it for this part of the discussion and return to it later.

To obtain KDM security for higher degree polynomials, we use a technique similar to that of [27]. We focus on quadratic functions for the sake of concreteness: We change our encryption algorithm so that encryption of a message m is performed in 2 stages: First, we compute a ciphertext as in our previous scheme $(a_1s + 2e_1 + m, -a_1)$, but then, rather than sending $-a_1$ as a part of the ciphertext, we encrypt it too and obtain $(a_2s + 2e_2 - a_1, -a_2)$. The final ciphertext will be $\mathbf{c} = (c_0, c_1, c_2) = (a_1s + 2e_1 + m, a_2s + 2e_2 - a_1, -a_2)$. To decrypt, we first extract $-a_1$ (plus some noise) from c_2, c_1, and then use this noisy $-a_1$ to decrypt c_0 and extract m. The decryption process here involves more noise than our standard scheme, but an appropriate choice of parameters enables correct decryption. Now let us consider $\mathbf{c} = (c_0, c_1, c_2) = (a_1s + 2e_1 + s^2, a_2s + 2e_2 - a_1, -a_2)$. Defining $a'_1 = a_1 + s$, $a'_2 = a_2 + 1$, we have that $\mathbf{c} = (a'_1s + 2e_1, a'_2s + 2e_2 - a'_1, -a'_2 + 1)$ Applying PLWE twice, we have that $\mathbf{c}$ is computationally indistinguishable from a tuple of uniform ring elements.

By repeating the above process, we can securely encrypt degree D polynomials using ciphertexts of length D. We notice that similar considerations apply in this case and in the case of the somewhat homomorphic scheme defined above, and indeed the decryption process in the two cases is very similar.

We further remark that our scheme can be proven to be $\mathrm{KDM}^{(\nu)}$ secure w.r.t. the same class of polynomial functions. Namely, even in the case where there is a polynomial number, ν, of users, encrypting functions of each other's secret keys, our scheme remains secure. However, this property is less relevant for homomorphism and we refer the reader to Section 4.1 for details. Achieving $\mathrm{KDM}^{(\nu)}$ security w.r.t. super-constant degree polynomials of the secret key was not known under any lattice assumption.

Full Homomorphism Using Squashing. One way to obtain a fully homomorphic scheme is to use Gentry's "bootstrapping" and "squashing". First, we notice that, as in all previously known somewhat homomorphic encryption schemes, the decryption circuit of our basic scheme has higher degree than can be homomorphically evaluated. Thus we use the by now established technique of posting, along with the public key, a sequence of elements that "hide" the secret key as a sparse subset sum. This enables reducing the complexity of decryption as we describe below.

We consider the vector $\mathbf{s} = (1, s, \ldots, s^D) \in R_q^{D+1}$ which contains all powers of the secret key s that are relevant for decryption. Note that to decrypt a ciphertext vector $\mathbf{c} = (c_0, \ldots, c_D)$, one needs to compute $\sum_{i=0}^{D} c_i s^i = \langle \mathbf{c}, \mathbf{s} \rangle$, where the "inner product" is over the ring R_q, and then take the result mod 2.

We post, along with the public key, a sequence of vectors $\mathbf{z}_1, \ldots, \mathbf{z}_m \in R_q^{D+1}$ that are uniformly sampled conditioned on the existence of a small set $L \subseteq [m]$ s.t. $|L| = n^\delta$ and $\sum_{\ell \in L} \mathbf{z}_i = \mathbf{s}$. The new secret key, therefore, is the set L (represented as a binary incidence vector over $\{0,1\}^m$). The encryptor then, along with the ciphertext vector $\mathbf{c}$, also posts $\tau_j = \langle \mathbf{c}, \mathbf{z}_\ell \rangle$, for all $\ell \in [m]$. The decryption task reduces to computing $\sum_{\ell \in L} \tau_\ell$ and taking the result modulo 2. As can be verified, this process, expressed as a polynomial over the bits of the incidence vector, has degree $\sim n^\delta$. If we choose $\delta < \epsilon$ (recall that n^ϵ is the maximal degree that can be evaluated), we have that the decryption circuit is shallow enough to be homomorphically evaluated. This is sufficient for the "bootstrapping" procedure a la Gentry.

As explained above, in order to use the bootstrapping method to obtain a public key whose size does not depend on the evaluated circuit, it is necessary to provide the evaluator with an encryption of the secret key of the somewhat homomorphic scheme. This requirement refers to the secret key of the *squashed scheme*, namely to the bits of the incidence vector of L. As we explained above, our KDM security proof does not extend to this case.

*Full Homomorphism Using Sparse-*PLWE. We mention a different way to achieve full homomorphism, as an alternative to squashing. In a followup of this paper, [10] introduced a "re-linearization" technique which they use to construct a fully homomorphic scheme based on the standard LWE assumption. One can verify that the re-linearization technique can be applied to PLWE as well – namely, a ciphertext $\mathbf{c} = (c_0, \ldots, c_D)$ can be "re-linearized" to a ciphertext $\mathbf{c} = (c_0, c_1)$ that is, in turn, decrypted in the same way. The decryption circuit thus becomes $c_0 + c_1 s \pmod 2$ and its complexity depends on the number of non-zero coefficients in the polynomial $s \in R_q$. If we sampled s from a distribution over n^δ*-sparse* polynomials, namely ones that have at most n^δ non-zero coefficients, then the decryption complexity will reduce in a similar manner to squashing. For this method to work, one needs to explicitly assume that PLWE is secure even when using such sparse s. Although the hardness of such assumption has not been thoroughly explored, we are not aware of an approach for breaking it either.

An Application: Private Information Retrieval. Our somewhat homomorphic encryption scheme can be used to construct a very efficient private information retrieval protocol [11,18] with almost logarithmic communication complexity, and security under worst-case hardness assumptions. While any (appropriate) somewhat homomorphic encryption scheme can be used to construct a PIR protocol (in particular, the scheme of [14]), our construction from ring LWE results in a particularly efficient and elegant PIR scheme. Due to space constraints, we do not provide a detailed explanation in this extended abstract.

1.2 Other Related Works

The only known candidate for fully homomorphic encryption, aside from Gentry's aforementioned scheme (and a variant thereof [38]), was presented by van Dijk, Gentry, Halevi and Vaikuntanathan [13]. Their scheme works over the integers and relies on a new assumption which roughly states that finding the greatest common divisor of many "noisy" multiples of a number is computationally hard. They cannot, however, reduce their assumption to worst case hardness.

The efficiency of implementing Gentry's scheme also gained much attention. Smart and Vercauteren [38], as well as Gentry and Halevi [16] conduct a study on reducing the complexity of implementing the scheme, specifically the key generation process. We note that the key generation process in this work is simpler and does not require generating lattice bases.

Candidate KDM secure encryption schemes in the standard model (i.e. without random oracles) started with the work of Boneh, Halevi, Hamburg and Ostrovsky [7] who presented a scheme based on the decisional Diffie-Hellman assumption that can securely encrypt linear combinations of the bits of its secret key. The aforementioned work of [2] showed a similar result based on LWE, where now the linear functions were over the components of the secret key, that reside in the space $\mathbb{Z}_p$ for some prime p. Later, Brakerski and Goldwasser [8] showed KDM security for linear functions based on a class of assumptions they refer to as "subgroup indistinguishability assumptions", which includes quadratic residuosity and decisional composite residuosity. The domain of the functions varied by the assumption. A number of works [4,9,1] showed that KDM w.r.t. linear functions can be extended to more complex functions. The work of [27] takes a different path by treating the secret key as an element in a ring (integers modulo a composite, in their case), an approach we adopt here as well.

In a followup work, [10] showed that fully homomorphic encryption can be achieved from the classical LWE assumption, without referring to ideal lattices and without squashing. Some of their techniques can also be applied to PLWE (e.g. to achieve full homomorphism via sparse polynomials as we describe above).

1.3 Notation

Let $\mathcal{D}$ denote a distribution over some finite set S. Then, $d \xleftarrow{\$} \mathcal{D}$ is used to denote the fact that d is chosen from the distribution $\mathcal{D}$. When we say $d \xleftarrow{\$} S$, we simply mean that d is chosen from the uniform distribution over S.

The ring of polynomials over the integers (i.e. symbolic polynomials with integer coefficients) is denoted $\mathbb{Z}[x]$. Given a degree n polynomial $f(x)$, the ring $\mathbb{Z}[x]/\langle f(x)\rangle$ is the ring of all polynomials modulo $f(x)$. The ring of polynomials with coefficients in $\mathbb{Z}_q$ is denoted $\mathbb{Z}_q[x]$ and $\mathbb{Z}_q[x]/\langle f(x)\rangle$ is defined analogously to above. For additional background in algebraic number theory, we refer the reader to to [40].

We denote scalars in plain (e.g. x) and vectors in bold (e.g. $\mathbf{v}$). A norm of a vector is denoted by $\|\mathbf{v}\|$ and always refers to ℓ_∞: $\|\mathbf{v}\| = \max_i |v_i|$. The norm of a polynomial $\|p(x)\|$ is the norm of its coefficient vector. More generally, we use the standard isomorphism between degree $(n-1)$ polynomials in $\mathbb{Z}[x]$ and vectors in $\mathbb{Z}^n$, given by the vector of coefficients, that allows to treat the two objects interchangeably: the vector $\mathbf{p}$ will indicate the vector of coefficients of $p(x)$. We explicitly mention when we use this isomorphism.

We let the distribution $D_{\mathbb{Z}^n,r}$ to indicate the *n-dimensional discrete Gaussian distribution*. To sample a vector $\mathbf{x} \in \mathbb{Z}^n$ from this distribution, sample $y_i \in \mathbb{R}$ from the Gaussian of standard deviation r and set $x_i := \lfloor y_i \rceil$, where $\lfloor \cdot \rceil$ represents rounding to the nearest integer. Using the isomorphism mentioned above, we treat $D_{\mathbb{Z}^n,r}$ as a distribution over integer degree n polynomials. Note that in this work we only need *spherical* Gaussian distributions, in which the standard deviation over each dimension is the same.

2 The Ring LWE Problem, and Variants

In this section, we describe a variant of the "ring learning with errors" (RLWE) assumption of Lyubaskevsky, Peikert and Regev [26], that we call *polynomial* LWE (or, PLWE). This assumption is in fact implicit in [26], and can be thought of as a special case of their general RLWE assumption. Fortunately, for the parameters of interest to us, it follows from [26] that breaking the PLWE assumption leads to an algorithm to solve worst-case ideal lattice problems. Our motivation in working with the PLWE assumption is due in part to our desire to keep the exposition *elementary*, but is also dictated by the particular choice of the message encoding in our encryption schemes. See Section 2.1 for a detailed comparison, and the statement of the worst-case to average-case reduction for PLWE.

The PLWE assumption is analogous to the (by now standard) "learning with errors" (LWE) assumption, defined by Regev [34,35] (generalizing the learning parity with noise assumption of Blum et al. [5]). In the PLWE assumption, we consider rings $R \doteq \mathbb{Z}[x]/\langle f(x)\rangle$ and $R_q \doteq R/qR$ for some degree n integer polynomial $f(x) \in \mathbb{Z}[x]$ and a prime integer $q \in \mathbb{Z}$. Note that $R_q \equiv \mathbb{Z}_q[x]/\langle f(x)\rangle$, i.e. the ring of degree $(n-1)$ polynomials with coefficients in $\mathbb{Z}_q$. Addition in these rings is done component-wise in their coefficients (thus, their additive group is isomorphic to $\mathbb{Z}^n$ and $\mathbb{Z}_q^n$ respectively). Multiplication is simply polynomial multiplication modulo $f(x)$ (and also q, in the case of the ring R_q).

Thus an element in R (or R_q) can be viewed as a degree $(n-1)$ polynomial over $\mathbb{Z}$ (or $\mathbb{Z}_q$). As we mentioned in Section 1.3, we represent such an element using the vector of its coefficients. For an element $a(x) = a_0 + a_1x + \ldots + a_{n-1}x^{n-1} \in R$, we let $\|a\| = \max |a_i|$ denote its ℓ_∞ norm.

The $\mathsf{PLWE}_{f,q,\chi}$ assumption is parameterized by an integer polynomial $f(x) \in \mathbb{Z}[x]$ of degree n (which defines the ring $R = \mathbb{Z}[x]/\langle f(x)\rangle$), a prime integer $q \in \mathbb{Z}$ and an error distribution χ over R.[6] We require that χ is efficiently sampleable in our representation, namely that it is efficient to sample the coefficients of the polynomial representing the sampled element.

Let $s \xleftarrow{\$} R_q$ be a uniformly random ring element. The assumption is that given any polynomial number of samples of the form $(a_i, b_i = a_i \cdot s + e_i) \in (R_q)^2$, where a_i is uniformly random in R_q and e_i is drawn from the error distribution χ, the b_i's are computationally indistinguishable from uniform in R_q. If the number of samples that the distinguisher obtains is limited by $\ell = \ell(\kappa)$, then we call this assumption $\mathsf{PLWE}^{(\ell)}_{f,q,\chi}$. Our formal definition below presents the *hermite normal form* of the assumption, where the secret s is sampled from the noise distribution χ rather than being uniform in R_q. This presentation is more useful for the purposes of this paper and it turns out that to be equivalent to the original one, up to obtaining one additional sample [2,26].

Definition 1 (The PLWE Assumption - Hermite Normal Form). *For all $\kappa \in \mathbb{N}$, let $f(x) = f_\kappa(x) \in \mathbb{Z}[x]$ be a polynomial of degree $n = n(\kappa)$, let*

[6] To be precise, $n(\kappa), q(\kappa)$ are functions of the security parameter κ and $\{f_\kappa(x)\}$ and $\{\chi_\kappa\}$ are ensembles of polynomials and distributions respectively.

$q = q(\kappa) \in \mathbb{Z}$ be a prime integer, let the ring $R \doteq \mathbb{Z}[x]/\langle f(x)\rangle$ and $R_q \doteq R/qR$, and let χ denote a distribution over the ring R.

The polynomial LWE assumption $\mathsf{PLWE}_{f,q,\chi}$ *states that for any $\ell = \mathrm{poly}(\kappa)$ it holds that*

$$\{(a_i, a_i \cdot s + e_i)\}_{i\in[\ell]} \stackrel{c}{\approx} \{(a_i, u_i)\}_{i\in[\ell]} ,$$

where s is sampled from the noise distribution χ, a_i are uniform in R_q, the "error polynomials" e_i are sampled from the error distribution χ, and finally, the ring elements u_i are uniformly random over R_q.

When we require the indistinguishability to hold given only ℓ samples (for some $\ell = \mathrm{poly}(\kappa)$), we denote the assumption by $\mathsf{PLWE}^{(\ell)}_{f,q,\chi}$.

Note that we define the PLWE assumption as a *decisional* assumption. One could also define the search assumption which requires an adversary to find $s \in R_q$, given any polynomial number of samples $(a_i, a_i \cdot s + e_i)$. The search and decisional assumptions are equivalent for some range of parameters, as shown by [26]. We focus here on the decisional assumption since that is the most natural for cryptographic applications.

Scaling the noise. It is very useful in our schemes to generate the PLWE samples as $(a_i, a_i \cdot s + t \cdot e_i)$, where a_i, s, e_i are as above and $t \in \mathbb{Z}_q^*$. This variant is equivalent to PLWE just by virtue of q and t being relatively prime as stated below. The proof is straightforward and is omitted.

Proposition 1. *Let $f(x), q$ and χ be as in Definition 1. Let $t = t(\kappa) \in \mathbb{Z}_q^*$ (thus t and q are relatively prime). Then for any $\ell = \mathrm{poly}(\kappa)$, the $\mathsf{PLWE}^{(\ell)}_{f,q,\chi}$ assumption implies that,*

$$\{(a_i, a_i \cdot s + t \cdot e_i)\}_{i\in[\ell]} \stackrel{c}{\approx} \{(a_i, u_i)\}_{i\in[\ell]} .$$

where a_i, s, e_i and u_i are as in Definition 1.

2.1 Choice of Parameters

Our results rely on a specific choices of the polynomial $f(x)$, the modulus q, and the error distribution χ. The parameter choices are dictated by the search-to-decision reduction of [26], as well as our choice of message encoding in the encryption scheme (which seems necessary to achieve homomorphic properties). In particular, setting κ as our security parameter, we assume that:

- We set $f(x) = x^n + 1$, where $n = 2^{\lfloor \log \kappa \rceil - 1}$ (this polynomial is also denoted $\Phi_m(x)$, where $m = 2n$; see below). Since $n \in (\kappa/4, \kappa]$, all asymptotics can be stated in terms of n. For the knowledgable reader we mention that $f(x) = \Phi_m(x)$ is the mth cyclotomic polynomial.
  In addition, the fact that $f(x) = x^n + 1$ means that multiplication of ring elements does not increase their norm by too much (see lemmas below).
- The error distribution χ is the discrete Gaussian distribution $D_{\mathbb{Z}^n,r}$ for some $r > 0$. A sample from this distribution defines a polynomial $e(x) \in R$.

Some Useful Facts. We present some elementary facts about the Gaussian error distribution, and multiplication over the ring $\mathbb{Z}[x]/\langle x^n+1\rangle$. The first fact bounds the (Euclidean and therefore, the ℓ_∞) length of a vector drawn from a discrete Gaussian of standard deviation r by $r\sqrt{n}$. The second says that the statistical distance between two Gaussian distributions with the same standard deviation r (but different centers) is proportional to Δ/r, where Δ is the distance between their centers. The third and final fact says that multiplication in the ring $\mathbb{Z}[x]/\langle\Phi_m(x)\rangle$ increases the norm of the constituent elements only by a modest amount.

Lemma 1 (see [29], Theorem 4.4). *Let $n\in\mathbb{N}$. For any real number $r=\omega(\sqrt{\log n})$, we have $\Pr_{x\leftarrow D_{\mathbb{Z}^n,r}}[||x||>r\sqrt{n}]\leq 2^{-n+1}$.*

Lemma 2 (see [20], Lemma 3). *Let $n\in\mathbb{N}$. For any real number $r=\omega(\sqrt{\log n})$, and any $\mathbf{c}\in\mathbb{Z}^n$, the statistical distance between the distributions $D_{\mathbb{Z}^n,r}$ and $D_{\mathbb{Z}^n,r,\mathbf{c}}$ is at most $||\mathbf{c}||/r$.*

Lemma 3 (see [23,14]). *Let $n\in\mathbb{N}$, $m=2n$, and let $f(x)=\Phi_m(x)=x^n+1$ and let $R=\mathbb{Z}[x]/\langle\Phi_m(x)\rangle$. For any $s,t\in R$, $||s\cdot t \pmod{\Phi_m(x)}||\leq\sqrt{n}\cdot||s||\cdot||t||$, and $||s\cdot t \pmod{\Phi_m(x)}||_\infty\leq n\cdot||s||_\infty\cdot||t||_\infty$.*

The Worst-case to Average-case Connection. We state a worst-case to average-case reduction from the shortest vector problem on ideal lattices to the PLWE problem for our setting of parameters. The reduction stated below is a special case of the results of [26].

Theorem 1 (A special case of [26]). *Let κ be the security parameter. Let $k\in\mathbb{N}$ and let $m=2^{\lceil\log\kappa\rceil}$ be a power of two. Let $\Phi_m(x)=x^n+1$ be the mth cyclotomic polynomial of degree $n=\varphi(m)=m/2$. Let $r\geq\omega(\sqrt{\log n})$ be a real number, and let $q\equiv 1 \pmod m$ be a prime integer. Let $R=\mathbb{Z}[x]/\langle\Phi_m(x)\rangle$. Then:*

- *There is a randomized reduction from $2^{\omega(\log n)}\cdot(q/r)$-approximate R-SVP to $\mathsf{PLWE}_{\Phi_m,q,\chi}$ where $\chi=D_{\mathbb{Z}^n,r}$ is the discrete Gaussian distribution. The reduction runs in time $\mathrm{poly}(n,q)$.*
- *There is a randomized reduction from $(n^2q/r)\cdot(n(\ell+1)/\log(n(\ell+1)))^{1/4}$-approximate R-SVP to $\mathsf{PLWE}^{(\ell)}_{\Phi_m,q,\chi}$ where $\chi=D_{\mathbb{Z}^n,r}$ is the discrete Gaussian distribution. The reduction runs in time $\mathrm{poly}(n,q,\ell)$.*[7]

3 A Somewhat Homomorphic Encryption Scheme

We present a somewhat homomorphic encryption scheme with message space $R_t=\mathbb{Z}_t[x]/\langle f(x)\rangle$ for some integer $t=t(\kappa)$. The homomorphism will be over this ring. For the sake of concreteness, we advise the reader to think of $t=2$

[7] For the interested reader, we remark that the term $(\ell+1)$ replaces the original ℓ of [26] due to our choice to define PLWE in hermite normal form.

as a running example. We describe our scheme in the symmetric case in Section 3.1 and then describe the public-key variant in Section 3.2. The transition to full homomorphism via squashing and bootstrapping is fairly standard and is omitted due to space limitations.

3.1 The Symmetric Scheme

Let κ denote the security parameter. Our scheme is parameterized by a prime number q and a prime $t \in \mathbb{Z}_q^*$, a degree n polynomial $f(x) \in \mathbb{Z}[x]$, and an error distribution χ over the ring $R_q \doteq \mathbb{Z}_q[x]/\langle f(x)\rangle$. The parameters n, f, q and χ are public and we assume that given κ, there are polynomial-time algorithms that output f and q, and sample from the error distribution χ. An additional parameter of the scheme is an integer $D \in \mathbb{N}$ that is related to the maximal degree of homomorphism allowed (and to the maximal ciphertext length). This is not a "free" parameter, and is determined by f, q, χ in the analysis of the scheme.

- $\mathsf{SH.Keygen}(1^\kappa)$: Sample a ring element $s \xleftarrow{\$} \chi$ and set the secret key $sk{:=}s$. Define the *secret key vector* as $\mathbf{s}{:=}(1, s, s^2, \ldots, s^D) \in R_q^{D+1}$, which is efficiently computable given s and will be used in the decryption process.
- $\mathsf{SH.Enc}(sk, m)$: Recall that our message space is R_t. Namely, we encode our message as a degree n polynomial with coefficients in $\mathbb{Z}_t$.
  To encrypt, sample $(a, b = as{+}te) \in R_q^2$, where $a \xleftarrow{\$} R_q$ and $e \xleftarrow{\$} \chi$. Compute

  $$c_0{:=}b + m \in R_q \qquad \text{and} \qquad c_1{:=} - a$$

  and output the ciphertext $\mathbf{c}{:=}(c_0, c_1) \in R_q^2$.
  An important note is that the encryptor only uses the key s in order to sample (a, b). This will be important when we present our public-key scheme, where we will show that the public key enables sampling from this distribution without direct access to s.
  While the encryption algorithm only generates ciphertexts $\mathbf{c} \in R_q^2$, homomorphic operations (described below) might add more elements to the ciphertext. Thus the most generic form of a decryptable ciphertext in our scheme is $\mathbf{c} = (c_0, \ldots, c_d)$ for $d \leq D$. We remark that, as we will show below, "padding with zeros" does not effect the ciphertext. Namely $(c_0, \ldots, c_d) \equiv (c_0, \ldots, c_d, 0, \ldots, 0)$.
- $\mathsf{SH.Eval}(p(\xi_1, \ldots, \xi_\ell), (\mathbf{c}_1, \ldots, \mathbf{c}_\ell))$: We show how to evaluate an ℓ-variate polynomial $p : R_t^\ell \to R_t$. To this end, we show how to homomorphically add and multiply two elements in R_t.
  - Given two ciphertexts $\mathbf{c} = (c_0, \ldots, c_d)$ and $\mathbf{c}' = (c'_0, \ldots, c'_d)$ (we assume w.l.o.g that they have the same length, e.g. by padding), output the ciphertext $\mathbf{c}_{\text{add}} = \mathbf{c} + \mathbf{c}' = (c_0 + c'_0, \ldots, c_d + c'_d) \in R_q^{d+1}$, as an encryption of the *sum* of the underlying messages. Namely, addition is done by coordinate-wise vector addition of the ciphertext vectors. Note that addition does not increase the number of elements in the ciphertext vectors.

- Given two ciphertexts $\mathbf{c} = (c_0, \ldots, c_d)$ and $\mathbf{c}' = (c'_0, \ldots, c'_{d'})$ (here we do not pad with zeros), an encryption of their *product* is computed as follows.
  Let v be a *symbolic* variable and consider the expression $\left(\sum_{i=0}^{d} c_i v^i\right) \cdot \left(\sum_{i=0}^{d'} c'_i v^i\right)$ (over R_q). We can (symbolically, treating v as an unknown variable) open the parenthesis to compute $\hat{c}_0, \ldots, \hat{c}_{d+d'} \in R_q$ such that for all $v \in R_q$

$$\left(\sum_{i=0}^{d} c_i v^i\right) \cdot \left(\sum_{i=0}^{d'} c'_i v^i\right) \equiv \sum_{i=0}^{d+d'} \hat{c}_i v^i \ .$$

  The output ciphertext is $\mathbf{c}_{\text{mult}} = (\hat{c}_0, \ldots, \hat{c}_{d+d'})$.

We claim that if $\mathbf{c}$ is an encryption of a message $m \in R_t$ and $\mathbf{c}'$ is an encryption of $m' \in R_t$, then these two operations generate ciphertexts that decrypt to $m + m'$ and mm', respectively, where arithmetics is over R_t.

– $\mathsf{SH.Dec}(sk, \mathbf{c})$: Recall that the general form of a decryptable ciphertext is w.l.o.g $\mathbf{c} = (c_0, c_1, \ldots, c_D) \in R_q^{D+1}$, e.g. by padding.
To decrypt, we first compute $\langle \mathbf{c}, \mathbf{s} \rangle \doteq \sum_{i=0}^{D} c_i s^i \in R_q$, which can be interpreted as inner product over R_q^{D+1}, and output $m = \langle \mathbf{c}, \mathbf{s} \rangle \pmod{t}$ as the message.
Note that the condition for correct decryption is that the ℓ_∞ norm of the polynomial $\langle \mathbf{c}, \mathbf{s} \rangle$ is smaller than $q/2$.

We note that for the sake of the symmetric key somewhat homomorphic encryption scheme, the secret key s can be chosen uniformly at random. Choosing s from the error distribution is important both in the public-key variant as well as for KDM security.

We state the correctness and security below. Proofs are omitted from this extended abstract.

Theorem 2. *Let $\chi = D_{\mathbb{Z}^n, r}$ be the discrete Gaussian noise distribution with standard deviation r. The scheme described above is a somewhat homomorphic encryption scheme capable of evaluating ℓ-variate degree-D polynomials over R_t, as long as $M \cdot (trn^{1.5})^D < q/2$, where M is the ℓ_∞ norm of the polynomial (i.e. its maximal coefficient).*

Theorem 3. *Let $n, q, f(x)$ be as in the scheme, let $r = \text{poly}(n)$ and $q = 2^{n^\epsilon}$ for some $0 < \epsilon < 1$. Then, the scheme allows evaluation of degree-$O(n^\epsilon / \log n)$ polynomials with at most $2^{O(n^\epsilon / \log n)}$ terms, and is secure under the worst-case hardness of approximating shortest vectors on ideal lattices to within a factor of $O(2^{n^\epsilon})$.*

3.2 Public-Key Encryption

There is a number of ways to go from symmetric-key to public-key somewhat homomorphism. The work of Rothblum [37] provides a generic though inefficient way to go from homomorphic symmetric to public key encryption. Alternatively, one can use re-randomization via the leftover hash lemma (as used in Regev's LWE based scheme). However, the greatest efficiency is achieved using a method that appears in the full version of [26] and due to space limitations will be the only one discussed here.

Recall that in order to encrypt with our symmetric scheme, we only need the ability to generate pairs of the form $(a, as+te)$. We show that given one such pair (with smaller noise parameter), it is easy to re-randomize and generate as many of them as we want.

Concretely, we show that given one sample $(a, b = as+te)$, where the noise e comes from a distribution χ, we can generate as many additional samples as we would like, without knowing s, but with noise coming from a distribution χ' of greater standard deviation. To be even more precise, we will generate samples that are computationally indistinguishable from the desired distribution, even given the secret key s, under the $\mathsf{PLWE}^{(1)}_{f,q,\chi}$ assumption. This is sufficient for all of our purposes since in all scenarios we consider (including KDM security) the randomness used to generate these samples is not revealed to any entity (including the decryptor). The re-randomization lemma follows.

Lemma 4. *Let $f, q, \chi = D_{\mathbb{Z}^n,r}$ be parameters for* PLWE *and let t be co-prime to q. Let $\chi' = D_{\mathbb{Z}^n,r'}$, with $r' \geq 2^{\omega(\log n)} \cdot r$. Let $s, v, e, e' \xleftarrow{\$} \chi$, $a, a' \xleftarrow{\$} R_q$, $e'' \xleftarrow{\$} \chi'$, $b \doteq as + te$, then under the* $\mathsf{PLWE}^{(1)}_{f,q,\chi}$ *assumption,*

$$(s, (a, b), (av + te', bv + te'')) \overset{c}{\approx} (s, (a, b), (a', a's + te'')) \ .$$

Proof. Denote $\alpha \doteq av + te'$ and $\beta \doteq bv + te''$. Then it holds that $\beta = (as + te)v + te'' = \alpha s + t(e'' + ev - e's)$. By Lemma 1 and Lemma 2, it holds that $e'' + ev - e's \overset{s}{\approx} e''$. Namely

$$(s, (a, b), (av + te', bv + te'')) \overset{s}{\approx} (s, (a, b), (\alpha, \alpha s + te'')) \ .$$

However, $(s, a, \alpha) = (s, a, av + te') \overset{c}{\approx} (s, a, a')$ by $\mathsf{PLWE}^{(1)}_{f,q,\chi}$ and the result follows.

Therefore, to achieve a public key scheme, the following changes need to be made in our scheme. Let χ, χ' be as above.

1. In the key generation, in addition to the secret key $sk = s \xleftarrow{\$} \chi$, a public key $pk \doteq (a_0, b_0 = a_0 s + te_0)$ is output. Where $a_0 \xleftarrow{\$} R_q$, $e_0 \xleftarrow{\$} \chi$.
2. In the encryption algorithm, instead of using $(a, as + te)$, the encryptor will use $(a_0 v + te', b_0 v + te'')$, where $v, e' \xleftarrow{\$} \chi$ and $e'' \xleftarrow{\$} \chi'$.

As a side note we remark that for all applications except KDM security, it is sufficient to generate $(a, as + te)$ where the error e is not distributed according to the correct error distribution. Generating such "skewed" samples is easier and improves the parameters of the scheme. We omit the details.

4 Key Dependent Message (Circular) Security

We show that our somewhat homomorphic scheme from Section 3 (using a sufficiently large parameter t) is KDM secure w.r.t. linear functions of the secret key, over the ring R_t. We further show that changing just the encryption algorithm, allows for KDM security w.r.t. degree-d polynomials. An interesting interplay between KDM security and somewhat homomorphism allows the key generation and even the decryption circuit to stay unchanged. We note that even though we change the encryption algorithm, the ciphertexts of the resulting scheme can still undergo homomorphic operations (although a little fewer than before). Let us elaborate a bit more about the connection between homomorphism and KDM security.

Assume that we can prove that our basic scheme, as is, is secure w.r.t. linear functions of the secret key, then somewhat homomorphism implies that we can generate encryptions of degree $d \leq D$ polynomials of the secret key, by taking encryptions of the secret key and e.g. multiplying them together to generate quadratic polynomials, multiply by constants, add more terms etc. The above implies a very weak form of KDM: that it is possible to generate secure ciphertexts that decrypt to the right function of the secret key. To show full KDM, we need to present an encryption algorithm that produces indistinguishable ciphertexts whether it encrypts functions of the secret key or the constant message 0. Not surprisingly, we accomplish this by modifying the encryption algorithm to generate ciphertexts that look a lot like the one generated by homomorphism. Specifically, to be secure against degree-d polynomials, our encryption algorithm will generate $(d+1)$-element ciphertexts (contrast this with the encryption algorithm of the somewhat homomorphic encryption scheme in Section 3, that generates ciphertexts with just two non-zero ring elements). Our techniques borrow from a recent work of Malkin et al. [27]. We describe our scheme in the symmetric key setting only, noting that the public key variant applies to here as well. Due to space limitations, we do not provide proofs in this section. The formal definition of KDM security is omitted as well (see e.g. [7]).

We define $\mathcal{P}_d = \mathcal{P}_d[R_t]$ to be the class of all degree d polynomials over R_t, i.e. all functions of the form $p(z) = \sum_{i=0}^{d} \alpha_i z^i$, where $\alpha_i \in R_t$ and arithmetics is over R_t as well.

The Scheme. Let κ denote the security parameter. Our scheme is parameterized by the same parameters as our somewhat homomorphic scheme from Section 3: The primes $q, t \in \mathbb{Z}_q^*$, a degree n polynomial $f(x) \in \mathbb{Z}[x]$, and an error distribution χ over the ring $R_q \doteq \mathbb{Z}_q[x]/\langle f(x) \rangle$. As before, there is the maximal degree parameter D. An additional parameter $d \leq D$ determines the class of functions for which KDM security holds. The message space is R_t.

- $\mathsf{KDM.Keygen}(1^\kappa)$: As we explained above, the key generation is identical to that of our basic scheme $\mathsf{SH.Keygen}$: The secret key is generated as $s \stackrel{\$}{\leftarrow} \chi$. (Note that we will set χ to be a Gaussian distribution with small enough parameter $r \ll t$ such that with all but negligible probability $s \in R_t$)

- $\mathsf{KDM.Enc}(sk, m)$: To encrypt a message $m \in R_t$, we generate the ciphertext $c = (c_0, \ldots, c_d) \in R_q^{d+1}$ as follows.
  We generate d pairs $\{(a_i, b_i = a_i s + te_i)\}_{i \in [d]}$, where $a_i \xleftarrow{\$} R_q$, $e_i \xleftarrow{\$} \chi'$ (χ' will be set with noise parameter much larger than χ) and set:
  $$c_0 = b_1 + m \ ; \qquad \forall_{i \in [d-1]}.\ c_i = b_{i+1} - a_i \ ; \qquad c_d = -a_d \ .$$
  This encryption algorithm coincides with our basic scheme when $d = 1$. While we analyze the correctness of the scheme separately, let us justify our encryption algorithm by noting that for our ciphertext it holds that
  $$\begin{aligned}\sum_{i=0}^{d} c_i s^i &= b_1 + m + \sum_{i \in [d-1]} (b_{i+1} - a_i)\, s^i \quad a_d s^d \\ &= m + \sum_{i \in [d]} b_i s^{i-1} - \sum_{i \in [d]} a_i s^i \\ &= m + \sum_{i \in [d]} (b_i - a_i s) s^{i-1} \\ &= m + t \cdot \sum_{i \in [d]} e_i s^{i-1} \ . \end{aligned} \tag{1}$$
- $\mathsf{KDM.Dec}(sk, \mathbf{c})$: The decryption algorithm is identical to the basic scheme and in fact is able to decrypt ciphertexts that were originally generated by the encryption algorithm of our KDM scheme above, and then underwent somewhat homomorphic operations. Recall that the general form of a decryptable ciphertext is, w.l.o.g, $\mathbf{c} = (c_0, c_1, \ldots, c_D) \in R_q^{D+1}$. To decrypt, we first compute $\langle \mathbf{c}, \mathbf{s} \rangle = \sum_{i=0}^{D} c_i s^i \in R_q$, and output $m = \langle \mathbf{c}, \mathbf{s} \rangle \pmod t$ as the message.

Parameter Setting. Let us now describe a plausible parameter setting for our scheme. As usual, we will set $f(x) = \Phi_m(x)$ (where $n = 2^{\lfloor \log \kappa \rceil - 1}$, and $m = 2n$) and our error distributions will be Gaussian $\chi = D_{\mathbb{Z}^n, r}$, $\chi' = D_{\mathbb{Z}^n, r'}$ for r, r' defined next. Our q needs to be super-polynomial as implied by the selection below.

We set $r = 2^{\omega(\log n)}$ to be a super-polynomial function, and $r' = 2^{\omega(\log n)} \cdot r^d$. We note that the r^d factor is so that χ' can "swallow" degree d polynomials over χ.

The parameter t is chosen so that $t > r\sqrt{n}$ (i.e. a sample from χ resides in R_t with all but negligible probability) and on the other hand, for correctness, $t < 2^{-\omega(\log n)} \cdot r^{-d} \cdot q$.

We conclude that in our parameter setting, the scheme supports KDM functions of degree $d \approx (\log q - \log t - O(1))/\omega(\log n) = \log q/\omega(\log n)$, for some super-logarithmic $\omega(\log n)$. Security will be based on $\mathsf{PLWE}_{\Phi_m, q, r}$.

We next state correctness in light of this parameter setting. (We remark that we can somewhat improve efficiency with a more aggressive parameter setting; we choose to present the concrete setting above for simplicity).

Lemma 5. *Consider the parameters of our scheme as defined above. Then for all $m \in R_t$ it holds that*

$$\Pr[\mathsf{KDM.Dec}(s, \mathsf{KDM.Enc}(s, m)) \neq m] = \text{negl}(\kappa) ,$$

where the probability is taken over the choice of s and over the randomness of KDM.Enc.

The $\text{KDM}^{(1)}$-security of our scheme is stated below.

Theorem 4. *Our scheme is $\text{KDM}^{(1)}_{\mathcal{P}_d}$-secure under the $\mathsf{PLWE}_{\Phi_m,q,\chi}$ assumption.*

4.1 $\text{KDM}^{(\nu)}$ Security

We proceed to show that our scheme is $\text{KDM}^{(\nu)}$-secure for any polynomial ν. We use a methodology introduced by [7] and used by all following $\text{KDM}^{(\nu)}$ constructions: The ν secret keys associated with the ν users are simulated by one "real" secret key. The secret key of each specific user is obtained by offsetting the "real" secret key by a known (to the challenger) amount. The offset can be done without knowing the real key and the offset keys look like appropriately generated keys. This enables using the same techniques as for $\text{KDM}^{(1)}$. We present a variant of this argument where the offset is drawn from a distribution that "swallows" the real secret key. A formal statement follows, the proof is omitted.

Towards formally stating our scheme, we introduce an additional distribution and parameter. For the purpose of the proof, we will need to sample our keys from an even narrower distribution as before. We denote this distribution by $\chi^* = D_{\mathbb{Z}_q^n, r^*}$, and require that $r^* = 2^{-\omega(\log n)} \cdot r$. Namely that it is "swallowed" by our "normal" secret key distribution. One can verify that such r^* can be chosen without affecting the other parameters of the scheme.

Theorem 5. *Our scheme is $\text{KDM}^{(\nu)}_{\mathcal{P}_d}$-secure under the $\mathsf{PLWE}_{\Phi_m,q,\chi^*}$ assumption, for any $\nu = \text{poly}(\kappa)$.*

Acknowledgments. We thank Chris Peikert for providing us with a full version of [26], and Nigel Smart and the anonymous CRYPTO reviewers for numerous insightful comments on the draft.

References

1. Applebaum, B.: Key-dependent message security: Generic amplification and completeness theorems. In: Paterson, K.G. (ed.) EUROCRYPT 2011. LNCS, vol. 6632, pp. 527–546. Springer, Heidelberg (2011)
2. Applebaum, B., Cash, D., Peikert, C., Sahai, A.: Fast cryptographic primitives and circular-secure encryption based on hard learning problems. In: Halevi, S. (ed.) CRYPTO 2009. LNCS, vol. 5677, pp. 595–618. Springer, Heidelberg (2009)
3. Applebaum, B., Harnik, D., Ishai, Y.: Semantic security under related-key attacks and applications. In: To Appear in Innovations in Computer Science, ICS (2011), `http://eprint.iacr.org/2010/544`

4. Barak, B., Haitner, I., Hofheinz, D., Ishai, Y.: Bounded key-dependent message security. In: Gilbert [19], pp. 423–444
5. Blum, A., Furst, M.L., Kearns, M.J., Lipton, R.J.: Cryptographic primitives based on hard learning problems. In: Stinson, D.R. (ed.) CRYPTO 1993. LNCS, vol. 773, pp. 278–291. Springer, Heidelberg (1994)
6. Boneh, D., Goh, E.-J., Nissim, K.: Evaluating 2-DNF formulas on ciphertexts. In: Kilian, J. (ed.) TCC 2005. LNCS, vol. 3378, pp. 325–341. Springer, Heidelberg (2005)
7. Boneh, D., Halevi, S., Hamburg, M., Ostrovsky, R.: Circular-secure encryption from decision diffie-hellman. In: Wagner, D. (ed.) CRYPTO 2008. LNCS, vol. 5157, pp. 108–125. Springer, Heidelberg (2008)
8. Brakerski, Z., Goldwasser, S.: Circular and leakage resilient public-key encryption under subgroup indistinguishability - (or: Quadratic residuosity strikes back). In: Rabin [33], pp. 1–20
9. Brakerski, Z., Goldwasser, S., Kalai, Y.: Balck-box circular secure encryption beyond affine functions. In: Ishai, Y. (ed.) TCC 2011. LNCS, vol. 6597, pp. 201–218. Springer, Heidelberg (2011)
10. Brakerski, Z., Vaikuntanathan, V.: Efficient fully homomorphic encryption from standard lwe (2011) (manuscript)
11. Cachin, C., Micali, S., Stadler, M.: Computationally private information retrieval with polylogarithmic communication. In: Stern, J. (ed.) EUROCRYPT 1999. LNCS, vol. 1592, pp. 402–414. Springer, Heidelberg (1999)
12. Cohen, J.D., Fischer, M.J.: A robust and verifiable cryptographically secure election scheme (extended abstract). In: FOCS, pp. 372–382. IEEE, Los Alamitos (1985)
13. Dijk, M., Gentry, C., Halevi, S., Vaikuntanathan, V.: Fully homomorphic encryption over the integers. In: Gilbert [19], pp. 24–43, Full Version in `http://eprint.iacr.org/2009/616.pdf`
14. Gentry, C.: Fully homomorphic encryption using ideal lattices. In: Mitzenmacher, M. (ed.) STOC, pp. 169–178. ACM, New York (2009)
15. Gentry, C.: Toward basing fully homomorphic encryption on worst-case hardness. In: Rabin [33], pp. 116–137
16. Gentry, C., Halevi, S.: Implementing gentry's fully-homomorphic encryption scheme. In: Paterson, K.G. (ed.) EUROCRYPT 2011. LNCS, vol. 6632, pp. 129–148. Springer, Heidelberg (2011)
17. Gentry, C., Peikert, C., Vaikuntanathan, V.: Trapdoors for hard lattices and new cryptographic constructions. In: Dwork, C. (ed.) STOC, pp. 197–206. ACM, New York (2008)
18. Gentry, C., Ramzan, Z.: Single-database private information retrieval with constant communication rate. In: Caires, L., Italiano, G.F., Monteiro, L., Palamidessi, C., Yung, M. (eds.) ICALP 2005. LNCS, vol. 3580, pp. 803–815. Springer, Heidelberg (2005)
19. Gilbert, H. (ed.): EUROCRYPT 2010. LNCS, vol. 6110. Springer, Heidelberg (2010)
20. Goldwasser, S., Kalai, Y.T., Peikert, C., Vaikuntanathan, V.: Robustness of the learning with errors assumption. In: Yao, A.C.-C. (ed.) ICS, pp. 230–240. Tsinghua University Press, Beijing (2010)
21. Goldwasser, S., Micali, S.: Probabilistic encryption and how to play mental poker keeping secret all partial information. In: STOC, pp. 365–377. ACM, New York (1982)

22. Hoffstein, J., Pipher, J., Silverman, J.H.: Ntru: A ring-based public key cryptosystem. In: Buhler, J.P. (ed.) ANTS 1998. LNCS, vol. 1423, pp. 267–288. Springer, Heidelberg (1998)
23. Lyubashevsky, V., Micciancio, D.: Generalized compact knapsacks are collision resistant. In: Bugliesi, M., Preneel, B., Sassone, V., Wegener, I. (eds.) ICALP 2006. LNCS, vol. 4052, pp. 144–155. Springer, Heidelberg (2006)
24. Lyubashevsky, V., Micciancio, D.: Asymptotically efficient lattice-based digital signatures. In: Canetti, R. (ed.) TCC 2008. LNCS, vol. 4948, pp. 37–54. Springer, Heidelberg (2008)
25. Lyubashevsky, V., Micciancio, D., Peikert, C., Rosen, A.: Swifft: A modest proposal for fft hashing. In: Nyberg, K. (ed.) FSE 2008. LNCS, vol. 5086, pp. 54–72. Springer, Heidelberg (2008)
26. Lyubashevsky, V., Peikert, C., Regev, O.: On ideal lattices and learning with errors over rings. In: Gilbert [19], pp. 1–23, Draft of full version was provided by the authors
27. Malkin, T., Teranishi, I., Yung, M.: Efficient circuit-size independent public key encryption with kdm security. In: Paterson, K.G. (ed.) EUROCRYPT 2011. LNCS, vol. 6632, pp. 507–526. Springer, Heidelberg (2011)
28. Micciancio, D.: Generalized compact knapsacks, cyclic lattices, and efficient one-way functions. Computational Complexity 16(4), 365–411 (2007)
29. Micciancio, D., Regev, O.: Worst-case to average-case reductions based on gaussian measures. SIAM J. Comput. 37(1), 267–302 (2007)
30. Paillier, P.: Public-key cryptosystems based on composite degree residuosity classes. In: Stern, J. (ed.) EUROCRYPT 1999. LNCS, vol. 1592, pp. 223–238. Springer, Heidelberg (1999)
31. Peikert, C., Rosen, A.: Efficient collision-resistant hashing from worst-case assumptions on cyclic lattices. In: Halevi, S., Rabin, T. (eds.) TCC 2006. LNCS, vol. 3876, pp. 145–166. Springer, Heidelberg (2006)
32. Peikert, C., Rosen, A.: Lattices that admit logarithmic worst-case to average-case connection factors. In: Johnson, D.S., Feige, U. (eds.) STOC, pp. 478–487. ACM, New York (2007)
33. Rabin, T. (ed.): CRYPTO 2010. LNCS, vol. 6223. Springer, Heidelberg (2010)
34. Regev, O.: On lattices, learning with errors, random linear codes, and cryptography. In: Gabow, H.N., Fagin, R. (eds.) STOC, pp. 84–93. ACM, New York (2005)
35. Regev, O.: On lattices, learning with errors, random linear codes, and cryptography. J. ACM 56(6) (2009)
36. Rivest, R., Adleman, L., Dertouzos, M.: On data banks and privacy homomorphisms. In: Foundations of Secure Computation, pp. 169–177. Academic Press, London (1978)
37. Rothblum, R.: Homomorphic encryption: From private-key to public-key. In: Ishai, Y. (ed.) TCC 2011. LNCS, vol. 6597, pp. 219–234. Springer, Heidelberg (2011)
38. Smart, N.P., Vercauteren, F.: Fully homomorphic encryption with relatively small key and ciphertext sizes. In: Nguyen, P.Q., Pointcheval, D. (eds.) PKC 2010. LNCS, vol. 6056, pp. 420–443. Springer, Heidelberg (2010)
39. Stehlé, D., Steinfeld, R., Tanaka, K., Xagawa, K.: Efficient public key encryption based on ideal lattices. In: Matsui, M. (ed.) ASIACRYPT 2009. LNCS, vol. 5912, pp. 617–635. Springer, Heidelberg (2009)
40. Stein, W.: A Brief Introduction to Classical and Adelic Algebraic Number Theory (2004)

Bi-Deniable Public-Key Encryption

Adam O'Neill[1,⋆], Chris Peikert[2,⋆⋆], and Brent Waters[1,⋆⋆⋆]

[1] University of Texas at Austin
[2] Georgia Institute of Technology

Abstract. In 1997, Canetti *et al.* (CRYPTO 1997) put forward the intruiging notion of *deniable encryption*, which (informally) allows a sender and/or receiver, having already performed some encrypted communication, to produce 'fake' (but legitimate-looking) random coins that open the ciphertext to another message. Deniability is a powerful notion for both practice and theory: apart from its inherent utility for resisting coercion, a deniable scheme is also noncommitting (a useful property in constructing adaptively secure protocols) and secure under selective-opening attacks on whichever parties can equivocate. To date, however, known constructions have achieved only limited forms of deniability, requiring at least one party to withhold its randomness, and in some cases using an interactive protocol or external parties.

In this work we construct *bi-deniable* public-key cryptosystems, in which both the sender and receiver can simultaneously equivocate; we stress that the schemes are noninteractive and involve no third parties. One of our systems is based generically on "simulatable encryption" as defined by Damgård and Nielsen (CRYPTO 2000), while the other is lattice-based and builds upon the results of Gentry, Peikert and Vaikuntanathan (STOC 2008) with techniques that may be of independent interest. Both schemes work in the so-called "multi-distributional" model, in which the parties run alternative key-generation and encryption algorithms for equivocable communication, but claim under coercion to have run the prescribed algorithms. Although multi-distributional deniability has not attracted much attention, we argue that it is meaningful and useful because it provides credible coercion resistance in certain settings, and suffices for all of the related properties mentioned above.

⋆ Supported in part by the grants of the third author. Most of this work was completed while at the Georgia Institute of Technology.

⋆⋆ This material is based upon work supported by the National Science Foundation under Grant CNS-0716786 and CAREER Award CCF-1054495, and by the Alfred P. Sloan Foundation. Any opinions, findings, and conclusions or recommendations expressed in this material are those of the authors and do not necessarily reflect the views of the National Science Foundation.

⋆⋆⋆ Department of Computer Science, University of Texas at Austin. Supported by NSF CNS-0915361 and CNS-0952692, AFOSR Grant No: FA9550-08-1-0352, DARPA PROCEED, DARPA N11AP20006, Google Faculty Research award, the Alfred P. Sloan Foundation, and Microsoft Faculty Fellowship.

P. Rogaway (Ed.): CRYPTO 2011, LNCS 6841, pp. 525–542, 2011.

1 Introduction

Suppose that Eve has two children: Alice, who is away at college, and a young Bob, who still lives at home. The siblings are planning a surprise party for Eve, so to keep their plans secret, they communicate using public-key encryption. Eve, however, has taken note of their encrypted communications and grows suspicious. Using her inherent parental authority, she demands that Alice and Bob reveal their secret decryption keys, as well as any of the encryption randomness they might have retained. Is there any way for Alice and Bob to comply, without spoiling the surprise? The answer seems to be obviously no: using the secret keys, Eve can simply decrypt their messages and learn about the party.

However, the above argument misses a subtle point: if Alice and Bob are able to produce *alternative* keys and randomness that are consistent with their ciphertexts so far, then they might be able to fool Eve into thinking that they are communicating about something else (or at least not alert her to the party). A scheme that makes this possible is said to be *deniable*, a notion formally introduced by Canetti, Dwork, Naor, and Ostrovsky [10].

In practice, deniable encryption has been sought by users whose legitimate activities may not always be protected from subpoenas or legal coercion, e.g., journalists and whistleblowers, or lawyers and activists in repressive regimes. Indeed, several commercial and open-source storage encryption products claim limited forms of deniability (see, for example, [1, 2], and further references in [24]), though without formal definitions or supporting security proofs. More worryingly, these products only allow for denying the *existence* of messages on a storage medium, not for *equivocating* those messages. This is insufficient in a communications setting, where the mere exchange of messages between parties indicates that they are communicating in some form.

Deniability is also a compelling property for theoretical reasons: in particular, deniable encryption schemes are *noncommitting* (a fundamental concept in the design of adaptively secure protocols) [11, 16, 14], secure against *selective-opening* attacks [19, 7], and imply incoercible multiparty computation [12]. We point out that deniable encryption is stronger than noncommitting encryption, because equivocable ciphertexts actually decrypt to the intended messages, and users of the system (not just a simulator) can themselves produce such ciphertexts.

Canetti *et al.* distinguish between two different models of deniability. The first is *full* deniability, in which the parties always run the prescribed key-generation and encryption algorithms, and can equivocate their messages later on if they so choose. The second model is called *multi-distributional* deniability, in which there exist alternative "deniable" algorithms whose outputs can be equivocated, so that it appears as if the *prescribed* algorithms had been used all along. Whether these models are useful in various settings has been the subject of some debate over the years; we discuss these issues in Section 1.2 below. We also discuss some recent developments and related work in Section 1.3.

Under standard assumptions, Canetti *et al.* construct a multi-distributional *sender*-deniable scheme (i.e., one that remains secure if only the sender is coerced), and give a *fully* sender-deniable scheme where the coercer's distinguishing advantage between a 'real' and 'fake' opening is an inverse polynomial that depends on the public key size. They also construct a receiver-deniable scheme that requires an additional round of interaction, and a sender- and receiver-deniable protocol that relies on third parties, at least one of whom must remain uncoerced. In particular, up to this point there have not been any noninteractive schemes offering receiver-deniability, nor any schemes (interactive or not) in which all the parties can be coerced simultaneously, in either of the two models (full or multi-distributional deniability).

1.1 Our Contributions

Bideniable public-key encryption. Our main results are the first known *bideniable* (that is, simultaneously sender- and receiver-deniable) public-key encryption schemes, in the multi-distributional model. We stress that the schemes are noninteractive, require no third parties, and are immediately noncommitting and secure under selective-opening attacks.

We give two qualitatively different constructions. The first is built generically, using a combinatorial construction with somewhat large overhead, from any "simulatable" encryption scheme in the sense of Damgård and Nielsen [16]. This shows (perhaps surprisingly) that simulatability is sufficient not only for noncommitting encryption, but for a form of deniability as well. The scheme is presented in Section 4.

Our second scheme is based on worst-case lattice problems via "learning with errors" [23], and builds upon the trapdoor function and identity-based encryption techniques of Gentry, Peikert, and Vaikuntanathan [21]. In particular, it exploits a unique property of the GPV IBE, namely its negligible chance of oblivious decryption error when the secret key vector and the error vector in the ciphertext are too highly "aligned." Our scheme relies on the ability of the receiver to resample a fresh secret key that is highly correlated with the ciphertext error term. Proving that this secret key "looks real" relies on a symmetry between correlated (discrete) Gaussians, which we believe may be of independent interest and application in lattice cryptography. Interestingly, the deniable scheme is essentially identical to the GPV cryptosystem, and it appears to be the first known cryptosystem that "naturally" supports receiver-deniability without any substantial changes. It is also essentially as efficient for the receiver as the GPV system, but it is less efficient for the sender because she must encrypt each bit separately (rather than amortizing). The details of the system are described in Section 6.

In addition to our public-key schemes, we also define notions of deniability for the identity-based setting, and show how our techniques immediately adapt to it as well. As we discuss below, multi-distributional deniability (especially for the receiver) may be more palatable in this setting because the receiver does not run a different key-generation algorithm (indeed, there is no such algorithm). We also remark that to be meaningful, the identity-based setting inherently requires any solution to be noninteractive.

Plan-ahead bideniability with short keys. A simple information-theoretic argument reveals that in any noninteractive receiver-deniable encryption scheme, the secret key must be at least as long as the message: a fixed ciphertext can only decrypt to N different plaintexts if there are at least N distinct secret keys for the public key (see also [22] and the recent work [9]). We circumvent this constraint by designing a scheme offering *"plan-ahead"* bideniability (a notion also introduced in [10]) that can encrypt arbitrarily long messages using *fixed-sized* keys. In plan-ahead deniability, the sender must choose at encryption time a bounded number of potential 'fake' messages, to which the parties may later equivocate. This may be seen as the deniable analogue of "somewhat noncommitting encryption," introduced by Garay, Wichs and Zhou [20]. In many cases, this model would seem to be sufficient for coercion resistance, since the sender can just include one "innocuous" message along with the real one.

Our plan-ahead scheme is a hybrid system that reduces the deniable encryption of an arbitrary-length message to that of a short symmetric key. For example, when combined with the moderately good efficiency of our GPV-style bideniable scheme, the overall system is potentially usable in practice. (Though as noted above, our scheme is not able to amortize many bits into one ciphertext as the GPV scheme does, so our bandwidth requirements are larger.) Due to space constraints, we defer the details of our plan-ahead scheme to the full version.

Relations among notions. We clarify and study relations among the various types of deniability introduced in [10]. Our main contribution is that any type of *multi-distributional* deniability suffices to obtain the corresponding type of *full* deniability, with an inverse-polynomial distinguishing advantage related to the size of the public key. This further reinforces the usefulness of multi-distributional deniability itself. We also observe that for multi-distributional schemes, bideniability implies sender-deniability, but perhaps surprisingly, it may not imply receiver-deniability alone. That is, bideniability relies on the sender to correctly run the deniable encryption algorithm.

1.2 Discussion

Is (multi-distributional) deniability useful? The ideal deniable encryption scheme would be noninteractive and fully deniable for both parties. Unfortunately, it has been recently shown [9] that these properties cannot all be achieved at once, even for receiver deniability alone (see related work below). So to obtain a noninteractive scheme we must work in the multi-distributional model.

A common objection to multi-distributional deniability is that, since there are alternative deniable algorithms (for encryption and key generation) that are strictly more powerful than the normal ones, why would anyone ever run the normal algorithms? And given this situation, why would a coercer ever accept a transcript corresponding to the normal algorithms? Whether this is a significant problem will depend on the setting in which coercion happens, and what recourse the coercer has in response to the users' claims.

For example, if there is a prescribed legal process (such as a subpoena or search warrant) by which parties are forced to reveal their transcripts, then multi-distributional deniability may be sufficient to protect the users. Even if the coercer asks for a transcript corresponding to the deniable algorithms, the users can simply assert that they did not run those algorithms, and so cannot produce coins for them. The users' claims might also gain in credibility via "safety in numbers," if a deployed implementation defaults to the normal algorithms — which do make up an operational cryptosystem, after all — or by formally standardizing on the normal algorithms within an organization. Since the coercer would only have *reason to believe* — but not any actual *evidence* — that the deniable algorithms were used in a particular instance, imposing a sanction seems fundamentally unjust, and might even be ruled out by the prescribed process. If, on the other hand, the coercer is able to punish the users until they "tell him what he wants to hear," then multi distributional deniability might not be enough to protect the users — but neither might full deniability! After all, in either model the coercer has no reason to believe what the users have revealed. Anticipating potential punishment, the users of a multi-distributional scheme might retain the coins of the deniable algorithms as a "backup plan," just in case they might later want to reveal a convincing proof for the true message (e.g., if the punishment becomes too severe). But even a fully deniable scheme allows for a similar backup plan and proof of the true message, by using "verifiably random" coins such as the digits of π or the output of a pseudorandom generator.

In the identity-based setting, multi-distributional deniability may be useful as well, especially for the receiver. Here the receiver does not run a key-generation algorithm at all, but instead gets his secret key from an authority who possesses a 'master' secret key for all users. Our model allows the receiver to ask the authority for a fake (but real-looking) secret key that causes a particular ciphertext to decrypt to any desired message. This could be useful if the authority is out of the coercer's jurisdiction (e.g., if the receiver is travelling in another country), or if it can argue that exposing its master secret key would harm the privacy of too many other users.

In summary, the purpose of deniability is not at all to 'convince' the coercer that the surrendered transcripts are real; indeed, it is common knowledge that they can easily be faked. Instead, the goal is to *preempt coercion* in the first place by making it useless, since parties who "stick to their stories" can never be pinned down to the real message. At the same time, neither form of deniability seems appropriate if a user might eventually want to convincingly reveal the true plaintext, e.g., to sell her vote in an election. The main significant difference we see between the two models relates not to security, but usability: multi-distributional deniability requires the users to know in advance which messages they might want to equivocate, whereas full deniability allows the user to decide afterward.

Why not erase? At first glance, erasures appear to provide a very simple way of achieving deniability: the parties can just tell the coercer that they deliberately erased their coins (perhaps according to a published schedule), and therefore cannot surrender them. For sender deniability, this claim might be credible, since there is no point in the sender keeping her ephemeral encryption coins. (And indeed, none of our results preclude using erasures on the sender side.) For receiver deniability, erasures seem much less credible, since the receiver must store some form of decryption key in order to decrypt messages, and at some point in time this key can be subject to coercion. Certain regulatory regimes (e.g., in the financial sector) might also mandate retention of all data for a certain time, for potential audit. The existence of deniable encryption means that such requirements would still not necessarily guarantee compliance with the intent of the regulations. In any case, we contend that there is a significant qualitative difference between deniable schemes that use erasures, and those that do not. In the former, the coerced parties must resort to the claim that they no longer have the desired evidence, even though they once did. In the latter, the parties can credibly claim to have provided the coercer with all the evidence they have ever had, and yet still equivocate.

1.3 Other Related Work

Subsequent to the initial dissemination of this work in 2010, there has been additional work on deniable encryption that complements our own. Dürmuth and Freeman [18] announced an interactive, fully sender-deniable encryption protocol (i.e., one with a single encryption protocol and negligible detection advantage). However, following its publication Peikert and Waters found a complete break of this system (and a corresponding flaw in the claimed security proof); see [17] for details. In particular, the problem of constructing a fully deniable encryption scheme remains an intriguing open question. Bendlin *et al.* [9] recently showed that any noninteractive public-key scheme having key size σ can be fully receiver-deniable (or bideniable) only with non-negligible $\Omega(1/\sigma)$ detection advantage. In particular, our use of the multi-distributional model is necessary to achieve a noninteractive receiver-deniable scheme.

Deniability can be seen as a type of security that holds true even when secret information is revealed to the adversary. In the case of break-ins, one relevant notion is "forward security" (see, e.g., [8, 13]), which relies on secure erasures to update secret state over time. In the case of side-channel or memory attacks, relevant notions include the "bounded retrieval model" and "leakage-resilient" cryptography, which limit the amount of secret information the adversary may learn (see the recent survey [5] and references therein). In contrast, deniability ensures security in contexts where the adversary obtains the *entire, unchanging* secret key and/or encryption randomness, but cannot tell whether those values came from the actual execution of the system.

2 Preliminaries

We denote the set of all binary strings by $\{0,1\}^*$ and the set of all binary strings of length i by $\{0,1\}^i$. The length of a string s is denoted by $|s|$. By $s_1 \| s_2$ we denote an encoding of strings s_1, s_2 from which the two strings are unambiguously recoverable. (If the lengths of s_1, s_2 are known, then concatenation suffices.) For $i \in \mathbb{N}$ we denote by $[i]$ the set $\{1, \ldots, i\}$. We use the standard definitions of negligible functions, and statistical and computational indistinguishability.

We say an algorithm A with input space $\mathcal{X}$ has *invertible sampling* [16] if there is an efficient inverting algorithm, denoted I_A, such that for all $x \in \mathcal{X}$ the outputs of the following two experiments are indistinguishable (either computationally, or statistically):

$y \leftarrow A(x; r)$	$y \leftarrow A(x; r)$
	$r' \leftarrow I_A(x, y)$
Return (x, y, r)	Return (x, y, r')

In other words, given just an input-output pair of A, it is possible to efficiently generate appropriately distributed randomness that "explains" it. It may also be the case that I_A requires some "trapdoor" information about x in order to do so. Namely, we say that A has *trapdoor* invertible sampling if we replace the second line in the right-hand experiment above with "$r' \leftarrow I_A(td_x, x, y)$," where td_x is a trapdoor corresponding to x (we think of x and td_x as being sampled jointly as the setup to both of the above experiments).

A *public-key cryptosystem* PKC with message space $\mathcal{M}$ consists of three algorithms: The *key generation algorithm* $\mathsf{Gen}(1^n; r_R)$ outputs a public key pk, and the randomness r_R is used as the associated secret decryption key. (This convention is natural in the context of deniability, where we might even consider coercing the receiver to reveal the random coins r_R it used to generate its public key. This is without loss of generality, since the stored "secret key," whatever its form, can always be computed from r_R.) The *encryption algorithm* $\mathsf{Enc}(pk, m; r_S)$ outputs a ciphertext c. The deterministic *decryption algorithm* $\mathsf{Dec}(pk, r_R, c)$ outputs a message m or $\perp$. For correctness, we require the probability that $m' \neq m$ be negligible for all messages $m \in \mathcal{M}$, over the experiment $pk \leftarrow \mathsf{Gen}(1^n; r_R)$, $c \leftarrow \mathsf{Enc}(pk, m)$, $m' \leftarrow \mathsf{Dec}(pk, r_R, c)$. To distinguish between the above notion and deniable encryption as defined in Section 3, we sometimes refer to the former as *normal* encryption.

3 Bideniable Encryption

Here we formally define bideniable encryption and its security properties, along with some weaker variants. (We use "bideniable" as shorthand for "sender-and-receiver deniable" in the language of Canetti *et al.* [10].) Following the definitions, we discuss further how our notion relates to the variants of deniable encryption introduced in [10].

As with normal encryption, bideniable encryption allows a sender in possession of the receiver's public key to communicate a message to the latter, confidentially. Additionally, if the parties are later coerced to reveal all their secret data — namely, the coins used by the sender to encrypt her message and/or those used by the receiver to generate her key — bideniable encryption allows them to do so as if *any desired message* (possibly chosen as late as at the time of coercion) had been encrypted.

In a *multi-distributional* deniable encryption scheme, there are 'normal' key generation, encryption, and decryption algorithms that can be run as usual

— though the resulting communication may not be equivocable later on. In addition, there are 'deniable' key generation and encryption algorithms that can be used for equivocable communication. Associated with these deniable algorithms are 'faking' algorithms, which can generate secret coins that open a deniably generated public key and ciphertext to any desired message, as if the *normal* algorithms had been used to generate them. Note that the ability to generate fake *random coins* for the parties yields the strongest definition, since such coins can be used to compute whatever locally stored values (e.g., a secret key of some form) the coercer might expect. We now give a formal definition.

Definition 1 (Deniable encryption). *A* multi-distributional sender-, receiver-, *or* bi-deniable *encryption scheme DEN with message space $\mathcal{M}$ is made up of the following algorithms:*

- *The* normal *key-generation, encryption, and decryption algorithms* $\mathsf{Gen}, \mathsf{Enc}, \mathsf{Dec}$ *are defined as usual for public-key encryption (see Section 2). These algorithms make up the* induced normal scheme.
- *The* deniable *key-generation algorithm* $\mathsf{DenGen}(1^n)$ *outputs* (pk, fk)*, where* fk *is the* faking key.[1] *We also extend* Dec *to so that it can decrypt using* fk *in lieu of the usual receiver randomness* r_R.
- *The* deniable *encryption algorithm* DenEnc *has the same interface as the normal encryption algorithm.*
- *The* sender faking algorithm $\mathsf{SendFake}(pk, r_S, m', m)$*, given a public key* pk*, original coins* r_S *and message* m' *of* DenEnc*, and desired message* m*, outputs faked random coins* r_S^* *for* Enc.
- *The* receiver faking algorithm $\mathsf{RecFake}(pk, fk, c, m)$*, given the public and faking keys* pk *and* fk *(respectively), a ciphertext* c*, and a desired message* m*, outputs faked random coins* r_R^* *for* Gen.

We require the following properties:

1. Correctness. *Any triplet* $(\mathsf{G}, \mathsf{E}, \mathsf{Dec})$*, where* $\mathsf{G} \in \{\mathsf{Gen}, \mathsf{DenGen}\}$ *and* $\mathsf{E} \in \{\mathsf{Enc}, \mathsf{DenEnc}\}$*, should form a correct public-key encryption scheme.*
2. Multi-distributional deniability. *Let* $m, m' \in \mathcal{M}$ *be arbitrary messages, not necessarily different. The appropriate experiment below, which represents equivocation (by the appropriate party/parties) of an encrypted* m' *as* m*, should be computationally indistinguishable from the following 'honest opening' experiment: let* $pk \leftarrow \mathsf{Gen}(1^n; r_R)$*,* $c \leftarrow \mathsf{Enc}(pk, m; r_S)$*, and output* pk*,* c*, and whichever of* r_R*,* r_S *are appropriate to the type of deniability under consideration.*

[1] Without loss of generality, we could replace fk with the randomness of DenGen, but since this randomness will never be exposed to the adversary, we elect to define a distinguished faking key.

Sender-Deniable	*Receiver-Deniable*	*Bi-Deniable*
$pk \leftarrow \mathsf{Gen}(1^n; r_R)$	$(pk, fk) \leftarrow \mathsf{DenGen}(1^n)$	$(pk, fk) \leftarrow \mathsf{DenGen}(1^n)$
$c \leftarrow \mathsf{DenEnc}(pk, m'; r_S)$	$c \leftarrow \mathsf{Enc}(pk, m'; r_S)$	$c \leftarrow \mathsf{DenEnc}(pk, m'; r_S)$
	$r_R^* \leftarrow \mathsf{RecFake}(pk, fk, c, m)$	$r_R^* \leftarrow \mathsf{RecFake}(pk, fk, c, b)$
$r_S^* \leftarrow \mathsf{SendFake}(pk, r_S, m', m)$		$r_S^* \leftarrow \mathsf{SendFake}(pk, r_S, m', m)$
Return (pk, c, r_S^*)	*Return* (pk, c, r_R^*)	*Return* (pk, c, r_R^*, r_S^*)

Multi-distributional bideniability is a particularly strong theoretical notion. For example, it immediately implies non-committing encryption as defined in [11] — but in addition, equivocable ciphertexts actually decrypt to the intended messages, and can be produced by the regular users of the scheme, not just by a simulator. Bideniability is also important in practice; in particular, each party's security does not depend upon the incoercibility of the other.

Note that we did not explicitly require DEN to satisfy the standard notion of indistinguishability under chosen-plaintext attack; this is because it is implied by any of the above notions of deniability.

Proposition 1. *Suppose that DEN satisfies any of sender-, receiver-, or bideniability. Then any triplet* $(\mathsf{G}, \mathsf{E}, \mathsf{Dec})$, *where* $\mathsf{G} \in \{\mathsf{Gen}, \mathsf{DenGen}\}$ *and* $\mathsf{E} \in \{\mathsf{Enc}, \mathsf{DenEnc}\}$, *is semantically secure.*

While our focus is on multi-distributional bideniability, we also briefly examine interrelations among the other types. We start with a basic question: for a particular deniable encryption scheme DEN, which notions of deniability imply which others? (This question is also important in practice, since an encryption scheme may not always be used in the ways it is intended.) First, we show that bideniablility implies sender deniability.

Proposition 2. *If DEN is bideniable, then DEN is sender-deniable.*

Proof. By assumption, we know that the distributions (pk, c, r_R, r_S) and (pk, c, r_R^*, r_S^*) are computationally indistinguishable, as produced by the two bideniability experiments. Clearly, the distributions (pk, c, r_S) and (pk, c, r_S^*) are indistinguishable when produced by the same two experiments, where we can now omit the generation of r_R^* in the faking experiment. This modified bideniable faking experiment producing (pk, c, r_S^*) differs from the sender-deniable faking experiment only its use of DenGen instead of Gen. Because neither experiment uses fk, all we need is for the pks output by DenGen and by Gen to be indistinguishable. But this follows directly from bideniability, by restricting the outputs of the two experiments to just pk.

One might also expect bideniability to imply receiver-deniability. Perhaps surprisingly, at least in the multi-distributional setting this appears not to be the case! For example, in our abstract scheme from Section 5, the receiver can equivocate a *normal* ciphertext in one direction (from 1 to 0), but apparently not from 0 to 1. In general, the problem is that to allow the receiver to equivocate a message, DenEnc may need to produce special ciphertexts that Enc would never (or very rarely) output. In other words, the receiver's ability to equivocate a message may depend crucially the sender's use of DenEnc.

In the reverse direction, it appears that a scheme that is both sender-deniable and (separately) receiver-deniable still might not be bideniable. Indeed, the fake randomness produced by SendFake and RecFake (which depend on the ciphertext c) might be obviously correlated in such a way that exposing both together is easily detected, whereas exposing only one is safe. Constructing a concrete example along these lines to demonstrate a formal separation remains an interesting problem.

4 Bideniable Encryption from Simulatable Encryption

Here we give a bideniable public-key encryption scheme from any *simulatable* one, in the sense of Damgård and Nielsen [16]. In particular, this shows that simulatable encryption suffices to realize not just a noncommitting encryption scheme, but also a (multi-distributional) bideniable one.

Simulatable public-key encryption. We recall the notion of a simulatable public-key encryption scheme from [16]. Intuitively, this is a scheme in which it is possible to 'obliviously' sample a public key without knowing the secret key, and to 'obliviously' sample the encryption of a random message without knowing the message.

Definition 2 (Simulatable PKE [16]). *A* simulatable public-key encryption scheme *PKC-SIM with message space $\mathcal{M}$ is made up of the following algorithms:*

- *The* normal *key-generation, encryption, and decryption algorithms* Gen, Enc, Dec *are defined as usual for a public-key encryption scheme (see Section 2).*
- *The* oblivious *key-generation algorithm* $\mathsf{OGen}(1^n; r_{\mathsf{OGen}})$ *outputs a public key opk, and has invertible sampling via algorithm I_{OGen}.*
- *The* oblivious *encryption algorithm* $\mathsf{OEnc}(pk)$ *outputs a ciphertext oc, and has invertible sampling via algorithm I_{OEnc}.*

We require the following properties:

1. Oblivious key generation. *The distribution of pk, where $pk \leftarrow \mathsf{Gen}(1^n; r_R)$, should be computationally indistinguishable from opk, where $opk \leftarrow \mathsf{OGen}(1^n; r_{\mathsf{OGen}})$.*
2. Oblivious ciphertext generation. *The output distributions of the following two experiments should be computationally indistinguishable:*

$pk \leftarrow \mathsf{Gen}(1^n; r_R)$	$pk \leftarrow \mathsf{Gen}(1^n; r_R)$
$m \leftarrow \mathcal{M}$	
$c \leftarrow \mathsf{Enc}(pk, m)$	$oc \leftarrow \mathsf{OEnc}(pk)$
Return (pk, r_R, c)	*Return* (pk, r_R, oc)

Note that the above conditions are non-trivial in light of the the fact that OGen, OEnc *are required to have invertible sampling. In particular, it follows by Condition 2 (after removing r_R from the output distributions) that any simulatable public-key encryption scheme is semantically secure.*

Simulatable encryption can be realized under a variety of standard computational assumptions such as DDH and RSA [16], as well as worst-case lattice assumptions [16, 21] (though we later show a more efficient bideniable encryption scheme based on the latter using an entirely different approach).

4.1 A "Coordinated" Scheme

Overview. To get at the technical core of our construction, we first present a *coordinated* scheme in which the faked random coins r_R^* for the receiver and r_S^* for the sender are outputs of the *same* algorithm $\mathsf{FakeCoins}(pk, fk, r_S, b', b, c)$ (we give the corresponding ciphertext c for convenience); i.e., the sender and receiver coordinate their faked coins upon being coerced. We later describe how the coordination can be removed for our specific scheme. Additionally, we present a scheme that encrypts one-bit messages, but a scheme that encrypts poly(n)-bit messages then follows generically by parallel repetition with independent keys.

The high-level idea for our scheme draws on and extends that of Choi *et al.* [14] (who generalized the approach of Damgård and Nielsen [16]) used to achieve noncommitting encryption. In our scheme, we run $5n$ instances of an underlying simulatable encryption scheme in parallel. In normal (non-deniable) operation, by using the oblivious key and ciphertext generation algorithms appropriately, the receiver "knows" the secret keys corresponding to a random size-n subset $\mathcal{S} \subset [5n]$ of indices and the sender "knows" the plaintexts and associated encryption randomness corresponding to the ciphertexts for an independent random size-n subset $\mathcal{R} \subset [5n]$. In particular, $\mathcal{S}$ corresponds to encryptions of the message bit b, and $[5n] \setminus \mathcal{S}$ corresponds to oblivious encryptions of random bits. To decrypt, the receiver decrypts ciphertexts corresponding to $\mathcal{R}$ and takes a majority vote. In deniable operation, the receiver knows *all* the secret keys and the sender knows *all* the plaintexts and associated encryption randomness. In particular, in addition to choosing a random size-n 'biasing' subset $\mathcal{Y} \subset [5n]$ corresponding to encryptions of the true message bit b', the sender also chooses two random pairwise disjoint size-n subsets $\mathcal{S}_0, \mathcal{S}_1 \subset [5n]$, also disjoint from $\mathcal{Y}$, corresponding to encryptions of 0s and 1s respectively; the ciphertexts in $[5n] \setminus \mathcal{Y} \cup \mathcal{S}_0 \cup \mathcal{S}_1$ encrypt random bits. To open a deniably encrypted message b' as b, the sender chooses $\mathcal{S}^* = \mathcal{S}_b$ and the receiver chooses $\mathcal{R}^*$ so that it has an appropriate random number of elements in common with $\mathcal{S}_b$, and then chooses the remainder of $\mathcal{R}^*$ as random indices from $[5n] \setminus (\mathcal{S}_0 \cup \mathcal{S}_1 \cup \mathcal{Y})$.

The scheme. To formally define our scheme, let PKC-SIM be simulatable public-key encryption scheme with message space $\{0, 1\}$. Below, for a set $\mathcal{X}$ we denote by $\mathcal{P}(\mathcal{X})$ the set of all subsets of $\mathcal{X}$ and by $\mathcal{P}_i(\mathcal{X})$ the set of all subsets of $\mathcal{X}$ of size i. Additionally, for nonnegative integers $x, y, N \geq M$, let $P_{\mathsf{HGD}}(x; N, M, y)$ denote the probability that exactly x values from $[M]$ are chosen after a total of y values are drawn uniformly at random from $[N]$, without replacement. We denote by $\mathsf{HGD}(N, M, y)$ the *hypergeometric distribution on* $[N]$ *with parameters* M, y that assigns to each $x \in \{0, \ldots, M\}$ the probability $P_{\mathsf{HGD}}(x; N, M, y)$. Below we will use parameters N, M, y polynomial in the security parameter, so we can sample efficiently from this distribution simply by running the sampling experiment. Define scheme BI-DEN[PKC-SIM] with message-space $\{0, 1\}$ as follows:

$\text{BI-DEN.}\mathsf{Gen}(1^n)$:
- $\mathcal{R} \leftarrow \mathcal{P}_n([5n])$
- For $i = 1$ to $5n$ do:
- If $i \in \mathcal{R}$ then
- $pk_i \leftarrow \mathsf{Gen}(1^n; r_{R,i})$
- Else $pk_i \leftarrow \mathsf{OGen}(1^n; r_{R,i})$
- EndFor
- $pk \leftarrow pk_1 \| \dots \| pk_{5n}$
- Return pk

$\text{BI-DEN.}\mathsf{Enc}(pk, b)$:
- $\mathcal{S} \leftarrow \mathcal{P}_n([5n])$
- For $i = 1$ to $5n$ do:
- If $i \in \mathcal{S}$ then
- $c_i \leftarrow \mathsf{Enc}(pk_i, b; r_{S,i})$
- Else $c_i \leftarrow \mathsf{OEnc}(pk_i; r_{S,i})$
- EndFor
- $c \leftarrow c_1 \| \dots \| c_{5n}$
- Return c

$\text{BI-DEN.}\mathsf{Dec}((\mathcal{R}, r_R), c)$:
- For all $i \in \mathcal{R}$ do:
- $d_i \leftarrow \mathsf{Dec}(r_{R,i}, c_i)$
- EndFor
- If most d_i's are 1 then
- Return 1
- Else return 0

$\text{BI-DEN.}\mathsf{DenGen}(1^n)$:
- $\mathcal{R} \leftarrow \mathcal{P}_n([5n])$
- For $i = 1$ to $5n$ do:
- $pk_i \leftarrow \mathsf{Gen}(1^n; r_{R,i})$
- EndFor
- $pk \leftarrow pk_1 \| \dots \| pk_{5n}$
- $r \leftarrow r_{R,1} \| \dots \| r_{R,5n}$
- Return $(pk, (\mathcal{R}, r))$

$\text{BI-DEN.}\mathsf{DenEnc}(pk, b')$:
- $\mathcal{S}_0 \leftarrow \mathcal{P}_n([5n])$
- $\mathcal{S}_1 \leftarrow \mathcal{P}_n([5n] \setminus \mathcal{S}_0)$
- $\mathcal{Y} \leftarrow \mathcal{P}_n([5n] \setminus (\mathcal{S}_0 \cup \mathcal{S}_1))$
- For $i = 1$ to $5n$ do:
- If $i \in \mathcal{S}_0$ then
- $c_i \leftarrow \mathsf{Enc}(pk_i, 0; r_{S,i})$
- If $i \in \mathcal{S}_1$ then
- $c_i \leftarrow \mathsf{Enc}(pk_i, 1; r_{S,i})$
- If $i \in \mathcal{Y}$ then
- $c_i \leftarrow \mathsf{Enc}(pk_i, b'; r_{S,i})$
- Else $c_i \leftarrow \mathsf{OEnc}(pk; r_{S,i})$
- EndFor
- $c \leftarrow c_1 \| \dots \| c_{5n}$
- Return c

$\text{BI-DEN.}\mathsf{FakeCoins}(pk, fk, r_S, b', b, c)$:
- $z \leftarrow \mathsf{HGD}(5n, n, n)$
- $\mathcal{Z} \leftarrow \mathcal{P}_z(\mathcal{S}_b)$
- $\mathcal{Z}' \leftarrow \mathcal{P}_{n-z}([5n] \setminus (\mathcal{S}_0 \cup \mathcal{S}_1 \cup \mathcal{Y}))$
- $\mathcal{R}^* \leftarrow \mathcal{Z} \cup \mathcal{Z}'$
- $\mathcal{S}^* \leftarrow \mathcal{S}_b$
- For $i = 1$ to $5n$ do:
- If $i \in \mathcal{S}^*$ then $r^*_{S,i} \leftarrow r_{S,i}$
- Else $r^*_{S,i} \leftarrow I_{\mathsf{OEnc}}(pk_i, c_i)$
- If $i \in \mathcal{R}^*$ then $r^*_{R,i} \leftarrow r_{R,i}$
- Else $r^*_{R,i} \leftarrow I_{\mathsf{OGen}}(pk_i)$
- EndFor
- $r^*_S \leftarrow r^*_{S,1} \| \dots \| r^*_{S,5n}$
- $r^*_R \leftarrow r^*_{R,1} \| \dots \| r^*_{R,5n}$
- Return (r^*_S, r^*_R)

Removing the coordination. To remove the coordination between the sender and receiver in the faking algorithm, the idea is for the sender to communicate its choices of $\mathcal{S}_0, \mathcal{S}_1, \mathcal{Y}$ to the receiver *in-band* by using a separate instance of the simulatable encryption scheme. Details are deferred to the full verison.

4.2 Correctness and Security

Theorem 1. *Let PKC-SIM be a simulatable public-key encryption scheme. Then BI-DEN[PKC-SIM] is correct.*

The proof is a simple argument that relies on a Chernoff-like tail inequality for the hypergeometric distribution, and the Chernoff bound. Due to space restrictions we leave it to the full version. We now turn to security.

Theorem 2. *Let PKC-SIM be a simulatable public-key encryption scheme. Then BI-DEN[PKC-SIM] satisfies bideniability under chosen-plaintext attacks.*

We sketch the proof outline, leaving details to the full version. We consider three hybrid experiments that transition from an 'honest' opening of an 'honest' encryption of b to a 'fake' opening of a deniably encrypted bit b' as b. The first two experiments differ only in how they choose the size-n subsets $\mathcal{Y}, \mathcal{S}_0, \mathcal{S}_1, \mathcal{R} \subset [5n]$ of indices (i.e., the corresponding key pairs and ciphertexts are generated identically; in fact, $\mathcal{S}_{1-b}, \mathcal{Y}$ are not used). Namely, the first experiment chooses $\mathcal{S}_b$ and $\mathcal{R}$ at random independently and then S_{1-b} and $\mathcal{Y}$ at random as pairwise disjoint and disjoint from $\mathcal{S}_b \cup \mathcal{R}$, while the second chooses S_b and S_{1-b} at random as pairwise disjoint, then $\mathcal{Y}$ at random as disjoint from $\mathcal{S}_b \cup \mathcal{S}_{1-b}$, and finally $\mathcal{R}$ to have an HGD-distributed random number of elements in common with $\mathcal{S}_b$ and the remainder at random disjoint from $\mathcal{S}_0 \cup \mathcal{S}_1 \cup \mathcal{Y}$. The outputs of these experiments are identically distributed by a combinatorial argument. Finally, the third experiment appropriately changes how the key pairs and ciphertexts corresponding to $\mathcal{Y}, \mathcal{S}_0, \mathcal{S}_1, \mathcal{R}$ are generated. The outputs of the last two experiments are computationally indistinguishable by simulatability of the underlying encryption scheme.

5 Bideniable Encryption from Bitranslucent Sets

Here we construct a bideniable encryption scheme based on a new primitive we call a *bitranslucent set*, which extends the notion of a translucent set from [10]. Whereas translucent sets can be constructed straightforwardly from any trapdoor permutation (and other specific assumptions) [10], bitranslucent sets appear much more challenging to realize. In Section 6 we give a novel construction based on lattices.

Bitranslucent sets. Informally, a *translucent set* is a subset P of a universe U, which can be sampled efficiently using only public information, and which is pseudorandom unless its associated secret key is known (in which case it is easily distinguishable from uniform over the universe).

Here we strengthen the notion of a translucent set in two ways. The first, main strengthening essentially preserves the pseudorandomness of P *even if the secret key is revealed.* Of course this is impossible as just stated, because the secret key makes P 'transparent' by design. Instead, we introduce an algorithm that, given a 'faking' key for the set system, and some c drawn from the pseudorandom set P, is able to *resample* a new, 'good-looking' secret key for which c appears uniformly random. We stress that such keys are necessarily very rare, because a typical secret key should correctly recognize P with high probability. Nevertheless, it can still be the case that for any $c \in P$, there are a few rare keys that misclassify c (without making it apparent that they are doing so). A bitranslucent set scheme is able to use the faking key find such keys.

Looking ahead to our bideniable encryption scheme, bitranslucency will allow a coerced sender and receiver both to plausibly claim that a value $c \in P$ is actually uniform: the sender simply declares that c was chosen uniformly from U (by claiming c itself as the random coins), and the receiver resamples a secret key that also makes c appear uniform.

The second strengthening, which yields qualitative improvements in efficiency and may have independent applications, allows for multiple translucent sets to share a single, fixed-size faking key. Essentially, this makes the bitranslucent set 'identity-based' (although we do not refer to identities explicitly): each translucent set has its own public and secret keys for generating and distinguishing P- and U-samples, and the master faking key makes it possible to generate a good-looking secret key that equivocates a given P-sample as a U-sample. Interestingly, this implies a bideniable encryption scheme in which the deniable key generator's faking key is a *fixed* size independent of the message length, despite the information-theoretic bound that *normal* secret keys must exceed the message length.

Definition 3 (Bitranslucent Set Scheme (BTS)). *A* bitranslucent set scheme *BTS is made up of the following algorithms:*

- *The* normal setup *procedure* $\mathsf{Setup}(1^n; r_{\mathsf{Setup}})$ *outputs a public parameter pp. We require that* Setup *has invertible sampling via an algorithm* I_{Setup}.
- *The* deniable setup *procedure* $\mathsf{DenSetup}(1^n)$ *outputs a public parameter pp and a* faking key fk.
- *The* key generator $\mathsf{Gen}(pp; r_R)$ *outputs a public key pk, whose associated secret key is the randomness* r_R *(without loss of generality).*
  We require that Gen *has trapdoor invertible sampling via an algorithm* I_{Gen} *and trapdoor* fk, *where* $(pp, fk) \leftarrow \mathsf{DenSetup}(1^n)$.
- *The* P- *and* U-samplers $\mathsf{SampleP}(pp, pk; r_S)$ *and* $\mathsf{SampleU}(pp, pk; r_S)$ *each output some* $c \in \{0,1\}^*$.
- *The* P-tester $\mathsf{TestP}(pp, r_R, c)$ *either accepts or rejects.*
- *The* sender-coins faker $\mathsf{FakeSCoins}(pp, pk, r_S)$ *outputs some coins* r_S^* *for the* U*-sampler.*[2]
- *The* receiver-coins faker $\mathsf{FakeRCoins}(pp, fk, pk, c)$ *outputs some coins* r_R^* *for the key generator.*

We require:

1. (Correctness.) *The following experiment should accept (respectively, reject) with overwhelming probability over all its randomness: let* $pp \leftarrow \mathsf{Setup}(1^n)$, $pk \leftarrow \mathsf{Gen}(pp; r_R)$, $c \leftarrow \mathsf{SampleP}(pp, pk; r_S)$ *(resp.,* $c \leftarrow \mathsf{SampleU}(pp, pk; r_S)$*), and output* $\mathsf{TestP}(pp, r_R, c)$.
   We also require correctness for the faking algorithms: with overwhelming probability over all the random choices, letting $(pp, fk) \leftarrow \mathsf{DenSetup}(1^n)$ *and* $pk \leftarrow \mathsf{Gen}(pp; r_R)$, *we should have*

$$\mathsf{SampleU}(pp, pk; \mathsf{FakeSCoins}(pk, r_S)) = \mathsf{SampleP}(pp, pk; r_S)$$
$$\mathsf{Gen}(pp; \mathsf{FakeRCoins}(pp, fk, pk, \mathsf{SampleP}(pp, pk; r_S))) = pk.$$

[2] In some realizations, including our own, FakeSCoins can directly compute coins r_S^* given just a ciphertext $c \leftarrow \mathsf{SampleP}(pp, pk; r_S)$, not r_S (or even pp, pk). We retain the more relaxed definition above for generality.

2. (Indistinguishable public parameters.) *The public parameters pp as produced by the two setup procedures* $pp \leftarrow \mathsf{Setup}(1^n; r_{\mathsf{Setup}})$ *and* $(pp, fk) \leftarrow \mathsf{DenSetup}(1^n)$ *should be indistinguishable (either statistically or computationally).*
3. (Bideniability.) *The following two experiments should be computationally indistinguishable:*

$(pp, fk) \leftarrow \mathsf{DenSetup}(1^n)$	$(pp, fk) \leftarrow \mathsf{DenSetup}(1^n)$
$pk \leftarrow \mathsf{Gen}(pp; r_R)$	$pk \leftarrow \mathsf{Gen}(pp; r_R)$
$c \leftarrow \mathsf{SampleU}(pp, pk; r_S)$	$c \leftarrow \mathsf{SampleP}(pp, pk; r_S)$
	$r_R^* \leftarrow \mathsf{FakeRCoins}(pp, fk, c)$
	$r_S^* \leftarrow \mathsf{FakeSCoins}(pk, r_S)$
Return (pp, r_R, r_S)	*Return* (pp, r_R^*, r_S^*)

Remark 1. For correctness, it suffices (and is more convenient) to require that when $c \leftarrow \mathsf{SampleU}(pp, pk)$, $\mathsf{TestP}(pp, r_R, c)$ just rejects with probability at least (say) 1/2. The error probability can then be made negligible by parallel repetition of Gen and SampleU, using the same public parameters *pp*.

The definition is designed to allow for the use of many public keys pk_i and associated secret keys $r_{R,i}$ under the same public parameter *pp*. This lets the sender and receiver exchange multiple bits more efficiently, and have shorter keys/randomness, than if the entire system were parallelized. In addition, in deniable mode, the receiver can perform all its secret-key operations (for multiple public keys pk_i) using just the single faking key fk, without keeping any of the random strings $r_{R,i}$ that were used to generate the pk_i, by just resampling $r_{R,i}$ as needed using $I_{\mathsf{Gen}}(fk, pp, pk_i)$. This trivially allows the receiver to decrypt, and to open P-samples and U-samples 'honestly' (without equivocation), in deniable mode.

Note that in the bideniability property above, both experiments use the *deniable* setup algorithm DenSetup, rather than using the normal setup in the left-hand experiment. However, the choice of setup in the left experiment is essentially arbitrary, and the definition would be equivalent if we replaced the first line of the left-hand experiment with a normal setup $pp \leftarrow \mathsf{Setup}(1^n; r_{\mathsf{Setup}})$. This is because the faking key fk is never used, the public parameters are indistinguishable, and Setup has invertible sampling. We chose the presentation above because it yields a more modular proof that one can securely equivocate many independent public key/ciphertext pairs (pk_i, c_i) under a single public parameter *pp*: first, the bideniability property allows us to replace each pair of calls to the faking algorithms, one by one, with their normal counterparts. Then, finally, the deniable setup can be replaced with a normal setup, as just described.

Construction of deniable encryption. Canetti *et al.* [10] described a simple encoding trick to construct a multi-distributional sender-deniable encryption scheme from a translucent set: the normal encryption algorithm encodes a 0 message as "UU" whereas the deniable encryption algorithm encodes it as "PP;" both algorithms encode 1 as "UP." Thus, the sender can always open a deniably generated ciphertext as any message bit, by equivocating zero or more Ps as Us.

The same encoding lets us construct a multi-distributional *bideniable* encryption scheme from a bitranslucent set, since now the sender and receiver are both able to produce 'fake' coins that simultaneously equivocate a P as a U. Using the security properties of BTS, the proof of the following theorem (which we given in the full version) is routine.

Theorem 3. *Existence of a bitranslucent set scheme implies existence of a bideniable encryption scheme, secure under chosen-plaintext attacks.*

Canetti *et al.* [10] also construct a *fully* sender-deniable scheme from a translucent set, where the 'fake' coins can be distinguished with an inverse-polynomial probability related to the public key size. We show that for bideniability an analogous construction from a bitranslucent set also works, but requires the sender and receiver to 'coordinate' their fake coins as described Section 4. However, this coordination can again be removed using simulatable encryption. Details are deferred to the full version.

6 Lattice-Based Bitranslucent Set

In this section we give an overview of a bideniable translucent set scheme based on lattice problems, and the intuition behind its security. Due to the large amount of background required to formalize the system, space restrictions require us to defer a complete description and security proof to the full version.

Here we describe the main ideas behind our construction. To start, it will help to consider a basic mechanism for a (sender-deniable) translucent set. The public key is the description of a lattice Λ, and the secret key is a *short* lattice vector $\mathbf{z} \in \Lambda$. (A lattice is a periodic "grid" of points — formally, a discrete additive subgroup — in $\mathbb{R}^m$.) A P-sample is a point very close to the *dual lattice* Λ^*, while a U-sample is a uniformly random point in a certain large region. (The dual lattice Λ^* is the set of points $\mathbf{v} \in \text{span}(\Lambda)$ for which $\langle \mathbf{z}, \mathbf{v} \rangle \in \mathbb{Z}$ for every $\mathbf{z} \in \Lambda$.) With knowledge of $\mathbf{z}$, it is possible to distinguish these two types of points $\mathbf{c}$ by computing the inner product $\langle \mathbf{z}, \mathbf{c} \rangle$: when $\mathbf{c}$ is close to some $\mathbf{v} \in \Lambda^*$, the inner product $\langle \mathbf{z}, \mathbf{c} \rangle \approx \langle \mathbf{z}, \mathbf{v} \rangle \in \mathbb{Z}$ is close to an integer. On the other hand, when $\mathbf{c}$ is uniform, the inner product is uniformly random modulo $\mathbb{Z}$. Moreover, distinguishing between P-samples and U-samples (given only the public information) is essentially the decision-LWE problem, when the lattice Λ is defined appropriately in relation to the LWE instance. All of this is so far very similar to the structure of lattice-based encryption going back to the seminal scheme of Ajtai and Dwork [4], but in that system, the public key uniquely defines a secret key while the ciphertext does not uniquely define the encryption randomness. By contrast, here we have essentially described the 'dual' cryptosystem of Gentry, Peikert, and Vaikuntanathan [21], where there are many short vectors in Λ and hence many valid secret keys, and the ciphertext uniquely specifies the encryption randomness.

To construct a *bitranslucent* set, we exploit and build upon additional techniques from [21]. The faking key for the scheme is a *short basis* $\mathbf{T}$ of Λ (constructed using techniques from [3, 6]). As shown in [21], such a basis acts as a 'master trapdoor' that allows for efficiently sampling secret keys under a Gaussian-like distribution, and for efficiently extracting the short offset vector $\mathbf{x}$ from a P-sample $\mathbf{c} = \mathbf{v} + \mathbf{x}$, where $\mathbf{v} \in \Lambda^*$.

Using the above facts, our receiver faking algorithm works as follows: given the short basis $\mathbf{T}$ and a P-sample $\mathbf{c} = \mathbf{v} + \mathbf{x}$ that needs to be 'faked' as a U-sample, the algorithm first extracts $\mathbf{x}$, and then samples a secret key $\mathbf{z}^* \in \Lambda$ that is *highly correlated* with $\mathbf{x}$. By this we mean that $\mathbf{z}^*$ comes from a Gaussian distribution (over Λ) centered at $u \cdot \mathbf{x}$, for some random and large enough correlation coefficient $u \in \mathbb{R}$. Modulo $\mathbb{Z}$, the inner product $\langle \mathbf{z}^*, \mathbf{c} \rangle \approx \langle \mathbf{z}^*, \mathbf{x} \rangle \approx u \cdot \|\mathbf{x}\|^2$, because $\mathbf{z}^* \in \Lambda$ is short. When u is chosen from a wide enough Gaussian distribution, this inner product is uniform (modulo $\mathbb{Z}$), hence $\mathbf{c}$ "looks like" a U-sample when tested with the fake secret key $\mathbf{z}^*$. The natural question at this point is, why should it be secure to release a $\mathbf{z}^*$ that is so correlated with the secret offset vector $\mathbf{x}$? That is, why do $\mathbf{z}^*$ and $\mathbf{c}$ 'look like' an honestly generated secret key and U-sample, respectively? The first key point is that as Gaussian random variables, the correlation between $\mathbf{x}$ and $\mathbf{z}^*$ is essentially *symmetric*. That is, we can consider an alternative experiment in which $\mathbf{z}^*$ is chosen first (according to the normal key generation algorithm), and then $\mathbf{x}$ is chosen from a Gaussian centered at $v \cdot \mathbf{z}^*$ (for some appropriate random $v \in \mathbb{R}$). In both experiments, the pair $(\mathbf{z}^*, \mathbf{x})$ is jointly distributed as a nonspherical Gaussian, with the same covariance. This allows us to switch from the faking experiment to one in which the 'faked' secret key $\mathbf{z}^*$ is generated normally, followed by $\mathbf{c} = \mathbf{v} + \mathbf{x}$ where $\mathbf{v} \in \Lambda^*$ and $\mathbf{x}$ is centered at $v \cdot \mathbf{z}^*$. The second key point is that when $\mathbf{x}$ is correlated with $\mathbf{z}^*$ as described above, it may be written as the sum of two terms: the component $v \cdot \mathbf{z}^*$, and an independent (spherical) Gaussian component $\mathbf{g}$. Under the LWE assumption, we can switch from $\mathbf{c} = (\mathbf{v} + \mathbf{g}) + v \cdot \mathbf{z}^*$ to $\mathbf{c}' = \mathbf{u} + v \cdot \mathbf{z}^*$, where $\mathbf{u}$ is uniformly random. Of course, the latter quantity $\mathbf{c}'$ is truly uniform and independent of the (normally generated) secret key $\mathbf{z}^*$. This coincides exactly with the generation and subsequent honest opening of a normal secret key and U-sample, as desired.

Acknowledgments. We thank Ran Canetti and Cynthia Dwork for suggesting this research topic, Josh Benaloh and Ari Juels for helpful discussions about deniability, and the anonymous reviewers for useful comments.

References

1. The rubberhose encryption system. Internet website (accessed February 9, 2010), `http://iq.org/~proff/marutukku.org/`
2. Truecrypt: Free open-source on-the-fly encryption. Internet website (accessed Feburary 9, 2010), `http://www.truecrypt.org`
3. Ajtai, M.: Generating hard instances of the short basis problem. In: Wiedermann, J., Van Emde Boas, P., Nielsen, M. (eds.) ICALP 1999. LNCS, vol. 1644, pp. 1–9. Springer, Heidelberg (1999)
4. Ajtai, M., Dwork, C.: A public-key cryptosystem with worst-case/average-case equivalence. In: STOC, pp. 284–293 (1997)

[5] Alwen, J., Dodis, Y., Wichs, D.: Survey: Leakage resilience and the bounded retrieval model. In: Kurosawa, K. (ed.) Information Theoretic Security. LNCS, vol. 5973, pp. 1–18. Springer, Heidelberg (2010)

[6] Alwen, J., Peikert, C.: Generating shorter bases for hard random lattices. In: STACS, pp. 75–86 (2009)

[7] Bellare, M., Hofheinz, D., Yilek, S.: Possibility and impossibility results for encryption and commitment secure under selective opening. In: Joux, A. (ed.) EUROCRYPT 2009. LNCS, vol. 5479, pp. 1–35. Springer, Heidelberg (2009)

[8] Bellare, M., Miner, S.K.: A forward-secure digital signature scheme. In: Wiener, M. (ed.) CRYPTO 1999. LNCS, vol. 1666, pp. 431–448. Springer, Heidelberg (1999)

[9] Bendlin, R., Nielsen, J.B., Nordholt, P.S., Orlandi, C.: Receiver-deniable public-key encryption is impossible. Cryptology ePrint Archive, Report 2011/046 (2011), `http://eprint.iacr.org/`

[10] Canetti, R., Dwork, C., Naor, M., Ostrovsky, R.: Deniable encryption. In: Kaliski Jr., B.S. (ed.) CRYPTO 1997. LNCS, vol. 1294, pp. 90–104. Springer, Heidelberg (1997)

[11] Canetti, R., Feige, U., Goldreich, O., Naor, M.: Adaptively secure multi-party computation. In: STOC, pp. 639–648 (1996)

[12] Canetti, R., Gennaro, R.: Incoercible multiparty computation (extended abstract). In: FOCS, pp. 504–513 (1996)

[13] Canetti, R., Halevi, S., Katz, J.: A forward-secure public-key encryption scheme. J. Cryptology 20(3), 265–294 (2007), Preliminary version in EUROCRYPT 2003.

[14] Choi, S.G., Dachman-Soled, D., Malkin, T., Wee, H.: Improved non-committing encryption with applications to adaptively secure protocols. In: Matsui, M. (ed.) ASIACRYPT 2009. LNCS, vol. 5912, pp. 287–302. Springer, Heidelberg (2009)

[15] Chvátal, V.: The tail of the hypergeometric distribution. Discrete Math. 25, 285–287 (1979)

[16] Damgård, I., Nielsen, J.B.: Improved non-committing encryption schemes based on a general complexity assumption. In: Bellare, M. (ed.) CRYPTO 2000. LNCS, vol. 1880, pp. 432–450. Springer, Heidelberg (2000)

[17] Duermuth, M., Freeman, D.M.: Deniable encryption with negligible detection probability: An interactive construction. Cryptology ePrint Archive, Report 2011/066 (2011), `http://eprint.iacr.org/`

[18] Dürmuth, M., Freeman, D.M.: Deniable encryption with negligible detection probability: An interactive construction. In: Paterson, K.G. (ed.) EUROCRYPT 2011. LNCS, vol. 6632, pp. 610–626. Springer, Heidelberg (2011)

[19] Dwork, C., Naor, M., Reingold, O., Stockmeyer, L.J.: Magic functions. J. ACM 50(6), 852–921 (2003); Preliminary version in FOCS 1999

[20] Garay, J.A., Wichs, D., Zhou, H.-S.: Somewhat non-committing encryption and efficient adaptively secure oblivious transfer. In: Halevi, S. (ed.) CRYPTO 2009. LNCS, vol. 5677, pp. 505–523. Springer, Heidelberg (2009)

[21] Gentry, C., Peikert, C., Vaikuntanathan, V.: Trapdoors for hard lattices and new cryptographic constructions. In: STOC, pp. 197–206 (2008)

[22] Nielsen, J.B.: Separating random oracle proofs from complexity theoretic proofs: The non-committing encryption case. In: Yung, M. (ed.) CRYPTO 2002. LNCS, vol. 2442, pp. 111–126. Springer, Heidelberg (2002)

[23] Regev, O.: On lattices, learning with errors, random linear codes, and cryptography. J. ACM 56(6), 1–40 (2009); Preliminary version in STOC 2005

[24] Wikipedia. Deniable encryption — Wikipedia, the free encyclopedia. Internet website (2010), `http://en.wikipedia.org/wiki/Deniable_encryption` (accessed February 9, 2010)

Better Security for Deterministic Public-Key Encryption: The Auxiliary-Input Setting

Zvika Brakerski[1] and Gil Segev[2]

[1] Department of Computer Science and Applied Mathematics
Weizmann Institute of Science, Rehovot 76100, Israel
zvika.brakerski@weizmann.ac.il

[2] Microsoft Research, Mountain View, CA 94043, USA
gil.segev@microsoft.com

Abstract. Deterministic public-key encryption, introduced by Bellare, Boldyreva, and O'Neill (CRYPTO '07), provides an alternative to randomized public-key encryption in various scenarios where the latter exhibits inherent drawbacks. A deterministic encryption algorithm, however, cannot satisfy any meaningful notion of security when the plaintext is distributed over a small set. Bellare et al. addressed this difficulty by requiring semantic security to hold only when the plaintext has high min-entropy from the adversary's point of view.

In many applications, however, an adversary may obtain auxiliary information that is related to the plaintext. Specifically, when deterministic encryption is used as a building block of a larger system, it is rather likely that plaintexts do not have high min-entropy from the adversary's point of view. In such cases, the framework of Bellare et al. might fall short from providing robust security guarantees.

We formalize a framework for studying the security of deterministic public-key encryption schemes with respect to auxiliary inputs. Given the trivial requirement that the plaintext should not be efficiently recoverable from the auxiliary input, we focus on *hard-to-invert* auxiliary inputs. Within this framework, we propose two schemes: the first is based on the decisional Diffie-Hellman (and, more generally, on the d-linear) assumption, and the second is based on a rather general class of subgroup indistinguishability assumptions (including, in particular, quadratic residuosity and Paillier's composite residuosity). Our schemes are secure with respect to any auxiliary input that is subexponentially hard to invert (assuming the *standard* hardness of the underlying computational assumptions). In addition, our first scheme is secure even in the multi-user setting where related plaintexts may be encrypted under multiple public keys. Constructing a scheme that is secure in the multi-user setting (even without considering auxiliary inputs) was identified by Bellare et al. as an important open problem.

1 Introduction

Public-key encryption is one of the most basic cryptographic tasks. A public-key encryption scheme consists of three algorithms: a key-generation algorithm that

P. Rogaway (Ed.): CRYPTO 2011, LNCS 6841, pp. 543–560, 2011.

produces a secret key and a corresponding public key, an encryption algorithm that uses the public key for mapping plaintexts into ciphertexts, and a decryption algorithm that uses the secret key for recovering plaintexts from ciphertexts. For modeling the security of public-key encryption schemes, the fundamental notion of *semantic security* was introduced in the seminal work of Goldwasser and Micali [20]. Semantic security asks that it should be infeasible to gain any effective information on the plaintext by seeing the ciphertext and the public key. More specifically, whatever can be computed efficiently from the ciphertext, the public key and possibly some auxiliary information, can essentially be computed efficiently from the public key and the auxiliary information alone.

Together with its rigorous, robust, and meaningful modeling of security, semantic security inherently carries the requirement for a randomized encryption algorithm. In some cases, however, a randomized encryption algorithm may suffer from various drawbacks. In terms of efficiency, ciphertexts are not length preserving (and might be significantly longer than their corresponding plaintexts), and are in general not efficiently searchable. These properties severely limit the deployment of public-key encryption schemes in applications involving, for example, massive data sets where the ciphertext expansion is crucial, or global deduplication-based storage systems where searches are highly frequent (e.g., [26]). In addition, in terms of security, the security guarantees provided by randomized public-key encryption, and by randomized cryptographic primitives in general, are typically highly dependant on the availability of true and fresh random bits (see, for example, [3] and the references therein).

Deterministic public-key encryption. For dealing with these kind of drawbacks, Bellare, Boldyreva, and O'Neill [2] initiated the study of *deterministic* public-key encryption schemes. These are public-key encryption schemes in which the encryption algorithm is deterministic[1]. In this setting, where full-fledged sematic security is out of reach, Bellare et al. put forward the goal of formalizing a notion of security that captures semantic security as much as possible. An immediate consequence of having a deterministic encryption algorithm, however, is that essentially no meaningful notion of security can be satisfied if the plaintext is distributed over a set of polynomial size. Indeed, in such a case an adversary who is given a public key pk and an encryption c of some plaintext m under the public key pk, can simply encrypt all possible plaintexts, compare each of them to the given ciphertext c, and thus recover the plaintext m.

Bellare et al. addressed this problem by requiring security to hold only when the plaintext is sampled from a distribution of high min-entropy. Subject to this restriction, they adapted semantic security to the setting of deterministic encryption: For any high-entropy plaintext distribution, whatever can be computed efficiently from the ciphertext and the public key, can also be computed efficiently from the public key alone. Constructions of deterministic public-key encryption schemes satisfying this and similar notions of security were proposed in the random oracle model by Bellare et al. [2], and then in the standard model

[1] Note that this is effectively a collection of injective trapdoor functions (assuming the decryption algorithm is deterministic as well).

by Bellare, Fischlin, O'Neill, and Ristenpart [4], by Boldyreva, Fehr, and O'Neill [5], and by O'Neill [23]. We refer the reader to Section 1.2 for an elaborated discussion of these constructions.

Security with respect to auxiliary information. In typical applications, a deterministic public-key encryption scheme is used as building block of a larger system. In such a setting, an adversary usually has additional information that it can use when trying to break the security of the scheme. This danger becomes even more critical when such additional information is related to the encrypted plaintext. In general, security with respect to auxiliary information is essential towards obtaining composable security (see, for example, [11] and the references therein). More closely related to our approach are the studies of security with respect to auxiliary information in the contexts of perfect one-way functions [10], program obfuscation [19], and leakage-resilient encryption [13,12,8].

For example, when using a deterministic public-key encryption scheme for enabling efficient searches on encrypted databases, as suggested by Bellare et al. [2], it is not unlikely that the same plaintext belongs to more than one database, and is therefore encrypted under several public keys; or that various statistics of the database are publicly available. A more acute example is when using a deterministic public-key encryption scheme for a key-encapsulation mechanism that "hedges against bad randomness" [3]. In such a case an adversary that observes the usage of the encapsulated key (say, as a key to a symmetric-key encryption scheme) may in fact obtain a huge amount of additional information on the encapsulated key.

In this light, the notion of security proposed by Bellare et al. [2] might fall short of capturing the likely case where auxiliary information is available. That is, although a plaintext may be sampled from a distribution with high min-entropy to begin with, it might still have no entropy, from the point of view of an adversary, in many realistic scenarios. We note that already in the setting of deterministic *symmetric-key* encryption of high-entropy messages, Dodis and Smith [14] observed that the main weakness of an approach that does not take into account auxiliary information, is the lack of composable security. It is thus a highly desirable task to model and to construct secure deterministic encryption schemes in the setting of auxiliary information, as a crucial and essential step towards obtaining more realistic security guarantees.

1.1 Our Contributions

In this paper we introduce a framework for modeling the security of deterministic public-key encryption schemes with respect to auxiliary inputs. Within this framework we propose constructions that are based on standard cryptographic assumptions in the standard model (i.e., without random oracles). Our framework is a generalization of the one formalized by Bellare et al. [2] (and further studied in [4,5,23]) to the auxiliary-input setting, in which an adversary possibly obtains additional information that is related to the encrypted plaintext, and might even fully determine the encrypted plaintext information theoretically.

Modeling auxiliary information. An immediate consequence of having a deterministic encryption algorithm is that no meaningful notion of security can be satisfied if the plaintext can be recovered from the adversary's auxiliary information (see Section 3 for a discussion of this inherent constraint[2]). Thus, we focus our attention on the case of *hard-to-invert* auxiliary inputs, where the source of hardness may be any combination of information-theoretic hardness (where the auxiliary-input function is many-to-one) and computational hardness (where the auxiliary input function is injective, but is hard to invert by efficient algorithms).

Notions of security. Following [2,4,5] we formalize three notions of security with respect to auxiliary inputs, and prove that all three are equivalent. The first is a simulation-based notion, capturing the intuitive meaning of semantic security: whatever can be computed efficiently given a public key, an encryption of a message, and *hard-to-invert auxiliary input*, can be computed efficiently given only the public key and the auxiliary input. The second is a comparison-based notion, which essentially serves as an intermediate notion towards an indistinguishability-based one that is somewhat easier to handle in proofs of security. The high-level approach of the equivalence proofs is motivated by those of [4,5], but the existence of auxiliary inputs that may fully determine the encrypted messages introduces various difficulties that our techniques overcome.

Constructions. We propose two constructions in the standard model satisfying our notions of security. At a first glance, one might hope that the constructions proposed in [2,4,5,23] can be naturally extended to the auxiliary-input setting by replacing the notion of statistical min-entropy with an appropriate notion of computational min-entropy. This, however, does not seem to be the case (at least without relying on random oracles), as these constructions seem to heavily rely on information-theoretic properties that might not have natural computational analogues[3].

Our first construction is based on the decisional Diffie-Hellman assumption, (and more generally, on any of the d-linear assumptions), and our second construction is based on a rather general class of subgroup indistinguishability assumptions as defined in [8] (including, in particular, the quadratic residuosity assumption, and Paillier's composite residuosity assumption [24]). The resulting schemes are secure with respect to any auxiliary input that is subexponentially hard to invert[4]. In addition, our first scheme is secure even in the multi-user setting where related messages may be encrypted under multiple public keys. In this setting we obtain security (with respect to auxiliary inputs) for any polynomial number of messages and users as long as the messages are related by invertible linear transformations. Constructing a scheme that is secure is the

---

[2] This is somewhat similar to the observation that security is impossible to achieve when the plaintext is distributed over a small set.

[3] A prime example is the generalized crooked leftover hash lemma [5], for which a computational analogue may seem somewhat challenging to devise.

[4] We emphasize that in this paper we rely on standard computational assumptions (i.e., d-linear or quadratic residuosity), and only the auxiliary inputs are assumed to have subexponential hardness.

multi-user setting (even without considering auxiliary inputs) was identified as an important open problem by Bellare et al. [2]. Finally, we note that this scheme also exhibits an interesting homomorphic property: it allows homomorphic additions and one multiplication, in the spirit of [6,16]. This property may be found especially useful in light of the possible applications of deterministic public-key encryption schemes in database systems [2].

1.2 Related Work

Exploiting the entropy of messages to prove otherwise-impossible security was first proposed by Russell and Wang [25], followed by Dodis and Smith [14]. These works achieved information-theoretic security for symmetric-key encryption with short keys.

In the setting of public-key encryption, deterministic encryption for high min-entropy messages was proposed by Bellare, Boldyreva, and O'Neill [2] who formalized a definitional framework, which was later refined and extended by Bellare, Fischlin, O'Neill, and Ristenpart [4], by Boldyreva, Fehr, and O'Neill [5], and by O'Neill [23]. Bellare et at. [2] presented two constructions in the random oracle model: The first relies on any semantically-secure public-key encryption scheme; whereas the second relies on the RSA function (and is in fact length preserving). Constructions in the standard model (i.e., without random oracles), were then presented in [4,5]. Bellare et al. [4] presented a construction based on trapdoor permutations, which is secure as long as the messages are (almost) uniformly distributed. Boldyreva et al. [5] presented a construction based on lossy trapdoor functions, which is secure as long as its n-bit messages have min-entropy at least n^ϵ for some constant $0 < \epsilon < 1$. These constructions, however, fall short in two interesting cases: In the multi-message setting, where arbitrarily related messages are encrypted under the same public key; and in the multi-user setting where the same message is encrypted under several (independently chosen) public keys. Recently, O'Neill [23] made a step towards addressing the former, by presenting a scheme that can securely encrypt any fixed number q of messages, but whose parameters depend polynomially on q. The latter case remained unexplored until this work.

Deterministic public-key encryption was used by Bellare et al. [3] who defined and constructed "hedged" public-key encryption schemes. These are schemes that are semantically secure in the standard sense, and maintain a meaningful and realistic notion of security even when "corrupt" randomness is used for the encryption, so long as the joint message-randomness pair has sufficient min-entropy. The definition of security in the latter case takes after that of deterministic public-key encryption.

The tools underlying our constructions in this paper are inspired by the line of research on "encryption in the presence of auxiliary input", initiated by Dodis, Kalai, and Lovett [13] in the context of symmetric-key encryption, and then extended in [12,8] to public-key encryption. These works consider encryption schemes where the adversary may obtain a hard-to-invert function of the secret key — extending the frameworks of "bounded leakage" [1] and "noisy leakage" [22].

1.3 Overview of Our Approach

In this section we provide a high-level overview of our approach and techniques. We begin with a brief description of the notions of security that we consider in the auxiliary-input setting, and then describe the main ideas underlying our two constructions. For simplicity, in what follows we consider the case where one message is encrypted under one public key, and refer the reader to the relevant sections for the more general case.

Defining security with respect to auxiliary inputs. Towards describing our notions of security, we first discuss our notion of hard-to-invert auxiliary inputs. We consider any auxiliary input $f(x)$ from which it is hard to recover the input x. The source of hardness may be any combination of information-theoretic hardness (where the function f is many-to-one), and computational hardness (where $f(x)$ fully determines x, but x is hard to recover by efficient algorithms). Informally, we say that a function f is ϵ-hard-to-invert with respect to a distribution $\mathcal{D}$, if for every efficient algorithm A it holds that $A(f(x)) = x$ with probability at most ϵ, over the choice of $x \leftarrow \mathcal{D}$ and the internal coin tosses of A.

As discussed in Section 1.1, we formalize three notions of security with respect to auxiliary inputs, and prove that all three are equivalent. For concreteness we focus here on the simulation-based definition, which captures the intuitive meaning of semantic security: Whatever can be computed efficiently given a public key, an encryption of a message, and *hard-to-invert auxiliary input*, can be computed efficiently given only the public key and the auxiliary input. A bit more formally, we say that a scheme is secure with respect to ϵ-hard-to-invert auxiliary inputs if for any probabilistic polynomial-time adversary A, and for any efficiently samplable plaintext distribution $\mathcal{M}$, there exists a probabilistic polynomial-time simulator S, such that for any efficiently computable function f that is ϵ-hard-to-invert with respect to $\mathcal{M}$, and for any function $g \in \{0,1\}^* \rightarrow \{0,1\}^*$, the probabilities of the events $A\left(pk, \mathsf{Enc}_{pk}(m), f(m)\right) = g(m)$ and $S\left(pk, f(m)\right) = g(m)$ are negligibly close, where $m \leftarrow \mathcal{M}$. We note that the functions f and g may be arbitrary related[5]. This is a generalization of the definitions considered in [2,4,5,23].

The scheme of Boldyreva et al. [5]. Our starting point is the scheme of Boldyreva et al. [5] that is based on lossy trapdoor functions. This is in fact the only known construction in the standard model (i.e., without random oracles) that is secure for arbitrary plaintext distributions with high (but not nearly full) min-entropy. In their construction, the public key consists of a function h that is sampled from the injective mode of the collection of lossy trapdoor functions, and a pair-wise independent permutation π. The secret key consists of the trapdoor for inverting h (we assume that π is efficiently invertible). The encryption of a message m is defined as $\mathsf{Enc}_{pk}(m) = h(\pi(m))$, and decryption is naturally defined.

[5] In fact, the "target" function g is allowed to take as input also the randomness that is used for sampling m, and any other public randomness – see Section 3.

In a high level, the proof of security in [5] considers the joint distribution of the public key and the ciphertext $(pk, \mathsf{Enc}_{pk}(m))$, and argues that it is computationally indistinguishable from a distribution that is independent of the plaintext m. This is done by considering a distribution of malformed public keys, that is computationally indistinguishable from the real distribution. Specifically, the injective function h is replaced with a lossy function $\tilde{h}$ to obtain an indistinguishable public key $\widetilde{pk}$. The next step is to show that the ciphertext $\tilde{c} = \mathsf{Enc}_{\widetilde{pk}}(m)$ can be described by the following two-step process. First, an analogue of a strong extractor is applied to m (where the seed is the permutation π that lies in $\widetilde{pk}$) to obtain $v = \mathsf{ext}_{\widetilde{pk}}(m)$. Then, the output of the extractor is used to compute the ciphertext $\tilde{c} = g(\widetilde{pk}, v)$. From this point of view, it is evident that so long as the plaintext m is drawn from a distribution with high min-entropy, it holds that $v = \mathsf{ext}_{\widetilde{pk}}(m)$ is statistically close to a uniform distribution (over some domain). This holds even given the malformed public key, and does not depend on the distribution of m. This methodology of using an analog of a strong extractor relies on the *crooked leftover hash lemma* of Dodis and Smith [14], that enables to base the construction on any collection of lossy trapdoor functions.

Our constructions. In our setting, we wish to adapt this methodology to rely on computational hardness instead of min-entropy. However, there is currently no known analog of the crooked leftover hash lemma in the computational setting. This is an interesting open problem. We overcome this difficulty by relying of specific collections of lossy trapdoor functions, for which we are in fact able to extract pseudorandomness from computational hardness. We do this by replacing the strong extractor component with a hard-core function of the message (with respect to the auxiliary input). Specifically, our encryption algorithm (when using the malformed public key) can be interpreted as taking an inner product between our message m (viewed as a vector of bits) and a random vector a, where the resulting ciphertext depends only on $(a, \langle m, a\rangle)$. This is similar to the Goldreich-Levin hard-core predicate [18], except that the vector a is not binary and the inner product is performed over some large $\mathbb{Z}$-module and not over the binary field. We thus require the generalized Goldreich-Levin theorem of Dodis et al. [12] to obtain that even given the auxiliary input, the distributions $(a, \langle m, a\rangle)$ and (a, u) are computationally indistinguishable, where u is uniformly distributed and does not depend on the distribution of m.

To be more concrete, let us consider our DDH-based scheme (formally presented in Section 4) which is based on the lossy trapdoor functions of Freeman et al. [15]. The scheme is instantiated by a DDH-hard group $\mathbb{G}$ of prime order q that is generated by g. The message space is $\{0,1\}^n$ (where n is polynomial in the security parameter) and the public key is $g^{\mathbf{A}}$, for a random $n \times n$ matrix $\mathbf{A}$ over $\mathbb{Z}_q$. Encryption is done by computing $\mathsf{Enc}_{g^{\mathbf{A}}}(\mathbf{m}) = g^{\mathbf{A}\cdot\mathbf{m}}$ and decryption is performed using $sk = \mathbf{A}^{-1}$.[6]

[6] We overload the notation g^x to matrices as follows: for $\mathbf{X} \in \mathbb{Z}_q^{k\times n}$, we let $g^{\mathbf{X}} \in \mathbb{G}^{k\times n}$ denote the matrix defined as $(g^{\mathbf{X}})_{i,j} = g^{(\mathbf{X})_{i,j}}$.

For analyzing the security of the scheme, we consider the joint distribution of the public key, ciphertext and auxiliary input $(pk, \mathsf{Enc}_{pk}(\mathbf{m}), f(\mathbf{m})) = (g^{\mathbf{A}}, g^{\mathbf{A}\cdot\mathbf{m}}, f(\mathbf{m}))$. The malformed distribution $\widetilde{pk}$ is obtained by taking $\mathbf{A}$ to be a random rank-1 matrix (rather than completely random). DDH implies that pk and $\widetilde{pk}$ are computationally indistinguishabile. Such a low-rank matrix takes the form $\mathbf{A} = \mathbf{r}\cdot\mathbf{b}^T$, and therefore $\mathbf{A}\cdot\mathbf{m} = \mathbf{r}\cdot\mathbf{b}^T\cdot\mathbf{m}$, for random vectors $\mathbf{r}$ and $\mathbf{b}$. Thus, our ciphertext depends only on $(\mathbf{b}, \langle\mathbf{b}, \mathbf{m}\rangle)$ which is indistinguishable from $(\mathbf{b}, u)$, for a uniformly random u, even given $f(\mathbf{m})$, by the generalized Goldreich-Levin theorem [12]. Our initial distribution is therefore indistinguishable from the distribution $(g^{\mathbf{r}\cdot\mathbf{b}^T}, g^{\mathbf{r}\cdot u}, f(\mathbf{m}))$ as required.

In the multi-user setting, we observe that any polynomial number of public keys $g^{\mathbf{A}_1}, \ldots, g^{\mathbf{A}_\ell}$ are computationally indistinguishable, by DDH, from having *joint rank*-1. Namely, in this case the distributions $(g^{\mathbf{A}_1}, \ldots, g^{\mathbf{A}_\ell})$ and $(g^{\mathbf{r}_1\cdot\mathbf{b}^T}, \ldots, g^{\mathbf{r}_\ell\cdot\mathbf{b}^T})$ are computationally indistinguishable, where the same vector $\mathbf{b}$ is used for all keys. Encrypting a message $\mathbf{m}$ under all such ℓ public keys results in a set of ciphertexts $(g^{\mathbf{r}_1\cdot\mathbf{b}^T\cdot\mathbf{m}}, \ldots, g^{\mathbf{r}_\ell\cdot\mathbf{b}^T\cdot\mathbf{m}})$, where all elements depend on $(\mathbf{b}, \langle\mathbf{b}, \mathbf{m}\rangle)$. This enables to apply the above approach, and we show that it in fact extends to linearly-related messages.

Our second scheme (based on subgroup indistinguishability assumptions) is analyzed quite similarly. We rely on the lossy trapdoor functions of [21] and can again show that our public key distribution is indistinguishable from one over rank-1 matrices. However, the groups under consideration might be non-cyclic. This adds additional complications into the analysis. In addition, this scheme does not seem to allow a "joint rank" argument as above, and we leave it as an open problem to construct an analogous scheme that is secure in the multi-user setting.

Paper organization. The remainder of this paper is organized as follows. In Section 2 we formalize a general notion for hard-to-invert auxiliary inputs. In Section 3 we introduce a framework for modeling the security of deterministic public-key encryption schemes with respect to auxiliary inputs, consisting of three main notions of security. In Section 4 we present a construction based on the decisional Diffie-Hellman assumption (and, more generally, on any of the d-linear assumptions), and in Section 5 we present a construction based on subgroup indistinguishability assumptions.

Due to space limitations, not all results and proofs appear in this extended abstract. We refer the reader to the full version of this paper [9] for more details.

Notation. Throughout the paper we denote scalars in plain lowercase letters ($x \in \{0, 1\}$). We use the term "vector" both in the algebraic sense, where it indicates an element in a vector space and denoted by bold lowercase letters ($\mathbf{x} \in \{0, 1\}^k$); and in the "combinatorial" sense, indicating an ordered set of elements (not necessarily having any algebraic properties) for which we use the notation $\vec{x}$. We denote a combinatorial vector whose elements are algebraic vectors by $\vec{\mathbf{x}}$, combinatorial vector of combinatorial vectors by $\vec{\vec{x}}$, and combinatorial vector of combinatorial vectors of algebraic vectors by $\vec{\mathbf{x}}$. Matrices (always algebraic) are denoted in bold uppercase ($\mathbf{X} \in \{0,1\}^{k \times n}$). The $k \times k$ identity matrix is denoted $\mathbf{I}_k$. All vectors are column vectors by default, and a row vector is denoted by $\mathbf{x}^T$.

2 Hard-to-Invert Auxiliary Inputs

In this work we consider any auxiliary input $f(x)$ from which it is hard to recover the input x. The source of hardness may be any combination of information-theoretic hardness (where the function f is many-to-one) and computational hardness (where $f(x)$ fully determines x, but x is hard to recover by efficient algorithms). Informally, we say that a function f is ϵ-hard-to-invert with respect to a distribution $\mathcal{D}$, if for every efficient algorithm A it holds that $A(f(x)) = x$ with probability at most ϵ over the choice of $x \leftarrow \mathcal{D}$ and the internal coin tosses of A.

For our purposes, we formalize a slightly more general notion in which $\mathcal{D}$ is a distribution over vectors of inputs $\vec{x} = (x_1, \ldots, x_t)$, and for every $i \in \{1, \ldots, t\}$ it should be hard to efficiently recover x_i when given $f(\vec{x})$. In addition, we also consider a *blockwise* variant of this notion, in which it should be hard to efficiently recover x_i when given $(x_1, \ldots, x_{i-1}, f(\vec{x}))$.

Definition 2.1. *An efficiently computable function $\mathcal{F} = \{f_k\}_{k \in \mathbb{N}}$ is* $\epsilon(k)$-hard-to-invert with respect to an efficiently samplable distribution $\mathcal{D} = \{\mathcal{D}_k\}_{k \in \mathbb{N}}$ *over vectors of $t(k)$ inputs, if for every probabilistic polynomial-time algorithm A and for every $i \in \{1, \ldots, t(k)\}$ it holds that*

$$\Pr\left[A\left(1^k, f_k(\vec{x})\right) = x_i\right] \leq \epsilon(k) ,$$

for all sufficiently large k, where the probability is taken over the choice of $\vec{x} = (x_1, \ldots, x_{t(k)}) \leftarrow \mathcal{D}_k$, and over the internal coin tosses of A.

Definition 2.2. *An efficiently computable function $\mathcal{F} = \{f_k\}_{k \in \mathbb{N}}$ is* $\epsilon(k)$-block-wise-hard-to-invert with respect to an efficiently samplable distribution $\mathcal{D} = \{\mathcal{D}_k\}_{k \in \mathbb{N}}$ *over vectors of $t(k)$ inputs, if for every probabilistic polynomial-time algorithm A and for every $i \in \{1, \ldots, t(k)\}$ it holds that*

$$\Pr\left[A\left(1^k, x_1, \ldots, x_{i-1}, f_k(\vec{x})\right) = x_i\right] \leq \epsilon(k) ,$$

for all sufficiently large k, where the probability is taken over the choice of $\vec{x} = (x_1, \ldots, x_{t(k)}) \leftarrow \mathcal{D}_k$, and over the internal coin tosses of A.

Note that Definition 2.1 implies in particular that the distribution $\mathcal{D}$ is such that each x_i has min-entropy at least $\log(1/\epsilon(k))$. Furthermore, Definition 2.2 implies that the distribution $\mathcal{D}$ is a block source in which each block x_i has (average) min-entropy at least $\log(1/\epsilon(k))$ conditioned on the previous blocks $(x_1, \ldots, x_{i-1})$.

3 Modeling Security in the Auxiliary-Input Setting

In this section we present a framework for modeling the security of deterministic public-key encryption schemes with respect to auxiliary inputs. Our framework is obtained as a generalization of those considered in [2,4,5] to a setting in which the encrypted plaintexts may be fully determined by some auxiliary information that is available to the adversary. Following [2,4,5] we formalize three notions of security with respect to auxiliary inputs, and prove that all three are equivalent. The first is a simulation-based semantic security notion (PRIV-SSS), capturing the intuitive meaning of semantic security: whatever can be computed given an encryption of a message and *auxiliary input*, can also be computed given only the auxiliary input. The second is a comparison-based semantic-security notion (PRIV-CSS), which essentially serves as an intermediate notion towards an indistinguishability-based one (PRIV-IND) that is somewhat easier to handle in proofs of security.

In the remainder of this paper we use the following notation. For a deterministic public-key encryption scheme $\Pi = (\mathsf{KeyGen}, \mathsf{Enc}, \mathsf{Dec})$, a public key pk, and a vector of messages $\vec{m} = (m_1, \ldots, m_t)$ we denote by $\vec{\mathsf{Enc}}_{pk}(\vec{m})$ the vector $(\mathsf{Enc}_{pk}(m_1), \ldots, \mathsf{Enc}_{pk}(m_t))$. When considering a distribution $\mathcal{M}$ over vectors of messages $\vec{m} = (m_1, \ldots, m_t)$ all of which are encrypted under the same public key, then for the case of hard-to-invert auxiliary inputs we make in this paper the simplifying assumption that $m_i \neq m_j$ for every $i \neq j$ (a bit more formally, one should require that all distributions have identical equality patterns – see [2]). In the case of blockwise-hard-to-invert auxiliary inputs this assumption is not necessary. In addition, for simplicity we present our definitions for the case of hard-to-invert auxiliary inputs, and note that they naturally extend to the case of blockwise-hard-to-invert auxiliary inputs.

Definition 3.1 (Simulation-based security). *A deterministic public-key encryption scheme* $\Pi = (\mathsf{KeyGen}, \mathsf{Enc}, \mathsf{Dec})$ *is* PRIV-SSS-secure with respect to ϵ-hard-to-invert auxiliary inputs *if for any probabilistic polynomial-time algorithm A and for any efficiently samplable distribution $\mathcal{M} = \{\mathcal{M}_k\}_{k \in \mathbb{N}}$, there exists a probabilistic polynomial-time algorithm S, such that for any efficiently computable function $\mathcal{F} = \{f_k\}_{k \in \mathbb{N}}$ that is ϵ-hard-to-invert with respect to $\mathcal{M}$, and for any function $g \in \{0,1\}^* \to \{0,1\}^*$, there exists a negligible function $\nu(k)$ such that*

$$\mathrm{Adv}^{\mathsf{PRIV-SSS}}_{\Pi,A,\mathcal{M},S,\mathcal{F},g}(k) \stackrel{\mathsf{def}}{=} \left|\mathrm{Real}^{\mathsf{PRIV-SSS}}_{\Pi,A,\mathcal{M},\mathcal{F},g}(k) - \mathrm{Ideal}^{\mathsf{PRIV-SSS}}_{\Pi,S,\mathcal{M},\mathcal{F},g}(k)\right| \leq \nu(k)$$

for all sufficiently large k, where

$$\begin{aligned} \mathrm{Real}^{\mathsf{PRIV-SSS}}_{\Pi,A,\mathcal{M},\mathcal{F},g}(k) &= \Pr\left[A\left(1^k, pk, \vec{\mathsf{Enc}}_{pk}(\vec{m}), f_k(\vec{m})\right) = g(\vec{m})\right] \\ \mathrm{Ideal}^{\mathsf{PRIV-SSS}}_{\Pi,S,\mathcal{M},\mathcal{F},g}(k) &= \Pr\left[S\left(1^k, pk, f_k(\vec{m})\right) = g(\vec{m})\right] \ , \end{aligned}$$

and the probability is taken over the choices of $\vec{m} \leftarrow \mathcal{M}_k$, $(sk, pk) \leftarrow \mathsf{KeyGen}(1^k)$, and over the internal coin tosses of A and S.

Definition 3.2 (Comparison-based security). *A deterministic public-key encryption scheme* $\Pi = (\mathsf{KeyGen}, \mathsf{Enc}, \mathsf{Dec})$ *is* PRIV-CSS-secure with respect to ϵ-hard-to-invert auxiliary inputs *if for any probabilistic polynomial-time algorithm A, for any efficiently samplable distribution $\mathcal{M} = \{\mathcal{M}_k\}_{k\in\mathbb{N}}$, for any efficiently computable function $\mathcal{F} = \{f_k\}_{k\in\mathbb{N}}$ that is ϵ-hard-to-invert with respect to $\mathcal{M}$, and for any function $g \in \{0,1\}^* \to \{0,1\}^*$, there exists a negligible function $\nu(k)$ such that*

$$\mathrm{Adv}^{\mathsf{PRIV-CSS}}_{\Pi,A,\mathcal{M},\mathcal{F},g}(k) \stackrel{\mathsf{def}}{=} \left|\mathrm{Adv}^{\mathsf{PRIV-CSS}}_{\Pi,A,\mathcal{M},\mathcal{F},g}(k,0) - \mathrm{Adv}^{\mathsf{PRIV-CSS}}_{\Pi,A,\mathcal{M},\mathcal{F},g}(k,1)\right| \leq \nu(k)$$

for all sufficiently large k, where

$$\mathrm{Adv}^{\mathsf{PRIV-CSS}}_{\Pi,A,\mathcal{M},\mathcal{F},g}(k,b) = \Pr\left[A\left(1^k, pk, \vec{\mathsf{Enc}}_{pk}(\vec{m}_b), f_k(\vec{m}_0)\right) = g(\vec{m}_0)\right] ,$$

and the probability is taken over the choices of $\vec{m}_0 \leftarrow \mathcal{M}_k$, $\vec{m}_1 \leftarrow \mathcal{M}_k$, $(sk, pk) \leftarrow$ $\mathsf{KeyGen}(1^k)$*, and over the internal coin tosses of A.*

Definition 3.3 (Indistinguishability-based security). *A deterministic public-key encryption scheme* $\Pi = (\mathsf{KeyGen}, \mathsf{Enc}, \mathsf{Dec})$ *is* PRIV-IND-secure with respect to ϵ-hard-to-invert auxiliary inputs *if for any probabilistic polynomial-time algorithm A, for any two efficiently samplable distributions $\mathcal{M}_0 = \{\mathcal{M}_{0,k}\}_{k\in\mathbb{N}}$ and $\mathcal{M}_1 = \{\mathcal{M}_{1,k}\}_{k\in\mathbb{N}}$, and for any efficiently computable function $\mathcal{F} = \{f_k\}_{k\in\mathbb{N}}$ that is ϵ-hard-to-invert with respect to both $\mathcal{M}_0$ and $\mathcal{M}_1$, there exists a negligible function $\nu(k)$ such that*

$$\mathrm{Adv}^{\mathsf{PRIV-IND}}_{\Pi,A,\mathcal{M}_0,\mathcal{M}_1,\mathcal{F}}(k) \stackrel{\mathsf{def}}{=} \left|\mathrm{Adv}^{\mathsf{PRIV-IND}}_{\Pi,A,\mathcal{M}_0,\mathcal{M}_1,\mathcal{F}}(k,0) - \mathrm{Adv}^{\mathsf{PRIV-IND}}_{\Pi,A,\mathcal{M}_0,\mathcal{M}_1,\mathcal{F}}(k,1)\right| \leq \nu(k)$$

for all sufficiently large k, where

$$\mathrm{Adv}^{\mathsf{PRIV-IND}}_{\Pi,A,\mathcal{M}_0,\mathcal{M}_1,\mathcal{F}}(k,b) = \Pr\left[A\left(1^k, pk, \vec{\mathsf{Enc}}_{pk}(\vec{m}_b), f_k(\vec{m}_0)\right) = 1\right] ,$$

and the probability is taken over the choices of $\vec{m}_0 \leftarrow \mathcal{M}_{0,k}$, $\vec{m}_1 \leftarrow \mathcal{M}_{1,k}$, $(sk, pk) \leftarrow$ $\mathsf{KeyGen}(1^k)$*, and over the internal coin tosses of A.*

The hard-to-invert requirement. We emphasize that in the setting of *deterministic* public-key encryption the requirement that the encrypted messages cannot be efficiently recovered from the auxiliary input is essential (unlike in the setting of *randomized* encryption, where the notion of semantic security takes into account any auxiliary input – see, for example, [17, Ch. 5]). This is easily observed using our indistinguishability-based formulation (Definition 3.3): an algorithm that on input $f_k(\vec{m}_0)$ (where $\vec{m}_0 = (m_{0,1}, \ldots, m_{0,t(k)})$) can recover one of the $m_{0,i}$ values can then encrypt this value under pk, compare the resulting ciphertext with the i-th component of $\vec{\mathsf{Enc}}_{pk}(\vec{m}_b)$, and thus learn the bit b.

Relation to previous notions. We note that any *constant function* is ϵ-hard-to-invert with respect to any message distribution of min-entropy at least

$\log(1/\epsilon)$. Thus, our notion of auxiliary-input security strictly generalizes previous security notions, in which auxiliary input is not considered, and the message distributions need to have sufficient min-entropy [2,4,5,23].

Access to the public key. As observed by Bellare et al. [2] it is essential that the "target" function g does not take the public key as input. Specifically, with a deterministic encryption algorithm the ciphertext itself is a non-trivial information that it leaked about the plaintext, and can clearly be computed efficiently using the public key. We refer the reader to [2] for a more elaborated discussion.

The randomness of sampling. For our notions of security we in fact allow the auxiliary-input function f and the "target" function g to take as input not only the vector of message $\vec{m}$, but also the random string $r \in \{0,1\}^*$ that was used for sampling $\vec{m}$ from the distribution $\mathcal{D}_k$. When this aspect plays a significant role we explicitly include r as part of the input for f and g, and denote by $\vec{m} \leftarrow \mathcal{D}_k(r)$ the fact that $\vec{m}$ is sampled using the random string r. When this aspect does not play a significant role we omit it for ease of readability (in particular, we omitted it from the above definitions).

PRIV1: focusing on a single message. As in [5] we also consider the PRIV1-variants of our notion of security that focus on a single message (instead of vectors of any polynomial number of messages). In the full version [9] we also provide proof that security for a vector of messages with respect to a blockwise-hard-to-invert auxiliary input is in fact equivalent to security for a single message with respect to a hard-to-invert auxiliary input.

The multi-user setting. So far our notions of security considered vectors of messages that are encrypted under the same public key. Our definitions in this section naturally generalize to the multi-user setting, where there are multiple public keys, each of which is used for encrypting a vector of messages. Due to space limitations, we refer the reader to the full version [9] for this generalization.

An even stronger notion of security. Note that in Definition 3.3 the algorithm A is given as input the vector $\left(1^k, pk, \vec{\mathsf{Enc}}_{pk}(\vec{m}_b), f_k(\vec{m}_0)\right)$, and that a seemingly stronger definition would even consider the vector

$$\left(1^k, pk, \vec{\mathsf{Enc}}_{pk}(\vec{m}_b), \vec{\mathsf{Enc}}_{pk}(\vec{m}_{1-b}), f_k(\vec{m}_0), f_k(\vec{m}_1)\right)$$

as its input. As indicated by the equivalence of our three definitions, such a stronger variant is not needed for capturing the intuitive meaning of semantic security as in Definition 3.1. Nevertheless, our schemes in this paper in fact satisfy this stronger variant. We refer to this notion as *strong indistinguishability* (PRIV-sIND), formally defined as follows:

Definition 3.4. *A deterministic public-key encryption scheme* $\Pi =$ (KeyGen, Enc, Dec) *is* PRIV-sIND-secure with respect to ϵ-hard-to-invert auxiliary inputs *if for any probabilistic polynomial-time algorithm A, for any two efficiently samplable distributions $\mathcal{M}_0 = \{\mathcal{M}_{0,k}\}_{k\in\mathbb{N}}$ and $\mathcal{M}_1 = \{\mathcal{M}_{1,k}\}_{k\in\mathbb{N}}$, and for any efficiently computable function $\mathcal{F} = \{f_k\}_{k\in\mathbb{N}}$ that is ϵ-hard-to-invert with respect to both $\mathcal{M}_0$ and $\mathcal{M}_1$, there exists a negligible function $\nu(k)$ such that*

$$\mathrm{Adv}^{\mathsf{PRIV-sIND}}_{\Pi,A,\mathcal{M}_0,\mathcal{M}_1,\mathcal{F}}(k) \stackrel{\mathsf{def}}{=} \left|\mathrm{Adv}^{\mathsf{PRIV-sIND}}_{\Pi,A,\mathcal{M}_0,\mathcal{M}_1,\mathcal{F}}(k,0) - \mathrm{Adv}^{\mathsf{PRIV-sIND}}_{\Pi,A,\mathcal{M}_0,\mathcal{M}_1,\mathcal{F}}(k,1)\right| \leq \nu(k)$$

for all sufficiently large k, where

$$\begin{aligned}&\mathrm{Adv}^{\mathsf{PRIV-sIND}}_{\Pi,A,\mathcal{M}_0,\mathcal{M}_1,\mathcal{F}}(k,b)\\ &\quad= \Pr\left[A\left(1^k, pk, \vec{\mathsf{Enc}}_{pk}(\vec{m}_b), \vec{\mathsf{Enc}}_{pk}(\vec{m}_{1-b}), f_k(\vec{m}_0), f_k(\vec{m}_1)\right) = 1\right] ,\end{aligned}$$

and the probability is taken over the choices of $\vec{m}_0 \leftarrow \mathcal{M}_{0,k}$, $\vec{m}_1 \leftarrow \mathcal{M}_{1,k}$, $(sk, pk) \leftarrow \mathsf{KeyGen}(1^k)$, and over the internal coin tosses of A.

4 A Scheme Based on the d-Linear Assumption

In this section we present our d-linear-based based deterministic encryption scheme and discuss its properties. We show that the d-linear-based lossy trapdoor function of Freeman et al. [15] is in fact a deterministic public-key encryption that is secure with respect to hard-to-invert auxiliary inputs.

The scheme $\Pi_{\mathbf{Lin}}$. Let $\mathsf{GroupGen}$ be a probabilistic polynomial-time algorithm that takes as input a security parameter 1^k, and outputs a triplet $(\mathbb{G}, q, g)$ where $\mathbb{G}$ is a group of prime order q that is generated by $g \in \mathbb{G}$, and q is a k-bit prime number. For describing the scheme we overload the notation g^x to matrices: for $\mathbf{X} \in \mathbb{M}^{k\times n}$, we let $g^{\mathbf{X}} \in \mathbb{G}^{k\times n}$ denote the matrix defined as $(g^{\mathbf{X}})_{i,j} = g^{(\mathbf{X})_{i,j}}$. The scheme is parameterized by the security parameter k and the message length $n = n(k)$.

- **Key generation.** The key-generation algorithm $\mathsf{KeyGen}(1^k)$ samples $(\mathbb{G}, q, g) \leftarrow \mathsf{GroupGen}(1^k)$, and a matrix $\mathbf{A} \leftarrow \mathbb{Z}_q^{n\times n}$. It then outputs $pk = (\mathbb{G}, q, g, g^{\mathbf{A}})$ and $sk = \mathbf{A}^{-1}$ (note that $\mathbf{A}$ is invertible with all but a negligible probability).
- **Encryption.** The encryption algorithm $\mathsf{Enc}_{pk}(\mathbf{m})$, where $\mathbf{m} \in \{0,1\}^n \subseteq \mathbb{Z}_q^n$, outputs the ciphertext $g^{\mathbf{c}} = g^{\mathbf{A}\cdot\mathbf{m}}$.
- **Decryption.** The decryption algorithm $\mathsf{Dec}_{sk}(g^{\mathbf{c}})$, where $g^{\mathbf{c}} \in \mathbb{G}^n$, first computes $g^{\mathbf{m}} = g^{\mathbf{A}^{-1}\cdot\mathbf{c}}$. Then, note that if $\mathbf{m} \in \{0,1\}^n$ then it can be efficiently extracted from $g^{\mathbf{m}}$. In such case it outputs $\mathbf{m}$, and otherwise it outputs $\perp$.

Correctness follows immediately as in [15]. We prove the following theorem:

Theorem 4.1. *Let $d \in \mathbb{N}$ be some integer. Then under the d-linear assumption, for any constant $0 < \mu < 1$ and for any sufficiently large message length $n = n(k)$, the scheme Π_{Lin} is PRIV-IND-secure with respect to 2^{-n^μ}-blockwise-hard-to-invert auxiliary inputs.*

Due to space limitations, we only describe the main ideas underlying the security of the scheme. The full proof can be found in the full version [9]. For simplicity, we focus here on the case $d = 1$ (i.e., we rely on the DDH assumption). Given a distribution $\mathcal{M}$ over messages $\mathbf{m} \in \{0,1\}^n$, and an auxiliary-input function f that is sub-exponentially hard to invert with respect to $\mathcal{M}$, we argue that an encryption of a messages $\mathbf{m}$ sampled from the distribution $\mathcal{M}$ is computationally indistinguishable from being completely independent of the public key pk and the auxiliary input $f(\mathbf{m})$. More specifically, we prove that $(pk, \mathsf{Enc}_{pk}(\mathbf{m}), f(\mathbf{m})) \stackrel{c}{\approx} (pk, g^{\mathbf{u}}, f(\mathbf{m}))$, for a uniformly chosen vector $\mathbf{u}$. Transforming this into either one of our notions of security from Section 3 is rather standard.

Consider the joint distribution $(pk, \mathsf{Enc}_{pk}(\mathbf{m}), f(\mathbf{m})) = (g^{\mathbf{A}}, g^{\mathbf{A}\cdot\mathbf{m}}, f(\mathbf{m}))$ of the public key, the ciphertext, and the auxiliary input. The DDH assumption implies that replacing the uniformly chosen matrix $\mathbf{A}$ with a random matrix of rank 1 results in a computationally indistinguishable distribution. Such a low-rank matrix can be written as $\mathbf{A} = \mathbf{r} \cdot \mathbf{b}^T$, for random vectors $\mathbf{r}$ and $\mathbf{b}$, and therefore $\mathbf{A} \cdot \mathbf{m} = \mathbf{r} \cdot \mathbf{b}^T \cdot \mathbf{m}$. However $\mathbf{b}^T \cdot \mathbf{m} = \langle \mathbf{b}, \mathbf{m} \rangle$ is indistinguishable from the uniform distribution, even given $\mathbf{b}$ and $f(\mathbf{m})$, according to the generalized Goldreich-Levin theorem of [12]. Our initial distribution is thus indistinguishable from the distribution $(g^{\mathbf{r}\cdot\mathbf{b}^T}, g^{\mathbf{r}\cdot\alpha}, f(\mathbf{m}))$.

Now, notice that the matrix $[\mathbf{r} \cdot \mathbf{b}^T \| \mathbf{r} \cdot \alpha] \in \mathbb{Z}_q^{n\times(n+1)}$ is essentially a random matrix of rank 1. Relying on the DDH assumption once again, it can be replaced with a completely random matrix while preserving computational indistinguishability. This yields the distribution $(g^{\mathbf{A}}, g^{\mathbf{u}}, f(\mathbf{m}))$, where $\mathbf{A}$ and $\mathbf{u}$ are chosen uniformly at random.

Homomorphic properties. The scheme naturally exhibits homomorphic properties w.r.t. multiplication by a scalar or addition of two ciphertexts over $\mathbb{Z}_q^n$. This follows from "arithmetics in the exponent". We stress, however, that the output of such homomorphic operations will be decryptable if it lies in the message space of our scheme, $\{0,1\}^n$, which is a proper subset of the domain $\mathbb{Z}_q^n$ on which these operations are performed. More generally, decryption is possible as long as each entry of the encrypted plaintext vector belongs to a predetermined set of logarithmic size.

In addition, if the underlying group $\mathbb{G}$ is associated with a *bilinear map*, then our scheme enjoys an additional homomorphism w.r.t. *one* matrix multiplication. This is similar to the homomorphism style achieved in [6] and in [16]. We stress that in such case we base the security of the scheme on the d-linear assumption for $d \geq 2$ (as the 1-linear, i.e. DDH, cannot hold in such a group). Formally, let $\mathbb{G}$, q, and g be as in the parameters of our scheme, and let $\mathbb{G}_T$ be a (different) group of order q. A bilinear map $e : \mathbb{G} \times \mathbb{G} \to \mathbb{G}_T$ has the following properties. *Bilinearity:* for all $x, y \in \mathbb{G}$, $a, b \in \mathbb{Z}$ it holds that $e(x^a, y^b) = e(x,y)^{ab}$; *Non-degeneracy:* $e(g,g) \neq 1$. It follows that $g_T \stackrel{\mathsf{def}}{=} e(g,g)$ generates $\mathbb{G}_T$.

Homomorphic matrix multiplication, thus, is performed in our scheme as follows: Given two ciphertexts $g^{\mathbf{A}\mathbf{m}_1}$ and $g^{\mathbf{A}\mathbf{m}_2}$, one can compute $e(g,g)^{\mathbf{A}\mathbf{m}_1\mathbf{m}_2^T\mathbf{A}^T}$.

This ciphertext can be decrypted by multiplying by $\mathbf{A}^{-1}$ from the left (in the exponent) and $\mathbf{A}^{-T}$ from the right (again, in the exponent) to obtain $e(g,g)^{\mathbf{m}_1 \cdot \mathbf{m}_2^T}$. Since $\mathbf{m}_1$ and $\mathbf{m}_2$ are binary, $\mathbf{m}_1 \cdot \mathbf{m}_2^T$ is binary as well and can be extracted from the exponent.

The multi-user setting. We now show that Π_{Lin} is secure (with respect to auxiliary inputs) even in the multi-user setting, where related messages may be encrypted under multiple public keys. We allow any polynomial number of users, and for simplicity we assume that each public key encrypts one message. As in the single-user setting, this natural extends to the case where several messages are encrypted under each public key with blockwise-hard-to-invert auxiliary input. In addition, we require that the messages to be encrypted come from an *affine distribution*, a term we define below. Intuitively, this means that there are publicly known invertible linear relations (over $\mathbb{Z}_q^n$) between the messages.

Definition 4.2 (Affine message distributions). *Let $n = n(k)$ and $\ell = \ell(k)$ be integer functions of the security parameter, and let $\mathcal{M} = \{\mathcal{M}\}_k \subseteq (\{0,1\}^n)^\ell$ be a distribution ensemble.*[7] *Then $\mathcal{M}$ is* affine *if there exist invertible and efficiently computable (given k) matrices $\mathbf{V}_2, \ldots, \mathbf{V}_\ell \subseteq \mathbb{Z}_q^{n\times n}$ and vectors $\mathbf{w}_2, \ldots, \mathbf{w}_\ell \in \mathbb{Z}_q^n$, such that for all $(\mathbf{m}_1, \ldots, \mathbf{m}_\ell)$ in the support of $\mathcal{M}$ and for all $i \in \{2, \ldots, \ell\}$ it holds that $\mathbf{m}_i = \mathbf{V}_i \cdot \mathbf{m}_1 + \mathbf{w}_i$ (where arithmetics is over $\mathbb{Z}_q$).*

Note that we require that messages are taken over the space $\{0,1\}^n$, and arithmetics is over $\mathbb{Z}_q$. In particular, this captures the case of "broadcast encryption" where encrypting the same message under many public keys. Furthermore, this also captures XORing with a constant vector over the binary field, or permuting the coordinates of a binary vector (a tool used, e.g., in [7]). The result is formally stated below. For proof, see full version [9].

Theorem 4.3. *Let $d \in \mathbb{N}$ be some integer. Then under the d-linear assumption, for any constant $0 < \mu < 1$ and for any sufficiently large message length $n = n(k)$, the scheme Π_{Lin} is PRIV1-IND-MU-secure with respect to 2^{-n^μ}-hard-to-invert auxiliary inputs.*

5 A Scheme Based on Subgroup Indistinguishability Assumptions

In this section we present our second deterministic encryption scheme, which is based on a rather general class of subgroup indistinguishability. For concreteness we first describe the scheme based on the quadratic residuosity assumption, and then describe the more general case. We show that (a slight generalization of) the QR-based lossy trapdoor function of Hemenway and Ostrovsky [21] is in fact a deterministic public-key encryption scheme that is secure against sub-exponentially hard-to-invert auxiliary inputs.

[7] To be absolutely precise should write that $\mathcal{M}_k$ is a distributions over $((\{0,1\}^n)^t)^\ell$ for $t = 1$, but this space is trivially isomorphic to the one we consider.

The scheme Π_{QR}. Let GroupGen be a probabilistic polynomial-time algorithm that takes as input a security parameter 1^k, and outputs an integer $N = PQ$, where P and Q are k-bit prime numbers, and $P \pmod 4 = Q \pmod 4 = 3$ (i.e., N is a Blum integer). In addition, recall that $y \leftarrow g^x$ denotes an application of an isomorphism transforming an element x in the module $\mathbb{M}_{\mathbb{QR}_N}$ into an element y in the group $\mathbb{QR}_N$ (since we will never express elements in the module explicitly, we do not care which isomorphism is used). We let $\hat{g}$ denote the isomorphism between the group $\mathbb{J}_N$ and the corresponding module, such that the generating set that corresponds to $\hat{g}$ is the same as that of g, appended with (-1). The scheme is parameterized by the security parameter k and the message length $n = n(k)$.

- **Key generation.** The key-generation algorithm $\mathsf{KeyGen}(1^k)$ samples $N \leftarrow \mathsf{GroupGen}(1^k)$, a vector $g^{\mathbf{w}^T} \leftarrow \mathbb{QR}_N^n$, and a vector $\mathbf{r} \leftarrow \left([N^2]\right)^n$. It then outputs $pk = (N, g^{\mathbf{w}^T}, (-1)^{\mathbf{I}_n} \cdot g^{\mathbf{r}\cdot\mathbf{w}^T})$ and $sk = \mathbf{r}$.
  The matrix dot product above refers to element-wise multiplication:
  $$\left((-1)^{\mathbf{I}_n} \cdot g^{\mathbf{r}\cdot\mathbf{w}^T}\right)_{i,j} = \left((-1)^{\mathbf{I}_n}\right)_{i,j} \cdot \left(g^{\mathbf{r}\cdot\mathbf{w}^T}\right)_{i,j} .$$
  To be completely explicit, we emphasize that $pk \in \mathbb{N} \times \mathbb{J}_N^{1\times N} \times \mathbb{J}_N^{n\times n}$ and $sk \in \mathbb{N}^n$.
- **Encryption.** The encryption algorithm $\mathsf{Enc}_{pk}(\mathbf{m})$, where $pk = (N, \hat{g}^{\mathbf{w}^T}, \hat{g}^{\mathbf{T}})$ and $\mathbf{m} \in \{0,1\}^n$, outputs the ciphertext $c = (\hat{g}^{\mathbf{w}^T\cdot\mathbf{m}}, \hat{g}^{\mathbf{T}\cdot\mathbf{m}})$. We note that this computation can be performed efficiently and that $c \in \mathbb{J}_N \times \mathbb{J}_N^n$.
  For a legally generated public key $pk = (N, g^{\mathbf{w}^T}, (-1)^{\mathbf{I}_n} \cdot g^{\mathbf{r}\cdot\mathbf{w}^T})$ and $sk = \mathbf{r}$, we get $c = (g^{\mathbf{w}^T\cdot\mathbf{m}}, (-1)^{\mathbf{m}} \cdot g^{\mathbf{r}\cdot\mathbf{w}^T\cdot\mathbf{m}})$.
- **Decryption.** The decryption algorithm $\mathsf{Dec}_{sk}(c)$, where $c = (\hat{g}^v, \hat{g}^{\mathbf{y}})$, first computes $\hat{g}^{(\mathbf{y}-\mathbf{r}\cdot v)}$. If the output is of the form $(-1)^{\mathbf{m}}$, for $\mathbf{m} \in \{0,1\}^n$, then it outputs $\mathbf{m}$ and otherwise it outputs $\perp$.

Correctness follows immediately by definition. Security is stated below. The proof appears in the full version [9].

Theorem 5.1. *Under the quadratic residuosity assumption, for any constant $0 < \mu < 1$ and for any sufficiently large message length $n = n(k)$, the scheme Π_{QR} is PRIV-IND-secure with respect to 2^{-n^μ}-blockwise-hard-to-invert auxiliary inputs.*

Extension to general subgroup indistinguishability. As mentioned above, this construction can be extended to general subgroup indistinguishability assumptions [8] (these include, in particular, Paillier's composite residuosity assumption [24]). These assumptions are defined in a setting where $\mathbb{G}_U = \mathbb{G}_M \times \mathbb{G}_L$ is a product group such that $\mathbb{G}_U$ and $\mathbb{G}_L$ are computationally indistinguishable (there are of course additional requirements, we refer the reader to [8] for details). Specifically, quadratic residuosity fits into this setting by letting $\mathbb{G}_U = \mathbb{J}_N$, $\mathbb{G}_L = \mathbb{QR}_N$, $\mathbb{G}_M = \{\pm 1\}$. To generalize our construction, we let $\mathbb{M}$ be the module that corresponds to $\mathbb{G}_L$ and replace (-1) with a generator h of $\mathbb{G}_M$. Namely, our keys become $pk = (g^{\mathbf{w}^T}, h^{\mathbf{I}_n} \cdot g^{\mathbf{r} \cdot \mathbf{w}^T})$ and $sk = \mathbf{r}$; encryption of a message $\mathbf{m} \in \{0,1\}^n$ is done by computing $c = (g^{\mathbf{w}^T \cdot \mathbf{m}}, h^{\mathbf{m}} \cdot g^{\mathbf{r} \cdot \mathbf{w}^T \cdot \mathbf{m}})$; and decryption of $c = (\hat{g}^v, \hat{g}^{\mathbf{y}})$ is done by computing $\hat{g}^{(\mathbf{y} - \mathbf{r} \cdot v)} = h^{\mathbf{m}}$ and extracting $\mathbf{m}$. The proof of security in this case is similar to that of the QR-based scheme.

References

1. Akavia, A., Goldwasser, S., Vaikuntanathan, V.: Simultaneous hardcore bits and cryptography against memory attacks. In: Reingold, O. (ed.) TCC 2009. LNCS, vol. 5444, pp. 474–495. Springer, Heidelberg (2009)
2. Bellare, M., Boldyreva, A., O'Neill, A.: Deterministic and efficiently searchable encryption. In: Menezes, A. (ed.) CRYPTO 2007. LNCS, vol. 4622, pp. 535–552. Springer, Heidelberg (2007)
3. Bellare, M., Brakerski, Z., Naor, M., Ristenpart, T., Segev, G., Shacham, H., Yilek, S.: Hedged public-key encryption: How to protect against bad randomness. In: Matsui, M. (ed.) ASIACRYPT 2009. LNCS, vol. 5912, pp. 232–249. Springer, Heidelberg (2009)
4. Bellare, M., Fischlin, M., O'Neill, A., Ristenpart, T.: Deterministic encryption: Definitional equivalences and constructions without random oracles. In: Wagner, D. (ed.) CRYPTO 2008. LNCS, vol. 5157, pp. 360–378. Springer, Heidelberg (2008)
5. Boldyreva, A., Fehr, S., O'Neill, A.: On notions of security for deterministic encryption, and efficient constructions without random oracles. In: Wagner, D. (ed.) CRYPTO 2008. LNCS, vol. 5157, pp. 335–359. Springer, Heidelberg (2008)
6. Boneh, D., Goh, E.-J., Nissim, K.: Evaluating 2-DNF formulas on ciphertexts. In: Kilian, J. (ed.) TCC 2005. LNCS, vol. 3378, pp. 325–341. Springer, Heidelberg (2005)
7. Boneh, D., Halevi, S., Hamburg, M., Ostrovsky, R.: Circular-secure encryption from decision Diffie-Hellman. In: Wagner, D. (ed.) CRYPTO 2008. LNCS, vol. 5157, pp. 108–125. Springer, Heidelberg (2008)
8. Brakerski, Z., Goldwasser, S.: Circular and leakage resilient public-key encryption under subgroup indistinguishability - (or: Quadratic residuosity strikes back). In: Rabin, T. (ed.) CRYPTO 2010. LNCS, vol. 6223, pp. 1–20. Springer, Heidelberg (2010)
9. Brakerski, Z., Segev, G.: Better security for deterministic public-key encryption: The auxiliary-input setting. Cryptology ePrint Archive, Report 2011/209 (2011)
10. Canetti, R.: Towards realizing random oracles: Hash functions that hide all partial information. In: Kaliski Jr., B.S. (ed.) CRYPTO 1997. LNCS, vol. 1294, pp. 455–469. Springer, Heidelberg (1997)
11. Canetti, R.: Universally composable security: A new paradigm for cryptographic protocols. In: Proceedings of the 42nd Annual IEEE Symposium on Foundations of Computer Science, pp. 136–145 (2001)
12. Dodis, Y., Goldwasser, S., Tauman Kalai, Y., Peikert, C., Vaikuntanathan, V.: Public-key encryption schemes with auxiliary inputs. In: Micciancio, D. (ed.) TCC 2010. LNCS, vol. 5978, pp. 361–381. Springer, Heidelberg (2010)
13. Dodis, Y., Kalai, Y.T., Lovett, S.: On cryptography with auxiliary input. In: Proceedings of the 41st Annual ACM Symposium on Theory of Computing, pp. 621–630 (2009)
14. Dodis, Y., Smith, A.: Entropic security and the encryption of high entropy messages. In: Kilian, J. (ed.) TCC 2005. LNCS, vol. 3378, pp. 556–577. Springer, Heidelberg (2005)

15. Freeman, D.M., Goldreich, O., Kiltz, E., Rosen, A., Segev, G.: More constructions of lossy and correlation-secure trapdoor functions. In: Nguyen, P.Q., Pointcheval, D. (eds.) PKC 2010. LNCS, vol. 6056, pp. 279–295. Springer, Heidelberg (2010)
16. Gentry, C., Halevi, S., Vaikuntanathan, V.: A simple BGN-type cryptosystem from LWE. In: Gilbert, H. (ed.) EUROCRYPT 2010. LNCS, vol. 6110, pp. 506–522. Springer, Heidelberg (2010)
17. Goldreich, O.: Foundations of Cryptography – Volume 2: Basic Applications. Cambridge University Press, Cambridge (2004)
18. Goldreich, O., Levin, L.A.: A hard-core predicate for all one-way functions. In: Proceedings of the 21st Annual ACM Symposium on Theory of Computing, pp. 25–32 (1989)
19. Goldwasser, S., Kalai, Y.T.: On the impossibility of obfuscation with auxiliary input. In: Proceedings of the 46th Annual IEEE Symposium on Foundations of Computer Science, pp. 553–562 (2005)
20. Goldwasser, S., Micali, S.: Probabilistic encryption. Journal of Computer and System Sciences 28(2), 270–299 (1984)
21. Hemenway, B., Ostrovsky, R.: Lossy trapdoor functions from smooth homomorphic hash proof systems. Electronic Colloquium on Computational Complexity, Report TR09-127 (2009)
22. Naor, M., Segev, G.: Public-key cryptosystems resilient to key leakage. In: Halevi, S. (ed.) CRYPTO 2009. LNCS, vol. 5677, pp. 18–35. Springer, Heidelberg (2009)
23. O'Neill, A.: Deterministic public-key encryption revisited. Cryptology ePrint Archive, Report 2010/533 (2010)
24. Paillier, P.: Public-key cryptosystems based on composite degree residuosity classes. In: Stern, J. (ed.) EUROCRYPT 1999. LNCS, vol. 1592, pp. 223–238. Springer, Heidelberg (1999)
25. Russell, A., Wang, H.: How to fool an unbounded adversary with a short key. IEEE Transactions on Information Theory 52(3), 1130–1140 (2006)
26. Zhu, B., Li, K., Patterson, R.H.: Avoiding the disk bottleneck in the data domain deduplication file system. In: Proceedings of the 6th USENIX Conference on File and Storage Technologies, pp. 269–282 (2008)

The Collision Security of Tandem-DM in the Ideal Cipher Model

Jooyoung Lee[1], Martijn Stam[2,⋆], and John Steinberger[3,⋆⋆]

[1] Faculty of Mathematics and Statistics, Sejong University, Seoul, Korea
jlee05@sejong.ac.kr

[2] Department of Computer Science, University of Bristol, Bristol, United Kingdom
stam@cs.bris.ac.uk

[3] Institute of Theoretical Computer Science, Tsinghua University, Beijing, China
jpsteinb@gmail.com

Abstract. We prove that Tandem-DM, which is one of the two "classical" schemes for turning a blockcipher of $2n$-bit key into a double block length hash function, has birthday-type collision resistance in the ideal cipher model. A collision resistance analysis for Tandem-DM achieving a similar birthday-type bound was already proposed by Fleischmann, Gorski and Lucks at FSE 2009 [3]. As we detail, however, the latter analysis is wrong, thus leaving the collision resistance of Tandem-DM as an open problem until now. Our analysis exhibits a novel feature in that we introduce a trick not used before in ideal cipher proofs.

1 Introduction

The Tandem-DM compression function is a $3n$-bit to $2n$-bit compression function based on two applications of a blockcipher of $2n$-bit key and n-bit word length (Fig. 1). While Tandem-DM was proposed by Lai and Massey in 1992 [8] the first proof of collision security for Tandem-DM (in the ideal cipher model, as is usual for all such proofs) was only proposed in 2009 by Fleischmann, Gorski and Lucks [3]. Unfortunately, as we detail in Section 3, the "FGL proof" (as we shall refer to it) has a number of serious flaws which make it false and nonobvious to repair. The purpose of this paper is to offer a correct collision resistance analysis of Tandem-DM. We show that, as claimed in [3], Tandem-DM does indeed have birthday-type collision security (necessitating at least $2^{120.8}$ queries to break when the output length is $2n = 256$ bits). A nice feature of our work is that the analysis is relatively simple compared to typical results in this area. This simplicity is afforded by a new trick we introduce, apparently not used before in ideal cipher analyses.

⋆ Part of the work performed while at LACAL, École Polytechnique Fédérale de Lausanne, Switzerland.

⋆⋆ Supported in part by the National Basic Research Program of China Grant 2007CB807900, 2007CB807901, the National Natural Science Foundation of China Grant 61033001, 61061130540, 61073174 and by NSF grant CNS 0904380.

P. Rogaway (Ed.): CRYPTO 2011, LNCS 6841, pp. 561–577, 2011.

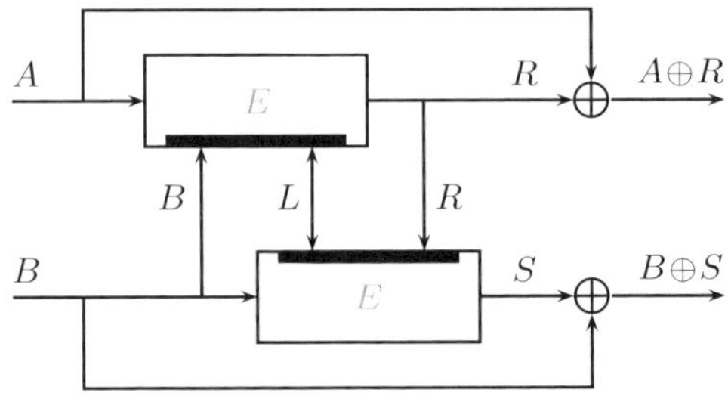

Fig. 1. The Tandem-DM compression function. All wires carry n-bit values. The top and bottom blockciphers are the same. Each has a $2n$-bit key and n-bit input/output. The wire marked L is an input to the compression function (along with A and B).

Related work on double block length constructions. Another classical scheme for turning a $2n$-bit key blockcipher into a $3n$-bit to $2n$-bit compression function is Abreast-DM, pictured in Fig. 2, which was proposed by Lai and Massey in the same paper as Tandem-DM [8]. The collision resistance of Abreast-DM was independently resolved by Fleischmann, Gorski and Lucks [4] and Lee and Kwon [9], who both showed birthday-type collision resistance for Abreast-DM. Before that, Hirose [5] had given a collision resistance analysis for a general class of compression functions that included Abreast-DM as a special case, but under the assumption that the top and bottom blockciphers of the diagram be distinct (this considerably simplifies the analysis). The work by Hirose was further generalized by Özen and Stam [14], who additionally discuss schemes that are only secure in the iteration.

Another $3n$-bit to $2n$-bit compression function making two calls to a blockcipher of $2n$-bit key was proposed by Hirose [6], who proved birthday-type collision resistance for his construction in the ideal cipher model. Hirose's construction (Fig. 3) is simpler than either Abreast-DM or Tandem-DM and in particular uses a single keying schedule for the top and bottom blockciphers. It is noteworthy that while Hirose introduced his construction over 10 years after Abreast-DM and Tandem-DM his collision resistance analysis pre-dates similar collision resistance analyses for Abreast-DM and Tandem-DM.

It is also possible to achieve birthday-type collision resistance for a $3n$-bit to $2n$-bit compression functions making only a single call to a $2n$-bit key blockcipher [22, 21, 13, 19, 20, 14, 10]; however these constructions have considerable overhead (typically comparable to a blockcipher call itself).

Comparison. Of the three well-known $3n$-bit to $2n$-bit compression functions making two calls to a $2n$-bit key blockcipher—those being Tandem-DM, Abreast-DM and Hirose's construction—the two constructions whose collision resistance has been successfully resolved (Hirose and Abreast-DM) share the feature that the inputs to the top and bottom blockcipher are bijectively related. For example, for Abreast-DM, if the top blockcipher call is $E_{B\|L}(A)$ then the bottom blockcipher call (for the same input $A\|B$) is $E_{L\|A}(\overline{B})$, where $\overline{B}$ denotes bit complementation of B; thus the inputs to the top and bottom blockciphers are

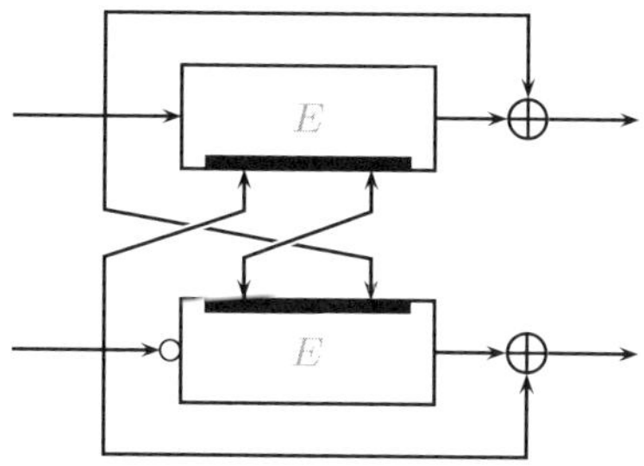

Fig. 2. The Abreast-DM compression function. The empty circle at bottom left denotes bit complementation.

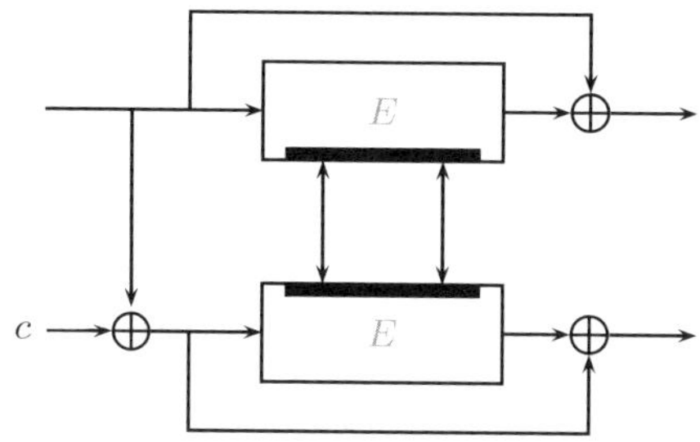

Fig. 3. Hirose's compression function. The bottom left-hand wire is not an input; it carries an arbitrary nonzero constant c.

related by the permutation $\pi : \{0,1\}^{3n} \to \{0,1\}^{3n}$, $\pi(X\|Y\|Z) = \overline{Y}\|Z\|X$. (Here the last $2n$ bits are the key.) In Hirose's construction, the inputs to the top and bottom blockciphers are related by the permutation $\pi' : \{0,1\}^{3n} \to \{0,1\}^{3n}$, $\pi'(X\|Y\|Z) = X \oplus c\|Y\|Z$.

By contrast, Tandem-DM exhibits a more subtle relationship between the inputs of the top and bottom blockciphers, as an *output* of the top blockcipher is used to key the bottom blockcipher. It is the presence of this "feedback" within the construction, it seems, that has complicated efforts to prove a collision resistance bound. On the other hand, Tandem-DM still has the agreeable feature that the top and bottom blockcipher calls uniquely determine each other in the following sense: given the key $B\|L$ and output R of the top cipher one can determine the key $L\|R$ and the input B of the bottom cipher, and vice-versa. This contrasts with constructions such as MDC-2 which use two calls to a blockcipher of n-bit key, and in which the top and bottom blockcipher calls do not uniquely determine each other. Typically, collision resistance analyses are much harder for the latter kind of compression functions. (MDC-2 can only be proved nontrivially collision resistant in the iteration, and the current best bound of $O(2^{\frac{3}{5}n})$ queries due to Steinberger [18] is undoubtedly suboptimal.)

We note that the permutations π and π' discussed above share the common feature of having *small cycle lengths*—all cycles of π have length 6 and all cycles of π' have length 2—which constitutes another strong similarity between Abreast-DM and Hirose's scheme. In fact, due to this reason, Hirose's collision resistance proof and the Abreast-DM collision resistance proof can be seen as special cases of the same framework, as noted in [4,9]. Building on this observation, Fleischmann et al. [4] defined a general class of compression functions called 'Cyclic-DM' that are amenable to collision resistance analyses and that include Hirose's scheme and Abreast-DM as special cases. Similarly, one can define collision-resistant generalizations of Tandem-DM by isolating those properties of Tandem-DM that are used in our proof. While defining the most all-encompassing possible collision resistant generalization of Tandem-DM is not the goal of our work, we do briefly discuss these key properties and the corresponding collision-resistant generalizations of Tandem-DM in the paper's full version [11].

We mention that Fleischmann et al. [2] also provided a comprehensive generalization of their earlier works [3,4]. In particular, a new and tighter collision resistance claim for Tandem-DM is made. Unfortunately, this second analysis has many similar flaws to the first, which are fatal to the integrity of the argument and to the final bound (in particular, the crucial "Argument B" of [2] is incorrect).

Full version contents. The proof of collision resistance that we provide in this paper is very slick, but slightly mysterious in its efficacy because it relies on a subtle trick that cuts out a large portion of the case analysis that "would have been there" in a more standard proof. As a kind of pedagogical bonus, the full version of this paper [11] contains a second collision resistance proof which does not utilize this trick. This proof is more straightforward but much longer, and the bound obtained is slightly worse (but still birthday).

Fleischmann et al. [3] also provide a preimage resistance proof for Tandem-DM which, unfortunately, suffers from similar flaws as their collision resistance proof. The full version of this paper [11] contains a description of these flaws, as well as a corrected preimage analysis. This preimage analysis shows $\Omega(2^n)$ queries are necessary to invert Tandem-DM on a random range point. We note, however, that dramatic progress has recently been made in this area, and it is now known that $\Omega(2^{2n})$ queries are necessary to invert Tandem-DM as well as Abreast-DM and Hirose's scheme [7,12].

Further possible improvements. We note that our collision resistance has the form $\tilde{O}(q/(2^n-q))$ rather than $\tilde{O}(q^2/(2^n-q)^2)$. Both bounds reach constant values when $q=\Omega(2^n)$, however $q^2/(2^n-q)^2$ grows slower than $q/(2^n-q)$ since our bound is (only) "linear birthday" rather than true "quadratic birthday". We leave it as an open problem to prove "quadratic birthday"-type collision resistance for Tandem-DM (as exists for Abreast-DM and Hirose's scheme).

2 Definitions

A blockcipher is a function $E:\{0,1\}^m \times \{0,1\}^n \rightarrow \{0,1\}^n$ such that $E(K,\cdot)$ is a permutation of $\{0,1\}^n$ for each $K \in \{0,1\}^m$. We call m the *key size* and

n the *word size* of the blockcipher. It is customary to write $E_K(X)$ instead of $E(K, X)$ for $K \in \{0,1\}^m$, $X \in \{0,1\}^n$. The function $E_K^{-1}(\cdot)$ denotes the inverse of $E_K(\cdot)$ (as $E_K(\cdot)$ is a permutation).

Given a blockcipher $E : \{0,1\}^{2n} \times \{0,1\}^n \to \{0,1\}^n$ we define the Tandem-DM compression function $TDM^E : \{0,1\}^{3n} \to \{0,1\}^{2n}$ by

$$TDM^E(A\|B\|L) = (A \oplus R)\|(B \oplus S)$$

where

$$R = E_{B\|L}(A) \qquad \text{and} \qquad S = E_{L\|R}(B).$$

In the collision resistance experiment, a computationally unbounded adversary $\mathcal{A}$ is given oracle access to a blockcipher E uniformly sampled among all blockciphers of key length $2n$ and word length n. We allow $\mathcal{A}$ to query both E and E^{-1}. After q queries to E, the *query history* of $\mathcal{A}$ is the set of triples $\mathcal{Q} = \{(X_i, K_i, Y_i)\}_{i=1}^q$ such that $E_{K_i}(X_i) = Y_i$ and $\mathcal{A}$'s i-th query is either $E_{K_i}(X_i)$ or $E_{K_i}^{-1}(Y_i)$ for $1 \le i \le q$. We let $\mathcal{Q}_i = \{(X_j, K_j, Y_j)\}_{j=1}^i$ be the first i elements of the query history; thus $\mathcal{Q} = \mathcal{Q}_q$. We say $\mathcal{A}$ *succeeds* or *finds a collision* after its first i queries if there exist distinct $3n$-bit values, $A\|B\|L$, $A'\|B'\|L'$ such that $TDM^E(A\|B\|L) = TDM^E(A'\|B'\|L')$ and such that $\mathcal{Q}_i$ contains both the queries necessary to compute $TDM^E(A\|B\|L)$ and $TDM^E(A'\|B'\|L')$. More formally—and see Fig. 4—we define this event by a predicate $\mathsf{Coll}(\mathcal{Q}_i)$, which is true if and only if there exist n-bit values A, B, L, R, S, A', B', L', R', S' such that

$$A\|B\|L \neq A'\|B'\|L', \qquad A \oplus R = A' \oplus R', \qquad B \oplus S = B' \oplus S' \tag{1}$$

and such that

$$(A, B\|L, R), (B, L\|R, S), (A', B'\|L', R'), (B', L'\|R', S') \in \mathcal{Q}_i. \tag{2}$$

We denote by

$$\mathbf{Adv}^{\mathrm{coll}}_{TDM}(q)$$

the maximum chance of an adversary making q queries causing $\mathsf{Coll}(\mathcal{Q})$ to become true. The probability occurs over the uniform choice of E and over $\mathcal{A}$'s coin tosses, if any. Also note that n is a hidden parameter.

The "XOR-output" of a query (X_i, K_i, Y_i) is the quantity $X_i \oplus Y_i$. Another predicate which plays an important part in both our proof and the FGL proof is the "many queries with the same XOR-output" predicate $\mathsf{Xor}(\mathcal{Q})$, defined on a query history $\mathcal{Q} = \{(X_i, K_i, Y_i)\}_{i=1}^q$ by

$$\mathsf{Xor}(\mathcal{Q}) \iff \max_{Z \in \{0,1\}^n} |\{i : X_i \oplus Y_i = Z\}| > \alpha.$$

Here α is a free parameter of the analysis which appears in the final collision resistance bound. (In [3] this predicate is named $\textsc{Lucky}(\mathcal{Q})$; in [18] a similar predicate is named $\mathsf{Win0}(\mathcal{Q})$.) Without going into details at this point, we mention that the FGL collision resistance proof—and ours, essentially, as well—upper bounds $\Pr[\mathsf{Coll}(\mathcal{Q})]$ by $\Pr[\mathsf{Xor}(\mathcal{Q})] + \Pr[\mathsf{Coll}(\mathcal{Q}) \wedge \neg\mathsf{Xor}(\mathcal{Q})]$. A larger α implies a lower value for $\Pr[\mathsf{Xor}(\mathcal{Q})]$ and a higher value for $\Pr[\mathsf{Coll}(\mathcal{Q}) \wedge \neg\mathsf{Xor}(\mathcal{Q})]$. The best value of α can be found numerically for a given value of n and q. Generally, readers may think of α as some small constant value (e.g. for $n = 128$ and $q = 2^{120.87}$, $\alpha = 16$).

So far, we have described "infrastructure" that is common to both proofs. We shall now introduce some material proper to our proof. Note a query history $\mathcal{Q} = \{(X_i, K_i, Y_i)\}_{i=1}^{q}$ does not record whether each triple (X_i, K_i, Y_i) was obtained by the adversary through a forward query $E_{K_i}(X_i)$ or a backward query $E_{K_i}^{-1}(Y_i)$. For this, we maintain two arrays $\mathsf{Fwd}[\cdot]$ and $\mathsf{Bwd}[\cdot]$ where $\mathsf{Fwd}[i] = 1$ if and only if the adversary's i-th query is a forward query and $\mathsf{Bwd}[i] = 1$ if and only if the adversary's i-th query is a backward query. We then define an additional predicate

$$\mathsf{FB}(\mathcal{Q}) \iff \max_{Z\in\{0,1\}^n} |\{i : (Y_i = Z \wedge \mathsf{Fwd}[i] = 1) \vee (X_i = Z \wedge \mathsf{Bwd}[i] = 1)\}| > \alpha. \tag{3}$$

('FB' stands for "Forward Backward".) Here α is the same free parameter as above. Note that $\neg\mathsf{FB}(\mathcal{Q})$ implies that

$$\max\nolimits_{Z\in\{0,1\}^n} |\{i : Y_i = Z \wedge \mathsf{Fwd}[i] = 1\}| \le \alpha, \tag{4}$$

$$\max\nolimits_{Z\in\{0,1\}^n} |\{i : X_i = Z \wedge \mathsf{Bwd}[i] = 1\}| \le \alpha. \tag{5}$$

It is really consequences (4) and (5) of $\neg\mathsf{FB}(\mathcal{Q})$ that interest us, though we define $\mathsf{FB}(\mathcal{Q})$ via (3) because this makes it slightly easier to bound $\Pr[\mathsf{FB}(\mathcal{Q})]$. We will use the bound

$$\begin{aligned}\Pr[\mathsf{Coll}(\mathcal{Q})] &\le \Pr[\mathsf{Xor}(\mathcal{Q})] + \Pr[\mathsf{Coll}(\mathcal{Q}) \wedge \neg\mathsf{Xor}(\mathcal{Q})] \\ &\le \Pr[\mathsf{Xor}(\mathcal{Q})] + \Pr[\mathsf{FB}(\mathcal{Q})] + \Pr[\mathsf{Coll}(\mathcal{Q}) \wedge \neg\mathsf{Xor}(\mathcal{Q}) \wedge \neg\mathsf{FB}(\mathcal{Q})]. \end{aligned} \tag{6}$$

One should thus think of $\mathsf{FB}(\mathcal{Q})$ and $\mathsf{Xor}(\mathcal{Q})$ as bad events whose nonoccurrence helps bound the probability of $\mathsf{Coll}(\mathcal{Q})$ occurring. We warn that (6) constitutes a slightly oversimplified encapsulation of our proof's high-level structure. We refer to Section 4 for more details.

3 The FGL Collision Resistance Proof

Since the interest of our paper would be substantially diminished (though not nullified, since our proof is much shorter) if the FGL collision resistance proof were correct, we detail here some of our objections to [3]. This material also serves as a good introduction to our own proof, and will give the reader more intuition about Tandem-DM.

Starting with a q-query collision-finding adversary $\mathcal{A}$, FGL first make the standard assumption that $\mathcal{A}$ never makes a query to which it already knows the answer (this could occur two ways: $\mathcal{A}$ could make the exact same query twice, or $\mathcal{A}$ could query (say) $E_K^{-1}(Y)$ after having received Y as an answer beforehand to a query $E_K(X)$). This ensures each answer $\mathcal{A}$ receives comes uniformly at

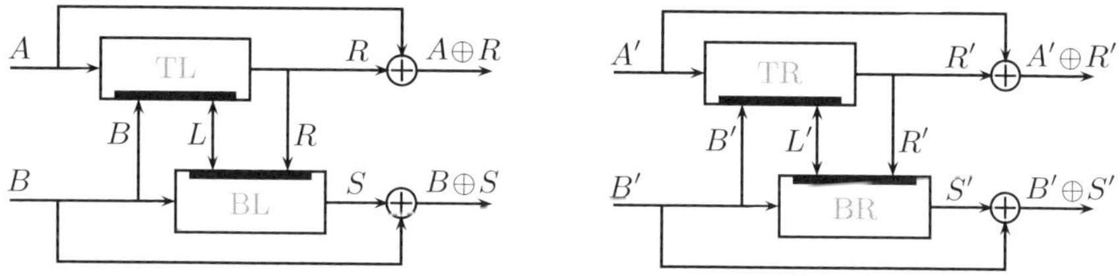

Fig. 4. The collision diagram for Tandem-DM. The adversary must find blockcipher queries to fit both sides of the diagram such that $A \oplus R = A' \oplus R'$, $B \oplus S = B' \oplus S'$ and $A\|B\|L \neq A'\|B'\|L'$. More precisely, the adversary must find four queries of the form $E_{B\|L}(A) = R$, $E_{L\|R}(B) = S$, $E_{B'\|L'}(A') = R'$, $E_{L'\|R'}(B') = S'$ such that the above conditions hold. Each query could either be learned through a forward query (to E) or through a backward query (to E^{-1}). The four queries in the diagram are labeled 'TL', 'BL', 'TR', 'BR' for 'Top Left', 'Bottom Left', etc.

random from a set of size at least $2^n - q$ (since $E_K(\cdot)$ is a random permutation for each K). Moreover, after $\mathcal{A}$ makes i queries its query history will contain exactly i distinct elements.

Say $\mathcal{A}$ *succeeds* at the i-th query if $\mathsf{Coll}(\mathcal{Q}_i)$ holds but neither $\mathsf{Coll}(\mathcal{Q}_{i-1})$ nor $\mathsf{Xor}(\mathcal{Q}_{i-1})$ holds. By upper bounding the probability that $\mathcal{A}$ ever succeeds we upper bound $\Pr[\mathsf{Coll}(\mathcal{Q}) \wedge \neg\mathsf{Xor}(\mathcal{Q})]$. (Upper bounding $\Pr[\mathsf{Xor}(\mathcal{Q})]$ is an easy probability exercise that we overlook for the purposes of this proof sketch.) A good analogy is to view $\mathcal{A}$ as trying to complete a puzzle where each element of its query history is a puzzle piece it can use to complete the collision diagram of Fig. 4. We use the expressions "$\mathcal{A}$ succeeds", "$\mathcal{A}$ finds a [puzzle] solution" or "$\mathcal{A}$ completes a collision" interchangeably (and we will rarely remind that the condition $\neg\mathsf{Xor}(\mathcal{Q}_{i-1})$ must hold for $\mathcal{A}$ to succeed). We refer to the four queries (in any hypothetical puzzle solution (a.k.a. collision)) as 'TL', 'BL', 'TR' and 'BR'; see Fig. 4.

Note the constraint $A\|B\|L \neq A'\|B'\|L'$ does not imply that the queries TL, BL, TR, BR are all distinct. For example, one could have TL = BR (in which case $(A, B\|L, R) = (B', L'\|R', S')$, so $A = B'$, $B = L'$, $L = R'$ and $R = S'$) or TL = BL (in which case we have the dramatic consequence that $A = B = L = R = S$, as is easy to check). This gives rise to several combinatorially distinct cases to consider; $\mathcal{A}$'s chance of obtaining a solution of each kind is upper bounded separately, and these probabilities are added together to form a final upper bound on $\mathcal{A}$'s chance of success. (Oddly, FGL include the cases TL = TR and BL = BR in their analysis, while these are impossible since they imply $A\|B\|L = A'\|B'\|L'$. This oversight, however, does not imply an incorrect proof in itself.)

We shall restrict our critique to FGL's analysis of the "generic" case when the queries TL, BL, TR, BR are all distinct. We note that, in these types of analyses, the generic case is usually the hardest to handle as $\mathcal{A}$'s job typically grows harder when additional constraints are placed on its solution. (The possibility of reusing the same query in two different positions of the collision diagram does, however, occasionally prove useful to $\mathcal{A}$, depending on the construction, so all cases must always be considered.) We call a puzzle solution in which TL, BL, TR, BR are distinct a "generic solution."

If $\mathcal{A}$ succeeds in finding a generic solution there is a smallest i such that a generic solution can be assembled from the queries in $\mathcal{Q}_i$. The i-th query is then called the "last query" of $\mathcal{A}$'s solution. To upper bound $\mathcal{A}$'s chance of obtaining a generic solution, FGL consider two cases. The first case is the event that $\mathcal{A}$'s last query can be used in position TL of the puzzle solution and the second case is the event that $\mathcal{A}$'s last query can be used in position BL (one of these two cases must occur). We shall focus on the first of these two cases, which is also the first analyzed in the order of the FGL proof. We call it the "TL generic" case.

One would typically consider two subcases for the TL generic case (or any other) depending on whether $\mathcal{A}$'s last query is a forward query to E or an inverse query to E^{-1}, but FGL lump their analysis into a single argument claiming that the two types of queries can be handled the same (in fact, they make this claim for every case in their proof, and never distinguish between forward and backward queries to E). For clarity, however, we shall restrict ourselves to considering the case of a forward query to E, and discuss how their argument specializes to that case. We also choose to specifically consider the forward query case because this is where FGL's analysis seems to be the most problematic.

The task at hand is thus to upper bound $\mathcal{A}$'s chance of completing a generic solution by making a forward query to E that can be used as query TL of such a solution. The usual approach for this, and the one used by FGL, is to consider any given forward query $E_{K_i}(X_i)$ made by $\mathcal{A}$ and to upper bound the probability that the answer Y_i to this query is such that the query history element (X_i, K_i, Y_i) can be used in the desired manner; one then multiplies this probability by q since $\mathcal{A}$ can make q queries total. With foresight on how we wish to use the query $E_{K_i}(X_i)$ it is convenient to rename K_i as $B\|L$ and X_i as A; thus the query is $E_{B\|L}(A)$. To proceed, one would typically upper bound the number of values $R \in \{0,1\}^n$ such that, if we had $E_{B\|L}(A) = R$, the query $(A, B\|L, R)$ could be used in position TL of a generic solution together with previous elements of the query history, and divide this number by $2^n - q$, since the answer to the query $E_{B\|L}(A)$ will come uniformly at random from a set of size at least $2^n - q$. In turn, the standard, formal way of bounding the number of such R's would be to upper bound the possible number of query triples (BL, BR, TR) already in the query history that could potentially be used with the query $E_{B\|L}(A)$ to form a generic solution, as the number of such triples is an upper bound for the number of R's. Note such a triple must have the form BL $= (B, L\|R, S)$, BR $= (B', L'\|R', S')$, TR $= (A', B'\|L', R')$ where $B \oplus S = B' \oplus S'$ (and note that A, B and L are fixed here by the last query).

FGL, however, do not adopt[1] this approach for bounding the number of good R's. Rather, they make the following argument: take the value of R, whatever it is, that is returned by the query $E_{B\|L}(A)$; because $\neg\mathsf{Xor}(\mathcal{Q}_{i-1})$ there will be

[1] Neither do we, in fact. Using a careful trick, we manage to upper bound the number of good R's by only considering the possibilities for the query BL rather than by considering the possible triples (BL, TR, BR). In the full version [11], however, we give for comparison the "brute force" proof which uses the method of upper bounding the number of triples (BL, TR, BR).

at most α queries TR $= (A', B'\|L', R')$ in the query history such that $A \oplus R = A' \oplus R'$; as the TR query uniquely determines the BR query, there are at most α possibilities for the BR query; now "give the query BL $= (B, L\|R, S)$ for free to the adversary"; then since there are at most α possibilities for the query BR $= (B', L'\|R', S')$ there is chance at most $\alpha/(2^n - q)$ that $B \oplus S = B' \oplus S'$ for one of the queries BR, so total chance at most $q\alpha/(2^n - q)$ that the adversary ever obtains a TL-generic solution with a forward query, there being at most q queries total.

The fallacy in the above argument can be succinctly summarized by pointing out that *the query* BL $= (B, L\|R, S)$ *may already be in the query history, in which case there is no randomness left in the value* $B \oplus S$. However, let us review in detail the argument in two different cases: when the query BL $= (B, L\|R, S)$ is already in the query history prior to the last query, and when it isn't. (Note that query BL only depends on R (besides B and L which are fixed by the last query), and not on which queries are "chosen" for TR and BR.) In the latter case, when BL $= (B, L\|R, S)$ is not yet in the query history at the i-th query, then A's last query can in any case not succeed in completing a generic TL collision since the query BL is missing; thus there is no need to bound anything (and no need even to "give the query BL for free"). In the case when query BL is already in the query history, on the other hand, all randomness is lost once R is revealed. FGL successfully argue that, for a given value of R, there will be at most α possibilities for the pair (TR , BR), but this does not in any way imply the *non-existence* of such queries TR, BR.

Note also that nothing in the FGL argument precludes the possibility that, when the adversary makes its i-th query $E_{B\|L}(A)$, there is not some very large number of distinct values of R—say $2^{0.5n}$—for which there exists a triplet of queries (BL, TR, BR) of the form BL $= (B, L\|R, S)$, BR $= (B', L'\|R', S')$, TR $= (A', B'\|L', R')$ where $B \oplus S = B' \oplus S'$, and such that R does not yet appear as the third coordinate of any query in the query history with key $B\|L$. Certainly, there being such a large number of values of R does not contradict $\neg\mathsf{Xor}(\mathcal{Q}_{i-1})$. Also certainly, the i-th query would have chance $2^{0.5n}/(2^n - q)$ of making the adversary succeed if such a large number of values of R existed, and not chance $\alpha/(2^n - q)$. In other words, one can infer something is wrong with the FGL argument because it simply does not address the main difficulty of the case at hand—that being the potential existence of a large number of triples (BL, BR, TR) that may fit with the query $E_{B\|L}(A)$.

Other issues are raised by FGL's casual comment that the query BL $= (B, L\|R, S)$ is simply "given for free" to the adversary. Indeed, if this query is not yet present, is it added to the query history before or after the i-th query itself? Is this query only made after the value of R is revealed, or is it somehow inserted into the query history before the value of R is revealed? The former might be all right; the latter not, since it would (drastically) alter R's distribution conditioned on the query history, i.e. R would no longer come uniformly at random from a set of size $\geq 2^n - q$. Most importantly, since this free query becomes part of the query history, one should account for the possibility that *this query* (not the i-th query) causes the adversary to succeed (and not necessarily by being used in position BL of a generic solution). Indeed, we are forced to give such credit to the adversary, since we have required the adversary never to make a query to which it already knows the answer, and since the adversary may have wished to subsequently make this query itself; this means the case analysis should be applied recursively to the free query, but if the case analysis requires other queries to be "given for free", then we bite our tail and end up giving an astronomical number of free queries to the adversary (e.g., nearly all possible queries).

While we singled out the TL generic case for examination, the same kinds of problems recur throughout the FGL case analysis, essentially invalidating the entire proof. Moreover, since the FGL proof sidesteps the most crucial challenges posed by an analysis of Tandem-DM (see the paragraph before last), it leaves little for any subsequent analysis to build on. We note that the FGL preimage resistance proof suffers from very similar flaws as the collision resistance proof, as discussed in the full version of this paper [11].

4 Main Result: Collision Resistance of Tandem-DM

It will be easier to explain the form of the probability bound in our main theorem if we explain a few high-level ideas from the proof beforehand. The proof starts by considering an arbitrary q-query collision-finding adversary $\mathcal{A}$ for Tandem-DM. We then construct an adversary $\mathcal{A}'$ as follows: $\mathcal{A}'$ simulates $\mathcal{A}$, but after each forward query $E_{V\|W}(U)$ made by $\mathcal{A}$, $\mathcal{A}'$ makes the backward query $E^{-1}_{U\|V}(W)$ *if* it does not already know[2] the answer to this query, and after each backward query $E^{-1}_{U\|V}(W)$ made by $\mathcal{A}$, $\mathcal{A}'$ makes the forward query $E_{V\|W}(U)$ *if* it does not already know[3] the answer to this query. (To better understand the relation of these instructions to Tandem-DM, view U, V, W as B, L, R.) Moreover if $\mathcal{A}$ ever makes a query to which $\mathcal{A}'$ already knows the answer from its query history, $\mathcal{A}'$ ignores this query. Thus $\mathcal{A}'$ never makes a query to which it knows the answer.

Let $\mathcal{Q}'$ be the query history of $\mathcal{A}'$ and $\mathcal{Q}$ be the query history of $\mathcal{A}$. Then $\mathcal{Q} \subseteq \mathcal{Q}'$ and $|\mathcal{Q}'| \leq 2q$. Since $\mathcal{Q} \subseteq \mathcal{Q}'$ we have

$$\begin{aligned} &\Pr[\mathsf{Coll}(\mathcal{Q})] \\ &\leq \Pr[\mathsf{Coll}(\mathcal{Q}')] \\ &\leq \Pr[\mathsf{Xor}(\mathcal{Q}')] + \Pr[\mathsf{FB}(\mathcal{Q}')] + \Pr[\mathsf{Coll}(\mathcal{Q}') \wedge \neg\mathsf{Xor}(\mathcal{Q}') \wedge \neg\mathsf{FB}(\mathcal{Q}')]. \end{aligned} \quad (7)$$

Our proof uses the inequality above to bound $\Pr[\mathsf{Coll}(\mathcal{Q})]$. We point out that if we construct an adversary $\mathcal{A}''$ from $\mathcal{A}'$ the same way $\mathcal{A}'$ is constructed from $\mathcal{A}$, then $\mathcal{A}''$ and $\mathcal{A}'$ will have the same query history, as is not difficult to see. In other words, *every* forward query $E_{V\|W}(U)$ made by $\mathcal{A}'$ (including its "own"

[2] More formally, if its query history does not contain any triple of the form $(\cdot, U\|V, W)$.

[3] More formally, if its query history does not contain any triple of the form $(U, V\|W, \cdot)$.

queries) is followed by the query $E^{-1}_{U\|V}(W)$ unless $\mathcal{A}'$ already knows this query, and likewise *every* backward query $E^{-1}_{U\|V}(W)$ made by $\mathcal{A}'$ is followed by the forward query $E_{V\|W}(U)$ unless $\mathcal{A}'$ already knows the answer to this query. The use of the augmented adversary $\mathcal{A}'$ may seem superficially similar to Fleischmann et al.'s idea of "giving away a query for free". However, it will become clear from our case analysis that we exploit the added structure of $\mathcal{Q}'$ entirely differently from the way Fleischmann et al. exploit their free queries. We also point out that the added structure of $\mathcal{Q}'$ enables the main interesting trick of our analysis, to be found in case 'TL Forward' of Proposition 3 below.

We can now more easily discuss our main result:

Theorem 1. *Let $N = 2^n$, $q < N/2$, $N' = N - 2q$ and let α be an integer, $1 \le \alpha \le 2q$. Then*

$$\mathbf{Adv}^{\mathrm{coll}}_{TDM}(q) \le 2N \left(\frac{2eq}{\alpha N'}\right)^{\alpha} + \frac{4q\alpha}{N'} + \frac{4q}{N'}.$$

The term $2N\left(\frac{2eq}{\alpha N'}\right)^{\alpha}$ in Theorem 1 is an upper bound for $\Pr[\mathsf{Xor}(\mathcal{Q}')]+\Pr[\mathsf{FB}(\mathcal{Q}')]$. In fact $\Pr[\mathsf{Xor}(\mathcal{Q}')] \le N\left(\frac{2eq}{\alpha N'}\right)^{\alpha}$ and $\Pr[\mathsf{FB}(\mathcal{Q}')] \le N\left(\frac{2eq}{\alpha N'}\right)^{\alpha}$. The two remaining terms $4q\alpha/N' + 4q/N'$ are an upper bound for $\Pr[\mathsf{Coll}(\mathcal{Q}') \wedge \neg\mathsf{Xor}(\mathcal{Q}') \wedge \neg\mathsf{FB}(\mathcal{Q}')]$. To bound $\mathbf{Adv}^{\mathrm{coll}}_{TDM}(q)$ for a given value of n and q one should optimize α numerically. For example, for $n = 128$, Theorem 1 yields that $\mathbf{Adv}^{\mathrm{coll}}_{TDM}(2^{120.87}) < \frac{1}{2}$ using $\alpha = 16$. Asymptotically, Theorem 1 yields the following result:

Corollary 1. $\lim_{n\to\infty} \mathbf{Adv}^{\mathrm{coll}}_{TDM}(N/n) = 0$.

Proof. Let $q = N/n$ and $\alpha = n/\log n$, where the logarithm takes base 2. Since $N' > N/2$ for $n > 4$, we have

$$\begin{aligned}\mathbf{Adv}^{\mathrm{coll}}_{TDM}(q) &\le 2N\left(\frac{2eq}{\alpha N'}\right)^{\alpha} + \frac{4q\alpha}{N'} + \frac{4q}{N'} \le 2N\left(\frac{4eq}{\alpha N}\right)^{\alpha} + \frac{8q\alpha}{N} + \frac{8q}{N} \\ &\le 2N\left(\frac{4e\log n}{n^2}\right)^{\frac{n}{\log n}} + \frac{8}{\log n} + \frac{8}{n} = 2\left(\frac{4e\log n}{n}\right)^{\frac{n}{\log n}} + \frac{8}{\log n} + \frac{8}{n}.\end{aligned}$$

The last expression obviously goes to zero as $n \to \infty$. □

In particular, $\lim_{n\to\infty} \mathbf{Adv}^{\mathrm{coll}}_{TDM}\left(2^{(1-\varepsilon)n}\right) = 0$ for any fixed $\varepsilon > 0$.

The proof of Theorem 1 uses refinements $\mathsf{Coll}_1(\mathcal{Q})$, $\mathsf{Coll}_2(\mathcal{Q})$, $\mathsf{Coll}_3(\mathcal{Q})$ of the collision predicate $\mathsf{Coll}(\mathcal{Q})$, defined as follows:

$\mathsf{Coll}_1(\mathcal{Q})$ occurs if $\mathcal{Q}$ contains a collision with TL, BL, TR, BR distinct.
$\mathsf{Coll}_2(\mathcal{Q})$ occurs if $\mathcal{Q}$ contains a collision with either TL = BL or TR = BR.
$\mathsf{Coll}_3(\mathcal{Q})$ occurs if $\mathcal{Q}$ contains a collision with either TL = BR or BL = TR.

For example, $\mathsf{Coll}_2(\mathcal{Q})$ occurs if there exist values $A, B, L, R, S, A', B', L', R', S'$ such that (1)–(2) hold and such that $(A, B\|L, R) = (B, L\|R, S)$. Since BL $\ne$ BR and TL $\ne$ TR in any collision, we have the following proposition.

Proposition 1. $\mathsf{Coll}(\mathcal{Q}) \implies \mathsf{Coll}_1(\mathcal{Q}) \vee \mathsf{Coll}_2(\mathcal{Q}) \vee \mathsf{Coll}_3(\mathcal{Q})$ *for any query history* $\mathcal{Q}$.

In view of proving Theorem 1, let $\mathcal{A}$ be an arbitrary q-query adversary for Tandem-DM, and let $\mathcal{A}'$ be obtained from $\mathcal{A}$ as outlined above; let $\mathcal{Q}$ be the query history of $\mathcal{A}$ and $\mathcal{Q}'$ be the query history of $\mathcal{A}'$. Then by (7) it suffices to show that

$$\Pr[\mathsf{Xor}(\mathcal{Q}')] \leq N\left(\frac{2eq}{\alpha N'}\right)^{\alpha}$$

$$\Pr[\mathsf{FB}(\mathcal{Q}')] \leq N\left(\frac{2eq}{\alpha N'}\right)^{\alpha}$$

$$\Pr[\mathsf{Coll}(\mathcal{Q}') \wedge \neg\mathsf{Xor}(\mathcal{Q}') \wedge \neg\mathsf{FB}(\mathcal{Q}')] \leq \frac{4q\alpha}{N'} + \frac{4q}{N'}$$

since the sum of the above probabilities is an upper bound for $\Pr[\mathsf{Coll}(\mathcal{Q})]$. Moreover, by Proposition 1, $\Pr[\mathsf{Coll}(\mathcal{Q}') \wedge \neg\mathsf{Xor}(\mathcal{Q}') \wedge \neg\mathsf{FB}(\mathcal{Q}')]$ can be upper bounded by finding upper bounds for $\Pr[\mathsf{Coll}_i(\mathcal{Q}') \wedge \neg\mathsf{Xor}(\mathcal{Q}') \wedge \neg\mathsf{FB}(\mathcal{Q}')]$ for $i = 1, 2, 3$ and taking the sum of these. We now upper bound these various probabilities in a series of propositions. For these propositions q, N and α are as in Theorem 1, and $\mathcal{Q}'$ is the query history of any adversary $\mathcal{A}'$ as just specified. We emphasize that $|\mathcal{Q}'| \leq 2q$ and that probabilities are taken over the random cipher E and over the coins of $\mathcal{A}'$, if any (it inherits these from $\mathcal{A}$).

Proposition 2. $\Pr[\mathsf{Xor}(\mathcal{Q}')] \leq N\left(\frac{2eq}{\alpha N'}\right)^{\alpha}$ *and* $\Pr[\mathsf{FB}(\mathcal{Q}')] \leq N\left(\frac{2eq}{\alpha N'}\right)^{\alpha}$.

Proof. Without loss of generality, we can assume that A' always makes exactly $2q$ queries. Let $\mathcal{Q}' = \{(X'_i, K'_i, Y'_i)\}_{i=1}^{2q}$ denote the query history of $\mathcal{A}'$. Since

$$\Pr[|\{i : X'_i \oplus Y'_i = Z\}| > \alpha] \leq \binom{2q}{\alpha}\left(\frac{1}{N'}\right)^{\alpha},$$

for each $Z \in \{0,1\}^n$, we have

$$\Pr[\mathsf{Xor}(\mathcal{Q}')] \leq N\binom{2q}{\alpha}\left(\frac{1}{N'}\right)^{\alpha} \leq N\left(\frac{2q\cdot e}{\alpha}\right)^{\alpha}\left(\frac{1}{N'}\right)^{\alpha} \leq N\left(\frac{2eq}{\alpha N'}\right)^{\alpha}.$$

$\Pr[\mathsf{FB}(\mathcal{Q}')]$ can be bounded similarly. □

Proposition 3. $\Pr[\mathsf{Coll}_1(\mathcal{Q}') \wedge \neg\mathsf{Xor}(\mathcal{Q}') \wedge \neg\mathsf{FB}(\mathcal{Q}')] \leq 4q\alpha/N'$.

Proof. Let

$$\mathsf{Success}_1(\mathcal{Q}'_i) = \mathsf{Coll}_1(\mathcal{Q}'_i) \wedge \neg\mathsf{Coll}_1(\mathcal{Q}'_{i-1}) \wedge \neg\mathsf{Xor}(\mathcal{Q}'_{i-1}) \wedge \neg\mathsf{FB}(\mathcal{Q}'_{i-1})$$

for $i = 1 \ldots 2q$. Then $\Pr[\mathsf{Coll}_1(\mathcal{Q}') \wedge \neg\mathsf{Xor}(\mathcal{Q}') \wedge \neg\mathsf{FB}(\mathcal{Q}')] \leq \sum_{i=1}^{2q} \Pr[\mathsf{Success}_1(\mathcal{Q}'_i)]$ and $\Pr[\mathsf{Success}_1(\mathcal{Q}'_i)] \leq \Pr[\mathsf{Coll}_1(\mathcal{Q}'_i) | \neg\mathsf{Coll}_1(\mathcal{Q}'_{i-1}) \wedge \neg\mathsf{Xor}(\mathcal{Q}'_{i-1}) \wedge \neg\mathsf{FB}(\mathcal{Q}'_{i-1})]$.

Fix a value of i, $1 \leq i \leq 2q$. We call the i-th query made by $\mathcal{A}'$ the *last query.* If $\mathsf{Success}_1(\mathcal{Q}'_i)$ occurs then either the adversary (i.e. $\mathcal{A}'$) can use its last query as query TL or as query BL of a collision in which TL, BL, TR and BR are distinct, by symmetry. Moreover the last query could either be a forward query or a backward query. This gives rise to four possible cases, and we bound $\Pr[\mathsf{Success}_1(\mathcal{Q}'_i)]$ for each separately. (We note the very first case, 'TL forward', is the case we discussed in Section 3.) For each case, we call the last query *successful* if this query completes a collision with TL, BL, TR, BR distinct and where the last query is used in the position stipulated by that case (e.g., for the case 'TL forward', the last query must be used in position TL).

TL forward: Let the last query be $E_{B\|L}(A)$. Call a value R *good* if there exists a query of the form $(B, L\|R, \cdot)$ in $\mathcal{Q}'$ that was obtained by $\mathcal{A}'$ as a backward query. We note that because of (5), $\neg\mathsf{FB}(\mathcal{Q}'_{i-1})$ implies there are at most α good R's.

We claim that for the last query to be successful the value R returned as an answer to the query must be good. Indeed, let R be the value returned; then a prerequisite for the query to be successful is that there be a query of the form $(B, L\|R, \cdot)$ in $\mathcal{Q}'_{i-1}$. We claim that this query must have been obtained as a backward query. Indeed, assume that the query $(B, L\|R, \cdot)$ was obtained as a forward query $E_{L\|R}(B)$ by $\mathcal{A}'$. Then, by construction, $\mathcal{A}'$ would have immediately followed this query by the query $E^{-1}_{B\|L}(R)$ unless $\mathcal{A}'$ already knew the answer to $E^{-1}_{B\|L}(R)$. Either way, $\mathcal{A}'$ would have the query $(A, B\|L, R)$ in its query history *prior* to the i-th (forward) query $E_{B\|L}(A)$, a contradiction since $\mathcal{A}'$ never makes a query to which it knows the answer. Thus the value R returned as an answer to the query $E_{B\|L}(A)$ must be good for the query to be successful.

Since there are at most α good values of R and since $\mathcal{A}'$ makes at most $2q$ queries, the probability that the last query is successful is therefore at most $\alpha/(2^n - 2q) = \alpha/N'$.

TL backward: Let the last query be $E^{-1}_{B\|L}(R)$. For the last query to be successful, there must be a (necessarily unique) query BL $= (B, L\|R, S) \in \mathcal{Q}'_{i-1}$, for some value $S \in \{0,1\}^n$. From the condition $B \oplus S = B' \oplus S'$ and from $\neg\mathsf{Xor}(\mathcal{Q}'_{i-1})$ there are at most α possibilities for the query BR. As each query BR uniquely determines the query TR, there are at most α possibilities for the query TR as well, and thus at most α possibilities for the value $A' \oplus R'$. Thus the value A returned by the last query has chance at most α/N' that $A \oplus R$ will be equal to $A' \oplus R'$ for one of these values $A' \oplus R'$, and so the last query has chance at most α/N' of being successful.

BL forward: A 180° rotation of the collision diagram shows this case is symmetric to the case TL backward. The chance of success in this case is therefore at most α/N'.

BL backward: A 180° rotation of the collision diagram shows this case is symmetric to the case TL forward. The chance of success in this case is therefore at most α/N'.

The chance a forward last query is successful is therefore at most $2\alpha/N'$ (adding the TL and BL forward cases) and likewise the chance that a backward last query is successful is at most $2\alpha/N'$. Thus $\Pr[\mathsf{Success}_1(\mathcal{Q}'_i)] \leq 2\alpha/N'$ for all i and $\sum_{i=1}^{2q} \Pr[\mathsf{Success}_1(\mathcal{Q}'_i)] \leq 4q\alpha/N'$. □

Proposition 4. $\Pr[\mathsf{Coll}_2(\mathcal{Q}') \wedge \neg\mathsf{Xor}(\mathcal{Q}') \wedge \neg\mathsf{FB}(\mathcal{Q}')] \leq 2q/N'$.

Proof. Note that when TL = BL, $B\|L = L\|R$, so $B = L = R$; moreover $R = S$ and $A = B$, so $A = B = L = R = S$. For the adversary to obtain a collision with TL = BL, therefore, it must obtain a query of the form $(U, U\|U, U)$. The same argument applies to the case TR = BR. The chance of a query $E_{U\|U}(U)$ or of a query $E^{-1}_{U\|U}(U)$ being answered by U is at most[4] $1/N'$. Thus, since $2q$ queries are made total, $\Pr[\mathsf{Coll}_2(\mathcal{Q}')] \leq 2q/N'$. □

Proposition 5. $\Pr[\mathsf{Coll}_3(\mathcal{Q}') \wedge \neg\mathsf{Xor}(\mathcal{Q}') \wedge \neg\mathsf{FB}(\mathcal{Q}')] \leq 2q\alpha/N' + 2q/N'$.

Proof. Note that in a collision with TL = BR we must have TL $\neq$ BL and $A \oplus R = B \oplus S$ (since $B \oplus S = B' \oplus S' = A \oplus R$, using TL = BR). Say the event $\mathsf{Coll}'_3(\mathcal{Q}'_i)$ occurs if there exist distinct queries $(A, B\|L, R)$, $(B, L\|R, S)$ in $\mathcal{Q}'_i$ such that $A \oplus R = B \oplus S$. With the same argument applied to the case BL = TR, we have $\mathsf{Coll}_3(\mathcal{Q}'_i) \implies \mathsf{Coll}'_3(\mathcal{Q}'_i)$. Therefore it suffices to show $\Pr[\mathsf{Coll}'_3(\mathcal{Q}') \wedge \neg\mathsf{Xor}(\mathcal{Q}') \wedge \neg\mathsf{FB}(\mathcal{Q}')] \leq 2q\alpha/N' + 2q/N'$.

The analysis now proceeds rather similarly to Proposition 3. Let

$$\mathsf{Success}'_3(\mathcal{Q}'_i) = \mathsf{Coll}'_3(\mathcal{Q}'_i) \wedge \neg\mathsf{Coll}'_3(\mathcal{Q}'_{i-1}) \wedge \neg\mathsf{Xor}(\mathcal{Q}'_{i-1}) \wedge \neg\mathsf{FB}(\mathcal{Q}'_{i-1}).$$

We have $\Pr[\mathsf{Coll}'_3(\mathcal{Q}') \wedge \neg\mathsf{Xor}(\mathcal{Q}') \wedge \neg\mathsf{FB}(\mathcal{Q}')] \leq \sum_{i=1}^{2q} \Pr[\mathsf{Success}'_3(\mathcal{Q}'_i)]$.

Fix a value of i, $1 \leq i \leq 2q$, and call the i-th query made by A' the *last query*. If $\mathsf{Success}'_3(\mathcal{Q}'_i)$ occurs then either the adversary (i.e. $\mathcal{A}'$) can use its last query as query TL or as query BL of its Coll'_3-solution. This gives rise to four possible cases given that the last query could be either forward or backward. In each case, we call the last query *successful* if $\mathsf{Success}'_3(\mathcal{Q}'_i)$ occurs and if the last query can be used in the position prescribed by that case (either TL or BL) in the Coll'_3-solution.

TL forward: We can use exactly the same analysis as in the case 'Forward TL' of Proposition 3. The probability that the last query is successful is therefore at most α/N'.

TL backward: Let $E^{-1}_{B\|L}(R)$ be the last query. For the last query to be successful, there must be a (necessarily unique) query of the form $(B, L\|R, S) \in \mathcal{Q}'_{i-1}$, for some $S \in \{0,1\}^n$. Since the answer A to the last query must be such that $A \oplus R = B \oplus S$ (as per the definition of Coll'_3) and $B \oplus S$ is uniquely determined, the last query has chance at most $1/N'$ of success.

[4] Since for each key there is only one relevant query, the tighter $1/N$ could be used as well.

BL forward: A 180° rotation of the collision diagram shows this case is symmetric to the case TL backward. The chance of success in this case is therefore at most $1/N'$.

BL backward: A 180° rotation of the collision diagram shows this case is symmetric to the case TL forward. The chance of success in this case is therefore at most α/N'.

The chance a forward last query is successful is therefore at most $(\alpha+1)/N'$ (adding the TL and BL forward cases) and likewise the chance that a backward last query is successful is at most $(\alpha+1)/N'$. Thus $\Pr[\mathsf{Success}'_3(\mathcal{Q}'_i)] \leq (\alpha+1)/N'$ for all i and $\sum_{i=1}^{2q} \Pr[\mathsf{Success}_1(\mathcal{Q}'_i)] \leq 2q\alpha/N' + 2q/N'$. (In fact, we even have $\Pr[\mathsf{Coll}_3(\mathcal{Q}') \wedge \neg\mathsf{FB}(\mathcal{Q}')] \leq 2q\alpha/N' + 2q/N'$ since $\neg\mathsf{Xor}(\mathcal{Q}')$ was never used in the above.) □

Taking the sum of the bounds of Propositions 3, 4 and 5 one obtains that

$$\Pr[\mathsf{Coll}(\mathcal{Q}') \wedge \neg\mathsf{Xor}(\mathcal{Q}') \wedge \neg\mathsf{FB}(\mathcal{Q}')] \leq \frac{6q\alpha}{N'} + \frac{4q}{N'}.$$

However, cases TL forward, BL backward and cases TL forward, BL backward of Propositions 3 and 5 reference the same events (the adversary is successful in case TL forward of Proposition 3 if and only if it is successful in case TL forward of Proposition 5, and likewise for the BL backward cases), which results in an "overcounting" of the adversary's probability of success by $2q\alpha/N'$. A more careful accounting of the adversary's probability of success thus shows

$$\Pr[\mathsf{Coll}(\mathcal{Q}') \wedge \neg\mathsf{Xor}(\mathcal{Q}') \wedge \neg\mathsf{FB}(\mathcal{Q}')] \leq \frac{4q\alpha}{N'} + \frac{4q}{N'}. \tag{8}$$

Here we have not established (8) entirely formally, though this is the bound we use for $\Pr[\mathsf{Coll}(\mathcal{Q}') \wedge \neg\mathsf{Xor}(\mathcal{Q}') \wedge \neg\mathsf{FB}(\mathcal{Q}')]$ in Theorem 1. Establishing (8) formally would require dividing the event $\mathsf{Coll}(\mathcal{Q})$ into a different, less intuitive set of events than $\mathsf{Coll}_1(\mathcal{Q})$, $\mathsf{Coll}_2(\mathcal{Q})$, $\mathsf{Coll}_3(\mathcal{Q})$, events that are directly based on those that occur in the case analyses of Propositions 3–5. (For example, one of these events would be the event that the adversary ever obtains a "good R" through a forward or backward query, as defined for forward queries in case TL forward of Proposition 3 and implicitly defined (by symmetry) for backward queries in case BL backward of Proposition 3; another event would cover the cases TL backward and BL forward of Proposition 5; and so on.) The current form of the proof is our best compromise between readability and formality. In any case, the difference between $4q\alpha/N'$ and $6q\alpha/N'$ is relatively minor.

Summing (8) with the bounds of Proposition 2 and using (7), we obtain

$$\Pr[\mathsf{Coll}(\mathcal{Q})] \leq 2N\left(\frac{2eq}{\alpha N'}\right)^{\alpha} + \frac{4q\alpha}{N'} + \frac{4q}{N'}. \tag{9}$$

Since (9) holds for an arbitrary q-query adversary $\mathcal{A}$, this establishes Theorem 1.

5 Conclusion

In this work, we have shown that an earlier work concerning the security of Tandem-DM was incorrect. However, with a new proof (exploiting new ideas) we have shown that, in the ideal-cipher model, Tandem-DM is collision resistant almost up to the birthday bound and (provably) preimage resistant essentially up to the birthday bound (leaving considerable room for improvement for the latter).

On a high level, our proof of collision resistance adheres to a (by now) standard framework. We first modify the collision-finding adversary by giving it several "free" queries and subsequently we bound the modified adversary's chance of success using a case analysis. This approach allows to easily bound both the number of free queries and the probability of a query (free or not) causing a collision.

In contrast, the FGL proof directly uses a case analysis and subsequently uses free queries within the case analysis. This ad hoc addition of free queries (and its binding to a particular case) is problematic, as it does not allow proper accounting of the free queries. In particular, if a free query is fresh it might cause a collision (or other bad event) elsewhere whereas if the free query has actually been asked before, no new randomness can be extracted from it.

Thus, apart from having established the security of Tandem-DM, we hope that our work also serves as a useful reminder to some of the subtleties involved in ICM proofs and as a guideline on how to avoid certain pitfalls.

References

1. Dodis, Y., Steinberger, J.: Message Authentication Codes from Unpredictable Block Ciphers. In: Halevi, S. (ed.) CRYPTO 2009. LNCS, vol. 5677, pp. 267–285. Springer, Heidelberg (2009), Full version available at `http://people.csail.mit.edu/dodis/ps/tight-mac.ps`
2. Fleischmann, E., Forler, C., Gorski, M., Lucks, S.: Collision resistant double-length hashing. In: Heng, S.-H., Kurosawa, K. (eds.) ProvSec 2010. LNCS, vol. 6402, pp. 102–118. Springer, Heidelberg (2010)
3. Fleischmann, E., Gorski, M., Lucks, S.: On the security of TANDEM-DM. In: Dunkelman, O. (ed.) FSE 2009. LNCS, vol. 5665, pp. 84–103. Springer, Heidelberg (2009)
4. Fleischmann, E., Gorski, M., Lucks, S.: Security of Cyclic Double Block Length Hash Functions. In: Parker, M.G. (ed.) Cryptography and Coding 2009. LNCS, vol. 5921, pp. 153–175. Springer, Heidelberg (2009)
5. Hirose, S.: Provably secure double-block-length hash functions in a black-box model. In: Park, C.-s., Chee, S. (eds.) ICISC 2004. LNCS, vol. 3506, pp. 330–342. Springer, Heidelberg (2005)
6. Hirose, S.: Some plausible constructions of double-block-length hash functions. In: Robshaw, M.J.B. (ed.) FSE 2006. LNCS, vol. 4047, pp. 210–225. Springer, Heidelberg (2006)
7. Krause, M., Armknecht, F., Fleischmann, E.: Preimage resistance beyond the birthday bound: double-length hashing revisited. Preprint, `http://eprint.iacr.org/2010/519`

8. Lai, X., Massey, J.: Hash functions based on block ciphers. In: Rueppel, R.A. (ed.) EUROCRYPT 1992. LNCS, vol. 658, pp. 55–70. Springer, Heidelberg (1993)
9. Lee, J., Kwon, D.: The security of Abreast-DM in the ideal cipher model. IEICE Transactions 94-A(1), 104–109 (2011), http://eprint.iacr.org/2009/225.pdf
10. Lee, J., Steinberger, J.: Multi-property-preserving Domain Extension Using Polynomial-Based Modes of Operation. In: Gilbert, H. (ed.) EUROCRYPT 2010. LNCS, vol. 6110, pp. 573–596. Springer, Heidelberg (2010)
11. Lee, J., Stam, M., Steinberger, J.: The collision security of Tandem-DM in the ideal-cipher model. Full version of this paper, http://eprint.iacr.org/2010/409
12. Lee, J., Stam, M., Steinberger, J.: The preimage security of double-block-length compression functions. Preprint, http://eprint.iacr.org/2011/210
13. Lucks, S.: A collision-resistant rate-1 double-block-length hash function. In: Symmetric Cryptography. Dagstuhl Seminar Proceedings, 07021 (2007)
14. Özen, O., Stam, M.: Another Glance at Double-Length Hashing. In: Parker, M.G. (ed.) Cryptography and Coding 2009. LNCS, vol. 5921, pp. 176–201. Springer, Heidelberg (2009)
15. Rogaway, P., Shrimpton, T.: Cryptographic hash-function basics: Definitions, implications, and separations for preimage resistance, second-preimage resistance, and collision resistance. In: Roy, B., Meier, W. (eds.) FSE 2004. LNCS, vol. 3017, pp. 371–388. Springer, Heidelberg (2004)
16. Rogaway, P., Steinberger, J.: Constructing cryptographic hash functions from fixed-key blockciphers. In: Wagner, D. (ed.) CRYPTO 2008. LNCS, vol. 5157, pp. 433–450. Springer, Heidelberg (2008)
17. Shrimpton, T., Stam, M.: Building a collision-resistant compression function from non-compressing primitives. In: Aceto, L., Damgård, I., Goldberg, L.A., Halldórsson, M.M., Ingólfsdóttir, A., Walukiewicz, I. (eds.) ICALP 2008, Part II. LNCS, vol. 5126, pp. 643–654. Springer, Heidelberg (2008)
18. Steinberger, J.P.: The Collision Intractability of MDC-2 in the Ideal-Cipher Model. In: Naor, M. (ed.) EUROCRYPT 2007. LNCS, vol. 4515, pp. 34–51. Springer, Heidelberg (2007)
19. Stam, M.: Beyond Uniformity: Better Security/Efficiency Tradeoffs for Compression Functions. In: Wagner, D. (ed.) CRYPTO 2008. LNCS, vol. 5157, pp. 397–412. Springer, Heidelberg (2008)
20. Stam, M.: Blockcipher-based hashing revisited. In: Dunkelman, O. (ed.) FSE 2009. LNCS, vol. 5665, pp. 67–83. Springer, Heidelberg (2009)
21. Wagner, D.: Cryptanalysis of the Yi-Lam Hash. In: Okamoto, T. (ed.) ASIACRYPT 2000. LNCS, vol. 1976, pp. 483–488. Springer, Heidelberg (2000)
22. Yi, X., Lam, K.-Y.: A new hash function based on block cipher. In: Mu, Y., Pieprzyk, J.P., Varadharajan, V. (eds.) ACISP 1997. LNCS, vol. 1270, pp. 139–146. Springer, Heidelberg (1997)

Order-Preserving Encryption Revisited: Improved Security Analysis and Alternative Solutions

Alexandra Boldyreva[1], Nathan Chenette[1], and Adam O'Neill[2,⋆]

[1] Georgia Institute of Technology
{sasha,nchenette}@gatech.edu
[2] University of Texas at Austin
adamo@cs.utexas.edu

Abstract. We further the study of order-preserving symmetric encryption (OPE), a primitive for allowing efficient range queries on encrypted data, recently initiated (from a cryptographic perspective) by Boldyreva et al. (Eurocrypt '09). First, we address the open problem of characterizing what encryption via a random order-preserving function (ROPF) leaks about underlying data (ROPF being the "ideal object" in the security definition, POPF, satisfied by their scheme.) In particular, we show that, for a database of randomly distributed plaintexts and appropriate choice of parameters, ROPF encryption leaks neither the precise value of any plaintext nor the precise distance between any two of them. The analysis here is quite technically non-trivial and introduces useful new techniques. On the other hand, we also show that ROPF encryption does leak both the value of any plaintext as well as the distance between any two plaintexts to within a *range* of possibilities roughly the square root of the domain size. We then study schemes that are not order-preserving, but which nevertheless allow efficient range queries and achieve security notions stronger than POPF. In a setting where the entire database is known in advance of key-generation (considered in several prior works), we show that recent constructions of "monotone minimal perfect hash functions" allow to efficiently achieve (an adaptation of) the notion of IND-O(rdered) CPA also considered by Boldyreva et al., which asks that *only* the order relations among the plaintexts is leaked. Finally, we introduce *modular* order-preserving encryption (MOPE), in which the scheme of Boldyreva et al. is prepended with a shift cipher. MOPE improves the security of OPE in a sense, as it does not leak any information about plaintext location. We clarify that our work should not be interpreted as saying the original scheme of Boldyreva et al., or the variants that we introduce, are "secure" or "insecure." Rather, the goal of this line of research is to help practitioners decide whether the options provide a suitable security-functionality tradeoff for a given application.

Keywords: Searchable encryption, symmetric encryption, hypergeometric distribution, range queries.

⋆ Part of the work done while at the Georgia Institute of Technology.

P. Rogaway (Ed.): CRYPTO 2011, LNCS 6841, pp. 578–595, 2011.

1 Introduction

Background and Motivation. An order-preserving symmetric encryption (or OPE) scheme is a deterministic symmetric encryption scheme whose encryption algorithm produces ciphertexts that preserve numerical ordering of the plaintexts. OPE was proposed in the database community by Agrawal et al. [1] in 2004 as a tool to support efficient range queries on encrypted data. (When encryption is done using an OPE scheme, a range query simply consists of the encryptions of the two end-points.) However, the first formal cryptographic treatment of OPE did not appear until recently, in the paper by Boldyreva et al. [8]. The authors formalized a security requirement for OPE and proposed an efficient blockcipher-based scheme provably meeting their security definition.

Yet despite having an OPE scheme that provably satisfies their security notion, the authors warn against its practical use before further studies of its security are performed. To explain this, consider the security notion (or "ideal object") from [8], called a pseudorandom order-preserving function (POPF).

Informally, the POPF notion calls an OPE scheme secure if oracle access to its encryption algorithm is indistinguishable from that to a *random* order-preserving function (ROPF), i.e., a random element of the set of all strictly-increasing functions on the same domain and range. This is a rather straight-forward adaptation of the classical notion of pseudorandom function (PRF)—which asks that oracle access to a function be indistinguishable from that to a truly random function on the same domain and range—to the order-preserving context, and it captures some intuition of what should be the "best possible" OPE scheme. However, the POPF definition is somewhat deceiving and confusing in terms of giving an idea of what kind of security it describes. A random function's behavior is well understood: on a new input the output is a random point in the range. Hence, an adversary seeing a function value learns absolutely no information about the pre-image, unless the former happens to coincide with one it has previously seen. But the situation with a random OPF is much harder to describe. It is clear that a random OPF cannot provide such strong security, but what exactly is leaked about the data and what is protected? The distribution of ciphertexts is known and it is not immediately clear if encryption is even one-way.

Despite its authors' warning of lingering unanswered questions, the OPE scheme from [8] immediately received attention from the applied community [21,20,18,17,14]. We agree that a secure OPE is better than no encryption at all and understand why the idea of its implementation may sound appealing. But practical use without a clear security understanding can be very dangerous and thus it is very important to clarify the security questions as soon as possible.

In this work we first address this open problem. We revisit the security of the "ideal object" ROPF introduced by [8] and provide results that help characterize what it leaks and what it protects about the underlying data. We then observe that it may be possible to achieve stronger security notions than POPF using schemes that fall outside the OPE class but nevertheless allow efficient range queries on encrypted data, and propose two such schemes. We now discuss our contributions in more detail.

New Definitions for Studying ROPF Security. As (perhaps surprisingly) pointed out by [8], a random order-preserving function—the ideal object in the POPF definition from that paper—itself requires a cryptographic treatment.

In order to better understand the strengths and limitations of encryption with an ROPF we first propose several security notions. One captures a basic one-wayness security and measures the probability that an adversary, given a set of ciphertexts of random messages, decrypts one of them. (The fact that messages are chosen uniformly at random we call the "uniformity assumption," and it will be discussed later.) We give the adversary multiple challenge ciphertexts because this corresponds to practical settings and because the ciphertexts are not independent from each other: learning more points of the OPE function may give the adversary additional information. We actually consider a more general security notion that asks the adversary given same inputs to guess an interval (window) within which the underlying challenge plaintext lies. This definition helps us get a better sense of how accurately the adversary can identify the location of a data point. The size of the window and the number of challenge ciphertexts are parameters of the definition. When the window size is one, the notion collapses to the case of simple one-wayness.

Our subsequent definitions address leakage of information not about the *location* of the data points but rather the *distances* between them, which seems crucial in other applications (e.g., a database of salaries). Indeed, [8] showed that an ROPF with a practical range size does not hide distances between plaintexts. We attempt to clarify this intuition. We consider a definition measuring the adversary's success in (precisely) guessing the distance[1] between the plaintexts corresponding to any two out of the set of ciphertexts of random messages given to the adversary. Again, we also consider a more general definition where the adversary is allowed to specify a window in which the distance falls.

We analyze security of an ROPF under these definitions as we believe this helps to understand secure pseudorandom OPE schemes' security guarantees and limitations, and also to evaluate the risk of their usage in various applications. (Indeed, we believe they capture the information about the data, namely location and relative distances, that practitioners are most likely to care about in applications.) However, especially in light of the uniformity assumption (which is unlikely to be satisfied in practice), we view our results as providing important steps in the direction of this understanding (as even under this assumption our results are challenging to prove) but still warn against practical usage of OPE based on current knowledge.

Analysis of an ROPF. We first give an upper bound on the one-wayness advantage of any adversary attacking an ROPF. The proof is quite involved (and is explained in detail in the full version [9]), but the result is a very concise, understandable bound that, under reasonable assumptions, does not even

[1] Technically, for purposes that will become clear in the paper, "distance" actually refers to "directed modular distance," i.e. the distance from one point "up" to the other point, possibly wrapping around the space. As such, distance in our context is non-commutative.

depend on the size of the ciphertext space. (Intuitively, an ROPF's one-wayness comes from the function's probability to deviate from points on the linear OPF $m \mapsto (N/M)m$. Increasing the ciphertext space size beyond a certain amount has little to no effect on these deviations.) We evaluate the bound for several parameters to get an idea of its quality. Our evaluation demonstrates that on practical parameters ROPF and POPF-secure OPEs significantly resist one-wayness attacks, i.e. the maximum one-wayness advantage of any adversary is quite low.

On the other hand, our ROPF analysis under the window one-wayness definition shows that a very efficient adversary can successfully break window one-wayness if the size of the window is not very small. In particular, for message space size M and arbitrary constant b, if the window size is approximately $b\sqrt{M}$, there exists an adversary A whose window one-wayness is at least $1 - 2e^{-b^2/2}$. Thus, for b large enough (say, $b \geq 8$), there exists an adversary with window one-wayness advantage very close to one.

We then extend our analysis of an ROPF to the distance one-wayness and window distance one-wayness definitions. Using similar techniques we show entirely analogous results, namely that the former is very small but the latter becomes large when the adversary is allowed to specify a window of size approximately $b\sqrt{M}$.

We conclude our ROPF analysis with several important supplemental remarks regarding the effect of known-plaintext attacks in the schemes, choosing an appropriate ciphertext space size, and the need to satisfy the uniformity assumption in practical implementations.

Achieving Stronger Security. We next consider the question of whether different types of schemes that support efficient range queries can achieve stronger security than POPF. To capture such schemes we introduce a general notion of *efficiently orderable encryption* (EOE), that covers all schemes supporting standard range queries by requiring a publicly computable function that determines order of the underlying plaintexts given any two ciphertexts. Since EOE leaks order of ciphertexts, IND-OCPA (which [8] showed is unachievable by OPE) is an ideal level of security for EOE schemes (although what information about the data can be inferred from it is outside the scope of the current paper).

An Optimally Secure Committed EOE Scheme. We focus on a scenario where we can show something like IND-OCPA security is possible. We define "committed" versions of EOE and IND-OCPA, called CEOE and IND-CCPA, corresponding to a setting where the database is static and completely known to the user in advance of encryption. Such a scenario is apparently important as it was considered in the first paper to propose an order-preserving scheme [1], and was also studied in several works including [13] for the case of exact-match queries. We observe that the more restrictive functionality in this setting allows one to achieve IND-CCPA. We propose a new scheme that uses a monotone minimal perfect hash function (MMPHF) directly as an "order preserving tagging algorithm" for the given message set, together with a secure encryption. The construction allows for easy implementation of range queries while also achieving the strongest security. Moreover, while MMPHFs are known to require long keys [4], recent constructions [4] are close to being space-optimal. Thus, this application of MMPHFs for tagging seems to be a novel, nearly efficient-as-possible way to support range queries, leaking nothing but the order of ciphertexts, when the database is fixed in advance.

A New Modular OPE Scheme and its Analysis. Finally, we propose a technique that improves on the security of any OPE scheme without sacrificing efficiency. Recall that our ROPF analysis reveals information leakage in OPE not alluded to by [8], namely about the *locations* of the data points rather than just the distances between them. We suggest a modification to (that can be viewed as a generalization of) an OPE scheme that overcomes this. The resulting scheme is not order-preserving per se, but still permits range queries—in this case, modular range queries. (When the left end of the queried range is greater than the right end, a modular range query returns the "wrap-around range," i.e. everything greater than the left end or less than the right end.) The modification to the scheme is simple and generic: the encryption algorithm just adds (modulo the size of the message space) a secret offset to the message before encryption. (The secret offset is the same for all messages.) We call a scheme obtained this way a modular OPE scheme, and generalize the security notion: the ideal object is now a random modular OPF (RMOPF), i.e. a random OPF applied to messages with a randomly picked offset. It is easy to see that any secure OPE scheme yields a secure modular OPE scheme using the above transformation.

We show that a random modular OPF, unlike a random OPF, completely hides the locations of the data points (but has the same leakage with respect to distance and window-distance one-wayness). On the other hand, if the adversary is able to recover a single known plaintext-ciphertext pair, security falls back to that of a random OPF.

We also note that the technique with a secret offset can be applied to the CEOE scheme to enhance its security even beyond IND-CCPA when support for modular range queries is sufficient.

Related Work. Efficient (sub-linear time) search on encrypted data for the case of simple exact-match queries has been addressed by [2] in the symmetric-key setting and [6,10,7] in the public-key setting. The work of [16] suggested enabling efficient range queries on encrypted data not by using OPE but so-called *prefix-preserving encryption* (PPE) [22,5]. But as discussed in [16,2], PPE schemes are subject to certain attacks. Allowing range queries on encrypted data in the public-key setting was studied in [11,19], but the solutions are not suitable for large databases, requiring to scan the whole database on every query. As we mentioned, order preserving encryption as an efficient solution for range queries has been proposed in [1], however, they do not provide any formal security analysis.

2 Preliminaries

Notation. If M is an integer, then $[M]$ denotes the set $\{1, \ldots, M\}$. For a set S and $n \leq |S|$, let Comb_n^S denote the set of n-element subsets of S. If $\mathcal{E}nc$ is an encryption function with key K, $\mathbf{x} = (x_1, \ldots, x_\ell)$ is a vector, and $X = \{x_1, \ldots, x_\ell\}$ is a set, then $\mathcal{E}nc(K, \mathbf{x})$ is shorthand for $(\mathcal{E}nc(K, x_1), \ldots, \mathcal{E}nc(K, x_\ell))$ and $\mathcal{E}nc(K, X)$ is shorthand for $\{\mathcal{E}nc(K, x_1), \ldots, \mathcal{E}nc(K, x_\ell)\}$. The same holds for decryption $\mathcal{D}ec$.

A Convention. For simplicity, in many cases we will assume a domain/plaintext space $[M]$ and range/ciphertext space $[N]$, for $N \geq M$. Naturally, all results for arbitrary spaces $\mathcal{D}$, $\mathcal{R}$ can be derived from those of $[|\mathcal{D}|]$, $[|\mathcal{R}|]$.

Range Queries. For fixed plaintext and ciphertext spaces $[M]$ and $[N]$, a range query *target* is a pair of plaintexts (m_L, m_R) that comes in two varieties: *standard* if $m_L \leq m_R$, or *wrap-around* if $m_L > m_R$. If (m_L, m_R) is a target, its associated *range* is $[m_L, m_R]$ in the standard case and $[m_L, M] \cup [1, m_R]$ in the wrap-around case.

To model the intended application, suppose a server has a database encrypted under a scheme $(\mathcal{K}, \mathcal{E}nc, \mathcal{D}ec)$ with key $K \xleftarrow{\$} \mathcal{K}$. In a *standard range query*, the user submits two unordered ciphertexts $\{c_1, c_2\}$ to the server. Let $(m_1, m_2) = \mathcal{D}ec(K, (c_1, c_2))$. Then the target is $(\min\{m_1, m_2\}, \max\{m_1, m_2\})$, and the server must return the set of ciphertexts in the database whose decryptions fall into the associated range. Notice that these targets are always standard.

In a *modular range query*, the user submits two ordered ciphertexts (c_L, c_R). Let $(m_L, m_R) = \mathcal{D}ec(K, (c_L, c_R))$. Then the range query target is (m_L, m_R), and the server must return the set of ciphertexts in the database whose decryptions fall into the associated range. Notice that these targets can be standard or wrap-around.

Order-Preserving Encryption (OPE). Following [8] we say that $\mathcal{SE}_{\mathcal{D},\mathcal{R}} = (\mathcal{K}, \mathcal{E}nc, \mathcal{D}ec)$ with associated *plaintext-space* $\mathcal{D}$ and *ciphertext-space* $\mathcal{R}$ is *deterministic* if the encryption algorithm $\mathcal{E}nc$ is deterministic. For $A, B \subseteq \mathbb{N}$ with $|A| \leq |B|$, a function $f\colon A \to B$ is *order-preserving* if for all $i, j \in A$, $f(i) > f(j)$ iff $i > j$. We say that deterministic encryption scheme $\mathcal{SE}_{\mathcal{D},\mathcal{R}} = (\mathcal{K}, \mathcal{E}nc, \mathcal{D}ec)$ is *order-preserving* if $\mathcal{E}nc(K, \cdot)$ is an order-preserving function from $\mathcal{D}$ to $\mathcal{R}$ for all K output by $\mathcal{K}$ (with elements of $\mathcal{D}, \mathcal{R}$ interpreted as numbers, encoded as strings).

Security of OPE. We recall the security definition for OPE from [8].[2] Informally (refer to [8] for the formal definition), it says that an OPE scheme is secure if oracle access to its encryption function is indistinguishable from oracle access to a random order-preserving function (ROPF) on the same domain and range. Any secure OPE scheme (including the only currently known blockcipher-based scheme from [8]) should "closely" imitate the behavior of an ROPF. Accordingly we focus in this paper on analyzing the ideal object, an ROPF.

---

[2] For simplicity, we do not discuss chosen-ciphertext attacks in detail. Note that symmetric schemes such as these can be made resistant to chosen-ciphertext attacks by implementing Encrypt-then-MAC with a MAC having strong unforgeability, preventing adversaries from even constructing valid ciphertexts.

An "Ideal" Scheme ROPF. We define the "ideal" ROPF scheme as follows. Let $\mathsf{OPF}_{\mathcal{D},\mathcal{R}}$ denote the set of all order-preserving functions from $\mathcal{D}$ to $\mathcal{R}$. Define $\mathsf{ROPF}_{\mathcal{D},\mathcal{R}} = (\mathcal{K}_{\mathrm{r}}, \mathcal{E}nc_{\mathrm{r}}, \mathcal{D}ec_{\mathrm{r}})$ as the following encryption scheme:

- $\mathcal{K}_{\mathrm{r}}$ returns a random element g of $\mathsf{OPF}_{\mathcal{D},\mathcal{R}}$.
- $\mathcal{E}nc_{\mathrm{r}}$ takes the key and a plaintext m to return $g(m)$.
- $\mathcal{D}ec_{\mathrm{r}}$ takes the key and a ciphertext c to return $g^{-1}(c)$.

Of course the above scheme is not computationally efficient, but our goal is its security analysis for the purpose of clarifying security of all POPF-secure constructions.

Most Likely Plaintext. Fix a symmetric encryption scheme $\mathcal{SE}_{\mathcal{D},\mathcal{R}} = (\mathcal{K}, \mathcal{E}nc, \mathcal{D}ec)$. For given $c \in \mathcal{R}$, if $m_c \in \mathcal{D}$ is a message such that

$$\Pr\left[K \xleftarrow{\$} \mathcal{K} \, : \, \mathcal{E}nc(K, m) = c \right]$$

achieves a maximum at $m = m_c$, then we call m_c a (if unique, "the") *most likely plaintext* for c.

Most Likely Plaintext Distance. Fix a symmetric encryption scheme $\mathcal{SE}_{[M],[N]} = (\mathcal{K}, \mathcal{E}nc, \mathcal{D}ec)$. For given $c_1, c_2 \in \mathcal{R}$, if $d_{c_1,c_2} \in \{0, 1, \ldots, M-1\}$ such that

$$\Pr\left[K \xleftarrow{\$} \mathcal{K} \, : \, (c_1, c_2) = \mathcal{E}nc(K, (m_1, m_2)) \, ; \, m_2 - m_1 \bmod M = d \right]$$

achieves a maximum at $d = d_{c_1,c_2}$, then we call d_{c_1,c_2} a (if unique, "the") *most likely plaintext distance* from c_1 to c_2.

3 New Security Definitions

As explained in the introduction, the "ideal" ROPF scheme defined in Section 2 itself requires a cryptographic treatment. Toward this end, we propose several generalized security definitions that help us understand its security.

Let $\mathcal{SE}_{[M],[N]} = (\mathcal{K}, \mathcal{E}nc, \mathcal{D}ec)$ be a deterministic symmetric encryption scheme.

Window One-Wayness. The most basic question left unanswered by [8] is whether a POPF-secure scheme is even one-way. Towards this end we start with the one-wayness definition. Our definition is a stronger and more general version of the standard notion of one-wayness. For $1 \leq r \leq M$ and $z \geq 1$, the adversary is given a set of z ciphertexts of (uniformly) random messages and is asked to come up with an interval of size r within which one of the underlying plaintexts lies. We call our notion r, z-window one-wayness (or r, z-WOW). Note that when $r = 1$, the definition collapses to the standard one-wayness definition (for multiple ciphertexts), and we will call it one-wayness for simplicity.

The *r, z-window one-wayness (r, z-WOW) advantage* of adversary A against $\mathcal{SE}_{[M],[N]}$ is

$$\mathbf{Adv}^{r,z\text{-wow}}_{[M],[N]}(A) \; = \; \Pr\left[\mathbf{Exp}^{r,z\text{-wow}}_{\mathcal{SE}_{[M],[N]}}(A) = 1 \right] ,$$

where

Experiment $\mathbf{Exp}^{r,z\text{-wow}}_{\mathcal{SE}_{[M],[N]}}(A)$
$K \overset{\$}{\leftarrow} \mathcal{K}$; $\mathbf{m} \overset{\$}{\leftarrow} \mathrm{Comb}^{[M]}_z$; $\mathbf{c} \leftarrow \mathcal{E}nc(K, \mathbf{m})$
$(m_L, m_R) \overset{\$}{\leftarrow} A(\mathbf{c})$
Return 1 if $(m_R - m_L) \bmod M + 1 \le r$ and there exists $m \in \mathbf{m}$ so that either $m \in [m_L, m_R]$ or ($m_L > m_R$ and $m \in [m_L, M] \cup [1, m_R]$)
Return 0 otherwise

Notice that the latter success condition allows the adversary to specify a window that "wraps around" the message space. Granting this extra power to the adversary will be useful in analyzing the MOPE scheme of Section 5.2.

Window Distance One-Wayness. To identify the extent to which an OPE scheme leaks distance between plaintexts, we also provide a definition in which the adversary attempts to guess the interval of size r in which the distance between any two out of z random plaintexts lies, for $1 \le r \le M$ and $z \ge 2$. We call the notion r, z-window distance one-wayness (r, z-WDOW). When $r = 1$, the adversary has to guess the exact distance between any two of z ciphertexts.

The *r, z-window distance one-way (r, z-WDOW) advantage* of adversary A against scheme $\mathcal{SE}_{[M],[N]}$ is

$$\mathbf{Adv}^{r,z\text{-wdow}}_{[M],[N]}(A) = \Pr\left[\mathbf{Exp}^{r,z\text{-wdow}}_{\mathcal{SE}_{[M],[N]}}(A) = 1\right],$$

where

Experiment $\mathbf{Exp}^{r,z\text{-wdow}}_{\mathcal{SE}_{[M],[N]}}(A)$
$K \overset{\$}{\leftarrow} \mathcal{K}$; $\mathbf{m} \overset{\$}{\leftarrow} \mathrm{Comb}^{[M]}_z$; $\mathbf{c} \leftarrow \mathcal{E}nc(K, \mathbf{m})$
$(d_1, d_2) \overset{\$}{\leftarrow} A(\mathbf{c})$
Return 1 if $d_2 - d_1 + 1 \le r$ and there exist distinct $m_i, m_j \in \mathbf{m}$ with $m_j - m_i \bmod M \in [d_1, d_2]$
Return 0 otherwise

4 One-Wayness of a Random OPF

This section is devoted to analyzing the "ideal" scheme $\mathsf{ROPF}_{[M],[N]}$ under the security definitions given in the previous section. The first result shows an upper bound on $1, z$-WOW advantage against the scheme. This demonstrates that on practical parameters, ROPF and POPF-secure OPEs significantly resist (size-1-window) one-wayness attacks. In contrast, the second result shows the ideal ROPF scheme is susceptible to an efficient large-window (a constant times $\sqrt{M}$) one-wayness attack, by constructing an adversary and lower-bounding its r, z-WOW advantage.

The analysis then proceeds similarly for window distance one-wayness definitions: we will show analogous contrasting results for small- versus large-window experiments. We now turn to the details of the analysis.

An Upper Bound on the $1, z$-WOW Advantage. The following theorem states an upper bound on the $1, z$-WOW advantage of any adversary against $\mathsf{ROPF}_{[M],[N]}$.

Theorem 1. *For any challenge set of size z and adversary A, if $N \geq 2M$ and $M \geq 15 + z$ then*

$$\mathbf{Adv}^{1,z\text{-wow}}_{\mathsf{ROPF}_{[M],[N]}}(A) < \frac{9z}{\sqrt{M - z + 1}} .$$

The formal proof is quite involved and is in the full version [9]. The idea is to first bound $1, z$-WOW security in terms of $1, 1$-WOW security; because ciphertexts are correlated, a simple hybrid argument does *not* work and our reduction uses new ideas. Then, to bound $1, 1$-WOW security, we take a combinatorial strategy, as follows. We define a ciphertext's most likely plaintext (m.l.p.) and recall the negative hypergeometric distribution (NHGD). We first relate the middle ciphertext's m.l.p.'s NHGD probability for a given plaintext/ciphertext space to that of a space twice the size; iterating this result produces a formula for the middle ciphertext's m.l.p.'s NHGD probability in a large space given the analogous value in a small space. We then relate *any* ciphertext's m.l.p.'s NHGD probability to that of the middle ciphertext in the space. Finally, we approximate the sum of m.l.p. NHGD probabilities over the ciphertext space in terms of that of the middle ciphertext, and hence to that of the middle ciphertext in a smaller space. Plugging in a value for the m.l.p. NHGD probability on the small space and simplifying yields the bound.

Evaluating the Bound. The bound of Theorem 1 is quite succinct—it does not even rely on N (as long as $N \geq 2M$). The result in essence shows that as long as the challenge set size z is small compared to M, the bound is a small constant times $z/\sqrt{M}$. This in turn is small as long as z is small compared to $\sqrt{M}$.

Plugging in some parameters, we can see some numerical bounds. (In all the following, we assume $N \geq 2M$.) For $M = 2^{80}$ and $z = 1$, the bound is $1.2 \cdot 2^{-37}$. For $M = 2^{80}$ and $z = 2^{20}$, the bound is $1.2 \cdot 2^{-17}$. For $M = 2^{80}$ and $z = 2^{38}$, the bound is no longer useful at 1.2.

We see that $\mathsf{ROPF}_{[M],[N]}$ has very good one-wayness security for reasonably-sized parameters. Given the results of [8] our bound for ROPF can be easily adjusted for their POPF construction, by taking into account pseudorandomness of an underlying blockcipher. But as we discussed in the introduction, standard one-wayness may not be sufficient in all applications and we have to also analyze the schemes under other security notions. Thus, we turn to the next result.

A Lower Bound on Large Window One-Wayness. Here we show that there exists a very efficient adversary attacking the window one-wayness of an ROPF for a sufficiently large window size. A more intuitive explanation of the result follows the theorem.

Theorem 2. *For any window size r and challenge set size z, there exists an adversary A such that*

$$\mathbf{Adv}^{r,z\text{-wow}}_{\mathsf{ROPF}_{[M],[N]}}(A) \geq \mathbf{Adv}^{r,1\text{-wow}}_{\mathsf{ROPF}_{[M],[N]}}(A) \geq 1 - 2e^{-\frac{(r-1)^2}{2}\frac{(M-1)}{M^2}}.$$

The proof is in the full version [9]. There, we construct a straightforward attack and demonstrate that it has the above probability of success, using some bounds by Chvátal on the tail probabilities of the hypergeometric distribution.

Intuitively, Theorem 2 implies that for $r \approx b\sqrt{M}$, where b is a large enough constant (say $b \geq 8$), there exists an adversary A whose r-window one-wayness is very close to 1. More precisely, let $r = b\frac{M}{\sqrt{M-1}} + 1$, and the theorem implies there exists an A such that

$$\mathbf{Adv}^{r,z\text{-wow}}_{\mathsf{ROPF}_{[M],[N]}}(A) \geq 1 - 2e^{-b^2/2}.$$

An Upper Bound on the $1, z$**-WDOW Advantage.** The following theorem, with the proof in the full version [9] , states an upper bound on the $1, z$-distance one-wayness of a random OPF that is very similar to the bound in Theorem 1.

Theorem 3. *For any challenge set size z and adversary A, if $N \geq 2M$ and $M \geq 16 + z$ then*

$$\mathbf{Adv}^{1,z\text{-wdow}}_{\mathsf{ROPF}_{[M],[N]}}(A) \leq \frac{9z(z-1)}{\sqrt{M-z+1}}.$$

Naturally, as this result looks very much like that of Theorem 1, the proof follows the same strategy and achieves similar results. The only differences are that the initial reduction relates r, z-WDOW security to $r, 2$-WDOW security, incurring a factor $z(z-1)$ advantage increase as opposed to just z, and the initial (tight) bound formula replaces parameters N, M with $N-1$, $M-1$. for proof details.

Thus, the $1, z$-window distance one-wayness of a random OPF is upper-bounded in a similar fashion as the $1, z$-window one-wayness, and we conclude that random OPFs have good $1, z$-WDOW security. Again, though, that is not the whole story, as we see next.

A Lower Bound on Window Distance One-Wayness of ROPF. Here, we derive a result similar to that of Theorem 2, but for the window distance one-wayness of a random OPF.

Theorem 4. *For any window size r and challenge set size z, there exists an efficient adversary A such that*

$$\mathbf{Adv}^{r,z\text{-wdow}}_{\mathsf{ROPF}_{[M],[N]}}(A) \geq \mathbf{Adv}^{r,1\text{-wdow}}_{\mathsf{ROPF}_{[M],[N]}}(A) \geq 1 - 2e^{-\frac{(r-1)^2}{2}\frac{(M-2)}{(M-1)^2}}.$$

The proof uses directly a result from the proof of Theorem 2 and appears in the full version [9].

Intuitively, Theorem 4 implies that for $r \approx b\sqrt{M}$, where b is a large enough constant (say, $b \geq 8$), there exists an efficient adversary A whose r-window distance one-wayness advantage is very close to 1. More precisely, let $r = b\frac{M-1}{\sqrt{M-2}}+1$, and the theorem implies there exists an A such that

$$\mathbf{Adv}^{r,z\text{-wow}}_{\mathsf{ROPF}_{[M],[N]}}(A) \geq 1 - 2e^{-b^2/2}.$$

4.1 Further Security Considerations for ROPFs

In this section, we explore several important questions regarding our ROPF security analysis.

Effect of Known-Plaintext Attacks. It is a natural question to ask what happens to the security of an ROPF scheme when the adversary knows a certain number of plaintext-ciphertext pairs. In general, we can answer this question for each definition of one-wayness using a simple extension of the arguments above.

In the scheme $\mathsf{ROPF}_{\mathcal{D},\mathcal{R}}$, known plaintext-ciphertext pairs split the plaintext and ciphertext spaces into subspaces. On each subspace, the analysis under each one-wayness definition reduces to that of an ROPF on the domain and range of the subspace. For instance, if (m_1, c_1) and (m_2, c_2) are known for $m_1 < m_2$, and no other known plaintext-ciphertext pairs occur between these two, then for $\mathcal{D}' = \{m \in \mathcal{D} \mid m_1 < m < m_2\}$ and $\mathcal{R}' = \{c \in \mathcal{R} \mid c_1 < c < c_2\}$, we analyze the behavior of the function on this subspace by considering the one-wayness bounds on $\mathsf{ROPF}_{\mathcal{D}',\mathcal{R}'}$.

This brings up an important issue. For much of our analysis to apply to a scheme, it must be the case that the ciphertext space is at least twice the size of the message space. Therefore, in order to make sure that our analysis will still apply to most subspaces once several plaintext-ciphertext pairs are discovered by the adversary, we would like to choose the initial parameters in such a way that subspaces are unlikely to violate this condition.

Choosing the Ciphertext Space Size. This brings us to the question posed in [8]: given a plaintext space of size M, what should be the size N of the ciphertext space? The recommendation and justification given in [8] was ad-hoc, necessarily so because the paper lacked a notion of security that would in any way depend on the size of N compared to M. Indeed, the choice of N has to do with the nature of the ideal object, an ROPF, while [8] was focused only on pseudorandomly sampling that ideal object, not analyzing it. Now that we have ways of characterizing the security of an ROPF using our one-wayness definitions, we can more justifiably discuss the question of what to choose for N.

For $g \in \mathsf{OPF}_{[M],[N]}$, if $m_1 < m_2 \in [M]$ exist such that $g(m_2) - g(m_1) < 2(m_2 - m_1)$, then we say that g is *shallow* on the ciphertext interval $[g(m_1), g(m_2)]$. The bounds found in the previous sections assume that $N \geq 2M$. Thus, any non-shallow interval can be analyzed through our theorems about one-wayness, and as a result we would like to choose N to avoid shallow intervals, both in the original space and in potential subspaces.

In particular, consider the following result, which bounds the probability that an interval between encryptions of two random plaintexts is shallow.

Proposition 1. *Let* $t = (N-1)/(M-1)$, *and assume* $t \geq 7$. *Let* $m_1 \xleftarrow{\$} [M]$, $m_2 \xleftarrow{\$} [M] \setminus \{m_1\}$, $K \xleftarrow{\$} \mathcal{K}_{\mathrm{r}}$, $\mathcal{E}nc_{\mathrm{r}}(K,(m_1,m_2)) = (c_1,c_2)$, $w = c_2 - c_1 \bmod M$, *and* $d = m_2 - m_1 \bmod M$. *Then over the choice of* m_1, m_2, K,

$$\Pr[\,2d > w\,] < \frac{3}{t}\frac{1}{\sqrt{(M-1)/\ln M}}\,.$$

The proof can be found in [9] and is mostly algebraic fiddling.

This bound gives us an idea of good values for $t \approx N/M$. In particular, it seems that choosing a constant for $t \geq 7$, that is, taking N to be a constant multiple of M, is sufficient in order to make the above probability negligible. Whether the constant should be large or small depends on one's tolerance for random intervals to be shallow.

On Implementing a Scheme to Support Range Queries using POPF. We stress that most of our analysis relies on the uniformity assumption assumption, namely that challenge messages come from a uniform distribution. (Intuitively, the we need this in our analysis so that the ciphertexts fall into a range subset of the range.) It is an open problem to extend our analysis to other input distributions, and until that is accomplished, we do not recommend practitioners draw any conclusions from the analysis.

5 Achieving Stronger Security

We study new ways to achieve better security than the OPE scheme of [8] while still allowing for efficient range queries on encrypted data. But first, we define a general primitive, Efficiently Orderable Encryption (EOE), that includes all schemes that support efficient standard range queries, including OPE. We show that IND-OCPA, defined and shown to be unachievable by OPE in [8], is the ideal security definition for such schemes.

We define "committed" analogues of EOE and IND-OCPA, namely CEOE and IND-CCPA, that apply to the practical scenario where the database to encrypt is pre-determined and static. Such a setting has been studied in several works on searchable encryption, including the first paper to propose an order-preserving scheme [1,13]. We then propose a new CEOE scheme that is CCPA-secure.

Finally, we develop a generic modification of an OPE that supports modular range queries (but not standard range queries) and overcomes some of the security weaknesses of any OPE that we studied in Section 4. The scheme is not EOE because it does not leak order; rather, it leaks only "modular" order.

Efficiently Orderable Encryption. We say that $\mathcal{EOE} = (\mathcal{K}, \mathcal{E}nc, \mathcal{D}ec, W)$ is an *efficiently-orderable encryption* (EOE) scheme if $\mathcal{K}, \mathcal{E}nc, \mathcal{D}ec$ are the algorithms of a symmetric encryption scheme, W is an efficient algorithm that takes two ciphertexts as input, and defining $C_K = \{\mathcal{E}nc(K,m) \mid m \in \mathcal{M}\}$ as the set of valid ciphertexts for key K,

$$W(c_0, c_1) = \begin{cases} 1 & \text{if } \mathcal{D}ec(K,c_0) < \mathcal{D}ec(K,c_1) \\ 0 & \text{if } \mathcal{D}ec(K,c_0) = \mathcal{D}ec(K,c_1) \\ -1 & \text{if } \mathcal{D}ec(K,c_0) > \mathcal{D}ec(K,c_1) \end{cases}$$

for any key K and all $c_0, c_1 \in C_K$. It is easy to see that such a scheme permits efficient standard range queries, as the server can keep the encrypted database sorted using W.

It is also clear that any OPE scheme $(\mathcal{K}, \mathcal{E}nc, \mathcal{D}ec)$ corresponds to an EOE scheme with the same key generation, encryption, and decryption algorithms, and $W(c_0, c_1)$ outputting 1, 0, or -1 if the relation between c_0 and c_1 is $<$, $=$, or $>$, respectively. But in general an EOE scheme does not have to be deterministic.

5.1 Committed Efficiently-Orderable Encryption

Range Queries on a Predetermined Static Database. Now we consider schemes for the settings when it is possible for the user to preprocess the whole data before encrypting and sending it to the server. For that we allow the key generation of an EOE scheme to take the message set as input, which we rename a *committed* EOE scheme.

Committed efficiently-orderable encryption. A *committed efficiently-orderable encryption* (CEOE) scheme on domain $\mathcal{D}$ is a tuple $(\mathcal{K}, \mathcal{E}nc, \mathcal{D}ec, W)$ satisfying the following.

- The randomized key generation algorithm $\mathcal{K}$ takes a message space $\mathcal{M} \subset \mathcal{D}$ (called the *committed* message space) as input and outputs a secret key K.
- For any committed message space $\mathcal{M} \subset \mathcal{D}$, $(\mathcal{K}(\mathcal{M}), \mathcal{E}nc, \mathcal{D}ec, W)$ is an EOE scheme on $\mathcal{M}$.

We will show that a CEOE scheme can achieve very strong security. In particular, it can achieve the "committed" adaptation of the IND-OCPA notion from [8], where the adversary outputs two vectors of plaintexts with the same order and equality pattern and is asked to guess whether it is given encryptions of the first or second vector. We define *indistinguishability under committed chosen plaintext attacks* (IND-CCPA). The definition mimics IND-OCPA except that the adversary chooses the challenge vectors (now viewed as message spaces) before key generation, and the scheme's key generation algorithm takes the appropriate message space as input.

IND-CCPA. Let $\mathcal{CEOE} = (\mathcal{K}, \mathcal{E}nc, \mathcal{D}ec, W)$ be a CEOE scheme on message space $\mathcal{M}$.

For an adversary $A = (A_1, A_2)$ and $b \in \{0,1\}$ consider the following experiment. (σ denotes a state.)

Experiment $\mathbf{Exp}^{\text{ind-ccpa-b}}_{\mathcal{CEOE}}(A)$

$(\mathcal{M}_0, \mathcal{M}_1, \sigma) \stackrel{\$}{\leftarrow} A_1$; If $|\mathcal{M}_0| \neq |\mathcal{M}_1|$ then output $\perp$.
Otherwise, let $l = |\mathcal{M}_0| = |\mathcal{M}_1|$
Let $m^j_1 < m^j_2 < \ldots < m^j_l$ be the elements of $\mathcal{M}_j$, for $j = 0, 1$
If there exist $1 \leq i \leq l$ so that $|m^0_i| \neq |m^1_i|$ then output $\perp$
$K \stackrel{\$}{\leftarrow} \mathcal{K}(\mathcal{M}_b)$; $c_j \leftarrow \mathcal{E}nc(K, m^b_j)$ for $j = 1, \ldots, l$
$d \stackrel{\$}{\leftarrow} A_2(\sigma, c_1, c_1, \ldots, c_l)$. Return d

For an adversary A, define its *ind-ccpa advantage* against $\mathcal{SE}$ as

$$\mathbf{Adv}^{\text{ind-ccpa}}_{\mathcal{CEOE}}(A) = \Pr\left[\mathbf{Exp}^{\text{ind-ccpa-1}}_{\mathcal{CEOE}}(A) = 1\right] - \Pr\left[\mathbf{Exp}^{\text{ind-ccpa-0}}_{\mathcal{CEOE}}(A) = 1\right].$$

We say that $\mathcal{CEOE}$ is IND-CCPA secure if the ind-ccpa advantage of any adversary against $\mathcal{CEOE}$ is small.

Our CEOE construction and its security. We now propose a CEOE scheme that will achieve IND-CCPA security. A ciphertext in our scheme consists of a semantically-secure ciphertext of the message concatenated with the tag, which indicates the order of the message in the ordered message list. As a building block for our scheme we use monotone minimal perfect hash functions, defined as follows.

Let $\mathcal{M}$ be a set with a total (lexicographical) order. h is a *monotone minimal perfect hash function* [4] (MMPHF) on $\mathcal{M}$ if h sends the ith largest element of $\mathcal{M}$ to i, for $i = 0, 1, \ldots, |\mathcal{M}| - 1$. Notice that the MMPHF on any given domain $\mathcal{M}$ is unique. So that we can use MMPHFs in the upcoming construction, let an *index tagging scheme* $(\mathcal{K}, \tau)$ be a pair of algorithms such that $\mathcal{K}$ takes a domain $\mathcal{M}$ and outputs a secret key $K_{\mathcal{M}}$ so that $\tau(K_{\mathcal{M}}, \cdot)$ is the (unique) MMPHF for $\mathcal{M}$, while $\tau(K, m) = \perp$ for any $m \notin \mathcal{M}$.

Our CEOE construction is based on two building blocks: MMPHF tagging and any symmetric encryption scheme.

Let $(\mathcal{K}_t, \tau)$ be an index tagging scheme. Fix a universe $\mathcal{D}$, and let $\mathcal{SE} = (\mathcal{K}', \mathcal{E}nc', \mathcal{D}ec')$ be any symmetric encryption scheme on $\mathcal{D}$. We construct a CEOE scheme $(\mathcal{K}, \mathcal{E}nc, \mathcal{D}ec, W)$ as follows.

- $\mathcal{K}$ takes $\mathcal{M} \subset \mathcal{D}$ as input, runs $K_t \leftarrow \mathcal{K}_t(\mathcal{M})$ and $K_e \leftarrow \mathcal{K}'$, and returns $K = K_t \| K_e$.
- $\mathcal{E}nc$ takes key $K = K_t \| K_e$ and message m as input, and computes $i = \tau(K_t, m)$. If $i = \perp$ then $\mathcal{E}nc$ returns $\perp$, otherwise it returns $i \| \mathcal{E}nc'(K_e, m)$.
- $\mathcal{D}ec$ takes key $K = K_t \| K_e$ and ciphertext $c = i \| c'$ as input, and returns $\mathcal{D}ec'(K_e, c')$.
- W takes ciphertexts $c_0 = i_0 \| c'_0$ and $c_1 = i_1 \| c'_1$ as input, and returns 1 if $i_0 < i_1$, 0 if $i_0 = i_1$, and -1 if $i_0 > i_1$.

We note that unlike the scheme with pre-processing for exact-match queries [13], when using the above scheme the server does indexing and query processing as for unencrypted data, which is a practical advantage. Also, as the following result shows, the scheme is secure under IND-CCPA.

Theorem 5. *The CEOE scheme defined above is IND-CCPA-secure provided the underlying symmetric encryption scheme is IND-CPA secure.*

The proof is in the full version [9].

Note that our secure CEOE construction relies on an efficient MMHPF implementation. Luckily, MMHPFs were studied recently by [4]. They showed that for a universe of size 2^w and for $n \geq \log w$, the shortest possible description of an MMPHF function (and thus, best possible key length for a tagging scheme) on n elements is unfortunately quite large at $\Omega(n)$ bits. This is somewhat disheartening, as a naive solution, in which the MMPHF key consists of an n-entry array whose ith entry is the ith largest element in the domain, has a key length of $O(nw)$. Nevertheless, the authors of [4] were able to generate MMPHF descriptions that are closer to the optimal bound: one construction uses $O(n \log \log w)$ bits and has query time $O(\log w)$, and the other uses $O(n \log w)$ bits and has constant query time. This is still large, but may be practical depending on the parameters involved.

5.2 Modular OPE and Analysis of an Ideal MOPE Scheme

Modular OPE. We propose a modification to (that can be viewed as a generalization of) an OPE scheme that improves the security performance of any OPE. The resulting scheme is no longer strictly order-preserving, but it still permits range queries. However, now the queries must be *modular* range queries. Standard range queries are not supported, as only "modular order" rather than order is leaked. The modification from OPE is simple, generic, and basically free computation-wise.

Let $\mathcal{SE}_{[M],[N]} = (\mathcal{K}, \mathcal{E}nc, \mathcal{D}ec)$ be an order-preserving encryption scheme. Define a *modular order-preserving encryption scheme* (MOPE) $\mathcal{SE}_{[M],[N]} = (\mathcal{K}_{\mathrm{m}}, \mathcal{E}nc_{\mathrm{m}}, \mathcal{D}ec_{\mathrm{m}})$ as follows.

- $\mathcal{K}_{\mathrm{m}}$ runs $\mathcal{K}$ to get K, picks $j \xleftarrow{\$} [M]$ and returns (K, j).
- $\mathcal{E}nc_{\mathrm{m}}$ on input (K, j) and m returns $\mathcal{E}nc(K, m - j \bmod M)$.
- $\mathcal{D}ec_{\mathrm{m}}$ on inputs (K, j) and c returns $\mathcal{D}ec(K, c) + j \bmod M$.

Notice that a MOPE is suitable for modular range query support as follows. To request the ciphertexts of the messages in the range $[m_1, m_2]$ (if $m_1 \leq m_2$), or $[m_1, M] \cup [1, m_2]$ (if $m_1 > m_2$), the user computes $c_1 \leftarrow \mathcal{E}nc_{\mathrm{m}}(K, m_1), c_2 \leftarrow \mathcal{E}nc_{\mathrm{m}}(K, m_2)$ and submits ciphertexts (c_1, c_2) as the query. The server returns the ciphertexts in the interval $[c_1, c_2]$ (if $c_1 \leq c_2$) or $[c_1, N] \cup [1, c_2]$ (if $c_1 > c_2$).

MOPE Security and Random MOPF. In order to define the security of an MOPE scheme, we introduce a generalization of OPFs. For $j \in [M]$, let $\phi_j : [M] \to [M]$ be the cyclic transformation $\phi_j(x) = (x - j - 1) \bmod M + 1$. We define the set of *modular order preserving functions* from $[M]$ to $[N]$ as

$$\mathsf{MOPF}_{[M],[N]} = \{f \circ \phi_j \mid f \in \mathsf{OPF}_{[M],[N]}, j \in [M]\} .$$

Note that all OPFs are MOPFs; on the other hand, most MOPFs are not OPFs. However, a MOPF g is "modular order-preserving" in that the function $g - g(0) \bmod N$ is order-preserving.

Now, define $\mathsf{RMOPF}_{[M],[N]} = (\mathcal{K}_{\mathrm{rm}}, \mathcal{E}nc_{\mathrm{rm}}, \mathcal{D}ec_{\mathrm{rm}})$, the *random modular order-preserving function* scheme, as the following (inefficient) encryption scheme:

- $\mathcal{K}_{\mathrm{rm}}$ returns a random instance g of $\mathsf{MOPF}_{[M][N]}$.
- $\mathcal{E}nc_{\mathrm{rm}}$ takes the key g and a plaintext m to return $g(m)$.
- $\mathcal{D}ec_{\mathrm{rm}}$ takes the key g and a ciphertext c to return $g^{-1}(c)$.

Note that an MOPF could alternatively be defined with a random ciphertext shift following the OPF rather than a random plaintext shift preceding it. The advantage of the above definition is that the map from (OPF, ciphertext offset) pairs to MOPFs is bijective whereas in the alternative it is not one-to-one.

We now are ready to define MOPE security. Fix an MOPE scheme $\mathcal{SE}_{[M],[N]} = (\mathcal{K}_{\mathrm{m}}, \mathcal{E}nc_{\mathrm{m}}, \mathcal{D}ec_{\mathrm{m}})$. Let $\mathsf{RMOPF}_{[M],[N]} = (\mathcal{K}_{\mathrm{rm}}, \mathcal{E}nc_{\mathrm{rm}}, \mathcal{D}ec_{\mathrm{rm}})$ be as defined above. For an adversary A, define its $\mathbf{Adv}^{\mathrm{pmopf}}_{\mathcal{SE}}(A)$, *pmopf-advantage* (or *pseudorandom modular order-preserving function advantage*) against $\mathcal{SE}$ as

$$\Pr\left[\, K \xleftarrow{\$} \mathcal{K}_{\mathrm{m}} \;:\; A^{\mathcal{E}nc_{\mathrm{m}}(K,\cdot)} = 1 \,\right] - \Pr\left[\, g \xleftarrow{\$} \mathsf{RMOPF}_{[M],[N]} \;:\; A^{g(\cdot)} = 1 \,\right] .$$

It is straightforward to show that the MOPE scheme obtained from any POPF-secure OPE scheme via the transformation defined in the beginning of Section 5.2 is PMOPF-secure, under the same assumption as the base scheme. We omit the details.

We now analyze the ideal object, RMOPF, under the one-wayness definitions.

Window One-Wayness of RMOPF. The following proposition, proved in [9], establishes that RMOPF is optimally r,z-window one-way (and hence optimally one-way, taking $r = 1$) in the sense that an adversary cannot do better than an adversary that outputs a random window independent of the challenge set. (Reminder: "window" includes windows that wrap around the edge of the space.)

Proposition 2. *Fix any window size r and challenge set size z. Let $A_{\mathrm{rand}}(r)$ be an r,z-WOW adversary that, on any input, outputs a random r-window from $[M]$. Then for any adversary A,*

$$\mathbf{Adv}^{r,z\text{-wow}}_{\mathsf{RMOPF}_{[M],[N]}}(A) \leq \mathbf{Adv}^{r,z\text{-wow}}_{\mathsf{RMOPF}_{[M],[N]}}(A_{\mathrm{rand}}(r)) \leq rz/M \;.$$

As one might surmise, the above "optimal" characterization of the one-wayness of a random MOPF fails to show a complete picture of the information a random MOPF leaks. To investigate further, we turn to distance one-wayness.

WDOW Advantage Bounds for RMOPF. We claim that the distance one-wayness analysis for RMOPF is exactly the same as for ROPF. To see this, consider the following proposition, whose (short) proof is in [9].

Proposition 3. *Let $c_1, c_2 \in [N]$. Then for any $d \in \{0, \dots, M-1\}$,*

$$\begin{aligned} &\Pr\left[\, \mathcal{D}ec_{\mathrm{r}}(K_1, c_2) - \mathcal{D}ec_{\mathrm{r}}(K_1, c_1) = d \,\right] \\ &= \Pr\left[\, \mathcal{D}ec_{\mathrm{rm}}(K_2, c_2) - \mathcal{D}ec_{\mathrm{rm}}(K_2, c_1) = d \,\right] , \end{aligned}$$

where the probabilities are over, respectively, $K_1 \xleftarrow{\$} \mathcal{K}_{\mathrm{r}}$ and $K_2 \xleftarrow{\$} \mathcal{K}_{\mathrm{rm}}$.

Therefore, the $1, z$-WDOW advantage upper bound of Theorem 3 and the r, z-WDOW advantage lower bound of Theorem 4 against ROPF schemes also apply to RMOPF schemes on the same parameters.

So, while an RMOPF has similar security to that of an ROPF for distance and window distance one-wayness, it is better in terms of one-wayness and window one-wayness. The analysis easily transfers to any secure MOPE scheme. We now discuss a few supplemental security considerations for RMOPF schemes.

Effect of a Known-Plaintext Attack on RMOPF. In the $\mathsf{RMOPF}_{[M],[N]}$ scheme, if the adversary learns a single plaintext-ciphertext pair, then the one-wayness analysis reduces to that of $\mathsf{ROPF}_{[M-1],[N-1]}$. To see this, note that if g is a random function in $\mathsf{MOPF}_{[M],[N]}$, and it is revealed that $g(m_0) = c_0$, then $f(m) = g(m + m_0 \bmod M) - c_0 \bmod N$ is a random function in $\mathsf{OPF}_{[M-1],[N-1]}$.

On Implementing a Scheme to Support Range Queries using PMOPF. We note that when a pseudorandom MOPF scheme is used to implement a range-query-supporting database, even wrap-around target range queries must be made, for otherwise an adversary may infer the secret offset of the MOPF scheme after observing many non-wrap-around target queries.

Remark. We finally note that the tagging scheme defined in Section 5.1 could be similarly modified so that its tag receives a secret offset. The resulting scheme would support modular range queries in predetermined static database scenario, and satisfy a stronger version of IND-CCPA, leaking only "modular" order.

Acknowledgements. We thank Nigel Smart, Abdullatif Shikfa and the anonymous reviewers for useful comments. We also thank Adam Smith and Brent Waters for useful discussions, and in particular Adam Smith for pointing out that ROPF encryption leaks the high-order bits of the plaintexts. Alexandra Boldyreva and Nathan Chenette are supported in part by Alexandra's NSF CAREER award 0545659 and NSF Cyber Trust award 0831184. Adam O'Neill was supported in part by Brent Waters grants NSF CNS-0915361 and CNS-0952692.

References

1. Agrawal, R., Kiernan, J., Srikant, R., Xu, Y.: Order-preserving encryption for numeric data. In: SIGMOD 2004, pp. 563–574. ACM, New York (2004)
2. Amanatidis, G., Boldyreva, A., O'Neill, A.: Provably-secure schemes for basic query support in outsourced databases. In: DBSec 2007, pp. 14–30. Springer, Heidelberg (2007)
3. Bauer, F.: Decrypted Secrets: Methods and Maxims of Cryptology. Springer, Heidelberg (2006)
4. Belazzougui, D., Boldi, P., Pagh, R., Vigna, S.: Monotone minimal perfect hashing: searching a sorted table with $o(1)$ accesses. In: SODA 2009, pp. 785–794. SIAM, Philadelphia (2009)
5. Bellare, M., Boldyreva, A., Knudsen, L.R., Namprempre, C.: Online ciphers and the hash-CBC construction. In: Kilian, J. (ed.) CRYPTO 2001. LNCS, vol. 2139, pp. 292–309. Springer, Heidelberg (2001)

6. Bellare, M., Boldyreva, A., O'Neill, A.: Deterministic and efficiently searchable encryption. In: Menezes, A. (ed.) CRYPTO 2007. LNCS, vol. 4622, pp. 535–552. Springer, Heidelberg (2007)
7. Bellare, M., Fischlin, M., O'Neill, A., Ristenpart, T.: Deterministic encryption: Definitional equivalences and constructions without random oracles. In: Wagner, D. (ed.) CRYPTO 2008. LNCS, vol. 5157, pp. 360–378. Springer, Heidelberg (2008)
8. Boldyreva, A., Chenette, N., Lee, Y., O'Neill, A.: Order-preserving symmetric encryption. In: Joux, A. (ed.) EUROCRYPT 2009. LNCS, vol. 5479, pp. 224–241. Springer, Heidelberg (2009)
9. Boldyreva, A., Chenette, N., O'Neill, A.: Order-Preserving Encryption Revisited: Improved Security Analysis and Alternative Solutions (2011) Full version of this paper, http://www.cc.gatech.edu/~aboldyre/publications.html
10. Boldyreva, A., Fehr, S., O'Neill, A.: On notions of security for deterministic encryption, and efficient constructions without random oracles. In: Wagner, D. (ed.) CRYPTO 2008. LNCS, vol. 5157, pp. 335–359. Springer, Heidelberg (2008)
11. Boneh, D., Waters, B.: Conjunctive, subset, and range queries on encrypted data. In: Vadhan, S.P. (ed.) TCC 2007. LNCS, vol. 4392, pp. 535–554. Springer, Heidelberg (2007)
12. Chvátal, V.: The tail of the hypergeometric distribution. Discrete Mathematics 25(3), 285–287 (1979)
13. Curtmola, R., Garay, J., Kamara, S., Ostrovsky, R.: Searchable symmetric encryption: Improved denitions and efficient constructions. In: CCS 2006, pp. 79–88. ACM, New York (2006)
14. Ding, Y., Klein, K.: Model-Driven Application-Level Encryption for the Privacy of E-health Data. In: International Conference on Availability, Reliability and Security, pp. 341–346 (2010)
15. Kershaw, D.: Some extensions of W. Gautschi's inequalities for the gamma function. Mathematics of Computation 41(164), 607–611 (1983)
16. Li, J., Omiecinski, E.: Efficiency and security trade-off in supporting range queries on encrypted databases. In: DBSec 2005, pp. 69–83. Springer, Heidelberg (2005)
17. Liu, H., Wang, H., Chen, Y.: Ensuring Data Storage Security against Frequency-Based Attacks in Wireless Networks. In: Rajaraman, R., Moscibroda, T., Dunkels, A., Scaglione, A. (eds.) DCOSS 2010. LNCS, vol. 6131, pp. 201–215. Springer, Heidelberg (2010)
18. Lu, W., Varna, A.L., Wu, M.: Security analysis for privacy preserving search of multimedia. In: Image Processing (ICIP), 2010, pp. 26–29 (2010)
19. Shi, E., Bethencourt, J., Chan, T.-H.H., Song, D., Perrig, A.: Multi-dimensional range query over encrypted data. In: Symposium on Security and Privacy 2007, pp. 350–364. IEEE, Los Alamitos (2007)
20. Tang, Q.: Privacy preserving mapping schemes supporting comparison. In: Proceedings of the ACM Workshop on Cloud Computing Security Workshop (CCSW 2010). ACM, New York (2010)
21. Wang, C., Cao, N., Li, J., Ren, K., Lou, W.: Secure Ranked Keyword Search over Encrypted Cloud Data. In: ICDCS 2010, pp. 253–262. IEEE, Los Alamitos (2010)
22. Xu, J., Fan, J., Ammar, M.H., Moon, S.B.: Prefix-preserving IP address anonymization: Measurement-based security evaluation and a new cryptography-based scheme. In: ICNP 2002, pp. 280–289. IEEE, Los Alamitos (2002)

A New Variant of PMAC: Beyond the Birthday Bound

Kan Yasuda

NTT Information Sharing Platform Laboratories,
NTT Corporation, Japan
yasuda.kan@lab.ntt.co.jp

Abstract. We propose a PMAC-type mode of operation that can be used as a highly secure MAC (Message Authentication Code) or PRF (Pseudo-Random Function). Our scheme is based on the assumption that the underlying n-bit blockcipher is a pseudo-random permutation. Our construction, which we call `PMAC_Plus`, involves extensive modification to PMAC, requiring three blockcipher keys. The `PMAC_Plus` algorithm is a first rate-1 (*i.e.*, one blockcipher call per n-bit message block) blockcipher-based MAC secure against $O(2^{2n/3})$ queries, increasing the $O(2^{n/2})$ security of PMAC at a low additional cost. Our analysis uses some of the security-proof techniques developed with the sum construction (Eurocrypt 2000) and with the encrypted-CBC sum construction (CT-RSA 2010).

Keywords: 64-bit blockcipher, PRP, sum construction, CBC vs. PMAC, game-playing technique.

1 Introduction

MACs (Message Authentication Codes) are frequently realized by making iterative use of blockciphers. They are called *blockcipher-based MACs* and specified in a large number of standardized documents including ISO 9797-1 [13].

The majority of blockcipher-based MACs iterate a blockcipher in the so-called *CBC* (*Cipher Block Chaining*) style [3,19,15], by xor-ing the current message block with the previous chaining value and then inputting the xor-ed result into the next blockcipher call. CBC-type MACs have a history of continuous updates, whose purpose is mainly to reduce the number of keys and to increase efficiency in the last message block. CMAC [18] *a.k.a.* OMAC [11], the first 1-key CBC MAC derived from XCBC [7], can be regarded as an outcome of such evolution.

On the other hand, PMAC (Parallelizable MAC) [8] is a distinctive, parallelizable blockcipher-based MAC. Its internal structure is completely different from CBC iteration, which is inherently sequential and not parallelizable. If sequentially implemented, then PMAC becomes slightly slower than CBC MACs, because PMAC requires an extra operation at every blockcipher call. The extra operation is typically a constant multiplication in the finite field, which is fast but still slower than a simple xor operation used by CBC MACs. However,

P. Rogaway (Ed.): CRYPTO 2011, LNCS 6841, pp. 596–609, 2011.

under parallel implementation, PMAC can possibly outperform CBC MACs significantly.

We believe it is worth re-evaluating PMAC-type constructions under the current trend of "parallelizable" (pipeline, superscalar, vector, and multi-core) CPUs (*e.g.*, see [14] for a state-of the-art implementation of AES in this direction). It seems that today PMAC is still not as widespread as CBC MACs, perhaps because most of the computational environments commercially available so far have been increasing the clock rate and hence the speed of sequential operations.

In this paper we look at another advantageous aspect of PMAC-type constructions. That is its proof of security. The parallel construction has a structure easy to analyze and to obtain better bounds. Intuitively, the difference between PMAC-type and CBC-type iterations lies in the "long-message attacks" noted by Preneel and Oorschot [20]: Suppose, for the moment, that we iterate an n-bit function rather than permutation. Then two messages $M00\cdots0$ and $M'00\cdots0$ would collide at some point of the CBC iteration if they are long enough—say 2^n blocks—and the collision would propagate through the output of the MAC algorithm. An event of this sort does not happen to PMAC. This appealing aspect of PMAC-type constructions is already pointed out, though implicitly, by Minematsu [17] in obtaining an $O(\ell q^2/2^n)$-type bound for PMAC (where ℓ is the maximum length of a message, q the maximum number of queries and n the block size). Recall that obtaining such a bound for CBC MACs seems more troublesome [4].

Besides PMAC, there have been a few proposals of parallelizable MACs, for which a blockcipher can be used. These include XOR MAC by Bellare *et al.* [2] and PCS by Bernstein [6]. There have been also some improvements or alternatives to PMAC. These include PMAC1 by Rogaway [21] and iPMAC by Sarkar [22].

Birthday-Bound Problems and Our Contributions. We take the advantage of the provable-security aspect of PMAC-type constructions in solving the so-called *birthday-bound problem* of iterative MACs [20]. That is, a MAC construction having an n-bit size of intermediate values cannot be secure against more than $O(2^{n/2})$ queries—a forgery becomes possible after so many queries. Typical MAC constructions iterating an n-bit blockcipher suffer from this problem.

This is a sever problem particularly for 64-bit blockciphers. Not to mention the fact that the legacy Triple-DES is still widely used (especially in financial services), we also have new 64-bit blockciphers such as HIGHT [10] and PRESENT [9], possibly due to industrial demands for "lightweight" algorithms. The birthday bound may not be a serious problem for 128-bit blockciphers at the current moment. However, it contributes not only to existing 64-bit blockciphers but also to the longevity of 128-bit blockciphers in future use to construct an efficient MAC mode which is free of the birthday-bound problem.

Table 1. Summary of our result and comparison with previous constructions

	Rate	# of keys	Parallelizable?	Security bound	Ref.
Alg. 6 of ISO 9797-1	1/2	6		$O(\ell^4q^3/2^{2n})$ or restricted $O(\ell^3q^3/2^{2n})$	[13]
SUM-ECBC	1/2	4		$O(\ell^4q^3/2^{2n})$ or restricted $O(\ell^3q^3/2^{2n})$	[23]
PMAC_Plus	1	3	✓	$O(\ell^3q^3/2^{2n} + \ell q/2^n)$	This work

The first attempt to solve this problem was made in ISO 9797-1 (without proofs of security).[1] Unfortunately, Algorithm 4 of ISO 9797-1 was attacked and shown insecure ("insecure" meaning secure only up to the $O(2^{n/2})$ birthday bound) by Joux *et al.* [12]. Algorithm 6 of ISO 9797-1, on the other hand, has been proven secure against $O(2^{2n/3})$ queries [23]. The $O(2^{2n/3})$ bound holds only with certain restrictions on the message length. The work [23] has also presented SUM-ECBC, the sum (xor) of two encrypted CBC MACs. SUM-ECBC has become a 4-key rate-1/2 (meaning two encryptions to process n-bits) blockcipher-based MAC having the same security bound as Algorithm 6 of ISO 9797-1. We improve over these MAC constructions by utilizing a PMAC-type iteration. We propose a new MAC algorithm `PMAC_Plus`. `PMAC_Plus` is an extensively modified version of PMAC. `PMAC_Plus` remains rate-1 but operates with three blockcipher keys. `PMAC_Plus` has an $O(2^{2n/3})$ bound without such a restriction on the message length as existed with the security bound for the previous constructions.[2] Table 1 summarizes our result.

Organization. Section 2 provides necessary background. In Section 3 we define our algorithm `PMAC_Plus` and state its security result. The entire Section 4 is devoted to proofs of security. The paper ends with brief discussion in Section 5.

2 Preliminaries

Symbols and Notation. We fix a block size n, which is typically 64 or 128. We write Perm(n) for the set of permutations $P : \{0,1\}^n \to \{0,1\}^n$. We also fix a key space $\mathscr{K}$. Usually $\mathscr{K} = \{0,1\}^\kappa$, where $\kappa = 80$, 128, 192 or 256. A blockcipher E is a function $E : \mathscr{K} \times \{0,1\}^n \to \{0,1\}^n$ such that for each key $K \in \mathscr{K}$ we have $E_K \in$ Perm(n), where $E_K : \{0,1\}^n \to \{0,1\}^n$ is defined as $E_K(X) := E(K, X)$. We can then write E_K^{-1} for the inverse permutation.

The set $\{0,1\}^n$ can be regarded as a set of integers $\{0, 1, \ldots, 2^n - 1\}$. This can be done by converting an n-bit string $a_{n-1} \cdots a_1 a_0 \in \{0,1\}^n$ to an

[1] See, for example, [1] for another direction (using random coins) of treating birthday-bound problems.

[2] Our new bound would become vacuous for very long messages, say $2^{2n/3}$ blocks.

integer $a_{n-1}2^{n-1} + \cdots + a_1 2 + a_0$, where multiplication and addition are integer arithmetic.

Let $GF(2^n)$ denote the finite field with 2^n elements. We regard $\{0,1\}^n$ as $GF(2^n)$. That is, we identify an n-bit string $a_{n-1} \cdots a_1 a_0 \in \{0,1\}^n$ with a formal polynomial $a_{n-1}\mathtt{x}^{n-1} + \cdots a_1\mathtt{x} + 1 \in GF(2)[\mathtt{x}]$. To do so we need to fix an irreducible polynomial $a(\mathtt{x}) = \mathtt{x}^n + a_{n-1}\mathtt{x}^{n-1} + \cdots + a_1\mathtt{x} + a_0 \in GF(2)[\mathtt{x}]$. We sometimes write $\oplus$ and $\odot$ to emphasize addition and multiplication in the field, respectively. So for example we have $2 \oplus 3 = \mathtt{x} + (\mathtt{x}+1) = 1$ and $3 \odot 3 = (\mathtt{x}+1)^2 = \mathtt{x}^2 + 1 = 5$ if $n \geq 3$.

We choose irreducible polynomials $a(\mathtt{x}) = \mathtt{x}^{64} + \mathtt{x}^4 + \mathtt{x}^3 + \mathtt{x} + 1$ for $n = 64$ and $a(\mathtt{x}) = \mathtt{x}^{128} + \mathtt{x}^7 + \mathtt{x}^2 + \mathtt{x} + 1$ for $n = 128$. These are actually *primitive* polynomials, meaning the element $2 = \mathtt{x}$ generates the entire multiplicative group $GF(2^n)^*$ of order $2^n - 1$.

Security Notions. An adversary $\mathscr{A}$ is an oracle machine. $\mathscr{A}$ has access to its oracle $O(\cdot)$ and, after interaction with the oracle, outputs a bit, 1 or 0. We write $\mathscr{A}^{O(\cdot)} = 1$ to denote the event that $\mathscr{A}$ outputs 1 after interacting with $O(\cdot)$. We measure the resources of $\mathscr{A}$ in terms of time and query complexities. We fix a model of computation and a method of encoding. The query complexity is measured in terms of the number of queries (usually denoted q) and also in terms of the maximum length of each query (denoted ℓ). The length of a query is measured in blocks (n bits).

We say that (informally) a block cipher E is a (*secure*) *pseudo-random permutation* (*PRP*) if it is indistinguishable from a random permutation $P \xleftarrow{\$} \mathrm{Perm}(n)$, where $\xleftarrow{\$}$ means uniformly random sampling. Specifically, we consider the advantage function

$$\mathbf{Adv}_E^{\mathrm{prp}}(\mathscr{A}) := \Pr\left[\mathscr{A}^{E_K(\cdot)} = 1; K \xleftarrow{\$} \mathscr{K}\right] - \Pr\left[\mathscr{A}^{P(\cdot)} = 1; P \xleftarrow{\$} \mathrm{Perm}(n)\right],$$

and if this quantity is "small enough" for a class of adversaries, then we say that E is a PRP. Here note that the probabilities are defined over internal coin tosses of $\mathscr{A}$, if any, as well as over the choices of K and P. We further define $\mathbf{Adv}_E^{\mathrm{prp}}(t,q) := \max_{\mathscr{A}} \mathbf{Adv}_E^{\mathrm{prp}}(\mathscr{A})$, where the max runs over all adversaries $\mathscr{A}$ whose running time is at most t, making at most q queries to its oracle.

With abuse of notation let $\{0,1\}^*$ denote the set of finite bit strings whose length is at most ℓq blocks. Let $\mathrm{Func}(*,n)$ denote the set of functions $G : \{0,1\}^* \to \{0,1\}^n$. Our goal is to construct a *pseudo-random function* (*PRF*) $F_K : \{0,1\}^* \to \{0,1\}^n$ having keys $K \in \mathscr{K}'$ (and preferably $\mathscr{K}'$ is not much larger than $\mathscr{K}$). Recall that any PRF can be used as a secure MAC. We say that F is a secure PRF if it is indistinguishable from a random function $G \xleftarrow{\$} \mathrm{Func}(*,n)$, or more precisely, we define

$$\mathbf{Adv}_F^{\mathrm{prf}}(\mathscr{A}) := \Pr\left[\mathscr{A}^{F_K(\cdot)} = 1; K \xleftarrow{\$} \mathscr{K}'\right] - \Pr\left[\mathscr{A}^{G(\cdot)} = 1; G \xleftarrow{\$} \mathrm{Func}(*,n)\right].$$

We also define $\mathbf{Adv}_F^{\mathrm{prf}}(t,q,\ell)$ to be the maximum advantage running over all adversaries $\mathscr{A}$ whose running time is at most t, making at most q queries to its oracle, each query being at most ℓ blocks.

Game-Playing Techniques. Our proofs of security largely depend on the so-called game-playing techniques [5]. In particular, we perform lazy sampling for a random permutation $P \xleftarrow{\$} \mathrm{Perm}(n)$. That is, P is initially set everywhere undefined, and when a value $P(X)$ becomes necessary at some point in the game, a corresponding range point Y is randomly sampled as $Y \xleftarrow{\$} \{0,1\}^n$, so that we have $P(X) = Y$. We implicitly maintain two sets, $\mathtt{Dom}\,P$ and $\mathtt{Ran}\,P$, which keep the record of already-defined domain points and that of range points, respectively.

3 PMAC_Plus: Specification and Security

There exist different versions of PMAC. The version that we use here is based on the discrete-log-based LFSR (linear feedback shift register) developed by Rogaway [21]. Our MAC construction PMAC_Plus is defined in Algorithm 1. It calls a subroutine Internal, which is described in Algorithm 2.

Algorithm 1. $\mathtt{PMAC\_Plus}[E_{K_1}, E_{K_2}, E_{K_3}](M)$

$(\Sigma, \Theta) \leftarrow \mathtt{Internal}[E_{K_1}](M)$
$T \leftarrow E_{K_2}(\Sigma) \oplus E_{K_3}(\Theta)$
return T

Algorithm 2. Subroutine $\mathtt{Internal}[E_K](M)$

$\Delta_0 \leftarrow E_K(0)$
$\Delta_1 \leftarrow E_K(1)$
$M \leftarrow M \| 10^*$
Partition M into $M[1] \| \cdots \| M[m]$
for $i = 1$ to m **do**
  $X[i] \leftarrow M[i] \oplus 2^i \cdot \Delta_0 \oplus 2^{2i} \cdot \Delta_1$
  $Y_i \leftarrow E_K(X[i])$
end for
$\Sigma \leftarrow Y_1 \oplus \cdots \oplus Y_m$
$\Theta \leftarrow 2^{m-1} \cdot Y_1 \oplus 2^{m-2} \cdot Y_2 \oplus \cdots \oplus Y_m$
return (Σ, Θ)

In Algorithm 2, by "$M[m] \,\|\, 10^*$" we mean appending a bit 1 and then an appropriate number of bits 0 so that the bit length of $M[m] \,\|\, 10^*$ becomes n. By "partition M" we mean $M = M[1] \,\|\, \cdots \,\|\, M[m]$ so that $|M[1]| = |M[m]| = n$. See Fig. 1 for a pictorial representation of our construction PMAC_Plus.

The security of our PMAC_Plus construction is as follows:

Theorem 1 (Security of PMAC_Plus). *We have*

$$\mathbf{Adv}^{\mathrm{prf}}_{\mathtt{PMAC\_Plus}}(t, q, \ell) \le \frac{27\ell^3 q^3}{2^{2n}} + \frac{3\ell q}{2^n} + 3\mathbf{Adv}^{\mathrm{prp}}_E(t', \ell q + 2),$$

where t' is about t plus a time complexity necessary to compute E for $\ell q + 2q + 2$ times.

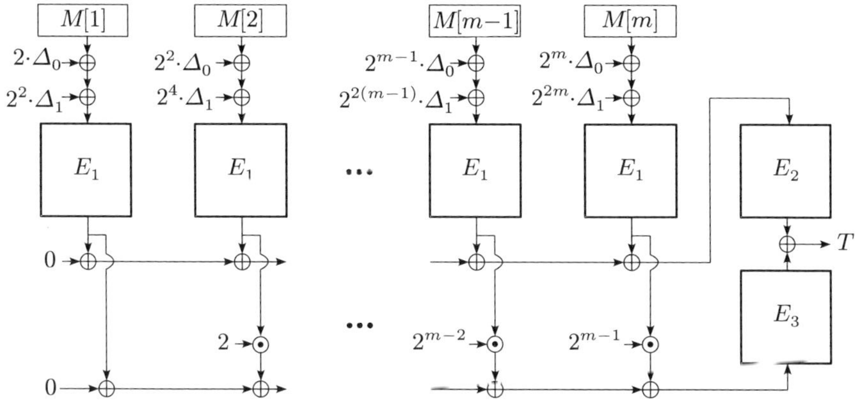

Fig. 1. Our PMAC_Plus algorithm using three blockcipher keys K_1, K_2 and K_3, where $E_1 = E_{K_1}$, $E_2 = E_{K_2}$, $E_3 = E_{K_3}$, $\Delta_0 = E_1(0)$ and $\Delta_1 = E_1(1)$

The additional term of $3\mathbf{Adv}_E^{\mathrm{prp}}(t', \ell q+2)$ comes from the standard argument of replacing actual blockciphers $E_1 = E_{K_1}$, $E_2 = E_{K_2}$ and $E_3 = E_{K_3}$ with random permutations P_1, P_2 and P_3, respectively.

4 Proofs of Security

We now prove that PMAC_Plus$[P_1, P_2, P_3]$ is an $O(2^{2n/3})$-secure PRF given random permutations P_1, P_2 and P_3.

4.1 Basic Ideas

Let $\mathscr{A}$ be an adversary that makes at most q queries, each query being at most ℓ blocks. The goal of $\mathscr{A}$ is to distinguish between the PMAC_Plus$[P_1, P_2, P_3](\cdot)$ oracle and a random function $G : \{0,1\}^* \to \{0,1\}^n$. We consider the game described in Fig. 2. In games, given a set $S \subset \{0,1\}^n$, we write for its complement $\overline{S} := \{0,1\}^n \setminus S$. Codes of subroutines are given in Figures 3, 4, 5 and 6. We observe that the game with single-boxed statements coincides with a random function G, whereas the game with double-boxed statements is exactly PMAC_Plus$[P_1, P_2, P_3]$. These two algorithms differ only when Bad events occur. Therefore, by the fundamental lemma of game-playing [5], we have

$$\Pr\big[\mathscr{A}^{\mathtt{PMAC\_Plus}[P_1,P_2,P_3](\cdot)} = 1\big] - \Pr\big[\mathscr{A}^{G(\cdot)} = 1\big] \le \Pr\big[\mathscr{A} \text{ sets } \mathtt{Bad}\big],$$

where the bad events are classified into five "winning" events as

$$\begin{aligned} &\Pr[\mathscr{A} \text{ sets } \mathtt{Bad}] \\ \le &\Pr[\mathscr{A} \text{ sets } \mathtt{Zero}^*, \mathtt{Unfair}^*, \mathtt{UpLow}^*, \mathtt{LowUp}^* \text{ or } \mathtt{Coll}^*] \\ \le &\Pr[\mathtt{Zero}^*] + \Pr[\mathtt{Unfair}^*] + \Pr[\mathtt{UpLow}^*] + \Pr[\mathtt{LowUp}^*] + \Pr[\mathtt{Coll}^*]. \end{aligned}$$

$\Delta_0, \Delta_1 \xleftarrow{\$} \{0,1\}^n$ // sampling $P_1(0)$ and $P_1(1)$
if $\Delta_0 = 0$ or $\Delta_1 = 0$ **then**
  $\mathtt{Zero}^* \leftarrow$ **true**
  $\mathtt{Bad} \leftarrow$ **true** $\boxed{\Delta_* \xleftarrow{\$} \overline{\{0\}}}$
end if

upon a query M **do**
$(\Sigma, \Theta) \leftarrow \mathtt{Internal}[P_1](M)$ // lazy sampling for P_1
if $\Sigma \notin \mathtt{Dom}\, P_2$ and $\Theta \notin \mathtt{Dom}\, P_3$ **then**
  go to **Case A** // lazy sampling for P_2 and P_3
end if
if $\Sigma \in \mathtt{Dom}\, P_2$ and $\Theta \notin \mathtt{Dom}\, P_3$ **then**
  go to **Case B** // lazy sampling for P_3
end if
if $\Sigma \notin \mathtt{Dom}\, P_2$ and $\Theta \in \mathtt{Dom}\, P_3$ **then**
  go to **Case C** // lazy sampling for P_2
end if
if $\Sigma \in \mathtt{Dom}\, P_2$ and $\Theta \in \mathtt{Dom}\, P_3$ **then**
  go to **Case D** // a bad event
end if
return T

Fig. 2. Main game

The first term can be easily bounded; it remains to bound the last four terms. These correspond to Cases A, B, C and D, respectively, and they are described as subroutines. Up to this point we essentially follow the same framework as the proof of SUM-ECBC [23].

Choose a fair set $R \subset \overline{\mathtt{Ran}\, P_2} \times \overline{\mathtt{Ran}\, P_3}$
$(U, L) \xleftarrow{\$} \overline{\mathtt{Ran}\, P_2} \times \overline{\mathtt{Ran}\, P_3}$
if $(U, L) \notin R$ **then**
  if $\neg\mathtt{Bad}$ **then**
    $\mathtt{Unfair}^* \leftarrow$ **true**
  end if
  $\mathtt{Bad} \leftarrow$ **true** $\boxed{(U, L) \xleftarrow{\$} R}$
end if
$T \leftarrow U \oplus L$

Fig. 3. Code for Case A

4.2 Bounding the Probability of Each Winning Event

Case A: `Unfair*`. This case can be essentially handled by the technique of fair sets developed by Lucks [16]. The proof is exactly the same as the case of SUM-ECBC [23].

```
U ← P2(Σ)
L ←$ {0,1}^n
if L ∈ Ran P3 then
    if ¬Bad then
        UpLow* ← true
    end if
    Bad ← true [L ←$ ~Ran P3]
end if
T ← U ⊕ L
```

Fig. 4. Code for Case B

```
L ← P3(Θ)
U ←$ {0,1}^n
if U ∈ Ran P2 then
    if ¬Bad then
        LowUp* ← true
    end if
    Bad ← true [U ←$ ~Ran P2]
end if
T ← U ⊕ L
```

Fig. 5. Code for Case C

```
if ¬Bad then
    Coll* ← true
end if
Bad ← true [T ←$ {0,1}^n] [T ← P2(Σ) ⊕ P3(Θ)]
```

Fig. 6. Code for Case D

Lemma 1 (Case A). *We have*

$$\Pr[\mathscr{A} \text{ sets } \texttt{Unfair}^*] \leq \frac{2q^3}{2^{2n}},$$

for $q \leq 2^{n-1}$.

Proof. The proof is given in [23], but for the sake of completeness we give one in Appendix. □

Case B: `UpLow*`**.** To treat this case we need the following:

Lemma 2. *For a pair of messages* (M, M') *such that* $M \neq M'$*, each being at most* ℓ *blocks, we have*

$$\Pr\big[\Sigma = \Sigma'; P \xleftarrow{\$} \mathrm{Perm}(n)\big] \leq \frac{8\ell}{2^n},$$

where $(\Sigma, \Theta) \leftarrow \texttt{Internal}[P](M)$ *and* $(\Sigma', \Theta') \leftarrow \texttt{Internal}[P](M')$.

Proof. Consider lazy sampling for P. First draw a range point $\Delta := P(0)$. We assume that $\Delta \neq 0$; the probability that $\Delta = 0$ occurs is $1/2^n$. We next assume that none of the input blocks $X[a]$ or $X'[a]$ is 0 or 1; the probability that $X[a] = 0, 1$ or $X'[a] = 0, 1$ occurs for some $a \leq \ell$ is at most $4\ell/2^n$.

Now without loss of generality we assume that $|M| \geq |M'|$. Let m, m' be the number of blocks in M and in M', respectively. Observe that we must have $m \geq 2$ for the above probability to be non-trivial. So assume $m \geq 2$.

Determine an index $i \in \{1, \ldots, m-1\}$ as follows. If $m > m'$, then set $i := m$. If $m = m'$ and the last blocks are the only blocks that differ, then the above probability becomes vacuous. So in the case of $m = m'$, let $i \le m$ be the maximum index such that $M[i] \ne M'[i]$.

We focus on the input block $X[i]$. Assume that $X[i]$ does not appear in any of $X[a]$ $(1 \le a \le m-1)$ or $X'[a']$ $(1 \le a \le m'-1)$ except for itself. The probability that $X[i] = X[a]$ or $X[i] = X'[a']$ occurs is at most $\big((\ell-1)+\ell\big)/2^n = (2\ell-1)/2^n$.

Finally, consider the condition $\Sigma = \Sigma'$. Now resume the lazy sampling for P. Sample the point $P(X[i])$ as $Y_i \xleftarrow{\$} \overline{\mathtt{Ran}\,P}$ after finishing sampling all other points; the point $P(X[i])$ gets always sampled according to the way of choosing the index i. In such a scenario the probability that $\Sigma = \Sigma'$ holds is at most $1/\big|\overline{\mathtt{Ran}\,P}\big| \le 1/\big(2^n - 1 - (\ell-2) - (\ell-1)\big)/2^n \le 1/(2^n - 2\ell) \le 2/2^n$, assuming $\ell \le 2^{n-2}$ (otherwise the desired inequality would become meaningless).

We sum up each terms. Overall, the probability can be bounded as

$$\frac{1}{2^n} + \frac{4\ell}{2} + \frac{2\ell-1}{2^n} + \frac{2}{2^n} \le \frac{8\ell}{2^n},$$

as desired. □

Now let $M^{(1)}, \ldots, M^{(q)}$ denote $\mathscr{A}$'s queries. Then the probability that $\mathscr{A}$ sets the $\mathtt{UpLow}^*$ flag can be bounded as

$$\sum_{i=2}^{q} \Pr\big[\Sigma^{(i)} \in \mathtt{Dom}\,P_2 \wedge L^{(i)} \in \mathtt{Ran}\,P_3; P_1, P_2, P_3 \xleftarrow{\$} \mathrm{Perm}(n)\big]$$

$$\le \sum_{i=2}^{q}\sum_{j=1}^{i-1} \Pr\big[(\Sigma^{(i)} = \Sigma^{(j)}); P_1 \xleftarrow{\$} \mathrm{Perm}(n)\big] \cdot \Pr\big[L^{(i)} \in \mathtt{Ran}\,P_3; L^{(i)} \xleftarrow{\$} \{0,1\}^n\big]$$

$$\le \sum_{i=2}^{q}\sum_{j=1}^{i-1} \frac{8\ell}{2^n} \cdot \frac{\big|\mathtt{Ran}\,P_3\big|}{2^n} \le \sum_{i=2}^{q}\sum_{j=1}^{i-1} \frac{8\ell}{2^n} \cdot \frac{2q}{2^n} \le \frac{q^2}{2} \cdot \frac{8\ell}{2^n} \cdot \frac{2q}{2^n} = \frac{8\ell q^3}{2^{2n}},$$

where we wrote $(\Sigma^{(i)}, \Theta^{(i)}) := \mathtt{Internal}[P_1]\big(M^{(i)}\big)$ and $L^{(i)}$ the sampling of L at the i-the query.

Case C: $\mathtt{LowUp}^*$. For this case we need the following lemma. Then the computation is similar to Case B, and we obtain the same bound of $8\ell q^3/2^{2n}$.

Lemma 3. *For a pair of messages (M, M') such that $M \ne M'$, each being at most ℓ blocks, we have*

$$\Pr\big[\Theta = \Theta'; P \xleftarrow{\$} \mathrm{Perm}(n)\big] \le \frac{8\ell}{2^n},$$

where $(\Sigma, \Theta) \leftarrow \mathtt{Internal}[P](M)$ and $(\Sigma', \Theta') \leftarrow \mathtt{Internal}[P](M')$.

Proof. Similar to Lemma 2. □

Case D: `Coll*`**.** Since P_1 is independent from P_2 and from P_3, we may fix $\mathscr{A}$'s (distinct) queries and let $M^{(1)}, \ldots, M^{(q)}$ denote them. We would like to compute the probability that at the i-th query $M^{(i)}$ we get $\Sigma^{(i)} \in \mathtt{Dom}\, P_2$ and $\Theta^{(i)} \in \mathtt{Dom}\, P_3$. The event implies that there exist some earlier queries $M^{(j)}$ and $M^{(k)}$ (j and k may be equal) such that $\Sigma^{(j)} = \Sigma^{(i)}$ and $\Theta^{(k)} = \Theta^{(i)}$.

Before evaluating the probability

$$\Pr\big[(\Sigma^{(j)} = \Sigma^{(i)}) \wedge (\Theta^{(k)} = \Theta^{(i)}); P_1 \stackrel{\$}{\leftarrow} \mathrm{Perm}(n)\big],$$

we first exclude the case that $Y_a \stackrel{\$}{\leftarrow} \overline{\mathtt{Ran}\, P_1}$ becomes zero (*i.e.*, $Y_a = 0$) in sampling range points of P_1 for messages $M^{(1)}, \ldots, M^{(q)}$. The overall probability that this event occurs is at most $\ell q/2^n$.

We then consider the case when an "input collision" occurs among $X^{(i)}[a]$, $X^{(j)}[a]$, $X^{(k)}[a]$. By an "input collision" we mean an event $X^{(*)}[a] = X^{(*)}[a']$ for some $a \neq a'$. If input collisions occur at indices $a < b < c$ such that $X^{(*)}[a] = X^{(*)}[b] = X^{(*)}[c]$, then this system of equations would determine the values Δ_0, Δ_1, so the probability that this occurs is at most $\binom{3\ell}{3} \cdot 1/2^n \leq 5\ell^3/2^n$.

Suppose no 3-collision occurs. The probability that a 2-collision happens upon sampling $\Delta \stackrel{\$}{\leftarrow} \{0,1\}^n$ for a fix set of $M^{(i)}$, $M^{(j)}$ and $M^{(k)}$ is at most $\binom{3\ell}{2} \cdot 1/2^n \leq 4.5\ell^2/2^n$. Under the event of an input (2-)collision, we focus on the equation $\Theta^{(i)} = \Theta^{(k)}$. We show that this provides a non-trivial equation for some random variable $Y_a^{(*)}$. Without loss of generality assume, for the moment, that $m^{(i)} \leq m^{(k)}$ where these are the number of blocks in the message $M^{(i)}$ and that in $M^{(k)}$, respectively. Let $\alpha \in \{1, \ldots, m^{(i)}\}$ be the largest index such that $M^{(i)}[\alpha] \neq M^{(k)}[\alpha]$, if such an index exists. Then we have $X^{(i)}[\alpha] \neq X^{(k)}[\alpha]$, so at least one of these input values is non-zero, which means that either $Y_\alpha^{(i)} = P(X^{(i)}[\alpha])$ or $Y_\alpha^{(k)} = P(X^{(k)}[\alpha])$ gets sampled. For that random variable the equation $\Theta^{(i)} = \Theta^{(k)}$ is non-trivial. On the other hand, if no such index α exists, then it means that $M^{(i)}[a] = M^{(k)}[a]$ for all $a \in \{1, \ldots, m^{(i)}\}$ and $m^{(i)}+1 \leq m^{(k)}$ (In such a case we say that $M^{(i)}$ is "contained" in $M^{(k)}$). Consider the values $X^{(k)}[m^{(i)}+1], \ldots, X^{(k)}[m^{(k)}]$. It cannot be the case that all of these input values are the same, as it would imply a zero value in the range of P_1. Therefore, we have $m^{(i)} + 2 \leq m^{(k)}$, and let $\beta \in \{m^{(i)} + 1, \ldots, m^{(k)}\}$ be the largest index such that $X^{(k)}[\beta] \neq 0$. Then we see that $Y_\beta^{(k)} = P(X^{(k)}[\beta])$ gets always sampled. Therefore, the probability that this equation is satisfied is at most $1/\big|\mathtt{Ran}\, P_1\big| \leq 1/(2^n - 2\ell) \leq 2/2^n$ assuming $\ell \leq 2^{n-2}$.

We can now bound the probability that, for a fix set of $M^{(i)}$, $M^{(j)}$ and $M^{(k)}$, an input collision occurs and the equation $\Theta^{(i)} = \Theta^{(k)}$ holds. It would be at most $4.5\ell^2/2^n \cdot 2 \cdot /2^n \leq 9\ell^2/2^{2n}$.

Consider the case when no input collision occurs among $X^{(i)}[a]$, $X^{(j)}[a]$, $X^{(k)}[a]$. We start with the case $j = k$. Without loss of generality assume, for the moment, that $m^{(i)} \leq m^{(j)}$ where these are the number of blocks in the message $M^{(i)}$ and that in $M^{(j)}$, respectively. It can be directly verified that we can choose indices $\alpha < \beta$ such that (a) $\beta \le m^{(i)}$ and $M^{(i)}[\alpha] \ne M^{(j)}[\alpha]$ and $M^{(i)}[\beta] \ne M^{(j)}[\beta]$, (b) $\alpha \le m^{(i)}$ and $M^{(i)}[\alpha] \ne M^{(j)}[\alpha]$ and $m^{(i)} + 1 \le \beta \le m^{(j)}$, or (c) $m^{(i)} + 1 \le \alpha < \beta \le m^{(j)}$. In any case, the system of equations $\Sigma^{(j)} = \Sigma^{(i)}$ and $\Theta^{(j)} = \Theta^{(i)}$ provides a unique solution set for the random variables $Y_\alpha^{(*)}$, $Y_\beta^{(*)}$. So the probability of this event is at most $1/|\mathtt{Ran}\,P_1| \cdot 1/|\mathtt{Ran}\,P_1| \le 1/(2^n - 2\ell)^2 \le 4/2^{2n}$, assuming $\ell \le 2^{n-2}$.

It remains to treat the case $j \ne k$. We consider the cases (a) $M^{(i)}$ is contained in $M^{(j)}$, (b) $M^{(j)}$ is contained in $M^{(i)}$, (c) $M^{(i)}$ is contained in $M^{(k)}$, or (d) $M^{(k)}$ is contained in $M^{(i)}$. For example, we discuss case (a). We can choose indices α, β such that (a1) $m^{(k)}+1 \le \alpha \le m^{(i)}$ and $m^{(i)}+1 \le \beta \le m^{(j)}$, (a2) $m^{(i)}+1 \le \alpha \le m^{(k)}$ and $m^{(k)}+1 \le \beta \le m^{(j)}$, (a3) $m^{(i)}+1 \le \alpha \le m^{(j)}$ and $m^{(j)}+1 \le \beta \le m^{(k)}$, (a4) $m^{(i)} + 1 \le \alpha \le m^{(j)}$ and $1 \le \beta \le m^{(i)}$ and $M^{(i)}[\beta] = M^{(j)}[\beta] \ne M^{(k)}[\beta]$, or (a5) $m^{(i)} + 1 \le \alpha \le m^{(j)}$ and $M^{(j)}[\alpha] \ne M^{(k)}[\alpha]$. In any case, the system of equations $\Sigma^{(j)} = \Sigma^{(i)}$ and $\Theta^{(j)} = \Theta^{(i)}$ provides a unique solution set for two random variables, so the probability of this event is at most $4/2^{2n}$, assuming $\ell \le 2^{n-2}$.

Lastly, assume that none of the containment (a) through (d) occurs. Then it means that there exist indices α, β such that $M^{(i)}[\alpha] \ne M^{(j)}[\alpha]$ and $M^{(i)}[\beta] \ne M^{(k)}[\beta]$. If $\alpha \ne \beta$, then we can simply choose $Y_\alpha^{(i)}$ and $Y_\beta^{(i)}$ to be the two variables. Suppose no such indices exist, that is, $\alpha = \beta$ and this is the only index that a difference occurs. If $M^{(j)}[\alpha] \ne M^{(k)}[\alpha]$, then we can choose two variables accordingly. If $M^{(j)}[\alpha] = M^{(k)}[\alpha]$, then since $M^{(j)} \ne M^{(k)}$, then $M^{(j)}$ is contained in $M^{(k)}$, or $M^{(k)}$ is contained in $M^{(j)}$, or there exists an index $\gamma > \alpha$ such that $M^{(j)}[\gamma] \ne M^{(k)}[\gamma]$. In any case, the system of equations $\Sigma^{(j)} = \Sigma^{(i)}$ and $\Theta^{(j)} = \Theta^{(i)}$ gives us a unique solution set for two random variables, and the probability of this event can be bounded as $4/2^{2n}$, again assuming $\ell \le 2^{n-2}$.

Now we are done with Case D. We just run indices i, j and k to get

$$\frac{\ell q}{2^n} + \sum_{i=2}^{q} \sum_{j=1}^{i-1} \sum_{k=1}^{i-1} \frac{5\ell^3 + 9\ell^2 + 4 + 4 + 4}{2^{2n}} \le \frac{\ell q}{2^n} + \frac{q^3}{3} \cdot \frac{26\ell^2}{2^{2n}} \le \frac{\ell q}{2^n} + \frac{9\ell^2 q^3}{2^{2n}}$$

for bounding the probability that case D happens.

4.3 Summing Up the Probabilities

We now bound the overall probability. The bound sums up to

$$\begin{aligned}
&\Pr[\mathscr{A} \text{ sets } \mathtt{Zero}^*, \mathtt{Unfair}^*, \mathtt{UpLow}^*, \mathtt{LowUp}^* \text{ or } \mathtt{Coll}^*] \\
\le& \frac{2}{2^n} + \frac{2q^3}{2^{2n}} + \frac{8\ell q^3}{2^{2n}} + \frac{8\ell q^3}{2^{2n}} + \frac{\ell q}{2^n} + \frac{9\ell^3 q^3}{2^{2n}} \\
\le& \frac{27\ell^2 q^3}{2^{2n}} + \frac{3\ell q}{2^n},
\end{aligned}$$

which completes the proof.

5 Discussion

We have presented a 3-key rate-1 MAC construction based on a PMAC-type iteration. This raises a challenge to come up with a 1-key rate-1 MAC construction which is secure beyond the birthday bound.

After beating the birthday bound of $O(2^{n/2})$, we seem to be encountering another "bound problem" at the query complexity of $O(2^{2n/3})$. To beat this new bound efficiently is also a challenge for blockcipher-based message authentication.

Acknowledgments. The author would like to thank CRYPTO 2011 program committee members and reviewers for valuable feedback.

References

1. Bellare, M., Goldreich, O., Krawczyk, H.: Stateless evaluation of pseudorandom functions: Security beyond the birthday barrier. In: Wiener, M. J. (ed.) CRYPTO 1999. LNCS, vol. 1666, pp. 270–287. Springer, Heidelberg (1999)
2. Bellare, M., Guérin, R., Rogaway, P.: XOR MACs: New Methods for Message Authentication Using Finite Pseudorandom Functions. In: Coppersmith, D. (ed.) CRYPTO 1995. LNCS, vol. 963, pp. 15–28. Springer, Heidelberg (1995)
3. Bellare, M., Kilian, J., Rogaway, P.: The Security of Cipher Block Chaining. In: Desmedt, Y. (ed.) CRYPTO 1994. LNCS, vol. 839, pp. 341–358. Springer, Heidelberg (1994)
4. Bellare, M., Pietrzak, K., Rogaway, P.: Improved Security Analyses for CBC MACs. In: Shoup, V. (ed.) CRYPTO 2005. LNCS, vol. 3621, pp. 527–545. Springer, Heidelberg (2005)
5. Bellare, M., Rogaway, P.: The Security of Triple Encryption and a Framework for Code-Based Game-Playing Proofs. In: Vaudenay, S. (ed.) EUROCRYPT 2006. LNCS, vol. 4004, pp. 409–426. Springer, Heidelberg (2006)
6. Bernstein, D.J.: How to stretch random functions: The security of Protected Counter Sums. J. Cryptology 12(3), 185–192 (1999)
7. Black, J., Rogaway, P.: CBC MACs for Arbitrary-Length Messages:The Three-Key Constructions. In: Bellare, M. (ed.) CRYPTO 2000. LNCS, vol. 1880, pp. 197–215. Springer, Heidelberg (2000)
8. Black, J., Rogaway, P.: A Block-Cipher Mode of Operation for Parallelizable Message Authentication. In: Knudsen, L.R. (ed.) EUROCRYPT 2002. LNCS, vol. 2332, pp. 384–397. Springer, Heidelberg (2002)
9. Bogdanov, A., Knudsen, L.R., Leander, G., Paar, C., Poschmann, A., Robshaw, M.J.B., Seurin, Y., Vikkelsoe, C.: PRESENT: An Ultra-Lightweight Block Cipher. In: Paillier, P., Verbauwhede, I. (eds.) CHES 2007. LNCS, vol. 4727, pp. 450–466. Springer, Heidelberg (2007)
10. Hong, D., Sung, J., Hong, S.H., Lim, J.-I., Lee, S.-J., Koo, B.-S., Lee, C.-H., Chang, D., Lee, J., Jeong, K., Kim, H., Kim, J.-S., Chee, S.: HIGHT: A New Block Cipher Suitable for Low-Resource Device. In: Goubin, L., Matsui, M. (eds.) CHES 2006. LNCS, vol. 4249, pp. 46–59. Springer, Heidelberg (2006)

11. Iwata, T., Kurosawa, K.: OMAC: One-Key CBC MAC. In: Johansson, T. (ed.) FSE 2003. LNCS, vol. 2887, pp. 129–153. Springer, Heidelberg (2003)
12. Joux, A., Poupard, G., Stern, J.: New Attacks against Standardized MACs. In: Johansson, T. (ed.) FSE 2003. LNCS, vol. 2887, pp. 170–181. Springer, Heidelberg (2003)
13. JTC1. ISO/IEC 9797-1:1999 Information technology—Security techniques—Message Authentication Codes (macs)—Part 1: Mechanisms using a block cipher (1999)
14. Käsper, E., Schwabe, P.: Faster and Timing-Attack Resistant AES-GCM. In: Clavier, C., Gaj, K. (eds.) CHES 2009. LNCS, vol. 5747, pp. 1–17. Springer, Heidelberg (2009)
15. Kurosawa, K., Iwata, T.: TMAC: Two-Key CBC MAC. In: Joye, M. (ed.) CT-RSA 2003. LNCS, vol. 2612, pp. 33–49. Springer, Heidelberg (2003)
16. Lucks, S.: The Sum of PRPs Is a Secure PRF. In: Preneel, B. (ed.) EUROCRYPT 2000. LNCS, vol. 1807, pp. 470–484. Springer, Heidelberg (2000)
17. Minematsu, K., Matsushima, T.: New Bounds for PMAC, TMAC, and XCBC. In: Biryukov, A. (ed.) FSE 2007. LNCS, vol. 4593, pp. 434–451. Springer, Heidelberg (2007)
18. NIST. Recommendation for block cipher modes of operation: The CMAC mode for authentication. SP 800-38B (2005)
19. Petrank, E., Rackoff, C.: CBC MAC for real-time data sources. J. Cryptology 13(3), 315–338 (2000)
20. Preneel, B., van Oorschot, P.C.: MDx-MAC and Building Fast MACs from Hash Functions. In: Coppersmith, D. (ed.) CRYPTO 1995. LNCS, vol. 963, pp. 1–14. Springer, Heidelberg (1995)
21. Rogaway, P.: Efficient Instantiations of Tweakable Blockciphers and Refinements to Modes OCB and PMAC. In: Lee, P.J. (ed.) ASIACRYPT 2004. LNCS, vol. 3329, pp. 16–31. Springer, Heidelberg (2004)
22. Sarkar, P.: Pseudo-random functions and parallelizable modes of operations of a block cipher. IEEE Transactions on Information Theory 56(8), 4025–4037 (2010)
23. Yasuda, K.: The Sum of CBC MACs Is a Secure PRF. In: Pieprzyk, J. (ed.) CT-RSA 2010. LNCS, vol. 5985, pp. 366–381. Springer, Heidelberg (2010)

A Proof Lemma 1

Proof. The proof is almost exactly the same as the one for Lucks' SUM^2 construction $P_2(X) \oplus P_3(X)$ [16]. The fact that we have $\Sigma \neq \Theta$ does not have much effect on the computation of the probability. Specifically, we consider the following simulation of $P_2(\Sigma) \oplus P_3(\Theta)$.

The code without the boxed statement corresponds with $P_2(\Sigma) \oplus P_3(\Theta)$. The code with the boxed statement corresponds with a random oracle $\mathscr{R}$, because the set R is fair; that is, R is chosen so that the number of pairs $(U, L) \in R$ such that

$$T = U \oplus L$$

is the same for each value $T \in \{0,1\}^n$. In the code, we choose a fair set R as follows. Enumerate $\mathtt{Ran}\, P_2$ as $\{U_1, \ldots, U_\alpha\}$ and $\mathtt{Ran}\, P_3$ as $\{L_1, \ldots, L_\beta\}$. For each

$Y \leftarrow \overline{\mathtt{Ran}\, P_2}$, $Z \leftarrow \overline{\mathtt{Ran}\, P_3}$
Choose a fair set $R \subset Y \times Z$
$(U, L) \xleftarrow{\$} Y \times Z$
if $(U, L) \notin R$ **then**
    $\mathtt{Bad} \leftarrow$ **true** $\boxed{(U, L) \xleftarrow{\$} R}$
end if
$T \leftarrow U \oplus L$
return T

i and j such that $1 \leq i \leq \alpha$ and $1 \leq j \leq \beta$ we choose arbitrarily representatives $(U'_i, L'_j) \in Y \times Z$ such that $U'_i \oplus L'_j = U_i \oplus L_j$. We then define $R \leftarrow Y \times Z \setminus \bigcup_{i,j}\{(U'_i, L'_j)\}$. We see that, for each value $T \in \{0,1\}^n$,

$$\left|\{(U, L) \in R \mid U \oplus L = T\}\right| = 2^n - \alpha - \beta,$$

so R is indeed a fair set.

After q queries, the overall probability that the bad event occurs becomes

$$\begin{aligned}
\Pr\left[\mathtt{Bad}\right] &\leq \sum_{i=1}^{q} \frac{\left|(Y \times Z) \setminus R\right|}{|Y \times Z|} \\
&= \sum_{i=1}^{q} \frac{\alpha\beta}{(2^n - \alpha)(2^n - \beta)} \\
&\leq \sum_{i=0}^{q-1} \frac{i^2}{(2^n - q)^2} \\
&\leq \frac{1}{(2^n - q)^2} \cdot \sum_{i=0}^{q-1} i^2 \\
&\leq \frac{1}{(2^{n-1})^2} \cdot \frac{q(q-1)(2q-1)}{6} \\
&\leq \frac{2q^3}{2^{2n}},
\end{aligned}$$

where we used the condition $q \leq 2^{n-1}$. □

Authenticated and Misuse-Resistant Encryption of Key-Dependent Data

Mihir Bellare and Sriram Keelveedhi

Department of Computer Science & Engineering, University of California San Diego, 9500 Gilman Drive, La Jolla, California 92093, USA
http://www.cs.ucsd.edu/users/{mihir,skeelvee}/

Abstract. This paper provides a comprehensive treatment of the security of authenticated encryption (AE) in the presence of key-dependent data, considering the four variants of the goal arising from the choice of universal nonce or random nonce security and presence or absence of a header. We present attacks showing that universal-nonce security for key-dependent messages is impossible, as is security for key-dependent headers, not only ruling out security for three of the four variants but showing that currently standarized and used schemes (all these target universal nonce security in the presence of headers) fail to provide security for key-dependent data. To complete the picture we show that the final variant (random-nonce security in the presence of key-dependent messages but key-independent headers) is efficiently achievable. Rather than a single dedicated scheme, we present a RO-based transform RHtE that endows *any* AE scheme with this security, so that existing implementations may be easily upgraded to have the best possible seurity in the presence of key-dependent data. RHtE is cheap, software-friendly, and continues to provide security when the key is a password, a setting in which key-dependent data is particularly likely. We go on to give a key-dependent data treatment of the goal of misuse resistant AE. Implementations are provided and show that RHtE has small overhead.

1 Introduction

The key used by BitLocker to encrypt your disk may reside on the disk. The key under which a secure filesystem is encrypted may itself be stored in a file on the same system. The result is encryption of key-dependent data.

There is growing recognition that security of key-dependent data, first defined to connect cryptography to formal methods [18] and provide anonymous credentials [24], is a more direct and widespread concern for secure systems. The problem is particularly acute when keys are passwords, for many of us store our passwords on our systems and systems store password hashes. If nothing else, one cannot expect applications to ensure or certify that their data is *not* key-dependent, making security for key-dependent data essential for easy-to-use, robust and misuse-resistant cryptography.

P. Rogaway (Ed.): CRYPTO 2011, LNCS 6841, pp. 610–629, 2011.

This paper provides a comprehensive treatment of security for key-dependent data for the central practical goal of symmetric cryptography, namely authenticated encryption. For each important variant of the goal we either show that it is impossible to achieve security or present an efficient solution. Our attacks rule out security for in-use and standardized schemes in their prescribed and common modes while our solutions show how to adapt them in minimal ways to achieve the best achievable security. Let us now look at all this more closely.

BACKGROUND. The standard IND-CPA and IND-CCA goals that our encryption schemes are proven to meet do not guarantee security when the message being encrypted depends on the key. (In the symmetric setting, we mean the single key used for both encryption and decryption.) Black, Rogaway and Shrimpton (BRS) [18] extend IND-CPA to allow key-dependent messages (KDMs). The adversary provides its encryption oracle with a function ϕ, called a message-deriving function, that the game applies to the target key K to get a message M, and the adversary is returned either an encryption of M under K or the encryption of $0^{|M|}$, and must be unable to tell which. (They, and we, actually consider a multi-key setting, but the single-key setting will simplify the current discussion.) They present a simple random-oracle (RO) model solution.

Post-BRS work has aimed mainly at showing existence of schemes secure against as large as possible a class of message deriving functions without random oracles [19,36,4,21,23,20,7,25,17,3,38]. The schemes suffer from one or more of the following: they are in the asymmetric setting while data encryption in practice is largely symmetric; they are too complex to consider usage; or security is provided for a limited, mathematical class of message-deriving functions which does not cover all key-dependencies in systems.

Backes, Pfitzmann and Scedrov (BPS) [6] define KDM-security for a basic form of authenticated encryption and and show that Encrypt-then-MAC [12] achieves it if the encryption scheme is KDM secure and the MAC is strongly unforgeable (remarkably, no KDM security is required from the MAC), resulting in RO model solutions via [18]. In this paper we will extend their treatment of AE in several directions.

SETTING. Privacy without authenticity, meaning plain (IND-CPA) encryption, is of limited utility. The most important symmetric primitive in practice is authenticated encryption (AE), which provides both privacy and integrity. This is evidenced by numerous standards and high usage: CCM [50,49] is in IEEE 802.11, IEEE 802.15.4, IPSEC ESP and IKEv2; GCM [39] is standardized by NIST as SP 800-38D; EAX [16] is in ANSI C12.22 and ISO/IEC 19772; OCB 2.0 [45,47] is in ISO 19772. Consideration of KDM security for these standards is compelling and urgent but has not been done. We seek to fill this gap.

Symmetric encryption schemes take as input a nonce, also called an IV. Classically — [9] following [30]— this was chosen at random by the encrypter. We call this random-nonce security (r). Later schemes targeted universal-nonce security (u) [44,46,48] where security must hold even when the adversary provides the nonce, as long as no nonce is re-used. This is adopted by the above-mentioned standards.

	(ki, ki)	(kd, kd)	(ki, kd)	(kd, ki)
u	Yes	No	No	No
r	Yes	No	No	Yes

Fig. 1. Each of message and header may be key-dependent (kd) or key-independent (ki), leading to the four choices naming the columns. Security could be universal-nonce (u) or random-nonce (r), leading to the two choices naming the rows. For each of the 8 possibilities, we indicate whether security is possible (Yes, meaning a secure scheme exists) or impossible (No, meaning there is an attack that breaks *any* scheme in this category). The first column reflects known results when inputs are not key-dependent.

Besides key, nonce and message, modern AE schemes, including the above standards, take input a header, or associated data [44]. The scheme must provide integrity but *not* privacy of the header. Thus we must consider that not just the message, but also the header, could be key-dependent.

Abbreviate key-dependent by kd and key-independent by ki. With two choices for nonce type —$\mathtt{nt} \in \{\mathtt{u}, \mathtt{r}\}$— two for message type —$\mathtt{mt} \in \{\mathtt{kd}, \mathtt{ki}\}$— and two for header type —$\mathtt{ht} \in \{\mathtt{kd}, \mathtt{ki}\}$— we have 8 variants of AE. The form of AE treated by Backes, Pfitzmann and Scedrov [6] is the special case of $(\mathtt{nt}, \mathtt{mt}, \mathtt{ht}) = (\mathtt{r}, \mathtt{kd}, \mathtt{ki})$ in which the header is absent.

DEFINITION. Our first contribution is a definition of security for AE under key-dependent inputs that captures all these 8 variants in a unified way. The encryption oracle takes functions ϕ_m, ϕ_h, and applies them to the key to get message and header respectively, and the adversary gets back either an encryption of these under the game-chosen target key, or a random string of the same length. The decryption oracle takes a ciphertext and, importantly, not a header but a function ϕ_h to derive it from the key, and either says whether or not decryption under the key is valid, or always says it is invalid. Varying the way nonces are treated and from what spaces ϕ_m, ϕ_h are drawn yields the different variants of the notion. A definition of MACs for key-dependent messages emerges as the special case of empty messages.

On a real system, the data may be a complex function of the key, such as a compressed (zipped) version of file containing, amongst other things, the key, or an error-corrected version of the key. If the key is a password the system will store its hash that will be encrypted as part of the disk, so common password-hashing functions must be included as message-deriving functions. All this argues for not restricting the types of message-deriving or header-deriving functions, and indeed, following [18,6], we allow any functions in this role. These functions are even allowed to call the RO, a source of challenges in proofs.

Underlying the above definition is a new one of the standard AE goal that simplifies that of [48] by having the decryption oracle turn into a verification oracle, returning, not the full decryption, but only whether it succeeded or not, along the lines of [12]. When data is key-independent, these and prior formulations [12,37] are equivalent, but the difference is important with key-dependent data.

IMPOSSIBILITY RESULTS. We present an attack that shows that *no* AE scheme can achieve universal-nonce security for key-dependent data. (Regardless of whether or not the header is key-dependent.) This explains the "No" entries in the first row of Fig. 1. The attack requires only that the nonce is predictable. Thus it applies even when the nonce is a counter, ruling out KDM security for counter-based AE schemes and showing that the standardized schemes (CCM, GCM, EAX, OCB) are all insecure for key-dependent messages in this case. The attack does not use the decryption oracle, so rules out even KDM universal-nonce CPA secure encryption. Thus, the universal-nonce security proven for the standardized schemes for key-independent messages fails to extend to key-dependent ones, demonstrating that security for key-dependent messages is a fundamentally different and stronger security requirement.

An attack aiming to show that no stateful scheme is KDM-CPA secure was described in [18] but the message-deriving functions execute a search and it is not clear how long this will take to terminate or whether it will even succeed. (In asymptotic terms, the attack is not proven to terminate in polynomial time.) Our attack extends theirs to use pairwise independent hash functions, based on which we prove that it achieves a constant advantage in a bounded (polynomial) amount of time. Interestingly, as a corollary of the bound proven on our *modified* attack, we are able to also prove a bound on the running time of the attack of [18], although it was not clear to us how to do this directly.

We also present an attack that shows that *no* AE scheme can achieve security for key-dependent headers. (Even for random, rather than universal, nonce security, and even for key-independent messages.) This explains the "No" entries in columns 2 and 3 of Fig. 1. This rules out security of the standardized schemes even with random nonces in a setting where headers may be key-dependent.

One might consider this trivial with the following reasoning: "Since the header is not kept private, the adversary sees it, and if it is key-dependent, it could for example just be the key, effectively giving the adversary the key." The fallacy is the assumption that the adversary sees the header. In our model, it is given a ciphertext but not directly given the header on which the ciphertext depends. This choice of model is not arbitrary but reflects applications, where a key-dependent header is present on the encrypting and decrypting systems (which may be the same system) but not visible to the adversary. Instead, the attack exploits the ability of the adversary to test validity of ciphertexts with implicitly specified headers.

RHtE. We turn to achieving security in the only viable, but still important setting, namely $(\mathtt{nt}, \mathtt{mt}, \mathtt{ht}) = (\mathtt{r}, \mathtt{kd}, \mathtt{ki})$. As background, recall that to achieve KDM-CPA security, BRS [18] encrypt message M by picking R at random and returning $H(K\|R)\oplus M)$ where H is a RO returning $|M|$ bits. (Here and below, it is assumed the decryptor and adversary also get the nonce R, but it is not formally part of the ciphertext.) We note that this is easily extended to achieve $(\mathtt{r}, \mathtt{kd}, \mathtt{ki})$-AE security. To encrypt header H and message M under key K, pick R at random and return (C, T) where $C = H_1(K\|R)\oplus M$ and $T = H_2(K\|R, H, C)$ and H_1, H_2 are ROs.

Randomized Hash then Encrypt (RHtE) is more practical. Unlike the above, it is not a dedicated scheme but rather transforms a standard (secure only for key-independent data) base AE scheme into a $(\mathtt{r}, \mathtt{kd}, \mathtt{ki})$-secure AE scheme. RHtE, given key L and randomness R, derives subkey $K = H(R\|L)$ via RO H and then runs the base scheme with key K on the header and message to get the ciphertext C. Only one-time security of the base scheme is required, so it could even be deterministic. The software changes are non-intrusive since the code of the base scheme is used unchanged. Thus RHtE can easily be put on top of existing standards like CCM, GCM, EAX, OCB to add security in the presence of key-dependent messages. As long as these base schemes transmit their nonce, RHtE has *zero overhead in bandwidth* because it can use the base scheme with some fixed, known nonce and use the nonce space for R. (It is okay to re-use the base-scheme nonce because this scheme is only required to be one-time secure. Its key is changing with every encryption.) The computational overhead of RHtE is independent of the lengths of header and message and hence becomes negligible as these get longer.

The proof of security is surprisingly involved due to a combination of three factors. First is that the message-deriving functions are allowed to call the RO. Second, while the BRS scheme and its extension noted above are purely information theoretic, the security of RHtE is computational due to the base scheme, and must be proven by reduction. Third, unlike BRS, we must deal with decryption queries. To handle all this we will need to invoke the security of the base scheme in multiple, inter-related ways, leading to a proof with two, interleaved hybrids that go in opposite directions.

Some indication of the complexity of the proof is provided by the fact that the bound we finally achieve in Theorem 5 is weaker than we would like. It is an interesting open problem to either prove a better bound for RHtE or provide an alternative scheme with such a bound.

EXTENSIONS. In filesystem encryption, as with most applications, security is likely to stem from your password pw. The system stores a hash $\overline{\mathsf{pw}} = h(\mathsf{pw})$ of it to authenticate you and an AE scheme must then encrypt or decrypt using pw. Key dependent data is now an even greater concern. One reason is that users tend to write their passwords in files in their filesystems. The other reasons is that $\overline{\mathsf{pw}}$ is a function of pw that must be stored on the system and thus will be encrypted with disk encryption. To address this, we show that RHtE is secure even when its starting key L is a password as long as the latter is drawn from a space that, asymptotically, has super-logarithmic min-entropy.

The security discussed so far relies crucially on using fresh randomness with each encryption. This is fine in theory but in real systems, failures of random-number generation (RNG) due to poorly gathered entropy or bugs are all too common and have lead to major security violations [29,33,22,42,41,1,51,27]. Simply asking that system designers get their RNGs "right" is unrealistic. Misuse-resistant [18] or hedged [8] encryption take a different approach, mitigating the damage caused by RNG failures by providing as much security as possible when randomness fails.

We extend this to the key-dependent data setting in the full version [10]. A misuse resistant AE scheme for key-dependent data provides two things. First, it must continue to provide (r, kd, kd)-security when the nonce is random. Second, even for nonces that are entirely adversary controlled (and may repeat), the scheme must meet a second condition that we define to capture its providing the security of deterministic AE in the presence of key-dependent data. In the latter case it is impossible to protect against certain classes of message-deriving functions. We show however that RHtE provides security against any class of functions satisfying the output-unpredictability and collision-resistance conditions of [11]. This is a fairly significant class, containing functions of pragmatic interest.

IMPLEMENTATION. We implemented RHtE for base schemes CCM, EAX and GCM, with SHA256 instantiating the RO. The results, provided in [10]show for example that with CCM the slowdown is 11% for 5KB messages and only 1% for 50KB messages. The implementations use the crypto++ library on a Intel Core i5 M460 CPU running at 2.53 GHz with code compiled using g++ -O3 for data sizes small enough to fit in the level 2 cache.

RELATED WORK. The issue (key-dependent messages) was pointed out as early as Goldwasser and Micali [30], and asymmetric encryption of decryption keys was treated by Camenisch and Lysyanskaya [24], but a full treatment of key-dependent message (KDM) encryption awaited BRS [18], who provided RO model KDM-CPA secure schemes. Researchers then asked for what classes of message-deriving functions one could achieve KDM security in the standard model, providing results for both symmetric and asymmetric encryption under different assumptions [19,36,4,21,23,20,7,25,17,3,38]. On the more practical side, Backes, Dürmuth and Unruh [5] show that RSA-OAEP [13,28] is KDM-secure in the RO model. Backes, Pfitzmann and Scedrov [6] treat active attacks and provide and relate a number of different notions of security.

By showing that IND-CPA security does not even imply security for the encryption of 2-cycles, Acar, Belenkiy, Bellare and Cash [2] and Green and Hohenberger [32] settled a basic question in this area and showed that achieving even weak KDM-security *requires* new schemes, validating previous efforts in that direction. Acar et. al. [2] also connect KDM secure encryption to cryptographic agility. Haitner and Holenstein [34] study the difficulty of proving KDM security by blackbox reduction to standard primitives.

Halevi and Krawczyk [35] consider blockciphers under key-dependent inputs. Muñiz and Steinwandt [40] study KDM secure signatures. González, in an unpublished thesis [31], studies KDM secure MACs.

Motivated by attacks on SSH, Paterson and Watson [43] consider notions of security (in the standard ki-data context) which allow the attacker to interact in a byte-by-byte manner with the decryption oracle. Our treatment does not encompass such attacks, and extending the model of [43] to allow key-dependent data is an interesting direction for future work.

2 Definitions

We provide a unified definition for universal and random nonce AE security and then extend this to definitions of universal and random nonce AE security in the presence of key-dependent messages and headers.

NOTATION. If S is a (finite) set then $s \leftarrow_{\$} S$ denotes the operation of picking s from S at random and $|S|$ is the size of S. Read the term "efficient" as meaning "polynomial-time" in the natural asymptotic extension of our concrete framework. If x is a string then $|x|$ denotes its length and $x[i]$ denotes its i-th bit. The empty string is denoted ε. By $a_1 \| \ldots \| a_n$, we denote the concatenation of strings $a_1, \ldots, a_n$. Unless otherwise indicated, an algorithm may be randomized. We denote by $y \leftarrow_{\$} A(x_1, x_2, \ldots)$ the operation of running A on the indicated inputs and fresh random coins to get an output denoted y. For integers k, w let $\mathsf{Fun}(k, w)$ be the set of all functions ϕ for which there exists an integer $\mathsf{ol}(\phi)$, called the output length of ϕ, such that ϕ: $(\{0,1\}^k)^w \to \{0,1\}^{\mathsf{ol}(\phi)}$. Input-deriving functions will be drawn from this set. Let $\mathsf{Cns}(k, w)$ be the subset of $\mathsf{Fun}(k, w)$ consisting of constant functions, restricting attention to which drops KDI (key-dependent input) notions of security down to their standard, non-KDI counterparts.

GAMES. Some of our definitions and proofs are expressed via code-based games [15]. Such a game —see Fig. 2 for an example— consists of procedures that respond to adversary oracle queries. In an execution of game G with an adversary A, the latter must make exactly one INITIALIZE query, this being its first oracle query, and exactly one FINALIZE query, this being its last oracle query. In between, it can query other game procedures. Each time it makes a query, the corresponding game procedure executes, and what it returns, if anything, is the response to A's query. The output of FINALIZE, denoted G^A, is called the output of the game, and we let "$\mathrm{G}^A \Rightarrow d$" denote the event that this game output takes value d. If FINALIZE is absent it is understood to be the identity function, so the game output is the adversary output. Boolean flags are assumed initialized to false and $\mathsf{BAD}(\mathrm{G}^A)$ is the event that the execution of game G with adversary A sets flag bad to true. The running time of an adversary by convention is the worst case time for the execution of the adversary with the game defining its security, so that the time of the called game procedures is included.

AE SYNTAX. A symmetric encryption scheme $\mathsf{SE} = (\mathcal{K}, \mathcal{E}, \mathcal{D})$ is specified by a key generation algorithm $\mathcal{K}$ that returns k-bit strings, an encryption function $\mathcal{E}$: $\{0,1\}^k \times \{0,1\}^n \times \{0,1\}^* \times \{0,1\}^* \to \{0,1\}^*$ and decryption function $\mathcal{D}$: $\{0,1\}^k \times \{0,1\}^n \times \{0,1\}^* \times \{0,1\}^* \to \{0,1\}^* \cup \{\bot\}$. Inputs to $\mathcal{E}$ are key, nonce, header and message, and output is a ciphertext. Inputs to $\mathcal{D}$ are key, nonce, header and ciphertext, and output is a message or $\bot$. We refer to k as the keylength and n as the noncelength. Both $\mathcal{E}$ and $\mathcal{D}$ are deterministic, it being the way nonces are handled by the games defining security that will distinguish universal-nonce and random-nonce security. We require that $\mathcal{D}(K, N, H, \mathcal{E}(K, N, H, M)) = M$ for all values of the inputs shown. We also require that $\mathcal{E}$ is length respecting in the sense that the length of a ciphertext depends only on the length of the message and header. Formally, there is a

proc INITIALIZE // $\mathrm{KIAE}_{\mathsf{SE},\mathtt{nt}}$	proc INITIALIZE(w) // $\mathrm{KDAE}_{\mathsf{SE},\mathtt{nt}}$
$K \leftarrow\!\$\; \mathcal{K}$	For $j = 1, \ldots, w$ do
$S \leftarrow \emptyset$	$\quad K_j \leftarrow\!\$\; \mathcal{K}$; $S_j \leftarrow \emptyset$
$b \leftarrow\!\$\; \{0,1\}$	$b \leftarrow\!\$\; \{0,1\}$
proc ENC(N, H, M) // $\mathrm{KIAE}_{\mathsf{SE},\mathtt{nt}}$	proc ENC(j, N, ϕ_h, ϕ_m) // $\mathrm{KDAE}_{\mathsf{SE},\mathtt{nt}}$
If ($\mathtt{nt} = \mathtt{r}$) then $N \leftarrow\!\$\; \{0,1\}^n$	$M \leftarrow \phi_m(K_1, \ldots, K_w)$; $H \leftarrow \phi_h(K_1, \ldots, K_w)$
If ($b = 1$) then $C \leftarrow \mathcal{E}(K, N, H, M)$	If ($\mathtt{nt} = \mathtt{r}$) then $N \leftarrow\!\$\; \{0,1\}^n$
Else $c \leftarrow \mathsf{cl}(\|M\|, \|H\|)$; $C \leftarrow\!\$\; \{0,1\}^c$	If ($b = 1$) then $C \leftarrow \mathcal{E}(K_j, N, H, M)$
$S \leftarrow S \cup \{(N, H, C)\}$	Else $c \leftarrow \mathsf{cl}(\mathsf{ol}(\phi_m), \mathsf{ol}(\phi_h))$; $C \leftarrow\!\$\; \{0,1\}^c$
Return (N, C)	$S_j \leftarrow S_j \cup \{(N, H, C)\}$
	Return (N, C)
proc DEC(N, H, C) // $\mathrm{KIAE}_{\mathsf{SE},\mathtt{nt}}$	proc DEC(j, N, ϕ_h, C) // $\mathrm{KDAE}_{\mathsf{SE},\mathtt{nt}}$
If $(N, H, C) \in S$ then return $\perp$	$H \leftarrow \phi_h(K_1, \ldots, K_w)$
If ($b = 1$) then $M \leftarrow \mathcal{D}(K, H, N, C)$	If $(N, H, C) \in S_j$ then return $\perp$
Else $M \leftarrow \perp$	If ($b = 1$) then $M \leftarrow \mathcal{D}(K_j, N, H, C)$
If $M = \perp$ then $V \leftarrow 0$ else $V \leftarrow 1$	Else $M \leftarrow \perp$
Return V	If $M = \perp$ then $V \leftarrow 0$ else $V \leftarrow 1$
	Return V
proc FINALIZE(b') // $\mathrm{KIAE}_{\mathsf{SE},\mathtt{nt}}$	proc FINALIZE(b') // $\mathrm{KDAE}_{\mathsf{SE},\mathtt{nt}}$
Return $(b' = b)$	Return $(b' = b)$

Fig. 2. On the left is game $\mathrm{KIAE}_{\mathsf{SE},\mathtt{nt}}$ defining AE-security of encryption scheme $\mathsf{SE} = (\mathcal{K}, \mathcal{E}, \mathcal{D})$, where $\mathtt{nt} \in \{\mathtt{u}, \mathtt{r}\}$ indicates universal or random nonce. On the right is game $\mathrm{KDAE}_{\mathsf{SE},w,\mathtt{nt}}$ defining KDI AE-security of SE.

function $\mathsf{cl}(\cdot, \cdot)$ called the ciphertextlength such that $|C| = \mathsf{cl}(|M|, |H|)$ for any C that may be output by $\mathcal{E}(\cdot, \cdot, H, M)$.

As in [46,48], $\mathcal{D}$ takes the nonce and header as an input. (In this view, the ciphertext in standard counter-mode encryption does not incude the counter. It is up to the application to transmit nonce and header if necessary, so the "ciphertext" in practice may be more than the output of $\mathcal{E}$, but in many settings the receiver gets nonce and header in out-of-band ways.) But our treatment differs from standard ones [9] in that the nonce must be explicitly provided to $\mathcal{D}$ even when it is random. This means that, for randomized schemes, we are limited to ones that make the randomness public, but this is typically true. The restriction is only to compact and unify the presentation. Otherwise we would have needed separate games to treat universal and random nonce security.

AE SECURITY. We now define standard (neither message nor header is key-dependent) AE security for $\mathsf{SE} = (\mathcal{K}, \mathcal{E}, \mathcal{D})$. Consider game $\mathrm{KIAE}_{\mathsf{SE},\mathtt{nt}}$ shown on the left side of Fig. 2. Define the advantage of adversary A via $\mathbf{Adv}^{\mathrm{ae}\text{-}\mathtt{nt}}_{\mathsf{SE}}(A) = 2\Pr[\mathrm{KIAE}^{A}_{\mathsf{SE},\mathtt{nt}} \Rightarrow \mathsf{true}] - 1$. When $\mathtt{nt} = \mathtt{u}$ the definition captures what we call universal-nonce security. (It is simply called nonce-based security in [46,44,48].)

It is understood that in this case we only consider A that is *unique-nonce*, meaning we have $N \neq N'$ for any two ENC queries N, H, M and N', H', M'. Security is thus required even for adversary-chosen nonces as long as no nonce is used for more than one encryption. When $\texttt{nt} = \texttt{r}$, the adversary-provided nonce in ENC is ignored, a random value being substituted by the game, and we have random-nonce security, in the classical spirit of randomized encryption [30,9]. The nonce returned by ENC is redundant in the u case but needed in the r case and thus always returned for uniformity.

Historically the first definitions of security for AE had separate privacy (IND-CPA) and integrity (INT-CTXT) requirements [12,37,14]. Our version is a blend of the single-game formulation of [48] and INT-CTXT. Privacy is in the strong sense of indistinguishability from random, meaning ciphertexts are indistinguishable from random strings, which implies the more common LR-style [9] privacy, namely that ciphertexts of different messages are indistinguishable from each other. (A subtle point is that the length-respecting property assumed of $\mathcal{E}$ is important for this implication.) The integrity is in the fact that the adversary can't create new ciphertexts with non-$\perp$ decryptions. ("New" means not output by ENC.) Unlike [48], oracle DEC does not return decryptions but only whether or not they succeed. This simpler version is nonetheless equivalent to the original. IND-CCA is implied by this definition of AE [12,44].

KDI SECURITY OF AE. We now extend the above along the lines of [18,6] to provide our definition of security for AE in the presence of key-dependent inputs, considering both key-dependent messages and key-dependent headers. Consider game $\text{KDAE}_{\mathsf{SE},\texttt{nt}}$ shown on the right side of Fig. 2. Define the advantage of adversary A via $\mathbf{Adv}^{\text{ae-nt}}_{\mathsf{SE}}(A) = 2\Pr[\text{KDAE}^{A}_{\mathsf{SE},\texttt{nt}} \Rightarrow \mathsf{true}] - 1$. The argument w to INITIALIZE is the number of keys; arguments ϕ_m, ϕ_h (message and header deriving functions, respectively) in the ENC, DEC queries must be functions in $\mathsf{Fun}(k, w)$; $\mathsf{ol}(\phi)$ is the output length of $\phi \in \mathsf{Fun}(k, w)$; and cl is the ciphertext length of SE. When $\texttt{nt} = \texttt{u}$ the definition again captures universal-nonce security. That A is unique-nonce (always assumed in this case) now means that for each $j \in [1..w]$ we have $N \neq N'$ for any two ENC queries j, N, ϕ_m, ϕ_h and j, N', ϕ'_m, ϕ'_h. When $\texttt{nt} = \texttt{r}$ we have random-nonce security.

Messages could be key-dependent or not, and so could headers, giving rise to four settings of interest. These are best captured by considering different classes of adversaries. For $\Phi_m, \Phi_h \subseteq \mathsf{Fun}(k, w)$ let $\mathcal{A}[\Phi_m, \Phi_h]$ be the class of all adversaries A for which ϕ_m in A's ENC queries is in Φ_m and ϕ_h in its ENC, DEC queries is in Φ_h. Let $\mathcal{A}[\texttt{mt}, \texttt{ht}] = \mathcal{A}[\Phi_m, \Phi_h]$ where the values of (Φ_m, Φ_h) corresponding to $(\texttt{mt}, \texttt{ht}) = (\texttt{kd}, \texttt{kd}), (\texttt{kd}, \texttt{ki}), (\texttt{ki}, \texttt{kd}), (\texttt{ki}, \texttt{ki})$ are, respectively, $(\mathsf{Fun}(k, w), \mathsf{Fun}(k, w))$, $(\mathsf{Fun}(k, w), \mathsf{Cns}(k, w))$, $(\mathsf{Cns}(k, w), \mathsf{Fun}(k, w))$, $(\mathsf{Cns}(k, w), \mathsf{Cns}(k, w))$. Say that $\mathsf{SE} = (\mathcal{K}, \mathcal{E}, \mathcal{D})$ is $(\texttt{nt}, \texttt{mt}, \texttt{ht})$-AE secure if $\mathbf{Adv}^{\text{ae-nt}}_{\mathsf{SE}}(A)$ is negligible for all efficient $A \in \mathcal{A}[\texttt{mt}, \texttt{ht}]$.

Now that the header may not be known to the adversary in a DEC query, it does not know in advance whether or not $(H, N, C) \in S_j$ and it deserves to know whether rejection took place due to this or due to unsuccessful decryption. This why we do not return $\perp$ for both but rather $\perp$ for one and 0 for the other. It was to disambiguate these that we found it convenient to modify the starting definition of AE. The issue is crucial when considering security with key-dependent headers.

In the RO model there is an additional procedure HASH representing the RO. As usual it may be invoked by the scheme algorithms and the adversary, but, importantly, also by the input-deriving functions ϕ_m, ϕ_h.

For input-deriving functions to be adversary queries it is assumed they are encoded in some way. Recall that, as per our convention, the running time of A is that of the execution of A with the game, so A pays in run time if it uses functions whose description or evaluation time is too long. In asymptotic terms, A is restricted to polynomial-time computable input-deriving functions, and their description could be set to the Turing-machine that computes them.

PASSWORDS AS KEYS. The key-generation algorithm $\mathcal{K}$ in our syntax $\mathsf{SE} = (\mathcal{K}, \mathcal{E}, \mathcal{D})$ does not have to output random k-bit strings but could induce an arbitrary distribution, allowing us to capture passwords. The metric of interest in this case is the min-entropy $\mathbf{H}_\infty(\mathcal{K}) = -\log_2(\mathbf{GP}(\mathcal{K}))$, where the guessing probability $\mathbf{GP}(\mathcal{K})$ is defined as the maximum, over all k-bit strings K, of the probability that $K' = K$ when $K' \leftarrow_{\$} \mathcal{K}$. We aim to provide security as long as the min-entropy of the key-generator is not too small.

Providing security when keys are passwords is crucial because key-dependent data is more natural and prevalent in this case. In practice, our keys are largely passwords. They may be stored on disk. Their hashes are stored on the disk by the system.

3 Impossibility Results

We rule out universal-nonce security for key-dependent messages as well as security for key-dependent headers.

3.1 Universal-Nonce Insecurity

Standardized schemes all achieve universal-nonce security for ki-messages. This is convenient because an application-setting often provides for free something that can play the role of a nonce, like a counter. It also increases resistance to misuse. We would like to maintain this type of security in the presence of key-dependent data. Unfortunately we show that this is impossible. We show that no scheme is $(\mathtt{u}, \mathtt{kd}, \mathtt{ki})$-AE secure:

Proposition 1. *Let* $\mathsf{SE} = (\mathcal{K}, \mathcal{E}, \mathcal{D})$ *be an encryption scheme. Then there is an efficient adversary* $A \in \mathcal{A}[\mathtt{kd}, \mathtt{ki}]$ *such that* $\mathbf{Adv}^{\text{ae-u}}_{\mathsf{SE}}(A) \geq 1/4$.

As the proof of the above will show, the attack we present is strong in that the adversary does not just distinguish real from random encryptions but recovers the key. (A simpler attack is possible if we only want to distinguish rather than recover the key.) Also the attack works even when the nonce is a counter rather than adversary controlled. And since the adversary does not use the decryption oracle we rule out even KDM-CPA security.

We begin with some background and an overview, then prove Proposition 1, and finally show how to apply an underlying lemma to provide the first analysis of an attack in BRS [18].

BACKGROUND AND OVERVIEW. BRS [18, Section 6] suggest an attack aimed at showing that no stateful symmetric encryption scheme is KDM-secure. For the purpose of our discussion we adapt it to an attack on universal-nonce security of an AE scheme $\mathsf{SE} = (\mathcal{K}, \mathcal{E}, \mathcal{D})$. Let k be the keylength of the scheme. We will use messages of length m. Let c denote the length of the resulting ciphertexts. Let $H_{\rm ip}(V, C) = V[1]C[1]+\cdots+V[c]C[c] \bmod 2$ denote the inner product modulo two of c-bit strings V, C. Let $\phi_{V,i}$ denote the message-deriving function that on input a key K returns the first m-bit message M such that $H_{\rm ip}(V, \mathcal{E}(K, i, \varepsilon, M)) = K[i]$, or 0^m if there is no such message. (Here we use i as the nonce and ε as the header.) The adversary can pick V (BRS do not say how, but the natural choice is at random), query $\phi_{V,i}$ to get (i, C), and then recover $K[i]$ as $H_{\rm ip}(V, C)$, repeating for $i = 1, \ldots, k$ to get K.

The difficulty is that $\phi_{V,i}$ must search the message space until it finds a message satisfying the condition, and it is unclear how long this will take. In asymptotic terms, this means there is no proof that the attack runs in polynomial time, meaning is a legitimate attack at all. This issue does not appear to be recognized by BRS, who provide no analysis or formal claims relating to the attack.

In order to have a polynomial time attack where the key-recovery probability is, say, a constant, one would need to show that there is a polynomial number l of trials in which the failure probability to recover a particular bit $K[i]$ of the key is $O(1/k)$. (A union bound will then give the desired result.) We did not see a direct way to show this. Certainly, for a particular i, the probability that the first message M fails to satisfy $H_{\rm ip}(V, \mathcal{E}(K, i, \varepsilon, M)) = K[i]$ is at most 1/2, but it is not clear what is the failure probability in multiple trials because they all use the same V. The first thought that comes to mind is to modify the attack so that $\phi_{V_1,\ldots,V_l,i}$ now depends on a sequence $V_1, \ldots, V_l$ of strings, chosen independently at random by the adversary. On input the key K, the function computes the smallest j such that $H_{\rm ip}(V_j, \mathcal{E}(K, i, \varepsilon, M_j)) = K[i]$, where $M_1, M_2, \ldots, M_l$ is a fixed sequence of messages, and returns M_j. Although one can prove that this "successful" j is quickly found, the attack fails to work, since, to recover $K[i] = H_{\rm ip}(V_j, C)$ from the ciphertext $C = \mathcal{E}(K, i, \varepsilon, M_j)$, the adversary needs to know j, and it is not clear how the ciphertext is to "communicate" the value of j to the adversary.

We propose a different modification, namely to replace the inner product function with a family $H\colon \{0,1\}^s \times \{0,1\}^c \to \{0,1\}$ of pairwise independent functions. The message-deriving function $\phi_{S,i}$, on input K, will now search for M such that $H(S, \mathcal{E}(K, i, \varepsilon, M)) = K[i]$. The adversary can pick S at random, query $\phi_{S,i}$ to get (i, C), and then recover $K[i]$ as $H(S, C)$, repeating for $i = 1, \ldots, k$ to get K. We will prove that $O(k)$ trials suffice for the search to have failure probability at most $O(1/k)$ for each i, and thus that the adversary gets a constant advantage in a linear number of trials.

This strategy can be instantiated by the pairwise independent family of functions H: $\{0,1\}^{c+1} \times \{0,1\}^c \rightarrow \{0,1\}$ defined by $H(S,C) = H_{\text{ip}}(S[1] \ldots S[c], C) + S[c+1] \bmod 2$ to get a concrete attack that is only a slight modification of the BRS one but is proven to work. Given this, the question of whether the original attack can be proven to work is perhaps moot, but we find it interesting for historical reasons. Our results would not at first appear to help to answer this because the inner product function is not pairwise independent. (For example, 0^c is mapped to 0 by all functions in the family.) But curiously, as a corollary of our proof that the attack works for the *particular* family H we just defined, we get a proof that the BRS attack works as well. This is because we show that the attack using H works for an overwhelming fraction of functions from H, and thus, with sufficient probability, even for functions drawn only from the subspace of inner-product functions. Let us now proceed to the details.

ATTACK AND ANALYSIS. We begin with a general lemma.

Lemma 2. *Let H: $\{0,1\}^s \times \{0,1\}^c \rightarrow \{0,1\}$ be a family of pairwise independent hash functions. Let $C_1, \ldots, C_l \in \{0,1\}^c$ be distinct and let $T \in \{0,1\}$. Then*

$$\Pr\left[\forall j \,:\, H(S, C_j) \neq T\right] \,\leq\, \frac{1}{l}$$

where the probability is over a random choice of S from $\{0,1\}^s$.

Proof (Lemma 2). For each $j \in \{1, \ldots, l\}$ define X_j: $\{0,1\}^s \rightarrow \{0,1\}$ to take value 1 on input S if $H(S, C_j) = T$ and 0 otherwise. Regard $X_1, \ldots, X_j$ as random variables over the random choice of S from $\{0,1\}^s$. Let $X = X_1 + \cdots + X_l$ and let $\mu = \mathbf{E}[X]$. By Chebyshev's inequality, the probability above is

$$\Pr\left[X = 0\right] \,\leq\, \Pr\left[|X - \mu| \geq \mu\right] \,\leq\, \frac{\mathbf{Var}[X]}{\mu^2} \,.$$

Since H is pairwise independent, so are $X_1, \ldots, X_l$ and hence $\mathbf{Var}[X] = \mathbf{Var}[X_1] + \cdots + \mathbf{Var}[X_l]$. But for each j we have $\mathbf{E}[X_j] = 1/2$ and $\mathbf{Var}[X_j] = 1/4$, so $\mu = l/2$ and $\mathbf{Var}[X] = l/4$. Thus the above is at most $(l/4)/(l/2)^2 = 1/l$ as desired. □

We now use this to prove Proposition 1.

Proof (Proposition 1). Let k be the keylength, n the noncelength and cl the ciphertextlength of SE. Let $l = 4k$. Let $\mathsf{NumToStr}(j)$ denote a representation of integer $j \in \{0, \ldots, l\}$ as a string of length exactly $m = \lceil \log_2(l+1) \rceil$ bits. Let H: $\{0,1\}^s \times \{0,1\}^{\mathsf{cl}(m,0)} \rightarrow \{0,1\}$ denote a family of pairwise independent hash functions with s-bit keys. We construct an adversary B that recovers the target key with probability at least 3/4 when playing the real game, meaning game $\text{KDAE}_{\mathsf{SE},\mathsf{u}}$ with challenge bit $b = 1$. From B it is easy to build A achieving advantage at least 1/4. Below we depict B and also define the message-deriving functions it uses. Nonces are given as integers and assumed encoded as n-bit strings:

Adversary B	Function $\phi_{\mathbf{m},S,i}(K)$

INITIALIZE(1)
For $j = 1, \ldots, l$ do
  $\mathbf{m}[j] \leftarrow \mathsf{NumToStr}(j)$
$S \leftarrow_{\$} \{0,1\}^s$
For $i = 1, \ldots, k$ do
  $(i, C) \leftarrow_{\$} \mathrm{ENC}(1, i, \phi_\varepsilon, \phi_{\mathbf{m},S,i})$; $L[i] \leftarrow H(S, C)$
Return L
	$M \leftarrow \mathsf{NumToStr}(0)$
	For $j = 1, \ldots, l$ do
	  $C_j \leftarrow \mathcal{E}(K, i, \varepsilon, \mathbf{m}[j])$
	  If $H(S, C_j) = K[i]$ then
	    $M \leftarrow \mathbf{m}[j]$
	Return M

Above $\mathbf{m}$ is a l-vector over $\{0,1\}^m$ and ϕ_ε is the constant function that returns the empty string on every input. In its first step, B initializes the game to play with $w = 1$, meaning a single target key. Function $\phi_{\mathbf{m},S,i}(K)$ returns a message from whose encryption under nonce i and empty header one can recover bit i of the key by encoding this bit as the result of $H(S, \cdot)$ on the ciphertext. For the analysis, Lemma 2 says that for each i, adversary B fails to recover $K[i]$ with probability at most $1/4k$. By the union bound B fails to recover K with probability at most $1/4$. □

ANALYSIS OF THE BRS ATTACK. As a corollary of Lemma 2 we not only show that the inner-product function works but that it is worse only by a factor of two:

Lemma 3. *Let* $H_{\mathrm{ip}}\colon \{0,1\}^c \times \{0,1\}^c \rightarrow \{0,1\}$ *be defined by* $H_{\mathrm{ip}}(V,C) = V[1]C[1] + \cdots + V[c]C[c] \bmod 2$. *Let* $C_1, \ldots, C_l \in \{0,1\}^c$ *be distinct and let* $T \in \{0,1\}$. *Then*

$$\Pr[\forall j : H_{\mathrm{ip}}(V, C_j) \neq T] \leq \frac{2}{l} \tag{1}$$

where the probability is over a random choice of V *from* $\{0,1\}^c$.

Proof (Lemma 3). Define $H\colon \{0,1\}^{c+1} \times \{0,1\}^c \rightarrow \{0,1\}$ by

$$H(S, C) = H_{\mathrm{ip}}(S[1] \ldots S[c], C) + S[c+1] \bmod 2 .$$

This family of functions is pairwise independent. Let G be the set of all $S \in \{0,1\}^{c+1}$ such that $H(S, C_j) = T$ for some j. For $b \in \{0,1\}$ let G_b be the set of all $S \in G$ with $S[c+1] = b$. Let $\epsilon = 1/l$. Lemma 2 says that $|G| \geq (1-\epsilon)2^{c+1}$. But $G = G_0 \cup G_1$ and G_0, G_1 are disjoint so

$$|G_0| = |G| - |G_1| \geq |G| - 2^c \geq (1-\epsilon)2^{c+1} - 2^c = (1-2\epsilon)2^c .$$

To conclude we note that the probability on the left of Equation (1) equals $1 - |G_0|/2^c$. □

With this in hand, one can substitute H by H_{ip} in the proof of Lemma 1. By also doubling the value of l, the analysis goes through and shows that the BRS attack terminates in a linear number of trials and achieves a constant advantage.

3.2 Header Insecurity

We would like to use schemes in such a way that headers are not key-dependent but it may not be under our control. Applications may create headers based on data present on the system in a way that results in their depending on the key. We would thus prefer to maintain security in the presence of key-dependent headers. We show that this, too, is impossible, even when messages are key-independent. For both $\mathtt{nt} = \mathtt{u}$ and $\mathtt{nt} = \mathtt{r}$, we present attacks showing no scheme is $(\mathtt{nt}, \mathtt{ki}, \mathtt{kd})$-secure.

Proposition 4. *Let* $\mathsf{SE} = (\mathcal{K}, \mathcal{E}, \mathcal{D})$ *be an encryption scheme. Then for any* $\mathtt{nt} \in \{\mathtt{u}, \mathtt{r}\}$ *there is an efficient adversary* $A \in \mathcal{A}[\mathtt{ki}, \mathtt{kd}]$ *such that* $\mathbf{Adv}^{\text{ae-}\mathtt{nt}}_{\mathsf{SE}}(A) \geq 1/2$.

Proof (Proposition 4). Let k be the keylength of SE. Again, we present an adversary B that recovers the key with probability 1, from which A is easily built. Below we depict B and also define the message-deriving functions it uses. Nonces are given as integers and assumed encoded as n-bit strings:

Adversary B	Function $\mathsf{bit}_i(K)$

INITIALIZE(1)
For $i = 1, \ldots, k$ do
    $(N_i, C_i) \leftarrow\$ \text{ENC}(1, i, \mathsf{bit}_i, \phi_0)$; $V_i \leftarrow \text{DEC}(1, N_i, \phi_0, C_i)$
    If $V_i = \perp$ then $L[i] \leftarrow 0$ else $L[i] \leftarrow 1$
Return L
	Return $K[i]$

Here ϕ_c denotes the constant function that returns $c \in \{0, 1\}$. The header computed and used by the game in response to the i-th ENC query is $K[i]$. The header computed and used by the game in response to the i-th DEC query is 0. Thus, DEC will return $\perp$ if $K[i] = 0$. Otherwise, it will most likely return 0 because the headers don't match, although it might return 1, but in either case we have learned that $K[i] = 1$.

The attack has been written so that it applies in both the universal and random nonce cases. In the first case we will have $N_i = i$. In the second case, N_i will be a random number independent of i chosen by the game.

3.3 Remarks

The message-deriving functions used by the adversary in the proof of Proposition 1 invoke the encryption algorithm, which is legitimate since any efficient function is allowed. Having encryption depend on a RO will not avoid the attack because the message-deriving functions are allowed to call the RO and can continue to compute encryptions. (In an instantiation the RO will be a hash function and the system may apply it to the key to get data that is later encrypted.)

We do not suggest that precisely these attacks may be mounted in practice. (The message-deriving functions in our attacks are contrived.) However, our attacks rule out the possibility of a proof of security and thus there may exist other, more practical attacks. Indeed, the history of cryptography shows that once an attack is uncovered, better and more practical ones often follow.

4 The RHtE Transform and Its Security

We describe our RHtE (Randomized Hash then Encrypt) transform and prove that it endows the base scheme to which it is applied with (r, kd, ki)-AE security.

THE TRANSFORM. Given a base symmetric encryption scheme $\mathsf{SE} = (\mathcal{K}, \mathcal{E}, \mathcal{D})$, a key-generation algorithm $\mathcal{L}$ returning l-bit strings, and an integer parameter r representing the length of the random seed used in the key-hashing, the RHtE transform returns a new symmetric encryption scheme $\overline{\mathsf{SE}} = \mathsf{RHtE}[\mathsf{SE}, \mathcal{L}, r] = (\mathcal{L}, \overline{\mathcal{E}}, \overline{\mathcal{D}})$. It has $\mathcal{L}$ as its key-generation algorithm, keylength l, noncelength r and the same ciphertextlength as the base scheme. Its encryption and decryption algorithms are defined as follows, where $\text{HASH}\colon \{0,1\}^{r+l} \to \{0,1\}^k$ is a RO, $L \in \{0,1\}^l$ is the key, $R \in \{0,1\}^r$ is the nonce (which in the security game will be random), H is the header and M is the message:

Algorithm $\overline{\mathcal{E}}(L, R, H, M)$	Algorithm $\overline{\mathcal{D}}(L, R, H, C)$
$K \leftarrow \text{HASH}(R \,\|\, L)$; $C \leftarrow \mathcal{E}(K, H, M)$ Return C	$K \leftarrow \text{HASH}(R \,\|\, L)$; $M \leftarrow \mathcal{D}(K, H, C)$ Return M

The base scheme $\mathsf{SE} = (\mathcal{K}, \mathcal{E}, \mathcal{D})$ is assumed to achieve standard (nt, ki, ki)-AE security, with nt being either u or r. It is assumed to be a standard (as opposed to RO) model scheme. This is not a restriction because for the type of security we assume of it (no key-dependent data) there is no need to use a RO and none of the standardized, in use schemes do, and in any case the assumption is only for simplicity. We are not concerned with keys of the base scheme being passwords because, in standard schemes, they aren't. (Most of the time the key is an AES key.) So it is assumed that $\mathcal{K}$ returns random strings of length k. We only require one-time security of the base scheme. Accordingly we assume it is nonceless and deterministic and drop the nonce input above for both encryption and decryption. One can obtain such a scheme from standard ones by fixing a single, public nonce and hardwiring it into the algorithm. The repeated use of the nonce causes no problems since the key K is different on each encryption.

We want the constructed scheme $\overline{\mathsf{SE}}$ to provide security not only when its keys are full-fledged cryptographic ones but also when they are passwords. Hence we view as given an (arbitrary) key-generation algorithm $\mathcal{L}$ returning l-bit strings under some arbitrary distribution, and design $\overline{\mathsf{SE}}$ to have $\mathcal{L}$ as its key-generation algorithm.

The ciphertext returned is a ciphertext of the base scheme but this is deceptive since in practice R will have to be transmitted too to enable decryption. Nonetheless, in common usage, there will be no bandwidth overhead. This is because we must compare to a standard use of the base scheme where it too uses and transmits a nonce. We have saved this space by fixing this nonce and can use it for R. However, if we are in a mode where the base scheme gets the nonce out-of-band, we have r bits of bandwidth overhead. The computational overhead is independent of the message size. Implementations with base schemes CCM, EAX and GCM (see Section 5) show that for the first the slowdown is 11% for 5KB messaegs and only 1% for 50KB messages.

The BRS scheme [18] is purely RO-based, and one needs ROs with outputs of length equal to the length of the message. In our scheme the RO is used only for key-derivation and its output length is independent of the length of the message to be encrypted. In this sense, the reliance on ROs is reduced.

SECURITY OF RHtE. The following theorem says that if the base scheme is secure for key-independent headers and messages then the constructed scheme is random-nonce secure for key-dependent messages and key-independent headers.

Theorem 5. *Let* $\mathsf{SE} = (\mathcal{K}, \mathcal{E}, \mathcal{D})$ *be a base symmetric encryption scheme as above. Let* $\mathcal{L}$ *be a key-generation algorithm with keylength* l *and let* r *be a positive integer. Let* $\overline{\mathsf{SE}} = \mathsf{RHtE}[\mathsf{SE}, \mathcal{L}, r]$ *be the RO model symmetric encryption scheme associated to* $\mathsf{SE}, \mathcal{L}, r$ *as above. Let* $A \in \mathcal{A}[\mathtt{kd}, \mathtt{ki}]$ *be an adversary making* $\mathsf{q_e}$ ENC *queries,* $\mathsf{q_d}$ DEC *queries and* $\mathsf{q_h}$ HASH *queries, and let* $w \leq 2^{\mathbf{H}_\infty(\mathcal{L})-1}$ *be the number of keys, meaning the argument of A's* INITIALIZE *query. Then there is an adversary D such that*

$$\mathbf{Adv}^{\text{ae-r}}_{\overline{\mathsf{SE}}}(A)$$

$$\leq (24\mathsf{q_e}^2 + 2\mathsf{q_d}) \cdot \mathbf{Adv}^{\text{ae}}_{\mathsf{SE}}(D) + \frac{8w\mathsf{q_e}\mathsf{q_h} + 2w(w-1)\mathsf{q_e}}{2^{\mathbf{H}_\infty(\mathcal{L})}} + \frac{2\mathsf{q_e}(\mathsf{q_h} + 2\mathsf{q_e}w)}{2^r} . \quad (2)$$

Adversary D makes only one ENC *query and has the same number of* DEC *queries and the same time complexity as A.* ∎

We have omitted the nt superscript in the advantage of D because SE is nonceless. That only one-time security is required of SE is reflected in the fact that D makes only one ENC query. We remark that the bound in Theorem 5 does not appear to be tight. It is an interesting open problem to either provide a proof with a better bound or an alternative scheme for which a tight bound can be proved.

PROOF OVERVIEW. As we noted in Section 1 the proof is surprisingly involved because message-deriving functions are allowed to query the RO and because the assumed security of the base scheme must be invoked in multiple, interrelated ways in different parts of the argument, leading to two hybrids in opposite directions, one, unusually, with steps that are differently weighted.

Assume for simplicity that $w = 1$, meaning there is a single target key, denoted L. Also assume A makes no DEC queries. Denote by $\phi_1, \ldots, \phi_{\mathsf{q_e}}$ the message-deriving functions in its ENC queries and ignore the corresponding headers. Picking index g at random we set up a hybrid in which the i-th ENC query ϕ_i is answered by encrypting message $\phi_i(L)$ under L as in the real game if $i < g$ and answered at random if $i > g$, the g-th query toggling between real and random to play the role of the challenge for an adversary B against the base scheme. Let $R_1, \ldots, R_{\mathsf{q_e}}$ denote the random nonces chosen by the game. The reduction B cannot answer hash oracle query $R_g \| L$ because the reply is its target key so a bad event is flagged if A either makes this query directly, or indirectly via a message-deriving function. But once query g has been answered, A has R_g and

Hash	Scheme	RHtE Relative Running Time			
		KeySetup	5KB	50KB	500KB
SHA256	CCM	2.73	1.11	1.01	1.00
	EAX	1.94	1.10	1.01	1.00
	GCM-2k	1.66	1.10	1.02	1.00
	GCM-64k	1.19	1.09	1.02	1.00

Fig. 3. Table showing relative slowdown of RHtE with SHA256 in Crypto++ for common AE schemes and different message sizes. KeySetup is the relative slowdown in the keysetup phase alone. GCM-2k and GCM-64k correspond to GCM implemented with tables of corresponding size.

thus for queries $i > g$, nothing can prevent ϕ_i from querying $R_g \| L$ to the RO, and how are these queries to be answered by B? Crucial to this was doing the hybrid top to bottom, meaning first real then random rather than the other way round. This enables us to avoid evaluating ϕ_i on L for post-challenge queries, so that its RO queries do not need to be answered at all. This leaves the possibility that A directly makes hash query $R_g \| L$ after it gets R_g. Intuitively this is unlikely because A does not know L. The subtle point is that this relies on the assumed security of the base scheme and hence must be proven by reduction. However, doing such a reduction means another hybrid and seems to simply shunt the difficulty to another query. To get around this circularity, we do the second hybrid in the opposite direction and also with different "weights" on the different steps. A full proof can be found in [10].

5 Implementation Results

We recall that RHtE works on an existing AE scheme and a hash function. We ran RHtE with common AE schemes like CCM, EAX and GCM (with tables of 2k and 64k entries) to measure the slowdown relative to the original schemes, using a truncated version of SHA256 as the hash function and setting $l = r = k = 128$. We ran these tests using Crypto++ [26], a standard cryptography library. The measurements in Fig. 3 correspond to a Intel Core i5 M460 64-bit CPU running at 2.53 GHz with code compiled using g++ -O3 for data sizes small enough to fit in the level 2 cache. For our purposes, the relative performance of these routines is of more importance. From Fig. 3, we can observe that even at modest message sizes of around 50KB, the slowdown due to RHtE is no more than 1%. Futhermore, if algorithms like GCM are implemented with large tables and in turn a lot of precomputation in the key-setup phase, the RHtE overhead is even less noticeable.

Acknowledgments. The authors are supported in part by NSF grants CNS-0904380 and CCF-0915675. We thank the Crypto 2011 reviewers for their comments.

References

1. Abeni, P., Bello, L., Bertacchini, M.: Exploiting DSA-1571: How to break PFS in SSL with EDH (July 2008), http://www.lucianobello.com.ar/exploiting_DSA-1571/index.html
2. Acar, T., Belenkiy, M., Bellare, M., Cash, D.: Cryptographic agility and its relation to circular encryption. In: Gilbert, H. (ed.) EUROCRYPT 2010. LNCS, vol. 6110, pp. 403–422. Springer, Heidelberg (2010)
3. Applebaum, B.: Key-dependent message security: Generic amplification and completeness. In: Paterson, K.G. (ed.) EUROCRYPT 2011. LNCS, vol. 6632, pp. 527–546. Springer, Heidelberg (2011)
4. Applebaum, B., Cash, D., Peikert, C., Sahai, A.: Fast cryptographic primitives and circular-secure encryption based on hard learning problems. In: Halevi, S. (ed.) CRYPTO 2009. LNCS, vol. 5677, pp. 595–618. Springer, Heidelberg (2009)
5. Backes, M., Dürmuth, M., Unruh, D.: OAEP is secure under key-dependent messages. In: Pieprzyk, J. (ed.) ASIACRYPT 2008. LNCS, vol. 5350, pp. 506–523. Springer, Heidelberg (2008)
6. Backes, M., Pfitzmann, B., Scedrov, A.: Key-dependent message security under active attacks - brsim/uc-soundness of dolev-yao-style encryption with key cycles. Journal of Computer Security 16(5), 497–530 (2008)
7. Barak, B., Haitner, I., Hofheinz, D., Ishai, Y.: Bounded key-dependent message security. In: Gilbert, H. (ed.) EUROCRYPT 2010. LNCS, vol. 6110, pp. 423–444. Springer, Heidelberg (2010)
8. Bellare, M., Brakerski, Z., Naor, M., Ristenpart, T., Segev, G., Shacham, H., Yilek, S.: Hedged public-key encryption: How to protect against bad randomness. In: Matsui, M. (ed.) ASIACRYPT 2009. LNCS, vol. 5912, pp. 232–249. Springer, Heidelberg (2009)
9. Bellare, M., Desai, A., Jokipii, E., Rogaway, P.: A concrete security treatment of symmetric encryption. In: 38th FOCS, pp. 394–403. IEEE Computer Society Press, Los Alamitos (1997)
10. Bellare, M., Keelveedhi, S.: Authenticated and misuse-resistant encryption of key-dependent data. Cryptology ePrint Archive, Report 2011/269 (2011), Full version of this paper, http://eprint.iacr.org/
11. Bellare, M., Kohno, T.: A theoretical treatment of related-key attacks: RKA-PRPs, RKA-PRFs, and applications. In: Biham, E. (ed.) EUROCRYPT 2003. LNCS, vol. 2656, pp. 491–506. Springer, Heidelberg (2003)
12. Bellare, M., Namprempre, C.: Authenticated encryption: Relations among notions and analysis of the generic composition paradigm. In: Okamoto, T. (ed.) ASIACRYPT 2000. LNCS, vol. 1976, pp. 531–545. Springer, Heidelberg (2000)
13. Bellare, M., Rogaway, P.: Optimal asymmetric encryption. In: De Santis, A. (ed.) EUROCRYPT 1994. LNCS, vol. 950, pp. 92–111. Springer, Heidelberg (1995)
14. Bellare, M., Rogaway, P.: Encode-then-encipher encryption: How to exploit nonces or redundancy in plaintexts for efficient cryptography. In: Okamoto, T. (ed.) ASIACRYPT 2000. LNCS, vol. 1976, pp. 317–330. Springer, Heidelberg (2000)
15. Bellare, M., Rogaway, P.: The security of triple encryption and a framework for code-based game-playing proofs. In: Vaudenay, S. (ed.) EUROCRYPT 2006. LNCS, vol. 4004, pp. 409–426. Springer, Heidelberg (2006)
16. Bellare, M., Rogaway, P., Wagner, D.: The EAX mode of operation. In: Roy, B. K., Meier, W. (eds.) FSE 2004. LNCS, vol. 3017, pp. 389–407. Springer, Heidelberg (2004)

17. Bitansky, N., Canetti, R.: On strong simulation and composable point obfuscation. In: Rabin, T. (ed.) CRYPTO 2010. LNCS, vol. 6223, pp. 520–537. Springer, Heidelberg (2010)
18. Black, J., Rogaway, P., Shrimpton, T.: Encryption-scheme security in the presence of key-dependent messages. In: Nyberg, K., Heys, H.M. (eds.) SAC 2002. LNCS, vol. 2595, pp. 62–75. Springer, Heidelberg (2003)
19. Boneh, D., Halevi, S., Hamburg, M., Ostrovsky, R.: Circular-secure encryption from decision diffie-hellman. In: Wagner, D. (ed.) CRYPTO 2008. LNCS, vol. 5157, pp. 108–125. Springer, Heidelberg (2008)
20. Brakerski, Z., Goldwasser, S.: Circular and leakage resilient public-key encryption under subgroup indistinguishability-(or: Quadratic residuosity strikes back). In: Rabin, T. (ed.) CRYPTO 2010. LNCS, vol. 6223, pp. 1–20. Springer, Heidelberg (2010)
21. Brakerski, Z., Goldwasser, S., Kalai, Y.T.: Black-box circular-secure encryption beyond affine functions. In: Ishai, Y. (ed.) TCC 2011. LNCS, vol. 6597, pp. 201–218. Springer, Heidelberg (2011)
22. Brown, D.R.: A weak randomizer attack on RSA-OAEP with e=3. IACR ePrint Archive (2005)
23. Camenisch, J., Chandran, N., Shoup, V.: A public key encryption scheme secure against key dependent chosen plaintext and adaptive chosen ciphertext attacks. In: Joux, A. (ed.) EUROCRYPT 2009. LNCS, vol. 5479, pp. 351–368. Springer, Heidelberg (2009)
24. Camenisch, J., Lysyanskaya, A.: An efficient system for non-transferable anonymous credentials with optional anonymity revocation. In: Pfitzmann, B. (ed.) EUROCRYPT 2001. LNCS, vol. 2045, pp. 93–118. Springer, Heidelberg (2001)
25. Canetti, R., Tauman Kalai, Y., Varia, M., Wichs, D.: On symmetric encryption and point obfuscation. In: Micciancio, D. (ed.) TCC 2010. LNCS, vol. 5978, pp. 52–71. Springer, Heidelberg (2010)
26. Dai, W.: Crypto++ library, `http://www.cryptopp.com`
27. Dorrendorf, L., Gutterman, Z., Pinkas, B.: Cryptanalysis of the windows random number generator. In: Ning, P., di Vimercati, S.D.C., Syverson, P.F. (eds.) ACM CCS 2007, pp. 476–485. ACM Press, New York (2007)
28. Fujisaki, E., Okamoto, T., Pointcheval, D., Stern, J.: RSA-OAEP is secure under the RSA assumption. Journal of Cryptology 17(2), 81–104 (2004)
29. Goldberg, I., Wagner, D.: Randomness in the Netscape browser. Dr. Dobb's Journal (January 1996)
30. Goldwasser, S., Micali, S.: Probabilistic encryption. Journal of Computer and System Sciences 28(2), 270–299 (1984)
31. González, M.: Cryptography in the Presence of Key Dependent Messages. PhD thesis, Florida Atlantic University (2009)
32. Green, M., Hohenberger, S.: CPA and CCA-secure encryption systems that are not 2-circular secure. Cryptology ePrint Archive, Report 2010/144 (2010), `http://eprint.iacr.org/`
33. Gutterman, Z., Malkhi, D.: Hold your sessions: An attack on java session-id generation. In: Menezes, A. (ed.) CT-RSA 2005. LNCS, vol. 3376, pp. 44–57. Springer, Heidelberg (2005)
34. Haitner, I., Holenstein, T.: On the (Im)Possibility of key dependent encryption. In: Reingold, O. (ed.) TCC 2009. LNCS, vol. 5444, pp. 202–219. Springer, Heidelberg (2009)

35. Halevi, S., Krawczyk, H.: Security under key-dependent inputs. In: Ning, P., di Vimercati, S.D.C., Syverson, P.F. (eds.) ACM CCS 2007, pp. 466–475. ACM Press, New York (2007)
36. Hofheinz, D., Unruh, D.: Towards key-dependent message security in the standard model. In: Smart, N.P. (ed.) EUROCRYPT 2008. LNCS, vol. 4965, pp. 108–126. Springer, Heidelberg (2008)
37. Katz, J., Yung, M.: Unforgeable encryption and chosen ciphertext secure modes of operation. In: Schneier, B. (ed.) FSE 2000. LNCS, vol. 1978, pp. 284–299. Springer, Heidelberg (2001)
38. Malkin, T., Teranishi, I., Yung, M.: Efficient circuit-size independent public key encryption with KDM security. In: Paterson, K.G. (ed.) EUROCRYPT 2011. LNCS, vol. 6632, pp. 507–526. Springer, Heidelberg (2011)
39. McGrew, D.A., Viega, J.: The security and performance of the galois/Counter mode (GCM) of operation. In: Canteaut, A., Viswanathan, K. (eds.) INDOCRYPT 2004. LNCS, vol. 3348, pp. 343–355. Springer, Heidelberg (2004)
40. Muñiz, M.G., Steinwandt, R.: Security of signature schemes in the presence of key-dependent messages. Tatra Mt. Math. Publ. 47, 15–29 (2010)
41. Mueller, M.: Debian OpenSSL predictable PRNG bruteforce SSH exploit (May 2008), `http://milw0rm.com/exploits/5622`
42. Ouafi, K., Vaudenay, S.: Smashing SQUASH-0. In: Joux, A. (ed.) EUROCRYPT 2009. LNCS, vol. 5479, pp. 300–312. Springer, Heidelberg (2009)
43. Paterson, K.G., Watson, G.J.: Plaintext-dependent decryption: A formal security treatment of SSH-CTR. In: Gilbert, H. (ed.) EUROCRYPT 2010. LNCS, vol. 6110, pp. 345–361. Springer, Heidelberg (2010)
44. Rogaway, P.: Authenticated-encryption with associated-data. In: Atluri, V. (ed.) ACM CCS 2002, pp. 98–107. ACM Press, New York (2002)
45. Rogaway, P.: Efficient instantiations of tweakable blockciphers and refinements to modes OCB and PMAC. In: Lee, P.J. (ed.) ASIACRYPT 2004. LNCS, vol. 3329, pp. 16–31. Springer, Heidelberg (2004)
46. Rogaway, P.: Nonce-based symmetric encryption. In: Roy, B., Meier, W. (eds.) FSE 2004. LNCS, vol. 3017, pp. 348–359. Springer, Heidelberg (2004)
47. Rogaway, P., Bellare, M., Black, J., Krovetz, T.: OCB: A block-cipher mode of operation for efficient authenticated encryption. In: ACM CCS 2001, pp. 196–205. ACM Press, New York (2001)
48. Rogaway, P., Shrimpton, T.: A provable-security treatment of the key-wrap problem. In: Vaudenay, S. (ed.) EUROCRYPT 2006. LNCS, vol. 4004, pp. 373–390. Springer, Heidelberg (2006)
49. Whiting, D., Housley, R., Ferguson, N.: Counter with CBC-MAC (CCM). Undated manuscript. Submission to NIST, available from their web page (June 2002)
50. Whiting, D., Housley, R., Ferguson, N.: Counter with CBC-MAC (CCM). RFC 3610 (Informational) (2003)
51. Yilek, S., Rescorla, E., Shacham, H., Enright, B., Savage, S.: When private keys are public: Results from the 2008 Debian OpenSSL vulnerability. In: IMC. ACM, New York (2009)

Round Optimal Blind Signatures

Sanjam Garg[1], Vanishree Rao[1], Amit Sahai[1], Dominique Schröder[2,⋆], and Dominique Unruh[3]

[1] University of California, Los Angeles, USA
[2] University of Maryland, USA
[3] University of Tartu, Estonia

Abstract. Constructing round-optimal blind signatures in the standard model has been a long standing open problem. In particular, Fischlin and Schröder recently ruled out a large class of three-move blind signatures in the standard model (Eurocrypt'10). In particular, their result shows that finding security proofs for the well-known blind signature schemes by Chaum, and by Pointcheval and Stern in the standard model via black-box reductions is hard. In this work we propose the first round-optimal, i.e., two-move, blind signature scheme in the standard model (i.e., without assuming random oracles or the existence of a common reference string). Our scheme relies on the Decisional Diffie Hellman assumption and the existence of sub-exponentially hard 1-to-1 one way functions. This scheme is also secure in the concurrent setting.

1 Introduction

Blind signature schemes [13, 14] provide the functionality of a carbon copy envelope: The user (receiver), puts his message into this envelope and hands it over to the signer (sender). The signer in return signs the envelope and gives it back to the user who uses the signed enelope to recover the original message together with a signature on it. The notion of security in this context entails (1) that the signer remains oblivious about the message (blindness), but at the same time, (2) the receiver cannot forge signatures for fresh messages (unforgeability).

Blind signatures are an important primitive, whose classical applications include e-cash, e-voting, and anonymous credentials [9, 10, 8]. Moreover, oblivious transfer can be built from *unique* blind signatures [12, 17]. The several known instantiations of blind signature schemes are based on security assumptions either in the random oracle model [35, 2, 6, 7, 5, 37], or in the standard model [11, 33, 24, 28, 3]. Constructions based on general assumptions are also known [25, 16, 23, 17].

One central measure of efficiency in these schemes is the round complexity of the signing protocol. This has been an explicit problem for at least a decade, since the work of Abe [2]. Currently, the best known blind signature scheme in terms of the round complexity in the standard model is due to Okamoto [33]. This scheme has *four* rounds.

⋆ Supported in part by a DAAD postdoctoral fellowship.

P. Rogaway (Ed.): CRYPTO 2011, LNCS 6841, pp. 630–648, 2011.

All round optimal solutions (the user sends a single message to the signer and gets a single response) rely either on the random oracle heuristic [14, 7], or they require a common reference string [16,3,22,31,5,4,19,27], and some instantiations even prove their security under an interactive assumption [6, 7, 22].

Many interesting impossibility results ruling out the existence of secure blind signatures have also been proposed. Most prominently, Fischlin and Schröder [18] provide a very general impossibility result for a large class of blind signature schemes. In their result, they rule out signing protocols of less than *four* rounds, but under some natural technical conditions on the protocols (motivated by existing blind signature schemes). Interestingly, most of the round optimal blind signature schemes known today have these properties [14, 7, 16]. In particular, this means that there is not much hope to instantiate one of the known schemes under weaker assumptions. Furthermore, Katz, Schröder and Yerukhimovich [26] rule out black-box constructions of blind signature schemes from one-way permutations. In light of these result, it seems clear that significant new ideas would be needed to construct round-optimal blind signatures.

Concurrently Secure Blind Signature Schemes. Another reason why round optimal blind signature schemes are desirable is that a solution would be concurrently secure. Concurrently secure blind signature schemes, however, are difficult to obtain. Juels, Luby, and Ostrovsky [25] explained why a straight forward approach does not work. The authors present a solution that is, according to Hazay et al. [23], only secure in the sequential setting. The reason is that the solution seems to require a *concurrently secure* protocol for two-party computation. Such a protocol, however, is a mayor open problem in the standard model [23].

Obtaining a concurrently secure protocol under simulation-based definition via black-box proofs is impossible as shown by Lindell [29]. Previous protocols overcome this impossibility result by assuming a common reference string and by relying on game-based definitions. The only exception is the protocol of Hazay et al. [23] that does not need a CRS. The authors build a blind signature scheme that uses the concurrent zero-knowledge protocol of Prabhakaran, Rosen, and Sahai [36] that has an (almost) logarithmic round complexity as a building block.

1.1 Our Contribution

In this work we give the *first* round-optimal, i.e., two-move, blind signature scheme in the standard model. Our scheme is also secure in the concurrent setting. This follows directly from the fact that any two round blind signature schemes is concurrently secure, as observed by [23]. In contrast to prior schemes, our solution does not need any setup assumption such as a common reference string.

This result is especially surprising in light of the recent impossibility result of Fischlin and Schröder [18]. They provide a very general impossibility result that rules out a large class of three (or less than three) round blind signature schemes in the setting of both statistical blindness and computational blindness. Specifically, they investigate the possibility of instantiating random oracles in the schemes by

Chaum [13] and by Pointcheval and Stern [35], and of giving a security proof based only on standard assumptions. Therefore, in order to make their Cproblem tractable they restrict themselves to those blind signature schemes which satisfy a few *technical conditions* which encompass most[1] known blind signature schemes. One of these conditions is that the reduction in the unforgeability proof needs to be efficient (since the reduction is transformed into an adversary against the blindness game). In fact, this is precisely the technical condition that we avoid in our scheme and overcome the impossibility result. We note that our scheme relies on the Decision Diffie Hellman (DDH) assumption[2] and the existence of sub-exponentially hard 1-to-1 one way functions. Further, we stress that our result is only a feasibility result, and it is not as efficient as the earlier constructions. However, our work opens doors to the possibility of constructing efficient round-optimal blind signature schemes in the standard model.

Besides being interesting in its own right, our construction is an example of a scenario in which known impossibility results for concurrently-secure 2-party computation [29, 30] can be avoided to achieve meaningful game-based security definitions. Finally, we note that in a recent result Pass [34] rules out the existence of unique blind signatures using super-polynomial reductions, as long as the blindness property holds for appropriately strong adversaries. In our case blindness holds against polynomial time adversaries only and hence our result is in some tight with respect this impossibility result.

The results presented in this paper are a merge between the following two publications [38, 20].

Notations. Before presenting our results we briefly recall some basic definitions. In what follows we denote by $\lambda \in \mathbb{N}$ the security parameter. We say that a function is *negligible* if it vanishes faster than the inverse of any polynomial. A function is non-negligible if it is not negligible. If S is a set, then $x \leftarrow S$ indicates that x is chosen uniformly at random over S (which in particular assumes that S can be sampled efficiently). We write $A(x; X)$ to indicate that A is an algorithm that takes as input a value x and uses randomness X. In general, we use capital letters for the randomness. W.l.o.g. we assume that X has bit length λ.

2 Blind Signatures and Their Security

By $(a, b) \leftarrow \langle \mathcal{X}(x), \mathcal{Y}(y) \rangle$ we denote interactive execution of algorithms $\mathcal{X}$ and $\mathcal{Y}$, where x (resp., y) is the private input of $\mathcal{X}$ (resp., $\mathcal{Y}$), and a (resp., b) is the private output of $\mathcal{X}$ (resp., $\mathcal{Y}$). We write $\mathcal{X}^{\langle \cdot, \mathcal{Y} \rangle^\infty}$ for $\mathcal{X}$ with oracle access two arbitrarily many interactions with $\mathcal{Y}$. And $\mathcal{X}^{\langle \cdot, \mathcal{Y} \rangle^1}$ for $\mathcal{X}$ with oracle access two arbitrarily a single interaction with $\mathcal{Y}$.

---

[1] Known blind signature schemes can indeed be modified in rather unnatural ways to construct blind signature schemes that do not satisfy their conditions.

[2] We also need a ZAP (a two round witness indistinguishable proof system) which can be constructed under the DDH assumption. We note that ZAPs can in fact be constructed from any one-way trapdoor permutation.

Experiment $\mathsf{Unforge}^{\mathsf{BS}}_{\mathcal{U}^*}(\lambda)$
  $(sk, vk) \leftarrow \mathsf{Gen}(1^\lambda)$
  $((m_1^*, \sigma_1^*), \ldots, (m_{k+1}^*, \sigma_{k+1}^*)) \leftarrow \mathcal{U}^{*\langle \mathcal{S}(sk), \cdot \rangle^\infty}(vk)$
  Return 1 iff
    $m_i^* \neq m_j^*$ for all i, j with $i \neq j$, and
    $\mathsf{Vrfy}(vk, m_i^*, \sigma_i^*) = 1$ for all i, and
    at most k interactions with $\mathcal{S}(sk)$
      were completed.

Experiment $\mathsf{Unblind}^{\mathsf{BS}}_{\mathcal{S}^*}(\lambda)$
  $(vk, m_0, m_1, \mathsf{st}_{\mathsf{find}}) \leftarrow \mathcal{S}^*(\mathsf{find}, 1^\lambda)$
  $b \leftarrow \{0, 1\}$
  $\mathsf{st}_{\mathsf{issue}} \leftarrow \mathcal{S}^{*\langle \cdot, \mathcal{U}(vk, m_b) \rangle^1, \langle \cdot, \mathcal{U}(vk, m_{\bar{b}}) \rangle^1}(\mathsf{issue}, \mathsf{st}_{\mathsf{find}})$
    and let $\sigma_b, \sigma_{\bar{b}}$ denote the
      (possibly undefined) local outputs
    of $\mathcal{U}(vk, m_b)$ resp. $\mathcal{U}(vk, m_{\bar{b}})$.
  set $(\sigma_0, \sigma_1) = (\bot, \bot)$ if $\sigma_0 = \bot$ or $\sigma_1 = \bot$
  $b^* \leftarrow \mathcal{S}^*(\mathsf{guess}, \sigma_0, \sigma_1, \mathsf{st}_{\mathsf{issue}})$
  return 1 iff $b = b^*$.

Fig. 1. Security games of blind signatures

Definition 1. *A* blind signature scheme BS *consists of PPT algorithms* Gen, Vrfy *along with interactive PPT algorithms* $\mathcal{S}, \mathcal{U}$ *such that for any* $\lambda \in \mathbb{N}$*:*

- $\mathsf{Gen}(1^\lambda)$ *generates a key pair* (sk, vk)*.*
- *The joint execution of* $\mathcal{S}(sk)$ *and* $\mathcal{U}(vk, m)$*, where* $m \in \{0,1\}^\lambda$*, generates an output* σ *for the user and no output for the signer. We write this as* $(\bot, \sigma) \leftarrow \langle \mathcal{S}(sk), \mathcal{U}(vk, m) \rangle$*.*
- *Algorithm* $\mathsf{Vrfy}(vk, m, \sigma)$ *outputs a bit* b*.*

We require completeness *i.e., for any* $m \in \{0,1\}^\lambda$*, and for* $(sk, vk) \leftarrow \mathsf{Gen}(1^\lambda)$*, and* σ *output by* $\mathcal{U}$ *in the joint execution of* $\mathcal{S}(sk)$ *and* $\mathcal{U}(vk, m)$*, it holds that* $\mathsf{Vrfy}(vk, m, \sigma) = 1$ *with overwhelming probability in* $\lambda \in \mathbb{N}$*.*

Note that it is always possible to sign messages of arbitrary length by applying a collision-resistant hash function to the message prior to signing.

Blind signatures must satisfy two properties: unforgeability and blindness [25, 35]. Notice that we can also achieve the stronger definition of unforgeability from [39] by applying their transformation which does not increase the round complexity.

For unforgeability we require that a user who runs k executions of the signature-issuing protocol should be unable to output $k+1$ valid signatures on $k+1$ distinct messages.

Definition 2. *A blind signature scheme* $\mathsf{BS} = (\mathsf{Gen}, \mathcal{S}, \mathcal{U}, \mathsf{Vrfy})$ *is* unforgeable *if for any PPT algorithm* $\mathcal{U}^*$ *the probability that experiment* $\mathsf{Unforge}^{\mathsf{BS}}_{\mathcal{U}^*}(\lambda)$ *defined in Figure 1 evaluates to* 1 *is negligible (as a function of* λ*).*

Blindness says that it should be infeasible for a malicious signer $\mathcal{S}^*$ to decide which of two messages m_0 and m_1 has been signed first in two executions with an honest user $\mathcal{U}$. This condition must hold, even if $\mathcal{S}^*$ is allowed to choose the public key maliciously [1]. If one of these executions has returned an invalid signature, denoted by $\bot$, then the signer is not informed about the other signature either.

Definition 3. *A blind signature scheme* $\mathsf{BS} = (\mathsf{Gen}, \mathcal{S}, \mathcal{U}, \mathsf{Vrfy})$ *satisfies* blindness *if for any efficient algorithm* $\mathcal{S}^*$ *(working in modes* find, issue, *and* guess*) the probability that experiment* $\mathsf{Unblind}^{\mathsf{BS}}_{\mathcal{S}^*}(\lambda)$ *defined in Figure 1 evaluates to* 1 *is negligibly bigger than* $1/2$*.*

A blind signature scheme is secure if it is unforgeable and blind.

3 Towards a Secure Construction

A central idea behind our work is to adapt techniques from secure two-party computation, despite the fact that we *cannot* achieve the traditional notions of secure two-party computation in the standard model with only 2 rounds. Indeed, unfortunately, very few two round protocol techniques are known at all.[3] The high-level idea of our construction is as follows: We employ a two-move secure function evaluation (SFE) protocol to let the signer and user compute a signature on a message chosen by the user. Using Yao's garbled circuits together with Naor-Pinkas OT [32], we get a two-move SFE protocol with the following properties: The user sends the first message, the signer replies, and only the user gets output. The user's input stays secret even in the case of an active malicious signer (as the user does not send any responses to the signer's messages, an active signer is no more powerful than a passive one). The signer's input stays secure against active malicious users. The correctness of the protocol's output is guaranteed against active users and against *passive* signers.

If we use this SFE protocol for signing, we face the following two problems:

(a) Although the signer does not learn the user's inputs, he could cheat in the SFE to make the signatures output by the two users in the blindness game depend on the message in different ways. Even if the SFE would have correctness against active signers, such cheating would still be possible by using different signing keys or randomnesses in the two interactions.

(b) To prove unforgeability of the blind signature scheme, we have to reduce it to the unforgeability of the underlying non-interactive signature scheme. To do so, we need to extract the messages sent by the user in order to feed them into a signing oracle. But in a two-round SFE protocol, we cannot use rewinding to extract the message.

To solve (a), we let the signer commit, as part of his public verification key, to the secret key, and to a random seed to be used when signing. In order to force the signer to actually use that key and randomness, we would like to use a zero-knowledge proof that the SFE was performed correctly and with the right inputs. Unfortunately, two-move zero-knowledge proofs do not exist in the standard model (without CRS). However, there are two-move *witness indistinguishable* proofs, so-called ZAPs. As these are not zero-knowledge, we cannot use them directly (witness indistinguishable proofs might still leak, e.g., the signing key). To make the ZAP "almost zero-knowledge", we introduce a trapdoor: We introduce the possibility of producing a fake proof by using the preimage under some one-way function of a value chosen by the user. A complexity-leveraged simulator can then use this trapdoor, and we can show that the ZAP does not reveal too much.

The solve (b), we again use complexity leveraging: Our SFE protocol only has computational security for the user, so the signer can extract the message m to

[3] We do know of two round witness indistinguishable protocols (ZAPs), which will be useful for us. But as we will see later, ZAPs by themselves are not sufficient for our purposes.

be signed in superpolynomial time T. Thus, we can transform the unforgeability game into one where the signer extracts m, sends it to a signing oracle, and re-inserts the resulting signature back into the SFE. This allows to reduce the unforgeability of the blind signature scheme against the unforgeability of the underlying non-interactive signature scheme. Note however, that the underlying scheme needs to be secure against T-time adversaries in this reduction. Also, standard properties of SFE do not seem to allow us to perform such an extraction and re-insertion. Thus, we define a new property called Alice-extraction-privacy for this purpose; fortunately, Yao's garbled circuits using Naor-Pinkas OTs have this property.

4 Required Primitives

Before presenting our generic construction, we review the required primitives. Most definitions are standard, except that in some cases we require security against superpolynomial adversaries. In these cases we write, e.g., T-one-wayness and mean one-wayness against adversaries running in time $T \cdot poly(\lambda)$. We will now review those primitives and security definitions that are not standard. Complete definitions are given in the full version [21].

ZAPs. A ZAP [15] is a two-round witness-indistinguishable proof system. That is, a ZAP for a language L consists of a prover $\mathcal{P}$ and a verifier $\mathcal{V}$. The first invocation of $\mathcal{V}(1^\lambda)$ is independent of the statement to be proven and outputs a message msg. Given that message, the prover $\mathcal{P}(1^\lambda, \mathsf{msg}, s, w)$ outputs a proof π for statement s with witness w. Finally the verifier $\mathcal{V}(\mathsf{msg}, s, \pi)$ checks the proof π. Notice that this verification only uses msg but no private state from the first invocation of $\mathcal{V}$. In particular, the prover can check on his own whether verification succeeds.

Two-party-computation. We will need a two-move two-party secure function evaluation protocol. Syntactically, such a protocol is described by three PPT algorithms $\mathsf{SFE}_1, \mathsf{SFE}_2, \mathsf{SFE}_3$. The intended use is as follows: Assume that Alice holds a circuit C and Bob holds an input m to that circuit. Then Bob first computes $(\mathsf{sfe}_1, \mathsf{sfest}) \leftarrow \mathsf{SFE}_1(1^\lambda, m)$ and sends sfe_1 to Alice (sfest is Bob's state). Then Alice computes $\mathsf{sfe}_2 \leftarrow \mathsf{SFE}_2(1^\lambda, \mathsf{sfe}_1, C)$ and sends sfe_2 to Bob. Finally, Bob computes the result σ of the computation via $\sigma \leftarrow \mathsf{SFE}_3(1^\lambda, \mathsf{sfe}_2, \mathsf{sfest})$. We require two standard properties, perfect completeness (an honest execution gives the right result with probability 1) and Bob-privacy (from Alice's point of view, different Bob-inputs are computationally indistinguishable). We also require one non-standard property for Alice's security:

Instead of requiring that Bob does not learn anything about the circuit C except for $C(m)$, we require the following: If Alice knows m (which we model by a superpolynomial-time extraction algorithm SFEExt), then instead of applying SFE_2 with circuit C, she can instead compute $\sigma := C(m)$ directly and hardcode the result into the function evaluation using a function $\mathsf{SFEFake}_2$. (This property will be needed so that we can outsource the evaluation of C to a signing oracle later in our proofs.)

Definition 4 (Non-uniform T-Alice-extraction-privacy). *There is a $(T \cdot poly(\lambda))$-time probabilistic algorithm* SFEExt *and a polynomial-time algorithm* $\mathsf{SFEFake}_2$ *such that the following holds:*

For an adversary $\mathcal{A}$, consider the following experiments:

Experiment 1	***Experiment*** 2
$(\mathsf{sfe}_1, C, \mathsf{sfest}) \leftarrow \mathcal{A}(1^\lambda)$	$(\mathsf{sfe}_1, C, \mathsf{sfest}) \leftarrow \mathcal{A}(1^\lambda)$
	$m \leftarrow \mathsf{SFEExt}(1^\lambda, \mathsf{sfe}_1)$
	$\sigma \leftarrow C(m)$
$\mathsf{sfe}_2 \leftarrow \mathsf{SFE}_2(1^\lambda, \mathsf{sfe}_1, C)$	$\mathsf{sfe}_2 \leftarrow \mathsf{SFEFake}_2(1^\lambda, \mathsf{sfe}_1, \sigma)$
$b \leftarrow \mathcal{A}(\mathsf{sfe}_2, \mathsf{sfest})$	$b \leftarrow \mathcal{A}(\mathsf{sfe}_2, \mathsf{sfest})$

Then for any PPT $\mathcal{A}$, $|\Pr[b = 1 : \text{Experiment 1}] - \Pr[b = 1 : \text{Experiment 2}]|$ is negligible in λ.

In the fullversion [21], we show that Yao's garbled circuits using Naor-Pinkas OTs [32] satisfies these conditions for any T such that the DDH problem can be decided in time T.

5 Construction and Security Proofs

In this section, we present our construction and show its security.

5.1 Construction

We proceed to define our blind signature scheme. Fix superpolynomial functions T and T' with $T' < T$. In our construction, we assume a one-way function f, a pseudorandom function F, commitment schemes $\mathcal{C}_R, \mathcal{C}_X$, a signature scheme $\mathsf{Sig} = (\mathsf{SigGen}, \mathsf{Sign}, \mathsf{SigVrfy})$, a ZAP $Z = (\mathcal{P}, \mathcal{V})$, and a two-message two-party secure function evaluation protocol $(\mathsf{SFE}_1, \mathsf{SFE}_2, \mathsf{SFE}_3)$. We then define the blind signature scheme $(\mathsf{Gen}, \mathcal{S}, \mathcal{U}, \mathsf{Vrfy})$ as follows:

Key Generation. $\mathsf{Gen}(1^\lambda)$ peforms the following steps:
- $R, S, T \leftarrow \{0,1\}^\lambda$
- $(ssk, svk) \leftarrow \mathsf{SigGen}(1^\lambda; S)$
- $\mathsf{com}_R \leftarrow \mathsf{Com}_R((R, ssk); T)$
- set $sk \leftarrow (svk, ssk, R, S, T)$ and $vk \leftarrow (svk, \mathsf{com}_R)$

Signing. The signature issue protocol between the signer $\mathcal{S}$ and the user $\mathcal{U}$ is as follows:
- $\mathcal{U}$ generates the first message of the SFE protocol $(\mathsf{sfe}_1, \mathsf{sfest}) \leftarrow \mathsf{SFE}_1(1^\lambda, m)$ and the challenge $\mathsf{msg} \leftarrow \mathcal{V}(1^\lambda)$ for the ZAP Z. It picks $x \leftarrow \{0,1\}^\lambda$ uniformly at random, sets $y \leftarrow f(x)$, and sends $(\mathsf{sfe}_1, \mathsf{msg}, y)$ to $\mathcal{S}$.
- $\mathcal{S}$ receives $(\mathsf{sfe}_1, \mathsf{msg}, y)$ from the user. If y is not a valid image of f, then $\mathcal{S}$ sends $\perp$. Otherwise, denote by $C_{ssk,R}(m)$ the circuit computing $\mathsf{Sign}(ssk, m; F_R(m))$. The signer $\mathcal{S}$ picks two random values V, X each of

Signer $\mathcal{S}(sk)$		**User** $\mathcal{U}(vk, m)$
parse $sk = (svk, ssk, R, S, T)$		parse $vk = (svk, \mathsf{com}_R)$
	// first message of the SFE	$(\mathsf{sfe}_1, \mathsf{sfest}) \leftarrow \mathsf{SFE}_1(1^\lambda, m)$
	//generate the challenge for the ZAP	$x \leftarrow \{0,1\}^\lambda, y \leftarrow f(x)$ $\mathsf{msg} \leftarrow \mathcal{V}(1^\lambda)$
if $y \in \mathsf{image}(f)$, then	$\xleftarrow{(\mathsf{sfe}_1, \mathsf{msg}, y)}$	
$V, X \leftarrow \{0,1\}^\lambda$ $\mathsf{sfe}_2 \leftarrow \mathsf{SFE}_2(1^\lambda, \mathsf{sfe}_1, C_{ssk,R}; V)$ $\mathsf{com}_x \leftarrow \mathsf{Com}_X(0^\lambda; X)$ $s \leftarrow (svk, \mathsf{com}_R, \mathsf{sfe}_1, \mathsf{sfe}_2, \mathsf{com}_x, y)$ $w \leftarrow (R, S, T, V, ssk)$ $\pi \leftarrow \mathcal{P}(\mathsf{msg}, s, w)$	*//sign the message*	
else $\mathsf{sfe}_2, \mathsf{com}_x, \pi \leftarrow \perp$	$\xrightarrow{(\mathsf{sfe}_2, \mathsf{com}_x, \pi)}$	$s \leftarrow (svk, \mathsf{com}_R, \mathsf{sfe}_1, \mathsf{sfe}_2, \mathsf{com}_x, y)$ if $\mathcal{V}(\mathsf{msg}, s, \pi) = 1$ $\sigma \leftarrow \mathsf{SFE}_3(1^\lambda, \mathsf{sfe}_2, \mathsf{sfest})$ output σ

Fig. 2. Issue protocol of the two move blind signature scheme

bit length λ, it computes $\mathsf{sfe}_2 \leftarrow \mathsf{SFE}_2(1^\lambda, \mathsf{sfe}_1, C_{ssk,R}; V)$ and $\mathsf{com}_x \leftarrow \mathsf{Com}_X(0^\lambda; X)$. Then, it generates a proof π (with respect to msg) for the statement $(svk, \mathsf{com}_R, \mathsf{sfe}_1, \mathsf{sfe}_2, \mathsf{com}_x, y) \in L$, where L contains tuples for which there exists either a witness $\omega_1 = (R, S, T, V, ssk)$ such that:

$$\mathsf{sfe}_2 = \mathsf{SFE}_2(1^\lambda, \mathsf{sfe}_1, C_{ssk,R}; V) \bigwedge$$
$$\mathsf{com}_R = \mathsf{Com}_R((R, ssk); T) \bigwedge (ssk, svk) = \mathsf{SigGen}(1^\lambda; S)$$

or there exists a witness $\omega_2 = (x, X)$ such that

$$\mathsf{com}_x = \mathsf{Com}_X(x; X) \bigwedge f(x) = y.$$

$\mathcal{S}$ then sends $(\mathsf{sfe}_2, \mathsf{com}_x, \pi)$ to $\mathcal{U}$.

- $\mathcal{U}$ verifies that π is a valid proof for the statement $(svk, \mathsf{com}_R, \mathsf{sfe}_1, \mathsf{sfe}_2, \mathsf{com}_x, y) \in L$ with respect to msg and the ZAP Z. If this proof fails, then $\mathcal{U}$ outputs $\perp$. Otherwise, it computes the signature $\sigma \leftarrow \mathsf{SFE}_3(1^\lambda, \mathsf{sfe}_2, \mathsf{sfest})$ and outputs σ.

Verification. $\mathsf{Vrfy}(vk, \sigma, m)$ returns $\mathsf{SigVrfy}(svk, \sigma, m)$.

5.2 Security

Theorem 5. *Assume that* Sig *is complete;* $(\mathsf{SFE}_1, \mathsf{SFE}_2, \mathsf{SFE}_3)$ *is correct;* $\mathcal{C}_R, \mathcal{C}_X$ *are complete; and* $Z = (\mathcal{P}, \mathcal{V})$ *is complete. Then the protocol from Section 5.1 is complete.*

Theorem 6. *Assume that f is invertible in time T and has efficiently decidable images; F is a T-pseudorandom function;* $(\mathsf{SFE}_1, \mathsf{SFE}_2, \mathsf{SFE}_3)$ *is non-uniformly T-Alice-extraction-private;* Sig *is T-unforgeable; $\mathcal{C}_R$ is T-hiding; $\mathcal{C}_X$ is non-uniformly hiding; the ZAP $Z = (\mathcal{P}, \mathcal{V})$ is non-uniformly computationally witness-indistinguishable.*

Then the blind signature scheme from Section 5.1 is unforgeable.

Theorem 7. *Assume that f is T'-one-way;* $(\mathsf{SFE}_1, \mathsf{SFE}_2, \mathsf{SFE}_3)$ *is perfectly correct and has Bob-privacy; $\mathcal{C}_R, \mathcal{C}_X$ are perfectly binding; $\mathcal{C}_X$ is T-extractable; ZAP $Z = (\mathcal{P}, \mathcal{V})$ is adaptively sound.*

Then the scheme from Section 5.1 is blind.

Theorem 5 is immediate from the construction of the protocol, and Theorems 6 and 7 will be shown in the next section.

Instantiating the Primitives. The primitives needed in our construction (see Theorems 5, 6, and 7) can be instantiated if the DDH problem is non-uniformly hard, and if subexponentially-hard[4] 1-1 one-way functions with efficiently decidable range exist: Given the one-way functions, we can construct a perfectly hiding non-uniformly T-hiding commitment $\mathcal{C}_R$, a T-unforgeable signature scheme Sig, and a T-pseudorandom function (for some $T \in 2^{\eta^{\Omega(1)}}$). By scaling of the security parameter (such that the used randomness is $< \log T$), we can produce a T-extractable non-uniformly commitment $\mathcal{C}_R$. By similar rescaling, we get a T'-one-way T-time invertible one-way function f. Non-uniform ZAPs can be instantiated from non-uniform DDH [15]. Finally, $(\mathsf{SFE}_1, \mathsf{SFE}_2, \mathsf{SFE}_3)$ can be instantiated given non-uniform one-way functions (for Yao's circuits) and DDH (for Naor-Pinkas OT), see the full version [21], as long as the underlying DDH can be broken in time T. This can again be achieved by rescaling the security parameter.

5.3 Security Proofs

Proof of Unforgeability. The proof idea is the following. We start with a game that corresponds to the unforgeability game of blind signatures and we then gradually change this game such that at the end we can build an adversary against the unforgeability of the underlying signature scheme. The main steps of the proof are the following:

- We apply a complexity leveraging argument. This technique allows us to extract the message m out of the first message of the secure function evaluation protocol. We also use this technique in order to invert the one-way function f and obtain $f^{-1}(y)$.
- We use the external signing oracle in the unforgeability game of the underlying signature scheme to sign the message m.

[4] By subexponentially-hard, we mean that a function $T \in 2^{\eta^{\Omega(1)}}$ exist, such that the primitive is hard against T-time adversaries.

– Instead of computing the second message of the SFE protocol honestly, we use the $\mathsf{SFEFake}_2$ algorithm. These modifications do not change the success probability of the adversary (against the unforgeability of the blind signature scheme) because:
  - the SFE protocol is extraction-private and thus, the attacker cannot tell the difference;
  - the ZAP remains valid as it now uses the previously computed preimage x of f as a witness.

Due to the complexity leveraging, we have to be careful which primitives need superpolynomial-hardness and which do not. In some cases, non-uniform security turns out to be sufficient, even though we use the security of a primitive in a game involving superpolynomial computations.

Proof (of Theorem 6). Assume towards contradiction that the construction from Section 5.1 is not unforgeable. Then there exists a PPT algorithm $\mathcal{U}^*$ that outputs $(\ell + 1)$ message/signature pairs (m_i, σ_i) after ℓ executions of the signature issue protocols. This adversary wins if all messages are distinct and all signatures verify under vk. Now, consider the following sequence of games, where the first game Game-0 is the unforgeability game in which we run the game with the forger $\mathcal{U}^*$. Within all games, the first digit of the line numbers is the number of the game (i.e., line 107 in Game-1 corresponds to 007 in Game-0).

Game-0

000 $R, S, T, V_i, X_i \leftarrow \{0,1\}^\lambda$
001 $(ssk, svk) \leftarrow \mathsf{SigGen}(1^\lambda; S)$
002 $\mathsf{com}_R \leftarrow \ldots$
003 $\mathsf{st}_0 \leftarrow \mathcal{U}^*(svk, \mathsf{com}_R)$
004 for $i = 1, \ldots, \ell$
005 $\quad (\mathsf{sfe}_{1,i}, \mathsf{msg}_i, y_i) \leftarrow \mathcal{U}^*(\mathsf{st}_{i-1})$
006 $\quad$ if $y_i \in \mathsf{image}(f)$ then
007 $\qquad \mathsf{sfe}_{2,i} \leftarrow \mathsf{SFE}_{2,i}(1^\lambda, \mathsf{sfe}_{1,i}, C_{ssk,R})$
008 $\qquad \mathsf{com}_{x,i} \leftarrow \mathsf{Com}_X(0^\lambda; X_i)$
009 $\qquad s_i \leftarrow (svk, \mathsf{com}_R, C, \mathsf{sfe}_{1,i}, \mathsf{sfe}_{2,i}, \mathsf{com}_{x,i}, y_i)$
010 $\qquad w_i \leftarrow (R, S, T, V_i, ssk)$
011 $\qquad \pi_i \leftarrow \mathcal{P}(\mathsf{msg}_i, s_i, w_i)$
012 $\quad$ else
013 $\qquad \mathsf{sfe}_{2,i}, \mathsf{com}_i^x, \pi_i \leftarrow \perp$
014 $\quad \mathsf{st}_i \leftarrow \mathcal{U}^*(\mathsf{sfe}_{2,i}, \mathsf{com}_{x,i}, \pi_i, \mathsf{st}_i)$
015 end for
016 $(m_1, \sigma_1, \ldots, m_{\ell+1}, \sigma_{\ell+1}) \leftarrow \mathcal{U}^*(\mathsf{st}_\ell)$
017 Return 1 iff $\mathsf{SigVrfy}(svk, m_i, \sigma_i) = 1$ for all $i = 1, \ldots, \ell + 1$ and $m_i \neq m_j$ for all $i \neq j$

Game-0 $\Rightarrow$ Game-1. We now modify the above game by letting the signer (after Step 005) extract the message $m_i \leftarrow \mathsf{SFEExt}(1^\lambda, \mathsf{sfe}_{1,i})$ from the first message of the SFE protocol $\mathsf{sfe}_{1,i}$ according to Definition 4. Analogously, $f^{-1}(y)$ is the algorithm that inverts the one-way function f. Both algorithms are running in time T.

Game-1

105a $\boldsymbol{x_i \leftarrow f^{-1}(y_i), m_i \leftarrow \mathsf{SFEExt}(1^\lambda, \mathsf{sfe}_{1,i})}$
108 $\mathsf{com}_{x,i} \leftarrow \mathsf{Com}_X(\boldsymbol{x_i}; X_i)$

The difficulty in showing that the adversary's success probability in both games is the same arises from the point that Step 105a cannot be computed efficiently. Nevertheless, we solve this issue by applying the following (standard) technique. The idea is to consider primitives that are secure against *non-uniform* adversaries. This allows us to perform computations *in advance* that are not feasible in polynomial time. More precisely, consider the commitment scheme $\mathcal{C}_X$ that is non-uniformly hiding. The adversary is allowed to be unbounded as long as it has not received the commitment (since the unbounded computation is captured by the auxiliary input). Once the attacker has obtained the commitment, it runs in polynomial time. The basic idea behind this approach is that it is possible to wire an advice into the circuit that is hard to compute. This observation allows us to perform a computation that is not feasible in polynomial time *before* seeing the commitment. During this step, we extract the message out of the encryption $m_i \leftarrow \mathsf{SFEExt}(1^\lambda, \mathsf{sfe}_{1,i})$ and we invert the one-way function $x_i \leftarrow f^{-1}(y_i)$. Then, we commit to x_i (instead of 0^λ). Note that this is only possible because Step 108 happens after Step 105. This, however, is not quite true because this step happens in a loop. Thus, at some point Step 108 happens before Step 105. To handle this issue, we refine our argument as follows: let Game-$\tilde{1}_i$ be the game where we applied the modifications during the first i iterations but not in iterations $i+1, \ldots, \ell$. Now, the same argument as above shows that Game-$\tilde{1}_i$ and Game-$\tilde{1}_{i+1}$ are indistinguishable for any i (even if i depends on the security parameter). This, however, also implies that Game-$\tilde{1}_0$ and Game-$\tilde{1}_\ell$ are indistinguishable. Furthermore Game-$\tilde{1}_0$ = Game-0 and Game-$\tilde{1}_\ell$ = Game-1, hence Game-0 $\approx$ Game-1 where $\approx$ indicates that the probability that both games output 1 is the same (except for a negligible amount).

Game-1 $\Rightarrow$ Game-2 $\Rightarrow$ Game-3. In the next game, Game-2, we first change the witness of the ZAP Z. That is, we use as a witness the pre-image x_i of the one-way function f that we have inverted in the previous step. Afterwards, in Game-3, we sign the message that was extracted in Game-1, and then use $\mathsf{SFEFake}_2$ in order to make the function evaluation output that signature σ_i.

Game-2

209 $s_i \leftarrow (svk, \mathsf{com}_R, \mathsf{sfe}_{1,i}, \mathsf{sfe}_{2,i}, \mathsf{com}_{x,i}, y_i)$
210 $w_i := \boldsymbol{(x_i, X_i)}$
211 $\pi_i \leftarrow \mathcal{P}(\mathsf{msg}, s_i, \boldsymbol{w_i})$

Game-3

307 $\boldsymbol{\sigma_i \leftarrow \mathsf{Sign}(ssk, m_i; F_R(m_i))}$
307a $\mathsf{sfe}_{2,i} \leftarrow \boldsymbol{\mathsf{SFEFake}_2(1^\lambda, \mathsf{sfe}_{1,i}, \sigma_i)}$

Now, we argue that both modifications do not change the success probability of the adversary $\mathcal{U}^*$ by more than a negligible amount and therefore, Game-0 $\approx$ Game-3. This should follow from the following two observations

- The one-way function f has efficiently decidable images and the signer checks if y_i is a valid image under f in step 006. Thus, according to our construction the witness w_i is a valid witness. Note that according to our construction the witness $w_i := (R, S, T, V_i, ssk)$ used in Game-1 is also valid. Since both witnesses are a valid witness and because we have assumed that the ZAP $\mathcal{Z}$ is non-uniformly witness indistinguishable, it follows that the success probability of $\mathcal{U}^*$ in both games is the same (except for a negligible amount).
- The secure function evaluation scheme is T-Alice-extraction-private. Thus, the adversary $\mathcal{U}^*$ does not notice the difference in the computation of $\mathsf{sfe}_{2,i}$.

We elaborate more on the second point: The only difference between Game-2 and Game-3 is that in the i-th iteration of the loop, in Game-3 we replace $\mathsf{sfe}_{2,i} \leftarrow \mathsf{SFE}_{2,i}(1^\lambda, \mathsf{sfe}_{1,i}, C_{ssk,R})$ (Step 207) by $\mathsf{sfe}_{2,i} \leftarrow \mathsf{SFEFake}_2(1^\lambda, \mathsf{sfe}_{1,i}, \sigma_i) = C_{ssk,R}(m_i)$, where $\sigma_i \leftarrow \mathsf{Sign}(ssk, m_i; F_R(m_i))$ (Steps 307, 307a). By definition of T-Alice-extraction-privacy, it follows that the games are indistinguishable. (Notice that we use *non-uniform* T-Alice-extraction-privacy, the fact that there are superpolynomial computations occurring before the SFE does not limit the applicability of T-Alice-extraction-privacy.) Unfortunately, we cannot apply the arguments justifying the transformations Game-1 $\Rightarrow$ Game-2 $\Rightarrow$ Game-3 directly. The reason is that these arguments are only applicable as long as the games run in polynomial time (or at least those steps of the games occurring after the modifications). In the previous step, however, we have inverted the one-way function and we have extracted the message from the first message of the SFE protocol. Both steps, however, are not computable in polynomial time. We handle this issue by carefully applying a hybrid argument. Now, we apply carefully a hybrid argument over all three games. We omit the details of this hybrid argument.

Game-3 $\Rightarrow$ Game-4. This game is identical to the prior one, but instead of committing to R and ssk, we commit to an all zero string.

Game-4	
401	$\mathsf{com}_R \leftarrow \mathsf{Com}_R(\mathbf{0}^\lambda; T)$

Since the commitment scheme $\mathcal{C}_R$ is T hiding, and since the commitment's randomness T is not used anywhere else, this modification changes the success probability of $\mathcal{U}^*$ only by a negligible amount and thus, Game-3 $\approx$ Game-4 and therefore Game-0 $\approx$ Game-4. (Remember that both f^{-1} and SFEExt and thus all steps in the game run in time T.)

Game-4 $\Rightarrow$ Game-5. In this game, we do not generate the signing key locally, but we build a forger $\mathcal{B}$ against the signature scheme Sig. The difference to the above described games is that it uses its external signing oracle in order to obtain the signature σ_i on the message m_i.

Game-5	
500	$x, T, V_i, X_i \leftarrow \{0,1\}^\lambda$ **(removed R, S)**
501	$\boldsymbol{svk} \leftarrow \widehat{\mathsf{SigGen}(1^\lambda)}, \mathsf{com}_R \leftarrow \mathsf{Com}_R(R, ssk; T)$
507	$\boldsymbol{\sigma_i} \leftarrow \widehat{\mathsf{Sign}(m_i)}$

Here $\widehat{\mathsf{SigGen}}$ and $\widehat{\mathsf{Sign}}$ constitute a signing oracle. $\widehat{\mathsf{SigGen}}$ produces a verification key and $\widehat{\mathsf{Sign}}$ signs messages, but whenever a message is submitted that was already signed, $\widehat{\mathsf{Sign}}$ returns the previously produced signature again.

Since F is a T-pseudorandom function, and since R and S are used in Game-4 only in the arguments of SigGen and Sign, it follows that Game-4 $\approx$ Game-5 and thus Game-0 $\approx$ Game-5.

Now, assume that the adversary $\mathcal{U}^*$ wins the unforgeability game Game-0 with non-negligible probability. Then, since Game-0 $\approx$ Game-5, $\mathcal{U}^*$ also wins with non-negligible probability in Game-5.

Then it returns $\ell + 1$ pairs (m_i, σ_i) such that $m_i \neq m_j$ for all $i \neq j$ and $\mathsf{SigVrfy}(vk, m_i, \sigma_i) = 1$ for all $i = 1, \ldots, \ell + 1$. We denote by $Q = (\hat{m}_1, \ldots, \hat{m}_\ell)$ the set of messages that have been asked to the external signing oracle $\widehat{\mathsf{Sign}}$. Since all messages m_i are distinct there exists at least one message $m_j \notin Q$. The forger $\mathcal{B}$ outputs (m_j, σ_j). Since $\mathsf{SigVrfy}(vk, m_i, \sigma_i) = 1$ for all i, we have in particular that the pair (m_j, σ_j) verifies and thus $\mathcal{B}$ succeeds with non-negligible probability. Since $\mathcal{B}$ runs in time $T \cdot poly(\lambda)$, this contradicts the assumption that Sig is T-unforgeable. This concludes the proof.

Proof of Blindness. Before proving the blindness property, we discuss the central points. In our protocol, the privacy of the user (blindness) is preserved by the fact that the secure function evaluation does not reveal the message to be signed (Bob-privacy). We cannot, however, directly apply Bob-privacy: Bob-privacy only guarantees that the users are unlinkable as long as the outputs of the SFE are not revealed. In our setting, however, the outputs are revealed. Thus, we first have to transform the blindness game into one where the signer does not get the signatures output by the users. To achieve this, we use a rewinding argument: Instead of using the signatures produced in the main execution of the blindness game, we rewind the blindness game and use the signatures from the rewound execution (called look-ahead threads). The ZAP sent by the signer guarantees that the signatures are always produced using the same randomness, hence the signatures in the look-ahead threads equal those of the main thread. Finally, we show that the signer can simulate the look-ahead threads on his own. Thus we have a game in which the output of the SFE in the main thread is not used, and we can apply Bob-privacy.

Care needs to be taken with the ZAPs: In our ZAP, one can fake the proof by committing to the preimage $f^{-1}(y)$. Since a ZAP is not a proof of knowledge, the mere fact that the signer does not know $f^{-1}(y)$ is not sufficient to show that the signer cannot fake the ZAP. A complexity-leveraging argument between Games 3 and 4 solves this issue.

Proof (of Theorem 7). We prove this theorem via a sequence of games. In order to facilitate notation, we assume that the malicious signer $\mathcal{S}^*$ is given by a deterministic, stateless algorithm $\mathcal{S}^*$. That is, in its first invocation, $\mathcal{S}^*$ expects an explicit random tape as argument and returns its new internal state st.

In further invocations, $\mathcal{S}^*$ expects the previous internal state st and returns a new internal state st'.

In the blindness game, $\mathcal{S}^*$ has the liberty to schedule the instances of the user in an arbitrary order. In case of a two-move scheme, however, we can fix the ordering. Since $\mathcal{S}^*$ does not receive any response to the message it sends to the user, we can assume that $\mathcal{S}^*$ sends these messages after having gathered all incoming messages from the user. Thus, without loss of generality, $\mathcal{S}^*$ first outputs the public key vk and the challenge messages m_0, m_1, then expects the two incoming blinded messages from the two user instances, and then outputs its responses to the user messages.

With these assumptions about $\mathcal{S}^*$, the blindness game can be reformulated as follows:

Game-0

000 $\mathsf{rand}_{\mathcal{S}^*} \leftarrow \{0,1\}^\infty,\ b \leftarrow \{0,1\}$
001 $(\mathsf{st}, vk, m_0, m_1) \leftarrow \mathcal{S}^*(1^\lambda, \mathsf{rand}_{\mathcal{S}^*})$
002 $\mathsf{usermsg}_0 \leftarrow \mathcal{U}(vk, m_b) \mid \mathsf{usermsg}_1 \leftarrow \mathcal{U}(vk, m_{\bar{b}})$
003 $(\mathsf{st}', \mathsf{signermsg}_0, \mathsf{signermsg}_1) \leftarrow \mathcal{S}^*(\mathsf{st}, \mathsf{usermsg}_0, \mathsf{usermsg}_1)$
004 if fail then $b' \leftarrow \mathcal{S}^*(\mathsf{st}', \bot, \bot)$
005 else $b' \leftarrow \mathcal{S}^*(\mathsf{st}', \sigma_b, \sigma_{\bar{b}})$

In Game-0 and in the following, we use the abbreviation fail for $\sigma_0 = \bot$ or $\sigma_1 = \bot$. Blindness is then equivalent to requiring that $\left|\Pr[b' = b : \text{Game-0}] - \frac{1}{2}\right|$ is negligible. For contradiction, we assume that blindness does not hold, i.e., that $\left|\Pr[b' = b : \text{Game-0}] - \frac{1}{2}\right|$ is non-negligible. Then there exists a polynomial $p = p(\lambda)$ such that $\Pr[b' = b : \text{Game-0}] \geq \frac{1}{2} + 1/p$ for infinitely many λ. (Or $\leq \frac{1}{2} - 1/p$, but in this case we can replace $\mathcal{S}^*$ by an adversary that outputs the negated guess $1 - b'$.)

Game-0 $\Rightarrow$ Game-1. Our first transformation is to make the definition of the user explicit in the blindness game. That is, we replace invocations to $\mathcal{U}$ by its program code. For notational convenience later on, we split the description of Game-0 into the actual game and a subroutine Thread that executes the interaction between $\mathcal{S}^*$ and the users.

Thread$(vk, m_0, m_1, b, \mathsf{st})$

T00 $K_0, K_1, X_0, X_1, E_0, E_1, x_0, x_1 \leftarrow \{0,1\}^\lambda$

T01 $(\mathsf{sfe}_{1,0}, \mathsf{sfest}_0) \leftarrow \mathsf{SFE}_1(1^\lambda, m_b)$	$(\mathsf{sfe}_{1,1}, \mathsf{sfest}_1) \leftarrow \mathsf{SFE}_1(1^\lambda, m_{\bar{b}})$
T07 $y_0 \leftarrow f(x_0)$	$y_1 \leftarrow f(x_1)$
T08 $\mathsf{msg}_0 \leftarrow \mathcal{V}(1^\lambda)$	$\mathsf{msg}_1 \leftarrow \mathcal{V}(1^\lambda)$

T09 $(\mathsf{st}', (\mathsf{sfe}_{2,0}, \mathsf{com}_{x,0}, \pi_0), (\mathsf{sfe}_{2,1}, \mathsf{com}_{x,1}, \pi_1)) \leftarrow \mathcal{S}^*(\mathsf{st}, (\mathsf{sfe}_{1,0}, \mathsf{msg}_0, y_0), (\mathsf{sfe}_{1,1}, \mathsf{msg}_1, y_1))$

T10 $s_0 \leftarrow (svk, \mathsf{com}_R, \mathsf{sfe}_{1,0}, \mathsf{sfe}_{2,0}, \mathsf{com}_{x,0}, y_0)$	$s_1 \leftarrow (svk, \mathsf{com}_R, \mathsf{sfe}_{1,1}, \mathsf{sfe}_{2,1}, \mathsf{com}_{x,1}, y_1)$
T11 if $\mathcal{V}(\mathsf{msg}_0, s_0, \pi_0) = 1$ then	if $\mathcal{V}(\mathsf{msg}_0, s_1, \pi_1) = 1$ then
T12 $\sigma_b \leftarrow \mathsf{SFE}_3(1^\lambda, \mathsf{sfe}_{2,0}, \mathsf{sfest}_0)$ else $\sigma_b \leftarrow \bot$	$\sigma_{\bar{b}} \leftarrow \mathsf{SFE}_3(1^\lambda, \mathsf{sfe}_{2,1}, \mathsf{sfest}_1)$ else $\sigma_{\bar{b}} \leftarrow \bot$

Return$(\sigma_0, \sigma_1, \mathsf{st}')$

Game-1

100 $b \leftarrow \{0,1\}$, $\mathsf{rand}_{\mathcal{S}^*} \leftarrow \{0,1\}^\infty$
101 $(\mathsf{st}, vk, m_0, m_1) \leftarrow \mathcal{S}^*(1^\lambda; \mathsf{rand}_{\mathcal{S}^*})$ with $vk = (svk, \mathsf{com}^R)$
102 $(\sigma_0, \sigma_1, \mathsf{st}') \leftarrow \mathsf{Thread}(vk, m_0, m_1, b, \mathsf{st})$
103 if fail then $b' \leftarrow \mathcal{S}^*(\mathsf{st}', \perp, \perp)$
104 else $b' \leftarrow \mathcal{S}^*(\mathsf{st}', \sigma_0, \sigma_1)$

Since we have only restructured the game, we have $\Pr[b' = b : \mathsf{Game\text{-}0}] = \Pr[b' = b : \mathsf{Game\text{-}1}]$.

Game-1 $\Rightarrow$ Game-2. Game-2 is identical to Game-1 except for the following modifications. Once both user instances have computed their signatures, i.e., right after Step 202, we reset the malicious signer $\mathcal{S}^*$ to the point where it has returned the two messages, i.e., to after Step 201. We repeat this procedure p times. Since $\mathcal{S}^*$ is deterministic and stateless, this can be modeled by running the subroutine Thread p times using the same initial state st for $\mathcal{S}^*$ in each thread. Each thread uses a fresh random bit $\hat{b}$ to assign the messages to the user instances. We refer to the first execution as the *main thread* (representing the original blindness game) and to the other p executions as *look-ahead threads*. Observe that this rewinding *only* involves Step T01 to T12. The other steps are part of the blindness game. In particular, $\mathcal{S}^*$ gets in Step 204 the signatures σ_0 and σ_1 from the main thread.

Game-2

200 $b \leftarrow \{0,1\}$, $\mathsf{rand}_{\mathcal{S}^*} \leftarrow \{0,1\}^\infty$, $\boldsymbol{\hat{b}_1, \ldots, \hat{b}_p \leftarrow \{0,1\}}$
201 $(vk, m_0, m_1, \mathsf{st}) \leftarrow \mathcal{S}^*(1^\lambda; \mathsf{rand}_{\mathcal{S}^*})$ with $vk = (svk, \mathsf{com}^R)$
202 $(\sigma_0, \sigma_1, \mathsf{st}') \leftarrow \mathsf{Thread}(vk, m_0, m_1, b, \mathsf{st})$
202a $\boldsymbol{(\sigma^{\mathsf{la}}_{0,i}, \sigma^{\mathsf{la}}_{1,i}, \mathsf{st}'_i) \leftarrow \mathsf{Thread}(vk, m_0, m_1, \hat{b}_i, \mathsf{st})}$ **for** $\boldsymbol{i = 1, \ldots, p}$
203 if fail then $b' \leftarrow \mathcal{S}^*(\mathsf{st}', \perp, \perp)$
204 else $b' \leftarrow \mathcal{S}^*(\mathsf{st}', \sigma_0, \sigma_1)$

The success probability of $\mathcal{S}^*$ in both games is clearly the same, because the results of the look-ahead threads are not used and Thread has no side effects. That is, $\Pr[b' = b : \mathsf{Game\text{-}1}] = \Pr[b' = b : \mathsf{Game\text{-}2}]$.

Now, we bound the probability that both user instances in the main thread return valid signatures, but at least one user algorithm in *each* look-ahead threads fails. Observe that this includes aborting parties, decryption attempts that fail, or the case where a certain ZAP may be invalid. Recall that $\mathsf{fail} = 1$ if $\sigma_0 = \perp$ or $\sigma_1 = \perp$. We define $\mathsf{fail}^{\mathsf{la}}_i$ analogously to fail, i.e., $\mathsf{fail}^{\mathsf{la}}_i = 1$ if $\sigma^{\mathsf{la}}_{0,i} = \perp$ or $\sigma^{\mathsf{la}}_{1,i} = \perp$.

Lemma 8. *Denote by* bad *the event that* $\mathsf{fail}^{\mathsf{la}}_i$ *holds for all* $i = 1, \ldots, p$ *but that* fail *does not hold. The probability that* bad *happens in* Game-2 *is less or equal* $\frac{1}{p+1}$.

Game-2 $\Rightarrow$ Game-3. In this game, we set $b' = 0$ whenever the the main thread produces valid signatures ($\neg\mathsf{fail}$) but the look-ahead threads do not (fail_1 and ... and fail_p).

```
Game-3
303  if fail then b' ← S*(st', ⊥, ⊥)
303a else if fail_1^la and ... and fail_p^la then b' ← 0
304  else b' ← S*(st', σ_0, σ_1)
```

Notice that the else-branch in line 303a is only taken if the event bad occurs. This happens with probability at most $\frac{1}{p+1}$ by Lemma 8. Thus, $\mathcal{S}^*$'s avantage reduces at most by $\frac{1}{p+1}$, i.e., $\Pr[b' = b : \mathsf{Game\text{-}3}] \geq \Pr[b' = b : \mathsf{Game\text{-}2}] - \frac{1}{p+1}$.

$\mathsf{Game\text{-}3} \Rightarrow \mathsf{Game\text{-}4}$. In this hybrid, we do not give the adversary $\mathcal{S}^*$ the signatures from the main thread, but the ones from one of the successful look-ahead threads. That is, we choose a random g with $\neg\mathsf{fail}_g^{\mathsf{la}}$ and we give the signatures $(\sigma_{0,g}, \sigma_{1,g})$ to $\mathcal{S}^*$. Notice that we only need to do this if $\mathsf{fail}_1^{\mathsf{la}} \wedge \cdots \wedge \mathsf{fail}_p^{\mathsf{la}}$ does not hold (otherwise line 404 would not have been reached), so there is always at least one such g.

```
Game-4
404 else g ← {i : ¬fail_i^la}, b' ← S*(st', σ_{0,g}, σ_{1,g})
```

We first argue that all messages $\sigma_0, \sigma_{0,g}$ have the same value (if defined), and analogously for $\sigma_1, \sigma_{1,g}$: Due to the adaptive soundness of the ZAP, we have that with overwhelming probability the statements proven by the ZAPs are true. That is, for each thread it holds that $\mathsf{com}_{x,0}$ contains a preimage of y under f or that the message $\mathsf{sfe}_{2,0}$ is constructed correctly. The first occurs only with negligible probability, otherwise by using the T'-extractability of $\mathcal{C}_X$ we could build a T'-time inverter for f, breaking the T'-onewayness of f. Thus $\mathsf{sfe}_{2,0}$ is constructed correctly in all thread with overwhelming probability. Analogously for $\mathsf{sfe}_{2,1}$. Due to the perfect correctness of SFE (and the perfect binding property of $\mathcal{C}_R$), this implies that all signatures $\sigma_{i,g}$ are produced by applying the same circuit $C_{ssk,R}$ with the same ssk and R to the message m_i. Thus, giving the malicious signer $\mathcal{S}^*$ the signatures returned from the gth look-ahead thread does not change its success probability by more than a negligible amount. That is, $|\Pr[b' = b : \mathsf{Game\text{-}3}] - \Pr[b' = b : \mathsf{Game\text{-}4}]|$ is negligible.

$\mathsf{Game\text{-}4} \Rightarrow \mathsf{Game\text{-}5}$. Now, we modify $\mathsf{Game\text{-}4}$ to $\mathsf{Game\text{-}5}$ by returning $(\perp, \perp)$ to $\mathcal{S}^*$ only if one of the ZAPs in the main thread failed. Formally, we check this by validating the ZAP π. In what follows, we denote by $\mathsf{msg}_0(\mathsf{st}'), s_0^{u'}(\mathsf{st}'), \pi_0(\mathsf{st}')$ the messages that are needed to verify the ZAP as seen by $\mathcal{S}^*$. We assume, w.l.o.g., that these messages are stored in the state st'.

```
Game-5
503 if V(msg_0(st'), s_0^u(st'), π_0(st')) = 0 or
                  V(msg_1(st'), s_1^u(st'), π_1(st')) = 0 then b' ← S*(st', ⊥, ⊥)
```

Due to the soundness of the ZAP, this modification does not change the success probability of $\mathcal{S}^*$ by more than a negligible amount.

$\mathsf{Game\text{-}5} \Rightarrow \mathsf{Game\text{-}6}$. In this game, we build an attacker $\mathcal{B}$ that simulates the look-ahead threads and the malicious signer $\mathcal{S}^*$ locally.

$\mathcal{B}(\mathsf{st}, \mathsf{st}')$

B00 $\hat{b}_1, \ldots, \hat{b}_p \leftarrow \{0,1\}$
B01 $(\sigma^{\mathsf{la}}_{0,i}, \sigma^{\mathsf{la}}_{1,i}, \mathsf{st}'_i) \leftarrow \mathsf{Thread}(vk, m_0, m_1, \hat{b}_i, \mathsf{st})$ for $i = 1, \ldots, p$
B02 if $\mathcal{V}(\mathsf{msg}_0(\mathsf{st}'), s_0^{u'}(\mathsf{st}'), \pi_0(\mathsf{st}')) = 0$ or
$\mathcal{V}(\mathsf{msg}_1(\mathsf{st}'), s_1^{u}(\mathsf{st}'), \pi_1(\mathsf{st}')) = 0$ then $b' \leftarrow \mathcal{S}^*(\mathsf{st}', \bot, \bot)$
B03 else if fail_1 and $\ldots$ and fail_p then $b' \leftarrow 0$
B04 else $g \leftarrow \{i : \neg\mathsf{fail}_i\}$, $b' \leftarrow \mathcal{S}^*(\mathsf{st}', \sigma_{0,g}, \sigma_{1,g})$

The game looks as follows (here, we show the whole Game-6, not just the lines changed with respect to Game-5):

Game-6

600 $b \leftarrow \{0,1\}$, $\mathsf{rand}_{\mathcal{S}^*} \leftarrow \{0,1\}^\infty$
601 $(vk, m_0, m_1, \mathsf{st}) \leftarrow \mathcal{S}^*(1^\lambda; \mathsf{rand}_{\mathcal{S}^*})$ with $vk = (svk, \mathsf{com}^R)$
602 $(\sigma_0, \sigma_1, \mathsf{st}) \leftarrow \mathsf{Thread}(vk, m_0, m_1, b, \mathsf{st})$
603 $\boldsymbol{b' \leftarrow \mathcal{B}(\mathsf{st}, \mathsf{st}')}$

Clearly, the success probability of $\mathcal{B}$ in this game is equal to the success probability of $\mathcal{S}^*$ in the previous game since we have just moved some of the computations of Game-5 into $\mathcal{B}$.

Game-6 $\Rightarrow$ Game-7. Now, we modify the algorithm Thread only for the main thread (recall that all other threads are computed by $\mathcal{B}$ locally). Our modification removes all dependencies on the input message m. That is, we let the user algorithm compute the first message of the SFE protocol on 0^λ instead of m. Additionally, we remove the computation of the signatures σ_0, σ_1 that are never used.

Thread'

T'01	$(\mathsf{sfe}_{1,0}, \mathsf{sfest}_0) \leftarrow \mathsf{SFE}_1(1^\lambda, \mathbf{0^\lambda})$	$(\mathsf{sfe}_{1,1}, \mathsf{sfest}_1) \leftarrow \mathsf{SFE}_1(1^\lambda, \mathbf{0^\lambda})$
T'11–T'12	**(removed)**	**(removed)**

Since the SFE scheme is Bob-private, the success probability of $\mathcal{B}$ remains the same (except for a negligible amount). This, however, means that the entire transcript is independent of the message.

Now, we obtain the following contradiction concluding that advantage of $\mathcal{S}^*$ in Game-0 is less or equal $\frac{1}{p+1} + \mathsf{negl}(\lambda)$, where $\mathsf{negl}(\lambda)$ is a negligible function. In Game-0, however, we assumed that $\mathcal{S}^*$ wins the blindness game with advantage at least $\nu \geq 1/p$ (infinitely often). Since $1/p > 1/(p+1) + \mathsf{negl}$ for sufficiently large λ we obtain a contradiction. Thus, our initial assumption that $\mathcal{S}^*$ succeeds with non-negligible probability was wrong and therefore, our construction is blind.

Acknowledgments. We thanks the anonymous reviewers for valuable comments and Masayuki Abe for his useful feedback on this merged paper. Dominique Unruh was supported by European Social Fund's Doctoral Studies and Internationalisation Programme DoRa. Dominique Schröder is also supported by the Emmy Noether Program Fi 940/2-1 of the German Research Foundation (DFG). Part of this work was supported by CASED (`www.cased.de`). Amit Sahai is supported in part from a DARPA/ONR PROCEED award, NSF grants 0916574 and 0830803, a

Xerox Foundation Award, a Google Faculty Research Award, an equipment grant from Intel, and an Okawa Foundation Research Grant.

References

1. Abdalla, M., Namprempre, C., Neven, G.: On the (im)possibility of blind message authentication codes. In: Pointcheval, D. (ed.) CT-RSA 2006. LNCS, vol. 3860, pp. 262–279. Springer, Heidelberg (2006)
2. Abe, M.: A secure three-move blind signature scheme for polynomially many signatures. In: Pfitzmann, B. (ed.) EUROCRYPT 2001. LNCS, vol. 2045, pp. 136–151. Springer, Heidelberg (2001)
3. Abe, M., Fuchsbauer, G., Groth, J., Haralambiev, K., Ohkubo, M.: Structure-preserving signatures and commitments to group elements. In: Rabin, T. (ed.) CRYPTO 2010. LNCS, vol. 6223, pp. 209–236. Springer, Heidelberg (2010)
4. Abe, M., Haralambiev, K., Ohkubo, M.: Signing on elements in bilinear groups for modular protocol design. IACR ePrint 2010/133 (2010)
5. Abe, M., Ohkubo, M.: A framework for universally composable non-committing blind signatures. In: Matsui, M. (ed.) ASIACRYPT 2009. LNCS, vol. 5912, pp. 435–450. Springer, Heidelberg (2009)
6. Bellare, M., Namprempre, C., Pointcheval, D., Semanko, M.: The one-more-RSA-inversion problems and the security of Chaum's blind signature scheme. Journal of Cryptology 16(3), 185–215 (2003)
7. Boldyreva, A.: Threshold signatures, multisignatures and blind signatures based on the gap-diffie-hellman-group signature scheme. In: Desmedt, Y.G. (ed.) PKC 2003. LNCS, vol. 2567, pp. 31–46. Springer, Heidelberg (2002)
8. Brands, S., Paquin, C.: U-prove cryptographic specification v1.0 (March 2010), http://connect.microsoft.com/site642/Downloads/DownloadDetails.aspx?Dow%nloadID=26953
9. Brands, S.A.: Rethinking Public Key Infrastructures and Digital Certificates: Building in Privacy. MIT Press, Cambridge (2000)
10. Camenisch, J., Groß, T.: Efficient attributes for anonymous credentials. In: ACM CCS 2008, pp. 345–356. ACM Press, New York (2008)
11. Camenisch, J., Koprowski, M., Warinschi, B.: Efficient blind signatures without random oracles. In: Blundo, C., Cimato, S. (eds.) SCN 2004. LNCS, vol. 3352, pp. 134–148. Springer, Heidelberg (2005)
12. Camenisch, J., Neven, G., Shelat, A.: Simulatable adaptive oblivious transfer. In: Naor, M. (ed.) EUROCRYPT 2007. LNCS, vol. 4515, pp. 573–590. Springer, Heidelberg (2007)
13. Chaum, D.: Blind signatures for untraceable payments. In: CRYPTO 1982, pp. 199–203. Plenum Press, New York (1983)
14. Chaum, D.: Blind signature system. In: CRYPTO 1983, p. 153. Plenum Press, New York (1984)
15. Dwork, C., Naor, M.: Zaps and their applications. SIAM J. Comput. 36(6), 1513–1543 (2007)
16. Fischlin, M.: Round-optimal composable blind signatures in the common reference string model. In: Dwork, C. (ed.) CRYPTO 2006. LNCS, vol. 4117, pp. 60–77. Springer, Heidelberg (2006)
17. Fischlin, M., Schröder, D.: Security of blind signatures under aborts. In: Jarecki, S., Tsudik, G. (eds.) PKC 2009. LNCS, vol. 5443, pp. 297–316. Springer, Heidelberg (2009)

18. Fischlin, M., Schröder, D.: On the impossibility of three-move blind signature schemes. In: Gilbert, H. (ed.) EUROCRYPT 2010. LNCS, vol. 6110, pp. 197–215. Springer, Heidelberg (2010)
19. Fuchsbauer, G.: Automorphic signatures in bilinear groups and an application to round-optimal blind signatures. IACR ePrint 2009/320 (2009)
20. Garg, S., Rao, V., Sahai Round, A.: optimal blind signatures in the standard model (2011) (manuscript)
21. Garg, S., Rao, V., Sahai, A., Schröder, D., Unruh, D.: Round optimal blind signatures. IACR ePrint (2011)
22. Ghadafi, E., Smart, N.: Efficient two-move blind signatures in the common reference string model. IACR ePrint 2010/568 (2010)
23. Hazay, C., Katz, J., Koo, C.Y., Lindell, Y.: Concurrently-secure blind signatures without random oracles or setup assumptions. In: Vadhan, S.P. (ed.) TCC 2007. LNCS, vol. 4392, pp. 323–341. Springer, Heidelberg (2007)
24. Horvitz, O., Katz, J.: Universally-composable two-party computation in two rounds. In: Menezes, A. (ed.) CRYPTO 2007. LNCS, vol. 4622, pp. 111–129. Springer, Heidelberg (2007)
25. Juels, A., Luby, M., Ostrovsky, R.: Security of blind digital signatures (extended abstract). In: Kaliski Jr., B.S. (ed.) CRYPTO 1997. LNCS, vol. 1294, pp. 150–164. Springer, Heidelberg (1997)
26. Katz, J., Schröder, D., Yerukhimovich, A.: Impossibility of blind signature from one-way permutation. In: Ishai, Y. (ed.) TCC 2011. LNCS, vol. 6597, pp. 615–629. Springer, Heidelberg (2011)
27. Kiayias, A., Zhou, H.S.: Concurrent blind signatures without random oracles. In: De Prisco, R., Yung, M. (eds.) SCN 2006. LNCS, vol. 4116, pp. 49–62. Springer, Heidelberg (2006)
28. Kiayias, A., Zhou, H.S.: Equivocal blind signatures and adaptive UC-security. In: Canetti, R. (ed.) TCC 2008. LNCS, vol. 4948, pp. 340–355. Springer, Heidelberg (2008)
29. Lindell, Y.: Bounded-concurrent secure two-party computation without setup assumptions. In: STOC 2003, pp. 683–692. ACM Press, New York (2003)
30. Lindell, Y.: Lower bounds for concurrent self composition. In: Naor, M. (ed.) TCC 2004. LNCS, vol. 2951, pp. 203–222. Springer, Heidelberg (2004)
31. Meiklejohn, S., Shacham, H., Freeman, D.M.: Limitations on transformations from composite-order to prime-order groups: The case of round-optimal blind signatures. In: Abe, M. (ed.) ASIACRYPT 2010. LNCS, vol. 6477, pp. 519–538. Springer, Heidelberg (2010)
32. Naor, M., Pinkas, B.: Efficient oblivious transfer protocols. In: SODA 2001, pp. 448–457 (2001)
33. Okamoto, T.: Efficient blind and partially blind signatures without random oracles. In: Halevi, S., Rabin, T. (eds.) TCC 2006. LNCS, vol. 3876, pp. 80–99. Springer, Heidelberg (2006)
34. Pass, R.: Limits of provable security from standard assumptions. In: STOC 2011. ACM Press, New York (to appear, 2011)
35. Pointcheval, D., Stern, J.: Security arguments for digital signatures and blind signatures. Journal of Cryptology 13(3), 361–396 (2000)
36. Prabhakaran, M., Rosen, A., Sahai, A.: Concurrent zero knowledge with logarithmic round-complexity. In: FOCS 2002, pp. 366–375. IEEE, Los Alamitos (2002)
37. Rückert, M.: Lattice-based blind signatures. In: Abe, M. (ed.) ASIACRYPT 2010. LNCS, vol. 6477, pp. 413–430. Springer, Heidelberg (2010)
38. Schröder, D., Unruh, D.: Round optimal blind signatures. IACR ePrint (2011)
39. Schröder, D., Unruh, D.: Security of blind signatures revisited. IACR ePrint (2011)

Optimal Structure-Preserving Signatures in Asymmetric Bilinear Groups

Masayuki Abe[1], Jens Groth[2,⋆], Kristiyan Haralambiev[3], and Miyako Ohkubo[4]

[1] Information Sharing Platform Laboratories, NTT Corporation, Japan
abe.masyuki@lab.ntt.co.jp
[2] University College London, UK
j.groth@ucl.ac.uk
[3] Computer Science Department, New York University, US
kkh@cs.nyu.edu
[4] National Institute of Information and Communications Technology, Japan
m.ohkubo@nict.go.jp

Abstract. Structure-preserving signatures are signatures defined over bilinear groups that rely on generic group operations. In particular, the messages and signatures consist of group elements and the verification of signatures consists of evaluating pairing product equations. Due to their purist nature structure-preserving signatures blend well with other pairing-based protocols.

We show that structure-preserving signatures must consist of at least 3 group elements when the signer uses generic group operations. Usually, the generic group model is used to rule out classes of attacks by an adversary trying to break a cryptographic assumption. In contrast, here we use the generic group model to prove a lower bound on the complexity of digital signature schemes.

We also give constructions of structure-preserving signatures that consist of 3 group elements only. This improves significantly on previous structure-preserving signatures that used 7 group elements and matches our lower bound. Our structure-preserving signatures have additional nice properties such as strong existential unforgeability and can sign multiple group elements at once.

Keywords: Structure-Preservation, Digital Signatures, Generic Group Model.

1 Introduction

Digital signatures are fundamental cryptographic primitives used as building blocks in countless scenarios. Often, signatures are combined with zero-knowledge (ZK) proof systems, for example when constructing privacy-preserving cryptographic protocols. While suitable signature schemes for such cases have long been known, e.g., the schemes of Camenisch and Lysyanskaya [CL02, CL04], they were constructed with the intent to be used with interactive ZK proofs. The reason was the absence of an efficient non-interactive zero-knowledge (NIZK) proof system. Moreover, the only way to construct efficient NIZK proofs was using certain heuristics, e.g., random oracles, which transform interactive ZK proofs into NIZK proofs. In [GS08], Groth and Sahai

⋆ Supported by EPSRC grant number EP/G013829/1.

P. Rogaway (Ed.): CRYPTO 2011, LNCS 6841, pp. 649–666, 2011.

presented the first practical NIZK proof system for a non-trivial class of languages which does not resort to such heuristics. It is based on bilinear maps and is designed to be used on certain satisfiable systems of equations. The most interesting type of equation is the so-called "pairing-product equation" for which the proofs are also fully extractable, and therefore the proof system yields NIZK proofs of knowledge.

As pointed out in [AFG$^+$10], many previous signatures scheme were not fully "compatible" with pairing-product equations. Even if the verification algorithm used pairing-product equations, the signatures and messages were not composed entirely of group elements and thus were not ideal counterparts for the pairing product equations of Groth-Sahai proofs. That is why [AFG$^+$10] defined the notion of structure-preserving signatures which requires verification keys, messages, and signatures to be composed entirely of group elements and the verification equations to use pairing-product equations. Equipped with such signatures, one can easily design modular cryptographic protocols which rely on signatures and NIZK proofs and instantiate them efficiently. Of course some cryptographic protocols find other ingenious efficient solutions but these are specific to their tasks. In contrast, modular design makes constructions easier to build, less prone to errors, and provide a good alternative for efficiency comparisons. Moreover, modular constructions can be realized under different assumptions by choosing appropriate instantiations of the building blocks. Applications of structure-preserving signatures combined with Groth-Sahai proofs are numerous: group signatures, blind signatures, delegatable credentials, oblivious transfer, credential-based identification/key-exchange with hierarchical certification, etc.

Efficient structure-preserving signatures were presented in [AFG$^+$10] and were applied to the construction of round-optimal blind signatures and fully-secure group signatures with concurrent join protocols. Although they were efficient, it was left as an open problem to find the optimal signature size and determine whether more efficient schemes can be constructed. These are the problems we consider in this work.

1.1 Our Contribution

Results. We prove lower bounds on the complexity of structure-preserving signatures based on asymmetric bilinear groups. As far as we are aware, this is the first non-trivial lower bound for the complexity of practical signature schemes. We also construct a structure-preserving signature scheme that matches the lower bounds giving an optimal solution in terms of efficiency.

We demonstrate that a structure-preserving signature scheme must use at least two pairing product equations to verify a signature. Any structure-preserving signature scheme where the verification only uses one pairing product equation can be broken with a random message attack.

We also give a lower bound on the size of a signature. A structure-preserving signature with less than 3 group elements is vulnerable to a random message attack. The lower bound holds even when the message is a single group element.

Finally, we prove that the lower bounds are optimal by presenting a structure-preserving signature scheme where the verification of signatures uses only two verification equations and the signatures consist of only 3 group elements.

Our signature scheme has several nice properties. First, it is structure preserving. Second, it is strongly existentially unforgeable against adaptive chosen message attacks. Third, messages to be signed can consist of many group elements, which can be drawn from both of the base groups of the bilinear map.

The existential unforgeability of our signature scheme against adaptive chosen message attacks corresponds to an interactive cryptographic assumption, which we prove is true when the adversary only uses generic group operations. By adding a few extra group elements to the signatures (1 or 3 depending on whether the messages only contain elements in one base group or contains a mix of elements from both base groups) we can base security on a non-interactive cryptographic assumption.

Techniques. The lower bound on the number of pairing product equations needed in the verification process follows from a demonstration that any two signatures on two different random messages can be combined to yield signatures on different messages.

However, when there are two or more verification equations, the analysis of the number of group elements involved in a signature becomes intricate. We base our analysis on the signer being a generic algorithm. This differs from the standard use of the generic group model to rule out classes of attacks on cryptographic assumptions since the analysis is based on what the signing algorithm can do, not what some arbitrary unknown adversary can do. Arguably this is a more compelling way to use the generic group model since the analysis only fails for signature schemes where the designer invents a non-generic signing algorithm.

A generic signer must create signatures that are related to the messages in a specific way. Furthermore, the correctness of the signature scheme implies that signatures created this way must be valid signatures. With these two facts in mind, we analyze the pairing product equations in the verification and show that all pairing product equations must be linearly related if the signatures consist of 1 or 2 group elements. We conclude that they can be replaced by an equivalent single verification equation. But that would make the signature scheme vulnerable to a random message attack.

Our work on lower bounds on structure-preserving signatures gives insight into what a structure-preserving signature with more group elements should look like. The verification equations must be organized such that a generic signer can use the secret signing key to solve them for arbitrary messages. A random choice of 2 or more verification equations is unlikely to be solvable for a generic signing algorithm. With signature sizes of 3 or more group elements, however, it is possible to carefully select the verification equations such that they are solvable by a generic signer. We find such a set of verification equations that are solvable by a generic signer and at the same time resists generic attackers with access to an adaptive chosen message attack.

1.2 Related Work

Lower bounds for cryptographic protocols have been studied extensively. For some tasks it is possible to give information-theoretic lower bounds; ciphertexts must, for instance, be longer than plaintexts to enable correct decryption. In the context of zero-knowledge proofs lower bounds on the round complexity [GO94] have been found by exploiting the tension between soundness and zero-knowledge. However, these lower bounds do not readily apply to digital signatures where the hash-and-sign paradigm rules out strong information-theoretic bounds on the size and the protocols are non-interactive by definition. Gennaro, Gertner and Katz [GGK03] instead investigated the complexity of digital signatures that only make black-box calls to a one-way permutation and found asymptotic lower bounds on the number of black-box queries. In contrast, our lower bounds apply to practical pairing-based signature schemes.

The generic group model [Nec94, Sho97] is widely used in pairing-based cryptography to rule out generic attacks on cryptographic assumptions. However, there has been little work on using the generic group model to prove lower bounds on the efficiency of cryptographic protocols except for Bangerter, Camenisch and Krenn [BCK10] that gave lower bounds on the knowledge error in certain Sigma-protocols and Ostrovsky and Skeith [OS08] that gave lower bounds on single-server private information retrieval protocols based on homomorphic encryption. The generic group model has not been used to give lower bounds for the complexity of signature schemes.

The first structure-preserving signatures were presented by Groth [Gro06] who used them to build group signatures. Groth's signature scheme is based on the decision linear assumption but consists of thousands of group elements and is therefore not practical.

Green and Hohenberger [GH08] presented a structure-preserving signature scheme that provides security against random-message attacks, which they used to build a universally composable adaptive oblivious transfer protocol.

Cathalo, Libert and Yung [CLY09] constructed a partially structure-preserving signature scheme which signs only a single group element and used it for the construction of a group-encryption scheme.

Fuchsbauer [Fuc09] presented a structure-preserving signature scheme for signing messages that are Diffie-Hellman pairs. Fuchsbauer's scheme is automorphic, i.e., the public verification keys belong to the message space. Automorphic signatures have several applications including blind signatures, group signatures, anonymous proxy signatures and anonymous delegatable credentials [Fuc09, FV10, Fuc11].

Abe, Haralambiev and Ohkubo [AHO10] presented several constructions of structure-preserving signatures and found applications to blind signatures and group signatures. A merged version of [Fuc09, AHO10, Gro09] first coined the term structure-preserving signatures. The most efficient structure-preserving signature scheme from [AFG$^+$10] can sign multiple group elements belonging to one of the base groups with signatures that consist of 7 group elements and use two pairing product equations in the verification. In comparison, we present a scheme that can sign messages that contain group elements from both base groups and only uses 3 group elements in the signatures.

2 Preliminaries

2.1 Bilinear Groups

Throughout the paper we let $\mathcal{G}$ be a bilinear group generator that on security parameter k returns $(p, \mathbb{G}, \mathbb{H}, \mathbb{T}, e, G, H) \leftarrow \mathcal{G}(1^k)$ with the following properties:

- $\mathbb{G}, \mathbb{H}, \mathbb{T}$ are groups of prime order p.
- $e : \mathbb{G} \times \mathbb{H} \rightarrow \mathbb{T}$ is a bilinear map such that $\forall U \in \mathbb{G}, \forall V \in \mathbb{H}, \forall a, b \in \mathbb{Z}$: $e(U^a, V^b) = e(U, V)^{ab}$.

- G generates $\mathbb{G}$, H generates $\mathbb{H}$ and $e(G, H)$ generates $\mathbb{T}$.
- There are efficient algorithms for computing group operations, evaluating the bilinear map, comparing group elements and deciding membership of the groups.

There are many ways to set up bilinear groups. We will work in what Galbraith, Paterson and Smart [GPS08] call type III groups, where there are no efficiently computable isomorphisms $\mathbb{G} \to \mathbb{H}$ or $\mathbb{H} \to \mathbb{G}$. We focus on type III groups here because they have the most efficient instantiations and therefore the highest relevance for cryptographic purposes.

In a group $(p, \mathbb{G}, \mathbb{H}, \mathbb{T}, e, G, H)$ generated by $\mathcal{G}$ we refer to deciding group membership, computing group operations in $\mathbb{G}, \mathbb{H}$ or $\mathbb{T}$, comparing group elements and evaluating the bilinear map as the generic group operations. In the signature schemes we construct we only use generic group operations.

As a matter of notation, we will mostly use capital letters A, G, M, R, S, U for group elements in $\mathbb{G}$ and capital letters B, H, N, T, V, W for group elements in $\mathbb{H}$ and capital letter Z for group elements in $\mathbb{T}$. We will use small letters $r, s, t, \ldots$ for discrete logarithms of group elements with respect to base G or base H. We use Greek letters $\alpha, \beta, \ldots$ for hidden field elements in $\mathbb{Z}_p$ chosen by algorithms as part of their operation.

2.2 Secure Signature Schemes

A digital signature scheme over groups generated by a bilinear group generator $\mathcal{G}$ is a triple of efficient algorithms $(\mathcal{K}, \mathcal{S}, \mathcal{V})$. The key generation algorithm $\mathcal{K}$ takes a description of the bilinear group as input and returns a public verification key VK and a secret signing key SK. The signing algorithm $\mathcal{S}$ takes a signing key SK and a message M in the message space $\mathcal{M}$ defined by GK and VK as input and returns a signature Σ. The verification algorithm $\mathcal{V}$ takes the verification key VK, a message M and the signature Σ and returns either 1 (accept) or 0 (reject).

Definition 1 (Correctness). *We say the signature scheme $(\mathcal{K}, \mathcal{S}, \mathcal{V})$ over bilinear group generator $\mathcal{G}$ is (perfectly) correct if for all security parameters $k \in \mathbb{N}$*

$$\Pr[GK \leftarrow \mathcal{G}(1^k); (VK, SK) \leftarrow \mathcal{K}(GK); M \leftarrow \mathcal{M}; \Sigma \leftarrow \mathcal{S}_{SK}(M) : \mathcal{V}_{VK}(M, \Sigma) = 1] = 1.$$

A signature scheme is said to be existentially unforgeable if it is hard to forge a signature on a new message that has not been signed before. The adversary may see signatures on other messages before making the forgery. We distinguish between a random message attack, where the adversary gets pairs of random messages and corresponding signatures, and an adaptive chosen message attack where the adversary can choose arbitrary messages and receive signatures on them. Our signatures will be secure against adaptive chosen message attack, but our lower bounds on the complexity of signature schemes will hold even for the weaker random message attacks. We now formally define existential unforgeability against an adaptive chosen message attacks.

Definition 2 (EUF-CMA). *A signature scheme $(\mathcal{K}, \mathcal{S}, \mathcal{V})$ over bilinear group generator $\mathcal{G}$ is existentially unforgeable against adaptive chosen message attacks if for all non-uniform polynomial time $\mathcal{A}$*

$$\Pr[GK \leftarrow \mathcal{G}(1^k); (VK, SK) \leftarrow \mathcal{K}(GK); (M, \Sigma) \leftarrow \mathcal{A}^{\mathcal{S}_{SK}(\cdot)}(VK) : \\ M \notin Q \ \wedge \ \mathcal{V}_{VK}(M, \Sigma) = 1] = \mathrm{negl}(k),$$

where Q is the set of queries made by $\mathcal{A}$ to the signing oracle.

Sometimes it is also useful to prevent the adversary from issuing a new signature for a message that has already been signed. A signature scheme is strongly existentially unforgeable if it is hard to find a signature on a message that has not been signed before and also hard to find a new signature for a message that has already been signed.

Definition 3 (sEUF-CMA). *A signature scheme $(\mathcal{K}, \mathcal{S}, \mathcal{V})$ over bilinear group generator $\mathcal{G}$ is strongly existentially unforgeable against adaptive chosen message attacks if for all non-uniform polynomial time $\mathcal{A}$*

$$\Pr[GK \leftarrow \mathcal{G}(1^k); (VK, SK) \leftarrow \mathcal{K}(GK); (M, \Sigma) \leftarrow \mathcal{A}^{\mathcal{S}_{SK}(\cdot)}(VK) : \\ (M, \Sigma) \notin Q \ \wedge \ \mathcal{V}_{VK}(M, \Sigma) = 1] = \mathrm{negl}(k),$$

where Q is the set of message-signature pairs from $\mathcal{A}$'s queries to the signing oracle.

2.3 Structure-Preserving Signature Schemes

In this paper, we study structure-preserving signature schemes [AFG+10]. In a structure preserving signature scheme the verification key, the messages and the signatures consist only of group elements and the verification algorithm evaluates the signature by deciding group membership of elements in the signature and by evaluating pairing product equations, which are equations of the form

$$\prod_i \prod_j e(A_i, B_j)^{a_{ij}} = Z,$$

where $A_1, A_2, \ldots \in \mathbb{G}, B_1, B_2, \ldots \in \mathbb{H}, Z \in \mathbb{T}$ are group elements appearing in GK, VK, M or Σ and $a_{11}, a_{12}, \ldots \in \mathbb{Z}$ are constants. Structure-preserving signatures are extremely versatile because they mix well with other pairing-based protocols. Groth-Sahai proofs [GS08] are for instance designed with pairing product equations in mind and can therefore easily be applied to structure-preserving signatures.

Definition 4 (Structure-preserving signatures). *A signature scheme $(\mathcal{K}, \mathcal{S}, \mathcal{V})$ over bilinear group generator $\mathcal{G}$ is said to be structure preserving if*

- *$\mathcal{G}$ generates a bilinear group $GK = (p, \mathbb{G}, \mathbb{H}, \mathbb{T}, e, G, H)$,*
- *the verification key consists of GK and group elements in $\mathbb{G}$ and $\mathbb{H}$,*
- *the messages consist of group elements in $\mathbb{G}$ and $\mathbb{H}$,*
- *the signatures consist of group elements in $\mathbb{G}$ and $\mathbb{H}$, and*
- *the verification algorithm evaluates membership in $\mathbb{G}$ and $\mathbb{H}$ and pairing product equations with $Z = 1$.*

Our signatures are structure-preserving as defined above. When proving our lower bounds, we will relax the definition of structure-preserving signatures to allow arbitrary target group elements $Z \in \mathbb{T}$ to be included in the verification key and to appear in the verification equations. This strengthens our results, getting lower bounds in a relaxed model of structure-preserving signatures and constructing signatures in a strict model of structure-preserving signatures.

Generic Signer. Abe et al. [AFG$^+$10] did not explicitly require the signing algorithm to only use generic group operations when they defined structure-preserving signatures. However, it would be a natural addition to the definition of structure-preserving signatures because otherwise the cryptographic designer would have to invent some non-generic operations to be used in the signature scheme and that would be a surprising result in itself. All our signature schemes have a generic signer; as do all earlier structure-preserving signatures in the literature.

3 Lower Bounds on Structure-Preserving Signatures

In this section, we will prove lower bounds on the complexity of structure-preserving signatures. We summarize our lower bounds in the following main theorem, which follows from Theorems 2, 3 and 4.

Theorem 1. *All generic-signer structure-preserving signature schemes that are existentially unforgeable against random message attacks must use at least two verification equations and have signatures consisting of at least three group elements drawn from both $\mathbb{G}$ and $\mathbb{H}$. This holds even when the messages are single group elements and even if we allow the verification key to contain elements of $\mathbb{T}$.*

3.1 Impossibility of One Verification Equation

Theorem 2. *There is no structure-preserving signature with a single verification equation that is existentially unforgeable against random message attacks.*

Proof. Consider a structure preserving signature scheme for messages $M \in \mathbb{G}$ with the verification key containing group elements $U_1, U_2, \ldots \in \mathbb{G}, V_1, V_2, \ldots \in \mathbb{H}, Z \in \mathbb{T}$. Signatures are of the form $(S_1, S_2, \ldots, T_1, T_2, \ldots)$ with $S_i \in \mathbb{G}$ and $T_j \in \mathbb{H}$ and are verified by the following verification equation

$$\prod_i \prod_j e(S_i, T_j)^{a_{ij}} \cdot \prod_i \prod_j e(S_i, V_j)^{b_{ij}} \cdot \prod_j e(M, T_j)^{c_j} \cdot \prod_j e(M, V_j)^{d_j} \cdot \prod_i \prod_j e(U_i, T_j)^{e_{ij}} = Z.$$

Please note there is no need for terms of the form $e(U_i, V_j)$ because without loss of generality they can be incorporated into $Z \in \mathbb{T}$.

Suppose we get a signature $(S_1, \ldots, T_1, \ldots)$ on a random message $M \in \mathbb{G}$. Isolating T_ℓ and M in the verification, define for every ℓ

$$A_\ell = \prod_i S_i^{a_{i\ell}} \cdot \prod_i U_i^{e_{i\ell}} \qquad B_\ell = \prod_{j \neq \ell} T_j^{c_j} \cdot \prod_j V_j^{d_j}.$$

Suppose there is an ℓ for which $A_\ell \neq M^{-c_\ell}$. We can rewrite the verification equation

$$e(M,T_\ell)^{c_\ell} e(A_\ell,T_\ell) e(M,B_\ell) \cdot \prod_i \prod_{j \neq \ell} e(S_i,T_j)^{a_{ij}} \cdot \prod_i \prod_j e(S_i,V_j)^{b_{ij}} \cdot \prod_i \prod_{j \neq \ell} e(U_i,T_j)^{e_{ij}} = Z.$$

If $c_\ell = 0$ then setting $T'_\ell = T_\ell B_\ell^{-1}$ while keeping the rest of the signature intact gives us a forged signature on $M' = MA_\ell$, where $A_\ell \neq M^{-c_\ell} = M^0 = 1$. If $c_\ell \neq 0$ then setting $T'_\ell = T_\ell^{-1} B_\ell^{-\frac{2}{c_\ell}}$ while keeping the rest of the signature intact gives us a forged signature on $M' = M^{-1} A_\ell^{-\frac{2}{c_\ell}} \neq M$, where the inequality follows from $A_\ell \neq M^{-c_\ell}$. To avoid forged signatures must therefore, with overwhelming probability, have $A_\ell = M^{-c_\ell}$ for all ℓ.

If there is overwhelming probability that $A_\ell M^{c_\ell} = 1$ for all ℓ, then each T_ℓ is cancelled out in the verification. We can therefore without loss of generality ignore $T_1, T_2, \ldots$ and look at the case where signatures are of the form $(S_1, S_2, \ldots)$ with $S_i \in \mathbb{G}$ and the verification equation is of the form

$$\prod_i \prod_j e(S_i, V_j)^{b_{ij}} \cdot \prod_j e(M, V_j)^{d_j} = Z.$$

Obtaining two signatures $(S_1, S_2, \ldots)$ and $(S'_1, S'_2, \ldots)$ on two random messages M and M' gives us a signature $(S_1^2/S'_1, S_2^2/S'_2, \ldots)$ on M^2/M'. With overwhelming probability $M^2/M' \notin \{M, M'\}$ and we have a forgery. □

3.2 Impossibility of Unilateral Signatures

Let us call a signature unilateral if it only contains group elements in $\mathbb{G}$ or only contains group elements in $\mathbb{H}$. In other words, a unilateral signature is either of the form $(S_1, S_2, \ldots)$ with $S_i \in \mathbb{G}$ or of the form $(T_1, T_2, \ldots)$ with $T_i \in \mathbb{H}$.

Theorem 3. *There is no unilateral generic-signer structure-preserving signature scheme that is existentially unforgeable against random message attacks.*

Proof. Let us without loss of generality look at a signature scheme for single group element messages $M \in \mathbb{G}$. The verification key contains group elements $U_1, U_2, \ldots \in \mathbb{G}, V_1, V_2, \ldots \in \mathbb{H}, Z_1, Z_2, \ldots \in \mathbb{T}$.

We first look at the case, where signatures are of the form $(S_1, S_2, \ldots)$ with $S_i \in \mathbb{G}$ and fit a number of verification equations of the form

$$\prod_i \prod_j e(S_i, V_j)^{b_{qij}} \cdot \prod_j e(M, V_j)^{d_{qj}} = Z_q.$$

Given two signatures $(S_1, \ldots)$ and $(S'_1, \ldots)$ on random messages M and M' we see that $(S_1^2/S'_1, \ldots)$ is a signature on M^2/M'. There is negligible probability of $M^2/M' \in \{M, M'\}$ so this gives us an existential forgery.

Next, consider the case where signatures are of the form $(T_1, T_2, \ldots)$ with $T_j \in \mathbb{H}$ and satisfy verification equations of the form

$$\prod_j e(M, T_j)^{c_{qj}} \cdot \prod_j e(M, V_j)^{d_{qj}} \cdot \prod_i \prod_j e(U_i, T_j)^{e_{qij}} = Z_q.$$

A generic signer chooses $(T_1, \ldots)$ independently of M because they belong to different groups. Generating the signature independently of M combined with correctness of the signature scheme means that the resulting signature must be valid for all messages M so it is trivial to find a selective forgery after a one-time random message attack. □

3.3 Impossibility of Signatures with 2 Group Elements

Theorem 4. *No generic-signer structure-preserving signature scheme with signatures having two group elements is existentially unforgeable against random message attacks.*

Proof. Theorem 3 ruled out the existence of unilateral generic-signer structure-preserving signatures. The remaining question is therefore, whether we can have signatures of the form (S, T) with $S \in \mathbb{G}$ and $T \in \mathbb{H}$. Suppose without loss of generality that we have a generic-signer structure-preserving signature scheme for messages $M \in \mathbb{G}$. The public verification key contains $U_1, \ldots \in \mathbb{G}, V_1, \ldots \in \mathbb{H}, Z_1, \ldots \in \mathbb{T}$ and a signature (S, T) on M satisfies a number of verification equations of the form

$$e(S,T)^{a_q} \cdot \prod_j e(S, V_j)^{b_{qj}} \cdot e(M,T)^{c_q} \cdot \prod_j e(M, V_j)^{d_{qj}} \cdot \prod_i e(U_i, T)^{e_{qi}} = Z_q.$$

Without loss of generality we may assume that the signer knows the discrete logarithms of all the elements in the public verification key. Using generic group operations it can only construct $S = M^\alpha G^\beta$ and $T = H^\tau$, where α, β, τ may be correlated to each other and the public verification key but are independent of M. Taking discrete logarithms of the verification equations, we get equations of the form

$$(\alpha m + \beta)\tau a_q + (\alpha m + \beta) \sum_j v_j b_{qj} + m\tau c_q + m \sum_j v_j d_{qj} + \tau \sum_i u_i e_{qi} = z_q.$$

The correctness of the signature scheme means that these equations are satisfied for any choice of m. Defining $b_q = \sum_j v_j b_{qj}, d_q = \sum_j v_j d_{qj}, e_q = \sum_i u_i e_{qi}$ this means that the choice of α, β and τ must satisfy pairs of equations of the form

$$a_q \alpha\tau + b_q \alpha + c_q \tau + d_q = 0 \qquad a_q \beta\tau + b_q \beta + e_q \tau = z_q.$$

By taking suitable non-trivial linear combinations of two such pairs of equations, say equation q_1 and q_2, we can eliminate the $\alpha\tau$ and $\beta\tau$ terms to get a pair of equations of the form

$$b\alpha + c\tau + d = 0 \qquad b\beta + e\tau = z.$$

If $b = 0$ and $c \neq 0$ or $b = 0$ and $e \neq 0$ we get a fixed τ and $T = H^\tau$ is uniquely determined. This T can therefore without loss of generality be published as part of the verification key making the signature scheme unilateral. Theorem 3 therefore tells us that if $b = 0$ then $c = 0$ and $e = 0$. This implies $d = 0$ and $z = 0$ as well, and we conclude that the two verification equations q_1 and q_2 are linearly related and one of them can without loss of generality be eliminated from the signature scheme. From Theorem 2 we deduce that there must be at least two verification equations that are not linearly related giving a linear combination with $b \neq 0$.

If $b \neq 0$ we have

$$\alpha = -\frac{c}{b}\tau - \frac{d}{b} \qquad \beta = -\frac{e}{b}\tau + \frac{z}{b}.$$

Plugging them into the verification equations gives us equations of the form

$$-a_q\frac{c}{b}\tau^2 + (c_q - a_q\frac{d}{b} - b_q\frac{c}{b})\tau = b_q\frac{d}{b} - d_q \qquad -a_q\frac{e}{b}\tau^2 + (e_q + a_q\frac{z}{b} - b_q\frac{e}{b})\tau = -b_q\frac{z}{b} + z_q.$$

If one of the quadratic equations in τ is non-trivial then T can take at most two possible values T_0 or T_1. After obtaining signatures on three random messages, two of them would be using the same T. The adversary would thus have signatures (S, T) and (S', T) on messages M and M' and this would give a signature $(S^2/S', T)$ on M^2/M', which with overwhelming probability gives an existential forgery.

If all the quadratic equations are trivial there are two possibilities. The first possibility is that $a_1 = 0, a_2 = 0, \ldots$ but then

$$c_q = b_q\frac{c}{b} \qquad d_q = b_q\frac{d}{b} \qquad e_q = b_q\frac{e}{b} \qquad z_q = b_q\frac{z}{b}$$

and it can be seen that all the verification equations are linearly related and can be replaced with a single verification equation. Theorem 2 rules out this possibility. The other possibility is that $c = 0$ and $e = 0$. This gives us

$$c_q = a_q\frac{d}{b} \qquad d_q = b_q\frac{d}{b} \qquad e_q = -a_q\frac{z}{b} \qquad z_q = b_q\frac{z}{b} \qquad \alpha = -\frac{d}{b} \qquad \beta = \frac{z}{b}.$$

Plugging $S = M^\alpha G^\beta$ into the verification equations shows the verification equations completely ignore T. With all verification equations ignoring T we are back in the unilateral case that we ruled out in Theorem 3. □

4 Minimal Structure-Preserving Signatures

We will now present a structure-preserving signature scheme that matches the lower bounds we found in Section 3. The signature scheme is strongly existentially unforgeable against adaptive chosen message attacks. We can simultaneously sign tuples of messages in $\mathbb{G}$ and tuples of messages in $\mathbb{H}$. A signature consists of three group elements and is verified using two verification equations.

Let us first discuss the case of signing a pair of group elements $(M, N) \in \mathbb{G} \times \mathbb{H}$. Working over a bilinear group $(p, \mathbb{G}, \mathbb{H}, \mathbb{T}, e, G, H)$ the verification key is of the form

$(U, V, W, Z) \in \mathbb{G} \times \mathbb{H}^3$. A signature on a message $(M, N) \in \mathbb{G} \times \mathbb{H}$ is of the form $(R, S, T) \in \mathbb{G}^2 \times \mathbb{H}$ and is verified by two verification equations

$$e(R, V)e(S, H)e(M, W) = e(G, Z) \qquad e(R, T)e(U, N) = e(G, H).$$

It is instructive to look at the verification equations from a generic signer's perspective in light of the same type of equations we used to prove the lower bounds in Section 3. Using $R = M^\alpha G^\beta, S = M^\gamma G^\delta$ and $T = N^\epsilon H^\eta$ we get after taking discrete logarithms of the verification equations

$$(\alpha m + \beta)v + (\gamma m + \delta) + mw = z \qquad (\alpha m + \beta)(\epsilon n + \eta) + un = 1.$$

The signer does not know the discrete logarithms of M and N so the verification equations should hold for all choices of m and n. The signer must therefore choose $\alpha, \beta, \gamma, \delta, \epsilon, \eta \in \mathbb{Z}_p$ such that the following equations are satisfied

$$v\alpha + \gamma + w = 0 \qquad \beta v + \delta = z \qquad \alpha\epsilon = 0 \qquad \alpha\eta = 0 \qquad \beta\epsilon + u = 0 \qquad \beta\eta = 1.$$

This gives six constraints on $\alpha, \beta, \gamma, \delta, \epsilon, \eta$. An arbitrary pair of equations could in contrast give eight constraints on the six variables and might not be solvable. Furthermore, if we pick $\alpha = 0$ we are left with only four constrains

$$\gamma = -w \qquad \beta v + \delta = z \qquad \beta\epsilon + u = 0 \qquad \beta\eta = 1$$

on the five variables $\beta, \gamma, \delta, \epsilon, \eta$. This makes it possible to have many different solutions to the equations and avoids R, S or T being constrained to a single fixed value, which would bring us into conflict with the lower bounds from Section 3.

We extend the signature scheme sketched above in a natural way to sign messages in $\mathbb{G}^{k_M} \times \mathbb{H}^{k_N}$. The full signature scheme can be found in Figure 1.

Key generation $\mathcal{K}(GK)$**:** Parse GK as $(p, \mathbb{G}, \mathbb{H}, \mathbb{T}, e, G, H)$.
Pick at random $u_1, \ldots, u_{k_N}, v, w_1, \ldots, w_{k_M}, z \leftarrow \mathbb{Z}_p^*$ and compute

$$U_i = G^{u_i} \qquad V = H^v \qquad W_i = H^{w_i} \qquad Z = H^z.$$

Return the verification key $VK = (GK, U_1, \ldots, U_{k_N}, V, W_1, \ldots, W_{k_M}, Z)$ and the signing key $SK = (VK, u_1, \ldots, u_{k_N}, v, w_1, \ldots, w_{k_M}, z)$.

Signing $\mathcal{S}_{SK}(M_1, \ldots, M_{k_M}, N_1, \ldots, N_{k_N})$**:**
Given $(M_1, \ldots, M_{k_M}, N_1, \ldots, N_{k_N}) \in \mathbb{G}^{k_M} \times \mathbb{H}^{k_M}$ pick $r \leftarrow \mathbb{Z}_p^*$ and compute

$$R = G^r \qquad S = G^{z - rv} \prod_i M_i^{-w_i} \qquad T = (H \prod_i N_i^{-u_i})^{\frac{1}{r}}.$$

Return the signature (R, S, T).

Verification $\mathcal{V}_{VK}((M_1, \ldots, M_{k_M}, N_1, \ldots, N_{k_N}), (R, S, T))$**:**
Accept if $M_1, \ldots, M_{k_M}, R, S \in \mathbb{G}$ and $N_1, \ldots, N_{k_N}, T \in \mathbb{H}$ and

$$e(R, V)e(S, H) \prod_i e(M_i, W_i) = e(G, Z) \quad \wedge \quad e(R, T) \prod_i e(U_i, N_i) = e(G, H).$$

Fig. 1. Structure-preserving signature scheme for messages in $\mathbb{G}^{k_M} \times \mathbb{H}^{k_N}$

Theorem 5. *The signature scheme $(\mathcal{K}, \mathcal{S}, \mathcal{V})$ described in Figure 1 is a structure-preserving signature scheme over $\mathcal{G}$ that is strongly existentially unforgeable against adaptive chosen message attacks in the generic group model.*

Proof. The verification key, the messages and the signatures consist of group elements in $\mathbb{G}$ and $\mathbb{H}$ and the verification consists of verifying two pairing product equations, so it is a structure-preserving scheme. Correctness follows from verifying that

$$e(G^r, H^v)e(G^{z-vr}\prod_i M_i^{-w_i}, H)\prod_i e(M_i, H^{w_i}) = e(G, H^z)$$

$$e(G^r, (H\prod_i N_i^{-u_i})^{\frac{1}{r}})\prod_i e(G^{u_i}, N_i) = e(G, H).$$

Lemma 1 shows that the signature scheme is secure in the generic group model for $k_M = 1$ and $k_N = 2$. We will show that if the signature scheme is secure for $k_M = 1$ and $k_N = 2$, then the signature scheme is also secure when using arbitrary constants $k_M \geq 1$ and $k_N \geq 2$. In the following we write $(\mathcal{K}, \mathcal{S}, \mathcal{V})$ and $(\mathcal{K}', \mathcal{S}', \mathcal{V}')$ to distinguish between the two settings. We will show that if there is an adversary $\mathcal{A}'$ that can break $(\mathcal{K}', \mathcal{S}', \mathcal{V}')$, then there is an adversary $\mathcal{A}$ that can break $(\mathcal{K}, \mathcal{S}, \mathcal{V})$.

The adversary $\mathcal{A}$ gets as input a verification key $VK = (GK, U_1, U_2, V, W_1, Z)$. It picks at random $\alpha_i, \beta_i \leftarrow \mathbb{Z}_p$ and $\gamma_i, \delta_i \leftarrow \mathbb{Z}_p$ and computes

$$U_1' = U_1^{\gamma_1}U_2^{\delta_1} \ldots U_{k_N}' = U_1^{\gamma_{k_N}}U_2^{\delta_{k_N}} \qquad W_1' = W_1^{\alpha_1}H^{\beta_1} \ldots W_{k_M}' = W_1^{\alpha_{k_M}}H^{\beta_{k_M}}.$$

It gives the verification key $VK' = (GK, U_1', \ldots, U_{k_N}', V, W_1', \ldots, W_{k_N}', Z)$ to $\mathcal{A}'$. Conditioned on the overwhelmingly likely event $U_i' \neq 1$ and $W_i' \neq 1$ this has the same distribution as a normal key produced by $(\mathcal{K}', \mathcal{S}', \mathcal{V}')$.

When $\mathcal{A}'$ asks for a signature on $(M_1', \ldots, M_{k_M}', N_1', \ldots, N_{k_N}') \in \mathbb{G}^{k_M} \times \mathbb{H}^{k_N}$ the adversary $\mathcal{A}$ computes

$$M = \prod_i (M_i')^{\alpha_i} \qquad N_1 = \prod_i (N_i')^{\gamma_i} \qquad N_2 = \prod_i (N_i')^{\delta_i}.$$

It asks the signing oracle for a signature (R, S, T) on (M, N_1, N_2). It then computes $S' = S\prod_i (M_i')^{-\beta_i}$. It returns the signature (R, S', T) to $\mathcal{A}$. It is straightforward to verify that a valid signature is returned to $\mathcal{A}'$. Furthermore, we observe that the returned signature is uniformly random over all possible solutions to the two verification equations just like a normal signature would be. It is therefore a good simulation.

Suppose $\mathcal{A}'$ produces a signature (R', S', T') on some $(M_1', \ldots, M_{k_M}', N_1', \ldots, N_{k_N}')$ satisfying the two verification equations using the key VK'. $\mathcal{A}$ can translate that into a valid signature (R', S, T') on a message (M, N_1, N_2) using VK by computing

$$S = S'\prod_i (M_i')^{\beta_i} \qquad M = \prod_i (M_i')^{\alpha_i} \qquad N_1 = \prod_i (N_i')^{\gamma_i} \qquad N_2 = \prod_i (N_i')^{\delta}.$$

We now have a strong existential forgery unless (R', S, T') has been used before in some query q to sign a message $(M^{(q)}, N_1^{(q)}, N_2^{(q)}) = (M, N_1, N_2)$. That would give

$$\prod_i (M_i')^{\alpha_i} = \prod_i (M_i^{(q)'})^{\alpha_i} \qquad \prod_i (N_i')^{\gamma_i} = \prod_i (N_i^{(q)'})^{\gamma_i}.$$

Observe that α_i and γ_i are information-theoretically hidden to $\mathcal{A}'$ who only sees $W_i' = W_1^{\alpha_i} H^{\beta_i}$ and $U_i' = U_1^{\gamma_i} U_2^{\delta_i}$. Furthermore, no matter the values of α_i, γ_i the adversary gets uniformly random signatures as answer to the chosen message attacks, so these signatures do not reveal anything about the α_i's and the γ_i's either. The only way the adversary can have more than negligible chance of success is by choosing $M_1' = M_1^{(q)'}, \ldots, M_{k_M}' = M_{k_M}^{(q)'}, N_1' = N_1^{(q)'}, \ldots, N_{k_N}' = N_{k_N}^{(q)'}$. This means $\mathcal{A}'$ has repeated the message from query q and some calculation shows that it has also repeated the signature $(R^{(q)'}, S^{(q)'}, T^{(q)'})$. We conclude that $\mathcal{A}'$ has negligible chance of breaking the strong existential unforgeability against chosen message attacks on the signature scheme with $k_M \geq 1$ and $k_N \geq 2$.

The remaining case to consider is $k_M = 0$ or $k_N \in \{0, 1\}$. Here it is easy to get a secure signature scheme, because we can simply require that the signer always uses $M = 1$ or $N_1 = 1$ or $N_2 = 1$, which can be checked in the verification step. Furthermore, when we always have $M = 1$ or $N_1 = 1$ or $N_2 = 1$ then the corresponding W_1 or U_1 or U_2 is not needed in the verification key. □

Lemma 1. *The signature scheme $(\mathcal{K}, \mathcal{S}, \mathcal{V})$ described in Figure 1 is strongly existentially unforgeable against adaptive chosen message attacks in the generic group model for messages $(M, N_1, N_2) \in \mathbb{G} \times \mathbb{H}^2$.*

Proof. Let us for ease of notation write W instead of W_1 and U, U' instead of U_1, U_2. We write $(M, N, N') \in \mathbb{G} \times \mathbb{H}^2$ for the messages we are signing. We consider an adversary that only uses generic group operations on the group elements it sees and is unaware of the random u, u', v, w, z used in the public key and is unaware of the randomness r_i used to form the signature in signing query number i. Seeing signatures (R_i, S_i, T_i) on queries (M_i, N_i, N_i') the generic adversary is restricted to picking $\rho, \rho_u, \rho_{u'}, \rho_1, \rho_1', \ldots, \sigma, \sigma_u, \sigma_{u'}, \sigma_1, \sigma_1', \ldots, \tau, \tau_v, \tau_w, \tau_z, \tau_1, \ldots \in \mathbb{Z}_p$ and computing

$$R = G^{\rho} U^{\rho_u} (U')^{\rho_{u'}} \prod_i R_i^{\rho_i} S_i^{\rho_i'}\ ,\ S = G^{\sigma} U^{\sigma_u} (U')^{\sigma_{u'}} \prod_i R_i^{\sigma_i} S_i^{\sigma_i'}\ ,\ T = H^{\tau} V^{\tau_v} W^{\tau_w} Z^{\tau_z} \prod_i T_i^{\tau_i}.$$

The queries (M_i, N_i, N_i') are computed as products of $G, U, U', R_1, S_1, \ldots, R_{i-1}, S_{i-1}$ and $H, V, W, Z, T_1, \ldots, T_{i-1}$ raised to exponents chosen by the adversary and the message (M, N, N') for which a forgery is obtained is computed similarly. Taking discrete logarithms we have

$$\begin{aligned}
m_i &= \text{linear combination of } 1, u, u', r_1, s_1, \ldots, r_{i-1}, s_{i-1} \\
m &= \text{linear combination of } 1, u, u', r_1, s_1, \ldots, r_q, s_q \\
r &= \rho + \rho_u u + \rho_{u'} u' + \sum_i \rho_i r_i + \sum_i \rho_i'(z - r_i v - m_i w) \\
s &= \sigma + \sigma_u u + \sigma_{u'} u' + \sum_i \sigma_i r_i + \sum_i \sigma_i'(z - r_i v - m_i w) \\
n_i, n_i' &= \text{linear combination of } 1, v, w, z, t_1, \ldots, t_{i-1} \\
n, n' &= \text{linear combination of } 1, v, w, z, t_1, \ldots, t_q \\
t &= \tau + \tau_v v + \tau_w w + \tau_z z + \sum_i \tau_i \frac{1 - u n_i - u' n_i'}{r_i}
\end{aligned}$$

We first consider elements formal polynomials in the variables $u, u', v, w, z, r_1, \ldots, r_q$ and show that the generic adversary cannot make an existential forgery when they are viewed as formal multi-variate polynomials. Later, we will then consider the risk of two different formal polynomials resulting in identical values when evaluated over concrete random choices of $u, u', v, w, z, r_1, \ldots, r_q \in \mathbb{Z}_p^*$.

Taking discrete logarithms of the first verification equation gives us $rv+s+mw = z$, which means

$$0 = \rho v + \rho_u uv + \rho_{u'} u'v + \sum_i \rho_i r_i v + \sum_i \rho_i'(vz - r_i v^2 - m_i vw) + \sigma + \sigma_u u + \sigma_{u'} u' + \sum_i \sigma_i r_i + \sum_i \sigma_i'(z - r_i v - m_i w) + mw - z.$$

Since $s_i = z - r_i v - m_i w$ we have that $m_1, \ldots, m_q$ and m are multi-variate polynomials in $u, u', v, w, z, r_1, \ldots, r_q$. Each m_i has degree at most i and m has degree at most $q+1$.

Looking at the coefficients for $1, u, u', r_i$ we see that $\sigma = 0, \sigma_u = 0, \sigma_{u'} = 0$ and $\sigma_i = 0$ giving us $s = \sum_i \sigma_i'(z - r_i v - m_i w)$. Looking at the coefficients for $v, uv, u'v, r_i v^2$ we get $\rho = 0, \rho_u = 0, \rho_{u'} = 0, \rho_i' = 0$ giving us $r = \sum_i \rho_i r_i$. The coefficients for $r_i v$ give us $\sigma_i' = \rho_i$ so $s = \sum_i \rho_i(z - r_i v - m_i v)$.

Switching to the second verification equation we have $rt + un + u'n' = 1$. Define $\pi = \prod_i r_i$ and $\pi_j = \prod_{i \neq j} r_i$ such that $\pi = \pi_j r_j$. Multiplying the equation on both sides with π we get $rt\pi + un\pi + u'n'\pi = \pi$ so

$$0 = \left(\sum_i \rho_i r_i\right)\left(\tau\pi + \tau_v v\pi + \tau_w w\pi + \tau_z z\pi + \sum_j \tau_j(\pi_j - un_j\pi_j - u'n_j'\pi_j)\right) + un\pi + u'n'\pi - \pi.$$

Observe, $n_1, n_1', \ldots, n_q, n_q', n, n'$ are polynomials in $u, u', v, w, z, r_1^{-1}, \ldots, r_q^{-1}$. Each r_i^{-1} has at most degree 1 and a closer inspection reveals that $n_1\pi_1, n_1'\pi_1, \ldots, n_q\pi_q, n_q'\pi_q$ and $n\pi, n'\pi$ are polynomials in $u, u', v, w, z, r_1, \ldots, r_q$ of degree at most $q+1$.

Looking at the coefficients for π we see that there must exist some ℓ such that $\rho_\ell \neq 0$ and $\tau_\ell \neq 0$. Looking at the coefficients for $r_\ell\pi, r_\ell v\pi, r_\ell w\pi, r_\ell z\pi$ we see that $\tau = 0, \tau_v = 0, \tau_w = 0, \tau_z = 0$. Looking at the coefficients for $r_\ell\pi_j$ we see that $\tau_j = 0$ for $j \neq \ell$. Looking at the coefficients for $r_i\pi_\ell$ we see that $\rho_i = 0$ for $i \neq \ell$. This means $r = \rho_\ell r_\ell$ and $t = \tau_\ell \frac{1 - un_\ell - u'n_\ell'}{r_\ell}$. We now have

$$\rho_\ell r_\ell \cdot \tau_\ell \frac{1 - un_\ell - u'n_\ell'}{r_\ell}\pi - un\pi - \pi = 0.$$

From the coefficient of π we deduce that $\tau_\ell = \frac{1}{\rho_\ell}$. The equation now reads

$$\pi - un_\ell\pi - u'n_\ell'\pi + un\pi + u'n'\pi - \pi = 0,$$

which implies $un_\ell\pi + u'n_{\ell'}\pi = un\pi + u'n_\ell'\pi$. Plugging in all the possible linear combinations of $1, v, w, z, \frac{1 - un_1 - u'n_1'}{r_1}, \ldots, \frac{1 - un_q - u'n_q'}{r_q}$ that can make n, n', n_ℓ, n_ℓ' in this equation, we get $n = n_\ell$ and $n' = n_\ell'$.

Going back to the first equation we now have $r = \rho_\ell r_\ell$ and therefore $s = \rho_\ell(z - r_\ell v - m_\ell v)$, which gives us the equality

$$\rho_\ell r_\ell v + \rho_\ell(z - r_\ell v - m_\ell v) + mv - z = 0.$$

Looking at the coefficient of z we conclude $\rho_\ell = 1$. That tells us $m = m_\ell$. The adversary has therefore reused $m = m_\ell$ and $n = n_\ell, n' = n'_\ell$ for some ℓ and not obtained an existential forgery. Furthermore, $r = r_\ell, s = s_\ell, t = t_\ell$ so the adversary cannot even find a new signature on the same message.

We have now seen that the adversary cannot make an existential forgery when viewing group elements as formal multi-variate polynomials. However, it may be the case that for concrete choices of variables, two formally different polynomials evaluate to the same value. In this case, we cannot simulate the generic group and it may be that the adversary can make an existential forgery. The verification equations can be evaluated using generic group operations, so without loss of generality we can assume the adversary knows it when it has made a successful forgery. Since the polynomials have degree $O(q)$ we get with a birthday paradox argument and the Schwartz-Zippel lemma that the probability of this type of error occurring in the generic group simulation is a negligible $O(\frac{q^3}{p})$ when the adversary makes $O(q)$ generic group operations. □

5 Other Aspects of Structure-Preserving Signatures

5.1 Strong One-Time Signatures Based on Standard Assumptions

We present below a strong one-time signature scheme for messages from $\mathbb{G}^{k_M} \times \mathbb{H}^{k_N}$ with signature size 5 group elements. If the message is one-sided, i.e., $(M_1, \ldots, M_{k_M}) \in \mathbb{G}^{k_M}$, there is a simpler signature with 2 group elements and a single verification equation $e(R, H)e(S, V) \prod_i e(M_i, V_i) = e(G, W)$ [AHO10]. These schemes complement the lower bounds in Section 3 where it was shown that structure-preserving signature schemes with a single verification equation or with unilateral signatures or with signatures with less than 3 group elements do not exist if the adversary gets access to signatures on *two* random messages.

Key generation $\mathcal{K}(GK)$**:** Parse GK as $(p, \mathbb{G}, \mathbb{H}, \mathbb{T}, e, G, H)$.
Pick $w, u, u_1, \ldots, u_{k_N}, v, z, v_1, \ldots, v_{k_M} \leftarrow \mathbb{Z}_p^*$ at random and compute

$$W = H^w, U = G^u, U_1 = G^{u_1}, \ldots, U_{k_N} = G^{u_{k_N}}, \quad \text{and}$$
$$Z = H^z, \quad V = H^v, V_1 = H^{v_1}, \ldots, V_{k_M} = H^{v_{k_M}}.$$

Return verification key $VK = (GK, U, U_1, \ldots, U_{k_N}, V, Z, V_1, \ldots, V_{k_M}, W)$ and signing key $SK = (VK, w, u, u_1, \ldots, u_{k_N}, v, z, v_1, \ldots, v_{k_M})$.

Signing $\mathcal{S}_{SK}(M_1, \ldots, M_{k_M}, N_1, \ldots, N_{k_N})$**:** Given $(M_1, \ldots, M_{k_M}, N_1, \ldots, N_{k_N}) \in \mathbb{G}^{k_M} \times \mathbb{H}^{k_N}$ pick at random $s_1, s_2, t \leftarrow \mathbb{Z}_p^*$ and compute

$$T = G^t, \qquad S_2 = H^{s_2}, \qquad R_2 = H^t S_2^{-u} \prod_i N_i^{-u_i}$$
$$S_1 = G^{s_1}, \qquad R_1 = G^w S_1^{-v} T^{-z} \prod_i M_i^{-v_i}$$

Verification $\mathcal{V}_{VK}((M_1, \dots, M_{k_M}, N_1, \dots, N_{k_N}), (R_1, S_1, T, R_2, S_2))$**:**
Accept if $M_1, \dots, M_{k_M}, R_1, S_1, T \in \mathbb{G}$ and $N_1, \dots, N_{k_N}, R_2, S_2 \in \mathbb{H}$ and

$$e(R_1, H)e(S_1, V)e(T, Z)\prod_i e(M_i, V_i) = e(G, W) \quad \wedge$$

$$e(G, R_2)e(U, S_2)\prod_i e(U_i, N_i) = e(T, H)$$

Theorem 6 (Full paper). *The signature scheme is strongly existentially unforgeable against one-time chosen message attacks if the DDH assumption holds in $\mathbb{G}$ and $\mathbb{H}$.*

5.2 Non-interactive Assumptions

The existential unforgeability of our signature scheme in Figure 1 against adaptive chosen message attacks corresponds to an interactive cryptographic assumption. It would be nice to base the security of the signature scheme on a non-interactive assumption but we do not know of any such security reduction.

By adding a few group elements to the signature it is possible to base the signature scheme on a non-interactive cryptographic assumption though. Consider the following variant of the signature scheme in Figure 1, where the signer picks $M_1 \leftarrow \mathbb{G}$ and $N_1, N_2 \leftarrow \mathbb{H}$ at random when making signatures. In other words, we can sign messages of the form $(M_2, \dots, M_{k_M}, N_3, \dots, N_{k_N}) \in \mathbb{G}^{k_M - 1} \times \mathbb{H}^{k_N - 2}$ and a signature consists of $(R, S, M_1, T, N_1, N_2) \in \mathbb{G}^3 \times \mathbb{H}^3$, which is verified by the verification equations

$$e(R, V)e(S, H)\prod_i e(M_i, W_i) = e(G, Z) \text{ and } e(R, T)\prod_i e(U_i, N_i) = e(G, H).$$

The signature scheme is strongly existentially unforgeable against adaptive chosen message attacks if the following non-interactive assumption holds for $\mathcal{G}$, which essentially says the signature scheme $(\mathcal{K}, \mathcal{S}, \mathcal{V})$ from Figure 1 is strongly existentially unforgeable against *random* message attacks for message space $\mathbb{G} \times \mathbb{H}^2$.

Assumption 1. *Given a random bilinear group $(p, \mathbb{G}, \mathbb{H}, \mathbb{T}, e, G, H) \leftarrow \mathcal{G}(1^k)$ and uniformly random group elements $(U, \hat{U}, V, W, Z) \in \mathbb{G}^2 \times \mathbb{H}^3$ and uniformly random $(R_1, S_1, M_1, T_1, N_1, \hat{N}_1), \dots, (R_q, S_q, M_q, T_q, N_q, \hat{N}_q) \in \mathbb{G}^3 \times \mathbb{H}^3$ such that*

$$e(R_j, V)e(S_j, H)e(M_j, W) = e(G, Z) \text{ and } e(R_j, T_j)e(U, N_j)e(\hat{U}, \hat{N}_j) = e(G, H)$$

a non-uniform polynomial time adversary has negligible probability of finding a different tuple $(R, S, M, T, N, \hat{N}) \in \mathbb{G}^3 \times \mathbb{H}^3$ satisfying the two pairing product equations.

Lemma 1 implies that Assumption 1 holds in the generic group model. Actually, a careful analysis of the proof of Lemma 1 shows that a generic adversary using $O(q)$ group operations has probability $O(\frac{q^2}{p})$ of breaking Assumption 1.

Theorem 7 (Full paper). *If Assumption 1 holds, then the signature scheme $(\mathcal{K}, \mathcal{S}, \mathcal{V})$ in Figure 1 is strongly existentially unforgeable against adaptive chosen message attacks when the signer always chooses $M_1 \leftarrow \mathbb{G}$ and $N_1, N_2 \leftarrow \mathbb{H}$ at random.*

The signature scheme we just described has signatures consisting of 6 group elements. By setting $U_1 = 1, \ldots, U_{k_N} = 1$ and dropping N_1 and N_2 from a signature, the scheme can be used to sign messages of the form $(M_2, \ldots, M_{k_M}) \in \mathbb{G}^{k_M - 1}$ using only 4 group elements. This variant is secure under a related non-interactive assumption.

5.3 Rerandomizable Signatures

The signature scheme in Figure 1 is strongly existentially unforgeable, so it is not possible even to forge a new signature on a message that has already been signed before. In some cases strong existential unforgeability is a useful feature, while in other cases standard existential unforgeability suffices. In this section, we present a rerandomizable signature scheme where a signature can be modified into a different signature for the same message. Rerandomizability may for instance be useful in settings where the signature has to be hidden. One might choose to hide the signature by encrypting it but if the signature is rerandomizable it may be possible to send part of the rerandomized signature in the clear. An additional advantage of the rerandomizable signature scheme we are about to present is that after rerandomization we may only need to hide elements in one of the groups $\mathbb{H}$. This makes it possible to use special purpose variants of Groth-Sahai proofs (they refer to it as the linear case) that are particularly efficient.

We do not know how to construct a rerandomizable signature scheme with 3 group elements that can simultaneously sign messages both in $\mathbb{G}$ and $\mathbb{H}$. But by setting $W_i = 1$ and $Z = 1$ in the signature scheme in Figure 1 we get an efficient rerandomizable signature scheme for messages containing group elements in $\mathbb{H}$. The full description of our rerandomizable signature scheme can be found below.

Key generator $\mathcal{K}(GK)$: Parse GK as $(p, \mathbb{G}, \mathbb{H}, \mathbb{T}, e, G, H)$.
Pick at random $u_1, \ldots, u_{k_N}, v \leftarrow \mathbb{Z}_p^*$ and compute

$$U_1 = G^{u_1} \quad \ldots \quad U_{k_N} = G^{u_{k_N}} \qquad V = H^v.$$

Return $VK = (GK, U_1, \ldots, U_{k_N}, V)$ and $SK = (VK, u_1, \ldots, u_{k_N}, v)$.

Signing $\mathcal{S}_{SK}(N_1, \ldots, N_{k_N})$: Given $(N_1, \ldots, N_{k_N}) \in \mathbb{H}^{k_N}$ pick $r \leftarrow \mathbb{Z}_p^*$ and set

$$R = G^r \qquad S = R^v \qquad T = (H \prod_i N_i^{-u_i})^{\frac{1}{r}}.$$

Rerandomization $R_{VK}(R, S, T)$:
Pick $r' \leftarrow \mathbb{Z}_p^*$ and return the rerandomized signature $(R', S', T') = (R^{r'}, S^{r'}, T^{\frac{1}{r'}})$.

Verification $\mathcal{V}_{VK}((N_1, \ldots, N_{k_N}), (R, S, T))$:
Accept if $R, S \in \mathbb{G}$ and $N_1, \ldots, N_{k_N}, T \in \mathbb{H}$ and

$$e(R, V) = e(S, H) \qquad \wedge \qquad e(R, T) \prod_i e(U_i, N_i) = e(G, H).$$

Theorem 8 (Full paper). *The signature scheme $(\mathcal{K}, \mathcal{S}, \mathcal{V})$ over $\mathcal{G}$ described above is a rerandomizable structure-preserving signature scheme that is existentially unforgeable against adaptive chosen message attacks in the generic group model.*

References

[AFG+10] Abe, M., Fuchsbauer, G., Groth, J., Haralambiev, K., Ohkubo, M.: Structure-preserving signatures and commitments to group elements. In: Rabin, T. (ed.) CRYPTO 2010. LNCS, vol. 6223, pp. 209–236. Springer, Heidelberg (2010)

[AHO10] Abe, M., Haralambiev, K., Ohkubo, M.: Signing on elements in bilinear groups for modular protocol design. Cryptology ePrint Archive, Report 2010/133 (2010)

[BCK10] Bangerter, E., Camenisch, J., Krenn, S.: Efficiency limitations for Σ-protocols for group homomorphisms. In: Micciancio, D. (ed.) TCC 2010. LNCS, vol. 5978, pp. 553–571. Springer, Heidelberg (2010)

[CL02] Camenisch, J., Lysyanskaya, A.: A signature scheme with efficient protocols. In: Cimato, S., Galdi, C., Persiano, G. (eds.) SCN 2002. LNCS, vol. 2576, pp. 268–289. Springer, Heidelberg (2003)

[CL04] Camenisch, J., Lysyanskaya, A.: Signature schemes and anonymous credentials from bilinear maps. In: Franklin, M. (ed.) CRYPTO 2004. LNCS, vol. 3152, pp. 56–72. Springer, Heidelberg (2004)

[CLY09] Cathalo, J., Libert, B., Yung, M.: Group encryption: Non-interactive realization in the standard model. In: Matsui, M. (ed.) ASIACRYPT 2009. LNCS, vol. 5912, pp. 179–196. Springer, Heidelberg (2009)

[Fuc09] Fuchsbauer, G.: Automorphic signatures in bilinear groups and an application to round-optimal blind signatures. Cryptology ePrint Archive, Report 2009/320 (2009)

[Fuc11] Fuchsbauer, G.: Commuting signatures and verifiable encryption. In: Paterson, K.G. (ed.) EUROCRYPT 2011. LNCS, vol. 6632, pp. 224–245. Springer, Heidelberg (2011)

[FV10] Fuchsbauer, G., Vergnaud, D.: Fair blind signatures without random oracles. In: Bernstein, D.J., Lange, T. (eds.) AFRICACRYPT 2010. LNCS, vol. 6055, pp. 16–33. Springer, Heidelberg (2010)

[GGK03] Gennaro, R., Gertner, Y., Katz, J.: Lower bounds on the efficiency of encryption and digital signature schemes. In: STOC, pp. 417–425 (2003)

[GH08] Green, M., Hohenberger, S.: Universally composable adaptive oblivious transfer. In: Pieprzyk, J. (ed.) ASIACRYPT 2008. LNCS, vol. 5350, pp. 179–197. Springer, Heidelberg (2008)

[GO94] Goldreich, O., Oren, Y.: Definitions and properties of zero-knowledge proof systems. Journal of Cryptology 7(1), 1–32 (1994)

[GPS08] Galbraith, S.D., Paterson, K.G., Smart, N.P.: Pairings for cryptographers. Discrete Applied Mathematics 156(16), 3113–3121 (2008)

[Gro06] Groth, J.: Simulation-sound NIZK proofs for a practical language and constant size group signatures. In: Lai, X., Chen, K. (eds.) ASIACRYPT 2006. LNCS, vol. 4284, pp. 444–459. Springer, Heidelberg (2006)

[Gro09] Groth, J.: Homomorphic trapdoor commitments to group elements. Cryptology ePrint Archive, Report 2009/007 (2009)

[GS08] Groth, J., Sahai, A.: Efficient non-interactive proof systems for bilinear groups. In: Smart, N.P. (ed.) EUROCRYPT 2008. LNCS, vol. 4965, pp. 415–432. Springer, Heidelberg (2008)

[Nec94] Nechaev, V.I.: Complexity of a determinate algorithm for the discrete logarithm. Mat. Zametki 55(2), 91–101 (1994)

[OS08] Ostrovsky, R., Skeith III, W.E.: Communication complexity in algebraic two-party protocols. In: Wagner, D. (ed.) CRYPTO 2008. LNCS, vol. 5157, pp. 379–396. Springer, Heidelberg (2008)

[Sho97] Shoup, V.: Lower bounds for discrete logarithms and related problems. In: Fumy, W. (ed.) EUROCRYPT 1997. LNCS, vol. 1233, pp. 256–266. Springer, Heidelberg (1997)

Constant-Rate Oblivious Transfer from Noisy Channels

Yuval Ishai[1,⋆], Eyal Kushilevitz[1,⋆⋆], Rafail Ostrovsky[2,⋆⋆⋆]
Manoj Prabhakaran[3,†], Amit Sahai[2,‡], and Jürg Wullschleger[4,§]

[1] Technion, Haifa, Israel
{yuvali,eyalk}@cs.technion.il
[2] University of California, Los Angeles
{rafail,sahai}@cs.ucla.edu
[3] University of Illinois, Urbana-Champaign
mmp@cs.uiuc.edu
[4] Université of Montréal and McGill University
juerg@wulli.com

Abstract. A *binary symmetric channel* (BSC) is a noisy communication channel that flips each bit independently with some fixed error probability $0 < p < 1/2$. Crépeau and Kilian (FOCS 1988) showed that oblivious transfer, and hence general secure two-party computation, can be *unconditionally* realized by communicating over a BSC. There has been a long line of works on improving the efficiency and generality of this construction. However, all known constructions that achieve security against *malicious* parties require the parties to communicate $poly(k)$ bits over the channel for each instance of oblivious transfer (more precisely, $\binom{2}{1}$-*bit*-OT) being realized, where k is a statistical security parameter. The question of achieving a constant (positive) rate was left open, even in the easier case of realizing a single oblivious transfer of a long string.

We settle this question in the affirmative by showing how to realize n independent instances of oblivious transfer, with statistical error that vanishes with n, by communicating just $O(n)$ bits over a BSC. As a corollary, any boolean circuit of size s can be securely evaluated by two parties with $O(s)+poly(k)$ bits of communication over a BSC, improving over the $O(s) \cdot poly(k)$ complexity of previous constructions.

⋆ Work done in part while visiting UCLA. Supported by ERC Starting Grant 259426, ISF grant 1361/10, and BSF grant 2008411.

⋆⋆ Work done in part while visiting UCLA. Supported by ISF grant 1361/10 and BSF grant 2008411.

⋆⋆⋆ Research supported in part by DARPA, IBM Faculty Award, Xerox Innovation Group Award, the Okawa Foundation Award, Intel, Teradata, NSF grants 0830803, 0916574, BSF grant 2008411 and U.C. MICRO grant.

† Supported by NSF grant CNS 07-47027.

‡ Research supported in part from a DARPA/ONR PROCEED award, NSF grants 0916574 and 0830803, a Xerox Foundation Award, a Google Faculty Research Award, an equipment grant from Intel, and an Okawa Foundation Research Grant.

§ Research supported by the Canada-France NSERC-ANR project FREQUENCY.

P. Rogaway (Ed.): CRYPTO 2011, LNCS 6841, pp. 667–684, 2011.

1 Introduction

One of the attractive features of modern cryptography is its ability to "turn lemons into lemonade." Indeed, traditional complexity-based cryptography turns *computational intractability*, a major obstacle tackled by Computer Science, into a blessing. The present work is concerned with a similar phenomenon in the context of information-theoretic cryptography: the ability to turn *noise*, a major obstacle tackled by Information Theory, into a blessing.

Originating from the seminal work of Wyner [38] on the usefulness of noise for secure communication, there has been a large body of work on basing various cryptographic primitives on different types of noisy communication channels. The most fundamental type of a noisy channel in information theory is the *binary symmetric channel* (BSC). A BSC with crossover probability p, where $0 < p < \frac{1}{2}$, flips each communicated bit independently with probability p.

In 1988, Crépeau and Kilian [10] showed that two parties can make use of a BSC to realize *oblivious transfer* (OT) [32,15] with unconditional security. By OT we refer by default to $\binom{2}{1}$-*bit*-OT, a protocol which allows a receiver to select exactly one of two bits held by a sender without revealing the identity of the received bit to the sender. We require by default that security hold even against *malicious* parties. It is known that OT on a pair of m-bit strings reduces to $O(m)$ instances of bit-OT [4]. Much more broadly, OT can be used as a basis for general secure two-party computation [39,19,26,24]. This settles the main *feasibility* question concerning the cryptographic power of a BSC.

In contrast to the basic feasibility question, the corresponding *efficiency* questions are far less understood. To explain the main relevant issues, it is instructive to draw an analogy with classical information theory. A naive approach to send n bits of information over a noisy channel is to do it bit-wise, by repeating every bit k times. A major breakthrough in information theory was the seminal result of Shannon [33] that by sending bits in blocks and by using the right encoding, one can achieve a *constant transmission rate,* namely use only a constant number of channel transmissions per information bit with error that vanishes with n. One can analogously define the notion of a *constant-rate protocol* for OT from BSC (or a *constant-rate reduction* of OT to BSC) as a protocol which realizes n independent instances of OT with negligible (in n) statistical error[1] by exchanging $O(n)$ bits over the channel.[2] In such a protocol, the amortized communication complexity for each instance of OT tends to a constant which is independent of the desired level of security.

The existence of a constant-rate protocol for OT from BSC has been a long-standing open question. The original protocol from [10] required $O(k^{11})$ bits of communication over a BSC to realize each instance of OT with error 2^{-k}. This communication overhead was subsequently improved by Crépeau [9] to $O(k^3)$.

[1] By the *error* of OT or other secure computation protocol we refer to the statistical simulation error under standard simulation-based definitions [5,6,18].

[2] This is the best one can hope for up to the exact constant. Indeed, it is known that $\Omega(n)$ bits over the BSC are necessary even if one additionally allows unlimited communication over a noiseless channel [35].

A major progress was made by Harnik et al. [20], who showed that constant rate can be achieved in the *semi-honest* model, in which parties do not deviate from the protocol except for trying to infer additional information from their view.

Constant-rate protocols for *string OT*, realizing a single selection between two n-bit strings by communicating $O(n)$ bits over the channel, are considerably easier to obtain. (Indeed, known reductions [4] can be used to get constant-rate string-OT from constant-rate bit-OT, but not the other way around.) Constant-rate string-OT protocols from an *erasure* channel, which erases every bit with probability $0 < p < 1$ and informs the receiver of the erasures, were presented in [30,22].

To summarize the prior state of the art, constant-rate protocols for bit-OT from BSC were only known in the semi-honest model, and constant-rate string-OT protocols could only be based on an erasure channel. The existence of constant-rate bit-OT protocols from a BSC (or even from an erasure channel) as well as the existence of constant-rate string-OT protocols from a BSC were left open.

1.1 Our Results

We settle the above questions in the affirmative by presenting a statistically secure protocol which realizes n independent instances of OT, with 2^{-k} error, in which the parties communicate only $O(n) + poly(k)$ bits over a BSC.[3] This should be compared to the $n \cdot poly(k)$ bits required by previous constructions.

Combining the above main result with known results for secure two-party computation based on OT [24] we get the following corollaries:

- Any boolean circuit of size s can be securely evaluated by two parties with $O(s) + poly(k)$ bits of communication over a BSC, improving over the $O(s) \cdot poly(k)$ complexity of previous constructions.
- Applying the previous corollary, any discrete memoryless channel (described by rational crossover probabilities) can be faithfully emulated by a BSC at a constant rate.

Our techniques can be used to get similar results based on any "non-trivial" channel rather than just a BSC. We defer this generalization to the full version of this paper.

1.2 Overview of Techniques

Our construction uses a novel combination of previous results on OT from BSC [10,20], recent techniques from the area of secure computation [8,24], and some new tools that may be of independent interest.

[3] The protocol also involves a similar amount of communication over a noiseless channel. This additional communication can be implemented using the BSC with a constant rate.

Among the new general-interest tools is a so-called "Statistical-to-Perfect Lemma," showing roughly the following. Given a 2-party functionality $\mathcal{F}_f$ for securely evaluating a function f and $0 \leq \delta \leq 1$, we define $\widetilde{\mathcal{F}}_f^{(\delta)}$ to be an "δ-faulty" version of $\mathcal{F}_f$ that with probability δ allows the adversary to learn the inputs and have full control over the outputs but otherwise behaves normally. The lemma says that *any* ϵ-secure protocol for $\mathcal{F}$ in a $\mathcal{G}$-hybrid model (i.e., using oracle access to $\mathcal{G}$) *perfectly* realizes the functionality $\widetilde{\mathcal{F}}_f^{(\delta)}$ in the $\mathcal{G}$-hybrid model, where δ tends to 0 with ϵ (but inherently grows with the size of the input domain). The above lemma allows one to take an *arbitrary* (and possibly inefficient) protocol for OT from a noisy channel, such as the one from [10], and use it with a sufficiently large security parameter to get a perfectly secure implementation of $\widetilde{\mathcal{F}}_{\mathsf{OT}}^{(\delta)}$, for an arbitrarily small constant $\delta > 0$, while communicating just a constant number of bits (depending on δ) over the channel.

This calls for the use of *OT combiners* [21,20,31], which combine n OT implementation candidates of which some small fraction may be faulty into $m < n$ good instances of OT. A similar high level approach was used in [20] to solve our main question in the semi-honest model. While in the semi-honest model there are constant-rate combiners (tolerating a constant fraction of faulty candidates with $m = \Omega(n)$) that make only a single use of each OT candidate [20], known constant-rate OT combiners in the malicious model require a large number of calls to each candidate, making them insufficient for our purposes. Instead, we take the following alternative approach.

1. We give a direct construction of a constant-rate protocol for *string-OT* from a BSC. (As discussed above, such a result was only known for the easier cases of an erasure channel or in the semi-honest model.) The protocol employs previous protocols for OT from BSC [10], the completeness of OT for secure two-party computation [26,24], techniques from secure multiparty computation (including the use of algebraic-geometric multiplicative secret sharing [8,25]), and privacy amplification techniques [3,2]. Its analysis relies on the Statistical-to-Perfect lemma discussed above.
2. We extend the IPS protocol compiler [24] to apply also when the so-called "inner protocol" can employ a BSC channel. The main difficulty is that even when being forced to reveal their secrets, parties can use the uncertainty of the channel to lie without taking the risk of being caught. We address this difficulty in a natural way by employing statistical tests to ensure that *significant* deviations are being caught with high probability. The extended protocol compiler requires the inner protocol to satisfy an intermediate notion of security, referred to as "error-tolerance," that is stronger than security in the semi-honest model and weaker than security in the malicious model.
3. We instantiate the ingredients required by the extended compiler from Step 2 as follows. The so-called "watchlists" are implemented using string-OTs obtained via the protocol described in Step 1 above. The outer protocol is an efficient honest-majority MPC protocol for n instances of OT (see [23], building on [13,8]). The error-tolerant inner protocol is based on an error-tolerant constant-rate OT combiner from [20].

1.3 Related Work

There is a very large body of related work on cryptography from noisy channels that was not accounted for in the above survey, and even here we can only give a very partial account. For the question of basing other cryptographic primitives (such as key agreement and commitment) on noisy channels see [2,3,28,14,36,37] and references therein. The question of characterizing the types of channels on which OT can be based was studied in [27,11,14,12,37]. A general approach for converting feasibility results for OT from noisy channels into constant-rate protocols in the *semi-honest* model was given in [25]. Our work introduces a similar conversion technique that can be applied in the malicious model.

2 Preliminaries

Some of our results and analysis (in particular Theorem 1) apply to general 2-party secure function evaluation (SFE) functionalities. Such a functionality is characterized by a pair of functions $f = (f_A, f_B)$, $f_A : \mathcal{X} \times \mathcal{Y} \rightarrow \mathcal{Z}_A$ and $f_B : \mathcal{X} \times \mathcal{Y} \rightarrow \mathcal{Z}_B$ for (often finite) domains $\mathcal{X}$, $\mathcal{Y}$ and ranges $\mathcal{Z}_A, \mathcal{Z}_B$. We will refer to such an f as a 2-pary function. We associate a functionality $\mathcal{F}_f$ with a 2-party function f, which behaves as follows: $\mathcal{F}_f$ waits for inputs from both parties, and computes the respective outputs. Then if either party is corrupted, it sends the corresponding output to that party. Then it waits for an instruction from the adversary to release the output(s) to the uncorrupted party (or parties). We shall refer to such a functionality $\mathcal{F}_f$ as a 2-party SFE functionality.

The two main functionalities in this work are $\mathcal{F}_{\mathsf{BSC}}$ and $\mathcal{F}_{\mathsf{OT}}$. The $\mathcal{F}_{\mathsf{BSC}}$ functionality (BSC stands for Binary Symmetric Channel) takes as input a bit x from one of the parties (Alice), and outputs a single bit z to the other party (Bob) such that $\Pr[x \neq z] = p$ for some fixed constant probability strictly less than half. (Note that this is a randomized functionality.) $\mathcal{F}_{\mathsf{OT}}$ is an SFE functionality, associated with a function defined by $f_B(x_0, x_1; b) = x_b$ where x_0, x_1, b are single bits each; for $\mathcal{F}_{\mathsf{OT}}$ f_A is a constant function. The functionality $\mathcal{F}_{\mathsf{string\text{-}OT}}$ is similar to $\mathcal{F}_{\mathsf{OT}}$, but the inputs from Alice x_0, x_1 are longer strings.

For every 2-party SFE functionality $\mathcal{F}_f$, we define a weakened variant $\widetilde{\mathcal{F}}_f^{(p)}$ where $0 \leq p \leq 1$ is a constant error probability in the following sense. When invoked, an instance of $\widetilde{\mathcal{F}}_f^{(p)}$ would first generate a random bit which is 1 with probability p. Note that the bit is sampled before receiving inputs from any party. If the bit is 0, then the functionality behaves exactly as $\mathcal{F}_f$. Otherwise, if the bit is 1, then the functionality yields itself to adversarial control: i.e., the input(s) it receives are passed on to the adversary, and the adversary specifies the outputs to be sent (and when they should be sent) to the honest party (parties). In this case, even if neither party interacting with the functionality is corrupt, the functionality will allow the adversary to control it.

The main security definition we use is of *statistical* Universally Composable (UC) security [6]. The level of security – called statistical error – is indicated by the maximum distinguishing advantage between the real execution of the protocol and a simulated execution involving the ideal functionality that any environment can get (the distinguishing advantage being the difference in probabilities of the environment outputting 1 when interacting with the two systems). We require that the statistical error goes down as $2^{-\Omega(k)}$, where k is the security parameter. The computational complexity of the protocols should be polynomial in k and the input size. For intermediate constructions (and in Theorem 1) we consider *perfect* security as well.

We say that a protocol Π is in the $\mathcal{G}$-hybrid model if the parties can initiate and interact with (any number of) instances of the ideal functionality $\mathcal{G}$. Our goal is to give a "constant-rate" protocol for $\mathcal{F}_{\mathsf{OT}}$ in the $\mathcal{F}_{\mathsf{BSC}}$-hybrid model. A protocol Π in the $\mathcal{G}$-hybrid model is said to be a constant-rate protocol for a functionality $\mathcal{F}$, if the total communication in Π (including communication with instances of $\mathcal{G}$) is $O(\ell) + \text{poly}(k)$, where ℓ is the total communication with $\mathcal{F}$. We will be interested in realizing *parallel* instances of a target functionality, given the number of instances as a parameter (during run-time). More formally we can define a functionality $\mathcal{F}^*$ which takes ℓ as an initial input from one of the parties, and then implements ℓ parallel copies of $\mathcal{F}$. Note that when $\mathcal{F}$ and $\mathcal{G}$ are finite functionalities (i.e., with the total communication with a single instance upperbounded by a constant, as is the case for $\mathcal{F}_{\mathsf{OT}}$ and $\mathcal{F}_{\mathsf{BSC}}$), to securely realize $\mathcal{F}^*$, a constant-rate protocol Π will instantiate only $O(\ell) + \text{poly}(k)$ instances of $\mathcal{G}$. (For simplicity, we shall refer to Π as a protocol for $\mathcal{F}$, rather than $\mathcal{F}^*$.)

An Arithmetic Encoding Scheme. Our protocol (particularly, the sub-protocol in Section 4.1) relies on an efficient secret-sharing scheme that supports entrywise addition and multiplication of shared vectors. Following the terminology of [7], we refer to such a scheme as an *arithmetic encoding scheme.* Our abstraction captures the useful features of algebraic-geometric secret-sharing, introduced in [8] (see [25,7] for related abstractions).

Our notion of arithmetic encoding is parameterized by a tuple $(\mathbb{F}, \rho, \delta, \delta')$ and is defined by three efficient algorithms $(\mathsf{Encode}, \mathsf{Encode}', \mathsf{Decode}')$. Here $\mathbb{F}$ is a constant-size finite field, ρ, δ, δ' are positive constants less than 1, and the three algorithms satisfy the following properties.

- Encode and Encode' define constant-rate, *probabilistic* encodings of vectors over $\mathbb{F}$. More precisely, for every integer $m > 0$, there is an n, with $m > \rho n$, such that Encode and Encode' probabilistically map vectors in $\mathbb{F}^m$ to $\mathbb{F}^n$. Further, Encode and Encode' are linear: i.e., each entry of $\mathsf{Encode}(x)$ (respectively, $\mathsf{Encode}'(x)$) is a linear function of the entries of x and a set of independent random elements.
- The joint distribution of any $\lfloor \delta n \rfloor$ entries of the output of $\mathsf{Encode}(x)$ is independent of the input x.
- Decode' is an efficient δ-error-correcting decoder for Encode'. More precisely, we require that if y has Hamming distance at most δn from a vector in the support of $\mathsf{Encode}'(x)$, then $\mathsf{Decode}'(y) = x$.

- We require the following "homomorphic" properties. For any X, Y, X', Y' in the support of $\mathsf{Encode}(x), \mathsf{Encode}(y), \mathsf{Encode}'(x'), \mathsf{Encode}'(y')$, respectively:
  - $X * Y$ is in the support of $\mathsf{Encode}'(x * y)$
  - $X' + Y'$ is in the support of $\mathsf{Encode}'(x' + y')$

  where $*$ and $+$ denote entrywise multiplication and addition over $\mathbb{F}$ respectively.
- We require Encode' to be sufficiently randomizing. Note that $\mathsf{Encode}'(x)$ is uniform over an affine subspace of $\mathbb{F}^n$ whose dimension is at most $n - m$. We require that this dimension be at least $n - m(1 + \delta')$.

An arithmetic encoding scheme with the above properties can be obtained from the classes of algebraic geometric codes used in [8]. See Appendix A for details.

3 Statistical Security to Perfect Security

A crucial ingredient in our constructions and analysis is the ability to consider a weakly secure protocol to be a perfectly secure protocol for a weaker variant of the functionality. More precisely, we show the following.

Theorem 1. *Let $f : \mathcal{X} \times Y \to \mathcal{Z}_A \times \mathcal{Z}_B$ be a 2-party function, and $\mathcal{F}_f$ the secure function evaluation functionality for f. Suppose $\mathcal{G}$ is a 2-party functionality and Π is a D-round protocol such that Π UC securely realizes $\mathcal{F}_f$ in the $\mathcal{G}$-hybrid model, with a statistical security error of ϵ. Then Π UC securely realizes $\widetilde{\mathcal{F}}_f^{(p)}$ in the $\mathcal{G}$-hybrid model with perfect security, where $p = D|\mathcal{X}||\mathcal{Y}|\epsilon$.*

Above, if Π is only standalone-secure for $\mathcal{F}_f$, then the same conclusion holds for standalone security of Π for $\widetilde{\mathcal{F}}_f^{(p)}$.

This result gives a powerful composition theorem when multiple instances of the protocol Π are used together. Note that by UC security, it is indeed the case that if k copies of Π are run, one could instead consider k copies of $\mathcal{F}$, with a statistical security error bounded by $k\epsilon$. However, if ϵ is not negligible, say $\epsilon > 1/k$, then this bound gives us no useful security guarantee. What the above result does is to give a strong security guarantee for the case when ϵ is non-negligible, or even when it is a constant. It says that when k copies of Π are run, it roughly yields $(1 - p)k$ copies of $\mathcal{F}$ (mixed with about pk copies under adversarial control). In fact, it is further guaranteed that which copies will be corrupted is not under adversarial control.

In the full version we show that it is unavoidable that p is bigger than ϵ by a factor that grows linearly with the domain size of the function.

We give a high-level idea of how we prove the above theorem. Given the systems corresponding to REAL and IDEAL executions, the overall approach is to decompose each of the REAL and IDEAL systems into two parts – $\text{REAL}_0, \text{REAL}_1$ and $\text{IDEAL}_0, \text{IDEAL}_1$ – so that REAL_0 and IDEAL_0 are identical and carry much of the "mass" of the systems; then we construct a new ideal system by combining IDEAL_0 and REAL_1, to get a system that is identical to the REAL system. Here, a combination of two systems means that with a *fixed* probability one of the two systems is chosen (corresponding to whether $\widetilde{\mathcal{F}}_f^{(p)}$ lets the adversary control it or not, corresponding to choosing IDEAL_1 and IDEAL_0 respectively): in particular, the simulator in the new system is not allowed to influence this choice. Further – and this is the main difficulty in the proof – we need to ensure that IDEAL_0 can be implemented by a simulator interacting with $\mathcal{F}_f$ (without access to an honest party's inputs or outputs); to implement REAL_1 the simulator may control the functionality as well.

Note that Theorem 1 is related to Lemma 5 in [29]. The main difference are the abovementioned restrictions on the system IDEAL_0 which require extra care in our proof.

Splitting the systems in this manner needs to be carefully defined, see full version for details. Here we illustrate this by a toy example, to give a sense of how the simulator for perfect security is derived from the simulator for statistical security. The protocol we consider is for a degenerate 2-party function f which provides a constant output to both parties. Further, it takes a fixed input from Alice ($|\mathcal{X}| = 1$) and takes a bit from Bob ($\mathcal{Y} = \{0, 1\}$). The protocol Π for our example consists of a single message z from Bob to Alice, which is equal to y with probability $\frac{1}{2}$ and $\perp$ otherwise. We shall consider the case when Alice is corrupt and Bob is honest. Further we need to consider only a "dummy adversary" who simply allows the environment to play the role of Alice in the (real) protocol. The simulator simulates a message from Bob, which is equal to $\perp$ with probability $\frac{1}{2}$, and a uniformly chosen bit otherwise. It is easy to see that this simulation is good up to a statistical distance of $\frac{1}{4}$.

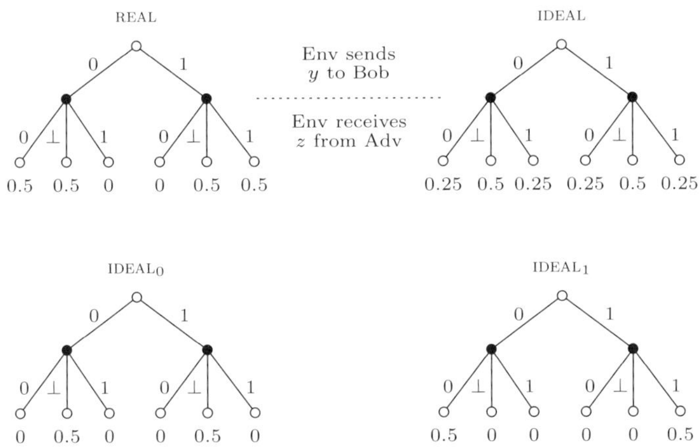

Fig. 1. An example to illustrate Theorem 1. The protocol used in the example (in which Bob sends a single message to Alice — please see text) is depicted as the interaction of a system REAL with the environment. The original simulated system is IDEAL. The modified simulation (for a functionality that yields to the adversary with probability 0.5) is obtained as the combination $\text{IDEAL}_0 + \text{IDEAL}_1$ which is exactly the same as the REAL system.

In Figure 1 we illustrate this example using what we call *interaction trees*, which capture an execution involving a system (REAL or IDEAL) and an environment. The edges from the top node in these trees correspond to the two possible inputs that the environment can give to Bob ($y = 0$ and $y = 1$). The edges out of the black nodes correspond to corrupt Alice reporting the (only) message it receives from Bob in the protocol: this can be one of 0, 1, or $\perp$. The leaves correspond to complete transcripts. The probabilities of the system reaching a leaf, provided the environment sends the messages (in our case, just y) that lead to that leaf is considered the "weight" assigned to that leaf by the system.

The top-left figure corresponds to the real execution of the protocol, and the top-right corresponds to the ideal execution. Note that in the simulation, the behavior of the simulator is independent of the input y. Then we obtain a "partial system" (with total weight only 0.5, for each value of y), IDEAL_0 by comparing the REAL and IDEAL systems. In this example, IDEAL_0 is obtained by retaining in each leaf the minimum of the weights assigned by the two systems, on that leaf, but for any choice of y. We will use the ideal functionality $\widetilde{\mathcal{F}}_f^{(p)}$, with $p = \frac{1}{2}$, since that is the weight not retained by IDEAL_0.

IDEAL_1 is obtained by "subtracting" IDEAL_0 from REAL, so that the combination of IDEAL_0 and IDEAL_1 is indeed REAL. In doing this we needed to ensure that weights induced by IDEAL_0 are no more than what REAL assigns (so that the system REAL $-$ IDEAL_0 does not have negative weights). Also we needed to ensure that IDEAL_0 can be implemented by a simulator which does not have access to y. Note that to implement IDEAL_1, the simulator will need to know y.

In going from this toy example to the general case poses several issues. Here the simulator for IDEAL_0 was determined without considering the interaction between the simulator and the functionality. (Indeed, there was little interaction between the two.) In general we cannot afford to do this. To properly take into account how the simulator's behavior depends on what it learns from the functionality, we consider a separate interaction which the simulator is the system and it interacts with an "enhanced environment" consisting of the original environment and the functionality. But the original statistical security guarantee is only against normal environments (and indeed, does not make sense against enhanced environments, since in the real execution there is no ideal functionality present). This requires us to relate the behavior of the enhanced environment to the behavior of the environment in the ideal world.

The final proof uses several carefully defined quantities for the three systems (the real and ideal executions, and the simulator system), and shows how one can define IDEAL_0 which can be implemented without using y, ensures that it can be extended to a perfect simulation (i.e., that the remainder of the simulation is a non-negative system), while retaining as much weight as possible (to keep p low as promised in Theorem 1).

4 A Constant-Rate OT Protocol

In this section we present our constant-rate protocol for $\mathcal{F}_{\mathsf{OT}}$ in the $\mathcal{F}_{\mathsf{BSC}}$-hybrid model. The construction follows the paradigm of the IPS compiler [24] of combining an outer protocol secure in the honest-majority setting, with an inner protocol secure in the passive corruption (semi-honest) setting, using "watchlists" implemented using string-OTs. For this we need to instantiate these components in the $\mathcal{F}_{\mathsf{BSC}}$-hybrid model, and also extend the IPS compiler so that it admits an inner protocol in the $\mathcal{F}_{\mathsf{BSC}}$-hybrid model. We outline these steps below, and present the details in the subsequent sections.

- In order to construct the inner protocol, we will need a constant-rate OT protocol using $\mathcal{F}_{\mathsf{BSC}}$, that is secure against adaptive *passive* corruption. However, since monitoring the use of a $\mathcal{F}_{\mathsf{BSC}}$ functionality (which inherently allows errors) is harder than monitoring the use of the $\mathcal{F}_{\mathsf{OT}}$ functionality we will need a somewhat stronger security guarantee from this protocol (namely, passive security should hold even when a small constant fraction of the $\mathcal{F}_{\mathsf{BSC}}$ instances can be corrupted). We shall formalize this notion of "error-tolerance" and observe that a protocol in [20] already has the requisite properties (Lemma 2).
- The next step is to construct a constant-rate *string-OT* protocol from $\mathcal{F}_{\mathsf{BSC}}$, with security against active corruption (Lemma 1). The protocol implements a single instance of string-OT (i.e., takes only one choice bit as input from the receiver), and the rate refers to the ratio of the length of the string to the number of instances of $\mathcal{F}_{\mathsf{BSC}}$ used. This crucially relies on Theorem 1 which states that a weakly secure protocol for a functionality is a perfectly secure protocol for a weak version of the same functionality.
- The final step involves an extension of the IPS compiler [24] wherein the "inner-protocol" is in the $\mathcal{F}_{\mathsf{BSC}}$-hybrid model (rather than in the $\mathcal{F}_{\mathsf{OT}}$-hybrid model) and enjoys error-tolerance (Lemma 3).

The extended IPS compiler from above is used to combine an appropriate constant-rate[4] outer protocol for $\mathcal{F}_{\mathsf{OT}}$ (based on [13,8], as used in [24]) with an error-tolerant inner protocol obtained from the first step, using watchlists implemented using string-OTs from the second step.

To implement n instances of $\mathcal{F}_{\mathsf{OT}}$, the resulting compiled protocol will invoke the string-OT protocols $O(k)$ times with $O(n/k)$ long strings. Since these string-OTs are implemented using the constant-rate protocol from the second step, the compiled protocol uses a total of $O(n)$ instances of $\mathcal{F}_{\mathsf{BSC}}$ for the watchlists.

Similarly the compiled protocol invokes k instances of the inner-protocol (for a functionality defined by the outer protocol). Originally, each instance

[4] Here the constant-rate refers to the total communication in the protocol, and the total computation of all the servers per instance of $\mathcal{F}_{\mathsf{OT}}$ produced. More precisely, regarding the computational complexity of the servers, we are interested in the complexity of a passive-secure protocol for implementing the server computations, and it is only the so-called "type II" computations of the servers that contribute to this. See [24] for details.

of this inner-protocol can be implemented using $O(n/k)$ instances of $\mathcal{F}_{\mathsf{OT}}$, and is passive-secure in the $\mathcal{F}_{\mathsf{OT}}$-hybrid model. We shall replace the $\mathcal{F}_{\mathsf{OT}}$ instances used by the inner protocol with the constant-rate error-tolerant protocol from the first step. This results in an error-tolerant inner protocol in the $\mathcal{F}_{\mathsf{BSC}}$-hybrid model (for the same functionality as the original inner-protocol), which uses $O(n/k)$ instances of $\mathcal{F}_{\mathsf{BSC}}$. Thus overall, for the inner-protocol instances too, the compiled protocol uses $O(n)$ instances of $\mathcal{F}_{\mathsf{BSC}}$.

In the following sub-sections we describe how the above three steps are carried out, and what precise security guarantees they provide. Then, by following the above sketched construction we obtain our main result.

Theorem 2. *There exists a UC-secure constant-rate protocol for $\mathcal{F}_{\mathsf{OT}}$ in the $\mathcal{F}_{\mathsf{BSC}}$-hybrid model. That is, there is a protocol that securely realizes n parallel, independent instances of $\mathcal{F}_{\mathsf{OT}}$ with statistical error 2^{-k}, with $O(n)+poly(k)$ bits of communication (including communication over $\mathcal{F}_{\mathsf{BSC}}$).*

An important corollary of implementing oblivious transfer is that it can be used to implement arbitrary function evaluation, quite efficiently [24]. Thus combined with the main result of [24] we have the following.

Corollary 1. *For any two party function f that can be computed using a boolean circuit of size s, there is a UC-secure protocol for $\mathcal{F}_f$ in the $\mathcal{F}_{\mathsf{BSC}}$-hybrid model, with $O(s)+\mathrm{poly}(k)$ bits of communication.*

4.1 A Constant-Rate String-OT Protocol

We denote by $\mathcal{F}_{\mathsf{string\text{-}OT}[\ell]}$ a string-OT functionality for which the sender's inputs are two strings from $\{0,1\}^{\ell}$. In this section we prove the following.

Lemma 1. *There exists a protocol which securely realizes a single instance of $\mathcal{F}_{\mathsf{string\text{-}OT}[\ell]}$ in the $\mathcal{F}_{\mathsf{BSC}}$-hybrid model, with total communication of $O(\ell)+\mathrm{poly}(k)$ bits.*

This constant-rate protocol for $\mathcal{F}_{\mathsf{string\text{-}OT}}$ in the $\mathcal{F}_{\mathsf{BSC}}$-hybrid model is constructed in three steps. The construction relies on an intermediate functionality, namely $\mathcal{F}_{\mathsf{OLE}}$ (or more precisely, $\widetilde{\mathcal{F}}_{\mathsf{OLE}}$). The $\mathcal{F}_{\mathsf{OLE}}$ functionality (OLE stands for Oblivious Linear function Evaluation) over the field $\mathbb{F}$ evaluates the following function: it takes $p, r \in \mathbb{F}$ from Alice and $q \in \mathbb{F}$ from Bob and outputs $pq+r$ to Bob (and sends an empty output to Alice). $\widetilde{\mathcal{F}}_{\mathsf{OLE}}$ is the error-prone version of $\mathcal{F}_{\mathsf{OLE}}$ as defined in Section 2. For simplicity we omit here the error parameter, which will be chosen as a sufficiently small constant.

Our protocol for $\mathcal{F}_{\mathsf{string\text{-}OT}}$ in the $\mathcal{F}_{\mathsf{BSC}}$-hybrid model is constructed by composing the following protocols:

1. $\mathcal{F}_{\mathsf{string\text{-}OT}}$ protocol in the $\widetilde{\mathcal{F}}_{\mathsf{OLE}}$-hybrid model, using a constant-rate protocol that relies on a constant-rate arithmetic encoding scheme as defined in Section 2.

2. $\widetilde{\mathcal{F}}_{\mathsf{OLE}}$ protocol in the $\widetilde{\mathcal{F}}_{\mathsf{OT}}$-hybrid model, and
3. $\widetilde{\mathcal{F}}_{\mathsf{OT}}$ protocol in the $\mathcal{F}_{\mathsf{BSC}}$-hybrid model.

The second step is obtained by applying Theorem 1 to any OT-based protocol for $\mathcal{F}_{\mathsf{OLE}}$ (e.g., [26,24]), where the latter is invoked with a sufficiently large (but *constant*) security parameter. The third step is obtained by similarly applying Theorem 1 to any protocol for $\mathcal{F}_{\mathsf{OT}}$ from $\mathcal{F}_{\mathsf{BSC}}$ (e.g., [10]). See full version for further details on the last two steps. In the rest of this section we focus on the first step.

Reducing $\mathcal{F}_{\mathsf{string\text{-}OT}}$ *to* $\widetilde{\mathcal{F}}_{\mathsf{OLE}}$. This construction uses an arithmetic encoding scheme as defined in Section 2, with parameters $0 < \rho, \delta, \delta' < 1$ and a constant-size $\mathbb{F}$. We point out that a given arithmetic encoding scheme can be considered to be a scheme with any smaller (positive) value of δ than originally specified. Hence, to suit the requirements of our protocol, we shall assume that $\delta < (1-\delta')\rho/6$. The protocol is in the $\widetilde{\mathcal{F}}^{(\phi)}_{\mathsf{OLE}}$-hybrid where $\phi \leq \delta/2$.

We shall also use a strong randomness extractor Ext in our construction – a family of pairwise independent hash functions suffices.

Suppose Alice's inputs are two strings s_0 and s_1 and Bob's input is a choice bit b. Then the $\mathcal{F}_{\mathsf{string\text{-}OT}}$ protocol for ℓ-bit strings proceeds as follows, where Encode and Encode' map strings in $\mathbb{F}^m$ to strings in $\mathbb{F}^n$ and $d = \lfloor \delta n \rfloor$. The parameter m (and the parameters of the extractor Ext) will be chosen such that for a string x distributed uniformly over any set of size $|\mathbb{F}|^{m(1-\delta')/2}$ or more, then $\mathsf{Ext}(x;h) \in \{0,1\}^\ell$ (where $\ell = \Omega(m)$ is the length of Alice's strings), and $(h, \mathsf{Ext}(x;h))$ (for a randomly chosen h) is almost uniformly random (up to a statistical distance of $2^{-\Omega(k)}$) over its range, even given up to $3d \log |\mathbb{F}| + \ell$ additional bits of information about x.

- Alice's input is (s_0, s_1) where $s_0, s_1 \in \{0,1\}^\ell$, and Bob's input is $b \in \{0,1\}$.
- Alice lets $X_0 = \mathsf{Encode}(x_0)$, and $X_1 = \mathsf{Encode}(x_1)$, where x_0, x_1 are randomly drawn from $\mathbb{F}^m$. She also sets $Z = \mathsf{Encode}'(0^m)$.
- Bob lets $B = \mathsf{Encode}(b^m)$ (the bit b is identified with 0 or 1 in $\mathbb{F}$).
- They invoke n instances of the $\widetilde{\mathcal{F}}^{(\phi)}_{\mathsf{OLE}}$ functionality, as follows. For each $i \in [n]$, Alice inputs $(p_i, r_i) = (X_1^{(i)} - X_0^{(i)}, X_0^{(i)} + Z^{(i)})$ and Bob inputs $q_i = B^{(i)}$ to an instance of $\widetilde{\mathcal{F}}^{(\phi)}_{\mathsf{OLE}}$, and Bob gets the output $y_i = p_i q_i + r_i$. ($X^{(i)}$ stands for the ith bit of the vector X.) The vector $y \in \mathbb{F}^n$ that Bob gets from this is a (possibly) noisy version of $X_0 * (1-B) + X_1 * B + Z$, which in turn is in the support of $\mathsf{Encode}'(x_b)$. Bob sets $x_b = \mathsf{Decode}'(y)$.
- Alice picks two seeds h_0, h_1 for Ext and lets $w_0 = s_0 \oplus \mathsf{Ext}(x_0; h_0)$ and $w_1 = s_1 \oplus \mathsf{Ext}(x_1; h_1)$. (The parameters of Ext are chosen as mentioned above.) She sends (h_0, h_1, w_0, w_1) to Bob.
- Bob obtains $s_b = w_b \oplus \mathsf{Ext}(x_b; h_b)$.

It is easy to see that the above protocol has a constant rate (since $\ell = \Omega(m)$). To prove the UC security of the protocol, we need to consider the case where both parties are honest, as well as when one of the parties is corrupt. In all cases, note that at most $2\phi n < d$ out of n instances of $\widetilde{\mathcal{F}}^{(\phi)}_{\mathsf{OLE}}$ will let the adversary control them, except with negligible probability. In the simulation for all three cases, the simulator starts off by faithfully simulating whether each instance of $\widetilde{\mathcal{F}}^{(\phi)}_{\mathsf{OLE}}$ lets the adversary control it or not. If more than d instances yield to adversarial control, the simulation aborts; as in the real execution, this happens with negligible probability. In the rest of the analysis, we condition on this not happening in the real execution as well as in the simulation.

When neither party is corrupted, security follows from the error-correcting property of Decode'. See full version for details.

Security When Neither Party Is Corrupt. In the simulation (i.e., ideal execution of $\mathcal{F}_{\mathsf{string\text{-}OT}}$), Bob's output is always s_b when Alice's inputs are (s_0, s_1) and Bob's input is b. In the real execution of the protocol (conditioned on less than d instances of $\widetilde{\mathcal{F}}^{(\phi)}_{\mathsf{OLE}}$ being under adversarial control) the vector y received by Bob has a Hamming distance of less than d from a vector in the support of $\mathsf{Encode}'(x_b)$. So, by the error-correcting guarantee of Decode' Bob recovers x_b, and hence outputs s_b correctly.

Security against Corrupt Alice. Here the simulation proceeds in two steps. First, Alice's view is completely straight-line simulated (if well-formed messages are not received from Alice in any round, then Bob aborting the protocol can be simulated). Next Bob's view is sampled for the two cases $b = 0$ and $b = 1$, conditioned on Alice's view, from which Bob's outputs for each case, denoted s_0 and s_1, respectively, are obtained. To complete the simulation the simulator sends (s_0, s_1) as the input to the ideal $\mathcal{F}_{\mathsf{string\text{-}OT}}$ functionality. Details of the two steps follow.

Conditioned on $\widetilde{\mathcal{F}}^{(\phi)}_{\mathsf{OLE}}$ yielding to the adversary at most d times, Alice sees at most d entries of the encoding of B and this can be perfectly simulated since they are independent of Bob's input. So first the simulator will sample the (at most) d entries for B and hands them over to the Alice as the message from the instances of $\widetilde{\mathcal{F}}^{(\phi)}_{\mathsf{OLE}}$ controlled by her. Then it receives from Alice the output for Bob from these instances. Further, the simulator receives all but d entries of (X_0, X_1, Z) from Alice as inputs to the instances of $\widetilde{\mathcal{F}}^{(\phi)}_{\mathsf{OLE}}$ not controlled by her. In the next round, the simulator receives (h_0, h_1, w_0, w_1) from Alice. This completes the first part of the simulation.

For the next part, the simulator picks $B_0 = \mathsf{Encode}(0^m)$ and $B_1 = \mathsf{Encode}(1^m)$ randomly, conditioned to match the d coordinates of B that were already simulated. Then, it computes s_0 as what Bob would compute in the protocol if it uses $B = B_0$ and receives the messages implied by what Alice sent to the simulator in the first part. Similarly, it computes s_1 if Bob used $B = B_1$. Then the simulator will send (s_0, s_1) to the functionality.

Given our conditioning the real and simulated executions on there being no more than d instances of $\widetilde{\mathcal{F}}^{(\phi)}_{\mathsf{OLE}}$ under adversarial control, this is a perfect simulation. Thus over all, this gives a statistically good simulation.

Security against Corrupt Bob. If Bob is corrupt, then he may not input a valid B in the range of $\mathsf{Encode}(0^m)$ or $\mathsf{Encode}(1^m)$. Nevertheless, we shall see that by using appropriate encodings and the extractor, there is a string s_b such that Bob learns a negligible amount of information about s_{1-b}.

Note that what Bob learns by Step 2 of the protocol is given by a system of n linear equations (defined by his input B to $\widetilde{\mathcal{F}}_{\mathsf{OLE}}$) over x_0, x_1 and the random elements used by Alice in forming the encodings X_0, X_1 and the entries of Z, and the values of X_0, X_1 and Z at no more than d randomly chosen entries. Alice's secrets at this point are two vectors x_0, x_1 of length only m, so it is non-trivial to ensure that the information that Bob learns (which is more than n field elements, and typically $n > 2m$) does not contain both x_0 and x_1. This is ensured by the blinding: intuitively, Z encodes at least $t' := n - m(1+\delta')$ random field elements, and so effectively Bob learns at most as much information about (x_0, x_1) as from $n - t' = m(1+\delta') < 2m$ linear equations.

More formally, let U denote the output vector in $\mathbb{F}^n$ obtained by Bob from $\widetilde{\mathcal{F}}_{\mathsf{OLE}}$, and let $\widetilde{U}$ denote the (at most $3d$) field elements learned by Bob from corrupted instances of $\widetilde{\mathcal{F}}_{\mathsf{OLE}}$. It can be shown (see full version) that for any possible value of Bob's $\widetilde{\mathcal{F}}_{\mathsf{OLE}}$ input B there is $c \in \{0, 1\}$ such that the distribution of x_c conditioned on B and any possible U is uniform in an affine space whose dimension is at least $\frac{m(1-\delta')}{2}$.

Note that c as above can be efficiently computed by solving for x_0 and x_1 from the equations defined by B and Encode', and checking the dimension of their solution spaces. The simulator first perfectly simulates $B, \widetilde{U}$, then uses B to compute c, and then sends $b = 1 - c$ to the functionality to obtain s_b. To complete the simulation, the simulator sets $s_c = 0$ and generates Alice's messages (to both Bob and $\widetilde{\mathcal{F}}_{\mathsf{OLE}}$) at random conditioned on the values of $B, \widetilde{U}$ that were already simulated. The correctness of the simulator follows from the fact that in the real protocol, conditioned on the choice of B, the string $\mathsf{Ext}(x_c; h_c)$ masking s_c is almost uniformly random even when further conditioned on the remaining view of Bob. This follows from the fact that $\widetilde{U}$ and $(h_b, \mathsf{Ext}(x_b; h_b))$ leak at most $3d \log |\mathbb{F}|$ and ℓ additional bits of information, respectively, which our choice of parameters for Ext tolerates. See full version for further details.

4.2 Error-Tolerant Protocol for $\mathcal{F}_{\mathsf{OT}}$ over $\mathcal{F}_{\mathsf{BSC}}$

Error-tolerance. We say that a protocol π is an *error-tolerant* protocol for $\mathcal{F}$ in the $\mathcal{G}$-hybrid model if it is secure against adaptive passive corruption, and in addition tolerates active corruption of a small constant fraction of $\mathcal{G}$ instances that it invokes. More formally, we can define a modified functionality $\mathcal{G}'$, which behaves exactly as $\mathcal{G}$ until a new command $\mathsf{corrupt}$ is received as input from the adversary; if this command is received, then this instance of $\mathcal{G}$ will yield to adversarial control – i.e., send its current view to the adversary, forward immediately any subsequent message that it receives, and send messages to other parties in the protocol as instructed by the adversary. π is called a ϵ_0-error-tolerant protocol for $\mathcal{F}$ in the $\mathcal{G}$-hybrid model if π is a secure protocol for $\mathcal{F}$ in the $\mathcal{G}'$-hybrid model against adaptive passive corruption, against the class of adversaries who send out the corrupt command to at most $\epsilon_0 T$ of $\mathcal{G}'$ instances, where T is (an upperbound on) the total number of $\mathcal{G}$ instances invoked by π. We will call π simply an error-tolerant protocol if it is ϵ_0-error-tolerant for any constant $\epsilon_0 > 0$.

As described above in the inner protocol in our construction, we will require a constant-rate error-tolerant protocol for $\mathcal{F}_{\mathsf{OT}}$ in the $\mathcal{F}_{\mathsf{BSC}}$-hybrid model.

We observe that such a protocol is implicit in a result in [20]. They present a constant-rate OT protocol in the $\mathcal{F}_{\mathsf{BSC}}$-hybrid model which is secure against adaptive passive adversaries. This construction starts with a simple passive-secure constant-rate protocol Φ for $\mathcal{F}_{\mathsf{OT}}$ in the $\mathcal{F}_{\mathsf{BSC}}$-hybrid model, with a small but constant probability of error, and then uses a constant-rate *combiner* to reduce the error to negligible. This combiner uses each candidate $\mathcal{F}_{\mathsf{OT}}$ instance once, and (passive-securely) realizes a constant fraction of $\mathcal{F}_{\mathsf{OT}}$ instances. As mentioned in [20], the "error-tolerant" version of their combiner allows a small fraction of the candidate $\mathcal{F}_{\mathsf{OT}}$ instances to be actively and adaptively corrupted, though requires the parties themselves to follow the combiner's protocol honestly. The combiner corresponds to a constant-rate protocol for $\mathcal{F}_{\mathsf{OT}}$ in the $\mathcal{F}_{\mathsf{OT}}$-hybrid model with error tolerance as we have defined above. By composing this protocol with Φ, we get a constant-rate protocol for $\mathcal{F}_{\mathsf{OT}}$ in the $\mathcal{F}_{\mathsf{BSC}}$-hybrid model, with the property that if a small constant fraction of the instances of $\mathcal{F}_{\mathsf{BSC}}$ are corrupted (resulting in the corruption of a small fraction of $\mathcal{F}_{\mathsf{OT}}$ instances used by the combiner protocol), security remains intact.

Lemma 2. *[20] There is a constant-rate, error-tolerant protocol for $\mathcal{F}_{\mathsf{OT}}$ in the $\mathcal{F}_{\mathsf{BSC}}$-hybrid model.*

4.3 An Extension to the IPS Compiler

IPS compiler requires a semi-honest inner protocol over $\mathcal{F}_{\mathsf{OT}}$. We need to extend this compiler to work with inner protocols over $\mathcal{F}_{\mathsf{BSC}}$. The IPS compiler depends on being able to monitor the use of $\mathcal{F}_{\mathsf{OT}}$ channels with a good probability of catching errors; however, one cannot monitor the $\mathcal{F}_{\mathsf{BSC}}$ functionality at the same level. Hence we shall depend on the slightly stronger error-tolerance guarantee of the inner protocol. Here we shall limit ourselves to the 2-party setting (since we are interested in a 2-party functionality, namely $\mathcal{F}_{\mathsf{OT}}$).

Below we state the extension of the IPS compiler (with the new elements underlined). See full version for a proof.

Lemma 3. *Suppose Π is a protocol among $n = \Theta(k)$ servers and 2 clients, for a 2-party functionality $\mathcal{F}$ between the clients, with UC-security against adaptive, active corruption of $\Omega(n)$ servers and adaptive, active corruption of (any number of) clients. Suppose $\rho^{\mathcal{F}_{\mathsf{BSC}}}$ is a 2-party protocol in the $\underline{\mathcal{F}_{\mathsf{BSC}}\text{-hybrid model}}$, that realizes the functionality of each server in the protocol Π, with $\underline{\text{error tolerance}}$. Then there is a 2-party protocol (compiled protocol) for the functionality $\mathcal{F}$ in the $\underline{(\mathcal{F}_{\mathsf{BSC}}, \mathcal{F}_{\mathsf{string\text{-}OT}})\text{-hybrid model}}$, with UC-security against adaptive, active* *adversaries. Further, if the (insecure) protocol obtained by directly implementing each server of Π using $\rho^{\mathcal{F}_{\text{BSC}}}$ has constant rate, then the compiled protocol has constant rate too.*

Putting Things Together. The final protocol is obtained from Lemma 3 by using the following outer and inner protocols. The outer protocol is the one used in Section 5.1 of [24] (based on [13,8]) applied to the functionality which realizes n instances of OT. The inner protocol is the standard OT-based implementation of the GMW protocol in the semi-honest OT-hybrid model [18], except that the OT instances consumed by this protocol are implemented using the error-tolerant protocol of Lemma 2. The watchlists are implemented using the protocol of Lemma 1.

References

1. Ahlswede, R., Csiszar, I.: On Oblivious Transfer Capacity. In: ISIT 2007, pp. 2061–2064 (2007)
2. Bennett, C.H., Brassard, G., Crépeau, C., Maurer, U.: Generalized privacy amplification. IEEE Transactions on Information Theory 41, 1915–1923 (1995)
3. Bennett, C.H., Brassard, G., Robert, J.-M.: Privacy Amplification by Public Discussion. SIAM J. Comput. 17(2), 210–229 (1988)
4. Brassard, G., Crépeau, C., Robert, J.-M.: All-or-nothing disclosure of secrets. In: Odlyzko, A.M. (ed.) CRYPTO 1986. LNCS, vol. 263, pp. 234–238. Springer, Heidelberg (1987)
5. Canetti, R.: Security and composition of multiparty cryptographic protocols. Journal of Cryptology 13(1), 143–202 (2000)
6. Canetti, R.: Universally Composable Security: A New Paradigm for Cryptographic Protocols. In: FOCS 2001, pp. 136–145 (2001)
7. Cascudo, I., Cramer, R., Xing, C.: The Torsion-Limit for Algebraic Function Fields and Its Application to Arithmetic Secret Sharing. In: Crypto 2011 (2011)
8. Chen, H., Cramer, R.: Algebraic geometric secret sharing schemes and secure multiparty computations over small fields. In: Dwork, C. (ed.) CRYPTO 2006. LNCS, vol. 4117, pp. 521–536. Springer, Heidelberg (2006)
9. Crépeau, C.: Efficient cryptographic protocols based on noisy channels. In: EUROCRYPT 1997, pp. 306–317 (1997)
10. Crépeau, C., Kilian, J.: Achieving oblivious transfer using weakened security assumptions. In: FOCS 1988, pp. 42–52 (1988)
11. Crépeau, C., Morozov, K., Wolf, S.: Efficient Unconditional Oblivious Transfer from Almost Any Noisy Channel. In: Blundo, C., Cimato, S. (eds.) SCN 2004. LNCS, vol. 3352, pp. 47–59. Springer, Heidelberg (2005)
12. Damgård, I., Fehr, S., Morozov, K., Salvail, L.: Unfair Noisy Channels and Oblivious Transfer. In: Naor, M. (ed.) TCC 2004. LNCS, vol. 2951, pp. 355–373. Springer, Heidelberg (2004)
13. Damgård, I., Ishai, Y.: Scalable Secure Multiparty Computation. In: Dwork, C. (ed.) CRYPTO 2006. LNCS, vol. 4117, pp. 501–520. Springer, Heidelberg (2006)
14. Damgård, I., Kilian, J., Salvail, L.: On the (Im)possibility of Basing Oblivious Transfer and Bit Commitment on Weakened Security Assumptions. In: Stern, J. (ed.) EUROCRYPT 1999. LNCS, vol. 1592, pp. 56–73. Springer, Heidelberg (1999)
15. Even, S., Goldreich, O., Lempel, A.: A randomized protocol for signing contracts. Communications of the ACM 28(6), 637–647 (1985)
16. Garcia, A., Stichtenoth, H.: On the asymptotic behavior of some towers of function fields over finite fields. Journal of Number Theory 61(2), 248–273 (1996)

17. Gemmell, P., Sudan, M.: Highly Resilient Correctors for Polynomials. Information Processing Letters 43(4), 169–174 (1992)
18. Goldreich, O.: Foundations of Cryptography, vol. 2. Cambridge University Press, Cambridge (2004)
19. Goldreich, O., Micali, S., Wigderson, A.: How to play any mental game. In: STOC 1987, pp. 218–229 (1987)
20. Harnik, D., Ishai, Y., Kushilevitz, E., Nielsen, J.B.: OT-Combiners via Secure Computation. In: Canetti, R. (ed.) TCC 2008. LNCS, vol. 4948, pp. 393–411. Springer, Heidelberg (2008)
21. Harnik, D., Kilian, J., Naor, M., Reingold, O., Rosen, A.: On tolerant combiners for oblivious transfer and other primitives. In: EUROCRYPT 2005, pp. 96–113 (2005)
22. Imai, H., Morozov, K., Nascimento, A.: Efficient Oblivious Transfer Protocols Achieving a Non-Zero Rate from Any Non-Trivial Noisy Correlation. In: Desmedt, Y. (ed.) ICITS 2007. LNCS, vol. 4883, pp. 183–194. Springer, Heidelberg (2009)
23. Ishai, Y., Kushilevitz, E., Ostrovsky, R., Sahai, A.: Zero-knowledge from secure multiparty computation. In: STOC 2007, pp. 21–30 (2007)
24. Ishai, Y., Prabhakaran, M., Sahai, A.: Founding Cryptography on Oblivious Transfer - Efficiently. In: Wagner, D. (ed.) CRYPTO 2008. LNCS, vol. 5157, pp. 572–591. Springer, Heidelberg (2008)
25. Ishai, Y., Kushilevitz, E., Ostrovsky, R., Sahai, A.: Extracting Correlations. In: FOCS 2009, pp. 261–270 (2009)
26. Kilian, J.: Founding cryptography on oblivious transfer. In: STOC 1988, pp. 20–31 (1988)
27. Kilian, J.: More general completeness theorems for secure two-party computation. In: STOC 2000, pp. 316–324 (2000)
28. Maurer, U.: Perfect Cryptographic Security from Partially Independent Channels. In: STOC 1991, pp. 561–571 (1991)
29. Maurer, U.M., Pietrzak, K., Renner, R.: Indistinguishability Amplification. In: Menezes, A. (ed.) CRYPTO 2007. LNCS, vol. 4622, pp. 130–149. Springer, Heidelberg (2007)
30. Nascimento, A., Winter, A.: On the Oblivious Transfer Capacity of Noisy Correlations. In: ISIT 2006, pp. 1871–1875 (2006)
31. Przydatek, B., Wullschleger, J.: Error-Tolerant Combiners for Oblivious Primitives. In: Aceto, L., Damgård, I., Goldberg, L.A., Halldórsson, M.M., Ingólfsdóttir, A., Walukiewicz, I. (eds.) ICALP 2008, Part II. LNCS, vol. 5126, pp. 461–472. Springer, Heidelberg (2008)
32. Rabin, M.O.: How to exchange secrets by oblivious transfer. In: TR 1981, Harvard (1981)
33. Shannon, C.E.: A mathematical theory of communication. Bell System Technical Journal 27, 379–423, 623-656 (1948)
34. Wiesner, S.: Conjugate coding. SIGACT News 15(1), 78–88 (1983)
35. Winkler, S., Wullschleger, J.: On the Efficiency of Classical and Quantum Oblivious Transfer Reductions. In: Rabin, T. (ed.) CRYPTO 2010. LNCS, vol. 6223, pp. 707–723. Springer, Heidelberg (2010)
36. Winter, A., Nascimento, A.C.A., Imai, H.: Commitment Capacity of Discrete Memoryless Channels. In: IMA Int. Conf. pp. 35–51 (2003)
37. Wullschleger, J.: Oblivious Transfer from Weak Noisy Channels. In: Reingold, O. (ed.) TCC 2009. LNCS, vol. 5444, pp. 332–349. Springer, Heidelberg (2009)
38. Wyner, A.D.: The wire-tap channel. Bell Cyst. Tech. J. 54, 1355–1387 (1975)
39. Yao, A.C.: How to generate and exchange secrets. In: FOCS 1986, pp. 162–167 (1986)

A Arithmetic Encoding from MPC-Friendly Codes

In this section, we briefly sketch how an implementation of our notion of an arithmetic encoding scheme $(\mathsf{Encode}, \mathsf{Encode}', \mathsf{Decode}')$ (as defined in Section 2) follows from the literature.

Below we recall (verbatim) the notion of MPC-friendly codes from [25], which *almost* have all the properties we need. (The parameter k in this section should not be confused with the use of k as a security parameter in the rest of the paper.)

Claim ([25], implicit in [8]). There exists a finite field $\mathbb{F}$ and an efficiently constructible family of linear error-correcting codes $C_k : \mathbb{F}^k \to \mathbb{F}^{n_k}$ with the following properties: (1) $n_k = O(k)$; (2) The dual distance of C_k is $\delta_k = \Omega(k)$; (3) The linear code C'_k spanned by all entrywise-products of pairs of codewords in C_k supports efficient decoding of up to $\mu_k = \Omega(k)$ errors. (The entrywise product of $(c_1, \ldots, c_n)$ and $(c'_1, \ldots, c'_n)$ is $(c_1c'_1, \ldots, c_nc'_n)$.)

A family of codes C_k as above can be obtained from the construction of Garcia and Stichtenoth [16]. An efficient decoder for C'_k can be obtained via the Gemmel-Sudan generalization of the Welch-Berlekamp decoder for Reed-Solomon codes [17].

The one stronger property that we need here (beyond what was needed in [25]), in order to guarantee that Encode' generates the amount of entropy that we need, is a "near-MDS" property of the code C'_k. Specifically, it suffices to ensure that, for a small enough constant $\delta_0 > 0$, we have:

$$(n_k - \dim(C'_k) - \Delta_k) < \delta_0 k.$$

Indeed, this follows immediately from the construction of [16], which in fact allows us to obtain $\delta_0 = o(1)$.

The code C_k corresponds to the Encode algorithm, and the code C'_k corresponds to the Encode' algorithm. In both cases, not just the message, but additional randomness would also be encoded (in the standard method for secret sharing). More specifically, it will suffice to have $\mathsf{Encode}(x)$ (respectively, $\mathsf{Encode}'(x)$) sample a random codeword y in the range of C_k (respectively, C'_k) such that y has x as a prefix, and set the encoding to be y modified by dropping this prefix. This can be done efficiently since the codes are linear (by solving a system of linear equations over x and randomly chosen field elements). In particular, if C_k and C'_k are systematic codes, then $\mathsf{Encode}(x)$ (respectively, $\mathsf{Encode}'(x)$) will pick a random vector r of the appropriate dimension, let $y = C_k(x||r)$ (respectively, $y = C'_k(x||r)$), and output the last $n - m$ entries of y.

It is instructive to note that Reed-Solomon codes satisfy all the properties we need, except that Reed-Solomon codes would require the size of the field $\mathbb{F}$ to grow linearly with k. One could use Reed-Solomon codes in our constructions (instead of algebraic geometric codes) at the cost of a polylogarithmic deterioration of the parameters.

The Torsion-Limit for Algebraic Function Fields and Its Application to Arithmetic Secret Sharing

Ignacio Cascudo[1], Ronald Cramer[2], and Chaoping Xing[3]

[1] CWI Amsterdam, The Netherlands
i.cascudo@cwi.nl
[2] CWI Amsterdam & Mathematical Institute, Leiden University, The Netherlands
cramer@cwi.nl, cramer@math.leidenuniv.nl
[3] Division of Mathematical Sciences, Nanyang Technological University, Singapore
xingcp@ntu.edu.sg

Abstract. An $(n, t, d, n-t)$-arithmetic secret sharing scheme (with uniformity) for $\mathbb{F}_q^k$ over $\mathbb{F}_q$ is an $\mathbb{F}_q$-linear secret sharing scheme where the secret is selected from $\mathbb{F}_q^k$ and each of the n shares is an element of $\mathbb{F}_q$. Moreover, there is t-privacy (in addition, any t shares are uniformly random in $\mathbb{F}_q^t$) and, if one considers the d-fold "component-wise" product of any d sharings, then the d-fold component-wise product of the d respective secrets is $(n-t)$-wise uniquely determined by it. Such schemes are a fundamental primitive in information-theoretically secure multi-party computation. Perhaps counter-intuitively, secure *multi-party* computation is a very powerful primitive for *communication-efficient two-party* cryptography, as shown recently in a series of surprising results from 2007 on. Moreover, the existence of *asymptotically good* arithmetic secret sharing schemes plays a crucial role in their communication-efficiency: for each $d \geq 2$, if $A(q) > 2d$, where $A(q)$ is Ihara's constant, then there exists an infinite family of such schemes over $\mathbb{F}_q$ such that n is unbounded, $k = \Omega(n)$ and $t = \Omega(n)$, as follows from a result at CRYPTO'06. Our main contribution is a novel paradigm for constructing asymptotically good arithmetic secret sharing schemes from towers of algebraic function fields. It is based on a new limit that, for a tower with a given Ihara limit and given positive integer ℓ, gives information on the cardinality of the ℓ-torsion sub-groups of the associated degree-zero divisor class groups and that we believe is of independent interest. As an application of the bounds we obtain, we relax the condition $A(q) > 2d$ from the CRYPTO'06 result substantially in terms of our torsion-limit. As a consequence, this result now holds over *nearly all finite fields* $\mathbb{F}_q$. For example, if $d = 2$, it is sufficient that $q = 8, 9$ or $q \geq 16$.

1 Introduction

An $(n, t, d, n-t)$-arithmetic secret sharing scheme (with uniformity) for $\mathbb{F}_q^k$ over $\mathbb{F}_q$ is an $\mathbb{F}_q$-linear secret sharing scheme where $k, n, t \geq 1, d \geq 2$, the secret is selected from $\mathbb{F}_q^k$ and each of the n shares is an element of $\mathbb{F}_q$. Moreover, there is t-privacy (in addition, any t shares are uniformly random in $\mathbb{F}_q^t$) and, if one

P. Rogaway (Ed.): CRYPTO 2011, LNCS 6841, pp. 685–705, 2011.

considers the d-fold "component-wise" product of any d sharings, then the d-fold component-wise product of the d respective secrets is $(n-t)$-wise uniquely determined by it.

Such schemes, first based on Shamir's scheme and later abstracted and generalized, are fundamental to (information-theoretically) secure multi-party computation [2,6,15,10]. Please refer to Section 5 for details about two main, well-known applications. Note that both concern protocols for "secure multiplication" and that the properties of arithmetic secret sharing are used somewhat differently from what their definition perhaps seems to suggest on first encounter. Secure multiplication is a fundamental primitive in its own right, as secure multi-party computation is often based on combinations of secure addition and secure multiplication, the latter typically being demanding and involved while the former is typically much more straightforward. Arithmetic secret sharing allows efficient recovery of the secret in the presence of faulty shares, by a generalization of a result from [12] (see Section 5) and also gives rise to verifiable secret sharing [10].

A series of surprising results, concerning zero-knowledge for circuit satisfiability ("MPC in the Head"), two-party secure computation, OT-combiners, correlation extractors, and OT from noisy channels [23,24,18,22,13,21], has caused nothing less than a paradigm shift that perhaps appears even as counter-intuitive: secure *multi-party* computation is a very powerful abstract primitive for *communication-efficient two-party cryptography*. All these results use arithmetic secret sharing schemes, typically with $d = 2$ (see also [11] for an application with $d > 2$). Note that both [22,21] are information-theoretic in nature, require the uniformity property, and also use the error correction procedure.

Also surprisingly, the existence of *asymptotically good* arithmetic secret sharing schemes plays a crucial role in the communication-efficiency of these recent, fundamental results on two-party cryptography: as follows from [7], for each $d \geq 2$, if $A(q) > 2d$ (where $A(q)$ is Ihara's constant from algebraic geometry, see Section 2) then there exists an infinite family of such schemes over $\mathbb{F}_q$ such that n is unbounded, $k = \Omega(n)$ and $t = \Omega(n)$. Using these schemes (for $d = 2$), "constant-rate communication" has been achieved in those results, due to the removal of logarithmic terms caused by approaches using (appropriate modes of) Shamir's scheme [32]. Note that the original motivation of [7] was to give a communication-efficient asymptotic version of the "fundamental theorem of perfect information-theoretically secure multi-party computation" of [2,6], by combining the asymptotically good scheme from [7] with the results from [10]. In particular, the field $\mathbb{F}_q$ of computation can be fixed, the number n of players is unbounded, and a malicious t-adversary is tolerated with $t = \Omega(n)$.

In [9], an extension of [7] is given where $\mathbb{F}_q^k$ is replaced by $\mathbb{F}_{q^k}$. It follows by the results of [4] that asymptotically good $(n, t, d, n-t)$-arithmetic secret sharing schemes for $\mathbb{F}_q^k$ over $\mathbb{F}_q$ exist over *any finite field*, with the caveat that the uniformity property does *not* hold after application of the dedicated descent technique to the result from [7]. Therefore, $A(q) > 2d$ is the weakest known condition under which asymptotically good $(n, t, d, n-t)$-arithmetic secret sharing schemes *with uniformity* (for $\mathbb{F}_q^k$ over $\mathbb{F}_q$) are known to exist. All these asymptotic results rely crucially on good towers of algebraic function fields, and currently *do not seem to admit more elementary proofs avoiding this.*[1]

The Drinfeld-Vlăduţ bound states that $A(q) \leq \sqrt{q} - 1$. By Ihara [20], $A(q) = \sqrt{q} - 1$ if q is a square. By Serre's Theorem [31], $A(q) \geq c \cdot \log q$ for some absolute real constant $c > 0$ (for which the current best lower bound is about $\frac{1}{96}$). If we take $d = 2$, for example, then the condition $A(q) > 4$ is satisfied if $q \geq 49$ is a square (alternatively, q is a large enough cubic [3,1]) or if q is *very* large. Thus, existence is unresolved for *many* values of q.

Our main contribution is a novel paradigm for constructing asymptotically good arithmetic secret sharing schemes from towers of algebraic function fields. It is based on a new limit that, for a tower with a given Ihara limit and given positive integer ℓ, gives information on the cardinality of the ℓ-torsion sub-groups of the associated degree-zero divisor class groups. Our "torsion limit," which we believe is of independent interest, can in general be upper bounded using Weil's classical theorem on torsion in Abelian varieties (and in many cases using the Weil-pairing). However, the resulting bound is far too pessimistic, as we present a tower for which our torsion limit is *considerably smaller*, yet it attains the Drinfeld-Vlăduţ bound.

By means of this paradigm, we *weaken* the condition $A(q) > 2d$ to $A(q) > 1 + J_d(q, A(q))$, where $J_d(q, A(q))$ upper-bounds a "d-torsion" rate (based on the logarithm of the cardinality of the d-torsion, divided by the genus) taken over all infinite families of curves defined over $\mathbb{F}_q$ such that the genus tends to infinity and such that the Drinfeld–Vlăduţ bound is attained.

More precisely, the bounds we obtain on this torsion limit allow us to show the existence of the claimed arithmetic secret sharing schemes by solving an appropriate system of "Riemann-Roch type of equations" over an algebraic function field (in fact, one such system for each algebraic function field in a given infinite family). Each such equation is of the form $\ell(\lambda_i X + Y_i) = 0$, where X is the divisor to be solved for, $\lambda_i \in \{-1, d\}$, Y_i is a given divisor, and $\ell(\cdot)$ denotes Riemann-Roch dimension. The solution X of such a system defines a certain AG-code with properties as claimed. The necessity of studying d-torsion arises from the fact that $\lambda_i = d$ does occur.

Concretely, for $d = 2$ we prove that for *all* finite fields $\mathbb{F}_q$ with $q \geq 16$ (as well as for $q = 8, 9$), asymptotically good $(n, t, d, n - t)$-arithmetic secret sharing schemes with uniformity (for $\mathbb{F}_q^k$ over $\mathbb{F}_q$) exist. This settles existence in the affirmative for *nearly all finite fields*. As an application, the results from [22,21] can in principle be based on smaller finite fields.

Finally, using our paradigm we also improve the explicit lower bounds on the asymptotic optimal normalized corruption tolerance $\widehat{\tau}(q)$ from [4] for all q with $q \leq 81$ and q square, as well as for all q with $q \leq 9$. For instance, $\widehat{\tau}(64) \geq 0.52$,

[1] The existence of asymptotically good $(n, t, 2, n)$-arithmetic secret sharing schemes for $\mathbb{F}_q$ over $\mathbb{F}_q$ (so $k = 1$!) can be shown by elementary means [8], with asymptotically good self-dual error correcting codes as a special case. But this is a *much weaker* class that neither supports the mentioned applications in two-party cryptography, nor the asymptotic version of the "fundamental MPC theorem" given in [7].

whereas previously the best known lower bound was 0.42. As an application, the asymptotic version of the Fundamental Theorem in principle tolerates a stronger adversary (by a constant factor). Our results also have a bearing on the study of the asymptotic complexity of multiplication in finite extension fields of $\mathbb{F}_q$, but we do not elaborate on this here.

This paper is organized as follows. Our main contributions are captured in Definition 2 (the torsion-limit), Theorem 1 (bounds for this limit), Theorem 6 (sufficient conditions for Riemann-Roch system solvability) and Main Theorems 1 and 2 (claimed arithmetic secret sharing schemes). After giving some preliminaries in Section 2, we introduce our torsion limit in Section 3 and show our bounds. In Section 4 we introduce Riemann-Roch systems of equations and show how these may be solved using the bounds from Section 3. In Section 5 we introduce an elementary framework in which our quantitative results can be conveniently stated and apply our bounds to obtain the claimed arithmetic secret sharing schemes. Efficiency issues are also discussed there.

2 Preliminaries

For a prime power q, let $\mathbb{F}_q$ be a finite field of q elements. An *algebraic function field* over $\mathbb{F}_q$ in one variable is a field extension $F \supset \mathbb{F}_q$ such that F is a finite algebraic extension of $\mathbb{F}_q(x)$ for some $x \in F$ that is transcendental over $\mathbb{F}_q$. $F/\mathbb{F}_q$ denotes a function field with full constant field $\mathbb{F}_q$; $g(F)$ and $N(F)$ are the genus and the number of rational places of F respectively. $\mathbb{P}(F)$ denotes the set of places of F, which is an infinite set, and $\mathbb{P}^{(k)}(F)$ is the (finite) subset consisting of the places of degree k of F. $N_i(F)$ is the number of rational places of the constant field extension $\mathbb{F}_{q^i}F$, i.e., $N_i(F) = N(\mathbb{F}_{q^i}F)$ (note that $N(F) = N_1(F)$); $\text{Div}(F)$ is the divisor group of F and $\text{Div}^0(F)$ its subset consisting of the divisors of degree 0; $\text{Prin}(F)$ is the principal divisor group of F; $\text{Cl}(F)$ is the divisor class group $\text{Div}(F)/\text{Prin}(F)$ of F and $\text{Cl}_0(F) = \mathcal{J}_F$ is the zero divisor class group $\text{Div}^0(F)/\text{Prin}(F)$ of F, which is a finite group of cardinality $h(F) = |\text{Cl}_0(F)|$ (the class number); $\mathcal{A}_r(F)$ is the set of effective divisors of degree $r \geq 0$, which is a finite set, and $A_r(F)$ denotes its cardinality; $\text{Cl}_r(F)$ is the set of $\{[D] : \deg(D) = r\}$, where $[D]$ stands for the divisor class containing D and $\text{Cl}_r^+(F)$ is the subset of $\text{Cl}_r(F)$ of classes which contain an effective divisor, i.e., $\{[D] : \deg(D) = r, D \geq 0\}$. In case there is no confusion, we omit F in some of the above notations. For instance, $A_r(F)$ is denoted by A_r if it is clear in the context. For a divisor G of F, $\mathcal{L}(G) := \{f \in F^* : \text{div}(f) + G \geq 0\} \cup \{0\}$ is its Riemann-Roch space. It is a finite dimensional space over $\mathbb{F}_q$. Its dimension $\ell(G)$ satisfies $\ell(G) = \deg(G) + 1 - g(F) + \ell(K - G)$ where K is a canonical divisor of degree $2g(F)-2$ (Riemann-Roch theorem). Therefore, $\ell(G) \geq \deg(G)+1-g(F)$, with equality if $\deg(G) \geq 2g(G) - 1$. The zeta function of F is defined by the following power series

$$Z_F(T) := \text{Exp}\left(\sum_{i=1}^{\infty} \frac{N_i(F)}{i} T^i\right) = \sum_{i=0}^{\infty} A_i(F) T^i.$$

In fact (Weil), $Z_F(T) = \frac{L_F(T)}{(1-T)(1-qT)}$ where $L_F(T)$ is a polynomial of degree $2g(F)$ in $\mathbb{Z}[T]$, called *L-polynomial* of F. Furthermore, $L_F(0) = 1$. If we factorize $L_F(T)$ into a linear product $\prod_{i=1}^{2g(F)}(w_iT - 1)$ in $\mathbb{C}[T]$, then Weil showed that $|w_i| = \sqrt{q}$ for all $1 \leq i \leq 2g(F)$. The Functional Equation of the L-polynomial states $L_F(T) = q^{g(F)}T^{2g(F)}L_F(1/qT)$. Finally we know $L_F(1) = h(F)$. All these facts about the L-polynomial can be found in [34]. Again, when F is clear from the context we write $Z(T)$ and $L(T)$ for its zeta function and L-polynomial, respectively. From the definition of zeta function, one obtains $N_m(F) = q^m + 1 - \sum_{i=1}^{2g(F)} w_i^m$ for all $m \geq 1$. This gives the Hasse-Weil bound $N(F) = N_1(F) \leq q + 1 + 2g(F)\sqrt{q}$. Define $N_q(g) = \max_F N(F)$, where F ranges over all function fields of genus g over $\mathbb{F}_q$, and define $A(q) := \limsup_{g\to\infty} N_q(g)/g$, Ihara's constant. Vladuţ and Drinfeld showed $A(q) \leq \sqrt{q}-1$. Ihara [20] showed that $A(q) \geq \sqrt{q}-1$ for any square power q. Hence, $A(q) = \sqrt{q} - 1$ for all square powers. Zink [42], and Bezerra et al. [3] (see also Bassa et al. [1]) showed that $A(q^3) \geq \frac{2(q^2-1)}{q+2}$. Serre showed that there is a absolute constant $c > 0$ such that $A(q) \geq c \cdot \log(q)$ for all prime powers q. In [41], Xing and Yeo showed that $A(2) \geq 0.258$. Very recently, Duursma and Mak have reported in [14] the stronger bound $A(2) \geq 0.316$. For a family $\mathcal{F} = \{F/\mathbb{F}_q\}$ of function fields with $g(F) \to \infty$ such that $\lim_{g(F)\to\infty} N(F)/g(F)$ exists, one can define this limit to be the *Ihara* limit, denoted by $A(\mathcal{F})$. It is clear that there exists a family $\mathcal{E} = \{E/\mathbb{F}_q\}$ of function fields such that $g(E) \to \infty$ and the Ihara limit $A(\mathcal{E})$ is equal to $A(q)$.

Remark 1. *In general, we can define the Ihara limit for any family $\mathcal{F} = \{F/\mathbb{F}_q\}$ of function fields with $g(F) \to \infty$ by $\limsup_{g(F)\to\infty} N(F)/g(F)$. However, for convenience of this paper, we define the Ihara limit only for those families $\{E/\mathbb{F}_q\}$ whose limit $\lim_{g(E)\to\infty} N(E)/g(E)$ exists.*

3 Torsion Point Limits

For the applications in this paper, we are interested in considering, in addition to the Ihara limit of a family of function fields, a limit for the number of torsion points of the zero divisor class groups of these function fields.

Let $F/\mathbb{F}_q$ be a function field. For a positive integer $r > 1$, we denote by $\mathcal{J}_F[r]$ the r-torsion point group in $\mathcal{J}_F$, i.e., $\mathcal{J}_F[r] := \{[D] \in \mathcal{J}_F : r[D] = 0\}$.

Definition 1. *For each family $\mathcal{F} = \{F/\mathbb{F}_q\}$ of function fields with $g(F) \to \infty$, we define*

$$J_r(\mathcal{F}) := \liminf_{F\in\mathcal{F}} \frac{\log_q |\mathcal{J}_F[r]|}{g(F)}.$$

We define an asymptotic notion involving both $J_r(\mathcal{F})$ and the Ihara limit $A(\mathcal{F})$.

Definition 2 (The torsion-limit). *For a prime power q, $r \in \mathbb{Z}_{>1}$ and $a \in \mathbb{R}$, let $\mathfrak{F}$ be the set of families $\{\mathcal{F}\}$ of function fields over $\mathbb{F}_q$ such that genus in each family tends to ∞ and the Ihara limit $A(\mathcal{F}) \geq a$ for every $\mathcal{F} \in \mathfrak{F}$. Then the asymptotic quantity $J_r(q,a)$ is defined by $J_r(q,a) = \liminf_{\mathcal{F}\in\mathfrak{F}} J_r(\mathcal{F})$.*

Thus, for a given family, the limit $J_r(\mathcal{F})$ measures the r-torsion against the genus. The corresponding constant $J_r(q,a)$ measures, for a given Ihara limit a and for given r, the "least possible r-torsion." Note that $A(q)$, Ihara's constant, is the supremum of $A(\mathcal{F})$ taken over all asymptotically good $\mathcal{F}$ over $\mathbb{F}_q$. Now we are ready to state the main results of this section.

THEOREM 1. *Let $\mathbb{F}_q$ be a finite field and let $r > 1$ be a prime.*

(i) *If $r \mid (q-1)$, then $J_r(q, A(q)) \leq \frac{2}{\log_r q}$.*
(ii) *If $r \nmid (q-1)$, then $J_r(q, A(q)) \leq \frac{1}{\log_r q}$*
(iii) *If q is square and $r \mid q$, then $J_r(q, \sqrt{q}-1) \leq \frac{1}{(\sqrt{q}+1)\log_r q}$.*

The *first* part of Theorem 1, as well as the second part when, additionally, $r|q$, is proved directly using a theorem of Weil [38,27] on torsion in Abelian varieties. [2] The second part, in the case $r \nmid q$ and $r \nmid (q-1)$, can be proved by using Weil pairing for abelian varieties and we will show it in Section 3.1 below. The most interesting is perhaps the bound in the *third part*, which is substantially smaller (we prove that bound in Section 3.2).

THEOREM 2. *Let $\mathbb{F}_q$ be a finite field of characteristic p.*

(i) *If $r \geq 2$ is an integer, then $J_r(q, A(q)) \leq \log_q(dr)$, where $d = \gcd(r, q-1)$.*
(ii) *Write r as $p^\ell m$ for some $\ell \geq 0$ and a positive integer m co-prime to p. If q is a square, then $J_r(q, \sqrt{q}-1) \leq \frac{\ell}{\sqrt{q}+1}\log_q(p) + \log_q(cm)$, where $c = \gcd(m, q-1)$.*

PROOF. The result follows quite directly from the case of prime r considered in Theorem 1 together with some observations about group torsion. We prove this formally in Section 3.3 below. △

Finally, we show existence of certain function field families that is essential for our applications in Section 5.

THEOREM 3. *For every $q \geq 8$ except for $q = 11$ or 13, there exists a family $\mathcal{F}$ of function fields over $\mathbb{F}_q$ such that the Ihara limit $A(\mathcal{F})$ exists and it satisfies $A(\mathcal{F}) > 1 + J_2(\mathcal{F})$.*

PROOF. We prove it in two steps. The first one is to prove that the result is true for all $q \geq 17$ by using class field theory. The second step is to show that the result holds for $q = 8, 9, 16$ by looking at each individual q. For $q \geq 17$, we prove the result only for odd q. For even q, we can similarly get it by considering the Artin-Schreier extensions. Choose 7 nonzero square elements $t_1, \ldots, t_7$ in $\mathbb{F}_q$ (this is possible since $(q-1)/2 \geq 7$). For each i, consider the extension $K_i = \mathbb{F}_q(x, y_i)$, where $y_i^2 = x + t_i$. Then the place x is completely splitting in

[2] If K is algebraically closed, then, for any $m \neq 0$, $A[m]$ is isomorphic to $(\mathbb{Z}/m\mathbb{Z})^{2g}$ if m is co-prime to the characteristic p of K; and $A[p]$ is isomorphic to $(\mathbb{Z}/p\mathbb{Z})^a$ for some $0 \leq a \leq g$, where g is the dimension of A. See also [30]. This implies upper bounds if K is not algebraically closed.

K_i. Let K be the field $\mathbb{F}_q(x,y)$, where $y^2=\prod_{i=1}^7(x+t_i)$. Then K is a subfield of $K_1\cdots K_7/\mathbb{F}_q(x)$ such that $[K:\mathbb{F}_q(x)]=2$ and $K_1\cdots K_7/K$ is an unramified abelian extension. The three places, ∞ and those lying above x, are completely splitting in $K_1\cdots K_7/K$. Since the 2-rank of the Galois group of $K_1\cdots K_7/K$ is 6 which is equal to $2+2\sqrt{3+1}$, K has an infinite $(2,S)$-Hilbert class field tower $\mathcal{F}$, where S consists of the three places ∞ and those lying above x. This yields $A(\mathcal{F})\ge 3/(g(K)-1)=3/2$ (see [31] or [29, Corollary 2.7.8]). Now we have $A(\mathcal{F})\ge 3/2>1+2/\log_2(17)\ge 1+2/\log_2 q\ge 1+J_2(\mathcal{F})$. For $q=8$, by the lower bound for cubics from Section 2 we know that there exists a family $\mathcal{F}$ over $\mathbb{F}_8$ such that $A(\mathcal{F})\ge 3/2$. Thus, $A(\mathcal{F})\ge\frac{3}{2}>1+\frac{1}{3}\ge 1+J_2(\mathcal{F})$. For $q=9$, by the result for squares from Section 2 we know that there exists a family $\mathcal{F}$ over $\mathbb{F}_9$ such that $A(\mathcal{F})=2$. Thus, $A(\mathcal{F})=2>1+\frac{2}{\log_2 9}\ge 1+J_2(\mathcal{F})$. For $q=16$, by the result for squares from Section 2 we know that there exists a family $\mathcal{F}$ over $\mathbb{F}_{16}$ such that $A(\mathcal{F})=3$. Thus, $A(\mathcal{F})=3>1+\frac{1}{4}\ge 1+J_2(\mathcal{F})$. △

3.1 Proof Theorem 1(ii)

For an abelian variety A defined over a field k and a positive integer m, the m-torsion point group, denoted by $A[m]$, is defined to be the set of the points over the algebraic closure $\bar{k}$ annihilated by m. We know that $A[m]$ is isomorphic to $(\mathbb{Z}/m\mathbb{Z})^{2g}$ if m is co-prime to the characteristic p of k; and $A[p]$ is isomorphic to $(\mathbb{Z}/p\mathbb{Z})^a$ for a non-negative integer $a\le g$, where g is the dimension of A (see [38,27]). We also denote by $A(k)$ the set of k-rational points. Thus, the set of m-torsion k-rational points is $A(k)[m]=A(k)\cap A[m]$. If m is co-prime with the characteristic of k, then we can define the Weil pairing to be a map e_m from $A[m]\times\hat{A}[m]$ to G_m, where $\hat{A}$ denotes the dual abelian variety of A and $G_m\simeq\mathbb{Z}/m\mathbb{Z}$ is the group of m-th roots of unity in $\bar{k}$. The Weil paring e_m has some properties such as bilinear, non-degenerate, commuting with the Galois action of $\mathrm{Gal}(\bar{k}/k)$ (see [26]), etc. More precisely:

(i) $e_m(S_1+S_2,T)=e_m(S_1,T)e_m(S_2,T)$; $e_m(S,T_1+T_2)=e_m(S,T_1)e_m(S,T_2)$;
(ii) If $e_m(S,T)=1$ for all $S\in A[m]$, then $T=0$;
(iii) $e_m(S^\sigma,T^\sigma)=e_m(S,T)^\sigma$.

If there is a polarization λ from A to $\hat{A}$, we get a pairing: e_m^λ from $A[m]\times A[m]$ to G_m defined by $e_m^\lambda(P,Q)=e_m(P,\lambda(Q))$. From now on, we assume that A is a Jacobian over k. Then there is a principal polarization λ from A to $\hat{A}$ which is an isomorphism. In this case, we denote e_m^λ by w_m, i.e., w_m is a pairing from $A[m]\times A[m]$ to G_m. It is clear that w_m satisfies all three properties above as well. From the bilinear property, we have $w_m(tP,Q)=w_m(P,Q)^t$ and $w_m(P,tQ)=w_m(P,Q)^t$ for any $t\ge 0$ and $P,Q\in A[m]$. To derive an upper bound on the size of r-torsion points, we need the following result which can be derived easily by using linear algebra.

LEMMA 1. *For a prime r, consider an $\mathbb{F}_r$-vector space W of dimension n and a non-degenerate bilinear map $e:W\times W\longrightarrow\mathbb{F}_r$, i.e., $e(\mathbf{x}+\mathbf{z},\mathbf{y})=e(\mathbf{x},\mathbf{y})+e(\mathbf{z},\mathbf{y})$,*

$e(\mathbf{x}, \mathbf{y}+\mathbf{z}) = e(\mathbf{x}, \mathbf{y}) + e(\mathbf{x}, \mathbf{z})$, *and if* $e(\mathbf{x}, \mathbf{u}) = 0$ *for all* $\mathbf{x} \in W$, *then* $\mathbf{u} = \mathbf{0}$. *If* V *is an* $\mathbb{F}_r$*-subspace of* W *with* $e(\mathbf{x}, \mathbf{y}) = 0$ *for all* $\mathbf{x}, \mathbf{y} \in V$, *then* $\dim_{\mathbb{F}_r} V \leq n/2$.

Applying Lemma 1 to the Weil paring w_r, we obtain the following:

Corollary 1. *If* V *is an* $\mathbb{F}_r$*-subspace of* $A[r]$ *such that* $w_r(P, Q) = 1$ *for all* $P, Q \in V$, *then* $\dim_{\mathbb{F}_r}(V) \leq g$.

Proof. Let ζ be a rth primitive root of unity and consider the bilinear map $(P, Q) \mapsto a \in \mathbb{Z}/r\mathbb{Z}$, where a satisfies $\zeta^a = w_r(P, Q)$. Now apply Lemma 1. △

Proposition 1. *Let* $k = \mathbb{F}_q$ *and assume that a prime* r *does not divide* $q-1$. *If* A *is a Jacobian variety over* k, *then* $\dim(A(k)[r]) \leq g$.

Proof. If r is the characteristic of k, then it follows from the Weil bound. Now assume that r is not the characteristic of k. It is easy to verify that $A(k)[r]$ is an $\mathbb{F}_r$-subspace of $A[r]$. For any σ in the Galois group $\mathrm{Gal}(\bar{k}/k)$, one has $w_r(P, Q) = w_r(P^\sigma, Q^\sigma) = w_r(P, Q)^\sigma$. This implies that $w_r(P, Q)$ is an element of k. However, the only r-th root of unity in k is 1. We get $w_r(P, Q) = 1$ for all $P, Q \in A(k)[r]$. Our desired result follows from Corollary 1. △

3.2 Proof of Theorem 1(iii)

Let $\mathbb{F}_q$ be a finite field. Write p for its characteristic. For a function field F over $\mathbb{F}_q$ denote by $\gamma(F)$ the p-rank of F. It holds that $\gamma(F) \geq \log_p(J_F[p])$. Assume q is a square. Consider the tower $\mathcal{F} = (F^{(0)} \subset F^{(1)} \subset \cdots)$ over F_q introduced in [16], recursively defined by $F^{(0)} = \mathbb{F}_q(x_0)$ and $F^{(n+1)} = F^{(n)}(x_{n+1})$, where $x_n^{\sqrt{q}-1} x_{n+1}^{\sqrt{q}} + x_{n+1} = x_n^{\sqrt{q}}$. The following facts can be found in [16].

1. The tower $\mathcal{F}$ attains Drinfeld-Vlăduţ bounds, i.e., its limit $A(\mathcal{F})$ is given by $A(\mathcal{F}) := \lim_{n\to\infty} \frac{N(F^{(n)})}{g(F^{(n)})} = \sqrt{q} - 1$.
2. For any place $P \in \mathbb{P}(F^{(n-1)})$ and any place $Q \in \mathbb{P}(F^{(n)})$ such that $Q|P$ we have $d(Q|P) = (\sqrt{q}+2)(e(Q|P) - 1)$, where $d(Q|P)$ and $e(Q|P)$ denote the different exponent and ramification index.
3. $g(F^{(n)}) = q^{\frac{n+1}{2}} + q^{\frac{n}{2}} - q^{\frac{n+2}{4}} - 2q^{\frac{n}{2}} + 1$ if $n \equiv 0 \pmod 2$ and $g(F^{(n)}) = q^{\frac{n+1}{2}} + q^{\frac{n}{2}} - \frac{1}{2}q^{\frac{n+3}{4}} - \frac{3}{2}q^{\frac{n+1}{4}} - q^{\frac{n-1}{4}} + 1$ if $n \equiv 1 \pmod 2$.

We will now show

Theorem 4. *It holds that* $\gamma(F^{(n)}) = (\sqrt{q}^{n/2} - 1)^2$ *if* $n \equiv 0 \pmod 2$ and $\gamma(F^{(n)}) = (\sqrt{q}^{(n-1)/2} - 1)(\sqrt{q}^{(n+1)/2} - 1)$ *if* $n \equiv 1 \pmod 2$.
In particular $\lim_{n\to\infty} \frac{g(F^{(n)})}{\gamma(F^{(n)})} = \sqrt{q} + 1$.

Then Theorem 1(iii) is a direct corollary of the above theorem.

Without loss of generality we can assume that the constant fields of the function fields are $\overline{\mathbb{F}}_q$. We will use the following theorem.

Theorem 5 (Deuring-Shafarevich (see e.g. [19])). *Let E/F be a Galois extension of function fields over $\overline{\mathbb{F}}_q$. Suppose its Galois group is a p-group.*

Then $\gamma(E) - 1 = [E:F](\gamma(F)-1) + \sum_{P\in\mathbb{P}(F)}\sum_{Q\in\mathbb{P}(E),Q|P}(e(Q|P)-1)$.

Proof of Theorem 4: Consider the extension $F^{(n)}/F^{(n-1)}$. As this is an Artin Schreier extension, it is Galois and its Galois-group is a p-group. By Riemann-Hurwitz (see e.g. [34] and fact 2. above), and by Deuring-Shafarevich, respectively,

$$2\cdot g(F^{(n)})-2 = \sqrt{q}\cdot(2g(F^{(n-1)})-2)+(\sqrt{q}+2)\cdot \sum_{P\in\mathbb{P}(F^{(n-1)})}\sum_{\substack{Q\in\mathbb{P}(F^{(n)})\\ Q|P}}(e(Q|P)-1),$$

$$\gamma(F^{(n)})-1 = \sqrt{q}\cdot(\gamma(F^{(n-1)})-1) + \sum_{P\in\mathbb{P}(F^{(n-1)})}\sum_{\substack{Q\in\mathbb{P}(F^{(n)})\\ Q|P}}(e(Q|P)-1).$$

Combining these two equations we find

$$\gamma(F^{(n)}) = \sqrt{q}\cdot\gamma(F^{(n-1)})+2\cdot(g(F^{(n)})-2\sqrt{q}\cdot g(F^{(n-1)})-\sqrt{q}^2+\sqrt{q})(\sqrt{q}+2)^{-1}.$$

Using the fact that $\gamma(F^{(0)})=0$ and applying induction, the result follows. △

3.3 Proof of Theorem 2

We first show how to lift the previous results from $A(k)[r]$ to $A(k)[r^t]$.

Lemma 2. *Let $k=\mathbb{F}_q$ and let r be a prime. If A is an Abelian variety over k with $|A(k)[r]|\le a$, then $|A(k)[r^t]|\le a^t$ for every $t\ge 1$.*

Proof. We prove it by induction. The case $t=1$ is the given condition. Assume it holds for $t-1$. Consider the map $[r]_k: A(k)[r^t]\to A(k)[r^{t-1}], P\mapsto rP$. Clearly the kernel of $[r]_k$ is $A(k)[r]$. Thus, $|A(k)[r^t]| = |\mathrm{Ker}([r]_k)|\times|\mathrm{Im}([r]_k)|\le a\times a^{t-1}=a^t$. The desired result follows. △

Proposition 2. *Let $k=\mathbb{F}_q$ and assume that a prime r does not divide $q-1$.*

1. *If A is a Jacobian variety over k, then $|A(k)[r^t]|\le r^{gt}$ for every $t\ge 1$.*
2. *If $m\ge 2$ is an integer, then $|A(k)[m]|\le (dm)^g$, where $d=\gcd(m,q-1)$.*

Proof. Part 1 is the direct result of Proposition 1 and Lemma 2. To prove Part 2, we factorize m into the product $\prod_p p^{s_p}\times\prod_\ell \ell^{s_\ell}$ of prime powers, where $d=\prod_p p^{s_p}$ is a factor of $q-1$ and $\prod_\ell \ell^{s_\ell}=m/d$. By Part 1 and the following isomorphism $A(k)[m]\simeq\prod_p A(k)[p^{s_p}]\times\prod_\ell A(k)[\ell^{s_\ell}]$, we have $|A(k)[m]| = |\prod_p A(k)[p^{s_p}]|\times\prod_\ell |A(k)[\ell^{s_\ell}]|\le d^{2g}\times(m/d)^g=(dm)^g$. △

Theorem 1(iii), Lemma 2, and Proposition 2 now imply Theorem 2.

4 Riemann Roch Systems of Equations

Let $F/\mathbb{F}_q$ be an algebraic function field.

DEFINITION 3. Let $L \in \mathbb{Z}_{>0}$ and let $Y_i \in \mathrm{Cl}(F)$, $d_i \in \mathbb{Z} \setminus \{0\}$ for $i = 1, \ldots, L$. The *Riemann-Roch system of equations* in the indeterminate X is the system $\{\ell(d_i X + Y_i) = 0\}_{i=1}^{L}$ determined by these data. A solution is some $[G] \in \mathrm{Cl}(F)$ which satisfies all equations when substituted for X.

It is often more convenient to define systems over $\mathrm{Div}(F)$ rather than $\mathrm{Cl}(F)$. The idea of using Riemann-Roch systems of equations was already present in some papers, e.g. [36], [39], [40]. However, those systems are less general, namely they have $d_i = \pm 1$ for all i.[3] The following theorem shows that a solution of degree s exists if a certain numerical condition is satisfied that involves the class number, the number A_{r_i} of effective divisors of degree r_i and the cardinality of the d_i-torsion subgroups of the degree-zero divisor class group, where the d_i are determined by the system and the r_i are determined by s and the d_i.

THEOREM 6. *Consider the Riemann-Roch system* $\{\ell(d_i X + Y_i) = 0\}_{i=1}^{L}$. *Write* $s_i = \deg Y_i$ *for* $i = 1, \ldots, L$. *Denote by* A_r *the number of effective divisors of degree* r *in* $Div(F)$ *for* $r \geq 0$, *and* 0 *for* $r < 0$. *Let* $s \in \mathbb{Z}$ *and define* $r_i = d_i s + s_i$ *for* $i = 1, \ldots, L$. *If* $h > \sum_{i=1}^{L} A_{r_i} \cdot |\mathcal{J}_F[d_i]|$, *then the Riemann-Roch system has a solution* $[G] \in Cl_s(F)$.

PROOF. Let S be the set $\{1 \leq i \leq L : r_i \geq 0\}$. For each $i \in S$, we argue as follows. Define the maps $\phi_i : \mathrm{Cl}_s(F) \to \mathrm{Cl}_{d_i s}(F)$, $X \mapsto d_i X$ and $\psi_i : \mathrm{Cl}_{d_i s}(F) \to \mathrm{Cl}_{r_i}(F)$, $X' \mapsto X' + Y_i$. Then ψ_i is an injection and each image under ϕ_i has exactly $|\mathcal{J}_F[d_i]|$ pre-images. Write $\sigma_i = \psi_i \circ \phi_i$. Then, for any element $Z \in \mathrm{Cl}_{r_i}^{+}(F)$, $|\sigma_i^{-1}(Z)| \leq |\mathcal{J}_F[d_i]|$. Hence, $|\sigma_i^{-1}(\mathrm{Cl}_{r_i}^{+}(F))| \leq A_{r_i} \cdot |\mathcal{J}_F[d_i]|$. Thus, $\bigcup_{i \in S} \sigma_i^{-1}(\mathrm{Cl}_{r_i}^{+}(F))| \leq \sum_{i \in S} A_{r_i} \cdot |\mathcal{J}_F[d_i]|$. Since by hypothesis we have $|\mathrm{Cl}_s(F)| = h > \sum_{i=1}^{L} A_{r_i} \cdot |\mathcal{J}_F[d_i]| = \sum_{i \in S} A_{r_i} \cdot |\mathcal{J}_F[d_i]|$, there is an element $[G] \in \mathrm{Cl}_s(F) \setminus \bigcup_{i \in S} \sigma_i^{-1}(\mathrm{Cl}_{r_i}^{+}(F))$. Since $\sigma_i([G]) \in \mathrm{Cl}_{r_i}(F)$ but $\sigma_i([G]) \notin \mathrm{Cl}_{r_i}^{+}(F)$, it follows that $\ell(\sigma_i([G])) = 0$ for $i \in S$, i.e., $[G]$ is a solution of the system $\{\ell(d_i X + Y_i + T_i) = 0\}_{i \in S}$. Finally $[G]$ is also a solution of $\{\ell(d_i X + Y_i) = 0\}_{i \notin S}$, because $\deg(d_i[G] + Y_i) = r_i < 0$ for all $i \notin S$. △

REMARK 2. ("Solving by taking any divisor X of large enough degree")

(i) If $r_i < 0$ for all $i = 1, \ldots, L$, then the inequality in Theorem 6 is automatically satisfied and hence the Riemann-Roch system always has a solution.
(ii) For instance, in [7], it was simply assumed that $r_i < 0$ to obtain $(n, t, d, n-t)$-arithmetic secret sharing schemes. But this does not always give the best results. In particular, in Section 5, we will show how we can employ Theorem 6 to get improvements, *especially for small finite fields.*

[3] In [33], the case $d_i = 2$ was considered but their result must be corrected for torsion.

5 Application to Arithmetic Secret Sharing

We first recall the results of [7], cast in a novel, technical framework.[4] This will make it possible to state our main quantitative results on arithmetic secret sharing in transparent language, which also facilitates easy comparison with earlier work. Let k, n be integers with $k, n \geq 1$. Consider the $\mathbb{F}_q$-vector space $\mathbb{F}_q^k \times \mathbb{F}_q^n$, where $\mathbb{F}_q$ is an arbitrary finite field.

Definition 4. *The $\mathbb{F}_q$-vector space morphism $\pi_0 : \mathbb{F}_q^k \times \mathbb{F}_q^n \to \mathbb{F}_q^k$ is defined by the projection $(s_1, \ldots, s_k, c_1, \ldots, c_n) \mapsto (s_1, \ldots, s_k)$. For each $i \in \{1, \ldots, n\}$, the $\mathbb{F}_q$-vector space morphism $\pi_i : \mathbb{F}_q^k \times \mathbb{F}_q^n \to \mathbb{F}_q$ is defined by the projection $(s_1, \ldots, s_k, c_1, \ldots, c_n) \mapsto c_i$. For $\emptyset \neq A \subset \{1, \ldots, n\}$, the $\mathbb{F}_q$-vector space morphism $\pi_A : \mathbb{F}_q^k \times \mathbb{F}_q^n \to \mathbb{F}_q^{|A|}$ is defined by the projection $(s_1, \ldots, s_k, c_1, \ldots, c_n) \mapsto (c_i)_{i \in A}$. For $\mathbf{v} \in \mathbb{F}_q^k \times \mathbb{F}_q^n$, it is sometimes convenient to denote $\pi_0(\mathbf{v}) \in \mathbb{F}_q^k$ by $\mathbf{v}_0$ and $\pi_A(\mathbf{v}) \in \mathbb{F}_q^{|A|}$ by $\mathbf{v}_A$. We write $\mathcal{I}^* = \{1, \ldots, n\}$. It is also sometimes convenient to refer to $\mathbf{v}_0$ as the* secret-component *of $\mathbf{v}$ and to $\mathbf{v}_{\mathcal{I}^*}$ as its* shares-component*.*

Definition 5. *An n-code for $\mathbb{F}_q^k$ (over $\mathbb{F}_q$) is an $\mathbb{F}_q$-vector space $C \subset \mathbb{F}_q^k \times \mathbb{F}_q^n$ such that $\pi_0(C) = \mathbb{F}_q^k$ and $(\text{Ker } \pi_{\mathcal{I}^*}) \cap C \subset (\text{Ker } \pi_0) \cap C$. For $\mathbf{c} \in C$, $\mathbf{c}_0 \in \mathbb{F}_q^k$ is the* secret *and $\mathbf{c}_{\mathcal{I}^*} \in \mathbb{F}_q^n$ the* shares.

The first condition means that, in C, the secret can take any value in $\mathbb{F}_q^k$. More precisely, for a uniformly random vector $\mathbf{c} \in C$, the secret $\mathbf{c}_0$ is uniformly random in $\mathbb{F}_q^k$. This follows from the fact that the projection $(\pi_0)_{|C}$ is regular (since it is a surjective $\mathbb{F}_q$-vector space morphism). The second condition means that the shares uniquely determine the secret. Indeed, the shares do not always determine the secret uniquely if and only if there are $\mathbf{c}, \mathbf{c}' \in C$ such that their shares coincide but not their secrets. Therefore, by linearity, the shares determine the secret uniquely if and only if the shares being zero implies the secret being zero. Moreover this condition implies that $k \leq n$. Note that an n-code with the stronger condition $(\text{Ker } \pi_{\mathcal{I}^*}) \cap C = (\text{Ker } \pi_0) \cap C$ is a k-dimensional error correcting code of length n.

Definition 6 (r-reconstructing). *An n-code C for $\mathbb{F}_q^k$ is* r-reconstructing *$(1 \leq r \leq n)$ if $(\text{Ker } \pi_A) \cap C \subset (\text{Ker } \pi_0) \cap C$ for each $A \subset \mathcal{I}^*$ with $|A| = r$.*

In other words, r-reconstructing means that any r shares uniquely determine the secret. Note that $r \leq n$ by definition of an n-code.

Definition 7 (t-Disconnected). *An n-code C for $\mathbb{F}_q^k$ is* t-disconnected *if $t = 0$ or else if $1 \leq t < n$, the projection $\pi_{0,A} : C \longrightarrow \mathbb{F}_q^k \times \pi_A(C)$, $\mathbf{c} \mapsto (\pi_0(\mathbf{c}), \pi_A(\mathbf{c}))$ is surjective for each $A \subset \mathcal{I}^*$ with $|A| = t$. If, additionally, $\pi_A(C) = \mathbb{F}_q^t$, we say C is* t-uniform.

[4] This is a special case of the notion of an *(arithmetic) codex* that we introduced in an invited talk at EUROCRYPT'11 and, earlier, at the IPAM workshop on Information-Theoretic Cryptography.

If $t > 0$, then t-disconnectedness means the following. Let $A \subset \mathcal{I}^*$ with $|A| = t$. Then, for uniformly randomly $\mathbf{c} \in C$, the secret $\mathbf{c}_0$ is independently distributed from the t shares $\mathbf{c}_A$. Indeed, for the same reason that the secret $\mathbf{c}_0$ is uniformly random in $\mathbb{F}_q^k$, it holds that $(\mathbf{c}_0, \mathbf{c}_A)$ is uniformly random in $\mathbb{F}_q^k \times \pi_A(C)$. Since the uniform distribution on the Cartesian-product of two finite sets corresponds to the uniform distribution on one set, and independently, the uniform distribution on the other, the claim follows. Uniformity means that, in addition, $\mathbf{c}_A$ is uniformly random in $\mathbb{F}_q^t$.

Definition 8 (Generator of an n-Code). *A generator (k_0, σ) of an n-code C consists of a positive integer k_0 and a surjective $\mathbb{F}_q$-vector space morphism $\sigma : \mathbb{F}_q^k \times \mathbb{F}_q^{k_0} \longrightarrow C$, such that $\pi_0(\sigma(\mathbf{s}, \mathbf{z})) = \mathbf{s}$ for all $(\mathbf{s}, \mathbf{z}) \in \mathbb{F}_q^k \times \mathbb{F}_q^{k_0}$.*

In particular, a generator selects an element of C with a prescribed secret. We can always assume $k_0 \leq n$. Note that a generator can be represented by a matrix defined by the columns (or rows, depending on one's view) $\sigma(\mathbf{e}_1, \mathbf{0}), \ldots, \sigma(\mathbf{e}_k, \mathbf{0})$, $\sigma(\mathbf{0}, \mathbf{e}'_1), \ldots, \sigma(\mathbf{0}, \mathbf{e}'_{k_0})$, where the $\mathbf{e}_i$'s are the standard unit-vectors in $\mathbb{F}_q^k$, and the $\mathbf{e}'_j$'s are the standard unit-vectors in $\mathbb{F}_q^{k_0}$. Given such a matrix-representation, selecting a uniformly random $\mathbf{c} \in C$ such that its secret equals some prescribed value, can be done efficiently. By elementary linear algebra, this also holds for r-reconstruction of a secret. Similarly, a generator can be computed efficiently from a basis of C.

Definition 9 (Powers of an n-Code). *Let $m \in \mathbb{Z}_{>0}$. For $\mathbf{x}, \mathbf{x}' \in \mathbb{F}_q^m$, their product $\mathbf{x} * \mathbf{x}' \in \mathbb{F}_q^m$ is defined as $(x_1 x'_1, \ldots, x_m x'_m)$. Let d be a positive integer. If C is an n-code for $\mathbb{F}_q^k$, then $C^{*d} \subset \mathbb{F}_q^k \times \mathbb{F}_q^n$ is the $\mathbb{F}_q$-linear subspace generated by all terms of the form $\mathbf{c}^{(1)} * \ldots * \mathbf{c}^{(d)}$ with $\mathbf{c}^{(1)}, \ldots, \mathbf{c}^{(d)} \in C$. For $d = 2$, we use the abbreviation $\widehat{C} := C^{*2}$.*

Remark 3 (Powering Need Not Preserve n-Code). *Suppose $C \subset \mathbb{F}_q^k \times \mathbb{F}_q^n$ is an n-code for $\mathbb{F}_q^k$. It follows immediately that the secret-component in C^{*d} takes any value in $\mathbb{F}_q^k$. However, the shares-component in C^{*d}* need not *determine the secret-component uniquely. Thus, C^{*d}* need not be an n-code for $\mathbb{F}_q^k$.

Remark 4. *Let C be an n-code for $\mathbb{F}_q^k$ and let (k_0, σ) be a generator. For an integer $d \geq 2$, suppose C^{*d} is an n-code. If $d = 2$, then it is easy to see that $\widehat{C}$ is generated by the vectors $\mathbf{x} * \mathbf{y} \in \mathbb{F}_q^k \times \mathbb{F}_q^n$, where $\mathbf{x}, \mathbf{y}$ range over all pairs of vectors selected from $\sigma(\mathbf{e}_1, \mathbf{0}), \ldots, \sigma(\mathbf{e}_k, \mathbf{0}), \sigma(\mathbf{0}, \mathbf{e}'_1), \ldots, \sigma(\mathbf{0}, \mathbf{e}'_{k_0})$. Since, for $i = 1, \ldots, k$, the vector $\sigma(\mathbf{e}_i, \mathbf{0}) * \sigma(\mathbf{e}_i, \mathbf{0})$ has the i-th unit vector as its secret-component, a generator for $\widehat{C}$ can be efficiently constructed from (k_0, σ). This generalizes to $d > 2$ in a straightforward way.*

Definition 10 (Arithmetic secret sharing scheme).
An (n, t, d, r)-arithmetic secret sharing scheme for $\mathbb{F}_q^k$ (over $\mathbb{F}_q$) *is an n-code C for $\mathbb{F}_q^k$ such that $t \geq 1$, $d \geq 2$, C is t-disconnected, C^{*d} is in fact an n-code for $\mathbb{F}_q^k$, and C^{*d} is r-reconstructing. C has* uniformity *if, in addition, it is t-uniform.*

For example, the case $k = 1$, $d = 2$, $n = 3t + 1$, $r = n - t$, $q > n$ obtained from Shamir's secret sharing scheme (taking into account that degrees sum up when taking products of polynomials) corresponds to the secret sharing scheme used in [2,6]. The properties are easily proved using Lagrange's Interpolation Theorem. The generalization to $k > 1$ of this Shamir-based approach is due to [15]. The abstract notion is due to [10], where also constructions for $d = 2$ were given based on general linear secret sharing. See also [7,8,9]. On the other hand the following limitations are easy to establish.

PROPOSITION 3. *Let C be an (n, t, d, r)-arithmetic secret sharing scheme for $\mathbb{F}_q^k$ over $\mathbb{F}_q$. As a linear secret sharing scheme for $\mathbb{F}_q^k$ over $\mathbb{F}_q$, C has t-privacy and $(r - (d - 1)t)$-reconstruction. Hence, $dt + k \leq r$. Particularly, if $k = 1$, $d = 2$, $r = n - t$, then $3t + 1 \leq n$.*

Stronger bounds are also known [5]. We note that arithmetic secret sharing schemes enjoy *efficient recovery of the secret in the presence of faulty shares.* The theorem below is a generalization of a result from [12].

THEOREM 7. *Let i, j be integers with $1 \leq i < j$. Suppose C and C^{*j} are n-codes for $\mathbb{F}_q^k$. If the n-code C^{*i} is t-disconnected and if C^{*j} has $(n - t)$-reconstruction, then, given a generator for C, there is an efficient algorithm for the n-code $C^{*(j-i)}$ that, on input $\tilde{\mathbf{a}} := \mathbf{c}_{\mathcal{I}^*} + \mathbf{e} \in \mathbb{F}_q^n$ (faulty shares) with $\mathbf{c} \in C^{*(j-i)}$ and $\mathbf{e} \in \mathbb{F}_q^n$ of Hamming-weight at most t, outputs $\mathbf{c}_0 \in \mathbb{F}_q^k$ (correct secret).*

PROOF. Note that if C and C^{*j} are an n-codes, then so is $C^{*j'}$ for any j' with $1 \leq j' \leq j$. Let $\mathbf{c} \in C^{*(j-i)}$. Let $(\mathbf{0}, \mathbf{e}) \in \mathbb{F}_q^k \times \mathbb{F}_q^n$ such that $\mathbf{e}$ has Hamming-weight at most t. Define $\tilde{\mathbf{c}} = \mathbf{c} + (\mathbf{0}, \mathbf{e})$ and let $\tilde{\mathbf{a}} = \tilde{\mathbf{c}}_{\mathcal{I}^*} = \mathbf{c}_{\mathcal{I}^*} + \mathbf{e}$. Write $\mathbf{u} = (1, \ldots, 1) \in \mathbb{F}_q^k$. Consider the system of equations $\{\tilde{\mathbf{a}} * \mathbf{x}_{\mathcal{I}^*} = \mathbf{y}_{\mathcal{I}^*},\ \mathbf{x}_0 = \mathbf{u},\ \mathbf{x} \in C^{*i},\ \mathbf{y} \in C^{*j}\}$, in the unknowns $\mathbf{x}, \mathbf{y}$. Note that this is in fact a *linear* system of equations (taking into account that, of course, membership of a subspace can be captured by a linear system of equations). We prove now that, first, this system has *some* solution $(\mathbf{x}, \mathbf{y})$, and that, second, *any* solution $(\mathbf{x}, \mathbf{y})$ satisfies $\mathbf{c}_0 = \mathbf{y}_0$. Efficiency then follows by linear algebra, in combination with the fact that a generator for C is given and that generators for the higher powers can be constructed efficiently from it. First, define $A \subset \{1, \ldots, n\}$ as the set of all i with $e_i \neq 0$. Since C^{*i} is $t-$disconnected, there is $\mathbf{z} \in C^{*i}$ such that $\mathbf{z}_0 = \mathbf{u}$ and $\mathbf{z}_A = \mathbf{0}$. Then $\mathbf{x} := \mathbf{z}$, $\mathbf{y} := \mathbf{c} * \mathbf{z}$ is a solution. Second, let $(\mathbf{x}, \mathbf{y})$ be any solution. Then, for at least $n - |A| \geq n - t$ indices within $\mathcal{I}^*$ it holds that the vectors $\mathbf{c} * \mathbf{x} \in C^{*j}$ and $\mathbf{y} \in C^{*j}$ coincide. Since C^{*j} has $(n - t)$-reconstruction and since $\mathbf{x}_0 = \mathbf{u}$, the claim follows. △

We briefly sketch two well-known applications. First, consider an $(n, t, 2, n)$-arithmetic secret sharing scheme C for $\mathbb{F}_q^k$ over $\mathbb{F}_q$. Such a scheme can be used to reduce n-party secure multiplication to secure addition in the case of a *honest-but-curious adversary*, at the cost of one round of interaction. From the definition, it follows there is an $\mathbb{F}_q$-vector space morphism $\psi : \mathbb{F}_q^n \longrightarrow \mathbb{F}_q^k$ such that, for all $\mathbf{c}, \mathbf{c}' \in C$, $\psi(c_1c_1', \ldots, c_nc_n') = \mathbf{c}_0 * \mathbf{c}_0'$. The reduction works as follows.[5] Let $\mathbf{c}_{\mathcal{I}^*}, \mathbf{c}'_{\mathcal{I}^*} \in \mathbb{F}_q^n$ be secret-sharings, with respective secrets $\mathbf{c}_0, \mathbf{c}_0' \in \mathbb{F}_q^k$ ($\mathbf{c}, \mathbf{c}' \in C$). Using a generator for C, player P_i secret-shares $(\lambda_{i1}c_ic_i', \ldots, \lambda_{ik}c_ic_i') \in \mathbb{F}_q^k$, where the coefficient vector is the "i-th row of the matrix representing ψ in the standard basis" ($i = 1, \ldots, n$). Next, player P_j sums the n received shares ($j = 1, \ldots, n$). This gives a secret-sharing of $\mathbf{c}_0 * \mathbf{c}_0'$ according to C (see e.g. [9]). This generalizes Shamir-based solutions from [2,6,15] (see also [10]).

Second, consider an $(n, t, 2, n-t)$-arithmetic secret sharing scheme C for $\mathbb{F}_q^k$ over $\mathbb{F}_q$. Such a scheme can be used for "zero-knowledge verification of secret multiplications." In a nutshell, the main idea is as follows. Suppose a prover puts forward commitments to secrets $\mathbf{x}_0, \mathbf{y}_0, \mathbf{z}_0 \in \mathbb{F}_q^k$, and claims that $\mathbf{x}_0 * \mathbf{y}_0 = \mathbf{z}_0$. To prove his claim, he gives ("coordinate-wise") commitments to random $\mathbf{x}, \mathbf{y} \in C$ where the respective secrets are the $\mathbf{x}_0, \mathbf{y}_0$ from the input, and a ("coordinate-wise") commitment to a random $z \in C^{*2}$ where the secret is the $\mathbf{z}_0 \in \mathbb{F}_q^k$ from the input. If the commitment scheme is $\mathbb{F}_q$-linear, then it is easy to enforce that indeed $\mathbf{x}, \mathbf{y} \in C$ and $\mathbf{z} \in C^{*2}$, and that the respective secrets are indeed the ones from the input. Now, if $\mathbf{z} = \mathbf{x} * \mathbf{y}$ (as an honest prover would choose), then indeed $\mathbf{x}_0 * \mathbf{y}_0 = \mathbf{z}_0$. In this case, inspection of any t "share-triples" (x_i, y_i, z_i) gives no information on the "secret-triple" $(\mathbf{x}_0, \mathbf{y}_0, \mathbf{x}_0\mathbf{y}_0)$. Yet, $z_i = x_iy_i$ for each of those t share-triples. On the other hand, suppose $\mathbf{z}_0 \neq \mathbf{x}_0 * \mathbf{y}_0$. Then there are at most $n - t - 1$ share-triples (x_i, y_i, z_i) such that $z_i = x_iy_i$, and hence there are at least $t + 1$ share-triples for which an inconsistency could show up. These facts together give a handle to checking that $\mathbf{z}_0 = \mathbf{x}_0\mathbf{y}_0$ in several different application scenarios, most notably perfect information-theoretically secure general multi-party computation, in the case of a *malicious adversary*. See [11] for an application with $d > 2$. The procedure above is essentially from [10] (which was inspired by ideas from [2,6]).

We are now ready to state the asymptotical results from [7] in full generality.[6] Let $F/\mathbb{F}_q$ be an algebraic function field (in one variable, with $\mathbb{F}_q$ as field of constants). Let g denote the genus of F. Let $k, t, n \in \mathbb{Z}$ with $n > 1$, $1 \leq t \leq n$, $1 \leq k \leq n$. Suppose $Q_1, \ldots, Q_k, P_1, \ldots, P_n \in \mathbb{P}^{(1)}(F)$ are *pairwise distinct* $\mathbb{F}_q$-rational places. Write $Q = \sum_{j=1}^k Q_j \in \text{Div}(F)$ and $D = Q + \sum_{i=1}^n P_i \in \text{Div}(F)$. Let $G \in \text{Div}(F)$ be such that supp $D \cap$ supp $G = \emptyset$, i.e, they have disjoint support. Consider the AG-code

$$C(G; D) = \{(f(Q_1), \ldots, f(Q_k), f(P_1), \ldots, f(P_n)) \mid f \in \mathcal{L}(G)\} \subset \mathbb{F}_q^k \times \mathbb{F}_q^n.$$

THEOREM 8. *(from [7]). Let $t \geq 1, d \geq 2$. Let $C = C(G; D)$ with deg $G \geq 2g + t + k - 1$. If $n > 2dg + (d+1)t + dk - d$, then C is an $(n, t, d, n-t)$-arithmetic sharing scheme for $\mathbb{F}_q^k$ over $\mathbb{F}_q$ with uniformity.*

THEOREM 9. *(from [7]). Fix $d \geq 2$ and a finite field $\mathbb{F}_q$. Suppose $A(q) > 2d$, where $A(q)$ is Ihara's constant. Then there is an infinite family of $(n, t, d, n-t)$-arithmetic secret sharing schemes for $\mathbb{F}_q^k$ over $\mathbb{F}_q$ with uniformity such that n*

[5] This "local share-multiplication plus re-sharing" simplification in the case of Shamir's scheme has been attributed to Michael Rabin

[6] In fact, we state a version that is proved by exactly the same arguments as in [7].

is unbounded, $k = \Omega(n)$ and $t = \Omega(n)$. Moreover, for every scheme C in the family, a generator for C is poly(n)-time computable and C^{*i} has poly(n)-time reconstruction of a secret in the presence of t faulty shares ($i = 1, \ldots, d-1$).

Since $A(q) = \sqrt{q} - 1$ if q is a square, it holds that $A(q) > 2d$ if q is a square with $q > (2d+1)^2$. Also, by Serre's Theorem, $A(q) > c \log q$ for some absolute constant $c > 0$. Therefore, $A(q) > 2d$ if q is (very) large.[7] The asymptotical result from [7] plays an important communication-saving role in *two-party cryptography*, see [23,24,18,22,13,21]. Often, the point is that terms in the communication analysis which would otherwise be logarithmic can be made constant using the [7] results. Note that [22,21] also use the efficient error correction. We will now apply our results on the torsion-limit in combination with appropriate Riemann-Roch systems in order to relax the condition $A(q) > 2d$ considerably. As a result, we attain the result of [7] but this time over *nearly all finite fields.*

THEOREM 10. *Let $t \geq 1, d \geq 2$. Define $\mathcal{I}^* = \{1, \ldots, n\}$. For $A \subset \mathcal{I}^*$ with $A \neq \emptyset$, define $P_A = \sum_{j \in A} P_j \in \mathrm{Div}(F)$. Let $K \in \mathrm{Div}(F)$ be a canonical divisor. If the system $\{\ell(dX - D + P_A + Q) = 0,\ \ell(K - X + P_A + Q) = 0\}_{A \subset \mathcal{I}^*, |A| = t}$ is solvable, then there is a solution $G \in \mathrm{Div}(F)$ such that $C(G; D)$ is an $(n, t, d, n-t)$-arithmetic secret sharing scheme for $\mathbb{F}_q^k$ over $\mathbb{F}_q$ (with uniformity).*

PROOF. First note that if the system is solvable, then the Weak Approximation Theorem guarantees that we can take a solution $G \in \mathrm{Div}(F)$ such that $\mathrm{supp}\ G \cap \mathrm{supp}\ D = \emptyset$. We claim that the condition that $\ell(K - G + P_A + Q) = 0$ for $A \subset \mathcal{I}^*$ with $|A| = t$ implies t-disconnection and uniformity on the code. Write $A = \{i_1, \ldots, i_t\}$. Consider the map $\phi : \mathcal{L}(G) \to \mathbb{F}_q^{k+t}$ given by $f \mapsto (f(Q_1), \ldots, f(Q_k), f(P_{i_1}), \ldots, f(P_{i_t}))$. Its kernel is $\mathcal{L}(G - Q - P_A)$. Consequently $\dim(\mathrm{Im}\ \phi) = \ell(G) - \ell(G - Q - P_A) = \ell(K - G) - \ell(K - G + Q + P_A) + \deg(Q + P_A)$, where the second equality follows by application of the Riemann-Roch theorem to G and to $G - Q - P_A$. Hence, $\ell(K - G) \leq \ell(K - G + Q + P_A) = 0$, where the inequality follows from the fact that $Q, P_A \geq 0$ and where the equality holds by assumption. Therefore, $\ell(K - G) = 0$ and $\dim(\mathrm{Im}\ \phi) = \deg(Q + P_A) = k + t$. We conclude that ϕ is surjective and this proves the claim. Finally we prove $(n-t)$-reconstruction in C^{*d}. Let $B = \{i_1, \ldots, i_{n-t}\}$ for distinct indices $i_1, \ldots, i_{n-t} \in \mathcal{I}^*$. Since $f_1, \ldots, f_d \in \mathcal{L}(G)$ implies $\prod_{i=1}^d f_i \in \mathcal{L}(dG)$, it is sufficient to prove that, for all $f \in \mathcal{L}(dG)$, the following holds: if the condition $f(P_i) = 0$ holds for all $i \in B$, then $f(Q_j) = 0$ for all $j \in \{1, \ldots, k\}$. Since $P_B = D - Q - P_A$ for some $A \subset \mathcal{I}^*$ with $|A| = t$, it holds that $\mathcal{L}(dG - P_B) = \mathcal{L}(dG - D + P_A + Q)$, which by assumption has dimension 0. Hence, $f \in \mathcal{L}(dG - P_B) = \{0\}$, and $f = 0$. △

And now as a corollary of Theorems 6 and 10 we get the following:

COROLLARY 2. *Let $F/\mathbb{F}_q$ be an algebraic function field. Let $d, k, t, n \in \mathbb{Z}$ with $d \geq 2$, $n > 1$ and $1 \leq t < n$. Suppose $Q_1, \ldots, Q_k, P_1, \ldots, P_n \in \mathbb{P}^{(1)}(F)$ are pairwise distinct. If there is $s \in \mathbb{Z}$ such that $h > \binom{n}{t}(A_{r_1} + A_{r_2}|\mathcal{J}_F[d]|)$ where $r_1 := 2g - s + t + k - 2$ and $r_2 := ds - n + t$, then there exists an $(n, t, d, n-t)$-arithmetic secret sharing scheme for $\mathbb{F}_q^k$ over $\mathbb{F}_q$ with uniformity.*

[7] The best known estimate for c is currently about $\frac{1}{96}$.

Main Theorem 1. *Let $\mathbb{F}_q$ be a finite field and $d \in \mathbb{Z}_{\geq 2}$. If there exists $0 < A \leq A(q)$ such that $A > 1 + J_d(q, A)$, then there is an infinite family of $(n, t, d, n-t)$-arithmetic secret sharing schemes for $\mathbb{F}_q^k$ over $\mathbb{F}_q$ with t-uniformity where n is unbounded, $k = \Omega(n)$ and $t = \Omega(n)$.*

This will follow from the more precise statement in Theorem 11 below. Combining Main Theorem 1 with Theorem 3 we obtain, in the special case $d = 2$:

Main Theorem 2. *For $q = 8, 9$ and for all prime powers $q \geq 16$ there is an infinite family of $(n, t, 2, n-t)$-arithmetic secret sharing schemes for $\mathbb{F}_q^k$ over $\mathbb{F}_q$ with t-uniformity where n is unbounded, $k = \Omega(n)$ and $t = \Omega(n)$.*

As to efficiency, when given a divisor that is a solution to these Riemann-Roch systems, it is efficient to compute a generator for the scheme C defined by this divisor. However, solving Riemann-Roch systems efficiently in full generality is subject of further research. In particular, for our strongest results to follow it is not known at present how to efficiently compute a generator. But of course, there exists a poly(n)-size description of generators, so overall, there is efficiency as before, but now in the weaker model where such description is given as advice.

More precisely, we have the following result (for $d > 2$ there is a similar analysis).

Theorem 11. *Let $\mathbb{F}_q$ be a finite field. Suppose $\kappa \in [0, \frac{1}{3})$ and $\tau \in (0, 1]$ and $0 < A \leq A(q)$ are real number such that $A > \frac{1+\kappa}{1-3\kappa}(1 + J_2(q, A))$ and*

$$\tau + \frac{H_2(\tau)}{\log q} < \frac{1}{3}\left(1 - 3\kappa - \frac{(1 + J_2(q, A))(1 + \kappa)}{A}\right)$$

Then there is an infinite family of $(n, t, 2, n-t)$-arithmetic secret sharing schemes for $\mathbb{F}_q^k$ over $\mathbb{F}_q$ with uniformity where n is unbounded, $k = \lfloor \kappa n \rfloor + 1$ and $t = \lfloor \tau n \rfloor$.

The proof of this fact relies on showing that the conditions in Corollary 2 are satisfied asymptotically for a family of function field with Ihara's limit A, if the requirements of Theorem 11 are met. It is easy to show why Theorem 11 implies Main Theorem 2: if $0 < A \leq A(q)$ is such that $A > 1 + J_2(q, A)$ we can always select $\kappa \in (0, \frac{1}{3})$ and $\tau \in (0, 1]$ satisfying the conditions in Theorem 11. Note that in order to obtain the result in Main Theorem 2 we require $\kappa > 0$.

We prove Theorem 11 formally below, but give here an an indication of how one would bound asymptotically each parameter in the inequality of Corollary 2. Of course $|\mathcal{J}_F[2]|$ is dealt with asymptotically with the torsion limit $J_2(q, A)$ which we have introduced in this paper. Stirling's Formula gives an asymptotical bound for the binomial coefficients $\binom{n}{t}$ when t is some fixed fraction of n. Finally the quotients A_r/h can be bounded by means of algebraic geometric techniques which have been used before in the code theoretic literature, for instance [25], [28], [39], [40]. We state now an upper bound of this type.

Proposition 4. *Let $F/\mathbb{F}_q$ be a function field with $g \geq 1$. Then, for any $r \in \mathbb{Z}$ with $0 \leq r \leq g - 1$, $A_r/h \leq \frac{g}{q^{g-r-1}(\sqrt{q}-1)^2}$.*

PROOF. For $i \geq 2g-1$, $A_i = \frac{h}{q-1}(q^{i+1-g}-1)$ (see Lemma 5.1.4 and Corollary 5.1.11 in [34]). This has been exploited in Lemma 3 (ii) from [28], to show that

$$\sum_{i=0}^{g-2} A_i T^i + \sum_{i=0}^{g-1} q^{g-1-i} A_i T^{2g-2-i} = \frac{L(T) - hT^g}{(1-T)(1-qT)}$$

where $L(T)$ is the L-polynomial associated to the zeta function of F.

The claim from Proposition 4 can be derived from a relation that is obtained by taking the limit as T tends to $1/q$ on both sides of the equation above, where l'Hôpital's Rule is applied on the right hand side, then finding an expression for $L'(1/q)$ (using the Functional Equation for L-polynomials and the fact that $L(1) = h$) and substituting that back in. This is similar to the proof of Proposition 2.5 (in the case $s = 0$) in [40].

△

PROOF OF THEOREM 11. Fix any A, κ, τ satisfying the conditions of the statement. Let $\mathcal{F} = \{F_m\}_{m>0}$ be an infinite family of algebraic function fields over $\mathbb{F}_q$ with $g(F_m) \to \infty$ such that $A(\mathcal{F}) \geq A$ and $J := J_2(\mathcal{F}) = J_2(q, A)$. Define $g_m = g(F_m)$, $h_m = h(F_m)$, $j_m = \log_q(|\mathcal{J}(F_m)[2]|)$. Let $n_m = \lfloor \frac{1}{1+\kappa}(N(F_m)-1) \rfloor$ and $k_m = \lfloor \kappa n_m \rfloor + 1$. Note $n_m + k_m \leq N(F_m)$ so we can pick $n_m + k_m$ distinct rational points in F_m. We set $t_m = \lfloor \tau n_m \rfloor$. We choose $d_m = \lfloor \delta g_m \rfloor$ where $\delta = 1 + \frac{A-1-J}{3}$. Define $(r_1)_m = 2g_m - d_m + t_m + k_m - 2$ and $(r_2)_m = 2d_m - n_m + t_m$. For m large enough we want to verify that we can apply Corollary 2 to F_m. We already noted we can take $n_m + k_m$ distinct points in $\mathbb{P}^{(1)}(F_m)$ so we now need to verify the condition

$$h_m > \binom{n_m}{t_m}(A_{(r_1)_m} + A_{(r_2)_m}|\mathcal{J}_{F_m}[2]|).$$

We will use Proposition 4. It is easy to see that $0 \leq (r_1)_m, (r_2)_m \leq g_m$ for large enough m for our selection of the parameters. Thus,

$$A_{(r_i)_m} \leq \frac{g_m h_m}{q^{g_m-(r_i)_m-1}(\sqrt{q}-1)^2}$$

for large enough m and $i = 1, 2$. Consequently it is sufficient to show that

$$\binom{n_m}{t_m} \frac{g_m q^{t_m}}{q^{g_m-1}(\sqrt{q}-1)^2} \left(q^{(r_1)_m - t_m} + q^{(r_2)_m - t_m}|\mathcal{J}_{F_m}[2]|\right) < 1$$

which is equivalent, taking logarithms, to

$$\log_q \binom{n_m}{t_m} + \log_q \left(\frac{g_m q^{t_m}}{q^{g_m-1}(\sqrt{q}-1)^2}\right) + \log_q \left(q^{(r_1)_m - t_m} + q^{(r_2)_m - t_m}|\mathcal{J}_{F_m}[2]|\right) < 0. \tag{1}$$

Take $\epsilon \in \mathbb{R}_{>0}$ such that $\tau + \frac{H_2(\tau)}{\log q} < \frac{1}{3}\left(1 - 3\kappa - \frac{(1+J)(1+\kappa)}{3A} - 3\epsilon\right)$, which exists by hypothesis. For large enough m, by definition of J, $j_m < (J+\epsilon)g_m$. Moreover by definition of A we have $(A-\epsilon)g_m < n_m + k_m \leq Ag_m$ for large enough m. Note that this implies $\frac{1}{1+\kappa}(A-\epsilon)g_m \leq n_m \leq \frac{1}{1+\kappa}Ag_m$ and $k_m \leq \frac{\kappa}{1+\kappa}Ag_m + 1$. We have the following observations: First, since $t_m \leq \tau n_m$, from Stirling's Formula $\binom{n_m}{t_m} \leq 2^{H_2(\tau)n_m}$, and hence $\log_q \binom{n_m}{t_m} \leq \frac{H_2(\tau)}{\log q} n_m \leq \frac{H_2(\tau)}{(1+\kappa)\log q}Ag_m$. Second, we have

$$\log_q \left(q^{(r_1)_m - t_m} + |\mathcal{J}(\mathbb{F}_m)[2]| q^{(r_2)_m - t_m} \right) \leq$$

$$\log_q 2 + max\{2g_m - d_m + k_m - 2, 2d_m - n_m + j_m\}.$$

Now for large enough m, the following two inequalities hold:

$$2g_m - d_m + k_m - 2 \leq \left(2 - \delta + \frac{\kappa}{1+\kappa}A\right) g_m = \left(1 + \frac{1}{3}(1+J) + \frac{2\kappa - 1}{3(1+\kappa)}A\right) g_m,$$

$$2d_m - n_m + j_m \leq \left(2\delta - \frac{1}{1+\kappa}(A-\epsilon) + (J+\epsilon)\right) g_m$$

$$\leq \left(1 + \frac{1}{3}(1+J) + \frac{2\kappa-1}{3(1+\kappa)}A + 2\epsilon\right) g_m.$$

Finally, for large enough m, using elementary calculus and noticing $t_m \leq \tau n_m$ we get

$$\log_q \left(\frac{g_m q^{t_m}}{q^{g_m - 1}(\sqrt{q}-1)^2} \right) \leq \left(\frac{\tau}{1+\kappa}A - 1 + \epsilon \right) g_m.$$

Putting all these observations together we obtain that the left part of Equation 1 is at most

$$\frac{H_2(\tau)}{(1+\kappa)\log q} A g_m + \left(\frac{\tau}{1+\kappa}A - 1 + \epsilon\right) g_m +$$

$$\log_q 2 + \left(1 + \frac{1}{3}(1+J) + \frac{2\kappa-1}{3(1+\kappa)}A + 2\epsilon\right) g_m.$$

Now using $\tau + \frac{H_2(\tau)}{\log q} < \frac{1}{3}\left(1 - 3\kappa - \frac{(1+J)(1+\kappa)}{3A} - 3\epsilon\right)$ one can see that this expression is at most $\log_q 2 - \frac{\kappa}{3(1+\kappa)}Ag_m$ and this is clearly smaller than 0 for large enough m. Therefore, we can apply Corollary 2 to F_m, for each $m > M_0$ (for some constant M_0), and we have an $(n_m, t_m, 2, n_m - t_m)$-arithmetic secret sharing scheme for $\mathbb{F}_q^{k_m}$ over $\mathbb{F}_q$ with uniformity, with $k_m = \lfloor \kappa n_m \rfloor + 1$ and $t_m = \lfloor \tau n_m \rfloor$. Since $N(F_m)$ tends to ∞ as m tends to ∞ (because $A(\mathcal{F}) \geq A > 0$) then the set $\mathcal{M} = \{n_m\}_{m \geq M_0}$ is infinite. This concludes the proof. △

Finally, using our paradigm we also improve the explicit lower bounds for the parameter $\widehat{\tau}(q)$ from [7] and [4] for all q with $q \leq 81$ and q square, as well as for all q with $q \leq 9$. Recall $\widehat{\tau}(q)$ is defined as the maximum value of $3t/(n-1)$ which can be obtained asymptotically (when n tends to infinity) when t, n are subject to the condition that an $(n, t, 2, n-t)$-arithmetic secret sharing for $\mathbb{F}_q$ over $\mathbb{F}_q$ exists (no uniformity required here). The new bounds are shown in the upper row of Table 1. All the new bounds marked with a star (*) are obtained by applying

Theorem 11 in the case $\kappa = 0$ and using the upper bounds given in Theorem 1 for the torsion limits. To obtain the rest of the new upper bounds, for each q, we apply the field descent technique in [4] to $\mathbb{F}_{q^2}$(in the special case of $\mathbb{F}_9$, even though Theorem 11 can be applied directly, as remarked in Main Theorem 2, it is better to apply Theorem 11 to $\mathbb{F}_{81}$ and then use the descent technique). These are compared with the previous bounds: the ones obtained in [7] (marked also with the symbol (*)), and the rest, which were obtained in [4] by means of the aforementioned field descent technique.

Table 1. Lower bounds for $\widehat{\tau}(q)$

q	2	3	4	5	7	8	9
New bounds	0.034	0.057	0.104	0.107	0.149	0.173(*)	0.173
Previous bounds	0.028	0.056	0.086	0.093	0.111	0.143	0.167

q	16	25	49	64	81
New bounds	0.298(*)	0.323(*)	0.448(*)	0.520(*)	0.520(*)
Previous bounds	0.244	0.278	0.333(*)	0.429(*)	0.500(*)

Acknowledgments. We are grateful for valuable contributions to the refinements on the bounds for the torsion-limit in Theorem 1. Bas Edixhoven and Hendrik Lenstra suggested the generic approach we used in its second part. Alp Bassa and Peter Beelen confirmed our hope that stronger bounds should be attainable from certain *specific* recursive towers, by contributing the proof of its third part. We also thank Hendrik for many helpful discussions, and for his encouragement since the paper was first circulated in the Fall of 2009. Part of this research was done when Cascudo was with University of Oviedo, partially supported by Spanish MEC project MTM2010-18370-C04-01. Cramer's research was supported by his NWO VICI project *Mathematical Foundations of Secure Computation*. Xing's research is partially supported by the Singapore National Research Foundation Competitive Research Program grant NRF-CRP2-2007-03 and the Singapore Ministry of Education under Research Grant T208B2206.

References

1. Bassa, A., Garcia, A., Stichtenoth, H.: A new tower over cubic finite fields. Moscow Mathematical Journal 8(3), 401–418 (2008)
2. Ben-Or, M., Goldwasser, S., Wigderson, A.: Completeness theorems for non-cryptographic fault-tolerant distributed computation. In: Proceedings of STOC 1988, pp. 1–10. ACM Press, New York (1988)
3. Bezerra, J., Garcia, A., Stichtenoth, H.: An explicit tower of function fields over cubic finite fields and Zink's lower bound. J. Reine Angew. Math. 589, 159–199 (2005)
4. Cascudo, I., Chen, H., Cramer, R., Xing, C.: Asymptotically Good Ideal Linear Secret Sharing with Strong Multiplication over *Any* Finite Field. In: Halevi, S. (ed.) CRYPTO 2009. LNCS, vol. 5677, pp. 466–486. Springer, Heidelberg (2009)

5. Cascudo, I., Cramer, R., Xing, C.: Upper Bounds on Asymptotic Optimal Corruption Tolerance in Strongly Multiplicative Linear Secret Sharing (2009) (manuscript)
6. Chaum, D., Crépeau, C., Damgaard, I.: Multi-party unconditionally secure protocols. In: Proceedings of STOC 1988, pp. 11–19. ACM Press, New York (1988)
7. Chen, H., Cramer, R.: Algebraic Geometric Secret Sharing Schemes and Secure Multi-Party Computations over Small Fields. In: Dwork, C. (ed.) CRYPTO 2006. LNCS, vol. 4117, pp. 516–531. Springer, Heidelberg (2006)
8. Chen, H., Cramer, R., Goldwasser, S., de Haan, R., Vaikuntanathan, V.: Secure Computation from Random Error Correcting Codes. In: Naor, M. (ed.) EUROCRYPT 2007. LNCS, vol. 4515, pp. 329–346. Springer, Heidelberg (2007)
9. Chen, H., Cramer, R., de Haan, R., Cascudo Pueyo, I.: Strongly multiplicative ramp schemes from high degree rational points on curves. In: Smart, N.P. (ed.) EUROCRYPT 2008. LNCS, vol. 4965, pp. 451–470. Springer, Heidelberg (2008)
10. Cramer, R., Damgaard, I., Maurer, U.: General secure multi-party computation from any linear secret sharing scheme. In: Preneel, B. (ed.) EUROCRYPT 2000. LNCS, vol. 1807, p. 316. Springer, Heidelberg (2000)
11. Cramer, R., Damgaard, I., Pastro, V.: On the Amortized Complexity of Zero Knowledge Protocols for Multiplicative Relations (2010) (manuscript)
12. Cramer, R., Daza, V., Gracia, I., Jiménez Urroz, J., Leander, G., Martí-Farré, J., Padró, C.: On codes, matroids and secure multi-party computation from linear secret sharing schemes. IEEE Transactions on Information Theory 54, 2644–2657 (2008); Earlier version: CRYPTO 2005
13. Damgaard, I., Ishai, Y., Krøigaard, M.: Perfectly Secure Multiparty Computation and the Computational Overhead of Cryptography. In: Gilbert, H. (ed.) EUROCRYPT 2010. LNCS, vol. 6110, pp. 445–465. Springer, Heidelberg (2010)
14. Duursma, I., Mak, K.-H.: On lower bounds for the Ihara constants A(2) and A(3). preprint (2011), `http://arxiv.org/abs/1102.4127`
15. Franklin, M., Yung, M.: Communication Complexity of Secure Computation. In: ACM STOC 1992, pp. 699–710
16. Garcia, A., Stichtenoth, H.: A tower of Artin-Schreier extensions of function fields attaining the Drinfeld-Vlăduţ bound. Invent. Math. 121, 211–222 (1995)
17. Garcia, A., Stichtenoth, H.: On the asymptotic behavior of some towers of function fields over finite fields. J. Number Theory 61, 248–273 (1996)
18. Harnik, D., Ishai, Y., Kushilevitz, E., Nielsen, J.: OT-Combiners via Secure Computation. In: Canetti, R. (ed.) TCC 2008. LNCS, vol. 4948, pp. 393–411. Springer, Heidelberg (2008)
19. Hirschfeld, J.W.P., Korchmáros, G., Torres, F.: Algebraic Curves of Finite Fields. Princeton Series in Applied Mathematics (2008)
20. Ihara, Y.: Some remarks on the number of rational points of algebraic curves over finite fields. J. Fac. Sci. Tokyo 28(3), 721–724 (1981)
21. Ishai, Y., Kushilevitz, E., Ostrovsky, R., Prabhakaran, M., Sahai, A., Wullschleger, J.: Constant-rate OT from Noisy Channels. These proceedings, CRYPTO (2011)
22. Ishai, Y., Kushilevitz, E., Ostrovsky, R., Sahai, A.: Extracting Correlations. In: Proc. 50th IEEE FOCS, pp. 261–270 (2009)
23. Ishai, Y., Kushilevitz, E., Ostrovsky, R., Sahai, A.: Zero-knowledge from secure multiparty computation. In: Proceedings of 39th STOC, San Diego, Ca., USA, pp. 21–30 (2007)
24. Ishai, Y., Prabhakaran, M., Sahai, A.: Founding Cryptography on Oblivious Transfer-Efficiently. In: Wagner, D. (ed.) CRYPTO 2008. LNCS, vol. 5157, pp. 572–591. Springer, Heidelberg (2008)

25. Lachaud, G., Martin-Deschamps, M.: Deschamps Nombre de points des jacobiennes sur un corps fini. Acta Arith. 56, 329–340 (1990)
26. Milne, J.S.: Abelian Varities. Online Lecture Notes (2009)
27. Mumford, D.: Abelian Varieties. Oxford University Press, Oxford (1970)
28. Niederreiter, H., Xing, C.: Low-Discrepancy Sequences and Global Function Fields with Many Rational Places. Finite Fields and Their Applications 2, 241–273 (1996)
29. Niederreiter, H., Xing, C.: Rational points on curves over finite fields-theory and applications, Cambridge (2000)
30. Rosen, M.: Number Theory in Function Fields. GTM, Springer (2001)
31. Serre, J.-P.: Rational points on curves over finite fields. Harvard University, Cambridge (1985)
32. Shamir, A.: How to share a secret. Comm. of the ACM 22(11), 612–613 (1979)
33. Shparlinski, I., Tsfasman, M., Vlăduţ, S.: Curves with many points and multiplication in finite fields. Lecture Notes in Math., vol. 1518, pp. 145–169. Springer, Berlin (1992)
34. Stichtenoth, H.: Algebraic function fields and codes. Springer, Heidelberg (1993) (new edition: 2009)
35. Tsfasman, M., Vlăduţ, S.: Modular curves, Shimura curves, and Goppa codes, better than Varshamov Gilbert bound. Math. Nachr. 109, 21–28 (1982)
36. Vlăduţ, S.G.: An exhaustion bound for algebro-geometric modular codes. Probl. Inf. Transm. 23, 22–34 (1987)
37. Vlăduţ, S.G., Drinfeld, V.G.: Number of points of an algebraic curves. Funct. Anal. Appl. 17, 53–54 (1983)
38. Weil, A.: Variétés Abéliennes et Courbes Algébriques. Hermann, Paris (1948)
39. Xing, C.: Algebraic geometry codes with asymptotic parameters better than the Gilbert-Varshamov and the Tsfasman-Vlăduţ-Zink bounds. IEEE Trans. on Inf. Theory 47(1), 347–352 (2001)
40. Xing, C.: Goppa Geometric Codes Achieving the Gilbert-Varshamov Bound. IEEE Trans. on Inf. Theory 51(1), 259–264 (2005)
41. Xing, C., Ling, Y.S.: Algebraic curves with many points over the binary field. J. Algebra 311, 775–780 (2007)
42. Zink, T.: Degeneration of Shimura surface and a problem in coding theory. In: Budach, L. (ed.) FCT 1985. LNCS, vol. 199, pp. 503–511. Springer, Heidelberg (1985)

Public-Key Identification Schemes Based on Multivariate Quadratic Polynomials

Koichi Sakumoto, Taizo Shirai, and Harunaga Hiwatari

Sony Corporation
5-1-12 Kitashinagawa Shinagawa-ku, Tokyo 141-0001, Japan
{Koichi.Sakumoto,Taizo.Shirai,Harunaga.Hiwatari}@jp.sony.com

Abstract. A problem of solving a system of multivariate quadratic polynomials over a finite field, which is called an MQ problem, is a promising problem in cryptography. A number of studies have been conducted on designing public-key schemes using the MQ problem, which are known as multivariate public-key cryptography (MPKC). However, the security of the existing schemes in MPKC relies *not* only on the MQ problem but *also* on an Isomorphism of Polynomials (IP) problem. In this paper, we propose public-key identification schemes based on the conjectured intractability of the MQ problem under the assumption of the existence of a non-interactive commitment scheme. Our schemes do *not* rely on the IP problem, and they consist of an identification protocol which is zero-knowledge argument of knowledge for the MQ problem. For a practical parameter choice, the efficiency of our schemes is highly comparable to that of identification schemes based on another problem including Permuted Kernels, Syndrome Decoding, Constrained Linear Equations, and Permuted Perceptrons. Furthermore, even if the protocol is repeated in parallel, our scheme can achieve the security under active attack with some additional cost.

Keywords: identification scheme, zero knowledge, MQ problem.

1 Introduction

A problem of solving a system of multivariate quadratic polynomials over a finite field, which is called an MQ problem, is a promising problem in cryptography. The associated decision problem is known to be NP-complete [24,40], and a random instance of the MQ problem is widely believed to be intractable. In contrast to factorization or a discrete logarithm problem, there is no known polynomial-time quantum algorithm to solve the MQ problem. A function consisting of multivariate quadratic polynomials, which we call an MQ function, can be used as a one-way function with short input and output. Complexity of generic attacks using Gröbner basis is known to be exponential in time and space [3,16], and the best known attack to break the MQ function over $\mathbb{F}_2$ with 84-bit input and 80-bit output requires $2^{88.7} (> 2^{80})$ bit operations [10].

P. Rogaway (Ed.): CRYPTO 2011, LNCS 6841, pp. 706–723, 2011.

A number of studies on designing primitives based on the MQ function have been conducted both in symmetric and in asymmetric cryptography. In symmetric cryptography, a stream cipher which is named QUAD is proposed by Berbain et al. [7]. The security of QUAD is provably reducible to the conjectured intractability of the MQ problem. In asymmetric cryptography, several public-key schemes have been proposed, which are known as multivariate public-key cryptography (MPKC) [30,35,39]. However, the security of the existing schemes in MPKC relies not only on the MQ problem but also on an Isomorphism of Polynomials (IP) problem. The IP problem consists of recovering a particular transformation between two sets of multivariate polynomials, and some cryptanalyses of the problem have been reported [11,17,22,41]. In fact, some schemes in MPKC have been already shown to be insecure [11,14,31,38].

In this paper, we propose public-key identification schemes based on the conjectured intractability of the MQ problem under the assumption of the existence of a non-interactive commitment scheme which is statistically-hiding and computationally-binding. We emphasize that our schemes do not rely on the IP problem. The assumption for the commitment scheme is natural, since it can be constructed from a collision resistant hash function [27]. Our identification protocols are non-trivial constructions of statistical zero-knowledge argument of knowledge for the MQ problem. Assuming the intractability of the MQ function, our identification schemes consisting of the *sequential* composition and the *parallel* composition of the protocols are secure against impersonation under *active* attack and *passive* attack, respectively. These security levels are the same as those of known identification schemes based on another problem including Permuted Kernels (PK) [46], binary Syndrome Decoding (SD) [47,49], Constrained Linear Equations (CLE) [48], Permuted Perceptrons (PP) [42,43], and q-ary SD [12].

For a practical parameter choice, the sizes of a public key, a secret key, and communication data of our schemes are comparable to those of the schemes based on PK, SD, CLE, PP, and q-ary SD. In particular, the sizes of a public key and a secret key of our 3-pass scheme are only 80 bits and 84 bits for 80-bit security, respectively. These are smaller than those of the known schemes [12,42,43,46,47,48,49]. This is due to the fact that the MQ function has short input and output. The size of communication data in our 3-pass protocol is 29,640 bits when the impersonation probability is less than 2^{-30}. This is also small compared to those of the existing 3-pass protocols [42,43,47,48,49], which are between 45,517 bits and 100,925 bits. Although the data size of system parameter of our scheme is relatively large, it can be reduced to some small seed, e.g. 128 bits, by employing a pseudo-random number generator. The technique is also used in the implementation of QUAD [2].

Furthermore, we consider the case that our scheme employs the MQ function which is substantially compressing (e.g., mapping 160 bits to 80 bits), although the sizes of the secret key and the communication data increase compared to those of the practical parameter choice. In this case, when such a function is preimage resistant, our scheme is secure under active attack even if the protocol is repeated in parallel. The proof of the security is non-trivial, since zero knowledge is not preserved under the parallel composition. Thus we prove the security by also showing that the MQ function is *second-preimage* resistant if such a function is *preimage* resistant, although the MQ function is known not to have the collision resistance [9].

Techniques for Our Constructions. Our protocols employ the cut-and-choose approach, where a prover first divides her secret into shares and then proves the correctness of some shares depending on the choice of a verifier without revealing the secret itself. The property of group homomorphism such as a modular exponentiation $x \mapsto g^x \mod p$ and a linear function $x \mapsto Mx$ is useful for this approach, since dividing a secret $s = r_0 + r_1$ simply corresponds to dividing its image $g^s = (g^{r_0})(g^{r_1})$ and $Ms = Mr_0 + Mr_1$, respectively. However, the MQ function $(x_1, \ldots, x_n) \mapsto (y_1, \ldots, y_m)$ where $y_l = \sum_{i,j} a_{l,i,j} x_i x_j + \sum_i b_{l,i} x_i$ does not seem to have such a property.

Therefore, we introduce new dividing techniques using the bilinearity of a *polar form* of the MQ function. The polar form $\mathbf{G}$ of the MQ function $\mathbf{F}$ is a function $\mathbf{G}(\mathbf{x}_1, \mathbf{x}_2) = \mathbf{F}(\mathbf{x}_1 + \mathbf{x}_2) - \mathbf{F}(\mathbf{x}_1) - \mathbf{F}(\mathbf{x}_2)$ and is known to be bilinear. It was introduced as the differential of the quadratic system, and has been used for cryptanalysis of MPKC so far [14,15,21,22]. To our knowledge, this is the first time that it is constructively used in a context of a public-key identification scheme.

Our dividing techniques are briefly described as follows. Let $\mathbf{s}$ and $\mathbf{v} = \mathbf{F}(\mathbf{s})$ be a secret key and a public key, respectively. When the secret key is divided as $\mathbf{s} = \mathbf{r}_0 + \mathbf{r}_1$, the public key $\mathbf{v} = \mathbf{F}(\mathbf{r}_0 + \mathbf{r}_1)$ can be represented as $\mathbf{v} = \mathbf{F}(\mathbf{r}_0) + \mathbf{F}(\mathbf{r}_1) + \mathbf{G}(\mathbf{r}_0, \mathbf{r}_1)$ by using the polar form $\mathbf{G}$ of $\mathbf{F}$. However, this representation still contains the term $\mathbf{G}(\mathbf{r}_0, \mathbf{r}_1)$ which depends on both $\mathbf{r}_0$ and $\mathbf{r}_1$. Consider that $\mathbf{r}_0$ and $\mathbf{F}(\mathbf{r}_0)$ are further divided as $\mathbf{r}_0 = \mathbf{t}_0 + \mathbf{t}_1$ and $\mathbf{F}(\mathbf{r}_0) = \mathbf{e}_0 + \mathbf{e}_1$, respectively. In this case, the public key can be divided into two parts $\mathbf{v} = \big(\mathbf{G}(\mathbf{t}_0, \mathbf{r}_1) + \mathbf{e}_0\big) + \big(\mathbf{F}(\mathbf{r}_1) + \mathbf{G}(\mathbf{t}_1, \mathbf{r}_1) + \mathbf{e}_1\big)$, due to the bilinearity of $\mathbf{G}$. Each of the two parts is represented by either a tuple $(\mathbf{r}_1, \mathbf{t}_0, \mathbf{e}_0)$ or a tuple $(\mathbf{r}_1, \mathbf{t}_1, \mathbf{e}_1)$, while no information on the secret key $\mathbf{s}$ can be obtained from one out of the two tuples.

Related Work. Identification schemes based on PK [46], SD [47,49], CLE [48], PP [42,43], and q-ary SD [12] have some features similar to our schemes as follows. First, these schemes rely on the hardness of a random instance of each of the problems whose associated decision version is known to be NP-complete. Second, their protocols have perfect correctness. Finally, assuming the existence of a non-interactive commitment scheme, the sequential version and the parallel version of the schemes are secure against impersonation under active attack and passive attack, respectively. However, it is not explicitly known that the parallel versions of these schemes achieve the security under active attack.

On the other hand, lattice-based schemes [29,33,34,36] have other features. They are based on an average-case problem which is as hard as worst-case problems, and some of them [29,33,34] are secure under active attack even if repeated in parallel. Lyubashevsky's scheme [34] is stated to be more practically efficient than the others [29,33,36]. The size of communication data of the scheme [34] for 80-bit security against impersonation is only about 65,000 bits, although the scheme has small correctness error 2^{-20}. Both of the sizes of the public key and the secret key are 16,000 bits.

In a context of post-quantum cryptography, Komano et al. proposed a signature scheme based on a section finding problem on algebraic surface [32]. Their construction, similarly to our schemes, does not rely on a property of homomorphism. However, their scheme is universally forgeable under key-only attack, and their technique turned out to be unsuccessful to realize a signature scheme [45].

Paper Organization. The remainder of this paper is organized as follows. In Section 2 we present several notions and tools that are used in our constructions. In Section 3 and Section 4, our 3-pass and 5-pass constructions are presented, respectively. In Section 5 we discuss their security and efficiency for a practical parameter choice. In Section 6 we study the security of the parallel composition of our scheme at the expense of the efficiency. In Section 7 we mention some extensions of our scheme.

2 Preliminaries

A finite field of order q is denoted by $\mathbb{F}_q$. If an element x is randomly chosen from a finite set S, it is expressed by $x \in_R S$. If A and B are sets, and $R \subset A \times B$ is a binary relation, then we define $R(x) = \{s : (x, s) \in R\}$. If $s \in R(x)$, then s is called a solution for x.

Identification Scheme. An identification scheme is a tuple of algorithms ($\mathtt{Setup}$, $\mathtt{Gen}, \mathtt{P}, \mathtt{V}$) defined as follows. $\mathtt{Setup}$ is a setup algorithm which takes a security parameter 1^λ and outputs a system parameter *param*. $\mathtt{Gen}$ is a key-generation algorithm which takes *param*, and outputs a public key and a secret key (pk, sk). A pair of a prover $\mathtt{P}$ and a verifier $\mathtt{V}$ is an interactive protocol where a common input is $(param, pk)$ and an auxiliary input of $\mathtt{P}$ is sk. After interactions, $\mathtt{V}$ outputs a bit as a verification result. The protocol $(\mathtt{P}, \mathtt{V})$ is called an identification protocol.

Security against impersonation under *passive*/*active* attacks considers an adversary whose goal is to impersonate the prover without the knowledge of the secret key. The adversary under *passive* attack has access to interactions between the real prover and an honest verifier. The adversary under *active* attack can interact with the prover. Requiring security against impersonation under active attack is stronger than under passive attack. The details are described in [1,18].

The definitions of zero knowledge, witness indistinguishability, and argument of knowledge are omitted. For formal definitions, refer to textbooks, e.g., [25].

String Commitment Scheme. A string commitment function is denoted by *Com*. The commitment scheme runs in two phases. In the first phase, the sender computes a commitment value $c \leftarrow Com(s; \rho)$ and sends c to the receiver, where s

is a string and ρ is a random string. In the second phase, the sender gives (s, ρ) to the receiver and the receiver verifies $c = Com(s; \rho)$. We require two security properties of *Com*, statistically hiding and computationally binding. Informally, the former means that, at the end of the first phase, no receiver can distinguish two commitment values generated from two distinct strings even if the receiver is computationally unbounded. The latter means that, no polynomial-time sender can change the committed string after the first phase. The formal definitions and a practical construction are given in [27]. Throughout this paper, we assume the existence of such a commitment scheme. The assumption is natural, since it can be constructed from a collision resistant hash function [27]. Note that such an *interactive* commitment scheme can be constructed from any one-way function including the MQ function [26].

The MQ Function. We denote by $\mathcal{MQ}(n, m, \mathbb{F}_q)$ a family of functions

$$\left\{ \mathbf{F}(\mathbf{x}) = (f_1(\mathbf{x}), \ldots, f_m(\mathbf{x})) \,\middle|\, \begin{array}{r} f_l(\mathbf{x}) = \sum_{i,j} a_{l,i,j} x_i x_j + \sum_i b_{l,i} x_i, \\ a_{l,i,j}, b_{l,i} \in \mathbb{F}_q \;\text{ for }\; l = 1, \ldots, m \end{array} \right\}$$

where $\mathbf{x} = (x_1, \ldots, x_n)$. For the simplicity, constant terms are omitted without any security loss. We call $\mathbf{F} \in \mathcal{MQ}(n, m, \mathbb{F}_q)$ an MQ function. A function $\mathbf{G}(\mathbf{x}, \mathbf{y}) = \mathbf{F}(\mathbf{x} + \mathbf{y}) - \mathbf{F}(\mathbf{x}) - \mathbf{F}(\mathbf{y})$ is called the polar form of $\mathbf{F}$. The function $\mathbf{G} = (g_1, \ldots, g_m)$ is bilinear, since $g_l(\mathbf{x}, \mathbf{y}) = \sum_{i,j} a_{l,i,j}(y_i x_j + x_i y_j)$ where $\mathbf{x} = (x_1, \ldots, x_n)$ and $\mathbf{y} = (y_1, \ldots, y_n)$. An intractability assumption for a random instance of $\mathcal{MQ}(n, m, \mathbb{F}_q)$ is defined as follows.

Definition 1. *For polynomially bounded functions $n = n(\lambda)$, $m = m(\lambda)$, and $q = q(\lambda)$, it is said that $\mathcal{MQ}(n, m, \mathbb{F}_q)$ is intractable if there is no polynomial-time algorithm that takes $(\mathbf{F}, \mathbf{v})$ generated via $\mathbf{F} \in_R \mathcal{MQ}(n, m, \mathbb{F}_q)$, $\mathbf{s} \in_R \mathbb{F}_q^n$, and $\mathbf{v} \leftarrow \mathbf{F}(\mathbf{s})$ and finds a preimage $\mathbf{s}' \in \mathbb{F}_q^n$ such that $\mathbf{F}(\mathbf{s}') = \mathbf{v}$ with non-negligible probability $\epsilon(\lambda)$.*

All the state-of-the-art solving techniques have exponential complexity to break the intractability [8,10,16]. In particular, it is known that complexity of generic attacks using Gröbner basis is exponential in time and space [3,16]. Bouillaguet et al. stated that it would not outperform exhaustive search in the practically interesting range $m = n \leq 200$ [10]. They proposed an improved exhaustive search algorithm to break $\mathcal{MQ}(n, m, \mathbb{F}_2)$ in $2^{n+2} \cdot \log_2 n$ bit operations, which is the best known algorithm [10].

In addition, for $\mathbf{F} \in \mathcal{MQ}(n, m, \mathbb{F}_q)$, we define a binary relation $R_{\mathbf{F}} = \{(\mathbf{v}, \mathbf{x}) \in \mathbb{F}_q^m \times \mathbb{F}_q^n : \mathbf{v} = \mathbf{F}(\mathbf{x})\}$. Given an instance $\mathbf{F} \in \mathcal{MQ}(n, m, \mathbb{F}_q)$ and a vector $\mathbf{v} \in \mathbb{F}_q^m$, the MQ problem is finding a solution $\mathbf{s} \in R_{\mathbf{F}}(\mathbf{v})$.

3 A 3-pass Identification Scheme

In this section, we construct an identification scheme which consists of a 3-pass statistical zero-knowledge argument of knowledge for $R_{\mathbf{F}}$ with knowledge error 2/3, assuming the existence of a non-interactive commitment scheme *Com* which is statistically hiding and computationally binding.

Prover's input: $((\mathbf{F}, \mathbf{v}), \mathbf{s})$		**Verifier**'s input: $(\mathbf{F}, \mathbf{v})$
Pick $\mathbf{r}_0, \mathbf{t}_0 \in_R \mathbb{F}_q^n$, $\mathbf{e}_0 \in_R \mathbb{F}_q^m$ $\mathbf{r}_1 \leftarrow \mathbf{s} - \mathbf{r}_0$, $\mathbf{t}_1 \leftarrow \mathbf{r}_0 - \mathbf{t}_0$ $\mathbf{e}_1 \leftarrow \mathbf{F}(\mathbf{r}_0) - \mathbf{e}_0$ $c_0 \leftarrow Com(\mathbf{r}_1, \mathbf{G}(\mathbf{t}_0, \mathbf{r}_1) + \mathbf{e}_0)$ $c_1 \leftarrow Com(\mathbf{t}_0, \mathbf{e}_0)$ $c_2 \leftarrow Com(\mathbf{t}_1, \mathbf{e}_1)$	$\xrightarrow{(c_0, c_1, c_2)}$	
	$\xleftarrow{Ch}$	Pick $Ch \in_R \{0, 1, 2\}$
If $Ch = 0$, $Rsp \leftarrow (\mathbf{r}_0, \mathbf{t}_1, \mathbf{e}_1)$ If $Ch = 1$, $Rsp \leftarrow (\mathbf{r}_1, \mathbf{t}_1, \mathbf{e}_1)$ If $Ch = 2$, $Rsp \leftarrow (\mathbf{r}_1, \mathbf{t}_0, \mathbf{e}_0)$	$\xrightarrow{Rsp}$	If $Ch=0$, parse $Rsp=(\mathbf{r}_0, \mathbf{t}_1, \mathbf{e}_1)$ and check $c_1 \stackrel{?}{=} Com(\mathbf{r}_0 - \mathbf{t}_1, \mathbf{F}(\mathbf{r}_0) - \mathbf{e}_1)$ $c_2 \stackrel{?}{=} Com(\mathbf{t}_1, \mathbf{e}_1)$ If $Ch=1$, parse $Rsp=(\mathbf{r}_1, \mathbf{t}_1, \mathbf{e}_1)$ and check $c_0 \stackrel{?}{=} Com(\mathbf{r}_1, \mathbf{v} - \mathbf{F}(\mathbf{r}_1) - \mathbf{G}(\mathbf{t}_1, \mathbf{r}_1) - \mathbf{e}_1)$ $c_2 \stackrel{?}{=} Com(\mathbf{t}_1, \mathbf{e}_1)$ If $Ch=2$, parse $Rsp=(\mathbf{r}_1, \mathbf{t}_0, \mathbf{e}_0)$ and check $c_0 \stackrel{?}{=} Com(\mathbf{r}_1, \mathbf{G}(\mathbf{t}_0, \mathbf{r}_1) + \mathbf{e}_0)$ $c_1 \stackrel{?}{=} Com(\mathbf{t}_0, \mathbf{e}_0)$

Fig. 1. Our 3-pass identification protocol

Key Generation. We begin with describing a setup algorithm and a key-generation algorithm. Let λ be a security parameter. Let $n = n(\lambda)$, $m = m(\lambda)$, and $q = q(\lambda)$ be polynomially bounded functions. The setup algorithm `Setup` takes 1^λ and outputs a system parameter $\mathbf{F} \in_R \mathcal{MQ}(n, m, \mathbb{F}_q)$ which consists of m-tuple of random multivariate quadratic polynomials. The key-generation algorithm `Gen` takes $\mathbf{F}$. After choosing a random vector $\mathbf{s} \in_R \mathbb{F}_q^n$, `Gen` computes $\mathbf{v} \leftarrow \mathbf{F}(\mathbf{s})$, then outputs $(pk, sk) = (\mathbf{v}, \mathbf{s})$.

An Identification Protocol. The basic idea for our 3-pass construction is that a prover proves that she has a tuple $(\mathbf{r}_0, \mathbf{r}_1, \mathbf{t}_0, \mathbf{t}_1, \mathbf{e}_0, \mathbf{e}_1)$ satisfying

$$\mathbf{G}(\mathbf{t}_0, \mathbf{r}_1) + \mathbf{e}_0 = \mathbf{v} - \mathbf{F}(\mathbf{r}_1) - \mathbf{G}(\mathbf{t}_1, \mathbf{r}_1) - \mathbf{e}_1 \tag{1}$$

$$\text{and} \quad (\mathbf{t}_0, \mathbf{e}_0) = (\mathbf{r}_0 - \mathbf{t}_1, \mathbf{F}(\mathbf{r}_0) - \mathbf{e}_1), \tag{2}$$

since if the tuple satisfies (1) and (2) then $\mathbf{v} = \mathbf{F}(\mathbf{r}_0 + \mathbf{r}_1)$. Note that $\mathbf{G}$ is the polar form of $\mathbf{F}$. In the concrete protocol, corresponding to a challenge $Ch \in \{0, 1, 2\}$ of a verifier, the prover reveals one out of three tuples $(\mathbf{r}_0, \mathbf{t}_1, \mathbf{e}_1)$, $(\mathbf{r}_1, \mathbf{t}_1, \mathbf{e}_1)$, and $(\mathbf{r}_1, \mathbf{t}_0, \mathbf{e}_0)$. The verifier can check each side of each equations (1) and (2) by using either of the three tuples. Such vectors $\mathbf{r}_0, \mathbf{r}_1, \mathbf{t}_0, \mathbf{t}_1, \mathbf{e}_0, \mathbf{e}_1$ are produced by using the dividing techniques described in Section 1. Thus, when $\mathbf{r}_0$, $\mathbf{t}_0$, and $\mathbf{e}_0$ are randomly chosen, the verifier can obtain no information on the secret key $\mathbf{s}$ from only one out of the three tuples.

The 3-pass identification protocol is described in Figure 1. For the simplicity, a random string ρ in *Com* is not written explicitly. The verifier finally outputs 1 if both the checks of "$\stackrel{?}{=}$" are passed, otherwise outputs 0. This is denoted by $0/1 \leftarrow Dec(\mathbf{F}, \mathbf{v}; (c_0, c_1, c_2), Ch, Rsp)$. It is easy to see that the verifier always accepts an interaction with the honest prover. Thus the 3-pass scheme has perfect correctness.

Now we show two properties of the protocol in Theorem 2 and Theorem 3 as follows.

Theorem 2. *The 3-pass protocol is statistically zero knowledge when the commitment scheme Com is statistically hiding.*

Proof sketch. Let $\mathcal{S}$ be a simulator which takes $\mathbf{F}$ and $\mathbf{v}$ without knowing $\mathbf{s}$, and interacts with a cheating verifier $\mathcal{CV}$. We show that the simulator $\mathcal{S}$ can impersonate the honest prover with probability 2/3. The simulator $\mathcal{S}$ randomly chooses a value $Ch^* \in_R \{0, 1, 2\}$ and vectors $\mathbf{s}', \mathbf{r}'_0, \mathbf{t}'_0 \in_R \mathbb{F}_q^n$, $\mathbf{e}'_0 \in_R \mathbb{F}_q^m$, where Ch^* is a prediction of what value the cheating verifier $\mathcal{CV}$ will *not* choose. Then, it computes $\mathbf{r}'_1 \leftarrow \mathbf{s}' - \mathbf{r}'_0$ and $\mathbf{t}'_1 \leftarrow \mathbf{r}'_0 - \mathbf{t}'_0$. If $Ch^* = 0$ then it computes $\mathbf{e}'_1 \leftarrow \mathbf{v} - \mathbf{F}(\mathbf{s}') + \mathbf{F}(\mathbf{r}'_0) - \mathbf{e}'_0$, else $\mathbf{e}'_1 \leftarrow \mathbf{F}(\mathbf{r}'_0) - \mathbf{e}'_0$. If $Ch^* = 2$ then it computes $c'_0 \leftarrow Com(\mathbf{r}'_1, \mathbf{v} - \mathbf{F}(\mathbf{r}'_1) - \mathbf{G}(\mathbf{t}'_1, \mathbf{r}'_1) - \mathbf{e}'_1)$, else $c'_0 \leftarrow Com(\mathbf{r}'_1, \mathbf{G}(\mathbf{t}'_0, \mathbf{r}'_1) + \mathbf{e}'_0)$. It computes $c'_1 \leftarrow Com(\mathbf{t}'_0, \mathbf{e}'_0)$ and $c'_2 \leftarrow Com(\mathbf{t}'_1, \mathbf{e}'_1)$ and sends (c'_0, c'_1, c'_2) to $\mathcal{CV}$. Due to the statistically hiding property of *Com*, a challenge *Ch* from $\mathcal{CV}$ is different from Ch^* with probability 2/3. If $Ch \neq Ch^*$ then $(\mathbf{r}'_0, \mathbf{t}'_1, \mathbf{e}'_1)$, $(\mathbf{r}'_1, \mathbf{t}'_1, \mathbf{e}'_1)$, and $(\mathbf{r}'_1, \mathbf{t}'_0, \mathbf{e}'_0)$ are accepted responses to $Ch = 0$, 1, and 2, respectively. Note that if $Ch^* = 0$ and $Ch = 1$ then it is seen that $\mathbf{v} - \mathbf{F}(\mathbf{r}'_1) - \mathbf{G}(\mathbf{t}'_1, \mathbf{r}'_1) - \mathbf{e}'_1 = \mathbf{G}(\mathbf{t}'_0, \mathbf{r}'_1) + \mathbf{e}'_0$, since $\mathbf{e}'_1 = \mathbf{v} - \mathbf{F}(\mathbf{s}') + \mathbf{F}(\mathbf{r}'_0) - \mathbf{e}'_0$, $\mathbf{F}(\mathbf{s}') = \mathbf{F}(\mathbf{r}'_0) + \mathbf{F}(\mathbf{r}'_1) + \mathbf{G}(\mathbf{r}'_0, \mathbf{r}'_1)$, and $\mathbf{r}'_0 - \mathbf{t}'_1 = \mathbf{t}'_0$.

The details of the proof are given in the full paper, where we formally construct a black-box simulator $\mathcal{S}$ which has oracle access to a cheating verifier $\mathcal{CV}$, and outputs a successful transcript with probability 2/3. Furthermore, the distribution of the output of $\mathcal{S}$ is shown to be statistically close to the distribution of the real transcript. □

Theorem 3. *The 3-pass protocol is argument of knowledge for* $R_{\mathbf{F}}$ *with knowledge error* 2/3 *when the commitment scheme Com is computationally binding.*

Proof sketch. Let $((c_0, c_1, c_2), Ch_0, Rsp_0)$, $((c_0, c_1, c_2), Ch_1, Rsp_1)$, and $((c_0, c_1, c_2), Ch_2, Rsp_2)$ be three transcripts such that $Ch_i = i$ and $Dec(\mathbf{F}, \mathbf{v}; (c_0, c_1, c_2), Ch_i, Rsp_i) = 1$ for $i \in \{0, 1, 2\}$. Then, by using the three transcripts, it is shown to be able to either break the binding property of *Com* or extract a solution for $\mathbf{v}$. Consider the situation where the responses are parsed as $Rsp_0 = (\tilde{\mathbf{r}}_0^{(0)}, \tilde{\mathbf{t}}_1^{(0)}, \tilde{\mathbf{e}}_1^{(0)})$, $Rsp_1 = (\tilde{\mathbf{r}}_1^{(1)}, \tilde{\mathbf{t}}_1^{(1)}, \tilde{\mathbf{e}}_1^{(1)})$, and $Rsp_2 = (\tilde{\mathbf{r}}_1^{(2)}, \tilde{\mathbf{t}}_0^{(2)}, \tilde{\mathbf{e}}_0^{(2)})$. Then, it is seen that

$$\begin{aligned} c_0 &= Com(\tilde{\mathbf{r}}_1^{(1)}, \mathbf{v} - \mathbf{F}(\tilde{\mathbf{r}}_1^{(1)}) - \mathbf{G}(\tilde{\mathbf{t}}_1^{(1)}, \tilde{\mathbf{r}}_1^{(1)}) - \tilde{\mathbf{e}}_1^{(1)}) \\ &= Com(\tilde{\mathbf{r}}_1^{(2)}, \mathbf{G}(\tilde{\mathbf{t}}_0^{(2)}, \tilde{\mathbf{r}}_1^{(2)}) + \tilde{\mathbf{e}}_0^{(2)}), \end{aligned} \tag{3}$$

$$c_1 = Com(\tilde{\mathbf{r}}_0^{(0)} - \tilde{\mathbf{t}}_1^{(0)}, \mathbf{F}(\tilde{\mathbf{r}}_0^{(0)}) - \tilde{\mathbf{e}}_1^{(0)}) = Com(\tilde{\mathbf{t}}_0^{(2)}, \tilde{\mathbf{e}}_0^{(2)}), \quad \text{and} \tag{4}$$

$$c_2 = Com(\tilde{\mathbf{t}}_1^{(0)}, \tilde{\mathbf{e}}_1^{(0)}) = Com(\tilde{\mathbf{t}}_1^{(1)}, \tilde{\mathbf{e}}_1^{(1)}). \tag{5}$$

If the two pairs of the arguments of *Com* are distinct on any one of the above equations, the binding property of *Com* is broken. Otherwise, the equation (3) yields $\mathbf{v} = \mathbf{F}(\tilde{\mathbf{r}}_1^{(2)}) + \mathbf{G}(\tilde{\mathbf{t}}_1^{(1)} + \tilde{\mathbf{t}}_0^{(2)}, \tilde{\mathbf{r}}_1^{(2)}) + \tilde{\mathbf{e}}_0^{(2)} + \tilde{\mathbf{e}}_1^{(1)}$. Combining it with the equations (4) and (5), it is seen that $\mathbf{v} = \mathbf{F}(\tilde{\mathbf{r}}_1^{(2)}) + \mathbf{G}(\tilde{\mathbf{r}}_0^{(0)}, \tilde{\mathbf{r}}_1^{(2)}) + \mathbf{F}(\tilde{\mathbf{r}}_0^{(0)}) = \mathbf{F}(\tilde{\mathbf{r}}_0^{(0)} + \tilde{\mathbf{r}}_1^{(2)})$. It means that a solution $\tilde{\mathbf{r}}_0^{(0)} + \tilde{\mathbf{r}}_1^{(2)}$ for $\mathbf{v}$ is extracted.

The details of the proof are given in the full paper, where we formally construct a knowledge extractor which has oracle access to a message specification function $P_{\mathbf{F},\mathbf{v},\mathbf{s},r}$, and either breaks the binding property of *Com* or outputs a solution for $\mathbf{v}$. □

Extension. The trick mentioned in [49] for saving one hash value can be applied to our 3-pass identification protocol as follows. In the first pass, by using a collision resistant hash function H, one hash value $c = H(c_0, c_1, c_2)$ instead of three commitments (c_0, c_1, c_2) is sent. In the third pass, for a challenge Ch of a verifier, a prover sends $c_i|_{i=Ch}$ in addition to Rsp. Consequently, a verifier computes $c_i|_{i \neq Ch}$ by using Rsp and checks $c = H(c_0, c_1, c_2)$. The modified version of 3-pass protocol is also shown to be zero-knowledge argument of knowledge with knowledge error $2/3$.

4 A 5-pass Identification Scheme

In this section, we construct a 5-pass identification protocol which is statistical zero-knowledge argument of knowledge for $R_{\mathbf{F}}$ with knowledge error $1/2 + 1/2q$, assuming the existence of a non-interactive commitment scheme *Com* which is statistically hiding and computationally binding. The knowledge error of the 5-pass protocol is smaller than that of the 3-pass protocol when $q \geq 4$. The setup algorithm and the key-generation algorithm of the 5-pass scheme are identical to those of the 3-pass scheme.

In the 5-pass protocol, a prover also divides the secret key $\mathbf{s}$ and the public key $\mathbf{F}(\mathbf{s})$ as $\mathbf{s} = \mathbf{r}_0 + \mathbf{r}_1$ and $\mathbf{F}(\mathbf{s}) = \mathbf{F}(\mathbf{r}_0 + \mathbf{r}_1) = \mathbf{F}(\mathbf{r}_0) + \mathbf{F}(\mathbf{r}_1) + \mathbf{G}(\mathbf{r}_0, \mathbf{r}_1)$, respectively. The difference from the 3-pass protocol is that $\mathbf{r}_0$ and $\mathbf{F}(\mathbf{r}_0)$ are divided as $\alpha\mathbf{r}_0 = \mathbf{t}_0 + \mathbf{t}_1$ and $\alpha\mathbf{F}(\mathbf{r}_0) = \mathbf{e}_0 + \mathbf{e}_1$ where $\alpha \in \mathbb{F}_q$ is a choice of a verifier. After sending $(\mathbf{t}_1, \mathbf{e}_1)$ to the verifier, corresponding to a challenge $Ch \in \{0, 1\}$ of the verifier, the prover reveals one out of two vectors $\mathbf{r}_0$ and $\mathbf{r}_1$. When $\mathbf{r}_0$, $\mathbf{t}_0$, and $\mathbf{e}_0$ are randomly chosen, the verifier can obtain no information on the secret key $\mathbf{s}$ from only one out of the two vectors $\mathbf{r}_0$ and $\mathbf{r}_1$. On the other hand, the argument-of-knowledge property comes from that, for more than one choice of $\alpha \in \mathbb{F}_q$, an impersonator cannot response both of verifier's challenges $Ch = 0$ and $Ch = 1$ unless the impersonator has a solution $\mathbf{s}$ for $\mathbf{v}$.

The 5-pass identification protocol is described in Figure 2 where $\mathbf{G}$ is the polar form of $\mathbf{F}$. For the simplicity, a random string ρ in *Com* is not written

Prover's input: $((\mathbf{F}, \mathbf{v}), \mathbf{s})$		**Verifier**'s input: $(\mathbf{F}, \mathbf{v})$
Pick $\mathbf{r}_0, \mathbf{t}_0 \in_R \mathbb{F}_q^n$, $\mathbf{e}_0 \in_R \mathbb{F}_q^m$		
$\mathbf{r}_1 \leftarrow \mathbf{s} - \mathbf{r}_0$		
$c_0 \leftarrow Com(\mathbf{r}_0, \mathbf{t}_0, \mathbf{e}_0)$		
$c_1 \leftarrow Com(\mathbf{r}_1, \mathbf{G}(\mathbf{t}_0, \mathbf{r}_1) + \mathbf{e}_0)$	$\xrightarrow{(c_0, c_1)}$	
	$\xleftarrow{\alpha}$	Pick $\alpha \in_R \mathbb{F}_q$
$\mathbf{t}_1 \leftarrow \alpha\mathbf{r}_0 - \mathbf{t}_0$		
$\mathbf{e}_1 \leftarrow \alpha\mathbf{F}(\mathbf{r}_0) - \mathbf{e}_0$	$\xrightarrow{(\mathbf{t}_1, \mathbf{e}_1)}$	
	$\xleftarrow{Ch}$	Pick $Ch \in_R \{0, 1\}$
If $Ch = 0$, $Rsp \leftarrow \mathbf{r}_0$		
If $Ch = 1$, $Rsp \leftarrow \mathbf{r}_1$	$\xrightarrow{Rsp}$	If $Ch = 0$, parse $Rsp = \mathbf{r}_0$ and check $c_0 \stackrel{?}{=} Com(\mathbf{r}_0, \alpha\mathbf{r}_0 - \mathbf{t}_1, \alpha\mathbf{F}(\mathbf{r}_0) - \mathbf{e}_1)$
		If $Ch = 1$, parse $Rsp = \mathbf{r}_1$ and check $c_1 \stackrel{?}{=} Com(\mathbf{r}_1, \alpha(\mathbf{v} - \mathbf{F}(\mathbf{r}_1)) - \mathbf{G}(\mathbf{t}_1, \mathbf{r}_1) - \mathbf{e}_1)$

Fig. 2. Our 5-pass identification protocol

explicitly. The verifier finally outputs 1 if the check of "$\stackrel{?}{=}$" is passed, otherwise outputs 0. This is denoted by $0/1 \leftarrow Dec(\mathbf{F}, \mathbf{v}; (c_0, c_1), \alpha, (\mathbf{t}_1, \mathbf{e}_1), Ch, Rsp)$. It is easy to see that the verifier always accepts an interaction with the honest prover. Thus the 5-pass scheme has perfect correctness.

Now we show two properties of the protocol in Theorem 4 and Theorem 5 as follows.

Theorem 4. *The 5-pass protocol is statistically zero knowledge when the commitment scheme Com is statistically hiding.*

Proof sketch. Let $\mathcal{S}$ be a simulator which takes $\mathbf{F}$ and $\mathbf{v}$ without knowing $\mathbf{s}$, and interacts with a cheating verifier $\mathcal{CV}$. We show that the simulator $\mathcal{S}$ can impersonate the honest prover with probability $1/2$. The simulator $\mathcal{S}$ randomly chooses a value $Ch^* \in_R \{0, 1\}$ and vectors $\mathbf{s}', \mathbf{r}'_0, \mathbf{t}'_0 \in_R \mathbb{F}_q^n$, $\mathbf{e}'_0 \in_R \mathbb{F}_q^m$, where Ch^* is a prediction of what value the cheating verifier $\mathcal{CV}$ *will* choose. Then, it computes $\mathbf{r}'_1 \leftarrow \mathbf{s}' - \mathbf{r}'_0$, $c'_0 \leftarrow Com(\mathbf{r}'_0, \mathbf{t}'_0, \mathbf{e}'_0)$, and $c'_1 \leftarrow Com(\mathbf{r}'_1, \mathbf{G}(\mathbf{t}'_0, \mathbf{r}'_1) + \mathbf{e}'_0)$. It sends (c'_0, c'_1) to $\mathcal{CV}$. Receiving a challenge α from $\mathcal{CV}$, it computes $\mathbf{t}'_1 \leftarrow \alpha\mathbf{r}'_0 - \mathbf{t}'_0$. If $Ch^* = 0$ then it computes $\mathbf{e}'_1 \leftarrow \alpha\mathbf{F}(\mathbf{r}'_0) - \mathbf{e}'_0$, else $\mathbf{e}'_1 \leftarrow \alpha(\mathbf{v} - \mathbf{F}(\mathbf{s}') + \mathbf{F}(\mathbf{r}'_0)) - \mathbf{e}'_0$. It sends $(\mathbf{t}'_1, \mathbf{e}'_1)$ to $\mathcal{CV}$. Due to the statistically hiding property of Com, a challenge Ch from $\mathcal{CV}$ is equal to Ch^* with probability $1/2$. If $Ch = Ch^*$ then $\mathbf{r}'_0$ and $\mathbf{r}'_1$ are accepted responses to $Ch = 0$ and 1, respectively. Note that the case of $\alpha = 0$ does not spoil the zero-knowledge property. The details of the proof are given in the full paper, where we formally construct a black-box simulator $\mathcal{S}$ which outputs a successful transcript with probability $1/2 + 1/2q$. $\square$

Theorem 5. *The 5-pass protocol is argument of knowledge for $R_{\mathbf{F}}$ with knowledge error $1/2 + 1/2q$ when the commitment scheme Com is computationally binding.*

Proof sketch. Let $((c_0, c_1), \alpha_i, (\tilde{\mathbf{t}}_1^{(i)}, \tilde{\mathbf{e}}_1^{(i)}), Ch_j, Rsp^{(i,j)})$ be four transcripts for $i, j \in \{0, 1\}$ such that $Dec(\mathbf{F}, \mathbf{v}; (c_0, c_1), \alpha_i, (\tilde{\mathbf{t}}_1^{(i)}, \tilde{\mathbf{e}}_1^{(i)}), Ch_j, Rsp^{(i,j)}) = 1$, $\alpha_0 \neq \alpha_1$, and $Ch_j = j$. Then, by using the four transcripts, it is shown to be able to either break the binding property of *Com* or extract a solution for $\mathbf{v}$. Consider that the responses are parsed as $Rsp^{(0,0)} = \tilde{\mathbf{r}}_0^{(0)}$, $Rsp^{(0,1)} = \tilde{\mathbf{r}}_1^{(0)}$, $Rsp^{(1,0)} = \tilde{\mathbf{r}}_0^{(1)}$, and $Rsp^{(1,1)} = \tilde{\mathbf{r}}_1^{(1)}$. Then, it is seen that

$$\begin{aligned} c_0 &= Com(\tilde{\mathbf{r}}_0^{(0)}, \alpha_0\tilde{\mathbf{r}}_0^{(0)} - \tilde{\mathbf{t}}_1^{(0)}, \alpha_0\mathbf{F}(\tilde{\mathbf{r}}_0^{(0)}) - \mathbf{e}_1^{(0)}) \\ &= Com(\tilde{\mathbf{r}}_0^{(1)}, \alpha_1\tilde{\mathbf{r}}_0^{(1)} - \tilde{\mathbf{t}}_1^{(1)}, \alpha_1\mathbf{F}(\tilde{\mathbf{r}}_0^{(1)}) - \tilde{\mathbf{e}}_1^{(1)}) \quad \text{and} \end{aligned} \tag{6}$$

$$\begin{aligned} c_1 &= Com(\tilde{\mathbf{r}}_1^{(0)}, \alpha_0(\mathbf{v} - \mathbf{F}(\tilde{\mathbf{r}}_1^{(0)})) - \mathbf{G}(\tilde{\mathbf{t}}_1^{(0)}, \tilde{\mathbf{r}}_1^{(0)}) - \tilde{\mathbf{e}}_1^{(0)}) \\ &= Com(\tilde{\mathbf{r}}_1^{(1)}, \alpha_1(\mathbf{v} - \mathbf{F}(\tilde{\mathbf{r}}_1^{(1)})) - \mathbf{G}(\tilde{\mathbf{t}}_1^{(1)}, \tilde{\mathbf{r}}_1^{(1)}) - \tilde{\mathbf{e}}_1^{(1)}). \end{aligned} \tag{7}$$

If the two tuples of the arguments of *Com* are distinct on either of the above equations, the binding property of *Com* is broken. Otherwise, it is seen that $(\alpha_0 - \alpha_1)(\mathbf{v} - \mathbf{F}(\tilde{\mathbf{r}}_1^{(0)})) = \mathbf{G}(\tilde{\mathbf{t}}_1^{(0)} - \tilde{\mathbf{t}}_1^{(1)}, \tilde{\mathbf{r}}_1^{(0)}) + \tilde{\mathbf{e}}_1^{(0)} - \tilde{\mathbf{e}}_1^{(1)}$ from the equation (7). Combining it with the equation (6) yields $(\alpha_0 - \alpha_1)(\mathbf{v} - \mathbf{F}(\tilde{\mathbf{r}}_1^{(0)})) = \mathbf{G}((\alpha_0 - \alpha_1)\tilde{\mathbf{r}}_0^{(0)}, \tilde{\mathbf{r}}_1^{(0)}) + (\alpha_0 - \alpha_1)\mathbf{F}(\tilde{\mathbf{r}}_0^{(0)})$. Thus, $\mathbf{v} = \mathbf{F}(\tilde{\mathbf{r}}_1^{(0)}) + \mathbf{G}(\tilde{\mathbf{r}}_0^{(0)}, \tilde{\mathbf{r}}_1^{(0)}) + \mathbf{F}(\tilde{\mathbf{r}}_0^{(0)}) = \mathbf{F}(\tilde{\mathbf{r}}_1^{(0)} + \tilde{\mathbf{r}}_0^{(0)})$ is obtained, since $\alpha_0 \neq \alpha_1$. It means that a solution $\tilde{\mathbf{r}}_1^{(0)} + \tilde{\mathbf{r}}_0^{(0)}$ for $\mathbf{v}$ is extracted. The details of the proof are given in the full paper, where a knowledge extractor is formally constructed. □

5 Security and Efficiency

In this section, we summarize the security which is easily derived from the properties of zero-knowledge argument of knowledge, and give a practical parameter choice for each of the 3-pass scheme and the 5-pass scheme.

5.1 Security of the Identification Schemes

Here we briefly mention the security of each of the sequential and the parallel compositions when $\mathcal{MQ}(n, m, \mathbb{F}_q)$ is intractable and the commitment scheme *Com* is statistically hiding and computationally binding. Let $(\mathtt{P}, \mathtt{V})$ be an identification protocol described in Section 3 or Section 4. Then identification protocols which consist of repeating $(\mathtt{P}, \mathtt{V})$ N-times in sequential and in parallel are denoted by $(\mathtt{P}_N^{(s)}, \mathtt{V}_N^{(s)})$ and $(\mathtt{P}_N^{(p)}, \mathtt{V}_N^{(p)})$, respectively. The security of our identification schemes $(\mathtt{Setup}, \mathtt{Gen}, \mathtt{P}_N^{(s)}, \mathtt{V}_N^{(s)})$ and $(\mathtt{Setup}, \mathtt{Gen}, \mathtt{P}_N^{(p)}, \mathtt{V}_N^{(p)})$ is evaluated as follows.

First, we consider $(\mathtt{Setup}, \mathtt{Gen}, \mathtt{P}_N^{(s)}, \mathtt{V}_N^{(s)})$. From Theorem 2 and the sequential composition lemma [25], $(\mathtt{P}_N^{(s)}, \mathtt{V}_N^{(s)})$ is statistically zero knowledge. Furthermore, it is directly shown that the sequential repetition reduces the knowledge error at an optimal rate in the same way as [48,49]. We note that Bellare and Goldreich showed the theorem for a general reduction of a knowledge error by the sequential repetition [4]. Therefore, the identification scheme $(\texttt{Setup}, \texttt{Gen}, \texttt{P}_N^{(s)}, \texttt{V}_N^{(s)})$ is secure against impersonation under *active* attack where $N = \omega(\log \lambda)$.

Second, consider $(\texttt{Setup}, \texttt{Gen}, \texttt{P}_N^{(p)}, \texttt{V}_N^{(p)})$. It is easy to see that the parallel repetition of $(\texttt{P}, \texttt{V})$ reserves zero-knowledge with respect to an *honest verifier*. Because if the simulator $\mathcal{S}$ knows a challenge Ch which $\mathcal{CV}$ will choose, then $\mathcal{S}$ can *always* output a successful transcript in both case of the 3-pass protocol and the 5-pass protocol. Furthermore, Pass and Venkitasubramaniam mentioned that the parallel repetition reduces a knowledge error in a constant-round public-coin argument of knowledge [37]. In particular, the error rate drops exponentially with the number of repetitions N. Therefore, the identification scheme $(\texttt{Setup}, \texttt{Gen}, \texttt{P}_N^{(p)}, \texttt{V}_N^{(p)})$ is secure against impersonation under *passive* attack where $N = \omega(\log \lambda)$. In addition, for a certain parameter choice, the parallel version of our scheme is also secure under *active* attack as shown in Section 6.

5.2 Efficiency

We estimate practical sizes of system parameters, a public key, a secret key, and a transcript of our schemes. The numbers of arithmetic operations, computing permutations, and computing hash functions are also estimated as computational cost. Almost all arithmetic operations are done in evaluations of $\mathbf{F}$ and $\mathbf{G}$. The efficiency is compared with that of the identification schemes based on binary SD, q-ary SD, CLE, PP, and PK. The key lengths of these schemes for around 80-bit security is estimated in [12,23]. In our evaluation, the key lengths given in [12] are used, where lengths of a hash value and a random seed are 160 bits and 128 bits, respectively.

First, we consider the 3-pass identification scheme employing $\mathcal{MQ}(84, 80, \mathbb{F}_2)$. Following the same way as [7,3], the time complexity of the F_5 algorithm to break $\mathcal{MQ}(84, 80, \mathbb{F}_2)$ is estimated to be more than 2^{80}. Furthermore, the complexity of the improved exhaustive search algorithm to break $\mathcal{MQ}(84, 80, \mathbb{F}_2)$ [10], which is stated as the best known algorithm, is $2^{88.7}$ and thus also more than 2^{80}. Table 1 shows comparison of the sequential version of our scheme and the 3-pass schemes based on binary SD, CLE, and PP when each protocol is repeated until impersonation probability is less than 2^{-30}. In the SD-based scheme, the CLE-based scheme, and ours, we consider the case that $H(c_0, c_1, c_2)$ is sent in the first pass instead of (c_0, c_1, c_2) as mentioned at the end of Section 3. In the PP-based scheme, we consider the efficient version using hash tree [43]. The sizes of public/secret keys and communication of our scheme are smaller than those of the others. Although the size of system parameter of our scheme is relatively large, it can be reduced to some small seed, e.g. 128 bits, if a pseudo-random number generator is used as the implementation of QUAD [2]. Although the cost of arithmetic operations of our scheme is relatively high, it is still reasonable. In particular, our scheme does not require random permutations.

Second, consider the 5-pass identification scheme. As the order q of a field becomes larger, the knowledge error of the 5-pass protocol $1/2+1/2q$ is smaller. Here we use $\mathcal{MQ}(45,30,\mathbb{F}_{2^4})$ which is one of the minimal recommended parameters given in [8] for 80-bit security. Table 2 shows efficiency of the sequential

Table 1. Comparison of 3-pass schemes on 80-bit security against key-recovery attack when the impersonation probability is less than 2^{-30}

	SD [47,49]	CLE [48]	PP [43]	Our
round	52	52	73	52
system parameter (bit)	122,500[*1]	4,608[*1]	28,497[*1]	285,600[*1]
public key (bit)	350	288[*2]	245	80
secret key (bit)	700	192	177	84
communication (bit)	59,800[*6]	45,517[*3*4*6]	100,925[*6]	29,640
arithmetic ops. (times/field)	2^{24} / $\mathbb{F}_2$	2^{16} / $\mathbb{F}_{257}$	2^{22} / $\mathbb{F}_{127}$	2^{26} / $\mathbb{F}_2$
permutations[*5] (times/size)	$2/S_{700}$	$4/S_{24}$	$2/S_{161},S_{177}$	NO
hash function (times)	4	4	8	4
best known key-recovery attack	2^{87}	2^{84}	$> 2^{74}$	2^{80}

Table 2. Comparison of 5-pass schemes on 80-bit security against key-recovery attack when the impersonation probability is less than 2^{-30}

	SD [47,49]	SD [12]	PK [46]	CLE [48]	PP [42,43]	Our
round	31	31	31	31	52	33
system parameter (bit)	122,500[*1]	32,768[*1]	4,608[*1]	4,608[*1]	28,497[*1]	259,200[*1]
public key (bit)	2450	512	384	288[*2]	245	120
secret key (bit)	4900	1024	203[*7]	192	177	180
communication (bit)	120,652[*6]	61,783[*6]	27,234[*6]	27,528[*3*6]	105,060[*6]	26,565
arithmetic ops. (times/field)	$2^{23}/\mathbb{F}_2$	$2^{18}/\mathbb{F}_{256}$	$2^{15}/\mathbb{F}_{251}$	$2^{15}/\mathbb{F}_{257}$	$2^{21}/\mathbb{F}_{127}$	$2^{22}/\mathbb{F}_{2^4}$
permutations[*5] (times/size)	$8/S_{700}$	$2/S_{128}$	$3/S_{48}$	$4/S_{24}$	$2/S_{161},S_{177}$	NO
hash function (times)	2	2	2	2	5	2
best known key-recovery attack	2^{87}	2^{87}	2^{85}	2^{84}	$> 2^{74}$	2^{83}

[*1] These values can be reduced to 128 bit if a pseudo-random number generator is used.

[*2] For the verification, only one vector P is required for the public key whose size is 96 bits. However, as mentioned in [48,12], zero-knowledge property of the scheme can only be stated if two quantities ($S\sigma$ and $T\tau$) are public in addition to the vector P.

[*3] It is estimated for the case where elements in $\mathbb{F}_{257}$ are regarded as 8 bits.

[*4] In the original paper [48], a prover sends $(U\sigma, V\tau, (U+S)\sigma, (V-T)\tau)$ in the third pass. However, if the prover sends $(U\sigma, V\tau, S\sigma, T\tau)$ instead of $(U\sigma, V\tau, (U+S)\sigma, (V-T)\tau)$, then the communication cost is reduced. Our estimation employs the efficient version.

[*5] This shows the number of times of computing permutations and the size of the permutation, where S_n means a permutation over $\{1,\ldots,n\}$.

[*6] By following [46,48], the data size of S_n is regarded as $\lceil\log_2(n!)\rceil$ bits. Furthermore, in the same way as [48,49,12], the data size of a random permutation or a random vector is estimated at the length of random seed as 128 bits if it is over 128 bits.

[*7] We follow the original paper [46] and estimate the length of the secret key as $\lceil\log_2(n!)\rceil$ bits, although it is regarded as the length of the random seed in [12].

version of our scheme and the 5-pass schemes based on binary SD, q-ary SD, CLE, PK, and PP when each protocol is repeated until impersonation probability is less than 2^{-30}. This table tells us some advantages of our 5-pass scheme, which are similar to those of our 3-pass scheme.

6 On the Security against Active Attack in Parallel Repetition

In this section, we focus on the case that the underlying MQ function is substantially compressing, in particular, mapping $\mathbb{F}_q^n$ to $\mathbb{F}_q^m$ where $n = m + k$ and $k = \omega(\log \lambda)$. For example, the MQ function $\mathbf{F} \in \mathcal{MQ}(2m, m, \mathbb{F}_q)$ satisfies the requirement where $m = \omega(\log \lambda)$. In this case, the parallel version ($\mathtt{Setup}, \mathtt{Gen}, \mathtt{P}_N^{(p)}, \mathtt{V}_N^{(p)}$) of our 3-pass scheme is shown to be secure against impersonation under *active* attack, although the sizes of the secret key and the communication data increase at most double compared to those of Section 5.2. This argument can also be applied to our 5-pass scheme.

First, we define the preimage resistance and the second-preimage resistance of the MQ function as follows. The preimage resistance is slightly different from the intractability assumption of Definition 1 in the distribution of the challenge $\mathbf{v}$, but is also widely believed.

Definition 6. *For polynomially bounded functions $n = n(\lambda)$, $m = m(\lambda)$, and $q = q(\lambda)$, it is said that $\mathcal{MQ}(n, m, \mathbb{F}_q)$ is preimage resistant if there is no polynomial-time algorithm that takes $(\mathbf{F}, \mathbf{v})$ generated via $\mathbf{F} \in_R \mathcal{MQ}(n, m, \mathbb{F}_q)$ and $\mathbf{v} \in_R \mathbb{F}_q^m$ and finds a preimage $\mathbf{s} \in \mathbb{F}_q^n$ such that $\mathbf{F}(\mathbf{s}) = \mathbf{v}$ with non-negligible probability $\epsilon(\lambda)$. On the other hand, it is said that $\mathcal{MQ}(n, m, \mathbb{F}_q)$ is second-preimage resistant if there is no polynomial-time algorithm that takes $(\mathbf{F}, \mathbf{x})$ generated via $\mathbf{F} \in_R \mathcal{MQ}(n, m, \mathbb{F}_q)$ and $\mathbf{x} \in_R \mathbb{F}_q^n$ and finds a second preimage $\mathbf{x}' \in \mathbb{F}_q^n$ such that $\mathbf{F}(\mathbf{x}') = \mathbf{F}(\mathbf{x})$ and $\mathbf{x}' \neq \mathbf{x}$ with non-negligible probability $\epsilon(\lambda)$.*

When a second-preimage resistant hash function is substantially compressing, it is known to be preimage resistant [44]. Conversely, with respect to the MQ function, the following lemma is also shown.

Lemma 7. *If $\mathcal{MQ}(n, m, \mathbb{F}_q)$ is preimage resistant, then $\mathcal{MQ}(n + 1, m, \mathbb{F}_q)$ is second-preimage resistant.*

Proof sketch. Given $\mathbf{F} = (f_1, \ldots, f_m) \in_R \mathcal{MQ}(n, m, \mathbb{F}_q)$ and $\mathbf{v} = (v_1, \ldots, v_m) \in_R \mathbb{F}_q^m$, we show that a preimage $\mathbf{x}$ satisfying $\mathbf{v} = \mathbf{F}(\mathbf{x})$ can be found by using an algorithm $\mathcal{A}$ that breaks the second-preimage resistance of $\mathcal{MQ}(n + 1, m, \mathbb{F}_q)$, where $f_l(x_1, \ldots, x_n) = \sum_{i=1}^{n} \sum_{j=1}^{n} a_{l,i,j} x_i x_j + \sum_{i=1}^{n} b_{l,i} x_i$. For the simplicity, suppose that the algorithm $\mathcal{A}$ takes $\tilde{\mathbf{F}} = (\tilde{f}_1, \ldots, \tilde{f}_m) \in \mathcal{MQ}(n + 1, m, \mathbb{F}_q)$ and $\mathbf{t} = (t_1, \ldots, t_{n+1}) \in \mathbb{F}_q^{n+1}$ and outputs a second preimage $\mathbf{t}+\Delta$ such that $\tilde{\mathbf{F}}(\mathbf{t}+\Delta) = \tilde{\mathbf{F}}(\mathbf{t})$ and $\Delta = (d_1, \ldots, d_n, 1)$, where $\tilde{f}_l(x_1, \ldots, x_{n+1}) = \sum_{i=1}^{n+1} \sum_{j=1}^{n+1} \tilde{a}_{l,i,j} x_i x_j$

$+\sum_{i=1}^{n+1}\tilde{b}_{l,i}x_i$. In this case, the equation $\tilde{\mathbf{F}}(\mathbf{t}+\Delta)-\tilde{\mathbf{F}}(\mathbf{t})=\mathbf{0}$ is expanded as follows:

$$\sum_{i=1}^{n}\sum_{j=1}^{n}\tilde{a}_{l,i,j}d_id_j+\sum_{i=1}^{n}\Big(\sum_{j=1}^{n+1}(\tilde{a}_{l,i,j}+\tilde{a}_{l,j,i})t_j+\tilde{b}_{l,i}+(\tilde{a}_{l,i,n+1}+\tilde{a}_{l,n+1,i})\Big)d_i$$
$$+\sum_{j=1}^{n+1}(\tilde{a}_{l,n+1,j}+\tilde{a}_{l,j,n+1})t_j+\tilde{b}_{l,n+1}+\tilde{a}_{l,n+1,n+1}=0$$

for $l=1,\dots,m$. From the above equation, we can see that the output $\mathbf{t}+\Delta$ of $\mathcal{A}$ satisfies $\mathbf{v}=\mathbf{F}(d_1,\dots,d_n)$ if the input $(\tilde{\mathbf{F}},\mathbf{t})$ of $\mathcal{A}$ is produced as follows.

- The vector $\mathbf{t}$ is generated via $\mathbf{t}\in_R\mathbb{F}_q^{n+1}$.
- For $1\le i\le n$ and $1\le j\le n$ do $\tilde{a}_{l,i,j}\leftarrow a_{l,i,j}$, otherwise $\tilde{a}_{l,i,j}\in_R\mathbb{F}_q$.
- For $1\le i\le n$ do $\tilde{b}_{l,i}\leftarrow b_{l,i}-(\tilde{a}_{l,n+1,i}+\tilde{a}_{l,i,n+1})-\sum_{j=1}^{n+1}(\tilde{a}_{l,i,j}+\tilde{a}_{l,j,i})t_j$, otherwise $\tilde{b}_{l,n+1}\leftarrow -v_l-\tilde{a}_{l,n+1,n+1}-\sum_{j=1}^{n+1}(\tilde{a}_{l,n+1,j}+\tilde{a}_{l,j,n+1})t_j$.

The details of the proof of Lemma 7 are described in the full paper. □

Moreover, the following lemma is also shown.

Lemma 8. *Let $n=m+k$, $k=\omega(\log\lambda)$, and $N=\omega(\log\lambda)$. Suppose that $\mathcal{MQ}(n,m,\mathbb{F}_q)$ is second-preimage resistant. Then, $(\mathtt{P}_N^{(p)},\mathtt{V}_N^{(p)})$ achieves the security against impersonation under* active *attack when Com is statistically hiding and computationally binding.*

Proof sketch. The proof of the lemma follows standard techniques used in [19,29,33]. We construct an algorithm $\mathcal{B}$ breaking the second-preimage resistance of $\mathcal{MQ}(n,m,\mathbb{F}_q)$ by using an impersonator $\mathcal{I}=(\mathcal{CP},\mathcal{CV})$ which succeeds impersonation under active attack. Given $(\mathbf{F},\mathbf{x})$, the algorithm $\mathcal{B}$ runs the cheating verifier $\mathcal{CV}$ on input $(\mathbf{F},\mathbf{v})$ where $\mathbf{v}=\mathbf{F}(\mathbf{x})$. Using the secret key $\mathbf{x}$, $\mathcal{B}$ can simulate the prover oracle perfectly. After obtaining a state for $\mathcal{CP}$ from $\mathcal{CV}$, $\mathcal{B}$ feeds the state to $\mathcal{CP}$ and acts as the legitimate verifier. By using standard rewinding techniques, $\mathcal{B}$ either breaks the binding property of *Com* or obtains $\mathbf{x}'$ satisfying $\mathbf{v}=\mathbf{F}(\mathbf{x}')$, in the same way as the proof of Theorem 3. Furthermore, the event $\mathbf{x}'\neq\mathbf{x}$ occurs with non-negligible probability, because of the following (1) and (2): (1) $(\mathtt{P}_N^{(p)},\mathtt{V}_N^{(p)})$ is statistically witness indistinguishable when *Com* is statistically hiding, due to Theorem 2. (2) The probability that there is not another $\mathbf{x}'\in\mathbb{F}_q^n\setminus\{\mathbf{x}\}$ such that $\mathbf{F}(\mathbf{x})=\mathbf{F}(\mathbf{x}')$ is at most q^{-k} which is negligible, since $k=\omega(\log(\lambda))$. In the case of $\mathbf{x}'\neq\mathbf{x}$, $\mathcal{B}$ finds a second preimage $\mathbf{x}'$. The details of the proof of Lemma 8 are described in the full paper. □

We note that the above proof can be extended into that of the security under *concurrent* attack [6] as in the proof of Kawachi et al. [29].

Finally, combining Lemma 7 and Lemma 8 yields the following theorem.

Theorem 9. *Let $n=m+k$, $k=\omega(\log\lambda)$, and $N=\omega(\log\lambda)$. Suppose that $\mathcal{MQ}(n-1,m,\mathbb{F}_q)$ is preimage resistant. Then, $(\mathtt{P}_N^{(p)},\mathtt{V}_N^{(p)})$ achieves the security against impersonation under* active *attack when Com is statistically hiding and computationally binding.*

7 Extensions of Our Scheme

In this section we mention the following two extensions of our scheme.

Slightly Efficient Parallelization. The trick mentioned in the end of Section 3 can also be applied into the parallel version of our 3-pass scheme ($\mathtt{Setup}, \mathtt{Gen}, \mathtt{P}_N^{(p)}, \mathtt{V}_N^{(p)}$) without losing the security. After that, a hash value of $3N$-tuple of commitments $c = H((c_{0,1}, c_{1,1}, c_{2,1}), \ldots, (c_{0,N}, c_{1,N}, c_{2,N}))$ is sent by a prover in the first pass, where $c_{i,j}$ is a commitment and H is a collision resistant hash function. The sizes of a public key, a secret key, and communication data of the modified scheme are only 80 bits, 84 bits, and $160 + 410N$ bits, respectively.

A Signature Scheme. The Fiat-Shamir method is a generic technique which transforms an identification scheme into a signature scheme [20]. The signature scheme is secure against chosen-message attack in the random oracle model if the underlying identification scheme is secure against impersonation under passive attack [1]. Thus the transform yields a signature scheme based on the conjectured intractability of the MQ problem from the parallel version of our 3-pass identification scheme. Using the signature scheme, our identification/signature scheme can also be extended to an identity-based one in a natural way [5].

8 Conclusion

We introduced the dividing techniques using bilinearity of the polar form of the MQ function and proposed public-key identification schemes consisting of a non-trivial construction of zero-knowledge argument of knowledge for the MQ problem, assuming the existence of a non-interactive commitment scheme. For a practical parameter choice, the efficiency of our schemes is highly comparable to identification schemes based on another problem including PK, SD, CLE, and PP. Furthermore, even if the protocol is repeated in parallel, our scheme can achieve the security under active attack with some additional cost.

References

1. Abdalla, M., An, J.H., Bellare, M., Namprempre, C.: From Identification to Signatures via the Fiat-Shamir Transform: Minimizing Assumptions for Security and Forward-Security. In: Knudsen, L. (ed.) EUROCRYPT 2002. LNCS, vol. 2332, pp. 418–433. Springer, Heidelberg (2002)
2. Arditti, D., Berbain, C., Billet, O., Gilbert, H.: Compact FPGA Implementations of QUAD. In: Bao, F., Miller, S. (eds.) ASIACCS, pp. 347–349. ACM, New York (2007)
3. Bardet, M., Faugère, J.-C., Salvy, B.: Complexity of Gröbner Basis Computation for Semi-regular Overdetermined Sequences over F_2 with Solutions in F_2. Research Report RR-5049, INRIA (2003)
4. Bellare, M., Goldreich, O.: On Defining Proofs of Knowledge. In: Brickell, E.F. (ed.) CRYPTO 1992. LNCS, vol. 740, pp. 390–420. Springer, Heidelberg (1993)

5. Bellare, M., Namprempre, C., Neven, G.: Security Proofs for Identity-Based Identification and Signature Schemes. J. Cryptology 22(1), 1–61 (2009)
6. Bellare, M., Palacio, A.: GQ and Schnorr Identification Schemes: Proofs of Security against Impersonation under Active and Concurrent Attacks. In: Yung, M. (ed.) CRYPTO 2002. LNCS, vol. 2442, pp. 162–177. Springer, Heidelberg (2002)
7. Berbain, C., Gilbert, H., Patarin, J.: A Practical Stream Cipher with Provable Security. In: Vaudenay [50], pp. 109–128
8. Bettale, L., Faugère, J.-C., Perret, L.: Hybrid Approach for Solving Multivariate Systems over Finite Fields. Journal of Mathematical Cryptology 3(3), 177–197 (2009)
9. Billet, O., Robshaw, M.J.B., Peyrin, T.: On Building Hash Functions from Multivariate Quadratic Equations. In: Pieprzyk, J., Ghodosi, H., Dawson, E. (eds.) ACISP 2007. LNCS, vol. 4586, pp. 82–95. Springer, Heidelberg (2007)
10. Bouillaguet, C., Chen, H.-C., Cheng, C.-M., Chou, T., Niederhagen, R., Shamir, A., Yang, B.-Y.: Fast Exhaustive Search for Polynomial Systems in F_2. In: Mangard, S., Standaert, F.-X. (eds.) CHES 2010. LNCS, vol. 6225, pp. 203–218. Springer, Heidelberg (2010)
11. Bouillaguet, C., Faugère, J.-C., Fouque, P.-A., Perret, L.: Practical Cryptanalysis of the Identification Scheme Based on the Isomorphism of Polynomial with One Secret Problem. Cryptology ePrint Archive, Report 2010/504 (2010)
12. Cayrel, P.-L., Véron, P., El Yousfi Alaoui, S.M.: A Zero-Knowledge Identification Scheme Based on the q-ary Syndrome Decoding Problem. In: Biryukov, A., Gong, G., Stinson, D.R. (eds.) SAC 2010. LNCS, vol. 6544, pp. 171–186. Springer, Heidelberg (2011)
13. Cramer, R. (ed.): EUROCRYPT 2005. LNCS, vol. 3494. Springer, Heidelberg (2005)
14. Dubois, V., Fouque, P.-A., Shamir, A., Stern, J.: Practical Cryptanalysis of SFLASH. In: Menezes, A. (ed.) CRYPTO 2007. LNCS, vol. 4622, pp. 1–12. Springer, Heidelberg (2007)
15. Dubois, V., Fouque, P.-A., Stern, J.: Cryptanalysis of SFLASH with Slightly Modified Parameters. In: Naor, M. (ed.) EUROCRYPT 2007. LNCS, vol. 4515, pp. 264–275. Springer, Heidelberg (2007)
16. Faugère, J.C.: A New Efficient Algorithm for Computing Gröbner Bases without Reduction to Zero (F5). In: Proceedings of the 2002 International Symposium on Symbolic and Algebraic Computation, ISSAC 2002, pp. 75–83. ACM, New York (2002)
17. Faugère, J.-C., Perret, L.: Polynomial Equivalence Problems: Algorithmic and Theoretical Aspects. In: Vaudenay [50], pp. 30–47
18. Feige, U., Fiat, A., Shamir, A.: Zero-Knowledge Proofs of Identity. J. Cryptology 1(2), 77–94 (1988)
19. Feige, U., Shamir, A.: Witness Indistinguishable and Witness Hiding Protocols. In: STOC, pp. 416–426. ACM, New Orleans (1990)
20. Fiat, A., Shamir, A.: How to Prove Yourself: Practical Solutions to Identification and Signature Problems. In: Odlyzko, A.M. (ed.) CRYPTO 1986. LNCS, vol. 263, pp. 186–194. Springer, Heidelberg (1987)
21. Fouque, P.-A., Granboulan, L., Stern, J.: Differential Cryptanalysis for Multivariate Schemes. In: Cramer [13], pp. 341–353
22. Fouque, P.-A., Macario-Rat, G., Stern, J.: Key Recovery on Hidden Monomial Multivariate Schemes. In: Smart, N.P. (ed.) EUROCRYPT 2008. LNCS, vol. 4965, pp. 19–30. Springer, Heidelberg (2008)

23. Gaborit, P., Girault, M.: Lightweight Code-Based Identification and Signature. In: IEEE International Symposium on Information Theory, ISIT, pp. 191–195 (2007)
24. Garey, M.R., Johnson, D.S.: Computers and Intractability; A Guide to the Theory of NP-Completeness. W. H. Freeman & Co., New York (1979)
25. Goldreich, O.: Foundations of Cryptography: Volume I. Basic Tools. Cambridge University Press, Cambridge (2001)
26. Haitner, I., Reingold, O.: Statistically-Hiding Commitment from Any One-Way Function. In: Johnson, Feige [28], pp. 1–10
27. Halevi, S., Micali, S.: Practical and Provably-Secure Commitment Schemes from Collision-Free Hashing. In: Koblitz, N. (ed.) CRYPTO 1996. LNCS, vol. 1109, pp. 201–215. Springer, Heidelberg (1996)
28. Johnson, D.S., Feige, U. (eds.): Proceedings of the 39th Annual ACM Symposium on Theory of Computing, San Diego, California, USA, June 11-13. ACM, New York (2007)
29. Kawachi, A., Tanaka, K., Xagawa, K.: Concurrently Secure Identification Schemes Based on the Worst-Case Hardness of Lattice Problems. In: Pieprzyk, J. (ed.) ASIACRYPT 2008. LNCS, vol. 5350, pp. 372–389. Springer, Heidelberg (2008)
30. Kipnis, A., Patarin, J., Goubin, L.: Unbalanced Oil and Vinegar Signature Schemes. In: Stern, J. (ed.) EUROCRYPT 1999. LNCS, vol. 1592, pp. 206–222. Springer, Heidelberg (1999)
31. Kipnis, A., Shamir, A.: Cryptanalysis of the Oil & Vinegar Signature Scheme. In: Krawczyk, H. (ed.) CRYPTO 1998. LNCS, vol. 1462, pp. 257–266. Springer, Heidelberg (1998)
32. Komano, Y., Akiyama, K., Hanatani, Y., Miyake, H.: ASS-CC: Provably Secure Algebraic Surface Signature Scheme. In: The 2010 Symposium on Cryptography and Information Security 4A2-4 (2010)
33. Lyubashevsky, V.: Lattice-Based Identification Schemes Secure Under Active Attacks. In: Cramer, R. (ed.) PKC 2008. LNCS, vol. 4939, pp. 162–179. Springer, Heidelberg (2008)
34. Lyubashevsky, V.: Fiat-Shamir with Aborts: Applications to Lattice and Factoring-Based Signatures. In: Matsui, M. (ed.) ASIACRYPT 2009. LNCS, vol. 5912, pp. 598–616. Springer, Heidelberg (2009)
35. Matsumoto, T., Imai, H.: Public Quadratic Polynominal-Tuples for Efficient Signature-Verification and Message-Encryption. In: Gunther, C.G. (ed.) EUROCRYPT 1988. LNCS, vol. 330, pp. 419–453. Springer, Heidelberg (1988)
36. Micciancio, D., Vadhan, S.P.: Statistical Zero-Knowledge Proofs with Efficient Provers: Lattice Problems and More. In: Boneh, D. (ed.) CRYPTO 2003. LNCS, vol. 2729, pp. 282–298. Springer, Heidelberg (2003)
37. Pass, R., Venkitasubramaniam, M.: An Efficient Parallel Repetition Theorem for Arthur-Merlin Games. In: Johnson, Feige [28], pp. 420–429
38. Patarin, J.: Cryptanalysis of the Matsumoto and Imai Public Key Scheme of Eurocrypt '88. In: Coppersmith, D. (ed.) CRYPTO 1995. LNCS, vol. 963, pp. 248–261. Springer, Heidelberg (1995)
39. Patarin, J.: Hidden Fields Equations (HFE) and Isomorphisms of Polynomials (IP): Two New Families of Asymmetric Algorithms. In: Maurer, U. (ed.) EUROCRYPT 1996. LNCS, vol. 1070, pp. 33–48. Springer, Heidelberg (1996)
40. Patarin, J., Goubin, L.: Trapdoor One-Way Permutations and Multivariate Polynominals. In: Han, Y., Okamoto, T., Qing, S. (eds.) ICICS 1997. LNCS, vol. 1334, pp. 356–368. Springer, Heidelberg (1997)
41. Perret, L.: A Fast Cryptanalysis of the Isomorphism of Polynomials with One Secret Problem. In: Cramer [13], pp. 354–370

42. Pointcheval, D.: A New Identification Scheme Based on the Perceptrons Problem. In: Santis, A.D. (ed.) EUROCRYPT 1995. LNCS, vol. 950, pp. 319–328. Springer-Verlag, Heidelberg (1995)
43. Pointcheval, D., Poupard, G.: A New NP-Complete Problem and Public-key Identification. Des. Codes Cryptography 28(1), 5–31 (2003)
44. Rogaway, P., Shrimpton, T.: Cryptographic Hash-Function Basics: Definitions, Implications, and Separations for Preimage Resistance, Second-Preimage Resistance, and Collision Resistance. In: Roy, B., Meier, W. (eds.) FSE 2004. LNCS, vol. 3017, pp. 371–388. Springer, Heidelberg (2004)
45. Sakumoto, K., Shirai, T., Hiwatari, H.: On the Security of the Algebraic Surface Signature Scheme. IEICE Technical Report ISEC2010-39 (2010-9) (2010)
46. Shamir, A.: An Efficient Identification Scheme Based on Permuted Kernels (Extended Abstract). In: Brassard, G. (ed.) CRYPTO 1989. LNCS, vol. 435, pp. 606–609. Springer, Heidelberg (1990)
47. Stern, J.: A New Identification Scheme Based on Syndrome Decoding. In: Stinson, D.R. (ed.) CRYPTO 1993. LNCS, vol. 773, pp. 13–21. Springer, Heidelberg (1994)
48. Stern, J.: Designing Identification Schemes with Keys of Short Size. In: Desmedt, Y.G. (ed.) CRYPTO 1994. LNCS, vol. 839, pp. 164–173. Springer, Heidelberg (1994)
49. Stern, J.: A New Paradigm for Public Key Identification. IEEE Transactions on Information Theory, 13–21 (1996)
50. Vaudenay, S. (ed.): EUROCRYPT 2006. LNCS, vol. 4004. Springer, Heidelberg (2006)

Inverting HFE Systems Is Quasi-Polynomial for All Fields

Jintai Ding[1,2] and Timothy J. Hodges[2]

[1] South China University of Technology, Guangzhou, China
[2] Department of Mathematical Sciences, University of Cincinnati, Cincinnati, OH 45221-0025, USA
jintai.ding@gmail.com, timothy.hodges@uc.edu

Abstract. In this paper, we present and prove the first closed formula bounding the degree of regularity of an HFE system over an arbitrary finite field. Though these bounds are not necessarily optimal, they can be used to deduce

1. *if D, the degree of the corresponding HFE polynomial, and q, the size of the corresponding finite field, are fixed, inverting HFE system is polynomial for all fields;*
2. *if D is of the scale $O(n^{\alpha})$ where n is the number of variables in an HFE system, and q is fixed, inverting HFE systems is quasi-polynomial for all fields.*

We generalize and prove rigorously similar results by Granboulan, Joux and Stern in the case when $q = 2$ that were communicated at Crypto 2006.

1 Introduction

The security of cryptosystems such as RSA, ECC, and Diffie-Hellman key exchange scheme, depends on assumptions about the hardness of certain number theory problems, such as the Integer Prime Factorization Problem or the Discrete Logarithm Problem. However, in 1994 Peter Shor [19] showed that quantum computers could break all public key cryptosystems that are based on these hard number theory problems. People realize that we need to look ahead to a possible future of quantum computers. In recent years significant effort has been put into the search for alternative public key cryptosystems, now called post-quantum cryptosystems, which would remain secure in an era of quantum computers. Multivariate public key cryptosystems (MPKC) [5] are one of the main families of cryptosystems that have the potential to resist quantum computer attacks.

An MPKC is a cryptosystem whose public key is given as a set of multivariate polynomials over a normally small finite field. The security of such systems is suggested by the fact that solving a system of multivariate polynomial equations over a finite field is in general NP-complete [11]. A quantum computer has not yet been shown to be efficient in solving this problem. Furthermore, computations in a finite field are more efficient than the manipulation of large integers

P. Rogaway (Ed.): CRYPTO 2011, LNCS 6841, pp. 724–742, 2011.

which is required by the systems based on hard number theory problems. Thus MPKC's can be less computationally intense than these systems and therefore have potential for application in small ubiquitous computing devices with limited resources.

Research into MPKC's started in the middle of 1980s in work of Diffie, Fell, Tsujii, Shamir. However the success of this work was limited and the real breakthrough in this direction was the cryptosystem proposed by Matsumoto and Imai [16]. Their scheme used a simple quadratic function on an extension field whose field structure was kept hidden. Unfortunately this efficient scheme was proved to be insecure by Patarin using his linearization equation attack [18]. Hidden Field Equation (HFE) cryptosystems are a family of cryptosystems proposed by Patarin based on the same fundamental idea of quadratic functions on extension fields [18].

Fixing a finite field $\mathbb{F}$ of characteristic 2 and cardinality q, Matsumoto and Imai suggested using a bijective map P defined over $\mathbb{K}$, an extension field of degree n over $\mathbb{F}$. By identifying $\mathbb{K}$ with $\mathbb{F}^n$, one sees that P induces a multivariate polynomial map $\tilde{P}\colon \mathbb{F}^n \longrightarrow \mathbb{F}^n$. One can "hide" this map by composing on the left by L_1 and on the right by L_2, where the $L_i : \mathbb{F}^n \longrightarrow \mathbb{F}^n$ are invertible affine maps. This composition gives a map $\bar{P} : \mathbb{F}^n \longrightarrow \mathbb{F}^n$ defined by

$$\bar{P}(x_1, \ldots, x_n) = L_1 \circ \tilde{P} \circ L_2 (x_1, \ldots, x_n) = (y_1, \ldots, y_n) \ .$$

For a Hidden Field Equation (HFE) system [18], P is given as a univariate polynomial in the form:

$$P(X) = \sum_{q^i+q^j \leq D} a_{ij} X^{q^i+q^j} + \sum_{q^i \leq D} b_i X^{q^i} + c \ ,$$

where the coefficients are randomly chosen. Here the total degree D of P should not be too large since the decryption process involves solving the system of single variable polynomial equations given by $P(X) = Y'$ for a given Y' using the Berlekamp-Massey algorithm.

Faugère and Joux showed that these systems can be broken rather easily in the case when $q = 2$ and D is small [10] using the Gröbner basis algorithm F_4. Furthermore the experimental results suggested that such algorithms will finish at degree of order $\log_q(D)$ (by which we mean that the highest degree polynomials encountered are of degree of order $\log_q(D)$) and, therefore, that the complexity of the algorithm is $O(n^{\log_q(D)})$.

A key concept in the analysis of the complexity of these algorithms is that of *degree of regularity*. The degree of regularity of the component functions of P, $p_1(x_1, \ldots, x_n), \ldots, p_n(x_1, \ldots, x_n)$ is the lowest degree at which non-trivial polynomial relations between the p_i occur (we also talk about this as the degree of regularity of P or of the associated HFE system). Experimental evidence has shown that this is the point at which the algorithm will terminate. Here we mean by "*termination at a certain degree*", that the large matrices, whose entries are coefficients of multivariate polynomials, and on which the algorithm performs Gaussian eliminations, contain polynomials at most of the designated degree.

The largest size of all such matrices essentially determines the complexity of the algorithm. Bardet, Faugère and Salvy defined (in a different notation) the degree of regularity of random or generic systems and found an asymptotic formula for this degree. However since the systems arising from HFE polynomials were far from generic, the BFS bound does not yield useful information about the complexity of HFE systems. Granboulan, Joux and Stern [12] outlined a new way to bound the degree of regularity in the case $q = 2$. Their approach was to lift the problem back up to the extension field $\mathbb{K}$, an idea that originated in the work of Kipnis and Shamir [13] and Faugère and Joux [10]. They sketched that one can connect the degree of regularity of an HFE system to the degree of regularity of a lifted system over the big field. Assuming this assertion, the semi-regularity of a subsystem of the lifted system, and that the degree of regularity of a subsystem is greater than that of the original system, and using some asymptotic analysis of the degree of regularity of random systems found in [1], they derived heuristic asymptotic bounds for the case $q = 2$, which implied that if D is chosen to be $O(n^\alpha)$ for $\alpha \geq 1$, then the complexity of Gröbner basis attacks is quasi-polynomial. While the results derived from this method match well with experimental results, the asymptotic bound formula has not yet been proven rigorously. It relies on a formula that holds for a class of overdetermined generic systems but it is not yet clear how to prove their systems belong to this class. Therefore to derive any definitive general bounds on the degree of regularity for general q and n, or on the asymptotic behavior of the degree of regularity remained an open problem.

The security of HFE systems in the case when the characteristic of the field is odd has been the subject of much less study. The notions of degree of regularity and semi-regularity in [1] can be generalized to the case when q is odd. However, the asymptotic analysis on which the results of [12] depend has not yet been generalized to this situation. The work in [8] seemed to suggest that HFE systems over a field of odd characteristic could resist the attack of Gröbner basis algorithms even when D is small. When q is large the field equations $X_1^q - X_2, \ldots, X_n^q - X_1$ cannot be used effectively and this limits the efficiency of the Gröbner basis algorithms. A breakthrough in the case of general q came in the recent work of Dubois and Gama [9]. They first refined and gave a rigorous mathematical foundation for the arguments in [12]. They then derived a new method to compute the degree of regularity over any field similar to that in [1]. This led to an algorithm that can be used to calculate a bound for the degree of regularity for any choice of q, n and D. However it is not clear how to derive a closed form for their bound from their algorithm and therefore they were not able to answer the question of whether the complexity was quasi-polynomial in this case.

The contributions of this Paper. In this article we answer the above questions by giving a global bound on the degree of regularity (in the sense of [9]) of an HFE system. We begin with a similar idea to that used in [12] - roughly that one can bound the degree of regularity of a system by finding a bound for certain simpler subsystems. However we obtain our bound using a very different approach. Previously all estimates on the degree of regularity were based on a dimension counting argument. At some point the assumption that there are no trivial relations would imply that the space of linear combinations of the functions $p_1, \ldots, p_n$ by polynomials of degree k would be greater than the dimension of the space of all polynomials of degree $k+2$. Dubois and Gama were able to improve on earlier bounds by observing that the subspace of linear combinations has to lie in a special subspace of the space of all polynomials of a fixed degree. Unfortunately the dimension of this space is given by a recursive formula which does not have a known closed form. In contrast our approach is to find explicit non-trivial relations. Surprisingly, it is enough to do this for the case of a single polynomial. Moreover, we can find an explicit formula for the degree in which these non-trivial relations occur. This gives the degree of regularity as an explicit function of q and D (it does not depend on n unless the degree D is a function of n). Such explicit formulas enable us to draw conclusions about the complexity of inverting the system using Gröbner basis methods. Our conclusions rely on no heuristic assumptions beyond the standard assumption that the Gröbner basis algorithms terminate at or shortly after the degree of regularity.

Specifically, we give a closed formula bound for the degree of regularity of a multivariate quadratic polynomial of the form

$$P(X) = \sum_{q^i+q^j \leq D} a_{ij} X^{q^i+q^j} + \sum_{q^i \leq D} b_i X^{q^i} + c \ .$$

This bound depends on the rank of P (in a sense defined below); since the restriction $q^i + q^j \leq D$ implies a strong restriction on the rank of P, we are able to deduce a sharp bound for the degree of regularity of P over a field of any order q. When q is odd, these bounds are comparable with those found computationally by Dubois and Gama when the block size $n \log_2(q)$ is less than 700. Interestingly this formula also yields the degree of regularity of the Matsumoto-Imai system when $q = 2$ to be 3. This is precisely the statement that Patarin's linearization attack works in this case. Thus we believe that the notion of rank is a key new ingredient in the analysis of multivariate quadratic cryptosystems.

A crucial step in our approach is to look at the single polynomial

$$P_0(X_1, ..., X_n) = \sum_{i,j} a_{ij} X_i X_j$$

considered as an element of the graded algebra $\mathbb{K}[X_1, \ldots, X_n]/(X_i^q)$. Using methods from [9], the degree of regularity of the whole system is bounded by the degree of regularity of this single polynomial. This problem was studied in detail by the authors and their collaborators in [6,7] in the cases $q = 2$ and 3. Drawing on the ideas from this work we are able to find explicit relations which give us bounds on the degree of regularity of P_0 for any q (which we believe are sharp when q is odd). Specifically we show that the degree of regularity of the system defined by P is bounded by

$$\frac{\text{Rank}(P_0)(q-1)}{2} + 2 \leq \frac{(q-1)(\lfloor \log_q(D-1) \rfloor + 1)}{2} + 2$$

if $\text{Rank}(P_0) > 1$. Here $\text{Rank}(P_0)$ is the rank of the quadratic form associated to P_0. by It is important to note that these are universal bounds that require no assumption that the polynomials are of "generic type".

There are two very critical points in the formula. First, the degree of regularity depends only on the rank of P_0, not the degree of P; while the rank is bounded by $\log_q D + 1$, there are many situations (such as the Matsumoto-Imai operators or sparse HFE polynomials) where the rank is much smaller than can be predicted by looking at the degree; thus we believe that the rank of a HFE operator to be a more important invariant than its degree. Second, while we do not expect our formula to give sharp bounds, it yields similar results to those obtained in [9] for prime q. It also explains many of the jump discontinuities in their data, since jumps in the degree of regularity should be expected to occur when the degree reaches a value which allows the rank of P to increase.

The formulas above enable us to draw the following conclusions about the complexity of inverting an HFE polynomial using a Gröbner basis algorithm.

1. If D, the degree of the corresponding HFE polynomial, and q, the size of the corresponding finite field, are fixed, then the degree of regularity is bounded by a fixed integer $(q-1)(\lfloor \log_q(D-1) \rfloor + 1)/2 + 2$ or q. Therefore inverting HFE systems is polynomial for all fields;
2. If q is fixed and D is of the scale $O(n^\alpha)$, then inverting HFE systems is quasi-polynomial.

If, on the other hand, q is of the scale $O(n)$, then our results are inconclusive and the possibility remains that inverting HFE systems is actually exponential with respect to this parameter. Comparisons with the results of [9,12], suggest that our formulas asymptotically may be proportional to the degree of regularity at least for the case where q is a prime.

This paper is organized as follows. We first briefly introduce HFE cryptosystems in the section below. In Section 3, we review the definition and basic properties of the degree of regularity from [9]. In Section 4, we show how the notion of rank can give a useful closed formula bound on the degree of regularity and apply this to the analysis of the complexity of the Gröbner basis attacks on HFE systems.

In the appendix, we develop the key ideas of [9] in a more abstract mathematical framework using the language of graded algebras. This allows us to create different but abstractly more transparent proofs for the main theorems that we use in the paper.

After the present paper was submitted, a paper by Bettale, Faugère and Perret [2] was published that has some commonality with ours. In this article, the authors come to similar conclusions on the security of the HFE systems, but with respect only to the Kipnis-Shamir attack. They conjecture that if D is fixed the complexity of the Kipnis-Shamir attack is polynomial in n, the degree of the extension. Their experimental data backs up this conjecture.

2 HFE Systems

2.1 Quadratic Operators

Denote by $\mathbb{F}$ a finite field of order q and let $\mathbb{K}$ be an extension of $\mathbb{F}$ of degree n. Any function from $\mathbb{K}$ to $\mathbb{K}$ can be expressed as a polynomial with coefficients in $\mathbb{K}$ and degree less than q^n. Thus it has the general form

$$P(X) = \sum_{i=0}^{q^n-1} a_i X^i, \quad a_i \in \mathbb{K} .$$

There are two distinct notions of degree for P, the degree over $\mathbb{K}$ and the degree over $\mathbb{F}$. The degree over $\mathbb{K}$, denoted by $\deg_{\mathbb{K}}(P)$ is the degree in the usual sense of degree of a polynomial function. On the other hand, the functions X^{q^i} are all linear over $\mathbb{F}$. Thus the degree of the monomial X^d will be the sum of the digits in the base q expansion of d; that is, if $d = \sum_i d_i q^i$, $\deg_{\mathbb{F}}(X^d) = \sum_i d_i$. When $q = 2$, this is the Hamming weight of the binary representation of d. The degree of P over $\mathbb{F}$, denoted $\deg_{\mathbb{F}}(P)$ is the maximum of the degree of the monomial terms.

An $\mathbb{F}$-quadratic function from $\mathbb{K}$ to $\mathbb{K}$ is thus a polynomial all of whose monomial terms have exponent $q^i + q^j$ or q^i for some i and j. The general form of an $\mathbb{F}$-quadratic function is

$$P(X) = \sum_{i,j=0}^{n-1} a_{ij} X^{q^i+q^j} + \sum_{i=0}^{n-1} b_i X^{q^i} + c .$$

2.2 HFE Systems

In an HFE cryptosystem, plaintext from $\mathbb{F}^n$ is encrypted using an identification of $\mathbb{F}^n$ with $\mathbb{K}$ and an $\mathbb{F}$-quadratic map P. The nature of P is further hidden by pre- and post-composition with invertible affine linear maps $L_1, L_2 \colon \mathbb{F}^n \to \mathbb{F}^n$. Thus if $\phi \colon \mathbb{F}^n \to \mathbb{K}$ is the chosen linear isomorphism, the encryption is performed by the function $\bar{P} = L_1 \circ \phi^{-1} \circ P \circ \phi \circ L_2$. Decryption is performed by inverting the maps L_1 and L_2 and applying a standard root-finding algorithm for P. The public key is the function $\bar{P}$ expressed in terms of its quadratic component functions $\bar{p}_1, \ldots, \bar{p}_n \colon \mathbb{F}^n \to \mathbb{F}$. Provided the degree of P is not too high the decryption process will be manageable. However the direct solution of a system of quadratic multivariate equations

$$\bar{p}_1 = b_1, \ldots, \bar{p}_n = b_n$$

is a hard problem which provides the system with a certain level of security.

2.3 Gröbner Basis Attacks

One of the most successful attacks on HFE systems is to apply the refined Gröbner basis algorithms F_4 (and maybe F_5 if we do know how to make it work as efficiently as claimed) to convert the system of equations $\bar{p}_1 = b_1, \ldots, \bar{p}_n = b_n$ to a simpler system that can be solved quickly. From the point of view of security analysis it is sufficient to consider the system $p_1 = 0, \ldots, p_n = 0$ where the p_i are the component functions of $\phi^{-1} \circ P \circ \phi$ with respect to the given basis. From this point on, we restrict our attention to this case.

Implementation of the Gröbner basis algorithm involves searching through combinations of multiples of the p_i by polynomials of a fixed degree for polynomials of smaller degree. If the combination $\sum_i g_i p_i$ has smaller degree then the corresponding combination of leading components $\sum_i g_i^h p_i^h$ is zero. Here, by the leading component of a multivariate polynomial g we mean the multivariate polynomial g^h derived from removing all the monomial terms of g with degree lower than the degree of g, or the highest homogeneous component of g. In general, the decisive moment in the calculation is when *non-trivial* such combinations occur. These non-trivial relations will very likely generate what is called mutants [3,4,15], which are instrumental in solving the system. Obviously the combinations $p_i^h p_j^h - p_j^h p_i^h$ are tautologically zero and the equation $((p_i^h)^{q-1} - 1)p_i^h = 0$ is a result of the identity $a^q = a$ in $\mathbb{F}$. A non-trivial relation is one that does not follow from these trivial identities. The degree at which the first non-trivial relation occurs is called the *degree of regularity*. Extensive experimental evidence has shown that the algorithm will terminate at or shortly after the degree of regularity. Thus the calculation of the degree of regularity is crucial to understanding the complexity of the algorithm.

3 Degree of Regularity

We begin with the formal definition of degree of regularity as given in [9] and we summarize the key results from that paper. More abstract versions of these results are also given in the appendix. Let

$$A = \mathbb{F}[X_1, \ldots, X_n] / \langle X_1^q - X_1, \ldots, X_n^q - X_n \rangle \ .$$

This is the algebra of functions from $\mathbb{F}^n$ to $\mathbb{F}$. Let $p_1, \ldots, p_n$ be quadratic elements of A. Denote by A_k the subspace of A consisting of functions representable by a polynomial of degree less than or equal to k. For all j we have a natural map $\psi_j \colon A_j^n \to \sum_i A_j p_i$ given by

$$\psi_j(a_1, \ldots, a_n) = \sum_i a_i p_i \ .$$

We are interested in 'degree falls'; a degree fall occurs when the a_i have degree j but $\sum_i a_i p_i$ has degree less than degree $j+2$. Obviously we can have trivial degree falls of the form $p_j p_i + (-p_i)p_j$ or $(p_i^{q-1} - 1)p_i$. The *degree of regularity* of the set $\{p_1, \ldots, p_n\}$ is the first degree (measured as $\deg a_i + \deg p_i$) at which a non-trivial degree fall occurs. Obviously we can restrict our attention to the highest degree terms in the a_i and work modulo terms of smaller degree. Mathematically this means working in the associated graded ring $B = \mathbb{F}[X_1, \ldots, X_n] / \langle X_1^q, \ldots, X_n^q \rangle$.

The degree of regularity of the $\{p_1, \dots, p_n\}$ in A will be the first degree at which we find non-trivial relations among the leading components $p_1^h, \dots, p_n^h$ (considered as elements of B).

Denote by B_k the subspace of B consisting of homogeneous elements of degree k. Consider an arbitrary set of homogeneous quadratic elements $\{\lambda_1, \dots, \lambda_n\} \in B_2$. For all j we have a natural map $\psi_j \colon B_j^n \to B_{j+2}$ given by

$$\psi_j(b_1, \dots, b_n) = \sum_i b_i \lambda_i \ .$$

Let $R_j(\lambda_1, \dots, \lambda_n) = \ker \psi_j$; this is the subspace of relations of the form $\sum_i b_i \lambda_i = 0$. Inside $R_j(\lambda_1, \dots, \lambda_n)$ is the subspace of trivial relations, $T_j(\lambda_1, \dots, \lambda_n)$ generated by elements of the form:

1. $b(0, \dots, 0, \lambda_j, 0, \dots, 0, -\lambda_i, 0 \dots, 0)$ for $1 \leq i < j \leq n$ and $b \in B_{j-2}$; where λ_j is in the i-th position and $-\lambda_i$ is in the j-th position;
2. $b(0, \dots, 0, \lambda_i^{q-1} - 1, 0 \dots, 0)$ for $1 \leq i \leq n$ and $b \in B_{j-2(q-1)}$; where λ_i^{q-1} is in the i-th position;

The space of non-trivial relations is the quotient space $R_j(\lambda_1, \dots, \lambda_n)/T_j(\lambda_1, \dots, \lambda_n)$.

Definition 3.1. *The* degree of regularity *of* $\{\lambda_1, \dots, \lambda_n\}$ *is defined by*

$$D_{\text{reg}}(\{\lambda_1, \dots, \lambda_n\}) = \min\{j \mid R_{j-2}(\{\lambda_1, \dots, \lambda_n\})/T_{j-2}(\{\lambda_1, \dots, \lambda_n\}) \neq 0\}$$

It turns out that the degree of regularity is dependent only on the subspace generated by the λ_i so we can simplify the notation a little by denoting the space generated by the λ_i by V and writing $D_{\text{reg}}(V)$ for $D_{\text{reg}}(\{\lambda_1, \dots, \lambda_n\})$.

Two important properties of the degree of regularity were observed in [9]. First, the degree of regularity of a space is less than or equal to the degree of regularity of a subspace.

Property 3.2. [9, Property 11] Let V' be a subspace of V. Then $D_{\text{reg}}(V) \leq D_{\text{reg}}(V')$.

Second, the degree of regularity is invariant under field extension. Let $\mathbb{K}$ be an extension of $\mathbb{F}$. Define $B_{\mathbb{K}} = \mathbb{K}[X_1, \dots, X_n] / \langle X_1^q, \dots, X_n^q \rangle$ and denote by $V_{\mathbb{K}}$ be the $\mathbb{K}$-subspace of $B_{\mathbb{K}}$ generated by the λ_i.

Property 3.3. [9, §4.4] Let $\mathbb{K}$ be an extension of $\mathbb{F}$. Then $D_{\text{reg}}(V_{\mathbb{K}}) = D_{\text{reg}}(V)$.

Returning to the situation where P is a quadratic map with component functions $p_1, \dots, p_n \in A$. Let V and V^h be the vector spaces generated by the $p_1, \dots, p_n$ and their leading components $p_1^h, \dots, p_n^h$ (considered as elements of B). Our goal is to find a bound for $D_{\text{reg}}(V^h)$. We begin by extending the base field to $\mathbb{K}$. When we extend the base field in A, we pass from functions from $\mathbb{F}^n$ to $\mathbb{F}$ to functions from $\mathbb{F}^n$ to $\mathbb{K}$. Via the linear isomorphism $\phi^{-1} \colon \mathbb{K} \to \mathbb{F}^n$, this algebra is isomorphic to the algebra of functions from $\mathbb{K}$ to $\mathbb{K}$ which is simply $\mathbb{K}[X] / \left\langle X^{q^n} - X \right\rangle$.

It follows from elementary Galois theory that the space $V_{\mathbb{K}}$ corresponds under this identification with the space generated by $P, P^q, \ldots, P^{q^{n-1}}$. If we filter the algebra $\mathbb{K}[X]/\left\langle X^{q^n} - X\right\rangle$ by degree of functions over $\mathbb{F}$, then the linear component is spanned by $X, X^q, \ldots, X^{q^{n-1}}$. The associated graded ring will then be the algebra $B_{\mathbb{K}} = \mathbb{K}[X_0, \ldots, X_{n-1}]$ where X_i corresponds to X^{q^i} and $X_i^q = 0$. This is naturally isomorphic to the algebra B with coefficients extended to $\mathbb{K}$ (proofs in Appendix B). Thus the processes of extending the base field and taking the associated graded ring commute.

Let P_i denote the leading component of P^{q^i} in $B_{\mathbb{K}}$. Thus for instance if P is defined as above, then

$$P_0 = \sum_{i,j=0}^{n-1} a_{ij} X_i X_j \ .$$

The space generated by the P_i is exactly $V_{\mathbb{K}}^h$, the subspace of $B_{\mathbb{K}}$ generated by the p_i^h. Putting all the above together we get the following important theorem. A brief proof is given in Appendix B.

Theorem 3.4 ([9])

$$D_{\text{reg}}(\{p_1, \ldots, p_n\}) = D_{\text{reg}}(\{p_1^h, \ldots, p_n^h\}) = D_{\text{reg}}(\{P_0, \ldots, P_{n-1}\})$$

Using Property 2, we get the following immediate corollary.

Corollary 3.5. $D_{\text{reg}}(\{p_1, \ldots, p_n\}) \leq \min\{D_{\text{reg}}(Q) \mid Q \in V_{\mathbb{K}}^h\}$

4 Bounding the Degree of Regularity Using Q-Rank

Up to this point we have been following the ideas of [12,9]. In particular, the proofs of all the above results are given in [9], which is the basis for this work. We now, however, take a significant change of direction. The bounds on the degree of regularity in [12,9] and previous authors are all found by counting dimensions. The basic idea going back to [21,1] is that if $\dim B_j^n - \dim T_j > \dim B_j^{n+2}$, then $R_j \supsetneq T_j$ and the degree of regularity has been reached. This approach was refined in [12] by using Property 1 to reduce to subsets $\{P_0, \ldots, P_s\}$ which, for HFE systems, involve significantly fewer variables. It was further refined in [9] where the authors observed that the space on the right hand side of the inequality could be replaced by a significantly smaller space, allowing a more accurate computational estimate of the degree of regularity. The disadvantage of the approach in [9] was that no general formula for the degree of regularity could be derived.

Our approach is completely different. Instead of counting and comparing dimensions we actually find non-trivial relations in specific dimensions. Surprisingly, an important bound can be found by restricting to the case of a single polynomial. We begin by giving a bound on the degree of regularity of a single polynomial in terms of its rank. Applying this formula to P_0 yields a bound on the degree of regularity of an HFE system in terms of its degree.

The degree of regularity of a single polynomial has been studied in great detail in the cases where $q = 2$ and 3 [6,7]. In order to obtain the desired bound, we do not need the kind of exact information found in those papers. We merely need to demonstrate the existence of non-trivial relations. This we can do explicitly using the classification of quadratic forms. Recall that P_0 is a homogeneous quadratic polynomial in the algebra $\mathbb{K}[X_0, \ldots, X_{n-1}]/\left\langle X_0^q, \ldots, X_{n-1}^q \right\rangle$. Using the classification theorem of quadratic forms over finite fields, we are able to explicitly construct nontrivial relations and hence derive a simple bound for the degree of regularity of P_0 in terms of its rank.

We now briefly review the classification of quadratic forms over a finite field. We begin with the case when q is odd. A quadratic form in n indeterminates is a homogeneous quadratic polynomial in the polynomial ring $\mathbb{K}[X_1, \ldots, X_n]$. Two forms P and Q are said to be equivalent (written $P \sim Q$) if there is an invertible linear change of variables L which transforms P into Q:

$$P \circ L(X_1, ..., X_n) = Q(X_1, ..., X_n).$$

Pick an element $c \in \mathbb{K}$ that is not a square. Then a quadratic form is equivalent to one of the two types

1. $X_1^2 + \cdots + X_{r-1}^2 + X_r^2$
2. $X_1^2 + \cdots + X_{r-1}^2 + cX_r^2$

for some $r \leq n$ [17, §62]. The same classification applies to quadratic elements of the quotient ring $\mathbb{K}[X_1, \ldots, X_n]/\langle X_1^q, \ldots, X_n^q \rangle$.

When q is even, the situation is complicated by the fact that X^2 is linear rather than quadratic when $q = 2$. It is shown in [14, Theorem 6.30] that a quadratic polynomial in the polynomial algebra $\mathbb{K}[X_1, \ldots, X_n]$ is equivalent to an polynomial of one of the following forms for some $r \leq n$:

1. $X_1X_2 + ... + X_{r-1}X_r$
2. $X_1X_2 + ... + X_{r-2}X_{r-1} + X_r^2$
3. $X_1X_2 + ... + X_{r-1}X_r + X_{r-1}^2 + cX_r^2$ where $c \in \mathbb{K}\backslash\{0\}$ satisfies $\mathrm{TR}_{\mathbb{K}}(c) = 1$.

For $q > 2$ this classification carries over to the quotient ring $\mathbb{K}[X_1, \ldots, X_n]/\langle X_1^q, \ldots, X_n^q \rangle$. When $q = 2$, all quadratic elements of the quotient ring are equivalent to an element of the first type. In all cases the number r is known as the rank of Q. Note that if $q = 2$, the rank of a quadratic element must be at least 2.

When $r = 1$ (in which case $q > 2$), Q is actually equal to aX_1^2 for some $a \in \mathbb{K}$. It is easily verified that the smallest non-trivial relation is $X^{q-2}(aX^2) = 0$ and hence that $D_{\mathrm{reg}}(Q) = q$. More generally we have the following inequality.

Theorem 4.1. *Let Q be quadratic of rank r. If $r > 1$,*

$$D_{\mathrm{reg}}(Q) \leq \frac{r(q-1)}{2} + 2 \ .$$

Proof. In the case of a single polynomial, the definition of degree of regularity can be expressed in terms of non-trivial annihilators. Let Q be an arbitrary quadratic element of $B = \mathbb{K}[X_1, \dots, X_n]/\langle X_1^q, \dots, X_n^q\rangle$. The annihilators of Q are the elements of $\mathrm{Ann}(Q) = \{f \in B \mid fQ = 0\}$. The trivial annihilators are the multiples of Q^{q-1}. The degree of regularity is the first k such that there is a non-trivial annihilator of Q of degree $k-2$. The degree of regularity is invariant under a linear change of variables, so it is sufficient to prove the result by exhibiting explicit non-trivial annihilators for each of the above types of quadratic elements.

Because of the different types of standard forms we need to consider separately the cases when q is odd and even. We also need to divide these cases into the cases when r is odd or even.

– Case 1: q odd, r even

Set $s = r/2$. In this case Q is of the form

$$Q = X_1^2 + X_2^2 + \cdots + X_{2s-1}^2 + aX_{2s}^2$$

for some $a \in \mathbb{K}$. Let

$$K_i = X_{2i-1}^{q-1} - X_{2i}^2 X_{2i-1}^{q-3} + X_{2i}^4 X_{2i-1}^{q-5} + \cdots + (-1)^{(q-1)/2} X_{2i}^{q-1}$$

for $i = 1, \dots, s-1$; and

$$K_i = X_{2s-1}^{q-1} - aX_{2s}^2 X_{2s-1}^{q-3} + a^2 X_{2s}^4 X_{2s-1}^{q-5} + \cdots + (-a)^{(q-1)/2} X_{2s}^{q-1}$$

Set

$$K = K_1 K_2 \dots K_s \ .$$

It is clear that

$$K_i(X_{2i-1}^2 + X_{2i}^2) = X_{2i-1}^{q+1} - (-1)^{(q+1)/2} X_{2i}^{q+1} = 0 \ ,$$

for $i = 1, ..., s-1$; and

$$K_s(X_{2s-1}^2 + aX_{2s}^2) = X_{2s-1}^{q+1} - (-a)^{(q+1)/2} X_{2s}^{q+1} = 0 \ .$$

Hence $KQ = 0$. Thus $K \in \mathrm{Ann}(Q) \cap B_{s(q-1)}$. We claim that $K \notin \langle Q^{q-1}\rangle$. Consider the quotient algebra

$$\bar{B} = B/\left\langle X_{2i-1}^2 + X_{2i}^2, i = 1, ..., s-1;\ X_{2s-1}^2 + aX_{2s}^2\right\rangle \ .$$

The algebra $\bar{B}$ has a basis consisting of monomials with the powers of the variables $X_2, X_4, \dots, X_{2s}$ at most 1. It is clear that the image of Q (and hence also Q^{q-1}) in $\bar{B}$ is zero, whereas the image of K is

$$\prod_i^s X_{2i-1}^{q-1} \left(\frac{q+1}{2}\right)^s$$

which is non-zero. Therefore K is not in the ideal generated by Q^{q-1}. Hence $D_{\mathrm{reg}}(Q) \leq r(q-1)/2 + 2$.

– Case 2: q odd, r odd

Set $s = (r-1)/2$. In this case Q is of the form

$$Q = X_1^2 + X_2^2 + \cdots + X_{2s-1}^2 + X_{2s}^2 + aX_{2s+1}^2$$

for some $a \in \mathbb{K}$. From the classification of quadratic forms [20] we have

$$X_{2s-1}^2 + X_{2s}^2 + aX_{2s+1}^2 \sim X_{2s-1}^2 - X_{2s}^2 - aX_{2s+1}^2 \sim X_{2s-1}X_{2s} - aX_{2s+1}^2$$

so Q can be taken to be of the form:

$$Q = X_1^2 + X_2^2 + \cdots + X_{2s-2}^2 + X_{2s-1}X_{2s} - aX_{2s+1}^2 \ .$$

Let

$$K_i = X_{2i-1}^{q-1} - X_{2i}^2 X_{2i-1}^{q-3} + X_{2i}^4 X_{2i-1}^{q-5} + \cdots + (-1)^{(q-1)/2} X_{2i}^{q-1}$$

for $i = 1, \ldots, s-1$; and

$$K' = \frac{(X_{2s-1}X_{2s})^{(q+1)/2} - a^{(q+1)/2} X_{2s+1}^{(q+1)}}{X_{2s-1}X_{2s} - aX_{2s+1}^2} X_{2s-1}^{(q-1)/2} \ .$$

Note that

$$K'(X_{2s-1}X_{2s} - aX_{2s+1}^2) = (X_{2s-1}X_{2s})^{(q+1)/2} X_{2s-1}^{(q-1)/2} = 0 \ .$$

Set

$$K = K_1 K_2 \ldots K_{s-1} K'.$$

Note that the degree of K is $(s-1)(q-1) + 3(q-1)/2 = r(q-1)/2$. Again we see that $KQ = 0$ and so $K \in \mathrm{Ann}(Q) \cap B_{r(q-1)/2}$. Consider the quotient algebra

$$\bar{B} = B / \left\langle X_{2i-1}^2 + X_{2i}^2, i = 1, ..., s-1;\ X_{2s-1}X_{2s} - aX_{2s+1}^2 \right\rangle \ ,$$

Then $\bar{B}$ has a basis consists of monomials in which the powers of the variables $X_2, X_4, \ldots, X_{2(s-1)}, X_{2s+1}$ are at most one. The image of Q in $\bar{B}$ is zero, but that of K is

$$\prod_{i=1}^{s-1} X_{2i-1}^{q-1} \left(\frac{q+1}{2}\right)^{s-1} X_{2s-1}^{q-1} X_{2s}^{(q-1)/2} \left(\frac{q+1}{2}\right)$$

which is non-zero. Hence K is not in the ideal generated by Q^{q-1} and $D_{\mathrm{reg}}(Q) \leq r(q-1)/2 + 2$.

– Case 3: q even, r even (Q of type (1) or (3))

First suppose that Q is of the form $Q = X_1X_2 + \cdots + X_{2s-1}X_{2s}$ where $r = 2s$. Set $H = X_1^{q-1} X_3^{q-1} \ldots X_{2s-1}^{q-1}$. Then it is easily seen that $H \in \mathrm{Ann}(Q) \cap B_{s(q-1)}$. Consider the quotient algebra

$$\bar{B} = B / \langle X_1 - X_2, \ldots, X_{2s-1} - X_{2s} \rangle \ .$$

The image of Q in $\bar{B}$ is

$$\bar{Q} = X_1^2 + X_3^2 + \cdots + X_{2s-1}^2 = (X_1 + X_3 + \cdots + X_{2n-1})^2$$

so the image of Q^{q-1} is $\bar{Q}^{q-1} = 0$. On the other hand the image of H is

$$X_1^{q-1} X_3^{q-1} \dots X_{2s-1}^{q-1}$$

which is non-zero. Thus $H \notin \langle Q^{q-1} \rangle$ Hence $D_{\text{reg}}(Q) \leq r(q-1)/2 + 2$.

Next suppose that Q is of the form $Q = X_1X_2 + \cdots + X_{2s-1}X_{2s} + X_{2s-1}^2 + \alpha X_{2s}^2$. Let $\mathbb{L}$ be a finite extension field of $\mathbb{K}$ in which the equation $1 + X + aX^2$ has a root. In $\mathbb{L}[X_1, \dots, X_r]$, Q is equivalent to $X_1X_2 + \cdots + X_{2s-1}X_{2s}$. Since the degree of regularity is invariant under extensions of the base field by Property 2, it follows from the first part that $D_{\text{reg}}(Q) \leq r(q-1)/2 + 2$.

– Case 4: q even, r odd (Q of type (2))

Note that in this case we must have $q > 2$. We may assume that Q is of the form $Q = X_1X_2 + \cdots + X_{2s-1}X_{2s} + X_{2s+1}^2$ where $r = 2s + 1$. Set

$$H = X_1^{q-1} X_3^{q-1} \dots X_{2s-3}^{q-1} X_{2s-1}^{q/2} (X_{2s-1}X_{2s} + X_{2s+1}^2)^{(q-2)/2} \ .$$

Note that $\deg H = r(q-2)/2$ and

$$HQ = (X_{2s-1}X_{2s} + X_{2s+1}^2)^{q/2} X_{2s-1}^{q/2} = X_{2s-1}^q X_{2s}^{q/2} + X_{2s+1}^q X_{2s-1}^{q/2} = 0 \ .$$

Consider the quotient algebra

$$\bar{B} = B / \langle X_1 - X_2, \dots, X_{2s-1} - X_{2s} \rangle \ .$$

The image of Q in $\bar{B}$ is $\bar{Q} = X_1^2 + X_3^2 + \cdots + X_{2s-1}^2 + X_{2s+1}^2 = (X_1 + X_3 + \cdots + X_{2s-1} + X_{2s+1})^2$ so the image of Q^{q-1} is $\bar{Q}^{q-1} = 0$. On the other hand the image of H is $X_1^{q-1} X_3^{q-1} \dots X_{2s-3}^{q-1} X_{2s-1}^{q/2} (X_{2s-1}^2 + X_{2s+1}^2)^{(q-2)/2}$ which is non-zero. Thus $H \notin \langle Q^{q-1} \rangle$ Hence $D_{\text{reg}}(Q) \leq r(q-1)/2 + 2$.

Note that for q odd, these bounds were conjectured to be optimal in [7]. Experimental evidence suggests that this is not the case when q is even.

Let us define the Q-Rank of a quadratic operator $P(X)$ to be the minimal rank of elements of the space $V_{\mathbb{K}}^h$ generated by $P_0, \dots, P_{n-1}$.

$$\text{Q-Rank}\, P = \min\{\text{Rank}\, Q \mid Q \in V_{\mathbb{K}}^h\}$$

Note in particular that $\text{Q-Rank}(P) \leq \text{Rank}(P_0)$.

Theorem 4.2. *Let P be a quadratic operator of degree D. If* Q-Rank$(P) > 1$, *the degree of regularity of the associated system is bounded by*

$$\frac{(q-1)\,\text{Q-Rank}(P)}{2} + 2 \ .$$

In particular, this is less than or equal to

$$\frac{(q-1)(\lfloor \log_q(D-1) \rfloor + 1)}{2} + 2 \ .$$

If Q-Rank$(P) = 1$, *then the degree of regularity is less than or equal to q.*

Proof. The first assertion follows from Theorem 4.1 and Corollary 3.5. Suppose that

$$P(X) = \sum_{q^i+q^j \leq D} a_{ij} X^{q^i+q^j} + \sum_{q^i \leq D} b_i X^{q^i} + c \ .$$

Then

$$P_0 = \sum_{q^i+q^j \leq D} a_{ij} X_i X_j \ .$$

Let k be the largest subscript of a variable X_k that occurs non-trivially in P_0 (that is, $a_{ik} \neq 0$ for some i). The rank of P_0 is bounded by the number of variables involved in its expression which is at most $k+1$. On the other hand, by our assumption on D, $D \geq q^k + 1$ or equivalently, $k \leq \lfloor \log_q(D-1) \rfloor$. Thus the rank of P_0 is at most $\lfloor \log_q(D-1) \rfloor + 1$.

Example 4.3. Consider a Matsumoto-Imai operator of the form $P(X) = X^{1+2^\theta}$ over the field $GF(2)$. Then $P_0 = X_0 X_\theta$ has rank 2. So our theorem implies that the degree of regularity is less than or equal to three. This is precisely the statement that linearization equations exist in this case [18]. On the other hand if we consider a Matsumoto-Imai operator over a field of order $q = 2^m$, then the degree of regularity remains 3 but our bound is $2^m + 1$. Therefore our estimate formula needs to be improved when q is not a prime.

For fixed q the degree of regularity is $O(\log_q D)$. Consider now a Gröbner basis attack on an HFE system of degree D. We continue to make the assumption that these algorithms will terminate at degree equal to the degree of regularity or shortly after this. The runtime of this algorithm will be $O(n^{3D_{\text{reg}}})$. Assuming that the security parameter is chosen in such a way that $D = O(n^\alpha)$, the runtime for the Gröbner basis attack on an HFE system over any base field will be $2^{O(\log(n)^2)}$; that is, it will be quasi-polynomial.

On the other hand, suppose that q itself is a component of the security parameter and is taken to be of scale $O(n)$ (this assumption is reasonable since it will only increase the computation complexity for HFE systems by the scale of $O(\log_2 n)$). If the bound above is asymptotically sharp then the degree of regularity will be at least of the scale $O(n)$, and therefore inverting HFE systems will be exponential.

We do not expect or believe the bound obtained in Theorem 4.2 to be optimal in any degree of generality. We compare the bound $(q-1)(\lfloor \log_q(D-1) \rfloor + 1)/2 + 2$ with that obtained by Dubois and Gama for a large number of values of n and D and prime q from [9]. The tables Appendix C give a detailed comparison of our bound with the bound calculated in [9]. As n becomes large relative to q, the two bounds appear to be getting closer, though ours are frequently slightly higher. It seems possible that there may be a tighter upper bound of the form $cq \log_q(D)$ for some scalar c when q is a prime. The discontinuities in the Dubois-Gama data are close to the discontinuities of $\lfloor \log_q(D-1) \rfloor$ and the jump size seems proportional to q.

5 Conclusion

By finding explicit non-trivial relations, we prove that the degree of regularity of a multivariate quadratic cryptosystem over a field of arbitrary characteristic q is bounded above by a simple linear function of its Q-Rank and q. These universal estimate formulas for the degree of regularity for HFE systems for all finite fields allow us to show that if the degree D of the HFE formula is fixed and the number of variables increased, the complexity of a Gröbner basis attack on this system will grow as a polynomial function in n; if, on the other hand, the degree of the HFE polynomial is $O(n^\alpha)$, then the algorithm will take quasi-polynomial time, as was observed in the case $q = 2$ in [12].

Our bounds on the degree of regularity are not likely to be optimal even for large n - we look for relations involving a single polynomial rather than the whole polynomial systems to prove our estimates. We expect in general that there will be some non-trivial relations resulting from relations between polynomials in subsystems, which have smaller degree than relations coming from single polynomials. On the other hand there is a surprising similarity between our bounds and those found by Dubois and Gama using a very different approach. Of course, it is possible that neither bound is close to being optimal and it would be interesting to run experimental trials for large values of n, D and q. However, memory limitations prevent us from being able to do this at sufficiently large values. Taking all this into account, we conjecture that our formulas should give a good asymptotic estimate (up to a linear factor) of the degree of regularity in the case q when is prime. If this is true, this would imply that inverting an HFE system with q of size $O(n)$ is actually exponential.

Acknowledgments. J. Ding would like to thank V. Dubois and N. Gama for sending him their paper [9] before its publication and for sending him their data and their program to compute the estimated bound of the degree of regularity in terms of their formulation. J. Ding would like to thank V. Dubois for many stimulating and insightful discussions, without which this paper will not be possible. J. Ding would like to thank C. Christensen and J. Buchmann for useful discussions. J. Ding would also like to thank many years' crucial support of the **Charles Phelps Taft Foundation** and the support of the **National Science Foundation of China** under the grant #60973131.

References

1. Bardet, M., Faugère, J.-C., Salvy, B.: On the complexity of Gröbner basis computation of semi-regular overdetermined algebraic equations. In: International Conference on Polynomial System Solving - ICPSS, pp. 71–75 (November 2004)
2. Bettale, L., Faugère, J.-C., Perret, L.: Cryptanalysis of Multivariate and Odd-Characteristic HFE Variants. In: Catalano, D., Fazio, N., Gennaro, R., Nicolosi, A. (eds.) PKC 2011. LNCS, vol. 6571, pp. 441–458. Springer, Heidelberg (2011)
3. Ding, J.: Mutants and its impact on polynomial solving strategies and algorithms. Privately distributed research note, University of Cincinnati and Technical University of Darmstadt (2006)

4. Ding, J., Buchmann, J., Mohamed, M., Mohamed, W., Weinmann, R.-P.: Mutant XL. In: First International Conference on Symbolic Computation and Cryptography – SCC (2008)
5. Ding, J., Gower, J., Schmidt, D.: Multivariate Public Key Cryptography. Advances in Information Security series. Springer, Heidelberg (2006)
6. Ding, J., Hodges, T.J., Kruglov, V.: Growth of the ideal generated by a quadratic boolean function. In: Sendrier, N. (ed.) PQCrypto 2010. LNCS, vol. 6061, pp. 13–27. Springer, Heidelberg (2010)
7. Ding, J., Hodges, T.J., Kruglov, V., Schmidt, D., Tohaneanu, S.: Growth of the ideal generated by a multivariate quadratic function over GF(3), preprint
8. Ding, J., Schmidt, D., Werner, F.: Algebraic attack on HFE revisited. In: Wu, T.-C., Lei, C.-L., Rijmen, V., Lee, D.-T. (eds.) ISC 2008. LNCS, vol. 5222, pp. 215–227. Springer, Heidelberg (2008)
9. Dubois, V., Gama, N.: The degree of regularity of HFE systems. In: Abe, M. (ed.) ASIACRYPT 2010. LNCS, vol. 6477, pp. 557–576. Springer, Heidelberg (2010)
10. Faugère, J.-C., Joux, A.: Algebraic cryptanalysis of hidden field equation (HFE) cryptosystems using gröbner bases. In: Boneh, D. (ed.) CRYPTO 2003. LNCS, vol. 2729, pp. 44–60. Springer, Heidelberg (2003)
11. Garey, M.R., Johnson, D.S.: Computers and intractability, A Guide to the theory of NP-completeness. W.H. Freeman, San Francisco (1979)
12. Granboulan, L., Joux, A., Stern, J.: Inverting HFE Is Quasipolynomial. In: Dwork, C. (ed.) CRYPTO 2006. LNCS, vol. 4117, pp. 345–356. Springer, Heidelberg (2006)
13. Kipnis, A., Shamir, A.: Cryptanalysis of the HFE Public Key Cryptosystem by Relinearization. In: Wiener, M. (ed.) CRYPTO 1999. LNCS, vol. 1666, pp. 19–30. Springer, Heidelberg (1999)
14. Lidl, R., Niederreiter, H.: Finite Fields, Encyclopedia of Mathematics and its Applications, vol. 20. Cambridge University Press, Cambridge (1997)
15. Mohamed, M., Cabarcas, D., Ding, J., Buchmann, J., Bulygin, S.: MXL_3: An Efficient Algorithm for Computing Gröbner Bases of Zero-Dimensional Ideals. In: Lee, D., Hong, S. (eds.) ICISC 2009. LNCS, vol. 5984, pp. 87–100. Springer, Heidelberg (2010)
16. Matsumoto, T., Imai, H.: Public quadratic polynomial-tuples for efficient signature-verification and message-encryption. In: Günther, C.G. (ed.) EUROCRYPT 1988. LNCS, vol. 330, pp. 419–453. Springer, Heidelberg (1988)
17. O'Meara, O.T.: Introduction to Quadratic Forms. Springer, Berlin (1963)
18. Patarin, J.: Cryptanalysis of the Matsumoto and Imai Public Key Scheme of Eurocrypt '88. In: Coppersmith, D. (ed.) CRYPTO 1995. LNCS, vol. 963, pp. 248–261. Springer, Heidelberg (1995)
19. Shor, P.: Polynomial-time algorithms for prime factorization and discrete logarithms on a quantum computer. SIAM Rev. 41(2), 303–332 (1999)
20. Wan, Z.-X.: Lectures on Finite Fields and Galois Rings. World Scientific Publishing, Singapore (2003)
21. Yang, B.-Y., Chen, J.-M.: Theoretical Analysis of XL over Small Fields. In: Wang, H., Pieprzyk, J., Varadharajan, V. (eds.) ACISP 2004. LNCS, vol. 3108, pp. 277–288. Springer, Heidelberg (2004)

A Degree of Regularity

Let $\mathbb{F}$ be a finite field with $|\mathbb{F}| = q$. Denote by $B = \bigoplus_{k=0}^{N} B_k$ a graded finite dimensional algebra over $\mathbb{F}$. Let $V \subset B_d$ be a homogeneous subspace. Then for all j we have a natural map $\phi_j : B_j \otimes V \to B_j V$ given by $\phi(\sum b_i \otimes v_i) = \sum b_i v_i$. Let $R_j(V) = \ker \phi_j$. Inside $R_j(V)$ there is a subspace of "trivial relations" $T_j(V)$ spanned by the elements

1. $b(v \otimes w - w \otimes v)$ where $v, w \in V$ and $b \in B_{j-d}$;
2. $b(v^{q-1} \otimes v)$ where $v \in V$ and $b \in B_{j-(q-1)d}$.

A similar basis-dependent definition of trivial relations was give in [9]. It can be shown that these two definitions coincide.

Following Dubois and Gama [9], we define the degree of regularity of V to be the degree of the first space $B_j V$ in which non-trivial relations occur.

Definition A.1. *For a homogeneous subspace $V \subset B_d$, the* degree of regularity *of V is defined to be*

$$D_{\text{reg}}(V) = \min\{j \mid T_{j-d}(V) \subsetneq R_{j-d}(V)\}$$

Let A be a filtered algebra over $\mathbb{F}$ and let $\text{Gr}A = \bigoplus_j A_j/A_{j+1}$. Let V be a subspace of A_j. We denote by $\bar{V}$ its image in $\text{Gr}A$; that is, $\bar{V} = V + A_{j-1} \subset A_j/A_{j-1}$.

Definition A.2. *For a subspace $V \subset A_d$, we define the* degree of regularity *of V by*

$$D_{\text{reg}}(V) = \begin{cases} d & \textit{if} \quad \dim \bar{V} < \dim V \\ D_{\text{reg}}(\bar{V}) & \textit{otherwise} \end{cases}$$

Extension of the base field does not affect the degree of regularity.

Theorem A.3. *Let B be a graded algebra over $\mathbb{F}$, let $\mathbb{K}$ be an extension field of $\mathbb{F}$ and let $\tilde{B} = \mathbb{K} \otimes_{\mathbb{F}} B$. Let $\tilde{V} = \mathbb{K} \otimes V \subset \tilde{B}$. Then $D_{\text{reg}}(V) = D_{\text{reg}}(\tilde{V})$.*

Secondly, the degree of regularity of a subspace is at least that of the original space.

Theorem A.4. *Let B be a graded algebra. Let V be a homogeneous subspace and let V' be a subspace of V. Then $D_{\text{reg}}(V) \leq D_{\text{reg}}(V')$.*

B Quadratic Operators

Let $\mathbb{K}$ be an extension of $\mathbb{F}$ of degree n; hence $|\mathbb{K}| = q^n$. An $\mathbb{F}$-quadratic function $P\colon \mathbb{K} \to \mathbb{K}$ takes the form

$$P(X) = \sum_{i,j} a_{ij} X^{q^i} X^{q^j} + \sum_i b_i X^{q^i} + c$$

for some $a_{ij}, b_i, c \in \mathbb{K}$.

Fix a dual basis $\{e_i \in \mathbb{K}, x_i \in \mathbb{K}^* = Hom_{\mathbb{F}}(\mathbb{K}, \mathbb{F})\}$ for $\mathbb{K}$ over $\mathbb{F}$. Define

$$A = \text{Fun}(\mathbb{K}, \mathbb{F}) = \mathbb{F}[x_1, \ldots, x_n]$$

Note that $x_i^q = x_i$ and A is naturally isomorphic to $\mathbb{F}[T_1, \ldots, T_n]/(T_1^q - T_1, \ldots, T_n^q - T_n)$. Note also that $\dim_{\mathbb{F}} A = q^n$ and a basis for A is given by all monomials of the form

$$x_1^{i_1} \ldots x_n^{i_n}, \quad \text{where } 0 \leq i_j \leq q-1.$$

Analogously we have that

$$\tilde{A} = \mathrm{Fun}(\mathbb{K}, \mathbb{K}) = \mathbb{K}[X]$$

where $X^{q^n} = X$. It can be easily verified that

$$\tilde{A} \cong \mathbb{K} \otimes_{\mathbb{F}} A = \mathbb{K}[x_1, \ldots, x_n]$$

This isomorphism identifies X with the element $\sum_i e_i x_i$.

We can filter both of these algebra by degree over $\mathbb{F}$. Thus for A we define

$$A_0 = \mathbb{F}, \quad A_1 = \sum_i \mathbb{F} x_i + \mathbb{F}, \quad A_{i+1} = A_1 A_i$$

For $\tilde{A}$ we note that the maps X^{q^i} are $\mathbb{F}$-linear and that they span the $\mathbb{K}$-space of $\mathbb{F}$-linear maps from $\mathbb{K}$ to $\mathbb{K}$. Thus we define

$$\tilde{A}_0 = \mathbb{K}, \quad \tilde{A}_1 = \sum_i \mathbb{K} X^{q^i} + \mathbb{K}, \quad \tilde{A}_{i+1} = \tilde{A}_1 \tilde{A}_i$$

Recall from basic Galois theory that $\sum_i \mathbb{K} X^{q^i} = \sum_i \mathbb{K} x_i$. From this it follows easily that $\tilde{A}_i \cong \mathbb{K} \otimes_{\mathbb{F}} A_i$ and that

$$\tilde{B} = \mathrm{Gr}\tilde{A} \cong \mathbb{K} \otimes_{\mathbb{F}} \mathrm{Gr}A = \mathbb{K} \otimes_{\mathbb{F}} B.$$

Define $\bar{X}_i = X^{q^i} + \tilde{A}_0 \in \tilde{B}_1$ and $\bar{x}_i = x_i + A_0 \in B_1$. Note that $\bar{X}^q = 0$ and $\bar{x}^q = 0$ for all i. Note also that $\bar{X}_i = \sum_i e_i^{q^i} \bar{x}_i$ and that $\tilde{B}_1 = \sum \mathbb{K} \bar{x}_i$. Hence

$$\tilde{B} = \mathbb{K}[\bar{X}_1, \ldots, \bar{X}_n] = \mathbb{K}[\bar{x}_1, \ldots, \bar{x}_n].$$

Now consider a quadratic operator P. Let $p_i = x_i \circ P$ and let $V = \sum \mathbb{F} p_i$.

Theorem B.1. $\sum \mathbb{K} p_i = \sum \mathbb{K} P^{q^i}$.

Proof. Note that

$$\begin{aligned} \sum \mathbb{K} p_i &= \{L \circ P \mid L \in \sum \mathbb{K} x_i\} \\ &= \{L \circ P \mid L \in \sum \mathbb{K} X^{q^i}\} \\ &= \sum \mathbb{K} P^{q^i} \end{aligned}$$

since $X^{q^i} \circ P = P^{q^i}$.

Let $\bar{P}_i$ be the image of P^{q^i} in $\tilde{B}$; that is, $\bar{P}_i = P^{q^i} + \tilde{A}_1$.

Corollary B.2. $\sum \mathbb{K} \bar{p}_i = \sum \mathbb{K} \bar{P}_i$.

Proof.

$$\sum \mathbb{K} \bar{p}_i = \sum \mathbb{K}(p_i + A_1) = \sum \mathbb{K} p_i + \tilde{A}_1 = \sum \mathbb{K}(P^{q^i} + \tilde{A}_1) = \sum \mathbb{K} \bar{P}_i.$$

C Comparison with Dubois-Gama Bounds

The following tables give a detailed comparison of our bound with the bound calculated in [9].

In these tables, the symbol $\mathcal{D}$ stands for the bound on the degree of the HFE polynomial used in [9]. Rather than restrict by the total degree D of the HFE operator, their restriction is given by

$$P(X) = \sum_{i,j \leq \mathcal{D}} a_{ij} X^{q^i+q^j} + \sum_{q^i \leq D} b_i X^{q^i} + c \ .$$

Thus $\mathcal{D}$ is one less than the number of variables involved in the polynomial P_0 and so Q-Rank $\leq \mathcal{D}+1$. In the same row as $\mathcal{D}$, DG Dreg stands for the bound on the degree of regularity given in [9], and DH Dreg stands for the bound obtained from Theorem 4.2 using $\mathcal{D}+1$ in place of Q-Rank. Thus DH Dreg $= (q-1)(\mathcal{D}+1)/2+2$.

The authors would like to thank Vivien Dubois and Nicolas Gama for providing the detailed data that made this comparison possible.

Table 1. Comparison with Dubois-Gama bound

	$q=3$			$q=5$			$q=7$			$q=11$			$q=13$			$q=17$			$q=19$			$q=23$		
		D_{reg}			D_{reg}			D_{reg}			D_{reg}			D_{reg}			D_{reg}			D_{reg}			D_{reg}	
n	$\mathcal{D}$	DG	DH	$\mathcal{D}$	DG	DH	$\mathcal{D}$	DG	DH	$\mathcal{D}$	DG	DH	$\mathcal{D}$	DG	DH	$\mathcal{D}$	DG	DH	$\mathcal{D}$	DG	DH	$\mathcal{D}$	DG	DH
8	3	5	6	2	6	8	2	6	11	2	6	17	2	6	20	2	6	26	2	6	29	2	6	35
12	3	5	6	3	7	10	2	7	11	2	7	17	2	7	20	2	7	26	2	7	29	2	7	35
16	3	6	6	3	9	10	2	9	11	2	9	17	2	9	20	2	9	26	2	9	29	2	9	35
20	4	7	7	3	10	10	3	12	14	2	11	17	2	11	20	2	11	26	2	11	29	2	11	35
24	4	7	7	3	10	10	3	13	14	2	13	17	2	13	20	2	13	26	2	13	29	2	13	35
28	4	7	7	3	10	10	3	14	14	2	14	17	2	14	20	2	14	26	2	14	29	2	14	35
32	4	7	7	3	10	10	3	14	14	2	16	17	2	16	20	2	16	26	2	16	29	2	16	35
36	4	7	7	3	10	10	3	14	14	3	21	22	2	18	20	2	18	26	2	18	29	2	18	35
40	4	7	7	3	10	10	3	14	14	3	22	22	2	20	20	2	20	26	2	20	29	2	20	35
44	4	7	7	3	10	10	3	14	14	3	22	22	2	21	20	2	21	26	2	21	29	2	21	35
48	4	7	7	3	10	10	3	14	14	3	22	22	3	25	26	2	23	26	2	23	29	2	23	35
52	5	8	8	3	10	10	3	14	14	3	22	22	3	25	26	2	24	26	2	25	29	2	25	35
56	5	8	8	4	12	12	3	14	14	3	22	22	3	25	26	2	24	26	2	25	29	2	25	35
60	5	8	8	4	12	12	3	14	14	3	22	22	3	25	26	2	24	26	2	25	29	2	25	35
64	5	8	8	4	12	12	3	14	14	3	22	22	3	25	26	2	24	26	2	25	29	2	25	35
68	5	8	8	4	12	12	3	14	14	3	22	22	3	25	26	2	24	26	2	25	29	2	25	35
72	5	8	8	4	12	12	3	14	14	3	22	22	3	25	26	2	24	26	2	25	29	2	25	35
76	5	8	8	4	12	12	3	14	14	3	22	22	3	25	26	3	32	34	2	25	29	2	25	35
80	5	8	8	4	12	12	3	14	14	3	22	22	3	25	26	3	32	34	2	25	29	2	25	35
84	5	8	8	4	12	12	3	14	14	3	22	22	3	25	26	3	32	34	2	25	29	2	25	35
88	5	8	8	4	12	12	3	14	14	3	22	22	3	25	26	3	32	34	3	36	38	2	25	35
92	5	8	8	4	12	12	3	14	14	3	22	22	3	25	26	3	32	34	3	36	38	2	25	35
96	5	8	8	4	12	12	3	14	14	3	22	22	3	25	26	3	32	34	3	36	38	2	25	35
100	5	8	8	4	12	12	3	14	14	3	22	22	3	25	26	3	32	34	3	36	38	2	25	35
104	5	8	8	4	12	12	3	14	14	3	22	22	3	25	26	3	32	34	3	36	38	2	25	35
108	5	8	8	4	12	12	3	14	14	3	22	22	3	25	26	3	32	34	3	36	38	2	25	35
112	5	8	8	4	12	12	3	14	14	3	22	22	3	25	26	3	32	34	3	36	38	2	25	35
116	5	8	8	4	12	12	3	14	14	3	22	22	3	25	26	3	32	34	3	36	38	2	25	35
120	5	8	8	4	12	12	3	14	14	3	22	22	3	25	26	3	32	34	3	36	38	2	25	35
124	5	8	8	4	12	12	3	14	14	3	22	22	3	25	26	3	32	34	3	36	38	3	42	46

Smaller Decoding Exponents: Ball-Collision Decoding

Daniel J. Bernstein[1], Tanja Lange[2], and Christiane Peters[2]

[1] Department of Computer Science
University of Illinois at Chicago, Chicago, IL 60607–7045, USA
djb@cr.yp.to

[2] Department of Mathematics and Computer Science
Technische Universiteit Eindhoven, P.O. Box 513, 5600 MB Eindhoven, Netherlands
tanja@hyperelliptic.org, c.p.peters@tue.nl

Abstract. Very few public-key cryptosystems are known that can encrypt and decrypt in time $b^{2+o(1)}$ with conjectured security level 2^b against conventional computers and quantum computers. The oldest of these systems is the classic McEliece code-based cryptosystem.

The best attacks known against this system are generic decoding attacks that treat McEliece's hidden binary Goppa codes as random linear codes. A standard conjecture is that the best possible w-error-decoding attacks against random linear codes of dimension k and length n take time $2^{(\alpha(R,W)+o(1))n}$ if $k/n \to R$ and $w/n \to W$ as $n \to \infty$.

Before this paper, the best upper bound known on the exponent $\alpha(R,W)$ was the exponent of an attack introduced by Stern in 1989. This paper introduces "ball-collision decoding" and shows that it has a smaller exponent for each (R,W): the speedup from Stern's algorithm to ball-collision decoding is exponential in n.

Keywords: McEliece cryptosystem, Niederreiter cryptosystem, post-quantum cryptography, attacks, information-set decoding, collision decoding.

1 Introduction

In 1978, McEliece introduced a code-based public-key cryptosystem that has maintained remarkable strength against every proposed attack. The top threats against McEliece's system have always been generic decoding algorithms that decode random linear codes. The standard conjecture is that the best possible generic decoding algorithm takes exponential time for any constant asymptotic code rate R and constant asymptotic error fraction W: i.e., time $2^{(\alpha(R,W)+o(1))n}$ for some positive real number $\alpha(R,W)$ if $k/n \to R$ and $w/n \to W$ as $n \to \infty$. Here n is the code length, k is the code dimension, and w is the number of errors.

Permanent ID of this document: 0e8c929565e20cf63e6a19794e570bb1. Date: 2011.05.27. This work was supported by the Cisco University Research Program, by the National Institute of Standards and Technology under grant 60NANB10D263, and by the European Commission under Contract ICT-2007-216646 ECRYPT II.

P. Rogaway (Ed.): CRYPTO 2011, LNCS 6841, pp. 743–760, 2011.

Two decades ago a flurry of fundamental algorithmic improvements produced a new upper bound on the optimal decoding exponent $\alpha(R, W)$. The upper bound is the exponent of a 1989 algorithm by Stern [50]. This upper bound arises from an asymptotic binomial-coefficient optimization and does not have a simple formula, but it can be straightforwardly computed to high precision for any particular (R, W). For example, for $W = 0.04$ and $R = 1 + W \log_2 W + (1 - W) \log_2(1 - W) = 0.7577\ldots$, Stern's algorithm shows that $\alpha(R, W) \leq 0.0809\ldots$.

There have also been many polynomial-factor speedups in generic decoding algorithms; there are dozens of papers on this topic, both inside and outside cryptography. Here is an illustration of the cumulative impact of many of the speedups. McEliece's original parameter suggestions ("$n = 1024, k = 524, t = 50$") take about $524^3\binom{1024}{50}/\binom{500}{50} \approx 2^{81}$ operations to break by the simple information-set-decoding attack explained in McEliece's original paper [41, Section 3]. (McEliece estimated the attack cost as $524^3(1 - 50/1024)^{-524} \approx 2^{65}$; this underestimate was corrected by Adams and Meijer in [2, Section 3].) The attack we presented in [8], thirty years after McEliece's paper, builds on several improvements and takes only about $2^{60.5}$ operations for the same parameters. That attack was carried out successfully, decrypting a challenge ciphertext. More recent improvements include [28] and [45]; see Section 4 for a more comprehensive discussion of the literature.

However, polynomial factors are asymptotically $2^{o(n)}$, and thus have no relevance to the exponent $\alpha(R, W)$ in $2^{(\alpha(R,W)+o(1))n}$. The best known upper bound on $\alpha(R, W)$ has been unchanged since 1989.

Contents of this paper. This paper presents smaller upper bounds on the decoding exponent $\alpha(R, W)$. Specifically, this paper introduces a generic decoding algorithm and demonstrates that this algorithm is, for every (R, W), faster than Stern's algorithm by a factor exponential in n. We call this algorithm "ball-collision decoding" because of a geometric interpretation explained in Section 4. The change in the exponent is not very large — for example, this paper uses ball-collision decoding to demonstrate that $\alpha(R, W) \leq 0.0807\ldots$ for the (R, W) given above — but it is the first exponential improvement in decoding complexity in more than twenty years.

This paper also evaluates the exact cost of ball-collision decoding, using the same bit-operation-counting rules as in the previous literature, and uses this evaluation to illustrate the impact of ball-collision decoding upon cryptographic applications. For example, the parameters (6624, 5129, 117) were proposed in [8, Section 7] at a 256-bit security level against a state-of-the-art refinement of Stern's algorithm; this paper shows that ball-collision decoding costs 2.6 times fewer bit operations. At a theoretical 1000-bit security level the improvement grows to 15.5. These concrete figures are consistent with the asymptotic analysis.

Of course, actually breaking these parameters remains very far out of reach, and these results should not be interpreted as damaging the viability of the McEliece cryptosystem. However, these results *do* raise new questions regarding the proper choice of parameters for the McEliece cryptosystem. Section 8 discusses the problem of parameter selection for code-based cryptography.

We also wrote a straightforward reference implementation of ball-collision decoding, and tested the implementation on a long series of random challenges at a much lower security level. The costs and success probabilities observed in these experiments matched the formulas shown in this paper.

Attack model. "Attacks" above refer only to passive single-target inversion attacks. The original McEliece cryptosystem, like the original RSA cryptosystem, is really just a trapdoor one-way function; when used naively as a public-key cryptosystem it is trivially broken by chosen-ciphertext attacks such as Berson's attack [11] and the Verheul–Doumen–van Tilborg attack [53].

Protecting the McEliece system against these attacks, to meet the standard notion of IND-CCA2 security for a public-key cryptosystem, requires appropriate padding and randomization, similar to RSA-OAEP. As shown by Kobara and Imai in [36], adding this protection does not significantly increase the cost of the McEliece cryptosystem.

2 Review of the McEliece Cryptosystem

The public key in the McEliece cryptosystem consists of a random-looking rank-k matrix $G \in \mathbf{F}_2^{k\times n}$. The sender encrypts a message m in $\mathbf{F}_2^k$ by first multiplying it with the matrix G, producing mG; choosing uniformly at random a word e in $\mathbf{F}_2^n$ of Hamming weight w; and adding e to mG, producing a ciphertext $mG+e$. The cryptosystem parameters are n, k, w.

The legitimate receiver decrypts $mG+e$ using a secret key which consists of a secret decoding algorithm producing the *error vector* e given $mG+e$. The details are not relevant to our attack and can be found in, e.g., [44].

An attacker is faced with the problem of determining e given G and $mG+e$. Note that finding e is equivalent to finding the message m: subtracting e from $mG+e$ produces mG, and then simple linear transformations produce m.

The set $\mathbf{F}_2^k G = \{mG : m \in \mathbf{F}_2^k\}$ is called a *linear code* of *length* n and *dimension* k, specifically the linear code *generated by* G. The matrix G is called a *generator matrix* for this code. The elements of $\mathbf{F}_2^k G$ are called *codewords*. If the linear code $\mathbf{F}_2^k G$ equals $\{c \in \mathbf{F}_2^n : Hc = 0\}$ then the matrix H is called a *parity-check matrix* for the code.

Without loss of generality one can assume that the matrix G in a CCA2-secure version of the McEliece cryptosystem is given in *systematic form* $G = (I_k|-A^T)$ where I_k is a $k\times k$ identity matrix and A an $(n-k)\times k$ matrix. Then the matrix $H = (A|I_{n-k})$ is a parity-check matrix for the code generated by G.

An *information set* Z for H is a set of k integers in $\{1, 2, \ldots, n\}$ for which the $n-k$ columns of H that are not indexed by Z are linearly independent. Applying Gaussian elimination to those $n-k$ columns shows that codewords are determined by their Z-indexed components. For example, $\{1, 2, \ldots, k\}$ is an information set for $H = (A|I_{n-k})$.

Fix $m \in \mathbf{F}_2^k$ and $e \in \mathbf{F}_2^n$ with $\mathrm{wt}(e) = w$. Write $c = mG$. By linearity one has $H(c+e) = Hc + He = He$ since $Hc = 0$. The result $s = He$ is called the *syndrome of* e. It is the sum of the w columns of H indexed by the positions of 1's in e. The attacker's task is equivalent to finding e given H and $s = He$.

3 The Ball-Collision-Decoding Algorithm

This section introduces ball-collision decoding. It first states the algorithm and then discusses various optimizations. Section 4 explains how this algorithm relates to previous algorithms.

The algorithm is given a parity-check matrix $H \in \mathbf{F}_2^{(n-k)\times n}$, a syndrome $s \in \mathbf{F}_2^{n-k}$, and a weight $w \in \{0, 1, 2, \ldots\}$. The goal of the algorithm is to find a corresponding error vector e: i.e., a vector $e \in \mathbf{F}_2^n$ of weight w such that $s = He$.

Ball-collision decoding has its roots in information-set decoding, which was used against the McEliece system in, e.g., [50], [17], [18], and [8]. The previous algorithms select a random information set in the parity-check matrix and then search for vectors having a particular pattern of non-zero entries. Ball-collision decoding is similar but searches for a more complicated, and more likely, pattern. See Section 4 for further discussion of the previous work.

The reader is encouraged to consider, while reading the algorithm, the case that the algorithm is given a matrix H already in systematic form and that it chooses $Z = \{1, 2, \ldots, k\}$ as information set. The matrix U in Step 4 is then the identity matrix I_{n-k}. The algorithm divides H into blocks, and divides the syndrome s into corresponding blocks, as specified by algorithm parameters ℓ_1, ℓ_2:

$$H = \begin{pmatrix} A_1 & I_1 & 0 \\ A_2 & 0 & I_2 \end{pmatrix}, \qquad s = \begin{pmatrix} s_1 \\ s_2 \end{pmatrix},$$

where $s_1 \in \mathbf{F}_2^{\ell_1+\ell_2}$, $s_2 \in \mathbf{F}_2^{n-k-\ell_1-\ell_2}$, $A_1 \in \mathbf{F}_2^{(\ell_1+\ell_2)\times k}$, $A_2 \in \mathbf{F}_2^{(n-k-\ell_1-\ell_2)\times k}$, and each I_i is an identity matrix.

One iteration of ball-collision decoding:

CONSTANTS: $n, k, w \in \mathbf{Z}$ with $0 \le w \le n$ and $0 \le k \le n$.
PARAMETERS: $p_1, p_2, q_1, q_2, k_1, k_2, \ell_1, \ell_2 \in \mathbf{Z}$ with $0 \le k_1$, $0 \le k_2$, $k = k_1 + k_2$,
$0 \le p_1 \le k_1$, $0 \le p_2 \le k_2$, $0 \le q_1 \le \ell_1$, $0 \le q_2 \le \ell_2$,
and $0 \le w - p_1 - p_2 - q_1 - q_2 \le n - k - \ell_1 - \ell_2$.
INPUT: $H \in \mathbf{F}_2^{(n-k)\times n}$ and $s \in \mathbf{F}_2^{n-k}$.
OUTPUT: Zero or more vectors $e \in \mathbf{F}_2^n$ with $He = s$ and $\mathrm{wt}(e) = w$.

1. Choose a uniform random information set Z. Subsequent steps of the algorithm write "$\mathbf{F}_2^Z$" to refer to the subspace of $\mathbf{F}_2^n$ supported on Z.
2. Choose a uniform random partition of Z into parts of sizes k_1 and k_2. Subsequent steps of the algorithm write "$\mathbf{F}_2^{k_1}$" and "$\mathbf{F}_2^{k_2}$" to refer to the corresponding subspaces of $\mathbf{F}_2^Z$.
3. Choose a uniform random partition of $\{1, 2, \ldots, n\} \setminus Z$ into parts of sizes ℓ_1, ℓ_2, and $n - k - \ell_1 - \ell_2$. Subsequent steps of the algorithm write "$\mathbf{F}_2^{\ell_1}$" and "$\mathbf{F}_2^{\ell_2}$" and "$\mathbf{F}_2^{n-k-\ell_1-\ell_2}$" to refer to the corresponding subspaces of $\mathbf{F}_2^{\{1,2,\ldots,n\}\setminus Z}$.

4. Find an invertible $U \in \mathbf{F}_2^{(n-k)\times(n-k)}$ such that the columns of UH indexed by $\{1, 2, \ldots, n\}\setminus Z$ are an $(n-k)\times(n-k)$ identity matrix. Write the columns of UH indexed by Z as $\begin{pmatrix} A_1 \\ A_2 \end{pmatrix}$ with $A_1 \in \mathbf{F}_2^{(\ell_1+\ell_2)\times k}$, $A_2 \in \mathbf{F}_2^{(n-k-\ell_1-\ell_2)\times k}$.
5. Write Us as $\begin{pmatrix} s_1 \\ s_2 \end{pmatrix}$ with $s_1 \in \mathbf{F}_2^{\ell_1+\ell_2}$, $s_2 \in \mathbf{F}_2^{n-k-\ell_1-\ell_2}$.
6. Compute the set S consisting of all triples $(A_1x_0+x_1, x_0, x_1)$ where $x_0 \in \mathbf{F}_2^{k_1}$, $\mathrm{wt}(x_0) = p_1$, $x_1 \in \mathbf{F}_2^{\ell_1}$, $\mathrm{wt}(x_1) = q_1$.
7. Compute the set T consisting of all triples $(A_1y_0 + y_1 + s_1, y_0, y_1)$ where $y_0 \in \mathbf{F}_2^{k_2}$, $\mathrm{wt}(y_0) = p_2$, $y_1 \in \mathbf{F}_2^{\ell_2}$, $\mathrm{wt}(y_1) = q_2$.
8. For each $(v, x_0, x_1) \in S$:
   For each y_0, y_1 such that $(v, y_0, y_1) \in T$:
   If $\mathrm{wt}(A_2(x_0 + y_0) + s_2) = w - p_1 - p_2 - q_1 - q_2$:
   Output $x_0 + y_0 + x_1 + y_1 + A_2(x_0 + y_0) + s_2$.

Note that Step 8 is a standard "join" operation between S and T; it can be implemented efficiently by sorting, by hashing, or by simple table indexing. In [8, Section 6] we describe an efficient implementation of essentially the same operation using only about $2^{\ell_1+\ell_2+1}$ bits of memory. See Sections 5 and 6 for further discussion of arithmetic costs and memory-access costs.

Theorem 3.1 (Correctness of ball-collision decoding). *The set of output vectors e of the ball-collision decoding algorithm is the set of vectors e that satisfy $He = s$ and have weights $p_1, p_2, q_1, q_2, w - p_1 - p_2 - q_1 - q_2$ in $\mathbf{F}_2^{k_1}$, $\mathbf{F}_2^{k_2}$, $\mathbf{F}_2^{\ell_1}$, $\mathbf{F}_2^{\ell_2}$, and $\mathbf{F}_2^{n-k-\ell_1-\ell_2}$ respectively.*

Proof. Each element $(v, x_0, x_1) \in S$ satisfies $x_0 \in \mathbf{F}_2^{k_1}$ with $\mathrm{wt}(x_0) = p_1$; $v = A_1x_0 + x_1$ and $x_1 \in \mathbf{F}_2^{\ell_1}$ with $\mathrm{wt}(x_1) = q_1$. Similarly each element $(v, y_0, y_1) \in T$ satisfies $y_0 \in \mathbf{F}_2^{k_2}$ with $\mathrm{wt}(y_0) = p_2$; $v = A_1y_0 + y_1 + s_1$; $y_1 \in \mathbf{F}_2^{\ell_2}$ with $\mathrm{wt}(y_1) = q_2$. Now, with Z-indexed columns visualized as coming before the remaining columns, we have

$$UHe = UH \begin{pmatrix} x_0+y_0 \\ x_1+y_1 \\ A_2(x_0+y_0)+s_2 \end{pmatrix} = \begin{pmatrix} A_1(x_0+y_0)+x_1+y_1 \\ A_2(x_0+y_0)+A_2(x_0+y_0)+s_2 \end{pmatrix} = \begin{pmatrix} s_1 \\ s_2 \end{pmatrix} = Us$$

so $He = s$. Furthermore, $x_0 + y_0 \in \mathbf{F}_2^{k_1+k_2}$ has weights p_1, p_2 in $\mathbf{F}_2^{k_1}, \mathbf{F}_2^{k_2}$; $x_1 + y_1 \in \mathbf{F}_2^{\ell_1+\ell_2}$ has weights q_1, q_2 in $\mathbf{F}_2^{\ell_1}, \mathbf{F}_2^{\ell_2}$; and $\mathrm{wt}(A_2(x_0+y_0)+s_2) = w-p_1-p_2-q_1-q_2$.

Conversely, the iteration finds every vector e having this weight distribution and satisfying $He = s$. Indeed, write e as $x_0 + y_0 + x_1 + y_1 + e_2$ with $x_0 \in \mathbf{F}_2^{k_1}$, $y_0 \in \mathbf{F}_2^{k_2}$, $x_1 \in \mathbf{F}_2^{\ell_1}$, $y_1 \in \mathbf{F}_2^{\ell_2}$, and $e_2 \in \mathbf{F}_2^{n-k-\ell_1-\ell_2}$. By hypothesis the weights of x_0, y_0, x_1, y_1, e_2 are $p_1, p_2, q_1, q_2, w - p_1 - p_2 - q_1 - q_2$, respectively. Now define $v = A_1x_0 + x_1$. The equation $UHe = Us$ implies $v = A_1y_0 + y_1 + s_1$; and $e_2 = A_2(x_0 + y_0) + s_2$. Hence $(v, x_0, x_1) \in S$ and $(v, y_0, y_1) \in T$. Finally $\mathrm{wt}(A_2(x_0 + y_0) + s_2) = \mathrm{wt}(e_2) = w - p_1 - p_2 - q_1 - q_2$ so the algorithm prints e as claimed. □

Finding an information set. The simplest way to choose a uniform random information set is to repeatedly choose a uniform random size-k subset $Z \subseteq \{1, 2, \ldots, n\}$ until the $n-k$ columns of H indexed by $\{1, 2, \ldots, n\} \setminus Z$ are linearly independent. Standard practice (see, e.g., Stern [50]) is to eliminate the fruitless Gaussian-elimination steps here, at the expense of negligible bias, by assembling the information set one column at a time, ensuring that each newly added column is linearly independent of the previously selected columns. After this optimization there is only one Gaussian-elimination step per iteration.

Reusing intermediate sums. Computing the vector $A_1 x_0$ for a weight-p_1 word x_0 in $\mathbf{F}_2^{k_1}$ can be done by adding the specified p_1 columns of A_1 in $p_1 - 1$ additions in $\mathbf{F}_2^{\ell_1+\ell_2}$.

Computing $A_1 x_0$ for *all* the $\binom{k_1}{p_1}$ vectors x_0 can be done more efficiently than repeating this process for each of them. Start by computing all $\binom{k_1}{2}$ sums of 2 columns of A_1; each sum costs one addition in $\mathbf{F}_2^{\ell_1+\ell_2}$. Then compute all $\binom{k_1}{3}$ sums of 3 columns of A_1 by adding one extra column to the previous results. Proceed in the same way until all $\binom{k_1}{p_1}$ sums of p_1 columns of A_1 are computed. This produces all required sums in only marginally more than one $\mathbf{F}_2^{\ell_1+\ell_2}$ addition per sum; see Section 5 for a precise operation count.

Early abort. The vector $A_2(x_0 + y_0) + s_2$ is computed as a sum of $p_1 + p_2 + 1$ vectors of length $n - k - \ell_1 - \ell_2$. Instead of computing the sum on all $n - k - \ell_1 - \ell_2$ positions one computes the sum row by row and simultaneously checks the weight. If the weight exceeds $w - p_1 - p_2 - q_1 - q_2$ one can discard this particular pair (x_0, y_0).

We comment that one can further reduce the cost of this step by precomputing sums of smaller sets of columns, but we do not use this idea in our analysis, because it is not critical for the algorithm's performance.

4 Relationship to Previous Algorithms

This section discusses the relationship of ball-collision decoding to previous information-set-decoding algorithms.

Collision decoding vs. ball-collision decoding. We use the name "collision decoding" for the special case $q_1 = q_2 = 0$ of ball-collision decoding. The idea of collision decoding is more than twenty years old: Stern's algorithm in [50] is, aside from trivial details, exactly the special case $q_1 = q_2 = 0$, $p_1 = p_2$, $k_1 \approx k_2$. Dumer in [26] independently introduced the core idea, although in a more limited form, and in [27] achieved an algorithm similar to Stern's.

All state-of-the-art decoding attacks since [50] have been increasingly optimized forms of collision decoding. Other approaches to decoding, such as "gradient decoding" ([4]), "supercode decoding" ([5]), and "statistical decoding" (see [3] and [43]), have never been competitive with Stern's algorithm. This does not mean that those approaches should be ignored; our generalization from collision decoding to ball-collision decoding is inspired by one of the steps in supercode decoding.

Collision decoding searches for collisions in $\mathbf{F}_2^{\ell_1+\ell_2}$ between points A_1x_0 and points $A_1y_0 + s_1$. Ball-collision decoding expands each point A_1x_0 into a small ball (in the Hamming metric), namely $\{A_1x_0 + x_1 : x_1 \in \mathbf{F}_2^{\ell_1}, \mathrm{wt}(x_1) = q_1\}$; similarly expands each point A_1y_0 into a small ball; and searches for collisions between these balls.

From the perspective of ball-collision decoding, the fundamental disadvantage of collision decoding is that errors are required to avoid an asymptotically quite large stretch of $\ell_1 + \ell_2$ positions. Ball-collision decoding makes a much more reasonable hypothesis, namely that there are asymptotically increasingly many errors in those positions. It requires extra work to enumerate the points in each ball, but the extra work is only about the square root of the improvement in success probability. The cost ratio is exponential when all parameters are optimized properly; see Section 7.

Collision decoding also has a secondary disadvantage compared to ball-collision decoding: its inner loop is slower, since computing A_1x_0 for a new x_0 is considerably more expensive than adding x_1 for a new x_1. The cost ratio here is only polynomial, and is not relevant to the exponents (see Section 7), but is accounted for in the bit-operation count (see Section 5). This disadvantage of collision decoding is also visible in the number of memory accesses to A_1 (see Section 6); however, standard practice in the literature on this topic is to count the number of bit operations involved in arithmetic and to ignore the cost of memory access.

Additional credits. The simplest form of information-set decoding, introduced by Prange in [47], did not allow errors in the information set. For asymptotic analyses see [41], [1], and [2].

The idea of allowing errors was published by Lee and Brickell in [38], by Leon in [39], and by Krouk in [37], but without Stern's collision idea; in the terminology of ball-collision decoding, with $p_2 = 0$, $q_1 = q_2 = 0$, and $\ell_2 = 0$. For each pattern of p_1 errors in k columns, Lee and Brickell checked the weight of the remaining $n - k$ columns; Leon and Krouk required ℓ_1 columns to have weight 0, and usually checked only those columns. For asymptotic analyses see [37], [23], and [24].

Overbeck and Sendrier [44] give a visual comparison of the algorithms by comparing to which interval they restrict how many errors. Figure 4.1 extends their picture to include ball-collision decoding. It shows that the new algorithm allows errors in an interval that had to be error-free in Leon's and Stern's algorithms.

The idea of allowing errors everywhere can be extracted, with considerable effort, from the description of supercode decoding in [5]. After a detailed analysis we have concluded that the algorithm in [5] is much slower than collision decoding. The same algorithm is claimed in [5] to have smaller exponents than collision decoding (with astonishing gaps, often 15% or more), but this claim is based on a chain of exponentially large inaccuracies in the algorithm analysis

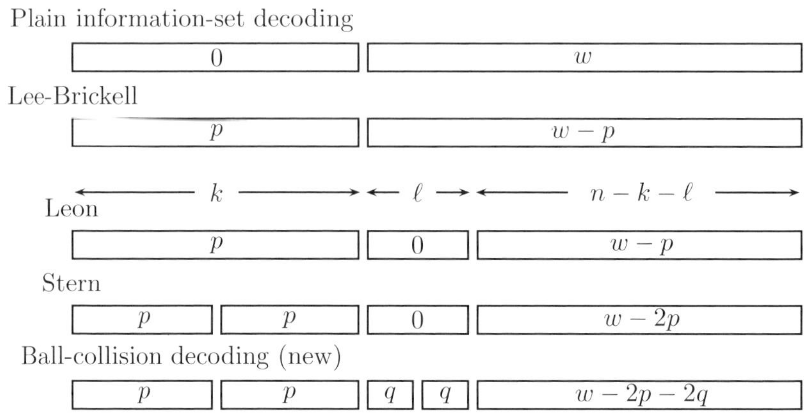

Fig. 4.1. Error positions hypothesized by various decoding algorithms

in [5]. The starting point of the chain is [5, "Corollary 12"], which claims size $\binom{k}{e_1}\binom{y}{e_2}/2^{by}$ for lists that actually have size $\binom{k}{e_1}\binom{y}{e_2}^b/2^{by}$.

The idea of allowing errors everywhere can also be found in the much more recent paper [28], along with a polynomial-factor "birthday" speedup obtained by dropping Stern's left-right separation. The algorithm analysis by Finiasz and Sendrier in [28] concludes that the overall "gain compared with Stern's algorithm" is a constant times "$\sqrt[4]{\pi p/2}$", which is bounded by a polynomial in n. Our own assessment is that if parameters had been chosen more carefully then the algorithm of [28] would have led to an exponential improvement over collision decoding, contrary to the conclusions in [28]. This algorithm would still have retained the secondary disadvantage described above, and therefore would not have been competitive with ball-collision decoding.

A more detailed analysis of the "birthday" speedup in collision decoding appeared in [45] along with an optimized generalization to $\mathbf{F}_q$. These modifications can be adapted to ball-collision decoding but would complicate the algorithm statement and analysis without changing the exponent of binary decoding; we have skipped these modifications for simplicity.

One way to speed up Gaussian elimination is to change only one information-set element in each iteration. This idea was introduced by Omura, according to [22, Section 3.2.4]. It was applied to increasingly optimized forms of information-set decoding by van Tilburg in [51] and [52], by Chabanne and Courteau in [19], by Chabaud in [20], by Canteaut and Chabanne in [16], by Canteaut and Chabaud in [17], and by Canteaut and Sendrier in [18]. In [8] we improved the balance between Gaussian-elimination cost and error-searching cost by changing c information-set elements in each iteration for an optimized value of c. The ideas of reusing sums and aborting weight calculations also appeared in [8], in the context of an improved collision-decoding algorithm.

5 Complexity Analysis

This section analyzes the complexity of ball-collision decoding. In particular, this section analyzes the success probability of each iteration and the number of bit operations needed for each iteration.

Success probability. Assume that e is a uniform random vector of weight w. One iteration of ball-collision decoding finds e exactly if it has the right weight distribution, namely weight p_1 in the first k_1 positions specified by the information set, weight p_2 in the remaining k_2 positions specified by the information set, weight q_1 on the first ℓ_1 positions outside the information set, and weight q_2 on the next ℓ_2 positions outside the information set.

The probability that e has this weight distribution is, by a simple counting argument, exactly

$$b(p_1, p_2, q_1, q_2, \ell_1, \ell_2) = \binom{n}{w}^{-1} \binom{n-k-\ell_1-\ell_2}{w-p_1-p_2-q_1-q_2} \binom{k_1}{p_1} \binom{k_2}{p_2} \binom{\ell_1}{q_1} \binom{\ell_2}{q_2}.$$

The expected number of iterations of the outer loop is, for almost all H, very close to the reciprocal of the success probability of a single iteration. We explicitly disregard, without further comment, the extremely unusual codes for which the average number of iterations is significantly different from the reciprocal of the success probability of a single iteration. For further discussion of this issue and how unusual it is see, e.g., [24] and [10].

Gaussian elimination. There are many ways to speed up Gaussian elimination, as discussed in Section 4; implementors are encouraged to use those optimizations. However, in this paper we will be satisfied with a quite naive form of Gaussian elimination, taking $(1/2)(n-k)^2(n+k)$ bit operations; our interest is in large input sizes, and elimination takes negligible time for those sizes.

Building the set S. The total cost of computing $A_1 x_0$ for all x_0 of Hamming weight p_1, using intermediate sums as explained in Section 3, is

$$(\ell_1+\ell_2)\left(\binom{k_1}{2} + \binom{k_1}{3} + \cdots \binom{k_1}{p_1}\right).$$

Using $L(k,p) = \sum_{i=1}^{p} \binom{k}{i}$ as a shorthand, the costs can be written as $(\ell_1 + \ell_2)\left(L(k_1, p_1) - k_1\right)$. The $\ell_1+\ell_2$ factor is the number of bit operations to compute $A_1 x_0$ from $A_1 x_0'$ where x_0 extends x_0' by a single bit.

Then for each x_0 all $\binom{\ell_1}{q_1}$ possible words x_1 in $\mathbf{F}_2^{\ell_1}$ of weight q_1 are added to $A_1 x_0$, producing $A_1 x_0 + x_1$. For x_1, as for x_0, we loop over the possible sets of indices, and reuse sums obtained from subsets. This slightly increases the number of sums up to $L(\ell_1, q_1)$, but decreases the cost of each sum down to a single bit operation, computing $A_1 x_0 + x_1$ from $A_1 x_0 + x_1'$. Overall this step takes $\min\{1, q_1\}\binom{k_1}{p_1} L(\ell_1, q_1)$ bit operations; note that for $q_1 = 0$ the cost of this step is indeed 0.

Each choice of (x_0, x_1) adds one element to S. Hence, the number of elements in S equals exactly the number of choices for x_0 and x_1, i.e. $\#S = \binom{k_1}{p_1}\binom{\ell_1}{q_1}$.

Building the set T. The set T is built similarly to the set S. The only difference is that the expression $A_1y_0 + y_1 + s_1$ involves adding s_1 and thus the single columns (corresponding to weight-1 words y_0) already cost $(\ell_1 + \ell_2)\binom{k_2}{1}$ bit operations. In total this step takes $(\ell_1 + \ell_2)L(k_2, p_2) + \min\{1, q_2\}\binom{k_2}{p_2}L(\ell_2, q_2)$.

The set T contains exactly $\#T = \binom{k_2}{p_2}\binom{\ell_2}{q_2}$ elements.

Checking collisions. The last step does one check for every (x_0, x_1, y_0, y_1) satisfying the equation $A_1x_0 + x_1 = A_1y_0 + y_1 + s_1$. There are $\binom{k_1}{p_1}\binom{k_2}{p_2}\binom{\ell_1}{q_1}\binom{\ell_2}{q_2}$ choices of (x_0, x_1, y_0, y_1).

If the vectors v appearing in S and T were uniformly distributed among the $2^{\ell_1+\ell_2}$ possible values then on average $\#S \cdot \#T \cdot 2^{-\ell_1-\ell_2}$ checks would be done. The expected number of checks is extremely close to this for almost all H; as above we disregard the extremely unusual codes with different behavior.

Each check consists of computing $\mathrm{wt}(A_2(x_0 + y_0) + s_2)$ and testing whether it equals $w - p_1 - p_2 - q_1 - q_2$. When using the early-abort weight calculation, on average only $2(w - p_1 - p_2 - q_1 - q_2 + 1)$ bits of the result are computed before the weight is found too high. Each bit of the result costs $p_1 + p_2$ bit operations because $x_0 + y_0$ has weight $p_1 + p_2$.

Cost of one iteration. To summarize, the total cost per iteration of the inner loop with parameters $p_1, p_2, q_1, q_2, \ell_1, \ell_2$ amounts to

$$
\begin{aligned}
&c(p_1, p_2, q_1, q_2, \ell_1, \ell_2)\\
&\quad = \frac{1}{2}(n-k)^2(n+k) + (\ell_1+\ell_2)\big(L(k_1,p_1) + L(k_2,p_2) - k_1\big)\\
&\qquad + \min\{1,q_1\}\binom{k_1}{p_1}L(\ell_1,q_1) + \min\{1,q_2\}\binom{k_2}{p_2}L(\ell_2,q_2)\\
&\qquad + 2(w - p_1 - p_2 - q_1 - q_2 + 1)(p_1+p_2)\binom{k_1}{p_1}\binom{k_2}{p_2}\binom{\ell_1}{q_1}\binom{\ell_2}{q_2}2^{-\ell_1-\ell_2}.
\end{aligned}
$$

6 Concrete Parameter Examples

This section considers concrete examples in order to show the speedup gained by ball-collision decoding in comparison to collision decoding. The first parameters were previously proposed to achieve 256-bit security against current attacks. We designed the second parameters according to similar rules to achieve a 1000-bit security level against current attacks. We do not mean to suggest that 1000-bit security is of any real-world relevance; we consider it to illustrate the asymptotic superiority of ball-collision decoding.

Finiasz and Sendrier in [28] presented "lower bounds on the effective work factor of existing real algorithms, but also on the future improvements that could be implemented"; and said that beating these bounds would require the introduction of "new techniques, never applied to code-based cryptosystems". For each set of parameters we evaluate the Finiasz–Sendrier lower bound and the costs of three algorithms:

(1) collision decoding ($q_1 = q_2 = 0$),
(2) collision decoding using the birthday trick from [28] as analyzed in [45], and
(3) ball-collision decoding.

Ball-collision decoding beats the Finiasz–Sendrier lower bound in both of these examples. The main reason for this is that ball-collision decoding dodges the secondary disadvantage described in Section 4; the lower bound assumes that each new vector requires $\ell_1 + \ell_2$ bit operations to update $A_1 x_0$, but in ball-collision decoding each new vector requires just 1 bit operation to update x_1.

We emphasize that all of these costs and bounds use the same model of computation, counting the number of bit operations for arithmetic and disregarding costs of memory access, copies, etc. A table-indexing join operation can easily be carried out for free in this model. We would prefer a more carefully defined model of computation that includes realistic memory-access costs, such as the Brent–Kung circuit model [13], but the bit-operation model is simpler and is standard in papers on this topic.

256-security revisited. According to [8, Section 7] a binary code with length $n = 6624$, $k = 5129$, $w = 117$ achieves 256-bit security. The best collision-decoding parameters are actually slightly below 2^{256} bit operations: they use $2^{181.4928}$ iterations (on average), each taking $2^{74.3741}$ bit operations, for a total of $2^{255.8669}$ bit operations.

Collision decoding with the birthday trick takes, with optimal parameters, $2^{255.54880}$ bit operations. The birthday trick increases the cost per iteration by a factor of 2.2420 compared to the classical collision-decoding algorithm, to $2^{75.5390}$ bit operations. However, the trick increases the chances of finding the desired error vector noticeably, reducing the number of iterations by a factor of 2.7951, to $2^{180.0099}$. Thus the birthday trick yields an overall 1.2467× speedup.

The Finiasz–Sendrier lower bound is $2^{255.1787}$ bit operations, 1.6112× smaller than the cost of collision decoding.

Ball-collision decoding with parameters $k_1 = 2565$, $k_2 = 2564$, $\ell_1 = \ell_2 = 47$, $p_1 = p_2 = 8$, and $q_1 = q_2 = 1$ needs only $2^{254.1519}$ bit operations to attack the same system. On average the algorithm needs $2^{170.6473}$ iterations each taking $2^{83.5046}$ bit operations.

Ball-collision decoding thus costs 3.2830× less than collision decoding, 2.6334× less than collision decoding with the birthday trick, and 2.0375× less than the Finiasz–Sendrier lower bound.

1000-bit security. Attacking a system based on a code of length $n = 30332$, $k = 22968$, $w = 494$ requires $2^{1000.9577}$ bit operations using collision decoding with the optimal parameters $k_1 = k_2 = 11484$, $\ell_1 = \ell_2 = 140$, $p_1 = p_2 = 27$ and $q_1 = q_2 = 0$.

The birthday trick reduces the cost by a factor of 1.7243, to $2^{1000.1717}$ bit operations. This means that this system offers 1000-bit security against all previously known attacks.

The Finiasz–Sendrier lower bound is $2^{999.45027}$ bit operations, 2.8430× smaller than the cost of collision decoding and 1.6488× smaller than the cost of collision decoding with the birthday trick.

Ball-collision decoding with parameters $k_1 = k_2 = 11484$, $\ell_1 = \ell_2 = 156$, $p_1 = p_2 = 29$, and $q_1 = q_2 = 1$ needs only $2^{996.21534}$ bit operations. This is $26.767\times$ smaller than the cost of collision decoding, $15.523\times$ smaller than the cost of collision decoding with the birthday trick, and $9.415\times$ smaller than the Finiasz–Sendrier lower bound.

7 Asymptotic Complexity of Ball-Collision Decoding

This section analyzes the asymptotic behavior of the cost of ball-collision decoding, and shows that it always has a smaller asymptotic exponent than the cost of collision decoding.

Input sizes. Fix a real number W with $0 < W < 1/2$, and fix a real number R with $-W \log_2 W - (1-W)\log_2(1-W) \le 1 - R < 1$.

Consider codes and error vectors of very large length n, where the codes have dimension $k \approx Rn$, and the error vectors have weight $w \approx Wn$. More precisely, fix functions $k, w : \{1, 2, \ldots\} \to \{1, 2, \ldots\}$ that satisfy $\lim_{n\to\infty} k(n)/n = R$ and $\lim_{n\to\infty} w(n)/n = W$; more concisely, $k/n \to R$ and $w/n \to W$.

Attack parameters. Fix real numbers P, Q, L with $0 \le P \le R/2$, $0 \le Q \le L$, and $0 \le W - 2P - 2Q \le 1 - R - 2L$. Fix ball-collision parameters $p_1, p_2, q_1, q_2, k_1, k_2, \ell_1, \ell_2$ with $p_i/n \to P$, $q_i/n \to Q$, $k_i/n \to R/2$, and $\ell_i/n \to L$.

We have also analyzed more general asymptotic parameter spaces, for example splitting P into P_1, P_2 where $p_i/n \to P_i$. Balanced parameters always turned out to be asymptotically optimal (as one would expect), so this section focuses on the parameter space (P, Q, L) stated above. Note that the asymptotic optimality of $P_1 = P_2$ does *not* imply the concrete optimality of $p_1 = p_2$; for example, $(p_1, p_2) = (2, 1)$ appears to be optimal for some small input sizes.

In the formulas below, expressions of the form $x \log_2 x$ are extended (continuously but not differentiably) to 0 at $x = 0$. For example, the expression $P \log_2 P$ means 0 if $P = 0$.

Success probability. We repeatedly invoke the standard asymptotic formula for binomial coefficients, namely

$$\frac{1}{n}\log_2\binom{(\alpha+o(1))n}{(\beta+o(1))n} \to \alpha\log_2\alpha - \beta\log_2\beta - (\alpha-\beta)\log_2(\alpha-\beta),$$

to compute the asymptotic exponent of the success probability of a single iteration of ball-collision decoding:

$$\begin{aligned}
B(P,Q,L) &= \lim_{n\to\infty}\frac{1}{n}\log_2\left(\binom{n}{w}^{-1}\binom{n-k-\ell_1-\ell_2}{w-p_1-p_2-q_1-q_2}\binom{k_1}{p_1}\binom{k_2}{p_2}\binom{\ell_1}{q_1}\binom{\ell_2}{q_2}\right)\\
&= W\log_2 W + (1-W)\log_2(1-W)\\
&\quad + (1-R-2L)\log_2(1-R-2L) - (W-2P-2Q)\log_2(W-2P-2Q)\\
&\quad - (1-R-2L-(W-2P-2Q))\log_2(1-R-2L-(W-2P-2Q))\\
&\quad + R\log_2(R/2) - 2P\log_2 P - (R-2P)\log_2(R/2-P)\\
&\quad + 2L\log_2 L - 2Q\log_2 Q - 2(L-Q)\log_2(L-Q).
\end{aligned}$$

The success probability of a single iteration is asymptotically $2^{n(B(P,Q,L)+o(1))}$.

Iteration cost. We similarly compute the asymptotic exponent of the cost of an iteration:

$$\begin{aligned}
&C(P,Q,L)\\
&= \lim_{n\to\infty}\frac{1}{n}\log_2\left(\binom{k_1}{p_1}\binom{\ell_1}{q_1}+\binom{k_2}{p_2}\binom{\ell_2}{q_2}+\binom{k_1}{p_1}\binom{\ell_1}{q_1}\binom{k_2}{p_2}\binom{\ell_2}{q_2}2^{-\ell_1-\ell_2}\right)\\
&= \max\{(R/2)\log_2(R/2) - P\log_2 P - (R/2-P)\log_2(R/2-P)\\
&\qquad + L\log_2 L - Q\log_2 Q - (L-Q)\log_2(L-Q),\\
&\qquad R\log_2(R/2) - 2P\log_2 P - (R-2P)\log_2(R/2-P)\\
&\qquad + 2L\log_2 L - 2Q\log_2 Q - 2(L-Q)\log_2(L-Q) - 2L\}.
\end{aligned}$$

The cost of a single iteration is asymptotically $2^{n(C(P,Q,L)+o(1))}$. Note that we have simplified the iteration cost to $\binom{k_1}{p_1}\binom{\ell_1}{q_1}+\binom{k_2}{p_2}\binom{\ell_2}{q_2}+\binom{k_1}{p_1}\binom{\ell_1}{q_1}\binom{k_2}{p_2}\binom{\ell_2}{q_2}2^{-\ell_1-\ell_2}$. The cost is actually larger than this, but only by a factor $\le \text{poly}(n)$, which we are free to disregard since $\frac{1}{n}\log_2 \text{poly}(n) \to 0$. We also comment that the bounds are valid whether or not $q_i = 0$.

Overall attack cost. The overall asymptotic ball-collision-decoding-cost exponent is the difference $D(P,Q,L)$ of the iteration-cost exponent $C(P,Q,L)$ and the success-probability exponent $B(P,Q,L)$, thus

$$\begin{aligned}
D(P,Q,L) = \max\{&-(R/2)\log_2(R/2)+P\log_2 P+(R/2-P)\log_2(R/2-P)\\
&-L\log_2 L+Q\log_2 Q+(L-Q)\log_2(L-Q), -2L\}\\
&-W\log_2 W-(1-W)\log_2(1-W)\\
&-(1-R-2L)\log_2(1-R-2L)+(W-2P-2Q)\log_2(W-2P-2Q)\\
&+(1-R-2L-(W-2P-2Q))\log_2(1-R-2L-(W-2P-2Q)).
\end{aligned}$$

Example: Take $W = 0.04$ and $R = 1 + W\log_2 W + (1-W)\log_2(1-W) = 0.7577078109\ldots$. Choose $P = 0.004203556640625$, $Q = 0.000192998046875$, and $L = 0.017429431640625$; we use very high precision here to simplify verification. The success-probability exponent is $-0.0458435310\ldots$, and the iteration-cost exponent is $0.0348588632\ldots$, so the overall cost exponent is $0.0807023942\ldots$. Ball-collision decoding with these parameters thus costs $2^{(0.0807023942\ldots+o(1))n}$ to correct $(0.04+o(1))n$ errors in a code of rate $0.7577078109\ldots+o(1)$.

Collision-decoding cost and the lower bound. Traditional collision decoding is the special case $p_1 = p_2$, $k_1 = k_2$, $\ell_1 = \ell_2$, $q_1 = q_2 = 0$ of ball-collision decoding. Its asymptotic cost exponent is the case $Q = 0$ of the ball-collision decoding exponent stated above.

Consider again $W = 0.04$ and $R = 1 + W\log_2 W + (1-W)\log_2(1-W)$. Choosing $P = 0.00415087890625$, $Q = 0$, and $L = 0.0164931640625$ achieves decoding exponent $0.0809085120\ldots$. We partitioned the (P, L) space into small intervals and performed interval-arithmetic calculations to show that $Q = 0$

cannot do better than 0.0809; ball-collision decoding therefore has a slightly smaller exponent than collision decoding in this case.

We performed similar calculations for other pairs (W, R) and in each case found that the infimum of all collision-decoding-cost exponents was beaten by a ball-collision-decoding-cost exponent. Ball-collision decoding therefore has a smaller exponent than collision decoding, as stated in the introduction of this paper.

The case $Q = 0$ is always suboptimal. The interval-arithmetic calculations described above are proofs of the suboptimality of $Q = 0$ for some specific pairs (W, R). These proofs have the advantage of computing explicit bounds on the collision-decoding-cost exponents for those pairs (W, R), but the proofs have two obvious disadvantages.

The first disadvantage is that these proofs do not cover *all* pairs (W, R); they leave open the possibility that ball-collision decoding has the same exponent as collision decoding for other pairs (W, R). The second disadvantage is that the proofs are much too long to verify by hand. The first disadvantage could perhaps be addressed by much more extensive interval-arithmetic calculations, partitioning the space of pairs (W, R) into boxes so small that, within each box, the ball-collision-decoding exponent is uniformly better than the minimum collision-decoding exponent; but this would exacerbate the second disadvantage.

To address both of these disadvantages we give, in the full version of this paper [9], a hand-verifiable proof that $Q = 0$ is always suboptimal: for *every* (W, R), ball-collision decoding has a smaller asymptotic cost exponent than collision decoding. Specifically, we prove the following theorem about the overall asymptotic cost exponent:

Theorem 7.1. *For each R, W it holds that*

$$\begin{aligned} &\min\{D(P, 0, L) : 0 \leq P \leq R/2,\ 0 \leq W-2P \leq 1-R-2L\} \\ > &\min\{D(P, Q, L) : 0 \leq P \leq R/2,\ 0 \leq Q \leq L,\ 0 \leq W-2P-2Q \leq 1-R-2L\}. \end{aligned}$$

Note that $\{(P, 0, L)\}$ and $\{(P, Q, L)\}$ are compact sets, and D is continuous, so we are justified in writing "min" rather than "inf". The proof strategy analyzes generic perturbations of D and combines all necessary calculations into a small number of elementary inequalities in the proofs in the full version of this paper [9].

8 Choosing McEliece Parameters

The traditional approach to selecting cryptosystem parameters is as follows:

- Consider the fastest known attacks against the system. For example, in the case of RSA, consider the latest refinements [35] of the number-field sieve.
- Restrict attention to parameters for which these attacks take time at least $2^{b+\delta}$. Here b is the desired security level, and δ is a "security margin" meant to protect against the possibility of further improvements in the attacks.

- Within the remaining parameter space, choose the most efficient parameters. The definition of efficiency depends on the target application: it could mean minimal key size, for example, or minimum decryption time.

This approach does not make clear how to choose the security margin δ. Some applications have ample time and space for cryptography, and can simply increase δ to the maximum value for which the costs of cryptography are still insignificant; but in some applications cryptography is an important bottleneck, and users insist on minimizing δ for the sake of performance.

Finiasz and Sendrier in [28] identified a bound on "future improvements" in attacks against the McEliece cryptosystem, and suggested that designers use this bound to "choose durable parameters". The general idea of identifying bottlenecks in any possible attack, and of using those bottlenecks to systematically choose δ, is quite natural and attractive, and has been used successfully in many contexts. However, as discussed in Section 6, ball-collision decoding disproves the specific bound in [28], violating one of the assumptions in [28] and raising the question of how many more assumptions can be violated.

We propose replacing the bound in [28] with the simpler bound

$$\min\left\{\frac{1}{2}\binom{n}{w}\binom{n-k}{w-p}^{-1}\binom{k}{p}^{-1/2} : p \geq 0\right\};$$

i.e., choosing the code length n, code rate k/n, and error fraction w/n so that this bound is at least 2^b. As usual, implementors can exploit the remaining flexibility in parameters to optimize decryption time, compressed key size $k(n-k)$, or efficiency in any other metric of interest.

This bound has several attractive features. It is easy to estimate via standard binomial-coefficient approximations. It is easy to compute exactly. It covers a very wide class of attacks, as explained in the full version [9] of this paper. It is nevertheless in the same ballpark as the cost of known attacks: for example, it is $2^{49.69}$ for the original parameters $(n, k, w) = (1024, 524, 50)$, and $2^{236.49}$ for $(n, k, w) = (6624, 5129, 117)$. Note that these numbers give lower bounds on the cost of the attack. Parameters protecting against this bound pay only about a 20% performance penalty at high security levels, compared to parameters that merely protect against known attacks.

The reader can easily verify that parameters $(n, k, w) = (3178, 2384, 68)$ achieve 128-bit security against this bound. For 256-bit security $(n, k, w) = (6944, 5208, 136)$ are recommended.

References

[1] Adams, C.M., Meijer, H.: Security-related comments regarding McEliece's public-key cryptosystem. In: Crypto'87 [46], pp. 224–228 (1987); See also newer version [2]; Citations in this document: §4

[2] Adams, C.M., Meijer, H.: Security-related comments regarding McEliece's public-key cryptosystem. IEEE Transactions on Information Theory 35, 454–455 (1988); See also older version [1]; Citations in this document: §1, §4

[3] Al Jabri, A.: A statistical decoding algorithm for general linear block codes. In: IMA 2001 [31], pp. 1–8 (2001); Citations in this document: §4
[4] Ashikhmin, A.E., Barg, A.: Minimal vectors in linear codes. IEEE Transactions on Information Theory 44, 2010–2017 (1998); Citations in this document: §4
[5] Barg, A., Krouk, E.A., van Tilborg, H.C.A.: On the complexity of minimum distance decoding of long linear codes. IEEE Transactions on Information Theory 45, 1392–1405 (1999); Citations in this document: §4, §4, §4, §4, §4, §4
[6] Batten, L., Safavi-Naini, R. (eds.): Information security and privacy: 11th Australasian conference, ACISP 2006, Melbourne, Australia, July 3–5, 2006, proceedings. LNCS, vol. 4058. Springer, Heidelberg (2006); See [43]
[7] Bernstein, D.J., Buchmann, J., Dahmen, E. (eds.): Post-quantum cryptography. Springer, Heidelberg (2009); See [44]
[8] Bernstein, D.J., Lange, T., Peters, C.: Attacking and defending the McEliece cryptosystem. In: PQCrypto 2008 [14], pp. 31–46 (2008), `http://eprint.iacr.org/2008/318`; Citations in this document: §1, §1, §3, §3, §4, §4, §6
[9] Bernstein, D.J., Lange, T., Peters, C.: Smaller decoding exponents: ball-collision decoding (full version) (2010), `http://eprint.iacr.org/2010/585`; Citations in this document: §7, §7, §8
[10] Bernstein, D.J., Lange, T., Peters, C., van Tilborg, H.C.A.: Explicit bounds for generic decoding algorithms for code-based cryptography. In: WCC 2009 (2009); Citations in this document: §5
[11] Berson, T.A.: Failure of the McEliece public-key cryptosystem under message-resend and related-message attack. In: Crypto '97 [33], pp. 213–220 (1997); Citations in this document: §1
[12] Blaum, M., Farrell, P.G., van Tilborg, H.C.A. (eds.): Information, coding and mathematics. Kluwer International Series in Engineering and Computer Science, vol. 687. Kluwer, Dordrecht (2002); See [53]
[13] Brent, R.P., Kung, H.T.: The area-time complexity of binary multiplication. Journal of the ACM 28, 521–534 (1981), `http://wwwmaths.anu.edu.au/~brent/pub/pub055.html`; Citations in this document: §6
[14] Buchmann, J., Ding, J. (eds.): Post-quantum cryptography, second international workshop, PQCrypto 2008, Cincinnati, OH, USA, October 17–19, 2008, proceedings. LNCS, vol. 5299. Springer, Heidelberg (2008); See [8]
[15] Camion, P., Charpin, P., Harari, S. (eds.): Eurocode '92: proceedings of the international symposium on coding theory and applications held in Udine, October 23–30, 1992. Springer, Heidelberg (1993); See [20]
[16] Canteaut, A., Chabanne, H.: A further improvement of the work factor in an attempt at breaking McEliece's cryptosystem. In: EUROCODE '94 [21] (1994), `http://www.inria.fr/rrrt/rr-2227.html`; Citations in this document: §4
[17] Canteaut, A., Chabaud, F.: A new algorithm for finding minimum-weight words in a linear code: application to McEliece's cryptosystem and to narrow-sense BCH codes of length 511. IEEE Transactions on Information Theory 44, 367–378 (1998), `ftp://ftp.inria.fr/INRIA/tech-reports/RR/RR-2685.ps.gz`; Citations in this document: §3, §4
[18] Canteaut, A., Sendrier, N.: Cryptanalysis of the original McEliece cryptosystem. In: Asiacrypt '98 [42], pp. 187–199 (1998); Citations in this document: §3, §4
[19] Chabanne, H., Courteau, B.: Application de la méthode de décodage itérative d'Omura à la cryptanalyse du système de McEliece. Université de Sherbrooke, Rapport de Recherche, number 122 (1993); Citations in this document: §4

[20] Chabaud, F.: Asymptotic analysis of probabilistic algorithms for finding short codewords. In: [15], pp. 175–183 (1993); Citations in this document: §4
[21] Charpin, P.(ed.): Livre des résumé — EUROCODE '94. Abbaye de la Bussière sur Ouche, France, October 1994 (1994); See [16]
[22] Clark Jr., G.C., Bibb Cain, J.: Error-correcting coding for digital communication. Plenum, New York (1981); Citations in this document: §4
[23] Coffey, J.T., Goodman, R.M.: The complexity of information set decoding. IEEE Transactions on Information Theory 35, 1031–1037 (1990); Citations in this document: §4
[24] Coffey, J.T., Goodman, R.M., Farrell, P.: New approaches to reduced complexity decoding. Discrete and Applied Mathematics 33, 43–60 (1991); Citations in this document: §4, §5
[25] Cohen, G.D., Wolfmann, J. (eds.): Coding theory and applications. LNCS, vol. 388. Springer, Heidelberg (1989); See [50]
[26] Dumer, I.I.: Two decoding algorithms for linear codes. Problemy Peredachi Informatsii 25, 24–32 (1989); Citations in this document: §4
[27] Dumer, I.I.: On minimum distance decoding of linear codes. In: [32], pp. 50–52 (1991); Citations in this document: §4
[28] Finiasz, M., Sendrier, N.: Security bounds for the design of code-based cryptosystems. In: Asiacrypt 2009 [40] (2009), `http://eprint.iacr.org/2009/414`; Citations in this document: §1, §4, §4, §4, §4, §6, §2, §8, §8, §8, §8
[29] Goldwasser, S. (ed.): Advances in cryptology — CRYPTO '88, proceedings of the conference on the theory and application of cryptography held at the University of California, Santa Barbara, California, August 21–25, 1988. LNCS, vol. 403. Springer, Heidelberg (1990); See [51]
[30] Günther, C.G. (ed.): Advances in cryptology — EUROCRYPT '88, proceedings of the workshop on the theory and application of cryptographic techniques held in Davos, May 25–27, 1988. LNCS, vol. 330. Springer, Heidelberg (1988); See [38]
[31] Honary, B. (ed.): Cryptography and coding: proceedings of the 8th IMA international conference held in Cirencester, December 17–19. LNCS, vol. 2260. Springer, Heidelberg (2001); See [3]
[32] Kabatianskii, G.A. (ed.): Fifth joint Soviet-Swedish international workshop on information theory, Moscow, 1991 (1991); See [27]
[33] Kaliski Jr., B.S. (ed.): Advances in cryptology — CRYPTO '97: 17th annual international cryptology conference, Santa Barbara, California, USA, August 17–21, 1997, proceedings. LNCS, vol. 1294. Springer, Heidelberg (1997); See[11]
[34] Kim, K. (ed.): Public key cryptography: proceedings of the 4th international workshop on practice and theory in public key cryptosystems (PKC 2001) held on Cheju Island, February 13–15, 2001. LNCS, vol. 1992. Springer, Heidelberg (2001); See [36]
[35] Kleinjung, T., Aoki, K., Franke, J., Lenstra, A.K., Thomé, E., Bos, J.W., Gaudry, P., Kruppa, A., Montgomery, P.L., Osvik, D.A., te Riele, H., Timofeev, A., Zimmermann, P.: Factorization of a 768-bit RSA modulus. In: Crypto 2010 [48], pp. 333–350 (2010), `http://eprint.iacr.org/2010/006`;
Citations in this document: §8
[36] Kobara, K., Imai, H.: Semantically secure McEliece public-key cryptosystems — conversions for McEliece PKC. In: PKC 2001 [34], pp. 19–35 (2001); Citations in this document: §1
[37] Krouk, E.A.: Decoding complexity bound for linear block codes. Problemy Peredachi Informatsii 25, 103–107 (1989); Citations in this document: §4, §4

[38] Lee, P.J., Brickell, E.F.: An observation on the security of McEliece's public-key cryptosystem. In: Eurocrypt '88 [30], pp. 275–280 (1988), http://dsns.csie.nctu.edu.tw/research/crypto/HTML/PDF/E88/275.PDF; Citations in this document: §4
[39] Leon, J.S.: A probabilistic algorithm for computing minimum weights of large error-correcting codes. IEEE Transactions on Information Theory 34, 1354–1359 (1988); Citations in this document: §4
[40] Matsui, M. (ed.): Advances in cryptology — ASIACRYPT 2009, 15th international conference on the theory and application of cryptology and information security, Tokyo, Japan, December 6–10, 2009, proceedings. LNCS, vol. 5912. Springer, Heidelberg (2009); See [28]
[41] McEliece, R.J.: A public-key cryptosystem based on algebraic coding theory. JPL DSN Progress Report 114–116 (1978), http://ipnpr.jpl.nasa.gov/progress_report2/42-44/44N.PDF; Citations in this document: §1, §4
[42] Ohta, K., Pei, D. (eds.): Advances in cryptology — ASIACRYPT'98: proceedings of the international conference on the theory and application of cryptology and information security held in Beijing. LNCS, vol. 1514. Springer, Heidelberg (1998); See [18]
[43] Overbeck, R.: Statistical decoding revisited. In: ACISP 2006 [6], pp. 283–294 (2006); Citations in this document: §4
[44] Overbeck, R., Sendrier, N.: Code-based cryptography. In: [7], pp. 95–145 (2009); Citations in this document: §2, §4
[45] Peters, C.: Information-set decoding for linear codes over $\mathbf{F}_q$. In: Post-Quantum Cryptography [49], pp. 81–94 (2010); Citations in this document: §1, §4, §2
[46] Pomerance, C. (ed.): Advances in cryptology — CRYPTO '87, proceedings of the conference on the theory and applications of cryptographic techniques held at the University of California, Santa Barbara, California, August 16–20, 1987. LNCS, vol. 293. Springer, Heidelberg (1987), http://dsns.csie.nctu.edu.tw/research/crypto/HTML/PDF/C87/224.PDF; See [1]
[47] Prange, E.: The use of information sets in decoding cyclic codes. IRE Transactions on Information Theory IT-8, S5–S9 (1962); Citations in this document: §4
[48] Rabin, T. (ed.): Advances in cryptology — CRYPTO 2010, 30th annual cryptology conference, Santa Barbara, CA, USA, August 15–19, 2010, proceedings. LNCS, vol. 6223. Springer, Heidelberg (2010); See [35]
[49] Sendrier, N. (ed.): Post-quantum cryptography, third international workshop, PQCrypto, Darmstadt, Germany, May 25–28, 2010, proceedings. LNCS, vol. 6061. Springer, Heidelberg (2010); See [45]
[50] Stern, J.: A method for finding codewords of small weight. In: [25], pp. 106–113 (1989); Citations in this document: §1, §3, §3, §4, §4
[51] van Tilburg, J.: On the McEliece public-key cryptosystem. In: Crypto '88 [29], pp. 119–131 (1990); Citations in this document: §4
[52] van Tilburg, J.: Security-analysis of a class of cryptosystems based on linear error-correcting codes. Ph.D. thesis, Technische Universiteit Eindhoven (1994); Citations in this document: §4
[53] Verheul, E.R., Doumen, J.M., van Tilborg, H.C.A.: Sloppy Alice attacks! Adaptive chosen ciphertext attacks on the McEliece public-key cryptosystem. In: [12], pp. 99–119 (2002); Citations in this document: §1

McEliece and Niederreiter Cryptosystems That Resist Quantum Fourier Sampling Attacks

Hang Dinh[1], Cristopher Moore[2,⋆], and Alexander Russell[3,⋆⋆]

[1] Indiana University South Bend
hdinh@cs.iusb.edu
[2] University of New Mexico, and Santa Fe Institute
moore@cs.unm.edu
[3] University of Connecticut
acr@cse.uconn.edu

Abstract. Quantum computers can break the RSA, El Gamal, and elliptic curve public-key cryptosystems, as they can efficiently factor integers and extract discrete logarithms. This motivates the development of *post-quantum* cryptosystems: classical cryptosystems that can be implemented with today's computers, that will remain secure even in the presence of quantum attacks.

In this article we show that the McEliece cryptosystem over *rational Goppa codes* and the Niederreiter cryptosystem over *classical Goppa codes* resist precisely the attacks to which the RSA and El Gamal cryptosystems are vulnerable—namely, those based on generating and measuring coset states. This eliminates the approach of strong Fourier sampling on which almost all known exponential speedups by quantum algorithms are based. Specifically, we show that the natural case of the Hidden Subgroup Problem to which McEliece-type cryptosystems reduce cannot be solved by strong Fourier sampling, or by any measurement of a coset state. To do this, we extend recent negative results on quantum algorithms for Graph Isomorphism to subgroups of the automorphism groups of linear codes.

This gives the first rigorous results on the security of the McEliece-type cryptosystems in the face of quantum adversaries, strengthening their candidacy for post-quantum cryptography. We also strengthen some results of Kempe, Pyber, and Shalev on the Hidden Subgroup Problem in S_n.

1 Introduction

If and when quantum computers are built, common public-key cryptosystems such as RSA, El Gamal, and elliptic curve cryptography will no longer be secure. Given that fact, the susceptibility or resistance of other well-studied public-key cryptosystems to quantum attacks is of fundamental interest. We present evidence for the strength of McEliece-type cryptosystems against quantum attacks, demonstrating that the quantum Fourier sampling attacks that cripple RSA and El Gamal do not apply to the McEliece or Niederreiter cryptosystems as long as the underlying code satisfies certain algebraic

⋆ This work was supported by the NSF under grants CCF-0829931, 0835735, and 0829917 and by the DTO under contract W911NF-04-R-0009.

⋆⋆ This work was supported by the NSF under grants 1117427 and 0835735 and by the DTO under contract W911NF-04-R-0009.

P. Rogaway (Ed.): CRYPTO 2011, LNCS 6841, pp. 761–779, 2011.

properties. While there are known classical attacks on these systems for the case of rational Goppa codes, our results also apply to the Niederreiter cryptosystem with classical Goppa codes, which to our knowledge is still believed to be classically secure. While our results do not rule out other quantum (or classical) attacks, they do demonstrate security precisely against the types of quantum algorithms that have proven so powerful for number-theoretic problems. We also strengthen some results of Kempe et al. [9] on subgroups of S_n reconstructible by Fourier sampling.

McEliece-type cryptosystems. The McEliece cryptosystem is a public-key cryptosystem proposed by McEliece in 1978 [13], conventionally built over Goppa codes. A dual variant of the system, proposed by Niederreiter [16], can provide slightly improved efficiency with equivalent security [10]. This dual system can additionally be used to construct a digital signature scheme [2], a shortcoming of the original system.

There are two basic types of attacks known against the McEliece-type cryptosystems: decoding attacks, and direct attacks on the private key. The former appears challenging, considering that the general decoding problem is NP-hard; indeed, historical confidence in the security of the McEliece system relies on the idea that this hardness can be retained for scrambled version of specific codes. This same intuition applies to quantum attacks: NP-hard problems are believed to be intractable, in general, for quantum computers and no significant quantum algorithmic developments appear to be directly relevant to these decoding problems. The latter—direct attacks on the key—can be successful on certain classes of linear codes, and is our focus. In a McEliece-type cryptosystem, the private key of a user Alice consists of three matrices: a $k \times n$ matrix M over a finite field $\mathbb{F}_{q^\ell}$, a $k \times k$ invertible matrix S over the field $\mathbb{F}_q$, and an $n \times n$ permutation matrix P. In the McEliece version, M is a generator matrix of a q-ary $[n,k]$-linear code (hence, $\ell = 1$), while in Niederreiter's dual system, M is a parity check matrix of a q-ary linear code of length n. The matrices S and P are selected randomly. Alice's public key consists of the matrix $M^* = SMP$. An adversary may attack the private key, attempting to recover the secret row "scrambler" S and the secret permutation P from M^* and M, assuming he already knew M.[1] As pointed out in [4], it crucial to keep S and P secret for the security of the McEliece system.

The security of these McEliece-type systems have received considerable attention in the literature, often focusing on particular choices for the underlying codes. Various classes of Goppa codes have received the greatest attention: along these lines, Sidelnokov and Shestakov's attack [23] can efficiently compute the matrices S and MP from the public matrix $M^* = SMP$ if the underlying code is a generalized Reed-Solomon code.[2] While this attack can reveal the structure of an alternative code, it does not reveal the secret permutation. An attack in which the secret permutation is revealed was proposed by Loidreau and Sendrier [11], using the Support Splitting Algorithm [21]. However, this attack only works with a very limited subclass of classical binary Goppa codes, namely those with a binary Goppa polynomial.

---

[1] Recovering the secret scrambler and the secret permutation is different from the Code Equivalence problem. The former finds a transformation between two equivalent codes, while the latter decides whether two linear codes are equivalent.

[2] We remark that the class of generalized Reed-Solomon codes is essentially equal to the class of rational Goppa codes.

Although the McEliece-type cryptosystems are efficient and still considered classically secure, at least with classical binary Goppa codes [4], they are rarely used in practice because of their comparatively large public key (see remark 8.33 in [14]). The discovery of successful quantum attacks on RSA and El Gamal, however, has changed the landscape: as suggested by Ryan [20] and Bernstein et al. [1], if "post-quantum" security guarantees can be made for the McEliece cryptosystem, this may compensate for its comparatively expensive computational demands.

Quantum Fourier sampling. Quantum Fourier Sampling (QFS) is the key ingredient in nearly all known efficient quantum algorithms for algebraic problems, including Shor's algorithms for factorization and discrete logarithm [22] and Simon's algorithm [24]. Shor's algorithm relies on quantum Fourier sampling over the cyclic group $\mathbb{Z}_N$, while Simon's algorithm uses quantum Fourier sampling over $\mathbb{Z}_2^n$. In general, these algorithms solve instances of the *Hidden Subgroup Problem* (HSP) over a finite group G. Given a function f on G whose level sets are left cosets of some unknown subgroup $H < G$, i.e., such that f is constant on each left coset of H and distinct on different left cosets, they find a set of generators for the subgroup H.

The standard approach to this problem treats f as a black box and applies f to a uniform superposition over G, producing the coset state $|cH\rangle = (1/\sqrt{|H|})\sum_{h\in H}|ch\rangle$ for a random c. We then measure $|cH\rangle$ in a Fourier basis $\{|\rho,i,j\rangle\}$ for the space $\mathbb{C}[G]$, where ρ is an irrep[3] of G and i,j are row and column indices of a matrix $\rho(g)$. In the *weak* form of Fourier sampling, only the representation name ρ is measured, while in the *strong* form, both the representation name and the matrix indices are measured, the latter in a chosen basis. This produces probability distributions from which classical information can be extracted to recover the subgroup H. Moreover, since $|cH\rangle$ is block-diagonal in the Fourier basis, the optimal measurement of the coset state can always be described in terms of strong Fourier sampling.

Understanding the power of Fourier sampling in nonabelian contexts has been an ongoing project, and a sequence of negative results [6, 15, 7] have suggested that the approach is inherently limited when the underlying groups are rich enough. In particular, Moore, Russell, and Schulman [15] showed that over the symmetric group, even the strong form of Fourier sampling cannot efficiently distinguish the conjugates of most order-2 subgroups from each other or from the trivial subgroup. That is, for any $\sigma \in S_n$ with large support, and most $\pi \in S_n$, if $H = \{1, \pi^{-1}\sigma\pi\}$ then strong Fourier sampling, and therefore any measurement we can perform on the coset state, yields a distribution which is exponentially close to the distribution corresponding to $H = \{1\}$. This result implies that GRAPH ISOMORPHISM cannot be solved by the naive reduction to strong Fourier sampling. Hallgren et al. [7] strengthened these results, demonstrating that even entangled measurements on $o(\log n!)$ coset states yield essentially no information.

Kempe and Shalev [8] showed that weak Fourier sampling of single coset states in S_n cannot distinguish the trivial subgroup from larger subgroups H with polynomial size and non-constant minimal degree.[4] They conjectured, conversely, that if a subgroup $H < S_n$ can be distinguished from the trivial subgroup by weak Fourier sampling, then

[3] Throughout the paper, we write "irrep" as short for "irreducible representation."

[4] The minimal degree of a permutation group H is the minimal number of points moved by a non-identity element of H.

the minimal degree of H must be constant. Their conjecture was later proved by Kempe, Pyber, and Shalev [9].

Our contributions. To state our results, we say that a subgroup $H < G$ is *indistinguishable by strong Fourier sampling* if the conjugate subgroups $g^{-1}Hg$ cannot be distinguished from each other (or from the trivial subgroup) by measuring the coset state in an arbitrary Fourier basis. A precise definition is presented in Section 3.2. Since the optimal measurement of a coset state can always be expressed as an instance of strong Fourier sampling, these results imply that no measurement of a single coset state yields any useful information about H. Based on the strategy of Moore, Russell, and Schulman [15], we first develop a general framework, formalized in Theorem 1, to determine indistinguishability of a subgroup by strong Fourier sampling. We emphasize that their results cover the case where the subgroup has order two. Our principal contribution is to show how to extend their methods to more general subgroups.

We then apply this general framework to a class of semi-direct products $(\mathsf{GL}_k(\mathbb{F}_q) \times S_n) \wr \mathbb{Z}_2$, bounding the distinguishability for the HSP corresponding to the private-key attack on a McEliece-type cryptosystem, i.e., the problem of determining a secret scrambler S and a secret permutation P from $M^* = SMP$ and M. Our bound, given in Corollary 1 of Theorem 4, depends on the column rank[5] of the matrix M as well as the minimal degree and the size of the *automorphism group* $\mathrm{Aut}(M)$, where $\mathrm{Aut}(M)$ is defined in Subsection 4.2 as the set of all permutations P on the columns of M such that $M = SMP$ for some $S \in \mathsf{GL}_k(\mathbb{F}_q)$. In general, our result indicates that McEliece-type cryptosystems resist known attacks based on strong Fourier sampling if M has column rank at least $k - o(\sqrt{n})/\ell$, and the automorphism group $\mathrm{Aut}(M)$ has minimal degree $\Omega(n)$ and size $\mathrm{e}^{o(n)}$. In particular, generator matrices of rational Goppa codes and canonical parity check matrices of classical Goppa codes have good values for these quantities (see Lemma 3). The result is most interesting for classical Goppa codes, which are considered classically secure; the McEliece system over *rational* Goppa codes is subject to the Sidelnokov-Shestakov [23] attack.

While our main application is the security of the McEliece cryptosystem, we show in addition that our general framework is applicable to other classes of groups with simpler structure, including the symmetric group and the finite general linear group[6] $\mathsf{GL}_2(\mathbb{F}_q)$. For the symmetric group, we extend the results of [15] to larger subgroups of S_n. Specifically, we show that any subgroup $H < S_n$ with minimal degree $m \geq \Theta(\log|H|) + \omega(\log n)$ is indistinguishable by strong Fourier sampling over S_n. This partially extends the results of Kempe et al. [9], which apply only to weak Fourier sampling.

Remark 1. Our results show that the natural reduction of McEliece to a hidden subgroup problem yields negligible information about the secret key. Thus they rule out the direct analogue of the quantum attack that breaks, for example, RSA. Of course, our results do not rule out other quantum (or classical) attacks. Neither do they establish that a quantum algorithm for the McEliece cryptosystem would violate a natural hardness

---

[5] The column rank of M is understood to be over the field F_{q^ℓ}. Recall that the entries of the matrix M are in F_{q^ℓ}.

[6] The case of $\mathsf{GL}_2(\mathbb{F}_q)$ is omitted in this version for lack of space.

assumption, as do recent lattice cryptosystem constructions whose hardness is based on the Learning With Errors problem (e.g. Regev [18]). Nevertheless, they indicate that any such algorithm would have to involve significant new ideas beyond than those that have been proposed so far.

Summary of technical ideas. Let G be a finite group. We wish to establish general criteria for indistinguishability of subgroups $H < G$ by strong Fourier sampling. We begin with the general strategy, developed in [15], that controls the resulting probability distributions in terms of the representation-theoretic properties of G. In order to handle richer subgroups, however, we have to overcome some technical difficulties. Our principal contribution here is a "decoupling" lemma that allows us to handle the cross terms arising from pairs of nontrivial group elements.

Roughly, the approach (presented in Section 3.2) identifies two disjoint subsets, SMALL and LARGE, of irreps of G. The set LARGE consists of all irreps whose dimensions are no smaller than a certain threshold D. While D should be as large as possible, we also need to choose D small enough so that the set LARGE is large. In contrast, the representations in SMALL must have small dimension (much smaller than $\sqrt{D}$), and the set SMALL should be small or contain few irreps that appear in the decomposition of the tensor product representation $\rho \otimes \rho^*$ for any $\rho \in$ LARGE. In addition, any irrep ρ outside SMALL must have small normalized character $|\chi_\rho(h)|/d_\rho$ for any nontrivial element $h \in H$. If two such sets exist, and if $|H|$ is sufficiently small, we establish that H is indistinguishable by strong Fourier sampling over G.

In the case $G = S_n$, as in [15] we define SMALL as the set Λ_c of all Young diagrams whose top row or left column has length at least $(1-c)n$, and define LARGE by setting $D = n^{dn}$, for appropriate constants $0 < c, d < 1$. We show that any irrep outside SMALL has large dimension and therefore small normalized characters.

For the case $G = (\mathsf{GL}_k(\mathbb{F}_q) \times S_n) \wr \mathbb{Z}_2$ corresponding to McEliece-type cryptosystems, the normalized characters on the hidden subgroup K depend on the minimal degree of the automorphism group $\mathrm{Aut}(M) < S_n$. If we choose SMALL as the set of all irreps constructed from tensor product representations $\tau \times \lambda$ of $\mathsf{GL}_k(\mathbb{F}_q) \times S_n$ with $\lambda \in \Lambda_c$, then the "small" features of Λ_c will induce the "small" features of this set SMALL. Finally, $|K|$ depends on $|\mathrm{Aut}(M)|$ and the column rank of M. When M is a generator matrix of a rational Goppa code or a canonical parity check matrix of a classical Goppa code, $\mathrm{Aut}(M)$ lies inside the automorphism group of a rational Goppa code, which can be controlled using Stichtenoth's Theorem [25].

2 Hidden Subgroup Attacks on McEliece-type Cryptosystems

As mentioned in the Introduction, we consider an attack attempting to recover the secret scrambler S and permutation P from M and M^*. We frame the problem such an attacker needs to solve as follows:

Definition 1 (Scrambler-Permutation Problem). *Given two $k \times n$ matrices M and M^* with entries in a finite field containing $\mathbb{F}_q$ such that $M^* = SMP$ for some $S \in \mathsf{GL}_k(\mathbb{F}_q)$ and some $n \times n$ permutation matrix P, find such a pair (S, P).*

In the case where the matrix M is a generator matrix of a linear code over $\mathbb{F}_q$, the decision version of this problem is known as the CODE EQUIVALENCE problem, which is at least as hard as GRAPH ISOMORPHISM, although it is unlikely to be NP-complete [17]. This problem can be immediately recast as a Hidden Subgroup Problem (described below). We begin with a presentation of the problem as a Hidden Shift Problem:

Definition 2 (Hidden Shift Problem). *Let G be a finite group and Σ be a finite set. Given two functions $f_0 : G \to \Sigma$ and $f_1 : G \to \Sigma$ with the promise that there is an element $s \in G$ for which $f_1(x) = f_0(sx)$ for all $x \in G$, the problem is to determine such s by making queries to f_0 and f_1. An element s with this property is called a* left shift *from f_0 to f_1 (or, simply, a* shift*).*

The Scrambler-Permutation Problem can be immediately reduced to the Hidden Shift Problem over the group $G = \mathsf{GL}_k(\mathbb{F}_q) \times S_n$ by defining functions f_0 and f_1 on $\mathsf{GL}_k(\mathbb{F}_q) \times S_n$ so that for all $(S,P) \in \mathsf{GL}_k(\mathbb{F}_q) \times S_n$,

$$f_0(S,P) = S^{-1}MP\,, \qquad f_1(S,P) = S^{-1}M^*P\,. \tag{1}$$

Here and from now on, we identify each $n \times n$ permutation matrix with its corresponding permutation in S_n. Evidently, $SMP = M^*$ if and only if (S^{-1},P) is a shift from f_0 to f_1.

Next, following the standard approach to developing quantum algorithms for such problems, we reduce this Hidden Shift Problem on a group G to the Hidden Subgroup Problem on the wreath product $G \wr \mathbb{Z}_2 = G^2 \rtimes \mathbb{Z}_2$. Given two functions f_0 and f_1 on G, we define the function $f : G \wr \mathbb{Z}_2 \to \Sigma \times \Sigma$ as follows: for $(x,y) \in G^2$ and $b \in \mathbb{Z}_2$,

$$f((x,y),b) \stackrel{\text{def}}{=} \begin{cases} (f_0(x), f_1(y)) & \text{if } b = 0 \\ (f_1(y), f_0(x)) & \text{if } b = 1 \end{cases} \tag{2}$$

Now we would like to see that the Hidden Shift Problem is equivalent to determining the subgroup whose cosets are distinguished by f. Recall that a function f on a group G *distinguishes the right cosets* of a subgroup $H < G$ if for all $x,y \in G$, $f(x) = f(y) \iff yx^{-1} \in H$.

Definition 3. *Let f be a function on a group G. We say that f is* injective under right multiplication *if for all $x,y \in G$, $f(x) = f(y) \iff f(yx^{-1}) = f(1)$. Define the subset $G|_f \subseteq G$ as the level set containing the identity,*

$$G|_f \stackrel{\text{def}}{=} \{g \in G \mid f(g) = f(1)\}\,.$$

Proposition 1. *Let f be a function on a group G. If f distinguishes the right cosets of a subgroup $H < G$, then f must be injective under right multiplication and $G|_f = H$. Conversely, if f is injective under right multiplication, then $G|_f$ is a subgroup and f distinguishes the right cosets of the subgroup $G|_f$.*

Hence, the function f defined in (2) can distinguish the right cosets of some subgroup if and only if it is injective under right multiplication.

Lemma 1. *The function f defined in* (2) *is injective under right multiplication if and only if (1) f_0 is injective under right multiplication and (2) $f_1(x) = f_0(sx)$ for some s.*

The proof of this lemma is straightforward, so we omit it here.

Proposition 2. *Assume f_0 is injective under right multiplication. Let $H_0 = G|_{f_0}$ and s be a shift. Then the function f defined in* (2) *distinguishes right cosets of the following subgroup of $G \wr \mathbb{Z}_2$:*

$$G \wr \mathbb{Z}_2|_f = \left((H_0, s^{-1}H_0 s), 0\right) \cup \left((H_0 s, s^{-1}H_0), 1\right),$$

which has size $2|H_0|^2$. The set of all shifts from f_0 to f_1 is $H_0 s$.

If we can determine the hidden subgroup $K = G \wr \mathbb{Z}_2|_f$, we can find a shift by selecting an element of the form $((g_1, g_2), 1)$ from K. Then g_1 must belong to $H_0 s$, and so is a shift from f_0 to f_1.

Application to the Scrambler-Permutation problem. Returning to the Hidden Shift Problem over $G = \mathsf{GL}_k(\mathbb{F}_q) \times S_n$ corresponding to the Scrambler-Permutation problem, it is clear that the function f_0 defined in (1) is injective under right multiplication, and that

$$H_0 = \mathsf{GL}_k(\mathbb{F}_q) \times S_n|_{f_0} = \left\{(S, P) \in \mathsf{GL}_k(\mathbb{F}_q) \times S_n \mid S^{-1}MP = M\right\}.$$

The automorphism group of M is the projection of H_0 onto S_n, i.e.,

$$\mathrm{Aut}(M) = \left\{P \in S_n \mid \exists S : S^{-1}MP = M\right\}.$$

Note that each $P \in \mathrm{Aut}(M)$ has the same number of preimages $S \in \mathsf{GL}_k(\mathbb{F}_q)$ in this projection.

3 Quantum Fourier sampling (QFS)

3.1 Preliminaries and Notation

Fix a finite group G, abelian or non-abelian, and let $\widehat{G}$ denote the set of irreducible unitary representations, or "irreps" for short, of G. For each irrep $\rho \in \widehat{G}$, let V_ρ denote a vector space over $\mathbb{C}$ on which ρ acts so that ρ is a group homomorphism from G to the general linear group over V_ρ, and let d_ρ denote the dimension of V_ρ. For each ρ, we fix an orthonormal basis $B_\rho = \left\{\mathbf{b}_1, \ldots, \mathbf{b}_{d_\rho}\right\}$ for V_ρ. Then we can represent each $\rho(g)$ as a $d_\rho \times d_\rho$ unitary matrix whose j^{th} column is the vector $\rho(g)\mathbf{b}_j$.

Viewing the vector space $\mathbb{C}[G]$ as the regular representation of G, we can decompose $\mathbb{C}[G]$ into irreps as the direct sum $\bigoplus_{\rho \in \widehat{G}} V_\rho^{\oplus d_\rho}$. This has a basis $\{|\rho, i, j\rangle : \rho \in \widehat{G}, 1 \le i, j \le d_\rho\}$, where $\{|\rho, i, j\rangle \mid 1 \le i \le d_\rho\}$ is a basis for the jth copy of V_ρ. Up to normalization, $|\rho, i, j\rangle$ corresponds to the i, j entry of the irrep ρ.

Definition 4. *The* Quantum Fourier transform *over G is the unitary operator, denoted F_G, that transforms a vector in $\mathbb{C}[G]$ from the point-mass basis $\{|g\rangle \mid g \in G\}$ into the basis given by the decomposition of $\mathbb{C}[G]$. For all $g \in G$,*

$$F_G\,|g\rangle = \sum_{\rho, i, j} \sqrt{\frac{d_\rho}{|G|}}\, \rho(g)_{i,j}\, |\rho, i, j\rangle\,,$$

where $\rho(g)_{ij}$ is the (i, j)-entry of the matrix $\rho(g)$. Alternatively, we can view $F_G\,|g\rangle$ as a block diagonal matrix consisting of the block $\sqrt{d_\rho/|G|}\,\rho(g)$ for each $\rho \in \widehat{G}$.

Notation. For each subset $X \subseteq G$, define $|X\rangle = (1/\sqrt{|X|})\sum_{x \in X}|x\rangle$, which is the uniform superposition over X. For each $X \subseteq G$ and $\rho \in \widehat{G}$, define the operator $\Pi_X^\rho \stackrel{\text{def}}{=} \frac{1}{|X|}\sum_{x \in X}\rho(x)$, and let $\widehat{X}(\rho)$ denote the $d_\rho \times d_\rho$ matrix block at ρ in the quantum Fourier transform of $|X\rangle$, i.e.,

$$\widehat{X}(\rho) \stackrel{\text{def}}{=} \sqrt{\frac{d_\rho}{|G||X|}}\sum_{x \in X}\rho(x) = \sqrt{\frac{d_\rho |X|}{|G|}}\Pi_X^\rho .$$

Fact. If X is a subgroup of G, then Π_X^ρ is a projection operator. That is, $(\Pi_X^\rho)^\dagger = \Pi_X^\rho$ and $(\Pi_X^\rho)^2 = \Pi_X^\rho$.

Quantum Fourier Sampling (QFS) is a standard procedure based on the Quantum Fourier Transform to solve the Hidden Subgroup Problem (HSP) (see [12] for a survey). An instance of the HSP over G consists of a black-box function $f : G \to \{0,1\}^*$ such that $f(x) = f(y)$ if and only if x and y belong to the same left coset of H in G, for some subgroup $H \leq G$. The problem is to recover H using the oracle $O_f : |x,y\rangle \mapsto |x, y \oplus f(x)\rangle$. The general QFS procedure for this is the following:

1. Prepare a 2-register quantum state, the first in a uniform superposition of the group elements and the second with the value zero: $|\psi_1\rangle = (1/\sqrt{|G|})\sum_{g \in G}|g\rangle|0\rangle$.
2. Query f, i.e., apply the oracle O_f, resulting in the state
$$|\psi_2\rangle = O_f|\psi_1\rangle = \frac{1}{\sqrt{|G|}}\sum_{g \in G}|g\rangle|f(g)\rangle = \frac{1}{\sqrt{|T|}}\sum_{\alpha \in T}|\alpha H\rangle|f(\alpha)\rangle$$
where T is a transversal of H in G.
3. Measure the second register of $|\psi_2\rangle$, resulting in the state $|\alpha H\rangle|f(\alpha)\rangle$ with probability $1/|T|$ for each $\alpha \in T$. The first register of the resulting state is then $|\alpha H\rangle$ for some uniformly random $\alpha \in G$.
4. Apply the quantum Fourier transform over G to the coset state $|\alpha H\rangle$ observed at step 3:
$$F_G|\alpha H\rangle = \sum_{\rho \in \widehat{G}, 1 \leq i,j \leq d_\rho} \widehat{\alpha H}(\rho)_{i,j}|\rho, i, j\rangle .$$
5. (Weak) Observe the representation name ρ. (Strong) Observe ρ and matrix indices i, j.
6. Classically process the information observed from the previous step to determine the subgroup H.

Probability distributions produced by QFS. For a particular coset αH, the probability of measuring the representation ρ in the state $F_G|\alpha H\rangle$ is

$$P_{\alpha H}(\rho) = \|\widehat{\alpha H}(\rho)\|_F^2 = \frac{d_\rho |H|}{|G|}\mathrm{Tr}\left((\Pi_{\alpha H}^\rho)^\dagger \Pi_{\alpha H}^\rho\right) = \frac{d_\rho |H|}{|G|}\mathrm{Tr}\left(\Pi_H^\rho\right)$$

where $\mathrm{Tr}(A)$ denotes the trace of a matrix A, and $\|A\|_F := \sqrt{\mathrm{Tr}(A^\dagger A)}$ is the Frobenius norm of A. The last equality is due to the fact that $\Pi_{\alpha H}^\rho = \rho(\alpha)\Pi_H^\rho$ and that Π_H^ρ is an orthogonal projector.

Since there is no point in measuring the rows [6], we are only concerned with measuring the columns. As pointed out in [15], the optimal von Neumann measurement on a coset state can always be expressed in this form for some basis B_ρ. Conditioned on observing ρ in the state $F_G|\alpha H\rangle$, the probability of measuring a given $\mathbf{b} \in B_\rho$ is $\|\widehat{\alpha H}(\rho)\mathbf{b}\|^2$. Hence the conditional probability that we observe the vector $\mathbf{b}$, given that we observe the representation ρ, is then

$$P_{\alpha H}(\mathbf{b} \mid \rho) = \frac{\|\widehat{\alpha H}(\rho)\mathbf{b}\|^2}{P_{\alpha H}(\rho)} = \frac{\|\Pi^\rho_{\alpha H}\mathbf{b}\|^2}{\mathrm{Tr}\left(\Pi^\rho_H\right)} = \frac{\|\Pi^\rho_H\mathbf{b}\|^2}{\mathrm{Tr}\left(\Pi^\rho_H\right)}$$

where in the last equality, we use the fact that as $\rho(\alpha)$ is unitary, it preserves the norm of the vector $\Pi^\rho_H\mathbf{b}$.

The coset representative α is unknown and is uniformly distributed in T. However, both distributions $P_{\alpha H}(\rho)$ and $P_{\alpha H}(\mathbf{b} \mid \rho)$ are independent of α and are the same as those for the state $F_G|H\rangle$. Thus, in Step 5 of the QFS procedure above, we observe $\rho \in \widehat{G}$ with probability $P_H(\rho)$, and conditioned on this event, we observe $\mathbf{b} \in B_\rho$ with probability $P_H(\mathbf{b} \mid \rho)$.

If the hidden subgroup is trivial, $H = \{1\}$, the conditional probability distribution on B_ρ is uniform,

$$P_{\{1\}}(\mathbf{b} \mid \rho) = \frac{\|\Pi^\rho_{\{1\}}\mathbf{b}\|^2}{\mathrm{Tr}\left(\Pi^\rho_{\{1\}}\right)} = \frac{\|\mathbf{b}\|^2}{d_\rho} = \frac{1}{d_\rho}.$$

3.2 Distinguishability by QFS

We fix a finite group G and consider quantum Fourier sampling over G in the basis given by $\{B_\rho\}$. For a subgroup $H < G$ and for $g \in G$, let H^g denote the conjugate subgroup $g^{-1}Hg$. Since $\mathrm{Tr}\left(\Pi^\rho_H\right) = \mathrm{Tr}\left(\Pi^\rho_{H^g}\right)$, the probability distributions obtained by QFS for recovering the hidden subgroup H^g are

$$P_{H^g}(\rho) = \frac{d_\rho|H|}{|G|}\mathrm{Tr}\left(\Pi^\rho_H\right) = P_H(\rho) \quad \text{and} \quad P_{H^g}(\mathbf{b} \mid \rho) = \frac{\|\Pi^\rho_{H^g}\mathbf{b}\|^2}{\mathrm{Tr}\left(\Pi^\rho_H\right)}.$$

As $P_{H^g}(\rho)$ does not depend on g, weak Fourier sampling can not distinguish conjugate subgroups. Our goal is to point out that for certain nontrivial subgroup $H < G$, strong Fourier sampling can not efficiently distinguish the conjugates of H from each other or from the trivial one. Recall that the distribution $P_{\{1\}}(\cdot \mid \rho)$ obtained by performing strong Fourier sampling on the trivial hidden subgroup is the same as the uniform distribution U_{B_ρ} on the basis B_ρ. Thus, our goal can be boiled down to showing that the probability distribution $P_{H^g}(\cdot \mid \rho)$ is likely to be close to the uniform distribution U_{B_ρ} in total variation, for a random $g \in G$ and an irrep $\rho \in \widehat{G}$ obtained by weak Fourier sampling.

Definition 5. *We define the* distinguishability *of a subgroup H (using strong Fourier sampling over G), denoted $\mathscr{D}_H$, to be the expectation of the squared L_1-distance between $P_{H^g}(\cdot \mid \rho)$ and U_{B_ρ}:*

$$\mathscr{D}_H \stackrel{\text{def}}{=} \mathbb{E}_{\rho,g}\left[\|P_{H^g}(\cdot \mid \rho) - U_{B_\rho}\|_1^2\right],$$

where ρ is drawn from $\widehat{G}$ according to the distribution $P_H(\rho)$, and g is chosen from G uniformly at random. We say that the subgroup H is indistinguishable *if $\mathscr{D}_H \leq \log^{-\omega(1)}|G|$.*

Note that if $\mathscr{D}_H$ is small, then the total variation distance between $P_{H^g}(\cdot \mid \rho)$ and U_{B_ρ} is small with high probability due to Markov's inequality: for all $\varepsilon > 0$,

$$\Pr_g\left[\|P_{H^g}(\cdot \mid \rho) - U_{B_\rho}\|_{t.v.} \geq \varepsilon/2\right] = \Pr_g\left[\|P_{H^g}(\cdot \mid \rho) - U_{B_\rho}\|_1^2 \geq \varepsilon^2\right] \leq \mathscr{D}_H/\varepsilon^2 .$$

In particular, if the subgroup H is indistinguishable by strong Fourier sampling, then for all constant $c > 0$,

$$\|P_{H^g}(\cdot \mid \rho) - U_{B_\rho}\|_{t.v.} < \log^{-c}|G|$$

with probability at least $1 - \log^{-c}|G|$ in both g and ρ. Our notion of indistinguishability is the direct analogue of that of Kempe and Shalev [8]. Focusing on weak Fourier sampling, they say that H is indistinguishable if $\|P_H(\cdot) - P_{\{1\}}(\cdot)\|_{t.v.} < \log^{-\omega(1)}|G|$.

Our main theorem below will serve as a general guideline for bounding the distinguishability of H. For this purpose we define, for each $\sigma \in \widehat{G}$, the *maximal normalized character of σ on H* as

$$\overline{\chi}_\sigma(H) \stackrel{\text{def}}{=} \max_{h \in H \setminus \{1\}} \frac{|\chi_\sigma(h)|}{d_\sigma} .$$

For each subset $S \subset \widehat{G}$, let

$$\overline{\chi}_{\overline{S}}(H) = \max_{\sigma \in \widehat{G} \setminus S} \overline{\chi}_\sigma(H) \quad \text{and} \quad d_S = \max_{\sigma \in S} d_\sigma .$$

In addition, for each reducible representation ρ of G, we let $I(\rho)$ denote the set of irreps of G that appear in the decomposition of ρ into irreps.

Theorem 1. (MAIN THEOREM) *Suppose S is a subset of $\widehat{G}$. Let $D > d_S^2$ and $L = L_D \subset \widehat{G}$ be the set of all irreps of dimension at least D. Let*

$$\Delta = \Delta_{S,L} = \max_{\rho \in L} \left| S \cap I(\rho \otimes \rho^*) \right| . \tag{3}$$

Then the distinguishability of H is bounded by $\mathscr{D}_H \leq 4|H|^2 \left(\overline{\chi}_{\overline{S}}(H) + \Delta \frac{d_S^2}{D} + \frac{|\overline{L}|D^2}{|G|} \right)$.

Intuitively, the set S consists of irreps of small dimension, and L consists of irreps of large dimension. Moreover, we wish to have that the size of S is small while the size of L is large, so that most irreps are likely in L. In the cases where there are relatively few irreps, i.e., $|S| \ll D$ and $|\widehat{G}| \ll |G|$, we can simply upper bound Δ by $|S|$ and upper bound $|\overline{L}|$ by $|\widehat{G}|$.

We discuss the proof of this theorem in Section 5. Most details are relegated to the Appendix A.

4 Applications of the Main Theorem

In this section, we present applications of Theorem 1 to analyze strong Fourier sampling over certain non-abelian groups, including the symmetric group and the wreath product corresponding to the McEliece-type cryptosystems. Another application to the HSP over the groups $\mathrm{GL}_2(\mathbb{F}_q)$ is omitted for lack of space.

4.1 Strong Fourier Sampling over S_n

We focus now on the case where G is the symmetric group S_n. Recall that each irrep of S_n is in one-to-one correspondence to an integer partition $\lambda = (\lambda_1, \lambda_2, \ldots, \lambda_t)$ of n often given by a *Young diagram* of t rows in which the ith row contains λ_i columns. The conjugate representation of λ is the irrep corresponding to the partition $\lambda' = (\lambda'_1, \lambda'_2, \ldots, \lambda'_{t'})$, obtained by flipping the Young diagram λ about the diagonal.

As in [15], we shall apply Roichman's upper bound [19] on normalized characters:

Theorem 2 (Roichman's Theorem [19]). *There exist constant $b > 0$ and $0 < q < 1$ so that for $n > 4$, for every $\pi \in S_n$, and for every irrep λ of S_n,*

$$\left| \frac{\chi_\lambda(\pi)}{d_\lambda} \right| \le \left(\max\left(q, \frac{\lambda_1}{n}, \frac{\lambda'_1}{n} \right) \right)^{b \cdot \mathrm{supp}(\pi)}$$

where $\mathrm{supp}(\pi) = \#\{k \in [n] \mid \pi(k) \neq k\}$ *is the support of π.*

This bound works well for unbalanced Young diagrams. In particular, for a constant $0 < c < 1/4$, let Λ_c denote the collection of partitions λ of n with the property that either $\frac{\lambda_1}{n} \ge 1 - c$ or $\frac{\lambda'_1}{n} \ge 1 - c$, i.e., the Young diagram λ contains at least $(1-c)n$ rows or contains at least $(1-c)n$ columns. Then, Roichman's upper bound implies that for every $\pi \in S_n$ and $\lambda \notin \Lambda_c$, and a universal constant $\alpha > 0$,

$$\left| \frac{\chi_\lambda(\pi)}{d_\lambda} \right| \le \mathrm{e}^{-\alpha \cdot \mathrm{supp}(\pi)}. \tag{4}$$

On the other hand, both $|\Lambda_c|$ and the maximal dimension of representations in Λ_c are small, as shown in the following Lemma of [15].

Lemma 2 (Lemma 6.2 in [15]). *Let $p(n)$ denote the number of integer partitions of n. Then $|\Lambda_c| \le 2cn \cdot p(cn)$, and $d_\mu < n^{cn}$ for any $\mu \in \Lambda_c$.*

To give a more concrete bound for the size of Λ_c, we record the asymptotic formula for the partition function [5, pg. 45]: $p(n) \approx \mathrm{e}^{\pi\sqrt{2n/3}}/(4\sqrt{3}n) = \mathrm{e}^{O(\sqrt{n})}/n$ as $n \to \infty$.

Now we are ready to prove the main result of this section, an application of Theorem 1.

Theorem 3. *Let H be a nontrivial subgroup of S_n with minimal degree m, i.e., $m = \min_{\pi \in H \setminus \{1\}} \mathrm{supp}(\pi)$. Then for sufficiently large n, $\mathscr{D}_H \le O(|H|^2 \mathrm{e}^{-\alpha m})$.*

Proof. Let $2c < d < 1/2$ be constants. We will apply Theorem 1 by setting $S = \Lambda_c$ and $D = n^{dn}$. By Lemma 2, we have $d_S \le n^{cn}$. Hence, the condition $2c < d$ guarantees that $D > d_S^2$. First, we need to bound the maximal normalized character $\overline{\chi}_{\overline{S}}(H)$. By (4), we have $\overline{\chi}_\mu(H) \le \mathrm{e}^{-\alpha m}$ for all $\mu \in \widehat{S_n} \setminus S$. Hence, $\overline{\chi}_{\overline{S}}(H) \le \mathrm{e}^{-\alpha m}$. To bound the second term in the upper bound of Theorem 1, as $\Delta \le |S|$, it suffices to bound:

$$|S| \cdot \frac{d_S^2}{D} \le 2cn \cdot p(cn) \cdot \frac{n^{2cn}}{n^{dn}} \le \mathrm{e}^{O(\sqrt{n})} \cdot n^{(2c-d)n} \le n^{-\gamma n}/2$$

for sufficiently large n, so long as $\gamma < d - 2c$. Now bounding the last term in the upper bound of Theorem 1: Since $|\overline{L_D}| \le |\widehat{S_n}| = p(n)$ and $n! > n^n \mathrm{e}^{-n}$ by Stirling's approximation,

$$\frac{|\overline{L_D}|D^2}{|S_n|} \le \frac{p(n)n^{2dn}}{n!} \le \frac{\mathrm{e}^{O(\sqrt{n})}n^{2dn}}{n^n \mathrm{e}^{-n}} \le \mathrm{e}^{O(n)} n^{(2d-1)n} \le n^{-\gamma n}/2$$

for sufficiently large n, so long as $\gamma < 1 - 2d$. By Theorem 1, $\mathscr{D}_H \le 4|H|^2(\mathrm{e}^{-\alpha m} + n^{-\gamma n})$.

Theorem 3 generalizes Moore, Russell, and Schulman's result [15] on strong Fourier sampling over S_n, which only applied in the case $|H| = 2$. To relate our result to the results of Kempe et al. [9], observe that since $\log|S_n| = \Theta(n \log n)$, the subgroup H is indistinguishable by strong Fourier sampling if $|H|^2 \mathrm{e}^{-\alpha m} \le (n \log n)^{-\omega(1)}$ or, equivalently, if $m \ge (2/\alpha)\log|H| + \omega(\log n)$.

4.2 Applications to McEliece-type Cryptosystems

Our main application of Theorem 1 is to show the limitations of strong Fourier sampling in attacking the McEliece-type cryptosystems. Throughout this section, we fix parameters n, k, q of a McEliece-type cryptosystem, and fix the underlying $k \times n$ matrix M of the system. Here, M can be a generator matrix or a parity check matrix of the q-ary linear code used in the cryptosystem. Note that the entries of M are in a finite field $\mathbb{F}_{q^\ell} \supset \mathbb{F}_q$ (when M is a generator matrix of a q-ary linear code, we must have $\ell = 1$).

Recall that the canonical quantum attack against this McEliece cryptosystem involves the HSP over the wreath product group $(\mathsf{GL}_k(\mathbb{F}_q) \times S_n) \wr \mathbb{Z}_2$; the hidden subgroup in this case is

$$K = ((H_0, s^{-1}H_0 s), 0) \cup ((H_0 s, s^{-1}H_0), 1) \tag{5}$$

for some hidden element $s \in \mathsf{GL}_k(\mathbb{F}_q) \times S_n$. Here, H_0 is a subgroup of $\mathsf{GL}_k(\mathbb{F}_q) \times S_n$ given by

$$H_0 = \left\{(A,P) \in \mathsf{GL}_k(\mathbb{F}_q) \times S_n \mid A^{-1}MP = M\right\}. \tag{6}$$

To understand the structure of the subgroup H_0, we define the *automorphism group* of M as $\mathrm{Aut}(M) \stackrel{\mathrm{def}}{=} \left\{P \in S_n \mid SMP = M \text{ for some } S \in \mathsf{GL}_k(\mathbb{F}_q)\right\}$. Note that $\mathrm{Aut}(M)$ is a subgroup of the symmetric group S_n and each element $(A,P) \in H_0$ must have $P \in \mathrm{Aut}(M)$. This allows us to control the maximal normalized characters on K through the minimal degree of $\mathrm{Aut}(M)$. Then applying Theorem 1, we show that

Theorem 4. *Assume $q^{k^2} \le n^{an}$ for some constant $0 < a < 1/4$. Let m be the minimal degree of the automorphism group* $\mathrm{Aut}(M)$. *Then for sufficiently large n, the subgroup K defined in* (5) *has $\mathscr{D}_K \le O(|K|^2 \mathrm{e}^{-\delta m})$, where $\delta > 0$ is a constant.*

The proof of Theorem 4 follows the technical ideas discussed in the Introduction. The details can be found in [3].

As $q^{k^2} \le n^{an}$, we have $\log\left|(\mathsf{GL}_k(\mathbb{F}_q) \times S_n) \wr \mathbb{Z}_2\right| = O(\log n! + \log q^{k^2}) = O(n \log n)$. Hence, the subgroup K is indistinguishable if $|K|^2 \mathrm{e}^{-\delta m} \le (n \log n)^{-\omega(1)}$. The size of the subgroup K is given by $|K| = 2|H_0|^2$, and $|H_0| = |\mathrm{Aut}(M)| \times |\mathrm{Fix}(M)|$, where

$\mathrm{Fix}(M) \overset{\text{def}}{=} \{S \in \mathsf{GL}_k(\mathbb{F}_q) \mid SM = M\}$ is the set of scramblers fixing M. To bound the size of $\mathrm{Fix}(M)$, we record an easy fact which can be obtained by the orbit-stabilizer formula:

Fact. Let r be the column rank of M. Then $|\mathrm{Fix}(M)| \leq q^{\ell k(k-r)}$.

Proof. WLOG, assume the first r columns of M are $\mathbb{F}_{q^\ell}$-linearly independent, and each remaining column is an $\mathbb{F}_{q^\ell}$-linear combination of the first r columns. Let N be the $k \times r$ matrix consisting of the first r columns of M. Then we can decompose M as $M = (N \mid NA)$, where A is an $r \times (n-r)$ matrix with entries in $\mathbb{F}_{q^\ell}$. Clearly, $\mathrm{Fix}(M) = \mathrm{Fix}(N)$. Consider the action of $\mathsf{GL}_k(\mathbb{F}_{q^\ell})$ on the set of $k \times r$ matrices over $\mathbb{F}_{q^\ell}$. Under this action, the stabilizer of N contains $\mathrm{Fix}(N)$, and the orbit of the matrix N, denoted $\mathrm{Orb}(N)$, consists of all $k \times r$ matrices over $\mathbb{F}_{q^\ell}$ whose columns are $\mathbb{F}_{q^\ell}$-linearly independent. Thus, $|\mathrm{Orb}(N)| = (q^{\ell k}-1)(q^{\ell k}-q^\ell)\ldots(q^{\ell k}-q^{\ell(r-1)})$. By the orbit-stabilizer formula, we have

$$|\mathrm{Fix}(N)| \leq \frac{|\mathsf{GL}_k(\mathbb{F}_{q^\ell})|}{|\mathrm{Orb}(N)|} = \frac{(q^{\ell k}-1)(q^{\ell k}-q^\ell)\ldots(q^{\ell k}-q^{\ell(k-1)})}{(q^{\ell k}-1)(q^{\ell k}-q^\ell)\ldots(q^{\ell k}-q^{\ell(r-1)})}$$
$$= (q^{\ell k}-q^{\ell r})(q^{\ell k}-q^{\ell(r+1)})\cdots(q^{\ell k}-q^{\ell(k-1)}) \leq q^{\ell k(k-r)}.$$

Corollary 1. *Assume $q^{k^2} \leq n^{0.2n}$ and the automorphism group $\mathrm{Aut}(M)$ has minimal degree $\Omega(n)$. Let r be the column rank of M. Then the subgroup K defined in* (5) *has $\mathscr{D}_K \leq |\mathrm{Aut}(M)|^4 q^{4\ell k(k-r)} \mathrm{e}^{-\Omega(n)}$. In particular, the subgroup K is indistinguishable if, further, $|\mathrm{Aut}(M)| \leq \mathrm{e}^{o(n)}$ and $r \geq k - o(\sqrt{n})/\ell$.*

The constraint $q^{k^2} \leq n^{0.2n}$ implies $\log|\mathsf{GL}_k(\mathbb{F}_q)| = O(n\log n)$, so Alice only needs to flip $O(n\log n)$ bits to pick a random S from $\mathsf{GL}_k(\mathbb{F}_q)$. Thus she needs only $O(n\log n)$ coin flips overall to generate her private key.

Application to the McEliece cryptosystem. Consider a McEliece cryptosystem using a q-ary linear $[n,k]$-code C, with parameters satisfying $q^{k^2} \leq n^{0.2n}$. Since the automorphism group of the code C equals the automorphism group of its generator matrix, we can conclude that this McEliece cryptosystem resists the standard quantum Fourier sampling attack if the code C is *(i) well-scrambled*, i.e., it has a generator matrix of rank at least $k - o(\sqrt{n})$, and is *(ii) well-permuted*, i.e., its automorphism group has minimal degree at least $\Omega(n)$ and has size at most $\mathrm{e}^{o(n)}$. Recall that in terms of security, the Niederreiter system using $(n-k) \times n$ parity check matrices over $\mathbb{F}_q$ of the same code C is equivalent to the McEliece system using the code C [10].

Application to Goppa codes. We would like to point out that if M is a generator matrix of a rational Goppa code or a canonical parity check matrix of a classical Goppa code, it will give good bounds in Corollary 1. Specifically, we consider a matrix M over a finite field $\mathbb{F}_{q^\ell} \supset \mathbb{F}_q$ of the following form:

$$M = \begin{pmatrix} v_1 f_1(\alpha_1) & v_2 f_1(\alpha_2) & \cdots & v_n f_1(\alpha_n) \\ v_1 f_2(\alpha_1) & v_2 f_2(\alpha_2) & \cdots & v_n f_2(\alpha_n) \\ \vdots & \vdots & \ddots & \vdots \\ v_1 f_k(\alpha_1) & v_2 f_k(\alpha_2) & \cdots & v_n f_k(\alpha_n) \end{pmatrix} \tag{7}$$

where $v_1, \ldots, v_n$ are nonzero elements in the field $\mathbb{F}_{q^\ell}$, $(\alpha_1, \ldots, \alpha_n)$ is a list of distinct points in the projective line $\mathbb{F}_{q^\ell} \cup \{\infty\}$, and $f_1, \ldots, f_k$ are $\mathbb{F}_{q^\ell}$-linearly independent polynomials in $\mathbb{F}_{q^\ell}[X]$ of degree less than k (by convention, $f_i(\infty)$ is the X^{k-1}-coefficient of $f_i(X)$). Note that such a matrix M is a generator matrix of a rational Goppa $[n,k]$-code over the field $\mathbb{F}_{q^\ell}$, and is also a parity check matrix of a classical Goppa $[n, \geq n-\ell k]$-code over $\mathbb{F}_q$. To apply Corollary 1, we show the following properties of the matrix M:

Lemma 3. *The matrix M in the form of* (7) *has full rank (i.e., its column rank equals k), and* $\mathrm{Aut}(M)$ *has minimal degree at least $n-2$, and* $|\mathrm{Aut}(M)| \leq q^{3\ell}$.

Proof. We can show that M has full rank directly by decomposing M as $M = AVD$, where $A = (a_{ij})$ is an $k \times k$ invertible matrix with entry a_{ij} being the the X^{j-1}-coefficient of polynomial $f_i(X)$; V is a $k \times n$ Vandermonde matrix with (i,j)-entry being α_j^{i-1}; and D is an $n \times n$ diagonal matrix with v_i in the (i,i)-entry. Then the rank of M equals the rank of the Vandermonde matrix V, which has full rank.

Now we can view M as a generator matrix of a rational Goppa $[n,k]$-code R over the field $\mathbb{F}_{q^\ell}$. Then we have $\mathrm{Aut}(M) \subset \mathrm{Aut}(R)$, where $\mathrm{Aut}(R)$ is the automorphism group of the code R, that is, $\mathrm{Aut}(R) = \left\{P \in S_n \mid SMP = M \text{ for some } S \in \mathrm{GL}_k(\mathbb{F}_{q^\ell})\right\}$. Now we can apply Stichtenoth's Theorem [25] to control the automorphism group $\mathrm{Aut}(R)$.

Theorem 5 (Stichtenoth [25]). *Let $2 \leq k \leq n-2$. Then the automorphism group of any rational Goppa $[n,k]$-code over a field F is isomorphic to a subgroup of* $\mathrm{Aut}(F(x)/F)$.

On the other hand, we also have the useful fact that $\mathrm{Aut}(F(x)/F) \simeq \mathrm{PGL}_2(F)$. Hence, $\mathrm{Aut}(R)$ is isomorphic to a subgroup of the projective linear group $\mathrm{PGL}_2(\mathbb{F}_{q^\ell})$, which implies that $|\mathrm{Aut}(M)| \leq |\mathrm{Aut}(R)| \leq |\mathrm{PGL}_2(\mathbb{F}_{q^\ell})| \leq q^{3\ell}$.

To show that the minimal degree of $\mathrm{Aut}(M)$ is at least $n-2$, we view $\mathrm{Aut}(M) \subset \mathrm{PGL}_2(\mathbb{F}_{q^\ell})$, and observe that any transformation in $\mathrm{PGL}_2(\mathbb{F}_{q^\ell})$ that fixes at least three distinct projective lines must be the identity. Q.E.D.

Hence, classical Goppa codes or rational Goppa codes are good choices for the security of McEliece-type cryptosystems against standard quantum Fourier sampling attacks. Since the rational Goppa codes are broken (classically) by the Sidelnokov-Shestakov [23] structural attack, we shall focus on the classical Goppa codes, which remain secure given suitable choice of parameters.

Application to Neiderreiter systems with classical Goppa codes. Consider a classical q-ary Goppa code C constructed by a support list of distinct points $\alpha_1, \ldots, \alpha_n \in \mathbb{F}_{q^\ell}$ and a Goppa polynomial $g(X) \in \mathbb{F}_{q^\ell}[X]$ of degree k. This code has dimension $k' \geq n-\ell k$. More importantly, it has $k \times n$ parity check matrices in the form of (7) in which $v_j = 1/g(\alpha_j)$ (see [26]), we refer to those matrices as *canonical parity check matrices* of the classical Goppa code C. By Corollary 1 and Lemma 3, *the Niederreiter cryptosystem using $k \times n$ canonical parity check matrices of this code C resists the known quantum attack, provided $q^{k^2} \leq n^{0.2n}$ and $q^{3\ell} \leq e^{o(n)}$.* As pointed out in [4], this Niederreiter system is secure under the Sidelnokov-Shestakov attack. We remark, however, that the security of this Niederreiter cryptosystem may *not* be equivalent to that of the McEliece cryptosystem using the same code C (unless $k' = n - \ell k$ as discussed below), since the equivalence showed in [10] only applies to the Niederreiter cryptosystem using a parity check matrix over the subfield $\mathbb{F}_q$.

Setting the parameters. We discuss the parameters for classical Goppa codes that meet our security requirement. Traditionally, the code length is chosen as $n = q^\ell$, then our parameter setting requires only one constraint, $k^2 \leq 0.2n\ell$, which imposes that the code C must have large dimension, i.e., $k' \geq n - \ell k \geq n - \sqrt{0.2n}(\log_q n)^{3/2}$.

Now we compare our parameter setting with practical parameters suggestion. In most McEliece cryptosystems considered in practice, classical binary Goppa codes are used, that is $q = 2$ and $n = 2^\ell$. The code is also designed so that it has dimension $k' = n - \ell k$ and minimal distance $d \geq 2t + 1$, where $t \ll n$ is a predetermined parameter indicating the number of errors the code can correct. For those systems, the original parameters suggested by McEliece were $(n = 1024, k' \geq 524, t = 50)$, which would meet our requirement as long as the dimension k' is chosen to be slightly larger ($k' \geq 572$). The parameters $(n = 1024, k' = 524, t = 50)$, which can be broken in just 7 days by a cluster of 200 CPUs under Bernstein et al.'s attack [1], clearly do not meet our requirement. An optimal choice of parameters for the Goppa code which maximizes the adversary's work factor was recommended to be $(n = 1024, k' \geq 644, t = 38)$ (see Note 8.32 in [14]). Bernstein et al. [1] suggested two other sets of parameters, $(n = 2048, k' = 1751, t = 27)$ and $(n = 1632, k' = 1269, t = 34)$, that achieve the standard security against all known (classical) attacks. All of these parameters meet our requirement. Well, of course, these parameters were recommended for the original McEliece, or for the equivalent Neiderreiter system that uses parity check matrices over the subfield $\mathbb{F}_2$ with $n - k' = \ell k$ rows. However, if we view each element in $\mathbb{F}_{q^\ell}$ as a vector of dimension ℓ over the subfield $\mathbb{F}_q$, then a $k \times n$ canonical parity check matrix over $\mathbb{F}_{q^\ell}$ can be viewed as a $\ell k \times n$ parity check matrix over $\mathbb{F}_q$.

5 Bounding Distinguishability

We now sketch the proof for the main theorem (Theorem 1). Fixing a nontrivial subgroup $H < G$, we want to upper bound $\mathscr{D}_H$. Let us start with bounding the expectation over the random group element $g \in G$, for a fixed irrep $\rho \in \widehat{G}$:

$$E_H(\rho) \stackrel{\text{def}}{=} \mathbb{E}_g\left[\|P_{H^g}(\cdot \mid \rho) - U_{B_\rho}\|_1^2\right].$$

Obviously we always have $E_H(\rho) \leq 4$. More interestingly, we have

$$\begin{aligned}
E_H(\rho) &= \mathbb{E}_g\left[\left(\sum_{\mathbf{b}\in B_\rho}\left|P_{H^g}(\mathbf{b} \mid \rho) - \frac{1}{d_\rho}\right|\right)^2\right] \\
&\leq \mathbb{E}_g\left[d_\rho \sum_{\mathbf{b}\in B_\rho}\left(P_{H^g}(\mathbf{b} \mid \rho) - \frac{1}{d_\rho}\right)^2\right] && \text{(by Cauchy-Schwarz)} \\
&= d_\rho \sum_{\mathbf{b}\in B_\rho} \mathrm{Var}_g[P_{H^g}(\mathbf{b} \mid \rho)] && \left(\text{since } \mathbb{E}_g[P_{H^g}(\mathbf{b} \mid \rho)] = \frac{1}{d_\rho}\right)
\end{aligned}$$

$$= \frac{d_\rho}{\mathrm{Tr}(\Pi_H^\rho)^2} \sum_{\mathbf{b}\in B_\rho} \mathrm{Var}_g\left[\|\Pi_{H^g}^\rho \mathbf{b}\|^2\right]. \tag{8}$$

The equation $\mathbb{E}_g[P_{H^g}(\mathbf{b} \mid \rho)] = 1/d_\rho$ can be shown using *Schur's lemma*.

From (8), we are motivated to bound the variance of $\|\Pi_{H^g}^\rho \mathbf{b}\|^2$ when g is chosen uniformly at random. We provide an upper bound that depends on the projection of the vector $\mathbf{b}\otimes\mathbf{b}^*$ onto irreducible subspaces of $\rho\otimes\rho^*$, and on maximal normalized characters of σ on H for all irreps σ appearing in the decomposition of $\rho\otimes\rho^*$. Recall that the representation $\rho\otimes\rho^*$ is typically reducible and can be written as an orthogonal direct sum of irreps $\rho\otimes\rho^* = \bigoplus_{\sigma\in\widehat{G}} a_\sigma\sigma$, where $a_\sigma \geq 0$ is the multiplicity of σ. Then $I(\rho\otimes\rho^*)$ consists of σ with $a_\sigma > 0$, and we let $\Pi_\sigma^{\rho\otimes\rho^*}$ denote the projection operator whose image is $a_\sigma\sigma$, that is, the subspace spanned by all copies of σ. Our upper bound given in Lemma 4 below generalizes the bound given in Lemma 4.3 of [15], which only applies to subgroups H of order 2.

Lemma 4. (DECOUPLING LEMMA) *Let ρ be an irrep of G. Then for any vector $\mathbf{b}\in V_\rho$,*

$$\mathrm{Var}_g\left[\|\Pi_{H^g}^\rho \mathbf{b}\|^2\right] \leq \sum_{\sigma\in I(\rho\otimes\rho^*)} \overline{\chi}_\sigma(H) \left\|\Pi_\sigma^{\rho\otimes\rho^*}(\mathbf{b}\otimes\mathbf{b}^*)\right\|^2.$$

Back to our goal of bounding $E_H(\rho)$ using the bound in Lemma 4, the strategy will be to separate irreps appearing in the decomposition of $\rho\otimes\rho^*$ into two groups, those with small dimension and those with large dimension, and treat them differently. If d_σ is large, we shall rely on bounding $\overline{\chi}_\sigma(H)$. If d_σ is small, we shall control the projection given by $\Pi_\sigma^{\rho\otimes\rho^*}$ using the following lemma which was proved implicitly in [15]:

Lemma 5. *For any irrep σ, we have $\sum_{\mathbf{b}\in B_\rho} \left\|\Pi_\sigma^{\rho\otimes\rho^*}(\mathbf{b}\otimes\mathbf{b}^*)\right\|^2 \leq d_\sigma^2$.*

The method discussed above for bounding $E_H(\rho)$ is culminated into Lemma 6 below.

Lemma 6. *Let $\rho\in\widehat{G}$ be arbitrary and $S\subset\widehat{G}$ be any subset of irreps that does not contain ρ. Then*

$$E_H(\rho) \leq 4|H|^2\left(\overline{\chi}_{\overline{S}}(H) + |S\cap I(\rho\otimes\rho^*)|\frac{d_S^2}{d_\rho}\right).$$

To apply this lemma, we should choose the subset S such that $d_S^2 \ll d_\rho$, that is, S should consist of small dimensional irreps. Then applying Lemma 6 for all irreps ρ of large dimension, we can prove our general main theorem straightforwardly.

The detailed proofs of the main theorem and the decoupling lemma are put in Appendix A. The proof for Lemma 6 is omitted for lack of space. See [3] for a full technical version.

References

1. Bernstein, D.J., Lange, T., Peters, C.: Attacking and defending the mcEliece cryptosystem. In: Buchmann, J., Ding, J. (eds.) PQCrypto 2008. LNCS, vol. 5299, pp. 31–46. Springer, Heidelberg (2008)

2. Courtois, N., Finiasz, M., Sendrier, N.: How to achieve a mcEliece-based digital signature scheme. In: Boyd, C. (ed.) ASIACRYPT 2001. LNCS, vol. 2248, pp. 157–174. Springer, Heidelberg (2001)
3. Dinh, H., Moore, C., Russell, A.: The McEliece cryptosystem resists quantum Fourier sampling attacks, preprint (2010), http://arxiv.org/abs/1008.2390
4. Engelbert, D., Overbeck, R., Schmidt, A.: A summary of McEliece-type cryptosystems and their security. J. Math. Crypt. 1, 151–199 (2007)
5. Fulton, W., Harris, J.: Representation Theory - A First Course. Springer-Verlag, New York Inc., Heidelberg (1991)
6. Grigni, M., Schulman, J., Vazirani, M., Vazirani, U.: Quantum mechanical algorithms for the nonabelian hidden subgroup problem. Combinatorica 24(1), 137–154 (2004)
7. Hallgren, S., Moore, C., Rötteler, M., Russell, A., Sen, P.: Limitations of quantum coset states for graph isomorphism. In: STOC 2006: Proceedings of the Thirty-Eighth Annual ACM Symposium on Theory of Computing, pp. 604–617 (2006)
8. Kempe, J., Shalev, A.: The hidden subgroup problem and permutation group theory. In: SODA 2005: Proceedings of the Sixteenth Annual ACM-SIAM Symposium on Discrete Algorithms, pp. 1118–1125 (2005)
9. Kempe, J., Pyber, L., Shalev, A.: Permutation groups, minimal degrees and quantum computing. Groups, Geometry, and Dynamics 1(4), 553–584 (2007), http://xxx.lanl.gov/abs/quant-ph/0607204
10. Li, Y.X., Deng, R.H., Wang, X.M.: On the equivalence of McElieces and Niederreiters public-key cryptosystems. IEEE Transactions on Information Theory 40(1), 271–273 (1994)
11. Loidreau, P., Sendrier, N.: Weak keys in the McEliece public-key cryptosystem. IEEE Transactions on Information Theory 47(3), 1207–1212 (2001)
12. Lomont, C.: The hidden subgroup problem - review and open problems (2004), http://arXiv.org:quantph/0411037
13. McEliece, R.J.: A public-key cryptosystem based on algebraic coding theory. JPL DSN Progress Report, 114–116 (1978)
14. Menezes, A.J., van Oorschot, P.C., Vanstone, S.A.: Handbook of applied cryptography. CRC Press, Boca Raton (1996)
15. Moore, C., Russell, A., Schulman, L.J.: The symmetric group defies strong quantum Fourier sampling. SIAM Journal of Computing 37, 1842–1864 (2008)
16. Niederreiter, H.: Knapsack-type cryptosystems and algebraic coding theory. Problems of Control and Information Theory. Problemy Upravlenija i Teorii Informacii 15(2), 159–166 (1986)
17. Petrank, E., Roth, R.M.: Is code equivalence easy to decide? IEEE Transactions on Information Theory 43(5), 1602–1604 (1997), doi:10.1109/18.623157
18. Regev, O.: On lattices, learning with errors, random linear codes, and cryptography. In: STOC 2005: Proceedings of the Thirty-Seventh Annual ACM Symposium on Theory of Computing, pp. 84–93 (2005)
19. Roichman, Y.: Upper bound on the characters of the symmetric groups. Invent. Math. 125(3), 451–485 (1996)
20. Ryan, J.A.: Excluding some weak keys in the McEliece cryptosystem. In: Proceedings of the 8th IEEE Africon, pp. 1–5 (2007)
21. Sendrier, N.: Finding the permutation between equivalent linear codes: the support splitting algorithm. IEEE Transactions on Information Theory 46(4), 1193–1203 (2000)
22. Shor, P.W.: Polynomial-time algorithms for prime factorization and discrete logarithms on a quantum computer. SIAM Journal on Computing 26, 1484–1509 (1997)
23. Sidelnikov, V.M., Shestakov, S.O.: On insecurity of cryptosystems based on generalized Reed-Solomon codes. Discrete Mathematics and Applications 2(4), 439–444 (1992)

24. Simon, D.R.: On the power of quantum computation. SIAM J. Comput. 26(5), 1474–1483 (1997)
25. Stichtenoth, H.: On automorphisms of geometric Goppa codes. Journal of Algebra 130, 113–121 (1990)
26. van Lint, J.H.: Introduction to coding theory, 2nd edn. Springer, Heidelberg (1992)

Appendix A Proofs for the Main Theorem

Proof of the Decoupling Lemma

Proof (Proof of Lemma 4). Fix a vector $\mathbf{b} \in V_\rho$. To simplify notations, we shall write Π_g as shorthand for $\Pi^\rho_{H^g}$, and write $g\mathbf{b}$ for $\rho(g)\mathbf{b}$. For any $g \in G$, we have

$$\begin{aligned}\|\Pi_g \mathbf{b}\|^2 &= \langle \Pi_g \mathbf{b}, \Pi_g \mathbf{b}\rangle = \langle \mathbf{b}, \Pi_g \mathbf{b}\rangle \\ &= \frac{1}{|H|}\left(\langle \mathbf{b}, \mathbf{b}\rangle + \sum_{h\in H\setminus\{1\}} \langle \mathbf{b}, g^{-1}hg\mathbf{b}\rangle\right).\end{aligned}$$

Let $S_g = \sum_{h\in H\setminus\{1\}} \langle \mathbf{b}, g^{-1}hg\mathbf{b}\rangle$. Then

$$\mathrm{Var}_g\left[\|\Pi_g \mathbf{b}\|^2\right] = \frac{\mathrm{Var}_g[S_g]}{|H|^2} = \frac{\mathbb{E}_g\left[S_g^2\right] - \mathbb{E}_g[S_g]^2}{|H|^2}.$$

To bound the variance, we upper bound S_g^2 for all $g \in G$. Since S_g is real, applying Cauchy-Schwarz inequality, we have

$$S_g^2 = \left|\sum_{h\in H\setminus\{1\}} \langle \mathbf{b}, g^{-1}hg\mathbf{b}\rangle\right|^2 \le (|H|-1)\left(\sum_{h\in H\setminus\{1\}} \left|\langle \mathbf{b}, g^{-1}hg\mathbf{b}\rangle\right|^2\right).$$

As in Lemma 4.2 of [15], one can express the second moment of the inner product $\langle \mathbf{b}, g^{-1}hg\mathbf{b}\rangle$ in terms of the projection of $\mathbf{b}\otimes\mathbf{b}^*$ into the irreducible constituents of the tensor product representation $\rho\otimes\rho^*$. Specifically, for any $h \in G$, we have

$$\mathbb{E}_g\left[|\langle \mathbf{b}, g^{-1}hg\mathbf{b}\rangle|^2\right] = \sum_{\sigma\in I(\rho\otimes\rho^*)} \frac{\chi_\sigma(h)}{d_\sigma} \left\|\Pi_\sigma^{\rho\otimes\rho^*}(\mathbf{b}\otimes\mathbf{b}^*)\right\|^2.$$

It follows that

$$\begin{aligned}\mathrm{Var}_g\left[\|\Pi^\rho_{H^g}\mathbf{b}\|^2\right] &\le \frac{|H|-1}{|H|^2} \sum_{h\in H\setminus\{1\}} \mathbb{E}_g\left[\left|\langle \mathbf{b}, g^{-1}hg\mathbf{b}\rangle\right|^2\right] \\ &\le \mathbb{E}_{h\in H\setminus\{1\}}\left[\sum_{\sigma\in I(\rho\otimes\rho^*)} \frac{\chi_\sigma(h)}{d_\sigma}\left\|\Pi_\sigma^{\rho\otimes\rho^*}(\mathbf{b}\otimes\mathbf{b}^*)\right\|^2\right] \\ &\le \sum_{\sigma\in I(\rho\otimes\rho^*)} \overline{\chi}_\sigma(H)\left\|\Pi_\sigma^{\rho\otimes\rho^*}(\mathbf{b}\otimes\mathbf{b}^*)\right\|^2.\end{aligned}$$

Proof of the Main Theorem

Proof (Proof of Theorem 1:). For any $\rho \in L$, since $d_\rho \geq D > d_S^2$, we must have $\rho \notin S$. By Lemma 6,

$$E_H(\rho) \leq 4|H|^2 \left(\overline{\chi}_{\overline{S}}(H) + \Delta \frac{d_S^2}{D} \right) \quad \text{for all } \rho \in L.$$

Combining this with the fact that $E_H(\rho) \leq 4$ for all $\rho \notin L$, we obtain

$$\mathscr{D}_H = \mathbb{E}_\rho[E_H(\rho)] \leq 4|H|^2 \left(\overline{\chi}_{\overline{S}}(H) + \Delta \frac{d_S^2}{D} \right) + 4\Pr_\rho[\rho \notin L].$$

To complete the proof, it remains to bound $\Pr_\rho[\rho \notin L]$. Since $\mathrm{Tr}(\Pi_H^\rho) \leq d_\rho$, we have

$$P(\rho) = \frac{d_\rho |H|}{|G|} \mathrm{Tr}(\Pi_H^\rho) \leq \frac{d_\rho^2 |H|}{|G|}.$$

Since $d_\rho < D$ for all $\rho \in \widehat{G} \setminus L$, it follows that

$$\Pr_\rho[\rho \notin L] = \sum_{\rho \notin L} P(\rho) \leq \frac{|\overline{L}| D^2 |H|}{|G|} \leq \frac{|\overline{L}| D^2 |H|^2}{|G|}.$$

Author Index